Cytomegaloviruses

From Molecular Pathogenesis to Intervention

Volume II

Edited by

Matthias J. Reddehase

with the assistance of

Niels A.W. Lemmermann

Institute for Virology
University Medical Center of the Johannes Gutenberg-University
Mainz
Germany

Caister Academic Press

Caister Academic Press
Norfolk, UK

www.caister.com

British Library Cataloguing-in-Publication Data
A catalogue record for this book is available from the British Library

ISBN for complete set of two volumes: 978-1-908230-18-8
ISBN of this volume: 978-1-908230-20-1

Cover image designed by Andrew Townsend, adapted from Figure I.6.1

Printed and bound in Malta by Gutenberg Press Ltd

Contents

Contents of Volume I

Contributors

Stuart P. Adler
Virginia Commonwealth University
Richmond, VA
USA

sadler@vcu.edu

Rebecca J. Aicheler
Department of Medical Microbiology
Cardiff Institute of Infection and Immunity
Cardiff University School of Medicine
Cardiff

UK

aichelerr@cardiff.ac.uk

Robin K. Avery
Transplant and Oncology Infectious Diseases Program
Division of Infectious Disease
Johns Hopkins University
Baltimore, MD
USA

averyr@ccf.org

Peter A. Barry
Center for Comparative Medicine
University of California Davis
Davis, CA
USA

pabarry@ucdavis.edu

Chris A. Benedict
La Jolla Institute For Allergy & Immunology
La Jolla, CA
USA

benedict@liai.org

Stephanie R. Bialek
National Center for Immunization and Respiratory Diseases
Centers for Disease Control and Prevention
Atlanta, GA
USA

zqg7@cdc.gov

Christine A. Biron
Department of Molecular Microbiology and Immunology and
The Division of Biology and Medicine
The Warren Alpert Medical School
Brown University
Providence, RI
USA

Christine_Biron@brown.edu

Michael Boeckh
Vaccine & Infectious Disease Division and Clinical Research Division
Fred Hutchinson Cancer Research Center
Seattle, WA
USA

mboeckh@fhcrc.org

Verena Böhm
Medical Clinic II
University of Kiel
Kiel
Germany

v.boehm@med2.uni-kiel.de

Suresh B. Boppana
Department of Pediatrics
University of Alabama
Birmingham, AL
USA

sboppana@peds.uab.edu

William Britt
Department of Pediatrics
University of Alabama
Birmingham, AL
USA

WBritt@peds.uab.edu

Michael J. Cannon
National Center on Birth Defects and Development Disabilities
Centers for Disease Control and Prevention
Atlanta, GA
USA

mcannon@cdc.gov

Djurdjica Cekinović
Faculty of Medicine
University of Rijeka
Rijeka
Croatia

durce@medri.hr

Meike Chevillotte
Rockefeller University
Center for the Study of Hepatitis C
New York, NY
USA

meike.chevillotte@rockefeller.edu

Karine Crozat
Centre d'Immunologie de Marseille Luminy (CIML)
Aix-Marseille University
INSERM
CNRS
Marseille
France

crozat@ciml.univ-mrs.fr

Marc Dalod
Centre d'Immunologie de Marseille Luminy (CIML)
Aix-Marseille University, UNIV UM2
Institut National de la Santé et de la Recherche Médicale (Inserm), UMR1104
Centre National de la Recherche Scientifique (CNRS), UMR 7280
CNRS
Marseille
France

dalod@ciml.univ-mrs.fr

Mariapia Degli-Esposti
Centre for Ophthalmology and Visual Science
Centre for Experimental Immunology
Lions Eye Institute
The University of Western Australia
Perth
Australia

mariapia@cyllene.uwa.edu.au

Sheila C. Dollard
National Center for Immunization and Respiratory Diseases
Centers for Disease Control and Prevention
Atlanta, GA
USA

sgd5@cdc.gov

Stefan Ebert
Institute for Virology
University Medical Center of the Johannes Gutenberg-University Mainz
Mainz
Germany

ebertst@uni-mainz.de

Vincent C. Emery
Department of Cellular and Microbial Sciences
University of Surrey
Guildford
UK

v.emery@medsch.ucl.ac.uk

Jakob Ettinger
Institute of Virology
Charité
Berlin
Germany

jakob.ettinger@charite.de

June Fang-Hoover
Cell and Tissue Biology Department
University of California San Francisco
San Francisco, CA
USA

june.fang-hoover@ucsf.edu

Annette Fink
Institute for Virology
University Medical Center of the Johannes Gutenberg-University Mainz
Mainz
Germany

finka@uni-mainz.de

Karen B. Fowler
Department of Pediatrics
University of Alabama
Birmingham, AL
USA

kfowler@uab.edu

Kirsten Freitag
Institute for Virology
University Medical Center of the Johannes Gutenberg-University Mainz
Mainz
Germany

kfreitag@uni-mainz.de

Klaus Früh
Vaccine and Gene Therapy Institute
Oregon Health and Science University
Beaverton, OR
USA

fruehk@ohsu.edu

Giuseppe Gerna
Laboratori Sperimentali di Ricerca
Area Trapiantologica
Fondazione IRCCS Policlinico San Matteo
Pavia
Italy

g.gerna@smatteo.pv.it

Paul D. Griffiths
Division of Infection and Immunity (Royal Free Campus)
University College London
London
UK

p.griffiths@ucl.ac.uk

Scott D. Grosse
National Center on Birth Defects and Development Disabilities
Centers for Disease Control and Prevention
Atlanta, GA
USA

sgg4@cdc.gov

Scott G. Hansen
Vaccine and Gene Therapy Institute
Oregon Health and Science University
Beaverton, OR
USA

hansensc@ohsu.edu

Ann B. Hill
Department of Molecular Microbiology and Immunology
Oregon Health and Science University
Portland, OR
USA

hillan@ohsu.edu

Rafaela Holtappels
Institute for Virology
University Medical Center of the Johannes Gutenberg-University Mainz
Mainz
Germany

r.holtappels@uni-mainz.de

Michael A. Jarvis
School of Biomedical and Biological Sciences
University of Plymouth
Plymouth
UK

michael.jarvis@plymouth.ac.uk

Stipan Jonjić
Faculty of Medicine
University of Rijeka
Rijeka
Croatia

stipan.jonjic@medri.uniri.hr

Astrid Krmpotić
Faculty of Medicine
University of Rijeka
Rijeka
Croatia

astrid.krmpotic@medri.uniri.hr

Niels A.W. Lemmermann
Institute for Virology
University Medical Center of the Johannes Gutenberg-University Mainz
Mainz
Germany

lemmermann@uni-mainz.de

Alistair McGregor
Division of Pediatric Infectious Diseases and Immunology
Center for Infectious Diseases and Microbiology Translational Research
University of Minnesota Medical School
Minneapolis, MN
USA

mcgre077@umn.edu

Michael Mach
Institut für klinische und molekulare Virologie
Friedrich-Alexander-Universität Erlangen-Nürnberg
Erlangen
Germany

michael.mach@viro.med.uni-erlangen.de

Michael McVoy
Division of Infectious Diseases
Virginia Commonwealth University School of Medicine
Richmond, VA
USA

mmcvoy@vcu.edu

Daniel Malouli
Vaccine and Gene Therapy Institute
Oregon Health and Science University
Beaverton, OR
USA

maloulid@ohsu.edu

Gavin M. Mason
Peter Gorer Department of Immunobiology
King's College London
Guy's Hospital
London
UK

gavin.mason@kcl.ac.uk

Thomas Mertens
Institute of Virology
Ulm University Medical Center
Ulm
Germany

thomas.mertens@uniklinik-ulm.de

Detlef Michel
Ulm University Medical Center
Institute of Virology
Ulm
Germany

detlef.michel@uniklinik-ulm.de

Richard S.B. Milne
Division of Infection and Immunity (Royal Free Campus)
University College London
London
UK

richard.milne@ucl.ac.uk

Jay A. Nelson
Vaccine and Gene Therapy Institute
Oregon Health and Science University
Beaverton, OR
USA

nelsonj@ohsu.edu

Giovanni Nigro
Pediatric Unit and School
San Salvatore Hospital
University of L'Aquila
L'Aquila
Italy

giovanni.nigro@univaq.it
nigriogio@libero.it

Ismael R. Ortega-Sanchez
National Center for Immunization and Respiratory Diseases
Centers for Disease Control and Prevention
Atlanta, GA
USA

iao8@cdc.gov

Kristie L. Oxford
Department of Medical Pathology & Laboratory Medicine
California National Primate Research Center
Center for Comparative Medicine
University of California Davis
Davis, CA
USA

kloxford@ucdavis.edu

Lenore Pereira
Cell and Tissue Biology Department
University of California San Francisco
San Francisco, CA
USA

lenore.pereira@ucsf.edu

Matthew Petitt
Cell and Tissue Biology Department
University of California San Francisco
San Francisco, CA
USA

matthew.petitt@ucsf.edu

Louis J. Picker
Vaccine and Gene Therapy Institute
Oregon Health and Science University
Beaverton, OR
USA

pickerl@ohsu.edu

Bodo Plachter
Institute for Virology
University Medical Center of the Johannes Gutenberg-University Mainz
Mainz
Germany

plachter@uni-mainz.de

Stanley A. Plotkin
University of Pennsylvania
Children's Hospital of Philadelphia
Philadelphia, PA
USA

stanley.plotkin@vaxconsult.com

Jürgen Podlech
Institute for Virology
University Medical Center of the Johannes Gutenberg-University Mainz
Mainz
Germany

podlech@uni-mainz.de

Michal Pyzik
Department of Human Genetics
McGill Life Sciences Complex
Montreal, QC
Canada

michal.pyzik@mail.mcgill.ca

Maria Grazia Revello
Clinica Ostetrico-Ginecologica
Fondazione IRCCS Policlinico San Matteo
Pavia
Italy

mg.revello@smatteo.pv.it

Matthias J. Reddehase
Institute for Virology
University Medical Center of the Johannes Gutenberg-University Mainz
Mainz
Germany

matthias.reddehase@uni-mainz.de

Angélique Renzaho
Institute for Virology
University Medical Center of the Johannes Gutenberg-University Mainz
Mainz
Germany

Renzaho@uni-mainz.de

Mark R. Schleiss
Division of Pediatric Infectious Diseases and Immunology
Center for Infectious Diseases and Microbiology Translational Research
University of Minnesota Medical School
Minneapolis, MN
USA

schleiss@umn.edu

Sachiko Seo
Vaccine & Infectious Disease Division and Clinical Research Division
Fred Hutchinson Cancer Research Center
Seattle, WA
USA

sseo-tky@umin.ac.jp

J.G. Patrick Sissons
Clinical School
Addenbrookes Hospital
University of Cambridge
Cambridge
UK

jgps10@medschl.cam.ac.uk

Nadja Spindler
Institut für klinische und molekulare Virologie
Friedrich-Alexander-Universität Erlangen-Nürnberg
Erlangen
Germany

nadja.spindler@viro.med.uni-erlangen.de

Daniel N. Streblow
Vaccine and Gene Therapy Institute
Oregon Health Sciences University
Beaverton, OR
USA

streblow@ohsu.edu

Takako Tabata
Cell and Tissue Biology Department
University of California San Francisco
San Francisco, CA
USA

takako.tabata@ucsf.edu

Doris Thomas
Institute for Virology
University Medical Center of the Johannes Gutenberg-University Mainz
Mainz
Germany

thomdo00@uni-mainz.de

Silvia Vidal
Department of Human Genetics
McGill Life Sciences Complex
Montreal, QC
Canada

silvia.vidal@mcgill.ca

Sebastian Voigt
Robert Koch Institute
Division of Viral Infections
Berlin
Germany

sebastian.voigt@charite.de

Eddie C.Y. Wang
Department of Medical Microbiology
Cardiff Institute of Infection and Immunity
Cardiff University School of Medicine
Cardiff
UK

wangec@cardiff.ac.uk

Anna-Katharina Wiegers
Institut für klinische und molekulare Virologie
Friedrich-Alexander-Universität Erlangen-Nürnberg
Erlangen
Germany

aawieger@viro.med.uni-erlangen.de

Gavin W.G. Wilkinson
Department of Medical Microbiology
Cardiff Institute of Infection and Immunity
Cardiff University School of Medicine
Cardiff
UK

wilkinsongw1@cardiff.ac.uk

Mark R. Wills
Department of Medicine
Addenbrookes Hospital
University of Cambridge
Cambridge
UK

mrw1004@cam.ac.uk

Thomas Winkler
Hematopoiesis Unit
Department of Biology
Nikolaus-Fiebiger Center for Molecular Medicine
Friedrich-Alexander-University Erlangen-Nürnberg
Erlangen
Germany

thomas.winkler@molmed.rz.uni-erlangen.de

Abbreviations

A

aa	amino acid(s)
Ab	antibody
ACV	aciclovir, acyclovir
AIDS	acquired immune deficiency syndrome
APC	antigen-presenting cell
APC	anaphase-promoting complex
AT	adoptive (cell) transfer
ATCC	American Type Culture Collection
ATG	anti-thymocyte globulin
ATP	adenosine triphosphate (also ADP, AMP, CTP, UDP, etc.)

B

BAC	bacterial artificial chromosome
BAL	bronchoalveolar lavage
BCG	bacillus Calmette–Guérin
BFU	burst-forming unit
BM	bone marrow
BMC	bone marrow cell
BMT	bone marrow (cell) transplantation
bp	base pair(s)

C

CAV	cardiac allograft vasculopathy
CCL	CC chemokine ligand
CCMV	chimpanzee CMV
CCR	CC chemokine receptor
CD	cluster of differentiation (e.g. CD8)
cDNA	complementary DNA
CDV	cidofovir
CFU	colony-forming unit
ChIP	chromatin immunoprecipitation
CI	confidence interval
CID	cytomegalic inclusion disease
CLP	common lymphoid progenitor
CLT	cytomegalovirus latency-associated transcript
CMI	cell-mediated immunity
CMP	common myeloid progenitor
CMV	cytomegalovirus (in general; also used as synonym for human CMV)
CNS	central nervous system
CPE	cytopathic effect
CSF	colony-stimulating factor
CSF	cerebrospinal fluid
CTL	cytolytic (or cytotoxic) T-lymphocyte(s)
CTLL	cytolytic T-lymphocyte line
CVD	cardiovascular disease

D

d	day(s)
D	donor (for transplantation)
DC	dendritic cell
DISC	death-inducing signalling complex
DLI	donor lymphocyte infusion
DN	dominant negative
DNA	deoxyribonucleic acid
dpi (d p.i.)	day(s) post infection

E

E	early (phase of cytomegaloviral gene expression)
EC	endothelial cell
ECM	extracellular matrix
EEC	early effector cell
EGFP	enhanced green fluorescent protein
ELISA	enzyme-linked immunosorbent assay
ELISPOT	enzyme-linked immunospot
EM	electron microscope (microscopy)
ER	endoplasmic reticulum
ERGIC	endoplasmic reticulum Golgi intermediate compartment
E/T ratio	effector–target (cell) ratio
EYFP	enhanced yellow fluorescent protein

F

FACS	fluorescence-activated cell sorting (sorter)
FasL:	Fas ligand
FcR	Fc receptor

FFU focus-forming unit
FL: fluorescence (intensity)
FLP flippase (recombination enzyme)
FOS foscarnet
FRT FLP recognition target

G
gB glycoprotein B (also gM, etc.)
GBM glioblastoma multiforme
gc glycoprotein complex(es)
G-CSF granulocyte colony-stimulating factor
GCV ganciclovir
GFP green fluorescent protein
GLP good laboratory practice
GM-CSF granulocyte/macrophage colony-stimulating factor
GOI gene of interest
gp glycoprotein
GPCMV guinea pig cytomegalovirus
GvH graft-versus-host
GvHD graft-versus-host disease

H
h hour(s)
H-2 major histocompatibility complex of the mouse
HAART highly active anti-retroviral therapy
HC haematopoietic cell(s)
HCMV human cytomegalovirus
HCT haematopoietic (stem) cell transplantation
HDAC histone deacetylase(s)
HFF human foreskin fibroblast(s)
HF human fibroblast(s)
HHV human herpesvirus
HLA (human) histocompatibility leucocyte antigen
HPLC high-performance liquid chromatography
HPS haematopoietic progenitor cell
HR hazard ratio
HSC (pluripotent, self-renewing) haematopoietic stem cell
HSCT haematopoietic stem cell transplantation
HSCTR HSCT recipient
HUVEC human umbilical vein endothelial cell
HvG host-versus-graft
HvGD host-versus-graft disease
HvGR host-versus-graft reaction

I
i.c. intracerebral(ly), intracranial(ly)
i.fp. intra-footpad, i.e. intraplantar(ily)
i.m. intramuscular(ly)
i.n. intranasal(ly)
i.p. intraperitoneal(ly)
i.v. intravenous(ly)
IDE immunodominant epitope(s)
IE immediate-early (phase of cytomegaloviral gene expression)
IFN interferon(s) (e.g. IFN-γ)
Ig immunoglobulin(s)
IHC immunohistochemistry
IL interleukin(s) (e.g. IL-2)
IM inflammatory monocyte
INM inner nuclear membrane
IRF IFN regulatory factor
ISG IFN-stimulated gene
ISH *in situ* hybridization
ISRE IFN-stimulated response element
ITAM immunoreceptor tyrosine-based activation motif
ITIM immunoreceptor tyrosine-based inhibitory motif
IVIG intravenous immunoglobulin (plasma protein replacement therapy)

K
Kan kanamycin
kb kilobase(s)
kbp kilobase pair(s)
kDa kilodalton
KIR killer cell Ig-like receptor(s)
KLH keyhole limpet haemocyanin
KO knockout

L
L late (phase of cytomegaloviral gene expression)
L ligand (in abbreviations)
LAT latency-associated transcript
LC Langerhans cell
LD lethal dose
LIR leucocyte Ig-like receptor
LN lymph node
LPS lipopolysaccharide
LSEC liver sinusoidal endothelial cell
LTR long terminal repeat

M
M molar
mAb monoclonal antibody
MAC macrophage
MBV maribavir
MCMV murine (or mouse) cytomegalovirus
MCP major capsid protein
mCP minor capsid protein
MDSC monocyte-derived suppressor cell

MEF mouse (murine) embryonic fibroblast(s)
MHC major histocompatibility complex
MHC-I MHC class I (genes or antigens)
MHC-II MHC class II (genes or antigens)
MIE major immediate-early
MIEP MIE promoter (also: promoter and enhancer)
min minute(s)
miRNA micro-RNA
mo month(s)
MOI multiplicity of infection
mRNA messenger RNA
MS mass spectrometry

N

n; *N* number in study or group
n.d. not determined
ND nuclear domain (e.g. ND10)
NF nuclear factor (e.g. NF-κB)
NGS next generation sequencing
NHP non-human primate
NK natural killer (e.g. NK cells or NKT-cells)
NLS nuclear localization signal
NOD non-obese diabetic
NPC nuclear pore complex
NPC neuronal precursor (stem) cell; neuroglial precursor stem cell
NPLC non-parenchymal liver cells
NSG NOD SCID gamma
nt nucleotide(s)
NT-Ab (virus) neutralizing antibody

O

ONM outer nuclear membrane
OR odds ratio
ORF open reading frame
Ori origin (e.g. ori*Lyt*)
OVA ovalbumin

P

p; *P* probability
PAA phosphonoacetic acid
PAGE polyacrylamide gel electrophoresis
PAMP pathogen-associated molecular pattern
profAPC professional antigen-presenting cell
PB peripheral blood
PBL peripheral blood lymphocyte/leucocyte
PBMC peripheral blood mononuclear cell
PCR polymerase chain reaction
pDC plasmacytoid DC
PEC peritoneal exudate cell
PFA phosphonoformic acid
PFU plaque-forming unit
p.i. post infection
PISH PCR *in situ* hybridization
PK protein kinase (e.g. PKC)
PLN popliteal lymph node
PM patrolling monocyte(s)
pMHC-I peptide-loaded major histocompatibility complex class I (protein)
PML promyelocytic leukaemia protein (e.g. PML bodies)
PMNL polymorphonuclear leucocyte(s)
POI protein of interest
poly(A)$^+$ polyadenylated (RNA)
pp phosphoprotein
PRR pattern recognition receptor

Q

qPCR quantitative polymerase chain reaction (equals real-time PCR)
qRT-PCR quantitative reverse transcription PCR (equals RT-qPCR)

R

r recombinant (e.g. rIFN-γ)
R receptor (e.g. IL-2R)
R recipient (of transplantation)
RAE-1 retinoic acid early inducible gene(s) 1
RCMVr at cytomegalovirus
RFLP restriction fragment length polymorphism
RFP red fluorescent protein
RhCMV rhesus (macaque) cytomegalovirus
RIP radioimmunoprecipitation
RLN regional lymph node
RNA ribonucleic acid
RNAi RNA interference
RT reverse transcription (transcriptase), as in RT-PCR
RT-qPCR reverse transcription quantitative PCR

S

s.c. subcutaneous(ly)
SCF stem cell factor
SCID severe combined immunodeficiency
SCT stem cell transplantation
SD standard deviation
SE standard error
SEM standard error of the mean
SES socioeconomic status
SG salivary gland(s)
siRNA short interfering RNA
SIV simian immunodeficiency virus
SLEC short-lived effector cell(s)
SLO secondary lymphoid organ(s)
SMC smooth muscle cell(s)

SOT	solid organ transplantation
SOTR	SOT recipient
STAT	signal transducer and activator of transcription

T

TAP	transporter associated with antigen processing
Tc	cytolytic (cytotoxic) T-cell
TC	tissue culture
TCID	tissue culture infective dose
T_{CM}	central memory T-cell
TCR	T-cell receptor for antigen
T_E	effector T-cell
T_{EM}	effector-memory T-cell
TEL	transcript(s) expressed in latency
TEM	transmission electron microscopy
TGF	transforming growth factor (e.g. TGF-β)
TGN	trans-Golgi network
Th	helper T-cell
TLR	toll-like receptor
T_M	memory T-cells
TMD	transmembrane domain
TNF	tumour necrosis factor (e.g. TNF-α)
TRAIL	TNF-related apoptosis-inducing ligand
Treg	regulatory T-cell(s)

U

U	unit(s)
UL	unique long (region)
US	unique short (region)
UTR	untranslated region
UV	ultraviolet (light)

V

V region	variable region of Ig or of TCR
valGCV	valganciclovir
VEGF	vascular endothelial cell growth factor
VGCV	valganciclovir
vol	volume
vRAP	viral regulator of (direct) antigen presentation

W

wt	weight
WT	wild-type (e.g. MCMV-WT)

Miscellaneous

β2m	β_2-microglobulin
γ_C	common γ-chain (of cytokine/interleukin receptors)
3D	three-dimensional

Current Books of Interest

Genome Analysis: Current Procedures and Applications	2014
Bacterial Toxins: Genetics, Cellular Biology and Practical Applications	2013
Cold-Adapted Microorganisms	2013
Fusarium: Genomics, Molecular and Cellular Biology	2013
Prions: Current Progress in Advanced Research	2013
RNA Editing: Current Research and Future Trends	2013
Real-Time PCR: Advanced Technologies and Applications	2013
Microbial Efflux Pumps: Current Research	2013
Oral Microbial Ecology: Current Research and New Perspectives	2013
Bionanotechnology: Biological Self-assembly and its Applications	2013
Real-Time PCR in Food Science: Current Technology and Applications	2013
Bacterial Gene Regulation and Transcriptional Networks	2013
Bioremediation of Mercury: Current Research and Industrial Applications	2013
Neurospora: Genomics and Molecular Biology	2013
Rhabdoviruses	2012
Horizontal Gene Transfer in Microorganisms	2012
Microbial Ecological Theory: Current Perspectives	2012
Two-Component Systems in Bacteria	2012
Malaria Parasites: Comparative Genomics, Evolution and Molecular Biology	2013
Foodborne and Waterborne Bacterial Pathogens	2012
Yersinia: Systems Biology and Control	2012
Stress Response in Microbiology	2012
Bacterial Regulatory Networks	2012
Systems Microbiology: Current Topics and Applications	2012
Quantitative Real-time PCR in Applied Microbiology	2012
Bacterial Spores: Current Research and Applications	2012
Small DNA Tumour Viruses	2012
Extremophiles: Microbiology and Biotechnology	2012
Bacillus: Cellular and Molecular Biology (Second edition)	2012
Microbial Biofilms: Current Research and Applications	2012
Bacterial Glycomics: Current Research, Technology and Applications	2012
Non-coding RNAs and Epigenetic Regulation of Gene Expression	2012
Brucella: Molecular Microbiology and Genomics	2012
Molecular Virology and Control of Flaviviruses	2012

www.caister.com

Synopsis of Clinical Aspects of Human Cytomegalovirus Disease

II.1

Suresh B. Boppana and William J. Britt

Abstract

The spectrum of clinical disease associated with infection with human cytomegalovirus (HCMV) ranges from severe multiorgan system disease with significant morbidity and mortality to nearly an asymptomatic infection. Populations susceptible to severe HCMV infections include transplant recipients undergoing immunosuppressive therapy, individuals with untreated HIV infection, and the developing fetus. The loss of adaptive immunity in transplant recipients and HIV-infected hosts represents a major risk factor for disseminated HCMV infection whereas the developmental immaturity of the immune system of the fetus has been proposed to predispose infants infected *in utero* to severe infection and disease. Specific immunological determinants that predispose infection and disease remain incompletely characterized but both CD8+ and CD4+ T-lymphocyte responses, antiviral antibodies, and natural cytotoxicity (natural killer cells, γδ T-cells) all have been shown to have a potential role in controlling HCMV replication. The clinical expression of HCMV infection appears directly related to the immune response of the host such that immunocompromised individuals are at risk for uncontrolled virus replication and dissemination. Paradoxically, under conditions in which the adaptive immune system is severely impaired and virus replication continues unabated, a combination of viral cytopathology and host immunopathological responses arising from residual immunity appear to contribute to disease. Under specific circumstances such as in solid organ allograft recipients, immunopathological responses appear to be a major determinant in the long-term outcome of allograft recipients. The mechanism(s) of disease in the HCMV-infected fetus remains poorly understood perhaps because of a lack of well-defined models of this infection and the hurdles inherent in the study of subclinical infections in pregnant women and silent transmission to the developing fetus. However, as in the immunocompromised host, maternal immunity and presumably, passively acquired immunity in the fetus, modify the infection and have prompted interests in developing prophylactic vaccines. However, recent studies have demonstrated that pre-existing maternal immunity is incomplete in the prevention of infection and disease suggesting that approaches that induce levels of immunity that are present in naturally immune individuals may be insufficient to prevent congenital HCMV infection and disease. Selective use of currently available antiviral therapies has provided a significant improvement in outcome of HCMV infections in immunocompromised hosts.

Introduction: spectrum of disease and role of virus replication

The clinical importance of human cytomegalovirus (HCMV) as the aetiological agent of cytomegalic inclusion disease (CID) (Jesionek and Kiolemenoglou, 1904) was unrecognized until this virus was isolated from infants in which the infection was present at birth (congenital) and therefore acquired *in utero* (Rowe *et al.*, 1956; Smith, 1956; Craig *et al.*, 1957; Weller, 1971). Although identification of HCMV and the development of tissue culture systems that enabled recovery of this virus from patient specimens accelerated studies of the natural history of this infection in the late 1960s and in the early 1970s, more contemporary technologies have not only extended the findings reported in these earlier studies but have resulted in reinterpretation of results from many of the original studies. The recognition of HCMV as a pathogen in populations of immunocompromised allograft recipients quickly established the importance of HCMV as a cause of disease in populations other than newborn infants (Rifkind, 1965; Ho, 1977; Rubin *et al.*, 1977). The recognition that immunosuppression was the most significant risk factor for HCMV associated disease also led to the early recognition of HCMV as an important cause of opportunistic infections in HIV-infected patients (Drew *et al.*, 1981; Gottlieb *et al.*, 1981; Drew and Mintz, 1984; Spector *et al.*, 1984, 1999; Gallant *et al.*, 1992). Even with the widespread use of highly active retroviral therapy in HIV-infected patients, HCMV continues to represent pathogen in these patients (Huppmann and Orenstein, 2010; Lemonovich and Watkins, 2011). Through studies in these populations that share the commonality of altered immune function, the broad spectrum of disease caused by HCMV became well described. The management of HCMV infection in many of these populations has been accomplished, though with varying degrees of success. These categories of disease associated with HCMV infection also exhibit common virological characteristics that include high levels of virus replication with easily detectable virus excretion and end-organ disease that can be attributed to virus replication and expression of lytic viral genes. More recently, new associations of human disease with HCMV infection have been proposed. These include (1) vascular disease in allograft recipients as well as in normal individuals (Ho *et al.*, 1991; Muhlestein *et al.*, 2000; Zhu *et al.*, 2000; Blankenberg *et al.*, 2001; Weis *et al.*, 2004), (2) chronic inflammatory diseases such as inflammatory bowel disease (Rahbar *et al.*, 2003; Hommes *et al.*, 2004; Xavier and Podolsky, 2007; Domenech *et al.*, 2008; Onyeagocha *et al.*, 2009; Lawlor and Moss, 2010), (3) accelerated immune senescence in the elderly (Almanzar *et al.*, 2005; Faist *et al.*, 2010; Brunner *et al.*, 2011; Pawelec and Derhovanessian, 2011) and (4) the development or phenotypic modification of the behaviour of malignant tumours (Ho *et al.*, 1991; Muhlestein *et al.*, 2000; Zhu *et al.*, 2000; Blankenberg *et al.*, 2001; Cobbs *et al.*, 2002; Rahbar *et al.*, 2003; Hommes *et al.*, 2004; Weis *et al.*, 2004; Almanzar *et al.*, 2005; Xavier and Podolsky, 2007; Domenech *et al.*, 2008; Mitchell *et al.*, 2008; Prins *et al.*, 2008; Onyeagocha *et al.*, 2009; Faist *et al.*, 2010; Lawlor and Moss, 2010; Brunner *et al.*, 2011; Dziurzynski *et al.*, 2011; Lucas *et al.*, 2011; Pawelec and Derhovanessian, 2011; Sampson and Mitchell, 2011). In contrast to diseases in individuals with altered immunological function, these later diseases are not associated with high levels of sustained virus replication nor the consistent expression of lytic viral genes. Thus, it is likely these diseases require a component of the host response to HCMV for manifestation of the disease phenotype. Recent terminology to describe HCMV disease have included the designation of 'direct' for diseases thought to arise secondary to virus replication and 'indirect' for those that are less convincingly associated with levels of virus replication (Fishman, 2007). Although perhaps useful as an operational definition, the term 'direct' suggests that disease can be attributed to virus -induced damage and the designation of 'indirect' implies that virus gene expression is not directly responsible for the phenotypes of disease. Neither of these inferences has been examined experimentally, except for a study providing evidence against an immunopathological cause of murine CMV interstitial pneumonia in a bone marrow transplantation model (Podlech *et al.*, 2000) and data from the overwhelming majority of clinical studies is observational. Thus, precise mechanisms of diseases in most human infections with HCMV is unknown but presumably consists of both virus induced host cellular dysfunction and components of host responses to this infection. Moreover, the features of the transcriptional programme of HCMV that contributes to disease phenotypes, particularly those that are not directly attributable to lytic virus replication, are not well understood. Further studies of viral gene expression and host responses under such conditions could identify targets for the development of new antiviral agents to also restrict expression of non-lytic viral genes.

For the purposes of this brief discussion of the manifestations of HCMV-associated disease we will instead describe disease syndromes associated with high levels of virus replication and that have been shown to be modified by treatment with antiviral agents that inhibit viral DNA replication. A second group of HCMV-associated disease syndromes include those in which the level of virus replication has not been closely correlated with the disease and for which antiviral agents have not been convincingly demonstrated to alter the course of disease.

Immunity and disease associated with HCMV infection

Perhaps the most consistent risk factor for clinically apparent disease associated with HCMV infection is a depression of adaptive immune responses. Although developmental immaturity of the adaptive immune system has been argued to explain the susceptibility of fetus and newborn infant to HCMV infection, there is little direct evidence to support this claim. Early findings of prolonged period of high levels of virus replication in congenitally infected infants were consistent with deficits in the adaptive immune responses in these infants. In addition, several early reports suggested that congenitally HCMV-infected infants lacked HCMV specific $CD4^+$ T-lymphocyte responses; however, these studies were carried out utilizing methodologies that likely underestimated the $CD4^+$ responses in these infants (Gehrz *et al.*, 1977; Starr *et al.*, 1979; Pass *et al.*, 1981a). More recent studies have suggested that infants with HCMV infection do indeed manifest some evidence of decreased $CD4^+$ T-lymphocyte responsiveness to HCMV antigens compared with adults, although this observation represents an association and not a mechanism of disease in these patients (Tu *et al.*, 2004; Miles *et al.*, 2008). Furthermore, findings from a novel study of congenitally infected infants demonstrated the presence of HCMV specific CD8$^+$ T-lymphocyte responses in preterm infants infected *in utero* with HCMV (Marchant *et al.*, 2003). More definitive studies in immunosuppressed transplant patients, some carried out decades ago, have clearly demonstrated that decreases in adaptive immune responses, particularly HCMV specific T-lymphocytes, represent an independent risk factor for the development of clinically important HCMV infection and end-organ disease (Schooley *et al.*, 1983; Reusser *et al.*, 1991, 1999; Li *et al.*, 1994; Walter *et al.*, 1995; Gallina *et al.*, 1996; Sester *et al.*, 2001; La Rosa *et al.*, 2007; Razonable, 2008; Pourgheysari *et al.*, 2009). Alternatively, the most direct positive evidence for the importance of adaptive immunity and resistance to disease in HCMV-infected hosts comes from original studies that demonstrated protective activity of passively transferred antiviral antibodies or *in vitro* expanded HCMV specific CD8$^+$ T-lymphocytes in allograft recipients (Riddell and Greenberg, 1994, 2000; Walter *et al.*, 1995). These studies complemented a large number of earlier studies that detailed the risk of HCMV disease in immunosuppressed patients as a function of the loss of T-lymphocyte responses to HCMV (Rubin *et al.*, 1979; Pass *et al.*, 1981b; Reusser *et al.*, 1991; Li *et al.*, 1994). Similarly, the loss of HCMV specific T-lymphocyte response was shown to be a key risk factor in the development of HCMV end-organ disease in HIV-infected patients (Autran *et al.*, 1997; Komanduri *et al.*, 1998; Bronke *et al.*, 2005). Interestingly, the loss of CD4$^+$ T-lymphocytes in HIV-infected patients was also associated with impaired anti-HCMV antibody responses, and limited data argued for an important role of anti-HCMV antibody activity and HCMV-associated disease progression in these patients (Dylewski *et al.*, 1983; Rasmussen *et al.*, 1994; Boppana *et al.*, 1995). In solid organ transplant recipients, the transplantation of an organ from a donor with HCMV into a non-infected recipient (D^+R^-) is associated with the highest risk of clinically significant infection, presumably because HCMV specific adaptive immune responses must develop in these patients in the face of potent immunosuppression, much of which targets adaptive immune functions. This observation provided further evidence for the critical role of adaptive immunity in the control of HCMV replication, particularly in HCMV infections characterized by high levels of virus replication.

In patients with compromised adaptive immune responses, HCMV infections can result in a variety of manifestations of end-organ damage. These include pneumonitis, hepatitis, colitis, retinitis, encephalitis, and bone marrow dysfunction. Disease is assumed to be secondary to the cytopathic effect of lytic virus replication, an assumption based on finding of viral inclusions and necrotic damage in infected organs. Treatment of patients with disease and active HCMV replication with antivirals such as ganciclovir results in decreased levels of virus and improvement in clinical symptoms and laboratory abnormalities. Together these observations strongly argue for a role of lytic virus gene expression in HCMV disease. In addition, these findings also provide convincing evidence of the role of the adaptive immune response in the control of virus replication in the normal host. However, it is also of interest that the clinical severity of disease often appears to exceed the degree of organ involvement even in patients with profound depression of adaptive immune responses and HCMV disease associated with high levels of virus replication (Myerson *et al.*, 1984; Arribas *et al.*, 1996; Evans *et al.*, 1999; Lautenschlager *et al.*, 2006). This observation suggests that residual host adaptive and/or innate responses contribute to disease in these patients. Future studies that dissect these responses could provide insight into the disease syndromes that are associated with low levels of virus replication and restricted viral gene expression.

Expression of disease independent of overt virus replication

In contrast to disease syndromes associated with high levels of virus replication, diseases that have been linked to low levels of virus replication and restricted viral gene expression often cannot be unambiguously related to specific organ dysfunction readily attributable to HCMV infection. Furthermore, it remains uncertain if antiviral therapy can impact the disease course. In many cases, HCMV appears to function as co-factor in the development of disease or alternatively, in disease progression. Because the contribution of HCMV to several diseases could be minor in comparison to the primary mechanism of disease, quantifying the role of HCMV in natural history of diseases in the normal host such as atherosclerosis (Adam *et al.*, 1987; Melnick *et al.*, 1993; Epstein *et al.*, 1999; Leinonen and Saikku, 2000; Streblow *et al.*, 2001; Libby, 2002; Cheng *et al.*, 2009; Gredmark-Russ *et al.*, 2009; Crumpacker, 2010), periodontitis (Contreras and Slots, 2000; Saygun *et al.*, 2004; Cappuyns *et al.*, 2005; Grenier *et al.*, 2009), and more recently, the morbidities related to normal ageing (Adam *et al.*, 1987; Melnick *et al.*, 1993; Epstein *et al.*, 1999; Contreras and Slots, 2000; Leinonen and Saikku, 2000; Streblow *et al.*, 2001; Libby, 2002; Saygun *et al.*, 2004; Cappuyns *et al.*, 2005; Cheng *et al.*, 2009; Gredmark-Russ *et al.*, 2009; Grenier *et al.*, 2009;

Crumpacker, 2010; Roberts *et al.*, 2010; Wang *et al.*, 2010; Moro-Garcia *et al.*, 2011; Wills *et al.*, 2011), has been difficult. Clearly, associations can be made, but definitive evidence is lacking. Perhaps a commonality that links many of these disease associations with HCMV is inflammation that underlies the primary disease. HCMV infection induces a vigorous innate and adaptive immune response and persistence of HCMV leads to chronic stimulation of the immune response. Thus, it is reasonable to suggest that a potential role of HCMV in disease is maintenance of the host inflammatory responses. The multitude of immune evasion functions encoded by HCMV that ensure its persistence in the infected host likely also contribute to its capacity to efficiently express a limited set of viral genes and in some cases, replicate in sites of inflammation. Consistent with this observation has been the finding that host inflammatory response can enhance HCMV replication, although the mechanism(s) associated with this interplay between virus and host has not been determined definitively (Prösch *et al.*, 2002; Benedict *et al.*, 2004; Caposio *et al.*, 2004, 2007; 2011; DeMeritt *et al.*, 2004, 2006; Simon *et al.*, 2005; Zhang *et al.*, 2009; Liu, 2011). A further understanding of the propensity for HCMV to persist in the presence of host inflammation could provide insight into the relative contribution of this virus to the phenotype of diseases that cannot be readily explained by lytic replication of virus. An important population of patients in which HCMV almost certainly plays a role in disease progression is solid organ transplant recipients. In addition to disease syndromes in the early period (acute) after transplantation, HCMV infection in allograft recipients has also been associated with graft loss, most commonly years after transplantation (chronic) (Tolkoff-Rubin *et al.*, 2001; Valantine, 2004; Fishman, 2007; Potena and Valantine, 2007; Dumortier *et al.*, 2008; Potena *et al.*, 2009; Erdbruegger *et al.*, 2011). In the case of renal allografts this takes the form of interstitial inflammation and damage to proximal tubules, changes that have been designated as IF/TA (interstitial fibrosis and tubular atrophy). These pathological changes lead to the loss of the functional capacity of the transplanted kidney whereas in the transplanted heart, HCMV is associated with a vasculopathy characterized by unrelenting concentric narrowing of the coronary arteries. In each of these diseases, HCMV is thought to accelerate ongoing processes and contribute to the loss of a functioning allograft. Understanding the role of HCMV in these diseases and more importantly, developing strategies to block its contribution to disease could represent a major advance in transplantation as it would extend lifetime of the functioning graft.

HCMV and cancer

The potential of HCMV to contribute to the malignant phenotype has been considered for well over 40 years (Geder *et al.*, 1977; Rapp, 1984). Because it lacks a defined transforming function, most investigators have remained sceptical of a possible association with human cancers. More recently, HCMV has been demonstrated in surgically resected malignant gliomas (Cobbs *et al.*, 2002; Mitchell *et al.*, 2008; Lucas *et al.*, 2011). Initially, these reports were met with concerns that detection of viral DNA and viral proteins were secondary to artefacts inherent in the improved sensitivity of the assays that were required for detection of HCMV. Subsequently, several other laboratories have confirmed these initial findings and have extended these results to include other CNS tumours (Baryawno *et al.*, 2011). Although the mechanism by which HCMV entry process could alter the malignant phenotype in human tumours is not understood, studies *in vitro* have suggested that HCMV interactions with cell surface molecules such as the PDGF receptor deliver proliferative or migratory signals to glioma cells, and that the reported induction of the IL-6/STAT3 signalling by the virus-encoded US28 or, alternatively, signalling triggered by HCMV could induce the production of molecules such as thrombospondin or survivin which could facilitate angiogenesis in the tumour (Soroceanu *et al.*, 2008; Maussang *et al.*, 2009; Slinger *et al.*, 2010; Botto *et al.*, 2011). To date, a cohesive mechanism that would explain the role of HCMV in the malignant behaviour of gliomas has not been put forth; however, there is a consensus of a growing number of investigators that HCMV during persistent infection could modify the environment of the tumour and promote the malignant phenotype of these lethal CNS tumours (Michaelis *et al.*, 2010). Alternatively, a definitive experiment has not ruled out the trivial explanation that HCMV is merely a passenger in these tumours as a result of either persistence in cells of this lineage or reactivation of latent virus secondary to the tumour milieu and that the persistent infection established by HCMV does not exert a measurable influence on the phenotype of tumours such as malignant gliomas.

Clinical disease associated with HCMV infection: congenital HCMV infection

Epidemiology of maternal and congenital HCMV infection

HCMV is an important cause of congenital infection and a leading cause of sensorineural hearing loss (SNHL) worldwide. Numerous cross-sectional serological studies dating back from the 1960s have demonstrated that HCMV infection is ubiquitous in humans. Increased rates of HCMV serological reactivity have been demonstrated in non-white and low-income populations in both developed and developing countries (Alford *et al.*, 1980; 1981). The patterns of HCMV acquisition vary greatly based on geographic, socioeconomic, and racial backgrounds and seroprevalence generally increases with age. Studies have shown that most preschool children (>90%) in South America, sub-Saharan Africa, East Asia and India are CMV antibody positive (Colugnati *et al.*, 2007; Staras *et al.*, 2008; Bate *et al.*, 2010; Britt, 2010). In contrast, seroepidemiological surveys in Great Britain and in the USA have found that less than 20% of children of similar age are seropositive (Britt, 2010). A recent study of CMV seroprevalence that utilized samples from the National Health and Examination Survey (NHANES), 1988–2004, showed that the overall age-adjusted CMV seroprevalence in the USA was 50.4% (Bate *et al.*, 2010). That study also showed that CMV seroprevalence was higher among non-Hispanic black children and Mexican-American children compared with non-Hispanic white children (Bate *et al.*, 2010).

Intrauterine transmission and congenital infection

The prevalence of maternal HCMV infection is an important determinant of the frequency and significance of vertical transmission of HCMV in a population. The prevalence of congenital HCMV infection is directly proportional to the maternal seroprevalence such that populations with higher seropositivity have higher rates of congenital infection (Alford *et al.*, 1981; Stagno *et al.*, 1982a; Wang *et al.*, 2011). The natural history of HCMV infection during pregnancy is complex and has not been defined completely. Unlike rubella and toxoplasmosis where intrauterine transmission is thought to occur as a result of primary infection acquired during pregnancy, congenital HCMV infection has been shown to also occur in children born to mothers who have had HCMV infection prior to pregnancy (non-primary infection) (Stagno *et al.*, 1977; Peckham *et al.*, 1983; Ahlfors *et al.*, 1984, 1999; Morris *et al.*, 1994; Gaytant *et al.*, 2003). In fact, congenital HCMV infection following a non-primary maternal infection has been shown to be common accounting for two-thirds to three-quarters of all congenital infections in highly seroimmune populations (Stagno *et al.*, 1977; 1982a; Schopfer *et al.*, 1978; Peckham *et al.*, 1983; Ahlfors *et al.*, 1984; Wang *et al.*, 2011). Studies of risk factors for congenital HCMV infection have documented the association between young maternal age and increased rates of congenital HCMV infection (White *et al.*, 1989; Fowler *et al.*, 1993). In addition, non-white race and single marital status were independently associated with increased risk of congenital HCMV infection (Preece *et al.*, 1986; Walmus *et al.*, 1988; White *et al.*, 1989; Fowler *et al.*, 1993; Kenneson and Cannon, 2007).

It has been shown that maternal immunity to HCMV can limit intrauterine transmission and fetal infection (Stagno *et al.*, 1982a,b; 1986; Preece *et al.*, 1983; Britt, 2010). This argument is supported by numerous studies that suggest that the rate of intrauterine infection is on the order of 1–1.5% in women with existing HCMV immunity (non-primary maternal infections), a rate about 20- to 30-fold less than in women with primary infection during pregnancy (Stagno *et al.*, 1977; 1982b; 1986; Britt, 2010). Yet, the interpretation of results from these studies has rested on the assumption that congenital HCMV infections following non-primary maternal infections result from a recurrence of infection in the mother and that risk for recurrence is similar for all women at all times during pregnancy. Although the later parameters of infections have not been well studied, recent studies have clearly demonstrated that reinfection following exposure to a new strain of virus can lead to intrauterine transmission and congenital infection (Boppana *et al.*, 2001; Yamamoto *et al.*, 2010). Because the incidence of reinfection remains undefined in maternal populations during pregnancy, it is unclear how frequently reinfection of previously HCMV immune women can lead to intrauterine transmission and congenital infection. Thus, it is unclear if maternal immunity can reduce intrauterine transmission rates by 20- to 30-fold as has been claimed.

Transmission to the developing fetus is believed to occur through haematogenous spread of infectious virus to fetal blood at the placental interface lying between maternal and fetal blood circulation (see also Chapter II.4). It is unknown if cell-free or cell-associated virus is transmitted to the developing fetus but because a primary function of the placenta is to limit transfer of maternal cells into the fetal circulation, it has been assumed that cell-free virus is transmitted to the fetus. Virus could initially infect the fetal liver and after amplification in the liver disseminate to other organs, or alternatively disseminate to fetal end-organs during the primary viraemia. Importantly, the key characteristics of this infection that result in infection of specific organs, particularly the fetal brain, are unknown. Whether widespread dissemination and replication in end organs is secondary to unique viral genotypes which exhibit extended tropism, the quantity of virus in the fetal blood, or the duration of a threshold level of viraemia required for efficient dissemination is unknown. Clearly, there are viral determinants of extended tropism but it must be assumed that such viral-encoded functions are present in the vast majority of viral strains that circulate within the community and recent findings have demonstrated that multiple viral genotypes are likely present within a single infection suggesting that it is unlikely that specific viral genotypes of HCMV can account for phenotypic variations that follow fetal infection (Renzette *et al.*, 2011; Ross *et al.*, 2011). Developing an understanding of the parameters for dissemination of virus to organs in the fetus, particularly the CNS, is critical for both understanding the pathogenesis of the fetal infection but perhaps more importantly, for the rational design of prophylactic vaccines and potentially other interventions such as targeted biologics. Animal models of neuroinvasion represent the only feasible approach to elucidate mechanisms of virus entry into the CNS as well as offering the only experimental systems available for testing potential therapies. As an example, recent studies in a murine model of neuroinvasion have provided some limited understanding into the mechanism of antiviral antibody protection from infection and clearance of replicating virus (see Chapter II.6). Antibodies passively transferred prior to virus dissemination to the brain dramatically altered the level of virus replication in the brain and decreased the number of virus-induced foci of inflammatory cells, a marker of encephalitis in this model (Cekinovic *et al.*, 2008). Perhaps more interesting were the results in animals receiving passively transferred antiviral antibodies after CNS infection was established. In these animals antiviral antibodies appeared to curtail virus replication in the CNS, suggesting a role of antibodies in both prevention of neuroinvasion and in clearance of the infection (Cekinovic *et al.*, 2008) and similar findings were reported for the role of antibodies in confining an established liver infection in an immunocompromised host mouse model (Wirtz *et al.*, 2008). Elucidation of the mechanism(s) of action of antiviral antibodies could provide insight into similar function(s) of passively acquired antiviral antibodies in the developing human fetus, thus identifying targets of protective responses and measures of protective antibody functions.

Autopsy findings in congenitally infected infants have included infection of all organ systems with characteristic HCMV inclusions in cells from the liver, spleen, lung, adrenals, pancreas, thyroid, colon, heart, and brain (Becroft, 1981; Gabrielli *et al.*, 2009; Britt, 2010). In most cases it can be assumed that these lesions developed *in utero*, an inference supported by (1) autopsy findings of stillborn infants, (2) ultrasound imaging findings of unborn infants and (3) the occurrence of congenital ascites in infants with congenital HCMV infection (Fowler *et al.*, 1992; Ancora *et al.*, 2007; Benoist *et al.*, 2008; Guerra *et al.*, 2008; Iwasenko *et al.*, 2011). However, similar findings can develop in the postnatal period in infants who were presumably infected late in gestation. These findings, including histopathological findings, are similar to those that have been described in murine and guinea pig models of congenital HCMV infection and strongly argue that these models faithfully recapitulate characteristic pathological features of the human disease (Ahlfors *et al.*, 1984; Harris *et al.*, 1984a; Woolf, 1991; Harrison *et al.*, 1995; Bourne *et al.*, 2001; Chatterjee *et al.*, 2001). Likewise, premature infants infected by ingestion of HCMV-containing breast milk or as the result of HCMV-positive blood products exhibit similar organ involvement with the exception of clinically apparent CNS disease (Adler *et al.*, 1983; Dworsky *et al.*, 1983; Yeager, 1983; Nankervis and Bhumbra, 1986; Maschmann *et al.*, 2001; Takahashi *et al.*, 2007; Hamprecht *et al.*, 2008; Hamele *et al.*, 2010). Interestingly, CNS involvement and long-term sequelae have not been reported in infants infected in the perinatal period, findings that have suggested that the developmental status of the CNS and its vascular supply could play a critical role in the susceptibility of the CNS to infection by HCMV (Vochem *et al.*, 1998; Hamprecht *et al.*, 2008). Whether such an explanation will be consistent with findings in extremely premature infants infected with HCMV following breast milk exposure awaits further study. From existing data, it is clear that the disease and long-term sequelae that follow intrauterine exposure to HCMV could be dependent on the highly complex interplay between the nature of the virus infection, the host (both maternal and fetal) immune responses, and the developmental status of fetal CNS that will determine the final phenotypic characteristics of the congenital infection. To model these variables in observational or even in targeted interventional studies in this relatively infrequent infection represents a challenging task.

Clinical findings at birth

The majority of the 20,000 to 30,000 children born with congenital HCMV infection each year in the USA (approximately 85–90%) do not exhibit clinical abnormalities at birth (asymptomatic congenital HCMV infection) (Peckham *et al.*, 1983; Yow *et al.*, 1988; Boppana *et al.*, 1992, 1999; Ahlfors *et al.*, 1999). The remaining 10–15% born with clinical abnormalities are categorized as having clinically apparent or symptomatic congenital infection. The infection involves multiple organ systems with particular predilection for the reticuloendothelial and central nervous system (Table II.1.1) (Boppana *et al.*, 1992). The most commonly observed physical signs are petechiae, jaundice and hepatosplenomegaly with neurological abnormalities such as microcephaly and lethargy affecting a significant proportion of symptomatic children. Ophthalmological examination is abnormal in approximately 10%, with chorioretinitis and/or optic atrophy most commonly observed (Conboy *et al.*, 1987; Boppana *et al.*, 1992; Kylat *et al.*, 2006). Approximately half of symptomatic children are small for gestational age and one-third are born prior to 38 weeks' gestational age. Early studies suggested that approximately 10% of symptomatic infants die in the newborn period. However, more recent data suggest that the mortality rate is probably less than 5% (Demmler, 1991; Boppana *et al.*, 1992; Ahlfors *et al.*, 1999; Dollard *et al.*, 2007).

However, the exact prevalence of symptomatic congenital HCMV infection and the disease severity are difficult to ascertain because of the lack of standard criteria for categorizing congenitally infected infants as symptomatic. For example, some studies have included infants with low birthweight or small for gestational age as the only clinical finding as symptomatic whereas other studies did not (Dworsky *et al.*, 1983; Demmler, 1991; Boppana *et al.*, 1992). In addition, children with symptomatic congenital HCMV infection identified on newborn screening generally have a milder newborn disease compared with those referred to studies based on clinical manifestations (Boppana *et al.*, 1992). Therefore, published studies that included large numbers of referral infants may have overestimated the proportion of symptomatic infants and the severity of newborn disease among children with congenital HCMV infection. Since symptomatic children are at much higher risk for long-term sequelae, it is important to have precise estimates of symptomatic infection for better counselling of parents and to target infected infants at increased risk for adverse long-term outcome for closer monitoring so that interventions can be provided during critical stages of development. The typical CID is seen less often in the screened population and therefore, the overall risk for permanent sequelae may be somewhat lower in symptomatic children identified on newborn screening than those referred to studies. Although it was thought based on reports during the 1980s and 1990s most (> 80%) symptomatic children will develop sequelae, the findings of recent studies that included more infants identified on newborn screening revealed that only about half of symptomatic children develop sequelae (McCracken *et al.*, 1969; Pass *et al.*, 1980; Conboy *et al.*, 1987; Ahlfors *et al.*, 1999; Boppana *et al.*, 1999; Grosse *et al.*, 2006). Hence, the lack of reliable estimates of both symptomatic infection and long-term sequelae in congenitally infected children identified on newborn HCMV screening is an important gap in our knowledge and need to be examined in future studies.

Table II.1.1 Clinical findings in 106 infants with symptomatic congenital HCMV infection in the newborn period

Abnormality	Positive/total examined (%)
Prematurity[a]	36/106 (34)
Small for gestational age[b]	56/106 (50)
Petechiae	80/106 (76)
Jaundice	69/103 (67)
Hepatosplenomegaly	63/105 (60)
Purpura	14/105 (13)
Microcephaly[c]	54/102 (53)
Lethargy/hypotonia	25/104 (27)
Poor suck	20/103 (19)
Seizures	7/105 (7)

Adapted from Boppana, S.B., Pass, R.F., Britt, W.J., Stagno, S., and Alford, C.A. (1992). Symptomatic congenital cytomegalovirus infection: neonatal morbidity and mortality. Pediatr. Infect. Dis. J. *11*, 93–99; with permission.

[a]Gestational age less than 38 weeks.
[b]Weight less than 10th percentile for gestational age.
[c]Head circumference less than 10th percentile.

Laboratory findings in children with symptomatic infection reflect the involvement of the hepatobiliary and reticuloendothelial systems and include conjugated hyperbilirubinaemia, thrombocytopenia and elevations of hepatic transaminases in over half of symptomatic newborns. Transaminases and bilirubin levels typically peak within the first 2 weeks of life and can remain elevated for several weeks thereafter while thrombocytopenia reaches its nadir by the second week of life and normalizes within 3–4 weeks of age (Conboy *et al.*, 1987; Boppana *et al.*, 1992; Kylat *et al.*, 2006). Radiographic imaging of the brain is abnormal in approximately 50–70% of children with symptomatic infection at birth. The most common finding is intracranial calcifications, with ventricular dilatation, cysts, and lenticulostriate vasculopathy.

Long-term outcome

A detailed discussion of long-term outcome in children with congenital HCMV infection is provided in the next chapters (II.2 and II.3). Briefly, studies have shown that approximately half of the infants with symptomatic infection will develop sensorineural hearing loss (SNHL), mental retardation with IQs less than 70, and microcephaly (Williamson *et al.*, 1982; Conboy *et al.*, 1987; Demmler, 1991; Boppana *et al.*, 1992, 1999; Fowler *et al.*, 1992; Ahlfors *et al.*, 1999; Dahle *et al.*, 2000; Kylat *et al.*, 2006; Ross *et al.*, 2006). Predictors of adverse neurological outcome in children with symptomatic congenital HCMV infection include microcephaly, chorioretinitis, the presence of other neurological abnormalities at birth or in early infancy, and cranial imaging abnormalities detected within the first month of life (Williamson *et al.*, 1982; Conboy *et al.*, 1987; Boppana *et al.*, 1997, 1999; Noyola *et al.*, 2001; Rivera *et al.*, 2002; Ancora *et al.*, 2007). A study examining the predictors of hearing loss in 190 children with symptomatic infection revealed that intrauterine growth retardation, petechiae, hepatosplenomegaly, hepatitis, thrombocytopenia and intracerebral calcifications were associated with hearing loss (Rivera *et al.*, 2002). However, only petechiae and intrauterine growth restriction were independently predictive of hearing loss on logistic regression analysis (Rivera *et al.*, 2002).

Although asymptomatic children have a better long-term prognosis than children with symptomatic congenital infection, approximately 7% of asymptomatic children will develop SNHL (Table II.1.2). Many prospective studies of children with asymptomatic CMV infection have been performed to define the natural history of hearing loss in this group (Harris *et al.*, 1984b; Williamson *et al.*, 1990, 1992; Fowler *et al.*, 1997; Ahlfors *et al.*, 1999; Dahle *et al.*, 2000). These studies show that approximately one-half of children with asymptomatic infection who develop hearing loss will have bilateral deficits which can vary from mild 'high-frequency hearing loss' to profound hearing impairment. In addition, hearing loss in these children is often progressive and/or of delayed onset requiring ongoing audiological evaluation (Williamson *et al.*, 1990, 1992; Fowler *et al.*, 1997; Dahle *et al.*, 2000). Other neurological complications can also occur in asymptomatic congenital CMV infection but at a lower frequency than in symptomatic infection (Williamson *et al.*, 1990).

Factors associated with disease in congenitally infected children

It has been thought that symptomatic infection with severe newborn disease and long-term sequelae are much more frequent in children born to mothers with primary HCMV infection during pregnancy. However, more recent data argues against this notion (Ahlfors *et al.*, 1999; Ross *et al.*, 2006; Mussi-Pinhata *et al.*, 2009). Data from Ahlfors and colleagues in Sweden, and studies in Alabama as well as in Brazil have shown that symptomatic infection occurs with similar frequency in children born to women with primary CMV infection during pregnancy and those born to women who were HCMV seroimmune prior to pregnancy (Ahlfors *et al.*, 1999; Ross *et al.*, 2006; Mussi-Pinhata *et al.*, 2009). In addition, the severity of newborn disease is also not different between primary and non-primary infection groups (Boppana *et al.*, 1999; Mussi-Pinhata *et al.*, 2009; Yamamoto *et al.*, 2011). As discussed in more detail in the next section, an accumulation of the data from recent studies of seropositive women suggest HCMV reinfection with a different virus strain is common and such reinfection could lead to intrauterine transmission and symptomatic congenital infection.

An early finding from natural history studies of congenital HCMV infections demonstrated that

Table II.1.2 Audiological results for children with congenital cytomegalovirus infection

	Symptomatic	Asymptomatic
No. (%) of children with SNHL	40.7%	7.4%
Bilateral loss	67.1%	47.9%
High-frequency loss only (4000–8000 Hz)	12.9%	37.5%
Delayed-onset loss	27.1%	37.5%
Median age (range) of delayed-onset loss	33 months (6–197)	44 months (24–182)
Progressive loss	54.1%	54.2%
Fluctuating loss	29.4%	54.1%
Improvement of loss	21.1%	47.9%

Adapted from Dahle, A.J., Fowler, K.B., Wright, J.D., Boppana, S.B, Britt, W.J., and Pass, R.F. (2000). Longitudinal investigations of hearing disorders in children with congenital cytomegalovirus. J. Am. Acad. Audiol. *11*, 283–290; with permission.

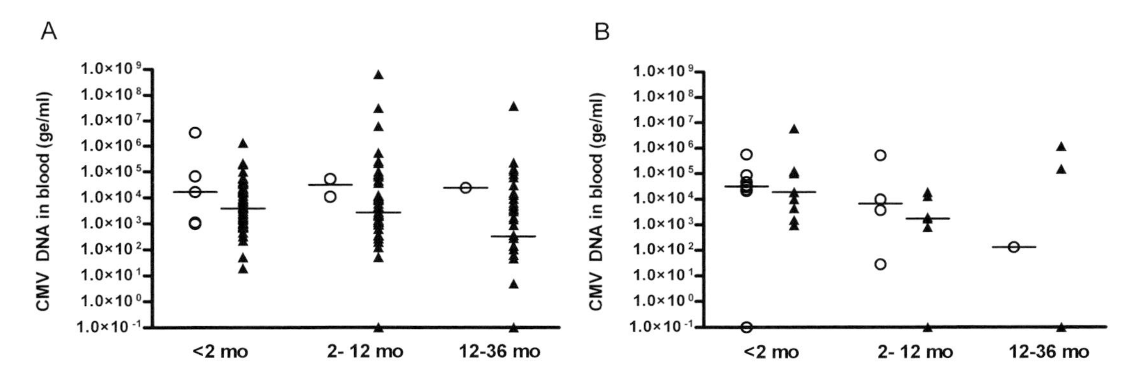

Figure II.1.1 Results of CMV DNA levels in peripheral blood at three different age ranges from children with congenital CMV infection with asymptomatic (A) and symptomatic (B) infection at birth that had hearing loss (○) and normal hearing (▲). The results are expressed as genomic equivalents per ml of blood (ge/ml). The horizontal bars represent median values. Adapted from Ross, S.A., Novak, Z., Fowler, K.B., Arora, N., Britt, W.J., and Boppana, S.B. (2009). Cytomegalovirus blood viral load and hearing loss in young children with congenital infection. Pediatr. Infect. Dis. J. *28*, 588–592; with permission.

infants with symptomatic infection excreted increased amounts of HCMV in their urine compared with infants with asymptomatic infection (Stagno *et al.*, 1983). In addition, monitoring of peripheral blood viral load has been shown to be useful in the clinical management of invasive HCMV infections in immunocompromised individuals including allograft recipients and HIV-infected patients (Gor *et al.*, 1998; Spector *et al.*, 1999; Emery *et al.*, 2000). Together, these findings suggested that quantitation of virus in various body fluids could identify infants with more significant infections as well as potentially identifying infants with a greater likelihood of developing sequelae. More recent studies utilizing quantitative PCR analysis of urine and blood have confirmed these early findings and demonstrated that infants with symptomatic infections have increased levels of viral DNA in these fluids compared with infants with asymptomatic infections (Boppana *et al.*, 2005; Lanari *et al.*, 2006; Ross *et al.*, 2009). Interestingly, the level of virus replication in symptomatic infants was most closely correlated with the presence of hepatic and splenic disease and not the CNS, a finding which raises further questions about the relationship between virus replication and long-term outcome in infants with congenital infection (Boppana *et al.*, 2005; Ross *et al.*, 2009). When a similar approach was used in an attempt to identify infants who would develop sequelae, the variability between the levels of viral DNA in individual patients limited the utility of the approach (Ross *et al.*, 2009). The most recent study which utilized peripheral blood samples from 135 children with congenital infection demonstrated no difference in CMV viral load levels in the first months of life and beyond among children with and without SNHL (Fig. II.1.1) (Ross *et al.*, 2009). As the frequency and natural history of SNHL in children with symptomatic and asymptomatic infection differ, data in the two groups of children were analysed independently (Table II.1.2). These data indicate that in individual children with congenital CMV infection, an elevated viral load measurement may not be useful in identifying a child at risk for CMV-related hearing loss. These findings argued that the development of sequelae following a congenital HCMV infection cannot be related to an absolute value of viral DNA in the blood or urine, a finding consistent with the recurring theme that the development of disease cannot be directly related to an absolute level of virus replication in an infant with congenital HCMV infection.

The gestational age at the time of intrauterine infection represents another characteristic of congenital HCMV infections that has been associated with more significant sequelae (Bodeus *et al.*, 1999; Enders *et al.*, 2001; Revello *et al.*, 2003; Gindes *et al.*, 2008). These studies suggested that maternal seroconversions that occurred in the late first and early second trimester were more often associated with more severe congenital HCMV infections with CNS disease than maternal seroconversions that occurred in the late second trimester and third trimester of pregnancy. These findings are consistent with the presumed vulnerability of the CNS during early development to a damaging congenital HCMV infection. However, the inability to define the timing of virus transmission to the fetus limits definitive interpretation of these studies. As noted previously, studies in the guinea pig model of congenital HCMV infection have suggested that intrauterine transmission of guinea pig CMV occurs during a secondary maternal viraemia and not at the time of the initial viraemia associated with maternal infection (Griffith *et al.*, 1986). Together, data from observational studies of infants with congenital HCMV infections have only hinted at the importance of the level of virus replication and the gestational age of the fetus at the time of infection in the pathogenesis of infection. Currently available data would not support either of these reported associations as being reliable surrogate markers of disease in infants with congenital HCMV infections.

Maternal reinfection and congenital HCMV infection

Soon after HCMV was discovered and CID in infants was described, investigators recognized that pre-existing maternal immunity (non-primary maternal infections) to HCMV did not prevent transmission to the fetus, suggesting that the natural history of congenital HCMV was unique amongst other perinatal infections (Ross *et al.*, 2009). In fact, studies in the USA, Sweden, and Africa all indicated that congenital infections following non-primary maternal infections were responsible for the majority of infants with congenital HCMV infections (Schopfer *et al.*, 1978; Ahlfors *et al.*, 1981; 1983; Stagno *et al.*, 1982a). The rate of congenital infection in these populations appeared to be directly related to the HCMV seroprevalence of the maternal population such that populations with high seroprevalence have higher rates of congenital HCMV infections (Table II.1.3) (Stagno *et al.*, 1982a; van der Sande *et al.*, 2007; Dar *et al.*, 2008; Mussi-Pinhata *et al.*, 2009). An initial explanation for this observation was that increased rates of HCMV infection in maternal populations would be consistent with an increased likelihood of infection during pregnancy of non-immune women and delivery of an infected infant. However, the increased rates of congenital infection in offspring from highly seroimmune women as have been reported in equatorial Africa, Brazil, and India occurred in maternal populations with near universal seroimmunity. This finding differs dramatically to other causes of congenital viral infections such as rubella (Ross *et al.*, 2009). In the case of rubella the incidence of congenital rubella syndrome fell dramatically when the rate of maternal seroimmunity was above 80–85%, suggesting that a sufficient pool of susceptible, seronegative women was necessary to result in maternal infection and transmission to the developing fetus (Stagno *et al.*, 1983). In contrast, the rates of congenital HCMV infection as noted above increase as the rates of maternal seroimmunity increase past 90%.

Although congenital HCMV infections were common in maternal populations with near universal seroimmunity, the prevailing view at that time was that congenital infections following non-primary maternal infections were rarely associated with disease and/or sequelae and thus of insignificant clinical importance than congenital infections associated with primary maternal infections. Yet, there were investigators who reported symptomatic infections in infants born to women with non-primary infections and significant rates of hearing loss in infants born to women with probable non-primary infections (Ahlfors *et al.*, 1981; 1983). When more rigorous epidemiological methodology was applied to the study of congenital HCMV infections (see also Chapter II.2), it became increasingly apparent that many of the early studies were heavily biased by patients referred to single centres and that the natural history of this infection was not derived from screened populations that would eliminate much of this bias

Table II.1.3 Rates of maternal CMV seroprevalence and congenital CMV infection in different populations

Location	Maternal CMV seroprevalence (%)	Congenital CMV infection (%)
Aarhus-Viborg, Denmark	52	0.4
Abidjan, Ivory Coast	100	1.4
Birmingham, USA		
Low income	77	1.25
Middle income	36	0.53
Hamilton, Ontario, Canada	44	0.42
London, UK	56	0.3
Seoul, South Korea	96	1.2
New Delhi, India	99	2.1
Ribeirão Preto, Brazil	96	1.1
Sukuta, The Gambia	96	5.4

(Demmler, 1991; Fowler *et al.*, 1992). From studies in screened populations, investigators have determined that congenitally infected infants born to women following non-primary infections actually account for the majority of HCMV-infected newborns and that the incidence of symptomatic disease and long-term sequelae is similar in infants born following primary and non-primary maternal infections (Ahlfors *et al.*, 1999; Boppana *et al.*, 1999; Mussi-Pinhata *et al.*, 2009; Yamamoto *et al.*, 2011). Some investigators continue to argue that infants infected following a primary maternal infection suffer more severe sequelae than those born to women with non-primary infection; however in many cases, these arguments rely on data from study populations contaminated with referral patients that were identified secondary to the severity of the congenital infection. Finally, recent studies from populations in which maternal seroimmunity exceeds 96% and almost all infected infants are born to women with non-primary infection have shown that the prevalence of symptomatic infection and the development of long-term sequelae in infected infants are very similar to those in reported studies of infants infected following primary maternal infections (Mussi-Pinhata *et al.*, 2009; Yamamoto *et al.*, 2011). Therefore, the contribution of primary maternal and non-primary maternal infections to disease burden appear similar in some populations in northern Europe and the USA but in most populations in the southern hemisphere, the near-universal prevalence of HCMV infection in women of child-bearing age suggests that the primary infections are rare and unlikely to contribute to the public health importance of congenital HCMV infections.

Non-primary maternal infections were initially believed to be secondary to recurrences of pre-existing maternal infection, presumably following reactivation of latent infections. This assumption was based on early studies of the genetic composition of viruses isolated from mother/newborn infant specimens and from the same mothers at time points prior to conception (Huang *et al.*, 1980). Using restriction fragment length polymorphisms of viral DNA, these early studies argued for identity between viral isolates, thus suggesting that non-primary maternal infection could follow reactivation of existing infections and ultimately lead to the infection of the fetus (Huang *et al.*, 1980). Although this approach included the most sophisticated methodology of its time, careful review of the findings from these studies suggests that the identity of viral isolates could be questioned. Secondly, these studies were carried out with labelled viral DNA and therefore required extensive passage *in vitro* to adapt the viral isolates to growth on human fibroblasts and the production of significant amounts of extracellular virus. Oftentimes this required passage of the viral isolates up to 25 times *in vitro*, a process that would put significant pressure on populations of viruses that may have been present in the initial inoculum. Thus, it is uncertain if definitive information was indeed derived from these studies. More recent studies utilizing direct sequencing, including next generation sequencing of viruses isolated from mothers and infants, has demonstrated remarkable genetic diversity, findings which further question the validity of the early findings (Renzette *et al.*, 2011; Ross *et al.*, 2011). Defining the sources of virus in infants infected in women undergoing a non-primary infection represents a challenge.

Two sources can be imagined. The first is recurrence of an existing infection and the second is reinfection. Notably, non-human primate (NHP) models have shown experimentally that reinfection is possible in the face of pre-existing immunity (see Chapters II.21 and II.22). Reinfection of previously immune individuals has been demonstrated in young children attending day-care, individuals with sexually acquired HCMV, transplant recipients, and in women undergoing non-primary infection (Chandler and McDougall, 1986; Bale *et al.*, 1999, 2001; Ishibashi *et al.*, 2007; Yamamoto *et al.*, 2010). Reinfection appears to be a relatively common occurrence in populations with significant exposure to HCMV (Chandler and McDougall, 1986; Bale *et al.*, 1999; Novak *et al.*, 2008). As has been noted previously, the importance of reinfection in the natural history of non-primary maternal infections and subsequent congenital infections has not been adequately defined. Our early studies clearly demonstrated that maternal reinfections can result in a damaging congenital infection, but the parameters of this infection in previously immune women remain poorly defined at present (Table II.1.4) (Boppana *et al.*, 2001). However, symptomatic congenital HCMV infections and the development of hearing loss occur in infants infected following a maternal reinfection, findings that are seemingly at odds with the concept of protective maternal immune responses (Table II.1.5) (Boppana *et al.*, 1999; Yamamoto *et al.*, 2010, 2011). In a recent study, we have shown mixed infection with more than one virus strain in approximately one-third of the infants with congenital HCMV infection and the presence of distinct virus strains in specimens from different compartments from the same child (Ross *et al.*, 2011). This finding demonstrates that there is great diversity in the CMV strains that cause congenital infection and that infection with multiple CMV strains occurs in congenital CMV infection. However, the relationship of specific genotypes and the implications of infection with multiple viral strains in the pathogenesis and long-term outcome in

Table II.1.4 Infection with multiple HCMV strains in mothers according to serological responses to two polymorphic determinants on gH and gB

Variable	Mothers of infected infants, number (%) ($n=40$)	Mothers of uninfected infants, number (%) ($n=109$)	*P*-value
Antibody reactivity against ≥2 HCMV strains at first prenatal visit	14 (35.0)	17 (15.6)	0.009
Seroconversion to new HCMV strain during pregnancy	7 (17.5)	5 (4.6)	0.02
Infection with ≥2 CMV strains before and/or during pregnancy	21 (52.5)	22 (20.2)	<0.0001

Table II.1.5 HCMV-related hearing loss according to type of maternal infection

	Maternal HCMV infection ($n=85$)		
Hearing Status	Primary ($n=3$)	Non-primary ($n=40$)	Indeterminate or samples not available ($n=42$) (95% CI)
Moderate to severe unilateral HL	0	4	1
Moderate to profound bilateral HL	1	2	2
Normal hearing	2	34	39

children with congenital CMV infection are not yet known, and remain an important area of investigation for future studies.

Clinical disease associated with HCMV infection: disease in allograft recipients

As noted in the previous sections, severe HCMV infections in immunosuppressed solid organ and haematopoietic cell transplant (HCT) recipients represent a major complication of transplantation (see also Chapters II.13, II.14 and II.16). Prior to the availability of effective antiviral therapy, mortality rates associated with HCMV pneumonia in some transplant recipients such haematopoietic or cardiac/lung allograft recipients ranged between 20% and 80% and donor/recipient mismatching for HCMV serostatus (D^+/R^-) precluded heart/lung transplantation in some centres (Myers *et al.*, 1975; Wreghitt *et al.*, 1988; Hutter *et al.*, 1989). In addition, the development of HCMV infections early in the post-transplant period in solid organ transplant (SOT) recipients resulted in clinically significant infection that not infrequently contributed to loss of the allograft (Rubin, 1990; Kidd *et al.*, 2000). HCMV clinical syndromes included fever, leucopenia, and non-specific findings (CMV syndrome) as well as a variety of invasive end-organ disease including pneumonitis, colitis, hepatitis, and haematological dysfunction. As noted above, clinical disease in these patients was heralded by increasing levels of virus replication measured by virus excretion in urine, viraemia, viral antigen expression in peripheral blood leucocytes, or perhaps more quantitatively, by PCR in peripheral blood (Schmidt *et al.*, 1991; Fox *et al.*, 1995; Cope *et al.*, 1997; Emery *et al.*, 2000). It is interesting that only ranges but not an absolute level of virus replication that predicts invasive infection has been defined presumably because of variability of patient populations and immunosuppressive protocols between transplant centres that seem to defy standardization. In contrast, increases in the level of virus replication between successive clinical specimens as measured by PCR forewarns of the loss of virus control by host adaptive immune responses and often dictates the treatment with antiviral agents (Emery *et al.*, 2000). This approach requires intensive monitoring of virus replication and institution of antiviral therapy based on evidence of increasing levels of virus or the detection of virus in specific tissues. This has been designated as pre-emptive therapy (Schmidt *et al.*, 1991; Boeckh, 1999; Emery, 2000, 2001; Simmen *et al.*, 2001; Singh, 2006). Importantly, this approach also demands accurate identification in timely fashion of patients whose infection will progress and prior to the development of significant disease. In HCT, the pre-emptive approach has become standard in most centres because of the adverse effects of ganciclovir on graft function in these patients. In contrast to the approach in HCT, many centres utilize antiviral prophylaxis during the first 3–12 months in the post-transplant period of SOT to eliminate significant infections during the period of maximal immunosuppression (Winston *et al.*, 1993;

Valantine *et al.*, 1999; Khoury *et al.*, 2006; Fishman, 2007). Recent studies have demonstrated the success of antiviral prophylaxis in lung transplant recipients and this approach reduces both the incidence and severity of HCMV infections (Palmer *et al.*, 2010) (see also Chapter II.13). The advantages and disadvantages of each approach has been detailed in several publications, including reports that suggest that routine prophylaxis is cost-effective and appears to be associated with improved graft survival (Khoury *et al.*, 2006; Fishman, 2007; Potena and Valantine, 2007; Potena *et al.*, 2009; Kotton *et al.*, 2010). Perhaps one of the most compelling arguments for pre-emptive approaches emerged from studies initially reported from HCT recipients and later from SOT recipients (Zaia *et al.*, 1997; Boeckh *et al.*, 2003; Limaye *et al.*, 2004; Singh, 2005). In these studies the authors suggested that although prophylaxis limited the onset of HCMV-associated disease in the early transplant period, a small but not insignificant percentage of patients developed HCMV disease in the late transplant period (late disease) when the antiviral prophylaxis was discontinued (Boeckh *et al.*, 2003). Risk factors for the development of late HCMV disease included (1) HCMV infection, (2) lymphopenia, (3) failure to reconstitute $CD4^+$ responses to HCMV and (4) graft versus host disease (Boeckh *et al.*, 2003; Lacey *et al.*, 2004). The susceptibility to disease in these patients was thought to be secondary to a failure to develop HCMV specific immunity, specifically T-lymphocyte responses, possibly secondary to the absence of virus replication during engraftment. It has been argued that the pre-emptive approach provides effective antiviral therapy while also allowing *de novo* immune stimulation from replicating HCMV. Yet more recent findings have argued that immune reconstitution of HCMV responses can occur in HCT in the absence of detectable HCMV infection, a finding that argues characteristics intrinsic to the transplanted cell population, efficiency of engraftment, and possibly thymic function of the recipient could contribute to the reconstitution of HCMV immune responses more significantly than the presence of an overt HCMV infection (Storek *et al.*, 1995, 2001, 2008; Hakki *et al.*, 2003; Lacey *et al.*, 2004; Gallez-Hawkins *et al.*, 2005; Clave *et al.*, 2009). Similar observations in SOT recipients has led to the description of late or delayed-onset CMV infection as a clinical entity in SOT recipients (Limaye *et al.*, 2004; Arthurs *et al.*, 2008; Husain *et al.*, 2009; Boudreault *et al.*, 2011). In studies of HCT recipients, the acquisition of HCMV specific $CD8^+$ T-lymphocyte responses (and probably more importantly, HCMV specific $CD4^+$ T-lymphocyte responses) have been correlated with resistance to disease but not infection. In one of the larger studies carried out in HCT recipients, investigators monitored the presence of HCMV specific $CD8^+$ T-lymphocyte reactivity using tetramers during the first year post transplantation and found that the delay in detection of tetramer positive cells by day 65 post transplantation was associated with an increased risk of HCMV disease (Gratama *et al.*, 2010). Similar findings were reported in SOT recipients when monitored with an assay of IFN-γ production by peptide stimulation of peripheral blood (Onyeagocha *et al.*, 2009). Monitoring these responses could allow withholding antiviral therapy while limiting the risk of invasive disease and permitting the recipient's immune system to establish protective immunity to HCMV. As an example of this approach, Gerna and colleagues have recently described their approach to the identification and treatment of cardiac transplant recipients at risk for invasive HCMV disease that includes precise monitoring of virus replication and the development of both $CD8^+$ and $CD4^+$ HCMV-specific T-lymphocyte responses (Gerna *et al.*, 2011). Their findings suggested that patients at risk for invasive disease could be identified using these criteria and therefore, candidates for antiviral therapy. It should be anticipated that these and other studies will lead to standardization of antiviral treatment only in patients who are at risk for invasive disease.

Although HCMV infection likely occurs in nearly all SOT recipients (> 85%) in which either the donor and/or recipient is infected prior to transplantation, clinically significant infections are more likely to develop in patients with identifiable risk factors. These include patients with mismatches between donor and recipient HCMV serology such that an organ from an HCMV-infected host is transplanted into a non-infected recipient and secondly, the degree of immunosuppression following transplantation. In seronegative SOT recipients, transplantation of an organ from a CMV-seropositive donor can result in a severe CMV infection in 60% of recipients in some clinical reports and almost always requires treatment with an antiviral. In this group of patients, development of an effective adaptive immune response, both T-lymphocyte and antiviral antibodies, can require weeks as well as be delayed by immunosuppression required for graft survival in the early post-transplant period. Thus, these patients are often at risk for clinically significant infections caused by CMV. In contrast, transplantation of an organ from a seropositive or seronegative donor into a seropositive recipient may result in infection as detected by PCR but no clinical disease. However, in the case of HCT recipients, the conditioning required for successful transplantation and/or ablative therapy for existing malignancies results in severely immunocompromised recipients who are at high risk for invasive HCMV infections. Similarly, the use of potent immunosuppressive medications including agents that deplete T-lymphocytes and block activation of adaptive responses to limit either graft versus host or host versus graft reactions, place allograft recipients at increased risk for invasive HCMV infection. This is particularly the case in transplant recipients that require depletion of T-lymphocytes and/or large doses of corticosteroids. Conversely, it has been argued that some immunosuppressive agents such as mTOR inhibitors could be associated with a decreased risk of significant HCMV infection, perhaps through intrinsic antiviral activity of these compounds (Moorman and Shenk, 2010). A recent report provided evidence that use of mTOR inhibitors in place of mycophenolate in immunosuppression protocols resulted in fewer CMV-associated events in all groups of renal allograft recipients except for CMV mismatched donor$^+$/recipient$^-$ transplants (Brennan *et al.*, 2011). Thus, when evaluating findings from clinical studies of HCMV infection and disease in transplant populations, careful attention must be paid to the immunosuppression regimen employed at individual transplantation centres.

The clinical syndromes associated with HCMV infection during the early post-transplant period are consistent with disease associated with high levels of virus replication that have been described above and extensive literature describing such infections in both haematopoietic and solid-organ transplant recipients is available. As noted above, treatment with effective antivirals decreases virus replication and leads to resolution of clinical signs and symptoms, which has clearly suggested a direct role for lytic viral gene expression in disease in these patients. In fact, although HCMV infection still represents an important cause of morbidity in the early post-transplant period, the evolving role of effective antiviral prophylaxis and more refined pre-emptive treatment protocols could suggest that at least in SOT, HCMV infection in the early transplant period represents a very manageable clinical problem. In contrast to its role in the early transplant period, chronic allograft rejection that is associated with HCMV infections during the post-transplant period has become of major importance and could eventually represent the major morbidity associated with HCMV in this population. Graft survival in the early post-transplant period has greatly improved with the availability of newer and more potent pharmaceuticals to suppress alloreactivity, but chronic allograft rejection leading to loss of functional graft continues to represent a major clinical problem in SOT.

A hallmark of chronic allograft rejection in cardiac allografts has been termed cardiac allograft vasculopathy (CAV) (see also Chapter II.13). HCMV infection has been associated with CAV in epidemiological studies and treatment of the cardiac allograft recipients with the antiviral ganciclovir has been reported to result in a reduction in the incidence of CAV (Grattan *et al.*, 1989; Colvin-Adams and Agnihotri, 2005; Potena and Valantine, 2007; Potena *et al.*, 2009). Together, these and other observations have argued for a role of HCMV in this post-transplant complication. More recently, small animal models of cardiac allograft vasculopathy have provided data consistent with the role of HCMV in this disease (Lemstrom *et al.*, 1993, 1995; Streblow *et al.*, 2005, 2008) (see also Chapter II.15). An important observation has suggested that prolonged prophylaxis with ganciclovir can reduce virus replication to minimal levels in cardiac allograft recipients and decrease the incidence of CAV, thus directly implicating HCMV in the development of CAV in cardiac allograft recipients (Potena *et al.*, 2009).

The importance of CAV as a complication in heart transplantation cannot be overstated in that it reportedly affects between 30% and 50% of cardiac allograft recipients during the first 10 years post transplantation and accounts for 32% of deaths by the fifth post-transplant year (Colvin-Adams and Agnihotri, 2005; Mehra, 2006). The pathogenesis of this disease is thought to include an initial step of endothelial damage and inflammation followed by intimal and myofibroblast proliferation and fibrosis leading to concentric narrowing of the coronary arteries of the allograft (Mehra, 2006). Endothelial damage and dysfunction appears to be a constant component of all proposed mechanisms of CAV. Mechanisms leading to endothelial damage range from antibody- and/or cellular immune-mediated damage (including CD8$^+$ T-lymphocytes and possibly NK cells) of the endothelium to mechanisms implicating HCMV in the inhibition of nitric oxide synthesis resulting in the loss of physiological vasodilatation of the coronary arteries (Fearon *et al.*, 2007). As with other investigations in humans with chronic disease, a precise contribution of HCMV to CAV cannot be easily assigned secondarily to the complexity of transplant recipients who often also have multiple underlying disease processes that could contribute to ongoing vasculature disease as well as the requirement of treatment with potent immunosuppressive drugs that could also contribute to this disease. Furthermore, CAV develops and progresses during an ongoing alloreaction. For these reasons it has been difficult to demonstrate a singular role of HCMV in this disease and similarly, difficult to dissect the mechanism(s) of disease associated with HCMV infection. Fortunately, studies in animal models of cardiac allograft transplantation have clearly demonstrated that infection with CMV leads to decreased graft survival and the development of histopathological changes that resemble findings reported in human cardiac allografts (Koskinen *et al.*, 1993; Lemstrom *et al.*, 1995; Streblow *et al.*, 2005, 2008). These models have already allowed a more controlled investigation of this particular disease. Generally, such model systems allow hypothesis testing that could never be accomplished in a clinical setting.

HCMV has also been shown to play a role in chronic renal allograft rejection (see also Chapter II.14), which has been more accurately described as interstitial inflammation and fibrosis (IF) with tubular atrophy (TA) leading to loss of proximal tubules, a collection of histological findings that are also described in animal models of renal allograft transplantation (Lautenschlager *et al.*, 1999; Krogerus *et al.*, 2001; Inkinen *et al.*, 2002, 2005a,b; Solez *et al.*, 2008; Mannon *et al.*, 2010; Racusen and Regele, 2010; Botto *et al.*, 2011). A discussion of the mechanisms that could account for IF/TA and chronic dysfunction in renal allografts is beyond the scope of this overview, but it should be noted that a specific role of HCMV in this chronic inflammatory process has not been universally accepted and continues to remain somewhat contentious (Erdbruegger *et al.*, 2011). It is important to note, however, that HCMV infection in allograft recipients has been consistently linked to an increased incidence of graft dysfunction in renal allograft recipients in both the early and late post-transplant period, findings that cannot be readily dismissed. Unfortunately, a coherent mechanism that could explain the role of HCMV in chronic allograft rejection has not emerged from studies in humans. As a result, it is likely that animal models of this disease will hold the key as they will allow hypothesis testing under more controlled experimental systems. Of the proposed mechanisms, persistent inflammation represents a commonality and at least at some level, viral gene expression appears to be required for continued inflammatory responses in renal allografts undergoing chronic rejection. Consistent with the latter have been studies demonstrating very restricted HCMV gene expression in animals models of cardiac allograft rejection and studies in humans that have demonstrated HCMV encoded proteins in endothelial cells of aorta (Streblow *et al.*, 2007; Gredmark-Russ *et al.*, 2009). Although the studies in animal models can be interpreted with some confidence, the findings in human must be viewed with the caveats that surround most studies in human allograft recipients and include obvious variables such as the heterogeneity of immune competence of the patient, the duration of HCMV infection, and the initial status of the allograft. Intuitively, the presence of HCMV in the allograft could contribute to the chronic inflammation associated with the alloreaction, even with only restricted expression of viral genes as has been demonstrated in small animal models of HCMV-associated graft dysfunction (Streblow *et al.*, 2007). Moreover, the presence of viral gene products could influence not only the intensity but the nature of the alloreaction by as yet undefined mechanisms. As an example, it has been shown recently that expression of the HCMV IE-1 protein in renal epithelial cells can lead to increased activity of TGF-β, resulting in an increase in epithelial–mesenchymal transition (EMT) of these cells (Shimamura *et al.*, 2010). Because EMT represents a phenotypic characteristic of epithelial cells that has been associated with fibrosis, these findings could provide a mechanism by which restricted gene expression in HCMV-infected renal tubular epithelium within an allograft could potentiate fibrosis associated with inflammation in the allograft. Furthermore, HCMV infection of allografts in small animal models has been shown to induce increased angiogenesis as well as transcription programmes that resemble that of wound healing (Streblow *et al.*, 2005, 2008; Dumortier *et al.*, 2008; Caposio *et al.*, 2011). Thus, there is a large body of evidence that points to HCMV as a proximal cause of chronic allograft dysfunction and IF/TA in renal allografts. Specific mechanisms leading to graft loss associated with HCMV infection will likely be defined in animal models and will await validation in human transplant populations.

Concluding remarks

The advent of effective antiviral therapy in the treatment of HCMV infection in immunocompromised hosts represented an unimaginable clinical advance for physicians responsible for care of these patients (see Chapter II.19). Over the ensuing decades, significant improvements in viral diagnostics have permitted rapid identification of patients at risk for invasive disease as well a monitoring of the efficacy of antiviral drug therapy in patients (see Chapter II.18). The formulation of ganciclovir into an oral preparation with satisfactory bioavailability has greatly expanded the potential to use this medication as prophylaxis in patients at risk for invasive HCMV disease. In addition, the availability of this medication has permitted extending its use into paediatric patients that may be at risk for hearing loss secondary to congenital HCMV infection. Several new agents currently under development and some that are in clinical trials suggest that the continued progress in effective antiviral therapy for invasive HCMV disease will continue, albeit perhaps not with the same trajectory that was seen with the introduction of ganciclovir and foscarnet into clinical medicine.

An advance in the prevention and treatment of congenital HCMV infections of similar size as that accomplished in transplant patients has not been reached. Ideally, an effective antiviral that could be safely administered to pregnant women and that would limit CNS damage associated with intrauterine HCMV infection could help reduce the incidence of disease associated with this infection. Perhaps such a quest will lead to an effective immunoglobulin preparation or alternatively, a biological such as a potent HCMV neutralizing monoclonal antibody that can be safely given to pregnant women who are at risk for delivering an infected infant with CNS disease. Alternatively, such products could be given to all women at risk for HCMV infection during pregnancy much as similar products were given to women at risk for Rh factor sensitization. However, each of these scenarios depend on accurate identification of pregnant women at risk for HCMV transmission to their fetus and possibly those women in whom intrauterine transmission would result in a clinically significant infection in the fetus leading to long-term sequelae. Since such cases occur in somewhere around 2–4% of infants with congenital HCMV infection (or alternatively in 1–2/10,000 live births based on prevalence of 5/1000 of congenital HCMV in the USA), it is imperative that screening tests identify women at risk with both high specificity and sensitivity. In addition, a significant number of infants with congenital HCMV and CNS disease are born to women with preconceptional immunity (non-primary infections); therefore such screening tests cannot rely entirely on the identification of women with primary exposure to HCMV. These features of the unique natural history of congenital HCMV infection present a challenge to the goal of limiting disease associated with this perinatal infection. Lastly, the promise of vaccine induced immunity to limit the disease burden that results from congenital HCMV infection remains unfulfilled. Although optimistic discussions have been and will continue to be written, this area of investigation appears to be in its infancy when one critically evaluates the progress that has been made (for the actual state of vaccine development, see Chapter II.20). Protective immunity to HCMV has been estimated from studies in experimental animal models and from extrapolation from unique clinical situations but none from studies in pregnant women. This coupled to the findings that reinfections are not infrequent and can result in fetal transmission and disease, suggests that fundamental knowledge about the role of maternal immunity in congenital HCMV infection must be secured before vaccinology can move forward.

References

Adam, E., Melnick, J.L., Probtsfield, J.L., Petrie, B.L., Burek, J., Bailey, K.R., McCollum, C.H., and DeBakey, M.E. (1987). High levels of cytomegalovirus antibody in patients requiring vascular surgery for atherosclerosis. Lancet *2*, 291–293.

Adler, S.P., Chandrika, T., Lawrence, L., and Baggett, J. (1983). Cytomegalovirus infections in neonates acquired by blood transfusions. Pediatr. Infect. Dis. J. *2*, 114–118.

Ahlfors, K., Ivarsson, S.A., Harris, S., Svanberg, L., Holmqvist, R., Lernmark, B., and Theander, G. (1984). Congenital cytomegalovirus infection and disease in Sweden and the relative importance of primary and secondary maternal infections. Scand. J. Infect. Dis. *16*, 129–137.

Ahlfors, K., Ivarsson, S., and Harris, S. (1999). Report on a long-term study of maternal and congenital cytomegalovirus infection in Sweden. Review of prospective studies available in the literature. Scand. J. Infect. Dis. *31*, 443–457.

Ahlfors, K., Harris, S., Ivarsson, S., and Svanberg, L. (1981). Secondary maternal cytomegalovirus infection causing symptomatic congenital infection. N. Engl. J. Med. *305*, 284.

Ahlfors, K., Forsgren, M., Ivarsson, S.A., Harris, S., and Svanberg, L. (1983). Congenital cytomegalovirus infection: on the relation between type and time of maternal infection and infant's symptoms. Scand. J. Infect. Dis. *15*, 129–138.

Alford, C.A., Stagno, S., and Pass, R.F. (1980). Natural history of perinatal cytomegalovirus infection. In Perinatal Infections (Excerpta Medica, Amsterdam), pp. 125–147.

Alford, C.A., Stagno, S., Pass, R.F., and Huang, E.S. (1981). Epidemiology of cytomegalovirus. In The Human Herpesviruses: An Interdisciplinary Perspective, Nahmais, A., Dowdle, W., and Schinazi, R., eds. (Elsevier, New York), pp. 159–171.

Almanzar, G., Schwaiger, S., Jenewein, B., Keller, M., Herndler-Brandstetter, D., Wurzner, R., Schonitzer, D., and Grubeck-Loebenstein, B. (2005). Long-term cytomegalovirus infection leads to significant changes in the composition of the CD8+ T-cell repertoire, which may be the basis for an imbalance in the cytokine production profile in elderly persons. J. Virol. *79*, 3675–3683.

Ancora, G., Lanari, M., Lazzarotto, T., Venturi, V., Tridapalli, E., Sandri, F., and Menarini, M. (2007). Cranial ultrasound scanning and prediction of outcome in newborns with congenital cytomegalovirus infection. J. Pediatr. *150*, 157–161.

Arribas, J.R., Storch, G.A., Clifford, D.B., and Tselis, A.C. (1996). Cytomegalovirus encephalitis. Ann. Int. Med. *125*, 577–587.

Arthurs, S.K., Eid, A.J., Pedersen, R.A., Kremers, W.K., Cosio, F.G., Patel, R., and Razonable, R.R. (2008). Delayed-onset primary cytomegalovirus disease and the risk of allograft failure and mortality after kidney transplantation. Clin. Infect. Dis. *46*, 840–846.

Autran, B., Carcelain, G., Li, T.S., Blanc, C., Mathez, D., Tubiana, R., Katlama, C., Debre, P., and Leibowitch, J. (1997). Positive effects of combined antiretroviral therapy on CD4+ T-cell homeostasis and function in advanced HIV disease. Science *277*, 112–116.

Bale, J.F., Zimmerman, B., Dawson, J., Souza, I., Petheram, B., and Murph, J.R. (1999). Cytomegalovirus transmission in child care homes. Arch. Pediatr. Adolesc. Med. *153*, 75–79.

Bale, J.F., Jr., Petheram, S.J., Robertson, M., Murph, J.R., and Demmler, G. (2001). Human cytomegalovirus a sequence and UL144 variability in strains from infected children. J. Med. Virol. *65*, 90–96.

Baryawno, N., Rahbar, A., Wolmer-Solberg, N., Taher, C., Odeberg, J., Darabi, A., Khan, Z., Sveinbjornsson, B., FuskevAg, O.M., Segerstrom, L., *et al.* (2011). Detection of human cytomegalovirus in medulloblastomas reveals a potential therapeutic target. J. Clin. Invest. *121*, 4043–4055.

Bate, S.L., Dollard, S.C., and Cannon, M.J. (2010). Cytomegalovirus seroprevalence in the USA: the national health and nutrition examination surveys, 1988–2004. Clin. Infect. Dis. *50*, 1439–1447.

Becroft, D.M. (1981). Prenatal cytomegalovirus infection: epidemiology, pathology and pathogenesis. Perspect. Pediatr. Pathol. *6*, 203–241.

Benedict, C.A., Angulo, A., Patterson, G., Ha, S., Huang, H., Messerle, M., Ware, C.F., and Ghazal, P. (2004). Neutrality of the canonical NF-kappaB-dependent pathway for human and murine cytomegalovirus transcription and replication *in vitro*. J. Virol. *78*, 741–750.

Benoist, G., Salomon, L.J., Mohlo, M., Saurez, B., Jacquemard, F., and Ville, Y. (2008). Cytomegalovirus-related fetal brain lesions: comparison between targeted ultrasound examination and magnetic resonance imaging. Ultrasound Obstet. Gynecol. *32*, 900–905.

Blankenberg, S., Rupprecht, H.J., Bickel, C., Espinola-Klein, C., Rippin, G., Hafner, G., Ossendorf, M., Steinhagen, K., and Meyer, J. (2001). Cytomegalovirus infection with interleukin-6 response predicts cardiac mortality in patients with coronary artery disease. Circulation *103*, 2915–2921.

Bodeus, M., Hubinont, C., Bernard, P., Bouckaert, A., Thomas, K., and Goubau, P. (1999). Prenatal diagnosis of human cytomegalovirus by culture and polymerase chain reaction: 98 pregnancies leading to congenital infection. Prenatal Diag. *19*, 314–317.

Boeckh, M. (1999). Management of cytomegalovirus infections in blood and marrow transplant recipients. Adv. Exp. Med. Biol. *458*, 89–109.

Boeckh, M., Leisenring, W., Riddell, S.R., Bowden, R.A., Huang, M.L., Myerson, D., Stevens-Ayers, T., Flowers, M.E., Cunningham, T., and Corey, L. (2003). Late cytomegalovirus disease and mortality in recipients of allogeneic hematopoietic stem cell transplants: importance of viral load and T-cell immunity. Blood *101*, 407–414.

Boppana, S.B., Pass, R.F., Britt, W.J., Stagno, S., and Alford, C.A. (1992). Symptomatic congenital cytomegalovirus infection: neonatal morbidity and mortality. Pediatr. Infect. Dis. J. *11*, 93–99.

Boppana, S.B., Polis, M.A., Kramer, A.A., Britt, W.J., and Koenig, S. (1995). Virus specific antibody responses to human cytomegalovirus (HCMV) in human immunodeficiency virus type 1-infected individuals with HCMV retinitis. J. Infect. Dis. *171*, 182–185.

Boppana, S.B., Fowler, K.B., Vaid, Y., Hedlund, G., Stagno, S., Britt, W.J., and Pass, R.F. (1997). Neuroradiographic findings in the newborn period and long-term outcome in children with symptomatic congenital cytomegalovirus infection. Pediatrics *99*, 409–414.

Boppana, S.B., Fowler, K.B., Britt, W.J., Stagno, S., and Pass, R.F. (1999). Symptomatic congenital cytomegalovirus infection in infants born to mothers with preexisting immunity to cytomegalovirus. Pediatrics *104*, 55–60.

Boppana, S.B., Rivera, L.B., Fowler, K.B., Mach, M., and Britt, W.J. (2001). Intrauterine transmission of cytomegalovirus to infants of women with preconceptional immunity. N. Engl. J. Med. *344*, 1366–1371.

Boppana, S.B., Fowler, K.B., Pass, R.F., Rivera, L.B., Bradford, R.D., Lakeman, F.D., and Britt, W.J. (2005). Congenital cytomegalovirus infection: The association between virus burden in infancy and hearing loss. J. Pediatr. *146*, 817–823.

Botto, S., Streblow, D.N., DeFilippis, V., White, L., Kreklywich, C.N., Smith, P.P., and Caposio, P. (2011). IL-6 in human cytomegalovirus secretome promotes angiogenesis and survival of endothelial cells through the stimulation of survivin. Blood *117*, 352–361.

Boudreault, A.A., Xie, H., Rakita, R.M., Scott, J.D., Davis, C.L., Boeckh, M., and Limaye, A.P. (2011). Risk factors for late-onset cytomegalovirus disease in donor seropositive /recipient seronegative kidney transplant recipients who receive antiviral prophylaxis. Transpl. Infect. Dis. *13*, 244–249.

Bourne, N., Schleiss, M., Bravo, J.J., and Bernstein, D.I. (2001). Preconception immunization with a cytomegalovirus (CMV) glycoprotein vaccine improves pregnancy outcome in a guinea pig model for congenital CMV infection. J. Infect. Dis. *183*, 59–64.

Brennan, D.C., Legendre, C., Patel, D., Mange, K., Wiland, A., McCague, K., and Shihab, F.S. (2011). Cytomegalovirus incidence between everolimus versus mycophenolate in de novo renal transplants: pooled analysis of three clinical trials. Am. J. Transplant. *11*, 2453–2462.

Britt, W.J. (2010). Cytomegalovirus. In Infectious Diseases of the Fetus and Newborn Infant, Remington, J.S., Klein, J.O., Wilson, C.B., Nizet, V., and Maldonaldo, Y., eds. (Elsevier Saunders, Philadelphia), pp. 706–755.

Bronke, C., Palmer, N.M., Jansen, C.A., Westerlaken, G.H., Polstra, A.M., Reiss, P., Bakker, M., Miedema, F., Tesselaar, K., and van Baarle, D. (2005). Dynamics of cytomegalovirus (CMV)-specific T-cells in HIV-1-infected individuals progressing to AIDS with CMV end-organ disease. J. Infect. Dis. *191*, 873–880.

Brunner, S., Herndler-Brandstetter, D., Weinberger, B., and Grubeck-Loebenstein, B. (2011). Persistent viral infections and immune aging. Ageing Res. Rev. *10*, 362–369.

Caposio, P., Dreano, M., Garotta, G., Gribaudo, G., and Landolfo, S. (2004). Human cytomegalovirus stimulates cellular IKK2 activity and requires the enzyme for productive replication. J. Virol. *78*, 3190–3195.

Caposio, P., Luganini, A., Hahn, G., Landolfo, S., and Gribaudo, G. (2007). Activation of the virus-induced IKK/NF-kappaB signalling axis is critical for the replication of human cytomegalovirus in quiescent cells. Cell Microbiol. *9*, 2040–2054.

Caposio, P., Orloff, S.L., and Streblow, D.N. (2011). The role of cytomegalovirus in angiogenesis. Virus Res. *157*, 204–211.

Cappuyns, I., Gugerli, P., and Mombelli, A. (2005). Viruses in periodontal disease – a review. Oral Dis. *11*, 219–229.

Cekinovic, D., Golemac, M., Pugel, E.P., Tomac, J., Cicin-Sain, L., Slavuljica, I., Bradford, R., Misch, S., Winkler, T.H., Mach, M., *et al.* (2008). Passive immunization reduces murine cytomegalovirus-induced brain pathology in newborn mice. J. Virol. *82*, 12172–12180.

Chandler, S.H., and McDougall, J.K. (1986). Comparison of restriction site polymorphisms among clinical isolates and laboratory strains of human cytomegalovirus. J. Gen. Virol. *67*, 2179–2192.

Chatterjee, A., Harrison, C.J., Britt, W.J., and Bewtra, C. (2001). Modification of maternal and congenital cytomegalovirus infection by anti-glycoprotein b antibody transfer in guinea pigs. J. Infect. Dis. *183*, 1547–1553.

Cheng, J., Ke, Q., Jin, Z., Wang, H., Kocher, O., Morgan, J.P., Zhang, J., and Crumpacker, C.S. (2009). Cytomegalovirus infection causes an increase of arterial blood pressure. PLoS Pathog. *5*, e1000427.

Clave, E., Busson, M., Douay, C., Peffault de Latour, R., Berrou, J., Rabian, C., Carmagnat, M., Rocha, V., Charron, D., Socie, G., *et al.* (2009). Acute graft-versus-host disease transiently impairs thymic output in young patients after allogeneic hematopoietic stem cell transplantation. Blood *113*, 6477–6484.

Cobbs, C.S., Harkins, L., Samanta, M., Gillespie, G.Y., Bharara, S., King, P.H., Nabors, L.B., Cobbs, C.G., and Britt, W.J. (2002). Human cytomegalovirus infection and expression in human malignant glioma. Cancer Res. *62*, 3347–3350.

Colugnati, F.A., Staras, S.A., Dollard, S.C., and Cannon, M.J. (2007). Incidence of cytomegalovirus infection among the general population and pregnant women in the USA. BMC Infect. Dis. *7*, 71.

Colvin-Adams, M., and Agnihotri, A. (2005). Cardiac allograft vasculopathy: current knowledge and future direction. Clin. Transplant. *25*, 175–184.

Conboy, T.J., Pass, R.F., Stagno, S., Alford, C.A., Myers, G.J., Britt, W.J., McCollister, F.P., Summers, M.N., McFarland, C.E., and Boll, T.J. (1987). Early clinical manifestations and intellectual outcome in children with symptomatic congenital cytomegalovirus infection. J. Pediatr. *111*, 343–348.

Contreras, A., and Slots, J. (2000). Herpesviruses in human periodontal disease. J. Periodontal Res. *35*, 3–16.

Cope, A.V., Sabin, C., Burroughs, A., Rolles, K., Griffiths, P.D., and Emery, V.C. (1997). Interrelationships among quantity of human cytomegalovirus (HCMV) DNA in blood, donor-recipient serostatus, and administration of methylprednisolone as risk factors for HCMV disease following liver transplantation. J. Infect. Dis. *176*, 1484–1490.

Craig, J.M., Macauley, J.C., Weller, T.H., and Wirth, P. (1957). Isolation of intranuclear inclusion producing agents from infants with illnesses resembling cytomegalic inclusion disease. Proc. Soc. Exp. Biol. Med. *94*, 4–12.

Crumpacker, C.S. (2010). Invited commentary: human cytomegalovirus, inflammation, cardiovascular disease, and mortality. Am. J. Epidemiol. *172*, 372–374.

Dahle, A.J., Fowler, K.B., Wright, J.D., Boppana, S.B., Britt, W.J., and Pass, R.F. (2000). Longitudinal investigation of hearing disorders in children with congenital cytomegalovirus. J. Am. Acad. Audiol. *11*, 283–290.

Dar, L., Pati, S.K., Patro, R.K., Deorari, A.K., Rai, S., Kant, S., Broor, S., Fowler, K.B., Britt, W.J., and Boppana, S.B. (2008). Congenital cytomegalovirus infection in a highly seropositive semi-urban population in India. Pediatr. Infect. Dis. J. *27*, 841–843.

DeMeritt, I.B., Milford, L.E., and Yurochko, A.D. (2004). Activation of the NF-kappaB pathway in human cytomegalovirus-infected cells is necessary for efficient transactivation of the major immediate-early promoter. J. Virol. *78*, 4498–4507.

DeMeritt, I.B., Podduturi, J.P., Tilley, A.M., Nogalski, M.T., and Yurochko, A.D. (2006). Prolonged activation of NF-kappaB by human cytomegalovirus promotes efficient viral replication and late gene expression. Virology *346*, 15–31.

Demmler, G.J. (1991). Infectious Diseases Society of America and Centers for Disease Control. Summary of a workshop on surveillance for congenital cytomegalovirus disease. Rev. Infect. Dis. *13*, 315–329.

Dollard, S.C., Grosse, S.D., and Ross, D.S. (2007). New estimates of the prevalence of neurological and sensory sequelae and mortality associated with congential cytomegalovirus infection. Rev. Med. Virol. *17*, 355–363.

Domenech, E., Vega, R., Ojanguren, I., Hernandez, A., Garcia-Planella, E., Bernal, I., Rosinach, M., Boix, J., Cabre, E., and Gassull, M.A. (2008). Cytomegalovirus infection in ulcerative colitis: a prospective, comparative study on prevalence and diagnostic strategy. Inflamm. Bowel Dis. *14*, 1373–1379.

Drew, W.L., and Mintz, L. (1984). Cytomegalovirus infection in healthy and immune-deficient homosexual men. In The Acquired Immune Deficiency Syndrome and Infections of Homosexual Men, Ma, P., and Armstrong, D., eds. (Yorke Medical Books, New York), pp. 117–123.

Drew, W.L., Mintz, L., Miner, R.C., Sands, M., and Ketterer, B. (1981). Prevalence of cytomegalovirus infection in homosexual men. J. Infect. Dis. *143*, 188–192.

Dumortier, J., Streblow, D.N., Moses, A.V., Jacobs, J.M., Kreklywich, C.N., Camp, D., Smith, R.D., Orloff, S.L., and Nelson, J.A. (2008). Human cytomegalovirus secretome contains factors that induce angiogenesis and wound healing. J. Virol. *82*, 6524–6535.

Dworsky, M., Yow, M., Stagno, S., Pass, R.F., and Alford, C.A. (1983). Cytomegalovirus infection of breast milk and transmission in infancy. Pediatrics *72*, 295–299.

Dylewski, J., Chou, S., and Merigan, T.C. (1983). Absence of detectable IgM antibody during cytomegalovirus disease in patients with AIDS [letter]. N. Engl. J. Med. *309*, 493.

Dziurzynski, K., Wei, J., Qiao, W., Hatiboglu, M.A., Kong, L.Y., Wu, A., Wang, Y., Cahill, D., Levine, N., Prabhu, S., *et al.* (2011). Glioma-associated cytomegalovirus mediates subversion of the monocyte lineage to a tumor propagating phenotype. Clin. Cancer Res. *17*, 4642–4649.

Emery, V.C. (2001). Prophylaxis for CMV should not now replace pre-emptive therapy in solid organ transplantation [comment]. Rev. Med. Virol. *11*, 83–86.

Emery, V.C., Sabin, C.A., Cope, A.V., Gor, D., Hassan-Walker, A.F., and Griffiths, P.D. (2000). Application of viral-load kinetics to identify patients who develop cytomegalovirus disease after transplantation. Lancet *355*, 2032–2036.

Enders, G., Bader, U., Lindemann, L., Schalasta, G., and Daiminger, A. (2001). Prental diagnosis of congenital cytomegalovirus infections in 189 pregnancies with known outcome. Prenat. Diag. *21*, 362–377.

Epstein, S.E., Zhou, Y.F., and Zhu, J. (1999). Infection and atherosclerosis: emerging mechanistic paradigms. Circulation *100*, e20–28.

Erdbruegger, U., Scheffner, I., Mengel, M., Schwarz, A., Verhagen, W., Haller, H., and Gwinner, W. (2011). Impact of CMV infection on acute rejection and long-term renal allograft function: a systematic analysis in patients with protocol biopsies and indicated biopsies. Nephrol. Dial. Transplant. *27*, 435–443.

Evans, P.C., Coleman, N., Wreghitt, T.G., Wight, D.G., and Alexander, G.J. (1999). Cytomegalovirus infection of bile duct epithelial cells, hepatic artery and portal venous endothelium in relation to chronic rejection of liver grafts. J. Hepatol. *31*, 913–920.

Faist, B., Fleischer, B., and Jacobsen, M. (2010). Cytomegalovirus infection- and age-dependent changes in human CD8+ T-cell cytokine expression patterns. Clin. Vaccine Immunol. *17*, 986–992.

Fearon, W.F., Potena, L., Hirohita, A., Sakurai, R., Yamasaki, M., Luikart, H., Lee, J., Vana, M.L., Cooke, J.P., Mocarski, E.S., *et al.* (2007). Changes in coronary arterial dimensions early after cardiac transplantation. Transplantation *83*, 700–705.

Fishman, J.A. (2007). Infection in solid-organ transplant recipients. N. Engl. J. Med. *357*, 2601–2614.

Fowler, K.B., Stagno, S., Pass, R.F., Britt, W.J., Boll, T.J., and Alford, C.A. (1992). The outcome of congenital cytomegalovirus infection in relation to maternal antibody status. N. Engl. J. Med. *326*, 663–667.

Fowler, K.B., Stagno, S., and Pass, R.F. (1993). Maternal age and congenital cytomegalovirus infection: screening of two diverse newborn populations, 1980–1990. J. Infect. Dis. *168*, 552–556.

Fowler, K.B., McCollister, F.P., Dahle, A.J., Boppana, S., Britt, W.J., and Pass, R.F. (1997). Progressive and fluctuating sensorineural hearing loss in children with asymptomatic congenital cytomegalovirus infection. J. Pediatr. *130*, 624–630.

Fox, J.C., Kidd, I.M., Griffiths, P.D., Sweny, P., and Emery, V.C. (1995). Longitudinal analysis of cytomegalovirus load in renal transplant recipients using a quantitative polymerase chain reaction: correlation with disease. J. Gen. Virol. *76*, 309–319.

Gabrielli, L., Bonasoni, M.P., Lazzarotto, T., Lega, S., Santini, D., Foschini, M.P., Guerra, B., Baccolini, F., Piccorilli, G., Chiereghin, A., *et al.* (2009). Histological findings in fetuses congenitally infected by cytomegalovirus. J. Clin. Virol. *46*, S16–21.

Gallant, J.E., Moore, R.D., Richman, D.D., Keruly, J., and Chaisson, R.E. (1992). Incidence and natural history of cytomegalovirus disease in patients with advanced human immunodeficiency virus disease treated with zidovudine. J. Infect. Dis. *166*, 1223–1227.

Gallez-Hawkins, G., Thao, L., Lacey, S.F., Martinez, J., Li, X., Franck, A.E., Lomeli, N.A., Longmate, J., Diamond, D.J., Spielberger, R., *et al.* (2005). Cytomegalovirus immune reconstitution occurs in recipients of allogeneic hematopoietic cell transplants irrespective of detectable cytomegalovirus infection. Biol. Blood Marrow Transplant. *11*, 890–902.

Gallina, A., Percivalle, E., Simoncini, L., Revello, M.G., Gerna, G., and Milanesi, G. (1996). Human cytomegalovirus protein pp65 lower matrix phosphoprotein harbours two transplantable nuclear localization signals. J. Gen. Virol. *77*, 1151–1157.

Gaytant, M.A., Rours, J.I.J.G., Steegers, E.A.P., Galama, J.M.D., and Semmekrot, B.A. (2003). Congenital cytomegalovirus infection after recurrent infection: case reports and review of the literature. Eur. J. Pediatr. *162*, 248–253.

Geder, L., Sanford, E.J., Rohner, T.J., and Rapp, F. (1977). Cytomegalovirus and cancer of the prostate: *in vitro* transformation of human cells. Cancer Treat. Rep. *61*, 139–146.

Gehrz, R.C., Marker, S.C., Knorr, S.O., Kalis, J.M., and Balfour, H.H., Jr. (1977). Specific cell-mediated immune defect in active cytomegalovirus infection of young children and their mothers. Lancet *2*, 844–847.

Gerna, G., Lilleri, D., Chiesa, A., Zelini, P., Furione, M., Comolli, G., Pellegrini, C., Sarchi, E., Migotto, C., Bonora, M.R., *et al.* (2011). Virologic and immunologic monitoring of cytomegalovirus to guide preemptive therapy in solid-organ transplantation. Am. J. Transplant. *11*, 2463–2471.

Gindes, L., Teperberg-Oikawa, M., Sherman, D., Pardo, J., and Rahav, G. (2008). Congenital cytomegalovirus infection following primary maternal infection in the third trimester. BJOG *115*, 830–835.

Gor, D., Sabin, C., Prentice, H.G., Vyas, N., Man, S., Griffiths, P.D., and Emery, V.C. (1998). Longitudinal fluctuations in cytomegalovirus load in bone marrow transplant patients: relationship between peak virus load, donor/recipient serostatus, acute GVHD and CMV disease. Bone Marrow Transplant. *21*, 597–605.

Gottlieb, M.S., Schroff, R., Schanker, H.M., Wesiman, J.D., Fan, P.T., Wofl, R.A., and Saxon, A. (1981). Pneumocystis carinii pneumonia and mucosal candidiasis in previously healthy homosexual men. Evidence of a new acquired cellular immunodeficiency. N. Engl. J. Med. *305*, 1425–1431.

Gratama, J.W., Boeckh, M., Nakamura, R., Cornelissen, J.J., Brooimans, R.A., Zaia, J.A., Forman, S.J., Gaal, K., Bray, K.R., Gasior, G.H., *et al.* (2010). Immune monitoring with iT Ag MHC tetramers for prediction of recurrent or persistent cytomegalovirus infection or disease in allogeneic hematopoietic stem cell transplant recipients: a prospective multicenter study. Blood *116*, 1655–1662.

Grattan, M.T., Moreno-Cabral, C.E., Starnes, V.A., Oyer, P.E., Stinson, E.B., and Shumway, N.E. (1989). Cytomegalovirus infection is associated with cardiac allograft rejection and atherosclerosis. JAMA *261*, 3561–3566.

Gredmark-Russ, S., Dzabic, M., Rahbar, A., Wanhainen, A., Bjorck, M., Larsson, E., Michel, J.B., and Soderberg-Naucler, C. (2009). Active cytomegalovirus infection in aortic smooth muscle cells from patients with abdominal aortic aneurysm. J. Mol. Med. *87*, 347–356.

Grenier, G., Gagnon, G., and Grenier, D. (2009). Detection of herpetic viruses in gingival crevicular fluid of patients suffering from periodontal diseases: prevalence and effect of treatment. Oral Microbiol. Immunol. *24*, 506–509.

Griffith, B.P., McCormick, S.R., Booss, J., and Hsiung, G.D. (1986). Inbred guinea pig model of intrauterine infection with cytomegalovirus. Am. J. Pathol. *122*, 112–119.

Grosse, S.D., Boyle, C.A., Kenneson, A., Khoury, M.J., and Wilfond, B.S. (2006). From public health emergency to public health service: the implications of evolving criteria for newborn screening panels. Pediatrics *117*, 923–929.

Guerra, B., Simonazzi, G., Puccetti, C., Lanari, M., Farina, A., Lazzarotto, T., and Rizzo, N. (2008). Ultrasound prediction of symptomatic congenital cytomegalovirus infection. Am. J. Obstet. Gynecol. *198*, 380.e1–7.

Hakki, M., Riddell, S.R., Storek, J., Carter, R.A., Stevens-Ayers, T., Sudour, P., White, K., Corey, L., and Boeckh, M. (2003). Immune reconstitution to cytomegalovirus after allogeneic hematopoietic stem cell transplantation: impact of host factors, drug therapy, and subclinical reactivation. Blood *102*, 3060–3067.

Hamele, M., Flanagan, R., Loomis, C.A., Stevens, T., and Fairchok, M.P. (2010). Severe morbidity and mortality with breast milk associated cytomegalovirus infection. Pediatr. Infect. Dis. J. *29*, 84–86.

Hamprecht, K., Maschmann, J., Jahn, G., Poets, C.F., and Goelz, R. (2008). Cytomegalovirus transmission to preterm infants during lactation. J. Clin. Virol. *41*, 198–205.

Harris, J.P., Woolf, N.K., Ryan, A.F., Butler, D.M., and Richman, D.D. (1984a). Immunologic and electrophysiological response to cytomegaloviral inner ear infection in the guinea pig. J. Infect. Dis. *150*, 523–530.

Harris, S., Ahlfors, K., Ivarsson, S., Lemmark, B., and Svanberg, L. (1984b). Congenital cytomegalovirus infection and sensorineural hearing loss. Ear Hearing *5*, 352–355.

Harrison, C.J., Britt, W.J., Chapan, N.M., Mullican, J., and Tracy, S. (1995). Reduced congenital cytomegalovirus (CMV) infection after maternal immunization with a guinea pig CMV glycoprotein before gestational primary CMV infection in the guinea pig model. J. Infect. Dis. *172*, 1212–1220.

Ho, K.L., Gottlieb, C., and Zarbo, R.J. (1991). Cytomegalovirus infection of cerebral astrocytoma in an AIDS patient. Clin. Neuropathol. *10*, 127–133.

Ho, M. (1977). Virus infections after transplantation in man. Arch. Virol. *55*, 1–24.

Hommes, D.W., Sterringa, G., van Deventer, S.J., Tytgat, G.N., and Weel, J. (2004). The pathogenicity of cytomegalovirus in inflammatory bowel disease: a systematic review and evidence-based recommendations for future research. Inflamm. Bowel Dis. *10*, 245–250.

Huang, E.S., Alford, C.A., Reynolds, D.W., Stagno, S., and Pass, R.F. (1980). Molecular epidemiology of cytomegalovirus infections in women and their infants. N. Engl. J. Med. *303*, 958–962.

Huppmann, A.R., and Orenstein, J.M. (2010). Opportunistic disorders of the gastrointestinal tract in the age of highly active antiretroviral therapy. Hum. Pathol. *41*, 1777–1787.

Husain, S., Pietrangeli, C.E., and Zeevi, A. (2009). Delayed onset CMV disease in solid organ transplant recipients. Transpl. Immunol. *21*, 1–9.

Hutter, J.A., Scott, J., Wreghitt, T., Higenbottam, T., and Wallwork, J. (1989). The importance of cytomegalovirus in heart-lung transplant recipients. Chest *95*, 627–631.

Inkinen, K., Soots, A., Krogerus, L., Bruggeman, C., Ahonen, J., and Lautenschlager, I. (2002). Cytomegalovirus increases collagen synthesis in chronic rejection in the rat. Nephrol. Dial. Transplant. *17*, 772–779.

Inkinen, K., Soots, A., Krogerus, L., Loginov, R., Bruggeman, C., and Lautenschlager, I. (2005a). Cytomegalovirus enhance expression of growth factors during the development of chronic allograft nephropathy in rats. Transplant Int. *18*, 743–749.

Inkinen, K.A., Soots, A.P., Krogerus, L.A., Lautenschlager, I.T., and Ahonen, J.P. (2005b). Fibrosis and matrix metalloproteinases in rat renal allografts. Transplant Int. *18*, 506–512.

Ishibashi, K., Tokumoto, T., Tanabe, K., Shirakawa, H., Hashimoto, K., Kushida, N., Yanagida, T., Inoue, O., Yamaguchi, H., Toma, T., *et al.* (2007). Association of the outcome of renal transplantation with antibody response to cytomegalovirus strain-specific glycoprotein H epitopes. J. Infect. Dis. *45*, 60–67.

Iwasenko, J.M., Howard, J., Arbuckle, S., Graf, N., Hall, B., Craig, M.E., and Rawlinson, W.D. (2011). Human cytomegalovirus infection is detected frequently in stillbirths and is associated with fetal thrombotic vasculopathy. J. Infect. Dis. *203*, 1526–1533.

Jesionek, A., and Kiolemenoglou, B. (1904). Uber einen befund von protozoenartigen gebilden in den organen eines heriditarluetischen fotus. Munch Med. Wochenschr. *51*, 1905–1907.

Kenneson, A., and Cannon, M.J. (2007). Review and meta-analysis of the epidemiology of congenital cytomegalovirus (CMV) infection. Rev. Med. Virol. *17*, 253–276.

Khoury, J.A., Storch, G.A., Bohl, D.L., Schuessler, R.M., Torrence, S.M., Lockwood, M., Gaudreault-Keener, M., Koch, M.J., Miller, B.W., Hardinger, K.L., *et al.* (2006). Prophylactic versus preemptive oral valganciclovir for the management of cytomegalovirus infection in adult renal transplant recipients. Am. J. Transplant. *6*, 2134–2143.

Kidd, I.M., Clark, D.A., Sabin, C.A., Andrew, D., Hassan-Walker, A.F., Sweny, P., Griffiths, P.D., and Emery, V.C. (2000). Prospective study of human betaherpesviruses after renal transplantation: association of human herpesvirus 7 and cytomegalovirus co-infection with cytomegalovirus disease and increased rejection. Transplantation *69*, 2400–2404.

Komanduri, K.V., Viswanathan, M.N., Wieder, E.D., Schmidt, D.K., Bredt, B.M., Jacobson, M., and McCune, J.M. (1998). Restoration of cytomegalovirus-specific CD4+ T-lymphocyte responses after ganciclovir and highly active antiretroviral therapy in individuals infected with HIV-1. Nat. Med. *4*, 953–956.

Koskinen, P.K., Nieminen, M.S., Krogerus, L.A., Lemstrom, K.B., Mattila, S.P., Hayry, P.J., and Lautenschlager, I.T. (1993). Cytomegalovirus infection and accelerated cardiac allograft vasculopathy in human cardiac allografts. J. Heart Lung Transplant. *12*, 724–729.

Kotton, C.N., Kumar, D., Caliendo, A.M., Asberg, A., Chou, S., Snydman, D.R., Allen, U., and Humar, A. (2010). International consensus guidelines on the management of cytomegalovirus in solid organ transplantation. Transplantation *89*, 779–795.

Krogerus, L., Soots, A., Loginov, R., Bruggeman, C., Ahonen, J., and Lautenschlager, I. (2001). CMV accelerates tubular apoptosis in a model of chronic renal allograft rejection. Transplant. Proc. *33*, 254.

Kylat, R.I., Kelly, E.N., and Ford-Jones, E.L. (2006). Clinical findings and adverse outcome in neonates with symptomatic congenital cytomegalovirus (SCCMV) infection. Eur. J. Pediatr. *165*, 773–778.

La Rosa, C., Limaye, A.P., Krishnan, A., Longmate, J., and Diamond, D.J. (2007). Longitudinal assessment of cytomegalovirus (CMV)-specific immune responses in liver transplant recipients at high risk for late CMV disease. J. Infect. Dis. *195*, 633–644.

Lacey, S.F., Diamond, D.J., and Zaia, J.A. (2004). Assessment of cellular immunity to human cytomegalovirus in recipients of allogeneic stem cell transplants. Biol. Blood Marrow Transplant. *10*, 433–447.

Lanari, M., Lazzarotto, T., Venturi, V., Papa, I., Gabrielli, L., Guerra, B., Landini, M.P., and Faldella, G. (2006). Neonatal cytomegalovirus blood load and risk of sequelae in symptomatic and asymptomatic congenitally infected newborns. Pediatrics *117*, e76-e83.

Lautenschlager, I., Soots, A., Krogerus, L., Inkinen, K., Kloover, J., Loginov, R., Holma, K., Kauppinen, H., Bruggeman, C., and Ahonen, J. (1999). Time-related effects of cytomegalovirus infection on the development of chronic renal allograft rejection in a rat model. Intervirology *42*, 279–284.

Lautenschlager, I., Halme, L., Hockerstedt, K., Krogerus, L., and Taskinen, E. (2006). Cytomegalovirus infection of the liver transplant: virological, histological, immunological, and clinical observations. Transpl. Infect. Dis. *8*, 21–30.

Lawlor, G., and Moss, A.C. (2010). Cytomegalovirus in inflammatory bowel disease: pathogen or innocent bystander? Inflamm. Bowel Dis. *16*, 1620–1627.

Leinonen, M., and Saikku, P. (2000). Infections and atherosclerosis. Scand. Cardiovasc. J. *34*, 12–20.

Lemonovich, T.L., and Watkins, R.R. (2012). Update on cytomegalovirus infections of the gastrointestinal system in solid organ transplant recipients. Curr. Infect. Dis. Rep. *14*, 33–40.

Lemstrom, K.B., Bruning, J.H., Bruggeman, C.A., Lautenschlager, I.T., and Hayry, P.J. (1993). Cytomegalovirus infection enhances smooth muscle cell proliferation and intimal thickening of rat aortic allografts. J. Clin. Invest. *92*, 549–558.

Lemstrom, K., Koskinen, P., Krogerus, L., Daemen, M., Bruggeman, C., and Hayry, P. (1995). Cytomegalovirus antigen expression, endothelial cell proliferation, and intimal thickening in rat cardiac allografts after cytomegalovirus infection. Circulation *92*, 2594–2604.

Li, C.R., Greenberg, P.D., Gilbert, M.J., Goodrich, J.M., and Riddell, S.R. (1994). Recovery of HLA-restricted cytomegalovirus (CMV)-specific T-cell responses after allogeneic bone marrow transplant: correlation with CMV disease and effect of ganciclovir prophylaxis. Blood *83*, 1971–1979.

Libby, P. (2002). Inflammation in atherosclerosis. Nature *420*, 868–874.

Limaye, A.P., Bakthavatsalam, R., Kim, H.W., Kuhr, C.S., Halldorson, J.B., Healey, P.J., and Boeckh, M. (2004). Late-onset cytomegalovirus disease in liver transplant recipients despite antiviral prophylaxis. Transplantation *78*, 1390–1396.

Liu, Y. (2011). Cellular and molecular mechanisms of renal fibrosis. Nat. Rev. Nephrol. *7*, 684–696.

Lucas, K.G., Bao, L., Bruggeman, R., Dunham, K., and Specht, C. (2011). The detection of CMV pp65 and IE1 in glioblastoma multiforme. J. Neuro-Oncol. *103*, 231–238.

McCracken, G.J., Shinefield, H.R., Cobb, K., Rausen, A.R., Dische, M.R., and Eichenwald, H.F. (1969). Congenital cytomegalic inclusion disease. A longitudinal study of 20 patients. Am. J. Dis. Child. *117*, 522–539.

Mannon, R.B., Matas, A.J., Grande, J., Leduc, R., Connett, J., Kasiske, B., Cecka, J.M., Gaston, R.S., Cosio, F., Gourishankar, S., *et al.* (2010). Inflammation in areas of tubular atrophy in kidney allograft biopsies: a potent predictor of allograft failure. Am. J. Transplant. *10*, 2066–2073.

Marchant, A., Appay, V., Van Der Sande, M., Dulphy, N., Liesnard, C., Kidd, M., Kaye, S., Ojuola, O., Gillespie, G.M., Vargas Cuero, A.L., *et al.* (2003). Mature CD8(+) T-lymphocyte response to viral infection during fetal life. J. Clin. Invest. *111*, 1747–1755.

Maschmann, J., Hamprecht, K., Dietz, K., Jahn, G., and Speer, C.P. (2001). Cytomegalovirus infection of extremely low-birth weight infants via breast milk. Clin. Infect. Dis. *33*, 1998–2003.

Maussang, D., Langemeijer, E., Fitzsimons, C.P., Stigter-van Walsum, M., Dijkman, R., Borg, M.K., Slinger, E., Schreiber, A., Michel, D., Tensen, C.P., *et al.* (2009). The human cytomegalovirus-encoded chemokine receptor US28 promotes angiogenesis and tumor formation via cyclooxygenase-2. Cancer Res. *69*, 2861–2869.

Mehra, M.R. (2006). Contemporary concepts in prevention and treatment of cardiac allograft vasculopathy. Am. J. Transplant. *6*, 1248–1256.

Melnick, J.L., Adam, E., and Debakey, M.E. (1993). Cytomegalovirus and atherosclerosis. Eur. Heart J. *14*, 30–38.

Michaelis, M., Baumgarten, P., Mittelbronn, M., Driever, P.H., Doerr, H.W., and Cinatl, J., Jr. (2010). Oncomodulation by human cytomegalovirus: novel clinical findings open new roads. Med. Microbiol. Immunol. *200*, 1–5.

Miles, D.J., Sande, M., Kaye, S., Crozier, S., Ojuola, O., Palmero, M.S., Sanneh, M., Touray, E.S., Waight, P., Rowland-Jones, S., *et al.* (2008). CD4(+) T-cell responses to cytomegalovirus in early life: a prospective birth cohort study. J. Infect. Dis. *197*, 658–662.

Mitchell, D.A., Xie, W., Schmittling, R., Learn, C., Friedman, A., McLendon, R.E., and Sampson, J.H. (2008). Sensitive detection of human cytomegalovirus in tumors and peripheral blood of patients diagnosed with glioblastoma. Neuro-Oncology *10*, 10–18.

Moorman, N.J., and Shenk, T. (2010). Rapamycin-resistant mTORC1 kinase activity is required for herpesvirus replication. J. Virol. *84*, 5260–5269.

Moro-Garcia, M.A., Alonso-Arias, R., Lopez-Vazquez, A., Suarez-Garcia, F.M., Solano-Jaurrieta, J.J., Baltar, J., and Lopez-Larrea, C. (2011). Relationship between functional ability in older people, immune system status, and intensity of response to CMV. Age (Dordr) *34*, 479–495.

Morris, D.J., Sims, D., Chiswick, M., Das, V.K., and Newton, V.E. (1994). Symptomatic congenital cytomegalovirus infection after maternal recurrent infection. Pediatr. Infect. Dis. J. *13*, 61–64.

Muhlestein, J.B., Horne, B.D., Carlquist, J.F., Madsen, T.E., Bair, T.L., Pearson, R.R., and Anderson, J.L. (2000). Cytomegalovirus seropositivity and C-reactive protein have independent and combined predictive value for mortality in patients with angiographically demonstrated coronary artery disease. Circulation *102*, 1917–1923.

Mussi-Pinhata, M.M., Yamamoto, A.Y., Moura Britto, R.M., de Lima Isaac, M., de Carvalho e Oliveira, P.F., Bopana, S., and Britt, W.J. (2009). Birth prevalence and natural history of congenital cytomegalovirus infection in a highly seroimmune population. Clin. Infect. Dis. *15*, 522–528.

Myers, J.D., Spencer, H.C., Jr., Watts, J.C., Gregg, M.B., Stewart, J.A., Troupin, R.H., and Thomas, E.D. (1975). Cytomegalovirus pneumonia after human marrow transplantation. Ann. Intern. Med. *82*, 181–188.

Myerson, D., Hackman, R.C., and Meyers, J.D. (1984). Diagnosis of cytomegaloviral pneumonia by *in situ* hybridization. J. Infect. Dis. *150*, 272–277.

Nankervis, G.A., and Bhumbra, N.A. (1986). Cytomegalovirus infections of the neonate and infant. Adv. Pediatr. Infect. Dis. *1*, 61–74.

Novak, Z., Ross, S.A., Patro, R.K., Pati, S.K., Kumbla, R.A., Brice, S., and Boppana, S.B. (2008). Cytomegalovirus strain diversity in seropositive women. J. Clin. Microbiol. *46*, 882–886.

Noyola, D.E., Demmler, G.J., Nelson, C.T., Griesser, C., Williamson, D., Atkins, J.T., Rozelle, J., Turcich, M., Llorente, A., Sllers-Vinson, S., *et al.* (2001).

Early predictors of neurodevelopmental outcome in symptomatic congenital CMV infection. J. Pediatr. *138*, 325–331.

Onyeagocha, C., Hossain, M.S., Kumar, A., Jones, R.M., Roback, J., and Gewirtz, A.T. (2009). Latent cytomegalovirus infection exacerbates experimental colitis. Am. J. Pathol. *175*, 2034–2042.

Palmer, S.M., Limaye, A.P., Banks, M., Gallup, D., Chapman, L., Lawrence, E.C., Dunitz, J., Milstone, A., Reynolds, J., Yung, G.L., *et al.* (2010). Extended valganciclovir prophylaxis to prevent cytomegalovirus after lung transplantation: a randomized, controlled trial. Ann. Intern. Med. *152*, 761–769.

Pass, R.F., Stagno, S., Myers, G.J., and Alford, C.A. (1980). Outcome of symptomatic congenital CMV infection: results of long-term longitudinal follow-up. Pediatrics *66*, 758–762.

Pass, R.F., Dworsky, M.E., Whitley, R.J., August, A.M., Stagno, S., and Alford, C.A., Jr. (1981a). Specific lymphocyte blastogenic responses in children with cytomegalovirus and herpes simplex virus infections acquired early in infancy. Infect. Immun. *34*, 166–170.

Pass, R.F., Reynolds, D.W., Welchel, J.D., Diethelm, A.G., and Alford, C.A. (1981b). Impaired lymphocyte transformation response to cytomegalovirus and phytohemagglutinin in recipients of renal transplants: association with antithymocyte globulin. J. Infect. Dis. *143*, 259–265.

Pawelec, G., and Derhovanessian, E. (2011). Role of CMV in immune senescence. Virus Res. *157*, 175–179.

Peckham, C.S., Chin, K.S., Coleman, J.C., Henderson, K., Hurley, R., and Preece, P.M. (1983). Cytomegalovirus infection in pregnancy: preliminary findings from a prospective study. Lancet *1*, 1352–1355.

Podlech, J., Holtappels, R., Pahl-Seibert, M.F., Steffens, H.P., and Reddehase, M.J. (2000). Murine model of interstitial cytomegalovirus pneumona in syngenic bone marrow transplantation: persistence of protective pulmonary CD8-T-cell infiltrates after clearance of acute infection. J. Virol. *74*, 7496–7507.

Potena, L., and Valantine, H.A. (2007). Cytomegalovirus-associated allograft rejection in heart transplant patients. Curr. Opin. Infect. Dis. *20*, 425–431.

Potena, L., Grigioni, F., Magnani, G., Lazzarotto, T., Musuraca, A.C., Ortolani, P., Coccolo, F., Fallani, F., Russo, A., and Branzi, A. (2009). Prophylaxis versus preemptive anti-cytomegalovirus approach for prevention of allograft vasculopathy in heart transplant recipients. J. Heart Lung Transplant. *28*, 461–467.

Pourgheysari, B., Piper, K.P., McLarnon, A., Arrazi, J., Bruton, R., Clark, F., Cook, M., Mahendra, P., Craddock, C., and Moss, P.A. (2009). Early reconstitution of effector memory CD4+ CMV-specific T-cells protects against CMV reactivation following allogeneic SCT. Bone Marrow Transplant. *43*, 853–861.

Preece, P.M., Blount, J.M., Glover, J., Fletcher, G.M., Peckham, C.S., and Griffiths, P.D. (1983). The consequences of primary cytomegalovirus infection in pregnancy. Arch. Dis. Child. *58*, 970–975.

Preece, P.M., Tookey, P., Ades, A., and Peckham, C.S. (1986). Congenital cytomegalovirus infection: predisposing maternal factors. J. Epidemiol. Comm. Health *40*, 205–209.

Prins, R.M., Cloughesy, T.F., and Liau, L.M. (2008). Cytomegalovirus immunity after vaccination with autologous glioblastoma lysate. N. Engl. J. Med. *359*, 539–541.

Prösch, S., Wuttke, R., Kruger, D.H., and Volk, H.D. (2002). NF-kappaB – a potential therapeutic target for inhibition of human cytomegalovirus (re)activation? Biol. Chem. *383*, 1601–1609.

Racusen, L.C., and Regele, H. (2010). The pathology of chronic allograft dysfunction. Kidney Int. *119*, S27–32.

Rahbar, A., Bostrom, L., Lagerstedt, U., Magnusson, I., Soderberg-Naucler, C., and Sundqvist, V.A. (2003). Evidence of active cytomegalovirus infection and increased production of IL-6 in tissue specimens obtained from patients with inflammatory bowel diseases. Inflamm. Bowel Dis. *9*, 154–161.

Rapp, F. (1984). Cytomegalovirus and carcinogenesis. J. Natl. Cancer I. *72*, 783–787.

Rasmussen, L., Morris, S., Wolitz, R., Dowling, A., Fessell, J., Holodniy, M., and Merigan, T.C. (1994). Deficiency in antibody response to human cytomegalovirus glycoprotein gH in human immunodeficiency virus-infected patients at risk for cytomegalovirus retinitis. J. Infect. Dis. *170*, 673–677.

Razonable, R.R. (2008). Cytomegalovirus infection after liver transplantation: current concepts and challenges. World J. Gastroentero. *14*, 4849–4860.

Renzette, N., Bhattacharjee, B., Jenson, J.D., Gibson, L., and Kowalik, T.F. (2011). Extensive genome-wide variability of human cytomegalovirus in congenitally infected infants. PLoS Pathog. *7*, e1001344.

Reusser, P., Riddell, S.R., Meyers, J.D., and Greenberg, P.D. (1991). Cytotoxic T-lymphocyte response to cytomegalovirus after human allogeneic bone marrow transplantation: pattern of recovery and correlation with cytomegalovirus infection and disease. Blood *78*, 1373–1380.

Reusser, P., Cathomas, G., Attenhofer, R., Tamm, M., and Thiel, G. (1999). Cytomegalovirus (CMV)-specific T-cell immunity after renal transplantation mediates protection from CMV disease by limiting the systemic virus load. J. Infect. Dis. *180*, 247–253.

Revello, M.G., Lilleri, D., Zavattoni, M., Furione, M., Middeldorp, J., and Gerna, G. (2003). Prenatal diagnosis of congenital human cytomegalovirus infection in amniotic fluid by nucleic acid sequence-based amplification assay. J. Clin. Microbiol. *41*, 1772–1774.

Riddell, S.R., and Greenberg, P.D. (1994). Therapeutic reconstruction of human viral immunity by adoptive transfer of cytotoxic T-lymphocyte clones. Curr. Top. Microbiol. Immunol. *189*, 9–34.

Riddell, S.R., and Greenberg, P.D. (2000). T-cell therapy of cytomegalovirus and human immunodeficiency virus infection. J. Antimicrob. Chemoth. *45*, 35–43.

Rifkind, D. (1965). Cytomegalovirus infection after renal transplantation. Arch. Intern. Med. *116*, 554–558.

Rivera, L.B., Boppana, S.B., Fowler, K.B., Britt, W.J., Stagno, S., and Pass, R.F. (2002). Predictors of hearing loss in children with symptomatic congenital cytomegalovirus infection. Pediatrics *110*, 762–767.

Roberts, E.T., Haan, M.N., Dowd, J.B., and Aiello, A.E. (2010). Cytomegalovirus antibody levels, inflammation, and mortality among elderly Latinos over 9 years of follow-up. Am. J. Epidemiol. *172*, 363–371.

Ross, S.A., Fowler, K.B., Guha, A., Stagno, S., Britt, W.J., Pass, R.F., and Boppana, S.B. (2006). Hearing loss in children with congential cytomegalovirus infection born to mothers with preexisting immunity. J. Pediatr. *148*, 332–336.

Ross, S.A., Novak, Z., Fowler, K.B., Arora, N., Britt, W.J., and Boppana, S.B. (2009). Cytomegalovirus blood viral load and hearing loss in young children with congenital infection. Pediatr. Infect. Dis. J. *28*, 588–592.

Ross, S.A., Novak, Z., Pati, S., Patro, R.K., Blumenthal, J., Danthaluri, V.R., Ahmed, A., Michaels, M.G., Sanchez, P.J., Bernstein, D.I., *et al.* (2011). Mixed infection and strain diversity in congenital cytomegalovirus infection. J. Infect. Dis. *204*, 1003–1007.

Rowe, W.P., Hartley, J.W., Waterman, S., Turner, H.C., and Huebner, R.J. (1956). Cytopathogenic agent resembling human salivary gland virus recovered from tissue cultures of human adenoids. Proc. Soc. Exp. Biol. Med. *92*, 418–424.

Rubin, R.H. (1990). Impact of cytomegalovirus infection on organ transplant recipients. Rev. Infect. Dis. *12*, S754-S766.

Rubin, R.H., Cosimi, A.B., Tolkoff-Rubin, N.E., Russell, P.S., and Hirsch, M.S. (1977). Infectious disease syndromes attributable to cytomegalovirus and their significance among renal transplant recipients. Transplantation *24*, 458–464.

Rubin, R.H., Russell, P.S., Levin, M., and Cohen, C. (1979). Summary of a workshop on cytomegalovirus infections during organ transplantation. J. Infect. Dis. *139*, 728–734.

Sampson, J.H., and Mitchell, D.A. (2011). Is cytomegalovirus a therapeutic target in glioblastoma? Clin. Cancer Res. *17*, 4619–4621.

van der Sande, M.A.B., Kaye, S., Miles, D.J., Waight, P., Jeffries, D.J., Ojoula, O.O., Palermo, M., Pinder, M., Ismaili, J., Flanagan, K.L., *et al.* (2007). Risk factors for and clinical outcome of congenital cytomegalovirus infection in a peri-urban West-African birth cohort. PLoS One *2*, e492.

Saygun, I., Kubar, A., Ozdemir, A., Yapar, M., and Slots, J. (2004). Herpesviral-bacterial interrelationships in aggressive periodontitis. J. Periodontal Res. *39*, 207–212.

Schmidt, G.M., Horak, D.A., Niland, J.C., Duncan, S.R., Forman, S.J., and Zaia, J.A. (1991). A randomized, controlled trial of prophylactic ganciclovir for cytomegalovirus pulmonary infection in recipients of allogeneic bone marrow transplants. N. Engl. J. Med. *324*, 1005–1011.

Schooley, R.T., Hirsch, M.S., Colvin, R.B., Cosimi, A.B., Tolkoff-Rubin, N.E., McCluskey, R.T., Burton, R.C., Russell, P.S., Herrin, J.T., Delmonico, F.L., *et al.* (1983). Association of herpes virus infection with T-lymphocyte subset alterations, glomerulopathy, and opportunistic infections after renal transplantation. N. Engl. J. Med. *308*, 307–313.

Schopfer, K., Lauber, E., and Krech, U. (1978). Congenital cytomegalovirus infection in newborn infants of mothers infected before pregnancy. Arch. Dis. Child. *53*, 536–539.

Sester, M., Sester, U., Gartner, B., Heine, G., Girndt, M., Mueller-Lantzsch, N., Meyerhans, A., and Kohler, H. (2001). Levels of virus-specific CD4 T-cells correlate with cytomegalovirus control and predict virus-induced disease after renal transplantation. Transplantation *71*, 1287–1294.

Shimamura, M., Murphy-Ullrich, J.E., and Britt, W.J. (2010). Human cytomegalovirus induces TGF-beta1 activation in renal tubular epithelial cells after epithelial-to-mesenchymal transition. PLoS Pathog. *6*, e1001170.

Simmen, K.A., Singh, J., Luukkonen, B.G., Lopper, M., Bittner, A., Miller, N.E., Jackson, M.R., Compton, T., and Fruh, K. (2001). Global modulation of cellular transcription by human cytomegalovirus is initiated by viral glycoprotein B. Proc. Natl. Acad. Sci. U.S.A. *98*, 7140–7145.

Simon, C.O., Seckert, C.K., Dreis, D., Reddehase, M.J., and Grzimek, N.K. (2005). Role for tumor necrosis factor alpha in murine cytomegalovirus transcriptional reactivation in latently infected lungs. J. Virol. *79*, 326–340.

Singh, N. (2005). Late-onset cytomegalovirus disease as a significant complication in solid organ transplant recipients receiving antiviral prophylaxis: a call to heed the mounting evidence. Clin. Infect. Dis. *40*, 704–708.

Singh, N. (2006). Antiviral drugs for cytomegalovirus in transplant recipients: advantages of preemptive therapy. Rev. Med. Virol. *16*, 281–287.

Slinger, E., Maussang, D., Schreiber, A., Siderius, M., Rahbar, A., Fraile-Ramos, A., Lira, S.A., Soderberg-Naucler, C., and Smit, M.J. (2010). HCMV-encoded chemokine receptor US28 mediates proliferative signaling through the IL-6-STAT3 axis. Sci. Signal. *3*, ra58.

Smith, M.G. (1956). Propagation in tissue cultures of cytopathogenic virus from human salivary gland (SVG) virus disease. Proc. Soc. Exp. Biol. Med. *92*, 424–430.

Solez, K., Colvin, R.B., Racusen, L.C., Haas, M., Sis, B., Mengel, M., Halloran, P.F., Baldwin, W., Banfi, G., Collins, A.B., *et al.* (2008). Banff 07 classification of renal allograft pathology: updates and future directions. Am. J. Transplant. *8*, 753–760.

Soroceanu, L., Akhavan, A., and Cobbs, C.S. (2008). Platelet-derived growth factor-alpha receptor activation is required for human cytomegalovirus infection. Nature *455*, 391–395.

Spector, S.A., Hirata, K.K., and Newman, T.R. (1984). Identification of multiple cytomegalovirus strains in homosexual men with acquired immunodeficiency syndrome. J. Infect. Dis. *150*, 953–956.

Spector, S.A., Hsia, K., Crager, M., Pilcher, M., Cabral, S., and Stempien, M.J. (1999). Cytomegalovirus (CMV) DNA load is an independent predictor of CMV disease and survival in advanced AIDS. J. Virol. *73*, 7027–7030.

Stagno, S., Reynolds, D.W., Huang, E.-S., Thames, S.D., Smith, R.J., and Alford, C.A. (1977). Congenital cytomegalovirus infection: occurrence in an immune population. N. Engl. J. Med. *296*, 1254–1258.

Stagno, S., Pass, R.F., Dworsky, M.E., and Alford, C.A. (1982a). Maternal cytomegalovirus infection and perinatal transmission. Clin. Obstet. Gynecol. *25*, 563–576.

Stagno, S., Pass, R.F., Dworsky, M.E., Henderson, R.E., Moore, E.G., Walton, P.D., and Alford, C.A. (1982b). Congenital cytomegalovirus infection: The relative importance of primary and recurrent maternal infection. N. Engl. J. Med. *306*, 945–949.

Stagno, S., Pass, R.F., Dworsky, M.E., and Alford, C.A. (1983). Congenital and perinatal cytomegaloviral infections. Semin. Perinatol. *7*, 31–42.

Stagno, S., Pass, R.F., Cloud, G., Britt, W.J., Henderson, R.E., Walton, P.D., Veren, D.A., Page, F., and Alford, C.A. (1986). Primary cytomegalovirus infection in pregnancy: Incidence, transmission to fetus and clinical outcome. JAMA *256*, 1904–1908.

Staras, S.A., Flanders, W.D., Dollard, S.C., Pass, R.F., McGowan, J.E., Jr., and Cannon, M.J. (2008).

Cytomegalovirus seroprevalence and childhood sources of infection: A population-based study among pre-adolescents in the USA. J. Clin. Virol. *43*, 266–271.

Starr, S.E., Tolpin, M.D., Friedman, H.M., Paucker, K., and Plotkin, S.A. (1979). Impaired cellular immunity to cytomegalovirus in congenitally infected children and their mothers. J. Infect. Dis. *140*, 500–505.

Storek, J., Witherspoon, R.P., and Storb, R. (1995). T-cell reconstitution after bone marrow transplantation into adult patients does not resemble T-cell development in early life. Bone Marrow Transplant. *16*, 413–425.

Storek, J., Dawson, M.A., Storer, B., Stevens-Ayers, T., Maloney, D.G., Marr, K.A., Witherspoon, R.P., Bensinger, W., Flowers, M.E., Martin, P., *et al.* (2001). Immune reconstitution after allogeneic marrow transplantation compared with blood stem cell transplantation. Blood *97*, 3380–3389.

Storek, J., Geddes, M., Khan, F., Huard, B., Helg, C., Chalandon, Y., Passweg, J., and Roosnek, E. (2008). Reconstitution of the immune system after hematopoietic stem cell transplantation in humans. Semin. Immunopathol. *30*, 425–437.

Streblow, D.N., Orloff, S.L., and Nelson, J.A. (2001). Do pathogens accelerate atherosclerosis? J. Nutr. *131*, 2798S–2804S.

Streblow, D.N., Kreklywich, C.N., Smith, P., Soule, J.L., Meyer, C., Yin, M., Beisser, P., Vink, C., Nelson, J.A., and Orloff, S.L. (2005). Rat cytomegalovirus-accelerated transplant vascular sclerosis is reduced with mutation of the chemokine-receptor R33. Am. J. Transplant. *5*, 436–442.

Streblow, D.N., van Cleef, K.W., Kreklywich, C.N., Meyer, C., Smith, P., Defilippis, V., Grey, F., Fruh, K., Searles, R., Bruggeman, C., *et al.* (2007). Rat cytomegalovirus gene expression in cardiac allograft recipients is tissue specific and does not parallel the profiles detected *in vitro*. J. Virol. *81*, 3816–3826.

Streblow, D.N., Dumortier, J., Moses, A.V., Orloff, S.L., and Nelson, J.A. (2008). Mechanisms of cytomegalovirus-accelerated vascular disease: induction of paracrine factors that promote angiogenesis and wound healing. Curr. Top. Microbiol. Immunol. *325*, 397–415.

Takahashi, R., Tagawa, M., Sanjo, M., Chiba, H., Ito, T., Yamada, M., Nakae, S., Suzuki, A., Nishimura, H., Naganuma, M., *et al.* (2007). Severe postnatal cytomegalovirus infection in a very premature infant. Neonatology *92*, 236–239.

Tolkoff-Rubin, N.E., Fishman, J.A., and Rubin, R.H. (2001). The bidirectional relationship between cytomegalovirus and allograft injury. Transplant. P. *33*, 1773–1775.

Tu, W., Chen, S., Sharp, M., Dekker, C., Manganello, A.M., Tongson, E.C., Maecker, H.T., Holmes, T.H., Wang, Z., Kemble, G., *et al.* (2004). Persistent and selective deficiency of CD4+ T-cell immunity to cytomegalovirus in immunocompetent young children. J. Immunol. *172*, 3260–3267.

Valantine, H.A. (2004). The role of viruses in cardiac allograft vasculopathy. Am. J. Transplant. *4*, 169–177.

Valantine, H.A., Gao, S.Z., Menon, S.G., Renlund, D.G., Hunt, S.A., Oyer, P., Stinson, E.B., Brown, B.W., Jr., Merrrigan, T.C., and Schroeder, J.S. (1999). Impact on prophylactic immediate posttransplant ganciclovir on development of transplant atherosclerosis: A post host analysis of a randomized, placebo-controlled study. Circulation *100*, 61–66.

Vochem, M., Hamprecht, K., Jahn, G., and Speer, C.P. (1998). Transmission of cytomegalovirus to preterm infants through breast milk. Pediatr. Infect. Dis. J. *17*, 53–58.

Walmus, B.F., Yow, M.D., Lester, J.W., Leeds, L., Thompson, P.K., and Woodward, R.M. (1988). Factors predictive of cytomegalovirus immune status in pregnant women. J. Infect. Dis. *157*, 172–177.

Walter, E.A., Greenberg, P.D., Gilbert, M.J., Finch, R.J., Watanabe, K.S., Thomas, E.D., and Riddell, S.R. (1995). Reconstitution of cellular immunity against cytomegalovirus in recipients of allogeneic bone marrow by transfer of T-cell clones from the donor. N. Engl. J. Med. *333*, 1038–1044.

Wang, C., Zhang, X., Bialek, S., and Cannon, M.J. (2011). Attribution of congenital cytomegalovirus infection to primary versus non-primary maternal infection. Clin. Infect. Dis. *52*, e11–13.

Wang, G.C., Kao, W.H., Murakami, P., Xue, Q.L., Chiou, R.B., Detrick, B., McDyer, J.F., Semba, R.D., Casolaro, V., Walston, J.D., *et al.* (2010). Cytomegalovirus infection and the risk of mortality and frailty in older women: a prospective observational cohort study. Am. J. Epidemiol. *171*, 1144–1152.

Weis, M., Kledal, T.N., Lin, K.Y., Panchal, S.N., Gao, S.Z., Valantine, H.A., Mocarski, E.S., and Cooke, J.P. (2004). Cytomegalovirus infection impairs the nitric oxide synthase pathway: role of asymmetric dimethylarginine in transplant arteriosclerosis. Circulation *109*, 500–505.

Weller, T.H. (1971). The cytomegaloviruses: ubiquitous agents with protean clinical manifestations. N. Engl. J. Med. *285*, 203–214.

White, N.H., Yow, M.D., Demmler, G.J., Norton, H.J., Hoyle, J., Pinckard, K., Mishaw, C., and Pokorny, S. (1989). Prevalence of cytomegalovirus antibody in subjects between the ages of 6 and 22 years. J. Infect. Dis. *159*, 1013–1017.

Williamson, W.D., Desmond, M.M., LaFevers, N., Taber, L.H., Catlin, F.I., and Weaver, T.G. (1982). Symptomatic congenital cytomegalovirus. Disorders of language, learning and hearing. Am. J. Dis. Child. *136*, 902–905.

Williamson, W.D., Percy, A.K., Yow, M.D., Gerson, P., Catlin, F.I., Koppelman, M.L., and Thurber, S. (1990). Asymptomatic congenital cytomegalovirus infection: audiologic, neuroradiologic, and neurodevelopmental abnormalities during the first year. Am. J. Dis. Child. *144*, 1365–1368.

Williamson, W.D., Demmler, G.J., Percy, A.K., and Catlin, F.I. (1992). Progressive hearing loss in infants with asymptomatic congenital cytomegalovirus infection. Pediatrics *90*, 862–866.

Wills, M., Akbar, A., Beswick, M., Bosch, J.A., Caruso, C., Colonna-Romano, G., Dutta, A., Franceschi, C., Fulop, T., Gkrania-Klotsas, E., *et al.* (2011). Report from the Second Cytomegalovirus and Immunosenescence Workshop. Immun. Ageing *8*, 10.

Winston, D.J., Ho, W.G., Bartoni, K., Du Mond, C., Ebeling, D.F., Buhles, W.C., and Champlin, R.E. (1993). Ganciclovir prophylaxis of cytomegalovirus infection and disease in allogeneic bone marrow transplant recipients. Results of a placebo-controlled, double-blind trial. Ann. Intern. Med. *118*, 179–184.

Wirtz, N., Schader, S.I., Holtappels, R., Simon, C.O., Lemmermann, N.A., Reddehase, M.J., and Podlech, J. (2008). Polycloncal cytomegalovirus-specific antibodies not only prevent virus dissemination from the portal of entry but also inhibit focal virus spread within target tissues. Med. Microbiol. Immunol. *197*, 151–158.

Woolf, N.K. (1991). Guinea pig model of congenital CMV-induced hearing loss: a review. Transplant. P. *23*, 32–34.

Wreghitt, T.G., Hakim, M., Gray, J.J., Kucia, S., Wallwork, J., and English, T.A. (1988). Cytomegalovirus infections in heart and heart and lung transplant recipients. J. Clin. Pathol. *41*, 660–667.

Xavier, R.J., and Podolsky, D.K. (2007). Unravelling the pathogenesis of inflammatory bowel disease. Nature *448*, 427–434.

Yamamoto, A.Y., Mussi-Pinhata, M.M., Boppana, S.B., Novak, Z., Wagatsuma, V.M., Oliviera, P.D., Duarte, G., and Britt, W.J. (2010). Human cytomegalovirus reinfection is associated with intrauterine transmission in a highly cytomegalovirus-immune maternal population. Am. J. Obstet. Gynecol. *202*, 297.e291–e298.

Yamamoto, A.Y., Mussi-Pinhata, M.M., Isaac, M.D.L., Amaral, F.R., Carvalheiro, C.G., Aragon, D.C., da Silva Mafredi, A.K., Boppana, S.B., and Britt, W.J. (2011). Congenital cytomegalovirus infection as a cause of sensorineural hearing loss in a highly seropositive population. Pediatr. Infect. Dis. J. *30*.1043–1046.

Yeager, A.S. (1983). Transmission of cytomegalovirus to mothers by infected infants: another reason to prevent transfusion-acquired infections. Pediatr. Infect. Dis. J. *2*, 295–297.

Yow, M.D., Williamson, D.W., Leeds, L.J., Thompson, P., Woodward, R.M., Walmus, B.F., Lester, J.W., Six, H.R., and Griffiths, P.D. (1988). Epidemiologic characteristics of cytomegalovirus infection in mothers and their infants. Am. J. Obstet. Gynecol. *158*, 1189–1195.

Zaia, J.A., Gallez-Hawkins, G.M., Tegtmeier, B.R., ter Veer, A., Li, X., Niland, J.C., and Forman, S.J. (1997). Late cytomegalovirus disease in marrow transplantation is predicted by virus load in plasma. J. Infect. Dis. *176*, 782–785.

Zhang, Z., Li, Z., Yan, S., Wang, X., and Abecassis, M. (2009). TNF-alpha signaling is not required for *in vivo* transcriptional reactivation of latent murine cytomegalovirus. Transplantation *88*, 640–645.

Zhu, J., Shearer, G.M., Norman, J.E., Pinto, L.A., Marincola, F.M., Prasad, A., Waclawiw, M.A., Csako, G., Quyyumi, A.A., and Epstein, S.E. (2000). Host response to cytomegalovirus infection as a determinant of susceptibility to coronary artery disease: sex-based differences in inflammation and type of immune response. Circulation *102*, 2491–2496.

The Epidemiology and Public Health Impact of Congenital Cytomegalovirus Infection

II.2

Michael J. Cannon, Scott D. Grosse and Karen B. Fowler[1]

Abstract

Congenital cytomegalovirus (CMV) is one of the most important causes of childhood disability in many countries. The dynamics of congenital CMV are driven by a complex interplay between factors that include the prevalence of CMV infection in the population, the frequency of viral shedding in bodily fluids, the relative importance of different transmission routes, and the incidence of CMV infection among women of reproductive age. Among children infected congenitally, many experience no disease sequelae, while others experience one or more disabilities which can vary in severity and which may be present at birth or may have a delayed onset. From a public health perspective, avenues for addressing congenital CMV include raising awareness, promoting primary prevention through behavioural change, and assessing the utility of prenatal and newborn screening.

Introduction

More than 40 years of research on the epidemiology of congenital CMV infection and disease has led to a clearer understanding of how mothers may become infected, how they transmit infection to their fetuses, and how congenital CMV can lead to hearing loss and developmental disabilities.

Prevalence of CMV infection

CMV infections are endemic and without seasonal variation and have been identified in every population that has been studied (Gold and Nankervis, 1976). The exact mode of CMV acquisition is unknown but is assumed to be through direct contact with body fluids from an infected person (Cannon *et al.*, 2011). Since most CMV infections are subclinical and therefore individuals do not know when they were infected with the virus, CMV infections are usually measured by the presence of anti-CMV antibodies in the blood. CMV prevalence (i.e. seroprevalence) is therefore a measure of the proportion of the population infected by CMV at some time in the past.

CMV seroprevalence has long been known to vary among differing geographic regions with developing countries in Africa, Asia and South America having very early acquisition of virus during early childhood (Alford and Pass, 1981; Gold and Nankervis, 1982; Kaye *et al.*, 2008). In contrast, studies in North America, Western Europe and Australia have shown lower CMV seroprevalence rates in children and young adults (Alford and Pass, 1981; Gold and Nankervis, 1982; Seale *et al.*, 2006; Staras *et al.*, 2006; Bate *et al.*, 2010; Lopo *et al.*, 2011). Essentially, CMV seroprevalence has maintained this similar geographic pattern over the decades (de Ory *et al.*, 2004; Marshall and Stout, 2005; Svahn *et al.*, 2006; Bate *et al.*, 2010).

CMV seroprevalence varies substantially by not just geographic location, but also demographic characteristics, behaviours, and exposure to infected persons. Therefore, due to the potential for uncontrolled confounding, it can be difficult to identify factors that are independently associated with CMV seropositivity. For example, if a study in one country reports seroprevalence in women while another reports seroprevalence in men, any geographic differences may be confounded by sex differences. Factors that are independently associated with CMV seropositivity may be identified by looking at trends across many studies and by using studies that attempt to control for confounding by using multivariate statistical methods.

1 The findings and conclusions in this report are those of the authors and do not necessarily represent the official position of the Centers for Disease Control and Prevention.

Geographic distribution

Studies of pregnant women or women of reproductive age may be the most useful for making geographic comparisons since their populations are similar by age and sex and such studies have been done in many countries. According to these studies, rates of seroprevalence in women are highest in Africa, South America and Asia, frequently reaching nearly 100% (Stagno *et al.*, 1982; Kamada *et al.*, 1983; Vial *et al.*, 1986; Schoub *et al.*, 1993; Yamamoto *et al.*, 2001; Adjei *et al.*, 2008; Chen, 2008; Dar *et al.*, 2008). With some exceptions, CMV seroprevalence is lower in North America, Western Europe and Australia, sometimes lower than 50% (Fig. II.2.1) (de Ory *et al.*, 2004; Seale *et al.*, 2006; Bate *et al.*, 2010; Cannon *et al.*, 2010). Many countries have no seroprevalence studies reported in the English-language literature or the studies are too small to constitute a reliable sample (Cannon *et al.*, 2010).

Demographic and maternal characteristics

Demographic differences in CMV seroprevalence help identify groups with higher infection risk and provide clues as to which behaviours and exposures may be responsible for viral transmission.

Age

Age is a strong predictor of CMV seropositivity in countries with moderate to low seroprevalence since seropositivity represents cumulative lifetime risk of infection. In North America, CMV seroprevalence typically rises steadily from childhood through age 60 and above (Fig. II.2.2). Seroprevalence among 6- to 11-year-olds in the USA is approximately 37% (Staras *et al.*, 2006; Bate *et al.*, 2010). Studies from Australia, New Zealand and Western Europe report similar results in young children (de Ory *et al.*, 2004; Seale *et al.*, 2006; Svahn *et al.*, 2006; O'Brien *et al.*, 2009). A similar pattern is found outside of North America and Western Europe, except that seroprevalence rises much more rapidly during childhood (Fig. II.2.2). In all populations, seroprevalences tend to be greater than 70–80% by age 50.

A nationally representative study in the USA found that seroprevalences tended not to change over time within age-groups, suggesting that the age-related upwards trend in CMV seroprevalence is due to the

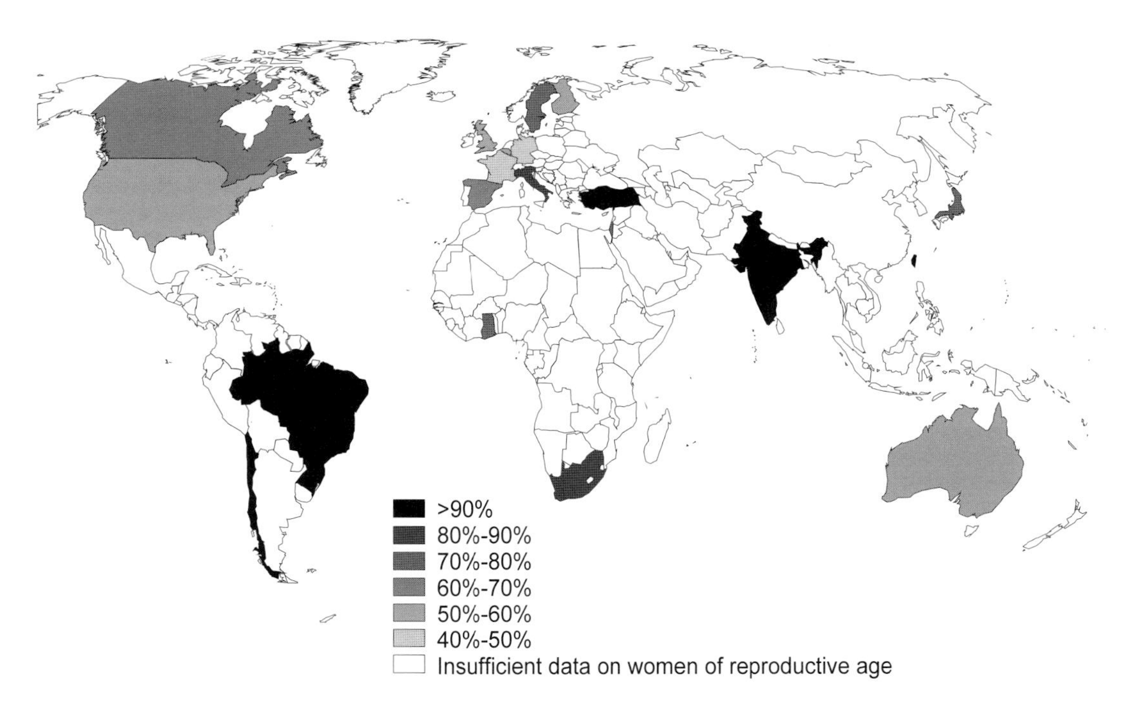

Figure II.2.1 Worldwide CMV seroprevalences among women of reproductive age (reproduced from Cannon *et al.*, 2010). Percentages were obtained by adding the number of seropositive women from all studies within a given country and dividing that number by the total number of women tested. Reproductive age was generally defined as between 12 and 49 years of age. To be included in the map, a country had to have a total of at least 500 women tested.

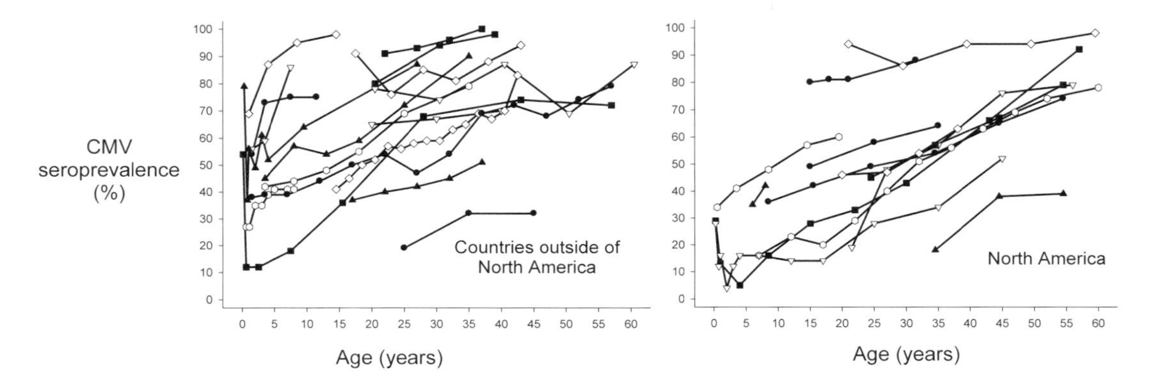

Figure II.2.2 CMV seroprevalence in various studies according to age (reproduced from Cannon *et al.*, 2010). Each line represents data from a single study. Data were presented in two panels (outside North America vs. North America) for better viewing clarity.

occurrence of new infections at each age rather than to a 'birth-cohort effect', where earlier cohorts had higher infection risk during childhood or adolescence (Bate *et al.*, 2010).

Race/ethnicity

Across many studies there is a consistent association within countries between non-White race/ethnicity and CMV seropositivity (Fig. II.2.3). On average, CMV seroprevalence in the USA and Europe is about 1.6 times higher in non-Whites compared to Whites (Cannon *et al.*, 2010). The reasons for these differences are unclear but may include differences in breastfeeding practices, child care practices, sexual behaviour and socioeconomic status (SES) (Cannon *et al.*, 2010).

Sex

In most studies, seroprevalence is slightly higher among women (Fig. II.2.3) (Cannon *et al.*, 2010). Seroprevalence sex differences do not appear in children but increase with age, suggesting that exposures during adolescence and adulthood are responsible for the higher seroprevalence among females (Bate *et al.*, 2010).

Socioeconomic status (SES)

SES, typically measured by income, is inversely associated with CMV seroprevalence. Across more than a dozen studies that measured this variable, CMV seroprevalence was 1.33 times higher among persons with lower SES (Fig. II.2.3) (Cannon *et al.*, 2010). Because CMV seroprevalence tends to be higher in developing countries and higher among lower SES persons within countries, crowding (i.e. number of persons per room) may facilitate transmission (Bate *et al.*, 2010).

Parity

Increasing parity in pregnant women has been independently associated with CMV seroprevalence (Chandler *et al.*, 1985a; Tookey *et al.*, 1992; Gratacap-Cavallier *et al.*, 1998; Knowles *et al.*, 2005). The relationship of parity with CMV seroprevalence could be accounted for by child-to-mother transmission in the household (Tookey *et al.*, 1992).

Other demographic characteristics

A number of other demographic characteristics have been evaluated but most are not measured in enough studies to support a reliable conclusion about associations with CMV seroprevalence. In two nationally representative samples in the USA, CMV seroprevalence decreased with increasing education, while it increased with increasing family size and crowding (Staras *et al.*, 2006; Bate *et al.*, 2010).

Multivariate analyses

In multivariate analyses in the USA, CMV seropositivity was associated with older age, black race, Mexican American ethnicity, female sex, lower income, lower education, and higher household crowding, although not all associations were statistically significant in all studies (Stover *et al.*, 2003; Staras *et al.*, 2006; Bate *et al.*, 2010; Stadler *et al.*, 2010).

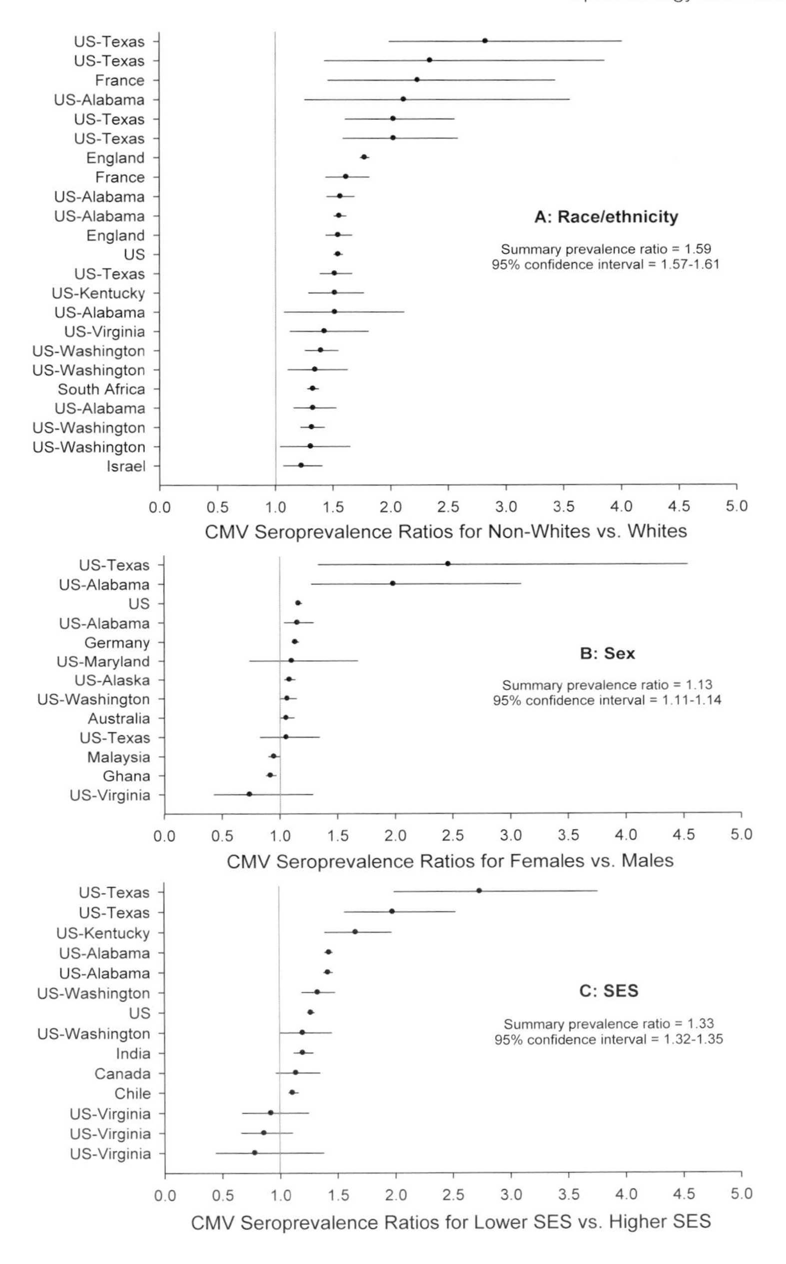

Figure II.2.3 CMV seroprevalence ratios comparing non-whites to whites, females to males, and lower socioeconomic status (SES) to higher SES (reproduced from Cannon *et al*., 2010). Each circle and associated 95% confidence interval comes from an individual study.

Behaviours

Associations of CMV seroprevalence with demographic characteristics indirectly suggest potential transmission routes. Studies of particular behaviours or behavioural risk factors are able to more directly identify transmission routes.

Sexual activity

A number of studies have demonstrated that CMV seroprevalence is higher in groups who on average report higher levels of sexual risk behaviours, such as STD clinic-attending men who have sex with men (Drew *et al.*, 1981; Mindel and Sutherland, 1984), women and adolescents who have multiple and new sexual partners

(Chandler *et al.*, 1985a,b; Collier *et al.*, 1990; Sohn *et al.*, 1991; Coonrod *et al.*, 1998), and persons who have other sexually transmitted diseases (Chandler *et al.*, 1985a,b; Collier *et al.*, 1990; Pereira *et al.*, 1990; Sohn *et al.*, 1991; Hyams *et al.*, 1993; Coonrod *et al.*, 1998) or who are HIV positive (Stover *et al.*, 2003). However, in studies of the general population, the association between CMV seropositivity and sexual behaviour is less pronounced and may only be found within some population subgroups such as non-Hispanic black and non-Hispanic white women (Staras *et al.*, 2008b). Furthermore, it is important to note that these studies are unable to identify particular transmission modes, although it seems plausible that CMV transmission may occur through several routes, such as kissing, oral sex or vaginal sex.

Caring for young children

Young children in the household often introduce CMV to susceptible parents (Spector and Spector, 1982; Dworsky *et al.*, 1983b; 1984; Taber *et al.*, 1985; Pass *et al.*, 1986; Adler, 1991b). Women who care for young children in the home or as day care providers are more likely to be CMV seropositive than women who do not (Adler, 1989; Pass *et al.*, 1990; Murph *et al.*, 1991; Ford-Jones *et al.*, 1996; Bale *et al.*, 1999; Fowler and Pass, 2006). However, these studies cannot indicate with certainty that CMV infections in day care providers were due to child-to-worker transmission.

Breastfeeding

Prevalence of infection is higher among babies who are breastfed and increases with duration of breastfeeding (Dworsky *et al.*, 1983a; Hamprecht *et al.*, 2001, 2008; Schleiss, 2006; Britt, 2011). Approximately 40% of all infants who are breast fed for at least 1 month by CMV-seropositive mothers become infected postnatally (Britt, 2011).

Exposure to persons known to be infected with CMV

Having older siblings and mothers who are known to be CMV seropositive is a very strong predictor of CMV seropositivity among children. In a large, nationally representative study in the USA, children who had an older sibling or a mother who was CMV seropositive had seroprevalences 30–50 percentage points higher than children who had only seronegative family members (Staras *et al.*, 2008a). Children in these households may have been exposed to CMV through breastfeeding or through contact with other children in the household.

Studies of discordant couples have shown that CMV-seropositive partners can transmit virus to their partners (Handsfield *et al.*, 1985; Chandler *et al.*, 1987). In a large study of couples at a fertility clinic, women who had a CMV-seropositive partner were several times more likely to be seropositive than women whose partner was seronegative (Francisse *et al.*, 2009).

Birth prevalence of congenital CMV infection

Prevalence of congenital CMV infection varies according to different factors but most typically is between about 0.4% and 1.0% in low-seroprevalence countries (Kenneson and Cannon, 2007). Recent US studies have reported estimates in the lower half of that range, between 0.4% and 0.7% (Boppana *et al.*, 2010; 2011; Kharrazi *et al.*, 2010; DePasquale *et al.*, 2011).

Studies that measure prevalence of fetal infection tend to report lower values than those that measure birth prevalence (Kenneson and Cannon, 2007); this might be because fetal ascertainment studies have typically been done in Western Europe where underlying prevalences are probably lower or because they select women who receive prenatal care and therefore may be higher SES than the general population.

Type of maternal infection

The natural history of CMV in pregnancy is complex. Primary maternal infections during pregnancy are more likely to be transmitted to the fetus than non-primary maternal infections (Stagno *et al.*, 1977; 1986; Fowler *et al.*, 2004). In almost 20 studies where the mother was diagnosed with primary infection during pregnancy, the average rate of congenital infection was 32.3% (Fig. II.2.4) (Kenneson and Cannon, 2007). In seven studies with possible primary maternal infection, the congenital rate averaged 20.7%. In 10 studies where maternal infection was shown to be non-primary, the rate of congenital infection was 1.4%, similar to the rate found in studies where type of maternal infection is unknown.

Maternal seroprevalence in the underlying population

There is evidence that congenital CMV infection occurs more frequently in populations that have higher

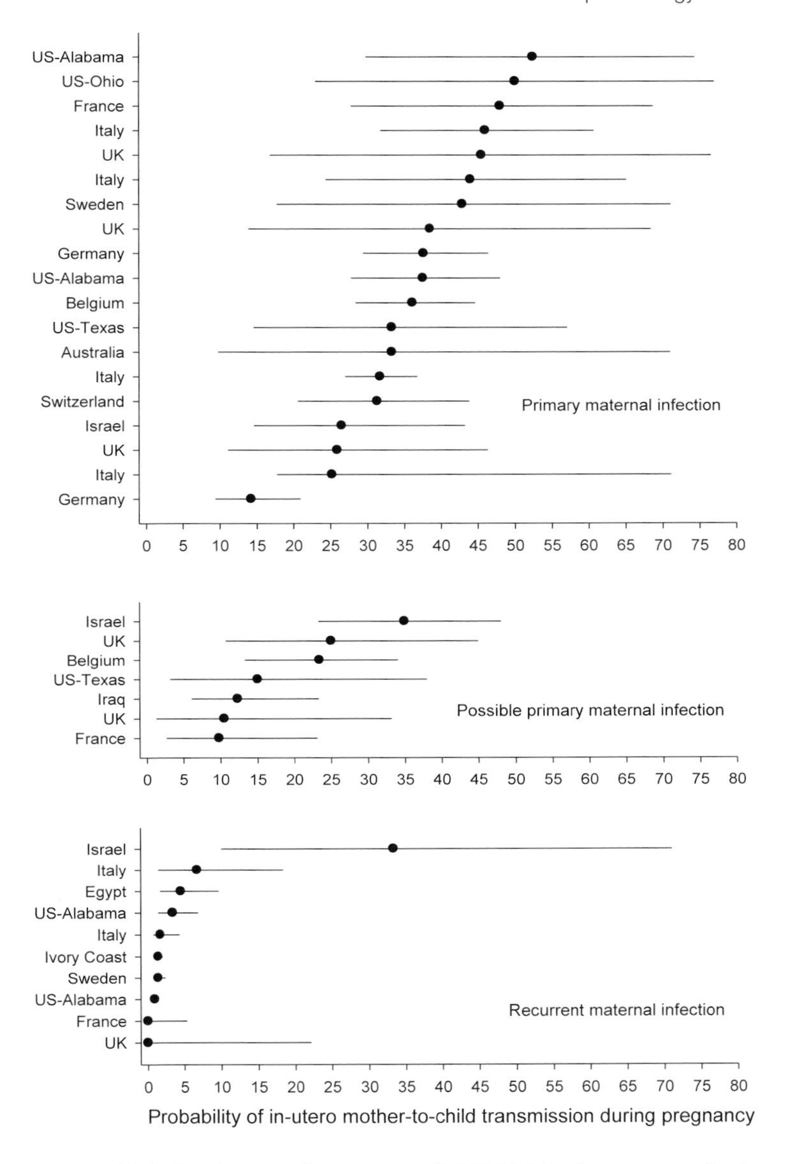

Figure II.2.4 Percentage of infants who experience congenital CMV infection as a result of a maternal primary infection during pregnancy (upper panel) (reproduced from Kenneson and Cannon, 2007). Percentage of infants who experience congenital CMV infection as a result of a possible maternal primary infection during pregnancy (centre panel). Percentage of infants who experience congenital CMV infection as a result of a maternal recurrent infection during pregnancy (lower panel). Lines represent 95% confidence intervals.

maternal seroprevalence. In an analysis that included 20 studies, every 10% increase in maternal seroprevalence resulted in a 0.26% increase in CMV birth prevalence (Kenneson and Cannon, 2007). Furthermore, maternal seroprevalence accounted for about 29% of the variability of CMV birth prevalence. This may seem somewhat paradoxical since primary maternal infections have such a high rate of transmission to the fetus. However, primary infections in seronegative women are relatively infrequent whereas more women are seropositive and they experience re-infections or reactivations that infect their fetuses. As a result, most congenital infections are associated with non-primary maternal infections (Stagno *et al.*, 1977; 1986; Yamamoto *et al.*, 2001; Mussi-Pinhata *et al.*, 2009; Wang *et al.*, 2011).

Demographic and maternal characteristics

Age

Young maternal age has been consistently reported to be linked to having an offspring with congenital CMV infection in differing populations (Larke *et al.*, 1980; Preece *et al.*, 1986; Fowler and Pass, 1991, 2006; Fowler *et al.*, 1993; Murph *et al.*, 1998; Mussi-Pinhata *et al.*, 2009). Risk of congenital CMV infection seems to have an inverse relationship with maternal age, with teen mothers (< 19 years) reported to have the highest rates of congenital CMV infections in their newborns (Preece *et al.*, 1986; Fowler *et al.*, 1993).

Race/ethnicity

Non-white race has been reported to be linked to congenital CMV infection in study populations of infants in which race was included as a parameter (Preece *et al.*, 1986; Fowler *et al.*, 1993; van der Sande *et al.*, 2007). Those studies which have included ethnicity as a parameter report that US Hispanic infants are at increased risk of congenital CMV infection relative to non-Hispanic white infants but lower than non-Hispanic black infants (Kharrazi *et al.*, 2010; DePasquale *et al.*, 2011). In Asian populations with high maternal CMV seroprevalence rates, congenital CMV infection rates are high, similar to rates observed in black infants in the USA and England (Sohn *et al.*, 1992; Tsai *et al.*, 1996; Dar *et al.*, 2008). One exception is Japan where even with high maternal CMV seroprevalence rates, congenital infection rates are similar to rates observed in white infants in North America and Western Europe (Kamada *et al.*, 1983; Preece *et al.*, 1986; Numazaki and Fujikawa, 2004).

Socioeconomic status (SES)

Lower SES has been consistently associated with higher rates of congenital CMV infection in differing populations. Although older studies often did not sort out the relative contribution of SES and race/ethnicity by adjusting for possible confounders, newer studies demonstrate an independent effect of SES controlling for race/ethnicity. One of the best predictors is low maternal education (DePasquale *et al.*, 2011).

Parity

In some studies parity has been associated with congenital CMV infection. Primiparous mothers appear to have an increased risk of offspring with congenital CMV infection (Larke *et al.*, 1980; Fowler and Pass, 1991, 2006; Noyola *et al.*, 2003). Although parity is linked to CMV seroprevalence and risk of primary infections during pregnancy, it is not clear that parity is an independent factor for congenital CMV infection.

Behaviours

Recent onset of sexual activity (< 2 years), sexually transmitted diseases during pregnancy, and history of sexually transmitted diseases have been independently associated with congenital CMV infection in the children of young urban women (Fowler and Pass, 1991, 2006). In the same population, caring for preschool children in the year before delivery and living in a household with more than three people were also risk factors for delivering a baby with congenital CMV infection (Fowler and Pass, 2006). In addition, women who both cared for preschool children in the year before delivery and also became sexually active < 2 years before delivery were at greatest risk (aOR = 7.2; 95% CI 3.2–16.1) of delivering a newborn with congenital CMV infection (Fowler and Pass, 2006). These data suggest that both sexual and young child exposures contribute to risk of congenital CMV infections in young urban populations.

Incidence of CMV infection

Whereas CMV seroprevalence studies measure cumulative lifetime risk of infection, CMV incidence studies assess the occurrence of new (i.e. primary) infections. Rates of primary CMV infection vary widely and are highly dependent on various risk factors.

General population

Most studies of CMV incidence have been done among particular risk groups rather than the general population (Hyde *et al.*, 2010). As a result, one of the best ways to assess incidence of CMV infection in the general population is through mathematical modelling using age-specific seroprevalence data. In a nationally representative study of 12- to 49-year-olds in the USA, the estimated annual rate of infection (which is approximated by a parameter called the force of infection) was 1.6% (Colugnati *et al.*, 2007). A similar study that used mathematical modelling found incidence of infection among pregnant women in England to be between 3.1% and 3.5% per year (Griffiths *et al.*, 2001).

Demographic characteristics

In the USA modelling study, annual incidence was significantly higher among non-Hispanic blacks (5.7%) and Mexican Americans (5.1%) than among non-Hispanic Whites (1.4%) (Fig. II.2.5) (Colugnati *et al.*, 2007; Cannon, 2009). Infection rates were also significantly higher in the low income group (3.5%) than in the middle (2.1%) and upper (1.5%) income groups. Age and race/ethnicity were also important predictors of incidence among seronegative persons. Compared to other age and race/ethnicity subgroups, incidence rates were highest among pre-adolescent Mexican-American girls (11.0%), adolescent non-Hispanic black girls (9.9%) and adolescent Mexican-American boys (8.7%) (Colugnati *et al.*, 2007).

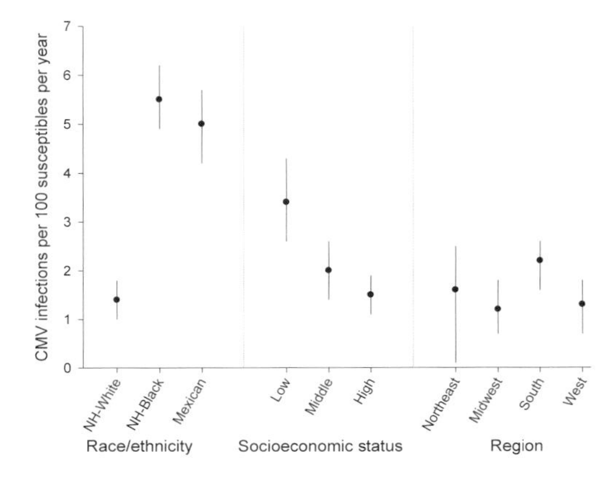

Figure II.2.5 Estimated CMV force of infection by race/ethnicity, socioeconomic status (SES), and region (adapted from Colugnati *et al.*, 2007). Data are from the Third National Health and Nutrition Examination Survey, 1988–1994.

Comparison with other infectious agents

Mathematical modelling studies allow for comparisons of incidence rates for different infectious agents. The force of infection (i.e. incidence) was much lower in similar age groups for CMV than for measles, mumps, rubella and varicella, but higher than for herpes simplex virus type 2 (HSV-2) and hepatitis B, which are frequently associated with sexual transmission (Table II.2.1) (Colugnati *et al.*, 2007).

Exposure to persons known to be infected with CMV

In a large cohort of women attending a fertility clinic, the risk of seroconversion was almost twice as high if their male sex partner was CMV seropositive rather than seronegative (Francisse *et al.*, 2009). Other studies found higher risk of seroconversion among those exposed to a CMV-seropositive husband (Numazaki *et al.*, 2000) or mother (Cabau *et al.*, 1979), a recently seroconverted family member (Taber *et al.*, 1985; Yow *et al.*, 1987), or children shedding CMV (Pass *et al.*, 1986; Adler, 1991a; Murph *et al.*, 1991).

Table II.2.1 Comparison of forces of infection for different viruses for selected[a] age ranges

Virus	Force of infection (per 100 persons per year)	Ages modelled	Study sample	Reference
Measles	20	11–17	Literature review – miscellaneous sources	Edmunds *et al.* (2000)
Mumps	12	11–17	Literature review – miscellaneous sources	Edmunds *et al.* (2000)
Rubella	10	11–17	Literature review – miscellaneous sources	Edmunds *et al.* (2000)
Varicella	6	≥10	Convenience sample	Gidding *et al.* (2003)
CMV[b]	3.1 and 3.5	16–40	Hospital-based	Griffiths *et al.* (2001)
CMV	1.8	12–49	Population-based	Colugnati *et al.* (2007)
HSV-2	0.84	≥12	Population-based	Armstrong *et al.* (2001)
Hepatitis B	0.15	6–39	Population-based	Coleman *et al.* (1998)

[a]Ages were selected to be roughly comparable with the ages modelled for CMV; in general, young children were not selected for comparison because they often had much higher forces of infection.
[b]Patients were recruited from two different hospitals.

Risk groups

Seroconversion studies have been able to detect considerable variation in rates of CMV infection among different risk groups. A review of various studies (Hyde *et al.*, 2010) found the highest annual seroconversion rates among parents of children who were shedding CMV (24%) and day care workers (8.5%), with much lower rates among pregnant women in the general population (2.3%), healthcare workers (2.3%), including those caring for infants and children, and parents whose child was not shedding CMV (2.1%) (Fig. II.2.6) (Hyde *et al.*, 2010). A few studies found elevated seroconversion rates among female minority adolescents (consistent with the modelling data) (Zanghellini *et al.*, 1999; Stanberry *et al.*, 2004) and women attending sexually transmitted disease clinics (Chandler *et al.*, 1985b; Coonrod *et al.*, 1998), but these studies had relatively small numbers of participants. Among HIV-infected persons, men who had sex with men had substantially higher seroconversion rates than heterosexual men and women (Robain *et al.*, 1998).

Prevalence of CMV in bodily fluids by risk group

CMV can be found in a wide range of bodily fluids, but shedding can be sporadic (Rosenthal *et al.*, 2009). This shedding pattern is consistent with latency and periodic reactivation, persistent low-level chronic infection, or re-infection with different CMV strains. Shedding in bodily fluids is important for understanding potential transmission modes. Shedding patterns have been characterized according to risk groups, age, and fluid type. Here we define shedding as the presence of CMV detected by culture techniques or CMV DNA detected by PCR.

Risk groups

In a review of many studies that reported CMV shedding, children who were born with congenital CMV were the most likely to subsequently be shedding. This makes sense because they were all, by definition, shedding when they were born. The next most frequent group of shedders is children who attend out-of-home child care. Their median shedding prevalence in any given fluid (usually urine or saliva) was 24% (Cannon *et al.*, 2011). Children who are not in out-of-home child care also shed frequently. In a number of studies their median shedding prevalence was 12% (Fig. II.2.7, reproduced from Cannon *et al.*, 2011). In multiple studies of children with medical conditions unrelated to CMV, shedding prevalences were similar to those of children not in out-of-home child care. Among adults, shedding prevalences are somewhat lower, approximately 7%, unless they had risk factors related to CMV infection, such as having a child with congenital CMV or being an STD clinic attendee, in which case the median shedding prevalence was 22% (Cannon *et al.*, 2011).

Seroconversion

Among adults, shedding is much more prevalent for several months following a seroconversion event than among adults in general (Cannon *et al.*, 2011).

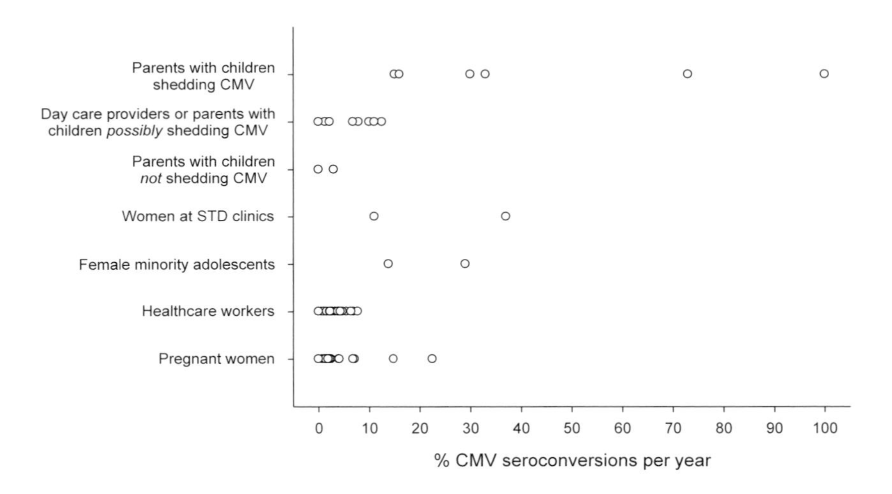

Figure II.2.6 CMV annual seroconversion rates according to selected population groups (reproduced from Hyde *et al.*, 2010). Each circle is the result for a single population from a single study.

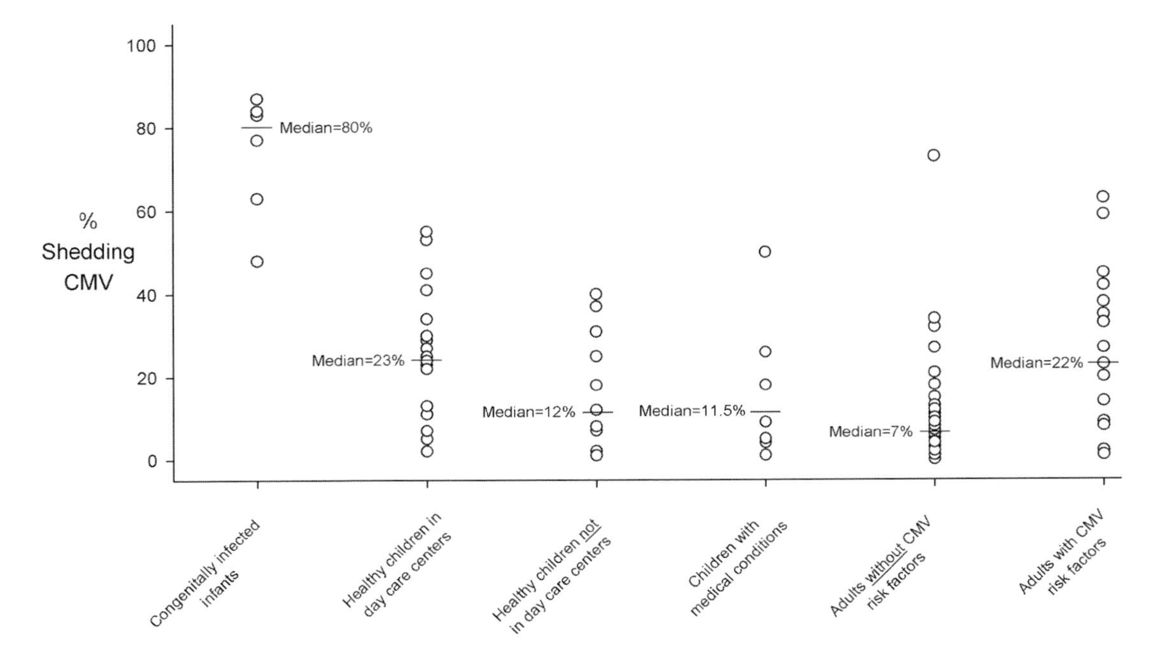

Figure II.2.7 Prevalences of CMV shedding in various studies that measured a single specimen type (e.g. urine, oral secretions) at a single point in time (reproduced from Cannon *et al*., 2011). Each circle represents the result from an individual study. Studies that measured more than one specimen type have more than one circle plotted above. Prevalences were not stratified by age.

Age

CMV shedding prevalence appears to be highly dependent on age. Shedding prevalences in children are, in general, higher than in adults. Studies that look at shedding prevalence among different age groups of children find that prevalences start relatively low in infants (i.e. < 1 year old), peak in 1- and 2-year-olds, and then begin to drop to approximately adult levels by the time the children are 5 years old (Fig. II.2.8) (Cannon *et al.*, 2011). Data on age-related shedding prevalences in adults are limited, although in one study it appeared that shedding decreased slightly with age among adolescents and women of reproductive age (Stagno *et al.*, 1975).

Fluid type or location

Within individuals, prevalence of CMV shedding differs by specimen type. In children, almost without exception, CMV is found more frequently in urine than in oral secretions, although the median difference is only about 10% (Fig. II.2.9) (Cannon *et al.*, 2011). CMV is less likely to be found on children's hands than in their urine or oral secretions. Among adults, the association with shedding and fluid type is less clear. Adults who did not have CMV risk factors were most likely to shed in genital secretions (i.e. cervical or vaginal swabs or lavages) than in urine, whereas adults with CMV risk factors were more likely to shed in urine (Fig. II.2.9) (Cannon *et al.*, 2011). However, shedding studies in adults are limited because they frequently did not collect oral secretions.

CMV transmission

Fetal transmission

The fetus becomes infected via the placenta and this transmission route will be discussed in more detail in other chapters (most specifically Chapter II.4). Such transmission is most likely to occur if the mother is experiencing a primary infection (30–40% chance) (Kenneson and Cannon, 2007). However, because primary infection is relatively infrequent among seronegative women and most women are seropositive before conception, the majority of congenital infections are due to maternal non-primary infections (Stagno *et al.*, 1977; 1986; Yamamoto *et al.*, 2001; Mussi-Pinhata *et al.*, 2009; Wang *et al.*, 2011). There is evidence that non-primary infections that result in congenital infection may be re-infections with a different strain of CMV rather than reactivations of latent virus (Boppana *et al.*, 2001; Ross *et al.*, 2010).

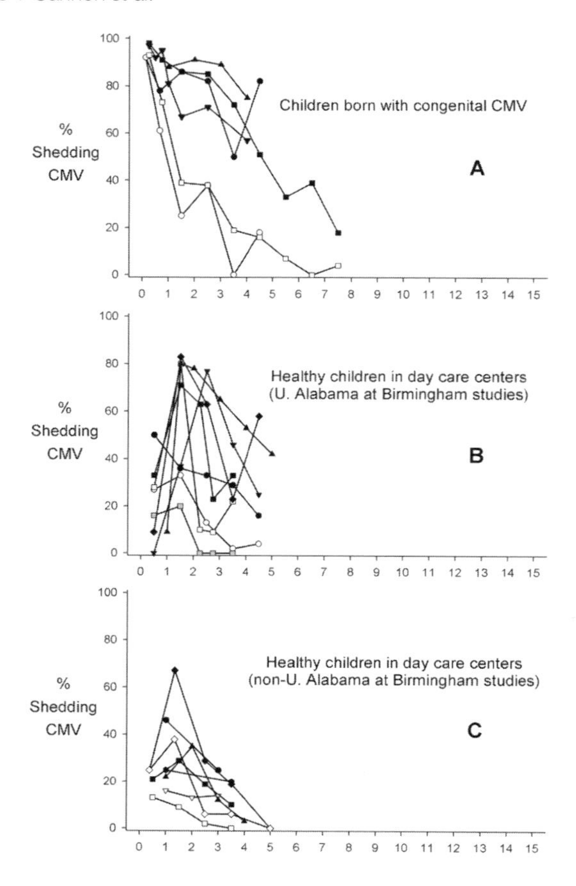

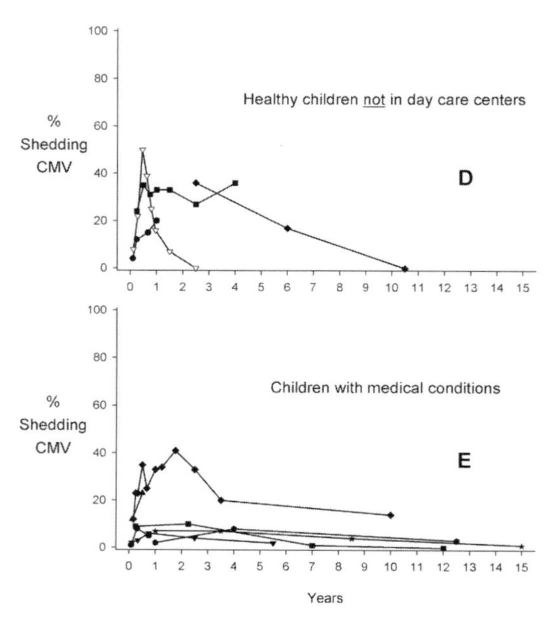

Figure II.2.8 CMV shedding as a function of age for different populations of children (reproduced from Cannon *et al.*, 2011). Each line represents results from a single study. Black symbols show shedding in urine, white symbols show shedding in oral secretions, and grey symbols show shedding in other specimens. Midpoints were used for ages when an age interval was reported.

Post-natal transmission

Based on data on CMV seroprevalence, seroconversion and shedding, it is possible to reach a number of conclusions about CMV transmission. Although CMV can be transmitted via multiple routes, it appears to be usually transmitted through direct contact with bodily fluids. Airborne transmission by aerosol or droplets is highly unlikely. The force of infection for CMV is much lower than for other viruses that have known airborne transmission (Colugnati *et al.*, 2007). Infection rates are not increased among healthcare workers, even though many care for sick children (Hyde *et al.*, 2010). In addition, despite CMV shedding being highly prevalent among young children, transmission to day care workers and parents occurs at much lower rates than ought to be expected if airborne transmission were occurring (Hyde *et al.*, 2010). CMV transmission via contaminated food or water also seems implausible since, as an enveloped virus, CMV does not survive well outside of the body (Stowell *et al.*, 2012) and is susceptible to detergents or drying out. Furthermore, there are no reports in the literature of point-source CMV outbreaks. Instead, it is believed that CMV transmission requires direct contact with bodily fluids, and that repeated exposures are generally needed for transmission to occur. Importantly, child-to-parent transmission must be highly inefficient because only one in four parents who have a child shedding CMV can expect to become infected over the course of a year, despite numerous exposures to child saliva and urine (Hyde *et al.*, 2010).

Children

The epidemiological data suggest that the primary routes of transmission for children are via breastfeeding (Dworsky *et al.*, 1983a; Hamprecht *et al.*, 2001, 2008; Schleiss, 2006; Britt, 2011) or direct contact with other fluids from their mothers or other young children, including siblings (Spector and Spector, 1982; Dworsky *et al.*, 1984; Taber *et al.*, 1985; Pass and Hutto, 1986; Adler, 1991b; Staras *et al.*, 2008a). Direct sources of contact may include kissing or sharing food with the mother or mouthing toys that other children have mouthed. The particular routes of child-to-child transmission have not been well-studied, perhaps because infection among children appears to cause little serious clinical disease.

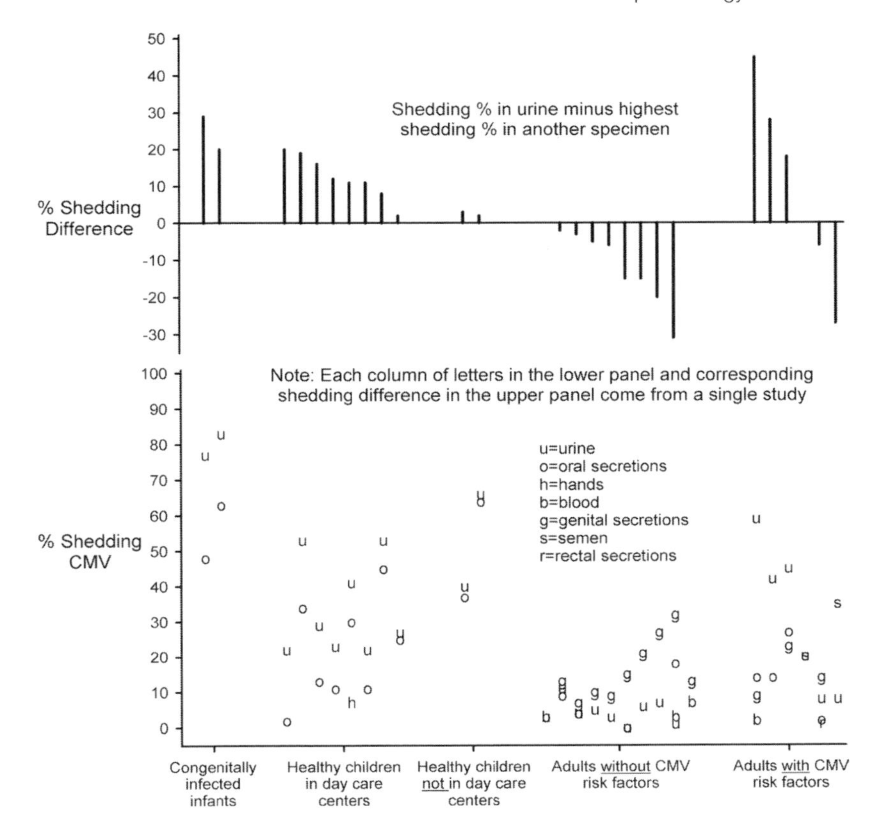

Figure II.2.9 Comparison of CMV shedding from different bodily fluids and locations according to risk group (reproduced from Cannon *et al.*, 2011). Studies were only included if they measured CMV shedding in multiple specimen types. The lower panel compares shedding in different specimens (e.g. urine vs. oral secretions). Each column of letters comes from a single study. Directly above each column of letters, in the top panel, the corresponding difference is shown between shedding percentage in urine and the highest shedding percentage in another specimen.

Women of reproductive age

The epidemiology of CMV infection in the USA during pregnancy is shown in Fig. II.2.10. It is likely that other developed countries share a similar epidemiological pattern. Parents with a child shedding CMV are ten times more likely to seroconvert than various comparison groups, including parents whose child is not shedding CMV (Fig. II.2.6) (Hyde *et al.*, 2010). These high rates of seroconversion are unlikely to be explained by differences in sexual activity, since there is no reason to believe that the sexual risks of parents in these groups differ substantially.

Women may also be infected through occupational exposure while working with young children in out-of-home child care (Adler, 1989; Murph *et al.*, 1991; Ford-Jones *et al.*, 1996; Bale *et al.*, 1999; Hyde *et al.*, 2010). However, health care workers, including those who care for children, do not appear to have elevated risk due to occupational exposure (Friedman *et al.*, 1984; Balfour and Balfour, 1986; Hatherley, 1986; Brady *et al.*, 1987; Balcarek *et al.*, 1990; Gerberding, 1994; Hyde *et al.*, 2010).

The seroconversion studies in women and shedding studies in young children suggest that the most important source of infection for women is contact with young children, presumably from exposure to their saliva and urine and that transmission likely occurs when they get these fluids in their eyes, nose, or mouth. Behaviours frequently practiced by women that are likely to expose them to these fluids include kissing young children on the lips and sharing food, cups, or utensils with young children (Cannon *et al.*, 2012). Women may also become infected after touching urine or nasal secretions and then touching their faces. However, CMV does not survive long on hands (J.D. Stowell, D. Forlin-Passoni, K. Radford, S.L. Bate, S.C. Dollard, S.R. Bialek, M.J. Cannon and D.S. Schmid, unpublished data), it is less likely to be infectious once a surface is not visibly wet

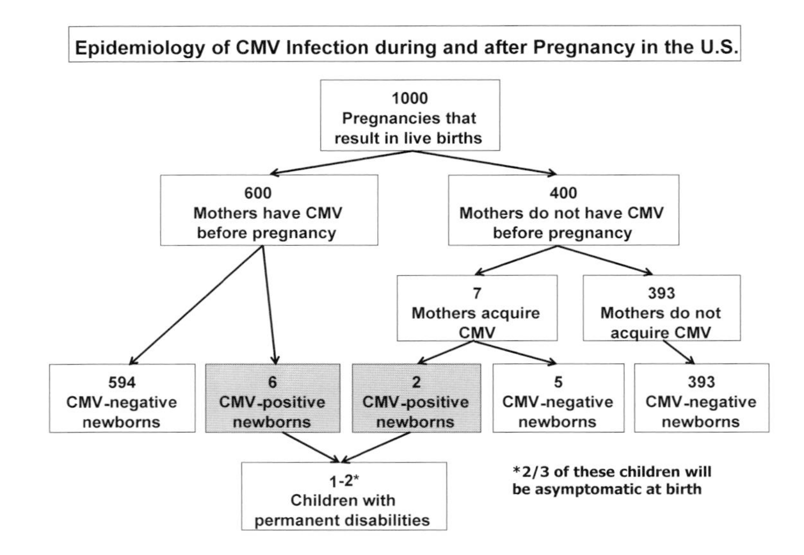

Figure II.2.10 Epidemiology of CMV infection in the USA during and after pregnancy (reproduced from Cannon, 2009). Highlighted boxes show that of the CMV-positive children, six of eight (75%) have mothers who are CMV seropositive before pregnancy. Data are from literature reviews on congenital CMV birth prevalence and outcomes with varying collection periods spanning multiple years.

(Stowell *et al.*, 2012), and it becomes non-infectious after hand cleansing (J.D. Stowell, D. Forlin-Passoni, K. Radford, S.L. Bate, S.C. Dollard, S.R. Bialek, M.J. Cannon and D.S. Schmid, unpublished data). Furthermore, women report frequently washing their hands after changing nappies (Cannon *et al.*, 2012). Taken together, these findings suggest that CMV transmission via contaminated hands, while plausible, may be inefficient.

To a lesser extent, activities associated with sex have a role in CMV transmission for women of reproductive age, especially among women with high-risk sexual behaviours. Since CMV can be found in both genital and oral secretions (Cannon *et al.*, 2011), there are probably multiple ways the virus can be transmitted, such as kissing, oral sex, and vaginal sex.

Women can be infected with CMV via blood transfusions or organ transplants. However, these exposures are uncommon and their associated CMV transmission rates are low, such that they would not have a significant population impact on CMV transmission among women.

Disease outcomes

Newborns with congenital CMV infections are classified as symptomatic or asymptomatic based on the presence of various characteristics at birth, most of which are not specific to CMV: petechiae, purpura, jaundice with associated hyperbilirubinaemia, hepatosplenomegaly, thrombocytopenia, chorioretinitis, seizures, microcephaly, and intracranial calcifications or fetal hydrops (see also Chapter II.3). Approximately 10% of newborns with congenital CMV infection will have signs and symptoms at birth that are consistent with the diagnosis of symptomatic congenital infection, with 5% of these having the typical generalized cytomegalic inclusion disease (CID) and the other 5% presenting with milder or atypical involvement (Britt, 2011). Approximately 90% of congenitally infected infants will have asymptomatic congenital infection at birth.

The most serious outcomes of congenital CMV infection are death, typically during infancy, sensorineural hearing loss (SNHL), cognitive impairment, and vision loss (see also Chapter II.3). Children may experience none of these outcomes, or one or more of them with varying degrees of severity. The strongest predictor for serious outcomes is symptomatic infection at birth (Dollard *et al.*, 2007; Rosenthal *et al.*, 2009).

Some early studies, whose populations included many children selected for the study because of severe disease, suggested relatively high mortality rates due to congenital CMV (Boppana *et al.*, 1992; Buchheit *et al.*, 1994). More recent studies of unselected populations suggest that death due to congenital CMV occurs in between 3% and 10% of infants with symptomatic infections, or 0.3–1.0% of all infants with congenital CMV (Fowler *et al.*, 1999a; Dollard *et al.*, 2007). For example, a study from Belgium which followed 29 infants with symptomatic infections reported two (6.9%) infant deaths (Pignatelli *et al.*, 2010). With approximately 30,000 infants born with CMV in the USA each year, 100–300 infant deaths are likely due to congenital CMV. An annual average of only 30 infant deaths are recorded as due to congenital CMV in the USA (Bristow *et al.*, 2011), which reflects the large degree of under-ascertainment of cases. There is also some evidence that CMV can be a cause of stillbirths (Iwasenko *et al.*, 2011).

The most common sequelae of congenital CMV is SNHL, occurring in up to 50% of children with symptomatic infections and up to 15% of those with asymptomatic infections (Harris *et al.*, 1984; Williamson *et al.*, 1992; Fowler *et al.*, 1999a; Dahle *et al.*, 2000; Foulon *et al.*, 2008; Britt, 2011; Liesbeth *et al.*, 2011). One-fourth to one-half of cases of SNHL that occur among children with congenital CMV in low-seroprevalence populations are late onset or progressive in nature and hence are not readily detectable through newborn hearing screening (Fowler *et al.*, 1999a; Grosse *et al.*, 2008; Korver *et al.*, 2009b). The proportion of hearing loss that is delayed may be lower in populations with high maternal seroprevalence. In a study in Brazil, only one of seven CMV-screened children with SNHL had delayed onset, although the sample size was two small to reach definitive conclusions (Yamamoto *et al.*, 2010). Documented cases of hearing loss diagnosed at 9 or 10 years of age in children with congenital CMV who had been monitored since birth are not uncommon (Fowler *et al.*, 1997, 1999a; Ikeno *et al.*, 2011).

Hearing loss varies across multiple dimensions, the most important of which are symmetry (bilateral vs. unilateral) and pure tone average (mild, moderate, severe, and profound). Mild bilateral or mild to moderate unilateral hearing losses may have limited implications for functioning. Clear evidence of adverse impacts on language development and school functioning is restricted to permanent SNHL of 40 decibels or greater in each ear, which is typically the cutoff used to qualify children for special education services. Among all children, 1–2 per 1000 have that level of permanent hearing loss. In contrast, it has been estimated that 3–5% of children with congenital CMV develop moderate to profound bilateral SNHL (Grosse *et al.*, 2008).

The proportion of cases of bilateral moderate to profound SNHL attributable to congenital CMV has been estimated as 15% overall and as high as 25% among those with permanent deafness (Grosse *et al.*, 2008). Recent estimates of the attributable fraction of all SNHL, including unilateral and mild losses, are in the range of 7–12%. The lower percentages in recent publications may be due to differences in methods of ascertainment of congenital CMV infection (i.e. use of stored dried blood spot specimens rather than urine or saliva), relatively short length of follow-up, and sampling from populations with relatively low rates of congenital CMV (a lower frequency of congenital CMV should be associated with a lower attributable fraction of SNHL). Furthermore, studies of children with SNHL who failed newborn hearing screening tests understate the aetiological fraction of CMV in hearing loss by excluding cases of hearing loss that are not detected by newborn hearing screening (Morton and Nance, 2006; Nance *et al.*, 2006).

Children with symptomatic congenital CMV are at risk of permanent neurological impairments that can cause both intellectual and physical disability. In some studies, up to 80% of children who are symptomatic at birth develop disabilities (Boppana *et al.*, 1992; Istas *et al.*, 1995; Noyola *et al.*, 2001). However, most of these studies include a large proportion of infants who were referred to the study because of their symptoms, and therefore the more severely affected symptomatic infants are overrepresented. Lower percentages of disability are reported among symptomatic infants identified through population-based, non-selective newborn screening (Dollard *et al.*, 2007).

Although asymptomatic congenital CMV infection is a strong risk factor for SNHL, it is not clear to what extent it is associated with other sequelae. Uncontrolled studies of children asymptomatic at birth have reported ocular defects (Boppana *et al.*, 1994; Anderson *et al.*, 1996; Coats *et al.*, 2000; Britt, 2011), developmental or motor delay (Preece *et al.*, 1983; Ahlfors *et al.*, 1984, 1999; Yow *et al.*, 1988; Ancora *et al.*, 2007), and low IQ (Britt, 2011). Among studies that had a comparison group, moderately lower average IQs have been reported in CMV-infected children less than 6 years of age (Williamson *et al.*, 1990; Kashden *et al.*, 1998; Fowler *et al.*, 1999b; Temple *et al.*, 2000; Zhang *et al.*, 2007), but no study appears to have found a significant difference in IQ scores among older children (Conboy *et al.*, 1986; Kashden *et al.*, 1998; Temple *et al.*, 2000; Cazacu *et al.*, 2004; Shan *et al.*, 2009). If there are cognitive differences associated with asymptomatic CMV infection, they are probably modest, affect a relatively small proportion of asymptomatic children, and may not persist into the school years.

Regrettably, limited information is available on the types and severities of impairments for population-based samples of affected children with either symptomatic or asymptomatic congenital CMV infection. Such information is needed for a quantitative assessment of the full public health burden of the disorder.

Awareness

The vast majority of women have never heard of congenital CMV. In three US studies that measured awareness, only 13%, 22%, and 14% had heard of the virus (Jeon *et al.*, 2006; Ross *et al.*, 2008; Cannon *et al.*, 2012), and among these women knowledge of transmission modes or prevention strategies was even lower (Jeon *et al.*, 2006). In addition, awareness for CMV was lower than for any other childhood condition included in the survey (Fig. II.2.11) (Jeon *et al.*, 2006; Cannon *et al.*, 2012). Furthermore, most obstetrician/gynaecologists do not counsel their patients about CMV or CMV prevention (CDC, 2008; Korver *et al.*, 2009a). So, few people are aware of congenital CMV, despite congenital CMV being one of the most common causes of childhood disabilities. If CMV prevention through behavioural change is going to occur, and if demand for a vaccine is going to be created, awareness among women and their healthcare providers must improve. Schleiss has noted that 'Increased public awareness about the risks of CMV is urgently needed: this in turn will drive the social, political, and economic forces necessary to increase the pace of progress of clinical trials.' (Schleiss, 2008).

Prevention

Prevention of CMV infection in pregnant women or fetal infection in their babies can take several forms. Vaccines are under development (Zhong and Khanna, 2007; Schleiss, 2008; Pass *et al.*, 2009) and there is some evidence that CMV hyperimmune globulin (HIG) may prevent fetal infection or disability (Nigro *et al.*, 2005). Because these topics are covered elsewhere in this book (in particular in Chapters II.3 and II.20), we focus the remainder of this section on prevention of maternal infection during pregnancy via reducing exposures to CMV.

Because exposure to viral shedding from young children is such a strong risk factor for CMV infection, reducing those exposures is essential. Two studies suggested that education and behavioural change reduced CMV infection rates among pregnant women who had young children. In a small study, 14 pregnant women who received a behavioural intervention related to avoiding oral exposures (e.g. kissing on the lips, sharing food, drink, or utensils) or hand transmission (e.g. via hand washing) had no seroconversions during pregnancy (Adler *et al.*, 1996). However, among the two non-pregnant intervention groups ($N=11$ and $N=11$), seroconversion rates were lower than those of the non-pregnant control group ($N=17$), but not significantly so. In a much larger study of 5000 pregnant women, rates of maternal CMV seroconversion were reduced by more than half the expected rate after women were tested for CMV, told they were CMV seronegative, and were educated about prevention via similar methods (Picone *et al.*, 2009; Vauloup-Fellous *et al.*, 2009). Developing and evaluating similar interventions to confirm their effectiveness and understand implementation issues in different populations and contexts, and identifying the key factors associated with behavioural changes should be high priorities. The success of the intervention probably speaks to the motivation of pregnant women to protect their developing babies when they are educated about health risks (see also Chapter II.3). Because CMV-seropositive women are also at risk for infection with a different CMV strain, behavioural

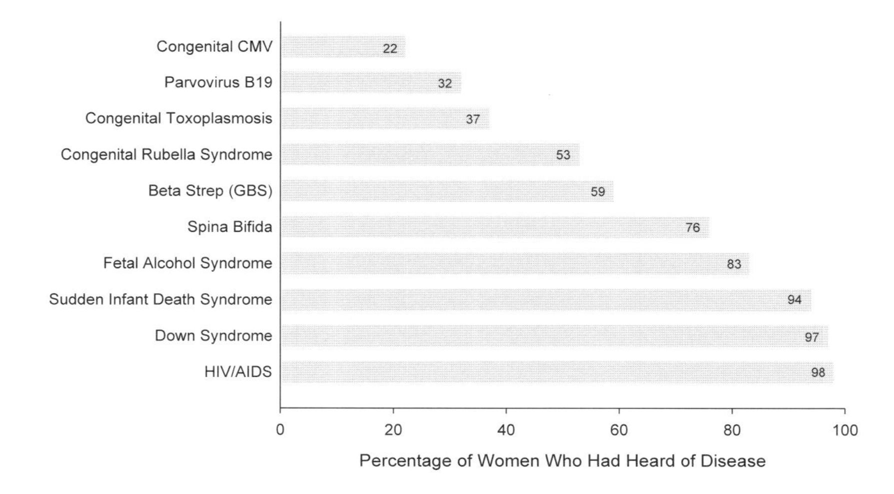

Figure II.2.11 Women's awareness of various diseases and conditions from a multisite survey conducted in 2005 (reproduced from Jeon *et al.*, 2006).

interventions may still play an important role for these women or for women who are vaccinated when a future vaccine becomes available.

Because CMV transmission is associated with sexual activity, pregnant women may also be able to lower their risk due to sexual exposures. However, prevention is complicated by the occurrence of CMV not just in genital secretions but in oral secretions as well. Thus, CMV transmission may occur among adults through kissing or oral sex in addition to vaginal or anal sex. At a minimum, it would be prudent to limit new sex partners during pregnancy and to use a condom with new partners, even though these practices would not eliminate risk due to oral exposures.

Prenatal screening

Routine CMV prenatal screening is not recommended by any medical association or public health body, although screening is commonly practiced in some countries such as Belgium (Naessens *et al.*, 2005), Italy (Revello *et al.*, 2011), and Israel (Schlesinger, 2007). CMV prenatal screening offers potential benefits but remains controversial because of lack of definitive benefit and risk of harms. In terms of benefit, screening early in pregnancy coupled with educational counselling about avoiding exposures has been reported to reduce CMV infection rates during pregnancy (Picone *et al.*, 2009; Vauloup-Fellous *et al.*, 2009). Screening would enable women to consider follow-up prenatal diagnosis (Revello and Gerna, 2002; Benoist *et al.*, 2008; Lazzarotto, 2010), although it is unclear whether that is a net benefit at present (see below). Further, screening would allow women and their providers to consider available treatments. Unfortunately, at present there are no treatments for CMV infection during pregnancy that have been approved for the prevention of fetal infection. The safety and efficacy of using CMV HIG for this purpose is currently being assessed in clinical trials. Although CMV HIG has an excellent safety profile and is currently being used in some instances among pregnant women in the USA and elsewhere (Adler, 2011), it is expensive and its effectiveness is unclear. As a result, CDC and ACOG do not have a formal recommendation for its use. In the absence of newborn CMV screening, prenatal screening and diagnosis may identify fetuses that are infected and may facilitate early detection and intervention for hearing loss and other developmental disabilities (Grosse *et al.*, 2009) and otherwise prepare families for possible adverse outcomes and treatment (see also Chapters II.3 and II.18).

One potential harm of CMV prenatal screening at present is the risk it might lead to termination of pregnancies in which the child would have been healthy. This is especially a risk when awareness and understanding among women and their health care providers are very low. In addition, prenatal screening can be costly and will lead to invasive diagnostic procedures for some uninfected children. A final challenge is that the uncertain prognosis for a fetus undergoing CMV prenatal screening and diagnosis can cause anxiety to parents. Although the prognostic value of laboratory tests has improved (Benoist *et al.*, 2008; Lazzarotto, 2010; Fabbri *et al.*, 2011), termination of pregnancy following a positive screen and without diagnostic testing remains common according to a recent Italian study (Revello *et al.*, 2011) (see also Chapter II.18).

Newborn screening

Screening newborns during the first week after birth for the presence of CMV infection has been proposed in order to detect asymptomatic children at risk of developing hearing loss not detected through newborn hearing screening. Even though at present antiviral treatment is not recommended for such children, regular monitoring to detect hearing loss may be justified in order to provide prompt intervention such as amplification or cochlear implants, which have been shown to improve language development.

The major challenge to newborn screening for CMV is the choice of an assay that is sufficiently sensitive, specific, affordable, and acceptable to health care providers and policy makers. Because dried blood spots (DBS) collected on filter paper cards is the standard specimen format used in almost all newborn screening programmes that employ laboratory testing, much work has been devoted to developing and evaluating DBS assays using polymerase chain reaction (PCR). The reported sensitivity of assays ranges from 34% to 100% (Barbi *et al.*, 1996, 2006, 2008; de Vries *et al.*, 2009; Boppana *et al.*, 2010; 2011; Leruez-Ville *et al.*, 2011). The rate of detection using DBS improves among infants infected as a result of primary maternal infection who present with higher viral loads, but the majority of congenital CMV infections result from non-primary maternal infections. Nevertheless, primary infection and higher viral loads may be associated with greater risk of sequelae (Fowler *et al.*, 1992; Cannon *et al.*, 2011), and thus DBS testing may detect most of the children who develop disabilities (Kharrazi *et al.*, 2010).

The current leading alternative to a DBS assay is a real-time PCR assay developed using either saliva swabs collected and stored in viral transport media kept at 4°C or a dried saliva swab collected and stored without media at room temperature (Boppana *et al.*, 2010).

Based on testing of more than 17,000 saliva specimens, the CHIMES study reported that the sensitivity of PCR for both the wet and dry saliva swabs is > 97% and the specificity is 99.9% as compared with standard saliva rapid culture (Boppana *et al.*, 2011). Those test characteristics are superior to many existing newborn screening assays for other conditions. Saliva swabs placed in sleeves or tubes kept at room temperature could be tested in the existing network of public health newborn screening laboratories. However, that would require a new infrastructure for specimen collection and transport.

Alternatively, testing urine collected with filter cards inserted into nappies has been shown to be feasible for large-scale newborn screening in Japan (Koyano *et al.*, 2011). In a study of 21,000 newborns enrolled at 25 sites in six geographically distinct areas of Japan, investigators found CMV screening to be effective without compromising detection sensitivity.

Newborn screening is not restricted to laboratory analysis of specimens but can employ point-of-care testing methods. Official US newborn screening recommendations since 2011 have included two point-of-care screening tests: audiometry for hearing loss and pulse oximetry for critical congenital heart disease (https://www.nbstrn.org/node/152). Although no point-of-care test for congenital CMV infection is currently available, this could be an option in the future.

An important limitation to the adoption of routine newborn screening for CMV is that most children with neonatal infection do not develop disabilities and would not benefit from screening and monitoring but their parents and health care providers would be burdened by the need for regular monitoring and might experience anxiety. A US survey indicates that approximately 85% of parents would favour newborn screening for CMV despite these limitations, but a significant minority would have concerns (Din *et al.*, 2011). In the absence of a safe, effective treatment, CMV would not fit the paradigm for routine public health newborn screening. One possible alternative would be voluntary newborn screening with informed decision-making (Grosse *et al.*, 2006) and perhaps alternative specimens such as saliva or urine. Such screening would require a new approach.

References

Adjei, A.A., Armah, H.B., Gbagbo, F., Boamah, I., du-Gyamfi, C., and Asare, I. (2008). Seroprevalence of HHV-8, CMV, and EBV among the general population in Ghana, West Africa. BMC. Infect. Dis. *8*, 111.

Adler, S.P. (1989). Cytomegalovirus and child day care. Evidence for an increased infection rate among day-care workers. N. Engl. J. Med. *321*, 1290–1296.

Adler, S.P. (1991a). Cytomegalovirus and child day care: risk factors for maternal infection. Pediatr. Infect. Dis. J. *10*, 590–594.

Adler, S.P. (1991b). Molecular epidemiology of cytomegalovirus: a study of factors affecting transmission among children at three day-care centers. Pediatr. Infect. Dis. J. *10*, 584–590.

Adler, S.P. (2011). Screening for cytomegalovirus during pregnancy. Infect. Dis. Obstetr. Gynecol. *2011*, 1–9.

Adler, S.P., Finney, J.W., Manganello, A.M., and Best, A.M. (1996). Prevention of child-to-mother transmission of cytomegalovirus by changing behaviors: a randomized controlled trial. Pediatr. Infect. Dis. J. *15*, 240–246.

Ahlfors, K., Ivarsson, S.A., Harris, S., Svanberg, L., Holmqvist, R., Lernmark, B., and Theander, G. (1984). Congenital cytomegalovirus infection and disease in Sweden and the relative importance of primary and secondary maternal infections. Preliminary findings from a prospective study. Scand. J. Infect. Dis. *16*, 129–137.

Ahlfors, K., Ivarsson, S.A., and Harris, S. (1999). Report on a long-term study of maternal and congenital cytomegalovirus infection in Sweden. Review of prospective studies available in the literature. Scand. J. Infect. Dis. *31*, 443–457.

Alford, C.A., and Pass, R.F. (1981). Epidemiology of chronic congenital and perinatal infections of man. Clinics Perinatol. *8*, 397–414.

Ancora, G., Lanari, M., Lazzarotto, T., Venturi, V., Tridapalli, E., Sandri, F., Menarini, M., Ferretti, E., and Faldella, G. (2007). Cranial ultrasound scanning and prediction of outcome in newborns with congenital cytomegalovirus infection. J. Pediatr. *150*, 157–161.

Anderson, K.S., Amos, C.S., Boppana, S., and Pass, R. (1996). Ocular abnormalities in congenital cytomegalovirus infection. J. Am. Optom. Assoc. *67*, 273–278.

Armstrong, G.L., Schillinger, J., Markowitz, L., Nahmias, A.J., Johnson, R.E., McQuillan, G.M., and St Louis, M.E. (2001). Incidence of herpes simplex virus type 2 infection in the USA. Am. J. Epidemiol. *153*, 912–920.

Balcarek, K.B., Bagley, R., Cloud, G.A., and Pass, R.F. (1990). Cytomegalovirus infection among employees of a children's hospital. No evidence for increased risk associated with patient care. JAMA *263*, 840–844.

Bale, J.F., Jr., Zimmerman, B., Dawson, J.D., Souza, I.E., Petheram, S.J., and Murph, J.R. (1999). Cytomegalovirus transmission in child care homes. Arch. Pediatr. Adolesc. Med. *153*, 75–79.

Balfour, C.L., and Balfour, H.H., Jr. (1986). Cytomegalovirus is not an occupational risk for nurses in renal transplant and neonatal units. Results of a prospective surveillance study. JAMA *256*, 1909–1914.

Barbi, M., Binda, S., Primache, V., Luraschi, C., and Corbetta, C. (1996). Diagnosis of congenital cytomegalovirus infection by detection of viral DNA in dried blood spots. Clin. Diagn. Virol. *6*, 27–32.

Barbi, M., Binda, S., and Caroppo, S. (2006). Diagnosis of congenital CMV infection via dried blood spots. Rev. Med. Virol. *16*, 385–392.

Barbi, M., MacKay, W.G., Binda, S., and van Loon, A.M. (2008). External quality assessment of cytomegalovirus DNA detection on dried blood spots. BMC. Microbiol. *8*, 2.

Bate, S.L., Dollard, S.C., and Cannon, M.J. (2010). Cytomegalovirus seroprevalence in the USA: the national health and nutrition examination surveys, 1988–2004. Clin. Infect. Dis. *50*, 1439–1447.

Benoist, G., Salomon, L.J., Jacquemard, F., Daffos, F., and Ville, Y. (2008). The prognostic value of ultrasound abnormalities and biological parameters in blood of fetuses infected with cytomegalovirus. BJOG *115*, 823–829.

Boppana, S., Amos, C., Britt, W., Stagno, S., Alford, C., and Pass, R. (1994). Late onset and reactivation of chorioretinitis in children with congenital cytomegalovirus infection. Pediatr. Infect. Dis. J. *13*, 1139–1142.

Boppana, S.B., Pass, R.F., Britt, W.J., Stagno, S., and Alford, C.A. (1992). Symptomatic congenital cytomegalovirus infection: neonatal morbidity and mortality. Pediatr. Infect. Dis. J. *11*, 93–99.

Boppana, S.B., Rivera, L.B., Fowler, K.B., Mach, M., and Britt, W.J. (2001). Intrauterine transmission of cytomegalovirus to infants of women with preconceptional immunity. N. Engl. J. Med. *344*, 1366–1371.

Boppana, S.B., Ross, S.A., Novak, Z., Shimamura, M., Tolan, R.W., Palmer, A.L., Ahmed, A., Michaels, M.G., Sanchez, P.J., Bernstein, D.I., *et al.* (2010). Dried blood spot real-time polymerase chain reaction assays to screen newborns for congenital cytomegalovirus infection. JAMA *303*, 1375–1382.

Boppana, S.B., Ross, S.A., Shimamura, M., Palmer, A.L., Ahmed, A., Michaels, M.G., Sanchez, P.J., Bernstein, D.I., Tolan, R.W., Novak, Z., *et al.* (2011). Saliva polymerase-chain-reaction assay for cytomegalovirus screening in newborns. N. Engl. J. Med. *364*, 2111–2118.

Brady, M.T., Demmler, G.J., and Anderson, D.C. (1987). Cytomegalovirus infection in pediatric house officers: susceptibility to and rate of primary infection. Infect. Control *8*, 329–332.

Bristow, B.N., O'Keefe, K.A., Shafir, S.C., and Sorvillo, F.J. (2011). Congenital cytomegalovirus mortality in the USA, 1990–2006. Plos Neglect. Trop. Dis. *5*, 5.

Britt, W.J. (2011). Cytomegalovirus. In Infectious Disease of the Fetus and Newborn Infant, Remington, J.S., Klein, J.O., Wilson, C.B., Nizet, V., and Maldonado, Y.A., eds. (Elsevier Saunders, Philadelphia, PA), pp. 706–755.

Buchheit, J., Marshall, G.S., Rabalais, G.P., and Dobbins, G.J. (1994). Congenital cytomegalovirus disease in the Louisville area: a significant public health problem. J. Ky. Med. Assoc. *92*, 411–415.

Cabau, N., Labadie, M.D., Vesin, C., Feingold, J., and Boue, A. (1979). Seroepidemiology of cytomegalovirus infections during the first years of life in urban communities. Arch. Dis. Child. *54*, 286–290.

Cannon, M.J. (2009). Congenital cytomegalovirus (CMV) epidemiology and awareness. J. Clin. Virol. *46S*, S6-S10.

Cannon, M.J., Schmid, D.S., and Hyde, T.B. (2010). Review of cytomegalovirus seroprevalence and demographic characteristics associated with infection. Rev. Med. Virol. *20*, 202–213.

Cannon, M.J., Hyde, T.B., and Schmid, D.S. (2011). Review of cytomegalovirus shedding in bodily fluids and relevance to congenital cytomegalovirus infection. Rev. Med. Virol. *21*, 240–255.

Cannon, M.J., Westbrook, K., Levis, D., Schleiss, M.R., Thackeray, R., and Pass, R.F. (2012). Awareness of and behaviors related to child-to-mother transmission of cytomegalovirus. Prevent. Med. *54*, 351–357.

Cazacu, A.C., Chung, S., Greisser, C., Sellers-Vinson, S., Williamson, W.D., Llorente, A.M., Rozelle, J., Turcich, M., Peters, S., and Demmler, G.J. (2004). Neurodevelopmental outcome at elementary school age of children born with asymptomatic congenital cytomegalovirus infection. Pediatr. Res. *55*, 319A–319A.

Centers for Disease Control and Prevention (2008). Knowledge and practices of obstetricians and gynecologists regarding cytomegalovirus infection during pregnancy – United States, 2007. MMWR Morb. Mortal. Wkly. Rep. *57*, 65–68.

Chandler, S.H., Alexander, E.R., and Holmes, K.K. (1985a). Epidemiology of cytomegaloviral infection in a heterogeneous population of pregnant women. J. Infect. Dis. *152*, 249–256.

Chandler, S.H., Holmes, K.K., Wentworth, B.B., and et.al (1985b). The epidemiology of cytomegaloviral infection in women attending sexually transmitted disease clinic. J. Infect. Dis. *152*, 597–605.

Chandler, S.H., Handsfield, H.H., and McDougall, J.K. (1987). Isolation of multiple strains of cytomegalovirus from women attending a clinic for sexually transmitted disease. J. Infect. Dis. *155*, 655–660.

Chen, M. (2008). High perinatal seroprevalence of cytomegalovirus in northern Taiwan. J. Paediatr. Child Health *44*, 166–169.

Coats, D.K., Demmler, G.J., Paysse, E.A., Du, L.T., and Libby, C. (2000). Ophthalmologic findings in children with congenital cytomegalovirus infection. J. AAPOS. *4*, 110–116.

Coleman, P.J., McQuillan, G.M., Moyer, L.A., Lambert, S.B., and Margolis, H.S. (1998). Incidence of hepatitis B virus infection in the USA, 1976–1994: estimates from the National Health and Nutrition Examination Surveys. J. Infect. Dis. *178*, 954–959.

Collier, A.C., Handsfield, H.H., Roberts, P.L., DeRouen, T., Meyers, J.D., Leach, L., Murphy, V.L., Verdon, M., and Corey, L. (1990). Cytomegalovirus infection in women attending a sexually transmitted disease clinic. J. Infect. Dis. *162*, 46–51.

Colugnati, F.A., Staras, S.A., Dollard, S.C., and Cannon, M.J. (2007). Incidence of cytomegalovirus infection among the general population and pregnant women in the USA. BMC Infect. Dis. *7*, 71.

Conboy, T.J., Pass, R.F., Stagno, S., Britt, W.J., Alford, C.A., McFarland, C.E., and Boll, T.J. (1986). Intellectual development in school-aged children with asymptomatic congenital cytomegalovirus infection. Pediatrics *77*, 801–806.

Coonrod, D., Collier, A.C., Ashley, R., DeRouen, T., and Corey, L. (1998). Association between cytomegalovirus seroconversion and upper genital tract infection among women attending a sexually transmitted disease clinic: a prospective study. J Infect. Dis. *177*, 1188–1193.

Dahle, A.J., Fowler, K.B., Wright, J.D., Boppana, S.B., Britt, W.J., and Pass, R.F. (2000). Longitudinal investigation of hearing disorders in children with congenital cytomegalovirus. J. Am. Acad. Audiol. *11*, 283–290.

Dar, L., Pati, S.K., Patro, A.R., Deorari, A.K., Rai, S., Kant, S., Broor, S., Fowler, K.B., Britt, W.J., and Boppana,

S.B. (2008). Congenital cytomegalovirus infection in a highly seropositive semi-urban population in India. Pediatr. Infect. Dis. J. *27*, 841–843.

DePasquale, J.M., Freeman, K., Amin, M.M., Park, S., Rivers, S., Hopkins, R., Cannon, M.J., Dy, B., and Dollard, S.C. (2011). Efficient linking of birth certificate and newborn screening databases for laboratory investigation of congenital cytomegalovirus infection and preterm birth: Florida, 2008. Matern. Child Health J.

Din, E.S., Brown, C.J., Grosse, S.D., Wang, C., Bialek, S.R., Ross, D.S., and Cannon, M.J. (2011). Attitudes toward newborn screening for cytomegalovirus infection. Pediatrics *128*, e1434–1442.

Dollard, S.C., Grosse, S.D., and Ross, D.S. (2007). New estimates of the prevalence of neurological and sensory sequelae and mortality associated with congenital cytomegalovirus infection. Rev. Med. Virol. *17*, 355–363.

Drew, W.L., Mintz, L., Miner, R.C., Sands, M., and Ketterer, B. (1981). Prevalence of cytomegalovirus infection in homosexual men. J. Infect. Dis. *143*, 188–192.

Dworsky, M., Yow, M., Stagno, S., Pass, R.F., and Alford, C. (1983a). Cytomegalovirus infection of breast milk and transmission in infancy. Pediatrics *72*, 295–299.

Dworsky, M.E., Welch, K., Cassady, G., and Stagno, S. (1983b). Occupational risk for primary cytomegalovirus infection among pediatric health-care workers. N. Engl. J. Med. *309*, 950–953.

Dworsky, M., Lakeman, A., and Stagno, S. (1984). Cytomegalovirus transmission within a family. Pediatr. Infect. Dis. J. *3*, 236–238.

Edmunds, W.J., Gay, N.J., Kretzschmar, M., Pebody, R.G., and Wachmann, H. (2000). The pre-vaccination epidemiology of measles, mumps and rubella in Europe: implications for modelling studies. Epidemiol. Infect. *125*, 635–650.

Fabbri, E., Revello, M.G., Furione, M., Zavattoni, M., Lilleri, D., Tassis, B., Quarenghi, A., Rustico, M., Nicolini, U., Ferrazzi, E., *et al.* (2011). Prognostic markers of symptomatic congenital human cytomegalovirus infection in fetal blood. BJOG *118*, 448–456.

Ford-Jones, L., Kitai, I.M., Davis, L., Corey, M., Farrell, H., Petric, M., Kyle, I., Beach, J., Yaffe, B., Kelly, E., *et al.* (1996). Cytomegalovirus infections in Toronto child-care centers: A prospective study of viral excretion in children and seroconversion among day-care providers. Pediatr. Infect. Dis. J. *15*, 507–514.

Foulon, I., Naessens, A., Foulon, W., Casteels, A., and Gordts, F. (2008). A 10-year prospective study of sensorineural hearing loss in children with congenital cytomegalovirus infection. J. Pediatr. *153*, 84–88.

Fowler, K.B., and Pass, R.F. (1991). Sexually transmitted diseases in mothers of neonates with congenital cytomegalovirus infection. J. Infect. Dis. *164*, 259–264.

Fowler, K.B., and Pass, R.F. (2006). Risk factors for congenital cytomegalovirus infection in the offspring of young women: exposure to young children and recent onset of sexual activity. Pediatrics *118*, e286–e292.

Fowler, K.B., Stagno, S., Pass, R.F., Britt, W.J., Boll, T.J., and Alford, C.A. (1992). The outcome of congenital cytomegalovirus infection in relation to maternal antibody status. N. Engl. J. Med. *326*, 663–667.

Fowler, K.B., Stagno, S., and Pass, R.F. (1993). Maternal age and congenital cytomegalovirus infection: screening of two diverse newborn populations, 1980–1990. J. Infect. Dis. *168*, 552–556.

Fowler, K.B., McCollister, F.P., Dahle, A.J., Boppana, S., Britt, W.J., and Pass, R.F. (1997). Progressive and fluctuating sensorineural hearing loss in children with asymptomatic congenital cytomegalovirus infection. J. Pediatr. *130*, 624–630.

Fowler, K.B., Dahle, A.J., Boppana, S.B., and Pass, R.F. (1999a). Newborn hearing screening: will children with hearing loss caused by congenital cytomegalovirus infection be missed? J. Pediatr. *135*, 60–64.

Fowler, K.B., Pass, R.F., Boppana, S., and Britt, W.J. (1999b). Neurodevelopmental deficits in school-aged children who have asymptomatic congenital cytomegalovirus infection. Paediatr. Perinat. Epidemiol. *13*, A19–20.

Fowler, K.B., Stagno, S., and Pass, R.F. (2004). Interval between births and risk of congenital cytomegalovirus infection. Clin. Infect. Dis. *38*, 1035–1037.

Francisse, S., Revelard, P., De Maertelaer, V., Strebelle, E., Englert, Y., and Liesnard, C. (2009). Human cytomegalovirus seroprevalence and risk of seroconversion in a fertility clinic population. Obstetr. Gynecol. *114*, 285–291.

Friedman, H.M., Lewis, M.R., Nemerofsky, D.M., and Plotkin, S.A. (1984). Acquisition of cytomegalovirus infection among female employees at a pediatric hospital. Pediatr. Infect. Dis. J. *3*, 233–235.

Gerberding, J.L. (1994). Incidence and prevalence of human immunodeficiency virus, hepatitis B virus, hepatitis C virus, and cytomegalovirus among health care personnel at risk for blood exposure: final report from a longitudinal study. J. Infect. Dis. *170*, 1410–1417.

Gidding, H.F., MacIntyre, C.R., Burgess, M.A., and Gilbert, G.L. (2003). The seroepidemiology and transmission dynamics of varicella in Australia. Epidemiol. Infect. *131*, 1085–1089.

Gold, E., and Nankervis, G.A. (1976). Cytomegalovirus. In Viral Infections of Humans: Epidemiology and Control, Evans, A.S., ed. (Plenum Press, New York), pp. 143–161.

Gold, E., and Nankervis, G.A. (1982). Cytomegalovirus. In Viral Infections of Humans: Epidemiology and Control, Evans, A.S., ed. (Plenum Press, New York), pp. 167–186.

Gratacap-Cavallier, B., Bosson, J.L., Morand, P., Dutertre, N., Chanzy, B., Jouk, P.S., Vandekerckhove, C., Cart-Lamy, P., and Seigneurin, J.M. (1998). Cytomegalovirus seroprevalence in French pregnant women: Parity and place of birth as major predictive factors. Eur. J. Epidemiol. *14*, 147–152.

Griffiths, P.D., McLean, A., and Emery, V.C. (2001). Encouraging prospects for immunisation against primary cytomegalovirus infection. Vaccine *19*, 1356–1362.

Grosse, S.D., Boyle, C.A., Kenneson, A., Khoury, M.J., and Wilfond, B.S. (2006). From public health emergency to public health service: the implications of evolving criteria for newborn screening panels. Pediatrics *117*, 923–929.

Grosse, S.D., Ross, D.S., and Dollard, S.C. (2008). Congenital cytomegalovirus (CMV) infection as a cause of permanent bilateral hearing loss: a quantitative assessment. J. Clin. Virol. *41*, 57–62.

Grosse, S.D., Dollard, S., Ross, D.S., and Cannon, M. (2009). Newborn screening for congenital cytomegalovirus:

Options for hospital-based and public health programs. J. Clin. Virol. *46*, S32-S36.

Hamprecht, K., Maschmann, J., Vochem, M., Dietz, K., Speer, C.P., and Jahn, G. (2001). Epidemiology of transmission of cytomegalovirus from mother to preterm infant by breastfeeding. Lancet *357*, 513–518.

Hamprecht, K., Maschmann, J., Jahn, G., Poets, C.F., and Goelz, R. (2008). Cytomegalovirus transmission to preterm infants during lactation. J. Clin. Virol. *41*, 198–205.

Handsfield, H.H., Chandler, S.H., Caine, V.A., Meyers, J.D., Corey, L., Medeiros, E., and McDougall, J.K. (1985). Cytomegalovirus infection in sex partners: evidence for sexual transmission. J. Infect. Dis. *151*, 344–348.

Harris, S., Ahlfors, K., Ivarsson, S., Lernmark, B., and Svanberg, L. (1984). Congenital cytomegalovirus infection and sensorineural hearing loss. Ear Hear. *5*, 352–355.

Hatherley, L.I. (1986). Is primary cytomegalovirus infection an occupational hazard for obstetric nurses? A serological study. Infect. Control *7*, 452–455.

Hyams, K.C., Krogwold, R.A., Brock, S., Wignall, F.S., Cross, E., and Hayes, C. (1993). Heterosexual transmission of viral hepatitis and cytomegalovirus infection among United States military personnel stationed in the Western Pacific. Sex. Transm. Dis. *20*, 36–40.

Hyde, T.B., Schmid, D.S., and Cannon, M.J. (2010). Cytomegalovirus seroconversion rates and risk factors: implications for congenital CMV. Rev. Med. Virol. *20*, 311–326.

Ikeno, M., Okumura, A., Ito, Y., Abe, S., Saito, M., and Shimizu, T. (2011). Late-onset sensorineural hearing loss due to asymptomatic congenital cytomegalovirus infection retrospectively diagnosed by polymerase chain reaction using preserved umbilical cord. Clin. Pediatr. *50*, 666–668.

Istas, A.S., Demmler, G.J., Dobbins, J.G., and Stewart, J.A. (1995). Surveillance for congenital cytomegalovirus disease: a report from the National Congenital Cytomegalovirus Disease Registry. Clin. Infect. Dis. *20*, 665–670.

Iwasenko, J.M., Howard, J., Arbuckle, S., Graf, N., Hall, B., Craig, M.E., and Rawlinson, W.D. (2011). Human cytomegalovirus infection is detected frequently in stillbirths and is associated with fetal thrombotic vasculopathy. J. Infect. Dis. *203*, 1526–1533.

Jeon, J., Victor, M., Adler, S., Arwady, A., Demmler, G., Fowler, K., Goldfarb, J., Keyserling, H., Massoudi, M., Richards, K., *et al.* (2006). Knowledge and awareness of congenital cytomegalovirus among women. Infect. Dis. Obstet. Gynecol. *2006*, 1–7.

Kamada, M., Komori, A., Chiba, S., and Nakao, T. (1983). A prospective study of congenital cytomegalovirus infection in Japan. Scand. J. Infect. Dis. *15*, 227–232.

Kashden, J., Frison, S., Fowler, K., Pass, R.F., and Boll, T.J. (1998). Intellectual assessment of children with asymptomatic congenital cytomegalovirus infection. J. Dev. Behav. Pediatr. *19*, 254–259.

Kaye, S., Miles, D., Antoine, P., Burny, W., Ojuola, B., Kaye, P., Rowland-Jones, S., Whittle, H., van der, S.M., and Marchant, A. (2008). Virological and immunological correlates of mother-to-child transmission of cytomegalovirus in The Gambia. J. Infect. Dis. *197*, 1307–1314.

Kenneson, A., and Cannon, M.J. (2007). Review and meta-analysis of the epidemiology of congenital cytomegalovirus (CMV) infection. Rev. Med. Virol. *17*, 253–276.

Kharrazi, M., Hyde, T., Young, S., Amin, M.M., Cannon, M.J., and Dollard, S.C. (2010). Use of screening dried blood spots for estimation of prevalence, risk factors, and birth outcomes of congenital cytomegalovirus infection. J. Pediatr. *157*, 191–197.

Knowles, S.J., Grundy, K., Cahill, I., Cafferkey, M.T., and Geary, M. (2005). Low cytomegalovirus sero-prevalence in Irish pregnant women. Irish Med. J. *98*, 210–212.

Korver, A.M.H., de Vries, J.J.C., de Jong, J.W., Dekker, F.W., Vossen, A., and Oudesluys-Murphy, A.M. (2009a). Awareness of congenital cytomegalovirus among doctors in the Netherlands. J. Clin. Virol. *46*, S11-S15.

Korver, A.M.H., de Vries, J.J.C., Konings, S., de Jong, J.W., Dekker, F.W., Vossen, A., Frijns, J.H.M., Oudesluys-Murphy, A.M., and Grp, D.C.S. (2009b). DECIBEL study: Congenital cytomegalovirus infection in young children with permanent bilateral hearing impairment in the Netherlands. J. Clin. Virol. *46*, S27-S31.

Koyano, S., Inoue, N., Oka, A., Moriuchi, H., Asano, K., Ito, Y., Yamada, H., Yoshikawa, T., Suzutani, T., and Japanese Congenital Cytomegalovirus Study, G. (2011). Screening for congenital cytomegalovirus infection using newborn urine samples collected on filter paper: feasibility and outcomes from a multicentre study. BMJ open *1*, e000118.

Larke, R.P., Wheatley, E., Saigal, S., and Chernesky, M.A. (1980). Congenital cytomegalovirus infection in an urban Canadian community. J. Infect. Dis. *142*, 647–653.

Lazzarotto, T. (2010). The best practices for screening, monitoring, and diagnosis of cytomegalovirus disease, Part II. Clin. Microbiol. Newsletter *32*, 9–15.

Leruez-Ville, M., Vauloup-Fellous, C., Couderc, S., Parat, S., Castel, C., Avettand-Fenoel, V., Guilleminot, T., Grangeot-Keros, L., Ville, Y., Grabar, S., *et al.* (2011). Prospective identification of congenital cytomegalovirus infection in newborns using real-time polymerase chain reaction assays in dried blood spots. Clin. Infect. Dis. *52*, 575–581.

Liesbeth, R., Christian, D., Frans, D., and Ermelinde, R. (2011). Hearing status in children with congenital cytomegalovirus: Up-to-6-years audiological follow-up. Int. J. Pediatr. Otorhinolaryngol. *75*, 376–382.

Lopo, S., Vinagre, E., Palminha, P., Paixao, M.T., Nogueira, P., and Freitas, M.G. (2011). Seroprevalence to cytomegalovirus in the Portuguese population, 2002–2003. Eurosurveillance *16*, 17–22.

Marshall, G.S., and Stout, G.G. (2005). Cytomegalovirus seroprevalence among women of childbearing age during a 10-year period. Am. J. Perinatol. *22*, 371–376.

Mindel, A., and Sutherland, S. (1984). Antibodies to cytomegalovirus in homosexual and heterosexual men attending an STD clinic. Brit. J. Vener. Dis. *60*, 189–192.

Morton, C.C., and Nance, W.E. (2006). Newborn hearing screening – a silent revolution. N. Engl. J. Med. *354*, 2151–2164.

Murph, J.R., Baron, J.C., Brown, C.K., Ebelback, C.L., and Bale, J.F. (1991). The occupational risk of cytomegalovirus infection among day-care providers. JAMA *265*, 603–608.

Murph, J.R., Souza, I.E., Dawson, J.D., Benson, P., Petheram, S.J., Pfab, D., Gregg, A., O'Neill, M.E., Zimmerman, B., and Bale, J.F., Jr. (1998). Epidemiology of congenital cytomegalovirus infection: maternal risk factors and molecular analysis of cytomegalovirus strains. Am. J. Epidemiol. *147*, 940–947.

Mussi-Pinhata, M.M., Yamamoto, A.Y., Brito, R.M.M., Isaac, M.D., Oliveira, P., Boppana, S., and Britt, W.J. (2009). Birth prevalence and natural history of congenital cytomegalovirus infection in a highly seroimmune population. Clin. Infect. Dis. *49*, 522–528.

Naessens, A., Casteels, A., Decatte, L., and Foulon, W. (2005). A serologic strategy for detecting neonates at risk for congenital cytomegalovirus infection. J. Pediatr. *146*, 194–197.

Nance, W.E., Lim, B.G., and Dodson, K.M. (2006). Importance of congenital cytomegalovirus infections as a cause for pre-lingual hearing loss. J. Clin. Virol. *35*, 221–225.

Nigro, G., Adler, S.P., La Torre, R., and Best, A.M. (2005). Passive immunization during pregnancy for congenital cytomegalovirus infection. N. Engl. J. Med. *353*, 1350–1362.

Noyola, D.E., Demmler, G.J., Nelson, C.T., Griesser, C., Williamson, W.D., Atkins, J.T., Rozelle, J., Turcich, M., Llorente, A.M., Sellers-Vinson, S., *et al.* (2001). Early predictors of neurodevelopmental outcome in symptomatic congenital cytomegalovirus infection. J. Pediatr. *138*, 325–331.

Noyola, D.E., Mejia-Elizondo, A.R., Canseco-Lima, J.M., Allende-Carrera, R., Hernandez-Salinas, A.E., and Ramirez-Zacarias, J.L. (2003). Congenital cytomegalovirus infection in San Luis Potosi, Mexico. Pediatr. Infect. Dis. J. *22*, 89–90.

Numazaki, K., and Fujikawa, T. (2004). Chronological changes of incidence and prognosis of children with asymptomatic congenital cytomegalovirus infection in Sapporo, Japan. BMC Infect. Dis. *4*, 22.

Numazaki, K., Fujikawa, T., and Chiba, S. (2000). Relationship between seropositivity of husbands and primary cytomegalovirus infection during pregnancy. J. Infect. Chemother. *6*, 104–106.

O'Brien, T.P., Thompson, J.M.D., Black, P.N., Becroft, D.M.O., Clark, P.M., Robinson, E., Wild, C., and Mitchell, E.A. (2009). Prevalence and determinants of cytomegalovirus infection in pre-school children. J. Paediatr. Child Health *45*, 291–296.

de Ory, F., Ramirez, R., Garcia, C.L., Leon, P., Sagues, M.J., and Sanz, J.C. (2004). Is there a change in cytomegalovirus seroepidemiology in Spain? Eur. J. Epidemiol. *19*, 85–89.

Pass, R.F., and Hutto, C. (1986). Group day care and cytomegaloviral infections of mothers and children. Rev. Infect. Dis. *8*, 599–605.

Pass, R.F., Hutto, C., Ricks, R., and Cloud, G.A. (1986). Increased rate of cytomegalovirus infection among parents of children attending day-care centers. N. Engl. J. Med. *314*, 1414–1418.

Pass, R.F., Hutto, C., Lyon, M.D., and Cloud, G. (1990). Increased rate of cytomegalovirus infection among day care center workers. Pediatr. Infect. Dis. J. *9*, 465–470.

Pass, R.F., Zhang, C., Evans, A., Simpson, T., Andrews, W., Huang, M.L., Corey, L., Hill, J., Davis, E., Flanigan, C., *et al.* (2009). Vaccine prevention of maternal cytomegalovirus infection. N. Engl. J. Med. *360*, 1191–1199.

Pereira, L.H., Embil, J.A., Haase, D.A., and Manley, K.M. (1990). Cytomegalovirus-infection among women attending a sexually-transmitted disease clinic – association with clinical symptoms and other sexually-transmitted diseases. Am. J. Epidemiol. *131*, 683–692.

Picone, O., Vauloup-Fellous, C., Cordier, A.G., Du Chatelet, I.P., Senat, M.V., Frydman, R., and Grangeot-Keros, L. (2009). A 2-year study on cytomegalovirus infection during pregnancy in a French hospital. BJOG-an Int. J. Obstetr. Gynaecol. *116*, 818–823.

Pignatelli, S., Lazzarotto, T., Gatto, M.R., Dal Monte, P., Landini, M.P., Faldella, G., and Lanari, M. (2010). Cytomegalovirus gN genotypes distribution among congenitally infected newborns and their relationship with symptoms at birth and sequelae. Clin. Infect. Dis. *51*, 33–41.

Preece, P.M., Blount, J.M., Glover, J., Fletcher, G.M., Peckham, C.S., and Griffiths, P.D. (1983). The consequences of primary cytomegalovirus infection in pregnancy. Arch. Dis. Child. *58*, 970–975.

Preece, P.M., Tookey, P., Ades, A., and Peckham, C.S. (1986). Congenital cytomegalovirus infection: predisposing maternal factors. J. Epidemiol. Comm. Health *40*, 205–209.

Revello, M.G., and Gerna, G. (2002). Diagnosis and management of human cytomegalovirus infection in the mother, fetus, and newborn infant. Clin. Microbiol. Rev. *15*, 680–715.

Revello, M.G., Fabbri, E., Furione, M., Zavattoni, M., Lilleri, D., Tassis, B., Quarenghi, A., Cena, C., Arossa, A., Montanari, L., *et al.* (2011). Role of prenatal diagnosis and counseling in the management of 735 pregnancies complicated by primary human cytomegalovirus infection: A 20-year experience. J. Clin. Virol. *50*, 303–307.

Robain, M., Carre, N., Dussaix, E., Salmon-Ceron, D., and Meyer, L. (1998). Incidence and sexual risk factors of cytomegalovirus seroconversion in HIV-infected subjects. The SEROCO Study Group. Sex. Transm. Dis. *25*, 476–480.

Rosenthal, L.S., Fowler, K.B., Boppana, S.B., Britt, W.J., Pass, R.F., Schmid, S.D., Stagno, S., and Cannon, M.J. (2009). Cytomegalovirus shedding and delayed sensorineural hearing loss: results from longitudinal follow-up of children with congenital infection. Pediatr. Infect. Dis. J. *28*, 515–520.

Ross, D.S., Victor, M., Sumartojo, E., and Cannon, M.J. (2008). Women's knowledge of congenital cytomegalovirus: results from the 2005 HealthStyles survey. J. Womens Health (Larchmt.) *17*, 849–858.

Ross, S.A., Arora, N., Novak, Z., Fowler, K.B., Britt, W.J., and Boppana, S.B. (2010). Cytomegalovirus reinfections in healthy seroimmune women. J. Infect. Dis. *201*, 386–389.

van der Sande, M.A.B., Kaye, S., Miles, D.J.C., Waight, P., Jeffries, D.J., Ojuola, O.O., Palmero, M., Pinder, M., Ismaili, J., Flanagan, K.L., *et al.* (2007). Risk factors for and clinical outcome of congenital cytomegalovirus

infection in a peri-urban West-African birth cohort. PLoS One 2.
Schleiss, M.R. (2006). Role of breast milk in acquisition of cytomegalovirus infection: recent advances. Curr. Opin. Pediatr. *18*, 48–52.
Schleiss, M.R. (2008). Cytomegalovirus vaccine development. Human Cytomegalovirus *325*, 361–382.
Schlesinger, Y. (2007). Routine screening for CMV in pregnancy: opening the Pandora box? Isr. Med. Assoc. J. *9*, 395–397.
Schoub, B.D., Johnson, S., McAnerney, J.M., Blackburn, N.K., Guidozzi, F., Ballot, D., and Rothberg, A. (1993). Is antenatal screening for rubella and cytomegalovirus justified? S. Afr. Med. J. *83*, 108–110.
Seale, H., MacIntyre, C.R., Gidding, H.F., Backhouse, J.L., Dwyer, D.E., and Gilbert, L. (2006). National serosurvey of cytomegalovirus in Australia. Clin. Vaccine Immunol. *13*, 1181–1184.
Shan, R.B., Wang, X.L., and Fu, P. (2009). Growth and development of infants with asymptomatic congenital cytomegalovirus infection. Yonsei Med. J. *50*, 667–671.
Sohn, Y.M., Oh, M.K., Balcarek, K.B., Cloud, G.A., and Pass, R.F. (1991). Cytomegalovirus infection in sexually active adolescents. J. Infect. Dis. *163*, 460–463.
Sohn, Y.M., Park, K.I., Lee, C., Han, D.G., and Lee, W.Y. (1992). Congenital cytomegalovirus infection in Korean population with very high prevalence of maternal immunity. J. Korean Med. Sci. *7*, 47–51.
Spector, S.A., and Spector, D.H. (1982). Molecular epidemiology of cytomegalovirus infections in premature twin infants and their mother. Pediatr. Infect. Dis. J. *1*, 405–409.
Stadler, L.P., Bernstein, D.I., Callahan, S.T., Ferreira, J., Simone, G.A.G., Edwards, K.M., Stanberry, L.R., and Rosenthal, S.L. (2010). Seroprevalence of cytomegalovirus (CMV) and risk factors for infection in adolescent males. Clin. Infect. Dis. *51*, E76–E81.
Stagno, S., Dworsky, M.E., Torres, J., Mesa, T., and Hirsh, T. (1982). Prevalence and importance of congenital cytomegalovirus infection in three different populations. J. Pediatr. *101*, 897–900.
Stagno, S., Pass, R.F., Cloud, G., Britt, W.J., Henderson, R.E., Walton, P.D., Veren, D.A., Page, F., and Alford, C.A. (1986). Primary cytomegalovirus infection in pregnancy. Incidence, transmission to fetus, and clinical outcome. JAMA *256*, 1904–1908.
Stagno, S., Reynolds, D., Tsiantos, A., Fuccillo, D.A., Smith, R., Tiller, M., and Alford, C.A., Jr. (1975). Cervical cytomegalovirus excretion in pregnant and nonpregnant women: suppression in early gestation. J. Infect. Dis. *131*, 522–527.
Stagno, S., Reynolds, D.W., Huang, E.S., Thames, S.D., Smith, R.J., and Alford, A.C. (1977). Congenital cytomegalovirus infection: occurance in an immune population. N. Engl. J. Med. *296*, 1254–1258.
Stanberry, L.R., Rosenthal, S.L., Mills, L., Succop, P.A., Biro, F.M., Morrow, R.A., and Bernstein, D.I. (2004). Longitudinal risk of herpes simplex virus (HSV) type 1, HSV type 2, and cytomegalovirus infections among young adolescent girls. Clin. Infect. Dis. *39*, 1433–1438.
Staras, S.A.S., Dollard, S.C., Radford, K.W., Flanders, W.D., Pass, R.F., and Cannon, M.J. (2006). Seroprevalence of cytomegalovirus infection in the USA, 1988–1994. Clin. Infect. Dis. *43*, 1143–1151.
Staras, S.A., Flanders, W.D., Dollard, S.C., Pass, R.F., McGowan, J.E., Jr., and Cannon, M.J. (2008a). Cytomegalovirus seroprevalence and childhood sources of infection: A population-based study among pre-adolescents in the USA. J. Clin. Virol. *43*, 266–271.
Staras, S.A., Flanders, W.D., Dollard, S.C., Pass, R.F., McGowan, J.E., Jr., and Cannon, M.J. (2008b). Influence of sexual activity on cytomegalovirus seroprevalence in the USA, 1988–1994. Sex. Transm. Dis. *35*, 472–479.
Stover, C.T., Smith, D.K., Schmid, D.S., Pellett, P.E., Stewart, J.A., Klein, R.S., Mayer, K., Vlahov, D., Schuman, P., and Cannon, M.J. (2003). Prevalence of and risk factors for viral infections among human immunodeficiency virus (HIV)-infected and high-risk HIV-uninfected women. J. Infect. Dis. *187*, 1388–1396.
Stowell, J.D., Forlin-Passoni, D., Din, E., Radford, K., Brown, D., White, A., Bate, S.L., Dollard, S.C., Bialek, S.R., Cannon, M.J., *et al.* (2012). Cytomegalovirus survival on common environmental surfaces: Opportunities for viral transmission. J. Infect. Dis. *205*, 211–214.
Svahn, A., Berggren, J., Parke, A., Storsaeter, J., Thorstensson, R., and Linde, A. (2006). Changes in seroprevalence to four herpesviruses over 30 years in Swedish children aged 9–12 years. J. Clin. Virol. *37*, 118–123.
Taber, L.H., Frank, A.L., Yow, M.D., and Bagley, A. (1985). Acquisition of cytomegaloviral infections in families with young children: a serological study. J. Infect. Dis. *151*, 948–952.
Temple, R.O., Pass, R.F., and Boll, T.J. (2000). Neuropsychological functioning in patients with asymptomatic congenital cytomegalovirus infection. J. Dev. Behav. Pediatr. *21*, 417–422.
Tookey, P.A., Ades, A.E., and Peckham, C.S. (1992). Cytomegalovirus prevalence in pregnant women: the influence of parity. Arch. Dis. Child. *67*, 779–783.
Tsai, C.H., Tsai, F.J., Shih, Y.T., Wu, S.F., Liu, S.C., and Tseng, Y.H. (1996). Detection of congenital cytomegalovirus infection in Chinese newborn infants using polymerase chain reaction. Acta Paediatr. *85*, 1241–1243.
Vauloup-Fellous, C., Picone, O., Cordier, A.-G., Parent-du-Chatelet, I., Senat, M.-V., Frydman, R., and Grangeot-Keros, L. (2009). Does hygiene counseling have an impact on the rate of CMV primary infection during pregnancy? Results of a 3-year prospective study in a French hospital. J. Clin. Virol. *46S*, S49-S53.
Vial, P., Torres-Pereyra, J., Stagno, S., Gonzalez, F., Donoso, E., Alford, C.A., Hirsch, T., and Rodriguez, L. (1986). Serologic screening for cytomegalovirus, rubella virus, herpes simplex virus, hepatitis B virus, and Toxoplasma gondii in two urban populations of pregnant women in Chile. Bull. Pan Am. Health Organ. *20*, 53–61.
de Vries, J.J.C., Claas, E.C.J., Kroes, A.C.M., and Vossen, A. (2009). Evaluation of DNA extraction methods for dried blood spots in the diagnosis of congenital cytomegalovirus infection. J. Clin. Virol. *46*, S37-S42.
Wang, C.B., Zhang, X.Y., Bialek, S., and Cannon, M.J. (2011). Attribution of congenital cytomegalovirus infection to primary versus non-primary maternal infection. Clin. Infect. Dis. *52*, E11-E13.
Williamson, W.D., Percy, A.K., Yow, M.D., Gerson, P., Catlin, F.I., Koppelman, M.L., and Thurber, S. (1990).

Asymptomatic congenital cytomegalovirus infection. Audiologic, neuroradiologic, and neurodevelopmental abnormalities during the first year. Am. J. Dis. Child. *144*, 1365–1368.

Williamson, W.D., Demmler, G.J., Percy, A.K., and Catlin, F.I. (1992). Progressive hearing loss in infants with asymptomatic congenital cytomegalovirus infection. Pediatrics *90*, 862–866.

Yamamoto, A.Y., Mussi-Pinhata, M.M., Cristina, P., Pinto, G., Moraes Figueiredo, L.T., and Jorge, S.M. (2001). Congenital cytomegalovirus infection in preterm and full-term newborn infants from a population with a high seroprevalence rate. Pediatr. Infect. Dis. J. *20*, 188–192.

Yamamoto, A.Y., Mussi-Pinhata, M.M., Boppana, S.B., Novak, Z., Wagatsuma, V.M., Oliveira, P.d.F., Duarte, G., and Britt, W.J. (2010). Human cytomegalovirus reinfection is associated with intrauterine transmission in a highly cytomegalovirus-immune maternal population. Am. J. Obstet. Gynecol. *202*, 297.e291–298.

Yow, M.D., White, N.H., Taber, L.H., Frank, A.L., Gruber, W.C., May, R.A., and Norton, H.J. (1987). Acquisition of cytomegalovirus infection from birth to 10 years: a longitudinal serologic study. J. Pediatr. *110*, 37–42.

Yow, M.D., Williamson, D.W., Leeds, L.J., Thompson, P., Woodward, R.M., Walmus, B.F., Lester, J.W., Six, H.R., and Griffiths, P.D. (1988). Epidemiologic characteristics of cytomegalovirus infection in mothers and their infants. Am. J. Obstet. Gynecol. *158*, 1189–1195.

Zanghellini, F., Boppana, S.B., Emery, V.C., Griffiths, P.D., and Pass, R.F. (1999). Asymptomatic primary cytomegalovirus infection: virologic and immunologic features. J. Infect. Dis. *180*, 702–707.

Zhang, X.W., Li, F., Yu, X.W., Shi, X.W., Shi, J., and Zhang, J.P. (2007). Physical and intellectual development in children with asymptomatic congenital cytomegalovirus infection: a longitudinal cohort study in Qinba mountain area, China. J. Clin. Virol. *40*, 180–185.

Zhong, J., and Khanna, R. (2007). Vaccine strategies against human cytomegalovirus infection. Expert Rev. Anti. Infect. Ther. *5*, 449–459.

II.2 – Addendum

The Economic Impact of Congenital CMV Infection: Methods and Estimates

Scott D. Grosse, Ismael R. Ortega-Sanchez, Stephanie R. Bialek and Sheila C. Dollard[1]

Abstract

Calculation of the economic burden of congenital cytomegalovirus (CMV) is needed to quantify the potential economic impact of preventative interventions. Although no preventative interventions are currently available other than hygiene education and avoidance of exposures by pregnant women, economic assessment may be needed in coming years for a CMV vaccine, prenatal screening followed by pharmacological treatment, or newborn screening and early intervention for hearing loss and other disabilities. Economic estimates of the costs associated with the long-term consequences of congenital CMV infections are important both to demonstrate the potential benefit from the development of preventative interventions and also to assess their expected cost-effectiveness or cost–benefit once they are ready to be implemented.

Introduction: concepts and definitions

Economic burden measures are of three types: direct costs (resources consumed in providing care), indirect costs (forgone economic productivity), and intangible costs or lost utility (quality-adjusted life-years or QALYs). All economic burden assessments include direct medical costs. Cost-of-illness (COI) analyses conducted from the societal perspective also assess non-medical direct costs and indirect costs, most commonly using a human capital approach to assess loss of life or work disability (Grosse and Krueger, 2011). Intangible costs can be included in cost–benefit or cost-effectiveness analyses of interventions such as vaccines. A cost-effectiveness analysis can include indirect costs or QALYs but should not include both measures (Gold *et al.*, 1996). The US Advisory Committee on Immunization Practices has guidance for cost-effectiveness or cost–benefit analyses of vaccines (Lieu *et al.*, 2008; Smith *et al.*, 2009).

In economic analyses costs are defined in terms of resources consumed. In the absence of direct data on costs, it is typical to use data on payments as a proxy for costs. Charges (list prices), however, are not valid estimates of costs in US health care because charges typically exceed actual payments (Gold *et al.*, 1996).

Direct costs of care for childhood-onset disability are of five major types: medical costs, special education services, developmental and habilitation services, residential care, and time costs for unpaid family caregivers. Although US guidelines call for the inclusion of unpaid family caregiving time cost (Gold *et al.*, 1996), such costs are typically not included because of lack of data. Also guidelines in other countries generally classify unpaid family caregiving time cost as an indirect cost, since it represents lost productivity (Drummond, 2005).

Incremental costs represent the difference in costs due to disease or disability. For example, the incremental cost of special education for children with a given condition is the difference in average total schooling costs for children with that condition relative to average costs for all children, not the average cost of special education services *per se* (CDC, 2004).

1 The findings and conclusions in this report are those of the authors and do not necessarily represent the official position of the Centers for Disease Control and Prevention.

The cost of a childhood-onset condition such as congenital CMV represents the present value of the stream of future costs of care discounted back to the time of birth. Costs in future years are projected on the basis of the assumption that current costs remain constant in inflation-adjusted currency. The use of a discount rate, typically 3% to 5%, is an adjustment for the time value of money, not inflation, and is equivalent to the long-term risk-free real rate of return on capital.

The QALY is a summary measure of population health that integrates premature mortality, morbidity, and disability (Gold *et al.*, 2002). A QALY is calculated as 1 year of life lived in a health state multiplied by a QALY weight between 0 and 1, where 1 is equivalent to perfect health and 0 is equal to death. QALY weights can be calculated through use of a generic preference-based instrument which asks respondents to rate themselves (or their child) on various attributes. Or respondents may be asked to trade off between dying prematurely and living with a less than optimal health state (Prosser *et al.*, 2012). For example, one study which used the latter approach with a sample of US parents found that the average weights for children with mild or severe intellectual disability were 0.84 and 0.59, respectively, and for mild or severe bilateral hearing loss were 0.92 and 0.86, respectively (Carroll and Downs, 2009; Grosse *et al.*, 2010).

Previous economic estimates of congenital CMV disease burden

Peer-reviewed publications

Existing cost estimates for congenital CMV are poorly documented. For example, Arvin *et al.* (2004) stated that in the 1990s the economic impact of congenital CMV in the USA was estimated to be $1.86 billion and that the lifetime cost of caring for a child with congenital CMV was $300,000. However, neither of the references cited provided cost estimates. Consequently, the methods used are unknown. Those cost estimates have been cited in subsequent publications on congenital CMV (Rahav, 2007; Ludwig and Hengel, 2009; Walter *et al.*, 2011).

A few estimates of the economic burden or impact of congenital CMV have been published as part of cost-effectiveness analyses of prevention strategies (Porath *et al.*, 1990; Stratton *et al.*, 2000; Cahill *et al.*, 2009). None of these studies estimated indirect costs or monetary estimates of intangible costs, which would be needed for the preparation of a cost–benefit analysis. Existing burden estimates for the USA are limited to direct costs and/or QALYs. One recent analysis from Germany estimated the sum of direct and indirect costs associated with congenital CMV and the cost–benefit of prenatal screening and treatment (Walter *et al.*, 2011).

Porath *et al.* (1990) appear to have been the first investigators to calculate what they referred to as the potential 'cost–benefit' of a CMV vaccine. However, it did not include indirect costs or any measure of health outcomes which are required for a cost–benefit analysis. Also, because their analysis was restricted to potentially preventable charges, not actual costs, of medical and educational services, estimates of costs were likely inflated. The authors assumed a lifetime direct charge of $127,000 for a child with non-fatal symptomatic CMV infection, using a 5% discount rate. They assumed that 50% of children would require special schooling for the blind or deaf and another 38% would require schooling or institutional care for intellectual disability. No sources were cited for those assumptions.

Cahill *et al.* (2009) published a cost-effectiveness analysis of prenatal screening for CMV in the USA which calculated the costs of screening pregnant women for primary maternal CMV infection followed by the treatment of those women who screen positive with CMV-intravenous immune globulin (IVIG), the averted costs of care for cases of congenital CMV prevented, and the numbers of QALYs gained by preventing infant deaths and cases of disability. The key economic assumptions about the societal burden of congenital CMV in the absence of intervention were the lifetime cost of care for a severely affected child, assumed to be $995,940 using a 3% discount rate, and the utility weights of 0.75 for a mildly affected child and 0.55 for a severely affected child. The key epidemiological assumptions were that more than 2000 infants would die each year and more than 8300 would experience severe disability in the absence of prenatal screening and treatment. Assumptions about the effectiveness of treatment are not the focus of this appendix on costs.

A serious limitation of the estimates in Cahill *et al.* is the lack of data to support both the epidemiological and economic assumptions. The assumptions of more than 2000 infant deaths and more than 8300 cases of severe disability each year in the USA caused by congenital CMV infection are roughly an order of magnitude greater than what could be justified based on population-based estimates. The one reference cited for the cost assumption was a study which reported two lifetime cost estimates for a child born with cerebral palsy, $1,067,000 using a 2% discount rate and $445,000 using a 5% discount rate, in 1988 dollars (Waitzman *et al.*, 1994). There are two problems with these cost assumptions. First, Cahill *et al.* did not explain how they adjusted for inflation and the difference in discount rates (3% vs. 2% or 5%). More seriously, the Waitzman *et al.* estimate was the sum of direct and indirect costs. US guidelines for cost-effectiveness analyses that use QALYs as the outcome measure require that only direct costs be included (Gold *et al.*, 1996). Approximately half of the total cost for children with cerebral palsy consisted of indirect costs (Waitzman *et al.*, 1994), which should have been excluded. A minor limitation of the Cahill analysis was that the assumed utility of mildly affected children was low relative to published estimates for children with moderate hearing loss (Grosse *et al.*, 2010).

A recently published German analysis by Walter *et al.* (2011) estimated the aggregate economic burden of congenital CMV infection to be 242.9 million euros. They assumed that a German birth cohort of 700,000 births has 6500 infants with congenital CMV infection (0.9% birth prevalence), of whom 40 die as infants and 1200 others experience disability. The lifetime cost per patient was reported to be 766,444 euros using a 5% discount rate. However, if one divides the aggregate cost estimate by the lifetime cost per patient, the implied number of patients is 317. It is unclear how either cost estimate was calculated. The source for the estimated numbers of infants with congenital CMV and outcomes was a webpage with no documentation (http://dgk.de/gesundheit/impfen-infektionskrankheiten/krankheiten-von-a-bis-z/cytomegalie-virus-cmv.html). The authors also cited a German text on CMV which reported birth prevalence estimates of 0.04–0.49% (Halwachs-Baumann and Benser, 2003).

Institute of Medicine report

The remainder of this section discusses the estimates of the preventable economic burden of CMV prepared by the US Institute of Medicine (IOM) Committee to Study Priorities for Vaccine Development (Stratton *et al.*, 2000). The IOM report estimated that a hypothetical vaccine that was 100% effective and had 100% uptake would result in the saving of 70,000 discounted QALYs and a reduction in direct cost of $4 billion (Stratton *et al.*, 2000). The report did not present cost estimates in the absence of a vaccine. The report included benefits of a CMV vaccine to immunocompromised adults, but almost all of the calculated benefit was attributable to prevention of congenital CMV infections.

The epidemiological assumptions in the IOM report need to be detailed in order to be able to follow the methods used to estimate economic costs and QALYs. The report assumed that 1.0% of children in a single year's US cohort of 4 million births are born with congenital CMV, of whom 10% are symptomatic at birth. Of symptomatic cases, 10% ($n=400$) were assumed to die during infancy and of survivors 90% ($n=3240$) would develop severe sequelae (severe intellectual disability) and the remaining 10% ($n=360$) would all have moderate sequelae (primarily moderate to profound bilateral hearing loss or impairment). It was assumed that 15% ($n=5400$) of cases of congenital CMV infection asymptomatic at birth would develop moderate sequelae. These assumptions resulted in estimates that each year 40,000 children are born with congenital CMV infection in the USA each year, 400 die during infancy, 3240 develop severe sequelae, and 5760 develop moderate sequelae.

The bulk of costs in the IOM calculations were attributed to lifelong residential care for all individuals with severe sequelae and for 10% of those with moderate sequelae of asymptomatic infections, It was assumed that children with severe sequelae would all require residential care at a cost of roughly $82,125 per year for 20 years followed by death. With a 3% discount rate, that is equivalent to 14.8 discounted years times $82,125, or roughly $1.2 million per child. The estimate is said to be based on a 1995 report, and it is not clear what currency year it applies to. For the 3240 children with symptomatic infections the discounted cost of institutional care was estimated at $3.9 billion.

For children with congenital CMV infections that were asymptomatic at birth, the IOM report assumed that 15% would develop neurological sequelae. Among those 15% with sequelae, 10% ($n=540$) of children were assumed to require life-long institutional care, at an aggregate cost of $1.3 billion. It was assumed that their life expectancy was 75.35 years. At a reported cost of $82,125 per year for institutional care, that is equivalent to $2.5 million per child, or an aggregate cost of $1.3 billion. Special education costs for those children added $300 million and medical follow-up added $100 million. The other 90% of children with sequelae of infections asymptomatic at birth were assumed to incur minimal economic costs.

The total implied economic cost for children with sequelae of both symptomatic and asymptomatic congenital CMV infections was $5.6 billion, in dollars of unstated currency year (Stratton *et al.*, 2000). In 2011 dollars, adjusted for inflation, the total cost would be much larger, of course.

The most serious limitation of the IOM estimates of economic costs is the unsupported assumptions of the prevalence of the disorder and its sequelae. Dollard *et al.* (2007) in a systematic review of findings of long-term outcomes in prospective cohorts based on unbiased ascertainment at birth concluded that only a minority of children with symptomatic infections had cognitive impairment. The largest such study reported that 2 of 11 (18%) children with neonatal symptoms of congenital CMV had intellectual disability at age 7, only one of whom had severe disability (Ahlfors *et al.*, 1999). Clinic-based studies, which are subject to referral bias and therefore likely to overstate the frequency of long-term sequelae, have reported rates of severe intellectual disability of 28% (Conboy *et al.*, 1987) to 56% (Bale *et al.*, 1990), much lower than the 90% assumed in the IOM report. Also, there is no evidence that school-age children with asymptomatic congenital CMV infections are at risk of serious non-auditory sequelae (Dollard *et al.*, 2007). In particular, the assumption that 10% of children with sequelae of asymptomatic CMV infections require lifelong institutional care has no empirical support.

Another major weakness of the IOM economic cost calculations is that the unit cost estimates were based on hypothetical patterns of care which were not supported by or consistent with empirical data. First, it was assumed that all children with severe disability receive institutional care. In fact, very few children with severe developmental disabilities in the US, regardless of aetiology, receive institutional care (Harrington *et al.*, 2009). Even among adults with moderate to profound intellectual disability (i.e. IQ < 50), who are more likely than children to be institutionalized, just half are in residential care (Harrington *et al.*, 2009). It might be argued that those data represent a milder spectrum of outcomes. However, a Dutch study of seven survivors of metabolic crises who experienced severe neurological disability, comparable to severe congenital CMV sequelae, found that just two (28%) required institutional care (Derks *et al.*, 2006). The assumption that all children with severe neurological disability require life-long institutional care has also been made in other economic analyses of vaccine-preventable diseases (e.g. Shepard *et al.*, 2005), despite an absence of data. On the other hand, the IOM understated costs by assuming a life expectancy of 20 years for severely disabled children. In fact, the vast majority of children with severe to profound intellectual disability and no physical disabilities survive to adulthood (Shavelle *et al.*, 2003). However, the latter understatement is much smaller in magnitude than the overstatement resulting from the assumption of universal institutionalization.

Another weakness of the IOM report was in the estimation of education costs for children with disability. The IOM report assumed that children with symptomatic infections would receive special education services for 10 years at a cost of $8000 per year regardless of whether they had severe or mild sequelae. In contrast, for children with sequelae of asymptomatic infections, 90% were assumed to have education costs of just $2000 per year for 10 years, and the remaining 10% had no education costs assigned. The average incremental cost of schooling a child with multiple handicaps in the USA in 2000 was approximately $13,500 per year (Chambers *et al.*, 2003). The incremental cost of education for children with hearing loss and no other impairment in the USA in 2000 was close to $10,000 (Chambers *et al.*, 2003), almost five times higher than the IOM assumption. Also, the IOM analysis made no allowance for the cost of assistive devices such as hearing aids or cochlear implants.

One strength of the IOM report lies in its methods for the calculation of QALYs. In addition to deaths, for which the QALY weight is 0, the report assumed a QALY weight of 0.48 for years lived with severe disability and 0.89 for years lived with moderate neurological sequelae, especially 'moderate deafness'. These QALY weights are consistent with the published literature for those endpoints (Carroll and Downs, 2009; Grosse *et al.*, 2010).

Need for better estimates

More data on congenital CMV outcomes based on the long-term follow-up of cohorts of universally screened infants are needed. The CMV & Hearing Multicenter Screening (CHIMES) Study at seven US sites is following children up to 4 years of age with congenital CMV detected through newborn screening to identify cases of hearing loss; the research protocol does not include assessments for other types of sequelae (http://main.uab.edu/Sites/chimes/about/). The Centers for Disease Control and Prevention (CDC) is currently working with the Baylor College of Medicine to retrospectively analyse outcomes data collected from a screened cohort.

Existing cost estimates for neurological impairments in children have serious limitations. One study reported cost estimates for intellectual disability without distinction by severity (Honeycutt *et al.*, 2003; CDC, 2004). All intellectual disability was treated as one category. Approximately 90% of all individuals with intellectual disability have mild impairment, classified as those with an IQ of between 55 and 70 (Shavelle *et al.*, 2003). Consequently, average cost estimates for people with intellectual disability as a group cannot be used to assess the economic costs of severe sequelae of congenital CMV.

Cost estimates should include all components of direct costs. Reasonable data exist for costs of educational services, for which year 2000 estimates are available which can be updated for inflation (Chambers *et al.*, 2003). For mild disability, medical costs are typically quite low, especially for hearing loss apart from other physical disabilities, and existing estimates could be adjusted for inflation and the use of cochlear implants.

The most important economic impact of CMV is severe neurological sequelae specific to symptomatic congenital CMV. Ideally, direct cost estimates would be based on long-term cost and outcomes data collected for a representative cohort of children with congenital CMV infections as has been done for medium chain acyl-coA dehydrogenase (MCAD) deficiency in the Netherlands (Derks *et al.*, 2006; van der Hilst *et al.*, 2007). Alternatively, modelling can be done based on information on survival and empirical patterns of age-specific use of services among persons with relatively severe intellectual and other developmental disabilities. For example, in California during 2004–2005, individuals receiving public services with a diagnosis of severe or profound intellectual disability had average annual expenditures of roughly $12,000, and those with a dual diagnosis with another condition had average annual expenditures of $16,000 (Kang and Harrington, 2008).

Conclusion

Existing US cost estimates for congenital CMV infection do not reflect current epidemiological and clinical information. Given the substantial limitations of existing estimates, there is limited value to be gained by updating previous cost estimates for inflation. Rather, new estimates must be calculated using population-based epidemiological data derived from long-term follow-up to newborn screening of unselected cohorts of infants to project costs for representative cohorts of children with congenital CMV.

Acknowledgements

Mike Cannon provided helpful comments on an earlier draft.

References

Ahlfors, K., Ivarsson, S.-A., and Harris, S. (1999). Report on a long-term study of maternal and congenital cytomegalovirus infection in Sweden. Review of prospective studies available in the literature. Scand. J. Infect. Dis. *31*, 443–457.

Arvin, A.M., Fast, P., Myers, M., Plotkin, S., and Rabinovich, R. (2004). Vaccine development to prevent cytomegalovirus disease: report from the National Vaccine Advisory Committee. Clin. Infect. Dis. *39*, 233–239.

Bale, J.F., Jr., Blackman, J.A., and Sato, Y. (1990). Outcome in children with symptomatic congenital cytomegalovirus infection. J. Child. Neurol. *5*, 131–136.

Cahill, A.G., Odibo, A.O., Stamilio, D.M., and Macones, G.A. (2009). Screening and treating for primary cytomegalovirus infection in pregnancy: where do we stand? A decision-analytic and economic analysis. Am. J. Obstet. Gynecol. *201*, 466.e1–e7.

Carroll, A.E., and Downs, S.M. (2009). Improving decision analyses: parent preferences (utility values) for pediatric health outcomes. J. Pediatr. *155*, 21–25.

Centers for Disease Control and Prevention (CDC). (2004). Economic costs associated with mental retardation, cerebral palsy, hearing loss, and vision impairment – United States, 2003. MMWR Morb. Mortal. Wkly. Rep. *53*, 57–59.

Chambers, J.G., Shkolnik, J., and Perez, M. (2003). Total expenditures for students with disabilities, 1999–2000: Spending variation by disability. Special Education Expenditure Project Report #5. Washington, DC: American Institutes for Research, 2003. Available at: http://csef.air.org/publications/seep/national/Final_SEEP_Report_5.PDF (accessed 20 February 2012)

Conboy, T.J., Pass, R.F., Stagno, S., Britt, W.J., Alford, C.A., McFarland, C.E., and Boll, T.J. (1987). Early clinical manifestations and intellectual outcome in children with symptomatic congenital cytomegalovirus infection. J. Pediatr. *111*, 343–348.

Derks, T.G., Reijngoud, D.J., Waterham, H.R., Gerver, W.J., van den Berg, M.P., Sauer, P.J., and Smit, G.P. (2006). The natural history of medium-chain acyl CoA dehydrogenase deficiency in the Netherlands: clinical presentation and outcome. J. Pediatr. *148*, 665–670.

Dollard, S.C., Grosse, S.D., and Ross, D.S. (2007). New estimates of the prevalence of neurological and sensory sequelae and mortality associated with congenital cytomegalovirus infection. Rev. Med. Virol. *17*, 355–363.

Drummond, M.F. (2005). Methods for the Economic Evaluation of Health Care Programmes (Oxford University Press, Oxford, UK).

Fowler, K.B., Dahle, A.J., Boppana, S.B., and Pass, R.F. (1999). Newborn hearing screening: will children with hearing loss caused by congenital cytomegalovirus infection be missed? J. Pediatr. *135*, 60–64.

Gold, M.R., Siegel, J.E., Russell, L.B., and Weinstein, M.C., eds. (1996). Cost-Effectiveness in Health and Medicine (Oxford University Press, New York).

Gold, M.R., Stevenson, D., and Fryback, D.G. (2002). HALYs and QALYs and DALYs, oh my: similarities and differences in summary measures of population health. Annu. Rev. Public Health *23*, 115–134.

Grosse, S.D., and Krueger, K.V. (2011). Income-based human capital estimation in public health economics and forensic economics. J. Forensic Econ. *22*, 43–57.

Grosse, S.D., Ross, D.S., and Dollard, S.C. (2008). Congenital cytomegalovirus (CMV) infection as a cause of permanent bilateral hearing loss: a quantitative assessment. J. Clin. Virol. *41*, 57–62.

Grosse, S.D., Prosser, L.A., Asakawa, K., and Feeny, D. (2010). QALY weights for neurosensory impairments in pediatric economic evaluations: case studies and a critique. Expert Rev. Pharmacoecon. Outcomes Res. *10*, 293–308.

Halwachs-Baumann, G., and Benser, B. (2003). Die konnatale Zytomegalievirusinfektion. Epidemiologie – Diagnose – Therapie. (SpringerWienNewYork, Vienna, Austria).

Harrington, C., Kang, T., and Chang, J. (2007). Factors associated with living in developmental centers in California. Intellect. Dev. Disabil. *47*, 108–124.

Honeycutt, A.A., Grosse, S.D., Dunlap, J.L., Schendel, D.E., Chen, H., Brann, E., and al Homsi, G. (2003). Economic costs of mental retardation, cerebral palsy, hearing loss, and vision impairment. In Using Survey Data to Study Disability, Altman, B.M., Barnartt, S.N., and Hendershot, G., eds. (Elsevier, London, UK), pp. 207–228.

Kang, T., and Harrington, C. (2008). Variation in types of service use and expenditures for individuals with developmental disabilities. Disabil. Health J. *1*, 30–41.

Lieu, T., Meltzer, M.I., and Messonnier, M.L. (2008). Guidance for health economics studies presented to the Advisory Committee on Immunization Practices (ACIP). Centers for Disease Control and Prevention, Atlanta, GA. Available at: www.cdc.gov/vaccines/recs/acip/economic-studies.htm.

Ludwig, A., and Hengel, H. (2009). Epidemiological impact and disease burden of congenital cytomegalovirus infection in Europe. Euro Surveill. *14*, 26–32.

Porath, A., McNutt, R.A., Smiley, L.M., and Weigle, K.A. (1990). Effectiveness and cost benefit of a proposed live cytomegalovirus vaccine in the prevention of congenital disease. Rev. Infect. Dis. *12*, 31–40.

Prosser, L.A., Grosse, S.D., and Wittenberg, E. (2012). Health utility elicitation: is there still a role for direct methods? Pharmacoeconomics *30*, 83–86.

Rahav, G. (2007). Congenital cytomegalovirus infection – a question of screening. Isr. Med. Assoc. J. *9*, 392–394.

Shavelle, R.M., Strauss, D.J., and Day, S.M. (2003). Comparative mortality of persons with mental retardation in California 1980–1999. J. Insur. Med. *35*, 5–8.

Shepard, C.W., Ortega-Sanchez, I.R., Scott, R.D., and Rosenstein, N.E. (2005). Cost-effectiveness of conjugate meningococcal vaccination strategies in the USA. Pediatrics. *115*, 1220–1232.

Smith, J.C., Snider, D.E., and Pickering, L.K. (2009). Immunization policy development in the USA: the role of the Advisory Committee on Immunization Practices. Ann. Intern. Med. *150*, 45–49.

Stratton, K.R., Durch, J.S., and Lawrence, R.S., eds. (2000). Appendix 4, Cytomegalovirus. In Vaccines for the 21st Century: A Tool for Decisionmaking. Committee to Study Priorities for Vaccine Development, Division of Health Promotion and Disease Prevention, Institute of Medicine (National Academy Press, Washington, DC), pp. 165–171. Available at: http://www.nap.edu/openbook.php?record_id=5501&page=165 (accessed 20 February 2012).

van der Hilst, C.S., Derks, T.G., Reijngoud, D.J., Smit, G.P., and TenVergert, E.M. (2007). Cost-effectiveness of neonatal screening for medium chain acyl-CoA dehydrogenase deficiency: the homogeneous population of The Netherlands. J. Pediatr. *151*, 115–120.

Waitzman, N.J., Romano, P.S., and Scheffler, R.M. (1994). Estimates of the economic costs of birth defects. Inquiry *31*, 188–205.

Walter, E., Brennig, C., and Schöllbauer, V. (2011). How to save money: congenital CMV infection and the economy. In Congenital Cytomegalovirus Infection: Epidemiology, Diagnosis, Therapy, Halwachs-Baumann, G., ed. (New York: SpringerWienNewYork), pp. 121–144.

Clinical Cytomegalovirus Research: Congenital Infection

II.3

Stuart P. Adler and Giovanni Nigro

Abstract

This chapter focuses on recent research developments related to CMV infections during pregnancy. The developments include an understanding of the pathogenesis of CMV infections, knowledge of high risk women, and potentially effective interventions to prevent maternal infections and fetal disease.

Introduction

The epidemiology and pathogenesis of CMV infections among pregnant women have been intensely studied over the last three decades (Colugnati, 2007; Kenneson and Cannon, 2007). A primary CMV infection during pregnancy is a frequent and serious threat to the fetus. Annually in the USA approximately 40,000 pregnant women acquire a primary CMV infection (seroconvert). Of the 40,000 women who seroconvert approximately 6000–8000 of their infants develop severe and permanent neurological damage from this infection (Fowler *et al.*, 1992). Fetal or neonatal death occurs in about 10% of fetuses or newborns after an intrauterine CMV infection. Neurological damage includes impaired development, mental retardation, and unilateral or bilateral neurosensory hearing loss.

The rate of susceptibility to a primary CMV infection during pregnancy is known. In the USA and Europe, among women of child-bearing age, between 40% and 80% are susceptible (seronegative) to CMV at conception. The rate of susceptibility at conception varies by country and by ethnic or racial group.

Vertical and horizontal transmission of CMV

How is CMV transmitted? In up to 2% of all pregnancies, transplacental transmission (vertical) occurs. In the majority of cases of transmission, the mother is seropositive at conception, and the fetus is congenitally infected after a recurrence of the mother's infection. Although primary maternal infection during pregnancy is responsible for only a minority of congenitally infected newborns, it is responsible for the majority of the symptomatic infections in the newborn and the severe handicaps caused by congenital infection (Grant *et al.*, 1981; Ahlors *et al.*, 1984; Griffiths and Baboonian, 1984; Nankervis *et al.*, 1984; Stagno *et al.*, 1986).

Prenatal and perinatal transmission frequently occurs via cervical and vaginal secretions (Dowrsky *et al.*, 1983). Postnatal CMV transmission is more frequent than congenital infection and occurs frequently via breast milk; CMV can also be acquired from other children, as in a day care. Intrafamilial transmission is frequent but slow following a primary infection in a single family member, with a rate of transmission of about 50% (Taber *et al.*, 1985).

CMV is frequently excreted in semen and cervical secretions and CMV infections are common among those with multiple sex partners. The frequency of sexual transmission of CMV is unknown, because the virus can be transmitted orally and by close and frequent contact.

Person-to-person (horizontal) transmission is very slow, even under optimal circumstances such as daily contact as occurs within a household or child care centre. In studies of CMV transmission among children in day care, children initially shed CMV at a concentration of about 10^4 plaque-forming units per millilitre of urine following a primary infection. This titre declines slowly thereafter (Murph *et al.*, 1988). Children less than 2 years of age shed CMV for between 6 and 40 months with a mean of about 2 years (Adler, 1991).

Over 3 years at three day care centres, 14%, 27% and 45% of the children became infected, with the most infected in the second year of life (Adler, 1991). Even at centres with the high rates of infection, on average only one child per month acquires a primary CMV infection. Therefore, even under ideal transmission conditions of close, intimate daily contact (i.e. children playing daily together in the same room or sharing toys), the virus is transmitted slowly.

The rate of CMV transmission from infected children to their mothers or caregivers is also very slow and depends on the age of an infected child (Adler, 1991). We observed that, among the seronegative mothers of infected children, 16 (57%) of 28 mothers with infected children 20 months of age or younger acquired CMV from their children, while only three (14%) of 22 mothers with infected children over 20 months of age acquired the infection ($P < 0.007$). In the group of mothers with infected children less than 20 months of age, the average interval between identification of the child's infection and transmission to the mother was 8 months (SD ±6 months).

Caregivers can also be infected with CMV from children (Adler, 1989; Pass *et al.*, 1990; Murph *et al.*, 1991). We studied 614 caregivers in Richmond, and the rate of CMV infections among caregivers was independently associated with the age and race of the caregiver and the ages of children for whom they cared. The highest rate of CMV infections occurred in women caring for children younger than 2 years independent of age and race (Adler, 1989; Pass *et al.*, 1990). For the caregivers in our study, the annual seroconversion rate was 11% for a group of 202 initially seronegative women, compared with a 2% rate for a control group.

CMV is in the urine and saliva of children and is also on environmental surfaces such as toys and even hands (Hutto *et al.*, 1986; Demmler *et al.*, 1987). CMV easily deactivates with soaps, detergents, and alcohol and is not very stable in the environment (Faix, 1985). CMV will also wash off surfaces with plain water (Faix, 1986). CMV has a half-life of 2 to 6 hours on surfaces, but low titres of virus may persist for 24 hours.

Primary versus secondary maternal infection

Fig. II.3.1 shows an algorithm which indicates that between 20% and 60% of pregnant women are susceptible to CMV at conception. Of these between 1% to 4% will acquire CMV during pregnancy and on average between 40% and 50% of infected women will transmit the virus to the fetus. The lowest transmission rate (35%) occurs if the maternal infection is in the first trimester and as pregnancy progresses the transmission rate increases to 73% for women who acquire CMV infections in the third trimester (Bodéus *et al.*, 2010). This increasing rate of transmission during gestation may be related to the transfer of maternal IgG to the fetus (see Chapter II.4). Of infants infected *in utero* in the first half of pregnancy approximately one third to one half will have symptoms or develop severe neurological impairment (Nigro *et al.*, 2005; Pass *et al.*, 2006). The neonatal disease rate is probably highest for children of women who have had a primary infection in early gestation, but definitive data on this point are lacking.

There is uncertainty about the role of maternal immunity to CMV prior to conception. Infants born of mothers with preconception immunity not only give birth to infected infants but also occasionally give birth to infants with symptoms at birth that may develop delayed sequelae, particularly hearing loss (Fowler *et al.*, 1992; Nigro *et al.*, 1993; Revello *et al.*, 2006; Ross *et al.*, 2006). Nevertheless, there is no uncertainty concerning the fact that the rate of congenital infection among women with preconception immunity is only between 0.5% and 2% compared with an average rate of 40% to 50% in women who have seroconverted during pregnancy. In one study,

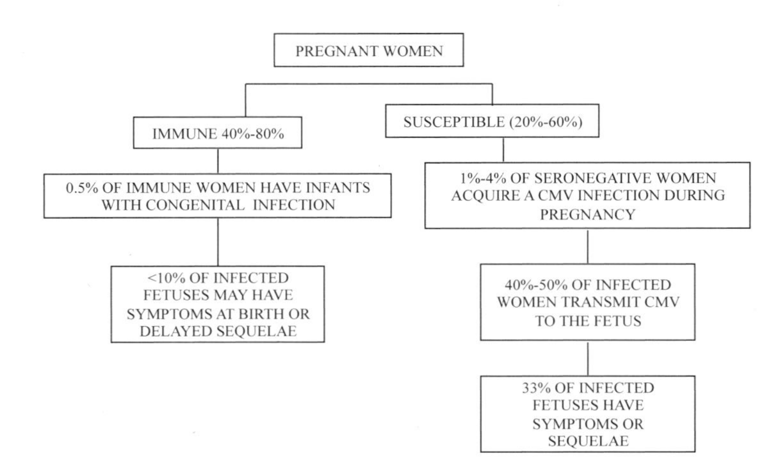

Figure II.3.1 Relationship of maternal immunity to disease caused by congenital CMV infection. Adapted from Stagno and Whitely (1985).

3% of pregnant women seronegative before conception had congenitally infected infants compared with 1% for women seropositive before conception (Fowler *et al.*, 2003).

The most severe infant sequelae occur only among women with a primary CMV infection during pregnancy (Fowler *et al.*, 1992) Infant hearing loss was observed in women who had a recurring infection, but it was not nearly as severe as among the children born of mothers without preconception immunity (Revello *et al.*, 2006; Ross *et al.*, 2006).

Immune adaptation to pregnancy

To avoid maternal immunological rejection of the fetus as a foreign tissue, the immune system of pregnant women adapts to coexist with the fetus (Moffett-King, 2002). The immunological relationship between the mother and the fetus is a bi-directional communication determined by fetal antigen presentation and by reaction to these antigens by the maternal immune response, which is biased towards humoral immunity and away from cell-mediated immunity that could be harmful to the fetus. Protective mechanisms for a normal pregnancy are a decrease of peripheral NK activity and progesterone-dependent immunomodulation (Szekeres-Bartho, 2002). The transient maternal immunodepression may enhance susceptibility to new pathogens or reactivation of latent infections. Women with impaired cellular immune responses to CMV (such as those with AIDS or those receiving immunosuppressive therapy) are more likely to transmit CMV to the fetus (Chandwani *et al.*, 1996; Doyle *et al.*, 1996). Regulation of the innate and adaptive immune responses plays a critical role in the balance between infection control and immunopathology (Cheeran *et al.*, 2009). In fact, neuroinflammation following fetal CMV infection can have neurotoxic effects due to an overexpression of cytokines. Neonatal disease may be also enhanced by transient cellular and humoral immunodepression occurring in the second and third trimesters. Like in other viral infections, both the rate of maternal–fetal CMV transmission and the subsequent disease may be associated with a high viral load either in fetal amniotic fluid or in the newborn's plasma (Lanari *et al.*, 2006). A protective role is displayed by high titres of maternal neutralizing antibodies and high IgG avidity to CMV antigens both against maternal–fetal transmission and viral pathogenicity (Boppana *et al.*, 2001; Nigro *et al.*, 2005).

Fetal infection and symptoms

Neuropathological processes are the most severe and characteristic features of congenital CMV infection, and include meningoencephalitis, periventricular calcifications, microcephaly, polymicrogyria and other migrational alterations. Viral predilection for the periventricular area may relate to the proximity to the cerebrospinal fluid (CSF) pathway, through which CMV probably spreads, and to the actively proliferating subependymal germinal matrix cells, which are particularly vulnerable to CMV. Microcephaly relates to encephaloclastic viral effects and possible neuroproliferative impairment due to CMV interference. Migrational alterations show that teratogenic effects of CMV may also occur in the second trimester of pregnancy, when neuronal migration takes place. In fact, CMV is the only congenitally transmitted pathogen causing altered gyral development, the pathogenesis of which includes both teratogenic and encephaloclastic mechanisms (Volpe, 2000).

The diagnosis of suspected or confirmed primary CMV infection in the first half of pregnancy, or the appearance of fetal abnormalities by ultrasound should be followed by testing the amniotic fluid for CMV DNA detection by PCR or nested PCR. Real-time PCR is currently the most sensitive test for prenatal diagnosis of CMV infection and is also highly specific since it avoids possible false-positive results due to carryover products (Revello *et al.*, 2009; see Chapter II.18). A rapid diagnosis of fetal infection is possible using monoclonal antibodies to the major IE protein p72 and the shell vial technique for CMV isolation and identification within 16–24 hours after sample collection (Gleaves *et al.*, 1984). Cordocentesis may complement amniocentesis, since fetal blood can be tested for detection of viral genome, IgM and culture together with haematological, immunological and enzymatic examinations. In particular, thrombocytopenia is a negative predictor for infant outcome (Liesnard *et al.*, 2000). However, umbilical cord sampling has an increased procedural morbidity and delays sampling until at least 20 weeks' gestation when the vein is large enough to puncture and the immune system is producing IgM antibodies. If CMV initially infects the placenta, chorionic villus sampling offers the potential for first-trimester diagnosis (Revello and Gerna, 2002). Prenatal diagnosis of CMV infection is most predictable at 20–21 weeks' gestation, since fetal diuresis is highly efficient in the second half of pregnancy. However, CMV DNA could be detected earlier, depending on the time of maternal and subsequent fetal infection as occurs in periconceptional or early in gestation CMV transmission (Nigro *et al.*, 2012a). According to Enders *et al.* (2011), the intrauterine transmission rates following primary CMV infection in the pre- and periconceptional period are about 17% and 34%, respectively. For the first, second and third trimester of pregnancy transmission rates are around 30%, 38% and 72% respectively.

CMV in the amniotic fluid does not predict an adverse fetal outcome. The prognostic value of quantification of viral DNA in the amniotic fluid has been evaluated as a prognostic marker of infant infection, in addition to the ultrasound abnormalities such as intrauterine growth retardation, ascites, placental enlargement, or central nervous system abnormalities. Guerra *et al.* (2000) reported that values greater than 1000 copies/ml were predictive of fetal infection and greater than 100,000 copies/mL were predictive of infant symptoms. However, although higher values were observed in symptomatic than in asymptomatic fetuses, a high degree of overlap was observed between the two groups (Revello and Gerna, 2000; Gouarin *et al.*, 2002). In addition, regardless of the fetal outcome, other variables such as the gestational age at time of amniocentesis and the time elapsed after maternal infection were related to viral load in amniotic fluid. Therefore, a possible correlation between viral load in amniotic fluid and clinical outcome should be considered with caution. Conversely, maternal viraemia was not correlated with symptoms in infants.

Fetal CMV infection should be closely monitored by ultrasound examination for evidence of intrauterine growth retardation, intestinal, periventricular and hepatic echodensities, hepatosplenomegaly, pyelectasis, oligohydramnios, polyhydramnios, ascites, placental enlargement and hydrops (Guerra *et al.*, 2008). Bowel hyperechodensity is a frequent first or single ultrasound abnormality of fetal CMV infection (Nigro *et al.*, 2012a). Other manifestations such as lung hypoplasia, ileus, pericardial and pleural effusions, intrahepatic calcifications, microcephaly, hydrocephalus, necrotic, cystic or calcified lesions in the brain, liver or placenta, and nephropathy can also occur (Stagno and Britt, 2006). In recent years, MRI appeared to be a very useful diagnostic and prognostic tool, more sensitive than ultrasound examination (Doneda *et al.*, 2010).

Symptoms at birth

CMV is the most common and serious congenital infection associated with severe neurological sequelae, although the majority of infants are asymptomatic at birth. Congenital CMV infection may occur in approximately 40% (range: 24% to 75%) of infants born to mothers with primary infection; of these, 10–12% are symptomatic at birth and will have a remote likelihood to develop normal intelligence and hearing (Ross and Boppana, 2005). Congenital infection due to recurrent maternal disease is considered far less severe, and less than 10% of infected infants have long-term sequelae, but the presence of maternal antibody could blunt the presentation and severity of fetal infection (Demmler, 1996). In symptomatic congenital infection, reticuloendothelial involvement, including jaundice, hepatosplenomegaly and thrombocytopenia, is predominant. Liver abnormalities generally disappear completely, although fatal cirrhosis can occur. Neurological syndromes, including seizures, hyper- or hypotonia, microcephaly and periventricular calcifications, is the most serious consequence of congenital CMV infection and occurs in 30–50% of the symptomatic infants (Istas *et al.*, 1996). Approximately 20% of these also develop sensorineural hearing loss, which probably is the most common handicap caused by congenital CMV infection, and is now considered one of the major causes of deafness in childhood. In some reports, hearing loss occurs in approximately half of the infants with symptoms at birth and about 15% of infants with asymptomatic congenital CMV infection (Fowler and Boppana, 2006). The majority of patients have CSF signs of encephalitis (e.g. pleocytosis, elevated protein concentration); diabetes insipidus may also occur.

Pneumonia and a purpuric rash may also be present. Ocular manifestations include chorioretinitis, which is prevalent in the premature infants, microphthalmia and cataract. Among other manifestations occurring in the infants with symptomatic congenital CMV infection, inguinal hernia is the most frequent, being present in approximately 25%. The combination of abnormal ultrasound findings and positive amniotic fluid culture appears to predict a worse neonatal outcome, although fetal signs do not always predict severe involvement at birth and do not necessarily predict an adverse long-term outcome (Stagno and Britt, 2006). In children with symptomatic congenital CMV infection, the outcome is closely related to the severity of neurological syndrome. In fact, approximately 95% of the infants with microcephaly, periventricular calcifications or chorioretinitis may have major neurological sequelae such as severe mental retardation, seizures, deafness, spasticity (Ross and Boppana, 2005). A CT scan or ultrasound examination of the brain is important to document the extension of neurological involvement. Children with microcephaly without calcifications may not have mental retardation. The clinical course is generally slow, but progressive encephaloclastic disease or hearing loss may occur (Volpe, 2000). In fact, viruria still persists in about 50% of patients at 5 years of age (Stagno and Britt, 2006). Below is a typical case report:

Among children with asymptomatic congenital

A 35-year-old woman with three previous children had a pregnancy complicated by a possible abortion at 11 weeks of gestation and excessive amniotic fluid at 32 weeks of gestation. During pregnancy the mother did not have serological tests for CMV or ultrasound examinations. At 36 weeks of gestation she delivered a CMV-infected boy who weighed 2350g at birth and who had respiratory distress. His head was small for his size (cranial circumference: 29.4cm), his liver and spleen were enlarged, and he had a hernia of the diaphragm. A computerized tomography of the baby revealed diffuse periventricular calcifications, and intraventricular haemorrhage. Visual and auditory tests showed bilateral retinitis, cataract, and deafness. At 3 months of age the infant developed enlarged ventricles in the brain and large head size (cranial circumference 37.2cm) as well as convulsions. Psychomotor development was very poor and the boy became tetraspastic (Fig. II.3.2). His urine contained CMV culture until 6 years of age. Other problems included pneumonias, thrombocytopenic purpura, and bone fractures. At 26 years of age he died of neurological complications associated with a flu vaccination. The child was never treated with antiviral drugs.

Figure II.3.2 Case report.

infections, approximately 10–15% have been reported to develop mental retardation or hearing loss, which may not be diagnosed until serious language impairment occurs. However, hearing loss may not be detectable before the first year of life, being related to direct and progressive lesions of the cochlear cells and neuronal cells of the eighth cranial nerve (Fowler and Boppana, 2006). Late onset and reactivation of chorioretinitis in children with congenital CMV infection may also occur. Therefore, clinical and laboratory follow-up is very important (Ross and Boppana, 2005).

Neonatal prognostic markers

In the neonate, predictors of long-term adverse outcomes are not well known. It has long been recognized that infants who are symptomatic at birth are at an increased risk of sequelae compared with those who are asymptomatic (Fowler *et al.*, 1992; Williamson *et al.*, 1992; Dahle *et al.*, 2000; Noyola *et al.*, 2001). Children with thrombocytopenia and intrauterine growth retardation, suggesting a more disseminated disease process, have an increased risk of sensorineural hearing loss, while microcephaly and abnormal cranial CT scan predict abnormal neurodevelopmental outcomes (Boppana *et al.*, 1997; Rivera *et al.*, 2002). Persistent CMV DNA detection in blood has been correlated with a poor audiological and neurological outcome (Dahle *et al.*, 2000; Lanari *et al.*, 2006). Alternatively, there is no evidence that isolated abnormal haematological or biochemical parameters are related to severe prognosis. CMV disease, once present, usually progresses: 30–80% of children with unilateral hearing loss will have hearing deterioration in the normal ear or experience progressive loss in the affected ear (Boppana *et al.*, 1997; Fowler *et al.*, 1997; Noyola *et al.*, 2000). This progression has been noted to occur as late as 6 years of age (Fowler *et al.*, 1992; Dahle *et al.*, 2000). In addition, children can have progressive motor or cognitive impairment. The precise pathogenesis of disease progression in congenital CMV is unclear. It is unknown if progressive disease is due to persistent viral activity, viral reactivation, abnormal immunological response of the host, or delayed clinical appearance of damage present at initial infection (Fowler *et al.*, 1997; Cheeran *et al.*, 2009). While neurological damage that occurred *in utero* may not be expected to reverse, it is possible that antiviral treatment may prevent disease progression.

Disease rates: neurological versus hearing deficit

Hearing loss and neurological impairment are common among infants born of mothers with a primary infection during pregnancy. We recently observed 27 congenitally infected infants with sequelae who were born of mothers with a primary CMV infection at less than 21 weeks of gestation (Nigro *et al.*, 2012). Of these

27 infants, 17 had both severe neurological damage and hearing deficit, five only hearing deficit and five only neurological damage.

If one considers only asymptomatic infants born to mothers with a non-primary infection during pregnancy then severe neurological damage is very rare (Ross *et al.*, 2006). In a study of 176 asymptomatic congenitally infected infants only 19 (11%) had only hearing loss as the only manifestation. This rate was the same as observed for 124 children born to mothers with known pre-existing immunity to CMV where 13 children (ca. 10%) developed hearing loss. However, significantly more children born of mothers with primary infection had progressive and severe/profound hearing loss compared with children born of mothers with pre-conceptual immunity. Severe to profound hearing loss was observed in only 3/13 (23%) children born to mothers with pre-conceptual immunity compared with 63% of the children born of mothers with a primary infection during pregnancy ($P=0.04$). The rates of bilateral, delayed onset, high-frequency, and fluctuating hearing loss were not different between pre-conceptually immune mothers versus mothers with a primary infection during pregnancy. The age at diagnosis of hearing loss occurs at an average age of 13 months among children born of mothers with a primary infection during pregnancy compared with an average age of 39 months among those born to mothers with pre-conceptual immunity.

Recurrent maternal infections

For pregnant women who are naturally CMV seropositive (i.e. those who were infected years prior to conception), the congenital infection rate is usually reported as < 2% (Stagno and Britt, 2006). Naturally CMV-seropositive women who give birth to congenitally infected infants are said to have 'recurrent infections', meaning that they were either reinfected with new CMV strains or a latent infection was reactivated. Given the reduction in disease severity in infants born to naturally CMV-seropositive mothers, and given that maternal antibodies – not immune cells – reach the fetus, it is presumed that, in fetal infections following recurrent maternal infection, pre-existing maternal antibodies reduce or eliminate maternal and fetal viraemia or viral load, thereby protecting the placenta and fetus. However, pre-existing humoral immunity does not completely protect seropositive women against reinfection or reactivation, which probably is common in pregnancy, similar to herpes simplex virus recurrence. In fact, higher rates, which range between 7% and 19.6% have been reported (Nigro *et al.*, 1999; Boppana *et al.*, 2001; Rahav *et al.*, 2007). Although recurrent infections occasionally cause mild-to-moderate neurosensory hearing deficits, they rarely result in severe sequelae following fetal microcephaly, ventriculomegaly, periventricular calcifications and cystic lesions, echogenic bowel, hydrops and hepatosplenomegaly (Zalel *et al.*, 2008). In highly CMV-immune maternal populations, reinfection by new CMV strains was a major source of congenital infection, being a significant cause of sensorineural hearing loss (Ross *et al.*, 2006; Dar *et al.*, 2008; Yamamoto *et al.*, 2010). However, the frequency and severity of fetal disease after maternal CMV reactivation is unknown (Boppana *et al.*, 2001).

An important question regarding recurrent maternal infections is: does a prior CMV infection protect healthy adults against reinfection? One theory holds the effectiveness of all licensed vaccines is due to a critical level of serum IgG against the microbe and that these antibodies act by inactivating the low dose inoculum of the pathogen, usually on mucosal surfaces (Robins *et al.*, 1995). This hypothesis does not exclude other immune mechanisms from being protective, such as T-cells.

A critical question for the feasibility of developing a CMV vaccine is to what extent antibodies induced by natural infection protect against reinfection with a second viral isolate. Considerable evidence indicates that seropositivity induced by a wild-type virus protects against reinfection (see Chapters II.10 and II.20):

1. Plotkin *et al.* (1989) showed that natural seropositivity and a low dose of Towne vaccine protected against reinfection via a subcutaneous challenge of 100 PFU of wild-type virus. Protection was lost if the inoculum size of the challenge virus (Toledo) was increased to 1000 PFU. These results indicated that protection against reinfection afforded by natural seropositivity depends on the route and dose of the challenge and on the host immune status. Natural seropositivity will not protect against reinfection when the challenge is via organ transplantation but does protect against transfusion acquisition (Adler *et al.*, 1985).
2. We showed that a primary infection with wild-type CMV is between 85% and 93% effective in preventing a second infection among adults exposed to children shedding CMV and protection was associated with a high level of neutralizing antibodies induced by wild-type infection (Adler *et al.*,1995). We have also detected IgG, serum IgA, and secretory IgA in the saliva and nasal washes obtained from women with wild-type CMV infections (Wang *et al.*, 1996). Serum IgG and IgA occur via transudation to mucosal surfaces. The

women we studied were challenged with a low dose inoculum by the mucosal route with protection occurring via serum neutralizing antibodies at the mucosal surfaces.

3 Fowler *et al.* (2003) reported that preconception maternal immunity resulted in 69% reduction in the risk of congenital CMV infection in future pregnancies, suggesting that seropositive women were protected against reinfection during pregnancy.
4 The gB/MF59 CMV vaccine trial found the vaccine was 50% effective in protecting women from acquiring a CMV infection (Pass *et al.*, 2009). This vaccine induces only neutralizing antibodies that block viral entry into fibroblasts but is a very poor inducer of antibodies that block viral entry into epithelial and endothelial cells. It does not induce broad based cellular responses.

Most recently, Ross *et al.* (2010) reported indirect data that contradicts the above observations. They observed 205 naturally seropositive women for over 3 years and, based on changes in serological responses to four polymorphic epitopes in CMV gB or gH, concluded that 29% of the women became reinfected with CMV. This observation, if correct, challenges the concept that natural infection induced a high level of protective immunity. This study however has limitations. First, it does not prove reinfection. It is possible that some portion of the serological response were induced either by strong anamnestic responses to gB or gH epitopes induced by reactivation of a latent virus, or that some women were initially infected with one or more genotypes, of which one was the predominant antibody inducer. Over time, the replication rate of one genotype may have increased and induced an anamnestic response to the gB or gH antigens of that genotype. A more important limitation of this study was that there was no comparison group of seronegative women allowing the rate of primary CMV infection to be compared to the observed rate of reinfection. That is, it remained unclear how much protection, if any, was afforded by natural immunity.

Maternal risk factors

The major risk factor for maternal acquisition of CMV during pregnancy is frequent and prolonged contact with a child less than 3 years of age (Hutto el al., 1985; Adler *et al.*, 1986, 1989, 1991, 1996, 2004; Pass *et al.*, 1986; Fowler *et al.*, 2006; Francisse *et al.*, 2009). This occurs among women with a child at home or among women employed in child care centres (Adler, 1989; Pass *et al.*, 1990; Murph *et al.*, 1991). CMV-seronegative health care workers, even those caring for hospitalized young children and infants, are not at an increased risk (Adler, 2004). Regardless of whether CMV was acquired *in utero*, via breast milk, or via contact with other children, unlike older children and adults, children who are less than age 2 years when they acquire CMV, shed CMV in urine and saliva for 6–42 months with a median of 18 months.

In the USA, an average 60% of the mothers of children in day care are CMV seronegative, and at least 25% of all young children attending large group child care centres are shedding CMV. Seropositivity rates are highest for women with an infected child and seronegative mothers with infected children acquire CMV at rates 10–25 times higher than other women in the population (Pass, 1986; Adler *et al.*, 1991). The annual infection rate for seronegative women without exposure to children is 2% (Adler, 1989).

To estimate the frequency of pregnancy and exposure to CMV among mothers contemplating a possible additional pregnancy and with a child less than 2 years of age in group day care, we recently performed a prospective study which included a demographic questionnaire and serological and virological monitoring of mothers and their children in day care (Marshall and Adler, 2009). Of 60 women, 62% were seronegative and 20% of seronegative women had a child shedding CMV. Of the 60 women, 23 women, or 38%, became pregnant on average 10 months after study enrolment. During pregnancy eight or 35% of these pregnant women had a child in day care who shed CMV. These results illustrate the potential magnitude of the public problem associated with exposure to a asymptomatic viral infection during pregnancy. Our data, when extrapolated to the US population, estimate that every 2 years between 31,000 and 168,000 susceptible pregnant women will be exposed to CMV by an infected child.

Another group of high risk women in the USA are those who are seronegative, young, poor, and predominantly African-American. For this group, contact with a young child is also an independent predictor of delivering a CMV congenitally infected infant, as is a history of frequent sexual activity (Fowler *et al.*, 2006). A recent study monitored CMV seroconversion among 1906 seronegative women attending a fertility clinic (Francisse *et al.*, 2009) and reported that seroconversion was associated not only with contact with a child younger than age 3 years but also with seropositivity of the sex partner. Because CMV is often present in semen, it may be prudent to include condom use as part of the hygienic precautions given to seronegative pregnant women.

Prevention of maternal infection via behaviour modification

Both the US Centers for Disease Control and American College of Obstetrics and Gynecology (ACOG) recommend that pregnant women be counselled on ways to reduce their risk of CMV acquisition during pregnancy (ACOG, 2000; Cannon and David, 2005; Anderson *et al.*, 2008). These simple hygienic precautions are summarized in Table II.3.1. Studies completed by us demonstrated that these hygienic measures, when provided to CMV seronegative pregnant women with a young child at home, were effective (Adler *et al.*, 1996, 2004). Women were educated about CMV, provided with written guidelines detailing hygienic precautions, watched a video on how to practice these precautions, and then a nurse answered questions the women asked. The precautions were well received by each pregnant woman and were easily accomplished. Based on interviews and a written survey done before enrolment and at the end of pregnancy, none of 130 seronegative pregnant women complained the precautions were burdensome or anxiety provoking during pregnancy. None of the pregnant women declined serotesting. To date, of 37 pregnant women with a child shedding CMV, we have observed only one who received hygienic precautions and nevertheless seroconverted to CMV during pregnancy, compared to infection rates of 42% for 64 of 154 non-pregnant women with a child shedding CMV, including seronegative women who were trying to conceive (Adler, 2004).

A French study has recently confirmed and expanded our observations. 5312 pregnant women were offered CMV serological screening during pregnancy (Vauloup-Fellous *et al.*, 2009). Of these women 97.4% agreed to screening and signed a consent. If a woman was seronegative at 12 weeks' gestation, detailed oral and written hygienic information was given to the woman and her spouse. Wearing protective gloves was not recommended. For 2595 seronegative women, the rate of maternal seroconversion during the first 12 weeks of gestation was compared with the rate between weeks 12 and 36. Prior to subject education and the receipt of hygienic precautions at 12 weeks the maternal seroconversion rate was 0.42%, compared with a rate of 0.19% for women from week 12 to 36 of the gestation. When adjusted for the number of woman-weeks observed, the rate prior to 12 weeks' gestation was 0.035% per woman-week with to a rate of 0.008% per woman-week after intervention ($P=0.0005$). Maternal primary infections and seroconversions were distributed evenly throughout gestation. A limitation of this study was that, for ethical practical reasons, women less than 12 weeks could not be given the hygienic information nor could be withheld from those between 12 and 36 weeks' gestations. This meant that a direct comparison of women at the same gestational times of pregnancy could not be made.

These studies provide compelling data on the simplicity and effectiveness of hygienic intervention to prevent CMV infection of high risk pregnant women. Seronegative health care workers do not have an increased risk but pregnant child care employees are at a significant risk for CMV acquisition. In 2008 the Centers for Disease Control's website recommended that pregnant child care employees be informed they could assess their risk by serological testing and if seronegative, avoid, if possible, caring for children less than 2 years of age for the duration of pregnancy.

Table II.3.1 Practices for seronegative pregnant women to reduce risk of CMV infection

1. Assume children under age 3 years in your care have CMV in their urine and saliva
2. Thoroughly wash hands with soap and warm water after:
   - diaper changes and handling child's dirty laundry
   - feeding or bathing child
   - wiping child's runny nose or drool
   - handling child's toys, pacifiers or toothbrushes
3. Do not:
   - share cups, plates, utensils, toothbrushes or food
   - kiss your child on or near the mouth
   - share towels or washcloths with your child
   - sleep in the same bed with your child

Pathophysiology of intrauterine infection

Serological testing for primary maternal CMV infection during pregnancy is not routine but ultrasound studies are routine and can show abnormalities of the placenta and fetus. We evaluated placental thickening in women with primary CMV infections during pregnancy (La Torre *et al.*, 2006). Ninety-two women with a primary CMV infection during pregnancy and 73 CMV-seropositive pregnant women without a primary CMV infection were studied. Thirty-two women were treated with CMV hyperimmune globulin to either prevent or treat intrauterine CMV infection. Maximal placental thickness was measured by longitudinal (*non oblique*) scanning, with ultrasound beam perpendicular to the chorial dish. Programmed placental ultrasound evaluations were performed from 16 to 36 weeks' gestation. At each placental measurement between 16 and 36 weeks' gestation, women with a primary CMV infection and a fetus and/or newborn with CMV disease had significantly ($P < 0.0001$) thicker placentas than women without diseased fetuses, who in turn had significantly ($P < 0.0001$) thicker placentas than seropositive controls. After a primary infection, for women both with and without diseased fetuses and/or newborns, receipt of CMV hyperimmune globulin was associated with significant ($P < 0.001$) reductions in placental thickness. Placental vertical thickness values, predictive of primary maternal infection, were observed at each measurement from 16 to 36 weeks' gestation and cut-off values ranged from 22mm to 35mm with the best sensitivity and specificity at 28 and 32 weeks.

We concluded that primary maternal CMV infections and fetal and/or neonatal disease are associated with sonographically thickened placentas which respond to CMV hyperimmune globulin.

In addition to placental thickening, several other lines of evidence suggest that most of the symptoms of a congenital CMV infection present at birth are not due to a direct effect of the virus on the fetus but rather due to a direct effect of the virus on the placenta (see Chapter II.4) which in turn becomes impaired in its ability to oxygenate and nourish the fetus.

First, most of all the manifestations of congenital infection resolve over the first weeks to months of life concurrent with adequate oxygenation and nutrition of the newborn. This includes weight gain, increased head size, as well as resolution of liver disease, of haematopoietic abnormalities, and of splenomegaly. The respective manifestations, plus neonatal neurological disease, are all associated with intrauterine hypoxaemia. For example, neonatal thrombocytopenia present at birth in 75% of symptomatic CMV-infected newborns is caused by impaired megakaryocytopoiesis and platelet production secondary to a pregnancy complicated by placental insufficiency and/or fetal hypoxia (Roberts *et al.*, 2003). Fetal hypoxia with cerebral hypoxia and ischaemia is a cause of perinatal brain injury and may be the cause of the periventricular calcifications common in affected newborns. During intrauterine hypoxia in premature infants, the cerebral white matter is the site of injury and hypoxia leads to a well-known histopathology called periventricular leucomalacia (Rees and Inder, 2005). This condition comprises both focal cystic infarcts adjacent to the lateral ventricles and a diffuse gliosis which extends throughout the cerebral white matter.

Second, that the symptoms of congenital disease are not due to the direct effects of CMV on the fetus is suggested by the fact that many infants, whether born of mothers with primary or recurrent infections, are asymptomatic and develop normally in spite of being viraemic and shedding high titres of virus in urine and saliva for 3–5 years after birth. In addition, infants acquiring CMV postnatally exhibit the similar patterns of viraemia and viral excretion but develop no symptoms. When CMV is acquired by transfusion or by breast milk, low-birthweight infants (< 1250 g) born of seronegative mothers may become ill, but do not develop any of the typical manifestations of disease acquired *in utero* after a primary maternal infection (Adler *et al.*, 1983; Hamprecht *et al.*, 2008).

Third, CMV infection is occasionally associated with a blueberry muffin syndrome, in which the purpura is caused by extramedullary haematopoiesis indicative of intrauterine hypoxia.

Fourth, among infants without the blueberry muffin rash the hepatomegaly is due to biliary obstruction secondary to extramedullary haematopoiesis. Extramedullary haematopoiesis and erythrocytic congestion also cause splenic enlargement seen in most symptomatic infants (Naeye, 1967).

Fifth, receipt of CMV hyperimmune globulin by pregnant women with a primary CMV infection and infected fetuses is associated with the resolution of the *in utero* signs of fetal disease and a decrease in placental size, as detected by ultrasound and the delivery of infants who develop normally (Nigro *et al.*, 2005; LaTorre *et al.*, 2006).

Sixth, this reversal of fetal disease as well as the prevention of the progression of fetal disease by CMV hyperimmune globulin suggests the effect of immunoglobulin is to improve placental function by reducing placental inflammation which occurs secondary to the viral infection of the placenta. A recent study performed immunohistological analysis of placental specimens from women with untreated congenital infection, CMV hyperimmune globulin treated congenital infection, and uninfected controls (Maidji *et al.*, 2010; see Chapter II.4). In untreated infection, viral replication proteins were found in trophoblasts and endothelial cells of chorionic villi and uterine arteries. Associated damage included extensive fibrinoid deposits, fibrosis, avascular villi and oedema, which could impair placental functions. With hyperimmune globulin treatment, placentas appeared uninfected and had an increase in the number of chorionic villi and blood vessels over that in controls suggesting compensation for a hypoxia-like condition. The results indicate that CMV hyperimmune globulin suppresses CMV replication in the placenta and prevents placental dysfunction, thus improving fetal outcome.

The frequent hearing loss caused by CMV among affected newborns has not been associated with intrauterine hypoxia and its pathogenesis is unknown.

Prevention of fetal infection by immunoglobulin

Routine antepartum and pregnancy serological screening should efficiently prevent or diagnose congenital CMV infection. However, it is still considered controversial. In fact, serial testing of CMV-specific IgG and IgM antibodies, at least in the first two trimesters for women at high risk like day care workers, could avoid difficulties in differentiating recurrent from primary infections and enhance patient counselling if IgM antibodies are detected. Serological testing will promptly reveal both primary (by seroconversion and high IgM levels) and recurrent (by significant IgG levels which increase concomitantly or not with IgM antibodies) infections. Available data indicate the majority of these fetal infections are either preventable or treatable. Given the high financial costs associated with these infections, this estimate is sufficiently high to justify appropriate interventions (Cahill *et al.*, 2009). In a prospective, non-randomized but controlled study, Nigro and colleagues (2005) reported positive results using passive immune globulin during pregnancy to prevent transmission of CMV from mother to fetus. Pregnant women with a primary CMV infection were offered amniocentesis and if positive they were offered treatment with CMV-specific hyperimmunoglobulin. The authors identified 55 women with primary CMV infection who had PCR-positive amniocentesis suggesting fetal infection. A total of 31 women elected treatment, 14 did not and ten chose terminate the pregnancy. Of the 31 treated women, 15 had fetal ultrasounds that were abnormal prior to treatment but at follow-up only one of these children had abnormalities. None of the infants with normal fetal ultrasounds developed symptoms later. These results contrasted with the non-treated women; 7 of the 14 had fetal abnormalities by ultrasounds, none of the infants were normal at birth and two died. Again, the infants who had normal fetal ultrasounds remained normal at follow-up at 2 years of age or older.

This study also described offering hyperimmune globulin as prophylaxis to women with primary infection but who did not undergo amniocentesis, so that fetal infection was not proven. Six of the 37 women (16%) who chose this prophylaxis therapy had infected infants, while 19 of the 47 women (40%) who did not agree to prophylaxis had infants who were infected at birth. Further work by the same group noted that the placentas of women with primary CMV which were enlarged compared with the women who had experienced a past infection, decreased in those women who received hyperimmune globulin. This led to the hypothesis that hyperimmune globulin might act by improving placental health (LaTorre *et al.*, 2006; Maidij *et al.*, 2010).

Treatment of fetal infection via immunoglobulin

Animal results

The guinea pig model of cytomegalovirus (GPCMV) infection is useful because GPCMV crosses the guinea pig placenta, causing infection *in utero*, and the structure of the guinea pig placenta is similar to the human placenta (Schleiss *et al.*, 2008; see Chapter II.5). In the guinea pig model fetal infection leads to fetal death, so pup survival is the endpoint used in guinea pig experiments. Thus, the guinea pig model has been used frequently to study the role of antibodies using either active or passive immunization to interrupt intrauterine transmission and/or protect the fetus. In these studies, pregnant guinea pigs were challenged with guinea pig CMV before or after passive administration of neutralizing anti-sera to either whole virus or gB, a glycoprotein that induces neutralizing antibodies (Bia *et al.*, 1980; Bratcher *et al.*, 2001; Chatterjee *et al.*, 2001).

In two separate reports passive administration of immune serum to whole virus significantly increased fetal survival even though it did not affect the rate of fetal infection, indicating that the immune serum was therapeutic. In other guinea pig experiments with immune sera to purified gB, there was reduced fetal infection, placental inflammation, fetal death, and enhanced fetal growth. In these experiments fetal survival was enhanced for dames treated with immune globulin to CMV gB compared with controls. This effect was independent of whether immune globulin was administered before or after the challenge virus (Bratcher *et al.*, 2001). Additional high-titre immune globulin given before or after maternal challenge significantly reduced the rate of fetal infection from 39% (9 of 23 fetuses infected) to 0% (0 of 18 fetuses infected). Immunoglobulin to gB, administered before or after maternal challenge, also significantly reduced placental inflammation and enhanced fetal growth, as measured by fetal weight. In the guinea pig model, there are several plausible mechanisms for the therapeutic efficacy of passive immunization including reduction of maternal viraemia and viral load, and/or decreased placental inflammation resulting in increased blood flow with enhanced fetal nutrition, oxygenation, and survival.

The brains of a high proportion of human fetuses infected with CMV *in utero* after a primary maternal infection contain CMV and an associated inflammatory response (Gabrielli *et al.*, 2010). Therefore of interest is a recent study of passive immunization that used a mouse model (Cekinovic *et al.*, 2008). Mouse CMV does not cross the mouse placenta, so mouse CMV was injected into the peritoneal cavity of newborn mice. This led to viral infection of the brains of the infant mice with associated inflammatory lesions in the brains. These lesions consisted of mononuclear cell infiltrations and prominent glial nodules. Treatment of the newborn mice with either immune sera or monoclonal antibodies directed against the mouse CMV gB resulted in markedly reduced amounts of virus in the brains as well as over 5-fold reductions in inflammatory lesions in the brain. These observations suggested that antiviral antibodies limited both viral replication in the brain and viral induced brain damage.

Effects in human pregnancy

Currently, there is no standard therapy for maternal CMV infections diagnosed during pregnancy, in spite of advances in the diagnosis of maternal–fetal CMV infection. Termination of pregnancy is often offered as the only option when the fetus becomes infected with CMV regardless of whether the disease is detected by ultrasound. Therefore, preventing transmission of CMV to the fetus by treating mothers experiencing an active CMV infection, mostly primary, would be ideal. In fact, it has long been recognized that the highest risk of transmission occurs with primary infection. For this reason, the strategy of screening women during pregnancy and treating those who develop a primary infection may be of great benefit.

There are four main reasons to consider passive immunization for treating or preventing fetal CMV infection:

1. Hyperimmunoglobulins have been shown to be efficacious in animal models.
2. Hyperimmunoglobulins are the most purified among the blood derivatives, including albumin, and are the only derivatives which can be pasteurized and are safe.
3. Hyperimmunoglobulins display immunomodulatory effects, which amplify their capacity of binding viral antigens.
4. Commercial preparations have been available since 1990.

In pregnant women with fetal CMV involvement after a primary infection, hyperimmunoglobulin was used to treat fetal disease (Nigro *et al.*, 2005). Twenty-two women with fetal CMV infection received one infusion of hyperimmunoglobulin (200 U/kg), and nine patients received two or three intravenous infusions and intra-amniotic infusions (400 U/kg of fetal weight) because of persistent fetal disease shown by ultrasound. 15 women had fetal ultrasound disease: of these 15 women, only one gave birth to an infant with severe encephalopathy. In three other fetuses, ultrasound examination showed after hyperimmunoglobulin the regression of cerebral involvement during pregnancy (Nigro *et al.*, 2008). Hyperimmunoglobulin was also associated with a reduction of placental enlargement induced by primary CMV infection (LaTorre *et al.*, 2006). On the other hand, of 14 women with CMV-positive amniotic fluid who refused hyperimmunoglobulin infusions and were followed as controls, seven had affected infants (50%). Another 10 women with infected amniotic fluid aborted, and three of them had fetuses with ultrasound disease. Immunological investigations, including antibody titres and avidity, T-cell subpopulations and NK cytotoxic activity, supported hyperimmunoglobulin efficacy.

These results are supported by a case–control study with 64 congenitally infected children (Nigro *et al.*, 2012a). Cases were 32 children with either hearing deficit and/or psychomotor retardation and whose mothers had a confirmed or probable primary CMV infection at ≤ 20 weeks' gestation. Controls were 32 congenitally infected children who were normal but whose mothers had a confirmed primary infection at ≤ 20 weeks' gestation. Case and controls were matched by the weeks of maternal gestation (± 1 week) at the mother's infection and by the child's age (± 1 year) at evaluation. Of the 32 cases, only four mothers received CMV immunoglobulin, compared with 27 of the 32 mothers of control infants (adjusted odds ratio 14; 95% CI 1.7–110). The only risk factor for an affected child was the mother not receiving immunoglobulin ($P = 0.001$).

Therefore, if ultrasound examinations show signs of fetal injury, or CMV is detected in the amniotic fluid, patients should be advised of the possible option of immunotherapy with hyperimmunoglobulin as an alternative to pregnancy termination (Nigro *et al.*, 2012b).

Treatment of fetal infection via antivirals

In theory, drugs which inhibit the replication of CMV should be as effective or even more effective than antibody treatment at either preventing fetal infection among mothers with a primary CMV infection during pregnancy or in treating fetuses infected *in utero*. That is, if antiviral therapy stops or reduces viral replication in the placenta the rate of fetal infection should be reduced. The problem is finding an antiviral agent which is active against CMV and can be used in pregnancy. Ganciclovir, a US Food and Drug Administration (FDA) approved drug for treating CMV infections, is not approved for use in pregnant women, mainly over concerns associated with drug induced neutropenia. Acyclovir does not cause neutropenia and the oral form (valacyclovir) has been evaluated during pregnancy.

Valacyclovir is active against CMV. Even though CMV does not have a thymidine kinase gene, valacyclovir at a dose of 8 g/day is effective. Peak blood levels are in the 13–20 μM range and IC_{50} are in the 1–3 μM range (Weller *et al.*, 1993). However, it is very difficult to measure IC_{50} values of clinical isolates of CMV (Landry *et al.*, 2000). This means one must rely on the results of clinical trials to assess the effectiveness of valacyclovir against CMV.

Valacyclovir is also clinically effective against CMV. Viral load seems to be correlated with the development of CMV disease in solid-organ transplant recipients, and treatment can prevent the occurrence of the disease in those with PCR positive viraemia. In numerous studies valacyclovir is as effective as ganciclovir or valganciclovir in treating transplant patients for CMV infections and reduced the risk of CMV disease and associated mortality in recipients of solid-organ transplants (Lowrance *et al.*, 1999; Winston *et al.*, 2003; Yango *et al.*, 2003; Emery *et al.*, 2005; Hodson *et al.*, 2005; Pavalopoulou *et al.*, 2005; Reishig *et al.*, 2005; see Chapters II.13 and II.14).

Valacyclovir has been used in pregnancy and has not been associated with adverse effects or teratogenicity. A recently published study (Pasternak *et al.*, 2010) used data from nationwide registries to conduct a historical cohort study including all infants born alive in Denmark from 1 January 1996 to 30 September 2008, and evaluated associations between exposure to oral acyclovir, valacyclovir, and famciclovir in the first trimester of pregnancy and major birth defects diagnosed within the first year of life. Their cohort included 837,950 live births (34,787 multiple births), among whom 19,960 (2.4%) were diagnosed with a major birth defect during the first year of life. Among 1804 pregnancies exposed to acyclovir (1561 infants), valacyclovir (229 infants) or famciclovir (26 infants) at any time in the first trimester, 40 infants (2.2%) had a diagnosis of a major birth defect, compared with 19,920 of 835,991 infants (2.4%) among the unexposed pregnancies (crude odds ratio, 0.93; 95% CI, 0.68–1.27). In a multivariate model, acyclovir, valacyclovir, or famciclovir exposure in the first trimester was not associated with increased rate of major birth defects (odds ratios 0.89; 95% CI, 0.65–1.22), as compared with pregnancies not exposed to these drugs. First-trimester use of acyclovir, the most commonly prescribed antiviral, was not associated with major birth defects [32 cases among 1561 exposed (2.0%) vs. 2.4% in the unexposed; adjusted odds ratio, 0.82; 95% CI 0.57–1.17]. Neither valacyclovir nor famciclovir was associated with major birth defects, although use of the latter was very uncommon. These results from Denmark support previously published data (Wilton *et al.*, 1998; Ratanajamit *et al.*, 2003; Stone *et al.*, 2004; US Food and Drug Administration, 2005 and 2008; Jacquemard *et al.*, 2007). Thus, acyclovir and related drugs appear safe for use in pregnancy.

Acyclovir levels of at least 20 μM in amniotic fluid and in maternal and fetal bloods have been achieved (Jacquemard *et al.*, 2007). Absorption and plasma levels in pregnant women are similar to those in non-pregnant women. A report of preliminary pharmacokinetics data from 20 pregnant women receiving therapy for herpes simplex virus infection found peak acyclovir plasma concentrations after oral administration of valacyclovir 500 mg three times a day were 13.9 ± 3 μM, with steady-state levels of 13.5 ± 4 μM (Kimberlin *et al.*, 1998). These concentrations are comparable with more recent data (Andrews *et al.*, 2006). The acyclovir and valacyclovir registry has no reports of developmental abnormalities in more than 600 embryos and fetuses exposed across all trimesters of pregnancy.

There is good placental transfer of acyclovir. It is concentrated in the amniotic fluid, most probably because of active transfer across the placenta. However, the drug does not accumulate, as levels seem to drop very rapidly after cessation of treatment. After maternal oral administration of valacyclovir drug concentrations within the expected IC_{50} are seen in maternal and fetal bloods, in most cases without evidence of maternal or fetal intolerance.

For all of the above reasons that indicate that valacyclovir is not harmful to the fetus valacyclovir is being evaluated in randomized double-blinded clinical trial being conducted in France.

Postnatal treatment of congenitally infected infants

Two forms of therapy are available: chemotherapy and passive immunization. Chemotherapy includes drugs inhibiting CMV DNA replication. Antiviral treatment of neonates or infants with congenital CMV infection is generally based on the use of ganciclovir, which has been given at different endpoints and regimens (both dosage and duration of therapy), and for different clinical manifestations. This drug is active only after phosphorylation to ganciclovir triphosphate, which substitutes for guanosine triphosphate in the viral DNA polymerase, with consequent inhibition of the CMV replication. Since the first phosphorylation requires the presence of the CMV-encoded (UL 97 gene) phosphotransferase, ganciclovir can display its activity mostly in the CMV-infected cells (Wutzler and Thust, 2001; see Chapter II.19). Ganciclovir stops or reduces the amount of virus excreted in urine and saliva and generally clears CMV from the blood. This antiviral effect occurs only when the patient is receiving the drug. After ganciclovir therapy, viral excretion and/or viraemia rapidly reappear. This is the reason a longer duration of antiviral therapy has been associated with a favourable outcome more frequently than shorter courses. Protracted and repeated courses of oral ganciclovir were useful for therapy of several CMV-related diseases.

A pilot study in 1994 comparing two regimens of intravenous ganciclovir treatment in 12 infants with severe neurological manifestations showed that an initial administration of high dose (15 mg/kg/day) followed by boluses of 10 mg/kg three times/week for 3 months was more efficacious than therapy with 10 mg/kg for only 2 weeks (Nigro *et al.*, 1994). A larger phase II study compared two 6-week regimens (8 mg/kg/day vs. 12 mg/kg/day, in 14 and 28 babies respectively) for toxicity, virological response and neurological outcome (Whitley *et al.*, 1997). The 12 mg/kg/day group showed a more pronounced antiviral effect in urine. In a subsequent randomized clinical trial in infants with congenital CMV and neurological involvement, treatment with intravenous ganciclovir for 6 weeks at a dose of 6 mg/kg/day resulted in improved hearing compared with controls after 6 months' follow-up (Kimberlin *et al.*, 2003). Further study from the same group of authors showed that infants with symptomatic congenital CMV cerebropathy receiving intravenous ganciclovir for 6 weeks had fewer developmental delays at 6 and 12 months compared with untreated infants (Oliver *et al.*, 2009). During therapy the viral excretion in urine decreased, but returned to near pre-treatment levels after cessation of therapy. Therefore, a 6-week course of intravenous ganciclovir suppresses virus replication temporarily but cannot prevent long-term sequelae due to persistent virus replication.

Oral administration of ganciclovir allowed for a longer duration of treatment avoiding the immediate CMV replication after stopping ganciclovir. In fact, prolonged courses of oral ganciclovir were associated with clinical improvement or recovery in children with neurological and auditory impairment or persistently severe enterocolitis requiring total parenteral or nasal gastric feeding (Michaels *et al.*, 2003; Nigro *et al.*, 2010). Although not yet licensed for therapy of congenital CMV infection, valganciclovir, a valine ester of ganciclovir, has been successfully used to treat children with congenital CMV disease. Initial dosage of 15 mg/kg given twice daily was suggested as the most suitable (Galli *et al.*, 2007). Pharmacokinetics showed ganciclovir stable and effective plasma antiviral concentrations: mean $C_{\text{trough}} = 0.51 \pm 0.3$ and $C_{2\text{h}} = 3.81 \pm 1.37$ µg/ml; no significant variability was seen neither intra-patient nor inter-patients (Lombardi *et al.*, 2009). Valganciclovir oral solution provides plasma concentrations of ganciclovir comparable to those achieved with administration of intravenous ganciclovir (Kimberlin *et al.*, 2008). Since ganciclovir has a potential toxicity, foscarnet, an organic analogue of inorganic pyrophosphate capable of blocking viral DNA polymerase, has been occasionally used for infants with congenital CMV disease (Nigro *et al.*, 2004). Cidofovir has also been used in the guinea pig model of congenital CMV showing potential benefit in preventing maternal transmission during pregnancy (Schleiss *et al.*, 2006). However, further data are lacking at this time. In conclusion, for infants with symptomatic infection a longer duration of antiviral therapy appears to be associated with better outcomes than shorter courses.

Hyperimmunoglobulin was not directly evaluated for therapy of neonates with congenital CMV infection but was efficient in infants with transfusion-acquired CMV disease (Yeager *et al.*, 1981; Adler *et al.*, 1983).

Concluding discussion: serological screening

Universal serological screening of pregnant women whether by an initial blood test or by serial testing during pregnancy for a primary CMV infection is controversial (Schlesinger, 2007; Yinon *et al.*, 2010). In the USA all women are screened via ultrasound at around 20 weeks' gestation. When CMV-associated fetal abnormalities are detected and an intrauterine CMV infection is confirmed by amniocentesis, data from the National CMV Registry for Pregnant Women (CMVregistry.org) indicates off-label immunoglobulin is often used. Of the first 48 women enrolled in the registry 23 had positive amniotic fluid and 18 of them (78%) were treated off-label with immunoglobulin. While apparently safe, this approach is far from ideal because treatment is given only to fetuses of mothers infected in early gestation and who have the poorest prognosis.

For those who advocate not screening or not using immunoglobulin for treatment until the results of a randomized prospective study, the fact is that for many reasons, there will probably never be a randomized prospective study of therapeutic passive immunization. No such study is either ongoing or contemplated. Barriers to such a study include (1) subject acceptance of placebo when the study drug appears safe and is readily available off-label, (2) the need to serologically test over

100,000 pregnant women to identify a few hundred with fetal infection, (3) the need for a large multicentre trial with costs in the tens of millions of dollars, and (4) a lack of enthusiasm for funding such a study both by industry and the government funding agencies.

A more rational approach to passive immunization is to prevent fetal infection following a primary maternal infection. Confirmatory trials of this approach are in their final stages in Europe, where universal screening permits identification of CMV-infected women during pregnancy and their potential enrolment in a randomized prophylactic trial.

Hygienic intervention for high-risk women is appropriate now. At least one-third of the pregnant women in the US and Europe are at high risk; that is, they have daily household or occupational contact with children less than 3 years old. Offering an initial blood test for CMV IgG antibodies in pregnancy, educating them about CMV, and providing simple hygienic precautions could be routine. Testing of household children for CMV excretion is not useful since they may not be shedding CMV initially but start shedding at any time during their mother's pregnancy (Adler *et al.*, 2004).

Whether to continue serial testing for seroconversion during pregnancy could initially be a decision for each woman and her obstetrician. In some countries the main rationale for almost universal serial screening was apparently to allow for an elective termination of pregnancy. In Israel approximately half of women who seroconvert to CMV electively terminate, although this rate is much lower in other countries (Guerra *et al.*, 2007; Rahav *et al.*, 2007).

An alternative to serological screening is to provide all high risk pregnant women the hygienic interventions. In the reported studies, however, women knew their serological status, so it is unclear if a pregnant women's perception of her risk (susceptible) would affect efficacy. Hygienic precautions do not work in non-pregnant women, suggesting seronegative pregnant women perceive a high risk and are more motivated to comply (Adler *et al.*, 1996, 2004). It is very likely that most high risk women would want to assess their risk by knowing their serological status given that serological testing is readily available.

Even limited serological screening has potential adverse effects such as false positive serological results which may lead to increased costs associated with unnecessary imaging and amniocentesis. The negative impact of these potential problems has not been reported, although this could be evaluated in countries where routine serological screening is frequent.

Regardless of the strategy used, no serological testing but only hygienic precautions, one time testing, or serial serological testing throughout the first two trimesters of pregnancy, education of pregnant women about CMV is necessary. With a rare exception, the reaction of women and especially of those who acquire a CMV infection during pregnancy is that they wish they had known about CMV and could have taken some measures to avoid infection.

Until further data become available one reasonable approach is to test all pregnant women at the beginning of pregnancy to determine their initial serological status. If the woman is seronegative and does not have a child under age 3 years in the home, she should be considered low risk and receive basic information about hygienic measures. If an ultrasound examination after the 20th of gestation is normal, no further intervention for CMV is necessary. If the ultrasound is abnormal, then maternal seroconversion to CMV should be confirmed and other causes of fetal ultrasound abnormalities excluded. Finally, fetal CMV infection should be confirmed via amniocentesis, and CMV hyperimmune globulin should be considered. At birth the baby should be evaluated for CMV disease.

If a pregnant woman is seronegative and at a high risk, that is she has household exposure to a child less than 3 years of age, she would receive the hygienic interventions and should be given the option for monthly serological testing until 20 weeks' gestation. If she does not seroconvert by 20 weeks no further intervention for CMV is necessary. If she does seroconvert to CMV during the first half of pregnancy then consideration should be given to using CMV hyperimmune globulin to prevent fetal infection. At birth the baby should be evaluated for CMV disease. This approach, given our limited knowledge, would at least offer pregnant women the opportunity to minimize their risk from having a child severely affected by CMV.

References

Adler, S.P. (1989). Cytomegalovirus and child day care. Evidence for an increased infection rate among day care workers. N. Engl. J. Med. *321*, 1290–1296.

Adler, S.P. (1991). Cytomegalovirus and child day care: risk factors for maternal infection. Pediatr. Infect. Dis J. *10*, 590–594.

Adler, S.P. (1991). Molecular epidemiology of cytomegalovirus: a study of factors affecting transmission among children at three day care centers. Pediatr. Infect. Dis. J. *10*, 584–590.

Adler, S.P. (2004). Cytomegalovirus. In Hospital Epidemiology and Infection Control, 3rd edn, Mayhall, G., ed. (Williams & Wilkens). pp.737–742.

Adler, S.P., Chandrika, T., Lawrence, L., and Baggett, J. (1983). Cytomegalovirus infections in neonates acquired by blood transfusions. Pediatr. Infect. Dis. J. *2*, 114–118.

Adler, S.P., Baggett, J., and McVoy, M. (1985). Transfusion associated cytomegalovirus infection among seropositive cardiac surgery patients. Lancet *2*, 743–746.

Adler, S.P., Starr, S.E., Plotkin, S.A., Hempfling, S.H., Buis, J., Manning, M.L., and Best, A.M. (1995). Immunity induced by a primary cytomegalovirus infection protects against secondary infection among women of childbearing age. J. Infect. Dis. *171*, 26–32.

Adler, S.P., Finney, J.W., Manganello, A.M., and Best, A.M. (1996). Prevention of child-to-mother transmission of cytomegalovirus by changing behaviors: A randomized controlled trial. Pediatr. Infect. Dis. J. *15*, 240–246.

Adler, S.P., Finney, J.W., Manganello, A.M., and Best, A.M. (2004). Prevention of child-to-mother transmission of cytomegalovirus among pregnant women. J. Pediatr. *145*, 485–491.

Ahlfors, K., Ivarsson, S.A., Harris, S., Svanberg, L., Holmqvist, R., Lernmark, B., and Theander, G. (1984). Congenital cytomegalovirus infection and disease in Sweden and the relative importance of primary and secondary maternal infections: preliminary findings from a prospective study. Scand. J. Infect. Dis. *16*, 129–137.

American College of Obstetricians and Gynecologists. (2000). Perinatal viral and parasitic infections. ACOG Practice Bulletin 20. 20th edn. (American College of Obstetricians and Gynecologists, Washington, DC).

Anderson, B., Schulkin, J., Ross, D.S., Rasmussen, S.A., Jones, J.L., and Cannon, M.J. (2008). Knowledge and practices of obstetricians and gynecologists regarding cytomegalovirus infection during pregnancy. MMWR Morb. Mortal. Wkly Rep. *57*, 65–68.

Andrews, W.W., Kimberlin, D.F., Whitley, R., Cliver, S., Ramsey, P.S., and Deeter, R. (2006). Valacyclovir therapy to reduce recurrent genital herpes in pregnant women. Am. J. Obstet. Gynecol. *194*, 774–781.

Bia, F.J., Griffith, B.P., Tarsio, M., and Hsiung, G.D. (1980). Vaccination for the prevention of maternal and fetal infection with guinea pig cytomegalovirus. J. Infect. Dis. *142*, 732–738.

Bodéus, M., Zech, F., Hubinont, C., Bernard, P., and Goubau, P. (2010). Human cytomegalovirus in utero transmission: Follow-up of 524 maternal seroconversions. J. Clin. Virol. *47*, 201–202.

Boppana, S.B., Fowler, K.B., Vaid, Y., Hedlund, G., Stagno, S., Britt, W.J., and Pass, R.F. (1997). Neuroradiographic findings in the newborn period and long-term outcome in children with symptomatic congenital cytomegalovirus infection. Pediatrics *99*, 409–414.

Boppana, S.B., Rivera, L.B., Fowler, K.B., Mach, M., and Britt, W.J. (2001). Intrauterine transmission of cytomegalovirus to infants of women with preconceptional immunity. N. Engl. J. Med. *344*, 1366–1371.

Bratcher, D.F., Bourne, N., Bravo, F.J., Schleiss, M.R., Slaoui, M., Myers, M.G., and Bernstein., D.I. (1995). Effect of passive antibody on congenital cytomegalovirus infection in guinea pigs. J. Infect. Dis. *172*, 944–950.

Cahill, A.G., Odibo, A.O., Stamilio, D.M., and Macones, G.A. (2009). Screening and treating for primary cytomegalovirus infection in pregnancy: where do we stand? A decision-analytic and economic analysis. Am. J. Obstet. Gynecol. *201*, 466.e1–7.

Cannon, M.J., and Davis, K.F. (2005). Washing our hands of the congenital disease epidemic. BMC Public Health *5*, 70.

Cekinović, D., Golemac, M., Pugel, E.P., Tomac, J., Cicin-Sain, L., Slavuljica, I., Bradford, R., Misch, S., Winkler, T.H., Mach, M., *et al.* (2008). Passive immunization reduces murine cytomegalovirus-induced brain pathology in newborn mice. J. Virol. *82*, 12172–12180.

Chandwani, S., Kaul, A., Bebenroth, D., Kim, M., John, D.D., Fidelia, A., Hassel, A., Borkowsky, W., and Krasinski, K. (1996). Cytomegalovirus infection in human immunodeficiency virus type 1-infected children. Pediatr. Infect. Dis. J. *15*, 310–314.

Chatterjee, A., Harrison, C.J., Britt, W.J., and Bewtra, C. (2001). Modification of maternal and congenital cytomegalovirus infection by anti-glycoprotein b antibody transfer in guinea pigs. J. Infect. Dis. *183*, 1547–1553.

Cheeran, M.C., Lokensgard, J.R., and Schleiss, M.R. (2009). Neuropathogenesis of congenital cytomegalovirus infection: disease mechanisms and prospects for intervention. Clin. Microbiol. Rev. *22*, 99–126.

Colugnati, F.A., Staras, S.A., Dollard, S.C., and Cannon, M.J. (2007). Incidence of cytomegalovirus infection among the general population and pregnant women in the USA. BMC Infect. Dis. *7*, 71.

Dahle, A.J., Fowler, K.B., Wright, J.D., Boppana, S.B., Britt, W.J., and Pass, R.F. (2000). Longitudinal investigation of hearing disorders in children with congenital cytomegalovirus. J. Am. Acad. Audiol. *11*, 283–290.

Dar, L., Pati, S.K., Patro, A.R., Deorari, A.K., Rai, S., Kant, S., Broor, S., Fowler, K.B., Britt, W.J., and Boppana, S.B. (2008). Congenital cytomegalovirus infection in a highly seropositive semi-urban population in India. Pediatr. Infect. Dis. J. *27*, 841–843.

Demmler, G.J. (1996). Congenital cytomegalovirus infection and disease. Adv. Pediatr. Infect. Dis. *12*, 35–62.

Demmler, G.J., Yow, M.D., Spector, S.A., Reis, S.G., Brady, M.T., Anderson, D.,C., and Taber, L.H. (1987). Nosocomial cytomegalovirus infections within two hospitals caring for infants and children. J. Infect. Dis. *156*, 9–16.

Doneda, C., Parazzini, C., Righini, A., Rustico, M., Tassis, B., Fabbri, E., Arrigoni, F., Consonni, D., and Triulzi, F. (2010). Early cerebral lesions in cytomegalovirus infection: prenatal MR imaging. Radiology *255*, 613–621.

Doyle, M., Atkins, J.T., and Rivera-Matos, I.R. (1996). Congenital cytomegalovirus infection in infants infected with human immunodeficiency virus type 1. Pediatr. Infect. Dis. J. *15*, 1102–1106.

Dworsky, M., Yow, M., Stagno, S., and Alford, C. (1983). Cytomegalovirus infection of breast milk and transmission in infancy. Pediatrics *72*, 295–299.

Emery, V.C., Sabin, C., Feinberg, J.E., Grywacz, M., Knight, S., and Griffiths, P.D. (1999). Quantitative effects of valacyclovir on the replication of cytomegalovirus (CMV) in persons with advanced human immunodeficiency virus disease: baseline CMV load dictates time to disease and survival. J. Infect. Dis. *180*, 695–701.

Faiz, R.G. (1986). Comparative efficacy of handwashing agents against cytomegalovirus. Pediatr. Res. *20*, 227A.

Faix, R.G. (1985). Survival of cytomegalovirus on environmental surfaces. J. Pediatr. *106*, 649–652.

Fowler, K.B., and Pass, R.F. (2006). Risk factors for congenital cytomegalovirus infection in the offspring of young women: exposure to young children and recent onset of sexual activity. Pediatrics *118*, 286–292.

Fowler, K.B., and Boppana, S.B. (2006). Congenital cytomegalovirus (CMV) infection and hearing deficit. J. Clin. Virol. *2*, 226 e31.

Fowler, K.B., Stagno, S., Pass, R.F., Britt, W.J., Boll, T.J., and Alford, C.A. (1992). The outcome of congenital cytomegalovirus infection in relation to maternal antibody status. N. Engl. J. Med. *326*, 663–667.

Fowler, K.B., McCollister, F.P., Dahle, A.J., Boppana, S.B., Britt, W.J., and Pass, R.F. (1997). Progressive and fluctuating sensorineural hearing loss in children with asymptomatic congenital cytomegalovirus infection. J. Pediatr. *130*, 624–630.

Fowler, K.B., Stagno, S., and Pass, R.F. (2003). Maternal immunity and prevention of congenital cytomegalovirus infection. JAMA *289*, 1008–1011.

Francisse, S., Revelard, P., De Maertelaer, V., Strebelle, E., Englert, Y., and Liesnard, C. (2009). Human cytomegalovirus seroprevalence and risk of seroconversion in a fertility clinic population. Obstet. Gynecol. *114*, 285–291.

Gabrielli, L., Bonasoni, P., Lazzarotto, T., Lega, S., Santini, D., Foschini, M.P., Guerra, B., Baccolini, F., Piccirilli, G., Chiereghin, A., *et al.* (2009). Histological findings in fetuses congenitally infected by cytomegalovirus. J. Clin. Virol. *46*, 16–21.

Galli, L., Novelli, A., Chiappini, E., Gervaso, P., Cassetta, M.I., Fallani, S., and de Martino, M. (2007). Valganciclovir for congenital CMV infection: A pilot study on plasma concentration in newborns and infants. Pediatr. Infect. Dis. J. *26*, 451–453.

Gleaves, C.A., Smith, T.F., Shuster, E.A., and Pearson, G.R. (1984). Rapid detection of cytomegalovirus in MRC-5 cells inoculated with urine specimens by using low speed centrifugation and monoclonal antibody to an early antigen. J. Clin. Microbiol. *19*, 917–919.

Gouarin, S., Gault, E., Vabret, A., Cointe, D., Rozenberg, F., Grangeot-Keros, L., Barjot, P., Garbarg-Chenon, A., Lebon, P., and Freymuth, F. (2002). Real time PCR quantification of human cytomegalovirus DNA in amniotic fluid samples from mothers with primary infection. J. Clin. Microbiol. *40*, 1767–1772.

Grant, S., Edmond, J.E., and Syme, J. (1981). A prospective study of cytomegalovirus infection in pregnancy. I. Laboratory evidence of congenital infection following maternal primary and reactivated infection. J. Infect. *3*, 24–31.

Griffiths, P.D., and Baboonian, C. (1984). A prospective study of primary cytomegalovirus infection during pregnancy: final report. Br. J. Obstet. Gynaecol. *9*, 307–315.

Guerra, B., Lazzarotto, T., Quarta, S., Lanari, M., Bovicelli, L., Nicolosi, A., and Landini, M.P. (2000). Prenatal diagnosis of symptomatic congenital cytomegalovirus infection. Am. J. Obstet. Gynecol. *183*, 476–482.

Guerra, B., Simonazzi, G., Puccetti, C., Lanari, M., Farina, A., Lazzarotto, T., and Rizzo, N. (2008). Ultrasound prediction of symptomatic congenital cytomegalovirus infection. Am. J. Obstet. Gynecol. *198*, 380 e 1–7.

Hamprecht, K., Maschmann, J., Jahn, G., Poets, C.F., and Goelz, R. (2008). Cytomegalovirus transmission to preterm infants during lactation. J. Clin. Virol. *41*,198–205.

Hodson, E.M., Barclay, P.G., Craig, J.C., Jones, C., Kable, K., Strippoli, G.F., Vimalachandra, D., and Webster, A.C. (2005). Antiviral medications to prevent cytomegalovirus disease and early death in recipients of solid-organ transplants: a systematic review of randomised controlled trials. Lancet *365*, 2105–2115.

Hutto, C., Ricks, R., Garvie, M., and Pass, R.F. (1985). Epidemiology of cytomegalovirus infections in young children: day care vs. home care. Pediatr. Infect. Dis. *4*, 149–152.

Hutto, C., Little, E.A., Ricks, R., Lee, J.D., and Pass, R.F. (1986). Isolation of cytomegalovirus from toys and hands in a day care center. J. Infect. Dis. *154*, 527–530.

Istas, A.S., Demmler, G.J., Dobbins, J.G., and Stewart, J.A. (1995). Surveillance for congenital cytomegalovirus disease: a report from the National Congenital Cytomegalovirus Disease Registry. Clin. Infect. Dis. *20*, 665–670.

Jacquemard, F., Yamamoto, M., Costa, J.M., Romand, S., Jaqz-Aigrain, E., Dejean, A., Daffos, F., and Ville, Y. (2007). Maternal administration of valaciclovir in symptomatic intrauterine cytomegalovirus infection. BJOG *114*, 1113–1121.

Kenneson, A., and Cannon, M.J. (2007). Review and meta-analysis of the epidemiology of congenital cytomegalovirus (CMV) infection. Rev. Med. Virol. *17*, 253–276.

Kimberlin, D.F., Weller, S., Whitley, R.J., Andrews, W.W., Hauth, J.C., Lakeman, F., and Miller, G. (1998). Pharmacokinetics of oral valacyclovir and acyclovir in late pregnancy. Am. J. Obstet. Gynecol. *179*, 846–851.

Kimberlin, D.W., Lin, C.Y., Sanchez, P.J., Demmler, G.J., Dankner, W., Shelton, M., Jacobs, R.F., Vaudry, W., Pass, R.F., Kiell, J.M., *et al.* (2003). Effect of ganciclovir on hearing in symptomatic congenital cytomegalovirus disease involving the central nervous system: a randomized, controlled trial. J. Pediatr. *143*, 16–25.

Kimberlin, D.W., Acosta, E.P., Sánchez, P.J., Sood, S., Agrawal, V., Homans, J., Jacobs, R.F., Lang, D., Romero, J.R., Griffin, J., *et al.* (2008). Pharmacokinetic and pharmacodynamic assessment of oral valganciclovir in the treatment of symptomatic congenital cytomegalovirus disease. J. Infect. Dis. *197*, 836–845.

La Torre, R., Nigro, G., Mazzocco, M., Best, A.M., and Adler, S.P. (2006). Placental enlargement in women with primary maternal cytomegalovirus infection is associated with fetal and neonatal disease. Clin Infect. Dis. *43*, 994–1000.

Lanari, M., Lazzarotto, T., Venturi, V., Papa, I., Gabrielli, L., Guerra, B., Landini, M.P., and Faldella, G. (2006). Neonatal cytomegalovirus blood load and risk of sequelae in symptomatic and asymptomatic congenitally infected newborns. Pediatrics *117*, e76–e83.

Landry, M.L., Stanat, S., Biron, K., Brambilla, D., Britt, W., Jokela, J., Chou, S., Drew, W.L., Erice, A., Gilliam, B., *et al.* (2000). A standardized plaque reduction assay for determination of drug susceptibilities of cytomegalovirus clinical isolates. Antimicrob. Agents Chemother. *44*, 688–692.

Liesnard, C., Donner, C., Bracart, F., Gosselin, F., Delforge, M.L., and Rodesch, F. (2000). Prenatal diagnosis of congenital cytomegalovirus infection: prospective study of 237 pregnancies at risk. Obstet. Gynecol. *95*, 881–888.

Lombardi, G., Garofoli, F., Villani, P., Tizzoni, M., Angelini, M., Cusato, M., Bollani, L., De Silvestri, A., Regazzi, M., and Stronati, M. (2009). Oral valganciclovir treatment in newborns with symptomatic congenital cytomegalovirus infection. Eur. J. Clin. Microbiol. Infect. Dis. *28*, 1465–1470.

Lowrance, D., Neumayer, H.H., Legendre, C.M., Squifflet, J.P., Kovarik, J., Brennan, P.J., Norman, D., Mendez, R., Keating, M.R., Coggon, G.L., *et al.* (1999). Valacyclovir for the prevention of cytomegalovirus disease after renal transplantation. International Valacyclovir Cytomegalovirus Prophylaxis Transplantation Study Group. N. Engl. J. Med. *340*, 1462–1470.

Maidji, E., Nigro, G., Tabata, T., McDonagh, S., Nozawa, N., Shiboski, S., Muci, S., Anceschi, M.M., Aziz, N., Adler, S.P., *et al.* (2010). Antibody treatment promotes compensation for human cytomegalovirus-induced pathogenesis and a hypoxia-like condition in placentas with congenital infection. Am. J. Pathol. *177*, 1298–1310.

Marshall, B.C., and Adler, S.P. (2009). The frequency of pregnancy and exposure to cytomegalovirus (CMV) infections among women with a young child in day care. Am. J. Obstet. Gynecol. *200*, 163.e1–163.e5.

Michaels, M.G., Greenberg, D.P., Sabo, D.L., and Wald, E.R. (2003). Treatment of children with congenital cytomegalovirus infection with ganciclovir. Pediatr. Infect. Dis. J. *22*, 504–509.

Moffett-King, A. (2002). Natural killer cells and pregnancy. Nat. Rev. Immunol. *2*, 656–663.

Murph, J.R., and Bale, J.F., Jr. (1988). The natural history of acquired cytomegalovirus infection among children in group day care. Am. J. Dis. Child. *142*, 843–846.

Murph, J.R., Baron, J.C., Brown, C.K., Ebelhack, C.L., and Bale, J.F., Jr. (1991). The occupational risk of cytomegalovirus infection among day-care providers. JAMA *265*, 603–608.

Naeye, R.L. (1967). Cytomegalic inclusion disease. Am. J. Clin. Path. *47*, 738–744.

Nankervis, G.A., Kumar, M.L., Cox, F.E., and Gold, E. (1984). A prospective study of maternal cytomegalovirus infection and its effect on the fetus. Am. J. Obstet. Gynecol. *149*, 435–440.

Nigro, G., Clerico, A., and Mondaini, C. (1993). Symptomatic congenital cytomegalovirus infection in two consecutive sisters. Arch. Dis. Child. *69*, 527–528.

Nigro, G., Scholz, H., and Bartmann, U. (1994). Ganciclovir therapy for symptomatic congenital cytomegalovirus infection in infants: a two-regimen experience. J. Pediatr. *124*, 318–322.

Nigro, G., Mazzocco, M., Anceschi, M.M., La Torre, R., Antonelli, G., and Cosmi, E.V. (1999). Prenatal diagnosis of fetal cytomegalovirus infection following primary or recurrent maternal infection. Obstet. Gynecol. *94*, 909–914.

Nigro, G., Sali, E., Anceschi, M.M., Mazzocco, M., Maranghi, L., Clerico, A., and Castello, M.A. (2004). Foscarnet therapy far congenital hepatic fibrosis by cytomegalovirus infection. J. Matern. Fetal Neonatal Med. *15*, 325–329.

Nigro, G., Adler, S.P., La Torre, R., and Best, A.M. (2005). Passive immunization during pregnancy for congenital cytomegalovirus infection. N. Engl. J. Med. *353*, 1350–1362.

Nigro, G., La Torre, R., Pentimalli, H., Taverna, P., Lituania, M., Martinez de Tejada, B., and Adler, S.P. (2008). Regression of fetal cerebral abnormalities by primary cytomegalovirus infection following hyperimmunoglobulin therapy. Prenat. Diagn. *28*, 512–517.

Nigro, G., Pietrobattista, A., Divito, S., and Gambarara, M. (2010). Oral ganciclovir therapy for immunocompetent infants with cytomegalovirus-associated hemorrhagic or intractable enterocolitis. J. Pediatr. Gastroenterol. Nutr. *50*, 111–113.

Nigro, G., Adler, S.P., Parruti, G., Anceschi, M.M., Coclite, E., Pezone, I., and Di Renzo, G.C. (2012a). Immunoglobulin therapy of fetal cytomegalovirus infection occurring in the first half of pregnancy – a case–control study of the outcome in children. J. Infect. Dis. *205*, 215–227.

Nigro, G., Adler, S.P., Gatta, E., Mascaretti, G., Megaloikonomou, A., La Torre, R., and Necozione, S. (2012b). Fetal hyperechogenic bowel may indicate congenital cytomegalovirus disease responsive to immunoglobulin therapy. J. Matern. Fetal Neonatal Med. *11*, 2202–2205.

Noyola, D.E., Demmler, G.J., Williamson, W.D., Griesser, C., Sellers, S., Llorente, A., Littman, T., Williams, S., Jarrett, L., and Yow, M.D. (2000). Cytomegalovirus urinary excretion and long-term outcome in children with congenital cytomegalovirus infection. Congenital CMV Longitudinal Study Group. Pediatr. Infect. Dis. J. *19*, 505–510.

Noyola, D.E., Demmler, G.J., Nelson, C.T., Griesser, C., Williamson, W.D., Atkins, J.T., Rozelle, J., Turcich, M., Llorente, A.M., Sellers-Vinson, S., *et al.* (2001). Early predictors of neurodevelopmental outcome in symptomatic congenital cytomegalovirus infection. J. Pediatr. *138*, 325–331.

Oliver, S.E., Cloud, G.A., Sánchez, P.J., Demmler, G.J., Dankner, W., Shelton, M., Jacobs, R.F., Vaudry, W., Pass, R.F., Soong, S.J., Whitley, R.J., *et al.* (2009). Neurodevelopmental outcomes following ganciclovir therapy in symptomatic congenital cytomegalovirus infections involving the central nervous system. J. Clin. Virol. *46*, 22–26.

Pass, R.F., Hutto, C., Ricks, R., and Cloud, G.A. (1986). Increased rate of CMV infection among parents of children attending day-care centers. N. Engl. J. Med. *314*, 1414–1416.

Pass, R.F., Hutto, C., Lyon, M.D., and Cloud, G. (1990). Increased rate of cytomegalovirus infection among day care center workers. Pediatr. Infect. Dis. J. *9*, 465–470.

Pass, R.F., Zhang, C., Evans, A., Simpson, T., Andrews, W., Huang, M.L., Corey, L., Hill, J., Davis, E., Flanigan, C., *et al.* (2009). Vaccine prevention of maternal cyotmegalovirus infection. New Eng. J. Med. *360*, 1191–1199.

Pasternak, B., and Hylid, A. (2010). Use of acyclovir, valacyclovir, and famciclovir in the first trimester of pregnancy and the risk of birth defects. JAMA *304*, 859–866.

Pavlopoulou, I.D., Syriopoulou,V.P., Chelioti, H., Daikos, G.L., Stamatiades, D., Kostakis, A., and Boletis, J.N. (2005). A comparative randomised study of valacyclovir vs. oral ganciclovir for cytomegalovirus prophylaxis in renal transplant recipients. Clin. Microbiol. Infect. *11*, 736–743.

Plotkin, S.P., Starr, S.E., Friedman, H.M., Gönczöl, E., and Weibel, R.E. (1989). Protective effects of Towne cytomegalovirus vaccine against low-passage cytomegalovirus administered as a challenge. J. Infect. Dis. *59*, 860–865.

Rahan, G., Gabbay, R., Ornoy, A., Shechtman, S., Arnon, J., and Diav-Citrin, D. (2007). Primary versus nonprimary cytomegalovirus infection during pregnancy, Israel. Emerg. Infect. Dis. *13*, 1791–1793.

Ratanajamit, C., Vinther Skriver, M., Nørgaard, M., Jepsen, P., Schønheyder, H.C., and Sørensen, H.T. (2003). Adverse pregnancy outcome in women exposed to acyclovir during pregnancy: a population-based observational study. Scand. J. Infect. Dis. *35*, 255–259.

Rees, S., and Inder, T., (2005). Fetal and neonatal origins of altered brain development. Early Hum. Dev. *81*, 753–761.

Reischig, T., Opatrný, Jr., K., Treska, V., Mares, J., Jindra, P., and Svecová, M. (2005). Prospective comparison of valacyclovir and oral ganciclovir for prevention of cytomegalovirus disease in high-risk renal transplant recipients. Kidney Blood Press. Res. *28*, 218–225.

Revello, M.G., and Gerna, G. (2002). Diagnosis and management of human cytomegalovirus infection in the mother, fetus, and newborn infant. Clin. Microbiol. Rev. *15*, 680–715.

Revello, M.G., Zavattoni, M., Furione, M., Fabbri, E., and Gerna, G. (2006). Preconceptional primary human cytomegalovirus infection and risk of congenital infection. J. Infect. Dis. *193*, 783–787.

Rivera, L.B., Boppana, S.B., Fowler, K.B., Britt, W.J., Stagno, S., and Pass, R.F. (2002). Predictors of hearing loss in children with symptomatic congenital cytomegalovirus infection. Pediatrics *110*, 762–767.

Robbins, J.B., Schneerson, R., and Szu, S.C. (1995). Perspective hypothesis: serum IgG antibody is sufficient to confer protection against infectious diseases by inactivating the inoculum. J. Infect. Dis. *171*, 1387–1398.

Roberts, I., and Murray, N.A. (2003). Neonatal thrombocytopenia: causes and management. Arch. Dis. Child. Fetal Neonatal Ed. *88*, F359–F364.

Ross, S.A., and Boppana, S.B. (2005). Congenital cytomegalovirus infection: outcome and diagnosis. Semin. Pediatr. Infect. Dis. *16*, 44–49.

Ross, S.A., Fowler, K.B., Ashrith, G., Stagno, S., Britt, W.J., Pass, R.F., and Boppana, S.B. (2006). Hearing loss in children with congenital cytomegalovirus infection born to mothers with preexisting immunity. J. Pediatr. *148*, 332–336.

Ross, S.A., Arora, N., Novak, Z., Fowler, K.B., Britt, W.J., and Boppana, S.B. (2010). Cytomegalovirus reinfections in healthy seroimmune women. J. Infect. Dis. *201*, 386–389.

Schleiss, M.R. (2008).Comparison of vaccine strategies against congenital CMV infection in the guinea pig model. J. Clin.Virol. *41*, 224–230.

Schleiss, M.R., Anderson, J.L., and McGregor, A. (2006). Cyclic cidofovir prevents congenital cytomegalovirus infection in a guinea pig model. Virol. J. *3*, 9.

Schlesinger, Y. (2007). Routine screening for CMV in pregnancy: Opening the pandora box? Isr. Med. Assoc. J. *9*, 395–397.

Stagno, S., and Britt, W. (2006). Cytomegalovirus Infections. In Infectious Diseases of the Fetus and Newborn Infant, 6th edn, Remington, J.S., Klein, J.O., Wilson, C.B., and Baker, C.J., eds. (W.B. Saunders, Philadelphia, PA), pp. 739–781.

Stagno, S., and Whitley, R.J. (1985). Herpesvirus infections of pregnancy. Part I: Cytomegalovirus and Epstein–Barr virus infections. N. Engl. J. Med. *313*, 1270–1274.

Stagno, S., Pass, R.F., Cloud, G., Britt, W.J., Henderson, R.E., Walton, P.D., Veren, D.A., Page, F., and Alford, C.A. (1986). Primary cytomegalovirus infection in pregnancy, incidence, transmission to fetus, and clinical outcome. JAMA *256*, 1904–1908.

Stone, K.M., Reiff-Eldridge, R., White, A.D., Cordero, J.F., Brown, Z., Alexander, E.R., and Andrews, E.B. (2004). Pregnancy outcomes following systemic prenatal acyclovir exposure: conclusions from the international acyclovir pregnancy registry, 1984–1999. Birth Defects Res A. Clin. Mol. Teratol. *70*, 201–207.

Szekeres-Bartho, J. (2002). Immunological relationship between the mother and the fetus. *21*, 471–495.

Taber, L.H., Frank, A.L., Yow, M.D., and Bagley, A. (1985). Acquisition of cytomegaloviral infections in families with young children: a serological study. J. Infect. Dis. *151*, 948–952.

US Food and Drug Administration (2005). Zovirax (acyclovir product information), GlaxoSmithKline. Available at: http://www.accessdata.fda.gov/drugsatfda_docs/label/2005/018828s030,020089s019,019909s020lbl.pd.

US Food and Drug Administration. (2008). Valtrex (valacyclovir product information), Glaxo Smith Kline. Available at: http://www.accessdata.fda.gov/drugsatfda_docs/label/2010/020487s016lbl.pdf.

Vauloup-Fellous, C., Picone, O., Cordier, A.G., Parent-du-Châtelet, I., Senat, M.V., Frydman, R., and Grangeot-Keros, L. (2009). Does hygiene counseling have an impact on the rate of CMV primary infection during pregnancy? Results of a 3-year prospective study in a French hospital. J. Clin. Virol. *46*, 49–53.

Volpe, J.J. (2000). Viral, protozoan, and related intracranial infections. In: Neurology of the Newborn, 3rd edition (WB Saunders Co., Philadelphia, PA), pp.675–729.

Wang, J.B., Adler, S.P., Hempfling, S., Burke, R.L., Duliège, A.M., Starr, S.E., and Plotkin, S.A. (1996). Mucosal antibodies to CMV glycoprotein B occur following both natural infection and immunization with human cytomegaloviral vaccines. J. Infect. Dis. *174*, 397–392.

Weller, S., Blum, M.R., Doucette, S., Burnette, T., Cederberg, D.M., de Miranda, P., and Smiley, M.L. (1993). Pharmacokinetics of the acyclovir pro-drug, valacyclovir after escalating single- and multiple-dose administration to normal volunteers. Clin. Pharmacol. Ther. *54*, 595–605.

Whitley, R.J., Cloud, G., Gruber, W., Storch, G.A., Demmler, G.J., Jacobs, R.F., Dankner, W., Spector, S.A., Starr, S., Pass, R.F., *et al.* (1997). Antiviral Study Group Ganciclovir treatment of symptomatic congenital cytomegalovirus infection: results of a phase II study. J. Infect. Dis. *75*, 1080–1086.

Williamson, W.D., Demmler, G.J., Percy, A.K., and Caitlin, F.I. (1992). Progressive hearing loss in infants with asymptomatic congenital cytomegalovirus infection. Pediatrics *90*, 862–868.

Wilton, L.V., Pearce, G.L., Martin, R.M., Mackay, F.J., and Mann, R.D. (1998). The outcomes of pregnancy in women exposed to newly marketed drugs in general practice in England. Br. J. Obstet. Gynaecol. *105*, 882–889.

Winston, D.J., Yeager, A.M., Chandrasekar, P.H., Snydman, D.R., Petersen, F.B., Territo, M.C., and Valacyclovir Cytomegalovirus Study Group (2003). Randomized comparison of oral valacyclovir and intravenous ganciclovir for prevention of cytomegalovirus disease after allogeneic bone marrow transplantation. Clin. Infect. Dis. *36*, 749–758.

Wutzler, P., and Thust, R. (2001). Genetic risks of antiviral nucleoside analogues – a survey. Antiviral Res. *49*, 55–74.

Yamamoto, A.Y., Mussi-Pinhata, M.M., Boppana, S.B., Novak, Z., Wagatsuma, V.M., Oliveira Pde, F., Duarte, G., and Britt, W.J. (2010). Human cytomegalovirus reinfection is associated with intrauterine transmission in a highly cytomegalovirus-immune maternal population. Am. J. Obstet. Gynecol. *202*, 297 e 1–8.

Yango, A., Morrissey, P., Zanabli, A., Beaulieu, J., Shemin, D., Dworkin, L., Monaco, A., and Gohh, R. (2003). Comparative study of prophylactic oral ganciclovir and valacyclovir in high-risk kidney transplant recipients. Nephrol. Dial. Transplant. *18*, 809–813.

Yeager, A.S., Grumet, F.C., Hafleigh, E.B., Arvin, A.M., Bradley, J.S., and Prober, C.G. (1981). Prevention of transfusion-acquired cytomegalovirus infections in newborn infants. J. Pediatr. 98, 281–287.

Yinon, Y., Farine, D., Yudin, M.H., Gagnon, R., Hudon, L., Basso, M., Bos, H., Delisle, M.F., Menticoglou, S., Mundle, W., *et al.* (2010). Cytomegalovirus infection in pregnancy. J. Obstet. Gynaecol. Can. 232, 348–354.

Cytomegalovirus Replication in the Developing Human Placenta

II.4

Lenore Pereira, Takako Tabata, Matthew Petitt and June Fang-Hoover

Abstract

During human pregnancy, HCMV spreads from the infected mother to the fetus, navigating the complex architecture of the human placenta, which anchors the fetus to the uterus. Primary sites of virus replication in the placenta include cytotrophoblast progenitor cells in chorionic villi and differentiating/invading cytotrophoblasts that breach uterine blood vessels and form a hybrid vasculature, increasing blood flow to the surface of the placenta. Focal virus replication and induction and release of paracrine factors result in pathology and a hypoxic intrauterine environment that stimulates compensatory development of vascularized floating villi. This chapter summarizes recent insights into the molecular changes that occur during virus replication, strategies to rescue development of the human placenta and the utility of placental villous explants and xenografts in SCID mice for quantifying infection *in vitro* and pathogenesis *in vivo*.

Introduction

HCMV is the leading cause of congenital viral infection, with an incidence in the USA of about 1–3% of live births (Fowler *et al.*, 2003) (for epidemiology, see also Chapter II.2). Primary maternal HCMV infection during gestation poses a 40–50% risk of transplacental transmission (Stagno *et al.*, 1986; Fowler *et al.*, 1992, 2003). In contrast, transmission and disease are considerably reduced in cases of recurrent infection in strongly seropositive mothers – highlighting the role of immunity in fetal protection. Symptomatic infants (25%) have intrauterine growth restriction (IUGR) and permanent birth defects, including neurological deficiencies, retinopathy and sensorineuronal deafness (Boppana *et al.*, 1992; Demmler, 1996; Noyola *et al.*, 2001) (see also Chapter II.3). Although virus transmission can occur throughout pregnancy, fetal disease is more severe when primary maternal infection occurs in the first trimester of gestation (Stagno *et al.*, 1986). IUGR and spontaneous abortion can occur in the absence of fetal infection, suggesting it can result from placental infection alone (Griffiths and Baboonian, 1984; Benirschke and Kaufmann, 2000). Leucocytic infiltration, oedema and villous fibrosis in infected placentas indicate that persistent replication and immune clearance can impair essential placental functions (Monif and Dische, 1972; Benirschke *et al.*, 1974; Garcia *et al.*, 1989; La Torre *et al.*, 2006). Recent reports suggest that 15% of stillbirths (i.e. fetal death *in utero* after 20 weeks' gestation) are associated with HCMV infection. The most prominent features in infected placentas from these stillbirths are fetal thrombotic vasculopathy, inflammation and oedema (Syridou *et al.*, 2008; Iwasenko *et al.*, 2011). This new evidence links congenital HCMV infection to significant morbidity from damaged placental and fetal blood vessels (Pereira, 2011).

Development of the haemochorial human placenta

Despite the prevalence and medical impact of congenital HCMV infection, the molecular mechanisms of virus replication, transmission and pathogenesis are still unresolved due to the complexity of human placental development and extreme species specificity of the virus. The embryo's acquisition of a supply of maternal blood is a critical hurdle in pregnancy maintenance (Cross *et al.*, 1994; Genbacev *et al.*, 1997; Zhou *et al.*, 1997; Norwitz *et al.*, 2001). The mechanics of this process are accomplished by cytotrophoblasts, which are specialized epithelial cells. In the first trimester, villous cytotrophoblasts of the placenta (i.e. trophoblast progenitor cells) form a polarized epithelium attached to the basement membrane that surrounds the stromal cores of chorionic villi, which contain the fetal blood vessels (Fig. II.4.1A, zone I). These cells may subsequently follow one of two developmental pathways. In one pathway, villous cytotrophoblasts detach from the

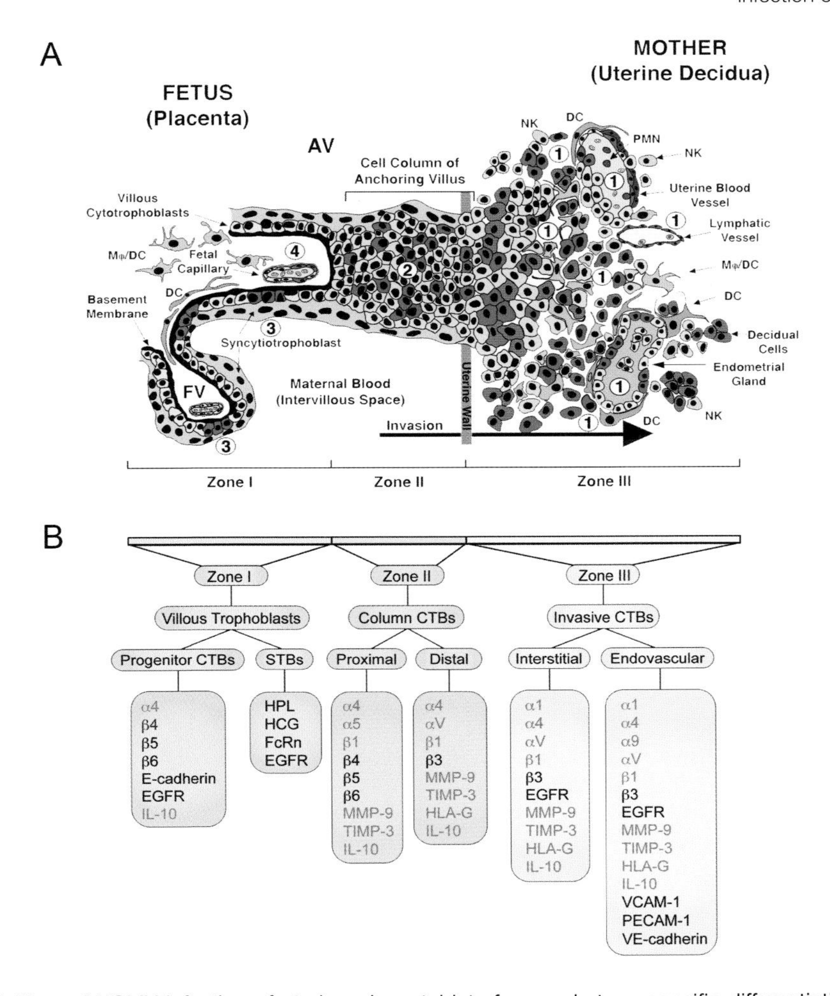

Figure II.4.1 Sites of HCMV infection of uterine–placental interface and stage-specific differentiation molecules altered. (A) The basic structural unit of the placenta is the chorionic villus, composed of a stromal core with blood vessels, surrounded by a basement membrane overlain by cytotrophoblast progenitor cells and syncytiotrophoblasts. The floating villus (FV) is in direct contact with maternal blood and anchoring villus (AV) attaches the fetus to the uterus by cytotrophoblast cell columns and conducts blood from maternal circulation into the intervillous blood space. As part of their differentiation programme, villous cytotrophoblasts detach from the basement membrane and adopt one of two lineage fates. They either fuse to form the syncytiotrophoblasts that cover floating villi or join a column of extravillous cytotrophoblasts at the tips of anchoring villi. The syncytial covering of floating villi mediates nutrient, gas and waste exchange and passive transfer of IgG from maternal blood to the fetus (zone 1). The anchoring villi, through the attachment of cytotrophoblast columns, establish physical connections between the fetus and the mother (zone II). Invasive cytotrophoblasts, blood vessels, lymphatic vessels and endometrial glands (Zone III). A portion of invasive extravillous cytotrophoblasts penetrate the uterine wall up to the first third of the myometrium, breach uterine spiral arterioles and remodel these blood vessels by destroying their muscular walls and replacing their endothelial linings. HCMV replication sites in uterine decidua and placenta are numbered: 1 (cytotrophoblasts, epithelial, blood and lymphatic endothelial cells); 2 (cell columns of AV), 3 and 4 (villous cytotrophoblasts, stromal fibroblasts and fetal capillaries in FV). Infected cells are coloured red (Fisher *et al.*, 2000; Pereira *et al.*, 2003). DC, dendritic cell; Mϕ, macrophage; NK, natural killer cell; PMN, polymorphonuclear neutrophil. (B) Zones I-III indicate the stage-specific antigens expressed by cytotrophoblasts. Proteins dysregulated in HCMV-infected cytotrophoblasts (red) include integrins (Damsky *et al.*, 1992; Zhou *et al.*, 1997; Fisher *et al.*, 2000; Yamamoto-Tabata *et al.*, 2004; Tabata *et al.*, 2007, 2008), IL-10 (Roth and Fisher, 1999; Yamamoto-Tabata *et al.*, 2004), matrix metalloproteinase 9 (MMP-9) (Librach *et al.*, 1991; Yamamoto-Tabata *et al.*, 2004), tissue inhibitor of metalloproteinases 3 (TIMP-3) (Bass *et al.*, 1997; Tabata and Pereira, unpublished), MHC class I HLA-G (McMaster *et al.*, 1995; Fisher *et al.*, 2000; Tabata *et al.*, 2007), epidermal growth factor receptor (EGFR) (Bass *et al.*, 1994), neonatal Fc receptor (FcRn) (Simister *et al.*, 1996), human placental lactogen (HPL) and human chorionic gonadotropin (HCG) (Kovalevskaya *et al.*, 2002). The diagram was modified from Zhou *et al.* (1997) and Pereira *et al.* (2005).

basement membrane and fuse to form multinucleated syncytiotrophoblasts, which transport substances between maternal blood in the intervillous space and the villus cores containing fetal blood vessels. In the pathway that leads to invasion, cytotrophoblasts form columns of non-polarized cells that attach to and then penetrate the uterine wall (Fig. II.4.1A, zone II). The columns terminate within the superficial endometrium, where they give rise to invasive (extravillous) cytotrophoblasts. During interstitial invasion, a subset of these cells, either individually or in small clusters, commingles with resident decidual, myometrial and immune cells (Fig. II.4.1A, zone III). During endovascular invasion, masses of cytotrophoblasts breach and plug the vessels. Subsequently, these cells create patent vessels by replacing the resident maternal endothelium and portions of the smooth muscle wall. These novel hybrid vessels divert uterine blood flow to the placenta.

As cytotrophoblasts differentiate they switch to an endothelial phenotype in a process akin to vasculogenesis (Damsky and Fisher, 1998) and up-regulate adhesion molecules and proteinases that enable their attachment to and invasion of the uterus (Fig. II.4.1B). Interstitial invasion requires down-regulation of integrin $\alpha6\beta4$, characteristic of epithelial cells, and *de novo* expression of the integrins $\alpha1\beta1$, $\alpha5\beta1$ and $\alpha v\beta3$ (Damsky *et al.*, 1994). Invasive cells up-regulate matrix metalloproteinase 9 (MMP-9) (Librach *et al.*, 1991, 1994), tissue inhibitor of metalloproteinase 3 (TIMP-3) (Bass *et al.*, 1997) and IL-10 (Roth and Fisher, 1999). Endovascular cytotrophoblasts that remodel uterine blood vessels transform their adhesion receptor phenotype to resemble that of endothelial cells, expressing VE-(endothelial) cadherin, platelet-endothelial adhesion molecule-1 and vascular endothelial adhesion molecule-1 (Zhou *et al.*, 1997; Damsky and Fisher, 1998). Like endothelial cells, cytotrophoblasts express proteins that influence vasculogenesis and angiogenesis, including the vascular endothelial growth factor (VEGF) family ligands VEGF-A and VEGF-C and receptors VEGFR-1 (fms-like tyrosine kinase 1, Flt-1) and VEGFR-3 (Zhou *et al.*, 2002, 2003; Red-Horse *et al.*, 2006). Expression of these molecules changes as the cells differentiate, and they come to regulate cytotrophoblast survival in the remodelled uterine vasculature. Finally, as hemi-allogeneic zygotic cells, invasive cytotrophoblasts must avoid maternal immune responses. It is thought that their expression of the non-classical MHC class I molecules HLA-G (Kovats *et al.*, 1990; McMaster *et al.*, 1995) and HLA-C, which have limited polymorphism (Hiby *et al.*, 2004, 2010), contributes to their lack of immunogenicity.

HCMV replicates at the uterine–placental interface

In the past decade, studies in naturally infected first-trimester placentas, primary placental cells and organ cultures of anchoring villous explants have been instrumental in understanding the detrimental effects of HCMV replication on differentiation. Virus transmission from mother to fetus is a stepwise process that unfolds in the complex architecture of the placenta, a site of replication. Analysis of paired samples from the uterine–placental interface revealed that HCMV first replicates in the decidua (Fig. II.4.1A, zone III) and infects glandular epithelial cells, endothelial cells and endovascular cytotrophoblasts (Fisher *et al.*, 2000; Pereira *et al.*, 2003; McDonagh *et al.*, 2004). Infection then spreads to invasive cytotrophoblasts that intercalate among decidual cells and form anchoring villi. In the otherwise immune-tolerant pregnant uterus, innate immune cells, e.g. macrophages and natural killer cells, limit replication in the uterine wall (Pereira *et al.*, 2003). Moreover, in seropositive women, high-avidity, HCMV-specific neutralizing IgG contributes to suppression of virus replication (Pereira *et al.*, 2003; Maidji *et al.*, 2006; McDonagh *et al.*, 2006; Nozawa *et al.*, 2009). Maternal IgG transcytosed across syncytiotrophoblasts by the neonatal Fc receptor reduces viral replication in villous cytotrophoblast progenitors that spread infection to placental fibroblasts and fetal capillaries (Fig. II.4.1A, azones I and II). In contrast, non-neutralizing low-avidity and high-avidity antibodies bind to virions and facilitate transport, but fail to prevent infection. Fig. II.4.1B lists stage-specific antigens expressed by villous cytotrophoblasts (zone I), cell column cytotrophoblasts (zone II) and differentiating invasive cytotrophoblasts (zone III). HCMV-infected cytotrophoblasts dysregulate the expression and functions of key proteins and impair cell migration and invasion (Fisher *et al.*, 2000; Yamamoto-Tabata *et al.*, 2004; Tabata *et al.*, 2007, 2008). Gene expression profiling has shown dramatic differences between second- and third-trimester placentas that could underlie differences in virus transmission in gestation (Pereira *et al.*, 2003; McDonagh *et al.*, 2006; Winn *et al.*, 2007; Nozawa *et al.*, 2009; Maidji *et al.*, 2010).

HCMV replication in cytotrophoblasts impairs differentiation

HCMV replicates in cytotrophoblasts, causing the release of factors with paracrine effects that contribute to pathogenesis and reduce expression of key differentiation molecules required for migration/

invasion. In particular, productive infection decreases the expression of integrin α1β1 (Halwachs-Baumann *et al.*, 1998; Hemmings *et al.*, 1998; Fisher *et al.*, 2000) and matrix metalloproteinase 9 (MMP-9) (Yamamoto-Tabata *et al.*, 2004), which are required for cytotrophoblast invasiveness, and alters other cell–cell and cell–matrix adhesion molecules, as well. In addition, cmvIL-10, a viral cytokine with immunosuppressive activities (Jones *et al.*, 2002; Spencer *et al.*, 2002; Chang *et al.*, 2004), binds the IL-10 receptor and, in a positive feedback manner, induces expression of IL-10, which suppresses MMP-9 and further reduces cytotrophoblast invasion (Roth *et al.*, 1996; Roth and Fisher, 1999; Yamamoto-Tabata *et al.*, 2004; Tabata *et al.*, 2007). HCMV also activates peroxisome proliferator-activated receptor (PPAR) gamma and its target genes, contributing to impaired invasion (Rauwel *et al.*, 2010; Fournier *et al.*, 2011). Finally, collagen deposition is increased through activation of TGF-β by integrin αvβ6 in HCMV-infected endothelial cells (Tabata *et al.*, 2008), potentially reducing blood flow in the uterine-placental vasculature. This constellation of molecular changes suggests multiple mechanisms by which direct HCMV infection and paracrine factors could impair cytotrophoblast invasion, cause inflammation and undermine placental and fetal development.

Maternal immunity to HCMV and fetal protection by passive immunity

How HCMV virions reach the fetus across the placenta is incompletely understood. FcRn expressed by syncytiotrophoblasts transports IgG for passive immunity (Fig. II.4.2) (Simister and Story, 1997). Immunohistochemical analysis of early-gestation biopsy specimens showed an unusual pattern of HCMV replication proteins in underlying villous cytotrophoblasts, whereas syncytiotrophoblasts

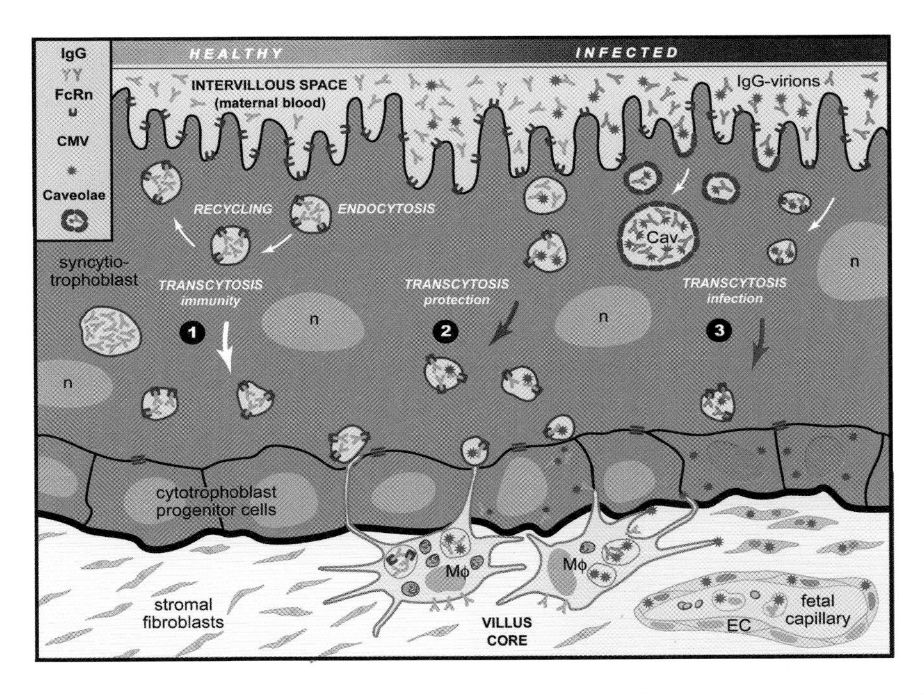

Figure II.4.2 Model illustrating transcytosis and sequestration of IgG-HCMV virion complexes in syncytiotrophoblasts. Pathway 1: Healthy placenta: IgG pinocytosed from maternal blood space binds to FcRn in endocytic vesicles in syncytiotrophoblasts and is recycled or transcytosed. Pathway 2: Suppressed placental infection in an immune mother; pinocytosed IgG–virion complexes of high-avidity, neutralizing IgG bind to FcRn in endocytic vesicles, are transcytosed and released into the intercellular space, where they contact cytotrophoblast progenitor cells, or are sequestered in caveolae. Virion complexes, internalized by immature Mϕs in the villous core that extend processes across cell–cell junctions, undergo rapid degradation in vacuoles. Pathway 3: Viral replication in the placenta; low- and high-avidity complexes of HCMV virions and IgG without neutralizing activity bind to FcRn in endocytic vesicles, are transcytosed and infect underlying cytotrophoblast progenitors. Focal infection spreads to stromal fibroblasts and fetal blood vessels in the villous core. Symbols: IgG, FcRn, HCMV virions and caveolae are shown at the left. Mϕ, fetal macrophages in the villous core. Model based on previous publications (Simister *et al.*, 1996; Fisher *et al.*, 2000; Pereira *et al.*, 2003; Maidji *et al.*, 2006).

were spared (Fisher *et al.*, 2000). Found in placentas with low to moderate HCMV neutralizing antibody titres, this pattern suggested virion transcytosis across syncytiotrophoblasts. In contrast, syncytiotrophoblasts from placentas with high neutralizing titres contained viral DNA and caveolin-1-positive vesicles in which IgG and HCMV gB colocalized (Maidji *et al.*, 2006). Fig. II.4.2 shows a model for dissemination of HCMV virion–IgG complexes across syncytiotrophoblasts in which the FcRn-mediated transport pathway for IgG is co-opted. These findings could offer an explanation for the efficacy of hyperimmune globulin for treatment of primary infection during gestation when low-avidity anti-HCMV IgG predominates in maternal circulation.

A recent study of placentas from healthy deliveries in HCMV-seropositive women showed that the concentration of virus-specific antibodies with neutralizing activity in the fetal bloodstream is equal to levels in maternal circulation (Nozawa *et al.*, 2009). Significantly increased levels of high-avidity, HCMV-neutralizing antibodies that suppress virus replication in the placenta are central to passive immunity and fetal protection (Pereira *et al.*, 2003; Yamamoto-Tabata *et al.*, 2004; McDonagh *et al.*, 2006; Nozawa *et al.*, 2009) and preclude symptomatic fetal infection (Fowler *et al.*, 1992; Grangeot-Keros *et al.*, 1997; Revello *et al.*, 2004). Passive immunity escalates in the third trimester of pregnancy (Malek *et al.*, 1996), mediated by the neonatal Fc receptor (Simister *et al.*, 1996; Simister and Story, 1997; Simister, 2003).

Recent evidence shows that supplementing maternal antibodies with HCMV-specific hyperimmune globulin (HIG) can alter the course of primary infection, reduce virus transmission and prevent congenital disease. In a clinical trial (Nigro *et al.*, 2005), early treatment with HIG to prevent transplacental transmission of infection to the fetus (i.e. prevention group) was associated with increased IgG avidity at delivery. Importantly, significantly fewer babies were infected in the prevention group and none had IUGR or permanent birth defects. Analysis of pathology in the placentas showed considerable differences between untreated infection, which correlated with sustained virus- and hypoxia-associated injury, and compensatory development with HIG administration, which prevented placental infection and damage from occurring (Maidji *et al.*, 2010). The results strongly support the possibility that administration of antiviral antibodies within a defined timeframe after maternal seroconversion reduces HCMV replication at the uterine–placental interface, reduces fetal infection and prevents ensuing pathology associated with congenital disease.

Congenital infection and associated injury at the uterine–placental interface leads to a hypoxic environment that stimulates compensatory development

VEGF-A, an angiogenic factor induced under both physiological and pathological conditions (Ferrara *et al.*, 2003), binds receptor tyrosine kinases that regulate cell proliferation, migration, survival and angiogenesis. The biological activity of VEGF-A is modulated by a soluble form of Flt-1 (sFlt1), which binds and reduces free circulating levels of VEGF and placental growth factor (PlGF) (Kendall and Thomas, 1993; Hornig *et al.*, 1999). Normal placental development depends on the appropriate balance between VEGF and PlGF at different stages of pregnancy. In the first trimester, physiological hypoxia favours VEGF-A expression and branching angiogenesis (Benirschke and Kaufmann, 2000). In the third trimester, normal oxygen levels correlate with strong PlGF expression and non-branching angiogenesis, whereas VEGF-A levels decline (Jackson *et al.*, 1994; Cooper *et al.*, 1996; Shiraishi *et al.*, 1996). Cytotrophoblasts cultured under hypoxic conditions, simulating early gestation, strongly up-regulate VEGF-A expression and down-regulate PlGF (Shore *et al.*, 1997). VEGF-A expression is dramatically up-regulated in placentas from women who smoke during gestation, and enormously enlarged blood vessels develop at the villous periphery to facilitate the transfer of oxygen to the fetus (Genbacev *et al.*, 2003; Zdravkovic *et al.*, 2005).

Congenitally infected placentas develop a hypoxia-like condition that stimulates compensatory development to increase the surface area in contact with maternal blood (Nigro *et al.*, 2005; Maidji *et al.*, 2010). VEGF-A immunostaining was found to increase in syncytiotrophoblasts, cytotrophoblasts, placental fibroblasts and fetal blood vessels in congenital infection. Strong up-regulation of VEGF-A parallels the increase in the number of blood vessels in untreated infection and the development of immature villi with HIG treatment. Under hypoxic conditions in the first trimester, villous cytotrophoblasts proliferate, differentiate and invade the uterine wall, regulated by oxygen tension (Genbacev *et al.*, 1996; Red-Horse *et al.*, 2004). Prolonged hypoxia from infection, inflammation and fibrosis at the uterine–placental interface could extend this process, i.e. villous cytotrophoblasts proliferate and new villi develop. As an example, in pregnancies from high altitudes with low oxygen, the relative volume of cytotrophoblasts increases (Mayhew *et al.*, 2002) and placentas weigh significantly more than they do at sea level, suggesting adaptation by development of a more extensive peripheral villous tree (Kruger and Arias-Stella, 1970). Placental weights also increase in heavy smokers (Williams *et al.*, 1997), but in this case the cytotrophoblast progenitor cell population is prematurely depleted, and the number of anchoring villi is reduced (Genbacev *et al.*, 2000). In addition, capillary densities increase within placental terminal convolutes, suggesting an adaptive angiogenic response in chorionic villi (Pfarrer *et al.*, 1999). Placentas with congenital HCMV infection increase in weight and thickness, and HIG treatment leads to some reduction in size (La Torre *et al.*, 2006), suggesting that irreversible enlargement could result from a combination of inflammation, oedema and development of chorionic villi that increases the surface area and compensates for uteroplacental hypoxia.

Explants of human placentas infected with pathogenic HCMV strains

The early stages of HCMV replication in the developing placenta are incompletely understood due to the virus's extreme host range restriction and the resulting inability to address early placental infection *in vivo*. Infection in fibroblasts and specialized epithelial and endothelial cells correlates with functions specified by gB (Navarro *et al.*, 1993, 1997; Tugizov *et al.*, 1994, 1996, 1999; Feire *et al.*, 2004) and a complex of gH/gL (Rasmussen *et al.*, 1984; Huber and Compton, 1997; Lopper and Compton, 2004) with pUL128-131A, highly conserved proteins in low-passage clinical isolates that promote infection of epithelial and endothelial cells and elicit neutralizing antibodies identified through isolation of human memory B-cells (Wang and Shenk, 2005; Baldanti *et al.*, 2006; Gerna *et al.*, 2008; Macagno *et al.*, 2010).

Organ cultures have been developed to quantify HCMV replication in the human placenta. With regard to infection in the maternal compartment, sections of decidua were cultured *ex vivo* to study virus infection and spread (Weisblum *et al.*, 2011). HCMV strain TB40/E expressing a UL83 (pp65)–GFP fusion was used, and the pattern of GFP-tagged-pp65 in the cytoplasm and levels of viral DNA were used to indicate infection at 6 and 10 days (see Chapters I.17 and II.10). Decidual fibroblasts, epithelial and endothelial cells and cytotrophoblasts could support virus replication. Vesicular patterns of GFP-tagged-pp65 in the cytoplasm of macrophages and dendritic cells suggested that virions were endocytosed. Neutralizing HCMV-specific HIG exhibited inhibitory activity against viral spread in decidua explants.

With regard to HCMV replication in the fetal compartment, we studied placental villous explants infected with the attenuated HCMV strain AD169 or the low-passage clinical isolate VR1814 (Tabata *et al.*, 2012). Significant differences were found in development of anchoring villi. Mock-infected controls and virus-infected explants are shown (Fig. II.4.3A–F). The controls developed robust cell columns and anchoring villi with cytotrophoblasts that aggregated and attached to the extracellular matrix (Fig. II.4.3A and D). AD169-infected explants also formed normal-size anchoring villi, indistinguishable from controls (Fig. II.4.3B and E). In contrast, VR1814-infected explants formed spindly cell columns composed largely of individual cytotrophoblasts migrating on the surface of the substrate (Fig. II.4.3C,F). Analysis of infected cytotrophoblasts within the placental villi revealed that few cells in AD169-infected explants expressed IE1 and 2 proteins (Fig. II.4.3G). In contrast, many VR1814-infected cytotrophoblasts expressed IE1 and 2 (Fig. II.4.3H, inset I) and gB (Fig. II.4.3J). Invasive cytotrophoblasts differentiated and expressed MHC class I HLA-G, but this was down-regulated in VR1814-infected cells (Fig. II.4.3J), as previously reported (Fisher *et al.*, 2000; Tabata *et al.*, 2007).

Since AD169 and VR1814 exhibited markedly distinct levels of infection and gB expression in placental explants over a 3-day interval – too short for titration of viral replication – the differences were quantified by counting the number of cytotrophoblasts expressing IE1 and IE2 proteins in the cell columns and anchoring villi (Fig. II.4.3K). AD169-infected explants contained a median of 2% infected cytotrophoblasts with a 5% maximum. In contrast, VR1814-infected placental villi contained a median of 26% infected cells with a 67% maximum. To quantify the effects on development of anchoring villi, the sizes of villi formed were determined by measuring the areas covered by the villous outgrowths at 3 days post infection (Fig. II.4.3L). Control and AD169-infected explants were comparable, with median sizes of 12 and 11 relative units for control and AD169, respectively, and maximums of about 45 relative units. In contrast, VR1814-infected explants formed significantly smaller villi, with a median size of 3.6 relative units (i.e. less than 10% the size of those in control and AD169-infected explants) and a maximum of 18 ($P < 0.0001$ compared with control). Together, the results showed that VR1814 infects cell column cytotrophoblasts of placental explants and impairs functions of the subpopulation of cells that contribute to forming anchoring villi, reducing their size.

Detailed analysis of intact placental villi showed surprising differences in the capacity of attenuated and pathogenic strains to infect cytotrophoblasts, replicate

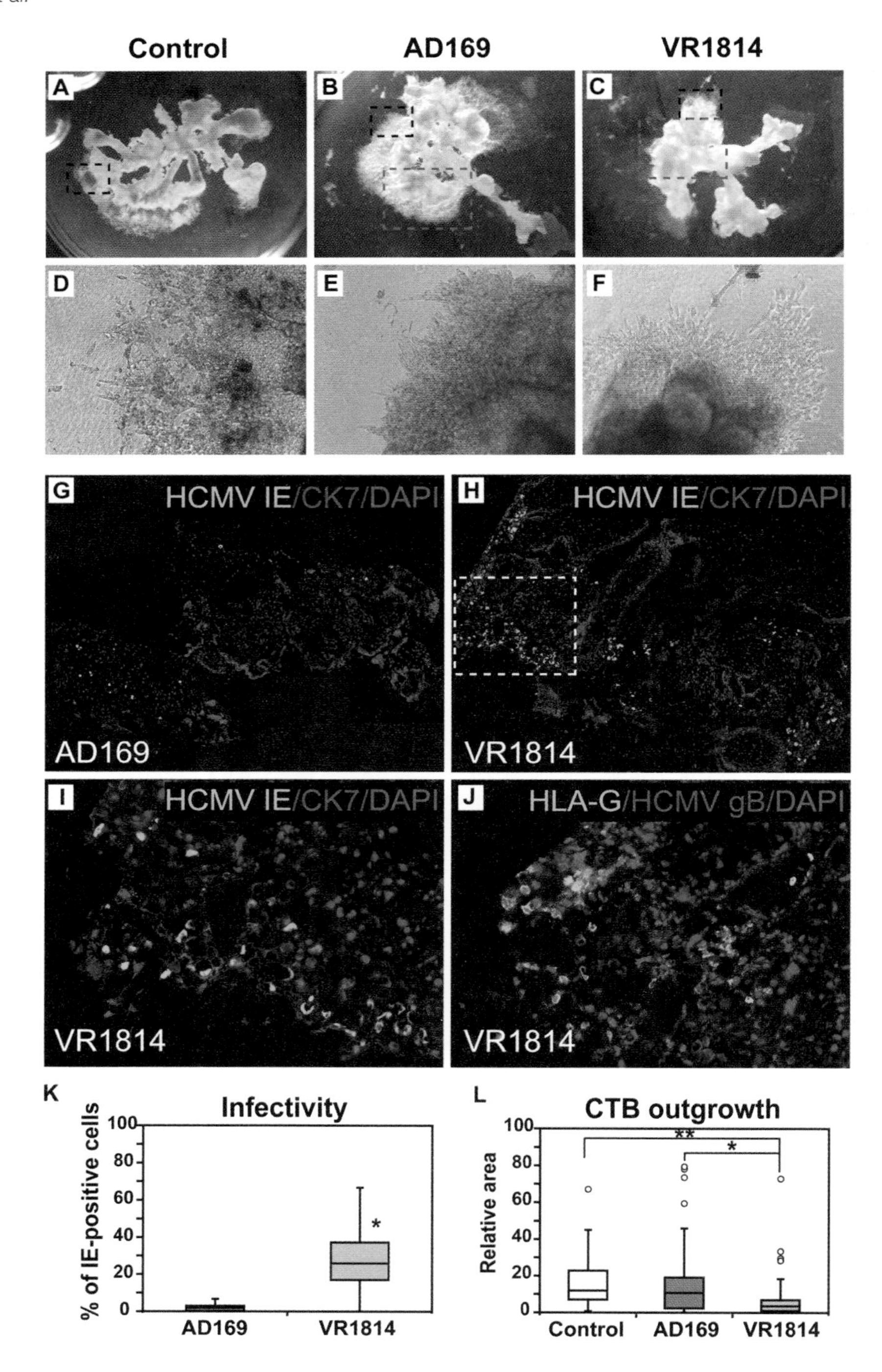

Figure II.4.3 A low-passage clinical HCMV strain infects cytotrophoblasts in placental chorionic villous explants and reduces anchoring villous development. (A–C) Villous explants with anchoring villi (2×) and (D–F) insets (black dashed box, 100×). (A and D) Mock-infected controls, (B and E) AD169 infection, and (C and F) VR1814 infection. Cytotrophoblasts (CTBs) immunostained for expression of cytokeratin 7 (CK7) and HCMV IE1 and IE2 in explants infected with (G) AD169 (blue dashed box from B) and (H) VR1814 (blue dashed box from C). Original magnification 40× (inset I, 200×). (J) Adjacent section was stained for HLA-G and gB (200×). Nuclei were counterstained with DAPI. (K) Box–whisker plots showing number of AD169- and VR1814-infected cytotrophoblasts in cell columns. (L) Measurements of anchoring villi of mock-infected control and AD169- and VR1814-infected explants. *$P<0.001$, **$P<0.0001$ (Student's *t*-test).

and progress to late infection, as judged by HCMV gB expression. HCMV strain Toledo infects villous cytotrophoblasts and differentiating cells that express EGFR and integrin α1β1 or αvβ3, but not cell column cytotrophoblasts (Maidji *et al.*, 2006, 2007). Like clinical HCMV isolates, strain Toledo carries additional DNA sequences that encode at least 19 genes but have rearrangements of the UL131A gene sequence, suggesting UL131A is required for infection (Cha *et al.*, 1996) and thus could be responsible for the deficiency. Cell columns bridge the gap between the placenta and the uterus and are suspended in the intervillous space by homotypic interactions. Cell columns express E-cadherin, as well as L-selectin and its carbohydrate ligand, part of a specialized adhesion system activated by shear stress that maintains column integrity during the early stages of placental development (Prakobphol *et al.*, 2006). This type of adhesion could also facilitate cytotrophoblast movement in and exit from cell columns, a prerequisite for uterine invasion. Virus replication could decrease the number of cytotrophoblasts that progress further down the differentiation pathway and could thereby perturb the specialized adhesive properties of cell columns or reduce the number of cytotrophoblasts that exit. Together these events would undermine development of anchoring villi in infected explants.

Human neutralizing antibodies rescue development of infected anchoring villi

Progression of congenital HCMV infection and placental fibrosis was prevented in pregnancies with early HIG treatment (Maidji *et al.*, 2010). This remarkable outcome suggested that human monoclonal antibodies (mAbs) with potent neutralizing activity could have comparable or even improved protective qualities. Antibodies to HCMV gB have long been known to neutralize virus infection and preclude virus entry (Britt, 1984; Navarro *et al.*, 1993, 1997; Tugizov *et al.*, 1994; Feire *et al.*, 2010) (for humoral immunity, see also Chapter II.10). Accordingly, mAbs to gB that neutralized VR1814 infection in retinal pigment epithelial cells, uterine vascular endothelial cells and primary cytotrophoblasts and fibroblasts isolated from the human placenta were evaluated in intact chorionic villous explants. The neutralizing activity of a commercial HIG preparation was compared with those of a panel of human mAbs. Efficacy was measured by counting the number of infected cytotrophoblasts in cell columns, identified by expression of CK7 (red) and HCMV IE1 and IE2 (green) (Fig. II.4.4A–C). Overall, mAbs to gB, gH/gL and UL128–131A (not shown) consistently had higher neutralizing titres than HIG products. These important results indicate that development of human placental villi can be rescued by neutralizing mAbs that preclude infection with a pathogenic HCMV strain.

HCMV replicates *in vivo* in xenografts of human placental villi

A successful approach to overcome the obstacle to studies of HCMV replication *in vivo* has been to infect mice with severe combined immunodeficiency (SCID) transplanted with xenografts of human tissues (for humanized mouse models, see Chapter I.23). Infection of human fetal thymus/liver (Thy/Liv) under the mouse kidney capsule showed that medullary epithelial cells are prominent targets of HCMV replication (Mocarski *et al.*, 1993). In this regard, dramatic interstrain differences were evident in replication of clinical isolates and laboratory strains in Thy/Liv xenografts *in vivo* (Mocarski *et al.*, 1993; Brown *et al.*, 1995). Toledo replicates to high titres in implants, whereas AD169 and Towne fail to propagate in tissues *in vivo* (Mocarski *et al.*, 1993). AD169 lacks a 15-kbp segment of the viral genome that encodes at least 19 ORFs present in the genomes of all pathogenic clinical strains (Cha *et al.*, 1996). A deletion mutant of Toledo lacking these sequences, although exhibiting only a minor growth defect in fibroblasts, fails to replicate in Thy/Liv implants in SCID mice, evidence that genes in this region are central to infection *in vivo* (Wang *et al.*, 2005).

In a recent study (Tabata *et al.*, 2012), HCMV pathogenesis was investigated by examining the vascular effects of fetal cytotrophoblasts in infected human placental villi transplanted beneath the kidney capsules of SCID mice (Red-Horse *et al.*, 2006). In this system, differentiating cytotrophoblasts invade the renal parenchyma, induce apoptosis of endothelial cells in resident arteries and establish remodelled blood vessels with large luminal diameter. The placental cells also induce formation of lymphatic vessels comparable to those of the decidua in human pregnancy. Xenografts were infected with VR1814 and AD169, and differentiation of the implants *in vivo* was quantified. VR1814 replicated in cytotrophoblasts, which had a severely diminished capacity to invade and remodel resident arteries, whereas AD169 replicated poorly and implants retained invasive capacity (Fig. II.4.5A–C). Quantification of implant development in the kidney parenchyma showed significant differences between uninfected control and AD169- and VR1814-infected implants (Fig. II.4.5D). Infiltrating lymphatic endothelial cells proliferated, aggregated and failed to form lymphatic vessels in VR1814-infected implants. In contrast, AD169 grew

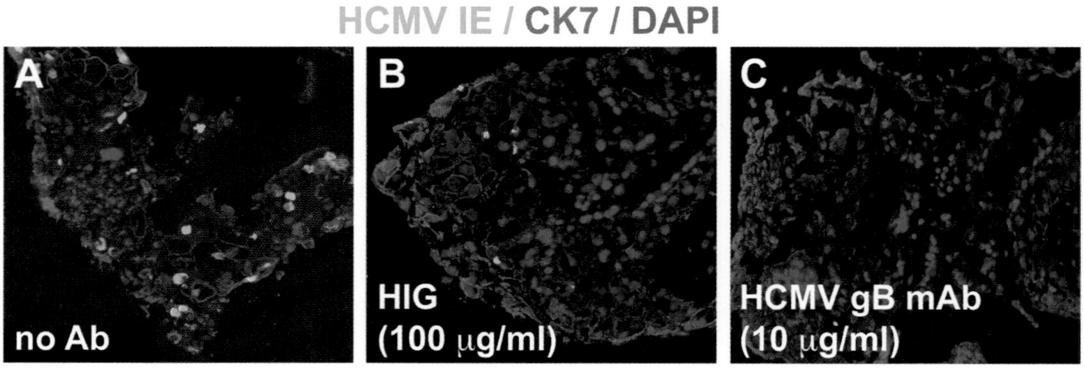

HCMV IE / CK7 / DAPI

A no Ab
B HIG (100 μg/ml)
C HCMV gB mAb (10 μg/ml)

Figure II.4.4 Neutralization of HCMV VR1814 infection in placental villous explants by hyperimmune globulin and a human mAb to gB. Chorionic villous explants showing cell column outgrowths of cytotrophoblasts infected with VR1814 and stained for IE1 and 2 protein expression and cytokeratin 7 (CK7) (A–C, 100x). Nuclei were counterstained with DAPI. Infected control without antibody (A) or with HCMV-specific HIG (B) or HCMV gB-specific mAb (C) that neutralize infection and rescue villous development.

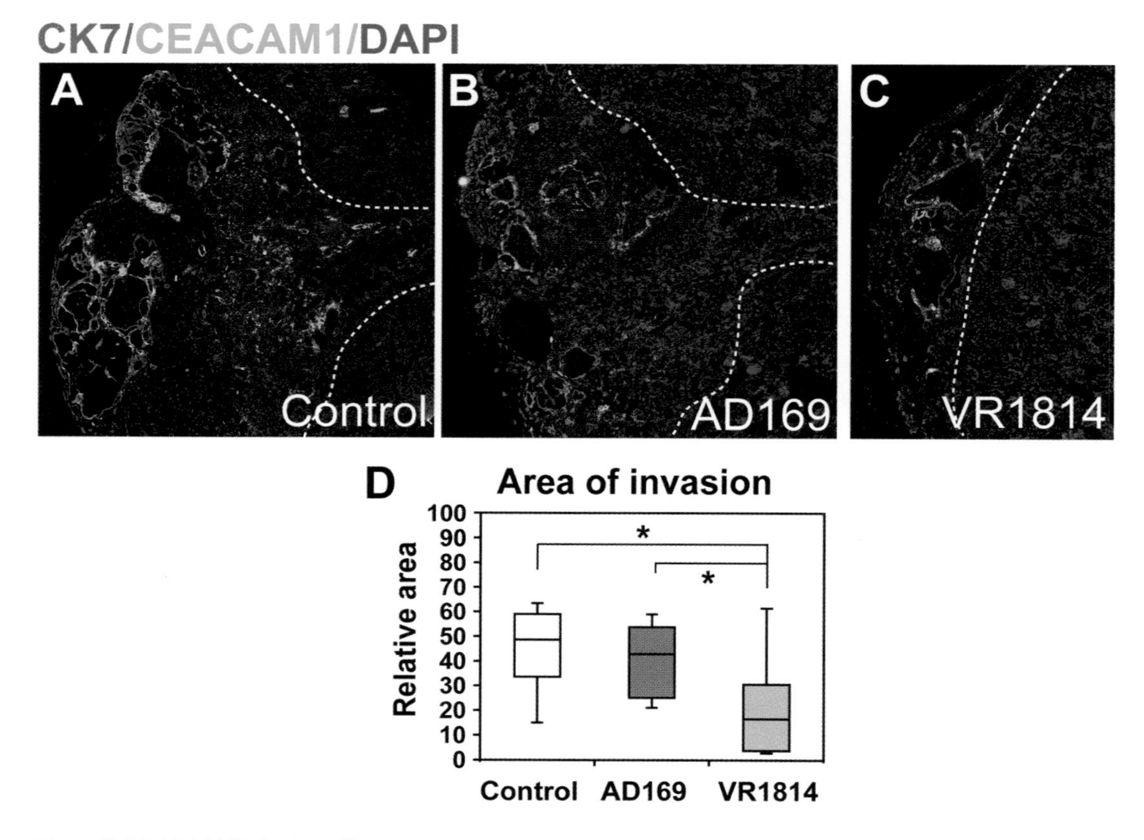

Figure II.4.5 VR1814 infection of human placental cells impairs cytotrophoblast invasion into the renal parenchyma in SCID mice. Human placental villi infected with AD169 and VR1814, transplanted under the kidney capsules and maintained for 3 weeks post infection. (A) Mock-infected control, (B) AD169-infected and (C) VR1814-infected implants. Cytotrophoblasts immunostained for cytokeratin 7 (CK7) or carcinoembryonic antigen-related cell adhesion molecule 1 (CEACAM1) and nuclei stained with DAPI. Original pictures were merged using Photoshop. Original magnification = 40×. (D) Box-whisker plots showing relative area of invasion, i.e. cytotrophoblast-occupied kidney parenchyma. Control implants ($n = 4$), AD169-infected implants ($n = 6$), VR1814-infected implants ($n = 14$), $P < 0.05$.

poorly and cytotrophoblasts retained invasive capacity, but some partially remodelled blood vessels incorporated lymphatic endothelial cells and were permeable to blood, suggesting paracrine factors could contribute to long-term viral pathogenesis *in vivo*.

Lymphangiogenic factors have autocrine and paracrine effects that transduce survival and migratory signals in lymphatic endothelial cells (Makinen *et al.*, 2001b; Red-Horse *et al.*, 2006) and maintain cytotrophoblast survival (Zhou *et al.*, 2002, 2003). Quantification of factors made in VR1814-infected cytotrophoblasts and explants *in vitro* showed VEGF-C and basic fibroblast growth factor levels were elevated as compared with mock-infected controls. In the model of cytotrophoblast-induced lymphangiogenesis *in vivo*, these factors and others promote migration of lymphatic endothelial cells (Red-Horse *et al.*, 2006). Several angiogenic factors are essential for the formation of functional vessels, and these must be expressed in a complementary and coordinated manner to balance stimulatory and inhibitory signals. Infected cytotrophoblasts could increase factors that promote cell proliferation, leading to aggregation that precludes formation of a functional lymphatic vasculature. In AD169-infected explants, there are elevated levels of both bioactive VEGF-A, which increases blood vessel permeability, and a soluble form of VEGFR-3 (sVEGFR-3), which binds to VEGF-C and prevents it from activating membrane-bound VEGFR-3 on lymphatic endothelium *in vitro* and lymphangiogenesis *in vivo* (Makinen *et al.*, 2001a). Moreover, VEGFR-3 antibodies and a sVEGFR-3–Ig fusion protein delivered by an adenovirus vector both inhibit angiogenesis by suppressing endothelial sprouting and vascular network formation (Tammela *et al.*, 2008). Dysregulation of these factors could result in blood vessels composed of both lymphatic endothelial cells and invasive cytotrophoblasts, as we observed in our implant model. This chimeric vasculature was hyperpermeable to blood, conceivably from incorporation of lymphatic endothelial cells that have junctions specialized for fluid uptake from surrounding tissue (Baluk *et al.*, 2007). It is likely that blood flow could exert pressure on the vessel wall, resulting in leakiness. In contrast to recent reports of an effect of IL-6 on lymphangiogenesis (Fiorentini *et al.*, 2010), endothelial cell survival and angiogenesis (Botto *et al.*, 2011), no significant differences were observed between AD169- and VR1814-infected cytotrophoblasts and villous explants. The results in infected placental xenografts also suggest that several paracrine factors could amplify pathology, mimicking effects seen *in vivo* that are not measurable in simple explant or cell culture systems, establishing the human placentation model as an experimental system with which to interrogate genetic determinants of viral pathogenesis. We also anticipate that this model will be invaluable for measuring neutralizing titres of monoclonal antibodies to HCMV and serological responses to novel vaccines to reduce congenital infection.

Perspectives

In this chapter, we described our current understanding of molecular mechanisms of pathogenesis in congenitally infected placentas and *in vitro* models of the differentiating human placenta, primary cytotrophoblasts and first trimester anchoring villous explants and xenografts infected with attenuated and low-passage pathogenic strains. Recently discovered is an important role for paracrine viral and cellular factors that suppress and/or stimulate cytotrophoblast differentiation, depending on ambient levels and temporal expression during gestation. New xenograft models utilizing intact tissues, such as the here discussed SCID mouse model with human placental villous implants, show tremendous promise for evaluating protection by antiviral antibodies and novel therapeutics.

Acknowledgements

Research in the Pereira laboratory was supported by US Public Health Service grants AI046657, AI073753, AI090200 (L.P.) and HD061890 (T.T.) and a grant from the International Aids Vaccine Initiative UCALI-FRSA1002.

References

Baldanti, F., Paolucci, S., Campanini, G., Sarasini, A., Percivalle, E., Revello, M.G., and Gerna, G. (2006). Human cytomegalovirus UL131A, UL130 and UL128 genes are highly conserved among field isolates. Arch. Virol. *151*, 1225–1233.

Baluk, P., Fuxe, J., Hashizume, H., Romano, T., Lashnits, E., Butz, S., Vestweber, D., Corada, M., Molendini, C., Dejana, E., *et al.* (2007). Functionally specialized junctions between endothelial cells of lymphatic vessels. J. Exp. Med. *204*, 2349–2362.

Bass, K.E., Morrish, D., Roth, I., Bhardwaj, D., Taylor, R., Zhou, Y., and Fisher, S.J. (1994). Human cytotrophoblast invasion is up-regulated by epidermal growth factor: evidence that paracrine factors modify this process. Dev. Biol. *164*, 550–561.

Bass, K.E., Li, H., Hawkes, S.P., Howard, E., Bullen, E., Vu, T.K., McMaster, M., Janatpour, M., and Fisher, S.J. (1997). Tissue inhibitor of metalloproteinase-3 expression is up-regulated during human cytotrophoblast invasion *in vitro*. Dev. Genet. *21*, 61–67.

Benirschke, K., and Kaufmann, P. (2000). Pathology of the Human Placenta, 4th edn (Springer, New York).

Benirschke, K., Mendoza, G.R., and Bazeley, P.L. (1974). Placental and fetal manifestations of cytomegalovirus infection. Virchows Arch. B. Cell Pathol. *16*, 121–139.

Boppana, S.B., Pass, R.F., Britt, W.J., Stagno, S., and Alford, C.A. (1992). Symptomatic congenital cytomegalovirus infection: neonatal morbidity and mortality. Pediatr. Infect. Dis. J. *11*, 93–99.

Botto, S., Streblow, D.N., DeFilippis, V., White, L., Kreklywich, C.N., Smith, P.P., and Caposio, P. (2011). IL-6 in human cytomegalovirus secretome promotes angiogenesis and survival of endothelial cells through the stimulation of survivin. Blood *117*, 352–361.

Britt, W.J. (1984). Neutralizing antibodies detect a disulfide-linked glycoprotein complex within the envelope of human cytomegalovirus. Virology *135*, 369–378.

Brown, J.M., Kaneshima, H., and Mocarski, E.S. (1995). Dramatic interstrain differences in the replication of human cytomegalovirus in SCID-hu mice. J. Infect. Dis. *171*, 1599–1603.

Cha, T.A., Tom, E., Kemble, G.W., Duke, G.M., Mocarski, E.S., and Spaete, R.R. (1996). Human cytomegalovirus clinical isolates carry at least 19 genes not found in laboratory strains. J. Virol. *70*, 78–83.

Chang, W.L., Baumgarth, N., Yu, D., and Barry, P.A. (2004). Human cytomegalovirus-encoded interleukin-10 homolog inhibits maturation of dendritic cells and alters their functionality. J. Virol. *78*, 8720–8731.

Cooper, J.C., Sharkey, A.M., Charnock-Jones, D.S., Palmer, C.R., and Smith, S.K. (1996). VEGF mRNA levels in placentae from pregnancies complicated by pre-eclampsia. Br. J. Obstet. Gynaecol. *103*, 1191–1196.

Cross, J.C., Werb, Z., and Fisher, S.J. (1994). Implantation and the placenta: key pieces of the development puzzle. Science *266*, 1508–1518.

Damsky, C.H., and Fisher, S.J. (1998). Trophoblast pseudo-vasculogenesis: faking it with endothelial adhesion receptors. Curr. Opin. Cell Biol. *10*, 660–666.

Damsky, C.H., Fitzgerald, M.L., and Fisher, S.J. (1992). Distribution patterns of extracellular matrix components and adhesion receptors are intricately modulated during first trimester cytotrophoblast differentiation along the invasive pathway, *in vivo*. J. Clin. Invest. *89*, 210–222.

Damsky, C.H., Librach, C., Lim, K.H., Fitzgerald, M.L., McMaster, M.T., Janatpour, M., Zhou, Y., Logan, S.K., and Fisher, S.J. (1994). Integrin switching regulates normal trophoblast invasion. Development *120*, 3657–3666.

Demmler, G.J. (1996). Congenital cytomegalovirus infection and disease. Adv. Pediatr. Infect. Dis. *11*, 135–162.

Feire, A.L., Koss, H., and Compton, T. (2004). Cellular integrins function as entry receptors for human cytomegalovirus via a highly conserved disintegrin-like domain. Proc. Natl. Acad. Sci. U.S.A. *101*, 15470–15475.

Feire, A.L., Roy, R.M., Manley, K., and Compton, T. (2010). The glycoprotein B disintegrin-like domain binds beta 1 integrin to mediate cytomegalovirus entry. J. Virol. *84*, 10026–10037.

Ferrara, N., Gerber, H.P., and LeCouter, J. (2003). The biology of VEGF and its receptors. Nat. Med. *9*, 669–676.

Fiorentini, S., Luganini, A., Dell'oste, V., Lorusso, B., Cervi, E., Caccuri, F., Bonardelli, S., Landolfo, S., Caruso, A., and Gribaudo, G. (2010). Human cytomegalovirus productively infects lymphatic endothelial cells and induces a secretome that promotes angiogenesis and lymphangiogenesis through interleukin-6 and granulocyte–macrophage colony-stimulating factor. J. Gen. Virol. *92*, 650–660.

Fisher, S., Genbacev, O., Maidji, E., and Pereira, L. (2000). Human cytomegalovirus infection of placental cytotrophoblasts *in vitro* and in utero: implications for transmission and pathogenesis. J. Virol. *74*, 6808–6820.

Fournier, T., Guibourdenche, J., Handschuh, K., Tsatsaris, V., Rauwel, B., Davrinche, C., and Evain-Brion, D. (2011). PPARgamma and human trophoblast differentiation. J. Reprod. Immunol. *90*, 41–49.

Fowler, K.B., Stagno, S., Pass, R.F., Britt, W.J., Boll, T.J., and Alford, C.A. (1992). The outcome of congenital cytomegalovirus infection in relation to maternal antibody status. N. Engl. J. Med. *326*, 663–667.

Fowler, K.B., Stagno, S., and Pass, R.F. (2003). Maternal immunity and prevention of congenital cytomegalovirus infection. JAMA *289*, 1008–1011.

Garcia, A.G., Fonseca, E.F., Marques, R.L., and Lobato, Y.Y. (1989). Placental morphology in cytomegalovirus infection. Placenta *10*, 1–18.

Genbacev, O., Joslin, R., Damsky, C.H., Polliotti, B.M., and Fisher, S.J. (1996). Hypoxia alters early gestation human cytotrophoblast differentiation/invasion *in vitro* and models the placental defects that occur in preeclampsia. J. Clin. Invest. *97*, 540–550.

Genbacev, O., Zhou, Y., Ludlow, J.W., and Fisher, S.J. (1997). Regulation of human placental development by oxygen tension. Science *277*, 1669–1672.

Genbacev, O., McMaster, M.T., Lazic, J., Nedeljkovic, S., Cvetkovic, M., Joslin, R., and Fisher, S.J. (2000). Concordant *in situ* and *in vitro* data show that maternal cigarette smoking negatively regulates placental cytotrophoblast passage through the cell cycle. Reprod. Toxicol. *14*, 495–506.

Genbacev, O., McMaster, M.T., Zdravkovic, T., and Fisher, S.J. (2003). Disruption of oxygen-regulated responses underlies pathological changes in the placentas of women who smoke or who are passively exposed to smoke during pregnancy. Reprod. Toxicol. *17*, 509–518.

Gerna, G., Sarasini, A., Patrone, M., Percivalle, E., Fiorina, L., Campanini, G., Gallina, A., Baldanti, F., and Revello, M.G. (2008). Human cytomegalovirus serum neutralizing antibodies block virus infection of endothelial/epithelial cells, but not fibroblasts, early during primary infection. J. Gen. Virol. *89*, 853–865.

Grangeot-Keros, L., Mayaux, M.J., Lebon, P., Freymuth, F., Eugene, G., Stricker, R., and Dussaix, E. (1997). Value of cytomegalovirus (CMV) IgG avidity index for the diagnosis of primary CMV infection in pregnant women. J. Infect. Dis. *175*, 944–946.

Griffiths, P.D., and Baboonian, C. (1984). A prospective study of primary cytomegalovirus infection during pregnancy: final report. Br. J. Obstet. Gynaecol. *91*, 307–315.

Halwachs-Baumann, G., Wilders-Truschnig, M., Desoye, G., Hahn, T., Kiesel, L., Klingel, K., Rieger, P., Jahn, G., and Sinzger, C. (1998). Human trophoblast cells are permissive to the complete replicative cycle of human cytomegalovirus. J. Virol. *72*, 7598–7602.

Hemmings, D.G., Kilani, R., Nykiforuk, C., Preiksaitis, J., and Guilbert, L.J. (1998). Permissive cytomegalovirus infection of primary villous term and first trimester trophoblasts. J. Virol. *72*, 4970–4979.

Hiby, S.E., Walker, J.J., O'Shaughnessy K.M., Redman, C.W., Carrington, M., Trowsdale, J., and Moffett, A. (2004). Combinations of maternal KIR and fetal HLA-C genes influence the risk of preeclampsia and reproductive success. J. Exp. Med. *200*, 957–965.

Hiby, S.E., Apps, R., Sharkey, A.M., Farrell, L.E., Gardner, L., Mulder, A., Claas, F.H., Walker, J.J., Redman, C.W., Morgan, L., *et al.* (2010). Maternal activating KIRs protect against human reproductive failure mediated by fetal HLA-C2. J. Clin. Invest. *120*, 4102–4110.

Hornig, C., Behn, T., Bartsch, W., Yayon, A., and Weich, H.A. (1999). Detection and quantification of complexed and free soluble human vascular endothelial growth factor receptor-1 (sVEGFR-1) by ELISA. J. Immunol. Methods *226*, 169–177.

Huber, M.T., and Compton, T. (1997). Characterization of a novel third member of the human cytomegalovirus glycoprotein H-glycoprotein L complex. J. Virol. *71*, 5391–5398.

Iwasenko, J.M., Howard, J., Arbuckle, S., Graf, N., Hall, B., Craig, M.E., and Rawlinson, W.D. (2011). Human cytomegalovirus infection is detected frequently in stillbirths and is associated with fetal thrombotic vasculopathy. J. Infect. Dis. *203*, 1526–1533.

Jackson, M.R., Carney, E.W., Lye, S.J., and Ritchie, J.W. (1994). Localization of two angiogenic growth factors (PDECGF and VEGF) in human placentae throughout gestation. Placenta *15*, 341–353.

Jones, B.C., Logsdon, N.J., Josephson, K., Cook, J., Barry, P.A., and Walter, M.R. (2002). Crystal structure of human cytomegalovirus IL-10 bound to soluble human IL-10R1. Proc. Natl. Acad. Sci. U.S.A. *99*, 9404–9409.

Kendall, R.L., and Thomas, K.A. (1993). Inhibition of vascular endothelial cell growth factor activity by an endogenously encoded soluble receptor. Proc. Natl. Acad. Sci. U.S.A. *90*, 10705–10709.

Kovalevskaya, G., Genbacev, O., Fisher, S.J., Caceres, E., and O'Connor, J.F. (2002). Trophoblast origin of hCG isoforms: cytotrophoblasts are the primary source of choriocarcinoma-like hCG. Mol. Cell. Endocrinol. *194*, 147–155.

Kovats, S., Main, E.K., Librach, C., Stubblebine, M., Fisher, S.J., and DeMars, R. (1990). A class I antigen, HLA-G, expressed in human trophoblasts. Science *248*, 220–223.

Kruger, H., and Arias-Stella, J. (1970). The placenta and the newborn infant at high altitudes. Am. J. Obstet. Gynecol. *106*, 586–591.

La Torre, R., Nigro, G., Mazzocco, M., Best, A.M., and Adler, S.P. (2006). Placental enlargement in women with primary maternal cytomegalovirus infection is associated with fetal and neonatal disease. Clin. Infect. Dis. *43*, 994–1000.

Librach, C.L., Werb, Z., Fitzgerald, M.L., Chiu, K., Corwin, N.M., Esteves, R.A., Grobelny, D., Galardy, R., Damsky, C.H., and Fisher, S.J. (1991). 92-kD type IV collagenase mediates invasion of human cytotrophoblasts. J. Cell Biol. *113*, 437–449.

Librach, C.L., Feigenbaum, S.L., Bass, K.E., Cui, T.Y., Verastas, N., Sadovsky, Y., Quigley, J.P., French, D.L., and Fisher, S.J. (1994). Interleukin-1 beta regulates human cytotrophoblast metalloproteinase activity and invasion *in vitro*. J. Biol. Chem. *269*, 17125–17131.

Lopper, M., and Compton, T. (2004). Coiled-coil domains in glycoproteins B and H are involved in human cytomegalovirus membrane fusion. J. Virol. *78*, 8333–8341.

Macagno, A., Bernasconi, N.L., Vanzetta, F., Dander, E., Sarasini, A., Revello, M.G., Gerna, G., Sallusto, F., and Lanzavecchia, A. (2010). Isolation of human monoclonal antibodies that potently neutralize human cytomegalovirus infection by targeting different epitopes on the gH/gL/UL128–131A complex. J. Virol. *84*, 1005–1013.

McDonagh, S., Maidji, E., Chang, H.T., and Pereira, L. (2006). Patterns of human cytomegalovirus infection in term placentas: a preliminary analysis. J. Clin. Virol. *35*, 210–215.

McDonagh, S., Maidji, E., Ma, W., Chang, H.T., Fisher, S., and Pereira, L. (2004). Viral and bacterial pathogens at the maternal–fetal interface. J. Infect. Dis. *190*, 826–834.

McMaster, M.T., Librach, C.L., Zhou, Y., Lim, K.H., Janatpour, M.J., DeMars, R., Kovats, S., Damsky, C., and Fisher, S.J. (1995). Human placental HLA-G expression is restricted to differentiated cytotrophoblasts. J. Immunol. *154*, 3771–3778.

Maidji, E., McDonagh, S., Genbacev, O., Tabata, T., and Pereira, L. (2006). Maternal antibodies enhance or prevent cytomegalovirus infection in the placenta by neonatal Fc receptor-mediated transcytosis. Am. J. Pathol. *168*, 1210–1226.

Maidji, E., Genbacev, O., Chang, H.T., and Pereira, L. (2007). Developmental regulation of human cytomegalovirus receptors in cytotrophoblasts correlates with distinct replication sites in the placenta. J. Virol. *81*, 4701–4712.

Maidji, E., Nigro, G., Tabata, T., McDonagh, S., Nozawa, N., Shiboski, S., Muci, S., Anceschi, M.M., Aziz, N., Adler, S.P., *et al.* (2010). Antibody treatment promotes compensation for human cytomegalovirus-induced pathogenesis and a hypoxia-like condition in placentas with congenital infection. Am. J. Pathol. *177*, 1298–1310.

Makinen, T., Jussila, L., Veikkola, T., Karpanen, T., Kettunen, M.I., Pulkkanen, K.J., Kauppinen, R., Jackson, D.G., Kubo, H., Nishikawa, S., *et al.* (2001a). Inhibition of lymphangiogenesis with resulting lymphedema in transgenic mice expressing soluble VEGF receptor-3. Nat. Med. *7*, 199–205.

Makinen, T., Veikkola, T., Mustjoki, S., Karpanen, T., Catimel, B., Nice, E.C., Wise, L., Mercer, A., Kowalski, H., Kerjaschki, D., *et al.* (2001b). Isolated lymphatic endothelial cells transduce growth, survival and migratory signals via the VEGF-C/D receptor VEGFR-3. EMBO J. *20*, 4762–4773.

Malek, A., Sager, R., Kuhn, P., Nicolaides, K.H., and Schneider, H. (1996). Evolution of maternofetal transport of immunoglobulins during human pregnancy. Am. J. Reprod. Immunol. *36*, 248–255.

Mayhew, T.M., Bowles, C., and Yucel, F. (2002). Hypobaric hypoxia and villous trophoblast: evidence that human pregnancy at high altitude (3600 m) perturbs epithelial turnover and coagulation-fibrinolysis in the intervillous space. Placenta *23*, 154–162.

Mocarski, E.S., Bonyhadi, M., Salimi, S., McCune, J.M., and Kaneshima, H. (1993). Human cytomegalovirus in a SCID-hu mouse: thymic epithelial cells are prominent targets of viral replication. Proc. Natl. Acad. Sci. U.S.A. *90*, 104–108.

Monif, G.R., and Dische, R.M. (1972). Viral placentitis in congenital cytomegalovirus infection. Am. J. Clin. Pathol. *58*, 445–449.

Navarro, D., Paz, P., Tugizov, S., Topp, K., La Vail, J., and Pereira, L. (1993). Glycoprotein B of human cytomegalovirus promotes virion penetration into cells, transmission of infection from cell to cell, and fusion of infected cells. Virology *197*, 143–158.

Navarro, D., Lennette, E., Tugizov, S., and Pereira, L. (1997). Humoral immune response to functional regions of human cytomegalovirus glycoprotein B. J. Med. Virol. *52*, 451–459.

Nigro, G., Adler, S.P., La Torre, R., and Best, A.M. (2005). Passive immunization during pregnancy for congenital cytomegalovirus infection. N. Engl. J. Med. *353*, 1350–1362.

Norwitz, E.R., Schust, D.J., and Fisher, S.J. (2001). Implantation and the survival of early pregnancy. N. Engl. J. Med. *345*, 1400–1408.

Noyola, D.E., Demmler, G.J., Nelson, C.T., Griesser, C., Williamson, W.D., Atkins, J.T., Rozelle, J., Turcich, M., Llorente, A.M., Sellers-Vinson, S., *et al.* (2001). Early predictors of neurodevelopmental outcome in symptomatic congenital cytomegalovirus infection. J. Pediatr. *138*, 325–331.

Nozawa, N., Fang-Hoover, J., Tabata, T., Maidji, E., and Pereira, L. (2009). Cytomegalovirus-specific, high-avidity IgG with neutralizing activity in maternal circulation enriched in the fetal bloodstream. J. Clin. Virol. *46*, S58–63.

Pereira, L. (2011). Have we overlooked congenital cytomegalovirus infection as a cause of stillbirth? J. Infect. Dis. *203*, 1510–1512.

Pereira, L., Maidji, E., McDonagh, S., Genbacev, O., and Fisher, S. (2003). Human cytomegalovirus transmission from the uterus to the placenta correlates with the presence of pathogenic bacteria and maternal immunity. J. Virol. *77*, 13301–13314.

Pereira, L., Maidji, E., McDonagh, S., and Tabata, T. (2005). Insights into viral transmission at the uterine–placental interface. Trends Microbiol. *13*, 164–174.

Pfarrer, C., Macara, L., Leiser, R., and Kingdom, J. (1999). Adaptive angiogenesis in placentas of heavy smokers. Lancet *354*, 303.

Prakobphol, A., Genbacev, O., Gormley, M., Kapidzic, M., and Fisher, S.J. (2006). A role for the L-selectin adhesion system in mediating cytotrophoblast emigration from the placenta. Dev. Biol. *298*, 107–117.

Rasmussen, L.E., Nelson, R.M., Kelsall, D.C., and Merigan, T.C. (1984). Murine monoclonal antibody to a single protein neutralizes the infectivity of human cytomegalovirus. Proc. Natl. Acad. Sci. U.S.A. *81*, 876–880.

Rauwel, B., Mariame, B., Martin, H., Nielsen, R., Allart, S., Pipy, B., Mandrup, S., Devignes, M.D., Evain-Brion, D., Fournier, T., *et al.* (2010). Activation of peroxisome proliferator-activated receptor gamma by human cytomegalovirus for de novo replication impairs migration and invasiveness of cytotrophoblasts from early placentas. J. Virol. *84*, 2946–2954.

Red-Horse, K., Zhou, Y., Genbacev, O., Prakobphol, A., Foulk, R., McMaster, M., and Fisher, S.J. (2004). Trophoblast differentiation during embryo implantation and formation of the maternal–fetal interface. J. Clin. Invest. *114*, 744–754.

Red-Horse, K., Rivera, J., Schanz, A., Zhou, Y., Winn, V., Kapidzic, M., Maltepe, E., Okazaki, K., Kochman, R., Vo, K.C., *et al.* (2006). Cytotrophoblast induction of arterial apoptosis and lymphangiogenesis in an *in vivo* model of human placentation. J. Clin. Invest. *116*, 2643–2652.

Revello, M.G., Gorini, G., and Gerna, G. (2004). Clinical evaluation of a chemiluminescence immunoassay for determination of immunoglobulin g avidity to human cytomegalovirus. Clin. Diagn. Lab. Immunol. *11*, 801–805.

Roth, I., and Fisher, S.J. (1999). IL-10 is an autocrine inhibitor of human placental cytotrophoblast MMP-9 production and invasion. Dev. Biol. *205*, 194–204.

Roth, I., Corry, D.B., Locksley, R.M., Abrams, J.S., Litton, M.J., and Fisher, S.J. (1996). Human placental cytotrophoblasts produce the immunosuppressive cytokine interleukin 10. J. Exp. Med. *184*, 539–548.

Shiraishi, S., Nakagawa, K., Kinukawa, N., Nakano, H., and Sueishi, K. (1996). Immunohistochemical localization of vascular endothelial growth factor in the human placenta. Placenta *17*, 111–121.

Shore, V.H., Wang, T.H., Wang, C.L., Torry, R.J., Caudle, M.R., and Torry, D.S. (1997). Vascular endothelial growth factor, placenta growth factor and their receptors in isolated human trophoblast. Placenta *18*, 657–665.

Simister, N.E. (2003). Placental transport of immunoglobulin G. Vaccine *21*, 3365–3369.

Simister, N.E., and Story, C.M. (1997). Human placental Fc receptors and the transmission of antibodies from mother to fetus. J. Reprod. Immunol. *37*, 1–23.

Simister, N.E., Story, C.M., Chen, H.L., and Hunt, J.S. (1996). An IgG-transporting Fc receptor expressed in the syncytiotrophoblast of human placenta. Eur. J. Immunol. *26*, 1527–1531.

Spencer, J.V., Lockridge, K.M., Barry, P.A., Lin, G., Tsang, M., Penfold, M.E., and Schall, T.J. (2002). Potent immunosuppressive activities of cytomegalovirus-encoded interleukin-10. J. Virol. *76*, 1285–1292.

Stagno, S., Pass, R.F., Cloud, G., Britt, W.J., Henderson, R.E., Walton, P.D., Veren, D.A., Page, F., and Alford, C.A. (1986). Primary cytomegalovirus infection in pregnancy. Incidence, transmission to fetus, and clinical outcome. JAMA *256*, 1904–1908.

Syridou, G., Spanakis, N., Konstantinidou, A., Piperaki, E.T., Kafetzis, D., Patsouris, E., Antsaklis, A., and Tsakris, A. (2008). Detection of cytomegalovirus, parvovirus B19 and herpes simplex viruses in cases of intrauterine fetal death: association with pathological findings. J. Med. Virol. *80*, 1776–1782.

Tabata, T., Petitt, M., Fang-Hoover, J., Rivera, J., Nozawa, N., Shiboski, S., Inoue, N., and Pereira, L. (2012). Cytomegalovirus impairs cytotrophoblast-induced lymphangiogenesis and vascular remodeling in an *in vivo* human placentation model. Am. J. Pathol. *18*, 1540–1559.

Tabata, T., McDonagh, S., Kawakatsu, H., and Pereira, L. (2007). Cytotrophoblasts infected with a pathogenic human cytomegalovirus strain dysregulate cell-matrix and cell–cell adhesion molecules: a quantitative analysis. Placenta *28*, 527–537.

Tabata, T., Kawakatsu, H., Maidji, E., Sakai, T., Sakai, K., Fang-Hoover, J., Aiba, M., Sheppard, D., and Pereira, L. (2008). Induction of an epithelial integrin alphavbeta6 in human cytomegalovirus-infected endothelial cells leads to activation of transforming growth factor-beta1 and increased collagen production. Am. J. Pathol. *172*, 1127–1140.

Tammela, T., Zarkada, G., Wallgard, E., Murtomaki, A., Suchting, S., Wirzenius, M., Waltari, M., Hellstrom, M., Schomber, T., Peltonen, R., *et al.* (2008). Blocking VEGFR-3 suppresses angiogenic sprouting and vascular network formation. Nature *454*, 656–660.

Tugizov, S., Navarro, D., Paz, P., Wang, Y., Qadri, I., and Pereira, L. (1994). Function of human cytomegalovirus glycoprotein B: syncytium formation in cells constitutively expressing gB is blocked by virus-neutralizing antibodies. Virology *201*, 263–276.

Tugizov, S., Maidji, E., and Pereira, L. (1996). Role of apical and basolateral membranes in replication of human cytomegalovirus in polarized retinal pigment epithelial cells. J. Gen. Virol. *77*, 61–74.

Tugizov, S., Maidji, E., Xiao, J., and Pereira, L. (1999). An acidic cluster in the cytosolic domain of human cytomegalovirus glycoprotein B is a signal for endocytosis from the plasma membrane. J. Virol. *73*, 8677–8688.

Wang, D., and Shenk, T. (2005). Human cytomegalovirus virion protein complex required for epithelial and endothelial cell tropism. Proc. Natl. Acad. Sci. U.S.A. *102*, 18153–18158.

Wang, W., Taylor, S.L., Leisenfelder, S.A., Morton, R., Moffat, J.F., Smirnov, S., and Zhu, H. (2005). Human cytomegalovirus genes in the 15-kilobase region are required for viral replication in implanted human tissues in SCID mice. J. Virol. *79*, 2115–2123.

Weisblum, Y., Panet, A., Zakay-Rones, Z., Haimov-Kochman, R., Goldman-Wohl, D., Ariel, I., Falk, H., Natanson-Yaron, S., Goldberg, M.D., Gilad, R., *et al.* (2011). Modeling of human cytomegalovirus maternal–fetal transmission in a novel decidual organ culture. J. Virol. *85*, 13204–13213.

Williams, L.A., Evans, S.F., and Newnham, J.P. (1997). Prospective cohort study of factors influencing the relative weights of the placenta and the newborn infant. BMJ *314*, 1864–1868.

Winn, V.D., Haimov-Kochman, R., Paquet, A.C., Yang, Y.J., Madhusudhan, M.S., Gormley, M., Feng, K.T., Bernlohr, D.A., McDonagh, S., Pereira, L., *et al.* (2007). Gene expression profiling of the human maternal–fetal interface reveals dramatic changes between midgestation and term. Endocrinology *148*, 1059–1079.

Yamamoto-Tabata, T., McDonagh, S., Chang, H.T., Fisher, S., and Pereira, L. (2004). Human cytomegalovirus interleukin-10 down-regulates metalloproteinase activity and impairs endothelial cell migration and placental cytotrophoblast invasiveness *in vitro*. J. Virol. *78*, 2831–2840.

Zdravkovic, T., Genbacev, O., McMaster, M.T., and Fisher, S.J. (2005). The adverse effects of maternal smoking on the human placenta: a review. Placenta *26*, 81–86.

Zhou, Y., Fisher, S.J., Janatpour, M., Genbacev, O., Dejana, E., Wheelock, M., and Damsky, C.H. (1997). Human cytotrophoblasts adopt a vascular phenotype as they differentiate. A strategy for successful endovascular invasion? J. Clin. Invest. *99*, 2139–2151.

Zhou, Y., Bellingard, V., Feng, K.T., McMaster, M., and Fisher, S.J. (2003). Human cytotrophoblasts promote endothelial survival and vascular remodeling through secretion of Ang2, PlGF, and VEGF-C. Dev. Biol. *263*, 114–125.

Zhou, Y., McMaster, M., Woo, K., Janatpour, M., Perry, J., Karpanen, T., Alitalo, K., Damsky, C., and Fisher, S.J. (2002). Vascular endothelial growth factor ligands and receptors that regulate human cytotrophoblast survival are dysregulated in severe preeclampsia and hemolysis, elevated liver enzymes, and low platelets syndrome. Am. J. Pathol. *160*, 1405–1423.

The Guinea Pig Model of Congenital Cytomegalovirus Infection

Alistair McGregor, Michael A. McVoy and Mark R. Schleiss

Abstract

In the study of the cytomegaloviruses of small mammals, the guinea pig cytomegalovirus (GPCMV) has distinctive advantages. These attributes are chiefly related to the ability of GPCMV to cross the placenta, causing infection *in utero*. For this reason, the model is well-suited to the study of vaccines and antiviral drugs for preventing or modifying the outcome of congenital CMV infection, and for the study of the role of viral genes in the pathogenesis of maternal, placental and fetal infection. Progress in GPCMV studies has been hampered by a lack of detailed characterization of the viral genome and gene products, and a lack of immunological reagents for animal study. However, recent efforts by several investigators have resulted in improved characterization of the GPCMV genome, and this information has in turn been applied to *in vivo* vaccine and pathogenesis studies. As is the case for human cytomegalovirus, the GPCMV glycoprotein B (gB) has proven to be a major target of humoral immune responses, and purified recombinant forms of gB have recently been shown to be effective vaccines in the guinea pig model. The study of viral genes has been facilitated by the availability of bacterial artificial chromosome (BAC) clones of the GPCMV genome in *E. coli*, which has enabled mutagenesis studies of the roles of specific viral genes in pathogenesis and immunity. Insights from the ongoing characterization of the GPCMV model should prove germane to the understanding of the correlates of protective immunity for the fetus. The study of protective immunity in this model as well as potential viral determinants of pathogenesis could facilitate a better understanding of congenital HCMV disease, and may help inform and direct future HCMV intervention trials designed to protect the fetus from the disabling effects of this infection.

Introduction

A vaccine for HCMV infection is a major public health priority, given the disability that congenital infection can cause in newborn infants (Plotkin, 1999; Arvin *et al.*, 2004). Ideally, HCMV vaccines would undergo preclinical evaluation for efficacy in small animal models of infection. Unfortunately, CMV infection is highly species specific, and given the divergence of CMVs at the molecular level, HCMV vaccines can be evaluated only for immunogenicity, and not for protective efficacy, in animal models (Staczek, 1990; Schleiss, 2002, 2006). Among the animal cytomegaloviruses, the rhesus macaque CMV (RhCMV) provides a highly relevant model of fetal and neonatal HCMV disease, characterized by placental infection and fetal injury (Tarantal *et al.*, 1998; Lockridge *et al.*, 1999, 2000; Barry *et al.*, 2006; Barry and William Chang, 2007; Yue and Barry, 2008; see also Chapter II.22), but the expense of these primates, and the difficulties in establishing RhCMV-seronegative animal colonies (Barry and Strelow, 2008), makes widespread use of this model impractical for vaccine studies. Mouse and rat cytomegaloviruses have been studied as models of CMV disease, but these CMVs generally do not cross the placentas of their host species, and are therefore not useful for vaccine studies that target prevention of congenital infection (Johnson, 1969; Woolf *et al.*, 2007). Fortunately, the guinea pig CMV (GPCMV) provides a uniquely useful model that is very relevant to HCMV vaccine studies. In contrast to the cytomegaloviruses of other small mammals, GPCMV is unique in its ability to cross the placenta and cause fetal infection. Thus, there has been extensive interest over many years in the development of this model for the study of immunity and pathogenesis. This chapter provides an overview of research on GPCMV, a summary of the uniquely valuable features of this model, and an update on recent advances in the study of this virus that have facilitated ongoing vaccine and pathogenesis study.

History of the GPCMV

GPCMV was first characterized approximately 85 years ago, and, interestingly, the history of the GPCMV parallels in many ways the description of the syndrome of congenital HCMV infection (Bia *et al.*, 1983; 1984a). GPCMV was first recognized by histopathological analysis, following identification of typical viral inclusions in the guinea pig salivary gland (Jackson, 1920). As was initially the case for HCMV (Vonglahn and Pappenheimer, 1925), these early studies mistakenly identified the virus as a protozoan (see Editor's Preface). By the late 1920s, a viral aetiology of these lesions was correctly postulated by Cole and Kuttner, when they noted that the infectious agent produced histological changes very similar to those observed in cells infected with herpes simplex virus: the cells were described as 'greatly swollen epithelial cells with nuclei having the same characters as the nuclei of the atypical cells in the lesions of herpes simplex' (Cole and Kuttner, 1926). These investigators also showed that injection of an emulsion of submaxillary glands of adult guinea pigs into the brains of young guinea pigs could produce a fatal meningitis, characterized by a 'large number of cells' possessing 'all the characteristics of the abnormal cells of herpes simplex'. It was later shown that salivary gland homogenates containing this agent were infectious, and could produce interstitial pneumonitis following intratracheal instillation in guinea pigs, establishing both the potential for this agent to cause disease in animals as well as the importance of the salivary gland as a reservoir for infection (Kuttner and T'ung, 1935). A series of subsequent reports identified this agent as a potential cause of systemic disease with attendant end-organ pathology, including brain (Hudson and Markham, 1932; Kuttner and Wang, 1934; Markham, 1938; Pappenheimer and Slanetz, 1942). Very early following the establishment of these animal models for the study of GPCMV, there was interest in the nature of the immune response to infection, foreshadowing later studies of protective immunity in this model. In what was in essence the first vaccine study in this model, Kuttner showed that young guinea pigs that received injections of heat-killed virus were protected against subsequent disease following intracerebral inoculation with the live agent (Kuttner, 1927). Andrewes showed in 1930 that mixing serum from infected animals with emulsions of infected salivary gland could neutralize the ability of these emulsions to produce salivary inclusions in recipients, and could protect them from the effects of intracerebral inoculation of virus (Andrewes, 1930).

Markham and Hudson were the first investigators to study GPCMV infection of the placenta. They found that intraplacental inoculation of pregnant dams led to fetal infection, characterized by generalized infection of brain, liver, lungs, spleen, and thymus. They also noted that fetal infection could lead to maternal infection and end-organ disease, and postulated that this was due to fetal macrophages entering the maternal circulation (Markham and Hudson, 1936). In a later paper describing the biology of GPCMV infection, Markham astutely noted the similarity between the pathology observed in the guinea pig and a disease that had been described in the human fetus (Farber and Wolbach, 1932; Markham, 1938). This human disease, which was associated with stillbirths, would later be more fully characterized as cytomegalic inclusion disease (CID) of the newborn. Demonstration of the permissiveness of the placenta for GPCMV infection as well as the ability of the virus to be transmitted to the fetus following direct inoculation were important milestones in the development of this model, as was the successful propagation of the virus, isolated from guinea pig salivary glands, in cell culture in 1957 (Hartley *et al.*, 1957). Hartley and colleagues provided the American Type Culture Collection with their strain of GPCMV (22122), which has been used in virtually all subsequent studies reported to date, in multiple laboratories, and remains the only fully characterized isolate.

Pathogenesis of GPCMV infection

Most publications of experimentally modelled GPCMV disease have used the 22122 strain of virus (Hartley *et al.*, 1957). Workpools for animal inoculation may be derived either from propagation in tissue culture (GPCMV-TC), or, more commonly, from homogenates of salivary glands (GPCMV-SG): either preparation is capable of inducing disease in guinea pigs, including congenital infection, but SG preparations are typically more virulent, and cause disease at a lower inoculum. The basis for the apparent increased virulence of SG passage is not completely characterized, but may relate to the maintenance in SG stocks of a 1.6-kbp locus from the viral genome that is preferentially deleted following serial TC passage in fibroblasts, reviewed later in this chapter (Nozawa *et al.*, 2008). Guinea pig pathogenesis experiments can be performed in outbred (Hartley) strain guinea pigs, and also in inbred (strain 2, JY-9) guinea pigs. An advantage of using inbred guinea pig strains is that this allows, in principle, more rigorous analysis of the cellular immune factors that participate in intrauterine CMV infections (Griffith *et al.*, 1986). The disease manifestations of GPCMV infection have been well-characterized in pregnant animals, but GPCMV is also useful for pathogenesis studies in non-pregnant guinea pigs (described below). Viral inoculation can be performed by subcutaneous (s.c.), intraperitoneal (i.p.), intracardiac (i.card.), intravaginal (i.vag.), or intranasal (i.n.) routes. Although infection at a mucosal surface may be the most relevant with respect to natural modes of acquisition, inoculation by the sc route is generally the most technically convenient approach, and is less stressful to the pregnant animal. In addition, although placental and fetal infections are known to occur after infection of pregnant animals by all of these routes, the transfer of GPCMV to placentas and fetuses appears to be more efficient in mothers inoculated by the s.c. route (Griffith *et al.*, 1990). However, this route bypasses the mucosal barrier that is present during natural routes of exposure, and thus fails to model the potential role of mucosal immunity in infection (including salivary neutralizing antibody responses, as well as submucosal innate and cellular responses).

Following experimental infection of guinea pigs with GPCMV, viraemia ensues, irrespective of the route of challenge, and this appears to be the mechanism by which end-organ disease (including of the placental-fetal unit) occurs. Viraemia persists for approximately 10 days post inoculation of non-pregnant animals, and generalized infection and disease involving the lungs, spleen, liver, kidney, thymus, pancreas, and brain can be demonstrated for approximately 3 weeks following acute infection (Hsiung *et al.*, 1978). Salivary gland titres of virus, first demonstrable at 7–10 days post infection, reach maximal levels at 3 weeks, and persist at high levels for at least 10 weeks. Salivary glands may be harvested during this time window for preparation of stocks for subsequent experiments.

Although the greatest value of the GPCMV model is the study of congenital infection, a number of other models of GPCMV-induced disease have been described in non-pregnant animals. Of considerable interest with respect to GPCMV pathogenesis is the description of a mononucleosis-like syndrome associated with acute infection, a syndrome of interest in light of the ability of HCMV to cause so-called 'heterophile negative' mononucleosis. Following sc inoculation of Hartley guinea pigs with GPCMV-SG, atypical lymphocytosis has been reported in peripheral blood, and a constellation of findings including splenomegaly, lymphadenopathy, anaemia, and neutropenia has been observed (Diosi and Georgescu, 1974; Griffith *et al.*, 1981). Interestingly, neutrophil dysfunction has been observed in this setting, suggesting an immunosuppressive effect of acute GPCMV infection (Yourtee *et al.*, 1982). Inbred strains of guinea pigs, in particular strain 2, are, in general, more prone to GPCMV disease than are outbred animals, and in particular are at risk to develop pneumonitis (Bia *et al.*, 1982). In guinea pigs that are pharmacologically immunosuppressed (e.g. agents such as cyclophosphamide or cyclosporine) extensive GPCMV-induced disease, including pneumonitis, is noted, establishing a model of disease that mimics the presentation of HCMV disease observed in immunosuppressed patients. This immunosuppression model provides a useful system in which to evaluate antiviral therapies (Aquino-de Jesus and Griffith, 1989). GPCMV has also been used as a model for CMV-induced sensorineural hearing loss (SNHL), following direct experimental inoculation of the cochlea (Harris *et al.*, 1984; Woolf *et al.*, 1985; Keithley *et al.*, 1988; White *et al.*, 2006; Schraff *et al.*, 2007a,b). SNHL has also been observed in the setting of congenital GPCMV transmission with associated labyrinthitis (Woolf *et al.*, 1989; Katano *et al.*, 2007; Park *et al.*, 2010). Interestingly, in one study, SNHL in guinea pigs was found to be both asymmetric and progressive (Park *et al.*, 2010), mimicking the pattern of hearing loss seen in many infants with congenital HCMV infection. Although congenital transmission with associated labyrinthitis has been described by several investigators, other studies have failed to reproduce these findings. In a study in which pregnant Hartley and Strain 2 guinea pigs were injected i.p. with GPCMV-SG during the first, second, or third trimesters, histological and immunocytochemical study of temporal bones in pups did not show evidence of labyrinthine infection (Strauss and Griffith, 1991). Further studies of congenital labyrinthitis in the GPCMV model are required to resolve this discrepancy. In a study where congenital GPCMV labyrinthitis could be experimentally engendered, ganciclovir (GCV) therapy of SNHL has been studied (Woolf *et al.*, 1988), although the study of this antiviral intervention is of limited translational relevance to HCMV, given the intrinsic resistance of GPCMV to this antiviral agent. However, GPCMV is very susceptible to several of the more recently developed experimental antivirals, and the evaluation of these agents, as well as strategies to improve the study of antiviral agents in the GPCMV model, is considered in greater detail later in this chapter.

Other models of non-congenital GPCMV-induced disease that have been described include a neuropathogenesis model, characterized by a glial nodule encephalitis (Booss *et al.*, 1988; 1989, 1990; Booss and Kim, 1989), and a model of experimental infection of newborn guinea pigs, characterized by features such as viraemia, hepatitis, pneumonitis, and brain involvement (Griffith *et al.*, 1985a; Zheng *et al.*, 1987; Bravo *et al.*, 2003). A GPCMV model of hepatobiliary injury, similar to the syndrome of biliary atresia observed in infants, has also been described (Wang *et al.*, 2011). Although these models are of interest and are useful in the study of antivirals, clearly the greatest strength of the GPCMV model is in the study of vaccines against congenital infection, reviewed in detail later in this chapter.

Immune response to GPCMV infection

Humoral immune response

As previously noted, the importance of immunity to GPCMV was first investigated over seventy years ago, when Andrewes assessed the ability of serum from immune animals to neutralize the infectivity of salivary gland homogenates prior to experimental inoculation (Andrewes *et al.*, 1930). In spite of considerable progress in characterization of the antibody response to GPCMV since these earliest reports, much still remains to be learned about the nature of humoral targets in the setting of acute infection. Early efforts to characterize the humoral response to GPCMV focused on the development of virus-specific polyclonal antibodies in rabbits, in order to reliably differentiate GPCMV in cell culture from a number of other related, endogenous guinea pig herpesviruses (Bia *et al.*, 1980b). Although these studies enabled the development of ELISA and immunofluorescence assays, there was no information available at this time about responses to specific GPCMV proteins. In a subsequent study, immunoblot analyses of the humoral response to infection in both pregnant and non-pregnant guinea pigs were performed, and defined the evolution of the antibody response to 12 GPCMV polypeptides during primary infection. Interestingly, immune responses were delayed, and of lower magnitude, in pregnant animals compared to non-pregnant animals, but the molecular identity of the polypeptides was not further studied (Bu and Griffith, 1990). In another immunoblot study of sera from GPCMV-infected, non-pregnant guinea pigs, at least 18 GPCMV polypeptides were identified using immune sera. Some of these polypeptides appeared to have immunological cross-reactivity with HCMV proteins, but the molecular identity of these proteins was not further explored (Kacica *et al.*, 1990). A study examining anti-gB antibodies to HCMV found no evidence of cross-reactivity with GPCMV proteins (Adler *et al.*, 1995). Some progress was made in the characterization of GPCMV proteins through the generation of monoclonal antibodies. This approach allowed characterization of a virion structural protein of ~ 160–180 kDa, a 50 kDa nuclear non-structural protein, and a 76 kDa matrix protein, although the precise identities of these proteins and their potential roles as immune targets was not examined (Tsutsui *et al.*, 1986; Nogami-Satake and Tsutsui, 1988; Jones *et al.*, 1994). Another murine mAb was described that was specific for a 60–90 kDa envelope glycoprotein complex; this antibody neutralized infectious virus in the presence of complement, and based on elucidation of intracellular processing pathways and evaluation of glycosylation patterns, was concluded to recognize the gB (gpUL55) homologue of GPCMV (Britt and Harrison, 1994). More recently, the use of recombinant technologies has allowed the characterization of specific GPCMV proteins, and this information has been applied to subunit vaccine studies (see below).

Innate and adaptive immune responses

As is the case for study of humoral responses, the contribution of cellular immune responses to infection with GPCMV has not been well-explored, due in part to the paucity of reagents for cell-mediated immunity (CMI) assays in guinea pigs. However, several lines of evidence support a role for CMI, including antiviral T-cells, in response to infection. An enhancement of cutaneous basophil responses was found after challenge with GPCMV in immune animals, and this occurred in association with stimulation of T-cell-dependent areas of lymph nodes, suggesting a role for T-cells in viral clearance (Griffith *et al.*, 1982a). In non-pregnant animals, cytofluorometric analysis of T-cells recovered following intradermal challenge of primed animals demonstrated $CD4^+$ and $CD8^+$ responses against the GPCMV pp65 homologue, GP83 (Schleiss *et al.*, 2007).

In addition to these studies, there is some information about the potential role of natural killer (NK) cells in the setting of acute GPCMV infection. In inbred strain 2 animals, peripheral blood mononuclear cell (PBMC)-mediated cytolytic activity, presumably representing the NK response, was described against GPCMV-infected syngeneic and allogeneic targets, and was unaffected by T-cell depletion (Harrison and Myers, 1988). These investigators also identified a potentially pathogenic effect of NK activity on the guinea pig fetus in the setting of GPCMV infection. In pregnant strain 2 guinea pigs, cytolysis of GPCMV-infected syngeneic fetal cells was associated with poor pregnancy outcomes, including pup mortality, runting, and conceptus loss (Harrison and Myers, 1989). Enhanced NK cell activity and increased tumour necrosis factor activity during early gestation of GPCMV-infected dams was also correlated with poor pregnancy outcomes (Harrison and Caruso, 2000).

Immune modulation following GPCMV infection

Acute GPCMV infection appears to induce diverse immunomodulatory and, potentially, immunoevasive effects on the guinea pig immune response. In a study of acute GPCMV mononucleosis, during which guinea pigs were noted to be viremic and neutropenic, mobilization of neutrophils to sites of inflammatory stimuli was impaired (Yourtee *et al.*, 1982). Neutrophils from infected animals were noted to have diminished killing in response to a bacterial stimulus. These alterations were further explored in a study of neutrophil migration, which demonstrated that acute infection resulted in abnormalities of neutrophil-directed chemotaxis (Tannous and Myers, 1983). Perturbation in macrophage function, as measured by H_2O_2 release, has been described in the setting of experimental GPCMV interstitial pneumonitis (Miller *et al.*, 1985), and was suggested as a predisposing factor for bacterial super-infection. In another study, the nonspecific functional capacity of spleen cells, taken from female guinea pigs with primary GPCMV infection, was assessed using lipopolysaccharide (LPS), a B-cell mitogen, and concanavalin A (Con A), a T-cell mitogen. Proliferative responses to the two mitogens were found to be significantly depressed in animals inoculated with GPCMV compared with controls (Griffith *et al.*, 1984). These differences were also found to persist in adult guinea pigs which had been infected as newborn pups, and were associated with significant depletion of the T-cell population in the thymus (Zheng *et al.*, 1987). GPCMV has also been shown to down-regulate cell surface expression of MHC class I, presumably to evade CTL recognition (Lacayo *et al.*, 2003). The viral gene(s) responsible for these effects remain to be fully defined.

Biology of GPCMV

GPCMV morphology and ultrastructure

The ultrastructure of the GPCMV, both in salivary gland and tissue culture, was analysed in detail by electron microscopy (EM) several decades ago (Middelkamp *et al.*, 1967). These studies demonstrated that GPCMV has a typical herpesvirus structure consisting of an icosahedral capsid core surrounded by a tegument layer and a viral membrane (see also Chapter I.6 and Chapter I.13). Morphological analyses at multiple time points post inoculation revealed the presence of tubular structures within nuclei of virus-infected cells, followed by the appearance of intranuclear inclusions containing virus nucleocapsids. Both intracellular and extracellular 'dense bodies' (DBs), representing enveloped tegument particles without a capsid core, were also identified (for the HCMV counterpart of DBs, see Chapter I.6). In the original ultrastructural analysis described by Fong and colleagues, it was noted that while some nucleocapsids were enveloped at the inner nuclear membrane, others were released into the cytoplasm where they were associated with, or within, dense matrix which was subsequently enveloped by cytoplasmic membranes to form what were described as 'enveloped dense virions' (Fig. II.5.1). Virus capsids were formed in the cytoplasm and enveloped in a similar manner. Immuno-EM revealed that enveloped virions and DBs share common envelope antigens (Fong *et al.*, 1979). Subsequently, it was shown that GPCMV DBs can be purified from enveloped virions using gradient centrifugation techniques and, as with HCMV DBs, a major component of the total tegument protein is the pp65 protein homologue, GP83 (Schleiss *et al.*, 1999). EM studies of the salivary glands from infected animals demonstrated that the virus was localized to ductal epithelial cells (Middelkamp *et al.*, 1967), an interesting observation in light of emerging observations about HCMV genes important in epithelial cell tropism. Representative EMs of GPCMV virions and DBs from infected cell cultures are demonstrated in Fig. II.5.1.

GPCMV DNA and genome configuration

The first inroads to a description of the molecular biology of GPCMV came in the early 1980s, when density gradient centrifugation of GPCMV DNA, and restriction endonuclease analysis of the viral genome, were first described (Bia *et al.*, 1980b). A comprehensive analyses of GPCMV genomic DNA resulted in the generation of *Hin*d III, *Eco*R I, and *Xba* I restriction maps of the viral genome, and a subset of these restriction fragments was cloned in plasmid pBR322 (Gao and Isom, 1984; Isom *et al.*, 1984; Yin *et al.*, 1990). Analyses of restriction endonuclease maps and cross-hybridization studies indicated that the viral genome consists of a long unique sequence, with terminal repeats. In contrast to HCMV (see Chapter I.1), the GPCMV genome does not contain internal repeat regions. Based on restriction endonuclease mapping of cloned fragments, the genome was initially estimated to be ~239 kbp (Gao and Isom, 1984), corresponding to a molecular mass of 158 MDa.

One novel aspect of GPCMV genome structure that was deduced from these studies was the finding that the genome exists as two isoforms in cell culture. In the predominant form, unit length genomes were noted to share sequence homology at the two termini, whereas in the less abundant genome isoform, one terminal fragment was noted to be smaller (by ~0.7 MDa), and lacked homology with the fragment from the other end of the genome (Gao and Isom, 1984). Cloning of the GPCMV genome as infectious BAC in *E. coli* generates a single genomic isoform variant, although regeneration of both genome isoforms (one variant with a single copy of the terminal sequence, and one variant with two copies) is observed when

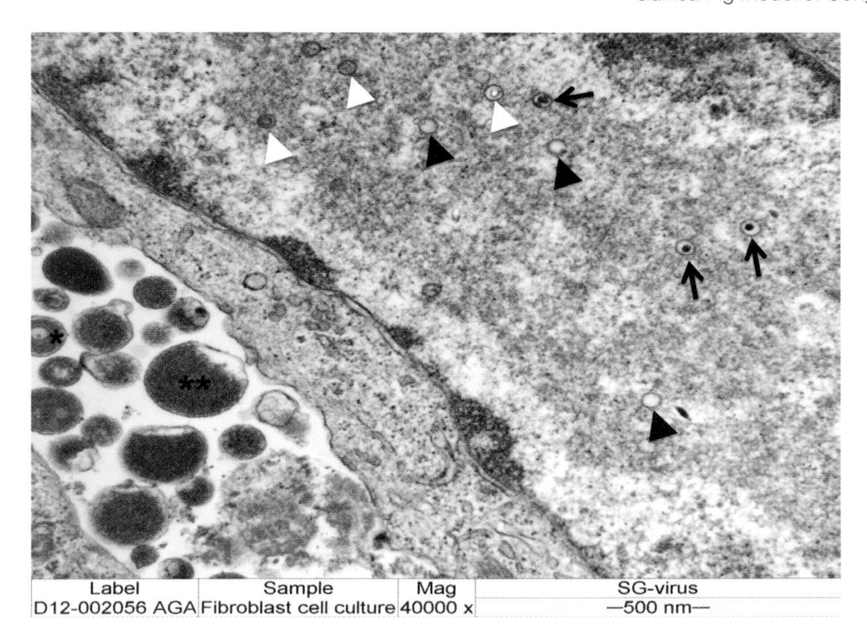

Figure II.5.1 Ultrastructural examination of GPCMV following infection of guinea pig lung fibroblasts using *in vivo* propagated virus purified from salivary gland homogenates. Cells were pelleted 1 week following infection in cell culture (MOI = 1) and subjected to transmission electron microscopy (TEM). In nucleus (upper portion of figure) nucleocapsids, including single-ring, type 'A' capsids (black arrowhead) and double-ring 'B' capsids (white arrowhead) are noted. Open arrows point to 'C' capsids containing whorls of DNA. Virions and dense bodies, electron-dense particles that do not contain capsids, are noted in lower left portion of figure, similar to findings observed for HCMV (see Chapter I.13). Enveloped dense virions, originally described by Fong and colleagues (Fong *et al.*, 1979), are also noted. TEM performed after phosphotungstic acid staining of infected cells following paraformaldehyde fixation, 40,000× magnification.

BAC DNA is transfected onto fibroblast cells to regenerate virus (McGregor and Schleiss, 2001a). The mechanisms by which these genomic isoforms occur were elucidated in a series of detailed studies (McVoy *et al.*, 1997). In circularized GPCMV DNA, two restriction fragments were identified which contained fused terminal sequences, and had sizes consistent with the presence of single or double terminal repeats. Cleavage to form the two genome types was found to occur at two sites. Viral mutants were generated at one site that authenticated the role in genomic cleavage of *cis* elements contained within a 64-bp region. In a follow-up study, it was demonstrated that the double repeats that are formed by circularization of infecting genomes are rapidly converted to single repeats, such that the junctions between genomes within replicative concatemers formed late in infection contain predominantly single copies of the terminal repeat. However, although each cleavage event begins with a single repeat within a concatemer, two repeats are produced, one at each of the resulting termini, demonstrating that terminal repeat duplication occurs in conjunction with cleavage, and that both duplicative and non-duplicative cleavage events may occur concurrently (Nixon and McVoy, 2002). Similar events involving duplication and deletion of terminal *a* sequence repeats and cleavage at alternative cleavage sites also occur in HCMV (Wang and McVoy, 2011). It remains to be demonstrated whether these variations in DNA replication have any implications for pathogenesis *in vivo*.

Analysis of GPCMV gene products and genomic sequence

Over the past decade, there have been significant advances in the molecular characterization of GPCMV. The first GPCMV gene to be identified was *GP55*, encoding glycoprotein B (gB). The approach of low-stringency Southern blot hybridization of HCMV gB probes with various combinations of restriction endonuclease-digested GPCMV DNA was used to pinpoint the region in the GPCMV genome which encoded the gB homologue. The GPCMV gB homologue was found to map to the *Hin*dIII 'K' and 'P' fragments of the genome. This data established the orientation of the GPCMV genome relative to other CMVs, and facilitated the identification of several other homologues. GPCMV gB is transcribed as an 'early' gene (6.8 kb mRNA), and encodes a 901 aa protein with 42% homology to HCMV gB (Schleiss, 1994). The protein is cleaved into amino- and carboxy-terminal subunits of approximately 90 and 58 kDa. GPCMV gB is a major target of the neutralizing antibody response following natural GPCMV infection, and antibodies to gB are uniformly found in GPCMV-seropositive animals (Schleiss and Jensen, 2003; Schleiss *et al.*, 2004), making this an ideal candidate protein for subunit vaccine studies in this animal model (see below).

The characterization of the GPCMV gB homologue defined the orientation of the viral genome relative to existing *Hin*dIII, *Eco*RI, and *Xba*I restriction maps, thus providing a framework for identification of other conserved genes. These included the homologues for glycoprotein H (pGP75, the pUL75 homologue), the glycoprotein L (pGP115, the pUL115 homologue), the DNA polymerase, pGP54, and the pUL97 phosphotransferase, pGP97 (Schleiss, 1995; Brady and Schleiss, 1996; Fox and Schleiss, 1997; Paglino *et al.*, 1999). The upper and lower matrix tegument protein homologues of pp71 (pUL82) and pp65 (pUL83) have also been characterized, and are encoded by homologue genes *GP82* and *GP83*, respectively (Schleiss *et al.*, 1999; McGregor *et al.*, 2004b). Of particular interest are conserved and novel genes with putative immunomodulatory functions. Homologues of the HCMV-encoded G-protein coupled receptors (GPCRs), pUL33 (pGP33) and pUL78 (pGP78) have been reported (Liu and Biegalke, 2001). The virus also encodes a protein having features of the CC (β) subfamily of chemotactic cytokines (chemokines). This protein, a homologue of macrophage inflammatory protein 1-α (also know as CCL-3), was designated GPCMV-MIP (Haggerty and Schleiss, 2002; Penfold *et al.*, 2003). The biological relevance of this putative chemokine was further explored using a recombinant form of the GPCMV-MIP protein, which was found to signal specifically through the CCR1 receptor in a series of *in vitro* assays (Penfold *et al.*, 2003).

An initial analysis of the GPCMV genomic sequence was assembled from DNA sequences obtained from a first-generation GPCMV BAC clone, a partial restriction library of viral subgenomic DNA restriction fragments, and fragments generated by PCR spanning uncloned regions (Schleiss, M.R., *et al.*, 2008). The G+C content of the genome determined by sequence analysis was ~55%, in good agreement with the estimation of 54.1% determined by CsCl buoyant gradient density (Isom *et al.*, 1984). This analysis indicated that the GPCMV genome is approximately 233 kbp in length, excluding the 1-kbp terminal repeat sequences. In total, 105 open reading frames (ORFs) of > 100 amino acids with sequence and/or positional homology to other CMV ORFs were annotated in this analysis, including positional and sequence homologues of HCMV ORFs *UL23* through *UL122*. Fig. II.5.2 illustrates co-linear and conserved homologues of GPCMV genes with HCMV genes involved in pathogenesis and immune response. Interestingly, evolutionary tree analysis suggested that the GPCMV genome has closer similarity to HCMV than to the true rodent CMVs, mouse CMV and rat CMV (Schleiss, M.R., *et al.*, 2008).

Subsequent to this report, which relied heavily on sequence assembly of restriction fragments and PCR-generated fragments in addition to first-generation BAC sequencing, additional studies demonstrated that this initial sequence report did not represent the *bona fide*, wild-type sequence of GPCMV (Schleiss and McVoy, 2010). More detailed analyses indicated that the first-generation GPCMV BAC construct contained deletions in the region of the BAC site of insertion, presumably introduced during the cloning process in *E. coli* (McGregor and Schleiss, 2001a), and also lacked a 1.6 kbp region of the viral genome that contained a number of previously unrecognized putative genes, including the presumptive GPCMV homologues of the HCMV *UL128–131* locus (Nozawa *et al.*, 2008). Although some of these sequences missing from the first-generation BAC were corrected in construction of a second generation BAC (Cui *et al.*, 2008), the latter was derived from a GPCMV-TC virus stock, and such stocks are known to be attenuated *in vivo* relative to salivary gland-adapted GPCMV-SG stocks. A re-evaluation of the GPCMV genome sequence recently reported by Inoue and colleagues, using ATCC-derived strain 21222 as a source of template, has provided a more complete representation of the genomic sequence (Kanai *et al.*, 2011). In addition, we are currently carrying out sequence analyses of ATCC-derived, BAC-derived, and SG virus, using 'deep' sequencing techniques (Dittmer *et al.*, 2011). The ability to sequence the GPCMV directly from salivary gland homogenates should help clarify the genetic differences between attenuated (BAC-derived/TC) and virulent (SG) viruses. To date, these analyses suggest that one potentially important difference between BAC virus sequence and the SG sequence affects a locus that is co-linear, and probably functionally homologous, to the region associated with epithelial and endothelial tropism in HCMV (discussed further below).

Targeted mutagenesis of a number of conserved gene homologues in the GPCMV genome have been carried out either by conventional or infectious BAC-mediated strategies to determine essential function

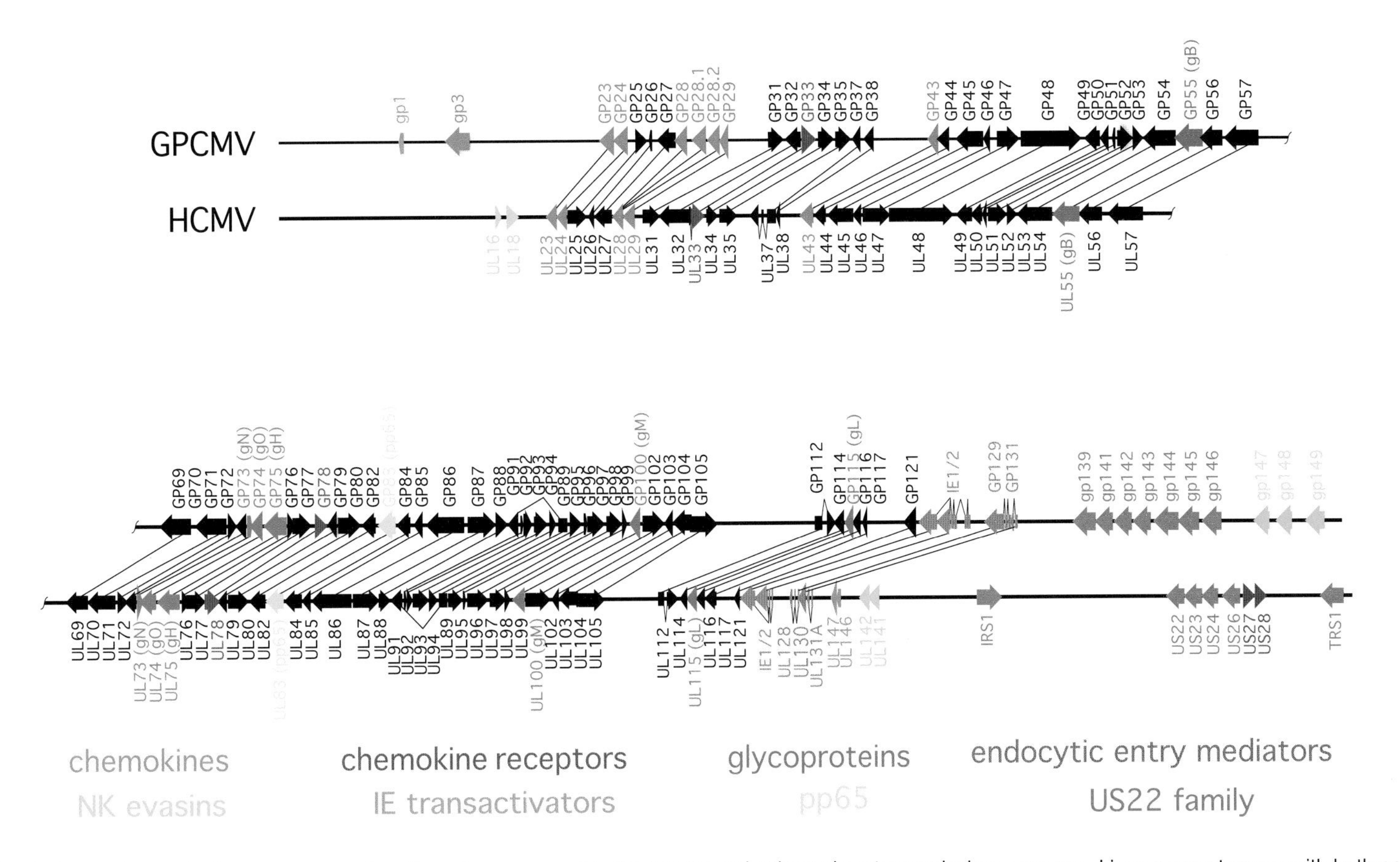

Figure II.5.2 Genes demonstrating sequence or functional homology within the human and guinea pig cytomegalovirus genomes. Lines connect genes with both positional and sequence conservation. Other genes that lack clear-cut positional homologues, but encode proteins with similar known or presumed functions (such as chemokines, chemokine receptors, and immunoevasins) are demonstrated in colour. The *US22* gene family and presumptive natural killer cell evasin family genes are present near the right-hand and the left-hand terminus of the GPCMV genome, respectively. Modified from Schleiss and McVoy (2010).

either *in vitro* or *in vivo* and this is discussed in a later section (GPCMV mutagenesis strategies). As with other CMVs, GPCMV gene expression can be conveniently categorized into immediate early (IE), early (E), or late (L) gene expression, using metabolic blockade with cycloheximide and phosphonoacetic acid (Yin *et al.*, 1990). Although a complete transcriptome map for the GPCMV genome remains to be established, the conserved HCMV gene homologues would appear to have the same class of expression kinetics relative to their counterparts in HCMV. One exception is GP55 (GPCMV gB), expressed as an early gene in GPCMV infection, but as a late gene in HCMV infection (Schleiss, 1994). GPCMV genes encoded at the respective termini of the genome appear to be species specific for the guinea pig host and are in general not homologous to HCMV genes. Notably, there is a plethora of potential ORFs at both termini with homology to the *US22* gene family. Many of these genes appear dispensable for replication; for example, the entire GPCMV *Hin*dIII 'D' locus (approximately 21 kbp), which does not encode any HCMV homologues, can be deleted from the virus genome without greatly impairing the ability of the virus to replicate in tissue culture. However, we have found that this virus is extremely impaired for replication and dissemination *in vivo*. Additional application of mutagenesis strategies to the study of the GPCMV genome will be discussed in a later section.

Tropisms, endocytic complex and 1.6 kbp region conferring pathogenesis

In HCMV the ability to efficiently enter epithelial and endothelial cells has been linked to three viral proteins, UL128, UL130, and UL131, that form a complex with the viral glycoproteins gH and gL (Hahn *et al.*, 2004; Wang and Shenk, 2005a,b; Ryckman *et al.*, 2006, 2008; see also Chapter I.17). At present it is not known whether GPCMV encodes proteins specifically required for epithelial/endothelial entry. Similarly, it is unclear if guinea pigs, like humans, engender high-level neutralizing antibodies specific for this entry pathway following infection, but preliminary evidence suggests that this may be the case. In both human and rhesus CMV, the ability to enter epithelial or endothelial cells appears to be detrimental to fibroblast replication, insofar as viruses passaged in fibroblasts rapidly acquire mutations that disable expression of one or more of the three epithelial/endothelial entry mediators (Hahn *et al.*, 2004; Oxford *et al.*, 2008; 2011). Interestingly, the ATCC stock of GPCMV strain 22122 was recently found in studies by Inoue and colleagues to consist of two genome variants. One variant, designated GPCMV/full, contains a region from which spliced transcripts were isolated that encode proteins, pGP129 and pGP131, that exhibit partial sequence homologies with pUL128 and pUL130, respectively (Nozawa *et al.*, 2008; Yamada *et al.*, 2009). The other variant, GPCMV/del, has a 1.6-kbp deletion in which these sequences are absent, possibly due to negative selection against these sequences upon culture in fibroblasts. In guinea pigs, GPCMV/full exhibited a growth advantage over GPCMV/del, while the opposite was the case *in vitro* in cultured fibroblasts (Nozawa *et al.*, 2008). Consistent with an important role for this region in pathogenesis *in vivo*, virulent SG passaged stocks have been found to retain the intact sequence found in GPCMV/full (Dettmer *et al.*, 2011).

We have recently independently confirmed these observations regarding the variations in the GPCMV genome that arise upon adaptation of GPCMV during cell culture (Fig. II.5.3). These genomic variants have been defined in BAC clones. It was noted that the first generation GPCMV BAC is similar to GPCMV/del, as it lacks the 1.6-kbp region, generated on the backdrop of high-passage ATCC virus lacking this locus (McGregor and Schleiss, 2001a). The second generation GPCMV BAC was generated in the backdrop of low pass ATCC stock (Cui *et al.*, 2008), and consequently contains the 1.6-kbp region, but also has a 4-bp deletion/frame shift impacting the sequences encoding pGP129, the homologue and putative orthologue of pUL128 (Yamada *et al.*, 2009). We have established that the pGP129 protein interacts with GPCMV gH by cellular co-localization and immunoprecipitation experiments of transiently expressed proteins that also include gL and pGP131, the pUL130 homologue. Inoue and colleagues further showed that pGP131 is present in GPCMV virions and, when co-expressed with GPCMV gH and gL, localizes to the cell surface (Yamada *et al.*, 2009). This suggests that there are pGP131 interactions with gH and gL that are similar to the HCMV gH/gL/UL130 interactions believed to be necessary for cell surface localization of pUL130 (Ryckman *et al.*, 2008). Passage of SG GPCMV on fibroblast cells rapidly results in a 1.6 kbp deletion which encompasses the GP129–131 locus in GPCMV, in a fashion analogous to the modifications observed in clinical strains of HCMV following passage in fibroblasts (Dargan *et al.*, 2010; Revello and Gerna, 2010; see also Chapter I.1). In tissue culture, we have noted that SG virus with an intact GP129–131 locus is capable of infecting a primary guinea pig epithelial cell line, unlike GPCMV/del virus (Fig. II.5.3).

Confirmation that GPCMV has a specialized method of epithelial/endothelial entry and that guinea pigs exhibit epithelial/endothelial entry-specific neutralizing antibodies awaits further study.

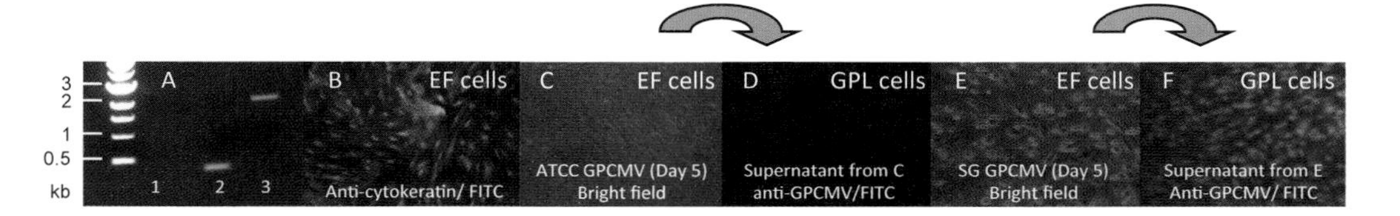

Figure II.5.3 Epithelial tropism of GPCMV-SG. (A) PCR analysis of the homologue *UL128–131* locus (*GP128–131*) in virulent GPCMV-SG (lane 3) and GPCMV adapted by passage on fibroblast cells (lane 2). Lane 1, kbp ladder. Common primers flanking the locus illustrate a deletion (1.6 kbp) in this locus similar to that identified in mixed strains of GPCMV identified in low pass ATCC stock (Nozawa *et al.*, 2008). (B) Immunofluorescence assay for characterization of epithelial cells. Clonal primary renal epithelial cell lines were established from Hartley guinea pigs. Immunofluorescence assay stained cells for cytokeratin (anti-pan-cytokeratin antibody; Applied Technology). Cytoplasmic green fluorescence (FITC) signal represents cytokeratin antigen. Nucleus was stained for nuclear proliferation antigen (orange). (C-F) GPCMV-TC or GPCMV-SG infection of epithelial cells. GPCMV was used to infect epithelial cells (EF cells) at an MOI of 0.5 pfu/cell, and after one week supernatant was removed and used to infect guinea pig lung (GPL) fibroblast cells. At 48 hours post infection GPL cells were fixed and stained for GPCMV by immunofluorescence assay (D and F) using polyclonal anti-GPCMV antibody as described in McGregor and Schleiss (2001). Green fluorescence (FITC) for GPCMV antigen (D and F). Only GPCMV-SG infected EF cells showed CPE (E) and only supernatant from GPCMV-SG infected EF cells produced subsequent infection of GPL cells (F).

However, in summary, several observations provide strong circumstantial evidence that GPCMV tropisms and entry mechanisms may closely parallel those of HCMV: (i) the GPCMV genome encodes proteins with amino acid similarities to HCMV epithelial/endothelial entry mediators; (ii) the genomic locations of these coding sequences are colinear with their counterparts in the HCMV genome (Fig. II.5.2); (iii) mutations impacting these coding sequences appear to have accumulated during fibroblast passage but not during *in vivo* passage and (iv) viruses lacking such mutations are more virulent. Thus, GPCMV may provide a powerful model for evaluation of vaccine strategies targeting the epithelial/endothelial entry pathway of viral entry.

GPCMV mutagenesis strategies

The development of strategies to generate recombinant CMVs has proven to be of tremendous benefit in animal model studies, by facilitating the study of the role of specific viral gene products in pathogenesis (Brune *et al.*, 2000; McGregor and Schleiss, 2001a,b; see also Chapter I.3). The xanthine–guanine phosphoribosyltransferase selection system was the first mutagenesis system employed for generation of GPCMV mutants, to enable the study of *cis* sequences in genomic replication, cleavage, and packaging (McVoy *et al.*, 1997). Subsequently, a similar strategy was successful in generating a mutant GPCMV with a GFP cassette inserted into the nonessential *Hin*dIII 'N' region of the viral genome. This mutant exhibited wild-type replication kinetics in cell culture, and was capable of inducing disease in animals (McGregor and Schleiss, 2001a). Subsequent to these studies, the GPCMV genome was cloned as an infectious BAC in *E. coli*, by insertion of an F plasmid into the *Hin*dIII 'N' region (McGregor and Schleiss, 2001a, 2004). This BAC, as with the parental ATCC plaque-purified virus, lacked the 1.6-kbp locus described by Inoue and colleagues (Nozawa *et al.*, 2008). Second generation GPCMV BACs, including a version in which the plasmid sequences are flanked with loxP sites for removal of prokaryotic sequence, have recently been generated on the backdrop of low-pass ATCC GPCMV stock and were found to contain the intact 1.6-kbp locus. The first generation BAC has the advantage of generating fully stable virus when transfected into guinea pig fibroblast cells where GFP reporter gene expression enables easy tracking of virus spread. In contrast, studies with the second generation BAC require the excision of the F plasmid encoding the GFP reporter gene to avoid spontaneous random deletions elsewhere in the genome because of constraints on genome over sizing (Cui *et al.*, 2009).

Both random transposon and targeted knockout mutagenic strategies have been employed to identify essential and non-essential genes for GPCMV replication in tissue culture but a global knockout map, as established for HCMV, has not yet been attained (McGregor *et al.*, 2004a,b, 2008, 2011a; Crumpler *et al.*, 2009). Importantly, GPCMV knockout mutants of HCMV-conserved homologue genes would appear to have the same ranking of essential, semi-essential or non-essential as their counterparts in HCMV. A typical analysis of essential function is illustrated in Fig. II.5.4, which demonstrates that the homologue of the *UL54* (*GP54*) gene that encodes the viral DNA

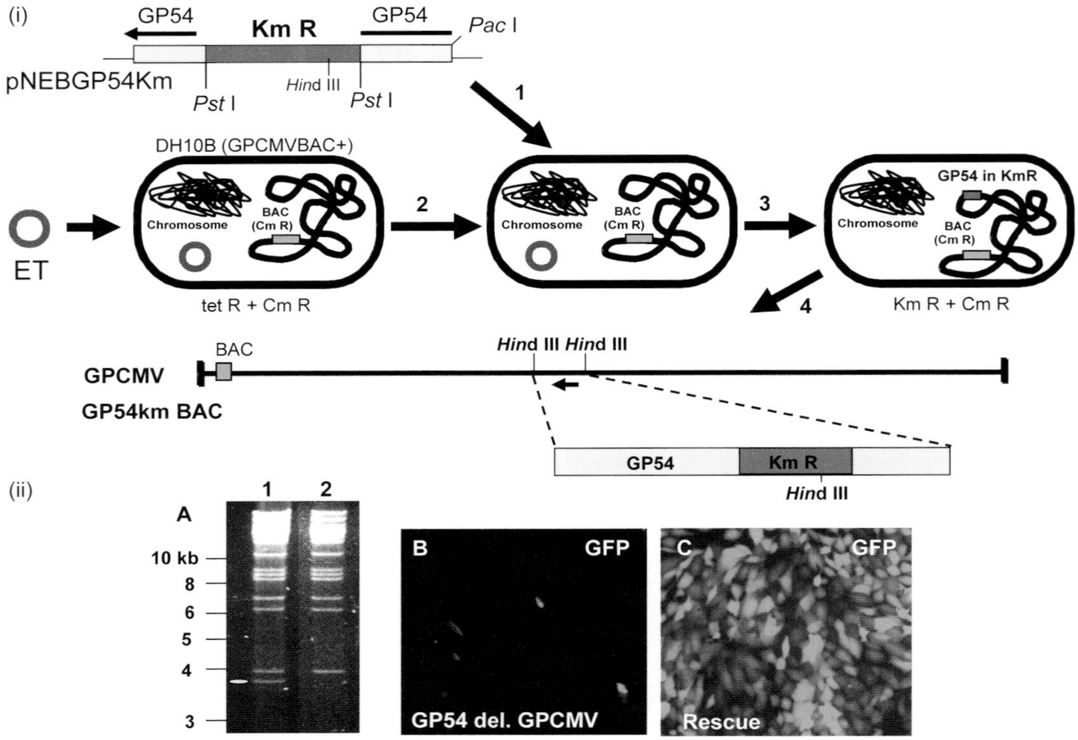

Figure II.5.4 Analysis of the essential nature of *GP54*. (i) Shown is a typical strategy employed for GPCMV gene knockout via BAC mutagenesis (see also Chapter I.3). Shuttle vector (pNEBGP54Km) encodes a 1 kbp deletion (and insertion of a kanamycin drug resistance marker, Km) in the *GP54*. Steps 1–2: The shuttle vector linearized (*Pac* I) and transformed into bacteria carrying the GPCMV BAC and temporally induced for recombination between shuttle vector and BAC. Step 3: Recombination culture plated out under selection for the BAC (chloramphenicol, Cm) and *GP54* mutation (Km) at 39°C to remove the temperature-sensitive plasmid encoding the red-ET recombination genes (for discussion of Red recombination, see also Chapter I.3). Step 4: Full-length GPCMV BAC mutants are characterized by DNA profile analysis and sequencing, and transfected onto fibroblasts to generate virus. (ii) (A) *Hin*dIII profile analysis of WT and *GP54* deletion GPCMV BACs. Km insertion into the GP54 gene introduces a novel *Hin*dIII site in the WT GPCMV subgenomic 3.7 kbp *Hin*dIII fragment (yellow dot). (B) The *GP54* gene is essential since transfection of *GP54* mutant BAC onto fibroblasts does not produce infectious virus unless co-transfected with rescue plasmid. (C) Virus spread tracked by GFP reporter gene encoded on the BAC plasmid. B and C, day 20 post transfection.

polymerase is demonstrated to be essential for virus replication. These data are directly relevant to antivirals that target the viral DNA replication pathway, as discussed in a later section.

Analyses of the viral genome (Cui *et al.*, 2008; M.R. Schleiss *et al.*, 2008; Kanai *et al.*, 2011) indicated that GPCMV encodes homologues to the HCMV glycoproteins (gB, gH, gL, gM, gN and gO) in genes co-linear with the HCMV genome (designated *GP55*, *GP75*, *GP115*, *GP100*, *GP73* and *GP74* respectively). Individual targeted knockouts of these various ORFs were carried out and, with the exception of the *GP74* gene (encoding the gO homologue), all genes were essential for virus replication in tissue culture, similar to what has been observed for HCMV (Hobom *et al.*, 2000, McGregor *et al.*, 2012). Fig. II.5.5 summarizes these studies and shows an example of the analysis of a gL (GP115) mutant. A GPCMV gO mutant produced characteristic small plaques and had modified levels of gH in isolated viral particles, as observed with an HCMV UL74 (gO) mutant (Ryckman *et al.*, 2010). We have also found that knockout of the *GP33* and *GP78* homologues encoding putative GPCRs does not impact on viral replication in tissue culture, but does impair viral dissemination *in vivo*. A recombinant virus with a 'knock-out' of the GPCMV-MIP gene induced less cochlear inflammation, and reduced hearing loss, compared to wild-type virus in a direct cochlear inoculation model of labyrinthitis (Schraff *et al.*, 2007a,b). A recombinant virus with a

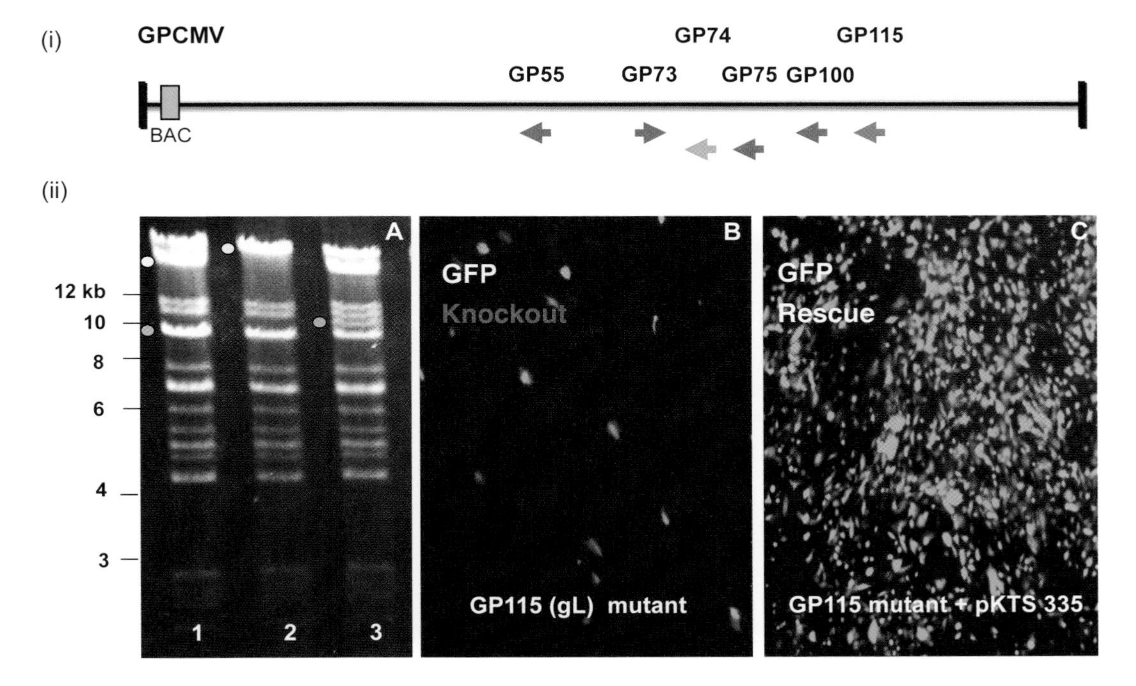

Figure II.5.5 Analysis of the essential nature of GPCMV homologue glycoprotein genes by knockout mutagenesis. The GPCMV genes encoding the homologue glycoproteins were individually knocked out via BAC mutagenesis in bacteria and DNA from confirmed mutant GPCMV BAC clones were transfected onto GPL fibroblast cells to determine their ability to produce infectious virus. (i) GPCMV genome showing position of the co-linear glycoprotein homologue genes: *GP55* (gB); *GP73* (gN); *GP74* (gO); *GP75* (gH); *GP100* (gM); *GP115* (gL). Genes are indicated by arrows and colour indicates essential (red) or non-essential (green). BAC plasmid insertion in the *Hin*dIII 'N' locus enables real time tracking of virus spread by GFP reporter gene expression. Only the *GP74* (gO) gene is non-essential for virus replication as is the case for HCMV. (ii) Analysis and rescue of a *GP115* (gL) mutant. (A) *Eco*RI restriction profile analysis of GPCMV wild-type and BAC glycoprotein mutant DNA. Gene knockout was by insertion of the Km resistance marker to disrupt the ORF, thus modifying the size of specific subgenomic fragments. Relevant band shifts are marked. Insertion of a Km marker into the *GP115* gene increases the size of the *EcoR*I 'G' band. (B) Transfection of *GP115* (gL) mutant onto GPL cells does not produce infectious virus and only single GFP positive cells are seen at 18 days post transfection. (C) Co-transfection rescue of mutant BAC with a rescue plasmid results in viable virus, detected by GFP expression in infected cells (day 18 post transfection).

targeted deletion of three MHC class I homologues, gp147–149, which attenuates the virus *in vivo* but not *in vitro*, has been utilized as a novel live vaccine strategy (Crumpler *et al.*, 2009) and will be discussed later. GPCMV encodes direct homologues to HCMV capsid genes and knockout of any of these genes is lethal for the virus (Dunn *et al.*, 2003; McGregor *et al.*, 2004a). A GPCMV minor capsid protein pUL85 homologue (pGP85) mutant grown on a supporting cell line has been investigated as a novel single cycle replication vaccine strategy and will be discussed in the vaccine section of this chapter.

One important aspect of GPCMV mutagenesis has been the delineation of function associated with the homologues of the *UL82-UL84* locus in GPCMV (*GP82-GP84*). In GPCMV, unlike MCMV, the encoded proteins would appear to be direct functional homologues to their HCMV counterparts (McGregor *et al.*, 2011a). This is particularly important in the context of the development of a valid model for the study of vaccine strategies against the pp65 homologue in GPCMV. The GP83 protein (pp65 homologue) has functions similar to pp65 in HCMV and a GP83 knockout mutant *in vivo* had dramatically impaired dissemination to target tissues in animals, but relatively normal growth kinetics in cell culture, as had previously been reported for an HCMV pp65 mutant (Schmolke *et al.*, 1995; McGregor *et al.*, 2004b). Similarly, a *GP97* deletion mutant was found to have an impaired growth phenotype comparable to that of a *UL97* knockout virus (Prichard *et al.*, 1999; McGregor *et al.*, 2008). The use of a 'chimeric' GPCMV containing the HCMV *UL97* gene in place of *GP97* in the study of antiviral therapy of GPCMV infection is further described later in this chapter (McGregor *et al.*, 2008).

GPCMV major immediate early locus

GPCMV has a number of immediate early (IE) transcripts that were originally assigned to three specific regions (Yin *et al.*, 1990) but we have found that those located in the *Hin*dIII 'D' locus towards the 5′ end of the viral genome can be deleted with only modest impairment of the ability of the virus to replicate in tissue culture (Cui *et al.*, 2009). The GPCMV major immediate early (MIE) locus of GPCMV is roughly co-linear with that of HCMV and has a very similar structure to HCMV (Yin *et al.*, 1990; Yamada *et al.*, 2009; see Chapter I.10). GPCMV encodes direct homologues to HCMV IE1 and IE2 that have recently been characterized by Inoue and colleagues. Both IE1 and IE2 transcripts are multiply spliced and share common exons but each encode a unique exon, *GP123* for IE1 and *GP122* for IE2, that are homologues to HCMV *UL123* and *UL122*, respectively (Yamada *et al.*, 2009; Fig. II.5.6).

We first reported functional analyses of the GPCMV MIE promoter/enhancer, based on evaluation of a series of promoter truncations fused with a betagalactosidase reporter gene, in 1996 (Sper-Whitis and Schleiss, 1996). These analyses mapped the MIE promoter/enhancer to a ~1.1 kbp region spanning the *Hin*dIII 'E' and 'B' regions of the genome. Detailed analysis of sequences upstream of the MIE TATA box and downstream into the first exon identified a number of features also observed in the HCMV MIE promoter/enhancer. The most notable of these is the location of multiple NF-κB binding sites (Fig. II.5.6). Interestingly, the GPCMV MIE promoter/enhancer also contains a large number of putative YY1 binding sites, especially downstream of the TATA box in the first exon, a feature also noted in the RhCMV MIE promoter/enhancer (Barry *et al.*, 1996). Several tandem reiterated sequences matched the consensus TTGGCN$_5$GCCAA NF-1 recognition element, and were contained within larger 30 bp direct repeats (Sper-Whitis and Schleiss, 1996). More recently, transient promoter expression studies in tissue culture have been performed with various truncations of the GPCMV MIE promoter cloned upstream of a luciferase reporter gene. In comparative studies, with a control plasmid encoding the HCMV MIE enhancer driving a luciferase reporter gene, the strength of GPCMV driven luciferase expression was similar to the HCMV MIE control plasmid (Fig. II.5.6). We have successfully used the GPCMV MIE promoter/enhancer to drive expression of essential GPCMV genes, such as *GP75*, in complementing cell lines.

GPCMV congenital transmission studies

As noted, the chief value of the GPCMV model is the ability of the virus to produce congenital infection, in contrast to other rodent CMVs. A number of aspects of guinea pig reproductive biology contribute to the usefulness of this model. Guinea pig gestational periods are fairly lengthy, ranging from 65 to 70 days, and can conveniently be divided into trimesters (Bia *et al.*, 1983). An important aspect of guinea pig reproductive biology facilitating congenital GPCMV transmission studies is the structure of the guinea pig placenta (Fig. II.5.7). The guinea pig placenta is haemomonochorial, containing a single trophoblast layer separating maternal and fetal circulation, and is histologically and biologically similar to the human placenta (Griffith *et al.*, 1985b; Mess, 2007; Mess *et al.*, 2007; see Chapter II.4). This structure differs from that of the mouse placenta, which contains three trophoblastic layers, a feature which possibly serves as a barrier to transmission of murine CMV (Medearis, 1964; Johnson, 1969). The ability of GPCMV to infect the guinea pig placenta following direct trans-uterine inoculation of virus was first described in the 1930s (Markham and Hudson, 1936), but recognition of viremic placental infection following a maternal systemic route of inoculation with GPCMV was noted in the late 1970s, when three groups independently described studies of fetal infection following infection of dams (Hsiung *et al.*, 1978; Kumar and Nankervis, 1978; Johnson and Connor, 1979). Since these initial reports on transplacental transmission, numerous investigations have employed the guinea pig model for vaccine, antiviral, and pathogenesis studies.

A variety of experimental endpoints are available in the GPCMV congenital infection model (Schleiss, 2006). Maternal and fetal outcomes following infection during pregnancy are dependent upon several variables. The timing of GPCMV inoculation is important. Viral inoculation in early pregnancy tends to lead to pup resorption, whereas challenge in late pregnancy tends to lead to pup mortality. Several strains of guinea pigs, including inbred (strain 2 and JY-9), and outbred (Hartley) strains, have been shown to be useful for congenital transmission studies. Typically, GPCMV-SG produces disease at lower inocula than does GPCMV-TC, although both sources of virus are capable of being congenitally transmitted. As discussed in detail elsewhere in this chapter, the decreased efficiency of transmission of GPCMV-TC is likely due to the fact that passage of virus in tissue culture results in a mixed population of viruses that includes a variant with a deletion of a 1.6 kbp genomic fragment that is essential for full virulence in animals (Nozawa *et al.*,

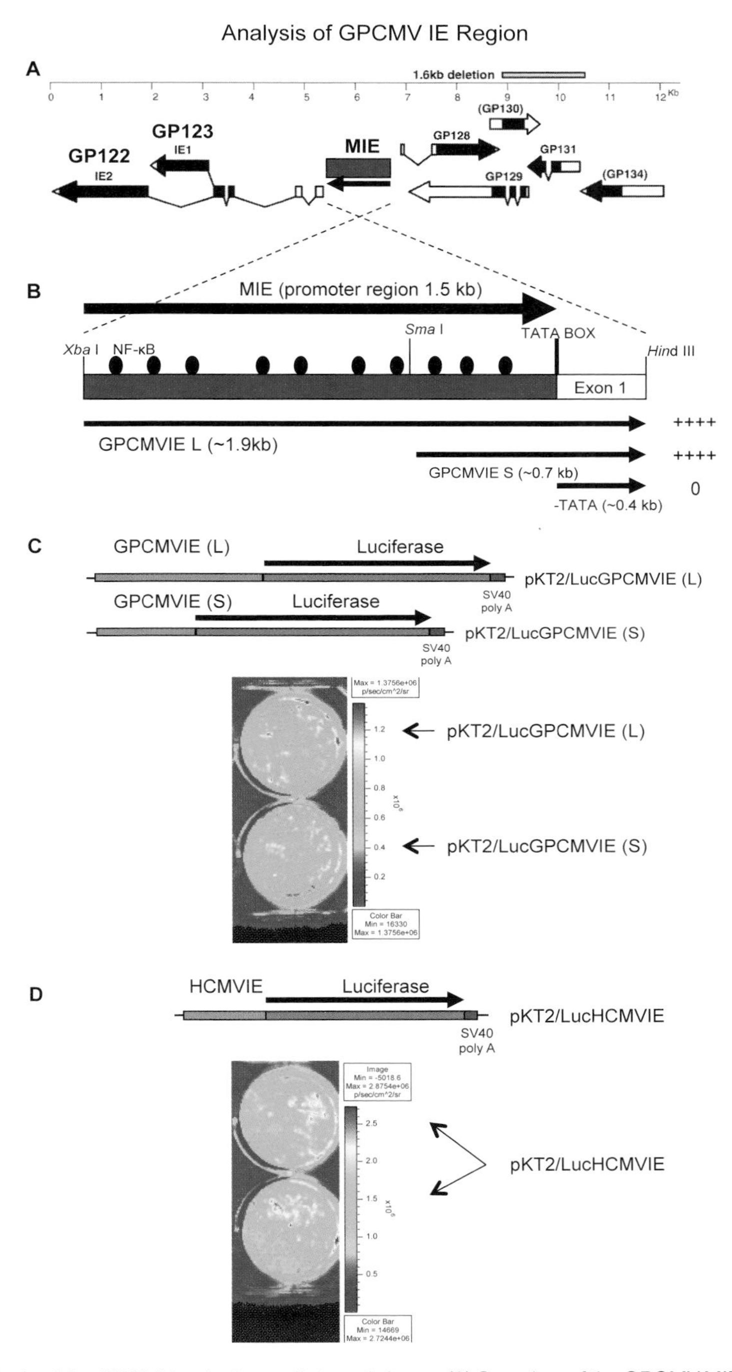

Figure II.5.6 Analysis of the GPCMV major immediate early locus. (A) Overview of the GPCMV MIE locus and flanking sequences. Shown are the multispliced transcripts that encode GPCMV IE1 and IE2. As in HCMV (see Chapter I.10), both share exons but each encodes a unique exon *GP122* (IE2) and *GP123* (IE1). The IE promoter/enhancer is indicated by a blue box. The adjacent region to the right of the IE genes encodes presumptive homologues of the *UL128–131* locus associated with epithelial and endothelial cell tropism in HCMV (see Fig. II.5.2; see also Chapter I.10). Figure was adapted from Yamada *et al.*, 2009. (B) Schematic representation of approximately 1-kbp region of the GPCMV MIE promoter/enhancer (*Xba* I/*Hin*d IIII) fragment. Indicated are presumptive NF-KB binding sites, and TATA box, and first exon. (C) Full length or truncated versions of the promoter drive equivalent levels of reporter gene expression (luciferase) in transient expression assays. Constructs lacking the TATA box exhibit no activity. (D) HCMV IE enhancer construct used as positive control for luciferase assays.

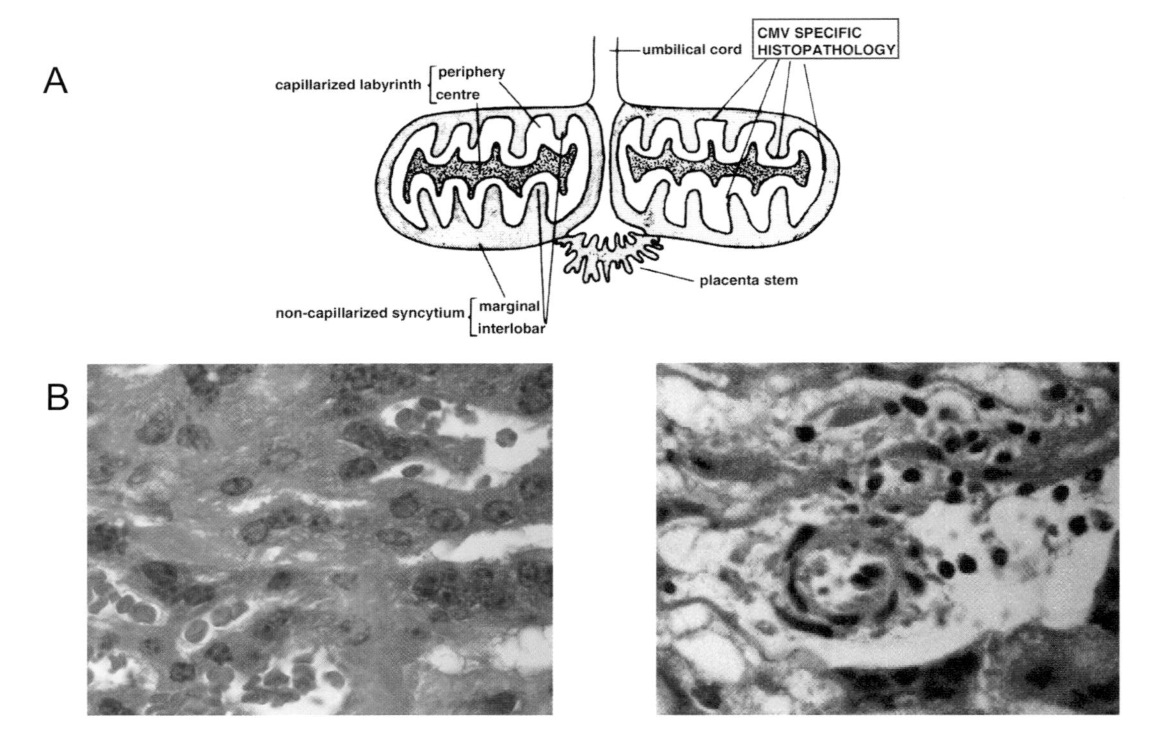

Figure II.5.7 Placental structure in guinea pigs and placental injury and inflammation in tissues obtained from GPCMV-inoculated dams. (A) Structure of guinea pig placenta and regions which harbour GPCMV-associated pathology. The mature guinea pig placenta has a lobular arrangement. In these lobules maternal blood circulates from centre to periphery through lacunae lined with syncytiotrophoblast. Two different zones have been described within this lobular arrangement: the capillarized labyrinth, containing maternal arterial blood and fetal vessels, and the non-capillarized syncytium (marginal or interlobar), which contains only maternal venous blood. GPCMV-induced lesions are found at the transitional zone between these two regions. Figure reproduced from Griffith *et al.* (1985a), with permission. (B) Histopathology of placenta following GPCMV infection during pregnancy. Following third trimester challenge with GPCMV, pup mortality rates range from 30% to 70%, depending upon dose of virus. Placenta obtained 14 days post maternal inoculation demonstrates infarction, and inflammatory injury is noted (haematoxylin–eosin stain). Left panel: healthy near-term placenta showing normal syncytiotrophoblast and vascularity. Right panel: placenta from pup GPCMV-infected *in utero*. Inflammatory cell infiltrate, particularly consisting of mononuclear cells, is noted. Placental and pup organs were positive for GPCMV by culture and quantitative PCR.

2008). Depending upon the strain of guinea pig, source of virus, and timing of inoculation during pregnancy, a variety of experimental endpoints can be modelled. These include mortality in dams, pup infection rate, pup mortality rate, maternal and pup weights, fetal resorption rate, total viral load in dams and pups (assessed by viral culture or PCR), and end-organ disease (including placenta), as assessed histologically and/or by viral load comparisons. Of particular relevance to human health, it has been observed that congenitally infected pups exhibit end-organ disease similar to that described in infants, including brain and visceral involvement, and inner ear pathology, providing additional useful endpoints for vaccine and pathogenesis studies (Griffith *et al.*, 1982b; Woolf *et al.*, 1989; Katano *et al.*, 2007).

Several studies have examined the impact of GPCMV infection during pregnancy on the guinea pig placenta. Following inoculation of pregnant dams at mid-gestation, virological and histological outcomes at the placental–fetal interface were assessed in a comprehensive study in Hartley guinea pigs (Griffith *et al.*, 1985b). Haematogenous spread of GPCMV from the mother to the placenta occurred early during the course of the infection, but GPCMV remained present in placental tissues long after CMV had been cleared from maternal blood, and, interestingly, was able to replicate in placental tissues even in the presence of specific maternal antibodies. This observation may be of particular relevance in light of the increased recognition of congenital HCMV transmission that occurs in the face of preconceptual maternal immunity (Ahlfors *et al.*, 1999; Boppana *et al.*, 2001; Dar *et al.*, 2008; Ross *et al.*, 2010; Yamamoto *et al.*, 2010; see also Chapters II.1, II.2, and II.3).

In the detailed analysis of placental infection conducted by Griffith, whenever GPCMV infection of the fetus occurred, virus was isolated from the associated placenta, but among placental–fetal units with GPCMV-infected placentas, only 27% of the fetuses were found to be infected, implying that the placenta, in addition to permitting access of virus to the fetus, may also, under some circumstances, limit it (Griffith *et al.*, 1985b). Histopathological and ultrastructural analyses of GPCMV-infected placentas in this study revealed a variety of forms of pathology, depending upon the timing of placental examination relative to maternal inoculation. Placentas were examined following mid-gestation inoculation with GPCMV. At early time points post infection (days 14 and 21), a number of placentas examined demonstrated ischaemic injury changes, ranging from foci of coagulative necrosis to large areas of frank infarction, but did not show areas of focal necrosis or viral inclusions. Ischaemic injury on days 14 and 21 post inoculation was predictive of pup mortality. Presumably, fetal loss in this setting is due to miscarriage/intrauterine death, reflecting maternal viraemia and placental insufficiency, rather than any specific effects of GPCMV on the fetus *per se*. Thus, GPCMV fetal mortality can occur in this model, even in the absence of fetal infection (for a discussion relating these findings to those of congenital HCMV-induced placental injury leading to fetal hypoxia, see Chapter II.3).

In contrast to the ischaemic injury seen early after maternal infection, typical CMV-specific histopathology, consisting of multiple areas of necrosis associated with acute and chronic inflammation, was frequently seen at the later time points, on days 21 and 28 post inoculation. Some lesions were associated with acute inflammatory cell infiltrates and prominent necrosis (Fig. II.5.7). Typical intranuclear cytomegalic inclusions were only observed in placentas examined 4 weeks post-maternal inoculation. By TEM, viral nucleocapsids were seen within nuclei of trophoblastic cells, and virions were present surrounding infected cells. Typical CMV-induced histopathological lesions bearing CMV antigens were consistently localized at the transitional zone between the capillarized labyrinth and the non-capillarized interlobium (Fig. II.5.7), a zone containing maternal venous lacunae and fetal arterioles. This zone represents the anatomical junction between the capillarized and non-capillarized trophoblastic tissue, and may function as a 'watershed' site for preferential localization of virus and viral antigen. Since the only continuous cellular layers separating the maternal and fetal blood circulations in the haemomonochorial placenta of the guinea pig are the syncytiotrophoblast and the endothelium of the fetal capillaries (Enders, 1965), it should be possible in future studies to pinpoint more precisely the mechanisms of transplacental transmission of GPCMV.

A number of studies have examined the patterns of GPCMV placental infection following maternal viral challenge very early during pregnancy. In one study, pregnant guinea pigs were inoculated with a low dose of virus during the first trimester (Goff *et al.*, 1987). Maternal viraemia, which had cleared by 2 weeks after initial inoculation, was found to reappear near the time of delivery in approximately one-third of dams, and in the last week of gestation (43–48 days after inoculation) GPCMV was detected at high titre in 95% of placentas. Fetal infection occurred in approximately 40% of pups in the last week of gestation, despite the presence of fetal antibody. These data suggested that the placenta may amplify CMV infection late in the third trimester, possibly secondary to changes in the hormonal milieu during pregnancy. Mechanisms by which GPCMV transmission occurs in spite of the presence of maternal and fetal antibody remain unclear, but these observations suggest that it may be possible to model GPCMV reinfection in pregnant animals. Another study evaluated the effect of inoculation with GPCMV in early gestation by subcutaneous, intracardiac, and intranasal route (Griffith *et al.*, 1990). Although placental and fetal infections occurred in all groups examined, transfer of GPCMV to placentas and fetuses was most efficient in mothers inoculated subcutaneously. Primary viraemia was followed by virus clearance from blood and by an episode of secondary viraemia in all groups. Notably, placental and fetal infections in animals increased during secondary viraemia. These results suggested that secondary maternal viraemia is associated with increased placental and fetal GPCMV infection rates. More recently, the route of intravaginal GPCMV challenge was compared in pregnant and non-pregnant animals (Olejniczak *et al.*, 2011). Pregnant animals had an earlier onset of DNAemia following intravaginal challenge than did non-pregnant animals, and the congenital transmission rate was 39%. Given the importance of sexual transmission for HCMV infection (Handsfield *et al.*, 1985; Collier *et al.*, 1990, 1995; Staras *et al.*, 2008), the establishment of this intravaginal challenge model may prove to be more relevant for future vaccine and antiviral studies, particularly those aimed at interrupting transmission at epithelial cell surfaces, than the less physiological s.c. challenge model.

GPCMV antiviral studies

This section provides an overview of CMV antiviral studies in the guinea pig and a summary of recently developed novel strategies to enhance the relevance and accessibility of this model. The utilization of animal models for CMV antiviral development is the subject of recent comprehensive reviews (Kern, 2006; McGregor and Choi, 2011).

GPCMV and licensed antivirals

There are currently three antivirals [ganciclovir (GCV), foscarnet (FOS) and cidofovir (CDV)] licensed for the treatment of systemic HCMV infection, and these function by interfering with viral DNA replication (see also Chapter II.19). Unfortunately, GPCMV is relatively resistant to GCV at clinically relevant dosage (Fong *et al.*, 1987), which has limited the usefulness of its study in guinea pigs. However, high-dose GCV therapy has been successfully used in preventing CMV related labyrinthitis in guinea pigs (Woolf *et al.*, 1988). GPCMV is susceptible to CDV and less susceptible to FOS (Lucia *et al.*, 1984; Li *et al.*, 1990). Unfortunately, CDV is extremely toxic in guinea pigs, causing kidney damage and death (Li *et al.*, 1990; Bravo *et al.*, 1993). Efforts to reduce the toxicity of CDV have focused on preclinical guinea pig studies with a cyclic congener of CDV (cCDV). This drug is better tolerated in guinea pigs with reduced toxicity while retaining antiviral activity. cCDV was highly effective against GPCMV in an immune compromised guinea pig model (Bourne *et al.*, 2000). Furthermore, cCDV prevented or reduces the rate of GPCMV congenital infection (Bravo *et al.*, 2006; Schleiss *et al.*, 2006a) and successfully prevented hearing loss following intracochlear challenge (White *et al.*, 2006). Recently, an orally bioavailable CDV prodrug, 1-*O*-hexadecyloxypropyl-cidofovir (also referred to as HPD-CDV or CMX001), has been evaluated as an antiviral in the guinea pig model. Oral therapy of GPCMV-infected pregnant dams with HPD-CDV resulted in increased pup survival in treated groups compared with a non-treated control group, but it did not completely prevent congenital CMV infection, in so far as virus could be detected in the tissue of some pups. In addition, at lower doses, although an effect on pup survival was still discernable, virus levels in the fetal tissues were similar to those in control tissues (Bravo *et al.*, 2011).

GPCMV and novel antiviral strategies

Antivirals that target the HCMV DNA terminase complex (pUL56 and pUL89) associated with DNA packaging are effective at blocking viral maturation and, in theory, GPCMV would be susceptible to these drugs since it encodes strongly conserved homologues to the target gene products. 2-bromo-5,6-dichloro-1-beta-D-ribofuranosyl benzimidazole (BDCRB) and a non-nucleoside 4-sulfonamide-substituted naphthalene derivative (BAY 38-4766) are effective against GPCMV (Nixon and McVoy, 2004; McGregor and Choi, 2011). The BAY compound was effective against systemic disease in immune compromised guinea pigs (Schleiss *et al.*, 2005) and resistant strains were isolated in tissue culture with sequence modifications to *GP89* similar to those observed for HCMV *UL89* mutants (Reefschlaeger *et al.*, 2001).

Protein kinase inhibitors of the chemical class of quinazolines, gefitinib (Iressa) and Ax7396 (RGB-315389) have also been evaluated as antivirals against GPCMV. Both compounds showed strong inhibitory effects *in vitro* against human and rodent CMVs with IC_{50} values in low micromolar ranges. The antiviral mode of action is based on the inhibition of protein kinase activity, mainly directed to the pUL97/GP97 kinases, in addition to cellular target genes. In guinea pigs, gefitinib showed inhibition of viral loads in blood and lung tissue, and reduction of mortality (Schleiss, M., *et al.*, 2008). Another novel antiviral strategy in development is based on the GPCMV *GP84* gene. The transdominant inhibitory activity of the HCMV UL84 protein has been explored as a potential antiviral strategy in cell culture (Gebert *et al.*, 1997; Kaiser *et al.*, 2009). The GPCMV homologue, GP84, exhibits transdominant activity against GPCMV when expressed in *trans* (McGregor *et al.*, 2011a). The localization of the inhibitory domain to a short peptide sequence in both HCMV and GPCMV potentially allows the development of a new peptide antiviral or development of a small molecule that mimics the inhibitory domain.

Development of a 'humanized' GPCMV for antiviral studies

As noted, GPCMV is resistant to GCV at medically relevant doses. Since resistance to GCV is mainly linked to codon modifications in the *UL97* kinase gene, and since the GPCMV homologue, pGP97, has multiple non-conserved amino acids (compared to pUL97) in potential kinase domains, including those important in phosphorylation of GCV (Fox and Schleiss, 1997), efforts were undertaken to substitute *GP97* with *UL97* coding sequence in a chimeric GPCMV generated by BAC mutagenesis (McGregor *et al.*, 2008). Fig. II.5.8 illustrates the selection strategy involved in the generation of the GPCMV chimera. A GFP reporter gene encoded in the recombinant virus enabled the tracking of virus in tissue culture. The chimeric GPCMV

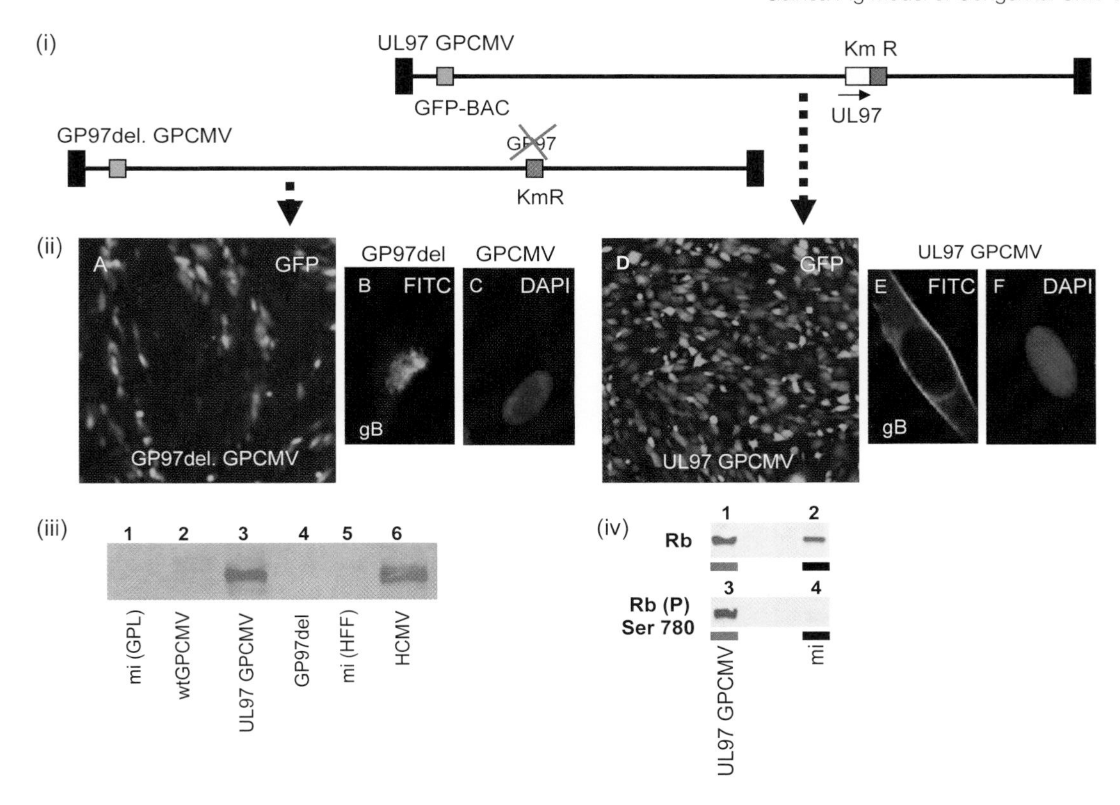

Figure II.5.8 Generation of a viable chimeric pUL97 GPCMV. (i) Overview of the strategy for targeted knockout of *GP97* or substitution of *UL97* in GPCMV BAC constructs. (ii) WT *GP97* or *UL97* chimeric GPCMV are both viable in tissue culture as shown by GFP-positive CPE (A and D) on cell monolayers (approximately 20 days post transfection). *GP97* knockout virus is greatly impaired for virus spread (A) and growth and demonstrates impairment in glycoprotein gB localization at late stage of virus infection (B), unlike chimeric virus (E). (B and E) anti-gB FITC conjugate; (C and F) DAPI stain. (iii) Western blot for HCMV pUL97 protein. Lane: (1) GPL cells, mock infected (mi); (2) GPCMV-infected GPL; (3) *UL97* GPCMV chimeric infected GPL; (4) *GP97* deletion GPCMV-infected GPL; (5) mock infected (mi) human fibroblast (HFF) cells; (6) HCMV (Towne) infected HFF cells. (iv) Western blots for detection of guinea pig retinoblastoma (Rb) protein and phosphorylated Rb protein (Ser_{780}) in *UL97* chimera GPCMV-infected cells at 48 hours post infection. Lanes 1 and 3, *UL97* chimeric GPCMV-infected GPLs; lanes 2 and 4, mock infected GPLs.

expressing pUL97 had similar growth kinetics to wild type GPCMV: in contrast, a *GP97* knockout mutant was impaired for growth and demonstrated impaired viral maturation. Characterization of the chimeric GPCMV verified that it expressed UL97 protein in virus-infected cells. In addition, the pUL97 kinase was capable of phosphorylating the guinea pig retinoblastoma protein when expressed in the context of infection with the GPCMV chimera (Fig. II.5.8).

Analysis of the chimeric virus by plaque reduction assay demonstrated that the *UL97* chimeric GPCMV had increased susceptibility to GCV (IC_{50} of 12.5 μM) compared to wild-type GPCMV (IC_{50} of 125 μM; Fig. II.5.9). In addition, the *UL97* chimeric GPCMV exhibited modified susceptibility to another antiviral that targets pUL97, maribavir, with an IC_{50} of 14 μM *UL97* chimeric GPCMV compared to 62 μM for wild-type virus (McGregor *et al.*, 2008). Antiviral susceptibility could be modified by introduction of specific codon changes found in the *UL97* sequence of resistant strains of HCMV, e.g. L595S (Lurain and Chou, 2010) without affecting viral growth kinetics. Preliminary studies have demonstrated that GCV was an effective antiviral against the *UL97* chimeric GPCMV at clinically relevant dosage in an immunosuppressed guinea pig model (McGregor *et al.*, 2010).

GPCMV detection by non-invasive bioluminescence imaging

As with other CMV models, detection of virus in target tissue of infected animals is either by co-culture of

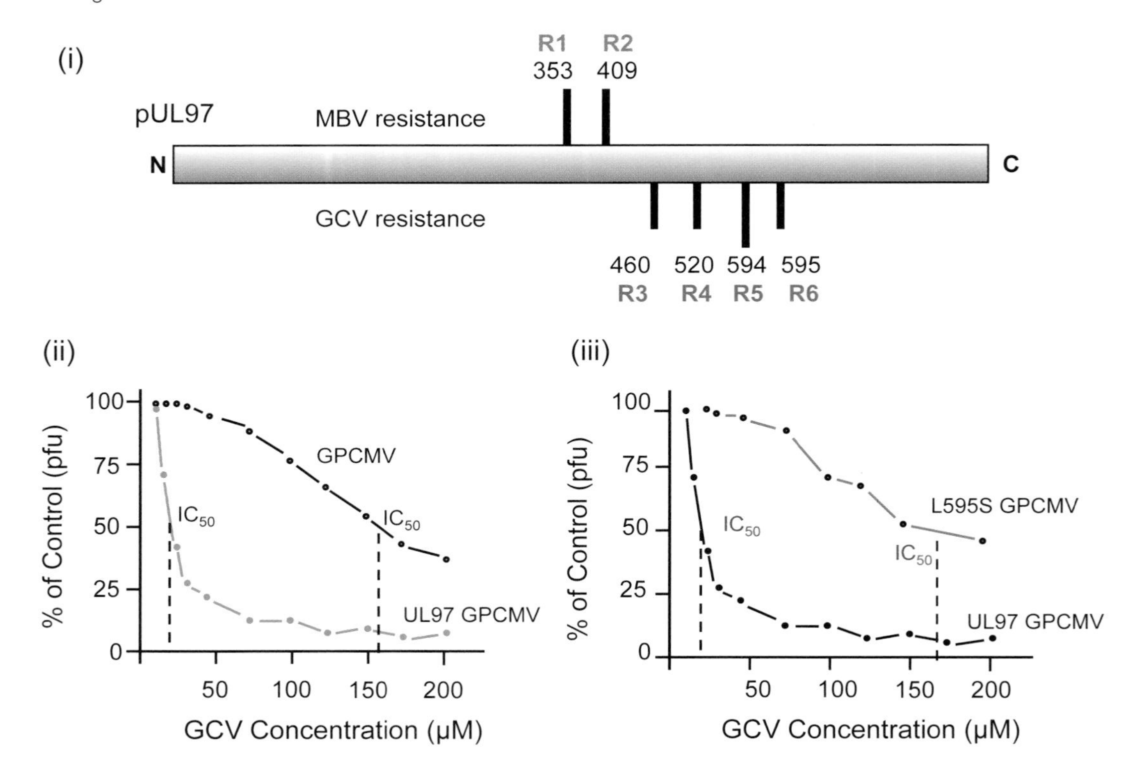

Figure II.5.9 Modification of GPCMV susceptibility to GCV by expression of wild-type or resistant mutant versions of *UL97* chimeric GPCMV. (i) The *UL97* ORF was modified by specific common codon changes found in HCMV GCV *UL97* mutants (R3–R6). In addition, codon mutations found in *UL97* mutants resistant to maribavir (MBV) were also introduced. (ii–iii) Wild type and *UL97* chimeric GPCMV were evaluated for susceptibility to GCV in plaque reduction assays (McGregor *et al.*, 2008). (ii) *UL97* chimeric GPCMV has improved GCV susceptibility relative to wild type GPCMV. (iii) Comparison of the susceptibility of WT *UL97* chimeric GPCMV and mutant *UL97* chimeric GPCMV (R6 mutant, L595S) demonstrates that a single codon change modifies susceptibility to GCV.

tissue homogenate (for virus) or by viral load by real time PCR assay of DNA extracted from target tissue (for viral DNA genome copies). These approaches are time consuming and typically require sacrifice of the animal. Additionally, except for evaluation of viral load in the blood, these protocols do not allow real time tracking of virus in a single animal. We have investigated the ability to track viral dissemination by whole animal bioluminescence imaging (BLI). In this approach recombinant viruses tagged with a firefly luciferase reporter gene can be tracked in real time in the same animal, provided the animal is injected with luciferase enzyme substrate (D-luciferin) prior to imaging.

In order to generate a luciferase-tagged GPCMV, a reporter gene under HCMV MIE promoter control was introduced into an intergenic locus between homologues of *GP25* and *GP26* genes via BAC mutagenesis. The luciferase-tagged virus (designated GPCMV LUC) grew with normal growth kinetics and produced high levels of bioluminescence activity upon infection of fibroblast cells (McGregor *et al.*, 2011b).

We investigated the ability to use the luciferase-tagged virus in both adult and neonatal guinea pigs. Infection via i.p. route allowed the tracking of the virus in animals in real time via BLI over a 2-week period. Virus disseminated to target organs (spleen, liver, lung) in adult animals. In neonates infection of the brain was also observed by approximately 5 days post challenge. Fig. II.5.10 demonstrates luciferase in neonate pups with varying BLI times to enable detection of 'hot spots' of infection. Currently, BLI strategy is being used to evaluate modified tropism of viral mutants in the GPCMV model. The use of this non-invasive imaging strategy for detection of virus dissemination has tremendous potential in aiding the development and evaluation of vaccines and new therapies.

GPCMV vaccine studies

For reasons outlined earlier in this chapter, the guinea pig provides an ideal small animal model for the study of CMV vaccines. Progress in GPCMV vaccine studies

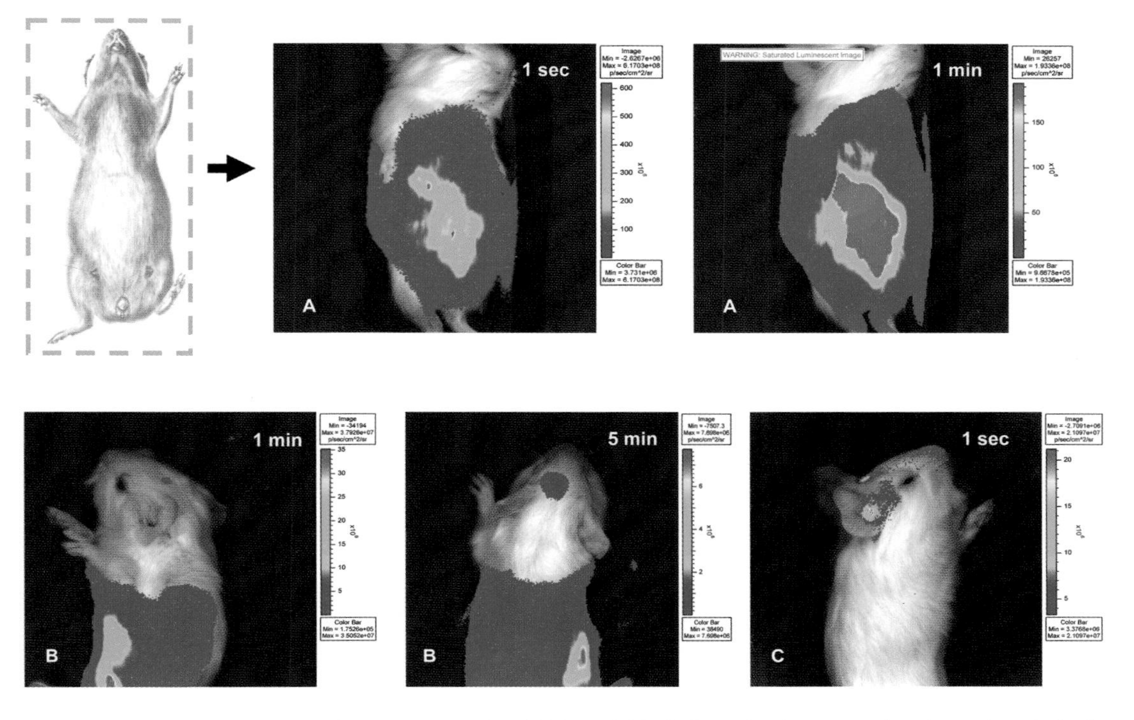

Figure II.5.10 Real time bioluminescence imaging (BLI) of GPCMV infection in neonatal guinea pigs. A firefly luciferase reporter gene under HCMV IE promoter/enhancer control was introduced into the *GP25-GP26* intergenic site. Recombinant virus (GPCMV LUC) grew with normal growth kinetics and exhibited high levels of bioluminescence in cell culture. Following i.p. inoculation (pups A and B) or intracardiac inoculation (pup C), animals were imaged for bioluminescence in a Xenogen IVIS imaging system. Shown are black and white images of pups with superimposed bioluminescence. BLI varied from 1 second to 5 min; the most intense signals are signified in red. Virus infection in the target organs (spleen, liver, and lungs) is indicated by high levels of bioluminescence in the animal torso and was verified by additional PCR studies quantifying viral load in target organs. Bioluminescence is demonstrated in brain by 4–6 days post infection. BLIs were obtained 5 days post infection for pups A and B and 3 days post infection for pup C.

mirrors strategies evaluated in clinical trials of HCMV vaccines (see Chapter II.20) and can be broadly subdivided into four general categories: live virus, attenuated vaccines; replication incompetent virus vaccines; subunit immunization using cloned, recombinant gene products; and passive immunization. Each of these strategies will be considered in some detail. Vaccination studies described to date in the GPCMV model are summarized and provided in Table II.5.1.

Live attenuated vaccination against congenital GPCMV infection

In the first active immunization study reported in the literature, a study in Hartley guinea pigs, it was found that vaccination by two different approaches was efficacious against congenital GPCMV infection (Bia *et al.*, 1980a). This study examined a live, attenuated GPCMV vaccine, and a partially purified, soluble envelope vaccine, administered with Freund's adjuvant. Live vaccine was prepared after 11 serial passages in tissue culture, whereas non-infectious envelope antigen vaccine was prepared by *n*-octyl glucoside treatment of GPCMV-derived dense bodies and virions, enriched for viral envelope antigens. After challenge with virulent GPCMV, animals previously inoculated with live virus vaccine were protected against acute viraemia and death, and a reduced incidence of generalized maternal and fetal infection was observed. Envelope antigen-vaccinated animals showed acute viraemia after similar challenge with virulent virus, but infection was less generalized than that in control animals, and GPCMV was not isolated from the fetuses of these vaccinated mothers. Virus-neutralizing titres were present in dams vaccinated by both strategies, but there was limited data concerning the qualitative aspects of the immune responses, because of the lack of any characterization of GPCMV structural proteins at the time these studies were performed. A subsequent vaccine study of live, attenuated tissue culture-passaged vaccine

Table II.5.1 Previous vaccines studies in the GPCMV congenital infection model

Vaccine approach	Guinea pig strain	Findings
Live, attenuated vaccine isolated following 11 serial passages of ATCC virus in tissue culture (Bia *et al.*, 1980a; 1982; 1984b)	Hartley strain	Reduced incidence of maternal and fetal infection Vaccine virus did not reactivate or produce fetal infection in subsequent pregnancy
	Strain 2	Protection against viraemia, pneumonitis and mortality in non-pregnant animals Reduction in end-organ recovery by culture of SG challenge virus in vaccinated animals
Envelope vaccine (dense bodies and virions) administered with Freund's adjuvant (Bia *et al.*, 1980a)	Hartley strain	Neutralizing antibody responses elicited Vaccinated animals demonstrated reduction in viraemia following SG virus challenge Protection against congenital infection was observed in vaccine group
Immunoaffinity purified envelope glycoprotein with Freund's adjuvant (Harrison *et al.*, 1995)	JY-9 strain	Neutralizing antibody and cell-mediated responses elicited Reduced maternal DNAemia Protection against GPCMV-associated pup mortality and congenital infection
Lectin-purified envelope glycoprotein(s) with Freund's adjuvant (Bourne *et al.*, 2001)	Hartley strain	Antibody responses to several glycoproteins including gB Neutralizing antibody responses Protection against GPCMV-associated pup mortality and congenital infection in live-born pups
DNA vaccines including gB/GP83 (Schleiss *et al.*, 2003) and replication-incompetent BAC DNA (Schleiss *et al.*, 2006b)	Hartley strain	gB DNA vaccine but not GP83 resulted in decreased pup mortality but not decreased vertical transmission rates Vaccination resulted in lower viral load in infected pups BAC DNA vaccine resulted in decreased pup mortality BAC DNA immunogenicity augmented by lipid adjuvant
Baculovirus-expressed, purified recombinant gB vaccine with Freund's and alum adjuvants (Schleiss *et al.*, 2004)	Hartley strain	Neutralizing antibody response adjuvant-dependent correlated with protection Decreased pup mortality and reduced viral load Magnitude of viral load reduction adjuvant-dependent
Single-cycle, RNA replicon-vectored vaccine based on virus-like replicon particles (VRPs) expressing GP83 (Schleiss *et al.*, 2007)	Hartley strain	Engineered using replication-deficient alphavirus Antibody, CD4 and CD8 responses elicited Reduced maternal and pup viral load Reduced pup mortality and improved weights Trend towards decreased congenital transmission
BAC-engineered, live, attenuated vaccine with genes encoding MHC class I homologues (presumed NK evasins; 3DX vaccine) deleted (Schleiss *et al.*, 2006b)	Hartley strain	3DX highly attenuated *in vivo* but not *in vitro* Vaccination with 3DX produced elevated cytokine levels and higher antibody titres than wild type (WT) virus while avidity and neutralizing titres were similar Decreased maternal viral loads and pup mortality noted in vaccinated animals following SG virus challenge

in non-pregnant strain 2 animals showed protection against pneumonitis, although once again the precise viral targets of protective immune responses were not known (Bia *et al.*, 1982).

BAC-based mutagenesis has allowed refinement of live, attenuated vaccine studies, and facilitated generation of vaccines with targeted 'knock-outs' of immune modulation genes. These viruses in turn can be analysed for protective efficacy in the congenital infection model, and the removal of such genes could potentially improve vaccine safety and efficacy. To test this hypothesis, three genes encoding MHC class I homologues (presumed NK evasins) were deleted from the GPCMV genome and the resulting virus, 3DX, was evaluated as a live attenuated vaccine. 3DX grew with wild-type kinetics in cell culture, but was attenuated in guinea pigs. In spite of attenuation, vaccination with 3DX produced elevated cytokine levels and higher antibody titres than wild-type virus, while avidity and neutralizing titres were similar. Protection, assessed by maternal viral loads and pup mortality following pathogenic viral challenge during pregnancy, was comparable in animals vaccinated with 3DX and wild-type virus, and statistically significant compared to control (non-vaccinated) animals (Crumpler *et al.*, 2009). These results suggest that the safety and perhaps efficacy of live attenuated HCMV vaccines could be enhanced by deletion of viral immunomodulatory genes, and evaluation of other live, attenuated vaccine candidates is ongoing.

Replication-incompetent, disabled virus vaccines against congenital GPCMV infection

In a variation of the live, attenuated vaccine strategy, an essential gene function was knocked out such that the virus could only replicate on a complementing cell line. In this approach, the disabled virus can enter a non-complementing cell line and go through an abortive round of replication and express a full array of viral genes, but lacks the ability to produce infectious virus because of a missing essential gene function. To test this approach, the GPCMV capsid gene *GP85* was mutated (Fig. II.5.11) to produce a disabled, infectious single-cycle (DISC) vaccine, designated as GP85 DISC. In preliminary studies this virus has been found to be highly immunogenic and produces neutralizing antibody responses and T-cell responses to GPCMV antigens (McGregor *et al.*, 2011c). This vaccine strategy is currently being evaluated for protection against congenital GPCMV infection.

Subunit native virion protein vaccination against congenital GPCMV infection

Initial evaluations of subunit approaches to vaccination against congenital GPCMV infection utilized lectin-purified or immuno-affinity purified, adjuvanted vaccines. In a vaccine study using an inbred model of congenital GPCMV infection, an immunoaffinity-purified glycoprotein was found to induce strong antibody and cell-mediated responses when administered with Freund's adjuvant, and newborn pups were protected against congenital infection and disease (Britt and Harrison, 1994; Harrison *et al.*, 1995). In another subunit vaccine study, immunity conferred by immunization with envelope glycoproteins resulted in substantial

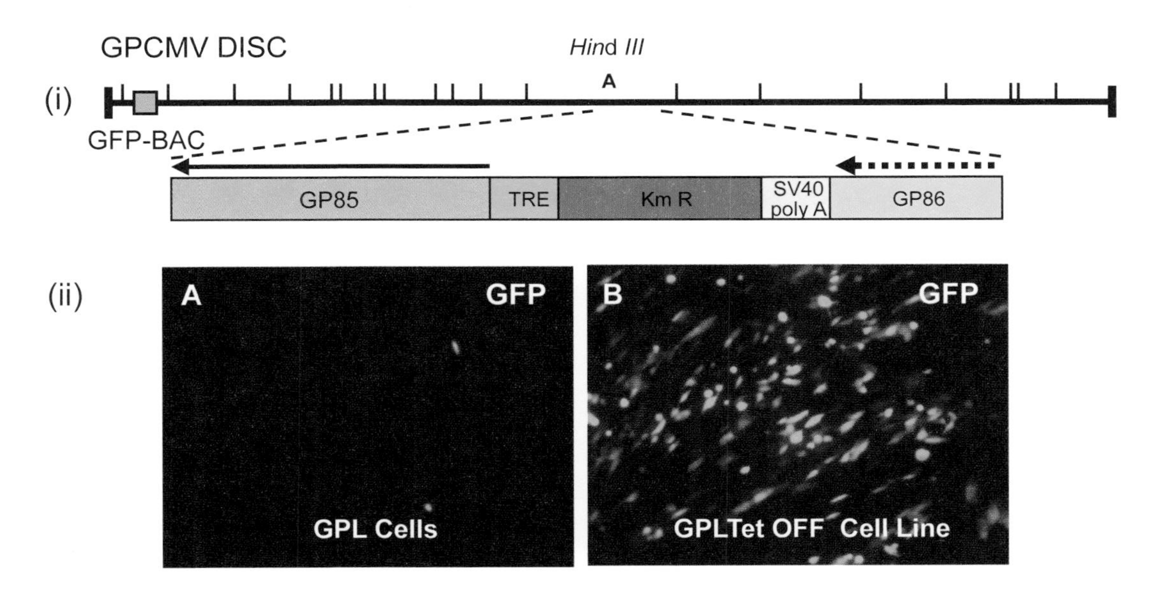

Figure II.5.11 Generation of a disabled infectious single cycle (DISC) vaccine strain for protection against experimental GPCMV infection. Recombinant virus capable of productive replication only on complementing cell line was established by placing the minor capsid gene (*GP85*) under Tet-off promoter control. (i) Overview of *GP85* under a TRE promoter that requires transactivating by the Tet-off transactivator, Tta. Construct was modified to remove all upstream non-coding intergenic sequence and to introduce an SV40 polyA sequence to terminate upstream transcripts. A Km drug resistance marker was also introduced to allow selection of recombinant BACs. (ii) (A) Transfection of *GP85* DISC BAC DNA onto GPL fibroblast cells did not produce infectious virus (lack of GFP virus spread). (B) Transfection of DISC mutant onto tet-off transactivator (Tta) GPL cell line results in viable virus being generated (as evidenced by GFP expression in the monolayer, indicating spread of viral infection). Clonal tet-off cell lines established by permanent transfection of Tta expression plasmid (Clonetech) into GPL cells under G418 selection.

protection against pup mortality, with a reduction in pup mortality from 56% to 14% in the immunized group compared to controls ($P < 0.001$). Among live-born pups in this study, glycoprotein immunization also substantially reduced infection rates. GPCMV was isolated from 24 of 54 live-born pups born to immunized mothers, compared with 16 of 20 live-born pups born to controls, indicating that immunization significantly reduced *in utero* transmission in surviving animals ($P < 0.01$). Since a number of GPCMV glycoproteins were present in the glycoprotein eluate, it was difficult to be certain which protein(s) were key in protection (Bourne *et al.*, 2001). Although both antibody and CMI responses have been noted in these adjuvanted glycoprotein studies, antibody appears to play the dominant role in protection of the fetus.

Cloned, recombinant subunit vaccines: DNA vaccines and recombinant protein expression-based vaccines

In order to optimize the translational relevance of vaccine studies in the GPCMV model, more recent efforts have focused on both molecular characterization of the viral genome, and applying this information to the generation of cloned, recombinant subunit vaccines. As noted, the first GPCMV homologue identified was gB (Schleiss, 1994). Subsequently, the GPCMV homologue of the HCMV pp65 tegument protein, GP83, was identified, and found to encode a 70-kDa phosphoprotein localized to virion and dense bodies (Schleiss *et al.*, 1999). In light of the critical role of immune responses to these two proteins in control of CMV infection, these genes have served as logical cornerstones of subunit vaccine studies, using DNA vaccines as well as recombinant protein expression techniques.

To examine the potential efficacy of DNA vaccines in the GPCMV model, the gB and GP83 ORFs were cloned in plasmid expression vectors, and immunogenicity was evaluated in both mice and guinea pigs (Schleiss *et al.*, 2000). These studies indicated optimal immunogenicity with a secreted form of gB, truncated upstream of the transmembrane domain. Plasmid DNA vaccine was administered epidermally, by 'gene gun' approach, in a protection study (Schleiss *et al.*, 2003). ELISA assay indicated that all animals ($n = 17$ with gB and $n = 14$ with GP83) seroconverted to GPCMV antigen following a 4-dose series of plasmid. In addition, all gB-immunized animals engendered antibody responses capable of precipitating gB species of ~150 kDa, ~90 kDa, and ~58 kDa by RIP-PAGE. Vaccination with gB reduced pup mortality in animals with high ELISA responses, and preconceptual vaccination with gB vaccine resulted in a significant decrease in the incidence of congenital CMV infection. Among 26 live-born pups in the control group, the total congenital infection rate was 77% (20/26). In contrast, in the gB group the total congenital infection rate was 41% (11/27; $P < 0.05$ vs. control group). Mean viral loads in infected pups born to gB and GP83-immunized dams were in the 1.3–1.8 $\log_{10}$ genomes/mg tissue range. In infected pups born to the control (unimmunized) dams, the mean viral load in liver was 3.8 $\log_{10}$ genomes/mg, and 4.0 $\log_{10}$ genomes/mg in spleen. Thus, both vaccines resulted in a reduced viral load in infected animals, compared to controls.

In a variation on the DNA immunization strategy, another study examined the immunogenicity and efficacy of a vaccine based on GPCMV genome cloned as a non-infectious BAC plasmid, modified by transposon insertion into an essential gene, *GP48* (Schleiss *et al.*, 2006b). Vaccinated animals engendered anti-GPCMV antibodies, including anti-gB antibody. Immunogenicity of BAC plasmid DNA was augmented by inclusion of the lipid adjuvant, a mixture of N-[1-(2, 3-dioleyloyx) propyl]-N-N-N-trimethyl ammonia chloride (DOTMA) and dioleoyl phosphatidylethanolamine (DOPE). Following vaccination of Hartley guinea pigs with BAC DNA, third-trimester GPCMV-SG challenge demonstrated that preconception immunization resulted in reduced pup mortality, to 10/34 pups (29%), compared to 23/35 (66%) in controls ($P < 0.005$, Fisher's exact test). In addition, vaccinated dams had reduced viral load, compared to controls, as assessed by quantitative, real-time PCR. This strategy may allow the elucidation of a broad-based immune response to multiple virally encoded immunogens, while at the same time ensuring safety, assuming appropriate mutations can be engineered in essential genes to preclude reconstitution of infectious virus.

For expression in recombinant baculovirus, a truncated, secreted form of gB was engineered, with a modified, insect cell-specific leader peptide (Schleiss and Jensen, 2003). Lectin column-purified gB was subsequently used in a vaccine study (Schleiss *et al.*, 2004). In order to compare and optimize adjuvant in our studies of purified recombinant gB, we compared immunogenicity, neutralizing titres, and protection against congenital GPCMV infection and disease following immunization with three 50 μg doses of vaccine administered over a six-month period with either Freund's adjuvant or alum. All vaccinated guinea pigs seroconverted by ELISA to GPCMV by the third dose. ELISA titres were significantly higher following vaccination with gB and Freund's adjuvant (mean titre, 4.9 log10), compared to gB with alum (mean titre, 4.5 log10; $P < 0.05$, Mann–Whitney test). Similarly, complement-dependent virus neutralizing titres were significantly higher in guinea pigs immunized with gB and Freund's (mean titre, 2.9 log10) compared to gB with alum (mean titre, 2.6 log10, $P < 0.05$, Mann–Whitney test). A total of 12 dams vaccinated with gB and Freund's adjuvant, and 14 dams immunized with gB and alum, became pregnant. A group of 14 unimmunized dams served as controls. Following third-trimester sc GPCMV-SG challenge, outcomes were compared. Among control dams ($n = 41$ pups), the pup mortality rate was 76% (31/41). In contrast, for pups born to gB- immunized dams, the overall pup mortality rate was 25% (23/91; $P < 0.0001$). There was a significant reduction in mortality both for the gB/Freund's group and the gB/alum group. For pups born to dams immunized with gB and alum, pup mortality was 35% (17/49; $P < 0.0001$ vs. control). Immunization with gB and Freund's adjuvant further reduced mortality to 14% (6/42; $P < 0.0001$ vs. control group, and $P < 0.05$ vs. the gB and alum group). Quantitative PCR indicated a significant reduction in maternal viral load conferred by gB/Freund's adjuvant, compared to alum.

To examine the role of a CMI target as a subunit vaccine against congenital GPCMV infection, a propagation-defective, single-cycle, RNA replicon vector system, derived from an attenuated strain of the alphavirus Venezuelan equine encephalitis virus, was used to produce virus-like replicon particles (VRPs) expressing GP83 (Schleiss *et al.*, 2007). Vaccination with VRP-GP83 induced antibodies and $CD4^+$ and $CD8^+$ T-cell responses. Furthermore, dams vaccinated with VRP-GP83 had improved pregnancy outcomes, compared with dams vaccinated with a control VRP expressing influenza haemagglutinin. For VRP-GP83-vaccinated dams, a 13% mortality was noted among ten evaluable litters, compared with 57% mortality among eight evaluable litters in the control group ($P < 0.001$, Fisher's exact test). Improved pregnancy outcome was accompanied by reductions in maternal blood viral load. These results indicate that CMI responses directed against a CMV matrix protein can also protect against congenital CMV infection and disease in the GPCMV model, and suggest that synergistic protection might be expected from a combination gB/GP83 vaccine.

In summary, these studies provide strong evidence for the importance of gB and GP83 in protective immunity against congenital CMV infection and disease in the guinea pig model. The magnitude of protection in these studies appears to depend on several factors including: the adjuvant used, the magnitude of the antibody response engendered, and the magnitude of reduction of maternal DNAemia. The extent of protection conferred by gB when administered with clinically relevant adjuvants warrants further exploration. It is also important to note that there are limitations to gB vaccines in the GPCMV model. Protection against congenital infection induced by gB, though significant, remains incomplete, particularly if a less efficacious adjuvant, alum, is utilized. These observations are clinically relevant, since new adjuvants, in addition to alum, are being introduced for human use. Even with Freund's adjuvant, 'sterilizing immunity' was not induced, and maternal viral loads were still significant. Other gB expression and immunization strategies, such as a combination of protein subunit with plasmid DNA vaccine in a 'prime-boost' approach, warrant future investigation, to identify strategies which optimize gB responses. A more diverse immune response may be required of a CMV vaccine to provide true sterilizing immunity, and a CMV vaccine may require other viral glycoproteins, or other T-cell targets, to confer optimal protection.

Passive immunization against congenital GPCMV infection

The GPCMV model has been used in the past to explore the potential for therapeutic administration of anti-CMV immune globulin (Ig), towards the goal of protecting the fetus against infection and disease. These studies are of considerable interest in light of recent clinical trials that have suggested a therapeutic role for HCMV-specific Ig in pregnant women with evidence of fetal infection (Nigro *et al.*, 2005, 2008; Adler and Nigro, 2008, 2009; Nigro and Adler, 2011). In one guinea pig study, administration of a pooled, polyclonal anti-GPCMV antibody administered prior to GPCMV challenge significantly increased pup survival at several different challenge doses, but when Ig was administered after infection pup survival was improved only after low-dose challenge (Bratcher *et al.*, 1995). Antibody did not protect against GPCMV infection of the pups. Thus, passive GPCMV antibody appeared to improve survival in some groups of congenitally infected pups, but did not affect vertical transmission.

In contrast, in a report in JY-9 animals (an inbred strain of guinea pigs), a hyperimmune anti-gB serum administered following GPCMV infection in early pregnancy was found to reduce the rate of congenital GPCMV transmission (Chatterjee *et al.*, 2001). In addition, a significantly shorter duration of primary maternal viraemia was observed, and fewer pregnancy losses occurred, in passively immunized dams compared with controls. Placentas from recipients of negative control serum were smaller and had cellular infiltrates, focal necrosis, and more viral foci than did those from recipients of anti-gB serum. These data not only support a role for therapeutic benefit of anti-CMV antisera in ameliorating CMV-mediated injury to the placenta and fetus, but also strongly suggest that the benefit is mediated through a specific antiviral effect, likely the virus-neutralizing function of IgG. However, another immunomodulatory function of IgG, such as ADCC or modulation of NK cell function, cannot at this time be excluded. Although anti-gB passive immunization appears to decrease fetal infection and mortality as well as reduce maternal viraemia, additional studies defining the protective response mediated by Ig specific to other glycoproteins, such as those associated with endocytic entry, are warranted, and may help inform and direct future human clinical trials of passive Ig therapy.

Summary and perspectives

Since its discovery nearly a century ago, the GPCMV has been utilized as a valuable small animal model of CMV pathogenesis, one that is uniquely well-suited to vaccine studies. This feature of the model has timely significance today, in light of the increasing recognition of the urgent need for a CMV vaccine. Strategies studied in the GPCMV model may help in developing therapeutic approaches for protecting women of child-bearing age and their newborns against the devastating effects of this congenital infection. Progress in this model should be facilitated by the improved molecular characterization of the viral genome, the ability to effect mutagenesis of the genome, and the recognition that cloned, recombinant subunit vaccine expression technologies can be used to evaluate candidates for efficacy against congenital infection and disease.

In vaccine studies, several areas of future research can be envisaged. Study of clinically relevant adjuvants is a high priority, given the important impact of adjuvant choice for vaccine-mediated protection already observed for recombinant gB. Another strategy which merits consideration is the testing of purified, recombinant forms of other envelope glycoproteins, in addition to gB, for vaccine efficacy. In addition to adjuvanted protein vaccines, strategies that utilize other expression modalities, including technologies that enhance the immunogenicity of DNA vaccines, warrant further testing. As reagents increasingly become available for detailed study of guinea pig cellular immune responses, a more complete examination of the role of CMI in protection of the maternal–placental–fetal unit will be possible. Delineation of the cellular targets for immune responses in this model will enable testing of vaccines primarily developed to elicit CMI responses, alone and in combination with glycoproteins. The ongoing characterization of the GPCMV genome, and the demonstrated ability to mutagenize the genome, now allows detailed study not only of the role of viral gene products in protective immunity but also in pathogenesis for the developing fetus. Study of other putative GPCMV immunomodulatory genes could have implications for future vaccine design. It is also important to emphasize future mucosal challenge studies in the GPCMV model. Mucosal infection represents a more physiological route of CMV entry and development of protocols using this route of challenge may have important translational relevance for future HCMV vaccine studies.

Acknowledgements

Supported by NIH grants HD038416, HD044864, AI083919, AI080972, AI090156 and grants from the March of Dimes Birth Defects Foundation. The assistance of Dr Y. Choi is gratefully acknowledged in establishing the bioluminescence imaging strategy in guinea pigs.

References

Adler, S.P., and Nigro, G. (2008). The importance of cytomegalovirus-specific antibodies for the prevention of fetal cytomegalovirus infection or disease. Herpes *15*, 24–27.

Adler, S.P., and Nigro, G. (2009). Findings and conclusions from CMV hyperimmune globulin treatment trials. J. Clin.Virol. *46 Suppl 4*, S54–57.

Adler, S.P., Shaw, K.V., McVoy, M., Burke, R.L., and Liu, H. (1995). Guinea pig and human cytomegaloviruses do not share cross-reactive neutralizing epitopes. J. Med. Virol. *47*, 48–51.

Ahlfors, K., Ivarsson, S.A., and Harris, S. (1999). Report on a long-term study of maternal and congenital cytomegalovirus infection in Sweden. Review of prospective studies available in the literature. Scand. J. Infect. Dis. *31*, 443–457.

Andrewes, C.H. (1930). Immunity of the salivary gland virus of guinea-pigs studied in the living animal, and in tissue culture. Br. J. Exp. Pathol. *11*, 23–34.

Aquino-de Jesus, M.J., and Griffith, B.P. (1989). Cytomegalovirus infection in immunocompromised guinea pigs: a model for testing antiviral agents *in vivo*. Antiviral Res. *12*, 181–193.

Arvin, A.M., Fast, P., Myers, M., Plotkin, S., and Rabinovich, R. (2004). Vaccine development to prevent cytomegalovirus disease: report from the National Vaccine Advisory Committee. Clin. Infect. Dis. *39*, 233–239.

Barry, P.A., and Strelow, L. (2008). Development of breeding populations of rhesus macaques (*Macaca mulatta*) that are specific pathogen-free for rhesus cytomegalovirus. Comp. Med. *58*, 43–46.

Barry, P.A., and William Chang, W. (2007). Primate betaherpesviruses. In Human Herpesviruses: Biology, Therapy, and Immunoprophylaxis, Arvin, A., Campadelli-Fiume, G., Mocarski, E., Moore, P.S., Roizman, B., Whitley, R., and Yamanishi, K., eds. (Cambridge University Press, Cambridge), pp. 1051–1075.

Barry, P.A., Alcendor, D.J., Power, M.D., Kerr, H., and Luciw, P.A. (1996). Nucleotide sequence and molecular analysis of the rhesus cytomegalovirus immediate-early gene and the UL121–117 open reading frames. Virology *215*, 61–72.

Barry, P.A., Lockridge, K.M., Salamat, S., Tinling, S.P., Yue, Y., Zhou, S.S., Gospe, S.M., Jr., Britt, W.J., and Tarantal, A.F. (2006). Nonhuman primate models of intrauterine cytomegalovirus infection. ILAR J. *47*, 49–64.

Bia, F.J., Griffith, B.P., Tarsio, M., and Hsiung, G.D. (1980a). Vaccination for the prevention of maternal and fetal infection with guinea pig cytomegalovirus. J. Infect. Dis. *142*, 732–738.

Bia, F.J., Summers, W.C., Fong, C.K., and Hsiung, G.D. (1980b). New endogenous herpesvirus of guinea pigs: biological and molecular characterization. J. Virol. *36*, 245–253.

Bia, F.J., Lucia, H.L., Fong, C.K., Tarsio, M., and Hsiung, G.D. (1982). Effects of vaccination on cytomegalovirus-associated interstitial pneumonia in strain 2 guinea pigs. J. Infect. Dis. *145*, 742–747.

Bia, F.J., Griffith, B.P., Fong, C.K., and Hsiung, G.D. (1983). Cytomegaloviral infections in the guinea pig: experimental models for human disease. Rev. Infect. Dis. *5*, 177–195.

Bia, F.J., Miller, S.A., and Davidson, K.H. (1984a). The guinea pig cytomegalovirus model of congenital human cytomegalovirus infection. Birth Defects Orig. Artic. Ser. *20*, 233–241.

Bia, F.J., Miller, S.A., Lucia, H.L., Griffith, B.P., Tarsio, M., and Hsiung, G.D. (1984b). Vaccination against transplacental cytomegalovirus transmission: vaccine reactivation and efficacy in guinea pigs. J. Infect. Dis. *149*, 355–362.

Booss, J., and Kim, J.H. (1989). Cytomegalovirus encephalitis: neuropathological comparison of the guinea pig model with the opportunistic infection in AIDS. Yale J. Biol. Med. *62*, 187–195.

Booss, J., Dann, P.R., Griffith, B.P., and Kim, J.H. (1988). Glial nodule encephalitis in the guinea pig: serial observations following cytomegalovirus infection. Acta Neuropathol. *75*, 465–473.

Booss, J., Dann, P.R., Griffith, B.P., and Kim, J.H. (1989). Host defense response to cytomegalovirus in the central nervous system. Predominance of the monocyte. Am. J. Pathol. *134*, 71–78.

Booss, J., Dann, P.R., Winkler, S.R., Griffith, B.P., and Kim, J.H. (1990). Mechanisms of injury to the central nervous system following experimental cytomegalovirus infection. Am. J. Otolaryngol. *11*, 313–317.

Boppana, S.B., Rivera, L.B., Fowler, K.B., Mach, M., and Britt, W.J. (2001). Intrauterine transmission of cytomegalovirus to infants of women with preconceptional immunity. N. Engl. J. Med. *344*, 1366–1371.

Bourne, N., Bravo, F.J., and Bernstein, D.I. (2000). Cyclic HPMPC is safe and effective against systemic guinea pig cytomegalovirus infection in immune compromised animals. Antiviral Res. *47*, 103–109.

Bourne, N., Schleiss, M.R., Bravo, F.J., and Bernstein, D.I. (2001). Preconception immunization with a cytomegalovirus (CMV) glycoprotein vaccine improves pregnancy outcome in a guinea pig model of congenital CMV infection. J. Infect. Dis. *183*, 59–64.

Brady, R.C., and Schleiss, M.R. (1996). Identification and characterization of the guinea-pig cytomegalovirus glycoprotein H gene. Arch. Virol. *141*, 2409–2424.

Bratcher, D.F., Bourne, N., Bravo, F.J., Schleiss, M.R., Slaoui, M., Myers, M.G., and Bernstein, D.I. (1995). Effect of passive antibody on congenital cytomegalovirus infection in guinea pigs. J. Infect. Dis. *172*, 944–950.

Bravo, F.J., Stanberry, L.R., Kier, A.B., Vogt, P.E., and Kern, E.R. (1993). Evaluation of HPMPC therapy for primary and recurrent genital herpes in mice and guinea pigs. Antiviral Res. *21*, 59–72.

Bravo, F.J., Bourne, N., Schleiss, M.R., and Bernstein, D.I. (2003). An animal model of neonatal cytomegalovirus infection. Antiviral Res. *60*, 41–49.

Bravo, F.J., Cardin, R.D., and Bernstein, D.I. (2006). Effect of maternal treatment with cyclic HPMPC in the guinea pig model of congenital cytomegalovirus infection. J. Infect. Dis. *193*, 591–597.

Bravo, F.J., Bernstein, D.I., Beadle, J.R., Hostetler, K.Y., and Cardin, R.D. (2011). Oral hexadecyloxypropyl-cidofovir therapy in pregnant guinea pigs improves outcome in the congenital model of cytomegalovirus infection. Antimicrob. Agents Chemother. *55*, 35–41.

Britt, W.J., and Harrison, C. (1994). Identification of an abundant disulfide-linked complex of glycoproteins in the envelope of guinea pig cytomegalovirus. Virology *201*, 294–302.

Brune, W., Messerle, M., and Koszinowski, U.H. (2000). Forward with BACs: new tools for herpesvirus genomics. Trends Genet. *16*, 254–259.

Bu, F.R., and Griffith, B.P. (1990). Immunoblot analysis of the humoral immune response to cytomegalovirus in non-pregnant and pregnant guinea pigs. Arch. Virol. *110*, 247–254.

Chatterjee, A., Harrison, C.J., Britt, W.J., and Bewtra, C. (2001). Modification of maternal and congenital cytomegalovirus infection by anti-glycoprotein b antibody transfer in guinea pigs. J. Infect. Dis. *183*, 1547–1553.

Cole, R., and Kuttner, A.G. (1926). A filterable virus present in the submaxillary glands of guinea pigs. J. Exp. Med. *44*, 855–873.

Collier, A.C., Handsfield, H.H., Roberts, P.L., DeRouen, T., Meyers, J.D., Leach, L., Murphy, V.L., Verdon, M., and Corey, L. (1990). Cytomegalovirus infection in women attending a sexually transmitted disease clinic. J. Infect. Dis. *162*, 46–51.

Collier, A.C., Handsfield, H.H., Ashley, R., Roberts, P.L., DeRouen, T., Meyers, J.D., and Corey, L. (1995). Cervical but not urinary excretion of cytomegalovirus is related to sexual activity and contraceptive practices in sexually active women. J. Infect. Dis. *171*, 33–38.

Crumpler, M.M., Choi, K.Y., McVoy, M.A., and Schleiss, M.R. (2009). A live guinea pig cytomegalovirus vaccine deleted of three putative immune evasion genes is highly attenuated but remains immunogenic in a vaccine/challenge model of congenital cytomegalovirus infection. Vaccine *27*, 4209–4218.

Cui, X., McGregor, A., Schleiss, M.R., and McVoy, M.A. (2008). Cloning the complete guinea pig cytomegalovirus genome as an infectious bacterial artificial chromosome with excisable origin of replication. J. Virol. Methods *149*, 231–239.

Cui, X., McGregor, A., Schleiss, M.R., and McVoy, M.A. (2009). The impact of genome length on replication and genome stability of the herpesvirus guinea pig cytomegalovirus. Virology *386*, 132–138.

Dar, L., Pati, S.K., Patro, A.R., Deorari, A.K., Rai, S., Kant, S., Broor, S., Fowler, K.B., Britt, W.J., and Boppana, S.B. (2008). Congenital cytomegalovirus infection in a highly seropositive semi-urban population in India. Pediatr. Infect. Dis. J. *27*, 841–843.

Dargan, D.J., Douglas, E., Cunningham, C., Jamieson, F., Stanton, R.J., Baluchova, K., McSharry, B.P., Tomasec, P., Emery, V.C., Percivalle, E., *et al.* (2010). Sequential mutations associated with adaptation of human cytomegalovirus to growth in cell culture. J. Gen. Virol. *91*, 1535–1546.

Diosi, P., and Georgescu, L. (1974). Degeneration of cytomegalovirus-infected cells in lymph nodes of naturally infected guinea-pigs. Rev. Roum. Morphol. Physiol. *20*, 91–94.

Dunn, C., Chalupny, N.J., Sutherland, C.L., Dosch, S., Sivakumar, P.V., Johnson, D.C., and Cosman, D. (2003). Human cytomegalovirus glycoprotein UL16 causes intracellular sequestration of NKG2D ligands, protecting against natural killer cell cytotoxicity. J. Exp. Med. *197*, 1427–1439.

Enders, A.C. (1965). A comparative study of the fine structure of the trophoblast in several hemochorial placentas. Am. J. Anat. *116*, 29–67.

Farber, S., and Wolbach, S.B. (1932). Intranuclear and cytoplasmic inclusions ('protozoan-like bodies') in the salivary glands and other organs of infants. Am. J. Pathol. *8*, 123–136 123.

Fong, C.K., Cohen, S.D., McCormick, S., and Hsiung, G.D. (1987). Antiviral effect of 9-(1,3-dihydroxy-2-propoxymethyl)guanine against cytomegalovirus infection in a guinea pig model. Antiviral Res. *7*, 11–23.

Fong, C.K., Bia, F., Hsiung, G.D., Madore, P., and Chang, P.W. (1979). Ultrastructural development of guinea pig cytomegalovirus in cultured guinea pig embryo cells. J. Gen. Virol. *42*, 127–140.

Fox, D.S., and Schleiss, M.R. (1997). Sequence and transcriptional analysis of the guinea pig cytomegalovirus UL97 homolog. Virus Genes *15*, 255–264.

Gao, M., and Isom, H.C. (1984). Characterization of the guinea pig cytomegalovirus genome by molecular cloning and physical mapping. J. Virol. *52*, 436–447.

Gebert, S., Schmolke, S., Sorg, G., Floss, S., Plachter, B., and Stamminger, T. (1997). The UL84 protein of human cytomegalovirus acts as a transdominant inhibitor of immediate-early-mediated transactivation that is able to prevent viral replication. J. Virol. *71*, 7048–7060.

Goff, E., Griffith, B.P., and Booss, J. (1987). Delayed amplification of cytomegalovirus infection in the placenta and maternal tissues during late gestation. Am. J. Obstet. Gynecol. *156*, 1265–1270.

Griffith, B.P., Lucia, H.L., Bia, F.J., and Hsiung, G.D. (1981). Cytomegalovirus-induced mononucleosis in guinea pigs. Infect. Immun. *32*, 857–863.

Griffith, B.P., Askenase, P.W., and Hsiung, G.D. (1982a). Serum and cell-mediated viral-specific delayed cutaneous basophil reactions during cytomegalovirus infection of guinea pigs. Cell. Immunol. *69*, 138–149.

Griffith, B.P., Lucia, H.L., and Hsiung, G.D. (1982b). Brain and visceral involvement during congenital cytomegalovirus infection of guinea pigs. Pediatr. Res. *16*, 455–459.

Griffith, B.P., Lavallee, J.T., Booss, J., and Hsiung, G.D. (1984). Asynchronous depression of responses to T- and B-cell mitogens during acute infection with cytomegalovirus in the guinea pig. Cell. Immunol. *87*, 727–733.

Griffith, B.P., Lavallee, J.T., Jennings, T.A., and Hsiung, G.D. (1985a). Transmission of maternal cytomegalovirus-specific immunity in the guinea pig. Clin. Immunol. Immunopathol. *35*, 169–181.

Griffith, B.P., McCormick, S.R., Fong, C.K., Lavallee, J.T., Lucia, H.L., and Goff, E. (1985b). The placenta as a site of cytomegalovirus infection in guinea pigs. J. Virol. *55*, 402–409.

Griffith, B.P., McCormick, S.R., Booss, J., and Hsiung, G.D. (1986). Inbred guinea pig model of intrauterine infection with cytomegalovirus. Am. J. Pathol. *122*, 112–119.

Griffith, B.P., Chen, M., and Isom, H.C. (1990). Role of primary and secondary maternal viremia in transplacental guinea pig cytomegalovirus transfer. J. Virol. *64*, 1991–1997.

Haggerty, S.M., and Schleiss, M.R. (2002). A novel CC-chemokine homolog encoded by guinea pig cytomegalovirus. Virus Genes *25*, 271–279.

Hahn, G., Revello, M.G., Patrone, M., Percivalle, E., Campanini, G., Sarasini, A., Wagner, M., Gallina, A., Milanesi, G., Koszinowski, U., *et al.* (2004). Human cytomegalovirus UL131–128 genes are indispensable for virus growth in endothelial cells and virus transfer to leukocytes. J. Virol. *78*, 10023–10033.

Handsfield, H.H., Chandler, S.H., Caine, V.A., Meyers, J.D., Corey, L., Medeiros, E., and McDougall, J.K. (1985). Cytomegalovirus infection in sex partners: evidence for sexual transmission. J. Infect. Dis. *151*, 344–348.

Harris, J.P., Woolf, N.K., Ryan, A.F., Butler, D.M., and Richman, D.D. (1984). Immunologic and electrophysiological response to cytomegaloviral inner ear infection in the guinea pig. J. Infect. Dis. *150*, 523–530.

Harrison, C.J., and Caruso, N. (2000). Correlation of maternal and pup NK-like activity and TNF responses against cytomegalovirus to pregnancy outcome in inbred guinea pigs. J. Med. Virol. *60*, 230–236.

Harrison, C.J., and Myers, M.G. (1988). Peripheral blood mononuclear cell-mediated cytolytic activity during cytomegalovirus (CMV) infection of guinea pigs. J. Med. Virol. *25*, 441–453.

Harrison, C.J., and Myers, M.G. (1989). Maternal cell-mediated cytolysis of CMV-infected fetal cells and the outcome of pregnancy in the guinea pig. J. Med. Virol. *27*, 66–71.

Harrison, C.J., Britt, W.J., Chapman, N.M., Mullican, J., and Tracy, S. (1995). Reduced congenital cytomegalovirus (CMV) infection after maternal immunization with a guinea pig CMV glycoprotein before gestational primary CMV infection in the guinea pig model. J. Infect. Dis. *172*, 1212–1220.

Hartley, J.W., Rowe, W.P., and Huebner, R.J. (1957). Serial propagation of the guinea pig salivary gland virus in tissue culture. Proc. Soc. Exp. Biol. Med *96*, 281–285.

Hobom, U., Brune, W., Messerle, M., Hahn, G., and Koszinowski, U.H. (2000). Fast screening procedures for random transposon libraries of cloned herpesvirus genomes: mutational analysis of human cytomegalovirus envelope glycoprotein genes. J. Virol. *74*, 7720–7729.

Hsiung, G.D., Choi, Y.C., and Bia, F. (1978). Cytomegalovirus infection in guinea pigs. I. Viremia during acute primary and chronic persistent infection. J. Infect. Dis. *138*, 191–196.

Hudson, N.P., and Markham, F.S. (1932). Brain to brain transmission of the submaxillary gland virus in young guinea pigs. J. Exp. Med. *55*, 405–415.

Isom, H.C., Gao, M., and Wigdahl, B. (1984). Characterization of guinea pig cytomegalovirus DNA. J. Virol. *49*, 426–436.

Jackson, L. (1920). An intracellular protozoan parasite of the ducts of the salivary glands of the guinea-pig. J. Infect. Dis. *26*, 347–351.

Johnson, K.P. (1969). Mouse cytomegalovirus: placental infection. J. Infect. Dis. *120*, 445–450.

Johnson, K.P., and Connor, W.S. (1979). Guinea pig cytomegalovirus: transplacental transmission. Brief report. Arch. Virol. *59*, 263–267.

Jones, C.T., Keay, S.K., and Swoveland, P.T. (1994). Identification of GPCMV infected cells *in vitro* and *in vivo* with a monoclonal antibody. J. Virol. Methods *48*, 133–144.

Kacica, M.A., Harrison, C.J., Myers, M.G., and Bernstein, D.I. (1990). Immune response to guinea pig cytomegalovirus polypeptides and cross reactivity with human cytomegalovirus. J. Med. Virol. *32*, 155–159.

Kaiser, N., Lischka, P., Wagenknecht, N., and Stamminger, T. (2009). Inhibition of human cytomegalovirus replication via peptide aptamers directed against the nonconventional nuclear localization signal of the essential viral replication factor pUL84. J. Virol. *83*, 11902–11913.

Kanai, K., Yamada, S., Yamamoto, Y., Fukui, Y., Kurane, I., and Inoue, N. (2011). Re-evaluation of the genome sequence of guinea pig cytomegalovirus. J. Gen. Virol. *92*, 1005–1020.

Katano, H., Sato, Y., Tsutsui, Y., Sata, T., Maeda, A., Nozawa, N., Inoue, N., Nomura, Y., and Kurata, T. (2007). Pathogenesis of cytomegalovirus-associated labyrinthitis in a guinea pig model. Microbes Infect. *9*, 183–191.

Keithley, E.M., Sharp, P., Woolf, N.K., and Harris, J.P. (1988). Temporal sequence of viral antigen expression in the cochlea induced by cytomegalovirus. Acta Otolaryngol. *106*, 46–54.

Kern, E.R. (2006). Pivotal role of animal models in the development of new therapies for cytomegalovirus infections. Antiviral Res. *71*, 164–171.

Kumar, M.L., and Nankervis, G.A. (1978). Experimental congenital infection with cytomegalovirus: a guinea pig model. J. Infect. Dis. *138*, 650–654.

Kuttner, A.G. (1927). Further studies concerning the filtrable virus present in the submaxillary glands of guinea pigs. J. Exp. Med. *46*, 935–956.

Kuttner, A.G., and T'ung, T. (1935). Further studies on the submaxillary gland viruses of rats and guinea pigs. J. Exp. Med. *62*, 805–822.

Kuttner, A.G., and Wang, S.H. (1934). The problem of the significance of the inclusion bodies found in the salivary glands of infants, and the occurrence of inclusion bodies in the submaxillary glands of hamsters, white mice, and wild rats (Peiping). J. Exp. Med. *60*, 773–791.

Lacayo, J., Sato, H., Kamiya, H., and McVoy, M.A. (2003). Down-regulation of surface major histocompatibility complex class I by guinea pig cytomegalovirus. J. Gen. Virol. *84*, 75–81.

Li, S.B., Yang, Z.H., Feng, J.S., Fong, C.K., Lucia, H.L., and Hsiung, G.D. (1990). Activity of (S)-1-(3-hydroxy-2-phosphonylmethoxypropyl)cytosine (HPMPC) against guinea pig cytomegalovirus infection in cultured cells and in guinea pigs. Antiviral Res. *13*, 237–252.

Liu, Y., and Biegalke, B.J. (2001). Characterization of a cluster of late genes of guinea pig cytomegalovirus. Virus Genes *23*, 247–256.

Lockridge, K.M., Sequar, G., Zhou, S.S., Yue, Y., Mandell, C.P., and Barry, P.A. (1999). Pathogenesis of experimental rhesus cytomegalovirus infection. J. Virol. *73*, 9576–9583.

Lockridge, K.M., Zhou, S.S., Kravitz, R.H., Johnson, J.L., Sawai, E.T., Blewett, E.L., and Barry, P.A. (2000). Primate cytomegaloviruses encode and express an IL-10-like protein. Virology *268*, 272–280.

Lucia, H.L., Griffith, B.P., and Hsiung, G.D. (1984). Effect of acyclovir and phosphonoformate on cytomegalovirus infection in guinea pigs. Intervirology *21*, 141–149.

Lurain, N.S., and Chou, S. (2010). Antiviral drug resistance of human cytomegalovirus. Clin. Microbiol. Rev. *23*, 689–712.

McGregor, A., and Choi, K.Y. (2011). Cytomegalovirus antivirals and development of improved animal models. Expert Opin. Drug Metab. Toxicol. 7, 1245–1265.

McGregor, A., and Schleiss, M.R. (2001a). Molecular cloning of the guinea pig cytomegalovirus (GPCMV) genome as an infectious bacterial artificial chromosome (BAC) in *Escherichia coli*. Mol. Genet. Metab. *72*, 15–26.

McGregor, A., and Schleiss, M.R. (2001b). Recent advances in herpesvirus genetics using bacterial artificial chromosomes. Mol. Genet. Metab. *72*, 8–14.

McGregor, A., and Schleiss, M.R. (2004). Herpesvirus genome mutagenesis by transposon-mediated strategies. Methods Mol. Biol. *256*, 281–302.

McGregor, A., Choi, K.Y., Cui, X., McVoy, M.A., and Schleiss, M.R. (2008). Expression of the human cytomegalovirus UL97 gene in a chimeric guinea pig cytomegalovirus (GPCMV) results in viable virus with increased susceptibility to ganciclovir and maribavir. Antiviral Res. *78*, 250–259.

McGregor, A., Liu, F., and Schleiss, M.R. (2004a). Identification of essential and non-essential genes of the guinea pig cytomegalovirus (GPCMV) genome via transposome mutagenesis of an infectious BAC clone. Virus Res. *101*, 101–108.

McGregor, A., Liu, F., and Schleiss, M.R. (2004b). Molecular, biological, and *in vivo* characterization of the guinea pig cytomegalovirus (CMV) homologs of the human CMV matrix proteins pp71 (UL82) and pp65 (UL83). J. Virol. *78*, 9872–9889.

McGregor, A., Choi, K.Y., and Schleiss, M.R. (2011a). Guinea pig cytomegalovirus GP84 is a functional homolog of the human cytomegalovirus (HCMV) UL84 gene that can complement for the loss of UL84 in a chimeric HCMV. Virology *410*, 76–87.

McGregor, A., Choi, Y., Schleiss, M., and Root, M. (2011b). Non-invasive bioluminescence imaging of cytomegalovirus (CMV) dissemination in real time in the guinea pig animal model. 30th American Society for Virology Meeting. Abstract W24–4, p. 147.

McGregor, A., Choi, Y., Schleiss, M., Kronneman, D., and Root, M. (2011c). Development of a DISC vaccine strategy against congenital CMV infection in the guinea pig model. 13th International CMV/Betaherpesvirus workshop. Abstract 4.11.

McGregor, A., Choi, K.Y., Schleiss, M., Bouska, C., Leviton, M., and Sweet, J. (2012). Analysis of the essential nature of viral proteins considered to be important target antigens against CMV in candidate vaccine strategies in the guinea pig model. Immunology *137* (Suppl. 1), 634–635.

McVoy, M.A., Nixon, D.E., and Adler, S.P. (1997). Circularization and cleavage of guinea pig cytomegalovirus genomes. J. Virol. *71*, 4209–4217.

Markham, F.S. (1938). A study of the submaxillary gland virus of the guinea pig. Am. J. Pathol. *14*, 311–322 311.

Markham, F.S., and Hudson, N.P. (1936). Susceptibility of the guinea pig fetus to the submaxillary gland virus of guinea pigs. Am. J. Pathol. *12*, 175–182.1.

Medearis, D.N., Jr. (1964). Mouse cytomegalovirus infection. 3. attempts to produce intrauterine infections. Am. J. Hyg. *80*, 113–120.

Mess, A. (2007). The Guinea pig placenta: model of placental growth dynamics. Placenta *28*, 812–815.

Mess, A., Zaki, N., Kadyrov, M., Korr, H., and Kaufmann, P. (2007). Caviomorph placentation as a model for trophoblast invasion. Placenta *28*, 1234–1238.

Middelkamp, J.N., Patrizi, G., and Reed, C.A. (1967). Light and electron microscopic studies of the guinea pig cytomegalovirus. J. Ultrastruct. Res. *18*, 85–101.

Miller, S.A., Bia, F.J., Coleman, D.L., Lucia, H.L., Young, K.R., Jr., and Root, R.K. (1985). Pulmonary macrophage function during experimental cytomegalovirus interstitial pneumonia. Infect. Immun. *47*, 211–216.

Nigro, G., and Adler, S.P. (2011). Cytomegalovirus infections during pregnancy. Curr. Opin. Obstet. Gynecol. *23*, 123–128.

Nigro, G., Adler, S.P., La Torre, R., and Best, A.M. (2005). Passive immunization during pregnancy for congenital cytomegalovirus infection. N. Engl. J. Med. *353*, 1350–1362.

Nigro, G., Torre, R.L., Pentimalli, H., Taverna, P., Lituania, M., de Tejada, B.M., and Adler, S.P. (2008). Regression of fetal cerebral abnormalities by primary cytomegalovirus infection following hyperimmunoglobulin therapy. Prenat. Diagn. *28*, 512–517.

Nixon, D.E., and McVoy, M.A. (2002). Terminally repeated sequences on a herpesvirus genome are deleted following circularization but are reconstituted by duplication during cleavage and packaging of concatemeric DNA. J. Virol. *76*, 2009–2013.

Nixon, D.E., and McVoy, M.A. (2004). Dramatic effects of 2-bromo-5,6-dichloro-1-beta-D-ribofuranosyl benzimidazole riboside on the genome structure, packaging, and egress of guinea pig cytomegalovirus. J. Virol. *78*, 1623–1635.

Nogami-Satake, T., and Tsutsui, Y. (1988). Identification and characterization of a 50K DNA-binding protein of guinea-pig cytomegalovirus. J. Gen. Virol. *69*, 2267–2276.

Nozawa, N., Yamamoto, Y., Fukui, Y., Katano, H., Tsutsui, Y., Sato, Y., Yamada, S., Inami, Y., Nakamura, K., Yokoi, M., *et al.* (2008). Identification of a 1.6 kb genome locus of guinea pig cytomegalovirus required for efficient viral growth in animals but not in cell culture. Virology *379*, 45–54.

Olejniczak, M.J., Choi, K.Y., McVoy, M.A., Cui, X., and Schleiss, M.R. (2011). Intravaginal cytomegalovirus (CMV) challenge elicits maternal viremia and results in congenital transmission in a guinea pig model. Virol. J. *8*, 89.

Oxford, K.L., Eberhardt, M.K., Yang, K.W., Strelow, L., Kelly, S., Zhou, S.S., and Barry, P.A. (2008). Protein coding content of the UL)b' region of wild-type rhesus cytomegalovirus. Virology *373*, 181–188.

Oxford, K.L., Strelow, L., Yue, Y., Chang, W.L., Schmidt, K.A., Diamond, D.J., and Barry, P.A. (2011). Open reading frames carried on UL/b' are implicated in shedding and horizontal transmission of rhesus cytomegalovirus in rhesus monkeys. J. Virol. *85*, 5105–5114.

Paglino, J.C., Brady, R.C., and Schleiss, M.R. (1999). Molecular characterization of the guinea-pig cytomegalovirus glycoprotein L gene. Arch. Virol. *144*, 447–462.

Pappenheimer, A.M., and Slanetz, C.A. (1942). A generalized visceral disease of guinea pigs, associated with intranuclear inclusions. J. Exp. Med. *76*, 299–306.

Park, A.H., Gifford, T., Schleiss, M.R., Dahlstrom, L., Chase, S., McGill, L., Li, W., and Alder, S.C. (2010). Development of cytomegalovirus-mediated sensorineural hearing loss in a Guinea pig model. Arch. Otolaryngol. Head Neck Surg. *136*, 48–53.

Penfold, M., Miao, Z., Wang, Y., Haggerty, S., and Schleiss, M.R. (2003). A macrophage inflammatory protein homolog encoded by guinea pig cytomegalovirus signals via CC chemokine receptor 1. Virology *316*, 202–212.

Plotkin, S.A. (1999). Vaccination against cytomegalovirus, the changeling demon. Pediatr. Infect. Dis. J. *18*, 313–325; quiz 326.

Prichard, M.N., Gao, N., Jairath, S., Mulamba, G., Krosky, P., Coen, D.M., Parker, B.O., and Pari, G.S. (1999). A recombinant human cytomegalovirus with a large deletion in UL97 has a severe replication deficiency. J. Virol. *73*, 5663–5670.

Reefschlaeger, J., Bender, W., Hallenberger, S., Weber, O., Eckenberg, P., Goldmann, S., Haerter, M., Buerger, I., Trappe, J., Herrington, J.A., *et al.* (2001). Novel non-nucleoside inhibitors of cytomegaloviruses (BAY 38-4766): *in vitro* and *in vivo* antiviral activity and mechanism of action. J. Antimicrob. Chemother. *48*, 757–767.

Revello, M.G., and Gerna, G. (2010). Human cytomegalovirus tropism for endothelial/epithelial cells: scientific background and clinical implications. Rev. Med. Virol. *20*, 136–155.

Ross, S.A., Arora, N., Novak, Z., Fowler, K.B., Britt, W.J., and Boppana, S.B. (2010). Cytomegalovirus reinfections in healthy seroimmune women. J. Infect. Dis. *201*, 386–389.

Ryckman, B.J., Jarvis, M.A., Drummond, D.D., Nelson, J.A., and Johnson, D.C. (2006). Human cytomegalovirus entry into epithelial and endothelial cells depends on genes UL128 to UL150 and occurs by endocytosis and low-pH fusion. J. Virol. *80*, 710–722.

Ryckman, B.J., Rainish, B.L., Chase, M.C., Borton, J.A., Nelson, J.A., Jarvis, M.A., and Johnson, D.C. (2008). Characterization of the human cytomegalovirus gH/gL/UL128–131 complex that mediates entry into epithelial and endothelial cells. J. Virol. *82*, 60–70.

Ryckman, B.J., Chase, M.C., and Johnson, D.C. (2010). Human cytomegalovirus TR strain glycoprotein O acts as a chaperone promoting gH/gL incorporation into virions but is not present in virions. J. Virol. *84*, 2597–2609.

Schleiss, M., Eickhoff, J., Auerochs, S., Leis, M., Abele, S., Rechter, S., Choi, Y., Anderson, J., Scott, G., Rawlinson, W., *et al.* (2008). Protein kinase inhibitors of the quinazoline class exert anti-cytomegaloviral activity *in vitro* and *in vivo*. Antiviral Res. *79*, 49–61.

Schleiss, M.R. (1994). Cloning and characterization of the guinea pig cytomegalovirus glycoprotein B gene. Virology *202*, 173–185.

Schleiss, M.R. (1995). Sequence and transcriptional analysis of the guinea-pig cytomegalovirus DNA polymerase gene. J. Gen. Virol. *76*, 1827–1833.

Schleiss, M.R. (2002). Animal models of congenital cytomegalovirus infection: an overview of progress in the characterization of guinea pig cytomegalovirus (GPCMV). J. Clin. Virol. *25(Suppl. 2)*, S37–S49.

Schleiss, M.R. (2006). Nonprimate models of congenital cytomegalovirus (CMV) infection: gaining insight into pathogenesis and prevention of disease in newborns. ILAR J. *47*, 65–72.

Schleiss, M.R., and Jensen, N.J. (2003). Cloning and expression of the guinea pig cytomegalovirus glycoprotein B (gB) in a recombinant baculovirus: utility for vaccine studies for the prevention of experimental infection. J. Virol. Methods *108*, 59–65.

Schleiss, M.R., McGregor, A., Jensen, N.J., Erdem, G., and Aktan, L. (1999). Molecular characterization of the guinea pig cytomegalovirus UL83 (pp65) protein homolog. Virus Genes *19*, 205–221.

Schleiss, M.R., Bourne, N., Jensen, N.J., Bravo, F., and Bernstein, D.I. (2000). Immunogenicity evaluation of DNA vaccines that target guinea pig cytomegalovirus proteins glycoprotein B and UL83. Viral Immunol. *13*, 155–167.

Schleiss, M.R., Bourne, N., and Bernstein, D.I. (2003). Preconception vaccination with a glycoprotein B (gB) DNA vaccine protects against cytomegalovirus (CMV) transmission in the guinea pig model of congenital CMV infection. J. Infect. Dis. *188*, 1868–1874.

Schleiss, M.R., Bourne, N., Stroup, G., Bravo, F.J., Jensen, N.J., and Bernstein, D.I. (2004). Protection against congenital cytomegalovirus infection and disease in guinea pigs, conferred by a purified recombinant glycoprotein B vaccine. J. Infect. Dis. *189*, 1374–1381.

Schleiss, M.R., Bernstein, D.I., McVoy, M.A., Stroup, G., Bravo, F., Creasy, B., McGregor, A., Henninger, K., and Hallenberger, S. (2005). The non-nucleoside antiviral, BAY 38-4766, protects against cytomegalovirus (CMV) disease and mortality in immunocompromised guinea pigs. Antiviral Res. *65*, 35–43.

Schleiss, M.R., Anderson, J.L., and McGregor, A. (2006a). Cyclic cidofovir (cHPMPC) prevents congenital cytomegalovirus infection in a guinea pig model. Virol. J. *3*, 9.

Schleiss, M.R., Stroup, G., Pogorzelski, K., and McGregor, A. (2006b). Protection against congenital cytomegalovirus (CMV) disease, conferred by a replication-disabled, bacterial artificial chromosome (BAC)-based DNA vaccine. Vaccine *24*, 6175–6186.

Schleiss, M.R., Lacayo, J.C., Belkaid, Y., McGregor, A., Stroup, G., Rayner, J., Alterson, K., Chulay, J.D., and Smith, J.F. (2007). Preconceptual administration of an alphavirus replicon UL83 (pp65 homolog) vaccine induces humoral and cellular immunity and improves pregnancy outcome in the guinea pig model of congenital cytomegalovirus infection. J. Infect. Dis. *195*, 789–798.

Schleiss, M.R., McGregor, A., Choi, K.Y., Date, S.V., Cui, X., and McVoy, M.A. (2008). Analysis of the nucleotide sequence of the guinea pig cytomegalovirus (GPCMV) genome. Virol. J. *5*, 139.

Schmolke, S., Kern, H.F., Drescher, P., Jahn, G., and Plachter, B. (1995). The dominant phosphoprotein pp65 (UL83) of human cytomegalovirus is dispensable for growth in cell culture. J. Virol. *69*, 5959–5968.

Schraff, S.A., Brown, D.K., Schleiss, M.R., Meinzen-Derr, J., Greinwald, J.H., and Choo, D.I. (2007a). The role of CMV inflammatory genes in hearing loss. Otol. Neurotol. *28*, 964–969.

Schraff, S.A., Schleiss, M.R., Brown, D.K., Meinzen-Derr, J., Choi, K.Y., Greinwald, J.H., and Choo, D.I. (2007b). Macrophage inflammatory proteins in cytomegalovirus-related inner ear injury. Otolaryngol. Head Neck Surg. *137*, 612–618.

Sper-Whitis, G.L., and Schleiss, M.R. (1996). DNA sequence and functional analysis of the putative guinea pig cytomegalovirus major immediate early promoter region. Abstracts of the 21st International Herpesvirus Workshop. Abstract 73.

Staczek, J. (1990). Animal cytomegaloviruses. Microbiol. Rev. *54*, 247–265.

Staras, S.A., Flanders, W.D., Dollard, S.C., Pass, R.F., McGowan, J.E., Jr., and Cannon, M.J. (2008). Influence of sexual activity on cytomegalovirus seroprevalence in the USA, 1988–1994. Sex Transm. Dis. *35*, 472–479.

Strauss, M., and Griffith, B.P. (1991). Guinea pig model of transplacental congenital cytomegaloviral infection with analysis for labyrinthitis. Am. J. Otol. *12*, 97–100.

Tannous, R., and Myers, M.G. (1983). Acquired chemotactic inhibitors during infection with guinea pig cytomegalovirus. Infect. Immun. *41*, 88–96.

Tarantal, A.F., Salamat, M.S., Britt, W.J., Luciw, P.A., Hendrickx, A.G., and Barry, P.A. (1998). Neuropathogenesis induced by rhesus cytomegalovirus in fetal rhesus monkeys (*Macaca mulatta*). J. Infect. Dis. *177*, 446–450.

Tsutsui, Y., Yamazaki, Y., Kashiwai, A., Mizutani, A., and Furukawa, T. (1986). Monoclonal antibodies to guinea-pig cytomegalovirus: an immunoelectron microscopic study. J. Gen. Virol. *67*, 107–118.

Vonglahn, W.C., and Pappenheimer, A.M. (1925). Intranuclear inclusions in visceral disease. Am. J. Pathol. *1*, 445–466 443.

Wang, D., and Shenk, T. (2005a). Human cytomegalovirus UL131 open reading frame is required for epithelial cell tropism. J. Virol. *79*, 10330–10338.

Wang, D., and Shenk, T. (2005b). Human cytomegalovirus virion protein complex required for epithelial and endothelial cell tropism. Proc. Natl. Acad. Sci. U.S.A. *102*, 18153–18158.

Wang, J.B., and McVoy, M.A. (2011). A 128-base-pair sequence containing the pac1 and a presumed cryptic pac2 sequence includes cis elements sufficient to mediate efficient genome maturation of human cytomegalovirus. J. Virol. *85*, 4432–4439.

Wang, W., Zheng, S., Shong, Z., and Zhao, R. (2011). Development of a guinea pig model of perinatal cytomegalovirus-induced hepatobiliary injury. Fetal Pediatr. Pathol. *30*, 301–311.

White, D.R., Choo, D.I., Stroup, G., and Schleiss, M.R. (2006). The effect of cidofovir on cytomegalovirus-induced hearing loss in a Guinea pig model. Arch. Otolaryngol. Head Neck Surg. *132*, 608–615.

Woolf, N.K., Harris, J.P., Ryan, A.F., Butler, D.M., and Richman, D.D. (1985). Hearing loss in experimental cytomegalovirus infection of the guinea pig inner ear: prevention by systemic immunity. Ann. Otol. Rhinol. Laryngol. *94*, 350–356.

Woolf, N.K., Ochi, J.W., Silva, E.J., Sharp, P.A., Harris, J.P., and Richman, D.D. (1988). Ganciclovir prophylaxis for cochlear pathophysiology during experimental guinea pig cytomegalovirus labyrinthitis. Antimicrob. Agents Chemother. *32*, 865–872.

Woolf, N.K., Koehrn, F.J., Harris, J.P., and Richman, D.D. (1989). Congenital cytomegalovirus labyrinthitis and sensorineural hearing loss in guinea pigs. J. Infect. Dis. *160*, 929–937.

Woolf, N.K., Jaquish, D.V., and Koehrn, F.J. (2007). Transplacental murine cytomegalovirus infection in the brain of SCID mice. Virol. J. *4*, 26.

Yamada, S., Nozawa, N., Katano, H., Fukui, Y., Tsuda, M., Tsutsui, Y., Kurane, I., and Inoue, N. (2009). Characterization of the guinea pig cytomegalovirus genome locus that encodes homologs of human cytomegalovirus major immediate-early genes, UL128, and UL130. Virology *391*, 99–106.

Yamamoto, A.Y., Mussi-Pinhata, M.M., Boppana, S.B., Novak, Z., Wagatsuma, V.M., Oliveira Pde, F., Duarte, G., and Britt, W.J. (2010). Human cytomegalovirus reinfection is associated with intrauterine transmission in a highly cytomegalovirus-immune maternal population. Am. J. Obstet. Gynecol. *202*, 297 e291–298.

Yin, C.Y., Gao, M., and Isom, H.C. (1990). Guinea pig cytomegalovirus immediate-early transcription. J. Virol. *64*, 1537–1548.

Yourtee, E.L., Bia, F.J., Griffith, B.P., and Root, R.K. (1982). Neutrophil response and function during acute cytomegalovirus infection in guinea pigs. Infect. Immun. *36*, 11–16.

Yue, Y., and Barry, P.A. (2008). Rhesus cytomegalovirus a nonhuman primate model for the study of human cytomegalovirus. Adv. Virus Res. *72*, 207–226.

Zheng, Z.M., Lavallee, J.T., Bia, F.J., and Griffith, B.P. (1987). Thymic hypoplasia, splenomegaly and immune depression in guinea pigs with neonatal cytomegalovirus infection. Dev. Comp. Immunol. *11*, 407–418.

Murine Model of Neonatal Cytomegalovirus Infection

William J. Britt, Djurdjica Cekinović and Stipan Jonjić

Abstract

Congenital HCMV infection of developing CNS is a major cause of long-term neuronal morbidity. Prevention of neurological damage due to CNS infection is considered as the primary goal of vaccine strategies and passively administered biologics. Owing to species specificity of CMVs animal models have been developed in order to paradigmatically analyse the course of CMV infections. While different animal models provide good tools for studies of HCMV infections in the immunocompromised host, no single animal model completely recapitulates the pathogenesis of congenital HCMV infection. The murine model of perinatal MCMV infection in newborn mice has been proven as powerful tool to analyse the pathogenesis of congenital HCMV infection and mechanisms of the immune response which control CNS infection in the developing brain. This chapter describes a model of MCMV infection in neonatal mice that represents various aspects of HCMV infection in neonates, and thus, could be highly predictive for possible antiviral interventions in humans.

Introduction – epidemiological overview of congenital HCMV infection

Chapter II.2 is dedicated to specifically detail epidemiological aspects of congenital HCMV infection and disease. Congenital (present at birth) HCMV infection is the most common viral infection that is transmitted *in utero* to the human fetus and occurs in about 0.4–1.2% of liveborn infants in most regions of the world (Kenneson and Cannon, 2007; Britt, 2010). HCMV infection in previously non-immune women results in what has been described as a primary infection. Primary maternal infections result in intrauterine transmission to the fetus in about 30% in the first and second trimesters and is increased later in pregnancy (Enders *et al.*, 2011). It is believed that maternal adaptive immunity plays a pivotal role in limiting transmission as evidenced in limited studies of the prevalence of congenital HCMV infections in offspring of women with HIV/AIDS (Mussi-Pinhata *et al.*, 1998; Guibert *et al.*, 2009; Duryea *et al.*, 2010). In contrast to other well studied congenital infections such as toxoplasmosis and rubella, congenital HCMV infections occur also in offspring of women with long standing immunity to the virus, infections that are classified as non-primary maternal infections (Britt, 2010). The rate of transmission following non-primary maternal infection is unknown but has been estimated at between 0.05–1.2% depending on the epidemiological characteristics of the maternal population (Kenneson and Cannon, 2007; Britt, 2010; Wang *et al.*, 2011). The primary determinant appears to be the HCMV seroprevalence in the maternal population (Stagno *et al.*, 1981; Britt, 2010). In some populations in Africa it has been estimated to be as high as 4–6% whereas in a population of higher socioeconomic demographics such as one studied decades ago in Birmingham, Alabama, the rate was estimated to be less than 0.02% (van der Sande *et al.*, 2007). Intrauterine transmission following maternal seroconversion occurs in the periconceptional period and in all three trimesters with an increased risk of transmission attributed to maternal infection in the third trimester of pregnancy in a series of recent studies (Revello *et al.*, 2002; Hadar *et al.*, 2010; Enders *et al.*, 2011). A critically important caveat of these data and one that must be considered in any discussion of the protective activity of maternal immunity or passively acquired immunity in the fetus is that although maternal seroconversion can be defined, the timing of transmission to the fetus is unknown in most cases of congenital infection unless the fetus is subjected to invasive sampling. This remains an unanswered question that could be of great importance when designing intervention strategies to prevent damaging infections following intrauterine transmission of this virus. Limited data (based on assumptions of the proximity of the time between transmission and maternal seroconversion) has suggested that maternal seroconversions in the first and early second trimester are more commonly associated with damaging fetal infections than infections in the third trimester (Stagno *et al.*, 1986; Enders *et al.*, 2011). However, third trimester maternal seroconversions can result in sequelae such as hearing loss (Pass *et al.*, 2006; Enders *et al.*, 2011). Studies in animal models have only further clouded this aspect of congenital HCMV infections in that GPCMV infection of pregnant guinea pigs (see also Chapter II.5) resulted in more frequent transmission events following the secondary viraemia associated with primary infections in the dams (Griffith and Aquino-de Jesus, 1991).

Congenital HCMV infection: symptoms and outcome

Chapter II.3 is dedicated to clinical aspects of congenital HCMV disease in general. Here we focus specifically on discussing issues of relevance for understanding CMV-associated CNS involvement. Newborn infants with congenital HCMV infections were initially classified as symptomatic (infants with physical findings recognizable on clinical examination) and asymptomatic (no physical findings noted on clinical examination). This early classification permitted epidemiological assignments of severe and less severe infections and remains useful in population studies and importantly, has been linked to long-term outcome. The original designation of symptomatic infection has evolved into a more operational definition of any evidence of end-organ system damage that can be documented by clinical or laboratory examination following this intrauterine infection. Interestingly, early virological monitoring of infants with congenital HCMV infections indicated the infants with more severe symptomatic infections excreted quantitatively more virus than those with asymptomatic infections, a finding that has been verified when quantitative PCR techniques have been used to monitor the virus burden in blood and other fluids from infected babies (Stagno *et al.*, 1980a; Lazzarotto *et al.*, 2003; Bradford *et al.*, 2005; Lanari *et al.*, 2006; Walter *et al.*, 2008; Ross *et al.*, 2009). Although viral load does correlate with severity of infection and outcome over relatively broad ranges, values from individual infants are in themselves poorly predictive of risk of long-term sequelae. Because of this variation, in many centres viral load measurements have not been routinely incorporated into clinical care of infected infants.

Symptomatic infections following congenital HCMV infection occur in about 10–12% of infected infants (Stagno *et al.*, 1982a; 1986; Fowler *et al.*, 1992; Ross and Boppana, 2005; Kenneson and Cannon, 2007). Symptoms include clinically obvious manifestations: (1) microcephaly or other neurological symptoms attributable to HCMV associated neurological disease, (2) hepatosplenomegaly with jaundice and (3) thrombocytopenia with petechial/purpuric rashes. Common laboratory findings include hepatitis with conjugated hyperbilirubinaemia, decreased number of platelets, and if imaging such as ultrasound, computerized tomography or magnetic resonance imaging is performed, intracranial calcifications and altered brain development can be documented (Boppana *et al.*, 1992a,b, 1997). Ophthalmological evaluation can reveal chorioretinitis in a minority (< 5%) of infants with symptomatic infections (Amos, 1977; Alford *et al.*, 1983; Anderson *et al.*, 1996). Other findings have also been reported to be associated with congenital HCMV infections, including dental enamel dysplasia and intrauterine growth restriction (Stagno *et al.*, 1982b). Severe congenital HCMV infection infrequently results in organ failure and death with most recent studies reporting mortality rates of around 10%, although these estimates include fatal cases from over two decades ago (Boppana *et al.*, 1992b). With improved care and the availability of effective antiviral therapies, organ failure and death in the perinatal period is infrequent and death in later infancy and childhood is often linked to significant neurological impairments associated with this intrauterine infection.

In contrast to the more dramatic clinical presentation of infants with severe congenital HCMV infection, over 90% of infants with congenital HCMV infection will have no demonstrable symptoms in the newborn period and thus are classified as asymptomatic infections. However, if carefully evaluated some infants with asymptomatic infections will have laboratory evidence of end-organ involvement. The percentage of infants with asymptomatic infections with end organ damage is unknown. This is likely the case in infants with subtle CNS damage that would be detectable only by imaging of the CNS as described above or in other cases by laboratory evaluation of hepatic functions because with the exception of hearing acuity testing, routine evaluation of infants with asymptomatic congenital HCMV infection is not performed. Thus, the interpretations of results from such evaluation that have been reported in the literature are, for the most part, limited by enrolment biases and by small numbers of subjects. Yet some studies have provided very provocative findings such as the autopsy study carried out in New Zealand over a period of 15 years (1964–1979) in which > 90% of infants that died in the perinatal period were autopsied and their CNS examined (Becroft, 1981). Unfortunately, this study took place at a time when viral diagnostics were not widely available and the diagnosis of HCMV was made based on the characteristic histological findings in sections from various organs. Therefore the sensitivity for the detection of HCMV in these autopsy specimens was considerably reduced compared with more recent series in adult patients (Arribas *et al.*, 1996). However, even with this limitation, the author noted that of a total of 16 infants with histopathological findings in various organ systems consistent with congenital HCMV, five (31%) had encephalitis and of these, four (80%) infants had no other histopathological or physical findings that would have identified them as infants with congenital HCMV infection (Table II.6.1) (Becroft, 1981). This report raises the intriguing possibility that infants with unrecognized, asymptomatic congenital HCMV infections could have CNS involvement more frequently than current literature would suggest.

Perinatal infections with HCMV can be acquired at the time of delivery through exposure to virus within genital tract, following exposure to blood products from HCMV-infected donors, and perhaps most commonly, from exposure to HCMV present in infected breast milk (Stagno *et al.*, 1980b; Dworsky *et al.*, 1983; Ahlfors and Ivarsson, 1985; Alford, 1991; Lawrence and Lawrence, 2004). Fortunately in most instances, infants infected in the perinatal period do not develop symptomatic infections and are not at risk for the long-term sequelae that follow intrauterine infection with this virus. Infrequently, preterm infants infected following ingestion of HCMV-infected breast milk provided by their seroimmune mothers can develop multiorgan disease and require antiviral therapy (Vochem *et al.*, 1998; Miron *et al.*, 2005; Hamele *et al.*, 2009). Although definitive data are limited, the current view of the mechanism of disease in premature infants is that they are essentially immunologically compromised secondary to the developmental immaturity of their adaptive immune system and limited concentrations of passively acquired maternal antiviral antibodies secondary to premature delivery. Together these factors are thought to contribute to the severity of the disease in these infants. Evidence derived over 30 years ago in a group of preterm infants who developed disease following transfusion of blood from HCMV-seropositive donors serves as the primary data for this proposed mechanism (Yeager *et al.*, 1981). As noted above, currently available evidence suggests that full-term infants acquiring HCMV in the perinatal period are at very limited risk for any long-term sequelae from this infection, although CNS involvement cannot be ruled out in most reported cases.

Long-term sequelae from congenital HCMV infection: the importance of central nervous system involvement

Although the vast majority of infants with congenital HCMV infection escape any permanent long-term sequelae from this intrauterine infection, somewhere between 12–24% can have CNS sequelae. CNS involvement can range from severe structural damage of the developing brain to minimal abnormalities of perceptual organ function such as hearing loss or chorioretinitis without visual impairment (Pass *et al.*, 1980; Williamson *et al.*, 1982; Conboy *et al.*, 1987; Bale *et al.*, 1990; McCollister *et al.*, 1996; Boppana *et al.*, 1997, 1999; Noyola *et al.*, 2001; Munro *et al.*, 2005). Long-term sequelae from end-organ damage in the liver or bone marrow has not been described in well studied patient cohorts although anecdotal descriptions of liver fibrosis have been circulated. However, clearly the most important sequelae associated with congenital HCMV infection results from CNS damage that follows intrauterine acquisition of this virus. Because of the complexity of the developmental programme of the CNS, including the susceptibility of the CNS to HCMV infection at different periods of development, it is unlikely that observational studies in infected human infants will provide definitive information on the mechanisms of this disease. Perhaps studies of the small number of infants with severe structural

Table II.6.1 CNS involvement in subclinical congenital HCMV infection

Total births (1964–1979)	Perinatal autopsy rate	Total number of congenital HCMV[a]	Number HCMV+ live births	CNS lesions	No. with encephalitis[b]
16 000	>90%	16	9	8	5

[a]Congenital infection diagnosed on autopsy by presence of cytomegalic inclusion and cells. In three cases viruses were isolated.
[b]Four out of five infants with focal encephalitis without other stigmata of congenital HCMV infection (Becroft *et al.*, 1981).

damage of the brain that can be related to morphological changes in regions of the brain will point to altered developmental programmes as mechanisms of disease. More subtle abnormalities of the CNS and associated clinical findings only suggest potential mechanisms. CNS imaging studies that have included magnetic resonance imaging have suggested mechanisms such as deficits in cellular positioning and loss of brain parenchyma that can lead to characteristic findings in infants with severe congenital HCMV infections and physical findings such as microcephaly. Such findings as lissencephaly, polymicrogyria, and cerebellar hypoplasia all point to deficits in cellular positioning in the CNS whereas periventricular calcifications argue for damage to the subventricular grey zone, an area of the brain that is thought to represent the source of neural progenitors that eventually populate the several regions of the brain including the neocortex (Table II.6.2). Definitive studies that could explain mechanisms of disease are not possible because of requirements for invasive sampling of surviving infants with congenital HCMV infections and the understandable reluctance of parents of deceased infants to grant permission for postmortem examination of the CNS. Thus, defining the mechanisms of spread to the CNS and the pathogenesis of CNS damage represent key questions that can only be definitively addressed in non-human models of congenital CMV infections.

Central nervous system disease associated with congenital HCMV infection: histopathological findings

As discussed above, the mechanisms associated with CNS disease in infants with congenital HCMV infection are largely unknown. Studies of infected fetuses and live born infants that have included magnetic resonance imagining and studies of autopsy tissue have defined CNS structural abnormalities that range from lissencephaly and near absence of cortical tissue with ventriculomegaly, pachygyria, polymicrogyria, widespread periventricular calcifications, and infiltration of mononuclear cells (encephalitis) (Table II.6.2). These patterns of disease have led investigators to suggest several mechanisms of disease including loss of neural progenitor cells in the periventricular grey

Table II.6.2 Findings from autopsy and imaging studies in infants with congenital HCMV infection and CNS disease

No. of cases	Source	Findings	Reference
9	Autopsy	Encephalitis (55%), cerebellar dysplasia (11%), micropolygyria (11%), calcifications (33%)	Becroft *et al*. (1981)
6	Autopsy	Encephalitis (100%), cerebellar dysplasia (66%), microgyria, calcifications	Marques Dias *et al*. (1984)
15	Autopsy	Encephalitis (80%), migration deficits (50%), cerebellar dysplasia (40%)	Perlman and Argyle (1992)
31	Autopsy	Encephalitis (55%), white matter involvement (13%)	Gabrielli *et al*. (2009)
8	Autopsy	Encephalitis (12%), calcifications (72%)	Benoist *et al*. (2008)
5	Imaging[b,c]	Lissencephaly, pachygyria (5%), calcifications	Hayward *et al*. (1991)
11	Imaging[b,c]	Lissencephaly (36%), cerebellar hypoplasia (64%), polymicrogyra (45%), calcifications (100%)	Barkovich and Lindan (1994)
11	Imaging[a,c]	Cerebellar hypoplasia (55%), polymicrogyria (45%)	de Vries *et al*. (2004)
56	Imaging	Calcifications (77%)	Boppana *et al*. (1996)
6	Imaging[c]	Cerebellar hypoplasia (33%), pachygyria(66%)	Sugita *et al*. (1991)
49	Imaging[c]	39% abnormal imaging: migration abnormalities (2%), calcifications (10%), cerebellar abnormalities (2%)	Benoist *et al*. (2008)
38	Imaging[c]	Ventriculomegaly (24%), calcifications (5%), cerebellar dysplasia (8%), cortical abnormality (13%), microcephaly (18%)	Doneda *et al*. (2010)
9	Imaging[a]	Lissencephaly (22%), pachy/polymicrogyria (77%), cerebellar hypoplasia	Bosnjak *et al*. (2011)
14	Imaging[c]	Polymicrogyria (43%), cerebellar hypoplasia (29%)	Manara *et al*. (2011)

[a]Ultrasound imaging
[b]Computed tomography imaging
[c]Magnetic resonance imaging

matter secondary to virus infection and subsequent abnormalities in cortical development. Findings in an RhCMV model of congenital HCMV infection were consistent with a loss of neuroprogenitor cells and cortical maldevelopment (Tarantal *et al.*, 1998). As will be discussed below, similar findings in mice inoculated intracranially (i.c.) with MCMV have been described suggesting that if neural progenitor cells are permissive for HCMV replication *in vivo*, the CNS structural abnormalities in infants with congenital HCMV infections could be attributed to the loss of neural progenitor cells (Naruse and Tsutsui, 1989; Shinmura *et al.*, 1997; Tsutsui *et al.*, 2005). An alternative possibility is that HCMV infects the endothelium of the developing vasculature of the fetal brain leading to loss of blood supply and the associated loss of neural cells and structures, a mechanism that would be consistent with symmetric loss of cortical parenchyma (Marques Dias *et al.*, 1984). This possibility is also consistent with findings in animal models of congenital CMV infection in that infection of brain endothelial cells by MCMV has been described (Koontz *et al.*, 2008). However, it is also important to note that infants presenting with severe CNS structural abnormalities secondary to congenital HCMV infection represent a very small fraction of infants with this infection. In fact, such cases represent only a minority of infants that have clinically apparent, symptomatic infections. More commonly, infants with manifestations of CNS disease such as hearing loss or isolated microcephaly do not exhibit such structural disease. Magnetic resonance imaging of infants in this category reveals more subtle findings such as cerebellar hypoplasia, abnormalities in the white matter, and focal calcifications (Table II.6.2). Autopsy studies of infants with congenital HCMV infections have confirmed that in many cases, imaging studies are reflective of the CNS disease but in addition, have also shown that most infants with CNS disease have mononuclear cell infiltrates consistent with histopathological descriptions of encephalitis (Table II.6.2). In contrast to the necrotizing encephalitis associated with herpes simplex virus infection in the newborn, congenital HCMV infection appears to be more frequently associated with focal encephalitis with little or no evidence of significant loss of brain parenchyma (Becroft, 1981; Perlman and Argyle, 1992; Gabrielli *et al.*, 2009; Teissier *et al.*, 2011). The foci contain virus infected cells, mononuclear lymphoid and myeloid cells, and reactive astrocytes. Microglial nodules and gliosis presumably represent the resolution of these areas of infection and inflammation. Careful characterization of the cell types infected with HCMV in these foci has not been accomplished in a sufficient number of specimens to permit a definitive assignment of cellular tropism in the CNS in animal models of CNS infection however, CMV do not exhibit a restricted tropism for specific resident cells of the CNS. The distribution of inflammatory foci is widespread and presumably secondary to the haematogenous spread of the virus to the CNS. Possible mechanisms of disease can be proposed based on pathological findings (Table II.6.3). However, when the descriptions of CNS disease following congenital HCMV infection are viewed collectively, it is difficult to ascribe a single disease pathway that would account for findings present in different cases. Perhaps this should be anticipated because it is almost a certainty that infection occurred at different times during CNS development in many of the infants described in clinical reports. The development of the CNS is extraordinarily complex and different regions mature at different times during fetal growth. Interruption of these pathways through direct virus-induced cell death, virus-induced cell dysfunction, loss of vascularization, or, as will be discussed below; virus-induced inflammation could lead to CNS damage and disease with unique features that could vary dramatically between individual cases. From this discussion, it should be obvious that dissection of disease processes will not be possible using human specimens because of the uncontrolled confounder of the timing of fetal infection together with other potential confounders such as the relationship between viral burden and fetal disease. Animal models will be essential for developing testable hypotheses to account for the manifestations of virus infection of the

Table II.6.3 Proposed mechanisms of HCMV-mediated disease in the CNS of congenitally infected newborns

Mechanism of infection	Pathological effect	Reference
Infection of endothelial cells and development of vasculitis	Insufficient blood supply to developing brain	Marques Dias *et al.* (1984), de Vries *et al.* (2004)
Infection of radial glial cells	Impaired neuronal migration	Kawasaki and Tsutsui (2003)
Infection and inflammation in the subventricular zone	NPCs death	Grassi *et al.* (1998)
Infection of glial cells	Impaired neuronal myelinization	van der Voorn *et al.* (2009)

CNS in human infants with congenital HCMV infections.

If the mechanism of disease in the brain can be viewed as poorly defined, then descriptions of the mechanisms leading to the hearing loss associated with congenital HCMV infection reflect an even more primitive level of understanding. The histopathology of the cochlea and related structures of the hearing apparatus in infants and children with congenital HCMV has been limited to a small number of temporal bones (Strauss, 1985, 1990; Boppana and Britt, 2006). In some infants included in these studies, hearing loss was not documented, thus it is unclear how the histopathology of the cochlea in these cases was related to hearing impairment (Boppana and Britt, 2006). Furthermore, some studies were done long after the perinatal period; those could have reflected ongoing disease and possibly a contribution of other ototoxic agents and/or other infections to the observed histopathology of the inner ear. Finally, it is highly unlikely that a significant number of autopsy specimens will ever be available to extend this data set because of (1) falling overall autopsy rates, (2) infrequent mortality of infants with congenital HCMV infections and (3) reluctance for caretakers to grant permission for recovery of temporal bones of fatal cases of congenital HCMV infections. Thus, studies in animal models offer the most obvious source of information for understanding this manifestation of the congenital HCMV infection.

With an understanding of the limitations of previous studies, the histopathological findings in the specimens from the inner ear of infants with congenital HCMV infection ranged from minimal evidence of infection to diffuse inflammation with hyalinization of the cochlea (Strauss, 1985, 1990; Boppana and Britt, 2006; Teissier *et al.*, 2011). Viral antigens were detected in the cochlea by immunofluorescence assays with polyvalent antisera derived from infected humans (Stagno *et al.*, 1977a). Histological evidence of viral inclusion containing cells was also reported (Davis *et al.*, 1977; Strauss, 1990; Teissier *et al.*, 2011). Thus, it appears that the virus infection of the cells of the inner ear takes place but it is uncertain whether infection of these cells is directly responsible for the development of hearing loss. Mononuclear cell infiltrates have been documented in the majority of specimens but the composition of these infiltrates is not completely defined (Strauss, 1990). A recent study of six terminated fetuses has provided confirmatory data but because it utilized more contemporary techniques, has extended our knowledge about the effects of congenital HCMV infection on the hearing apparatus (Teissier *et al.*, 2011). These authors documented inflammation with infiltration of $CD8^+$ T-lymphocytes and macrophages in the inner structures as well as virus infected cells in the cochlea (Teissier *et al.*, 2011). Specifically, virus and inflammatory cells were found in the stria vascularis and Reissner's membrane, and inflammatory cells in the spiral ganglion (Teissier *et al.*, 2011). In addition, virus and inflammatory cells were found in the vestibular system with involvement of the endolymphatic compartments (Teissier *et al.*, 2011). As has been noted above, there was no correlation between the viral load in the amniotic fluid or the level of placental involvement of the affected fetuses and histological patterns and inner ear lesions (Teissier *et al.*, 2011). Animal models of hearing loss in which virus is injected directly into the cochlea have suggested that the inflammatory infiltrates associated with CMVs in the inner ear correlated with hearing loss and that treatment with anti-inflammatory drugs could decrease the occurrence of hearing loss (Keithley *et al.*, 1988; Woolf *et al.*, 1989, 1990; Katano *et al.*, 2007; Li *et al.*, 2008; Park *et al.*, 2010). Thus, dissection of the nature of inflammatory response in the infected cochlea could offer invaluable insight into the pathogenesis of hearing loss in infants with congenital HCMV infections.

Mechanisms of CNS disease associated with congenital HCMV infections: the importance of inflammation

Mechanisms involving direct virus-induced cytopathology or cellular dysfunction following infection of resident cells of the CNS could explain the manifestations of CNS disease associated with congenital HCMV infections. However, more common findings of focal areas of involvement together with evidence of more global damage from magnetic resonance imaging studies suggesting more diffuse and symmetric involvement of regions of the brain such as the cerebellum have argued that soluble mediators likely play a more important role in CNS disease in infants with congenital HCMV infections. It is somewhat surprising that there is a paucity of studies that have explored the role of inflammation in the pathogenesis of congenital HCMV infection, perhaps because initial studies argued that the level of virus replication was a strong correlate of disease and subsequent studies have attempted to further this correlation. Although subsequent studies have confirmed the early observations correlating the level of virus replication with disease, it is important to also note that this correlation remains only a correlation and levels of viral replication are highly variable between patients precluding the assignment of an absolute level of virus replication associated with CNS disease (Ross *et al.*, 2009). Finally, there has been a general ignorance of the contribution of placental dysfunction and CNS disease in infants with congenital HCMV infection (see also Chapter II.4). Although in infants with congenital HCMV infection placental involvement is focal, the associated inflammatory infiltrate could impact placental function as well as fetal CNS development (La Torre *et al.*, 2006; Maidji *et al.*, 2006; Pereira and Maidji, 2008; Iwasenko *et al.*, 2011; see also Chapter II.4). Recent studies in the guinea pig model of congenital HCMV infection (see also Chapter II.5) have consistently demonstrated placental abnormalities and findings in pups derived from these pregnancies are consistent with placental dysfunction rather than infection of the CNS (Schleiss *et al.*, 2007). These and other findings in animals models as well as the presence of inflammatory cells in the parenchyma and cerebrospinal fluid of infants with CNS involvement together suggest that inflammatory mediators could adversely impact brain development in infected infants.

The nature of the inflammatory response associated with abnormal brain development in infants with congenital HCMV infections remains undefined. Studies in animal models of congenital HCMV infection have contributed the beginnings of an understanding of the role of inflammation in the manifestations of CNS disease in infected infants. CNS infection in murine models of congenital HCMV infection have pointed to a role of inflammation associated with the innate and adaptive immune responses to infection with murine CMV (Cheeran *et al.*, 2004; van den Pol, 2006, 2007; Koontz *et al.*, 2008). The non specificity of the innate response suggests several potential pathways that could alter CNS development. Similarly, inflammatory responses in the CNS of congenitally infected guinea pigs have been described, although in these reports disease was not associated with a specific inflammatory response (Booss *et al.*, 1988).

Prevention of congenital HCMV infection: maternal CMV-specific immunity and intrauterine transmission of HCMV infection

The rate of intrauterine transmission in women undergoing primary infection has been estimated to be about 20-fold higher than the estimated transmission rate of women with non-primary infections, suggesting that pre-existing maternal immunity can limit intrauterine transmission of HCMV following a maternal infection (Stagno *et al.*, 1977b; 1982a; Kenneson and Cannon, 2007; Britt, 2010; Wang *et al.*, 2011). However, it should also be stressed that the true rate of transmission following non-primary maternal infection is undefined as the often cited rates of transmission following non-primary maternal infections are based on assumptions that congenital HCMV infections in offspring of women with non-primary infection arise from recurrent infections and not exposure to a new strain of virus. Should reinfections with new strains of virus be responsible for the majority of congenital HCMV infections following non-primary maternal infections, the rate of intrauterine transmission could conceivably be as high as rates of transmission in women with primary infection. Although difficult to study, this possibility remains an unanswered question in the natural history of congenital HCMV infection. Primary maternal infections leading to intrauterine infection of the developing fetus can occur at any point during pregnancy but seem to increase with increasing gestation as has been observed in other congenital infections (Stagno *et al.*, 1986; Enders *et al.*, 2011). Investigators have speculated that primary maternal infections shortly before conception could also lead to intrauterine transmission at frequencies approaching infections occurring during pregnancy; however, this possibility was not confirmed in a prospective study (Revello *et al.*, 2002; Hadar *et al.*, 2010; Enders *et al.*, 2011).

The apparent difference in rates of intrauterine transmission between women with primary versus those with non-primary infection has prompted the search for a vaccine to prevent congenital HCMV infection (see also Chapter II.20). Furthermore, several investigators have argued that induction of adaptive immune responses that approximate immunity following natural infection could reduce the rate of congenital infection in some populations. The composition of such proposed vaccines remains ill-defined and importantly, studies in women undergoing primary infection have not convincingly identified specific immunological deficits or virological markers that are predictive of transmission to the developing fetus. Associations between intrauterine transmission and the titres of neutralizing antibodies and virus specific $CD4^+$ T-lymphocyte responses have been reported (Alford *et al.*, 1988; Boppana and Britt, 1995; Lilleri *et al.*, 2008). However, if such responses are protective as claimed by some investigators, it remains unclear why such immune responses routinely fail to prevent transmission of virus to the offspring of women with non-primary infections during pregnancy. Perhaps, immunity is waning in these specific women and boosting such responses to protective levels could represent an approach that would limit intrauterine transmission of HCMV. Yet there is little or no definitive data that supports such as hypothesis and immunity to HCMV appears to be boosted continually by reactivations/reinfections as illustrated by the stability of antiviral antibody titres and HCMV specific $CD8^+$ T-lymphocytes in most individuals (Sylwester *et al.*, 2005). In addition, women with pre-existing immunity can transmit virus to their offspring following reinfection with a genetically unrelated strain of HCMV and within highly seroimmune populations, reinfection with new strains of HCMV occurs not infrequently (Boppana *et al.*, 2001; Yamamoto *et al.*, 2010). This observation raised the troubling possibility that protective immunity is strain dependent, a possibility that could limit successful development of protective vaccines. Alternatively it could be argued that specific immunity to HCMV that can prevent infection is extremely limited or possibly even non-existent, an interpretation that can be readily appreciated from studies in non-human primates that have demonstrated repeated reinfection with related RhCMVs (Hansen *et al.*, 2009) (see also Chapter II.22). Thus, it should be obvious that the complexity of the biology of this maternal/fetal infection would suggest that simple approaches such as the induction of adaptive immune responses that mirror those that follow natural infection will be of little value in the prevention of congenital HCMV, particularly in many populations in the world where HCMV seroimmunity approaches 100% in women of child-bearing age.

The development of prophylactic (and potentially therapeutic) vaccines to prevent congenital HCMV infections and/or limit the morbidity of these infections has been a target of biomedical research for nearly four decades (Elek and Stern, 1974; Plotkin *et al.*, 1984). A variety of approaches have been studied which included attenuated replicating viral vaccines, adjuvanted subunit protein vaccines, vectored subunit vaccines, DNA vaccines, and killed virus vaccines. A review of the performance of various vaccines within these broad categories is beyond the scope of this chapter (for a comprehensive overview, see Chapter II.20), but suffice it to say that continued development and testing of vaccines would suggest that earlier candidates have proven unacceptable for routine use in humans. It is important to note that candidate preparations from most of these candidate vaccines could induce adaptive immune responses that in some cases were similar to selected responses in non-immune individuals undergoing infection with HCMV. Unfortunately, informative surrogates of protective immunity remain undefined. Therefore, it is unclear if protective responses were being generated in many of these studies and in cases where there were claims that protective responses were induced, this protection was remarkably short lived and outcome was only marginally significant (Pass *et al.*, 2009). What is more vexing for vaccine programmes is the observation that non-primary maternal infections account for the majority of congenitally infected infants in most populations in the USA and northern Europe and based on extrapolation of studies from Africa, South America, and Asia for almost all the cases of congenital HCMV infection in many countries in the southern hemisphere. In fact, even in the USA, as the number of women susceptible to primary infection with HCMV increases, the incidence of congenital HCMV infection falls dramatically (Britt, 2010). This relationship between maternal seroprevalence and the incidence of congenital HCMV infection suggests that an effective vaccine could have only a limited market unless such a vaccine could prevent intrauterine transmission from women with existing seroimmunity. Finally, it is important to remind the reader that the incidence of sequelae following congenital HCMV infection in offspring of women with non-primary infection is similar to that observed in offspring of women with primary infection during pregnancy (Boppana *et al.*, 1999; Ahlfors *et al.*, 2001; Ross *et al.*, 2006; Mussi-Pinhata *et al.*, 2009). Thus, it remains to be determined if candidate vaccines that induce responses similar to those following natural infection will limit the incidence of long-term sequelae in most populations of the world.

Congenital HCMV infection: treatment

Treatment of congenital HCMV includes both prophylactic immunotherapy and the use of antiviral agents in the postnatal period. Treatment of pregnant women undergoing primary infection with pooled intravenous immune globulin selected for increased titres reactive with HCMV was claimed to be potentially beneficial in terms of improved outcome of congenitally infected infants (Nigro *et al.*, 2005). However, it is almost universally agreed that the study design and patient inclusion/exclusion criteria rendered this study almost uninterpretable. Yet this study was provocative and animal studies clearly supported the rationale for this approach (Bratcher *et al.*, 1995; Chatterjee *et al.*, 2001; Cekinovic *et al.*, 2008). The results from a larger and a possibly better controlled study using this approach will be available within the very near future. Well defined human monoclonal antibodies with extraordinary *in vitro* antiviral activity are being developed for human studies such as those described above. Indeed if these biologics are shown to have similar activity *in vivo* and can be shown to be broadly reactive with clinical isolates of HCMV, they could offer a potential immunotherapeutic to limit both infection and disease in the fetuses of women undergoing primary HCMV infection during pregnancy. Whether the rationality for their use could be extended to women with non-primary infection is unclear and even if efficacious, the current cost of production would be prohibitive for widespread usage.

Treatment of congenital HCMV has been primarily targeted at congenitally infected infants identified in the newborn period. Infants with significant end-organ disease such as hepatitis, pneumonitis, and thrombocytopenia following perinatal infections such as those acquired through ingestion of HCMV-infected breast milk have been successfully treated with antivirals such as GCV. Treatment with GCV was also reported to limit the development and progression of hearing loss in a group of congenitally infected infants when their outcomes were compared to untreated historical controls (Kimberlin *et al.*, 2003). The use of antivirals such as GCV in the treatment of intrauterine infection would seem to represent a potentially therapeutic intervention. However, the unknown toxicity to the developing fetus and the known teratogenicity of this drug has prevented most investigators from considering prenatal treatment with this agent. Overall, findings from studies of infants with postnatal infections and infants with congenital infections are consistent with the capacity of antivirals to inhibit virus replication and limit organ system damage that is associated with virus replication. What remains to be determined is the effect of antivirals on long-term sequelae such as hearing loss that perhaps develops secondary to host derived responses with more limited contribution from productive virus replication. These questions can only be mechanistically defined in animal model systems and then confirmed in more well controlled human studies.

Animal models of congenital CMV infection: lessons from studies of CNS infection in murine models of congenital CMV infection

Species specificity of CMV precludes studying the pathogenesis of congenital HCMV infection in animal models. Among seven established animal models of CMV infection (chimpanzee, rhesus macaque, tupaia, pig, rat, mouse and guinea pig) the most extensively studied is the murine model. Advantages of the MCMV model include well defined viral genome and proteome, similar pathogenesis of infection and the availability of numerous viral mutants for *in vivo* studies. A mouse model is particularly suitable due to the genetic diversities between the various inbred mouse strains, especially in terms of differential response to MCMV infection. However, a significant but not insurmountable disadvantage of the murine model of congenital CMV infection is the inability of MCMV to cross the placental barrier and infect mouse embryos. To overcome this shortfall, investigators have developed alternative routes of infection including infection of either mouse embryos or newborn mice. MCMV infection of developing CNS can be established by direct virus inoculation into cerebral hemispheres or lateral ventricles of either newborn mice or mouse embryos (Ishiwata *et al.*, 2006; Li *et al.*, 2008), intraplacental inoculation of the virus (Tsutsui, 2009), i.p. infection of newborn mice (Koontz *et al.*, 2008) or infection of suckling mice by milk extracted from MCMV-infected dams (Wu *et al.*, 2011).

Intracranial (i.c.) inoculation of the virus requires cryoanesthesia of newborn mice prior to intraventricular inoculation of the virus (Kosugi *et al.*, 2002) or virus inoculation in lateral ventricles of mouse embryos in anesthetized pregnant mice at later stages of gestation, either of which could result in unintended injury to the CNS (Shinmura *et al.*, 1997). Intraplacental inoculation of the virus does not infect all offspring and it is unknown whether all offspring are infected with the same inoculum of virus (Li and Tsutsui, 2000). Furthermore, reports that have described findings from experiments utilizing this strategy have also reported the need for additional pre-treatments such as intraplacental inoculation of TNF-α (Tsutsui, 2009). Intraperitoneal inoculation of MCMV into newborn mice several hours after birth does not require any pre-treatment anaesthesia and additional techniques, while assuring an identical viral inoculum in all animals that leads to consistent systemic infection and predictable virus spread into CNS (Koontz *et al.*, 2008) (Fig. II.6.1).

Intraventricular virus inoculation has been reported to result in localized infection of periventricular glia cells, endothelial cells of the choroid plexus and neuroblasts in the subventricular area in close proximity to the site of virus inoculation. Infection of neural cells in subventricular areas is followed by infection of hippocampus or cortical neurons, most probably via continuous route, spreading from cell-to-cell or by radial migration of undifferentiated, developing neurons (Kosugi *et al.*, 2002). In adult mice, inoculation of MCMV by direct i.c. injection was associated with disruption of the blood brain barrier and infection along the track of the needle, findings which raise concerns about the validity of results from studies using this mode of infection (van Den Pol *et al.*, 1999). In the model utilizing i.p. infection of newborn mice, virus enters the CNS during haematogenous spread and replicates in the brain parenchyma without any predilection for specific regions of the brain or resident cell types of the CNS (Cekinovic *et al.*, 2008; Koontz *et al.*, 2008). Of available murine models of congenital CMV infection, this model more closely recapitulates the presumed route of HCMV entry into the developing CNS during systemic viraemia (Stagno and Britt, 2006). In fact, following i.p. inoculation, virus can be detected in both plasma and blood cells of newborn

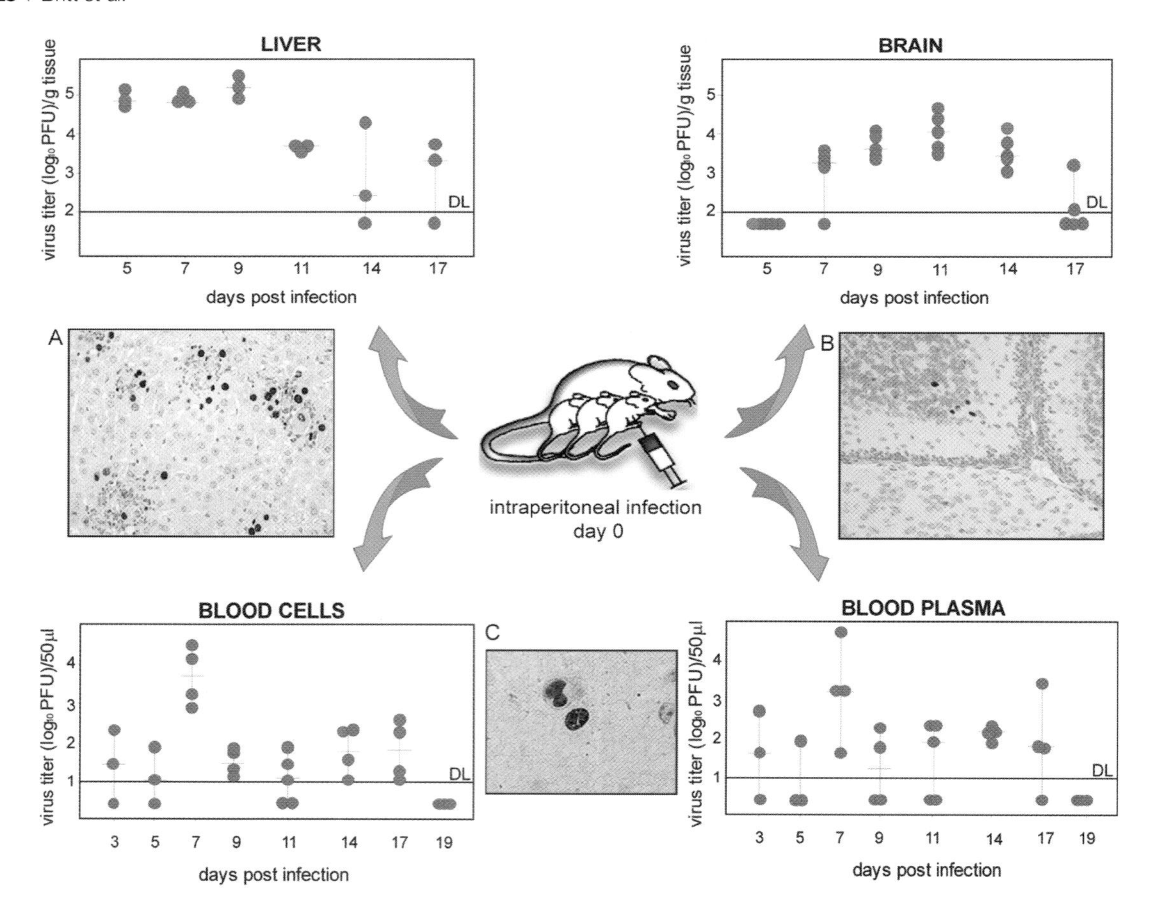

Figure II.6.1 Systemic MCMV infection in newborn mice. Following i.p. inoculation of the virus. MCMV can be detected in various organs, blood cells and plasma. IHC analysis using antibodies specific for MCMV IE-1-encoded protein reveals virus-infected cells in (A) liver, (B) cerebellum and (C) brain parenchymal blood vessels. Magnifications 10× (B), 20× (A) and 100× (C). Reproduced from Cekinović *et al*. (2008) and amended with permission from American Society for Microbiology.

mice prior to peak replication in the brain suggesting that virus enters the CNS either as cell-free virus or as cell-associated virus (Cekinovic *et al.*, 2008; Fig. II.6.1). Although the precise mechanism of spread into the CNS is unknown, preliminary findings suggest the possibility that it enters the CNS via mononuclear cells that populate developing CNS in the early post-natal period (Ling and Wong, 1993; Saederup *et al.*, 1999). Finally, it should be noted that direct i.c. virus inoculation bypasses the peripheral innate and adaptive immune responses of the developing mouse and this is in obvious contrast to responses observed in the model of i.p. infection of newborn mice (Bantug *et al.*, 2008).

CMV infection of the CNS: differential susceptibility of cells to CMV infection

In vitro analysis of infection of both human and murine cell lines has revealed that virtually all cell types present in the CNS are susceptible to infection by respectable CMV as measured by virus entry and transcription of IE genes. However, there exist remarkable difference in the ability of different cell types to support productive infection (Cheeran *et al.*, 2009). Astrocytes, cells which represent the most predominant cell type in the CNS, as well as ECs of the cerebral blood vessels both support virus replication (Jarvis and Nelson, 2002). *In vitro* analysis of HCMV infection of astrocyte cell cultures showed that virus replication can be abolished when cells are pre-treated with proinflammatory cytokines TNF-α, IL-1β or IFN-γ (Cheeran *et al.*, 2000). However, this finding has not been confirmed using recent isolates of clinical strains of HCMV. Importantly, astrocytes and ECs in the CNS form the blood–brain barrier (BBB), a selective barrier which insures nutrition but selectively limits passage of molecules and cells of the peripheral blood into the CNS (Abbott *et al.*, 2006). Productive CMV infection in ECs induces monocyte activation and recruitment, as well as infection of these cells, that could contribute to CNS infection during development (Bentz *et al.*, 2006). Infection with neurotropic viruses has been shown to disrupt the integrity of the BBB (Wang *et al.*, 2004). In the model of brain infection that we have described, we have not observed increased leakage of Evans blue dye through the BBB suggesting that the integrity of the BBB is unchanged during MCMV infection of the brain in newborn mice (Koontz *et al.*, 2008). However, this data does not exclude the possibility of transcytosis of cell-free virus into the CNS or cell-mediated entry during the early postnatal period, a time when monocytes actively populate the CNS pool of microglia cells (Ling and Wong, 1993). In contrast to astrocytes, in another glial cell type present in the CNS, microglia cells, productive infection has not been reported and although HCMV DNA can be detected in these cells, there is no expression of IE-encoded proteins (Lokensgard *et al.*, 1999). In recent studies, microglia cells have been infected *in vitro* as determined by IE-1 staining and viral RNA detection by RT-PCR suggesting that there is not an intrinsic block to infection in cells of this lineage (Britt, unpublished). Microglia progeny are thought to be derived from myeloid cells in the bone marrow and migrate into developing CNS; however more recent studies have argued this interpretation of existing data may be simplistic (Ling and Wong, 1993). In contrast to the reported phenotype of microglia cells, myeloid precursors at different stages of development and blood-derived monocytes support productive HCMV infection (Chan *et al.*, 2010) and are considered to contribute to virus spread within infected animals (Mocarski, 2002). Thus, these and other studies argue that the importance of microglial cells in the replication and spread of MCMV in the CNS remains to be defined.

The capacity of CMV to infect neurons appears to depend on the developmental state of these cells. While mouse embryonic stem cells are not permissive for MCMV infection (Kawasaki *et al.*, 2011), neural stem cells, which are precursors of both neuronal and glial cells, and localize to the subventricular area of the brain, fully support productive MCMV infection (Tsutsui *et al.*, 2008). It has been argued that replication of MCMV in these cells inhibits their ability to proliferate and differentiate (Cheeran *et al.*, 2005). MCMV infection in the developing brain induces the loss of CD133^{+} neural stem cells, specific cell clones which are involved in the formation of cerebral neurospheres, important for early postnatal neuronal development (Barraud *et al.*, 2007; Mutnal *et al.*, 2011). In contrast, primary differentiated neural cells, as well as mature neurons are refractory to HCMV infection (Lokensgard *et al.*, 1999), although viral entry into neurons is readily observed (Lokensgard *et al.*, 1999). Interestingly, although neural stem cells represent a very small population of cells in mature brain, they are preserved for the lifetime in the CNS, providing a population of susceptible cells for CMV infection even in the adult host. The block of virus replication in differentiated neural cells and end-stage neurons is observed at the level of transcription from the MIE promoter (MIEP), resulting in abortive infection (Cheeran *et al.*, 2005). One of the proposed mechanisms leading to abortive CMV infection of neurons is the presence of a dominant negative isoform of transcription factor C/EBP in differentiated neural cells. This molecule is proposed to inhibit transcription from the HCMV MIEP by binding to regions immediately downstream from its NF-kB binding sequence leading to inhibition of transcription (Ossipow *et al.*, 1993; Prösch *et al.*, 2001). More recent data by Luo and colleagues (Luo *et al.*, 2008) revealed *in vitro* susceptibility of neural precursor cells (NPCs) to HCMV infection and this permissiveness to infection is preserved in differentiated neurons or astroglia cells derived from cultured NPCs. This result can argue for neurons as cells that preserve virus in the CNS.

In models using i.c. virus inoculation to deliver MCMV to the CNS, infected neuronal cells, compared with uninfected neurons, exhibit decreased susceptibility to excitatory toxicity and to glutamate-induced cell death, and are resistant to apoptosis (Kosugi *et al.*, 1998). These results may imply MCMV latency in neurons (Tsutsui *et al.*, 1995). Interestingly, in contrast to MCMV infection, *in vitro* analysis in a model of HIV-associated neuronal death showed increased glutamate release from astrocytes leading to induced neuronal apoptosis (Bezzi *et al.*, 2001). In our model of perinatal MCMV infection in the developing brain we also have not observed an increased percentage of apoptotic cells as measured by caspase-3 expression in neurons (Koontz *et al.*, 2008). In contrast, results from a limited number of cases of HCMV-induced encephalitis in newborns showed significant numbers of apoptotic neurons and glial cells in areas of productive virus infection suggesting that apoptosis is a result from direct viral injury to neurons and not dependent on inflammatory response (DeBiasi *et al.*, 2002).

Moreover, human NPCs, when infected with HCMV have reduced differentiation ability and increased apoptotic index (Odeberg *et al.*, 2006).

In vitro analysis of HCMV infection revealed at least two genes in the HCMV genome that exhibit anti-apoptotic activity when expressed in the late phase of HCMV replication (see also Chapter I.15). The UL36 gene encodes the viral inhibitor of caspase-8 activation and the UL37 gene encodes the viral mitochondria-localized inhibitor of apoptosis (vMIA) (Goldmacher, 2005). More recent data show that HCMV can inhibit apoptosis in the early phase of infection by up-regulating the PI3K pathway (Chan *et al.*, 2010). The MCMV homologue of UL36, M36, also has a proven anti-apoptotic capacity (Menard *et al.*, 2003) and has a role in MCMV replication *in vivo* (McCormick *et al.*, 2003; Cicin-Sain *et al.*, 2008). Genes M41 and m38.5 of the MCMV genome have also been identified as genes encoding anti-apoptotic proteins either by interfering with resident apoptosis in the Golgi compartment or by inhibiting apoptosis induction at the mitochondrial checkpoint (Brune *et al.*, 2003; Jurak *et al.*, 2008). However, these mechanisms of action have been defined in murine fibroblast cell lines and further investigations are needed to evaluate a possible role of these genes in the control of neuronal survival following MCMV infection.

Developmental abnormalities in the CMV-infected brain

Mechanisms of HCMV-mediated neural disease are insufficiently defined and proposed mechanisms include (1) viral interference with normal neural cell growth and differentiation, suggesting a 'teratogenic' role of CMV during congenital infection (de Vries *et al.*, 2004); (2) compromise of the vascular supply and cerebral ischaemia as a consequence of endothelial cell infection and inflammation in the vessel wall (Marques Dias *et al.*, 1984); (3) impaired migration of differentiated neurons due to neural glial cell infection (Kawasaki and Tsutsui, 2003); and (4) subventricular neural stem cell loss as a result of periventriculitis (Grassi *et al.*, 1998) and loss of myelination of developing neurons (van der Voorn *et al.*, 2009; summarized in Table II.6.3). However, none of the proposed models can fully explain the mechanisms of neural maldevelopment during congenital HCMV infection. In our model of perinatal MCMV infection, we have described profound defects in the postnatal cerebellum development, a finding that is consistently described in studies utilizing either cranial ultrasound or magnetic resonance imaging of the brain of infants with congenital HCMV infection (de Vries *et al.*, 2004; Koontz *et al.*, 2008; Fig. II.6.2). Granule neuron cells in external granular layer (EGL) of the cerebellar cortex show proliferation and differentiation impairments in terms of low proliferation index, impaired expression of a differentiation specific protein neural cell adhesion molecule known as TAG-1 or contactin-3. As a consequence, migration of these cells into deeper parts of the cerebellar cortex is delayed, resulting in a characteristic phenotype of developing murine cerebellum in neonatally infected newborn mice: reduction in cerebellar foliation, decreased cerebellar area and increased thickness of the EGL of the cerebellum between the 9th and the 17th postnatal day (Koontz *et al.*, 2008). Purkinje cells in cerebella of infected mice also showed impairments in dendrite arborization resulting in decreased thickness of the cerebellar molecular layer, and inadequate alignment within the stratum gangliosum, leading to Purkinje neuron ectopic position (Koontz *et al.*, 2008; Fig. II.6.2). Although the exact mechanism of observed malformations has not been elucidated, impaired response of developing neurons to neurotrophins and inflammation with expression of proinflammatory cytokines and chemokines are thought to be possible explanations. Although the abnormalities temporally correlated with virus replication in the CNS, they were symmetrical and global in nature, and there was no obvious co-localization of virus replication and described abnormalities. This focal nature of the brain involvement argued that an indirect effect of virus infection was the more probable mechanism of the disease observed in these animals. Moreover, increased expression of a number of proinflammatory genes is readily observed in brains of MCMV-infected newborn mice (see below), while receptors specific for brain-derived neurotrophins are poorly expressed (Koontz *et al.*, 2008). In contrast to cerebellum, cerebrum volume of infected newborn mice corresponded to that of uninfected mice, indicating that the impact of MCMV infection is limited to regions of the brain that are actively developing in the early postnatal period (Koontz *et al.*, 2008). As in HCMV infection, long-term neurological sequelae in infected newborn mice comprise auditory and motor consequences in terms of profound uni- or bilateral hearing deficit or vestibular dysfunction (unpublished data). Interestingly, although few virus-infected cells are observed in the inner ear of infected newborn mice, it seems that MCMV infection induces a decrease in spiral ganglion cellular densities that could contribute to development of sensorineural hearing loss (unpublished data). It is worth mentioning that, despite the lack of virus replication in the inner ear, a strong inflammatory response characterized by mononuclear cell infiltrates is observable in cochlea, prominently in the stria vascularis and

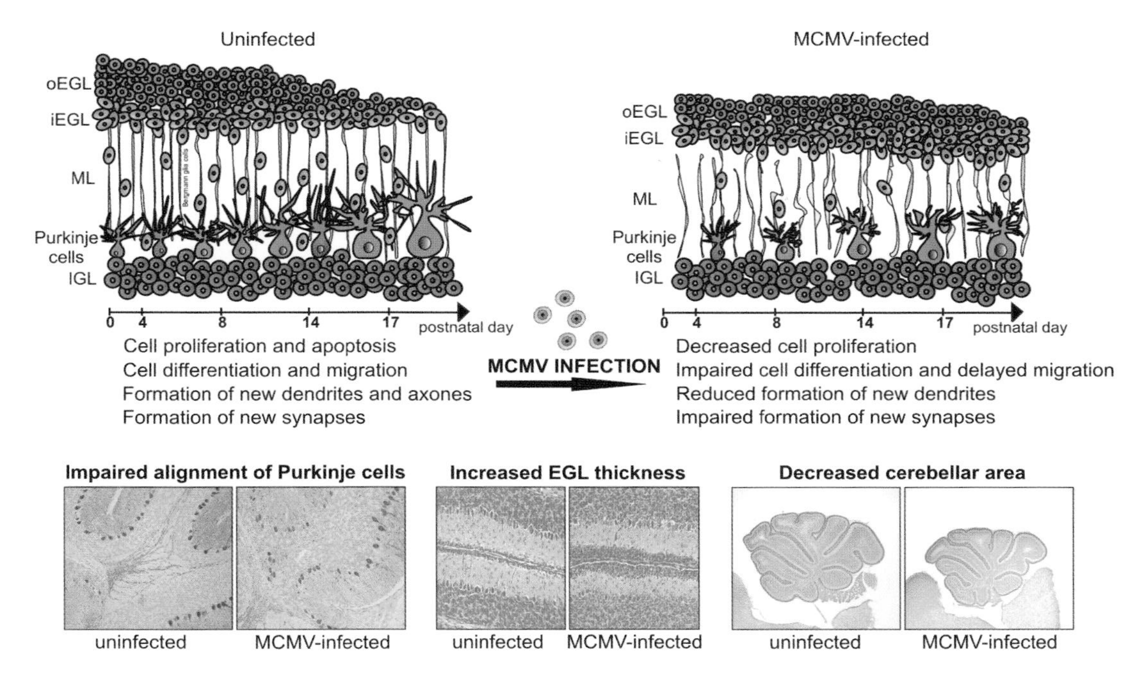

Figure II.6.2 MCMV infection in newborn mice influences the course of postnatal cerebellum development. Decreased proliferation of neurons in the outer external granular layer (oEGL) of the cerebellar cortex and impaired differentiation of proliferated neurons in the inner external granular layer (iEGL) are observed in MCMV-infected newborn mice. Differentiated neurons are delayed in migration from EGL into the internal granular layer (IGL). In parallel, Purkinje neurons fail to arborize dendrites in the molecular layer (ML) as efficiently as do those in uninfected newborn mice. Histochemical and IHC analysis of cerebellum slices reveals impaired alignment of Purkinje cells in stratum gangliosum, increased thickness of the EGL and decreased cerebellar area in infected newborn mice compared with healthy controls. Magnifications 4× (right panel), 10× (left panel) and 20× (centre panel).

in other perineural tissue (unpublished data). Other groups have also reported similar observations with the exception that the cell type undergoing cell death during MCMV infection in newborn mice were cochlear hair cells without any sign of virus replication in dying cells, but with a robust inflammatory response which comprises infiltrating macrophages (Schachtele *et al.*, 2011). Although this latter observation is consistent with the role of inflammation in the development of long-term neurological sequelae in MCMV-infected newborn mice, it should be noted that this study relied on direct i.c. inoculation of MCMV into newborn mice, a strategy that resulted in widespread brain involvement and limited survival of infected animals.

Host response and virus control in the brain of newborn mice

Following MCMV infection in the brain, the initial innate immune response is mediated by resident glial cells (astrocytes and activated microglia cells) and by infiltrating macrophages and NK cells. Components of the innate immune response are the primary source of proinflammatory cytokines, primarily IFN-γ and IFN-γ-induced chemokines CXCL9 and CXCL10 (Kosugi *et al.*, 2002). Numerous soluble mediators are activated in the brains of infected newborn mice in the early postinfection period, and are largely directed towards activation of adaptive immunity (Koontz *et al.*, 2008). Microarray analysis on brain tissue of infected newborn mice revealed increased expression of genes associated with the induction of interferon production (IRF-1, IRF-7, USP18 and LRG-47), the secretion of chemokines (CXCL3, CCL5, CCL21 and CXCL10) and the synthesis of TNF-α, TLR1, TLR2, TLR3 and MHC-I molecules (Koontz *et al.*, 2008). The most probable source of these molecules is activated glial cells and possibly infiltrating peripheral blood mononuclear cells. In the brains of MCMV-infected newborn mice, there is an obvious activation of microglia cells, coupled with infiltration of peripheral blood monocytes that exhibit phenotypic markers of tissue (brain) macrophages (Fig. II.6.3). Interestingly, activated microglia cells are co-localized within the sites of productive infection, which argues for an active role

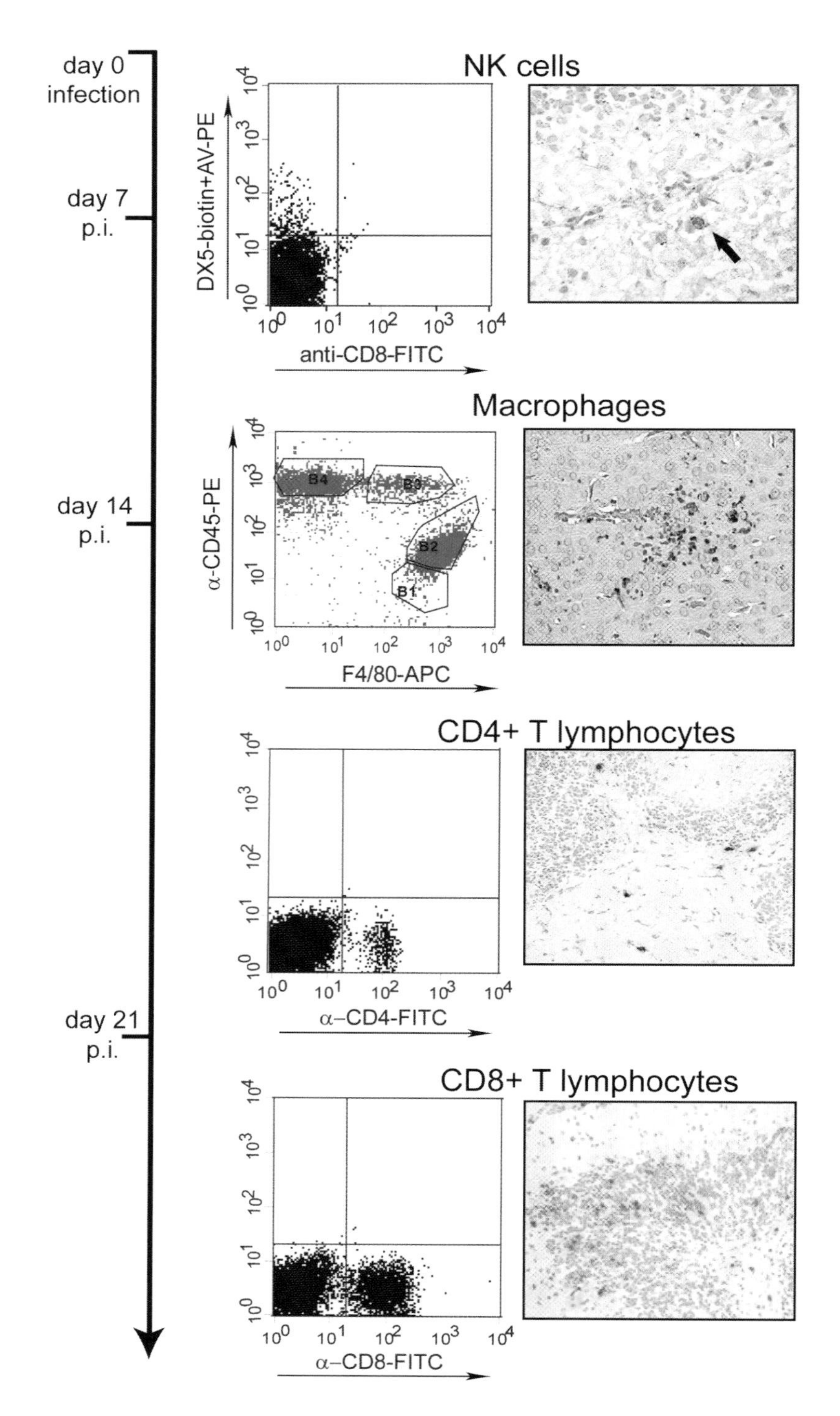

Figure II.6.3 Components of both innate and adaptive immune response infiltrate into the CNS of MCMV-infected newborn mice. NK cells can be isolated from infected brain already at day 7 p.i. Macrophage subpopulations can be distinguished in brains during the second week p.i.: activated microglia cells (B2 area) and infiltrating macrophages (B3 area) predominate over silent microglia (B1 area). $CD45^+$ cells (B4 area) present infiltrating lymphocytes. By d21 p.i. T-lymphocytes predominate in the MCMV-infected newborn brain with $CD8^+$ T-lymphocytes outnumber $CD4^+$ T-lymphocytes. Modified from Bantug *et al.* (2008). Copyright 2008. The American Association of Immunologists, Inc.

of these cells in the early control of infection (Bantug *et al.*, 2008). Although the essential role of NK cells in the control of MCMV replication at an early stage of infection has been defined in adult mice, the role of these cells in the control of MCMV infection in the CNS of newborn mice is still insufficiently defined (Jonjić *et al.*, 2005). Inflammatory cytokines appear to have an opposite effect on neural cells in terms of their proliferation and differentiation. Mouse NPCs in hippocampal slice cultures migrate into sites of neuroinflammation following TNF-α, IFN-γ or MCP-1 stimulation (Belmadani *et al.*, 2006), but these molecules appear to reduce the proliferation capacity of these cells. Similar effects have been observed following IFN-γ treatment of neuroblasts in terms of impaired proliferation of these cells in MCMV-infected mouse embryos (Cheeran *et al.*, 2008). In a model of CNS infection with neurotropic viruses, cytokine secretion was suggested to be a trigger of autoimmune activity of T-lymphocytes. In mouse hepatitis virus infection, both CD4+ and CD8+ T-cell-mediated immunopathological effects in the CNS were suggested to be dependent on RANTES-mediated activation of macrophages and trafficking of these cells into the CNS (Lane *et al.*, 2000), while others have reported IFN-γ dependent demyelination mediated by CD8+ T-lymphocytes (Pewe and Perlman, 2002). However, herpes simplex virus (HSV) infection in brains of neonatal mice is strictly controlled with the predominant CD4+ T-cell subpopulation being regulatory CD25+ CD4+ T-cells. Depletion of these cells resulted in increased CD8+ T-cell-mediated cytotoxicity and cytokine production *in vitro* (Fernandez *et al.*, 2008). However, a detrimental influence of RANTES and IFN-γ on neuronal proliferation in MCMV-infected newborn mice *in vivo* remains to be proven.

Strong inflammatory responses have also been observed in peripheral organs of MCMV-infected newborn mice. Systemic infection affects nearly all tissues and organ systems and inflammatory lesions were found to be comprised of both polymorphonuclear and mononuclear cell infiltrates (Trgovcich *et al.*, 1998). Focal hypoxic necrosis, numerous cytomegalic cells and cells with eosinophilic inclusions can be observed in specific organs (most commonly, the liver). Sera of these mice contain elevated levels of TNF-α that do not correlate temporally with the peak of virus replication in various organs, suggesting a cytokine-mediated pathogenesis of tissue damage in peripheral organs of MCMV-infected newborn mice (Trgovcich *et al.*, 1998). In contrast to peripheral organs, in brains of infected newborn mice, although elevated expression of TNF-α is observed, there was no sign of necrosis and tissue damage. This absence of histopathological lesions can be attributed to efficient regulation of immune response within the CNS (Koontz *et al.*, 2008; see below).

Role of CD8+ T-cell response in brains of MCMV-infected newborn mice

CD8+ T-cell-mediated immune response in HCMV-congenitally infected children can be detected and exhibits a full expression of molecules typical of terminally differentiated effector cells (HLA-DR+, CD95+, CD45RA+ and CD28−), but have low IFN-γ production capacity (Hassan *et al.*, 2007; Miles *et al.*, 2008). Interestingly, similar responses were detected from congenitally infected preterm infants, a result suggesting that intrauterine virus-specific CD8+ T-cell responses develop following intrauterine HCMV infections (Marchant *et al.*, 2003). In murine models of CMV infection, activation of T-cells is readily observed, and in fact, CD8+ T-cell control of MCMV infection in newborn brain has been suggested (Bantug *et al.*, 2008). The influx of CD8+ T-cells into the MCMV-infected brain correlated with significant decrease of virus replication in the CNS, whereas in neonatal mice depleted of CD8+ T-lymphocytes viral genome copy numbers in the CNS significantly increased and these mice succumbed to infection by postnatal day 15 (Bantug *et al.*, 2008). At an early stage of infection the surface phenotype of CD8+ T-cells isolated from brains of infected newborn mice reveals increased expression of CD44 and CD69 molecules, markers of T-lymphocyte activation (Sancho *et al.*, 2005). Expression of these activation markers is coupled with the increased expression of the CD49d molecule, necessary for T-lymphocyte infiltration into the CNS (Nandi *et al.*, 2004) and decreased expression of CD62L, a molecule whose expression diminishes following T-lymphocyte extravasation into tissue (Parish and Kaech, 2009). By the time virus replication is no longer detectable in the CNS of infected newborn mice, CD8+ T-cells present in the brain no longer express the CD62L molecule and have decreased expression of the CD69 molecule which indicates an activated effector phenotype of CD8+ T-cells (Bantug *et al.*, 2008). The drop in CD69 expression and the recovery of CD8+ T-cells from the brain tissue even months after the end of productive infection may indicate the switch of these cells from an activated effector phenotype into cells of the effector memory phenotype (Sierro *et al.*, 2005). Nevertheless, CD8+ T-lymphocytes isolated from brains of infected newborn mice at a later stage after virus clearance from the CNS are functional and can limit virus replication in γ-irradiated, immunocompromised animals upon adoptive cell transfer (Bantug et al., 2008), a finding that is strongly reminiscent of CD8+ T-cells accumulating in latently infected lungs (Holtappels et al., 2000; Podlech et al., 2000; see also Chapter I.22). As in congenital HCMV infection (Gibson et al., 2004), a significant proportion of CD8+ T-cells isolated from brains of infected mice shows specificity to IE-1 protein (Bantug et al., 2008). Although additional MCMV-derived peptides also induce the expansion of virus-specific CD8+ T-cells in newborn brain, IE-1 derived peptide proved to be predominant in the infected CNS in terms of eliciting specific immune response (Bantug et al., 2008). Following IE-1 derived peptide stimulation, CD8+ T-cells isolated from newborn brain secrete IFN-γ, whereas CD107a expression in these cells is like in a non-stimulated population of brain-isolated CD8+ T-cells, suggesting non-cytolytic mechanisms of virus control in the developing mouse CNS (Bantug et al., 2008). While the CD8+ T-cell-mediated antiviral response in brains of mice infected with West Nile virus (WNV) is dependent on perforin secretion (Shrestha et al., 2006), a similar pattern of CD8+ T-cell expansion and effector phenotype is observed in a model of CNS infection with this virus and other flaviviruses (Japanese encephalitis virus and tick-borne encephalitis virus), with diverse TCR phenotype patterns depending on the virus (Kitaura et al., 2011). This result would suggest that the CD8+ T-cell response in the CNS is highly oligoclonal. However, in a model of MCMV infection in brain this pattern still has to be determined.

Recent studies have revealed an active role of CD8+ T-lymphocytes in brains of MCMV-infected newborn mice in terms of constant activation of microglia cells via IFN-γ secretion (Mutnal et al., 2011). Our studies on brains of perinatally infected newborn mice revealed that a significant proportion of brain-isolated CD8+ T-lymphocytes express the PD-1 molecule on their surface, while brain-derived macrophages express high amounts of PD-1 ligands; PD-L1 (unpublished data). PD-1, a member of the CD28 family of molecules that negatively regulates CD8+ T-cell function, when engaged by its ligands PD-L1 and PD-L2, inhibits T-cell proliferation and cytokine production and is known to be up-regulated on CD8+ T-cells during other chronic infections like LCMV or HIV (Velu et al., 2007; Blattman et al., 2009). PD-1–PD-L1 interaction is very well described as being necessary for the induction and maintenance of CD8+ T-cell tolerance, a mechanism through which cells and tissues are protected from T-cell-mediated autoimmune activity (Sharpe et al., 2007). Possible engagement of these molecules in the regulation of CD8+ T-cell immune response in brains of MCMV-infected newborn mice remains to be elucidated.

Antibody-mediated immune responses to MCMV infection in the newborn mouse

Although dispensable for the resolution of acute CMV infections, antibodies have proven to be essential for the prevention of virus spread following reactivation (Jonjic et al., 1994; Polic et al., 1998). For a general and comprehensive discussion of the role of antibodies in CMV infections, see also Chapter II.10. The role of virus-specific antibodies is most evident in their effects on the risk of congenital HCMV infection. Pre-conceptual immunity has a demonstrable impact on the rate of virus transmission from mother to child but fails to provide complete protection to the fetus (Fowler et al., 1992; Adler et al., 1995). Several studies have proven the inability of antibodies to completely prevent transplacental virus transfer from mother to child (Boppana et al., 1993, 2001). Infection in this setting can be considered as reactivation of an endogenous virus or reinfection with a new serotype(s) of exogenous virus(es) (Boppana et al., 2001; Ross et al., 2011). In either case, it could be argued that the host is exposed to a genetic variant of HCMV (see also Chapter I.1) that was not recognized by existing anti-HCMV antibodies. Interestingly, high levels of antibodies specific for HCMV gB have been detected in sera of infected infants who developed profound sensorineural hearing loss, as well as in sera of their mothers (Boppana et al., 1999, 2001). This observation has been correlated with the degree of viraemia in infants, again suggesting only limited control of virus by antibody; however, as noted in previous sections, the amount of virus in the blood of women transmitting virus to their offspring is too variable to allow a direct correlation with rates of transmission and by inference, with the level of α-CMV antibody. In a model of perinatal MCMV infection antibodies have proven to play a role in the control and prevention of infection in the developing CNS (Cekinovic et al., 2008). Treatment of infected newborn mice with either immune sera or antibodies specific for gB prior to virus spread to the CNS provided protection and limited virus replication in the brain parenchyma to undetectable levels, while treatment of mice with antibodies even at the time of peak virus infection in the brain still significantly reduced virus titres in the CNS (Cekinovic et al., 2008). Undetectable virus replication in brains of newborn mice which received antibodies during an early p.i. period suggested that antibodies could prevent virus spread to the CNS by either direct neutralization of cell-free virus or prevention of cell-associated virus infection by mechanisms such as antibody-dependent cellular cytotoxicity (ADCC). On the other hand, virus clearance from brains of infected newborn mice that received antibodies in a period of peak virus replication in the brain implies that antibodies could be functioning even within the CNS to limit virus spread and/or replication. This is in accordance with a report showing that MCMV-specific antibodies control virus spread in the liver even when administered during an already established infection of the liver (Wirtz *et al.*, 2008), thus indicating that this antiviral property of antibodies is not restricted to the CNS. Virus-specific antibodies have proven to function in virus elimination from the CNS of mice infected with neurovirulent viruses (Griffin, 2003). B-lymphocytes have recently been proven to have a protective capacity towards MCMV replication even in immunodeficient mice. In contrast to previous data, memory B-cells were shown to proliferate and secrete antibodies following MCMV infection (Klenovsek *et al.*, 2007). In the model of Sindbis virus infection in the CNS, B-lymphocytes can be isolated from brains of infected mice during acute virus infection, and are retained in the CNS following the end of virus replication with capacity to secrete antibodies at times of RNA persistence (Metcalf and Griffin, 2011). When administered to MCMV-infected newborn mice, virus-specific antibodies can be detected in the CNS, surrounding blood vessels and neural cells (Cekinovic *et al.*, 2008), however, the role of B-lymphocytes in the control of MCMV infection in the developing brain and the mechanism of antibody-mediated virus clearance from the brain are still insufficiently defined. Based on the proven antiviral activity of antibodies in this model, a new vaccine approach for the prevention of CMV infection was defined. Infection of newborn mice with a recombinant mutant virus carrying a ligand that is recognized by the activating receptor NKG2D present on both NK cells and $CD8^+$ T-lymphocytes (see also Chapter II.9) was shown to be significantly attenuated compared with infection with WT virus (Slavuljica *et al.*, 2010). Moreover, infection of adult female mice with this virus elicited an efficient B-cell-mediated immune response. Antiviral antibodies generated prior to pregnancy were shown to protect offspring from virus dissemination and infection in the early postnatal period (Slavuljica *et al.*, 2010). This promising result encourages the development of new strategies for designing therapeutic and preventative vaccines that could limit disease associated with congenital CMV infection in the human infant.

Conclusion

The pathogenesis of congenital HCMV infection is mainly extrapolated from histopathological findings on limited numbers of autopsied cases in children who succumbed to infection or cases in which HCMV infection was retrospectively diagnosed. However, long-term neurological impairments develop also in a significant percentage of children with asymptomatic infection. Such findings have required the development of informative animal models of this human infection. When studied under the framework conditions of the natural history of congenital HCMV infections, models such as discussed in the above sections can be used to dissect critical parameters of protection from damaging CNS infections in congenitally infected infants for subsequently validating the experimental findings in humans by clinical trials. From such studies, rational strategies for the production of protective vaccines and biologics can be designed and tested.

Acknowledgements

William J. Britt is supported by NIH grants 1R01AI089956-01A1 and 1R01NS065845-01.

References

Abbott, N.J., Ronnback, L., and Hansson, E. (2006). Astrocyte–endothelial interactions at the blood–brain barrier. Nat. Rev. Neurosci. *7*, 41–53.

Adler, S.P., Starr, S.E., Plotkin, S.A., Hempfling, S.H., Buis, J., Manning, M.L., and Best, A.M. (1995). Immunity induced by primary human cytomegalovirus infection protects against secondary infection among women of childbearing age. J. Infect. Dis. *171*, 26–32.

Ahlfors, K., and Ivarsson, S.A. (1985). Cytomegalovirus in breast milk of Swedish milk donors. Scand. J. Infect. Dis. *17*, 11.

Ahlfors, K., Ivarsson, S.A., and Harris, S. (2001). Secondary maternal cytomegalovirus infection – a significant cause of congenital disease. Pediatrics *107*, 1227–1228.

Alford, C. (1991). Breast milk transmission of cytomegalovirus (CMV) infection. Adv. Exp. Med. Biol. *310*, 293–299.

Alford, C.A., Pass, R.F., and Stagno, S. (1983). Chronic congenital infections: common environmental causes for severe and subtle birth defects. Birth Defects Orig. Artic. Ser. *19*, 87–96.

Alford, C.A., Hayes, K., and Britt, W.J. (1988). Primary cytomegalovirus infection in pregnancy: comparison of antibody responses to virus encoded proteins between women with and without intrauterine infection. J. Infect. Dis. *158*, 917–924.

Amos, C.S. (1977). The ocular manifestations of congenital infections produced by toxoplasma and cytomegalovirus. J. Am. Optom. Assoc. *48*, 532–538.

Anderson, K.S., Amos, C.S., Boppana, S., and Pass, R. (1996). Ocular abnormalities in congenital cytomegalovirus infection. J. Am. Optom. Assoc. *67*, 273–278.

Arribas, J.R., Storch, G.A., Clifford, D.B., and Tselis, A.C. (1996). Cytomegalovirus encephalitis. Ann. Inter. Med. *125*, 577–587.

Bale, J.F., Blackman, J.A., and Sato, Y. (1990). Outcome in children with symptomatic congenital cytomegalovirus infection. J. Child. Neurol. *5*, 131–136.

Bantug, G.R., Cekinovic, D., Bradford, R., Koontz, T., Jonjic, S., and Britt, W.J. (2008). CD8+ T-lymphocytes control murine cytomegalovirus replication in the central nervous system of newborn animals. J. Immunol. *181*, 2111–2123.

Barkovich, A.J., and Lindan, C.E. (1994). Congenital cytomegalovirus infection of the brain: imaging analysis and embryologic considerations. AJNR Am. J. Neuroradiol. *15*, 703–715.

Barraud, P., Stott, S., Mollgard, K., Parmar, M., and Bjorklund, A. (2007). *In vitro* characterization of a human neural progenitor cell coexpressing SSEA4 and CD133. J Neurosci. Res. *85*, 250–259.

Becroft, D.M., Rainer, S.P., and Tonkin, S. (1981). Causes of infant deaths in New Zealand and Sweden. NZ Med. J. *93*, 316.

Becroft, D.M.O. (1981). Prenatal cytomegalovirus infection: epidemiology, pathology, and pathogenesis. In Perspective in Pediatric Pathology, Rosenberg, H.S., and Bernstein, J., eds. (Masson Press, New York), pp. 203–241.

Belmadani, A., Tran, P.B., Ren, D., and Miller, R.J. (2006). Chemokines regulate the migration of neural progenitors to sites of neuroinflammation. J. Neurosci. *26*, 3182–3191.

Benoist, G., Salomon, L.J., Mohlo, M., Suarez, B., Jacquemard, F., and Ville, Y. (2008). Cytomegalovirus-related fetal brain lesions: comparison between targeted ultrasound examination and magnetic resonance imaging. Ultrasound Obstet. Gynecol. *32*, 900–905.

Bentz, G.L., Jarquin-Pardo, M., Chan, G., Smith, M.S., Sinzger, C., and Yurochko, A.D. (2006). Human cytomegalovirus (HCMV) infection of endothelial cells promotes naïve monocyte extravasation and transfer of productive virus to enhance hematogenous dissemination of HCMV. J. Virol. *80*, 11539–11555.

Bezzi, P., Domercq, M., Brambilla, L., Galli, R., Schols, D., De Clercq, E., Vescovi, A., Bagetta, G., Kollias, G., Meldolesi, J., *et al.* (2001). CXCR4-activated astrocyte glutamate release via TNFalpha: amplification by microglia triggers neurotoxicity. Nat. Neurosci. *4*, 702–710.

Blattman, J.N., Wherry, E.J., Ha, S.J., van der Most, R.G., and Ahmed, R. (2009). Impact of epitope escape on PD-1 expression and CD8 T-cell exhaustion during chronic infection. J. Virol. *83*, 4386–4394.

Booss, J., Dann, P.R., Griffith, B.P., and Kim, J.H. (1988). Glial nodule encephalitis in the guinea pig: serial observations following cytomegalovirus infection. Acta Neuropathol. *75*, 465–473.

Boppana, S., and Britt, W. (2006). Cytomegalovirus. In Infection and Hearing Impairment, Newton, V.E., and Vallely, P.J., eds. (John Wiley and Sons, Sussex, UK), pp. 67–93.

Boppana, S.B., and Britt, W.J. (1995). Antiviral antibody responses and intrauterine transmission after primary maternal cytomegalovirus infection. J. Infect. Dis. *171*, 1115–1121.

Boppana, S.B., Fowler, K.B., Pass, R.F., Britt, W.J., Stagno, S., and Alford, C.A. (1992a). Newborn findings and outcome in children with symptomatic congenital CMV infection. Pediatr. Res. *31*, 158A.

Boppana, S.B., Pass, R.F., Britt, W.J., Stagno, S., and Alford, C.A. (1992b). Symptomatic congenital cytomegalovirus infection: neonatal morbidity and mortality. Pediatr. Infect. Dis. J. *11*, 93–99.

Boppana, S.B., Pass, R.F., and Britt, W.J. (1993). Virus-specific antibody responses in mothers and their newborn infants with asymptomatic congenital cytomegalovirus infections. J. Infect. Dis. *167*, 72–77.

Boppana, S.B., Miller, J., and Britt, W.J. (1996). Transplacentally acquired antiviral antibodies and outcome in congenital human cytomegalovirus infection. Viral Immunol. *9*, 211–218.

Boppana, S.B., Fowler, K.B., Vaid, Y., Hedlund, G., Stagno, S., Britt, W.J., and Pass, R.F. (1997). Neuroradiographic findings in the newborn period and long-term outcome in children with symptomatic congenital cytomegalovirus infection. Pediatrics 99, 409–414.

Boppana, S.B., Fowler, K.B., Britt, W.J., Stagno, S., and Pass, R.F. (1999). Symptomatic congenital cytomegalovirus infection in infants born to mothers with preexisting immunity to cytomegalovirus. Pediatrics *104*, 55–60.

Boppana, S.B., Rivera, L.B., Fowler, K.B., Mach, M., and Britt, W.J. (2001). Intrauterine transmission of cytomegalovirus to infants of women with preconceptional immunity. N. Engl. J. Med. *344*, 1366–1371.

Bosnjak, V.M., Dakovic, I., Duranovic, V., Lujic, L., Krakar, G., and Marn, B. (2011). Malformations of cortical development in children with congenital cytomegalovirus infection – a study of nine children with proven congenital cytomegalovirus infection. Coll. Antropol. *35(Suppl. 1)*, 229–234.

Bradford, R.D., Cloud, G., Lakeman, A.D., Boppana, S., Kimberlin, D.W., Jacobs, R., Demmler, G., Sanchez, P., Britt, W., Soong, S.J., *et al.* (2005). Detection of cytomegalovirus (CMV) DNA by polymerase chain reaction is associated with hearing loss in newborns with symptomatic congenital CMV infection involving the central nervous system. J. Infect. Dis. *191*, 227–233.

Bratcher, D.F., Bourne, N., Bravo, F.J., Schleiss, M.R., Slaoui, M., Myers, M.G., and Bernstein, D.I. (1995). Effect of passive antibody on congenital cytomegalovirus infection in guinea pigs. J. Infect. Dis. *172*, 944–950.

Britt, W. (2010). Cytomegalovirus. In Infectious Diseases of the Fetus and Newborn Infant, Remington, J.S., Klein, J.O., Wilson, C., Nizet, V., and Maldonado, Y.A., eds. (Elsevier, Philadelphia, PA), pp. 706–756.

Brune, W., Nevels, M., and Shenk, T. (2003). Murine cytomegalovirus m41 open reading frame encodes a Golgi-localized antiapoptotic protein. J. Virol. *77*, 11633–11643.

Cekinovic, D., Golemac, M., Pugel, E.P., Tomac, J., Cicin-Sain, L., Slavuljica, I., Bradford, R., Misch, S., Winkler, T.H., Mach, M., *et al.* (2008). Passive immunization reduces murine cytomegalovirus-induced brain pathology in newborn mice. J. Virol. *82*, 12172–12180.

Chan, G., Nogalski, M.T., Bentz, G.L., Smith, M.S., Parmater, A., and Yurochko, A.D. (2010). PI3K-dependent up-regulation of Mcl-1 by human cytomegalovirus is mediated by epidermal growth factor receptor and inhibits apoptosis in short-lived monocytes. J. Immunol. *184*, 3213–3222.

Chatterjee, A., Harrison, C.J., Britt, W.J., and Bewtra, C. (2001). Modification of maternal and congenital cytomegalovirus infection by anti-glycoprotein b antibody transfer in guinea pigs. J. Infect. Dis. *183*, 1547–1553.

Cheeran, M.C., Hu, S., Gekker, G., and Lokensgard, J.R. (2000). Decreased cytomegalovirus expression following proinflammatory cytokine treatment of primary human astrocytes. J. Immunol. *164*, 926–933.

Cheeran, M.C., Gekker, G., Hu, S., Min, X., Cox, D., and Lokensgard, J.R. (2004). Intracerebral infection with murine cytomegalovirus induces CXCL10 and is restricted by adoptive transfer of splenocytes. J. Neurovirol. *10*, 152–162.

Cheeran, M.C., Hu, S., Ni, H.T., Sheng, W., Palmquist, J.M., Peterson, P.K., and Lokensgard, J.R. (2005). Neural precursor cell susceptibility to human cytomegalovirus diverges along glial or neuronal differentiation pathways. J. Neurosci. Res. *82*, 839–850.

Cheeran, M.C., Jiang, Z., Hu, S., Ni, H.T., Palmquist, J.M., and Lokensgard, J.R. (2008). Cytomegalovirus infection and interferon-gamma modulate major histocompatibility complex class I expression on neural stem cells. J. Neurovirol. *14*, 437–447.

Cheeran, M.C., Lokensgard, J.R., and Schleiss, M.R. (2009). Neuropathogenesis of congenital cytomegalovirus infection: disease mechanisms and prospects for intervention. Clin. Microbiol. Rev. *22*, 99–126.

Cicin-Sain, L., Ruzsics, Z., Podlech, J., Bubic, I., Menard, C., Jonjic, S., Reddehase, M.J., and Koszinowski, U.H. (2008). Dominant-negative FADD rescues the *in vivo* fitness of a cytomegalovirus lacking an antiapoptotic viral gene. J. Virol. *82*, 2056–2064.

Conboy, T.J., Pass, R.F., Stagno, S., Alford, C.A., Myers, G.J., Britt, W.J., McCollister, F.P., Summers, M.N., McFarland, C.E., and Boll, T.J. (1987). Early clinical manifestations and intellectual outcome in children with symptomatic congenital cytomegalovirus infection. J. Pediatr. *111*, 343–348.

Davis, G.L., Spector, G.J., Strauss, M., and Middlekamp, J.N. (1977). Cytomegalovirus endolabyrinthitis. Arch. Pathol. Lab. Med. *101*, 118–121.

DeBiasi, R.L., Kleinschmidt-DeMasters, B.K., Richardson-Burns, S., and Tyler, K.L. (2002). Central nervous system apoptosis in human herpes simplex virus and cytomegalovirus encephalitis. J. Infect. Dis. *186*, 1547–1557.

van Den Pol, A.N., Mocarski, E., Saederup, N., Vieira, J., and Meier, T.J. (1999). Cytomegalovirus cell tropism, replication, and gene transfer in brain. J. Neurosci. *19*, 10948–10965.

Doneda, C., Parazzini, C., Righini, A., Rustico, M., Tassis, B., Fabbri, E., Arrigoni, F., Consonni, D., and Triulzi, F. (2010). Early cerebral lesions in cytomegalovirus infection: prenatal MR imaging. Radiology 255, 613–621.

Duryea, E.L., Sanchez, P.J., Sheffield, J.S., Jackson, G.L., Wendel, G.D., McElwee, B.S., Boney, L.F., Mallory, M.M., Owen, K.E., and Stehel, E.K. (2010). Maternal human immunodeficiency virus infection and congenital transmission of cytomegalovirus. Pediatr. Infect. Dis. J. *29*, 915–918.

Dworsky, M., Yow, M., Stagno, S., Pass, R.F., and Alford, C. (1983). Cytomegalovirus infection of breast milk and transmission in infancy. Pediatrics 72, 295–299.

Elek, S.D., and Stern, H. (1974). Development of a vaccine against mental retardation caused by cytomegalovirus infection in utero. Lancet *1*, 1–5.

Enders, G., Daiminger, A., Bader, U., Exler, S., and Enders, M. (2011). Intrauterine transmission and clinical outcome of 248 pregnancies with primary cytomegalovirus infection in relation to gestational age. J. Clin. Virol. 52, 244–246.

Fernandez, M.A., Puttur, F.K., Wang, Y.M., Howden, W., Alexander, S.I., and Jones, C.A. (2008). T regulatory cells contribute to the attenuated primary CD8+ and CD4+ T-cell responses to herpes simplex virus type 2 in neonatal mice. J. Immunol. *180*, 1556–1564.

Fowler, K.B., Stagno, S., Pass, R.F., Britt, W.J., Boll, T.J., and Alford, C.A. (1992). The outcome of congenital cytomegalovirus infection in relation to maternal antibody status. N. Engl. J. Med. *326*, 663–667.

Gabrielli, L., Bonasoni, M.P., Lazzarotto, T., Lega, S., Santini, D., Foschini, M.P., Guerra, B., Baccolini, F., Piccirilli, G., Chiereghin, A., *et al.* (2009). Histological findings in fetuses congenitally infected by cytomegalovirus. J. Clin. Virol. *46(Suppl. 4)*, 16–21.

Gibson, L., Piccinini, G., Lilleri, D., Revello, M.G., Wang, Z., Markel, S., Diamond, D.J., and Luzuriaga, K. (2004). Human cytomegalovirus proteins pp65 and immediate early protein 1 are common targets for CD8+ T-cell responses in children with congenital or postnatal human cytomegalovirus infection. J. Immunol. *172*, 2256–2264.

Goldmacher, V.S. (2005). Cell death suppression by cytomegaloviruses. Apoptosis *10*, 251–265.

Grassi, M.P., Clerici, F., Perin, C., D'Arminio Monforte, A., Vago, L., Borella, M., Boldorini, R., and Mangoni, A. (1998). Microglial nodular encephalitis and ventriculoencephalitis due to cytomegalovirus infection in patients with AIDS: two distinct clinical patterns. Clin. Infect. Dis. *27*, 504–508.

Griffin, D.E. (2003). Immune responses to RNA-virus infections of the CNS. Nat. Rev. Immunol. *3*, 493–502.

Griffith, B.P., and Aquino-de Jesus, M.J. (1991). Guinea pig model of congenital cytomegalovirus infection. Transplant. Proc. *23*, 29–31.

Guibert, G., Warszawski, J., Le Chenadec, J., Blanche, S., Benmebarek, Y., Mandelbrot, L., Tubiana, R., Rouzioux, C., and Leruez-Ville, M. (2009). Decreased risk of congenital cytomegalovirus infection in children born to HIV-1-infected mothers in the era of highly active antiretroviral therapy. Clin. Infect. Dis. *48*, 1516–1525.

Hadar, E., Yogev, Y., Melamed, N., Chen, R., Amir, J., and Pardo, J. (2010). Periconceptional cytomegalovirus infection: pregnancy outcome and rate of vertical transmission. Prenat. Diagn. *30*, 1213–1216.

Hamele, M., Flanagan, R., Loomis, C.A., Stevens, T., and Fairchok, M.P. (2009). Severe morbidity and mortality with breast milk associated cytomegalovirus infection. Pediatr. Infect. Dis. J. *29*, 84–86.

Hassan, J., Dooley, S., and Hall, W. (2007). Immunological response to cytomegalovirus in congenitally infected neonates. Clin. Exp. Immunol. *147*, 465–471.

Hansen, S.G., Vieville, C., Whizin, N., Coyne-Johnson, L., Siess, D.C., Drummond, D.D., Legasse, A.W., Axthelm, M.K., Oswald, K., Trubey, C.M., *et al.* (2009). Effector memory T-cell responses are associated with protection of rhesus monkeys from mucosal simian immunodeficiency virus challenge. Nat. Med. *15*, 293–299.

Hayward, J.C., Titelbaum, D.S., Clancy, R.R., and Zimmerman, R.A. (1991). Lissencephaly-pachygyria associated with congenital cytomegalovirus infection. J. Child. Neurol. *6*, 109–114.

Holtappels, R., Pahl-Seibert, M.F., Thomas, D., and Reddehase, M.J. (2000). Enrichment of immediate-early 1 (m123/pp89) peptide-specific CD8 T-cells in a pulmonary CD62L(lo) memory-effector cell pool during latent murine cytomegalovirus infection of the lungs. J. Virol. *74*, 11495–11503.

Ishiwata, M., Baba, S., Kawashima, M., Kosugi, I., Kawasaki, H., Kaneta, M., Tsuchida, T., Kozuma, S., and Tsutsui, Y. (2006). Differential expression of the immediate-early 2 and 3 proteins in developing mouse brains infected with murine cytomegalovirus. Arch. Virol. *151*, 2181–2196.

Iwasenko, J.M., Howard, J., Arbuckle, S., Graf, N., Hall, B., Craig, M.E., and Rawlinson, W.D. (2011). Human cytomegalovirus infection is detected frequently in stillbirths and is associated with fetal thrombotic vasculopathy. J. Infect. Dis. *203*, 1526–1533.

Jarvis, M.A., and Nelson, J.A. (2002). Human cytomegalovirus persistence and latency in endothelial cells and macrophages. Curr. Opin. Microbiol. *5*, 403–407.

Jonjic, S., Pavic, I., Polic, B., Crnkovic, I., Lucin, P., and Koszinowski, U.H. (1994). Antibodies are not essential for the resolution of primary cytomegalovirus infection but limit dissemination of recurrent virus. J. Exp. Med. *179*, 1713–1717.

Jonjić, S., Bubić, I., and Krmpotić, A. (2005). Innate immunity to cytomegalovirus. In Cytomegaloviruses: Molecular Biology and Immunology, Reddehase, M.J., ed. (Caister Academic Press, Norfolk, UK), pp. 286–319.

Jurak, I., Schumacher, U., Simic, H., Voigt, S., and Brune, W. (2008). Murine cytomegalovirus m38.5 protein inhibits Bax-mediated cell death. J. Virol. *82*, 4812–4822.

Katano, H., Sato, Y., Tsutsui, Y., Sata, T., Maeda, A., Nozawa, N., Inoue, N., Nomura, Y., and Kurata, T. (2007). Pathogenesis of cytomegalovirus-associated labyrinthitis in a guinea pig model. Microbes Infect. *9*, 183–191.

Kawasaki, H., and Tsutsui, Y. (2003). Brain slice culture for analysis of developmental brain disorders with special reference to congenital cytomegalovirus infection. Congenit. Anom. (Kyoto) *43*, 105–113.

Kawasaki, H., Kosugi, I., Arai, Y., Iwashita, T., and Tsutsui, Y. (2011). Mouse embryonic stem cells inhibit murine cytomegalovirus infection through a multistep process. PLoS One *6*, e17492.

Keithley, E.M., Sharp, P., Woolf, N.K., and Harris, J.P. (1988). Temporal sequence of viral antigen expression in the cochlea induced by cytomegalovirus. Acta Otolaryngol. *106*, 46–54.

Kenneson, A., and Cannon, M.J. (2007). Review and meta-analysis of the epidemiology of congenital cytomegalovirus (CMV) infection. Rev. Med. Virol. *17*, 253–276.

Kimberlin, D.W., Lin, C.Y., Sanchez, P.J., Demmler, G.J., Dankner, W., Shelton, M., Jacobs, R.F., Vaudry, W., Pass, R.F., Kiell, J.M., *et al.* (2003). Effect of ganciclovir therapy on hearing in symptomatic congenital cytomegalovirus disease involving the central nervous system: a randomized, controlled trial. J. Pediat. *143*, 16–25.

Kitaura, K., Fujii, Y., Hayasaka, D., Matsutani, T., Shirai, K., Nagata, N., Lim, C.K., Suzuki, S., Takasaki, T., Suzuki, R., *et al.* (2011). High clonality of virus-specific T-lymphocytes defined by TCR usage in the brains of mice infected with West Nile virus. J. Immunol. *187*, 3919–3930.

Klenovsek, K., Weisel, F., Schneider, A., Appelt, U., Jonjic, S., Messerle, M., Bradel-Tretheway, B., Winkler, T.H., and Mach, M. (2007). Protection from CMV infection in immunodeficient hosts by adoptive transfer of memory B cells. Blood *110*, 3472–3479.

Koontz, T., Bralic, M., Tomac, J., Pernjak-Pugel, E., Bantug, G., Jonjic, S., and Britt, W.J. (2008). Altered development of the brain after focal herpesvirus infection of the central nervous system. J. Exp. Med. *205*, 423–435.

Kosugi, I., Shinmura, Y., Li, R.Y., Aiba-Masago, S., Baba, S., Miura, K., and Tsutsui, Y. (1998). Murine cytomegalovirus induces apoptosis in non-infected cells of the developing mouse brain and blocks apoptosis in primary neuronal culture. Acta Neuropathol. *96*, 239–247.

Kosugi, I., Kawasaki, H., Arai, Y., and Tsutsui, Y. (2002). Innate immune responses to cytomegalovirus infection in the developing mouse brain and their evasion by virus-infected neurons. Am. J. Pathol. *161*, 919–928.

La Torre, R., Nigro, G., Mazzocco, M., Best, A.M., and Adler, S.P. (2006). Placental enlargement in women with primary maternal cytomegalovirus infection is associated with fetal and neonatal disease. Clin. Infect. Dis. *43*, 994–1000.

Lanari, M., Lazzarotto, T., Venturi, V., Papa, I., Gabrielli, L., Guerra, B., Landini, M.P., and Faldella, G. (2006). Neonatal cytomegalovirus blood load and risk of sequelae in symptomatic and asymptomatic congenitally infected newborns. Pediatrics *117*, e76–e83.

Lane, T.E., Liu, M.T., Chen, B.P., Asensio, V.C., Samawi, R.M., Paoletti, A.D., Campbell, I.L., Kunkel, S.L., Fox, H.S., and Buchmeier, M.J. (2000). A central role for CD4(+) T-cells and RANTES in virus-induced central nervous system inflammation and demyelination. J. Virol. *74*, 1415–1424.

Lawrence, R.M., and Lawrence, R.A. (2004). Breast milk and infection. Clin. Perinatol. *31*, 501–528.

Lazzarotto, T., Gabrielli, L., Foschini, M.P., Lanari, M., Guerra, B., Eusebi, V., and Landini, M.P. (2003). Congenital cytomegalovirus infection in twin pregnancies: viral load in the amniotic fluid and pregnancy outcome. Pediatrics *112*, e153–157.

Li, L., Kosugi, I., Han, G.P., Kawasaki, H., Arai, Y., Takeshita, T., and Tsutsui, Y. (2008). Induction of cytomegalovirus-infected labyrinthitis in newborn mice by lipopolysaccharide: a model for hearing loss in congenital CMV infection. Lab. Invest. *88*, 722–730.

Li, R.Y., and Tsutsui, Y. (2000). Growth retardation and microcephaly induced in mice by placental infection with murine cytomegalovirus. Teratology *62*, 79–85.

Lilleri, D., Fornara, C., Revello, M.G., and Gerna, G. (2008). Human cytomegalovirus-specific memory CD8+ and CD4+ T-cell differentiation after primary infection. J. Infect. Dis. *198*, 536–543.

Ling, E.A., and Wong, W.C. (1993). The origin and nature of ramified and amoeboid microglia: a historical review and current concepts. Glia *7*, 9–18.

Lokensgard, J.R., Cheeran, M.C., Gekker, G., Hu, S., Chao, C.C., and Peterson, P.K. (1999). Human cytomegalovirus replication and modulation of apoptosis in astrocytes. J. Hum. Virol. *2*, 91–101.

Luo, M.H., Schwartz, P.H., and Fortunato, E.A. (2008). Neonatal neural progenitor cells and their neuronal and glial cell derivatives are fully permissive for human cytomegalovirus infection. J. Virol. *82*, 9994–10007.

McCollister, F.P., Simpson, L.C., Dahle, A.J., Pass, R.F., Fowler, K.B., Amos, C.S., and Boll, T.J. (1996). Hearing loss and congenital symptomatic cytomegalovirus infection: a case report of multidisciplinary longitudinal assessment and intervention. J. Am. Acad. Audiol. *7*, 57–62.

McCormick, A.L., Skaletskaya, A., Barry, P.A., Mocarski, E.S., and Goldmacher, V.S. (2003). Differential function and expression of the viral inhibitor of caspase 8-induced apoptosis (vICA) and the viral mitochondria-localized inhibitor of apoptosis (vMIA) cell death suppressors conserved in primate and rodent cytomegaloviruses. Virology *316*, 221–233.

Maidji, E., McDonagh, S., Genbacev, O., Tabata, T., and Pereira, L. (2006). Maternal antibodies enhance or prevent cytomegalovirus infection in the placenta by neonatal Fc receptor-mediated transcytosis. Am. J. Pathol. *168*, 1210–1226.

Manara, R., Balao, L., Baracchini, C., Drigo, P., D'Elia, R., and Ruga, E.M. (2011). Brain magnetic resonance findings in symptomatic congenital cytomegalovirus infection. Pediatr. Radiol. *41*, 962–970.

Marchant, A., Appay, V., Van Der Sande, M., Dulphy, N., Liesnard, C., Kidd, M., Kaye, S., Ojuola, O., Gillespie, G.M., Vargas Cuero, A.L., *et al.* (2003). Mature CD8(+) T-lymphocyte response to viral infection during fetal life. J. Clin. Invest. *111*, 1747–1755.

Marques Dias, M.J., Harmant-van Rijckevorsel, G., Landrieu, P., and Lyon, G. (1984). Prenatal cytomegalovirus disease and cerebral microgyria: evidence for perfusion failure, not disturbance of histogenesis, as the major cause of fetal cytomegalovirus encephalopathy. Neuropediatrics *15*, 18–24.

Menard, C., Wagner, M., Ruzsics, Z., Holak, K., Brune, W., Campbell, A.E., and Koszinowski, U.H. (2003). Role of murine cytomegalovirus US22 gene family members in replication in macrophages. J. Virol. *77*, 5557–5570.

Metcalf, T.U., and Griffin, D.E. (2011). Alphavirus-induced encephalomyelitis: antibody-secreting cells and viral clearance from the nervous system. J. Virol. *85*, 11490–11501.

Miles, D.J., van der Sande, M., Jeffries, D., Kaye, S., Ojuola, O., Sanneh, M., Cox, M., Palmero, M.S., Touray, E.S., Waight, P., *et al.* (2008). Maintenance of large subpopulations of differentiated CD8 T-cells two years after cytomegalovirus infection in Gambian infants. PLoS One *3*, e2905.

Miron, D., Brosilow, S., Felszer, K., Reich, D., Halle, D., Wachtel, D., Eidelman, A.I., and Schlesinger, Y. (2005). Incidence and clinical manifestations of breast milk-acquired Cytomegalovirus infection in low birth weight infants. J. Perinatol. *25*, 299–303.

Mocarski, E.S., Jr. (2002). Immunomodulation by cytomegaloviruses: manipulative strategies beyond evasion. Trends Microbiol. *10*, 332–339.

Munro, S.C., Trincado, D., Hall, B., and Rawlinson, W.D. (2005). Symptomatic infant characteristics of congenital cytomegalovirus disease in Australia. J. Paediatr. Child Health *41*, 449–452.

Mussi-Pinhata, M.M., Yamamoto, A.Y., Figueiredo, L.T., Cervi, M.C., and Duarte, G. (1998). Congenital and perinatal cytomegalovirus infection in infants born to mothers infected with human immunodeficiency virus. J. Pediatr. *132*, 285–290.

Mussi-Pinhata, M.M., Yamamoto, A.Y., Moura-Britto, R.M., Lima-Issacs, M., Boppana, S., and Britt, W.J. (2009). Birth prevalence and natural history of congenital cytomegalovirus (CMV) infection in highly seroimmune population. Clin. Infect. Dis. *49*, 522–528.

Mutnal, M.B., Cheeran, M.C., Hu, S., and Lokensgard, J.R. (2011). Murine cytomegalovirus infection of neural stem cells alters neurogenesis in the developing brain. PLoS One *6*, e16211.

Nandi, A., Estess, P., and Siegelman, M. (2004). Bimolecular complex between rolling and firm adhesion receptors required for cell arrest; CD44 association with VLA-4 in T-cell extravasation. Immunity *20*, 455–465.

Naruse, I., and Tsutsui, Y. (1989). Brain abnormalities induced by murine cytomegalovirus injected into the cerebral ventricles of mouse embryos exo utero. Teratology *40*, 181–189.

Nigro, G., Adler, S.P., La Torre, R., and Best, A.M. (2005). Passive immunization during pregnancy for congenital cytomegalovirus infection. N. Engl. J. Med. *353*, 1350–1362.

Noyola, D.E., Demmler, G.J., Nelson, C.T., Griesser, C., Williamson, W.D., Atkins, J.T., Rozelle, J., Turcich, M., Llorente, A.M., Sellers-Vinson, S., *et al.* (2001). Early predictors of neurodevelopmental outcome in symptomatic congenital cytomegalovirus infection. J. Pediatr. *138*, 325–331.

Odeberg, J., Wolmer, N., Falci, S., Westgren, M., Seiger, A., and Soderberg-Naucler, C. (2006). Human cytomegalovirus inhibits neuronal differentiation and induces apoptosis in human neural precursor cells. J. Virol. *80*, 8929–8939.

Ossipow, V., Descombes, P., and Schibler, U. (1993). CCAAT/enhancer-binding protein mRNA is translated into multiple proteins with different transcription activation potentials. Proc. Natl. Acad. Sci. U.S.A. *90*, 8219–8223.

Parish, I.A., and Kaech, S.M. (2009). Diversity in CD8(+) T-cell differentiation. Curr. Opin. Immunol. *21*, 291–297.

Park, A.H., Gifford, T., Schleiss, M.R., Dahlstrom, L., Chase, S., McGill, L., Li, W., and Alder, S.C. (2010). Development of cytomegalovirus-mediated sensorineural hearing loss in a Guinea pig model. Arch. Otolaryngol. Head. Neck. Surg. *136*, 48–53.

Pass, R.F., Stagno, S., Myers, G.J., and Alford, C.A. (1980). Outcome of symptomatic congenital CMV infection: results of long-term longitudinal follow-up. Pediatrics *66*, 758–762.

Pass, R.F., Fowler, K.B., Boppana, S.B., Britt, W.J., and Stagno, S. (2006). Congenital cytomegalovirus infection following first trimester maternal infection: symptoms at birth and outcome. J. Clin. Virol. *35*, 216–220.

Pass, R.F., Zhang, C., Evans, A., Simpson, T., Andrews, W., Huang, M.L., Corey, L., Hill, J., Davis, E., Flanigan, C., *et al.* (2009). Vaccine prevention of maternal cytomegalovirus infection. N. Engl. J. Med. *360*, 1191–1199.

Pereira, L., and Maidji, E. (2008). Cytomegalovirus infection in the human placenta: maternal immunity and developmentally regulated receptors on trophoblasts converge. Curr. Top. Microbiol. Immunol. *325*, 383–395.

Perlman, J.M., and Argyle, C. (1992). Lethal cytomegalovirus infection in preterm infants: clinical, radiological, and neuropathological findings. Ann. Neurol. *31*, 64–68.

Pewe, L., and Perlman, S. (2002). Cutting edge: CD8 T-cell-mediated demyelination is IFN-gamma dependent in mice infected with a neurotropic coronavirus. J. Immunol. *168*, 1547–1551.

Plotkin, S.A., Friedman, H.M., Fleisher, G.R., Dafoe, D.C., Grossman, R.A., Smiley, M.L., Starr, S.E., Wlodaver, C., Friedman, A.D., and Barker, C.F. (1984). Towne-vaccine induced prevention of cytomegalovirus disease after renal transplants. Lancet *1*, 528–530.

Podlech, J., Holtappels, R., Pahl-Seibert, M.F., Steffens, H.P., and Reddehase, M.J. (2000). Murine model of interstitial cytomegalovirus pneumonia in syngeneic bone marrow transplantation: persistence of protective pulmonary CD8-T-cell infiltrates after clearance of acute infection. J. Virol. *74*, 7496–7507.

van den Pol, A.N. (2006). Viral infections in the developing and mature brain. Trends Neurosci. *29*, 398–406.

van den Pol, A.N., Robek, M.D., Ghosh, P.K., Ozduman, K., Bandi, P., Whim, M.D., and Wollmann, G. (2007). Cytomegalovirus induces interferon-stimulated gene expression and is attenuated by interferon in the developing brain. J. Virol. *81*, 332–348.

Polic, B., Hengel, H., Krmpotic, A., Trgovcich, J., Pavic, I., Luccaronin, P., Jonjic, S., and Koszinowski, U.H. (1998). Hierarchical and redundant lymphocyte subset control precludes cytomegalovirus replication during latent infection. J. Exp. Med. *188*, 1047–1054.

Prosch, S., Heine, A.K., Volk, H.D., and Kruger, D.H. (2001). CCAAT/enhancer-binding proteins alpha and beta negatively influence the capacity of tumor necrosis factor alpha to up-regulate the human cytomegalovirus IE1/2 enhancer/promoter by nuclear factor kappaB during monocyte differentiation. J. Biol. Chem. *276*, 40712–40720.

Revello, M.G., Zavattoni, M., Furione, M., Lilleri, D., Gorini, G., and Gerna, G. (2002). Diagnosis and outcome of preconceptional and periconceptional primary human cytomegalovirus infections. J. Infect. Dis. *186*, 553–557.

Ross, S.A., and Boppana, S.B. (2005). Congenital cytomegalovirus infection: outcome and diagnosis. Semin. Pediatr. Infect. Dis. *16*, 44–49.

Ross, S.A., Fowler, K.B., Ashrith, G., Stagno, S., Britt, W.J., Pass, R.F., and Boppana, S.B. (2006). Hearing loss in children with congenital cytomegalovirus infection born to mothers with preexisting immunity. J. Pediatr. *148*, 332–336.

Ross, S.A., Novak, Z., Fowler, K.B., Arora, N., Britt, W.J., and Boppana, S.B. (2009). Cytomegalovirus blood viral load and hearing loss in young children with congenital infection. Pediatr. Infect. Dis. J. *28*, 588–592.

Ross, S.A., Novak, Z., Pati, S., Patro, R.K., Blumenthal, J., Danthuluri, V.R., Ahmed, A., Michaels, M.G., Sanchez, P.J., Bernstein, D.I., *et al.* (2011). Mixed infection and strain diversity in congenital cytomegalovirus infection. J. Infect. Dis. *204*, 1003–1007.

Saederup, N., Lin, Y.C., Dairaghi, D.J., Schall, T.J., and Mocarski, E.S. (1999). Cytomegalovirus-encoded beta chemokine promotes monocyte-associated viremia in the host. Proc. Natl. Acad. Sci. U.S.A. *96*, 10881–10886.

Sancho, D., Gomez, M., and Sanchez-Madrid, F. (2005). CD69 is an immunoregulatory molecule induced following activation. Trends Immunol. *26*, 136–140.

van der Sande, M.A., Kaye, S., Miles, D.J., Waight, P., Jeffries, D.J., Ojuola, O.O., Palmero, M., Pinder, M., Ismaili, J., Flanagan, K.L., *et al.* (2007). Risk factors for and clinical outcome of congenital cytomegalovirus infection in a peri-urban West-African birth cohort. PLoS ONE *2*, e492.

Schachtele, S.J., Mutnal, M.B., Schleiss, M.R., and Lokensgard, J.R. (2011). Cytomegalovirus-induced sensorineural hearing loss with persistent cochlear inflammation in neonatal mice. J. Neurovirol. *17*, 201–211.

Schleiss, M.R., Lacayo, J.C., Belkaid, Y., McGregor, A., Stroup, G., Rayner, J., Alterson, K., Chulay, J.D., and Smith, J.F. (2007). Preconceptual administration of an alphavirus replicon UL83 (pp65 homolog) vaccine induces humoral and cellular immunity and improves pregnancy outcome in the guinea pig model of congenital cytomegalovirus infection. J. Infect. Dis. *195*, 789–798.

Sharpe, A.H., Wherry, E.J., Ahmed, R., and Freeman, G.J. (2007). The function of programmed cell death 1 and its ligands in regulating autoimmunity and infection. Nat. Immunol. *8*, 239–245.

Shinmura, Y., Kosugi, I., Aiba-Masago, S., Baba, S., Yong, L.R., and Tsutsui, Y. (1997). Disordered migration and loss of virus-infected neuronal cells in developing mouse brains infected with murine cytomegalovirus. Acta Neuropathol. *93*, 551–557.

Shrestha, B., Samuel, M.A., and Diamond, M.S. (2006). CD8+ T-cells require perforin to clear West Nile virus from infected neurons. J. Virol. *80*, 119–129.

Sierro, S., Rothkopf, R., and Klenerman, P. (2005). Evolution of diverse antiviral CD8+ T-cell populations after murine cytomegalovirus infection. Eur. J. Immunol. *35*, 1113–1123.

Slavuljica, I., Busche, A., Babic, M., Mitrovic, M., Gasparovic, I., Cekinovic, D., Markova Car, E., Pernjak Pugel, E., Cikovic, A., Lisnic, V.J., *et al.* (2010). Recombinant mouse cytomegalovirus expressing a ligand for the NKG2D receptor is attenuated and has improved vaccine properties. J. Clin. Invest. *120*, 4532–4545.

Stagno, S., and Britt, W.J. (2006). Cytomegalovirus. In Diseases of the Fetus and Newborn Infan, Remington, J.S., and Klein, J.O., eds. (Saunders, W.B., Philadelphia, PA), pp. 203–241.

Stagno, S., Reynolds, D.W., Amos, C.S., Dahle, A.J., McCollister, F.P., Mohindra, I., Ermocilla, R., and Alford, C.A. (1977a). Auditory and visual defects resulting from symptomatic and subclinical congenital cytomegaloviral and toxoplasma infections. Pediatrics *59*, 669–678.

Stagno, S., Reynolds, D.W., Huang, E.S., Thames, S.D., Smith, R.J., and Alford, C.A. (1977b). Congenital cytomegalovirus infection: occurrence in an immune population. N. Engl. J. Med. *296*, 1254–1258.

Stagno, S., Pass, R.F., Reynolds, D.W., Moore, M., Nahmias, A.J., and Alford, C.A. (1980a). Comparative study of diagnostic procedures for congenital cytomegalovirus infection. Pediatrics *65*, 251–257.

Stagno, S., Reynolds, D.W., Pass, R.F., and Alford, C.A. (1980b). Breast milk and the risk of cytomegalovirus infection. N. Engl. J. Med. *302*, 1073–1076.

Stagno, S., Pass, R.F., and Alford, C.A. (1981). Perinatal infections and maldevelopment. Birth. Defects. Orig. Artic. Ser. *17*, 31–50.

Stagno, S., Pass, R.F., Dworsky, M.E., Henderson, R.E., Moore, E.G., Walton, P.D., and Alford, C.A. (1982a). Congenital cytomegalovirus infection: The relative importance of primary and recurrent maternal infection. N. Eng. J. Med. *306*, 945–949.

Stagno, S., Pass, R.F., Thomas, J.P., Navia, J.M., and Dworsky, M.E. (1982b). Defects of tooth structure in congenital cytomegalovirus infection. Pediatrics *69*, 646–648.

Stagno, S., Pass, R.F., Cloud, G., Britt, W.J., Henderson, R.E., Walton, P.D., Veren, D.A., Page, F., and Alford, C.A. (1986). Primary cytomegalovirus infection in pregnancy. Incidence, transmission to fetus, and clinical outcome. JAMA *256*, 1904–1908.

Strauss, M. (1985). A clinical pathologic study of hearing loss in congenital cytomegalovirus infection. Laryngoscope *95*, 951–962.

Strauss, M. (1990). Human cytomegalovirus labyrinthitis. Am. J. Otolaryngol. *11*, 292–298.

Sugita, K., Ando, M., Makino, M., Takanashi, J., Fujimoto, N., and Niimi, H. (1991). Magnetic resonance imaging of the brain in congenital rubella virus and cytomegalovirus infections. Neuroradiology *33*, 239–242.

Sylwester, A.W., Mitchell, B.L., Edgar, J.B., Taormina, C., Pelte, C., Ruchti, F., Sleath, P.R., Grabstein, K.H., Hosken, N.A., Kern, F., *et al.* (2005). Broadly targeted human cytomegalovirus-specific CD4+ and CD8+ T-cells dominate the memory compartments of exposed subjects. J. Exp. Med. *202*, 673–685.

Tarantal, A.F., Salamat, M.S., Britt, W.J., Luciw, P.A., Hendrickx, A.G., and Barry, P.A. (1998). Neuropathogenesis induced by rhesus cytomegalovirus in fetal rhesus monkeys (*Macaca mulatta*). J. Infect. Dis. *177*, 446–450.

Teissier, N., Delezoide, A.L., Mas, A.E., Khung-Savatovsky, S., Bessieres, B., Nardelli, J., Vauloup-Fellous, C., Picone, O., Houhou, N., Oury, J.F., *et al.* (2011). Inner ear lesions in congenital cytomegalovirus infection of human fetuses. Acta Neuropathol. *122*, 763–774.

Trgovcich, J., Pernjak Pugel, E., Tomac, J., Koszinowski, U., and Jonjic, S. (1998). Pathogenesis of murine cytomegalovirus infection in neonatal mice. In CMV-related Immunopathology Monographs in Virology, Scholz, M., Rabenau, H.F., Doerr, H.W., and Cinatl, J.J., eds. (Karger, Basel, Switzerland), pp. 42–53.

Tsutsui, Y. (2009). Effects of cytomegalovirus infection on embryogenesis and brain development. Congenit. Anom. (Kyoto) *49*, 47–55.

Tsutsui, Y., Kashiwai, A., Kawamura, N., Aiba-Masago, S., and Kosugi, I. (1995). Prolonged infection of mouse brain neurons with murine cytomegalovirus after pre- and perinatal infection. Arch. Virol. *140*, 1725–1736.

Tsutsui, Y., Kosugi, I., and Kawasaki, H. (2005). Neuropathogenesis in cytomegalovirus infection: indication of the mechanisms using mouse models. Rev. Med. Virol. *15*, 327–345.

Tsutsui, Y., Kosugi, I., Kawasaki, H., Arai, Y., Han, G.P., Li, L., and Kaneta, M. (2008). Roles of neural stem progenitor cells in cytomegalovirus infection of the brain in mouse models. Pathol. Int. *58*, 257–267.

Velu, V., Kannanganat, S., Ibegbu, C., Chennareddi, L., Villinger, F., Freeman, G.J., Ahmed, R., and Amara, R.R. (2007). Elevated expression levels of inhibitory receptor programmed death 1 on simian immunodeficiency virus-specific CD8 T-cells during chronic infection but not after vaccination. J. Virol. *81*, 5819–5828.

Vochem, M., Hamprecht, K., Jahn, G., and Speer, C.P. (1998). Transmission of cytomegalovirus to preterm infants through breast milk. Pediatr. Infect. Dis. J. *17*, 53–58.

van der Voorn, J.P., Pouwels, P.J., Vermeulen, R.J., Barkhof, F., and van der Knaap, M.S. (2009). Quantitative MR imaging and spectroscopy in congenital cytomegalovirus infection and periventricular leukomalacia suggests a comparable neuropathological substrate of the cerebral white matter lesions. Neuropediatrics *40*, 168–173.

de Vries, L.S., Gunardi, H., Barth, P.G., Bok, L.A., Verboon-Maciolek, M.A., and Groenendaal, F. (2004). The spectrum of cranial ultrasound and magnetic resonance imaging abnormalities in congenital cytomegalovirus infection. Neuropediatrics *35*, 113–119.

Walter, S., Atkinson, C., Sharland, M., Rice, P., Raglan, E., Emery, V.C., and Griffiths, P.D. (2008). Congenital cytomegalovirus: association between dried blood spot viral load and hearing loss. Arch. Dis. Child. Fetal Neonatal. Ed. *93*, 280–285.

Wang, C., Zhang, X., Bialek, S., and Cannon, M.J. (2011). Attribution of congenital cytomegalovirus infection to primary versus non-primary maternal infection. Clin. Infect. Dis. *52*, e11–13.

Wang, T., Town, T., Alexopoulou, L., Anderson, J.F., Fikrig, E., and Flavell, R.A. (2004). Toll-like receptor 3 mediates West Nile virus entry into the brain causing lethal encephalitis. Nat. Med. *10*, 1366–1373.

Williamson, W.D., Desmond, M.M., LaFevers, N., Taber, L.H., Catlin, F.I., and Weaver, T.G. (1982). Symptomatic congenital cytomegalovirus: disorders of language, learning and hearing. Am. J. Dis. Child. *136*, 902–905.

Wirtz, N., Schader, S.I., Holtappels, R., Simon, C.O., Lemmermann, N.A., Reddehase, M.J., and Podlech, J. (2008). Polyclonal cytomegalovirus-specific antibodies not only prevent virus dissemination from the portal of entry but also inhibit focal virus spread within target tissues. Med. Microbiol. Immunol. *197*, 151–158.

Woolf, N.K. (1990). Experimental congenital cytomegalovirus labyrinthitis and sensorineural hearing loss. Am. J. Otolaryngol. *11*, 299–303.

Woolf, N.K., Koehrn, F.J., Harris, J.P., and Richman, D.D. (1989). Congenital cytomegalovirus labyrinthitis and sensorineural hearing loss in guinea pigs. J. Infect. Dis. *160*, 929–937.

Wu, C.A., Paveglio, S.A., Lingenheld, E.G., Zhu, L., Lefrancois, L., and Puddington, L. (2011). Transmission of murine cytomegalovirus in breast milk: a model of natural infection in neonates. J. Virol. *85*, 5115–5124.

Yamamoto, A.Y., Mussi-Pinhata, M.M., Boppana, S., Novak, Z., Wagatsuma, V., Oliveria, P., Duarte, G., and Britt, W.J. (2010). Human cytomegalovirus reinfection is associated with intrauterine transmission in a highly cytomegalovirus-immune maternal population. Am. J. Obstet. Gynecol. *202*, 297.e291–e298.

Yeager, A.S., Grumet, F.C., Hafleigh, E.B., Arvin, A.M., Bradley, J.S., and Prober, C.G. (1981). Prevention of transfusion-acquired cytomegalovirus infections in newborn infants. J. Pediatr. *98*, 281–287.

Adaptive Cellular Immunity to Human Cytomegalovirus

II.7

Mark R. Wills, Gavin M. Mason and J. G. Patrick Sissons

Abstract

Primary HCMV infection induces robust CD8$^+$ cytotoxic, and CD4$^+$ helper, T-cell-mediated immune responses, which are associated with the resolution of acute primary infection: these responses are maintained at high frequency in long-term memory as the virus establishes persistent infection, with latency and periodic reactivation. Many of these T-cells are specific for epitopes in the pp65 and IE1 HCMV proteins, but it is apparent that many other viral proteins can also be T-cell targets, and in some individuals pp65 and IE1 responses are not immunodominant. During long-term carriage of the virus a balance is established between the T-cell-mediated immune response and viral reactivation: the T-cell response controls viral spread following reactivation, but the virus encodes multiple genes that interfere with MHC-I antigen processing (US 2, 3, 6 and 11), and with MHC-II processing and NK cell killing, allowing limited viral evasion of the response. Loss of this balance is most evident in the immunocompromised host in whom reactivation of latent virus or primary infection can lead to unchecked viral replication, with consequent disease and mortality. This chapter describes current understanding of CD8$^+$ and CD4$^+$ T-cell responses to HCMV, and how these responses reconstitute after bone marrow transplantation and might be used as therapies to protect against HCMV disease in immunocompromised subjects. The phenomenon of 'memory T-cell inflation' associated with HCMV and its relationship to immunosenescence is also discussed.

Introduction

HCMV infection elicits high-frequency virus-specific CD4$^+$ and CD8$^+$ T-cell responses to numerous viral proteins: evidence suggesting that these responses are protective is inferential, derived from the increased rates of HCMV disease in subjects with impaired T-cell immunity and from data obtained from patients undergoing reconstitution of their immune systems following bone marrow transplantation or haematopoietic stem cell transplantation (BMT/HSCT). Further direct evidence has also been provided by the murine model of CMV; mice are protected from lethal MCMV challenge by CD8$^+$ T-cells specific for immediate early antigens (Reddehase *et al.*, 1987; Holtappels *et al.*, 2008). In B-cell deficient mice it has been demonstrated that CD4$^+$ T-cells and NK cells can substitute for CD8$^+$ T-cells in controlling MCMV reactivation (Polic *et al.*, 1998). However in murine models of BMT, removal of reconstituted CD8$^+$ T-cells leads to lethal disease (Podlech *et al.*, 1998) and reconstituted CD8$^+$ T-cells transferred to immunocompromised mice could prevent disease (Podlech *et al.*, 2000; see also Chapter II.17). Following BMT in humans there is a strong correlation between the recovery of cytolytic T-cell activity and recovery from HCMV infection (Reusser *et al.*, 1991). Subsequent studies using virus-specific MHC-I tetramers to quantify HCMV-specific CD8$^+$ T-cell recovery estimated that a recovery of virus-specific cells of > 10 per µl of blood was protective against serious HCMV disease (Cwynarski *et al.*, 2001).

This chapter reviews the current data available on the specificity, frequency, function and phenotype of CD4$^+$ and CD8$^+$ T-cell responses to HCMV. The frequency of virus-specific T-cells in peripheral blood responding to HCMV has been shown to progressively increase with time following primary infection, and the same phenomenon has also been recorded in the MCMV model and described as 'memory inflation': we discuss memory inflation and its relationship to 'immunosenescence' in the long-term carriage of virus. We also summarize current knowledge about the reconstitution of HCMV-specific immunity following BMT/HSCT. A practical consequence of all of this knowledge should be to inform the design of interventions to prevent or control HCMV reactivation and disease in both normal subjects and immunocompromised patients:

vaccine development for HCMV is the subject of a separate chapter (Chapter II.20), but we also review here the current state of adoptive immunotherapy to treat immunosuppressed patients.

The establishment and sites of HCMV latency are extensively reviewed in Chapters I.19 and I.20. A key tenet of this body of work is that HCMV establishes latency in CD34$^+$ bone marrow progenitor cells, carriage of viral genome persists in cells of the myeloid lineage and that reactivation is triggered following terminal differentiation of these myeloid cells (i.e. macrophages and dendritic cells). Thus, it should be remembered that in all phases of the HCMV life cycle the virus is intimately associated with professional antigen-presenting cells. As a consequence, it is likely that the evolution of multiple sophisticated T-cell immune evasion mechanisms encoded by the virus are driven by this association. Despite this, in healthy individuals primary HCMV infection is controlled by the immune response and long-term carriage of the virus is asymptomatic – however, the immune system never eliminates the virus.

The human CD8$^+$ T-cell response to HCMV

Human CMV specific CD8$^+$ T-cells that can mediate cytotoxicity against HCMV-infected cells have been isolated from the PBMCs of normal healthy HCMV carriers and established in *in vitro* culture since the early 1980s (Borysiewicz *et al.*, 1983). Since then considerable effort has been expended in determining their frequency, function, phenotype and specificity for HCMV proteins. Detailed studies of some of these proteins have identified the minimal T-cell epitopes and their MHC-I and II restrictions. Over the last 30 years new technologies for determining these parameters have been developed including cytokine ELISPOT assays, peptide loaded MHC-I multimers for use in flow cytometry and intracellular cytokine staining (ICS) to identify multiple cytokine secretion profiles in individual cells. Advances in flow cytometry instruments that can now readily measure eight parameters or more, simultaneously coupled with antigen specificity determined by ICS, have greatly increased our understanding of the CD8$^+$ T-cell response to HCMV following primary infection, in long-term carriage and following transplantation.

Antigen specificity

Initial analyses of antigen specificity utilized bulk T-cell cultures stimulated with HCMV-infected autologous fibroblast cells, and classic chromium release cytotoxicity assays as readouts, against target cells infected with recombinant vaccinia viruses expressing individual HCMV proteins. Following the identification of the major immediate early protein as a major T-cell target in murine CMV, it was subsequently established that human CD8$^+$ T-cells could also recognize the IE1 protein in HCMV (Borysiewicz *et al.*, 1983). Riddell and colleagues, based on their analysis of human T-cell clones derived for adoptive transfer, first suggested a large proportion of the CD8$^+$ T-cell response to HCMV was specific for the lower matrix tegument protein pp65 (McLaughlin Taylor *et al.*, 1994; Walter *et al.*, 1995) – an observation subsequently confirmed by numerous groups. pp65 is arguably the most studied HCMV T-cell antigen: it has been extensively mapped and numerous minimal T-cell epitopes restricted by a wide range of MHC-I alleles have been described (Table II.7.1). It was clear that the magnitude of the T-cell response to pp65 could be very large and was detectable in many, but significantly, not all donors. HCMV encodes over 165 ORFs (see Chapter I.1) and it seemed likely that other viral proteins would be potential CD8$^+$ T-cell targets.

Stimulation of PBMCs from healthy HCMV carriers with HCMV-infected autologous fibroblasts was used to identify CD8$^+$ T-cell responses to HCMV pp150 (UL32), pp50 (UL44), gB (UL55), gH (UL75) and pp28 (UL98) (Boppana and Britt, 1996). A combination of bioinformatics and synthetic peptides used in IFN-γ ELISPOT assays was also used to predict CD8$^+$ T-cell epitopes derived from the published sequence of 14 HCMV encoded proteins (pp28, pp50, pp65, pp150, pp71, gH, gB, IE1, US2, US3, US11, UL16 and UL18) and subsequently verified using an ELISPOT assay measuring IFN-γ production (Elkington *et al.*, 2003). The observation that HCMV is able to interfere with normal MHC-I processing and presentation, and that this interference is abrogated by a deletion mutant of HCMV which lacks the genes encoding US2 to US11 (see CD8 immune evasion section later in this chapter) led to this deletion mutant being used to infect autologous fibroblasts and stimulate PBMCs. This analysis revealed high frequency responses to pp65, IE1, pp150 and gB, in agreement with previous studies. However, T-cells with many other specificities were also generated to immediate early or early viral proteins, although individual viral proteins were not defined in this study (Manley *et al.*, 2004). A subsequent study using the complete HCMV proteome represents the most comprehensive analysis of the spectrum of HCMV-specific T-cell specificities to date; 13,687 overlapping peptides covering 213 predicted HCMV encoded ORFs were used to determine IFN-γ responses by intracellular cytokine analysis from both CD4$^+$ and CD8$^+$ T-cells

Table II.7.1 Class I MHC peptide epitopes recognized by human CD8+ T-cells derived from HCMV proteins

ORF	Protein	Peptide	MHC class I molecule	Assay[a]	Reference
UL32	pp150	945-TTVYPPSSTAK-955	HLA-A3	I M C	Longmate *et al.* (2001)
		792-QTVTSTPVQGR-802	HLA-A68	I C	Longmate *et al.* (2001)
UL44	pp50	245-VTEHDTLLY-253	HLA-A1	I M C	Elkington *et al.* (2003)
UL55	gB	618-(F)IAGNSAYEYV-628	HLA-A2	I C	Parker *et al.* (1992), Utz and Biddison (1992)
		731-AVGGAVASV-739	HLA-A2	I C	Elkington *et al.* (2003)
UL83	pp65	363-YSEHPTFTSQY-373	HLA-A1	I M C	Longmate *et al.* (2001), Hebart *et al.* (2002)
		14-VLGPISGHV-22	HLA-A2	I C	Solache *et al.* (1999)
		120-MLNIPSINV-128	HLA-A2	I C	Solache *et al.* (1999)
		490-ILARNLVPM-498	HLA-A2	I C	Elkington *et al.* (2003)
		495-NLVPMVATV-503	HLA-A2	I C	Wills *et al.* (1996), Diamond *et al.* (1997), Gillespie *et al.* (2000)
		522-RIFAELEGV-530	HLA-A0207	I C	Kondo *et al.* (2004)
		16-GPISGHVLK-24	HLA-A11	I C	Longmate *et al.* (2001), Hebart *et al.* (2002)
		501-ATVQGQNLK-509	HLA-A11	I C	Kondo *et al.* (2004)
		248-AYAQKIFKIL-257	HLA-A23/24	I M	Elkington *et al.* (2003)
		113-VYALPLKML-121	HLA-A24	I M C	Masuoka *et al.* (2001)
		341-QYDPVAALF-349	HLA-A24	I M C	Kuzushima *et al.* (2001)
		369-FTSQYRIQGKL-37	HLA-A24	I C	Longmate *et al.* (2001)
		186-FVFPTKDVALR-196	HLA-A68	I C	Longmate *et al.* (2001)
		265-RPHERNGFTV-274	HLA-B7	I M C	Weekes *et al.* (1999)
		417-(T)PRVTGGGAM-426	HLA-B7	I M C	Wills *et al.* (1996), Kern *et al.* (1998), Weekes *et al.* (1999)
		215-KMQVIGDQY-223	HLA-B15	I C	Kondo *et al.* (2004)
		103-CPSQEPMSIYVY-114	HLA-B35	I C	Rist *et al.* (2005)
		123-IPSINVHHY-131	HLA-B35	I M C	Gavin *et al.* (1993), Hassan-Walker *et al.* (2001)
		187-VFPTKDVAL-195	HLA-B35	I C	Wills *et al.* (1996)
		174-NQWKEPDVY-182	HLA-B35	I C	Kern *et al.* (2002)
		283-KPGKISHIMLDVA-295	HLA-B35	I	Elkington *et al.* (2003)
		367-PTFTSQYRIQGKL-379	HLA-B38	I C	Longmate *et al.* (2001)
		232-CEDVPSGKL-240	HLA-B40	I C	Kondo *et al.* (2004)
		267-HERNGFTVL-275	HLA-B40	I C	Kondo *et al.* (2004)
		525-AELEGVWQPA-534	HLA-B40	I C	Kondo *et al.* (2004)
		364-SEHPTFTSQY-373	HLA-B44	I C	Kondo *et al.* (2004)
		512-EFFWDANDIY-521	HLA-B44	I C	Wills *et al.* (1996)
		545-DALPGPCI-552	HLA-B51	I C	Kondo *et al.* (2004)
		155-QMWQARLTV-163	HLA-B52	I C	Kern *et al.* (2002)
		211-TRATKMQVI-219	HLA-B57/58	I M	Elkington *et al.* (2003)
		7-RCPEMISVL-15	HLA-Cw1	I C	Kondo *et al.* (2004)
		341-QYDPVAALF-349	HLA-Cw4	I C	Kondo *et al.* (2004)
		198-VVCAHELVC-206	HLA-Cw8	I C	Kern *et al.* (2002), Kondo *et al.* (2004)
		294-VAFTSHEHF-302	HLA-Cw12	I C	Kondo *et al.* (2004)
		198-VVCAHELVC-206	HLA-Cw15	I C	Kondo *et al.* (2004)
UL98		277-ARVYEIKCR-285	HLA-B27	I C	Elkington *et al.* (2003)
UL123, UL122	IE1, pp72, IE2	81-VLAELVKQI-89	HLA-A2	I M C	Elkington *et al.* (2003)
		315-Y(V/I)LEETSVM-323	HLA-A2	I C	Retiere *et al.* (2000), Khan *et al.* (2002)
		316-VLEETSVML-324	HLA-A2	I M C	Khan *et al.* (2002)
		354-YILGADPLRV-363	HLA-A2	I C	Frankenberg *et al.* (2002)
		184-KLGGALQAK-192	HLA-A3	M	[b]
		309-CRVLCCYVL-317	HLA-B7	I C	Kern *et al.* (1999), Wills *et al.* (2002)
		88-QIKVRVDMV-96	HLA-B8	I M C	Elkington *et al.* (2003)
		198-(D)ELRRKMMYM-207	HLA-B8	I C	Kern *et al.* (1999), Wills *et al.* (2002)
		199-ELKRKMIYM-207	HLA-B18	I M C	Retiere *et al.* (2000)
		279-CVETMCNEY-287	HLA-B18	I C	Retiere *et al.* (2000)
		379-DEEDAIAAY-387	HLA-B18	I C	Retiere *et al.* (2000)
		381-FEQPTETPP-389	HLA-B41	I C	Rist *et al.* (2005)
US2		190-SMMWMRFFV-198	HLA-A2	I C	Elkington *et al.* (2003)

[a]Epitopes listed elicit either a functional IFN-γ response (I) or have been established as functional cytotoxic T-cell lines or clones (C). Peptides that have been used to generate MHC Class I tetramers or multimers (M) are also listed with a reference to the first paper to generate them.
[b]Listed as commercially available from ProImmune UK.

in a cohort of 33 seropositive donors with disparate MHC-I and II types to provide a broad MHC polymorphism coverage. 151 ORFs were shown to elicit a CD4$^+$ or CD8$^+$ T-cell response in at least one donor. Three ORFs were recognized by more than half of the cohort, UL48, UL83 (pp65), and UL123 (IE1). CD8$^+$ T-cell responses from HCMV-seropositive donors recognized a median of eight ORFs: however responses were highly heterogeneous between individuals with some recognizing only a single ORF and others as many as 39 (Sylwester *et al.*, 2005). This analysis was performed on healthy HCMV-seropositive donors and as such reflects the range of specificities in the memory repertoire, it is possible that the spectrum maybe broader during acute primary infection.

During lytic infection all of the HCMV proteins are expressed in a temporal cascade of gene expression, but during latency the transcription of viral genes is highly restricted. A number of genes have been shown to be expressed during latency: these include UL111.5 a viral IL-10 homologue (Cheung *et al.*, 2009), UL138 (Goodrum *et al.*, 2007), and UL8182AS a viral transcript found antisense to UL81–82 latent undefined nuclear antigen (LUNA) (Bego *et al.*, 2005). UL138 and a truncated sequence of UL111.5 were included in the T-cell proteome screen; UL138 was both a CD4 and a CD8 target in just one donor (Sylwester *et al.*, 2005). More recently T-cell responses to UL138 and LUNA were examined in detail in 22 individuals: a CD8 response to a single peptide was seen exclusively in donors with HLA-B3501 and this was mapped to the same peptide in each case (Tey *et al.*, 2010).

This detailed knowledge of immunodominant peptides and their restricting MHC-I alleles has been put to a number of highly practical uses, including a rational vaccine design strategy which couples the extracellular part of HCMV gB (to elicit neutralizing antibodies) with multiple CD8$^+$ and CD4$^+$ T-cell epitopes to induce CTL and IFN-γ/TNF-α producing T-cells in a mouse model. The construct was also able to induce the expected CD4$^+$ and CD8$^+$ T-cell responses in restimulation assays on PBMCs from HCMV-seropositive donors (Zhong *et al.*, 2008). Adoptive immune therapy approaches utilizing specific MHC-I multimers to purify HCMV-specific T-cells, that have the benefit of not requiring extensive *in vivo* cell culture, are being explored (Cobbold *et al.*, 2005).

HCMV encoded interference with MHC class I processing and presentation

It has been recognized since the early 1990s that MHC-I expression on the surface of HCMV-infected cells progressively diminishes with increasing time after infection. The disruption of components of the MHC-I processing pathway is a property that many viruses display in an attempt to evade virus-specific CD8$^+$ T-cells (Petersen *et al.*, 2003).

HCMV can disrupt the normal MHC-I antigen processing pathways by a number of mechanisms, principally mediated by genes in the US2–11 region of the genome. These include the degradation of newly synthesized class I heavy chains mediated by US2 (Wiertz *et al.*, 1996a) and US11 (Jones *et al.*, 1995; Wiertz *et al.*, 1996b), which dislocate the nascent heavy chains from the ER into the cytoplasm where they undergo proteasomal degradation. The US3 gene product binds to tapasin in the ER and subsequently causes retention of MHC-I (Jones *et al.*, 1996; Lee *et al.*, 2000; Park *et al.*, 2004). The US6 gene product blocks peptide translocation into the ER by binding to the cytosolic face of the TAP (transporter of antigenic peptides) heterodimer (Ahn *et al.*, 1997; Lehner *et al.*, 1997) as summarized in Fig. II.7.1. US8 and US10 have been shown to physically interact with MHC-I (Furman *et al.*, 2002; Tirabassi and Ploegh, 2002), although US8 does not affect MHC-1 levels. More recently US10 has been shown to down-regulate surface expression of non-classical HLA-G which is targeted for degradation (Park *et al.*, 2010). Numerous reviews on the subject of immune evasion have been published (Reddehase, 2002; Powers *et al.*, 2008; Hansen and Bouvier, 2009).

The number of viral genes apparently dedicated to interfering with normal peptide presentation seems surprising. However the human immune system and HCMV have co-evolved, and it is possible that this diversity of immune evasion mechanisms reflects this co-evolution, and that some of the immune evasion mechanisms may be redundant. It may also be the case that different mechanisms working together may be a very efficient way of down-regulating MHC-I. More recent evidence suggests that these 'immune evasion' genes may not completely protect cells from CD8$^+$ T-cell recognition, and this is dependent on the antigen specificity of the T-cell. US2–11 could prevent any presentation of IE antigen in HCMV-infected cells: however the tegument and late antigen pp65 was still presented as evidenced by pp65 specific T-cell recognition (Besold *et al.*, 2007). The mechanism for this is unclear, it might be related to the overall 'avidity' of TCR for its peptide-MHC with high avidity clones able to recognize cells with only a low level of occupied MHC-I rather than the antigen specificity in terms of which viral gene product is recognized. There is also some evidence that the genes down-regulating MHC class I are not redundant, as recombinant HCMV engineered to express US2 and US11 alone incompletely protected HCMV-infected fibroblasts from both IE and

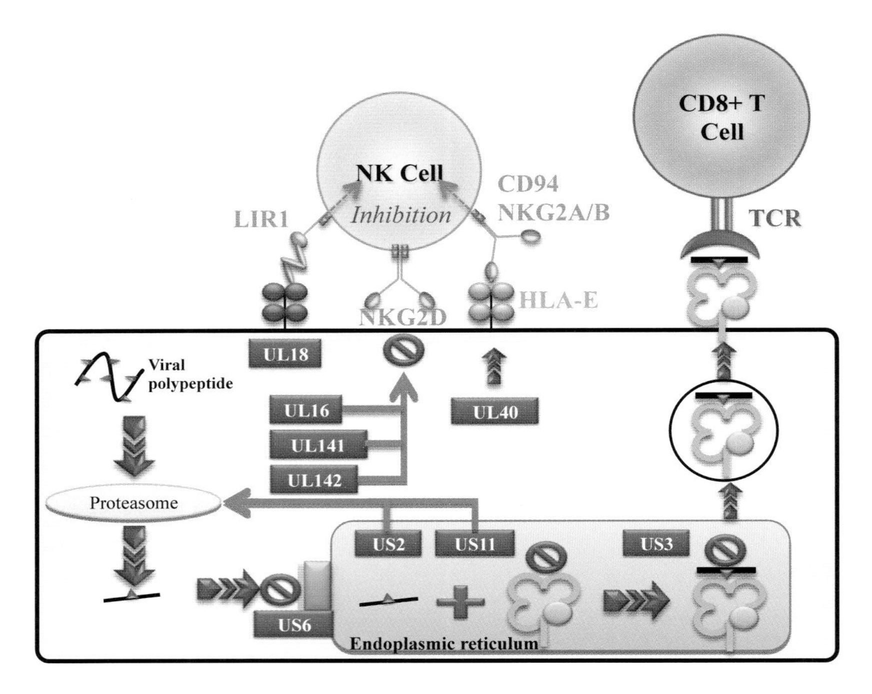

Figure II.7.1 CD8+ T-cell and NK cell evasion by HCMV-infected cells. Disruption of MHC-I processing and presentation to CD8+ T-cells by HCMV proteins US2, 3, 6 and 11 causes the inhibition of expression of cell surface MHC-I loaded with viral peptides and prevents CTL recognition. The lack of cell surface MHC-I decreases inhibitory signalling to NK cells which is substituted by UL18 and UL40 expression. Stress-induced ligands of NK cell receptors are prevented from cell surface expression by HCMV expression of UL16, UL141 and UL142 in order to inhibit NK cell activation.

pp65 specific T-cell recognition (Besold *et al.*, 2009). HCMV infects a number of different cell types *in vivo* and it remains possible that some of these viral gene products may be more efficient in some cell types than others.

These decreased levels of surface MHC-I on HCMV-infected cells might be predicted to render the cells more susceptible to lysis by NK cells: according to the 'missing self' hypothesis NK cells are normally inhibited by surface MHC-I molecules, but would be less inhibited by CMV-infected cells and would consequently lyse the cells (Karre *et al.*, 1986). However, it has been demonstrated in a number of experimental systems that HCMV-infected cells are in fact relatively resistant to NK-mediated lysis (Reyburn *et al.*, 1997; Fletcher *et al.*, 1998; Cerboni *et al.*, 2000; Wang *et al.*, 2002; Vales-Gomez *et al.*, 2003). This suggests that in addition to genes to evade T-cell surveillance, HCMV may also have evolved viral encoded functions to evade NK surveillance.

NK cells are subject to both activating and inhibitory signals when interacting with cells in the body that may be potential targets due to virus infection or transformation. The inhibitory signals are delivered via inhibitory receptors interacting with normal levels of MHC-I molecules on the surface of cells. Since HCMV-infected cells present substantially reduced levels of surface MHC-I, this would be predicted to reduce NK inhibitory signals (the missing self hypothesis), and consequently render the infected cells susceptible to NK cell-mediated cytotoxicity (Karre *et al.*, 1986; Ravetch and Lanier, 2000; Biassoni *et al.*, 2001). It was thus hypothesized that HCMV must encode mechanisms to inhibit NK cell recognition. Numerous groups have indeed observed that HCMV encodes multiple genes controlling NK cell activation and cytotoxicity both by the provision of inhibitory signals and suppression of activating signals (Wilkinson *et al.*, 2008; see also Chapter II.8).

HCMV-infected cells can induce inhibitory receptor signalling in NK cells by two mechanisms. These include the host HLA-E pathway to inhibit NK cells via the CD94/NKG2A heterodimeric inhibitory receptor (Braud *et al.*, 1998; Posch *et al.*, 1998). Viral UL40 protein contains a nonameric peptide which is processed in the ER and binds HLA-E promoting its cell surface expression (Tomasec *et al.*, 2000; Ulbrecht *et al.*, 2000). Secondly, HCMV expresses a viral homologue of cellular MHC-I, UL18, at the infected cell surface (Beck and Barrell, 1988). UL18 binds the inhibitory NK cell receptor, LILRB1 (LIR-1) with higher affinity than cellular MHC-I (Cosman *et al.*, 1997; Chapman *et al.*, 1999) inhibiting LILRB1$^+$ NK cell activation (Prod'homme *et al.*, 2007).

HCMV encodes four genes and a microRNA which prevent activating NK cell receptor signalling. The pp65 tegument protein (UL83) dissociates CD3ζ signalling from NKp30 (Arnon *et al.*, 2005). UL141 mediates the intracellular retention of CD155 and CD112 ligands for the NK activating receptors CD226 and CD96 (Tomasec *et al.*, 2005; Prod'homme *et al.*, 2010). The remaining viral proteins UL16 and UL142 interfere with NKG2D-mediated NK cell activation. NKG2D, a major activating receptor expressed on all human NK cells (Bauer *et al.*, 1999), is engaged by eight different NKG2D ligands (NKG2DL) induced by HCMV infection (Eagle *et al.*, 2006): MICA/B, ULBP1–3, RAET1E (ULBP4), RAET1G (ULBP5) and RAET1L (ULBP6). UL16 protein binds ULBP1, 2 and 6, as well as MICB (but not ULBP3/MICA) (Cosman *et al.*, 2001; Eagle *et al.*, 2009) mediating their intracellular retention (Dunn *et al.*, 2003; Rolle *et al.*, 2003). Down-regulated surface expression of the closely related MICA has also been demonstrated in infected cells to be mediated by UL142 (Wills *et al.*, 2005; Chalupny *et al.*, 2006; Ashiru *et al.*, 2009), and an as yet unidentified viral gene (Zou *et al.*, 2005). UL142 is a late viral gene and a member of the UL18 gene family (Dolan *et al.*, 2004): it localizes to the ER and cis-Golgi and prevents cell surface expression of both MICA and ULBP3 (Wills *et al.*, 2005; Ashiru *et al.*, 2009; Bennett *et al.*, 2010). An HCMV encoded microRNA, UL112–1, reduces translation of MICB mRNA expression (Stern-Ginossar *et al.*, 2008). Innate immunity and NK cell evasion is reviewed in more detail in Chapter II.8.

One may question whether this immune evasion is at all effective given that it is clear that upon primary HCMV infection the host mounts a strong T-cell response composed of both CD8$^+$ CTL and CD4$^+$ helper T-cells which produce antiviral cytokines, as well as the clear evidence of the wide extent of antigens recognized by this response. The outcome of acute primary infection in normal healthy individuals is control of viral replication, although HCMV is not cleared from the host but becomes latent with periodic reactivation and production of new virions (presumably to enable host to host transmission).

It seems plausible that the immune evasion mechanisms gives the virus a 'window of protection' during reactivation from latency in the face of a fully primed adaptive immune system, enabling the virus to complete its life cycle to produce new virions. Thus, the real functional significance *in vivo* of the viral immune evasion genes which target normal MHC-I antigen processing and presentation in HCMV, remains to be determined. However the murine model of MCMV infection (see Chapter II.17), and the more recently described RhCMV model (see Chapter II.22) do much to inform us of the role of MHC-I modulation *in vivo*.

It might be predicted that viruses that lack the MHC-I modulation genes would be better at priming T-cell responses (giving increased frequencies) in a naïve animal or indeed elicit a broader T-cell response. The results from experiments from a number of groups in fact demonstrated neither of these predictions to be true, (reviewed in Lemmermann *et al.*, 2011). MCMV infection of C57BL/6 mice with wild type (WT) virus actually elicited the same magnitude of response to an epitope in M45 as a virus which did not express one of the MHC-I evasion genes (Gold *et al.*, 2002). Later analysis of the T-cell response elicited to subdominant epitopes by a WT virus and a virus lacking all MHC-I down-regulation genes concluded that the immune evasion genes had little impact on the specificity of the response or the overall magnitude. (Munks *et al.*, 2007). For IE1 specific priming in BALB/c mice, Reddehase and colleagues have described better priming by WT virus and provided evidence to conclude that this paradoxical result occurs as a consequence of antigen-presenting cells infected with MCMV lacking the MHC-I down-regulation genes being rapidly cleared by newly produced virus-specific T-cells, thus reducing the available antigen and never stimulating the full potential magnitude of the T-cell response. APC infected with WT virus on the other hand persist much longer, producing more antigen which is available for uninfected APC to present (cross presentation), thus eliciting a stronger T-cell response (Böhm *et al.*, 2008). This group have suggested that one role for the immune evasion genes is to delay the rate of viral clearance and thus increase the chances of host to host transmission (Lemmermann *et al.*, 2011).

Do the immune evasion genes have an effect either on the establishment of latency or the efficiency with which reactivation can occur? In MCMV the latent viral load in the lungs is reduced when a virus lacking the MHC-I evasion genes is used for the infections: these authors suggest that the overall viral load is reduced with this mutant virus, leading to a reduction in the establishment of latency as the spread of the virus to the potential sites of latency is less efficient (Böhm *et al.*, 2009).

It has been suspected for some time that HCMV has the ability to super-infect a host which already has a well developed humoral and T-cell immune response as evidenced by the existence of multiple HCMV gB genotypes in a single normal host (Chou, 1990; Fries *et al.*, 1994). Cytomegalovirus super-infection has been studied experimentally using the RhCMV model, where the data show that the RhCMV homologues of HCMV US2, 3, 6 and 11 are required for super-infection. The study also demonstrated that the immune evasion genes were not required for primary infection, nor for the establishment or maintenance of persistent infection as the virus could be isolated from the urine of animals infected with the deletion mutant more than 400 days post infection (Hansen *et al.*, 2010).

HCMV-specific CD8+ T-cell phenotype and memory

The availability of HCMV-specific MHC-I multimers, monoclonal antibodies against cell surface molecules and multiparameter flow cytometry, as well as techniques for T-cell receptor analysis, have made possible the detailed characterization of antigen-experienced HCMV-specific CD8+ T-cells. The expression of different isoforms of the leucocyte common antigen CD45 was previously thought to distinguish naïve CD45RA^{hi} from memory CD45RO^{hi} T-cells. However in healthy carriers it has now been comprehensively shown that HCMV (pp65 and IE1) specific T-cells are present in the CD45RO^{hi} subpopulation but also that a large proportion of pp65 specific CD8+ memory T-cells have re-expressed CD45RA and have direct effector function as they contain preformed perforin and granzyme – these cells have subsequently been termed T_{EMRA} (effector memory expressing CD45RA) cells (Wills *et al.*, 1999, 2002; Gillespie *et al.*, 2000; Champagne *et al.*, 2001; Appay *et al.*, 2002a; Khan *et al.*, 2002b; Komatsu *et al.*, 2006; Miles *et al.*, 2007; Sauce *et al.*, 2007; van de Berg *et al.*, 2008; Harari *et al.*, 2009). In addition it has also been demonstrated that they are derived from the same clones that expressed CD45RO previously (Wills *et al.*, 1999; Iancu *et al.*, 2009). These T_{EMRA} cells retain functionality, in that they are still cytotoxic expressing both perforin and granzyme and readily express IFN-γ upon stimulation (Gillespie *et al.*, 2000; Appay *et al.*, 2002a; Khan *et al.*, 2002b; Wills *et al.*, 2002; Boutboul *et al.*, 2005; Libri *et al.*, 2008; van de Berg *et al.*, 2008; Harari *et al.*, 2009; Makedonas *et al.*, 2010). In post haematopoietic stem cell transplant studies it has been shown that reconstitution of T_{EMRA} cells correlated with better control of HCMV reactivation (Moins-Teisserenc *et al.*, 2008; Luo *et al.*, 2010). This CD45RA+ HCMV-specific memory T-cell population is also present from a very young age (Komatsu *et al.*, 2006; Miles *et al.*, 2007) and notably expands with old age (Komatsu *et al.*, 2006) (see discussion on 'memory inflation' later in this chapter).

A more detailed phenotypic analysis of the T_{EMRA} HCMV-specific T-cells, using markers of cell surface adhesion, co-stimulation and chemokine receptor molecules, has shown that they have usually lost expression of the co-stimulatory molecule CD28, have variable expression of CD27, and generally do not express the chemokine receptor CCR7, which is associated with homing to lymph nodes (Kern *et al.*, 1999; Gillespie *et al.*, 2000; Khan *et al.*, 2002a,b; Gamadia *et al.*, 2003; Komatsu *et al.*, 2006; Miles *et al.*, 2007; van de Berg *et al.*, 2008; Pita-Lopez *et al.*, 2009; Scheinberg *et al.*, 2009). However HCMV-specific memory cells do express other chemokine receptors, including CCR5 which enables migration to sites of inflammation and may be associated with the migration phenotype enabling homing to the bone marrow, where HCMV-specific memory CD8+ T-cell populations have been found (Letsch *et al.*, 2007; Palendira *et al.*, 2008; Melenhorst *et al.*, 2009).

Many HCMV-specific T_{EMRA} cells have also lost expression of IL-7Rα and express CD57, a marker which is associated with highly differentiated T-cells (Gillespie *et al.*, 2000; Khan *et al.*, 2002a,b; van Leeuwen *et al.*, 2005; Day *et al.*, 2007). HCMV-specific T_{EMRA} cells also express receptors found on NK cells, both inhibitory and activating receptors (van de Berg *et al.*, 2008; van Stijn *et al.*, 2008), including CD85j, CD244 (Pita-Lopez *et al.*, 2009) and KLRG1 (Ouyang *et al.*, 2003). In part this may explain why earlier reports of the highly differentiated phenotype of this population concluded that the cells were terminally differentiated (Champagne *et al.*, 2001) due to the expression of inhibitory receptors. It has been shown that KLRG1 signalling prevents Akt phosphorylation with a subsequent loss of proliferative capacity (Henson *et al.*, 2009) and that the transcription factor BMI-1 which is important for replicative competence in CD8+ T-cells following T-cell receptor ligation is not expressed in murine KLRG1+ T-cells (Heffner and Fearon, 2007). However, there is evidence from patients following stem cell transplantation that expression of inhibitory KIR NK receptors and loss of T-cell function is only observed following KIR ligation, whereas, in the absence of the corresponding KIR ligands, CMV-specific CD8+ T-cells retained functionality (van der Veken *et al.*, 2009).

CD8+ CMV specific memory T-cell populations also mediate effector function by cytokine secretion. So-called polyfunctional CMV specific CD8+ T-cells secrete a combination of TNF-α, MIP1-β, and IFN-γ and up-regulate CD107a, a marker of degranulation, (Gillespie *et al.*, 2000; Scheinberg *et al.*, 2009; Zhou *et al.*, 2009; Makedonas *et al.*, 2010). In post transplant patients the presence of polyfunctional CMV specific CD8^{+} T-cells was associated with more efficient prevention of CMV reactivation than if CD8^{+} CMV specific T-cells secreting only IFN-γ were detected (Zhou *et al.*, 2009).

Whether CD8^{+} T-cells that lack CD28 expression have a defect in proliferation has been controversial. Non-specific stimulation using anti-CD3, anti-CD2, PHA or PMA + Ionomycin has shown CD28^{-} T-cells have a lesser capacity for proliferation than CD28^{+} T-cells (Azuma *et al.*, 1993; Borthwick *et al.*, 1994; Brinchmann *et al.*, 1994; Lewis *et al.*, 1999; Scheuring *et al.*, 2002). However this has not been replicated in other studies using plate-bound anti-CD3 stimulation (Vingerhoets *et al.*, 1995) or autologous PHA-stimulated blast cells as stimulators in HIV-infected subjects (Fiorentino *et al.*, 1996).

A number of studies have suggested that CD8^{+} CD45RA^{+} CD28^{-} T-cells have a proliferation defect (Geginat *et al.*, 2003; Dunne *et al.*, 2005), including one using HCMV peptide stimulation (Champagne *et al.*, 2001). In contrast other studies dispute that there is a proliferation defect in this population; CD8^{+} CD45RA^{+} CD28^{-} CCR7^{-} T-cells have been shown to proliferate in response to HCMV peptide (Wills *et al.*, 2002). The cloning efficiency of CD8^{+} CD45RA^{+} CD28^{-} T-cells is not dissimilar to that of HCMV-specific central memory CD8^{+} CD45RO^{+} T-cells which are believed to be fully functional. CD8^{+} CD45RA^{+} CD28^{-} T-cells require professional antigen presentation as autologous fibroblasts pulsed with HCMV peptides are unable to activate these memory cells. While these T-cells have transient re-expression of CD28 they also up-regulate CD137 (4-1BB) and fibroblasts transfected with 4-1BBL are then able to support antigen-driven proliferation of CD45RA^{+} CD28^{-} T-cells. This activation also up-regulates the chemokine receptor CCR7 (lymph node homing) and later CXCR3 and CCR5 (inflamed peripheral tissue homing) (Waller *et al.*, 2007). Thus, the evidence is that HCMV-specific, CD45RA^{+} CD28^{-} T-cells are not terminally differentiated, are able to proliferate and acquire heightened effector function following restimulation, if they are provided with the necessary co-stimulatory signals (Fig. II.7.2).

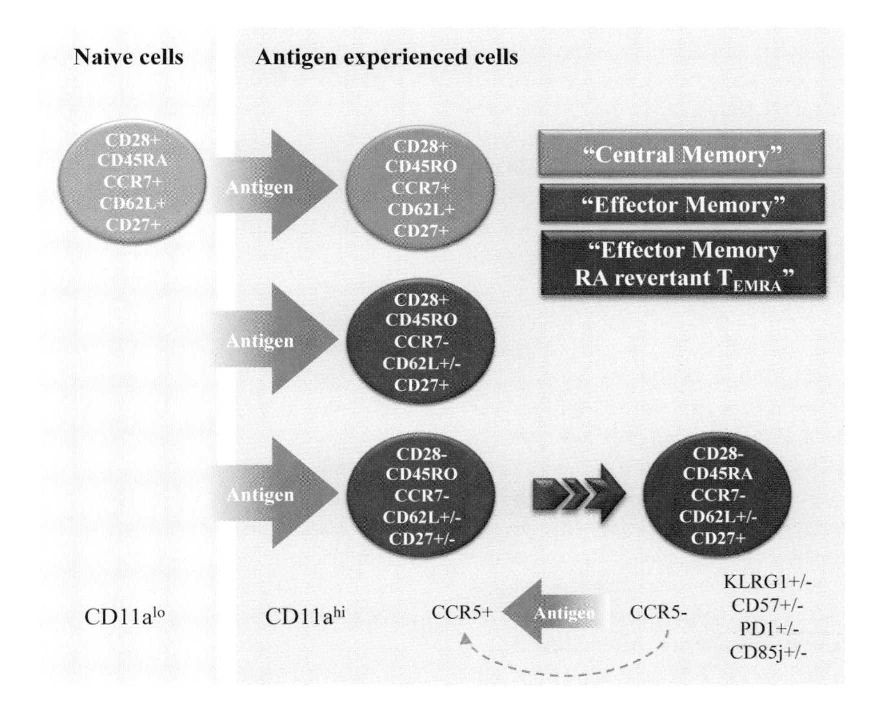

Figure II.7.2 A model of the differentiation of human CD8^{+} T-cells. Following activation of a CD28^{+}CD45RA^{hi} naïve cell by antigen, CD11a is up-regulated permanently and CD45RO is initially. Following resolution of primary infection some CD28^{-} CD45RO^{+} cells revert to CD28^{-}CD45RA^{hi} T effector memory RA cells T$_{EMRA}$. *In vitro* restimulation of these CD28^{-}CD45RA^{hi} CCR7^{-} cells induces them to proliferate and give rise to activated and cytotoxic CD28^{-} CD45RO^{+} cells. T$_{EMRA}$ cells often express CD57 as well as inhibitory NK cell receptors such as KLRG1, CD85J and other inhibitory receptors such as PD1.

CD8$^+$ T-cell clonality

Analysis of the clonal composition of the memory CD8$^+$ T-cells specific for defined pp65 and IE1 epitopes, by sequencing of the TCRs of multiple independently derived epitope-specific CTL clones, reveals a high degree of clonal focusing. In many donors, all of the CTL clones specific to a defined peptide use only one or two different TCRs as defined by their nucleotide sequence (Weekes *et al.*, 1999a; Khan *et al.*, 2002b). This implies that in many donors the circulating HCMV peptide-specific CD8$^+$ T-cells are composed of only a few individual CD8$^+$ clones that have undergone extensive clonal expansion *in vivo*.

As the extensive number of new HCMV proteins that can be recognized by a diverse population are defined at the level of individual peptides and their MHC-I restriction, it will be of interest to determine whether these T-cells show similar clonal TCR focusing and phenotype to those observed with pp65 and IE1 specific T-cells. These expanded individual HCMV-specific T-cell clones persist for years, as the same clones can be repeatedly isolated over time. HCMV pp65-specific CD8$^+$ T-cell clones show the property of containing public epitopes, a phenomenon described previously also in EBV specific T-cells: clones obtained from unrelated subjects that recognize the same defined peptide–MHC complex often use the same TCR Vβ segment, and have similar amino acid sequences within the hypervariable VDJ region of the TCR that binds to the viral peptide (Argaet *et al.*, 1994; Weekes *et al.*, 1999b). Furthermore, these oligoclonal responses can increase over time to form a very high proportion of all CD8$^+$ T-cells (a single clone forming up to 25% of all CD8$^+$ T-cells in an individual), as demonstrated by an analysis of elderly donors (Khan *et al.*, 2002).

CD8$^+$ T-cell response in primary infection and selection into memory

The functional and surface marker phenotypes of HCMV-specific T-cells described so far have been derived from normal healthy donors who are persistently/latently infected with HCMV, and thus do not inform us of how these phenotypes arose following primary infection, nor if intermediate phenotypes occur (Fig. II.7.2).

In long-term seropositive virus carriers CD8$^+$ T-cell responses to both IE and pp65 antigens are often highly focused with T-cells utilizing few or just a single TCR Vβ family and the sequence diversity within these families of T-cell clones is often highly conserved (Weekes *et al.*, 1999b). This may be because the primary T-cell response becomes focused with differential selection of certain clones into the memory T-cell pool, in contrast to the random, non-selective retention of the broad repertoire of the primary response reported in a number of model murine virus infection systems (Sourdive *et al.*, 1998; Blattman *et al.*, 2000; Karrer *et al.*, 2003; Pahl-Seibert *et al.*, 2005) or because long-term carriage of the virus with periodic reactivation could lead to a progressive focusing of the T-cell response with time. The evolution of the memory response to two different pp65 epitopes has been followed in otherwise normal healthy people who acquired natural HCMV infection. HCMV DNAemia in plasma peaked 2 to 3 weeks after infection and was at background levels within 4 to 5 weeks following the onset of symptoms. HCMV-specific T-cells peaked at 10–20% of total CD8$^+$ cells to a single pp65 peptide suggesting that the total HCMV-specific T-cell frequency in peripheral blood during primary infection was likely to be very large. The initial CD8$^+$ T-cell receptor repertoire was very diverse but rapidly focused as certain families of responding T-cells become undetectable. The cells that were selected into the memory pool were a subset of the initially responding CD8$^+$ T-cell population. Focusing onto particular TCR usage occurred during the resolution of primary infection and was much more rapid than expected, indicating that it is not long-term carriage and reactivation of the virus that drives clonal focusing (Day *et al.*, 2007).

Analyses of the phenotypic markers on HCMV-specific CD8$^+$ T-cells (it must be stressed that dominant pp65 and IE1 peptide specificities only have been studied) show that during primary HCMV infection, all the highly activated HCMV-specific effector T-cells express CD45RO. These activated CD45RO(high) effector CTL are already CD28 and CCR7 negative and remain negative as they enter the memory pool, but many accumulate in the CD45RA^{hi} subpopulation from 5–8 weeks following acute infection (Wills *et al.*, 1999) (Fig. II.7.3).

T-cells in the CD28$^-$, CD45RA^{hi}, CCR7$^-$ subpopulation have previously been described as 'terminally differentiated' (Champagne *et al.*, 2001). However, following stimulation *in vitro* with specific HCMV peptide, and the requisite co-stimulatory signals, these cells can undergo sustained clonal proliferation, up-regulate CD45RO and CCR5, and show strong peptide-specific cytotoxic activity (Wills *et al.*, 2002) (Fig. II.7.3). Thus, CD8$^+$ T-cell memory to HCMV is maintained by cells of expanded HCMV-specific clones that show heterogeneity of activation state and co-stimulation molecule expression, within both CD45RO^{hi} and CD28$^-$, CD45RA^{hi} T-cell pools. Similar techniques have been used by a number of groups to examine the memory T-cells generated in response to EBV: it is interesting to note that T-cells specific for lytic EBV

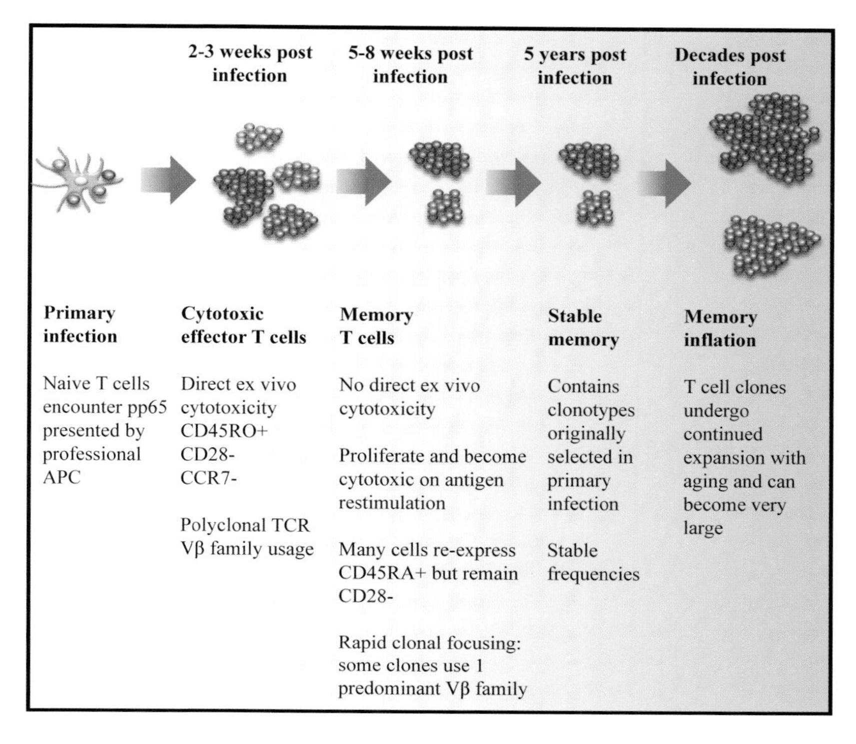

Figure II.7.3 Generation, selection and maintenance of HCMV specific CD8$^+$ T-cells following primary HCMV. Naïve T-cells expressing multiple Vβ TCR (multicoloured cells) respond to HCMV pp65 antigen during primary infection. Clonal focusing occurs rapidly during and after resolution of primary infection, with T-cells bearing particular Vβ TCR being deleted from the repertoire. The pp65-specific clones established in primary infection persist for years, maybe for life. These stable memory clones increase in frequency over a period of years and often become very large (representing a significant percentage of total CD8$^+$ cells), a phenomenon termed memory inflation.

antigens display very similar phenotypic profiles to HCMV pp65-specific T-cells, but also of interest that this does not hold for all persistent virus infections, e.g. hepatitis C virus (Faint *et al.*, 2001; Appay *et al.*, 2002a). van Lier and colleagues have attempted to unify these complex phenotypic differences in a variety of viral systems including HCMV (van Lier *et al.*, 2003).

Memory T-cell inflation

Following resolution of primary infection the memory HCMV-specific CD8$^+$ T-cell frequency that is established is large in comparison to other viruses, however in elderly HCMV-seropositive donors the HCMV-specific memory CD8$^+$ T-cells frequencies can be extremely high, more than 5% (of all peripheral blood CD8$^+$ T-cells) for a single epitope and in one case greater than 20% (Khan *et al.*, 2002a). It is presumed that the frequency of HCMV-specific T-cells increases with time and reflects long-term carriage of the virus with periodic reactivation and boosting of the T-cell response. The phenomenon was originally noted in MCMV infection of mice and has been termed 'memory inflation' (Holtappels *et al.*, 2000; Karrer *et al.*, 2003; Munks *et al.*, 2006).

In humans the expanded cells are often described as highly differentiated as they do not express CD27 or CD28 but many express CD57 and the inhibitory NK receptor CD85J (Northfield *et al.*, 2005) and have re-expressed CD45RA (Khan *et al.*, 2004; Wallace *et al.*, 2004). These T-cells are apoptosis resistant (Dunne *et al.*, 2002): they are often oligoclonal and the clones are stable over many years. In addition, comparison of CD8$^+$ cells from young and old donors showed no difference in either proliferative activity or cytotoxic function. Assessment of *in vivo* turnover rate by deuterated glucose labelling experiments suggested that the cells have a low proliferative rate *in vivo* and thus probably accumulate with time rather than by accelerated proliferation (Wallace *et al.*, 2004; 2011).

The phenomenon has been seen with T-cell responses to other viruses (e.g. HIV and parvovirus B19 infections), but not all T-cell epitopes are inflationary, even from the same virus. In MCMV certain antigens cause inflationary T-cell responses while others do not and the biological explanation underlying this phenomenon is unclear (for a discussion of possible explanations, see Chapter I.22). Using MCMV it has been shown that epitopes that showed inflation had a reduced dependence on the immunoproteasome (Hutchinson *et al.*, 2011). This observation is supported by a study that shows that cells of haematopoietic origin (professional antigen-presenting cells) are not required for memory inflation to occur and that epitope expression in MHC-I positive non-haematopoietic tissue was sufficient (Seckert *et al.*, 2011).

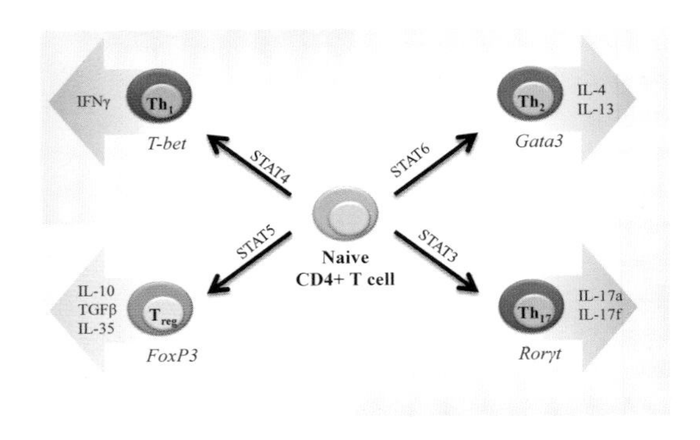

Figure II.7.4 The CD4$^+$ T-cell compartment is composed of distinct functional subsets. The diverse helper and direct effector functions performed by CD4$^+$ T-cells are mediated by different subsets of helper cells. These subsets are characterized by the transcription factors they express, and by the cytokines they secrete upon engagement of the T-cell receptor and activation.

The human CD4$^+$ T-cell response to HCMV

CD4$^+$ helper T-cells are presented with antigen by MHC-II molecules on the surface of APCs such as dendritic cells and macrophages. In contrast to MHC-I peptide presentation, antigen is classically derived exogenously and processed intracellularly, loaded onto the MHC-II molecules and expressed at the cell surface so that peptide can be presented to CD4$^+$ T-cells. CD4$^+$ T-cell responses play an essential and diverse role during virus infection. Classically, activation of the CD4$^+$ T-cell response leads to the activation of dendritic cells through ligation of CD40 on the dendritic cell surface by the T-cell associated surface molecule CD40L (CD154), the induction of B-cell immunoglobulin gene somatic hypermutation and class switching following B-cell CD40 ligation, and also the production of cytokines to promote proliferation and differentiation of virus-specific CD8$^+$ T-cells.

CD4$^+$ T-cells comprise a number of functional subsets, with diverse effector functions, creating a broad repertoire of T helper cell (Th) functions summarized in Fig. II.7.4. These subsets can be broadly defined by the cytokines they produce in response to antigen recognition and include: Th1 cells which provide help to CD8$^+$ T-cells, Th2 cells which provide B-cell help, Treg cells which provide a suppressive regulatory function, and further specialist subsets such as the highly proinflammatory Th17 cells that are implicated in autoimmunity and are essential in the inflammatory response to a range of microbial pathogens (Murphy and Reiner, 2002; Zenewicz *et al.*, 2009).

The importance of the CMV specific CD4$^+$ T-cell response is evident in MCMV infection, where long-term selective depletion of CD4$^+$ T-cells *in vivo* was associated with persistent viral replication at specific anatomical locations (Jonjic *et al.*, 1989). Furthermore, during MCMV infection in mice depleted of CD8$^+$ T-cells prior to infection, CD4$^+$ T-cells were shown to be critically involved in the control virus replication during primary infection (Jonjic *et al.*, 1990).

During CMV infection in the human host, in otherwise healthy children, an impaired CD4$^+$ T-cell response was associated with extended periods of virus shedding in both urine and saliva (Tu *et al.*, 2004). Asymptomatic infection during the course of primary infection of renal transplant patients was associated with the appearance of IFN-γ producing CD4$^+$ T-cells before the emergence of the HCMV-specific CD8$^+$ T-cell response (Gamadia *et al.*, 2003). In the same study, a delay in the generation of the HCMV-specific CD4$^+$ T-cell response was associated with extended virus replication and symptomatic disease. Other studies in the renal transplant setting, have shown that a decrease in the levels of HCMV-specific CD4$^+$ T-cells, preceded an increase in the viral load and the onset of clinical disease, possibly suggesting the HCMV CD4$^+$ T-cell response as a prognostic factor for clinical disease within this setting (Sester *et al.*, 2001). Furthermore, reduced HCMV-specific CD4$^+$ T-cells were also correlated with clinical disease amongst lung transplant recipients (Sester *et al.*, 2005). Maintenance of HCMV-specific CD8$^+$ T-cells infused following bone marrow transplantation has also been shown to require the presence of HCMV-specific CD4$^+$ T-cells, suggesting a further essential role for the CD4$^+$ T-cell response during HCMV infection (Walter *et al.*, 1995; Einsele *et al.*, 2002). Taken together these reports indicate that the HCMV-specific CD4$^+$ T-cell response is pivotal to providing long-term protection from HCMV replication and disease in both healthy seropositive hosts and the immunocompromised transplant populations (see also specific transplantation, Chapters II.13, II.14 and II.16).

CD4+ T-cell antigen specificity

The functional analysis of HCMV-specific $CD4^+$ T-cells has lagged behind that of the $CD8^+$ responses. However, the production of synthetic overlapping peptide libraries spanning the entire HCMV proteome, coupled with intracellular cytokine detection by flow cytometric methods have begun to address this situation (Waldrop *et al.*, 1997, 1998). In particular, the whole genome approach recently adopted by Picker and colleagues (Sylwester *et al.*, 2005) and described in greater detail in the CD8 section of this chapter has identified many ORFs which elicit $CD4^+$ responses: in due course these will no doubt be refined to minimal peptides and mapped to restricting MHC-II alleles. The definition of minimal peptides that bind MHC-I alleles, combined with tetramer technology, has greatly accelerated direct *ex vivo* analysis of HCMV-specific $CD8^+$ T-cells, but it has so far proved difficult to produce MHC-II peptide tetramers to identify HCMV-specific $CD4^+$ T-cells.

It has been recognized for some time that crude lysates of HCMV-infected cells contain virus encoded proteins which can elicit $CD4^+$ proliferative responses (measured by tritiated thymidine uptake): although these types of assays do not give any information about the properties of the responding $CD4^+$ T-cells, they do prove that they are generated in response to HCMV infection *in vivo* and maintained into long-term memory.

HCMV-infected cell lysates have also been used to restimulate PBMCs from seropositive donors and the responding $CD4^+$ T-cells detected using specific monoclonal antibodies in conjunction with detection of intracellular cytokine expression by flow cytometry. These experiments estimate the frequency of $CD4^+$ T-cells specific for HCMV in healthy HCMV carriers as 1–2% of all circulating $CD4^+$ T-cells (Waldrop *et al.*, 1998; Rentenaar *et al.*, 2000) with one report of much higher frequencies (Sester *et al.*, 2002). The HCMV used to prepare the viral lysates are not the strains of HCMV isolated from the donors tested and as such it is possible that some responses are not seen due to variation in protein sequences between the viruses.

Paralleling the specificities of $CD8^+$ HCMV-specific T-cells, $CD4^+$ T-cells specific for pp65 and IE1 can be readily demonstrated in approximately 60% of HCMV-seropositive individuals: in addition, proliferative $CD4^+$ responses to gB, gH, IE72, IE86 and UL69 have been detected in a number of donors, although no one donor had all of these specificities. (Beninga *et al.*, 1995; Davignon *et al.*, 1995; Kern *et al.*, 2002; Fuhrmann *et al.*, 2008).

In part because fewer peptide binding motifs are known for MHC-II alleles than for MHC-I alleles, relatively few MHC-II restricted epitopes have been identified within either pp65 (Khattab *et al.*, 1997; Kern *et al.*, 2002) or IE1 (Alp *et al.*, 1991; Li Pira *et al.*, 2004) (Table II.7.2). The approach used by Sylwester and colleagues using peptide libraries spanning the entire proteome indicated, importantly, that the $CD4^+$ T-cell response to HCMV, in contrast to what was originally believed, was in fact very broad, with HCMV-seropositive subjects responding to a median of 12 ORFs amongst a cohort of 33 HLA disparate donors (Sylwester *et al.*, 2005). In this same study five ORFs: UL55, UL83 (pp65), UL86, UL99 and UL122/123 (IE) were found to be recognized by more than half the donors enrolled for the study. The authors also showed that a total of 44 viral ORFs were recognized by $CD4^+$ T-cells alone and 81 ORFs were recognized by both the $CD4^+$ and $CD8^+$ T-cell response. As with the $CD8^+$ T-cell response to HCMV, the hierarchy of the $CD4^+$ T-cell response to individual ORFs was complex and showed no obvious pattern of immunodominance, as with all epitope screening with synthetic peptides, some peptides might not be represented and the processing of the overlong synthetic peptides may produce final peptides that are not optimal compared to the processing of the natural protein.

CD4+ T-cell surface phenotype and function

The lack of MHC-II tetramers makes it difficult to determine the surface phenotype of directly isolated HCMV-specific $CD4^+$ T-cells: however, multicolour flow cytometry can be combined with short-term peptide or lysate stimulation and intracellular cytokine detection, so that the phenotype of responding $CD4^+$ T-cells can be determined. It should be noted that cell surface molecules maybe rapidly modulated after stimulation and as such the results might differ compared with cells directly *ex vivo*.

Upon antigenic stimulation with whole HCMV antigen, the activated HCMV-specific $CD4^+$ T-cell population is enriched for subpopulations expressing CD45RO and $CD11a^{hi}$, whilst lacking the expression of CD27, CD62L and CCR7 (Rentenaar *et al.*, 2000; Bitmansour *et al.*, 2002; Sester *et al.*, 2002). A usually rare subset of peripheral blood $CD4^+$ T-cells which lack the CD28 co-stimulatory receptor have been found in high frequencies amongst HCMV-seropositive individuals (van Leeuwen *et al.*, 2004). These HCMV-specific $CD4^+CD28^-$ T-cells can proliferate and produce IFN-γ

Table II.7.2 MHC-II peptide epitopes recognized by human CD4+ T-cells from a number of HCMV ORFs

Gene	Product	Peptide	MHC class II restriction	Reference
UL55	gB	DYSNTHSTRYV	DR7	Elkington *et al.* (2004)
UL75	gH	HELLVLVKKAQL	DR11	Elkington *et al.* (2004)
UL123	IE1	91-VRVDMVRHRIKEHMLKKYTQ-110	DR3	Davignon *et al.* (1996)
		96-VRHRIKEHMLKKYTQTTEEKF-115	DR13	Davignon *et al.* (1996)
		162-DKREMWMACIKELH-175	DR8	Gautier *et al.* (1996)
UL83	pp65	41-LLQTGIHVRVSQPSLILVSQ-60	DQ6	Weekes *et al.* (2004)
	pp65	117-PLKMLNIPSINVHHY-131	DR1 DR3	Li Pira *et al.* (2004)
	pp65	169-TRQQNQWKEPDVYYT-183	DR1	Li Pira *et al.* (2004)
	pp65	177-EPDVYYTSAFVFPTK-191	DR7	Li Pira *et al.* (2004)
	pp65	225-KVYLESFCEDVPSGK-239	DR15	Li Pira *et al.* (2004)
	pp65	245-TLGSDVEEDLTMTRN-259	DR3	Li Pira *et al.* (2004)
	pp65	261-QPFMRPHERNGFTVL-275	DR13	Li Pira *et al.* (2004), Weekes *et al.* (2004)
	pp65	281-IIKPGKISHIMLDVA-295	DR4 DR7DR53	Kern *et al.* (2002), Li Pira *et al.* (2004)
	pp65	365-EHPTFTSQYRIQGKL-379	DR11	Kern *et al.* (2002) Li Pira *et al.* (2004)
	pp65	373-YRIQGKLEYRHTWDR-387	DR3	Li Pira *et al.* (2004)
	pp65	413-TERKTPRVTGGGAMA-427	DR14	Li Pira *et al.* (2004)
	pp65	429-ASTSAGRKRKSASSA-443	DR11	Li Pira *et al.* (2004)
	pp65	445-ACTSGVMTRGRLKAE-459	DR1	Li Pira *et al.* (2004)
	pp65	489-AGILARNLVPMVATV-503	DR11DR3	Kern *et al.* (2002), Li Pira *et al.* (2004)
	pp65	509-KYQEFFWDANDIYRI-523	DR1 DR3	Bitmansour *et al.* (2002), Li Pira *et al.* (2004)
	pp65	513-FFWDANDIYRI-523	DR1	Trivedi *et al.* (2005)
	pp65	166-LAWTRQQNQWKEPDV-180	DR1	Kern *et al.* (2002), Li Pira *et al.* (2004)
	pp65	510-YQEFFWDANDIYRIF-524	DR1DR3	Gallott *et al.* (2001), Kern *et al.* (2002), Li Pira *et al.* (2004)
	pp65	512-EFFWDANDIYRIF-524	DR1 DR3	Gallott *et al.* (2001), Kern *et al.* (2002), Li Pira *et al.* (2004)
	pp65	250-VEEDLTMTRNPQPFM-264	DR3	Li Pira *et al.* (2004)
	pp65	283-KPGKISHIMLDVAFTSH-299	DR4 DR7	Kern *et al.* (2002), Li Pira *et al.* (2004)
	pp65	370-TSQYRIQGKLEYRHT-384	DR4 DR13	Li Pira *et al.* (2004)
	pp65	180-VYYTSAFVFPTKDVA-194	DR7	Li Pira *et al.* (2004)
	pp65	360-GPQYSEHPTFTSQYRI-375	DR11	Khattab *et al.* (1997), Gallott *et al.* (2001) Kern *et al.* (2002), Li Pira *et al.* (2004)
	pp65	366-HPTFTSQYRIQGKLE-380	DR11 DR13	Khattab *et al.* (1997), Gallott *et al.* (2001) Kern *et al.* (2002), Li Pira *et al.* (2004)
	pp65	39-TRLLQTGIHVRVSQP-53	DR15	Gallott *et al.* (2001), Kern *et al.* (2002), Li Pira *et al.* (2004)
	pp65	109-MSIYVYALPLKMLNI-123	DR7	Nastke *et al.* (2005)
	pp65	339-LRQYDPVAALFFFDI353	DR7	Nastke *et al.* (2005)

upon activation: furthermore they express the cytotoxic granule components perforin and granzyme B, and interestingly, could also mediate MHC-II restricted cytotoxicity (van Leeuwen *et al.*, 2004, 2006). An independent study assessing the HCMV-specific TCR also identified this HCMV-specific CD4+

CD28⁻ population of cells and further showed the re-expression of CD45RA (Weekes *et al.*, 2004). Further studies of the HCMV-specific CD4⁺ T-cell response in both young and old subjects has confirmed high frequencies of CD4⁺ T-cells lacking CD28 expression in addition to the loss of CD27: it was noted in elderly populations that this subset of cells exhibited very low proliferative capacity (Fletcher *et al.*, 2005). These cells can be detected immediately after the disappearance of viral DNAemia in kidney transplant patients with active HCMV infection, and thus HCMV infection may trigger the formation of this particular CD4⁺ T-cell subset (van Leeuwen *et al.*, 2004; van de Berg *et al.*, 2008).

The majority of peripheral blood HCMV-specific memory CD4⁺ T-cells secrete IFN-γ upon stimulation with antigen; a subpopulation also secrete TNF-α, both of which are associated with Th1 type CD4⁺ T-cell effector function (Rentenaar *et al.*, 2000; Bitmansour *et al.*, 2002). In the same studies a further subpopulation showed the ability to secrete IL-2. Very few CD4⁺ T-cells in these studies were associated with IL-4 production and Th2 type effector function. A number of groups are accumulating evidence that in addition to providing classic T helper cell functions, HCMV-specific CD4⁺ T-cells can act as effector cells directly upon HCMV-infected cells (Rentenaar *et al.*, 2000; Appay *et al.*, 2002b; Gamadia *et al.*, 2004; van Leeuwen *et al.*, 2004). Individuals with higher levels of HCMV-specific CD4⁺ T-cells that secrete IFN-γ control virus replication better, clearing the virus quicker and exhibiting fewer symptoms *in vivo* (Sester *et al.*, 2001; Gamadia *et al.*, 2003). Furthermore, whilst T-cell-mediated cytotoxicity is usually associated with CD8⁺ T-cell effector function, a number of antigens from HCMV proteins including pp65, IE1, gB and gH, have been shown to induce the expansion of HCMV-specific CD4⁺ T-cells capable of mediating MHC-II restricted cytotoxicity *in vitro* (Hopkins *et al.*, 1996; Elkington *et al.*, 2004; Weekes *et al.*, 2004; Hegde *et al.*, 2005; van Leeuwen *et al.*, 2006). The majority of gB specific CD4⁺ T-cells express granzyme B when examined directly *ex vivo*, which in this study correlated with the ability of a range of CD4⁺ T-cell clones specific to gB to elicit cytotoxic effector function against target cells *in vitro* (Hegde *et al.*, 2005). Further studies involving gB specific cytotoxic CD4⁺ T-cells have identified a HLA-DRB*0701 restricted immunodominant peptide (DYSNTHSTRYV), showing that T-cell receptor usage for both the TCRα and TCRβ chains exhibited extreme conservation, reflecting previous findings in studies using CD8⁺ T-cells (Crompton *et al.*, 2008).

While HCMV-specific CD4⁺ T-cells have been shown to be predominantly Th1 type cells, with a low frequency of Th2 type cells (Rentenaar *et al.*, 2000; Bitmansour *et al.*, 2002), there is currently no report of HCMV-specific CD4⁺ T-cell responses being mediated by either Th17 or regulatory (Treg) CD4⁺ T-cells. Unpublished studies from our laboratory have detected IL-10 and TGF-β producing CD4⁺ T-cells which were specific to the latency associated gene products UL138 and LUNA (UL81–82as). This subset of the UL138 and LUNA specific CD4⁺ T-cells also had a regulatory T-cell phenotype being CD4⁺ CD25^{hi} and FoxP3⁺. A subpopulation of UL138 and LUNA specific CD4⁺ T-cells also produce IFN-γ (Th1 type). We have also screened 12 HCMV open reading frames including IE1/2, pp65 and gB for CD4⁺ T-cells that secrete IL-17 (Th17 cells) and have no evidence that such cells are elicited by HCMV infection.

It has been suggested that the ability of individual CD8⁺ or CD4⁺ T-cells to produce multiple cytokines (also termed polyfunctionality) is a useful measure of the 'quality' of a T-cell response (Boaz *et al.*, 2002; Younes *et al.*, 2003; Emu *et al.*, 2005; Harari *et al.*, 2005; Darrah *et al.*, 2007). A longitudinal study of HCMV infection in liver transplant patients suggests that this polyfunctionality of the CD4⁺ T-cell response is of critical importance during HCMV infection. The polyfunctional CD4⁺ T-cell response in this study was associated with protection against high-level viral replication, whilst significantly lower frequencies of HCMV-specific polyfunctional CD4⁺ T-cells (secreting IFN-γ and IL-2) were associated with patients exhibiting high levels of HCMV DNAemia (Nebbia *et al.*, 2008).

CD4⁺ T-cell clonality

The similarities of the pp65 specific CD4⁺ T-cell population to the pp65 specific CD8⁺ T-cell population also extends to the use of predominant clonotypes. In healthy HCMV carriers, the circulating population of HCMV-specific CD4⁺ T-cells is oligoclonal. Following *in vitro* stimulation with HCMV lysate, responding HCMV-specific CD4⁺ T-cells showed striking focussing of TCR Vβ segment usage, as detected by clonotypic monoclonal antibodies and RT-PCR analysis: these TCR Vβ expansions were also composed of a limited number of clonotypes in which one to three clones dominated the response together with numerous minor clones. The same dominant clonotypes were identified when CD4⁺ T-cells were stimulated with individual pp65 peptides (Bitmansour *et al.*, 2002; Weekes *et al.*, 2004). These studies noted that Vβ expansions were composed of few clonotypes, dominated by one to three individual clones in association with other subdominant and minor clones. Upon stimulation with a mixture of synthetic peptides spanning pp65, the same dominant CD4$^+$ T-cell clonotypes were detected and identified. Individual clones, upon *in vitro* expansion displayed a range of triggering thresholds, consistent with the idea of TCR-affinity independent but clonotypic threshold regulation, including threshold differences for the induction of both the IFN-γ and IL-2 response as well as differences in TCR cell surface expression densities that determine interaction avidity. No difference however, was seen in triggering thresholds between the CD27$^+$ and CD27$^-$ subpopulations of a given clonotype. Taken together, a particular antigen experienced CD4$^+$ T-cell clone can give rise to a qualitatively distinct functional response, dependent upon factors including both epitope dose and availability of costimulatory signals. It is unclear, however, how soon after infection clonal focusing of the CD4$^+$ T-cell response occurs and whether changes in clonal dominance occur in the host during long-term carriage of HCMV (Bitmansour *et al.*, 2002; Weekes *et al.*, 2004).

CD4$^+$ response during primary HCMV infection

The CD4$^+$ T-cell response induced in the course of natural primary HCMV infection in immunocompetent normal donors has not been described. However, valuable insight into the CD4$^+$ response can be gained from longitudinal studies of primary HCMV infection in the setting of HCMV-seronegative individuals who received a renal transplant from an HCMV-seropositive donor (D$^+$/R$^-$) without prophylactic Ganciclovir treatment, – although the concurrent use of immunosuppressive therapy (to prevent graft rejection) may modify the kinetics of infection and of the host T-cell response (Rentenaar *et al.*, 2000; Gamadia *et al.*, 2003).

HCMV DNAemia was first detected at a median of 25 days (range 18–29 days) after transplantation, and HCMV-specific IFN-γ producing CD4$^+$ T-cells were first detected (by intracellular cytokine staining following *in vitro* restimulation with whole HCMV antigen) at a median of 7 days (range 4–14 days) after first detection of HCMV DNAemia. In four of five subjects, HCMV-specific CD4$^+$ T-cells were detected 3–10 days before anti-HCMV IgM. The HCMV-specific CD4$^+$ T-cell response developed rapidly, reached peak frequencies of 0.46–2.5% of peripheral blood CD4$^+$ T-cells, and decreased steadily to a low level over at least the next 10 weeks. In contrast to the phenotype observed in long-term virus carriers, during primary infection HCMV-specific CD4$^+$ T-cells were predominantly CD28$^+$, many were CD27$^+$, and most showed co-expression of CD45RO and CD45RA. During primary infection a proportion of HCMV-specific CD4$^+$ T-cells were in cell cycle, as indicated by Ki67 expression (Rentenaar *et al.*, 2000).

In a second study, compared to asymptomatic primary HCMV infection, symptomatic primary infection was associated with the delayed appearance of HCMV-specific CD4$^+$ T-cells in peripheral blood. Among nine subjects who developed primary HCMV infection following D$^+$/R$^-$ renal transplantation, four had symptomatic infection that required Ganciclovir treatment, while the other five subjects had asymptomatic infection. In the symptomatic subjects the time at which HCMV DNAemia was first detected after transplantation (median 27 days) was similar to that in asymptomatic subjects, but the peak viral load and duration of HCMV DNAemia was greater in the symptomatic subjects. The time from first detection of HCMV DNAemia to first detection of HCMV-specific CD4$^+$ T-cells by intracellular cytokine staining was significantly longer in symptomatic infection (median 39 days, range 28–53) compared to asymptomatic infection (median 10 days, range 0–17). In symptomatic subjects, HCMV-specific CD4$^+$ T-cells only became detectable after starting Ganciclovir therapy, and reached peak frequencies of 0.36% to 1.42% of peripheral blood CD4$^+$ T-cells. Surprisingly, there were no differences observed in the kinetics of antibody or CD8$^+$ T-cell responses between symptomatic and asymptomatic subjects; the time from first detection of HCMV DNAemia to first detection of either anti-HCMV antibody (15 days vs. 17, days respectively) or HCMV-specific CD8$^+$ T-cells by peptide MHC class I tetramer staining (21 days vs. 24 days, respectively) was similar in symptomatic and asymptomatic subjects (Gamadia *et al.*, 2003).

HCMV mediated immune evasion of CD4$^+$ T-cell responses

The engagement of the TCR on CD4$^+$ cells by MHC-II molecules loaded with HCMV peptides is critical for CD4$^+$ T-cell activation. As discussed above, HCMV encodes functions which interfere with the MHC-I processing pathway, and which counteract the loss of negative signalling to NK cells consequent on MHC-I down-regulation. It might not thus be surprising that HCMV has also been shown to be capable of interfering with CD4$^+$ T-cell antigen recognition, as summarized in Fig. II.7.5.

MHC-II expression is usually restricted to professional APCs such as macrophages and dendritic cells, however, IFN-γ stimulation of certain cell types such as endothelial cells and epithelial cells results in expression of MHC-II on the cell surface via the JAK/STAT

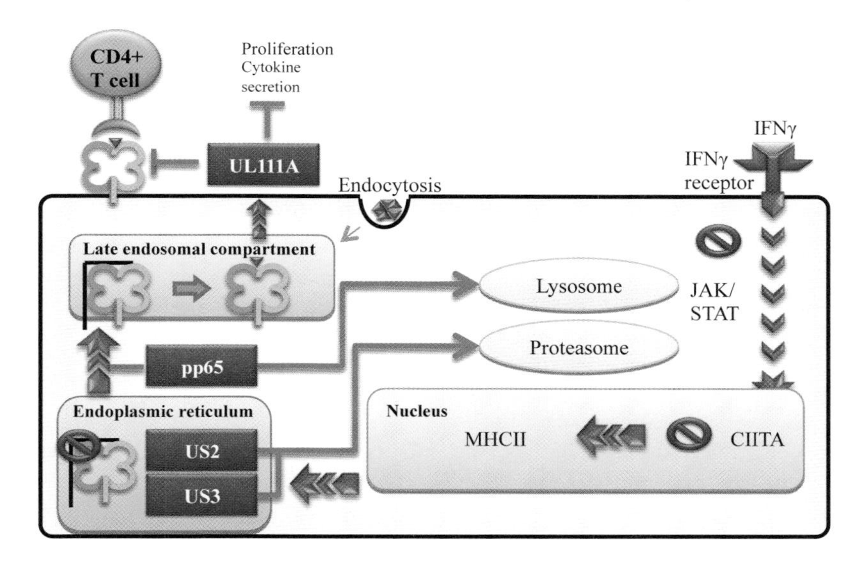

Figure II.7.5 HCMV induced evasion of the CD4+ T-cell response and MHC-II antigen presentation pathway. HCMV infection interferes with the IFN-γ induced JAK/STAT and CIITA up-regulation of MHC-II. The viral proteins US2 and US3 also disrupt the MHC-II pathway resulting in proteasome degradation of the MHC–II complex. Antigen presentation is further perverted by expression of the viral protein pp65 resulting in the inhibition of MHC-II peptide loading in the late endosomal compartment and redirecting to the lysosome. UL111A, a virally encoded homologue of the immunomodulatory cytokine IL-10 restricts CD4+ T-cell proliferation and proinflammatory cytokine production, whilst also resulting in reduced MHC-II expression on antigen-presenting cells.

pathway and the CIITA transactivator (Boss, 1997). This IFN-γ induced expression of MHC-II molecules is targeted during HCMV infection in these cell types, including fibroblasts, by two different mechanisms during immediate-early and late phases of HCMV infection respectively.

During immediate-early times of infection, repression of CIITA mRNA has been observed, which, results in inhibition of MHC-II expression, whilst the JAK/STAT signalling pathway remained intact (Le Roy *et al.*, 1999). Conversely, during later times of infection, inhibition of the JAK/STAT signalling pathway is achieved by reduced JAK expression, thus preventing STAT1 phosphorylation, dimerization, nuclear translocation and subsequent CIITA transactivation of MHC-II expression (Miller *et al.*, 1998, 2001).

US2 as mentioned previously in this chapter, represses MHC-I expression. Interestingly, US2 also appears to reduce the expression of HLA-DR, preventing CD4+ T-cell recognition (Tomazin *et al.*, 1999). US2 mediated redirecting of the HLA-DRα and HLA-DMα chains resulted in proteasomal degradation. US3 in a similar manner, inhibits both the MHC-I and MHC-II presentation pathways. US3 binds to MHC-II α/β complexes in the endoplasmic reticulum, compromising their interaction with Li chain, leading to inefficient sorting of MHC-II molecules into the peptide loading compartment, resulting in reduced peptide presentation to CD4+ T-cells (Hegde *et al.*, 2002). HLA-DR expression was shown to be inhibited during HCMV infection and was attributed to the expression of pp65, which mediated the transport and accumulation and eventual degradation in lysosomes of the HLA-DRα chain (Odeberg *et al.*, 2003).

The HCMV encoded gene product of UL111A is a homologue of the immunomodulatory cytokine IL-10, which interestingly, is expressed during both lytic and latent infection, although via differential splicing (Spencer *et al.*, 2002; Jenkins *et al.*, 2004, 2008). Both the lytic associated (cmv-IL-10) and latency associated (cmv-LA-IL-10) gene products of UL111A have been shown to down-regulate the expression of both MHC class I and II molecules, inhibit the proliferation of peripheral blood mononuclear cells and inhibit the production of inflammatory cytokines (Spencer *et al.*, 2002; Jenkins *et al.*, 2004, 2008). Furthermore using a model of HCMV latency in myeloid progenitor cells, the expression of cmv-LA-IL-10 has been associated with the suppression of both allogeneic and autologous CD4+ T-cell activation and the maintenance of the latent carriage of HCMV (Cheung *et al.*, 2009; see also Chapter I.20).

Taken together these reports show that HCMV dedicates significant resources to target the MHC-II presentation pathway, and hence the host CD4+ T-cell response. However, as for CD8+ T-cells, the T-cell response is nevertheless generated. The production of immune evasion molecules that target the MHC-II pathway may represent a mechanism, during early times of infection, reactivation or both, to create a window of opportunity for the virus to disseminate and establish latency before the inevitable induction of the CD4+ T-cell response and the control of infection.

HCMV-specific T-cell reconstitution following haematopoietic stem cell transplantation

The use of bone marrow transplantation (BMT) and more recently peripheral stem cell transplantation (SCT) as part of the treatment for haematological malignancies has steadily grown over the last few decades. It is well documented that primary infection or reactivation of latent HCMV can lead to serious morbidity and mortality in these patients (for specific clinical aspects, see also Chapter II.16). It has also been recognized for a considerable period of time that there is a direct correlation between effective HCMV-specific T-cell cytotoxicity and recovery from CMV infection or reactivation (Quinnan *et al.*, 1982). The basic research reviewed earlier in this chapter on the function, specificity, frequency and phenotype of HCMV-specific CD4+ and CD8+ T-cells in normal immunocompetent individuals provides the essential tools and baselines in order to understand the reconstitution of HCMV-specific T-cell responses following engraftment. It could be hoped that this knowledge would better inform clinicians as to when antiviral or adoptive immunotherapy interventions would be most appropriate.

Analysis of the generation or reconstitution of HCMV-specific CD4+ and CD8+ T-cells following BMT or SCT is complicated by a number of important factors. The donor (D) and recipient (R) may be matched siblings, but this is in a minority of transplants and the donor and recipient are more usually matched but unrelated (allo-SCT). In addition, the HCMV serostatus of the donor and recipient needs to be considered with there being four possible combinations: D+/R+, D−/R−, D+/R−, and D−/R+. Other variables to be considered include (1) the source of the stem cells, either bone marrow or peripheral stem cells, and in the case of peripheral stem cells whether the graft has been manipulated to remove mature T-cells, (2) the use of antiviral drugs such as Ganciclovir to suppress HCMV replication, and (3) the use of corticosteroid therapy to combat graft versus host disease (GvHD) which will of course act to globally impair CD4+ and CD8+ T-cells including those specific for CMV.

The techniques used to assess reconstitution inevitably mirror those used in more basic research on normal HCMV-specific T-cell responses and initially employed proliferation assays and the generation of CTL. More recently, pp65 specific MHC-I tetramers have been used to study the kinetics and frequency of CD8+ T-cell reconstitution (Aubert *et al.*, 2001; Cwynarski *et al.*, 2001; Gratama *et al.*, 2001) in conjunction with functional assays, such as IFN-γ production, which are appropriate for assessing a wider range of HCMV epitopes and MHC-I alleles, applied to CD8+ T-cell function, as well as Th1 CD4+ T-cells (Hebart *et al.*, 2002).

Donor and recipient HCMV serostatus

In solid organ transplantation (liver, kidney, heart or lung), transfer of HCMV in the graft from a seropositive donor into a seronegative recipient who also receives immunosuppressive treatment frequently leads to serious HCMV disease (see also Chapters II.13 and II.14). In haematopoietic transplantation, it should be noted that the serostatus of the graft donor indicates whether there are antigen experienced HCMV-specific donor T-cells that are transferred within the allograft (in addition to whether latent virus is present in the donor); the serostatus of the recipient indicates whether there is latent virus in the recipient. The serostatus of both donor and recipient therefore affect the reconstitution of donor-derived HCMV-specific CD8+ T-cells in the recipient (whose own CD8+ T-cells have of course been deliberately ablated prior to allografting). In the absence of HCMV infection, there is no antigen to stimulate a primary T-cell response; in two studies which included ten D−/R− transplantations, there was no detectable pp65 tetramer specific CD8+ T-cell reconstitution. In contrast, studies consistently report that almost all D+/R+ transplantations show early and sustained reconstitution of pp65 tetramer positive T-cells (Aubert *et al.*, 2001; Cwynarski *et al.*, 2001; Gratama *et al.*, 2001; Gandhi *et al.*, 2003).

The greatest risk of HCMV disease in haematopoietic transplantation thus occurs in D−/R+ transplantation, because the graft from the seronegative donor does not contain antigen-experienced HCMV-specific T-cells and the recipient has latent virus and simultaneously receives immunosuppressive treatment to prevent GvHD. Of eleven D−/R+ transplantations, seven patients failed to develop HCMV-specific tetramer positive T-cells, and the other four had very low levels of tetramer positive cells that were not maintained, results which are in agreement with previous studies not based on tetramers (Aubert *et al.*, 2001; Cwynarski *et al.*, 2001; Gratama *et al.*, 2001). Following D+/R− transplantation, most of the recipients have very low or undetectable HCMV-specific T-cells, however, 4/5 recipients did not reactivate CMV and the authors suggested that an episode of CMV replication is required in order to allow HCMV specific T-cells to become detectable and persist. In the one donor that did have detectable CMV they also developed high grade GvHD (Cwynarski *et al.*, 2001; Gandhi *et al.*, 2003). It is unclear why GvHD might promote expansion of HCMV specific T-cells. It maybe that the inflammatory environment promotes CMV reactivation and a subsequent T-cell response, or an issue of cross-reactivity, or that the highly inflammatory environment promotes an increase in frequency of the HCMV specific T-cells that were transplanted at low levels from the original donor graft.

Source of graft and manipulation

In recipients of sibling allografts where both the donor and recipient were HCMV-seropositive (D^+/R^+) before allo-SCT, recovery of HCMV-specific $CD8^+$ T-cells was rapid. However, early reconstitution of tetramer-positive cells was not observed in recipients of matched unrelated donor (MUD) allo-SCT (Aubert *et al.*, 2001; Cwynarski *et al.*, 2001; Gratama *et al.*, 2001). In order to reduce GvHD, MUD allo-SCT recipients receive T-cell depleted grafts with the intention to avoid alloreactive T-cells: however this will also deplete antigen experienced T-cells including mature HCMV-specific cells and in part may account for the delay in reconstitution.

The origin of the HCMV-specific clones that reconstitute these transplant patients is of interest: does reconstitution occur from the expansion of donor-derived antigen experienced T-cells or as the result of *de novo* naïve T-cell production and the mounting of a primary response? Staining with HCMV MHC-I tetramers does not distinguish between $CD8^+$ T-cells originating from the donor or from recipient, nor can tetramers distinguish the individual T-cell clones that comprise the tetramer positive population. These issues have been addressed by studying the reconstitution of HCMV-specific $CD8^+$ T-cells following allo-SCT at the level of individual clones, using clonotypic probing (Gandhi *et al.*, 2003). Following D^+/R^+ allo-SCT, immunodominant donor HCMV-specific clones transferred in the allograft underwent early expansion in the recipient and were maintained long-term, and retained *in vitro* proliferative and cytotoxic function. In some cases, the HCMV-seropositive recipients showed delayed diversification in the repertoire of clones that recognized specific HCMV peptides, with the delayed but persistent expansion of HCMV-specific clones that were not previously detected in the donor or in the recipient, during the first 5 months after transplantation. These clones contained donor DNA and may represent expansion of novel clones generated from donor-derived progenitor cells.

Effect of antiviral therapy on HCMV-specific T-cell reconstitution

In a randomized study of ganciclovir prophylaxis following allo-SCT, it was found that the majority of recipients did not have demonstrable functional T-cell responses before day 40. Between 40 and 90 days, HCMV-specific $CD4^+$ and $CD8^+$ responses became detectable in the majority of patients receiving placebo but in only a minority who received ganciclovir (Li *et al.*, 1994). A plausible explanation for this delay in reconstitution of HCMV-specific immunity in ganciclovir-treated recipients is that inhibition of viral replication may prevent antigen-driven expansion of HCMV-specific T-cells *in vivo*.

Effect of GvHD treatments

It might be expected that treatment of patients with high-dose steroids in order to control GvHD would globally suppress T-cell responses, and that even if HCMV-specific $CD4^+$ and $CD8^+$ T-cells were present in the graft or had reconstituted after allografting, their function would be compromised. This effect has indeed been directly demonstrated by the decline in tetramer-binding cells and/or reduction in the frequency of IFN-γ-producing cells in patients receiving high-dose steroids, associated with an increased risk of CMV disease (Einsele *et al.*, 2000; Aubert *et al.*, 2001; Cwynarski *et al.*, 2001; Gratama *et al.*, 2001; Hebart *et al.*, 2002).

Phenotype of reconstituted $CD8^+$ T-cells

Using MHC-I tetramer staining in conjunction with immunostaining for CD27, CD28, CD11b and CCR7 the phenotype of pp65 HCMV-specific $CD8^+$ T-cells in long-term surviving allo-SCT recipients was shown to be predominantly $CD27^-$, $CD28^-$, $CD11b^+$, $CCR7^-$ (Gandhi *et al.*, 2003): the phenotypic distribution of circulating HCMV-specific $CD8^+$ T-cells is similar in renal transplant recipients (Gamadia *et al.*, 2001). The $CCR7^-$ and $CD11b^+$ phenotype is consistent with the tissue homing 'effector-memory' phenotype. The CD45RA expression of HCMV-specific $CD8^+$ T-cells is heterogeneous (Wills *et al.*, 2002) but as in healthy HCMV carriers, HCMV pp65-specific $CD8^+$ T-cells in D^+/R^+ allo-SCT recipients are present in the $CD28^-$,

CD45RA^{hi} population, and very rare in the CD28^{+}, CD45RA^{hi} population (Gandhi *et al.*, 2003).

Reconstitution of HCMV-specific CD4^{+} T-cells

As discussed in an earlier section, it is clear that normal healthy seropositive donors have a high frequency of HCMV-specific CD4^{+} T-cells, a large proportion of which are pp65 specific, although many other CMV protein specificities are also detectable. The importance of the CMV specific CD4^{+} responses is highlighted by a number of observations in reconstituting transplant patients. The failure to reconstitute CD4^{+} T-cells ($< 100/\mu l$) is associated with impaired recovery of CMV specific CD8^{+} T-cells (Aubert *et al.*, 2001). Further evidence of the importance of CMV specific CD4^{+} T-cells is provided by adoptive immunotherapy studies as discussed in the next section.

Adoptive immunotherapy using HCMV-specific T-cells

Adoptive cellular immunotherapy of CMV infection in humans was first reported by Riddell and Greenberg in 14 patients (five D^{+}/R^{+} and nine D^{+}/R^{-}): HCMV-specific CD8^{+} T-cell clones were generated from seropositive allo-BMT donors by stimulation of PBMCs with HCMV-infected autologous fibroblasts. Each patient received four intravenous infusions of T-cell clones from their donors beginning 30–40 days after BMT. No toxic effects related to the infusions were observed, no prophylactic Ganciclovir was given, and HCMV-specific CD8^{+} T-cells were reconstituted in all patients. CMV-specific T-cell cytotoxicity was subsequently demonstrated in 11 patients in whom it was previously deficient. Analysis of rearranged T-cell-receptor genes in T-cells obtained from two recipients indicated that the transferred clones persisted for at least 12 weeks. However, cytotoxic CD8^{+} T-cell activity declined in those recipients who lacked CD4^{+} T-helper cells specific for HCMV, although it is unclear how many of these recipients actually had HCMV infection. Specifically, nine of the recipients were seronegative for HCMV prior to transplant, and HCMV viraemia and/or disease did not occur in any of the 14 patients (Riddell *et al.*, 1992; Walter *et al.*, 1995).

The observation that in this trial HCMV-specific CD8^{+} T-cell numbers declined in the absence of CD4^{+} T-cell help could also be explained by the absence of HCMV antigen in these recipients, with consequent failure to maintain transferred donor clones, as most of the recipients were HCMV-negative and no instances of HCMV viraemia occurred. However, more recent data reviewed in the previous section have provided us with direct evidence that HCMV-specific CD4^{+} T-cell help is desirable for the induction of effective CD8^{+} T-cell function. This is reflected in more recent attempts to refine the generation of HCMV-specific T-cells for adoptive immunotherapy, not only to include HCMV-specific CD4^{+} T-cell help but also to remove live HCMV from the culture systems, and reduce the risk of GvHD by directly selecting for HCMV-specific CD4^{+} and CD8^{+} T-cells using direct selection techniques based on peptide-MHC-I multimers or IFN-γ capture technologies – thus reducing the amount of time-consuming *ex vivo* manipulation of cells prior to their infusion into patients (Cobbold *et al.*, 2005; Einsele *et al.*, 2008; Peggs, 2009). The proof of concept for these antigen specific selection and adoptive transfer experiments was first demonstrated in the MCMV model (Reddehase, 2002; Pahl-Seibert *et al.*, 2005; see also Chapter II.17).

The ability to generate professional autologous antigen-presenting cells from SCT donors to be used to stimulate HCMV-specific T-cell responses ex vivo has led to their use in immununotherapy (Peggs *et al.*, 2001; Szmania *et al.*, 2001). The system developed by Peggs and colleagues utilizes donor monocyte-derived dendritic cells (by incubation of adherent PBMCs with IL-4 and GM-CSF) which are subsequently pulsed with a commercial CMV antigen preparation followed by co-culture with peripheral blood lymphocytes, as the source of HCMV-specific memory T-cells. This method generates polyclonal lines within 14 days with only a single round of antigen restimulation. The lines produced tended to be enriched in pp65 CD8^{+} T-cells (Peggs *et al.*, 2001, 2002) and the technique has been used to treat 16 HCMV-seropositive patients following allo-SCT (Peggs *et al.*, 2003). HCMV-specific CD8^{+} T-cells were quantified following infusion in five of these patients using an HLA-A2 restricted pp65 specific tetramer: four of the five patients showed large HCMV-specific T-cell expansions (the other patient still had cytotoxic levels of T-cell depleting CD52-specific mAb Alemtuzumab/CAMPATH-1H). Of the 16 patients treated, half required no antiviral drug intervention. A more recent phase II trial using a similar approach has been performed on 30 allogeneic transplant recipients and again showed reconstitution of CD4^{+} and CD8^{+} HCMV-specific T-cells preventing recurrent viral infection and late cytomegalovirus disease (Peggs *et al.*, 2009).

The importance of the HCMV-specific CD4^{+} T-cell response is highlighted by analysis of the reconstitution of CD8^{+} T-cell responses following allo-SCT, and showing their dependence on concomitant reconstitution of specific CD4^{+} T-cells. Further evidence for its importance is provided by the striking finding that infusion of T-cell lines that were predominantly HCMV-specific CD4$^+$ T-cells led to the control of HCMV replication as indicated by clearance of CMV DNA from the blood in five out of seven patients infused (Einsele *et al.*, 2002).

A potentially important advance in technology which reduces the time that cells are manipulated *in vitro* and targets HCMV-specific T-cells thus reducing the risk of GvHD is direct selection. Two methodologies have been used; purifying HCMV-specific CD8$^+$ T-cells using peptide-folded MHC-I tetramers binding to viral epitope-specific TCRs and short term stimulation of whole PBMCs with viral antigen or peptide pools followed by capturing cells which respond by IFN-γ production.

Cobbold used MHC-I tetramers specific for pp65 and IE1 peptides to purify HCMV-specific T-cells from the donor which were later infused into the SCT recipients. The results showed that HCMV-specific T-cells were detected in all recipients after adoptive transfer. A reduction in CMV viraemia was noted in eight patients including one patient who had become refractory to antiviral drug treatment and had high-level viral replication (Cobbold *et al.*, 2005). Peggs and colleagues have used a different approach in which both CD4$^+$ and CD8$^+$ T-cells isolated from donor PBMCs are incubated either with pp65 protein for 16–20 hours or a pool of overlapping peptides for 6 hours. Antigen specific cells are identified by their ability to produce IFN-γ, which is captured by specific antibody and cells then selected. Twenty-five patients were adoptively transferred and expansions of both CD4$^+$ and CD8$^+$ T-cells were seen *in vivo* in all recipients: the study provided evidence of protection from re-infection in the majority of patients (Peggs *et al.*, 2011).

It is clear from all these studies that if measures aimed at reconstituting HCMV-specific T-cell responses are to be routinely applicable and effective, the techniques used need to be rapid, induce high frequencies of HCMV-specific T-cells which contain both CD4$^+$ and CD8$^+$ T-cell populations, and in order to cover as wide a range of patient MHC-I and II haplotypes as possible, should not just rely on the very well characterized pp65 and IE1 peptides specific for the HLA-A201 and B7 alleles.

HCMV and T-cell immunosenescence

Together with the phenomenon of memory HCMV-specific T-cell inflation, a number of other lines of evidence have been drawn upon to suggest that long-term carriage of HCMV may be detrimental to the immune system: these include the suggestion that these large oligoclonal expansions of HCMV-specific T-cells have become dysfunctional which might lead to loss of control of HCMV in later life (Akbar and Fletcher, 2005; Koch *et al.*, 2006; Moss, 2010). Furthermore, it is suggested that the impact of HCMV on the immune system can cause wider dysfunction, such that immune responses to neo-antigens in the elderly are impaired (Goronzy *et al.*, 2001; Saurwein-Teissl *et al.*, 2002; Trzonkowski *et al.*, 2003). In one study HCMV seropositivity has also been shown to be associated with impairment of the immune response to co-existing EBV infection (Khan *et al.*, 2004). Taken together these observations have led to the suggestion that HCMV might be a driver of 'immunosenescence', an ultimate consequence of this being that long-term carriage of HCMV correlates with an increase in all-cause mortality. There is a growing body of literature which examines these questions, but it is also fair to say that the postulated causal association of HCMV with immunosenescence and decreased life expectancy is controversial, and that a similar weight of evidence probably exists against these arguments. In this section we try to summarize the arguments both for and against HCMV driving 'immunosenescence'.

The 'immune risk profile', HCMV and all-cause mortality

In studies of Swedish octogenarians and nonagenarians (OCTO and NONA) HCMV seropositivity has been correlated with a decrease in life expectancy in those over 85, in association with a number of other parameters which include; reduced B-cell numbers, an inverted CD4$^+$:CD8$^+$ T-cell ratio, increased frequency of CD8$^+$ CD28$^-$ T-cell populations and poor proliferative response to mitogen stimulation. This has been termed the 'immune risk profile' (IRP) (Olsson *et al.*, 2000; Wikby *et al.*, 2002, 2005). However, it should be noted that not all HCMV-seropositive individuals display this IRP and neither do all population groups – the reason for this is unknown.

Since these observations, a number of other studies from different geographical locations and population groups have been undertaken to address the question of whether long-term carriage of HCMV is associated with an increased likelihood of all-cause mortality.

Mortality measured over a 7-year period in 348 community-dwelling donors with stable cardiovascular disease showed a higher mortality rate in those individuals with the highest quartile of CMV specific IgG antibodies (Strandberg *et al.*, 2009). A study of 635 women aged 70–79 years living in the community also demonstrated that women with the highest anti CMV antibody titre had a significantly higher incidence of frailty compared with HCMV-seronegative women at baseline. Furthermore, CMV IgG concentration in the highest quartile also increased the risk of 5-year mortality, independent of potential confounders including cardiovascular disease (Wang *et al.*, 2010). An all-cause mortality study in a cohort of 1468 older Latinos in California again showed that the individuals in the highest quartile for CMV IgG titre had an increased mortality over 9 years (Roberts *et al.*, 2010). Data from the National Health and Nutrition Examination Study (NHANES) III also concluded that CMV seropositivity was associated with an increased risk of all-cause mortality, especially in individuals with a raised inflammatory marker (C-reactive protein, CRP) who were at a higher risk of both all-cause mortality and cardiovascular-associated mortality (Simanek *et al.*, 2011).

However, not all studies show such an association: cohorts comprising the Leiden 85-plus study, Danish Twin study, Leiden Longevity Study and the PROSPER study with a mortality follow-up of 6–12 years, did not support a relationship with CMV serostatus or antibody levels in all-cause mortality. A recent prospective longitudinal study with an 18-year follow-up of 915 patients undergoing coronary angiography, concluded that neither CMV seropositivity nor antibody levels were associated with longevity irrespective of the presence or absence of coronary heart disease (Wills *et al.*, 2011). This crucial question therefore remains to be resolved.

HCMV immunosenescence and T-cell function

Two studies have suggested that the proportion of HCMV-specific T-cells (as measured by MHC-I HLA-A2 NLV-pp65 tetramer stains) that are able to produce IFN-γ following peptide stimulation is reduced in the elderly (Ouyang *et al.*, 2003, 2004). This uses a single HCMV derived peptide and it is thus difficult to know if this is a more global phenomenon or one peculiar to this particular peptide and the highly oligoclonal T-cells that have been selected to recognize it. Function of these T-cells in this context is also only measured by IFN-γ; no assessment of other functional abilities such as cytotoxicity or proliferative capacity was determined.

As previously discussed in this chapter a large fraction of HCMV-specific T-cells re-express CD45RA, lose CD28, CD27 and gain CD57 and KLRG1, and had been reported to have poor replicative capacity to HCMV peptide stimulation (Champagne *et al.*, 2001). However, other studies show there is no proliferative defect in this population, particularly if appropriate systems of *in vitro* antigen presentation are employed (van Leeuwen *et al.*, 2002; Wills *et al.*, 2002; Waller *et al.*, 2007). Nevertheless, the term 'senescent' is still being applied to these late differentiated T-cells with the implication that they are non-functional: this is a clear misconception as there is solid evidence from humans and rhesus monkeys that shows such late-differentiated cells do maintain function. A comparison of the proliferative capacity and the cytotoxic function of HCMV-specific CD8$^+$ T-cells isolated from young and old HCMV-seropositive donors showed no difference in either. As others have shown, these HCMV-specific T-cells were often oligoclonal and the clones were stable over many years (Wallace *et al.*, 2011). Using the rhesus CMV (RhCMV) model, RhCMV CD4$^+$ and CD8$^+$ T-cell responses in adult animals, defined as aged 7–10 years, and aged animals, defined as 19–26 years old, were compared. The study concluded that there was no difference in the phenotype, numbers, proliferative capacity or functionality of the T-cells between the two groups of animals and that there was no evidence for either a functional decline or exhaustion in aged animals (Cicin-Sain *et al.*, 2010; see also Chapter II.22).

It might be predicted that if HCMV-specific T-cells did become dysfunctional this would result in viral replication following reactivation becoming less effectively suppressed, with consequent evidence of this, such as HCMV DNAemia and viral shedding. There are very few studies that have addressed this important prediction. One study has reported HCMV present in the urine of older individuals but not younger, although no virus was detected in the blood (Stowe *et al.*, 2007). In agreement with this a further more recent study in a small cohort of 40 elderly people categorized as either frail or not frail showed no HCMV DNA in serum in either group, with the exception of one frail individual who had eight copies/ml (this would be on the very edge of detection) (Wills *et al.*, 2011). Owing to the very limited studies this question thus still remains to be answered.

HCMV and the response to vaccination in the elderly

The possibility has also been raised that there is a generalized deleterious effect on the immune system associated with long-term carriage of HCMV. Some studies have correlated HCMV seropositivity, or a high frequency of CD8$^+$ CD28$^-$ T-cells in elderly individuals, with poor humoral responses to influenza vaccination (Goronzy *et al.*, 2001; Saurwein-Teissl *et al.*, 2002; Trzonkowski *et al.*, 2003).

A more recent study on the efficacy of influenza vaccination in a cohort of elderly individuals living in care facilities concluded that in this particular study group previous CMV infection did not explain poor responsiveness to influenza vaccination, and does not support CMV infection as being associated with dysfunction of the immune system of older people (den Elzen *et al.*, 2011). In this more recent study only the dominant H3N2 strain of influenza virus was used in comparison to the study by Trzonkowski *et al.* (2003) which used three seasonal strains, and the weakest correlation was between HCMV seropositivity and seroconversion to H3N2 in the latter. Thus, again, the issue is not clear cut and this important question deserves further study – if HCMV does drive immunosenescence, and a poor vaccination response, the evidence could support a rationale for a modified vaccination protocol in elderly HCMV-seropositive individuals.

Whether and how long-term carriage of HCMV might have a more global effect on the immune system is not known. One suggestion is that the accumulation of oligoclonal expansions of HCMV-specific T-cells of differentiated phenotype might 'fill up' the $CD8^+$ T-cell pool, reducing the size of the naïve pool and other memory T-cells, subsequently reducing recall responses and impairing the ability to mount new responses (Akbar and Fletcher, 2005; Koch *et al.*, 2006; Moss, 2010). A recent study comparing the frequency of EBV- and HCMV-specific T-cells in the blood and lymph nodes (LN) of donors showed that while the frequencies of EBV-specific T-cells were comparable between blood and LN, HCMV had a much lower frequency in LN than blood, including the late-differentiated cells as determined by $CD45RA^+$ and $CD27^-$ phenotyping. This study concluded that strong immune responses to HCMV were unlikely to restrict 'space' for naïve or other memory T-cells in LN (Remmerswaal *et al.*, 2012). It should be noted that lower frequencies of HCMV-specific T-cells might also be because the cells are preferentially recruited to organ sites.

It has also been noted that the absolute number of $CD8^+$ T-cells as well as the number and proportion of differentiated $CD45RA^+$ or $CD28^-$ T-cells is increased in HCMV seropositive, compared to seronegative, elderly or young individuals (Looney *et al.*, 1999): the absolute, as opposed to the relative, number of naïve cells may thus not differ dramatically.

It therefore remains, in the authors' view, unproven that HCMV is causally associated with systemic immunosenescence or whether the HCMV-specific T-cell population is dysfunctional in very elderly subjects – clearly these are important questions which require a definitive answer.

Conclusion

HCMV, as all persistent viruses, has to persist in the face of a powerful host immune response: antibody and, as discussed here, probably T-cells in particular, contain the infection in the normal host whilst impaired T-cell immunity is associated with HCMV disease. The virus encodes functions which can counter this immune response and there is evidence it also uses antigen-presenting cells as sites of latency. Although our knowledge of many aspects of the virus/host relationship is still incomplete, studies on HCMV over the past 25 years have given insight into how a large DNA virus achieves this coexistence with the normal immune response. It is also true that HCMV is becoming something of a paradigm for studying the way in which T-cell memory develops and is maintained against persistent virus infections, and provides insights which contribute to the basic understanding of T-cell memory.

Acknowledgement

Research was funded by the British Medical Research Council and the Wellcome Trust. I would like to thank all past and present members of the laboratory for their contributions to our understanding of HCMV immunobiology.

References

Ahn, K., Gruhler, A., Galocha, B., Jones, T.R., Wiertz, E.J., Ploegh, H.L., Peterson, P.A., Yang, Y., and Früh, K. (1997). The ER-luminal domain of the HCMV glycoprotein US6 inhibits peptide translocation by TAP. Immunity *6*, 613–621.

Akbar, A.N., and Fletcher, J.M. (2005). Memory T-cell homeostasis and senescence during aging. Curr. Opin. Immunol. *17*, 480–485.

Alp, N.J., Allport, T.D., Van Zanten, J., Rodgers, B., Sissons, J.G., and Borysiewicz, L.K. (1991). Fine specificity of cellular immune responses in humans to human cytomegalovirus immediate-early 1 protein. J. Virol. *65*, 4812–4820.

Appay, V., Dunbar, P.R., Callan, M., Klenerman, P., Gillespie, G.M., Papagno, L., Ogg, G.S., King, A., Lechner, F., Spina, C.A., *et al.* (2002a). Memory CD8+ T-cells vary in differentiation phenotype in different persistent virus infections. Nat. Med. *8*, 379–385.

Appay, V., Zaunders, J.J., Papagno, L., Sutton, J., Jaramillo, A., Waters, A., Easterbrook, P., Grey, P., Smith, D., McMichael, A.J., *et al.* (2002b). Characterization of CD4(+) CTLs ex vivo. J. Immunol *168*, 5954–5958.

Argaet, V.P., Schmidt, C.W., Burrows, S.R., Silins, S.L., Kurilla, M.G., Doolan, D.L., Suhrbier, A., Moss, D.J., Kieff, E.,

Suclley, T.B., *et al.* (1994). Dominant selection of an invariant T-cell antigen receptor in response to persistent infection by Epstein–Barr virus. J. Exp. Med. *180*, 2335–2340.

Arnon, T.I., Achdout, H., Levi, O., Markel, G., Saleh, N., Katz, G., Gazit, R., Gonen-Gross, T., Hanna, J., Nahari, E., *et al.* (2005). Inhibition of the NKp30 activating receptor by pp65 of human cytomegalovirus. Nat. Immunol. *6*, 515–523.

Ashiru, O., Bennett, N.J., Boyle, L.H., Thomas, M., Trowsdale, J., and Wills, M.R. (2009). NKG2D ligand MICA is retained in the cis-Golgi apparatus by human cytomegalovirus protein UL142. J. Virol. *83*, 12345–12354.

Aubert, G., Hassan-Walker, A.F., Madrigal, J.A., Emery, V.C., Morte, C., Grace, S., Koh, M.B., Potter, M., Prentice, H.G., Dodi, I.A., *et al.* (2001). Cytomegalovirus-specific cellular immune responses and viremia in recipients of allogeneic stem cell transplants. J. Infect. Dis. *184*, 955–963.

Azuma, M., Phillips, J.H., and Lanier, L.L. (1993). CD28- T-lymphocytes. Antigenic and functional properties. J. Immunol. *150*, 1147–1159.

Bauer, S., Groh, V., Wu, J., Steinle, A., Phillips, J.H., Lanier, L.L., and Spies, T. (1999). Activation of NK cells and T-cells by NKG2D, a receptor for stress-inducible MICA. Science *285*, 727–729.

Beck, S., and Barrell, B.G. (1988). Human cytomegalovirus encodes a glycoprotein homologous to MHC class-I antigens. Nature *331*, 269–272.

Bego, M., Maciejewski, J., Khaiboullina, S., Pari, G., and St Jeor, S. (2005). Characterization of an antisense transcript spanning the UL81–82 locus of human cytomegalovirus. J. Virol. *79*, 11022–11034.

Beninga, J., Kropff, B., and Mach, M. (1995). Comparative analysis of fourteen individual human cytomegalovirus proteins for helper T-cell response. J. Gen. Virol. *76*, 153–160.

Bennett, N.J., Ashiru, O., Morgan, F.J., Pang, Y., Okecha, G., Eagle, R.A., Trowsdale, J., Sissons, J.G., and Wills, M.R. (2010). Intracellular sequestration of the NKG2D ligand ULBP3 by human cytomegalovirus. J. Immunol. *185*, 1093–1102.

van de Berg, P.J., van Stijn, A., Ten Berge, I.J., and van Lier, R.A. (2008). A fingerprint left by cytomegalovirus infection in the human T-cell compartment. J. Clin. Virol. *41*, 213–217.

Besold, K., Frankenberg, N., Pepperl-Klindworth, S., Kuball, J., Theobald, M., Hahn, G., and Plachter, B. (2007). Processing and MHC class I presentation of human cytomegalovirus pp65-derived peptides persist despite gpUS2–11-mediated immune evasion. J. Gen. Virol. *88*, 1429–1439.

Besold, K., Wills, M., and Plachter, B. (2009). Immune evasion proteins gpUS2 and gpUS11 of human cytomegalovirus incompletely protect infected cells from CD8 T-cell recognition. Virology *391*, 5–19.

Biassoni, R., Cantoni, C., Pende, D., Sivori, S., Parolini, S., Vitale, M., Bottino, C., and Moretta, A. (2001). Human natural killer cell receptors and co-receptors. Immunol. Rev. *181*, 203–214.

Bitmansour, A.D., Douek, D.C., Maino, V.C., and Picker, L.J. (2002). Direct ex vivo analysis of human CD4(+) memory T-cell activation requirements at the single clonotype level. J. Immunol. *169*, 1207–1218.

Blattman, J.N., Sourdive, D.J., Murali-Krishna, K., Ahmed, R., and Altman, J.D. (2000). Evolution of the T-cell repertoire during primary, memory, and recall responses to viral infection. J. Immunol. *165*, 6081–6090.

Boaz, M.J., Waters, A., Murad, S., Easterbrook, P.J., and Vyakarnam, A. (2002). Presence of HIV-1 Gag-specific IFN-gamma+IL-2+ and CD28+IL-2+ CD4 T-cell responses is associated with nonprogression in HIV-1 infection. J. Immunol. *169*, 6376–6385.

Böhm, V., Simon, C.O., Podlech, J., Seckert, C.K., Gendig, D., Deegen, P., Gillert-Marien, D., Lemmermann, N.A., Holtappels, R., and Reddehase, M.J. (2008). The immune evasion paradox: immunoevasins of murine cytomegalovirus enhance priming of CD8 T-cells by preventing negative feedback regulation. J. Virol. *82*, 11637–11650.

Böhm, V., Seckert, C.K., Simon, C.O., Thomas, D., Renzaho, A., Gendig, D., Holtappels, R., and Reddehase, M.J. (2009). Immune evasion proteins enhance cytomegalovirus latency in the lungs. J. Virol. *83*, 10293–10298.

Boppana, S.B., and Britt, W.J. (1996). Recognition of human cytomegalovirus gene products by HCMV-specific cytotoxic T-cells. Virology *222*, 293–296.

Borthwick, N.J., Bofill, M., Gombert, W.M., Akbar, A.N., Medina, E., Sagawa, K., Lipman, M.C., Johnson, M.A., and Janossy, G. (1994). Lymphocyte activation in HIV-1 infection. II. Functional defects of CD28-T-cells. AIDS *8*, 431–441.

Borysiewicz, L.K., Morris, S., Page, J.D., and Sissons, J.G. (1983). Human cytomegalovirus-specific cytotoxic T-lymphocytes: requirements for *in vitro* generation and specificity. Eur. J. Immunol. *13*, 804–809.

Boss, J.M. (1997). Regulation of transcription of MHC class II genes. Curr. Opin. Immunol. *9*, 107–113.

Boutboul, F., Puthier, D., Appay, V., Pelle, O., Ait-Mohand, H., Combadiere, B., Carcelain, G., Katlama, C., Rowland-Jones, S.L., Debre, P., *et al.* (2005). Modulation of interleukin-7 receptor expression characterizes differentiation of CD8 T-cells specific for HIV, EBV and CMV. AIDS *19*, 1981–1986.

Braud, V.M., Allan, D.S., O'Callaghan, C.A., Soderstrom, K., D'Andrea, A., Ogg, G.S., Lazetic, S., Young, N.T., Bell, J.I., Phillips, J.H., *et al.* (1998). HLA-E binds to natural killer cell receptors CD94/NKG2A, B and C. Nature *391*, 795–799.

Brinchmann, J.E., Dobloug, J.H., Heger, B.H., Haaheim, L.L., Sannes, M., and Egeland, T. (1994). Expression of costimulatory molecule CD28 on T-cells in human immunodeficiency virus type 1 infection: functional and clinical correlations. J. Infect. Dis. *169*, 730–738.

Cerboni, C., Mousavi-Jazi, M., Linde, A., Soderstrom, K., Brytting, M., Wahren, B., Karre, K., and Carbone, E. (2000). Human cytomegalovirus strain-dependent changes in NK cell recognition of infected fibroblasts. J. Immunol. *164*, 4775–4782.

Chalupny, N.J., Rein-Weston, A., Dosch, S., and Cosman, D. (2006). Down-regulation of the NKG2D ligand MICA by the human cytomegalovirus glycoprotein UL142. Biochem. Biophys. Res. Commun. *346*, 175–181.

Champagne, P., Ogg, G.S., King, A.S., Knabenhans, C., Ellefsen, K., Nobile, M., Appay, V., Rizzardi, G.P., Fleury, S., Lipp, M., *et al.* (2001). Skewed maturation of memory HIV-specific CD8 T-lymphocytes. Nature *410*, 106–111.

Chapman, T.L., Heikeman, A.P., and Bjorkman, P.J. (1999). The inhibitory receptor LIR-1 uses a common binding interaction to recognize class I MHC molecules and the viral homolog UL18. Immunity *11*, 603–613.

Cheung, A.K., Gottlieb, D.J., Plachter, B., Pepperl-Klindworth, S., Avdic, S., Cunningham, A.L., Abendroth, A., and Slobedman, B. (2009). The role of the human cytomegalovirus UL111A gene in down-regulating CD4+ T-cell recognition of latently infected cells: implications for virus elimination during latency. Blood *114*, 4128–4137.

Chou, S.W. (1990). Differentiation of cytomegalovirus strains by restriction analysis of DNA sequences amplified from clinical specimens. J. Infect. Dis. *162*, 738–742.

Cicin-Sain, L., Sylwester, A.W., Hagen, S.I., Siess, D.C., Currier, N., Legasse, A.W., Fischer, M.B., Koudelka, C.W., Axthelm, M.K., Nikolich-Zugich, J., *et al.* (2010). Cytomegalovirus-specific T-cell immunity is maintained in immunosenescent rhesus macaques. J. Immunol. *187*, 1722–1732.

Cobbold, M., Khan, N., Pourgheysari, B., Tauro, S., McDonald, D., Osman, H., Assenmacher, M., Billingham, L., Steward, C., Crawley, C., *et al.* (2005). Adoptive transfer of cytomegalovirus-specific CTL to stem cell transplant patients after selection by HLA-peptide tetramers. J. Exp. Med. *202*, 379–386.

Cosman, D., Fanger, N., Borges, L., Kubin, M., Chin, W., Peterson, L., and Hsu, M.L. (1997). A novel immunoglobulin superfamily receptor for cellular and viral MHC class I molecules. Immunity *7*, 273–282.

Cosman, D., Mullberg, J., Sutherland, C.L., Chin, W., Armitage, R., Fanslow, W., Kubin, M., and Chalupny, N.J. (2001). ULBPs, novel MHC class I-related molecules, bind to CMV glycoprotein UL16 and stimulate NK cytotoxicity through the NKG2D receptor. Immunity *14*, 123–133.

Crompton, L., Khan, N., Khanna, R., Nayak, L., and Moss, P.A. (2008). CD4+ T-cells specific for glycoprotein B from cytomegalovirus exhibit extreme conservation of T-cell receptor usage between different individuals. Blood *111*, 2053–2061.

Cwynarski, K., Ainsworth, J., Cobbold, M., Wagner, S., Mahendra, P., Apperley, J., Goldman, J., Craddock, C., and Moss, P.A. (2001). Direct visualization of cytomegalovirus-specific T-cell reconstitution after allogeneic stem cell transplantation. Blood *97*, 1232–1240.

Darrah, P.A., Patel, D.T., De Luca, P.M., Lindsay, R.W., Davey, D.F., Flynn, B.J., Hoff, S.T., Andersen, P., Reed, S.G., Morris, S.L., *et al.* (2007). Multifunctional TH1 cells define a correlate of vaccine-mediated protection against Leishmania major. Nat. Med. *13*, 843–850.

Davignon, J.L., Clement, D., Alriquet, J., Michelson, S., and Davrinche, C. (1995). Analysis of the proliferative T-cell response to human cytomegalovirus major immediate-early protein (IE1): phenotype, frequency and variability. Scand. J. Immunol. *41*, 247–255.

Day, E.K., Carmichael, A.J., Ten Berge, I.J., Waller, E.C., Sissons, J.G., and Wills, M.R. (2007). Rapid CD8+ T-cell repertoire focusing and selection of high-affinity clones into memory following primary infection with a persistent human virus: human cytomegalovirus. J. Immunol. *179*, 3203–3213.

Dolan, A., Cunningham, C., Hector, R.D., Hassan-Walker, A.F., Lee, L., Addison, C., Dargan, D.J., McGeoch, D.J., Gatherer, D., Emery, V.C., *et al.* (2004). Genetic content of wild-type human cytomegalovirus. J. Gen. Virol. *85*, 1301–1312.

Dunn, C., Chalupny, N.J., Sutherland, C.L., Dosch, S., Sivakumar, P.V., Johnson, D.C., and Cosman, D. (2003). Human cytomegalovirus glycoprotein UL16 causes intracellular sequestration of NKG2D ligands, protecting against natural killer cell cytotoxicity. J. Exp. Med. *197*, 1427–1439.

Dunne, P.J., Faint, J.M., Gudgeon, N.H., Fletcher, J.M., Plunkett, F.J., Soares, M.V., Hislop, A.D., Annels, N.E., Rickinson, A.B., Salmon, M., *et al.* (2002). Epstein–Barr virus-specific CD8(+) T-cells that re-express CD45RA are apoptosis-resistant memory cells that retain replicative potential. Blood *100*, 933–940.

Dunne, P.J., Belaramani, L., Fletcher, J.M., Fernandez de Mattos, S., Lawrenz, M., Soares, M.V., Rustin, M.H., Lam, E.W., Salmon, M., and Akbar, A.N. (2005). Quiescence and functional reprogramming of Epstein–Barr virus (EBV)-specific CD8+ T-cells during persistent infection. Blood *106*, 558–565.

Eagle, R.A., Traherne, J.A., Ashiru, O., Wills, M.R., and Trowsdale, J. (2006). Regulation of NKG2D ligand gene expression. Hum. Immunol. *67*, 159–169.

Eagle, R.A., Traherne, J.A., Hair, J.R., Jafferji, I., and Trowsdale, J. (2009). ULBP6/RAET1L is an additional human NKG2D ligand. Eur. J. Immunol. *39*, 3207–3216.

Einsele, H., Hebart, H., Kauffmann-Schneider, C., Sinzger, C., Jahn, G., Bader, P., Klingebiel, T., Dietz, K., Loffler, J., Bokemeyer, C., *et al.* (2000). Risk factors for treatment failures in patients receiving PCR-based preemptive therapy for CMV infection. Bone Marrow Transplant. *25*, 757–763.

Einsele, H., Roosnek, E., Rufer, N., Sinzger, C., Riegler, S., Loffler, J., Grigoleit, U., Moris, A., Rammensee, H.G., Kanz, L., *et al.* (2002). Infusion of cytomegalovirus (CMV)-specific T-cells for the treatment of CMV infection not responding to antiviral chemotherapy. Blood *99*, 3916–3922.

Einsele, H., Kapp, M., and Grigoleit, G.U. (2008). CMV-specific T-cell therapy. Blood Cells Mol. Dis. *40*, 71–75.

Elkington, R., Walker, S., Crough, T., Menzies, M., Tellam, J., Bharadwaj, M., and Khanna, R. (2003). Ex vivo profiling of CD8+-T-cell responses to human cytomegalovirus reveals broad and multispecific reactivities in healthy virus carriers. J. Virol. *77*, 5226–5240.

Elkington, R., Shoukry, N.H., Walker, S., Crough, T., Fazou, C., Kaur, A., Walker, C.M., and Khanna, R. (2004). Cross-reactive recognition of human and primate cytomegalovirus sequences by human CD4 cytotoxic T-lymphocytes specific for glycoprotein B and H. Eur. J. Immunol. *34*, 3216–3226.

den Elzen, W.P., Vossen, A.C., Cools, H.J., Westendorp, R.G., Kroes, A.C., and Gussekloo, J. (2011). Cytomegalovirus infection and responsiveness to influenza vaccination in elderly residents of long-term care facilities. Vaccine *29*, 4869–4874.

Emu, B., Sinclair, E., Favre, D., Moretto, W.J., Hsue, P., Hoh, R., Martin, J.N., Nixon, D.F., McCune, J.M., and Deeks, S.G. (2005). Phenotypic, functional, and kinetic parameters associated with apparent T-cell control of human immunodeficiency virus replication in individuals with and without antiretroviral treatment. J. Virol. *79*, 14169–14178.

Faint, J.M., Annels, N.E., Curnow, S.J., Shields, P., Pilling, D., Hislop, A.D., Wu, L., Akbar, A.N., Buckley, C.D., Moss,

P.A., *et al.* (2001). Memory T-cells constitute a subset of the human CD8(+)CD45RA(+) pool with distinct phenotypic and migratory characteristics. J. Immunol. *167*, 212–220.

Fiorentino, S., Dalod, M., Olive, D., Guillet, J.G., and Gomard, E. (1996). Predominant involvement of CD8+CD28- lymphocytes in human immunodeficiency virus-specific cytotoxic activity. J. Virol. *70*, 2022–2026.

Fletcher, J.M., Prentice, H.G., and Grundy, J.E. (1998). Natural killer cell lysis of cytomegalovirus (CMV)-infected cells correlates with virally induced changes in cell surface lymphocyte function-associated antigen-3 (LFA-3) expression and not with the CMV-induced down-regulation of cell surface class I HLA. J. Immunol. *161*, 2365–2374.

Fletcher, J.M., Vukmanovic-Stejic, M., Dunne, P.J., Birch, K.E., Cook, J.E., Jackson, S.E., Salmon, M., Rustin, M.H., and Akbar, A.N. (2005). Cytomegalovirus-specific CD4+ T-cells in healthy carriers are continuously driven to replicative exhaustion. J. Immunol. *175*, 8218–8225.

Fries, B.C., Chou, S., Boeckh, M., and Torok-Storb, B. (1994). Frequency distribution of cytomegalovirus envelope glycoprotein genotypes in bone marrow transplant recipients. J. Infect. Dis. *169*, 769–774.

Fuhrmann, S., Streitz, M., Reinke, P., Volk, H.D., and Kern, F. (2008). T-cell response to the cytomegalovirus major capsid protein (UL86) is dominated by helper cells with a large polyfunctional component and diverse epitope recognition. J. Infect. Dis. *197*, 1455–1458.

Furman, M.H., Dey, N., Tortorella, D., and Ploegh, H.L. (2002). The human cytomegalovirus US10 gene product delays trafficking of major histocompatibility complex class I molecules. J. Virol. *76*, 11753–11756.

Gamadia, L.E., Rentenaar, R.J., Baars, P.A., Remmerswaal, E.B., Surachno, S., Weel, J.F., Toebes, M., Schumacher, T.N., ten Berge, I.J., and van Lier, R.A. (2001). Differentiation of cytomegalovirus-specific CD8(+) T-cells in healthy and immunosuppressed virus carriers. Blood *98*, 754–761.

Gamadia, L.E., Remmerswaal, E.B., Weel, J.F., Bemelman, F., van Lier, R.A., and Ten Berge, I.J. (2003). Primary immune responses to human CMV: a critical role for IFN-gamma-producing CD4+ T-cells in protection against CMV disease. Blood *101*, 2686–2692.

Gamadia, L.E., Rentenaar, R.J., van Lier, R.A., and ten Berge, I.J. (2004). Properties of CD4(+) T-cells in human cytomegalovirus infection. Hum. Immunol. *65*, 486–492.

Gandhi, M.K., Wills, M.R., Okecha, G., Day, E.K., Hicks, R., Marcus, R.E., Sissons, J.G., and Carmichael, A.J. (2003). Late diversification in the clonal composition of human cytomegalovirus-specific CD8+ T-cells following allogeneic hemopoietic stem cell transplantation. Blood *102*, 3427–3438.

Geginat, J., Lanzavecchia, A., and Sallusto, F. (2003). Proliferation and differentiation potential of human CD8+ memory T-cell subsets in response to antigen or homeostatic cytokines. Blood *101*, 4260–4266.

Gillespie, G.M., Wills, M.R., Appay, V., O'Callaghan, C., Murphy, M., Smith, N., Sissons, P., Rowland-Jones, S., Bell, J.I., and Moss, P.A. (2000). Functional heterogeneity and high frequencies of cytomegalovirus-specific CD8(+) T-lymphocytes in healthy seropositive donors. J. Virol. *74*, 8140–8150.

Gold, M.C., Munks, M.W., Wagner, M., Koszinowski, U.H., Hill, A.B., and Fling, S.P. (2002). The murine cytomegalovirus immunomodulatory gene m152 prevents recognition of infected cells by M45-specific CTL but does not alter the immunodominance of the M45-specific CD8 T-cell response *in vivo*. J. Immunol. *169*, 359–365.

Goodrum, F., Reeves, M., Sinclair, J., High, K., and Shenk, T. (2007). Human cytomegalovirus sequences expressed in latently infected individuals promote a latent infection *in vitro*. Blood *110*, 937–945.

Goronzy, J.J., Fulbright, J.W., Crowson, C.S., Poland, G.A., O'Fallon, W.M., and Weyand, C.M. (2001). Value of immunological markers in predicting responsiveness to influenza vaccination in elderly individuals. J. Virol. *75*, 12182–12187.

Gratama, J.W., van Esser, J.W., Lamers, C.H., Tournay, C., Lowenberg, B., Bolhuis, R.L., and Cornelissen, J.J. (2001). Tetramer-based quantification of cytomegalovirus (CMV)-specific CD8+ T-lymphocytes in T-cell-depleted stem cell grafts and after transplantation may identify patients at risk for progressive CMV infection. Blood *98*, 1358–1364.

Hansen, S.G., Powers, C.J., Richards, R., Ventura, A.B., Ford, J.C., Siess, D., Axthelm, M.K., Nelson, J.A., Jarvis, M.A., Picker, L.J., *et al.* (2010). Evasion of CD8+ T-cells is critical for superinfection by cytomegalovirus. Science *328*, 102–106.

Hansen, T.H., and Bouvier, M. (2009). MHC class I antigen presentation: learning from viral evasion strategies. Nat. Rev. Immunol. *9*, 503–513.

Harari, A., Vallelian, F., Meylan, P.R., and Pantaleo, G. (2005). Functional heterogeneity of memory CD4 T-cell responses in different conditions of antigen exposure and persistence. J. Immunol. *174*, 1037–1045.

Harari, A., Enders, F.B., Cellerai, C., Bart, P.A., and Pantaleo, G. (2009). Distinct profiles of cytotoxic granules in memory CD8 T-cells correlate with function, differentiation stage, and antigen exposure. J. Virol. *83*, 2862–2871.

Hebart, H., Daginik, S., Stevanovic, S., Grigoleit, U., Dobler, A., Baur, M., Rauser, G., Sinzger, C., Jahn, G., Loeffler, J., *et al.* (2002). Sensitive detection of human cytomegalovirus peptide-specific cytotoxic T-lymphocyte responses by interferon-gamma-enzyme-linked immunospot assay and flow cytometry in healthy individuals and in patients after allogeneic stem cell transplantation. Blood *99*, 3830–3837.

Heffner, M., and Fearon, D.T. (2007). Loss of T-cell receptor-induced Bmi-1 in the KLRG1(+) senescent CD8(+) T-lymphocyte. Proc. Natl. Acad. Sci. U.S.A. *104*, 13414–13419.

Hegde, N.R., Tomazin, R.A., Wisner, T.W., Dunn, C., Boname, J.M., Lewinsohn, D.M., and Johnson, D.C. (2002). Inhibition of HLA-DR assembly, transport, and loading by human cytomegalovirus glycoprotein US3: a novel mechanism for evading major histocompatibility complex class II antigen presentation. J. Virol. *76*, 10929–10941.

Hegde, N.R., Dunn, C., Lewinsohn, D.M., Jarvis, M.A., Nelson, J.A., and Johnson, D.C. (2005). Endogenous human cytomegalovirus gB is presented efficiently by MHC class II molecules to CD4+ CTL. J. Exp. Med. *202*, 1109–1119.

Henson, S.M., Franzese, O., Macaulay, R., Libri, V., Azevedo, R.I., Kiani-Alikhan, S., Plunkett, F.J., Masters, J.E., Jackson, S., Griffiths, S.J., *et al.* (2009). KLRG1 signaling induces defective Akt (ser473) phosphorylation and proliferative

dysfunction of highly differentiated CD8+ T-cells. Blood *113*, 6619–6628.
Holtappels, R., Pahl-Seibert, M.F., Thomas, D., and Reddehase, M.J. (2000). Enrichment of immediate-early 1 (m123/pp89) peptide-specific CD8 T-cells in a pulmonary CD62L(lo) memory-effector cell pool during latent murine cytomegalovirus infection of the lungs. J. Virol. *74*, 11495–11503.
Holtappels, R., Böhm, V., Podlech, J., and Reddehase, M.J. (2008). CD8 T-cell-based immunotherapy of cytomegalovirus infection: 'proof of concept' provided by the murine model. Med. Microbiol. Immunol. *197*, 125–134.
Hopkins, J.I., Fiander, A.N., Evans, A.S., Delchambre, M., Gheysen, D., and Borysiewicz, L.K. (1996). Cytotoxic T-cell immunity to human cytomegalovirus glycoprotein B. J. Med. Virol. *49*, 124–131.
Hutchinson, S., Sims, S., O'Hara, G., Silk, J., Gileadi, U., Cerundolo, V., and Klenerman, P. (2011). A dominant role for the immunoproteasome in CD8+ T-cell responses to murine cytomegalovirus. PLoS One *6*, e14646.
Iancu, E.M., Corthesy, P., Baumgaertner, P., Devevre, E., Voelter, V., Romero, P., Speiser, D.E., and Rufer, N. (2009). Clonotype selection and composition of human CD8 T-cells specific for persistent herpes viruses varies with differentiation but is stable over time. J. Immunol. *183*, 319–331.
Jenkins, C., Abendroth, A., and Slobedman, B. (2004). A novel viral transcript with homology to human interleukin-10 is expressed during latent human cytomegalovirus infection. J. Virol. *78*, 1440–1447.
Jenkins, C., Garcia, W., Godwin, M.J., Spencer, J.V., Stern, J.L., Abendroth, A., and Slobedman, B. (2008). Immunomodulatory properties of a viral homolog of human interleukin-10 expressed by human cytomegalovirus during the latent phase of infection. J. Virol. *82*, 3736–3750.
Jones, T.R., Hanson, L.K., Sun, L., Slater, J.S., Stenberg, R.M., and Campbell, A.E. (1995). Multiple independent loci within the human cytomegalovirus unique short region down-regulate expression of major histocompatibility complex class I heavy chains. J. Virol. *69*, 4830–4841.
Jones, T.R., Wiertz, E., Sun, L., Fish, K.N., Nelson, J.A., and Ploegh, H.L. (1996). Human cytomegalovirus US3 imparis transport and maturation of major histocompatability complex class I heavy chains. Proc. Natl. Acad. Sci. U.S.A. *93*, 11327–11333.
Jonjic, S., Mutter, W., Weiland, F., Reddehase, M.J., and Koszinowski, U.H. (1989). Site-restricted persistent cytomegalovirus infection after selective long-term depletion of CD4+ T-lymphocytes. J. Exp. Med. *169*, 1199–1212.
Jonjic, S., Pavic, I., Lucin, P., Rukavina, D., and Koszinowski, U.H. (1990). Efficacious control of cytomegalovirus infection after long-term depletion of CD8+ T-lymphocytes. J. Virol. *64*, 5457–5464.
Karre, K., Ljunggren, H.G., Piontek, G., and Kiessling, R. (1986). Selective rejection of H-2-deficient lymphoma variants suggests alternative immune defence strategy. Nature *319*, 675–678.
Karrer, U., Sierro, S., Wagner, M., Oxenius, A., Hengel, H., Koszinowski, U.H., Phillips, R.E., and Klenerman, P. (2003). Memory inflation: continuous accumulation of antiviral CD8+ T-cells over time. J. Immunol. *170*, 2022–2029. [Correction published in J. Immunol. (2003) *171*, 3895]
Kern, F., Khatamzas, E., Surel, I., Frommel, C., Reinke, P., Waldrop, S.L., Picker, L.J., and Volk, H.D. (1999). Distribution of human CMV-specific memory T-cells among the CD8pos. subsets defined by CD57, CD27, and CD45 isoforms. Eur. J. Immunol. *29*, 2908–2915.
Kern, F., Bunde, T., Faulhaber, N., Kiecker, F., Khatamzas, E., Rudawski, I.M., Pruss, A., Gratama, J.W., Volkmer-Engert, R., Ewert, R., *et al.* (2002). Cytomegalovirus (CMV) phosphoprotein 65 makes a large contribution to shaping the T-cell repertoire in CMV-exposed individuals. J. Infect. Dis. *185*, 1709–1716.
Khan, N., Shariff, N., Cobbold, M., Bruton, R., Ainsworth, J.A., Sinclair, A.J., Nayak, L., and Moss, P.A. (2002a). Cytomegalovirus seropositivity drives the CD8 T-cell repertoire toward greater clonality in healthy elderly individuals. J. Immunol. *169*, 1984–1992.
Khan, N., Cobbold, M., Keenan, R., and Moss, P.A. (2002b). Comparative analysis of CD8+ T-cell responses against human cytomegalovirus proteins pp65 and immediate early 1 shows similarities in precursor frequency, oligoclonality, and phenotype. J. Infect. Dis. *185*, 1025–1034.
Khan, N., Hislop, A., Gudgeon, N., Cobbold, M., Khanna, R., Nayak, L., Rickinson, A.B., and Moss, P.A. (2004). Herpesvirus-specific CD8 T-cell immunity in old age: cytomegalovirus impairs the response to a coresident EBV infection. J. Immunol. *173*, 7481–7489.
Khattab, B.A., Lindenmaier, W., Frank, R., and Link, H. (1997). Three T-cell epitopes within the C-terminal 265 amino acids of the matrix protein pp65 of human cytomegalovirus recognized by human lymphocytes. J. Med. Virol. *52*, 68–76.
Koch, S., Solana, R., Dela Rosa, O., and Pawelec, G. (2006). Human cytomegalovirus infection and T-cell immunosenescence: a mini review. Mech. Ageing. Dev. *127*, 538–543.
Komatsu, H., Inui, A., Sogo, T., Fujisawa, T., Nagasaka, H., Nonoyama, S., Sierro, S., Northfield, J., Lucas, M., Vargas, A., *et al.* (2006). Large scale analysis of pediatric antiviral CD8+ T-cell populations reveals sustained, functional and mature responses. Immun. Ageing *3*, 11.
Le Roy, E., Muhlethaler-Mottet, A., Davrinche, C., Mach, B., and Davignon, J.L. (1999). Escape of human cytomegalovirus from HLA-DR-restricted CD4(+) T-cell response is mediated by repression of gamma interferon-induced class II transactivator expression. J. Virol. *73*, 6582–6589.
Lee, S., Yoon, J., Park, B., Jun, Y., Jin, M., Sung, H.C., Kim, I.H., Kang, S., Choi, E.J., Ahn, B.Y., *et al.* (2000). Structural and functional dissection of human cytomegalovirus US3 in binding major histocompatibility complex class I molecules. J. Virol. *74*, 11262–11269.
van Leeuwen, E.M., Gamadia, L.E., Baars, P.A., Remmerswaal, E.B., ten Berge, I.J., and van Lier, R.A. (2002). Proliferation requirements of cytomegalovirus-specific, effector-type human CD8+ T-cells. J. Immunol. *169*, 5838–5843.
van Leeuwen, E.M., Remmerswaal, E.B., Vossen, M.T., Rowshani, A.T., Wertheim-van Dillen, P.M., van Lier, R.A., and ten Berge, I.J. (2004). Emergence of a CD4+CD28-granzyme B+, cytomegalovirus-specific T-cell subset after recovery of primary cytomegalovirus infection. J. Immunol. *173*, 1834–1841.

van Leeuwen, E.M., de Bree, G.J., Remmerswaal, E.B., Yong, S.L., Tesselaar, K., ten Berge, I.J., and van Lier, R.A. (2005). IL-7 receptor alpha chain expression distinguishes functional subsets of virus-specific human CD8+ T-cells. Blood *106*, 2091–2098.

van Leeuwen, E.M., Remmerswaal, E.B., Heemskerk, M.H., ten Berge, I.J., and van Lier, R.A. (2006). Strong selection of virus-specific cytotoxic CD4+ T-cell clones during primary human cytomegalovirus infection. Blood *108*, 3121–3127.

Lehner, P.J., Karttunen, J.T., Wilkinson, G.W., and Cresswell, P. (1997). The human cytomegalovirus US6 glycoprotein inhibits transporter associated with antigen processing-dependent peptide translocation. Proc. Natl. Acad. Sci. U.S.A. *94*, 6904–6909.

Lemmermann, N.A., Böhm, V., Holtappels, R., and Reddehase, M.J. (2011). *In vivo* impact of cytomegalovirus evasion of CD8 T-cell immunity: facts and thoughts based on murine models. Virus. Res. *157*, 161–174.

Letsch, A., Knoedler, M., Na, I.K., Kern, F., Asemissen, A.M., Keilholz, U., Loesch, M., Thiel, E., Volk, H.D., and Scheibenbogen, C. (2007). CMV-specific central memory T-cells reside in bone marrow. Eur. J. Immunol. *37*, 3063–3068.

Lewis, D.E., Yang, L., Luo, W., Wang, X., and Rodgers, J.R. (1999). HIV-specific cytotoxic T-lymphocyte precursors exist in a CD28-CD8+ T-cell subset and increase with loss of CD4 T-cells. AIDS *13*, 1029–1033.

Li, C.R., Greenberg, P.D., Gilbert, M.J., Goodrich, J.M., and Riddell, S.R. (1994). Recovery of HLA-restricted cytomegalovirus (CMV)-specific T-cell responses after allogeneic bone marrow transplant: correlation with CMV disease and effect of ganciclovir prophylaxis. Blood *83*, 1971–1979.

Li Pira, G., Bottone, L., Ivaldi, F., Pelizzoli, R., Del Galdo, F., Lozzi, L., Bracci, L., Loregian, A., Palu, G., De Palma, R., *et al.* (2004). Identification of new Th peptides from the cytomegalovirus protein pp65 to design a peptide library for generation of CD4 T-cell lines for cellular immunoreconstitution. Int. Immunol. *16*, 635–642.

Libri, V., Schulte, D., van Stijn, A., Ragimbeau, J., Rogge, L., and Pellegrini, S. (2008). Jakmip1 is expressed upon T-cell differentiation and has an inhibitory function in cytotoxic T-lymphocytes. J. Immunol. *181*, 5847–5856.

van Lier, R.A., ten Berge, I.J., and Gamadia, L.E. (2003). Human CD8(+) T-cell differentiation in response to viruses. Nat. Rev. Immunol. *3*, 931–939.

Looney, R.J., Falsey, A., Campbell, D., Torres, A., Kolassa, J., Brower, C., McCann, R., Menegus, M., McCormick, K., Frampton, M., *et al.* (1999). Role of cytomegalovirus in the T-cell changes seen in elderly individuals. Clin. Immunol. *90*, 213–219.

Luo, X.H., Huang, X.J., Liu, K.Y., Xu, L.P., and Liu, D.H. (2010). Protective immunity transferred by infusion of cytomegalovirus-specific CD8(+) T-cells within donor grafts: its associations with cytomegalovirus reactivation following unmanipulated allogeneic hematopoietic stem cell transplantation. Biol. Blood Marrow Transplant. *16*, 994–1004.

McLaughlin Taylor, E., Pande, H., Forman, S.J., Tanamachi, B., Li, C.R., Zaia, J.A., Greenberg, P.D., and Riddell, S.R. (1994). Identification of the major late human cytomegalovirus matrix protein pp65 as a target antigen for CD8+ virus-specific cytotoxic T-lymphocytes. J. Med. Virol. *43*, 103–110.

Makedonas, G., Hutnick, N., Haney, D., Amick, A.C., Gardner, J., Cosma, G., Hersperger, A.R., Dolfi, D., Wherry, E.J., Ferrari, G., *et al.* (2010). Perforin and IL-2 up-regulation define qualitative differences among highly functional virus-specific human CD8 T-cells. PLoS Pathog. *6*, e1000798.

Manley, T.J., Luy, L., Jones, T., Boeckh, M., Mutimer, H., and Riddell, S.R. (2004). Immune evasion proteins of human cytomegalovirus do not prevent a diverse CD8+ cytotoxic T-cell response in natural infection. Blood *104*, 1075–1082.

Melenhorst, J.J., Scheinberg, P., Chattopadhyay, P.K., Gostick, E., Ladell, K., Roederer, M., Hensel, N.F., Douek, D.C., Barrett, A.J., and Price, D.A. (2009). High avidity myeloid leukemia-associated antigen-specific CD8+ T-cells preferentially reside in the bone marrow. Blood *113*, 2238–2244.

Miles, D.J., van der Sande, M., Jeffries, D., Kaye, S., Ismaili, J., Ojuola, O., Sanneh, M., Touray, E.S., Waight, P., Rowland-Jones, S., *et al.* (2007). Cytomegalovirus infection in Gambian infants leads to profound CD8 T-cell differentiation. J. Virol. *81*, 5766–5776.

Miller, D.M., Rahill, B.M., Boss, J.M., Lairmore, M.D., Durbin, J.E., Waldman, J.W., and Sedmak, D.D. (1998). Human cytomegalovirus inhibits major histocompatibility complex class II expression by disruption of the Jak/Stat pathway. J. Exp. Med. *187*, 675–683.

Miller, D.M., Cebulla, C.M., Rahill, B.M., and Sedmak, D.D. (2001). Cytomegalovirus and transcriptional down-regulation of major histocompatibility complex class II expression. Semin. Immunol. *13*, 11–18.

Moins-Teisserenc, H., Busson, M., Scieux, C., Bajzik, V., Cayuela, J.M., Clave, E., de Latour, R.P., Agbalika, F., Ribaud, P., Robin, M., *et al.* (2008). Patterns of cytomegalovirus reactivation are associated with distinct evolutive profiles of immune reconstitution after allogeneic hematopoietic stem cell transplantation. J. Infect. Dis. *198*, 818–826.

Moss, P. (2010). The emerging role of cytomegalovirus in driving immune senescence: a novel therapeutic opportunity for improving health in the elderly. Curr. Opin. Immunol. *22*, 529–534.

Munks, M.W., Cho, K.S., Pinto, A.K., Sierro, S., Klenerman, P., and Hill, A.B. (2006). Four distinct patterns of memory CD8 T-cell responses to chronic murine cytomegalovirus infection. J. Immunol. *177*, 450–458.

Munks, M.W., Pinto, A.K., Doom, C.M., and Hill, A.B. (2007). Viral interference with antigen presentation does not alter acute or chronic CD8 T-cell immunodominance in murine cytomegalovirus infection. J. Immunol. *178*, 7235–7241.

Murphy, K.M., and Reiner, S.L. (2002). The lineage decisions of helper T-cells. Nat. Rev. Immunol. *2*, 933–944.

Nebbia, G., Mattes, F.M., Smith, C., Hainsworth, E., Kopycinski, J., Burroughs, A., Griffiths, P.D., Klenerman, P., and Emery, V.C. (2008). Polyfunctional cytomegalovirus-specific CD4+ and pp65 CD8+ T-cells protect against high-level replication after liver transplantation. Am. J. Transplant. *8*, 2590–2599.

Northfield, J., Lucas, M., Jones, H., Young, N.T., and Klenerman, P. (2005). Does memory improve with age? CD85j (ILT-2/LIR-1) expression on CD8 T-cells correlates with 'memory inflation' in human cytomegalovirus infection. Immunol. Cell Biol. *83*, 182–188.

Odeberg, J., Plachter, B., Branden, L., and Soderberg-Naucler, C. (2003). Human cytomegalovirus protein pp65 mediates accumulation of HLA-DR in lysosomes and destruction of the HLA-DR alpha-chain. Blood *101*, 4870–4877.

Olsson, J., Wikby, A., Johansson, B., Lofgren, S., Nilsson, B.O., and Ferguson, F.G. (2000). Age-related change in peripheral blood T-lymphocyte subpopulations and cytomegalovirus infection in the very old: the Swedish longitudinal OCTO immune study. Mech. Ageing Dev. *121*, 187–201.

Ouyang, Q., Wagner, W.M., Voehringer, D., Wikby, A., Klatt, T., Walter, S., Muller, C.A., Pircher, H., and Pawelec, G. (2003). Age-associated accumulation of CMV-specific CD8+ T-cells expressing the inhibitory killer cell lectin-like receptor G1 (KLRG1). Exp. Gerontol. *38*, 911–920.

Ouyang, Q., Wagner, W.M., Zheng, W., Wikby, A., Remarque, E.J., and Pawelec, G. (2004). Dysfunctional CMV-specific CD8(+) T-cells accumulate in the elderly. Exp. Gerontol. *39*, 607–613.

Pahl-Seibert, M.F., Juelch, M., Podlech, J., Thomas, D., Deegen, P., Reddehase, M.J., and Holtappels, R. (2005). Highly protective *in vivo* function of cytomegalovirus IE1 epitope-specific memory CD8 T-cells purified by T-cell receptor-based cell sorting. J. Virol. *79*, 5400–5413.

Palendira, U., Chinn, R., Raza, W., Piper, K., Pratt, G., Machado, L., Bell, A., Khan, N., Hislop, A.D., Steyn, R., *et al.* (2008). Selective accumulation of virus-specific CD8+ T-cells with unique homing phenotype within the human bone marrow. Blood *112*, 3293–3302.

Park, B., Kim, Y., Shin, J., Lee, S., Cho, K., Früh, K., Lee, S., and Ahn, K. (2004). Human cytomegalovirus inhibits tapasin-dependent peptide loading and optimization of the MHC class I peptide cargo for immune evasion. Immunity *20*, 71–85.

Park, B., Spooner, E., Houser, B.L., Strominger, J.L., and Ploegh, H.L. (2010). The HCMV membrane glycoprotein US10 selectively targets HLA-G for degradation. J. Exp. Med. *207*, 2033–2041.

Peggs, K., Verfuerth, S., and Mackinnon, S. (2001). Induction of cytomegalovirus (CMV)-specific T-cell responses using dendritic cells pulsed with CMV antigen: a novel culture system free of live CMV virions. Blood *97*, 994–1000.

Peggs, K., Verfuerth, S., Pizzey, A., Ainsworth, J., Moss, P., and Mackinnon, S. (2002). Characterization of human cytomegalovirus peptide-specific CD8(+) T-cell repertoire diversity following *in vitro* restimulation by antigen-pulsed dendritic cells. Blood *99*, 213–223.

Peggs, K.S. (2009). Adoptive T-cell immunotherapy for cytomegalovirus. Expert. Opin. Biol.Ther. *9*, 725–736.

Peggs, K.S., Verfuerth, S., Pizzey, A., Khan, N., Guiver, M., Moss, P.A., and Mackinnon, S. (2003). Adoptive cellular therapy for early cytomegalovirus infection after allogeneic stem-cell transplantation with virus-specific T-cell lines. Lancet *362*, 1375–1377.

Peggs, K.S., Verfuerth, S., Pizzey, A., Chow, S.L., Thomson, K., and Mackinnon, S. (2009). Cytomegalovirus-specific T-cell immunotherapy promotes restoration of durable functional antiviral immunity following allogeneic stem cell transplantation. Clin. Infect. Dis. *49*, 1851–1860.

Peggs, K.S., Thomson, K., Samuel, E., Dyer, G., Armoogum, J., Chakraverty, R., Pang, K., Mackinnon, S., and Lowdell, M.W. (2011). Directly selected cytomegalovirus-reactive donor T-cells confer rapid and safe systemic reconstitution of virus-specific immunity following stem cell transplantation. Clin. Infect. Dis. *52*, 49–57.

Petersen, J.L., Morris, C.R., and Solheim, J.C. (2003). Virus evasion of MHC class I molecule presentation. J. Immunol. *171*, 4473–4478.

Pita-Lopez, M.L., Gayoso, I., DelaRosa, O., Casado, J.G., Alonso, C., Munoz-Gomariz, E., Tarazona, R., and Solana, R. (2009). Effect of ageing on CMV-specific CD8 T-cells from CMV seropositive healthy donors. Immun. Ageing *6*, 11.

Podlech, J., Holtappels, R., Wirtz, N., Steffens, H.P., and Reddehase, M.J. (1998). Reconstitution of CD8 T-cells is essential for the prevention of multiple-organ cytomegalovirus histopathology after bone marrow transplantation. J. Gen. Virol. *79*, 2099–2104.

Podlech, J., Holtappels, R., Pahl-Seibert, M.F., Steffens, H.P., and Reddehase, M.J. (2000). Murine model of interstitial cytomegalovirus pneumonia in syngeneic bone marrow transplantation: persistence of protective pulmonary CD8-T-cell infiltrates after clearance of acute infection. J. Virol. *74*, 7496–7507.

Polic, B., Hengel, H., Krmpotic, A., Trgovcich, J., Pavic, I., Lucin, P., Jonjic, S., and Koszinowski, U.H. (1998). Hierarchical and redundant lymphocyte subset control precludes cytomegalovirus replication during latent infection. J. Exp. Med. *188*, 1047–1054.

Posch, P.E., Borrego, F., Brooks, A.G., and Coligan, J.E. (1998). HLA-E is the ligand for the natural killer cell CD94/NKG2 receptors. J. Biomed. Sci. *5*, 321–331.

Powers, C., DeFilippis, V., Malouli, D., and Früh, K. (2008). Cytomegalovirus immune evasion. Curr. Top. Microbiol. Immunol. *325*, 333–359.

Prod'homme, V., Griffin, C., Aicheler, R.J., Wang, E.C., McSharry, B.P., Rickards, C.R., Stanton, R.J., Borysiewicz, L.K., Lopez-Botet, M., Wilkinson, G.W., *et al.* (2007). The human cytomegalovirus MHC class I homolog UL18 inhibits LIR-1+ but activates LIR-1-NK cells. J. Immunol. *178*, 4473–4481.

Prod'homme, V., Sugrue, D.M., Stanton, R.J., Nomoto, A., Davies, J., Rickards, C.R., Cochrane, D., Moore, M., Wilkinson, G.W., and Tomasec, P. (2010). Human cytomegalovirus UL141 promotes efficient down-regulation of the natural killer cell activating ligand CD112. J. Gen. Virol. *91*, 2034–2039.

Quinnan, G.V., Jr., Kirmani, N., Rook, A.H., Manischewitz, J.F., Jackson, L., Moreschi, G., Santos, G.W., Saral, R., and Burns, W.H. (1982). Cytotoxic T-cells in cytomegalovirus infection: HLA-restricted T-lymphocyte and non-T-lymphocyte cytotoxic responses correlate with recovery from cytomegalovirus infection in bone-marrow-transplant recipients. N. Engl. J. Med. *307*, 7–13.

Ravetch, J.V., and Lanier, L.L. (2000). Immune inhibitory receptors. Science *290*, 84–89.

Reddehase, M.J. (2002). Antigens and immunoevasins: opponents in cytomegalovirus immune surveillance. Nat. Rev. Immunol. *2*, 831–844.

Reddehase, M.J., Mutter, W., Munch, K., Buhring, H.J., and Koszinowski, U.H. (1987). CD8-positive T-lymphocytes specific for murine cytomegalovirus immediate-early antigens mediate protective immunity. J. Virol. *61*, 3102–3108.

Remmerswaal, E.B., Havenith, S.H., Idu, M.M., van Leeuwen, E.M., van Donselaar, K.A., Ten Brinke, A., van der Bom-Baylon, N., Bemelman, F.J., van Lier, R.A., and Ten Berge, I.J. (2012). Human virus-specific effector-type T-cells accumulate in blood but not in lymph nodes. Blood. *119*, 1702–1712.

Rentenaar, R.J., Gamadia, L.E., van DerHoek, N., van Diepen, F.N., Boom, R., Weel, J.F., Wertheim-van Dillen, P.M., van Lier, R.A., and ten Berge, I.J. (2000). Development of virus-specific CD4(+) T-cells during primary cytomegalovirus infection. J. Clin. Invest. *105*, 541–548.

Reusser, P., Riddell, S.R., Meyers, J.D., and Greenberg, P.D. (1991). Cytotoxic T-lymphocyte response to cytomegalovirus after human allogeneic bone marrow transplantation: pattern of recovery and correlation with cytomegalovirus infection and disease. Blood *78*, 1373–1380.

Reyburn, H.T., Mandelboim, O., Vales-Gomez, M., Davis, D.M., Pazmany, L., and Strominger, J.L. (1997). The class I MHC homologue of human cytomegalovirus inhibits attack by natural killer cells. Nature *386*, 514–517.

Riddell, S.R., Watanabe, K.S., Goodrich, J.M., Li, C.R., Agha, M.E., and Greenberg, P.D. (1992). Restoration of viral immunity in immunodeficient humans by the adoptive transfer of T-cell clones. Science *257*, 238–241.

Roberts, E.T., Haan, M.N., Dowd, J.B., and Aiello, A.E. (2010). Cytomegalovirus antibody levels, inflammation, and mortality among elderly Latinos over 9 years of follow-up. Am. J. Epidemiol. *172*, 363–371.

Rolle, A., Mousavi-Jazi, M., Eriksson, M., Odeberg, J., Soderberg-Naucler, C., Cosman, D., Karre, K., and Cerboni, C. (2003). Effects of human cytomegalovirus infection on ligands for the activating NKG2D receptor of NK cells: up-regulation of UL16-binding protein (ULBP)1 and ULBP2 is counteracted by the viral UL16 protein. J. Immunol. *171*, 902–908.

Sauce, D., Larsen, M., Leese, A.M., Millar, D., Khan, N., Hislop, A.D., and Rickinson, A.B. (2007). IL-7R alpha versus CCR7 and CD45 as markers of virus-specific CD8+ T-cell differentiation: contrasting pictures in blood and tonsillar lymphoid tissue. J. Infect. Dis. *195*, 268–278.

Saurwein-Teissl, M., Lung, T.L., Marx, F., Gschosser, C., Asch, E., Blasko, I., Parson, W., Bock, G., Schonitzer, D., Trannoy, E., *et al.* (2002). Lack of antibody production following immunization in old age: association with CD8(+) CD28(-) T-cell clonal expansions and an imbalance in the production of Th1 and Th2 cytokines. J. Immunol. *168*, 5893–5899.

Scheinberg, P., Melenhorst, J.J., Brenchley, J.M., Hill, B.J., Hensel, N.F., Chattopadhyay, P.K., Roederer, M., Picker, L.J., Price, D.A., Barrett, A.J., *et al.* (2009). The transfer of adaptive immunity to CMV during hematopoietic stem cell transplantation is dependent on the specificity and phenotype of CMV-specific T-cells in the donor. Blood *114*, 5071–5080.

Scheuring, U.J., Sabzevari, H., and Theofilopoulos, A.N. (2002). Proliferative arrest and cell cycle regulation in CD8(+)CD28(-) versus CD8(+)CD28(+) T-cells. Hum. Immunol. *63*, 1000–1009.

Seckert, C.K., Schader, S.I., Ebert, S., Thomas, D., Freitag, K., Renzaho, A., Podlech, J., Reddehase, M.J., and Holtappels, R. (2011). Antigen-presenting cells of haematopoietic origin prime cytomegalovirus-specific CD8 T-cells but are not sufficient for driving memory inflation during viral latency. J. Gen. Virol *92*, 1994–2005.

Sester, M., Sester, U., Gartner, B., Heine, G., Girndt, M., Mueller-Lantzsch, N., Meyerhans, A., and Kohler, H. (2001). Levels of virus-specific CD4 T-cells correlate with cytomegalovirus control and predict virus-induced disease after renal transplantation. Transplantation *71*, 1287–1294.

Sester, M., Sester, U., Gartner, B., Kubuschok, B., Girndt, M., Meyerhans, A., and Kohler, H. (2002). Sustained high frequencies of specific CD4 T-cells restricted to a single persistent virus. J. Virol. *76*, 3748–3755.

Sester, U., Gartner, B.C., Wilkens, H., Schwaab, B., Wossner, R., Kindermann, I., Girndt, M., Meyerhans, A., Mueller-Lantzsch, N., Schafers, H.J., *et al.* (2005). Differences in CMV-specific T-cell levels and long-term susceptibility to CMV infection after kidney, heart and lung transplantation. Am. J. Transplant. *5*, 1483–1489.

Simanek, A.M., Dowd, J.B., Pawelec, G., Melzer, D., Dutta, A., and Aiello, A.E. (2011). Seropositivity to cytomegalovirus, inflammation, all-cause and cardiovascular disease-related mortality in the USA. PLoS One *6*, e16103.

Sourdive, D.J., Murali-Krishna, K., Altman, J.D., Zajac, A.J., Whitmire, J.K., Pannetier, C., Kourilsky, P., Evavold, B., Sette, A., and Ahmed, R. (1998). Conserved T-cell receptor repertoire in primary and memory CD8 T-cell responses to an acute viral infection. J. Exp. Med. *188*, 71–82.

Spencer, J.V., Lockridge, K.M., Barry, P.A., Lin, G., Tsang, M., Penfold, M.E., and Schall, T.J. (2002). Potent immunosuppressive activities of cytomegalovirus-encoded interleukin-10. J. Virol. *76*, 1285–1292.

Stern-Ginossar, N., Gur, C., Biton, M., Horwitz, E., Elboim, M., Stanietsky, N., Mandelboim, M., and Mandelboim, O. (2008). Human microRNAs regulate stress-induced immune responses mediated by the receptor NKG2D. Nat. Immunol. *9*, 1065–1073.

van Stijn, A., Rowshani, A.T., Yong, S.L., Baas, F., Roosnek, E., ten Berge, I.J., and van Lier, R.A. (2008). Human cytomegalovirus infection induces a rapid and sustained change in the expression of NK cell receptors on CD8+ T-cells. J. Immunol. *180*, 4550–4560.

Stowe, R.P., Kozlova, E.V., Yetman, D.L., Walling, D.M., Goodwin, J.S., and Glaser, R. (2007). Chronic herpesvirus reactivation occurs in aging. Exp. Gerontol. *42*, 563–570.

Strandberg, T.E., Pitkala, K.H., and Tilvis, R.S. (2009). Cytomegalovirus antibody level and mortality among community-dwelling older adults with stable cardiovascular disease. JAMA *301*, 380–382.

Sylwester, A.W., Mitchell, B.L., Edgar, J.B., Taormina, C., Pelte, C., Ruchti, F., Sleath, P.R., Grabstein, K.H., Hosken, N.A., Kern, F., *et al.* (2005). Broadly targeted human cytomegalovirus-specific CD4+ and CD8+ T-cells dominate the memory compartments of exposed subjects. J. Exp. Med. *202*, 673–685.

Szmania, S., Galloway, A., Bruorton, M., Musk, P., Aubert, G., Arthur, A., Pyle, H., Hensel, N., Ta, N., Lamb, L., Jr., *et al.* (2001). Isolation and expansion of cytomegalovirus-specific cytotoxic T-lymphocytes to clinical scale from a single blood draw using dendritic cells and HLA-tetramers. Blood *98*, 505–512.

Tey, S.K., Goodrum, F., and Khanna, R. (2010). CD8+ T-cell recognition of human cytomegalovirus latency-associated determinant pUL138. J. Gen. Virol. *91*, 2040–2048.

Tirabassi, R.S., and Ploegh, H.L. (2002). The human cytomegalovirus US8 glycoprotein binds to major histocompatibility complex class I products. J. Virol. *76*, 6832–6835.

Tomasec, P., Braud, V.M., Rickards, C., Powell, M.B., McSharry, B.P., Gadola, S., Cerundolo, V., Borysiewicz, L.K., McMichael, A.J., and Wilkinson, G.W. (2000).

Surface expression of HLA-E, an inhibitor of natural killer cells, enhanced by human cytomegalovirus gpUL40. Science *287*, 1031.

Tomasec, P., Wang, E.C., Davison, A.J., Vojtesek, B., Armstrong, M., Griffin, C., McSharry, B.P., Morris, R.J., Llewellyn-Lacey, S., Rickards, C., *et al.* (2005). Downregulation of natural killer cell-activating ligand CD155 by human cytomegalovirus UL141. Nat. Immunol. *6*, 181–188.

Tomazin, R., Boname, J., Hegde, N.R., Lewinsohn, D.M., Altschuler, Y., Jones, T.R., Cresswell, P., Nelson, J.A., Riddell, S.R., and Johnson, D.C. (1999). Cytomegalovirus US2 destroys two components of the MHC class II pathway, preventing recognition by CD4+ T-cells. Nat. Med. *5*, 1039–1043.

Trzonkowski, P., Mysliwska, J., Szmit, E., Wieckiewicz, J., Lukaszuk, K., Brydak, L.B., Machala, M., and Mysliwski, A. (2003). Association between cytomegalovirus infection, enhanced proinflammatory response and low level of anti-hemagglutinins during the anti-influenza vaccination--an impact of immunosenescence. Vaccine *21*, 3826–3836.

Tu, W., Chen, S., Sharp, M., Dekker, C., Manganello, A.M., Tongson, E.C., Maecker, H.T., Holmes, T.H., Wang, Z., Kemble, G., *et al.* (2004). Persistent and selective deficiency of CD4+ T-cell immunity to cytomegalovirus in immunocompetent young children. J. Immunol. *172*, 3260–3267.

Ulbrecht, M., Martinozzi, S., Grzeschik, M., Hengel, H., Ellwart, J.W., Pla, M., and Weiss, E.H. (2000). Cutting edge: the human cytomegalovirus UL40 gene product contains a ligand for HLA-E and prevents NK cell-mediated lysis. J. Immunol. *164*, 5019–5022.

Vales-Gomez, M., Browne, H., and Reyburn, H.T. (2003). Expression of the UL16 glycoprotein of human cytomegalovirus protects the virus-infected cell from attack by natural killer cells. BMC Immunol. *4*, 4.

van der Veken, L.T., Campelo, M.D., van der Hoorn, M.A., Hagedoorn, R.S., van Egmond, H.M., van Bergen, J., Willemze, R., Falkenburg, J.H., and Heemskerk, M.H. (2009). Functional analysis of killer Ig-like receptor-expressing cytomegalovirus-specific CD8+ T-cells. J. Immunol. *182*, 92–101.

Vingerhoets, J.H., Vanham, G.L., Kestens, L.L., Penne, G.G., Colebunders, R.L., Vandenbruaene, M.J., Goeman, J., Gigase, P.L., De Boer, M., and Ceuppens, J.L. (1995). Increased cytolytic T-lymphocyte activity and decreased B7 responsiveness are associated with CD28 down-regulation on CD8+ T-cells from HIV-infected subjects. Clin. Exp. Immunol. *100*, 425–433.

Waldrop, S.L., Pitcher, C.J., Peterson, D.M., Maino, V.C., and Picker, L.J. (1997). Determination of antigen-specific memory/effector CD4+ T-cell frequencies by flow cytometry: evidence for a novel, antigen-specific homeostatic mechanism in HIV-associated immunodeficiency. J. Clin. Invest. *99*, 1739–1750.

Waldrop, S.L., Davis, K.A., Maino, V.C., and Picker, L.J. (1998). Normal human CD4+ memory T-cells display broad heterogeneity in their activation threshold for cytokine synthesis. J. Immunol. *161*, 5284–5295.

Wallace, D.L., Zhang, Y., Ghattas, H., Worth, A., Irvine, A., Bennett, A.R., Griffin, G.E., Beverley, P.C., Tough, D.F., and Macallan, D.C. (2004). Direct measurement of T-cell subset kinetics *in vivo* in elderly men and women. J. Immunol. *173*, 1787–1794.

Wallace, D.L., Masters, J.E., De Lara, C.M., Henson, S.M., Worth, A., Zhang, Y., Kumar, S.R., Beverley, P.C., Akbar, A.N., and Macallan, D.C. (2011). Human cytomegalovirus-specific CD8(+) T-cell expansions contain long-lived cells that retain functional capacity in both young and elderly subjects. Immunology *132*, 27–38.

Waller, E.C., McKinney, N., Hicks, R., Carmichael, A.J., Sissons, J.G., and Wills, M.R. (2007). Differential co-stimulation through CD137 (4-1BB) restores proliferation of human virus-specific 'effector memory' (CD28-CD45RAhi) CD8+ T-cells. Blood *110*, 4360–4366.

Walter, E.A., Greenberg, P.D., Gilbert, M.J., Finch, R.J., Watanabe, K.S., Thomas, E.D., and Riddell, S.R. (1995). Reconstitution of cellular immunity against cytomegalovirus in recipients of allogeneic bone marrow by transfer of T-cell clones from the donor. N. Engl. J. Med. *333*, 1038–1044.

Wang, E.C., McSharry, B., Retiere, C., Tomasec, P., Williams, S., Borysiewicz, L.K., Braud, V.M., and Wilkinson, G.W. (2002). UL40-mediated NK evasion during productive infection with human cytomegalovirus. Proc. Natl. Acad. Sci. U.S.A. *99*, 7570–7575.

Wang, G.C., Kao, W.H., Murakami, P., Xue, Q.L., Chiou, R.B., Detrick, B., McDyer, J.F., Semba, R.D., Casolaro, V., Walston, J.D., *et al.* (2010). Cytomegalovirus infection and the risk of mortality and frailty in older women: a prospective observational cohort study. Am. J. Epidemiol. *171*, 1144–1152.

Weekes, M.P., Carmichael, A.J., Wills, M.R., Mynard, K., and Sissons, J.G. (1999a). Human CD28-CD8+ T-cells contain greatly expanded functional virus-specific memory CTL clones. J. Immunol. *162*, 7569–7577.

Weekes, M.P., Wills, M.R., Mynard, K., Carmichael, A.J., and Sissons, J.G. (1999b). The memory cytotoxic T-lymphocyte (CTL) response to human cytomegalovirus infection contains individual peptide-specific CTL clones that have undergone extensive expansion *in vivo*. J. Virol. *73*, 2099–2108.

Weekes, M.P., Wills, M.R., Sissons, J.G., and Carmichael, A.J. (2004). Long-term stable expanded human CD4+ T-cell clones specific for human cytomegalovirus are distributed in both CD45RA^{high} and CD45RO^{high} populations. J. Immunol. *173*, 5843–5851.

Wiertz, E.J., Tortorella, D., Bogyo, M., Yu, J., Mothes, W., Jones, T.R., Rapoport, T.A., and Ploegh, H.L. (1996a). Sec61-mediated transfer of a membrane protein from the endoplasmic reticulum to the proteasome for destruction. Nature *384*, 432–438.

Wiertz, E.J., Jones, T.R., Sun, L., Bogyo, M., Geuze, H.J., and Ploegh, H.L. (1996b). The human cytomegalovirus US11 gene product dislocates MHC class I heavy chains from the endoplasmic reticulum to the cytosol. Cell *84*, 769–779.

Wikby, A., Johansson, B., Olsson, J., Lofgren, S., Nilsson, B.O., and Ferguson, F. (2002). Expansions of peripheral blood CD8 T-lymphocyte subpopulations and an association with cytomegalovirus seropositivity in the elderly: the Swedish NONA immune study. Exp. Gerontol. *37*, 445–453.

Wikby, A., Ferguson, F., Forsey, R., Thompson, J., Strindhall, J., Lofgren, S., Nilsson, B.O., Ernerudh, J., Pawelec, G., and Johansson, B. (2005). An immune risk phenotype, cognitive impairment, and survival in very late life: impact of allostatic load in Swedish octogenarian and nonagenarian humans. J. Gerontol. A. Biol. Sci. Med. Sci. *60*, 556–565.

Wilkinson, G.W., Tomasec, P., Stanton, R.J., Armstrong, M., Prod'homme, V., Aicheler, R., McSharry, B.P., Rickards, C.R., Cochrane, D., Llewellyn-Lacey, S., *et al.* (2008). Modulation of natural killer cells by human cytomegalovirus. J. Clin. Virol. *41*, 206–212.

Wills, M., Akbar, A., Beswick, M., Bosch, J.A., Caruso, C., Colonna-Romano, G., Dutta, A., Franceschi, C., Fulop, T., Gkrania-Klotsas, E., *et al.* (2011). Report from the second cytomegalovirus and immunosenescence workshop. Immun. Ageing *8*, 10.

Wills, M.R., Carmichael, A.J., Weekes, M.P., Mynard, K., Okecha, G., Hicks, R., and Sissons, J.G. (1999). Human virus-specific CD8+ CTL clones revert from $CD45RO^{high}$ to $CD45RA^{high}$ *in vivo*: $CD45RA^{high}CD8+$ T-cells comprise both naïve and memory cells. J. Immunol. *162*, 7080–7087.

Wills, M.R., Okecha, G., Weekes, M.P., Gandhi, M.K., Sissons, P., and Carmichael, A.J. (2002). Identification of naïve or antigen-experienced human CD8(+) T-cells by expression of costimulation and chemokine receptors: analysis of the human cytomegalovirus-specific CD8(+) T-cell response. J. Immunol. *168*, 5455–5464.

Wills, M.R., Ashiru, O., Reeves, M.B., Okecha, G., Trowsdale, J., Tomasec, P., Wilkinson, G.W., Sinclair, J., and Sissons, J.G. (2005). Human cytomegalovirus encodes an MHC class I-like molecule (UL142) that functions to inhibit NK cell lysis. J. Immunol. *175*, 7457–7465.

Younes, S.A., Yassine-Diab, B., Dumont, A.R., Boulassel, M.R., Grossman, Z., Routy, J.P., and Sekaly, R.P. (2003). HIV-1 viremia prevents the establishment of interleukin 2-producing HIV-specific memory CD4+ T-cells endowed with proliferative capacity. J. Exp. Med. *198*, 1909–1922.

Zenewicz, L.A., Antov, A., and Flavell, R.A. (2009). CD4 T-cell differentiation and inflammatory bowel disease. Trends Mol. Med. *15*, 199–207.

Zhong, J., Rist, M., Cooper, L., Smith, C., and Khanna, R. (2008). Induction of pluripotent protective immunity following immunisation with a chimeric vaccine against human cytomegalovirus. PLoS One *3*, e3256.

Zhou, W., Longmate, J., Lacey, S.F., Palmer, J.M., Gallez-Hawkins, G., Thao, L., Spielberger, R., Nakamura, R., Forman, S.J., Zaia, J.A., *et al.* (2009). Impact of donor CMV status on viral infection and reconstitution of multifunction CMV-specific T-cells in CMV-positive transplant recipients. Blood *113*, 6465–6476.

Zou, Y., Bresnahan, W., Taylor, R.T., and Stastny, P. (2005). Effect of human cytomegalovirus on expression of MHC class I-related chains A. J. Immunol. *174*, 3098–3104.

Natural Killers Cells and Human Cytomegalovirus

Gavin W. G. Wilkinson, Rebecca J. Aicheler and Eddie C. Y. Wang

Abstract

The efficient down-regulation of HLA-I by HCMV has the clear potential to render infected cells extremely vulnerable to NK cells. Moreover, the major IE genes activate cell responses that stimulate efficient transcription of multiple ligands for NK cell activating receptors. The capacity of HCMV to persist *in vivo* can clearly be ascribed to its ability to modulate NK cell responses. To date seven functions encoded by HCMV have been formally demonstrated to suppress NK cell activation. UL18 is an MHC-I homologue that binds the inhibitory receptor LIR-1, while UL40 rescues expression of HLA-E, a ligand for the inhibitory receptor CD94-NKG2A. UL16, UL142 and miR-UL112 target multiple ligands for the ubiquitous NK activating receptor NKG2D, while UL141 targets ligands for the ubiquitous activating receptors DNAM-1 and TACTILE. The UL83-encoded major tegument protein (pp65) is unique in that it binds directly to inhibit the activating receptor NKp30. It is becoming evident that a substantial proportion of the remarkable coding capacity of this virus is directed at systematically addressing the NK cell response. Outwith the immediate goal of understanding HCMV pathogenesis, research on these immunomodulatory functions are providing remarkable insights into the mechanisms that regulate human NK cell responses. Recent studies demonstrate that during its lifelong persistent/latent infection, HCMV induces dramatic changes in the NK cell repertoire leading specifically to expansions of NK cell subsets expressing CD94-NKG2C, LIR-1 and CD57. There is growing interest in these changes in the NK cell response as they potentially contribute to an emerging paradox: an 'adaptive' response by a supposedly innate arm of the immune system. This amplification of specific NK cell subsets to the virus may be instrumental in controlling infections, and may also be disrupted by immunosuppression. NK cells undoubtedly play a crucial role in controlling HCMV infections. There is a compelling need to understand the mechanisms by which HCMV evades, modulates, and ultimately is recognized by 'innate' defence systems.

Introduction

NK cells play a crucial role in controlling cytomegalovirus infections. Individuals with genetic defects in their NK cell response are particularly vulnerable to herpesvirus infections and to HCMV in particular (Biron *et al.*, 1989; Orange, 2002; Gazit *et al.*, 2004). NK cells can control HCMV infection in the absence of T-cells (Kuijpers *et al.*, 2008), while recovery of NK cell function in haematopoietic stem cell transplant recipients correlates with protection against HCMV viraemia (Barron *et al.*, 2009). Consequently, there is a compelling need to investigate mechanisms by which NK cells recognize and eliminate HCMV-infected target cells. In recent years it has become apparent that a substantial proportion of HCMV's coding capacity has evolved to prevent NK cell activation. The task facing NK cells is formidable. This review examines an emerging story of how HCMV seeks to systematically modulate this crucial innate defence, and how the host immune system adapts to combat infection.

NK cell biology

NK cells constitute a heterogeneous population of large granular lymphocytes that exhibit an extraordinary capacity to kill tumour cell lines *in vitro*, and play a crucial role in controlling certain virus infections (Kiessling *et al.*, 1975; Biron *et al.*, 1999). As a key constituent of the innate immune response, they were differentiated from T and B cells functionally by their capacity to act independently of specific antigen recognition. Our appreciation of the mechanisms regulating NK cell function has been transformed in recent years with the identification and characterization of not only their inhibitory and activating receptors but also their associated ligands (Vivier *et al.*, 2008). NK cell function is regulated by integrating a combination of activating and inhibitory signals received from ligands on their targets. When an NK cell surveys a potential target, receptors bind their bespoke ligands at a highly structured immunological synapse organized at the interface. In addition to receptors that promote cell signalling pathways, the stability of the immunological synapse is modulated by adhesion molecules and membrane components that influence the underlying actin cytoskeleton (Davis, 2009). The complexity of the interaction is further enhanced by the existence of splice variants whose signalling moieties may be deleted or substituted.

HLA-I molecules act as ligands for the chief inhibitory receptors, which include a wide range of immunoglobulin-like (KIRs, LIR-1) and lectin-like receptors (e.g. CD94-NKG2A). NK cells are thus exquisitely sensitive to HLA-I down-regulation (Table II.8.1). *In vitro*, the level of HLA-I on tumour cell lines often determines their susceptibility to NK cells. To promote a response, however, NK cells require an activating signal. Many of these activating signals arise from proteins that are induced as a cellular response to stress, and/or are adhesion proteins that become exposed as a consequence of a breakdown of normal intercellular communications (Table II.8.2). Importantly, NK cells also have a key effector function in antibody dependent cellular cytoxicity (ADCC) where killing of target cells is directed by antibody binding to the Fc receptor CD16.

Somewhat surprisingly, a subset of NK cell ligands exhibit a capacity to bind both activating and inhibitory receptors. The existence of 'paired receptors' may be counterintuitive, yet it has been recognized for some time. The situation was first appreciated with the observation that both inhibitory and activating KIRs can recognize HLA-I molecules, but the phenomenon extends to other receptors (Tables II.8.1 and II.8.2). The foremost function of inhibitory receptors must primarily be to prevent self harm; NK cells are selected not to kill host cells through recognition of endogenous HLA-I molecules. However, the combination of inhibitory and activating receptors provides an NK cell with an inherent threshold that must be reached before NK cells can be activated. Indeed, signalling through the inhibitory receptor is thought to play an important role in 'educating' or 'licensing' NK cells (Yokoyama and Kim, 2006; Jonsson and Yokoyama, 2009). The presence of an inhibitory threshold allows an NK cell to be more responsive to activating signals. How quantitative and dynamic a process this is remains an area of increasing interest (Brodin and Hoglund, 2008; Brodin *et al.*, 2009). There will be examples later in this review of how this complex situation may impact in the context of HCMV and how it targets NK immunity.

Human NK cells are most readily analysed in blood, where they comprise between 5–20% of peripheral blood lymphocytes. Circulating NK cells are recognized to exhibit functional heterogeneity: classically $CD56^{dim}CD16^{+}$ are associated with high-level expression of cytolytic proteins (perforin, granzyme) required for direct target cell killing, whereas $CD56^{bright}CD16^{-}$ are associated with high-level cytokine (notably IFN-γ and TNF-α) secretion (Cooper *et al.*, 2001). The $CD56^{dim}CD16^{+}$ population comprises 90% of the

Table II.8.1 Inhibitory NK receptors

Receptor(s)	Ligand
KIRs	HLA-A, -B and -C
CD94-NKG2A	HLA-E
LIR-1 (ILT-2, CD85j, LILRB1)	HLA-I
CD161 (NKR-P1A, KLRB1)	LLT-1 (CLEC2D)
LAIR-1 (CD305)	Collagen
SIGLEC 3 (CD33), 7, 9	Sialic acid
TIGIT (VSTM3)	CD155, CD112
KLRG1 (CLEC15A)	Cadherins
CD200R	CD200
CEACAM	CEACAM
PD-1	PD-1L

Table II.8.2 Activating NK receptors

Receptor(s)	Ligand
KIR(s)	HLA-C
CD94-NKG2C/E	HLA-E
NKG2D	MICA, MICB, ULBP1, RAETII
CD226 (DNAM-1)	CD155 (PVR, necl-5) CD112 (nectin-2)
CD96 (TACTILE)	CD155 (PVR, necl-5) CD112 (PVRL2, nectin-2)
CRTAM	Necl2 (TSLC-1)
NKp30	BAT-3, B7-H6
NKp44	Viral?
NKp46	Viral?
NKp80	AICL (CLEC2B)
2B4	CD48
CRACC	CRACC
NTB-A	NTB-A
KLRG1 (CLEC15A)	–
CD16 (FcgRIIIA)	IgG1 and IgG3

circulating NK pool, but it is worth considering that this ratio is reversed in lymphoid organs, such that overall the cytokine-producing CD56^{bright}CD16^{-} population are the more numerous in the human body. For *in vitro* assays, NK cells are commonly first primed by treatment with a type I interferon, IL-2 and/or IL-15. This process mimics the situation *in vivo* where the basal state of NK cell activation is regulated by the cytokine/chemokine environment, TLR activation and cross talk with other immune cells. NK cells can sense pathogens directly through TLR 3, 7 and 8 (Alexopoulou *et al.*, 2001; Diebold *et al.*, 2004; Heil *et al.*, 2004; Lund *et al.*, 2004; Hart *et al.*, 2005), all of which have been implicated in recognition of virus (Akira and Sato, 2003). In addition to promoting innate and adaptive immunity, NK cells act to avert immunopathology by directly killing activated antigen-presenting cells and T-cells and by secretion of immunosuppressive cytokines.

Susceptibility of HCMV-infected cells to NK cells

A series of early studies observed that HCMV-infected cells generally exhibited enhanced sensitivity to NK cell-mediated cytolysis in chromium release assays (Thong *et al.*, 1976; Kirmani *et al.*, 1981; Starr and Garrabrant, 1981; Borysiewicz *et al.*, 1985; Gehrz and Rutzick, 1985). Cells were most sensitive to NK cell-mediated cytolysis early in infection (prior to 36 hours), while the later stages were characterized by enhanced resistance (Fig. II.8.1A) (Wang *et al.*, 2002). However, these studies utilized laboratory-adapted HCMV strains and consequently do not provide a complete picture. The laboratory strains AD169 and Towne lack 15 and 13 kbp of genomic DNA respectively from the right hand end of the U_L region. When low passage viruses were deployed, infection conferred much more robust protection against NK cell-mediated cytotoxicity (Waner and Nierenberg, 1985; Cerboni *et al.*, 2000; Wang *et al.*, 2002), while repair of the 13 kbp deletion in the Towne strain was associated with greatly enhanced resistance to NK cells (Tomasec *et al.*, 2005). This scenario is illustrated in Fig. II.8.1B where strains AD169 and Towne each have a large deletion in the U_L region, strain Toledo is a low passage virus while the recombinant Tx4 virus is a chimera in which the large deletion in strain Towne has been repaired (by E.S. Mocarski) with sequence from strain Toledo. When evaluating NK responses to HCMV-infected cells, notice clearly needs to be taken of the HCMV strain being deployed and the timing of infection. One further consideration is the use of allogeneic or autologous targets. As HLA is a common ligand for an array of NK activating and inhibitory receptors, the presence of mismatched HLA types could potentially alter the normal behaviour of primary NK cells. With this is in mind, the body of scientific research indicates that the virus is capable of taking on and comprehensively defeating the NK cell cytolytic response, at least *in vitro*. The only clear example of NK cells controlling a low passage HCMV strain *in vitro* was observed during long-term co-culture where lymphotoxin-dependent induction of

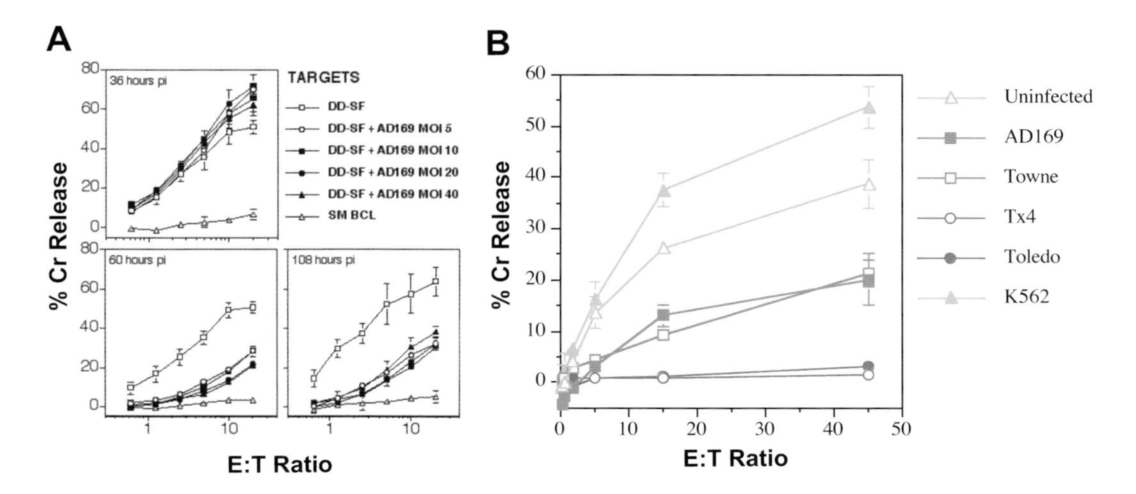

Figure II.8.1 Sensitivity of HCMV-infected cells to NK killing. (A) Killing of HCMV AD169-infected skin fibroblasts by the NK cell line, DEL, at the indicated times post infection. Fibroblasts were infected at a range of MOI as indicated and NK cells were used at the E/T ratios shown. (B) Killing of skin fibroblasts infected with the HCMV strains indicated. K562 is a HLA-I negative, positive control cell line for NK killing. Figure adapted with permission from (Wang *et al.*, 2002) and (Tomasec *et al.*, 2005) respectively.

interferon restricted replicative spread of the virus by a non-cytolytic mechanism (Iversen *et al.*, 2005).

Down-regulation of HLA-I

The primary role of classical HLA-I in antiviral immunity is to present foreign peptides such as those derived from viral proteins to CD8$^+$ cytotoxic T-cells, thereby allowing recognition and destruction of virus-infected cells. However, HCMV encodes multiple genes that inhibit surface expression of HLA-I. US3 is an IE gene that retains newly synthesized HLA-I molecules in the ER, US6 prevents peptide translocation by inhibiting TAP function, while US2 and US11 promote retrograde translocation of newly synthesized heavy chains from the ER to the cytosol where they is rapidly degraded by the proteasome (Wiertz *et al.*, 1996; Jones and Sun, 1997) (Fig. II.8.2A). US10 also delays trafficking of HLA-I molecules but preferentially targets HLA-G to enhance NK cell sensitivity, although the rationale for this remains unclear (Furman *et al.*, 2002; Park *et al.*, 2010), In addition, US8 binds HLA-I without an appreciable outcome, while miR-US4-1 suppresses expression of an endoplasmic reticulum aminopeptidase (ERAP1) responsible for trimming peptide epitopes (Kim *et al.*, 2011). Efficient down-regulation of multiple HLA-I alleles requires the HCMV immune evasion functions work in combination.

Since HLA-I molecules predominantly bind inhibitory receptors, their down-regulation would be expected to render infected cells more sensitive to NK cell recognition, and this is observed following infection with strain AD169 at least at early times after infection (Fig. II.8.1) (Falk *et al.*, 2002; Wang *et al.*, 2002). However, the HLA-I evasion functions all act post-translationally, and HCMV infection actually stimulates HLA-I transcription. Consequently, the rate of HLA-I expression is significantly increased following infection, although this only becomes apparent when the US2–11 functions are deleted from the virus (Fig. II.8.2B) (Warren *et al.*, 1994; Jones and Sun, 1997). The up-regulation of HLA-I induced by the HCMV US2-11 mutant during lytic infection is associated with enhanced protection against NK cell lysis (Fig. II.8.2B), consequently it can be concluded that the HLA-I evasion functions do render cells more susceptible to NK attack (Fig. II.8.1B) (Falk *et al.*, 2002).

It should be borne in mind that the down-regulation

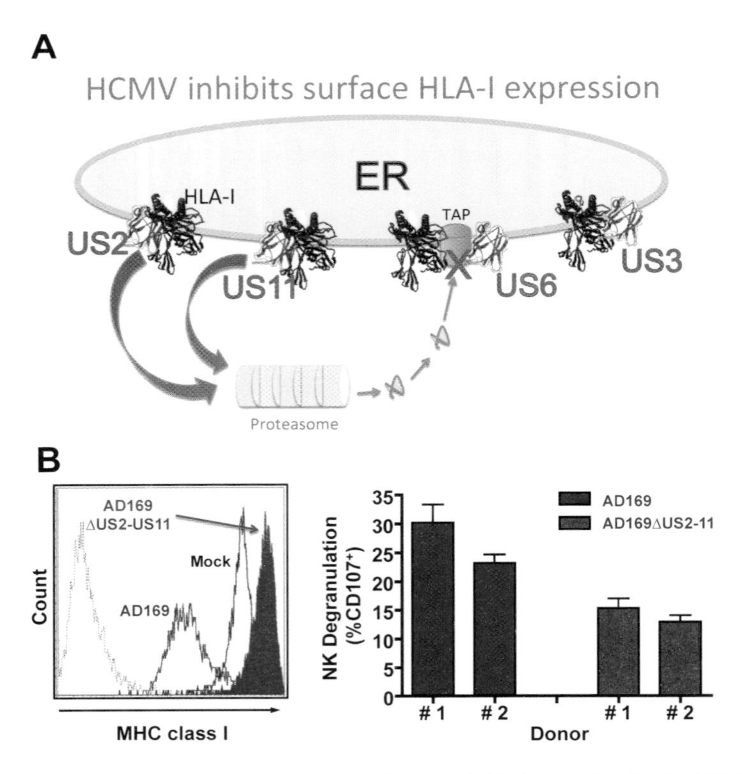

Figure II.8.2 Down-regulation of HLA-I and effects on killing of HCMV-infected cells. (A) HCMV encodes at least four genes involved in down-regulation of HLA-I; US2, US3, US6 and US11. (B) Removal of the US2–US11 region in AD169 results in up-regulation of HLA-I above normal levels and inhibition of NK degranulation.

of HLA-I expression, normally observed during lytic infection, can be overcome if cells are exposed to interferon prior to infection (Hengel *et al.*, 1995). It would be interesting to extend studies of this phenomenon *in vivo*. Theoretically at least, when an HCMV primary infection or reactivation is proceeding in a pro-inflammatory environment, endogenous HLA-I molecules may well be expressed, even up-regulated and capable of interacting with receptors on both T and NK cells. From an NK cell perspective, this means that the NK cell receptor recognition of HLA-I molecules can be expected to have a role in the response to HCMV-infected cells.

Compensating for HLA-I down-regulation

UL18

UL18 merits a special status as the first microbial gene implicated in NK cell modulation. The gene was identified as an HLA-I homologue during the sequencing of strain AD169 (Beck and Barrell, 1988) and in follow-up studies the prediction that it was highly glycosylated (13 consensus N-linked glycosylation sites) was confirmed. gpU18 forms a trimeric complex with β_2-microglobulin and peptide (Browne *et al.*, 1990; Fahnestock *et al.*, 1995; Griffin *et al.*, 2005); peptide loading is facilitated by UL18 acting directly on the TAP complex to overcome its inhibition by US6 (Kim *et al.*, 2008). Moreover, the molecular structure of gpUL18 (solved to 2.2 Å resolution) is similar to its cellular homologue (Yang and Bjorkman, 2008). A screen to find an interacting host cell protein successfully identified the inhibitory receptor LIR-1 (Cosman *et al.*, 1997). LIR-1, also known as LILRB1, ILT-2 (Colonna *et al.*, 1997) and CD85j, is expressed most efficiently on myeloid cells, but crucially is also present on NK and T-cell subsets. The normal function of LIR-1 is to engage with endogenous HLA-I (HLA-A, -B, -C, -F and -G) to provide an inhibitory signal through SHP1 (Wagner *et al.*, 2008), however, LIR-1 binds to gpUL18 with 1000-fold higher affinity than to HLA-1 (Chapman and Bjorkman, 1998). This high affinity interaction involves the D1 and D2 domains on the extended LIR-I molecule and a peptide in the α3 domain (strongly conserved in HLA heavy chains) and β_2-microglobulin (Chapman *et al.*, 1999, 2000; Willcox *et al.*, 2003; Wagner *et al.*, 2007; Occhino *et al.*, 2008; Yang and Bjorkman, 2008). The efficiency of this interaction implies that even low levels of gpUL18 expression should be able to promote a strong inhibitory signal, although to achieve this it has to be expressed on the cell surface – a precarious strategy as the protein then becomes a potential target for ADCC (Park *et al.*, 2002; Griffin *et al.*, 2005; Prod'homme *et al.*, 2007).

Working on gpUL18 has not proved straightforward: the gene is transcribed in low abundance (Hassan-Walker *et al.*, 2001; Gatherer *et al.*, 2011), a substantial proportion of newly synthesized protein is degraded (Tomasec, unpublished) and toxicity issues at high expression levels have hindered the establishment of stable cell lines. Fortunately, it can be produced in a soluble form where peptide-binding enhances its stability (Chapman *et al.*, 1999). However, results of early functional studies were conflicting, showing UL18 to be capable of both suppressing and stimulating NK cell activation (Reyburn *et al.*, 1997; Leong *et al.*, 1998; Odeberg *et al.*, 2002; Kim *et al.*, 2004). Resolution came with the advent of the CD107 mobilization assay, which measures NK cell degranulation by flow cytometry. Crucially, this technique permits phenotyping of responsive cells, thus allowing LIR-1^+ NK cells to be interrogated specifically. UL18 was indeed found to inhibit the activation of LIR-1^+ NK cells, but unexpectedly also stimulated degranulation of LIR-1^- NK cells (Prod'homme *et al.*, 2007). The net effect of UL18 expression depends on the proportion of LIR-1^+ NK cells in the donor. The result implies that gpUL18 is also recognized by an NK cell-activating receptor that is broadly expressed. The identity of this putative activating receptor is being sought. As an HLA-I homologue, it is natural to speculate that gpUL18 may be the target of an activating KIR, CD94-NKG2C/E, or an activating member of the LIR family. Yang and Bjorkman (2008) cautioned that 'most of the available surface area on the UL18 heavy chain was predicted to be covered with carbohydrate, particularly the α1–α2 peptide binding platform'. Glycosylation presumably helps protect against ADCC, but is a substantial issue when considering receptor binding. Nevertheless, the glycosylated form of gpUL18 was clearly demonstrated to preferentially bind the NK cell activating receptor CD94-NKG2C by surface plasmon resonance experiments (Kaiser *et al.*, 2008). At the time of writing, this biochemical finding awaits support from functional studies.

UL40

Overall cell surface levels of HLA-I can be monitored by the immune system through the non-classical MHC-I molecule HLA-E. Cell surface expression of HLA-E requires binding of a conserved nonameric peptide that is normally derived from the signal peptide of HLA-A, -B, -C and -G (Braud *et al.*, 1997). Although the HLA-E-binding peptide is highly conserved, some variation is compatible with binding and is observed amongst the different HLA-I alleles. HLA-E recognizes the inhibitory receptor CD94-NKG2A with high affinity and the paired activating receptor CD94-NKG2C with low affinity (Borrego *et al.*, 1998; Braud *et al.*, 1998, 2002; Lee *et al.*, 1998). During translation of ER-targeted proteins the hydrophobic signal peptide is cleaved by the signal peptidase (SP) to be released into the ER membrane. The signal peptide is further processed by a signal peptide peptidase (SPP). In the processing of the signal peptide for most HLA heavy chains, SPP releases the HLA-E-binding peptide into the cytosol. Trafficking of the HLA-E-binding peptide to the lumen of the ER is absolutely dependent on TAP (Fig. II.8.3). HLA-E maturation is therefore a sensor of both endogenous HLA-I expression and TAP function; this signal

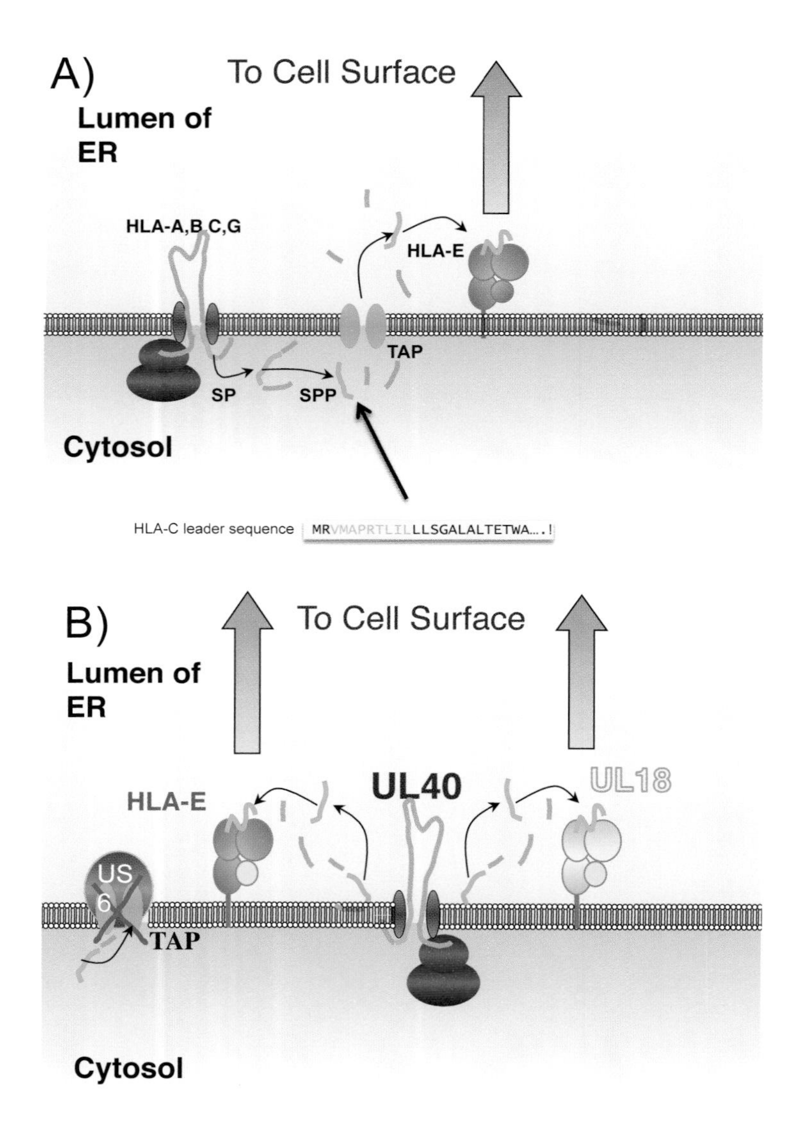

Figure II.8.3 Surface expression of HLA-E and gpUL18. (A) Nonameric peptides from the signal sequence of HLA-A, -B, -C and –G are generated by the action of signal peptidase (SP) and signal peptide peptidase (SPP) and transported from the cytosol to the ER by TAP. These peptides are then bound by HLA-E and allow transport of HLA-E to the cell surface. NK cells recognize HLA-E through their inhibitory CD94-NKG2A receptor complex. B) US6 inhibition of TAP prevents loading of the conserved HLA-I derived nonameric peptide on to HLA-E. An identical peptide encoded within the UL40 signal peptide (SP^{UL40}) is delivered directly to the ER to promote maturation of both HLA-I and gpUL18.

is monitored by NK cells. HCMV up-regulates endogenous transcription and translation of HLA-I heavy chains, and then acts post-translationally to impede their surface expression. Critically, the inhibition of TAP by US6 disrupts the supply of stabilizing peptides and thereby impeded HLA-E cell surface expression. HCMV UL40 overcomes this problem in an extremely elegant manner.

The signal peptide of UL40 (SP^{UL40}) contains an exact copy of a nine amino acid HLA-E binding peptide, but significantly this peptide is delivered to HLA-E in a TAP-independent manner, thus bypassing the HLA-I down-regulatory effect of US6 (Tomasec *et al.*, 2000; Ulbrecht *et al.*, 2000; Millo *et al.*, 2007). The *cis*-acting element responsible for bestowing TAP independence resides within the N-terminal 12 amino acids and is predicted to act by flipping the orientation of the signal peptide within the ER membrane so that SPP cleavage releases the HLA-E binding peptide directly into the lumen of the ER (Fig. II.8.4) (Prod'homme *et al.*, 2012). SP^{UL40}-mediated up-regulation of HLA-E preferentially elicits protection against $CD94^{+}$-$NKG2A^{+}$ NK cells when expressed in isolation or in the context of a productive virus infection (Wang *et al.*, 2002).

The low affinity interaction between HLA-E and CD94-NKG2C merits special consideration because of the observation that this subset is increased in HCMV-infected individuals (Guma *et al.*, 2004). Indeed, HCMV-infected fibroblasts are capable of driving the expansion of $CD94^{+}NKG2C^{+}$ NK cells *in vitro*, though this seems unaffected by the presence of UL40 (Guma *et al.*, 2006). Intriguingly, our data suggests there is also an inverse correlation with high CD94-NKG2C expression and low CD94-NKG2A expression. Not all HCMV-seropositive individuals show $CD94^{+}NKG2C^{+}$ expansions and this is in keeping with the presence of non-responders in *in vitro* expansion assays (Guma *et al.*, 2006). Non-responding HCMV-seropositive individuals mirror the CD94-NKG2A/NKG2C patterns observed in HCMV seronegative individuals (Fig. II.8.5). Further work is required to see whether there is potential conversion or adaptation between $NKG2A^{+}$ and $NKG2C^{+}$ NK subsets, and whether this relates to the length of HCMV infection or other as yet undefined factors. Moreover, the proportion of $CD94^{+}NKG2C^{+}$ NK cells expands markedly in patients experiencing acute HCMV infection (Kuijpers *et al.*, 2008; Lopez-Verges *et al.*, 2011) and in patients experiencing an acute infection with Puumala hantavirus; in the latter case this expansion correlates with HCMV-seropositivity (Bjorkstrom *et al.*, 2010). In the context of hantavirus infection, the expanded $CD94^{+}NKG2C^{+}$ subset was

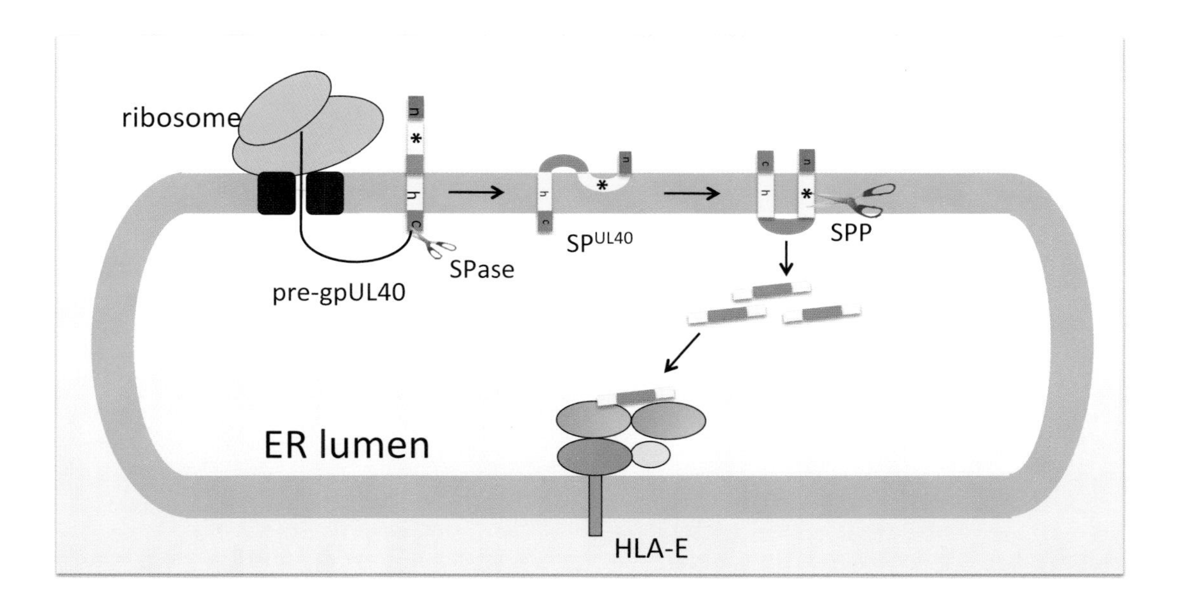

Figure II.8.4 Mechanism of TAP-independent peptide transfer by UL40. The signal peptide of UL40 allows insertion through the membrane. Signal peptides consist of a hydrophobic core region (*h-region*) flanked by an N- and C- terminal amino acids; termed the *n-* and *c-regions* and coloured red and green respectively. In SP^{UL40}, the *h-region* (yellow) is unusual in that it is interrupted by a hydrophilic spacer (green). SP^{UL40} must be configured so that hydrophobic sequences lie within the ER membrane. Our model implies SP^{UL40} re-orientates (flips) in the membrane to expose the HLA-E-binding peptide to the lumen of the ER. SP^{UL40} is then further processed by an SPP-type intramembrane protease, releasing the HLA-E-binding peptide within the lumen of the ER.

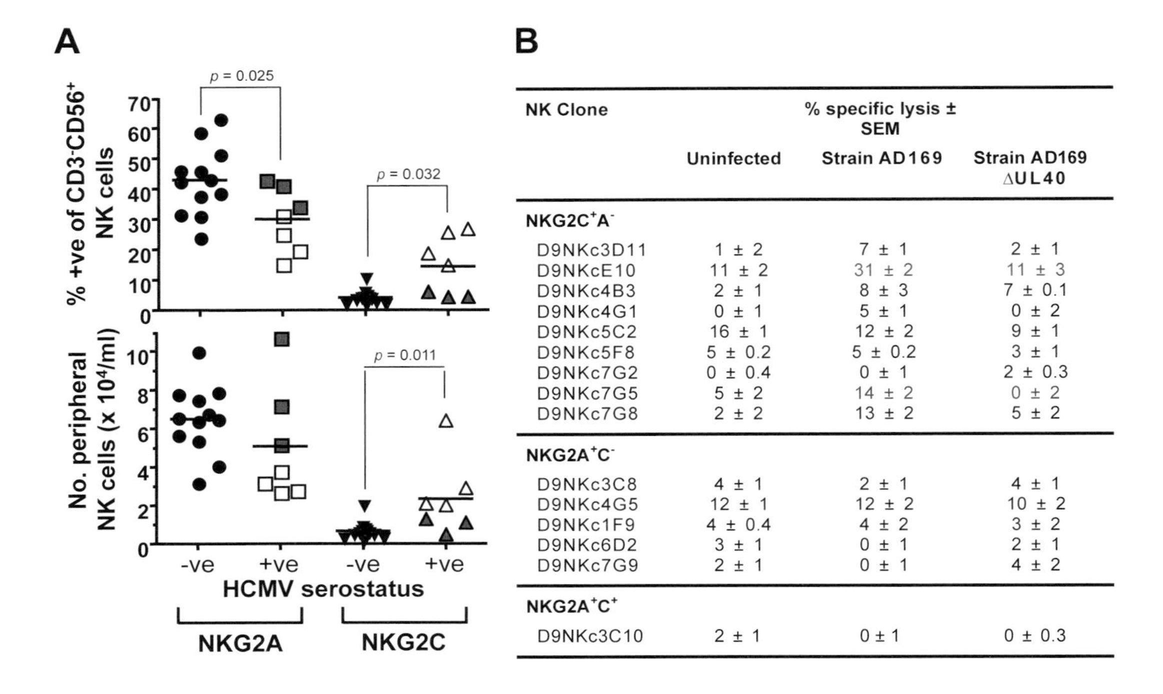

NK Clone	% specific lysis ± SEM		
	Uninfected	Strain AD169	Strain AD169 ΔUL40
NKG2C⁺A⁻			
D9NKc3D11	1 ± 2	7 ± 1	2 ± 1
D9NKcE10	11 ± 2	31 ± 2	11 ± 3
D9NKc4B3	2 ± 1	8 ± 3	7 ± 0.1
D9NKc4G1	0 ± 1	5 ± 1	0 ± 2
D9NKc5C2	16 ± 1	12 ± 2	9 ± 1
D9NKc5F8	5 ± 0.2	5 ± 0.2	3 ± 1
D9NKc7G2	0 ± 0.4	0 ± 1	2 ± 0.3
D9NKc7G5	5 ± 2	14 ± 2	0 ± 2
D9NKc7G8	2 ± 2	13 ± 2	5 ± 2
NKG2A⁺C⁻			
D9NKc3C8	4 ± 1	2 ± 1	4 ± 1
D9NKc4G5	12 ± 1	12 ± 2	10 ± 2
D9NKc1F9	4 ± 0.4	4 ± 2	3 ± 2
D9NKc6D2	3 ± 1	0 ± 1	2 ± 1
D9NKc7G9	2 ± 1	0 ± 1	4 ± 2
NKG2A⁺C⁺			
D9NKc3C10	2 ± 1	0 ± 1	0 ± 0.3

Figure II.8.5 Relationship between NKG2A, NKG2C and HCMV infection. (A) Proportions and numbers of CD94⁺NKG2A⁺ and CD94⁺NKG2C⁺ NK cells in peripheral blood of healthy HCMV seronegative and seropositive individuals. HCMV-seropositive subjects have increased proportions and numbers of CD94⁺NKG2C⁺ and decreased proportions and numbers of CD94⁺NKG2A⁺ NK cells. A proportion of HCMV seropositives were apparent non-responders (filled red symbols) and have a seronegative phenotype. (B) CD94⁺NKG2C⁺ NK clones can be expanded that recognize strain AD169-infected fibroblasts but only in the presence of UL40.

activated by cells expressing HLA-E, thus providing the first evidence that this NK cell population is functionally significant *in vivo*. Elevation of HLA-E expression levels by a combination of SP^{UL40} and HLA-E transcriptional up-regulation during HCMV infection has the potential to provide susceptibility to the expanded CD94⁺NKG2C⁺ NK cell subset. However, while we have generated CD94⁺NKG2C⁺ NK clones that are preferentially activated by strain AD169 and impaired by strain AD169ΔUL40 (Fig. II.8.5), we are not aware of any demonstration that the broad CD94⁺NKG2C⁺ response is preferentially activated by HCMV-infected targets.

HLA-I molecules are conventionally associated with promoting T-cell responses. Indeed, UL18 has been associated with the induction of an allospecific T-cell response (Wronska *et al.*, 1993). The sequence of the classical HLA-E-binding peptide exhibits a level of heterogeneity; the archetypical VMAPRTLIL epitope is encoded by multiple HLA-C alleles and by UL40 from strain AD169, while VMAPRTLVL is present in some HLA-A alleles and by UL40 from strain Toledo. The affinity of the interaction between HLA-E and CD94-NKG2A is affected by the peptide (Akira and Sato, 2003; Kaiser *et al.*, 2005; Sullivan *et al.*, 2007; Hoare *et al.*, 2008), but most significantly, if the host HLA haplotype does not encode a variant that exactly matches that of the infecting HCMV strain, then the potential exists to generate an HLA-E-restricted T-cell response. Such a response can be triggered by the single I/V difference at position 8 in the nonameric peptide (Pietra *et al.*, 2003; Hoare *et al.*, 2006), and is capable of killing HCMV-infected cells (Mazzarino *et al.*, 2005). While the value of SP^{UL40} to HCMV in promoting NK cell evasion is self evident, in certain circumstances or individuals the tables may turn with CD94⁺NKG2C⁺ NK or HLA-E restricted cytotoxic T-cells being able to preferentially target HCMV-infected cell as a consequence of the SP^{UL40} function.

SP^{UL40} and UL18 cross-talk

Up until now, SP^{UL40} and UL18 have been considered independently. However, their functions became linked by the discovery that the consensus HLA-E-binding peptide also stabilizes and thereby enhances cell surface expression of gpUL18. This finding is supported by three independent observations: (i) HLA-E-binding peptides were eluted from soluble UL18 expressed in transfected cells (Hugh Reyburn, personal communication); (ii) UL40 up-regulates gpUL18 expression (Prod'homme *et al.*, 2012) and; (iii) the UL40-derived peptide was identified *in silico* as a potential gpUL18-binding peptide (Yang and Bjorkman, 2008). Moreover, SP^{UL40}–mediated enhancement of gpUL18 also operates in the context of productive HCMV infection (Prod'homme *et al.*, 2012).

While gpUL18 and HLA-E bind an overlapping set of peptides, more detailed analyses indicate that the peptide-binding properties of HLA-E and gpUL18 are not identical, with gpUL18 appearing to be less fastidious. Most notably, gpUL18 appears able to tolerate the binding site for the first amino acid of the peptide being left vacant, whereas HLA-E does not. UL40 utilizes at least two transcriptional start sites: the upstream site provides for the coding of the expression of full length SP^{UL40}, whereas the prediction for the downstream site is that translation is actually initiated at P2 (position 2; second amino acid) of the HLA-E-binding epitope within SP^{UL40}. Full-length SP^{UL40} can up-regulate HLA-E and gpUL18, whereas the N-terminally truncated version of SP^{UL40} up-regulates UL18 only. Consequently, differential transcriptional initiation of UL40 results in differential regulation of gpUL18 and HLA-E expression.

Activating ligands

Induction and suppression of NKG2DLs

The activating receptor NKG2D is exceptional in that it is expressed on all NK cells and recognizes eight ligands: the polymorphic MHC-I-like molecules MICA and MICB, ULBP1-3, ULBP4/RAET-E, ULBP5/RAET-G, ULBP6/RAET1L (Eagle and Trowsdale, 2007; Eagle *et al.*, 2009b). NKG2D ligands (NKG2DLs) are considered to be stress proteins as their expression is induced by heat, genotoxic shock, oncogenic transformation or by virus infection. The existence of multiple NKG2DLs provides the opportunity for the expression of each ligand to be sensitive to distinct stimuli (Bauer *et al.*, 1999). During HCMV infections, transcription of all but ULBP4/RAET-E were up-regulated within 24 hours in fibroblasts (Eagle *et al.*, 2006), with IE1 and IE2 being specifically implicated in the activation of MICA, MICB and ULBP2 (Andresen *et al.*, 2007; Venkataraman *et al.*, 2007). While it needs to be ascertained whether the act of virion uptake alone would be sufficient to trigger NKG2DLs, the activation of multiple NKG2DLs with early phase kinetics necessitates a rapid response by HCMV to prevent immune elimination of the infected cell.

miR-UL112

The 3′ untranslated regions (3′UTR) of the MICA and MICB transcripts contain within them binding sites for multiple host miRNAs. These miRNAs act to suppress MICA/B expression, and thus impose a threshold level of transcription that must be surpassed before MICA and MICB expression can be activated (Stern-Ginossar *et al.*, 2008). Remarkably, a viral miRNA (miR-UL112) also binds to a miRNA seed sequence in the MICB 3′UTR, and acts synergistically with host miRNAs to suppress MICB gene expression (Stern-Ginossar *et al.*, 2007; Nachmani *et al.*, 2010). miR-UL112 has been shown to suppress expression from a number of key HCMV genes, including IE1 and IE2 (Grey *et al.*, 2007). During HCMV infection, MICB and IE transcripts will thus compete for, and be regulated by miR-UL112. Stimulating transcription from either one will relieve repression of the other. Consequently, activation of IE gene transcription would be expected to sequester miR-UL112 blocking MICB expression. The major IE gene transcriptional unit is autoregulatory. The major IE promoter is efficiently activated by the major IE protein itself until the product of the IE2 splice variant feeds back to specifically limit transcription of both IE1 and IE2. A stress response that efficiently activated MICB transcription could hence also be expected also trigger IE-1 expression and potentially HCMV reactivation.

UL16 and UL142

Following the identification of LIR-1 as a receptor for gpUL18, Cosman's group turned their attention to UL16, and in the process identified the ULBP family as a novel group of NK activating ligands (Cosman *et al.*, 2001; Kubin *et al.*, 2001). UL16 binds directly with three NKG2DLs (MICB, ULBP1 and ULBP2) to promote their intracellular retention within the ER/*cis*-Golgi and thereby preventing expression on the surface of HCMV-infected cells (Dunn *et al.*, 2003; Rolle *et al.*, 2003; Vales-Gomez *et al.*, 2003; Welte *et al.*, 2003; Wu *et al.*, 2003). The acronym ULBP stands for UL16-binding protein, even though ULBP3 and the mature form of ULBP5 were known not to bind gpUL16. Over time additional family members were characterized: ULBP4/RAET1E, ULBP6/RAET1L and the immature form of ULBP5/RAET1G were all found to bind gpUL16 (Bacon *et al.*, 2004; Eagle *et al.*, 2009a; Wittenbrink *et al.*, 2009). The capacity of gpUL16 to interact with 5/6 different NKG2DLs is remarkable. The ectodomains of the ULBP subfamily consist exclusively of MHC class I-like α1 α2 folds and thus have some structural similarity with MICA/B, which also have MHC class I-like α1 α2 folds, but MICA/B are distanced from the plasma membrane by an additional Ig-like α3 domain (Sutherland *et al.*, 2001; Wittenbrink *et al.*, 2009). A molecular understanding of why gpUL16 fails to bind MICA, ULBP3 and mature ULBP5 has come with a resolution of the gpUL16 structure and relates to a single incompatible amino acid in the host protein; consequently it has been postulated that these proteins have been selected on the basis of their resistance to down-regulation of UL16, thereby enabling recognition of HCMV-infected cells by NKG2D$^+$ effector cells (Wittenbrink *et al.*, 2009; Muller *et al.*, 2010).

UL142 was identified as an UL18 and MHC-I homologue (Davison *et al.*, 2003; Wills *et al.*, 2005), but its mechanism of action as an NK evasion function was unexpected (Wills *et al.*, 2005). Rather than acting as an inhibitory ligand, gpUL142 targeted the UL16-resistant NKG2D ligands MICA and ULBP3, interacting directly with both proteins to sequester them in the ER/*cis*-Golgi (Chalupny *et al.*, 2006; Ashiru *et al.*, 2009; Bennett *et al.*, 2010).

UL141

UL141 prevents activation of NK cells by down-regulating cell surface expression of CD155 and acting in combination with an additional undefined HCMV-encoded function(s) to degrade CD112 (Tomasec *et al.*, 2005; Prod'homme *et al.*, 2010). UL141, CD112 and CD155 are all members of the Ig superfamily. CD155 is also known as nectin-like molecule-5 (necl-5) or poliovirus receptor (PVR), and can be detected in complexes with the platelet derived growth factor (PDGF) receptor and αvβ3 integrin at the leading edge of proliferating or migrating cells. When cell–cell contact is initiated, CD155 will bind nectin-3 to promote the formation of adhesion junctions (Fujito *et al.*, 2005). On binding nectin-3, CD155 is normally internalized; a process that has been implicated in 'contact inhibition'. The suppression of CD155 expression can also be expected to impact on cell mobility and proliferation.

CD112 is also referred to as nectin2 or poliovirus receptor related 2 (PVRL2). Both nectins and necl proteins are homodimeric transmembrane proteins, but nectins differ in that they are attached to the actin cytoskeleton through an afadin bridge. The ectodomain of CD112 actively participates in adhesion junctions by binding to itself in neighbouring cells (heterophillic transdimerisation). CD112 and CD155 are therefore both constitutively expressed, but are normally preoccupied with intercellular adhesion junctions. Following virus infection, or cell transformation, intercellular contacts break down to render adhesion molecules more accessible to NK cells (Takai *et al.*, 2008). With respect to NK cell recognition, CD155 is a ligand for the NK cell receptors DNAM-1, CD96 (Tactile, T-cell-activated increased late expression) and TIGIT (T-cell immunoreceptor with Ig and ITIM domains) while CD112 is a ligand for DNAM-1 and TIGIT (Bottino *et al.*, 2003; Fuchs *et al.*, 2004; Stanietsky *et al.*, 2009). In functional assays on NK cell clones, UL141 clearly acts as a major suppressor of NK function, inhibiting two thirds of NK clones tested (Tomasec *et al.*, 2005).

While UL141 down-regulates cell surface CD155 expression efficiently, it is not absolute (Tomasec *et al.*, 2005). TIGIT is an inhibitory receptor that has high affinity for CD155 and CD122 and is ubiquitously expressed on all NK and T-cells (Stanietsky *et al.*, 2009). Stanietsky and Mandelboim propose that when CD155 is present at low levels, which is normal for healthy cells seeking to establish adhesion junctions, it will preferentially bind to TIGIT and will provide a dominant signal to the NK cell. However, following virus infection or cell transformation, the abundance of CD155 or CD112 will rise to a point that it will trigger DNAM-I or CD96 function (Stanietsky and Mandelboim, 2010). By maintaining CD155 and CD112 expression at a low level, UL141 will thus preserve inhibitory signalling by TIGIT.

Other mechanisms

Ongoing studies indicate that our understanding of the role and control of NK cell function during HCMV infection is far from complete. The UL83 gene encoding major tegument protein pp65 has been shown to bind directly with NKp30 to inhibit transmission of an activating signal (Arnon *et al.*, 2005). NKp30 is a member of the natural cytotoxicity receptor (NCR) family, which also includes NKp44 and NKp46. The mechanism by which pp65 gains access to directly bind the NK receptor is not clear, although models can be envisaged that involve pp65 release by lysis of infected cells or dense bodies. With the identification of the NKp30 ligands B7H6 and BAT-3 (Pogge von Strandmann *et al.*, 2007; Brandt *et al.*, 2009), it now becomes possible to initiate studies to directly examine the effect of virus infection on NCR ligands.

It is considered likely that additional HCMV functions exist that target the NKG2DLs; more specifically it is claimed a second uncharacterised HCMV gene also targets ULBP3 (Bennett *et al.*, 2010). The extreme polymorphism associated with MICA (80 alleles) and

MICB (33 alleles) may also pose problems for the virus. Most significantly, the unusual MICA*008 allele has a stop codon that results in a truncation of the cytosolic domain, and this allele has been shown to be UL142-resistant (Chalupny *et al.*, 2006). The MICA*008 allele is the most common allele in American Caucasians (42%), and is up-regulated during HCMV infection (Zou *et al.*, 2005). This surprising chink in HCMV's armour warrants further investigation.

It is arguable that the efficient down-regulation of HLA-I obviates a need to consider the role of activating and inhibitory KIRs. However, interferon stimulates HLA-I expression and may overcome the inhibitory effects of the HCMV HLA-I down-regulatory functions (Hengel *et al.*, 1995). Furthermore, the ligands for some KIRs are yet to be defined. In this context, it is interesting to note that the expression of activating KIRs has a positive impact on the clinical outcome of bone marrow recipients with HCMV infection (Cook *et al.*, 2006). Furthermore, a patient with aberrant expression of KIR2DL1 on all NK cells exhibited an immunodeficiency that manifested as recurrent overt HCMV infections (Gazit *et al.*, 2004), suggesting uniform expression of inhibitory KIRs can result in an easier evasion of NK immunity by HCMV. Together, these reports imply that HCMV infection may be expected to impact on KIR recognition.

Beyond fibroblasts

HCMV infects an extremely wide range of cell types *in vivo* yet to date the vast majority of NK studies have involved *in vitro* cytolysis or CD107 mobilization assays using human fibroblasts as targets. Most cultured HCMV strains will only replicate efficiently in fibroblasts, hence they are the cell of choice. Moreover, from an NK perspective it is possible to establish autologous assays using fibroblasts cultured from skin biopsies. However, it is appreciated that fibroblasts do not express the full gamut of ligands recognized by NK cell receptors. Wild-type HCMV does not replicate in any cell type efficiently *in vitro*, but can adapt to growth in fibroblasts *in vitro* by acquisition of mutations at two distinct loci (Dargan *et al.*, 2010; Stanton *et al.*, 2010). There is a clear need to progress NK cell studies to other physiologically relevant cells types such as endothelial, epithelial and myeloid cells; particularly with the need to evaluate the role played by cross-talk between NK cells and dendritic cells (DCs) in orchestrating the immune response. DCs are the primary cell type involved in priming the NK response. Depending on the maturation and differentiation state of the DCs, IL-15 trans-signalling can trigger NK proliferation (Chijioke and Munz, 2011), while type I IFNs, IL-2, IL-12 and IL-18 released from DCs can all induce IFN-γ secretion by NK cells, reviewed in da Silva and Munz (2011). Indeed for MCMV, IL-12 has been reported as an absolute requirement for NK cell IFN-γ production and viral control (Biron *et al.*, 1996). Other pathways also impact on IFN-γ secretion, such as the glucocorticoid-induced TNF-related protein (GITR), the ligand of which is induced by TLR9 signalling in human plasmacytoid DCs (Hanabuchi *et al.*, 2006). Moreover, cells of the myeloid lineage are sites of HCMV latency and reactivation.

Lopez-Botet and colleagues have pioneered research investigating the effect of HCMV infection on monocyte-derived DCs. Experiments were performed using the TB40/E strain, which is unusual in retaining a capacity to infect endothelial cells and DCs. Most remarkably, HCMV-infected cells were found to be sensitive to NK cells primarily via DNAM-1 and NKp46 ligands. Moreover, HLA-E was not up-regulated and NK cell activation was preferentially observed in $CD94^{+}NKG2A^{+}$ NK cells. Some care has to be taken in interpreting these results as the strain TB40/E virus used is both heavily passaged and not clonal. Indeed, the sequence corresponding to the consensus HLA-E-binding epitope within SP^{UL40} appeared to have acquired a mutation that would be expected to ablate its function. Nevertheless, HLA-E was also down-regulated by a clinical isolate whose SP^{UL40} sequence appeared intact. While it will be important to repeat these studies with additional DC-tropic HCMV strains as they become available, this landmark study suggests that the HCMV-encoded NK evasion functions may prove to be less effective in DCs than fibroblasts, and thus potentially identifies a point in the HCMV life cycle where infected cells may have heightened vulnerability to NK attack.

Changes in immune repertoire, NK cell adaptation, memory and HCMV selection

HCMV establishes a lifelong persistent infection. During clinical disease, expansion and enhanced killing by NK cells has been reported (Hokland *et al.*, 1988; Venema *et al.*, 1994). While the underlying mechanisms are not fully understood, it is possible that HCMV immune evasion functions could provide an engine to drive selection of NK and T-cells with enhanced responsiveness, either through influencing NK activation thresholds thereby affecting licensing (Joncker *et al.*, 2009) and/or through selection and expansion of NK cell populations that are more efficient at detecting and responding to HCMV infection. Although a key component of the innate immune response, there is increasing interest in the potential of NK cells to exhibit properties indicative of education and memory. Recently examples have arisen of NK cell subsets with properties associated with both adaptive responses and immunological memory that potentially have implications for HCMV biology (Paust *et al.*, 2010; Paust and von Andrian, 2011).

We have already described above how HCMV-infected individuals exhibit elevated proportions of LIR-1$^+$ and CD94$^+$NKG2C$^+$ NK cells (Berg *et al.*, 2003; Guma *et al.*, 2004). During acute HCMV infection, the CD94$^+$NKG2C$^+$ NK cells proliferate and acquire expression of CD57 (Lopez-Verges *et al.*, 2011), a marker that has previously been associated with expanded, highly differentiated, oligoclonal HCMV-specific CD4$^+$ and CD8$^+$ effector memory T-cells (Gratama *et al.*, 1989; Wang *et al.*, 1993, 1995; Kern *et al.*, 1999; Weekes *et al.*, 1999; Pourgheysari *et al.*, 2007). There have been suggestions that CD57 on T-cells is a marker for senescence as the cells were reported to be impaired in their proliferative ability (Brenchley *et al.*, 2003) and more sensitive to apoptosis (Shinomiya *et al.*, 2005). This remains controversial as there is also evidence for great proliferative capacity by CD8$^+$CD57$^+$ T-cells when given the right stimulation (Chong *et al.*, 2008). With regard to NK cells, CD57 is also associated with greater cytotoxic potential, but reduced responsiveness to cytokines mirroring early observations in T-cells (Lopez-Verges *et al.*, 2010). Whether this is a truly terminally differentiated subset or just highly differentiated but requiring as yet undiscovered stimulation for expansion remains to be seen. Irrespective of this, the expansion of CD94$^+$NKG2C$^+$ NK cells observed in HCMV-seropositive individuals may represent an adaptive antiviral NK response with associated characteristics of memory. Definition of the exact function of CD57 may help with our understanding this, though it has been difficult to dissect as CD57 is itself not a protein but is instead a sulfoglucoronyl carbohydrate moiety that is found on a range of different glycolipids and glycoproteins (Kruse *et al.*, 1984; Chou *et al.*, 1986). The glycoproteins on which CD57 is expressed on lymphocytes remain poorly defined and as a result, the underlying reasons why HCMV infection is associated with expanded CD57$^+$ NK and T-lymphocytes remain unknown.

A similar situation may yet unfold with gpUL18 when the nature of its putative activating receptor has yet to be defined. We have already described how NK cell activating and inhibitory receptors are capable of recognizing HLA-E and gpUL18 on HCMV-infected targets. Moreover, HLA-E-specific T-cells can be induced that are capable of targeting HCMV-infected cells depending on a mismatch between the HLA-E binding epitopes encoded by the host and the viral SP^{UL40} sequence. The full significance of gpUL18 binding the SP^{UL40}-derived peptide is not clear, yet it is likely that binding to both LIR-1 and the putative gpUL18 activating receptor may be influenced by the exact nature of the peptide present in the gpUL18 cleft, be it through a direct interaction with the peptide or a conformational effect on gpUL18 structure. In naïve populations, SP^{UL40} will be expected to act as an efficient immune evasion function conferring a clear selective advantage. Potentially, this situation can change in individuals whose T and/or NK cells adapt to target HLA-E stabilized with an SP^{UL40}-derived peptide (Sullivan *et al.*, 2007; Lopez-Verges *et al.*, 2011). Adaptive NK cell responses have been described in murine models, but it is clearly much more difficult to acquire definitive evidence in a human setting (Paust and von Andrian, 2011; Sun *et al.*, 2011). HCMV evasion functions, and SP^{UL40} in particular, may prove invaluable in providing systems to explore the phenomenon of NK cell 'memory' in man.

A natural consequence of the emergence of adaptive responses (T or NK) to SP^{UL40} is that the function could become a liability for the virus. In such circumstances, natural selection could be expected to favour the outgrowth of escape mutants. SP^{UL40} sequence variation is present in cultured HCMV strains, including a subset with variants that would be predicted to ablate its immune evasion function. However to date it has not been determinable whether these variants arose *in vitro* or *in vivo* (Tomasec *et al.*, 2000; Cerboni *et al.*, 2001; Magri *et al.*, 2011). SP^{UL40} mutations have recently been shown to arise in serial clinical isolates, however, their effect on function and pathogenesis is not known (Garrigue *et al.*, 2008). We identified a natural HCMV isolate (3157) from a congenital infection that has a mutation in the upstream UL40 translational initiation site (Cunningham *et al.*, 2010) that is also present in the clinical sample. SP^{UL40} in this virus retains a capacity to up-regulate gpUL18, but has lost the capacity to rescue HLA-E (Prod'homme *et al.*, 2012). HCMV strain 3157 is the first natural isolate that has been shown experimentally to lack a specific NK cell evasion function.

In conclusion

To gain a 'handle' to analyse an entity as complex as HCMV, immunomodulatory functions have been characterized individually and in isolation. To date seven HCMV genes have been defined in publications as NK cell evasion functions (Table II.8.3), yet it is clear that the picture is far from complete. Only a proportion of the genome has been systematically interrogated and

Table II.8.3 NK cell modulatory functions

Gene/locus	Comment/ligand	NK cell receptor[a]
UL18	MHC-I homologue	LIR1 (ILT2) unknown
SP^{UL40}	Upregulates HLA-E	CD94-NKG2A CD94-NKG2C
UL83	Direct binding	NKp30
UL16	MICB, ULBP1,2,4,5 (immature form) and 6	NKG2D
miR-UL112	MICB	NKG2D
UL142	MICA, ULBP3	NKG2D
UL141	PVR (CD155), Nectin-2	DNAM-1 (CD226) TACTILE (CD96) TIGIT

[a]Colour code: red and green inhibitory and activating receptors respectively.

additional HCMV-encoded functions that modulate recognized NK cell ligands are in the process of being characterized. In certain cases, multiple HCMV genes are required to control ligands for a single receptor (e.g. NKG2D, DNAM-1). On the other hand, an individual immune evasion gene (e.g. UL141, SP^{UL40}) can simultaneously impact on multiple ligands for multiple receptors. It is clear from ongoing studies that the virus has evolved to integrate its defences.

A picture is emerging of how the host immune response adapts over time to combat the NK cell evasion by HCMV. The up-regulation of HLA-E mediated by SP^{UL40} may be countered by expansion of CD94-NKG2C^{+} NK cell populations and, in some individuals, by HLA-E specific T-cells. The question naturally arises whether such adaptive responses could be exploited therapeutically, particularly because HCMV can acquire resistance rapidly *in vivo* to antiviral therapeutic agents. A remarkably high level of sequence variation exists among HCMV clinical isolates, patients are often infected with multiple strains and recombination between them is common. The biology of the virus is consistent with an inbuilt capacity to change, and thus potentially mutate to evade an effective host immune response. Indeed, *in vivo* mutations in SP^{UL40} already provides evidence of an immune evasion function being lost. During the course of lifelong human infection, HCMV may adapt its immunomodulatory functions to suit its immune environment; even to the point of dispensing with an NK cell evasion function if it no longer confers an immunological advantage for the virus.

This review has focussed on HCMV NK modulatory functions. However, the reader should be aware that many activating and inhibitory receptors detailed in Tables II.8.1–II.8.3 are not exclusive to NK cells, e.g. DNAM-1, NKG2D and LIR-1 can be found on T and myeloid cells. Consequently, HCMV NK modulatory functions can be expected to have important functions elsewhere in, and potentially even outwith the immune system.

HCMV has evolved a comprehensive strategy to evade NK cell recognition, yet *in vivo* NK cells are crucial in reaching effective control of infection. Despite its impressive counter-mechanisms, NK cells find a way to control HCMV infection. We provide evidence that this may be a dynamic relationship in which the host's NK response must *adapt* to control the virus, and in turn provides a powerful selection pressure for HCMV mutants/variants that avoid the immune control. By cataloguing and characterizing HCMV's arsenal of immune evasion functions, a better understanding of the complex mechanisms controlling human NK cell recognition is emerging.

Acknowledgements

The authors would like to thank Peter Tomasec, Richard Stanton, Ian Humphreys, Virginie Prod'homme, Veronique Braud, Marius Lemberg, Bruno Martoglio, Daniel Sugrue, Ofer Mandelboim, Hugh Reyburn and Andrew Davison for helpful discussions. Thanks to Hugh Reyburn for allowing findings to be cited prior to publication. Work in the laboratory was supported by funding from the Wellcome Trust (WT094210/Z/10/Z), BBSRC (BBF0098361) and MRC (G1000236, G0901119).

References

Akira, S., and Sato, S. (2003). Toll-like receptors and their signaling mechanisms. Scand. J. Infect. Dis. 35, 555–562.

Alexopoulou, L., Holt, A.C., Medzhitov, R., and Flavell, R.A. (2001). Recognition of double-stranded RNA and activation of NF-kappaB by Toll-like receptor 3. Nature *413*, 732–738.

Andresen, L., Jensen, H., Pedersen, M.T., Hansen, K.A., and Skov, S. (2007). Molecular regulation of MHC class I chain-related protein A expression after HDAC-inhibitor treatment of Jurkat T-cells. J. Immunol. *179*, 8235–8242.

Arnon, T.I., Achdout, H., Levi, O., Markel, G., Saleh, N., Katz, G., Gazit, R., Gonen-Gross, T., Hanna, J., Nahari, E., *et al.* (2005). Inhibition of the NKp30 activating receptor by pp65 of human cytomegalovirus. Nat. Immunol. *6*, 515–523.

Ashiru, O., Bennett, N.J., Boyle, L.H., Thomas, M., Trowsdale, J., and Wills, M.R. (2009). NKG2D ligand MICA is retained in the cis-Golgi apparatus by human cytomegalovirus protein UL142. J. Virol. *83*, 12345–12354.

Bacon, L., Eagle, R.A., Meyer, M., Easom, N., Young, N.T., and Trowsdale, J. (2004). Two human ULBP/RAET1 molecules with transmembrane regions are ligands for NKG2D. J. Immunol. *173*, 1078–1084.

Barron, M.A., Gao, D., Springer, K.L., Patterson, J.A., Brunvand, M.W., McSweeney, P.A., Zeng, C., Baron, A.E., and Weinberg, A. (2009). Relationship of reconstituted adaptive and innate cytomegalovirus (CMV)-specific immune responses with CMV viremia in hematopoietic stem cell transplant recipients. Clin. Infect. Dis. *49*, 1777–1783.

Bauer, S., Groh, V., Wu, J., Steinle, A., Phillips, J.H., Lanier, L.L., and Spies, T. (1999). Activation of NK cells and T-cells by NKG2D, a receptor for stress-inducible MICA. Science *285*, 727–729.

Beck, S., and Barrell, B.G. (1988). Human cytomegalovirus encodes a glycoprotein homologous to MHC class-I antigens. Nature *331*, 269–272.

Bennett, N.J., Ashiru, O., Morgan, F.J., Pang, Y., Okecha, G., Eagle, R.A., Trowsdale, J., Sissons, J.G., and Wills, M.R. (2010). Intracellular sequestration of the NKG2D ligand ULBP3 by human cytomegalovirus. J. Immunol. *185*, 1093–1102.

Berg, L., Riise, G.C., Cosman, D., Bergstrom, T., Olofsson, S., Karre, K., and Carbone, E. (2003). LIR-1 expression on lymphocytes, and cytomegalovirus disease in lung-transplant recipients. Lancet *361*, 1099–1101.

Biron, C.A., Su, H.C., and Orange, J.S. (1996). Function and Regulation of Natural Killer (NK) Cells during Viral Infections: Characterization of Responses *in vivo*. Methods *9*, 379–393.

Biron, C.A., Byron, K.S., and Sullivan, J.L. (1989). Severe herpesvirus infections in an adolescent without natural killer cells. N. Engl. J. Med. *320*, 1731–1735.

Biron, C.A., Nguyen, K.B., Pien, G.C., Cousens, L.P., and Salazar-Mather, T.P. (1999). Natural killer cells in antiviral defense: function and regulation by innate cytokines. Annu. Rev. Immunol. *17*, 189–220.

Bjorkstrom, N.K., Lindgren, T., Stoltz, M., Fauriat, C., Braun, M., Evander, M., Michaelsson, J., Malmberg, K.J., Klingstrom, J., Ahlm, C., *et al.* (2010). Rapid expansion and long-term persistence of elevated NK cell numbers in humans infected with hantavirus. J. Exp. Med. *208*, 13–21.

Borrego, F., Ulbrecht, M., Weiss, E.H., Coligan, J.E., and Brooks, A.G. (1998). Recognition of human histocompatibility leukocyte antigen (HLA)-E complexed with HLA class I signal sequence-derived peptides by CD94/NKG2 confers protection from natural killer cell-mediated lysis. J. Exp. Med. *187*, 813–818.

Borysiewicz, L.K., Rodgers, B., Morris, S., Graham, S., and Sissons, J.G. (1985). Lysis of human cytomegalovirus infected fibroblasts by natural killer cells: demonstration of an interferon-independent component requiring expression of early viral proteins and characterization of effector cells. J. Immunol. *134*, 2695–2701.

Bottino, C., Castriconi, R., Pende, D., Rivera, P., Nanni, M., Carnemolla, B., Cantoni, C., Grassi, J., Marcenaro, S., Reymond, N., *et al.* (2003). Identification of PVR (CD155) and Nectin-2 (CD112) as cell surface ligands for the human DNAM-1 (CD226) activating molecule. J. Exp. Med. *198*, 557–567.

Brandt, C.S., Baratin, M., Yi, E.C., Kennedy, J., Gao, Z., Fox, B., Haldeman, B., Ostrander, C.D., Kaifu, T., Chabannon, C., *et al.* (2009). The B7 family member B7-H6 is a tumor cell ligand for the activating natural killer cell receptor NKp30 in humans. J. Exp. Med. *206*, 1495–1503.

Braud, V., Jones, E.Y., and McMichael, A. (1997). The human major histocompatibility complex class Ib molecule HLA-E binds signal sequence-derived peptides with primary anchor residues at positions 2 and 9. Eur. J. Immunol. *27*, 1164–1169.

Braud, V.M., Allan, D.S., O'Callaghan, C.A., Soderstrom, K., D'Andrea, A., Ogg, G.S., Lazetic, S., Young, N.T., Bell, J.I., Phillips, J.H., *et al.* (1998). HLA-E binds to natural killer cell receptors CD94/NKG2A, B and C. Nature *391*, 795–799.

Braud, V.M., Tomasec, P., and Wilkinson, G.W. (2002). Viral evasion of natural killer cells during human cytomegalovirus infection. Curr. Top. Microbiol. Immunol. *269*, 117–129.

Brenchley, J.M., Karandikar, N.J., Betts, M.R., Ambrozak, D.R., Hill, B.J., Crotty, L.E., Casazza, J.P., Kuruppu, J., Migueles, S.A., Connors, M., *et al.* (2003). Expression of CD57 defines replicative senescence and antigen-induced apoptotic death of CD8+ T-cells. Blood *101*, 2711–2720.

Brodin, P., and Hoglund, P. (2008). Beyond licensing and disarming: a quantitative view on NK-cell education. Eur. J. Immunol. *38*, 2934–2937.

Brodin, P., Lakshmikanth, T., Johansson, S., Karre, K., and Hoglund, P. (2009). The strength of inhibitory input during education quantitatively tunes the functional responsiveness of individual natural killer cells. Blood *113*, 2434–2441.

Browne, H., Smith, G., Beck, S., and Minson, T. (1990). A complex between the MHC class I homologue encoded by human cytomegalovirus and beta 2 microglobulin. Nature *347*, 770–772.

Cerboni, C., Mousavi-Jazi, M., Linde, A., Soderstrom, K., Brytting, M., Wahren, B., Karre, K., and Carbone, E. (2000). Human cytomegalovirus strain-dependent changes in NK cell recognition of infected fibroblasts. J. Immunol. *164*, 4775–4782.

Cerboni, C., Mousavi-Jazi, M., Wakiguchi, H., Carbone, E., Karre, K., and Soderstrom, K. (2001). Synergistic effect of IFN-gamma and human cytomegalovirus protein UL40 in the HLA-E-dependent protection from NK cell-mediated cytotoxicity. Eur. J. Immunol. *31*, 2926–2935.

Chalupny, N.J., Rein-Weston, A., Dosch, S., and Cosman, D. (2006). Down-regulation of the NKG2D ligand MICA by the human cytomegalovirus glycoprotein UL142. Biochem. Biophys. Res. Commun. *346*, 175–181.

Chapman, T.L., and Bjorkman, P.J. (1998). Characterization of a murine cytomegalovirus class I major histocompatibility complex (MHC) homolog: comparison to MHC molecules and to the human cytomegalovirus MHC homolog. J. Virol. 72, 460–466.

Chapman, T.L., Heikeman, A.P., and Bjorkman, P.J. (1999). The inhibitory receptor LIR-1 uses a common binding interaction to recognize class I MHC molecules and the viral homolog UL18. Immunity *11*, 603–613.

Chapman, T.L., Heikema, A.P., West, A.P., Jr., and Bjorkman, P.J. (2000). Crystal structure and ligand binding properties of the D1D2 region of the inhibitory receptor LIR-1 (ILT2). Immunity *13*, 727–736.

Chijioke, O., and Munz, C. (2011). Interactions of human myeloid cells with natural killer cell subsets *in vitro* and *in vivo*. J. Biomed. Biotechnol. *2011*, 251679.

Chong, L.K., Aicheler, R.J., Llewellyn-Lacey, S., Tomasec, P., Brennan, P., and Wang, E.C. (2008). Proliferation and interleukin 5 production by CD8hi CD57+ T-cells. Eur. J. Immunol. *38*, 995–1000.

Chou, D.K., Ilyas, A.A., Evans, J.E., Costello, C., Quarles, R.H., and Jungalwala, F.B. (1986). Structure of sulfated glucuronyl glycolipids in the nervous system reacting with HNK-1 antibody and some IgM paraproteins in neuropathy. J. Biol. Chem. *261*, 11717–11725.

Colonna, M., Navarro, F., Bellon, T., Llano, M., Garcia, P., Samaridis, J., Angman, L., Cella, M., and Lopez-Botet, M. (1997). A common inhibitory receptor for major histocompatibility complex class I molecules on human lymphoid and myelomonocytic cells. J. Exp. Med. *186*, 1809–1818.

Cook, M., Briggs, D., Craddock, C., Mahendra, P., Milligan, D., Fegan, C., Darbyshire, P., Lawson, S., Boxall, E., and Moss, P. (2006). Donor KIR genotype has a major influence on the rate of cytomegalovirus reactivation following T-cell replete stem cell transplantation. Blood *107*, 1230–1232.

Cooper, M.A., Fehniger, T.A., and Caligiuri, M.A. (2001). The biology of human natural killer-cell subsets. Trends Immunol. *22*, 633–640.

Cosman, D., Fanger, N., Borges, L., Kubin, M., Chin, W., Peterson, L., and Hsu, M.L. (1997). A novel immunoglobulin superfamily receptor for cellular and viral MHC class I molecules. Immunity *7*, 273–282.

Cosman, D., Mullberg, J., Sutherland, C.L., Chin, W., Armitage, R., Fanslow, W., Kubin, M., and Chalupny, N.J. (2001). ULBPs, novel MHC class I-related molecules, bind to CMV glycoprotein UL16 and stimulate NK cytotoxicity through the NKG2D receptor. Immunity *14*, 123–133.

Cunningham, C., Gatherer, D., Hilfrich, B., Baluchova, K., Dargan, D.J., Thomson, M., Griffiths, P.D., Wilkinson, G.W., Schulz, T.F., and Davison, A.J. (2010). Sequences of complete human cytomegalovirus genomes from infected cell cultures and clinical specimens. J. Gen. Virol. *91*, 605–615.

Dargan, D.J., Douglas, E., Cunningham, C., Jamieson, F., Stanton, R.J., Baluchova, K., McSharry, B.P., Tomasec, P., Emery, V.C., Percivalle, E., *et al.* (2010). Sequential mutations associated with adaptation of human cytomegalovirus to growth in cell culture. J. Gen. Virol. *91*, 1535–1546.

Davis, D.M. (2009). Mechanisms and functions for the duration of intercellular contacts made by lymphocytes. Nat. Rev. Immunol. *9*, 543–555.

Davison, A.J., Akter, P., Cunningham, C., Dolan, A., Addison, C., Dargan, D.J., Hassan-Walker, A.F., Emery, V.C., Griffiths, P.D., and Wilkinson, G.W. (2003). Homology between the human cytomegalovirus RL11 gene family and human adenovirus E3 genes. J. Gen. Virol. *84*, 657–663.

Diebold, S.S., Kaisho, T., Hemmi, H., Akira, S., and Reis e Sousa, C. (2004). Innate antiviral responses by means of TLR7-mediated recognition of single-stranded RNA. Science *303*, 1529–1531.

Dunn, C., Chalupny, N.J., Sutherland, C.L., Dosch, S., Sivakumar, P.V., Johnson, D.C., and Cosman, D. (2003). Human cytomegalovirus glycoprotein UL16 causes intracellular sequestration of NKG2D ligands, protecting against natural killer cell cytotoxicity. J. Exp. Med. *197*, 1427–1439.

Eagle, R.A., Traherne, J.A., Ashiru, O., Wills, M.R., and Trowsdale, J. (2006). Regulation of NKG2D ligand gene expression. Hum. Immunol. *67*, 159–169.

Eagle, R.A., and Trowsdale, J. (2007). Promiscuity and the single receptor: NKG2D. Nat. Rev. Immunol. *7*, 737–744.

Eagle, R.A., Flack, G., Warford, A., Martinez-Borra, J., Jafferji, I., Traherne, J.A., Ohashi, M., Boyle, L.H., Barrow, A.D., Caillat-Zucman, S., *et al.* (2009a). Cellular expression, trafficking, and function of two isoforms of human ULBP5/RAET1G. PLoS One *4*, e4503.

Eagle, R.A., Traherne, J.A., Hair, J.R., Jafferji, I., and Trowsdale, J. (2009b). ULBP6/RAET1L is an additional human NKG2D ligand. Eur. J. Immunol. *39*, 3207–3216.

Fahnestock, M.L., Johnson, J.L., Feldman, R.M., Neveu, J.M., Lane, W.S., and Bjorkman, P.J. (1995). The MHC class I homolog encoded by human cytomegalovirus binds endogenous peptides. Immunity *3*, 583–590.

Falk, C.S., Mach, M., Schendel, D.J., Weiss, E.H., Hilgert, I., and Hahn, G. (2002). NK cell activity during human cytomegalovirus infection is dominated by US2–11-mediated HLA class I down-regulation. J. Immunol. *169*, 3257–3266.

Fuchs, A., Cella, M., Giurisato, E., Shaw, A.S., and Colonna, M. (2004). Cutting edge: CD96 (tactile) promotes NK cell-target cell adhesion by interacting with the poliovirus receptor (CD155). J. Immunol. *172*, 3994–3998.

Fujito, T., Ikeda, W., Kakunaga, S., Minami, Y., Kajita, M., Sakamoto, Y., Monden, M., and Takai, Y. (2005). Inhibition of cell movement and proliferation by cell–cell contact-induced interaction of Necl-5 with nectin-3. J. Cell. Biol. *171*, 165–173.

Furman, M.H., Dey, N., Tortorella, D., and Ploegh, H.L. (2002). The human cytomegalovirus US10 gene product delays trafficking of major histocompatibility complex class I molecules. J. Virol. *76*, 11753–11756.

Garrigue, I., Faure-Della Corte, M., Magnin, N., Recordon-Pinson, P., Couzi, L., Lebrette, M.E., Schrive, M.H., Roncin, L., Taupin, J.L., Dechanet-Merville, J., *et al.* (2008). UL40 human cytomegalovirus variability evolution patterns over time in renal transplant recipients. Transplantation *86*, 826–835.

Gatherer, D., Seirafian, S., Cunningham, C., Holton, M., Dargan, D.J., Baluchova, K., Hector, R.D., Galbraith, J., Herzyk, P., Wilkinson, G.W., *et al.* (2011). High-resolution human cytomegalovirus transcriptome. Proc. Natl. Acad. Sci. U.S.A. *108*, 19755–19760.

Gazit, R., Garty, B.Z., Monselise, Y., Hoffer, V., Finkelstein, Y., Markel, G., Katz, G., Hanna, J., Achdout, H., Gruda, R., *et al.* (2004). Expression of KIR2DL1 on the entire NK cell population: a possible novel immunodeficiency syndrome. Blood *103*, 1965–1966.

Gehrz, R.C., and Rutzick, S.R. (1985). Cytomegalovirus (CMV)-specific lysis of CMV-infected target cells can be mediated by both NK-like and virus-specific cytotoxic T-lymphocytes. Clin. Exp. Immunol. *61*, 80–89.

Gratama, J.W., Langelaar, R.A., Oosterveer, M.A., van der Linden, J.A., den Ouden-Noordermeer, A., Naipal, A.M., Visser, J.W., de Gast, G.C., and Tanke, H.J. (1989). Phenotypic study of CD4+ and CD8+ lymphocyte subsets in relation to cytomegalovirus carrier status and its correlate with pokeweed mitogen-induced B-lymphocyte differentiation. Clin. Exp. Immunol. 77, 245–251.

Grey, F., Meyers, H., White, E.A., Spector, D.H., and Nelson, J. (2007). A human cytomegalovirus-encoded microRNA regulates expression of multiple viral genes involved in replication. PLoS Pathog. 3, e163.

Griffin, C., Wang, E.C., McSharry, B.P., Rickards, C., Browne, H., Wilkinson, G.W., and Tomasec, P. (2005). Characterization of a highly glycosylated form of the human cytomegalovirus HLA class I homologue gpUL18. J. Gen. Virol. 86, 2999–3008.

Guma, M., Angulo, A., Vilches, C., Gomez-Lozano, N., Malats, N., and Lopez-Botet, M. (2004). Imprint of human cytomegalovirus infection on the NK cell receptor repertoire. Blood 104, 3664–3671.

Guma, M., Budt, M., Saez, A., Brckalo, T., Hengel, H., Angulo, A., and Lopez-Botet, M. (2006). Expansion of CD94/NKG2C+ NK cells in response to human cytomegalovirus-infected fibroblasts. Blood 107, 3624–3631.

Hanabuchi, S., Watanabe, N., Wang, Y.H., Ito, T., Shaw, J., Cao, W., Qin, F.X., and Liu, Y.J. (2006). Human plasmacytoid predendritic cells activate NK cells through glucocorticoid-induced tumor necrosis factor receptor-ligand (GITRL). Blood 107, 3617–3623.

Hart, O.M., Athie-Morales, V., O'Connor, G.M., and Gardiner, C.M. (2005). TLR7/8-mediated activation of human NK cells results in accessory cell-dependent IFN-gamma production. J. Immunol. 175, 1636–1642.

Hassan-Walker, A.F., Mattes, F.M., Griffiths, P.D., and Emery, V.C. (2001). Quantity of cytomegalovirus DNA in different leukocyte populations during active infection *in vivo* and the presence of gB and UL18 transcripts. J. Med. Virol. 64, 283–289.

Heil, F., Hemmi, H., Hochrein, H., Ampenberger, F., Kirschning, C., Akira, S., Lipford, G., Wagner, H., and Bauer, S. (2004). Species-specific recognition of single-stranded RNA via toll-like receptor 7 and 8. Science 303, 1526–1529.

Hengel, H., Esslinger, C., Pool, J., Goulmy, E., and Koszinowski, U.H. (1995). Cytokines restore MHC class I complex formation and control antigen presentation in human cytomegalovirus-infected cells. J. Gen. Virol. 76, 2987–2997.

Hoare, H.L., Sullivan, L.C., Pietra, G., Clements, C.S., Lee, E.J., Ely, L.K., Beddoe, T., Falco, M., Kjer-Nielsen, L., Reid, H.H., *et al.* (2006). Structural basis for a major histocompatibility complex class Ib-restricted T-cell response. Nat. Immunol. 7, 256–264.

Hoare, H.L., Sullivan, L.C., Clements, C.S., Ely, L.K., Beddoe, T., Henderson, K.N., Lin, J., Reid, H.H., Brooks, A.G., and Rossjohn, J.
(2008). Subtle changes in peptide conformation profoundly affect recognition of the non-classical MHC class I molecule HLA-E by the CD94-NKG2 natural killer cell receptors. J. Mol. Biol. 377, 1297–1303.

Hokland, M., Jacobsen, N., Ellegaard, J., and Hokland, P. (1988). Natural killer function following allogeneic bone marrow transplantation. Very early reemergence but strong dependence of cytomegalovirus infection. Transplantation 45, 1080–1084.

Iversen, A.C., Norris, P.S., Ware, C.F., and Benedict, C.A. (2005). Human NK cells inhibit cytomegalovirus replication through a noncytolytic mechanism involving lymphotoxin-dependent induction of IFN-beta. J. Immunol. 175, 7568–7574.

Joncker, N.T., Fernandez, N.C., Treiner, E., Vivier, E., and Raulet, D.H. (2009). NK cell responsiveness is tuned commensurate with the number of inhibitory receptors for self-MHC class I: the rheostat model. J. Immunol. 182, 4572–4580.

Jones, T.R., and Sun, L. (1997). Human cytomegalovirus US2 destabilizes major histocompatibility complex class I heavy chains. J. Virol. 71, 2970–2979.

Jonsson, A.H., and Yokoyama, W.M. (2009). Natural killer cell tolerance licensing and other mechanisms. Adv. Immunol. 101, 27–79.

Kaiser, B.K., Barahmand-Pour, F., Paulsene, W., Medley, S., Geraghty, D.E., and Strong, R.K. (2005). Interactions between NKG2x immunoreceptors and HLA-E ligands display overlapping affinities and thermodynamics. J. Immunol. 174, 2878–2884.

Kaiser, B.K., Pizarro, J.C., Kerns, J., and Strong, R.K. (2008). Structural basis for NKG2A/CD94 recognition of HLA-E. Proc. Natl. Acad. Sci. U.S.A. 105, 6696–6701.

Kern, F., Khatamzas, E., Surel, I., Frommel, C., Reinke, P., Waldrop, S.L., Picker, L.J., and Volk, H.D. (1999). Distribution of human CMV-specific memory T-cells among the CD8pos. subsets defined by CD57, CD27, and CD45 isoforms. Eur. J. Immunol. 29, 2908–2915.

Kiessling, R., Klein, E., and Wigzell, H. (1975). 'Natural' killer cells in the mouse. I. Cytotoxic cells with specificity for mouse Moloney leukemia cells. Specificity and distribution according to genotype. Eur. J. Immunol. 5, 112–117.

Kim, J.S., Choi, S.E., Yun, I.H., Kim, J.Y., Ahn, C., Kim, S.J., Ha, J., Hwang, E.S., Cha, C.Y., Miyagawa, S., *et al.* (2004). Human cytomegalovirus UL18 alleviated human NK-mediated swine endothelial cell lysis. Biochem. Biophys. Res. Commun. 315, 144–150.

Kim, S., Lee, S., Shin, J., Kim, Y., Evnouchidou, I., Kim, D., Kim, Y.K., Kim, Y.E., Ahn, J.H., Riddell, S.R., *et al.* (2011). Human cytomegalovirus microRNA miR-US4-1 inhibits CD8(+) T-cell responses by targeting the aminopeptidase ERAP1. Nat. Immunol. 12, 984–991.

Kim, Y., Park, B., Cho, S., Shin, J., Cho, K., Jun, Y., and Ahn, K. (2008). Human cytomegalovirus UL18 utilizes US6 for evading the NK and T-cell responses. PLoS Pathog. 4, e1000123.

Kirmani, N., Ginn, R.K., Mittal, K.K., Manischewitz, J.F., and Quinnan, G.V., Jr. (1981). Cytomegalovirus-specific cytotoxicity mediated by non-T-lymphocytes from peripheral blood of normal volunteers. Infect. Immun. 34, 441–447.

Kruse, J., Mailhammer, R., Wernecke, H., Faissner, A., Sommer, I., Goridis, C., and Schachner, M. (1984). Neural cell adhesion molecules and myelin-associated glycoprotein share a common carbohydrate moiety recognized by monoclonal antibodies L2 and HNK-1. Nature 311, 153–155.

Kubin, M., Cassiano, L., Chalupny, J., Chin, W., Cosman, D., Fanslow, W., Mullberg, J., Rousseau, A.M., Ulrich, D., and Armitage, R. (2001). ULBP1, 2, 3: novel MHC class I-related molecules that bind to human cytomegalovirus glycoprotein UL16, activate NK cells. Eur. J. Immunol. *31*, 1428–1437.

Kuijpers, T.W., Baars, P.A., Dantin, C., van den Burg, M., van Lier, R.A., and Roosnek, E. (2008). Human NK cells can control CMV infection in the absence of T-cells. Blood *112*, 914–915.

Lee, N., Llano, M., Carretero, M., Ishitani, A., Navarro, F., Lopez-Botet, M., and Geraghty, D.E. (1998). HLA-E is a major ligand for the natural killer inhibitory receptor CD94/NKG2A. Proc. Natl. Acad. Sci. U.S.A. *95*, 5199–5204.

Leong, C.C., Chapman, T.L., Bjorkman, P.J., Formankova, D., Mocarski, E.S., Phillips, J.H., and Lanier, L.L. (1998). Modulation of natural killer cell cytotoxicity in human cytomegalovirus infection: the role of endogenous class I major histocompatibility complex and a viral class I homolog. J. Exp. Med. *187*, 1681–1687.

Lopez-Verges, S., Milush, J.M., Pandey, S., York, V.A., Arakawa-Hoyt, J., Pircher, H., Norris, P.J., Nixon, D.F., and Lanier, L.L. (2010). CD57 defines a functionally distinct population of mature NK cells in the human CD56dimCD16+ NK-cell subset. Blood *116*, 3865–3874.

Lopez-Verges, S., Milush, J.M., Schwartz, B.S., Pando, M.J., Jarjoura, J., York, V.A., Houchins, J.P., Miller, S., Kang, S.M., Norris, P.J., *et al.* (2011). Expansion of a unique CD57+NKG2Chi natural killer cell subset during acute human cytomegalovirus infection. Proc. Natl. Acad. Sci. U.S.A. *108*, 14725–14732.

Lund, J.M., Alexopoulou, L., Sato, A., Karow, M., Adams, N.C., Gale, N.W., Iwasaki, A., and Flavell, R.A. (2004). Recognition of single-stranded RNA viruses by Toll-like receptor 7. Proc. Natl. Acad. Sci. U.S.A. *101*, 5598–5603.

Magri, G., Muntasell, A., Romo, N., Saez-Borderias, A., Pende, D., Geraghty, D.E., Hengel, H., Angulo, A., Moretta, A., and Lopez-Botet, M. (2011). NKp46 and DNAM-1 NK-cell receptors drive the response to human cytomegalovirus-infected myeloid dendritic cells overcoming viral immune evasion strategies. Blood *117*, 848–856.

Mazzarino, P., Pietra, G., Vacca, P., Falco, M., Colau, D., Coulie, P., Moretta, L., and Mingari, M.C. (2005). Identification of effector-memory CMV-specific T-lymphocytes that kill CMV-infected target cells in an HLA-E-restricted fashion. Eur. J. Immunol. *35*, 3240–3247.

Millo, E., Pietra, G., Armirotti, A., Vacca, P., Mingari, M.C., Moretta, L., and Damonte, G. (2007). Purification and HPLC-MS analysis of a naturally processed HCMV-derived peptide isolated from the HEK-293T/HLA-E+/Ul40+ cell transfectants and presented at the cell surface in the context of HLA-E. J. Immunol. Methods *322*, 128–136.

Muller, S., Zocher, G., Steinle, A., and Stehle, T. (2010). Structure of the HCMV UL16–MICB complex elucidates select binding of a viral immunoevasin to diverse NKG2D ligands. PLoS Pathog. *6*, e1000723.

Nachmani, D., Lankry, D., Wolf, D.G., and Mandelboim, O. (2010). The human cytomegalovirus microRNA miR-UL112 acts synergistically with a cellular microRNA to escape immune elimination. Nat. Immunol. *11*, 806–813.

Occhino, M., Ghiotto, F., Soro, S., Mortarino, M., Bosi, S., Maffei, M., Bruno, S., Nardini, M., Figini, M., Tramontano, A., *et al.* (2008). Dissecting the structural determinants of the interaction between the human cytomegalovirus UL18 protein and the CD85j immune receptor. J. Immunol. *180*, 957–968.

Odeberg, J., Cerboni, C., Browne, H., Karre, K., Moller, E., Carbone, E., and Soderberg-Naucler, C. (2002). Human cytomegalovirus (HCMV)-infected endothelial cells and macrophages are less susceptible to natural killer lysis independent of the down-regulation of classical HLA class I molecules or expression of the HCMV class I homologue, UL18. Scand. J. Immunol. *55*, 149–161.

Orange, J.S. (2002). Human natural killer cell deficiencies and susceptibility to infection. Microbes Infect. *4*, 1545–1558.

Park, B., Oh, H., Lee, S., Song, Y., Shin, J., Sung, Y.C., Hwang, S.Y., and Ahn, K. (2002). The MHC class I homolog of human cytomegalovirus is resistant to down-regulation mediated by the unique short region protein (US)2, US3, US6, and US11 gene products. J. Immunol. *168*, 3464–3469.

Park, B., Spooner, E., Houser, B.L., Strominger, J.L., and Ploegh, H.L. (2010). The HCMV membrane glycoprotein US10 selectively targets HLA-G for degradation. J. Exp. Med. *207*, 2033–2041.

Paust, S., and von Andrian, U.H. (2011). Natural killer cell memory. Nat. Immunol. *12*, 500–508.

Paust, S., Gill, H.S., Wang, B.Z., Flynn, M.P., Moseman, E.A., Senman, B., Szczepanik, M., Telenti, A., Askenase, P.W., Compans, R.W., *et al.* (2010). Critical role for the chemokine receptor CXCR6 in NK cell-mediated antigen-specific memory of haptens and viruses. Nat. Immunol. *11*, 1127–1135.

Pietra, G., Romagnani, C., Mazzarino, P., Falco, M., Millo, E., Moretta, A., Moretta, L., and Mingari, M.C. (2003). HLA-E-restricted recognition of cytomegalovirus-derived peptides by human CD8+ cytolytic T-lymphocytes. Proc. Natl. Acad. Sci. U.S.A. *100*, 10896–10901.

Pogge von Strandmann, E., Simhadri, V.R., von Tresckow, B., Sasse, S., Reiners, K.S., Hansen, H.P., Rothe, A., Böll, B., Simhadri, V.L., Borchmann, P., *et al.* (2007). Human leukocyte antigen-B-associated transcript 3 Is Released from tumor cells and engages the NKp30 receptor on natural killer cells. Immunity *27*, 965–974.

Pourgheysari, B., Khan, N., Best, D., Bruton, R., Nayak, L., and Moss, P.A. (2007). The cytomegalovirus-specific CD4+ T-cell response expands with age and markedly alters the CD4+ T-cell repertoire. J. Virol. *81*, 7759–7765.

Prod'homme, V., Griffin, C., Aicheler, R.J., Wang, E.C., McSharry, B.P., Rickards, C.R., Stanton, R.J., Borysiewicz, L.K., Lopez-Botet, M., Wilkinson, G.W., *et al.* (2007). The human cytomegalovirus MHC class I homolog UL18 inhibits LIR-1+ but activates LIR-1- NK cells. J. Immunol. *178*, 4473–4481.

Prod'homme, V., Sugrue, D.M., Stanton, R.J., Nomoto, A., Davies, J., Rickards, C.R., Cochrane, D., Moore, M., Wilkinson, G.W., and Tomasec, P. (2010). Human cytomegalovirus UL141 promotes efficient down-regulation of the natural killer cell activating ligand CD112. J. Gen. Virol. *91*, 2034–2039.

Prod'homme, V., Tomasec, P., Cunningham, C., Lemberg, M.K., Stanton, R.J., McSharry, B.P., Wang, E.C., Cuff, S., Martoglio, B., Davison, A.J., *et al.* (2012). Human cytomegalovirus UL40 signal peptide regulates cell surface expression of the NK cell ligands HLA-E and gpUL18. J. Immunol. *188*, 2794–2804.

Reyburn, H.T., Mandelboim, O., Vales-Gomez, M., Davis, D.M., Pazmany, L., and Strominger, J.L. (1997). The class I MHC homologue of human cytomegalovirus inhibits attack by natural killer cells. Nature *386*, 514–517.

Rolle, A., Mousavi-Jazi, M., Eriksson, M., Odeberg, J., Soderberg-Naucler, C., Cosman, D., Karre, K., and Cerboni, C. (2003). Effects of human cytomegalovirus infection on ligands for the activating NKG2D receptor of NK cells: up-regulation of UL16-binding protein (ULBP)1 and ULBP2 is counteracted by the viral UL16 protein. J. Immunol. *171*, 902–908.

Shinomiya, N., Koike, Y., Koyama, H., Takayama, E., Habu, Y., Fukasawa, M., Tanuma, S., and Seki, S. (2005). Analysis of the susceptibility of CD57 T-cells to CD3-mediated apoptosis. Clin. Exp. Immunol. *139*, 268–278.

da Silva, R.B., and Munz, C. (2011). Natural killer cell activation by dendritic cells: balancing inhibitory and activating signals. Cell. Mol. Life Sci. *68*, 3505–3518.

Stanietsky, N., and Mandelboim, O. (2010). Paired NK cell receptors controlling NK cytotoxicity. FEBS Lett. *584*, 4895–4900.

Stanietsky, N., Simic, H., Arapovic, J., Toporik, A., Levy, O., Novik, A., Levine, Z., Beiman, M., Dassa, L., Achdout, H., *et al.* (2009). The interaction of TIGIT with PVR and PVRL2 inhibits human NK cell cytotoxicity. Proc. Natl. Acad. Sci. U.S.A. *106*, 17858–17863.

Stanton, R.J., Baluchova, K., Dargan, D.J., Cunningham, C., Sheehy, O., Seirafian, S., McSharry, B.P., Neale, M.L., Davies, J.A., Tomasec, P., *et al.* (2010). Reconstruction of the complete human cytomegalovirus genome in a BAC reveals RL13 to be a potent inhibitor of replication. J. Clin. Invest. *120*, 3191–3208.

Starr, S.E., and Garrabrant, T. (1981). Natural killing of cytomegalovirus-infected fibroblasts by human mononuclear leucocytes. Clin. Exp. Immunol. *46*, 484–492.

Stern-Ginossar, N., Elefant, N., Zimmermann, A., Wolf, D.G., Saleh, N., Biton, M., Horwitz, E., Prokocimer, Z., Prichard, M., Hahn, G., *et al.* (2007). Host immune system gene targeting by a viral miRNA. Science *317*, 376–381.

Stern-Ginossar, N., Gur, C., Biton, M., Horwitz, E., Elboim, M., Stanietsky, N., Mandelboim, M., and Mandelboim, O. (2008). Human microRNAs regulate stress-induced immune responses mediated by the receptor NKG2D. Nat. Immunol. *9*, 1065–1073.

Sullivan, L.C., Clements, C.S., Beddoe, T., Johnson, D., Hoare, H.L., Lin, J., Huyton, T., Hopkins, E.J., Reid, H.H., Wilce, M.C., *et al.* (2007). The heterodimeric assembly of the CD94-NKG2 receptor family and implications for human leukocyte antigen-E recognition. Immunity *27*, 900–911.

Sun, J.C., Lopez-Verges, S., Kim, C.C., DeRisi, J.L., and Lanier, L.L. (2011). NK cells and immune 'memory'. J. Immunol. *186*, 1891–1897.

Sutherland, C.L., Chalupny, N.J., and Cosman, D. (2001). The UL16-binding proteins, a novel family of MHC class I-related ligands for NKG2D, activate natural killer cell functions. Immunol. Rev. *181*, 185–192.

Takai, Y., Miyoshi, J., Ikeda, W., and Ogita, H. (2008). Nectins and nectin-like molecules: roles in contact inhibition of cell movement and proliferation. Nat. Rev. Mol. Cell. Biol. *9*, 603–615.

Thong, Y.H., Hensen, S.A., Vincent, M.M., Fuccillo, D.A., Stiles, W.A., and Bellanti, J.A. (1976). Use of cryopreserved virus-infected target cells in a lymphocytotoxicity 51Cr release microassay for cell-mediated immunity to cytomegalovirus. Infect. Immun. *13*, 643–645.

Tomasec, P., Braud, V.M., Rickards, C., Powell, M.B., McSharry, B.P., Gadola, S., Cerundolo, V., Borysiewicz, L.K., McMichael, A.J., and Wilkinson, G.W. (2000). Surface expression of HLA-E, an inhibitor of natural killer cells, enhanced by human cytomegalovirus gpUL40. Science *287*, 1031.

Tomasec, P., Wang, E.C., Davison, A.J., Vojtesek, B., Armstrong, M., Griffin, C., McSharry, B.P., Morris, R.J., Llewellyn-Lacey, S., Rickards, C., *et al.* (2005). Down-regulation of natural killer cell-activating ligand CD155 by human cytomegalovirus UL141. Nat. Immunol. *6*, 181–188.

Ulbrecht, M., Martinozzi, S., Grzeschik, M., Hengel, H., Ellwart, J.W., Pla, M., and Weiss, E.H. (2000). Cutting edge: the human cytomegalovirus UL40 gene product contains a ligand for HLA-E and prevents NK cell-mediated lysis. J. Immunol. *164*, 5019–5022.

Vales-Gomez, M., Browne, H., and Reyburn, H.T. (2003). Expression of the UL16 glycoprotein of Human Cytomegalovirus protects the virus-infected cell from attack by natural killer cells. BMC Immunol. *4*, 4.

Venema, H., van den Berg, A.P., van Zanten, C., van Son, W.J., van der Giessen, M., and The, T.H. (1994). Natural killer cell responses in renal transplant patients with cytomegalovirus infection. J. Med. Virol. *42*, 188–192.

Venkataraman, G.M., Suciu, D., Groh, V., Boss, J.M., and Spies, T. (2007). Promoter region architecture and transcriptional regulation of the genes for the MHC class I-related chain A and B ligands of NKG2D. J. Immunol. *178*, 961–969.

Vivier, E., Tomasello, E., Baratin, M., Walzer, T., and Ugolini, S. (2008). Functions of natural killer cells. Nat. Immunol. *9*, 503–510.

Wagner, C.S., Rolle, A., Cosman, D., Ljunggren, H.G., Berndt, K.D., and Achour, A. (2007). Structural elements underlying the high binding affinity of human cytomegalovirus UL18 to leukocyte immunoglobulin-like receptor-1. J. Mol. Biol. *373*, 695–705.

Wagner, C.S., Walther-Jallow, L., Buentke, E., Ljunggren, H.G., Achour, A., and Chambers, B.J. (2008). Human cytomegalovirus-derived protein UL18 alters the phenotype and function of monocyte-derived dendritic cells. J. Leukoc. Biol. *83*, 56–63.

Waner, J.L., and Nierenberg, J.A. (1985). Natural killing (NK) of cytomegalovirus (CMV)-infected fibroblasts: a comparison between two strains of CMV, uninfected fibroblasts, and K562 cells. J. Med. Virol. *16*, 233–244.

Wang, E.C., Taylor-Wiedeman, J., Perera, P., Fisher, J., and Borysiewicz, L.K. (1993). Subsets of CD8+, CD57+ cells in normal, healthy individuals: correlations with human cytomegalovirus (HCMV) carrier status, phenotypic and functional analyses. Clin. Exp. Immunol. *94*, 297–305.

Wang, E.C., Moss, P.A., Frodsham, P., Lehner, P.J., Bell, J.I., and Borysiewicz, L.K. (1995). CD8highCD57+ T-lymphocytes in normal, healthy individuals are oligoclonal and respond to human cytomegalovirus. J. Immunol. *155*, 5046–5056.

Wang, E.C., McSharry, B., Retiere, C., Tomasec, P., Williams, S., Borysiewicz, L.K., Braud, V.M., and Wilkinson, G.W. (2002). UL40-mediated NK evasion during productive infection with human cytomegalovirus. Proc. Natl. Acad. Sci. U.S.A. *99*, 7570–7575.

Warren, A.P., Ducroq, D.H., Lehner, P.J., and Borysiewicz, L.K. (1994). Human cytomegalovirus-infected cells have unstable assembly of major histocompatibility complex class I complexes and are resistant to lysis by cytotoxic T-lymphocytes. J. Virol. *68*, 2822–2829.

Weekes, M.P., Wills, M.R., Mynard, K., Hicks, R., Sissons, J.G., and Carmichael, A.J. (1999). Large clonal expansions of human virus-specific memory cytotoxic T-lymphocytes within the CD57+ CD28- CD8+ T-cell population. Immunology *98*, 443–449.

Welte, S.A., Sinzger, C., Lutz, S.Z., Singh-Jasuja, H., Sampaio, K.L., Eknigk, U., Rammensee, H.G., and Steinle, A. (2003). Selective intracellular retention of virally induced NKG2D ligands by the human cytomegalovirus UL16 glycoprotein. Eur. J. Immunol. *33*, 194–203.

Wiertz, E.J., Tortorella, D., Bogyo, M., Yu, J., Mothes, W., Jones, T.R., Rapoport, T.A., and Ploegh, H.L. (1996). Sec61-mediated transfer of a membrane protein from the endoplasmic reticulum to the proteasome for destruction. Nature *384*, 432–438.

Willcox, B.E., Thomas, L.M., and Bjorkman, P.J. (2003). Crystal structure of HLA-A2 bound to LIR-1, a host and viral major histocompatibility complex receptor. Nat. Immunol. *4*, 913–919.

Wills, M.R., Ashiru, O., Reeves, M.B., Okecha, G., Trowsdale, J., Tomasec, P., Wilkinson, G.W., Sinclair, J., and Sissons, J.G. (2005). Human cytomegalovirus encodes an MHC class I-like molecule (UL142) that functions to inhibit NK cell lysis. J. Immunol. *175*, 7457–7465.

Wittenbrink, M., Spreu, J., and Steinle, A. (2009). Differential NKG2D binding to highly related human NKG2D ligands ULBP2 and RAET1G is determined by a single amino acid in the alpha2 domain. Eur. J. Immunol. *39*, 1642–1651.

Wronska, D., Jones, J., Browne, H., Wilkinson, G., Minson, A.C., Sissons, J.G.P., and Borysiewicz, L.K. (1993). The human MHC class I homologue induces an allospecific T-cell response. In Multidisciplinary Approach to Understanding Cytomegalovirus Disease, Michelson, S., and Plotkin, S.A., eds. (Elsevier Science Publishers, Amsterdam), pp. 321–326.

Wu, J., Chalupny, N.J., Manley, T.J., Riddell, S.R., Cosman, D., and Spies, T. (2003). Intracellular retention of the MHC class I-related chain B ligand of NKG2D by the human cytomegalovirus UL16 glycoprotein. J. Immunol. *170*, 4196–4200.

Yang, Z., and Bjorkman, P.J. (2008). Structure of UL18, a peptide-binding viral MHC mimic, bound to a host inhibitory receptor. Proc. Natl. Acad. Sci. U.S.A. *105*, 10095–10100.

Yokoyama, W.M., and Kim, S. (2006). Licensing of natural killer cells by self-major histocompatibility complex class I. Immunol. Rev. *214*, 143–154.

Zou, Y., Bresnahan, W., Taylor, R.T., and Stastny, P. (2005). Effect of human cytomegalovirus on expression of MHC class I-related chains A. J. Immunol. *174*, 3098–3104.

Innate Immunity to Cytomegalovirus in the Murine Model

Silvia Vidal, Astrid Krmpotić, Michal Pyzik and Stipan Jonjić

Abstract

Cells of the innate immune system, including macrophages, DCs and NK cells play an important role in the control of viral infection before the induction of a specific immune response, of which generation they are also crucial. The infection of mice with MCMV as a model of HCMV infection has been particularly informative in elucidating the role of innate and adaptive immune response mechanisms during infection. NK cells are considered the most important effector cells in early CMV surveillance. An evolutionary struggle between NK cells and CMVs can be inferred from the existence of a broad range of viral mechanisms designed to compromise NK-cell function. This chapter describes major innate immune response mechanisms involved in control of MCMV, with an emphasis on NK-cell-mediated viral detection as well as virally encoded immune evasion mechanisms.

Introduction

The immune system protects the host from numerous external and internal threats, such as pathogenic microorganisms or tumours. In order to mount an appropriate response against them, a plethora of leucocytes, epithelial cells, and soluble factors work in concerted fashion to recognize, contain, and possibly eradicate such threats. Innate immunity is the first and most ancient line of defence; it is immediate, but also non-specific. Indeed, innate effector cells, such as macrophages, dendritic cells (DCs), and natural killer (NK) cells, recognize structurally conserved, self or non-self molecular patterns with the help of their germline encoded receptors. However, it has recently come to light that infection with certain pathogens can induce an oddly 'specific' innate immune response. The case of mouse cytomegalovirus (MCMV) is a particularly cogent example. In fact, mouse NK cells can specifically detect MCMV infection.

Methods to study the mouse host

Over the course of millions of years of co-evolution, CMV has developed various evasion mechanisms against the immune responses devised by its vertebrate host. These have been studied from a balanced viewpoint or one focusing either on the host or the pathogen.

A mouse model is particularly well suited to the study of host resistance to CMV. Indeed, enough genetic diversity exists between the various, phylogenetically diverse, inbred mouse strains to allow a differential response to MCMV infection (Vidal *et al.*, 2008). Moreover, they can be tested *ad infinitum* to dissect virus pathogenesis as well as their own immune response to the infection. If more genetic variability is required, wild-derived mouse strains or wild MCMV strains are also available (see Chapter I.2). Two major approaches are used to identify host genes involved in the immune response against MCMV: forward and reverse genetics (for viral genes see Chapter I.3).

Forward genetics

Forward genetics was the first approach used to dissect the genetic basis of host resistance; it can be carried out in two ways. In the first method, investigators begin by infecting a number of naturally different mouse strains with MCMV in an attempt to identify those whose response to the infection sets them apart from the other strains (i.e. a unique pathology or an extreme phenotype). Common variables used to assess the immune response in mice following infection include: viral titres in target organs at different time-points p.i., serum cytokine levels, survival rates, and weight loss. The strain under question (strain A) is then crossed with another, well-characterized strain (strain B) in order to dissect the genetic basis of the observed phenotype; this entails determining the inheritance pattern of trait (dominant vs. recessive) as well as its complexity (monogenic vs. polygenic). This is followed by a genome-wide scan or quantitative trait loci analysis to identify and delimit the interval(s) of interest; additional crosses or exome sequencing can later on be used to further reduce the interval. Once the gene(s) underlying the phenotype is identified, strain A mice can be backcrossed to strain B mice until a pure background is achieved, creating a congenic mouse strain or chromosome substitution strain (i.e. differing only in one known locus or chromosome, respectively). This allows investigators to study the differential response of these mice in greater detail. Numerous genes and gene families have been identified in the aforementioned manner. Among the first MCMV susceptibility loci identified, we can name the *H-2* locus on mouse chromosome 17 or the non-*H-2* locus on mouse chromosome 6 (now known as the natural killer cell complex or NKC) (Grundy *et al.*, 1981; Shellam *et al.*, 1981).

Natural variation in mice is limited; that is why researchers have turned to chemically induced mutagenesis using different methods like the alkylating agent N-ethyl-*N*-nitrosourea (ENU). ENU introduces random mutations into the host genome; in theory any gene can be targeted without prior knowledge of the gene function. Thereby, the spectrum of variation is greatly expanded even to those mutations that might not be tolerated in natural settings. To start, the mutagenizing agent is administered to male mice before they are crossed to WT females of the same strain. The subsequent second and third generation of mice that carry heterozygous or homozygous ENU-induced mutations are then infected with MCMV to screen for mice that present new resistance or susceptibility phenotypes. Once a mutant is identified, positional cloning or, more recently, exome sequencing are used to isolate the gene mutation underlying the phenotype. The function of several genes was revealed by ENU mutagenesis, such as those involved in TLR signalling (e.g. *Lps2* mutant) (Hoebe *et al.*, 2003).

Reverse genetics

In this second approach, knock-out, knock-in, conditional or inducible transgenic mice are generated by the deletion or insertion of a given gene through homologous recombination. Transgenic animals are subsequently infected with MCMV to assess the effect of the presence or absence of the gene in question on their response to infection (see also Chapter II.12). Excellent examples of transgenic mice include animals engineered to express a particular subset of activating NK cell receptors, specific H-2 molecules or even viral genes (Lee *et al.*, 2003; Tripathy *et al.*, 2008; Xie *et al.*, 2010; Fodil-Cornu *et al.*, 2011). However, although this approach can yield rich data, it requires prior knowledge about the gene to be inserted or deleted and may cause embryonic lethality. Other limitations include, in the case of deletions, absence of phenotype due to redundancy or compensatory mechanisms; in the case of insertions, the insertion site or the strength of expression can be problematic, as well as the possible modulation of the phenotype due to background effects. Therefore, while reverse genetics is an excellent validation tool, forward genetics allows novel gene identification in an unbiased manner and without prior assumptions.

Innate immunity

Myeloid cells

Several innate immune cells are involved in the response against MCMV infection, namely, DCs, monocytes/macrophages, and $\gamma\delta$ T-cells (Fig. II.9.1). However, myeloid cells such as DCs and monocytes/macrophages are the primary sensors of MCMV infection via a number of innate immune receptors.

DCs constitute a heterogeneous cell population. They are dispersed throughout the body and are mainly known for their potent, professional antigen-presenting capacities, which are essential for the induction of immune responses and tolerance (Steinman, 1991). Based on their origin and the expression of surface proteins, DCs can be divided into several subsets: plasmocytoid DCs (pDCs), of lymphoid origin, and conventional DCs (cDCs), of myeloid origin. The latter group can be further subdivided into CD8$\alpha\alpha^+$ DCs and CD8$\alpha\alpha^-$CD11b$^+$CD4$^{+/-}$ DCs (Ardavin *et al.*, 2001; Arase *et al.*, 2002; Shortman and Liu, 2002). DCs are found in lymphoid organs as well as peripheral tissues like the skin, lungs, liver, kidneys, and intestinal tract. Immature DCs have a great capacity for endocytosis and antigen degradation, but they express low levels of MHC-I, MHC-II, and co-stimulatory molecules (Cooper *et al.*, 2004). Therefore, they make poor antigen-presenting cells. As for pDCs, they are known as the main producers of IFN-α/β during viral infection (Asselin-Paturel *et al.*, 2001). Although CD8$\alpha\alpha^+$ DCs can also produce type I IFNs (IFN-I), their specialty lies in the presentation of exogenous antigens within the context of MHC-I molecules (cross-presentation), which is of particular importance for CD8$^+$ T-cell priming (den Haan *et al.*, 2000; Pooley *et al.*, 2001; Iyoda *et al.*, 2002). In contrast, CD8$^-$CD11b$^+$ DCs have been shown to produce IL-12p70, and IFN-γ; they are more effective in stimulating CD4$^+$ T-cells (Hochrein *et al.*, 2001).

DC activation is primarily mediated by TLR ligation and proinflammatory cytokines such as TNF-α

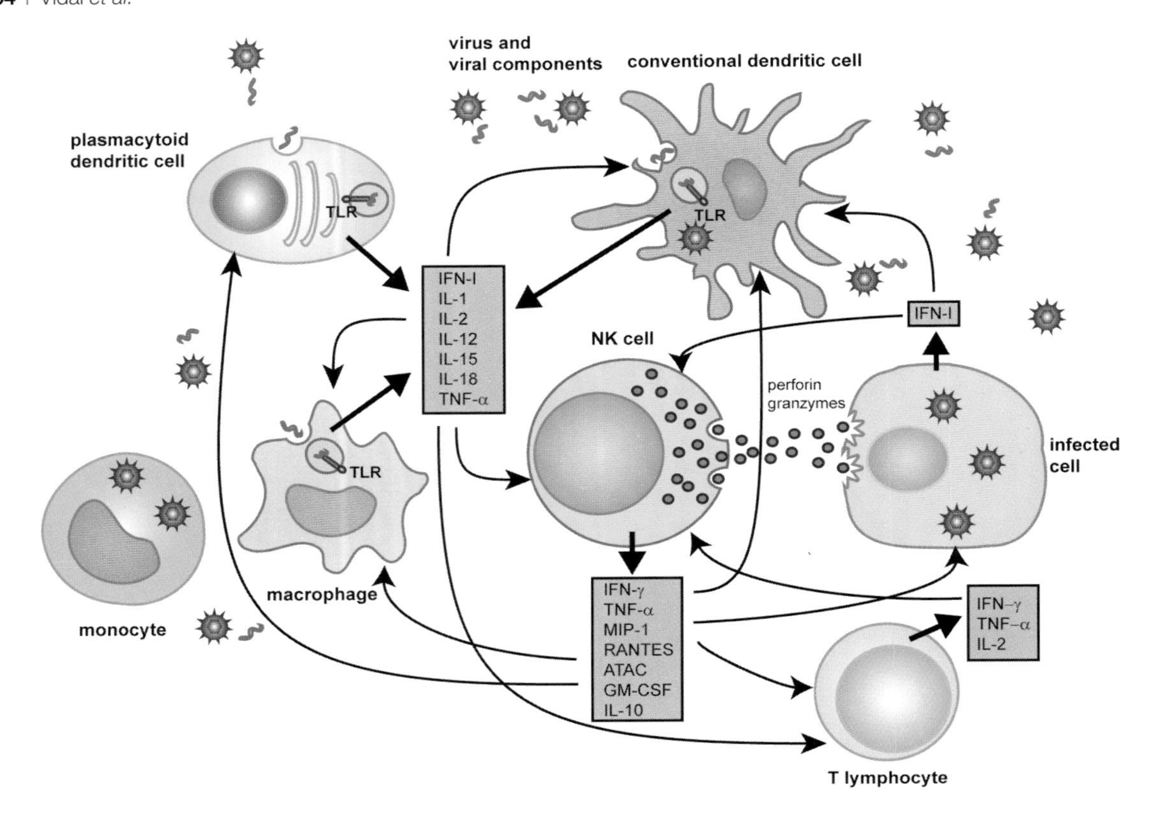

Figure II.9.1 Cellular and soluble components of innate immune response to MCMV infection. Following MCMV infection, DCs and macrophages secrete chemokines and cytokines, including IFN-I, which are important for the activation of NK cells. In turn, NK cells are important for protection of DCs. NK cells exert their antiviral activity through secretion of different cytokines and by direct cytolysis of infected cells. Both DCs and NK cells play a role in induction and regulation of subsequent adaptive immune response. MIP-1, macrophage inflammatory protein-1; RANTES, regulated upon activation, normal T-cell expressed, and secreted, also known as CCL5; ATAC, activation-induced, T-cell-derived and chemokine-related molecule, also known as lymphotactin.

and IL-1β (Zitvogel, 2002). These signals, together with the engagement of the CD40 receptor, stimulate DC maturation and migration to secondary lymphoid organs. For their part, mature DCs are less endocytotic, but become more potent in antigen presentation to T-lymphocytes. Besides their potent antigen-presenting capabilities, which enable the initiation of adaptive immunity, and secretion of IFN-α/β (Krug *et al.*, 2002), DCs produce proinflammatory cytokines (IL-2, IL-12, IL-15, IL-18) needed for optimal NK-cell activity and for the regulation of adaptive immune responses (Zitvogel, 2002) (Fig. II.9.1). CMVs can infect cDCs and modulate their function (Riegler *et al.*, 2000; Andrews *et al.*, 2001; Raftery *et al.*, 2001; Dalod *et al.*, 2003). In infected DCs, antigen uptake and degradation, as well as maturation and migration, is impaired. Infection also reduces the expression of MHC-I and co-stimulatory molecules, as well as cytokine secretion (Loewendorf *et al.*, 2011). Notably, DC and NK cell cross-talk was demonstrated during the earliest stages of MCMV infection (Walzer *et al.*, 2005).

Monocytes are also an important component of innate immunity against CMVs (Gordon, 1998). Both HCMV and MCMV use monocytes for their dissemination throughout the body (Collins *et al.*, 1994; Stoddart *et al.*, 1994; Hanson *et al.*, 1999; Noda *et al.*, 2006). Monocytes are recruited to sites of inflammation upon sensing a chemokine gradient (CCL2, CCL7, and CCL12), molecules whose expression is stimulated by infection. Once inside the tissue, these monocytes differentiate into macrophages or DCs. Macrophages produce inflammatory cytokines, possess a great ability for pathogen recognition, and play a role in antigen presentation. They are able to identify foreign antigens via TLRs and other innate receptors. In addition to phagocytosis and antigen presentation, macrophages also influence the development of innate and acquired immune responses by their secretion of pro-inflammatory cytokines such as TNF-α, IL-1 and IL-12. When infected with MCMV, their expression of MHC-II molecules decreases making them poor antigen-presenting cells (Hengel *et al.*, 2000). MHC-II down-regulation is probably triggered by the immunosuppressive cytokine IL-10, which is secreted in response to MCMV infection (Redpath *et al.*, 1999).

The type I IFN response

Effective control of MCMV replication has long been associated with an increase in type I IFN secretion in the spleen, liver, and serum during the first 48 h p.i. (Orange and Biron, 1996). Type I IFNs can be produced by almost all nucleated cells; they belong to a large cytokine family, which includes type II IFN (IFN-γ) and type III IFNs (IFN-λ1, IFN-λ2, IFN-λ3). Type I IFNs themselves include 16 members to date: IFN-β, IFN-ε, IFN-κ, IFN-ω and 12 subtypes of IFN-α (Borden *et al.*, 2007). Thus far, virus-infected cells have been shown to preferentially produce IFN-α, IFN-β, and IFN-ω. All type I IFNs bind the same heterodimeric receptor complex, made up of interferon alpha receptor 1 and 2 (IFNAR1 and IFNAR2). Following receptor engagement, Janus family kinases, namely Tyk2 and Jak1, become activated and phosphorylate tyrosine residues on the cytoplasmic tail of the receptor. This recruits STAT1 to the receptor, where it is also phosphorylated by Tyk2 and Jak1, along with the receptor-bound STAT2. Phosphorylated STAT1 and STAT2 dimerize before migrating to the nucleus, where they activate the transcription of IFN-inducible genes with the help of some other transcription factors (TFs) (see also Chapter I.16).

Upon intraperitoneal infection, the virus spread in the spleen begins by infecting cells of the marginal zone as early as 6 h p.i.; by 17 h p.i., it has spread to the red pulp. Although the spread is hastened during intravenous infection virus dissemination is similar. This viral localization coincides with the first wave of IFN secretion, though not much is known about the series of molecular events that leads from one to the other (see also Chapter II.11).

Type I IFN secretion is biphasic. The first wave peaks at 8 h p.i., while the second does so at approximately 36 h p.i. Each wave is associated with a particular cell type that recognizes the infection via specific receptors.

First wave of IFN-α/β secretion

From the evidence available, investigators were able to implicate lymphotoxin α and lymphotoxin β in this first wave of IFN-I secretion. These lymphotoxins are surface molecules belonging to the TNF family; they are involved in proper structural development of lymphoid tissues (Ware, 2005). Upon ligation with their cognate receptor, lymphotoxin β receptor (LTβR), the non-canonical NF-κB pathway is activated in the LTβR-bearing cell. Lymphotoxin α/β (LTα/β) expression is restricted to activated T-lymphocytes, NK cells, and a subset of follicular B-cells, whereas LTβR expression can been detected on a variety of cell types, including follicular DCs, high endothelial venules (HEVs), conventional DCs (cDCs), and macrophages (Ware, 2005). The importance of LTα/β for type I IFN production was first suggested by the increased sensitivity of LTα/β deficient mice to MCMV infection (Benedict *et al.*, 2001; Banks *et al.*, 2005). Furthermore, Schneider *et al.*, (2008) demonstrated that it was the interaction between LTα/β$^+$ B-cells and LTβR$^+$ MCMV-infected stromal cells that led to production of type I IFNs in the early hours following infection.

Second wave of IFN-α/β secretion

The second wave of type I IFN production is better characterized than the first, as precise causality has already been established. Indeed, this second wave mainly rests upon the recognition of viral PAMPs by TLRs expressed on pDCs. These pDCs (also known as natural IFN-producing cells) are morphologically similar to plasma cells; they represent ~ 1% of peripheral DCs, yet they secrete most of type I IFN following viral infection. They preferentially express the transcription factor IRF7 as well as toll-like receptors 7 and 9 (Krug *et al.*, 2004). Immediately after intraperitoneal infection, pDCs, which are not infected, probably endocytose virus infected apoptotic cells or virions and become activated through TLR9 and release IFN-α/β and IL-12. IFN-α/β increases the resistance of bystander uninfected cells to infection and stimulates NK-cell cytotoxicity, while IL-12 stimulates the secretion of IFN-γ. Reciprocally, activated NK cells promote maturation of activated DCs by TNF-α and IFN-γ.

Altogether, it appears that DCs enhance the antiviral activity of NK cells in the earliest stage of infection, mainly by providing IFN-α/β and IL-12 (Andoniou *et al.*, 2005; Andrews *et al.*, 2005). Effective virus control by NK cells limits the production of IFN-α/β and IL-12 by pDCs, which, if released in large amounts, can be immunosuppressive due to the loss of CD8αα$^+$ DCs in the spleen (Robbins *et al.*, 2007).

Receptors of innate immunity

Pathogen-associated molecular pattern (PAMP) recognition is critical in the early hours following MCMV infection. To do so, the host immune system relies almost exclusively on germline encoded pattern-recognition receptors (PRRs) to detect the virus-derived PAMPs. Toll-like receptors (TLRs), C-type lectin like receptors (CLRs), 'retinoic acid-inducible gene' (RIG)-I-like receptors (RLR), and the 'nucleotide-binding domain, leucine-rich repeat'-containing (NBD-LRR) proteins receptors (NLRs) are employed as PRRs. Virus-derived PAMP recognition by PRRs results in the early production of type I interferons (IFN-I: IFN-α/β), IL-12, IL-1β, IL-6, IL-18, and IL-15, as well as numerous other cytokines and chemokines that build on each other in order to limit the initial wave of viral replication until more targeted immune responses develop.

TLR sensing of MCMV infection

TLRs are type I transmembrane proteins involved in the recognition of conserved structures of fungal, bacterial, viral, parasitic or self origin. Up to 13 TLR genes have been identified in mammals (Casanova *et al.*, 2011). Their numerous ligands include components of the Gram-negative bacterial cell wall LPS (TLR4), Gram-positive bacterial cell wall, peptidoglycan (TLR2), bacterial flagellin (TLR5), multiple diacyl or triacyl lipopeptides (TLR6 or TLR1), double stranded RNA (TLR3), single-stranded RNA (TLR7), double stranded DNA, including CpG-rich DNA (TLR9) and others (Beutler, 2009; Blasius and Beutler, 2010). TLR receptors are expressed in different cell compartments: TLR1, TLR2, TLR4, TLR5, TLR6 and TLR11 are expressed at the cell surface while TLR3, TLR7, TLR8 and TLR9 are expressed intracellularly in the membranes of endosomes and the ER.

TLRs possess an extracellular 'leucine-rich repeat' domain (LRR), as well as an intracellular 'TLR and interleukin-1 receptor' (TIR) domain. Because of their complexity, the signalling pathways used by TLR receptors are usually divided into 'myeloid differentiation primary response 88' (MyD88)/'Toll-interleukin 1 receptor domain containing adapter protein' (TIRAP)-dependent, which is used by all TLRs except TLR3, and 'TIR-domain-containing adapter-inducing interferon-β' (TRIF)-dependent, which is used by TLR3 and TLR4 (TLR4 actually uses both signal paths), leading to the activation of the transcription factors IRF3, IRF7, and NF-κB (Barbalat *et al.*, 2009; Dietrich *et al.*, 2010). All of these transcription factors travel to the nucleus, where they promote the expression of pro-inflammatory cytokines such as IL-6, IL-1β, TNFα, and, most importantly, type I IFNs.

The role of TLR3, TLR9, MyD88 or TRIF in the innate immune response against MCMV infection has been well demonstrated. In one study, TLR3-deficient mice were found to be more susceptible to MCMV infection than WT mice, suggesting that dsRNA may be formed during MCMV infection. Moreover, ENU-induced mutation in *Lps2/Trif* showed that *Trif* mutant mice do not recognize dsRNA and are significantly more susceptible to MCMV infection (Hoebe *et al.*, 2003). In a similar fashion, survival of TLR9- or MyD88-deficient mice is very poor after MCMV infection due to significantly reduced production of IFN-α/β, IL-12 and lower NK-cell activity (Krug *et al.*, 2004; Tabeta *et al.*, 2004).

Recent data have underlined the importance of another molecular player in TLR9-induced IFN secretion. Indeed, in unstimulated pDCs, inactive TLR9 localizes predominantly to the endoplasmic reticulum (ER). Upon pDC stimulation, TLR9 movement from ER to the endosome is controlled by UNC93B (Kim *et al.*, 2008). As such, ENU mutations in the *Unc93b* gene greatly impaired TLR9 trafficking to endosomes, thus increasing host susceptibility to MCMV (Brinkmann *et al.*, 2007). Further data suggest that proteolytic cleavage of TLR9 in the lysosome is also necessary for it to recognize its cognate ligand, which might be an additional precautionary measure to prevent self-DNA recognition (Park *et al.*, 2008; Ewald *et al.*, 2011). Since TLR9 shares its endosomal localization with TLR3, the latter may similarly depend on UNC93B.

Although pDCs are essential, other cell types also contribute to this second wave of type I IFN production (Swiecki *et al.*, 2010). For instance, macrophages and cDCs detect MCMV infection via TLR3 or TLR2, the latter of which recognizes numerous PAMPs (e.g. peptidoglycan, zymosan, etc.). Interestingly, TLR2-deficient mice display increased viral titres and impaired NK cell activation following MCMV infection (Szomolanyi-Tsuda *et al.*, 2006). This suggests that an MCMV-derived molecule, which may be a structural component of the envelope, is triggering TLR2, though it remains unknown for the moment.

Yet, though type I IFN production is dramatically reduced in animals with deficient TLRs, it is not abrogated. This implies that additional mechanisms of viral recognition exist.

Other cytosolic sensors of MCMV infection

Other receptors have been implicated in the sensing of the viral DNA, though specific evidence of MCMV recognition by these sensors is not always available. In particular, CMV DNA is compartmentalized (both in entry and assembly/exit processes) so that viral DNA is never unpackaged in the cytoplasm/cytosol of an infected cell. Nevertheless, in uninfected DCs, which take up and degrade virions or fragments of infected cells for priming by cross-presentation, this recognition might be occurring. For instance, the helicases encoded by the 'retinoic acid-inducible gene I' (RIG-I) and the 'melanoma differentiation-associated gene 5′ (MDA5) were shown to be specifically responsible for type I IFN secretion in all cell types upon exposure to polyA:T (Loo and Gale, 2011). This was surprising, since RIG-I has been previously shown to interact with 5′-triphosphate RNA (Pichlmair *et al.*, 2006). Subsequent studies revealed that dsDNA is transcribed into RNA by RNA polymerase III; this RNA can then be sensed by either RIG-I or MDA5 (Ablasser *et al.*, 2009; Chiu *et al.*, 2009). Both of these proteins contain N-terminal caspase activation and recruitment domains (CARDs) and an internal DExD/H-box RNA helicase domain. Normally, RIG-I exists in an inactive, monomeric state due to its C-terminal repressor domain (RD). After binding its ligand, RIG-I dimerizes; through its CARD domains, the dimer can then interact with the signalling-adaptor molecule IPS-1/Cardiff/VISA/MAVS, which is bound to the mitochondrial membrane. This triggers a signalling cascade involving FADD, RIP1, and TRAF3, ultimately leading to the activation of NF-κB-dependent pathways, or to the activation of IRF3- and IRF7-dependent pathways.

Another cytosolic DNA sensor is the stimulator of interferon genes (STING). STING localizes to the ER membrane in numerous cell types, including macrophages, DCs, endothelial cells and epithelial cells. Initial studies identified STING as an RNA sensor; indeed, *sting* knock-out mice were found very sensitive to vesicular stomatitis virus (VSV, an RNA virus) infection (Ishikawa *et al.*, 2009). Yet, MEF cells isolated from *sting* knock-out mice failed to produce type I IFNs following infection with herpes simplex virus 1 (HSV-1, a DNA virus) or the bacteria *L. monocytogenes*. In the same vein, STING-deficient animals were shown to succumb to HSV-1 infection due to their lack of type I IFN production p.i. (Ishikawa *et al.*, 2009). Furthermore, the ENU induced mutation in *sting*, called *Goldenticket*, was found to be susceptible to *L. monocytogenes* infection and unresponsive to priming by cyclic dinucleotides (Sauer *et al.*, 2011). Various studies also revealed that STING enhances RIG-I-mediated signalling, thus promoting NF-κB activation (Barber, 2011).

Another known sensor is the IFN-inducible protein called 'DNA-dependent activator of IFN-regulatory factors' (DAI, also known as the Z-DNA binding protein ZBP1). It is expressed in numerous cell types, namely stromal cells, lymphocytes, and macrophages. This protein contains three putative DNA binding domains: Za, Zb, and D3 (Takaoka *et al.*, 2007). After binding to cytoplasmic DNA, DAI is thought to interact with STING and feed into its downstream signalling pathway. It has also been shown to intersect with the signalling cascade downstream of RIG-I, as it can bind RIP1 via its 'RIP homotypic interaction motif' (RHIM) domain and lead to NF-κB activation (Rebsamen *et al.*, 2009). Of note, the MCMV-encoded M45 protein has been shown to inhibit the interaction between DAI and RIP1 (Upton *et al.*, 2010). Though the precise nature of the interactions between DAI and other proteins involved in signalling pathways downstream of cytoplasmic DNA sensing is unclear, the end result of DAI activation remains the expression of type I IFNs and the subsequent induction of a potent antiviral state. However, cells isolated from DAI knock-out mice were not found to be more susceptible to infection with DNA viruses (HSV-1) than those taken from WT animals (Ishii *et al.*, 2008; Wang *et al.*, 2008).

The AIM2 inflammasome

The Nod-like receptors (NLR) are also involved in the innate immune recognition of infection. Activation of these receptors leads to the assembly of high-molecular-mass complexes called inflammasomes involved in the production of active caspase-1 and the production of mature IL-1β. Four inflammasome sub-families exist: 'NLR with PYRIN domain 1′ (NLRP1), 'NLR with PYRIN domain 3′ (NLRP3), 'NLR with CARD domains 4′ (NLRC4)/'IPAF and Absent in Melanoma 2′ (AIM2). Among these, only the latter has been found to directly detect viral DNA in the cytoplasm (Burckstummer *et al.*, 2009; Fernandes-Alnemri *et al.*, 2009; Hornung *et al.*, 2009; Roberts *et al.*, 2009).

AIM2 is a cytosolic dsDNA sensor, 'member of the PYRIN and HIN domain-containing protein' (PYHIN) family. It becomes activated after binding to its ligand, which in turn triggers inflammasome assembly: its pyrin domain associates with the apoptosis-related speck-like protein containing a CARD (ASC), resulting in the activation of both NF-κB and caspase-1. The activated caspase-1 cleaves pro-IL-1β and pro-IL-18 before the mature cytokines are released. It is the absence of self DNA in the cytoplasm (under normal conditions) which confers on AIM2 its ability to specifically recognize pathogen DNA.

The importance of the AIM2 inflammasome in the response to MCMV infection has been demonstrated in a study of AIM2$^{-/-}$ mice (Rathinam *et al.*, 2010). Upon infection with either vaccinia virus or MCMV, BM-derived macrophages isolated from AIM2$^{-/-}$ mice displayed reduced production of IL-1β and IL-18, as well as decreased activation of caspase-1. This was not the case following HSV-1 infection; perhaps because HSV-1 forms less cell-associated particles and debris than MCMV which can trigger the cytoplasmic receptors of phagocytes. Furthermore, in the case of MCMV infection, IL-18 production in AIM2$^{-/-}$ mice was found to be drastically reduced; this affected NK cell IFN-γ secretion and, thus, prevented efficient NK-cell control of viral replication compared with WT controls. These results suggested that AIM2/ASC-dependent induction of IL-18 contributes to the early control of MCMV infection by promoting NK-cell function.

In summary, several methods of cytoplasmic DNA sensing have been revealed in recent years. Though each sensor induces type I IFN or IL-1β/IL-18 secretion through a specific signalling cascade, it is obvious that these various pathways intersect and amplify each other, the end result being improved detection of infection by the host. Moreover, additional, putative cytoplasmic DNA sensors are currently under investigation, namely high mobility group box proteins, IFI16, DEAH/RNA helicase A and DDX41 (Sharma and Fitzgerald, 2011; Zhang *et al.*, 2011). This underlines the redundant and collaborative nature of these various mechanisms of cytoplasmic DNA sensing.

PAMP recognition, although essential, is an inherently unspecific mechanism of immune detection. As the infection progresses and the virus spreads further, more specific mechanisms of viral recognition are triggered, mechanisms which have arisen over eons of host–pathogen co-evolution.

The natural killer cell response

The crucial role of NK cells in controlling MCMV infection is best illustrated by the numerous evasion mechanisms devised by the virus to escape the NK-cell response. Studies with MCMV infection have shown that NK-cell dysfunction results in increased susceptibility to infection (Shellam *et al.*, 1981; Bukowski *et al.*, 1984; Shanley, 1990; Welsh *et al.*, 1991). Adoptive transfer of NK cells into immunocompromised adult or newborn mice protects the recipients from MCMV disease (Bukowski *et al.*, 1985; 1988). In addition, *beige* mice, which lack effective NK-cell effector functions, are susceptible to MCMV; in humans, NK cell deficiencies result in numerous primary and recurrent herpesviral infections (Shellam *et al.*, 1981; Biron *et al.*, 1989), including HCMV infection (see Chapter II.8).

Not much was known about NK cells at the time, apart from their ability to kill infected cells without prior stimulation and to produce type II IFN; in fact, it was their cytotoxic nature, initially described in the context of Moloney leukaemia virus infection, which earned them the name 'Natural Killer' (Kiessling *et al.*, 1975). Since then, NK-cell functions have been investigated more thoroughly. Today, they are known to be crucial effectors for tumour surveillance, viral control, placentation, and the regulation of adaptive immune response initiation; surprisingly, they also share some features of adaptive immunity. NK cells are large granular cells representing a minor lymphocyte population in spleen (approximately 5% of splenocytes), lymph nodes (less than 1%) and circulation (approximately 15%) (Vivier *et al.*, 2011). One can also discriminate several populations of NK cells based on their localization in lymphoid and other tissues and organs (reviewed in: Spits and Di Santo, 2011).

NK cells mediate their function by the secretion of cytokines (IFN-γ, TNF-α, GM-CSF, IL-3, M-CSF, CCL3/MIP-1α, CCL4/MIP-1β, CCL5/RANTES), the induction of target cell apoptosis via cell surface receptors (CD16, TRAIL), and the release of cytotoxic granules (granzymes and perforin) (Fig. II.9.1). NK cell activation depends on the integration of various signals arising from the ligation of their large array of surface receptors, which can either bind self peptides, non-self peptides or soluble factors. Indeed, type I IFN and pro-inflammatory cytokines (e.g. IL-15, IL-12, IL-18, etc.) produced by PRR triggering potentiates their effector abilities.

NK cell receptors

The most remarkable feature of NK cells is their ability to distinguish self from non-self. This distinction depends on germline encoded NK-cell receptors (NKRs), many of which recognize MHC class I or class I-related molecules (Fig. II.9.2). NKRs are classified into two main families according to their extracellular domain: (1) the C-type lectin like family, type II glycoproteins with a C-type lectin-like scaffold; and (2) the immunoglobulin like receptors, type I glycoproteins of the immunoglobulin superfamily. Both families can be further subdivided into activating or inhibitory receptors. Although the extracellular domains of the various NKRs are extremely diverse, their intracellular domains are mostly conserved so that inhibitory or activating receptors share common inhibitory or activating signalling pathways.

Inhibitory receptors contain a tyrosine-based inhibitory motif (ITIM) in their intracellular domain. Receptor ligation triggers tyrosine phosphorylation by Src family kinases. This recruits 'SH2-domain-containing inositol-5-phosphatase' (SHIP-1) to the membrane, which then degrades phosphatidylinositol-3,4,5-triphosphate to phosphatidylinositol-3,4-bisphosphate; 'SH2-domain-containing protein tyrosine phosphatase 1 or 2' (SHP-1, SHP-2) can also be recruited to the membrane after receptor ligation. Once these phosphatases dephosphorylate the protein substrates of

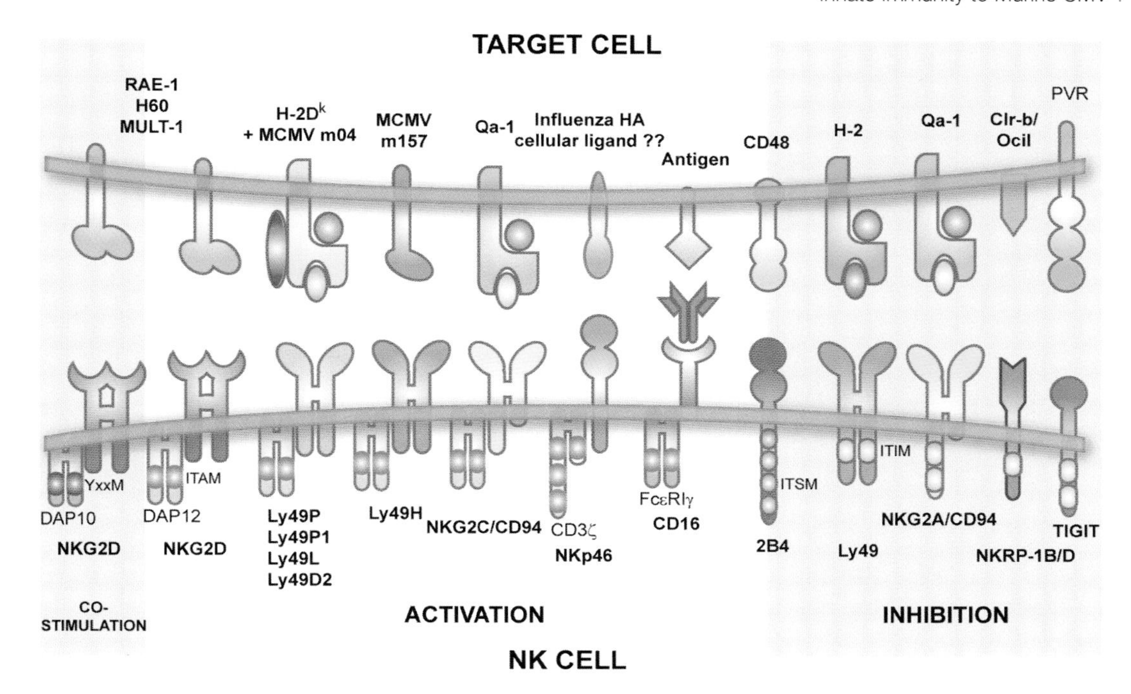

Figure II.9.2 Major NK cell receptors and their ligands.

tyrosine kinases, which are linked to activating NKRs, the activation signals are abrogated resulting in NK cell unresponsiveness. Activating NKRs lack ITIMs, but posses a positively charged arginine residue in their transmembrane domain, which allows them to interact with adaptor proteins such as DAP10, DAP12, FcεRI-γ or CD3-ξ (Lanier, 2008). These adaptors bear tyrosine-based activation motifs (ITAMs), which are phosphorylated upon receptor engagement, also by Src family kinases. Syk, ZAP-70, and PI3K or Grb2 are then recruited to the membrane, where they mediate actin cytoskeleton reorganization, cell polarization, release of cytolytic granules, and the transcription of many cytokine and chemokine genes.

Genetics of NK cell receptors and their ligands in mice

Early genetic studies into MCMV resistance revealed the dominant impact of two loci: the *H-2* locus on mouse chromosome 17 and a non-*H-2* locus on mouse chromosome 6 (Grundy *et al.*, 1981), within a region named the 'Natural Killer gene Complex' (NKC) (Fig. II.9.3).

The *H-2* locus spans about 2 Mbp of genomic DNA encoding the mouse major histocompatibility antigens that are the self ligands recognized by many NKRs. The *H-2* locus contains multiple, highly polymorphic genes and gene families that are involved in immune and non-immune functions and are inherited jointly as a haplotype; most importantly, it encodes MHC class Ia, class Ib and class II molecules. Three MHC class Ia genes exist (*H-2D*, *H-2K* and *H-2L*), all of which are multiallelic; however, a given inbred mouse strain may not necessarily possess all three of these genes. Further, MHC class Ib encodes, amongst others, the conserved monomorphic Qa-1 molecules which display restricted peptides derived from the leader sequence (Qdm) of MHC-I proteins. Qa-1 is the ligand to the heterodimeric CD94 and NKG2A or C or E receptors (Vance *et al.*, 1998).

The NKC spans ~8.7 Mbp encoding several multigene families that are inherited as a haplotype. It encodes amongst others many C-type lectin-like NKR families, as well as ligands to some of those NKRs. These include the killer lectin-like family 'a' (encoded by gene family *Klra*) [also known as 'lymphocyte antigen 49' (Ly49) receptors, members A to V], *Klrc* [also known as 'natural killer group 2' (NKG2) receptors, members A, C and E], *Klrk* (or NKG2D), *Klrd* (or CD94), CD69, *Klrb* [also known as 'natural killer cell receptor proteins' (NKRP1), members 'a' to 'g'], *Clr* (members 'a' to 'h'), and *Klrg1* (or MAFA1) (Yokoyama and Plougastel, 2003). Conversely, immunoglobulin-like NKRs map to several distinct chromosomes. For example, the 'natural cytotoxicity receptor 1' gene (*Ncr1*, also

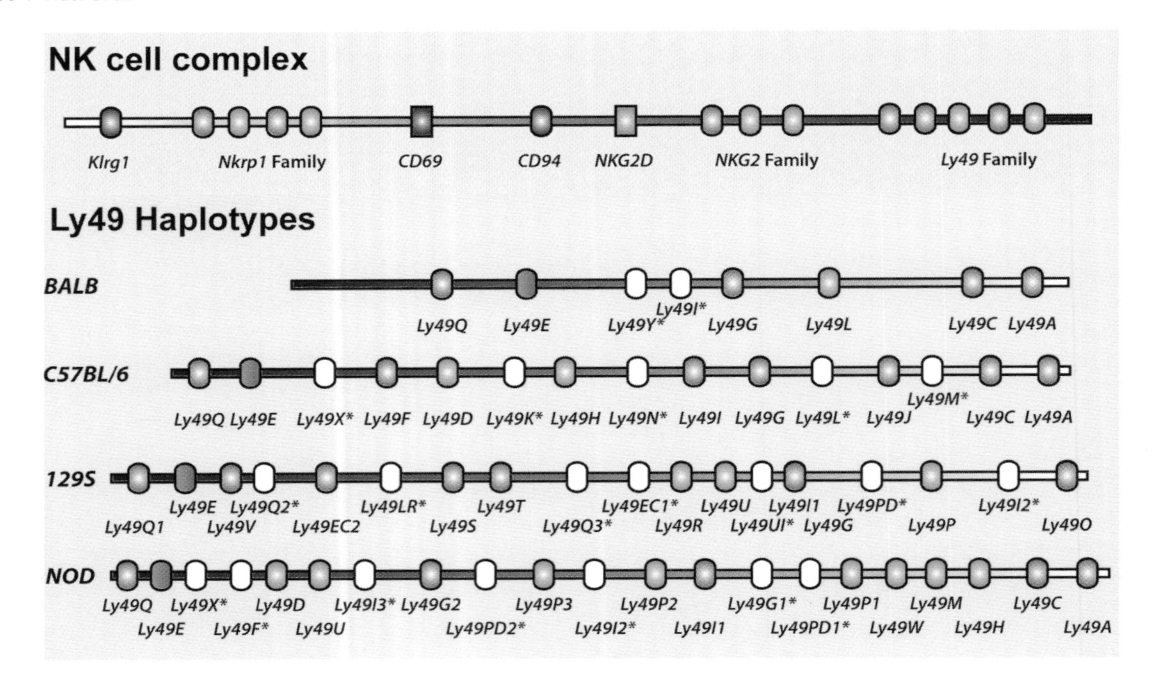

Figure II.9.3 Mouse NK cell complex and different *Ly49* haplotypes. Schematic organization of the NK cell complex (upper panel) and the *Ly49* gene cluster in BALB/c, C57BL/6, 129S6 and NOD strains (lower panel). Activating receptors are shown in green, inhibitory in red, and pseudogenes are shown in white and are asterisked. Ly49E has the potential to transduce both activating and inhibitory signals.

known as *Nkp46)* is encoded in the leucocyte receptor complex on mouse chromosome 7, proximal to DAP10 and DAP12; on the other hand, the 2B4 receptor maps to chromosome 1. Furthermore, expression of many NKRs is stochastic; with each NKR having limited ligand specificity, this entails that each NK cell is unique with regards to the repertoire of NKRs it expresses.

Ly49 haplotypes and receptor repertoire

The number of *Ly49* genes varies among different mouse strains with only four main haplotypes fully described (Fig. II.9.3). C57BL/6 mice posses 15 *Ly49* genes of which two (*Ly49d, h*) encode stimulatory and eight (*Ly49q, e, f, i, g, j, c, a*) inhibitory receptors. This haplotype is also shared with C56BL/10 mice. BALB/c mice have a minimal haplotype of seven genes (*Ly49q, e, i, g, l, c, a* genes) of which only *Ly49l* encodes an activating receptor. The BALB haplotype is shared among the DBA, C3H/He, CBA/J and A/J mice. A third haplotype was identified in the 129S6 strain and appears to be shared between 129P3, 129X1, 129S1, C57L/J, C57BR/cdJ, SJL/J, FVB/N and MA/My strains. The 129 haplotype encodes for three (Ly49r, u, p) activating receptors and nine (Ly49q_1, e, v, ec_2, s, t, i_1, g, o) inhibitory receptors. The fourth and largest *Ly49* haplotype was identified in NOD/Ltj animals. It comprises 21 genes seven of which (*Ly49d, u,* p_3, p_1, *w, m, h* genes) code for activating receptors and six (*Ly49q, e,* i_1, g_1, *c, a* genes) for inhibitory receptors, whereas eight being pseudogenes (Carlyle *et al.*, 2008). In addition to gene and functional diversity, variation of the Ly49 receptor repertoire also depends on their stochastic expression in overlapping NK cell subpopulations. In addition to NK cells, inhibitory Ly49 receptors are also expressed on monocytes, NKT cells, γδT cells and some effector T-cells (Coles *et al.*, 2000; Vogler and Steinle, 2011).

Missing-self recognition by inhibitory receptors

Inhibitory NKR function and ligand binding specificity have been extensively studied. The majority of NK cells express at least one of the inhibitory Ly49 receptors (Ly49A, Ly49C, Ly49G2, Ly49I); however, their contribution to the inhibitory signalling is strain specific. Ly49A is expressed on 5–20% of NK cells (Ortaldo *et al.*, 1999). Its natural ligands are H-2D^d and H-2D^k MHC-I molecules (Karlhofer *et al.*, 1992; Tormo *et al.*, 1999). Ly49G2 is expressed on 20–40% of NK cells and recognizes H-2D^k, H-2D^d and H-2L^d (Silver *et al.*, 2002). Ly49C is the most promiscuous and abundant receptor, since it binds H-2D^d, H-2K^d, H-2D^b, H-2K^b and H-2D^k and is expressed on about 50% of NK cells. It shares high homology with the Ly49I receptor in 129/J mice, which binds to H-2K^b, H-2^s and H-2^q molecules (Brennan *et al.*, 1996; Dimasi *et al.*, 2002). In general, those inhibitory receptors possess a high binding affinity for self MHC-I molecules. Since MHC-I molecules are expressed on virtually all nucleated cells, it is not surprising that the default state of NK cells is repression. Infection or tumour growth can perturb MHC-I expression, leading to down-regulation of the molecules. In the absence of their ligand, inhibitory NKRs can no longer suppress NK-cell activity; if the NK cell in question receives signals from its activating NKRs, NK-cell activation will be able to proceed unhindered (Fig. II.9.4A). This is termed 'missing-self' recognition or 'stressed-self' (in the case of NKG2D ligands) or 'non-self' (in the case of activating Ly49 ligands).

Inhibitory NKR engagement is also necessary for the maturing NK cell to develop proper effector functions. Indeed, although the underlying mechanism remains unclear, it has been demonstrated that the binding affinity of inhibitory NKRs is directly proportional

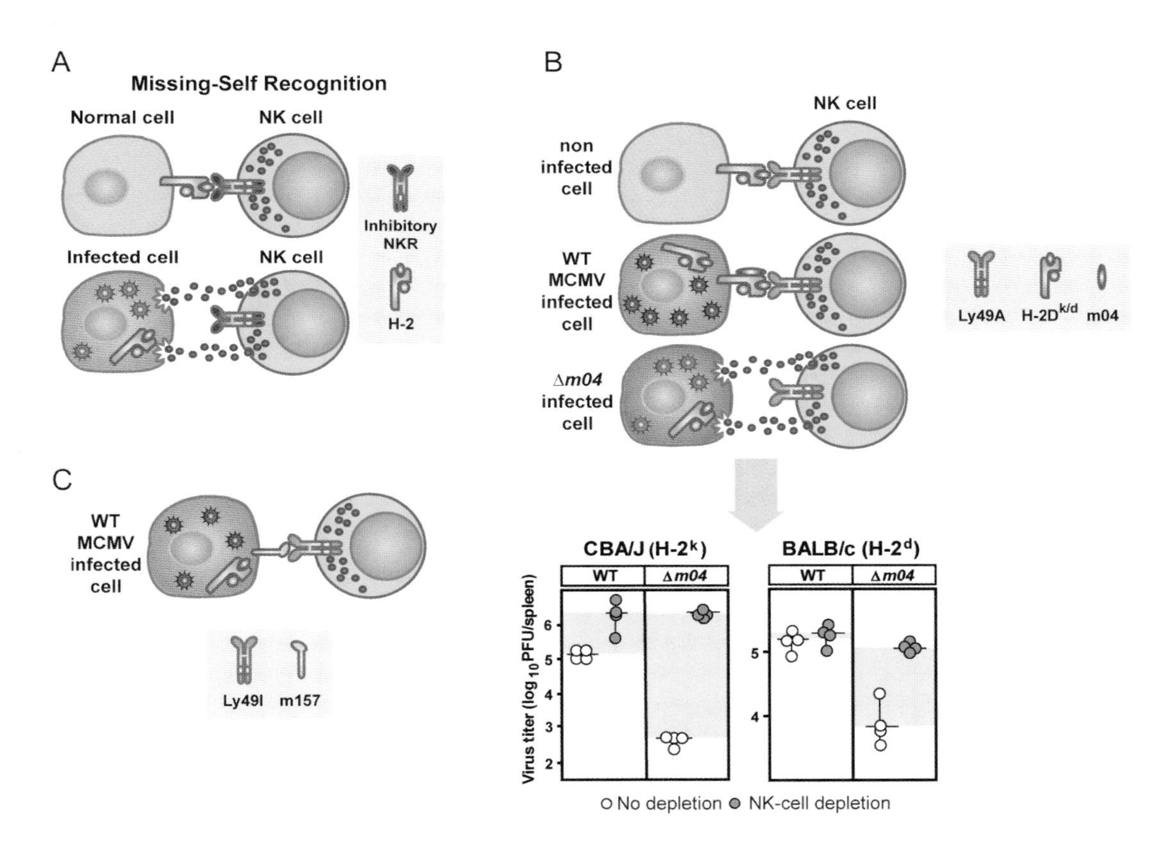

Figure II.9.4 MCMV evasion of NK cell activation via 'missing-self' mechanism. (A) The 'missing-self' hypothesis states that under normal conditions inhibitory NK-cell receptors (NKR, e.g. Ly49A) are engaged by surface MHC-I molecules and NK-cell activation is prevented. In WT MCMV-infected cells, MHC-I molecules are down-regulated, leaving inhibitory Ly49 receptors unengaged, which results in NK cell activation. (B) Upper panel: To prevent NK cell activation via 'missing-self', MCMV *m04*/gp34 protein binds to MHC-I molecules and escorts them in a complex to the cell surface to engage inhibitory Ly49 receptors. Infection with virus lacking m04 (Δ*m04*), generally down-modulates surface expression of MHC-I molecules, leaving inhibitory Ly49 receptors unengaged. Consequently, the activating signals in NK cells prevail, resulting in destruction of infected cells. Lower panel: Viral titres 3 days post infection in CBA/J (H-2^k) and BALB/c (H-2^d) mice which were either NK cell depleted (grey circle) or not (white circle) and subsequently infected with equivalent dose of WT or Δ*m04* MCMV. Absence of NK cells through their depletion renders mice unable to control viral replication demonstrating the non-redundant role of NK cells in MCMV resistance which is both H-2 and *m04*/gp34 dependent. Modified from Babic *et al.*, 2010. Originally published in J. Exp. Med. *207*, 2663–2673. (C) MCMV encodes several MHC-I homologues, amongst them m144 and m157, whose recognition by inhibitory NKR during infection (e.g. Ly49I) can potentially prevent NK-cell activation and target cell killing.

to NK-cell secretion of cytolytic granules and Type II IFN (Hoglund and Brodin, 2010). Thus, the greater the affinity of an inhibitory receptor for a given H-2 molecule, the more potent the subsequent NK-cell response will be. Interesting enough, this NK-cell education/licensing process is flexible, and the impact of MCMV infection on licensing is not completely understood.

Inhibitory Ly49–MHC-I axis

Until very recently, the impact of inhibitory Ly49 receptors on MCMV resistance had not been investigated *in vivo*. However, new studies in susceptible mice of BALB background carrying different H-2 haplotypes have revealed the importance of inhibitory NKR during infection. These mice possess the smallest Ly49 repertoire of all inbred mouse strains tested. Four receptors are mainly expressed on mature NK cells in these mice: inhibitory Ly49C, Ly49G, and Ly49A, as well as activating Ly49L. Depending on their H-2 haplotype, they display a varying range of MCMV susceptibility: H-2^k (BALB.K) mice are the most resistant to the infection and H-2^b (BALB.By) mice are the most susceptible, while H-2^d (BALB/c) mice have an intermediate phenotype. Regardless, all BALB strains show high viral titres in the spleen at day 3 p.i., which implies they are initially unable to control MCMV infection (Grundy *et al.*, 1981; Babic *et al.*, 2010; Pyzik *et al.*, 2011a,b).

Modulation of MHC-I expression

The virus can avoid cytotoxic cell detection by modulating the expression of MHC-I molecules on the surface of infected cells (reviewed by Hengel *et al.*, 1999). Nevertheless, the changes in MHC-I expression (Lemmermann *et al.*, 2010) should be sensed by NK cells and in the case of down-regulation of MHC-I molecules should render the infected cells sensitive to NK-cell control via the 'missing self'-mediated mechanism. MCMV encodes three proteins that tamper with the expression of MHC-I molecules. While the role of *m152*/gp40 and *m06*/gp48 in down-regulation of MHC-I molecules is well defined, the role of *m04*/gp34 protein, however, was not. The *m04*/gp34 protein is a type I- integral membrane glycoprotein that tightly associates with MHC-I molecules (Kleijnen *et al.*, 1997). Some data implicated *m04*/gp34 in escaping from MCMV-specific $CD8^+$ T-cell recognition (Kavanagh *et al.*, 2001), while other data suggested that *m04*/gp34 might facilitate the export of MHC-I molecules to the cell surface thereby antagonizing the effect of *m152*/gp40 (Kleijnen *et al.*, 1997; Wagner *et al.*, 2002; Holtappels *et al.*, 2006). Therefore, it was reasonable to propose that gp34/MHC-I complexes may silence NK cells (Kleijnen *et al.*, 1997) by serving as ligands for inhibitory Ly49 receptor.

m04/gp34 evasion of missing-self

When infected with a *m04*/gp34 knock-out virus (Δ*m04* MCMV), viral titres in the spleen of BALB mice decreased significantly compared with infection with WT virus (Fig. II.9.4B); conversely, WT and Δ*m04* MCMV infection did not differ in resistant B6 mice (Babic *et al.*, 2010). Moreover, the extent of viral attenuation depended on the H-2 haplotype: H-2^b (least), H-2^d (intermediate) or H-2^k (greatest). Furthermore, NK-cell proliferation was greater in H-2^k mice than H-2^d mice, which was in turn greater than in H-2^b mice.

The *m04*/gp34 viral protein associates with MHC-I molecules to maintain a basal level of expression during MCMV infection. Its absence during Δ*m04* MCMV infection leads to the reduction of MHC-I expression, thus decreasing inhibitory Ly49 triggering. This suggested that MCMV prevents the destruction of its host cell by circumventing missing-self recognition. Moreover, Ly49A – which can bind to H-$2D^d$ and H-$2D^k$ – was found to better recognize cells infected with WT MCMV as opposed to Δ*m04* MCMV in a reporter cell assay. This suggested that this inhibitory Ly49 was triggered not only by H-$2D^k$, but also by the complex of H-$2D^k$ and the *m04*/gp34 protein. Yet, when mice of BALB background carrying the H-2^d haplotype were infected with Δ*m04* MCMV, subsequent depletion of $Ly49G^+$ NK cells restored viral titres to WT infection levels; this also occurred when either $Ly49C^+$ or $Ly49A^+$ NK cells were depleted, but not to the same extent. These results suggest that Ly49G is the dominant inhibitory signal acting on the NK-cell response of these mice.

Missing-self independent inhibitory Ly49 receptor mechanisms

Surprisingly, inhibitory Ly49s have also been associated with resistance to MCMV in the MA/My strain. MA/MyJ resistance has been linked to a genetic interaction between the NKC and the *H-2^k* loci or to the *H-2^k* locus itself (Desrosiers *et al.*, 2005; Dighe *et al.*, 2005; Xie *et al.*, 2007). Indeed, following MCMV infection, $Ly49G^+$ NK cells are preferentially activated before they expand and secrete IFN-γ. Furthermore, depletion of this NK-cell fraction has been shown to reduce host resistance (Xie *et al.*, 2009). Since Ly49G recognizes H-$2D^k$, it was suggested that adequate education of $Ly49G^+$ NK cells in H-2^k mice was responsible for their improved NK-cell response to MCMV infection.

However, Ly49G$^+$ NK-cell expansion following infection with other viral pathogens or exposure to various stimuli rather suggests that Ly49G is simply an activation marker of mouse NK cells (Daniels *et al.*, 2001; Barao *et al.*, 2011). Therefore, the reduced resistance of MA/My mice, observed when their Ly49G$^+$ NK cells are depleted prior to MCMV infection, might be explained by the removal of the effector population that was activated through other receptors (Xie *et al.*, 2009; Babic *et al.*, 2010).

Viral encoded MHC-I homologs

The *m145* region of the MCMV genome encodes several MHC-I-like molecules, namely m144, m145, m152, m153 and m157 proteins (Mans *et al.*, 2009). m144, m157, and m153 possess the same protein fold as MHC-I molecules yet they cannot present peptides (Mans *et al.*, 2007). The function of most of these molecules is not well understood, but it is generally accepted that they contribute to viral evasion from NK-cell or CD8$^+$ T-cell detection (Fig. II.9.4C).

The first described MHC-I homologue encoded by MCMV is a product of the *m144* gene whose cellular ligand has not yet been identified (Farrell *et al.*, 1997). A recombinant virus lacking *m144* gene shows reduced pathogenesis that can be reversed by NK-cell depletion. In addition, the expression of m144 on tumour cells has an inhibitory effect on NK cells (Cretney *et al.*, 1999) and cell transfection of *m144* inhibits ADCC (Kubota *et al.*, 1999). The homologue of m144 in the RCMV has also been identified and the deletion mutant virus was found to be attenuated *in vivo* (Kloover *et al.*, 2002).

In recent years, m157 in particular has become the focus of much investigation. This GP-linked glycoprotein can be expressed as several differentially N-glycosylated isoforms on the surface of infected cells (Guseva *et al.*, 2010). Sequence analysis of m157 derived from MCMV samples isolated from the wild showed that, under these conditions, its sequence is highly variable (see also Chapter I.2). In fact, this great variety of sequences clusters in four classes, two of which are represented by the laboratory strains Smith and K181 (Voigt *et al.*, 2003; Smith *et al.*, 2006). Most important, several inhibitory Ly49 receptors have been shown to recognize m157, such as Ly49I from 129/J mice (Fig. II.9.4C). However, the contribution of this receptor to MCMV susceptibility could not be assessed due to an NK-cell signalling defect in this strain (Arase *et al.*, 2002; Patel *et al.*, 2010). In addition, the Ly49C receptor from the C57BL/6, BALB/c, and NZB strains was also shown to recognize the m157 derived from wild MCMV isolates, although it could not suppress the NK-cell activation induced by activating receptors (Corbett *et al.*, 2010). Other studies demonstrated that Ly49C$^-$ NK cells secreted more IFN-γ and proliferated better than their Ly49C$^+$ counterparts following MCMV infection of H-2^b mice; likewise, the adoptive transfer of Ly49C$^-$ NK cells conferred better protection from MCMV lethality than the Ly49C$^+$ fraction (Orr *et al.*, 2010). The cause of this difference remains uncertain, though possible explanations may be considered: 1) Ly49C-mediated inhibition dampens NK-cell effector functions and decreases host survival, or 2) Given its great binding affinity to H-2D^b and H-2K^b, Ly49C is responsible for NK-cell education in H-2^b mice, an education which makes them less apt to adequately respond to infection.

Inhibitory Nkrp1–Clr axis

Though 'missing-self' recognition classically involves the absence of MHC-I molecules, this mechanism has also been shown to occur in an MHC-independent manner in a model of RCMV infection. Typically, NKR-P1 receptors bind to molecules of the Clr family, which are ubiquitously expressed on healthy cells (Iizuka *et al.*, 2003). Each of these receptors displays selective binding affinity. For instance, the inhibitory NKR-P1B receptor from WAG rats is only triggered by Clr-b (Mesci *et al.*, 2006). However, RCMV infection blocks Clr-b expression. Therefore, the lack of NKR-P1B signalling in this strain can then potentially lead to NK-cell activation via 'missing-self' recognition. To avoid this, the RCMV genome encodes the *Clr-b* homologue RCTL; infection of WAG rats with an RCMV mutant lacking RCTL leads to decreased viral virulence (Voigt *et al.*, 2007; see also Chapter II.15).

As is the case in mice, rat NKR-P receptors display strain specific polymorphisms, such that different rat strains possess different NKR-P haplotypes. Moreover, a given NKR-P receptor may differ in its ligand binding affinity from one strain to the other. Indeed, while NKR-P1B strongly recognizes RCTL in the WAG strain, NKR-P1B from SD rats recognizes it only weakly. Taking this into consideration, susceptibility or resistance to RCMV becomes a strain-specific feature: strong RCTL recognition by NKR-P1B in one strain leads to RCMV susceptibility, while weak recognition in another strain leads to RCMV resistance (Voigt *et al.*, 2007). Given this differential recognition of viral proteins by inhibitory receptors, one might wonder if a similar phenomenon can be observed with activating NKR-P receptors, or even activating NKRs at large.

Regardless, inhibitory NKRs can specifically recognize CMVs, either through viral proteins (direct) or virus-induced modulation of MHC-I expression. The end result of inhibitory NKR triggering remains poor control of viral spread by the innate immune system.

Activating Ly49 receptor axis

Over the course of evolution, the capture of *H-2* genes by MCMV allowed it to produce MHC-I homologues for the purpose of immune-evasion, resulting in improved viral fitness. Many believe this exerted an evolutionary pressure on the host, culminating in the emergence of activating NKRs capable of MCMV recognition.

Ly49H–m157 axis

C57BL/6 mice are resistant to MCMV due to their expression of the activating receptor Ly49H on the surface of their NK cells (Daniels *et al.*, 2001; Dokun *et al.*, 2001; Lee *et al.*, 2002). Ly49H is known to bind m157 exclusively (Arase *et al.*, 2002; Smith *et al.*, 2002; Adams *et al.*, 2007) (Fig. II.9.5A). This dominant and direct ligation initiates an activating signalling cascade, resulting in clonal proliferation of Ly49H$^+$ NK cells, killing of infected cells, and secretion of cytokines (IFN-γ, MIP-1α, TNF-α, etc.); all of this culminates in the rapid control of viral replication, such that very little virus remains in the spleen at day 3 p.i.. C57BL/6 mice can be rendered susceptible if Ly49H:m157 signalling

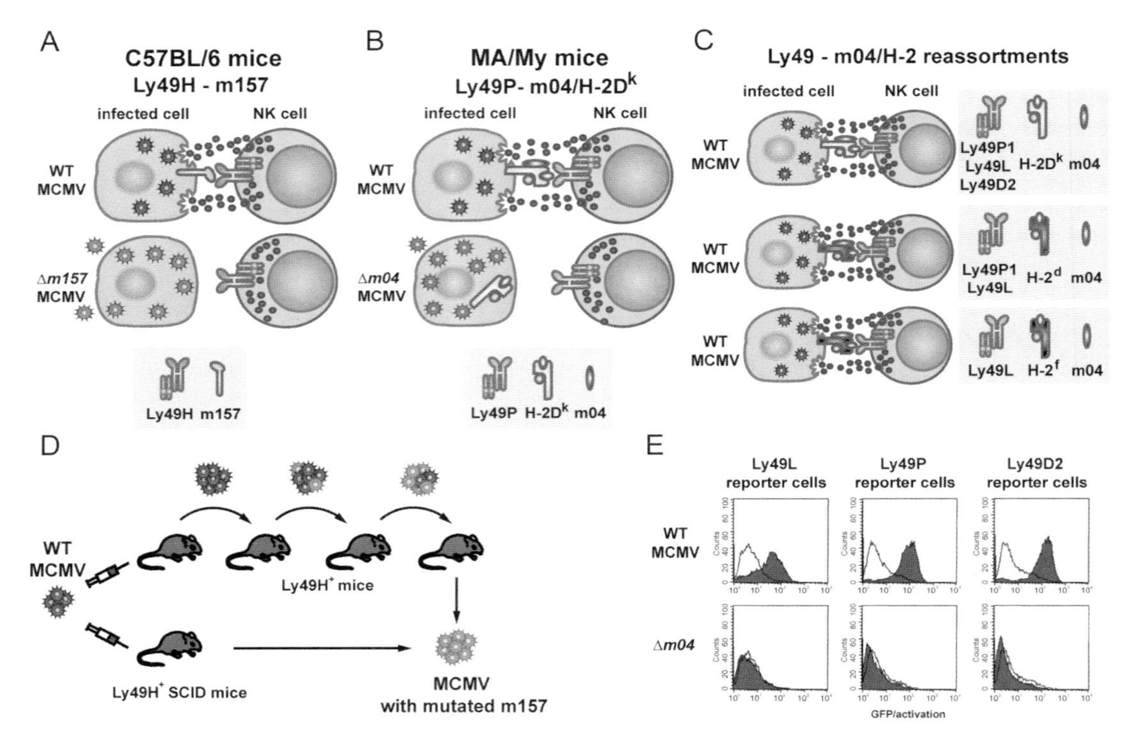

Figure II.9.5 Specific recognition of MCMV-infected cells by activating Ly49 receptors. (A) Ly49H NK-cell receptor specifically recognizes viral protein encoded by gene *m157*. As a consequence, mice expressing Ly49H receptor (C57BL/6) successfully control WT MCMV. Deletion of gene *m157* (Δ*m157* MCMV) results in inability of infected cells to activate Ly49H$^+$ NK cells. (B) Ly49P NK-cell receptor specifically recognizes a complex consisting of MCMV m04 protein associated with H-2D^k MHC-I molecule on the surface of infected cells. MA/My mice, which possess both Ly49P receptor and H-2D^k control WT MCMV infection. Deletion of gene *m04* (Δ*m04* MCMV) results in inability of infected cells to activate Ly49P$^+$ NK cells. (C) Ly49L, Ly49P1 and Ly49D2 receptors also recognize MCMV m04 protein, however, in association with particular H-2D^k (upper panel) or H-2^d (middle panel) or H-2^f (lower panel) MHC-I molecules. (D) Selective pressure in Ly49H$^+$ mice results in the emergence of m157 viral mutants which are no longer recognized by Ly49H receptor. This can be achieved either through serial passages of virus through normal Ly49H$^+$ mice, or by single passage in Ly49H$^+$ SCID mice. (E) Reporter cell assay involving the co-culture of H-2^k MEF cells infected with WT or Δ*m04* MCMV in presence of reporter cells stably transduced with specific Ly49 activating receptor and GFP expressed under the control of the Nuclear factor of activated T-cells (NFAT) promoter. Upon recognition of the ligand, activation of the NFAT pathway results in GFP expression, which can be detected by cytofluorometric analysis. Modified from Pyzik *et al.* (2011a). Originally published in J. Exp. Med. *208*, 1105–1117.

is abrogated in one of three ways: 1) infection with an *m157*-deleted virus (Δ*m157* MCMV); 2) deletion of the *Ly49h* gene in the host; or 3) a defect in DAP12 signalling (Sjolin *et al.*, 2002; Bubic *et al.*, 2004; Cheng *et al.*, 2008; Fodil-Cornu *et al.*, 2008). Conversely, Ly49H transgenesis into susceptible strains causes resistance to MCMV (Lee *et al.*, 2003).

Protein structure analysis has revealed that the Ly49H homodimerization domain is critical for m157 binding by this receptor; similarly, buried residues within m157 that are required for proper protein folding are also necessary for this viral molecule to be recognized by Ly49H (Kielczewska *et al.*, 2007; Davis *et al.*, 2008). Furthermore, m157 was found to be expressed on the surface of MCMV-infected cells in a variety of glycosylated isoforms, all of which can be recognized by Ly49H. Specifically, m157 binding stability and half-life were increased by glycosylation (Guseva *et al.*, 2010). Overall, the dissection of the interaction between Ly49H and m157 uncovered an extremely tolerant recognition mechanism, where several modifications to the receptor or the ligand were necessary to disrupt binding. Ly49H insensitivity to ligand modification is thought to maintain m157 recognition despite constant mutations in the latter, which occur as MCMV attempts to escape immune recognition; it might also allow the receptor to recognize newly emerged viral ligands with an m157-like fold. Yet, the driving force behind m157 mutations might not be to escape Ly49H detection, but rather to improve recognition by inhibitory Ly49 receptors.

Ly49P–m04–H-2^k axis

Another mechanism of MCMV recognition by NK cells was described in MA/My mice (Fig. II.9.5B). In this strain, an activating Ly49 receptor can recognize modified self MHC-I molecules rather than a viral MHC-I homologue. This line of investigation began when MA/MyJ mice were found to control MCMV replication at early time points p.i., even though they lack Ly49H (Rodriguez *et al.*, 2004). Later on, linkage analysis studies found that MA/MyJ resistance depended only on two genomic regions: the NKC$^{MA/My}$ and *H-2^k* loci (Desrosiers *et al.*, 2005; Dighe *et al.*, 2005). Among the three activating Ly49 receptors present in the MA/My repertoire, only Ly49P can recognize MCMV-infected cells (Desrosiers *et al.*, 2005). However, further investigation revealed that this recognition only occurs in the presence of both the viral *m04*/gp34 protein and H-2D^k (Kielczewska *et al.*, 2009). Yet, more precise analyses showed that Ly49P reporter cells could not recognize H-2D^k-bearing target cells that artificially expressed *m04*/gp34; infection of these target cells with Δ*m04* MCMV restored Ly49P recognition. This suggested that another, unknown host or viral factor is necessary for Ly49P/m04/H-2D^k-mediated recognition of MCMV-infected cells (Kielczewska *et al.*, 2009).

The importance of H-2D^k in MA/MyJ resistance was also demonstrated *in vivo*: 1) by the generation of an H-2^k congenic mouse panel; and 2) by H-2D^k transgenesis into susceptible H-2^b or H-2^q mice. Although mice became resistant to MCMV in both cases, small differences existed between experiments depending on the method used (Xie *et al.*, 2007, 2009; 2010; Fodil-Cornu *et al.*, 2011). The expression of H-2D^k in an endogenous H-2^b context was sufficient to produce resistance; however, this resistance was only partial in the H-2^q context (Xie *et al.*, 2010; Fodil-Cornu *et al.*, 2011). This suggested that the effect of an activating NKR can be masked by strong inhibitory NK-cell pathways.

Ly49P1/Ly49L/Ly49D2–m04-H-2 axis

Apart from Ly49P, several other activating Ly49 receptors can recognize MCMV-infected cells in an *m04*-specific and *H-2*-dependent manner: the Ly49P1 receptor from NOD/Ltj mice, Ly49D2 from PWK/Pas mice, and Ly49L from BALB/c mice (Pyzik *et al.*, 2011a) (Fig. II.9.5C,E). Each of these receptors can recognize MCMV infection in a different array of *H-2* contexts: Ly49D2$^{PWK/Pas}$ in H-2^k, Ly49P1^{NOD} in H-2^k and H-2^d, and Ly49L^{BALB} in H-2^k, H-2^f, and H-2^d. Among these receptors, Ly49L has been more characterized.

To begin with, *H-2^k* BALB mice were found to be more resistant to MCMV infection than *H-2^d* or *H-2^b* BALB animals; this improved control of viral spread was then found to be NK-cell dependent. Ly49L, the only activating Ly49 receptor in the BALB repertoire, was the candidate most likely to underlie this increased resistance. Indeed, in *H-2^k* BALB mice, the greater control of viral replication correlated with the specific expansion of Ly49L$^+$ NK cells and their secretion of IFN-γ. Moreover, adoptive transfer of Ly49L$^+$ NK cells into neonate mice increased their survival following MCMV infection (Pyzik *et al.*, 2011a). Further testing ascertained that Ly49L recognition of MCMV-infected cells was *m04*-dependent.

Because in these mice the signalling via inhibitory receptors prevails over the activating signals during the early days p.i., the effect of specific engagement of Ly49L receptor is evident only at later time points of infection. This is in contrast to immediate activation and proliferation of NK cells in C57BL/6 mice after engagement of Ly49H receptor with viral m157 protein. These results suggest that *m04*-specific, *H-2*-dependent detection of infected cells by activating Ly49 receptors is another mechanism of host defence.

Numerous activating and inhibitory Ly49 receptors share a similar mechanism of MCMV recognition. Given the speed at which the Ly49 locus evolves, this might suggest that an inhibitory receptor could have been converted to an activating one. However, a strong activating NKR might be detrimental if the pathogen is cleared completely before an effective adaptive immune response develops. This NK cell-mediated sterilizing immunity could in fact increase the individual risk of re-infection.

Virus escape: m157 and m04 variants

As described above, several MCMV encoded proteins serve as ligands for activating NK-cell receptors, either directly as in the case of m157 or in the MHC restricted manner, as shown for m04. Irrespective of the mode of receptor engagement, the recognition leads to NK-cell proliferation, lysis of infected cells and control of viral spread. Why should the virus preserve the genes encoding the ligand for a receptor that threatens its survival? The most plausible explanation is that m157 and m04 proteins have evolved to serve as ligands for inhibitory NK-cell receptors and that evolution of activating Ly49 receptors was the host response to selective pressure by the virus. Indeed, m04, as described above, is essential for evasion of 'missing-self' NK-cell activation, by escorting sufficient MHC-I molecules to ligate inhibitory Ly49 receptors.

Regardless of possible explanations of the origin of those genes, a key question involved the drive behind the viral retention of the *m157* gene upon exposure to the selective pressure of Ly49H$^+$ NK cells *in vivo*. It has been shown that MCMV can overall respond rapidly to the immune pressure of NK cells (Voigt *et al.*, 2003; French *et al.*, 2004) (Fig. II.9.5D). For instance, serial passage of WT MCMV in Ly49H$^+$ mice led to emergence of mutations within *m157* resulting in inability to activate Ly49H or control viral titre (Voigt *et al.*, 2003). Furthermore, the same phenomenon is observed in mice lacking adaptive immune responses (B6-SCID), as a large number of *m157* escape mutants were rapidly selected without need of serial passage (French *et al.*, 2004). The escape mutant viruses isolated from these mice were resistant to Ly49H$^+$ NK cells to the same extent as the mutant virus with the deliberate mutation of *m157* gene (Bubic *et al.*, 2004; French *et al.*, 2004). By contrast, the *m157* mutations did not occur after MCMV passage in susceptible mice, confirming that the selective presence of Ly49H$^+$ NK cells was essential for occurrence of escape mutants (French *et al.*, 2004). Therefore, how and why *m157* gene was preserved during the evolution still remains unknown. It is worth mentioning that most of wild derived MCMV isolates encode m157 proteins that do not activate NK cells, not even via Ly49H, yet were shown to trigger inhibitory receptors (Voigt *et al.*, 2003; Corbett *et al.*, 2010).

Sequence analysis of the *m04* gene variants obtained from MCMV isolates from wild mice also demonstrated significant sequence diversity, mostly in its extracellular domain (Corbett *et al.*, 2007). Compared to other MCMV genes belonging to the *m02* family (*m02*, *m03*, *m03.5*, *m04*, *m05*), the sequence identity of *m04* ranged from 40% to 95%, while for the other members it was found to exceed 85%. On the one hand, since *m04*/gp34 associates with MHC-I molecules it is possible that such variation was necessary to match the allelic diversity of MHC-I. However, the H-2-association regions mapped to the transmembrane domain of *m04*, which remains relatively conserved (Lu *et al.*, 2006). On the other hand, *m04* variability might result from the strong selective pressure exerted by Ly49 receptors, especially given the fact that numerous activating and inhibitory Ly49 receptors recognize the MHC-I–m04 complex (Kielczewska *et al.*, 2009; Babic *et al.*, 2010; Pyzik *et al.*, 2011a,b). Therefore, in a strain specific manner, *m04* variants might be selected for interaction with inhibitory receptors while in yet another strain they might be selected against the interaction with activating receptors. Still, it remains to be established whether the other m04 forms identified from wild-MCMV isolates are able to trigger the same receptors as the Smith-m04 and in the same H-2 contexts.

Induced self-recognition by NKG2D

A member of the NKR family of receptors, NKG2D, is undoubtedly one of the most valuable activating NKRs (Ogasawara *et al.*, 2005) (Fig. II.9.2). Unlike other members of its family, however, NKG2D does not associate with CD94. NKG2D is expressed on CD8$^+$ T-cells, $\gamma\delta$ T-cells, and some CD4$^+$ T-cells, as well as on all NK cells. While all CD8$^+$ T-cells in human peripheral blood express the receptor, this is true only of activated CD8$^+$ T-lymphocytes in mice. In mice, two variants of this receptor are expressed due to alternative splicing, NKG2D-L (long) and NKG2D-S (short), while only the NKG2D-L variant is present in humans (Carayannopoulos and Yokoyama, 2004). NKG2D-S associates with the adaptor molecules DAP10 or DAP12, while NKG2D-L associates only with DAP10.

NKG2D recognizes a wide array of ligands, characterized by a striking diversity in structure, expression pattern, and regulation. These ligands are distantly related to MHC-I molecules and expressed at low levels in normal cells. Indeed, NKG2D ligand expression is driven by cellular stress, such as infection or tumour transformation (Cerwenka and Lanier, 2003). Mouse NKG2D ligands include: (1) five members of the retinoic acid early transcript 1 family (RAE-1α-ε), which are differentially expressed in various mouse strains; (2) the three members of the H60 family (a, b, and c); and (3) murine UL16 protein-like transcript 1 (MULT-1), which is related to human NKG2D ligands, ULBPs (Cerwenka *et al.*, 2000; Diefenbach *et al.*, 2001; Carayannopoulos *et al.*, 2002). While on NK cells NKG2D acts as major activating receptor, on $CD8^+$ T-cells it is a co-stimulatory receptor (Vivier *et al.*, 2002). Binding of NKG2D to any of its ligands induces NK-cell cytotoxicity as well as production of IFN-γ and TNF-α.

Escape from NKG2D-mediated recognition

The importance of NKG2D in the immune response against CMV infection is best illustrated by the fact that HCMV and MCMV have developed numerous immune evasion mechanisms against this receptor (Jonjic *et al.*, 2008). Indeed, four MCMV genes (*m138, m152, m155, m145*) encode proteins that down-regulate the expression of NKG2D ligands (Lisnic *et al.*, 2010) (Fig. II.9.6). The precise mechanism of action of these viral proteins is not yet fully understood, but substantial information is already available.

MCMV *m152* encodes the gp40 protein, a molecule that is known to retain immature RAE-1 proteins in the endoplasmic reticulum-Golgi intermediate compartment (ERGIC) *in vitro* (Lodoen *et al.*, 2003; Arapovic *et al.*, 2009b). However, the various RAE-1 proteins show differential sensitivities to gp40-mediated ERGIC retention. Following MCMV infection of a variety of mouse cell lineages, surface expression of RAE-1α, β, γ, and ε could not be detected; in contrast, RAE-1δ surface expression was maintained (Arapovic *et al.*, 2009b). Sequence alignment of the RAE-1 proteins revealed that RAE-1α, β, and γ share the presence of a PLWY motif, which is deleted from RAE-1δ and mutated in RAE-1ε. This motif was thought to underlie this differential *m152*/gp40 sensitivity; indeed, upon insertion of this PLWY motif into RAE-1δ and transfection of the mutant protein into NIH-3T3 cells, MCMV infection caused a down-regulation of RAE-1δ-PLWY surface expression (Arapovic *et al.*, 2009b). Another study found that RAE-1β and RAE-1γ bind with high affinity to *m152*/gp40, while RAE-1δ interacts with it only weakly. Yet, whereas deletion of the PLWY motif from RAE-1β greatly decreased its affinity for *m152*/gp40, its deletion from RAE-1γ and its insertion into RAE-1δ had little effect on the binding affinity of either protein for *m152*/gp40; this suggests that other determinants are involved in the interaction of these two proteins with the viral immunoevasin (Zhi *et al.*, 2010).

As for the other MCMV-encoded immunoevasins

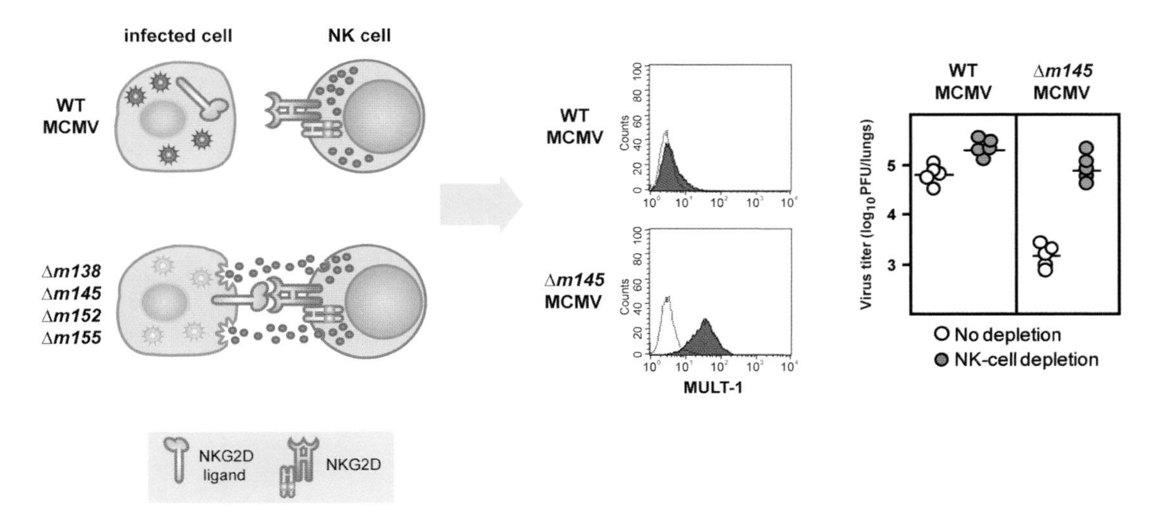

Figure II.9.6 MCMV evasion of NKG2D-dependant virus control. Left and middle panels: To prevent activation of NK cells via NKG2D, MCMV encodes four proteins involved in down-regulation of NKG2D ligands. Therefore, in spite of strong induction of these ligands in infected cells, their surface expression is compromised. Right panel: Viral titres 3 days post infection in BALB/c mice which were either NK cell depleted (grey circle) or not (white circle) and subsequently infected with equivalent dose of WT or *Δm145* MCMV. Deletion of any of four NKG2D inhibitors (e.g. m145) results in expression of corresponding ligand on infected cells and virus attenuation *in vivo*. Modified from Krmpotic *et al.* (2005). Originally published in J. Exp. Med. *201*, 211–220.

targeting NKG2D ligands, the m138 protein has diverse effects which complete those of the other immunoevasins. Indeed, presence of the m145 and m155 proteins can significantly reduce MULT-1 and H60 surface expression respectively; in both cases, the viral immunoevasins exert their effect after either MULT-1 or H60 exit the ERGIC-*cis*-Golgi (Hasan *et al.*, 2005; Krmpotic *et al.*, 2005). The m138 protein has been shown to interfere with the recycling of the intracellular portion of MULT-1, resulting in its subsequent lysosomal degradation; H60 surface expression is also down-regulated by the m138 protein, though by a different mechanism than that of MULT-1. In fact, it seems that different N-terminal ectodomains of the m138 protein are involved in the regulation of these two NKG2D ligands (Lenac *et al.*, 2006). Moreover, while *m152*/gp40 is essential for RAE-1 retention in the ER, it is not entirely responsible for the lack of RAE-1ε surface expression. Indeed, the m138 protein induces clathrin-dependent endocytosis of the surface-resident RAE-1ε, resulting in its down-regulation (Arapovic *et al.*, 2009a). Furthermore, the m138 protein can also inhibit the expression of the costimulatory molecule CD80 (B7-1) on DCs, leading to reduced $CD8^+$ T-cell priming (Mintern *et al.*, 2006).

The importance of MCMV-encoded NKG2D immunoevasins has been validated *in vivo*. Indeed, deletion of any of the four aforementioned genes decreases viral titres in target organs of infected animals. The replication of the attenuated virus can be reversed by NK-cell depletion or by NKG2D blocking (reviewed in: Lisnic *et al.*, 2010) (Fig. II.9.6). Interestingly, it is the deletion of the *m138* gene which results in the strongest MCMV attenuation; this is to be expected considering the variety of immunosubversive functions of the protein it encodes.

Conclusions and future perspective

Our current understanding of the principles of early immune surveillance, particularly the role of NK cells in virus control, was greatly refined by the use of a mouse model of CMV infection. On one hand, the roles of several activating NKRs and their ligands have been identified. Indeed, work on Ly49H and its virally encoded ligand has helped to better grasp NK-cell responses; it has also highlighted the strong selective pressure exerted on the virus, pushing it to develop NK-cell evasion mechanisms. Secondly, studying MCMV evasion of 'missing self' recognition and host responses to these viral immunoevasion strategies has served as undeniable evidence for the continued evolutionary struggle between NK cells and CMVs, as shown by the function of *m04*/gp34. In fact, while *m04*/gp34 prevents NK-cell activation by escaping 'missing-self' recognition, several activating NKRs depend on its presence for their own recognition of infected cells.

Of course, the study of such fundamental immune mechanisms gives rise to several questions. T-cell mechanisms were already well characterized when researchers began to investigate NK cells. It is entirely possible that the experimental design used to study T-cells has greatly coloured our study of NK cells. Indeed, the specific recognition of m157 by Ly49H, which results in NK-cell cytotoxicity, proliferation, contraction, and recall, fits the T-cell paradigm: that of strong, dominant activation. Yet, Ly49H is not protective for MCMV infection in any other strain than C57BL/6. In fact, integration of multiple activating and inhibitory signals from the various NKRs seems to be the primary mechanism determining NK-cell function. Moreover, NK cells are mostly subjected to inhibition, which can even be considered their default state. Why is it that we often study NK cells from the viewpoint of what the activation does to them? Inhibitory triggering might have other impacts than just blocking activation. Specific mechanisms of NK-cell inhibition should be investigated, especially since this is what many viruses target. Then again, why is NK cell inhibition so widespread? A weak NK-cell response might be preferable for the induction of stronger adaptive immunity. Conversely, a robust innate NK-cell response might lead to autoimmunity or sterilizing immunity, preventing the formation of memory. Evidently, the field is wrought with future research perspectives, concerning either the host immune system or CMV pathogenesis.

Acknowledgements

We are grateful to our laboratory members for their critical comments and to Eve-Marie Gendron-Pontbriand for editorial assistance. We apologize to our colleagues whose work was not referenced due to space limitations.

References

Ablasser, A., Bauernfeind, F., Hartmann, G., Latz, E., Fitzgerald, K.A., and Hornung, V. (2009). RIG-I-dependent sensing of poly(dA:dT) through the induction of an RNA polymerase III-transcribed RNA intermediate. Nat. Immunol. *10*, 1065–1072.

Adams, E.J., Juo, Z.S., Venook, R.T., Boulanger, M.J., Arase, H., Lanier, L.L., and Garcia, K.C. (2007). Structural elucidation of the m157 mouse cytomegalovirus ligand for Ly49 natural killer cell receptors. Proc. Natl. Acad. Sci. U.S.A. *104*, 10128–10133.

Andoniou, C.E., van Dommelen, S.L., Voigt, V., Andrews, D.M., Brizard, G., Asselin-Paturel, C., Delale, T., Stacey,

K.J., Trinchieri, G., and Degli-Esposti, M.A. (2005). Interaction between conventional dendritic cells and natural killer cells is integral to the activation of effective antiviral immunity. Nat. Immunol. *6*, 1011–1019.

Andrews, D.M., Andoniou, C.E., Granucci, F., Ricciardi-Castagnoli, P., and Degli-Esposti, M.A. (2001). Infection of dendritic cells by murine cytomegalovirus induces functional paralysis. Nat. Immunol. *2*, 1077–1084.

Andrews, D.M., Andoniou, C.E., Scalzo, A.A., van Dommelen, S.L., Wallace, M.E., Smyth, M.J., and Degli-Esposti, M.A. (2005). Cross-talk between dendritic cells and natural killer cells in viral infection. Mol. Immunol. *42*, 547–555.

Arapovic, J., Lenac Rovis, T., Reddy, A.B., Krmpotic, A., and Jonjic, S. (2009a). Promiscuity of MCMV immunoevasin of NKG2D: m138/fcr-1 down-modulates RAE-1epsilon in addition to MULT-1 and H60. Mol. Immunol. *47*, 114–122.

Arapovic, J., Lenac, T., Antulov, R., Polic, B., Ruzsics, Z., Carayannopoulos, L.N., Koszinowski, U.H., Krmpotic, A., and Jonjic, S. (2009b). Differential susceptibility of RAE-1 isoforms to mouse cytomegalovirus. J. Virol. *83*, 8198–8207.

Arase, H., Mocarski, E.S., Campbell, A.E., Hill, A.B., and Lanier, L.L. (2002). Direct recognition of cytomegalovirus by activating and inhibitory NK cell receptors. Science *296*, 1323–1326.

Ardavin, C., Martinez del Hoyo, G., Martin, P., Anjuere, F., Arias, C.F., Marin, A.R., Ruiz, S., Parrillas, V., and Hernandez, H. (2001). Origin and differentiation of dendritic cells. Trends Immunol. *22*, 691–700.

Asselin-Paturel, C., Boonstra, A., Dalod, M., Durand, I., Yessaad, N., Dezutter-Dambuyant, C., Vicari, A., O'Garra, A., Biron, C., Briere, F., *et al.* (2001). Mouse type I IFN-producing cells are immature APCs with plasmacytoid morphology. Nat. Immunol. *2*, 1144–1150.

Babic, M., Pyzik, M., Zafirova, B., Mitrovic, M., Butorac, V., Lanier, L.L., Krmpotic, A., Vidal, S.M., and Jonjic, S. (2010). Cytomegalovirus immunoevasin reveals the physiological role of 'missing self' recognition in natural killer cell dependent virus control *in vivo*. J. Exp. Med. *207*, 2663–2673.

Banks, T.A., Rickert, S., Benedict, C.A., Ma, L., Ko, M., Meier, J., Ha, W., Schneider, K., Granger, S.W., Turovskaya, O., *et al.* (2005). A lymphotoxin-IFN-beta axis essential for lymphocyte survival revealed during cytomegalovirus infection. J. Immunol. *174*, 7217–7225.

Barao, I., Alvarez, M., Ames, E., Orr, M.T., Stefanski, H.E., Blazar, B.R., Lanier, L.L., Anderson, S.K., Redelman, D., and Murphy, W.J. (2011). Mouse Ly49G2+ NK cells dominate early responses during both immune reconstitution and activation independently of MHC. Blood *117*, 7032–7041.

Barbalat, R., Lau, L., Locksley, R.M., and Barton, G.M. (2009). Toll-like receptor 2 on inflammatory monocytes induces type I interferon in response to viral but not bacterial ligands. Nat. Immunol. *10*, 1200–1207.

Barber, G.N. (2011). STING-dependent signaling. Nat. Immunol. *12*, 929–930.

Benedict, C.A., Banks, T.A., Senderowicz, L., Ko, M., Britt, W.J., Angulo, A., Ghazal, P., and Ware, C.F. (2001). Lymphotoxins and cytomegalovirus cooperatively induce interferon-beta, establishing host-virus detente. Immunity *15*, 617–626.

Beutler, B. (2009). Microbe sensing, positive feedback loops, and the pathogenesis of inflammatory diseases. Immunol. Rev. *227*, 248–263.

Biron, C.A., Byron, K.S., and Sullivan, J.L. (1989). Severe herpesvirus infections in an adolescent without natural killer cells. N. Engl. J. Med. *320*, 1731–1735.

Blasius, A.L., and Beutler, B. (2010). Intracellular toll-like receptors. Immunity *32*, 305–315.

Borden, E.C., Sen, G.C., Uze, G., Silverman, R.H., Ransohoff, R.M., Foster, G.R., and Stark, G.R. (2007). Interferons at age 50: past, current and future impact on biomedicine. Nat. Rev. Drug. Discov. *6*, 975–990.

Brennan, J., Mahon, G., Mager, D.L., Jefferies, W.A., and Takei, F. (1996). Recognition of class I major histocompatibility complex molecules by Ly-49: specificities and domain interactions. J. Exp. Med. *183*, 1553–1559.

Brinkmann, M.M., Spooner, E., Hoebe, K., Beutler, B., Ploegh, H.L., and Kim, Y.M. (2007). The interaction between the ER membrane protein UNC93B and TLR3, 7, and 9 is crucial for TLR signaling. J. Cell. Biol. *177*, 265–275.

Bubic, I., Wagner, M., Krmpotic, A., Saulig, T., Kim, S., Yokoyama, W.M., Jonjic, S., and Koszinowski, U.H. (2004). Gain of virulence caused by loss of a gene in murine cytomegalovirus. J. Virol. *78*, 7536–7544.

Bukowski, J.F., Woda, B.A., and Welsh, R.M. (1984). Pathogenesis of murine cytomegalovirus infection in natural killer cell-depleted mice. J. Virol. *52*, 119–128.

Bukowski, J.F., Warner, J.F., Dennert, G., and Welsh, R.M. (1985). Adoptive transfer studies demonstrating the antiviral effect of natural killer cells *in vivo*. J. Exp. Med. *161*, 40–52.

Bukowski, J.F., Yang, H., and Welsh, R.M. (1988). Antiviral effect of lymphokine-activated killer cells: characterization of effector cells mediating prophylaxis. J. Virol. *62*, 3642–3648.

Burckstummer, T., Baumann, C., Bluml, S., Dixit, E., Durnberger, G., Jahn, H., Planyavsky, M., Bilban, M., Colinge, J., Bennett, K.L., *et al.* (2009). An orthogonal proteomic-genomic screen identifies AIM2 as a cytoplasmic DNA sensor for the inflammasome. Nat. Immunol. *10*, 266–272.

Carayannopoulos, L.N., and Yokoyama, W.M. (2004). Recognition of infected cells by natural killer cells. Curr. Opin. Immunol. *16*, 26–33.

Carayannopoulos, L.N., Naidenko, O.V., Fremont, D.H., and Yokoyama, W.M. (2002). Cutting edge: murine UL16-binding protein-like transcript 1: a newly described transcript encoding a high-affinity ligand for murine NKG2D. J. Immunol. *169*, 4079–4083.

Carlyle, J.R., Mesci, A., Fine, J.H., Chen, P., Belanger, S., Tai, L.H., and Makrigiannis, A.P. (2008). Evolution of the Ly49 and Nkrp1 recognition systems. Semin. Immunol. *20*, 321–330.

Casanova, J.L., Abel, L., and Quintana-Murci, L. (2011). Human TLRs and IL-1Rs in host defense: natural insights from evolutionary, epidemiological, and clinical genetics. Annu. Rev. Immunol. *29*, 447–491.

Cerwenka, A., and Lanier, L.L. (2003). NKG2D ligands: unconventional MHC class I-like molecules exploited by viruses and cancer. Tissue Antigens *61*, 335–343.

Cerwenka, A., Bakker, A.B., McClanahan, T., Wagner, J., Wu, J., Phillips, J.H., and Lanier, L.L. (2000). Retinoic acid early inducible genes define a ligand family for the activating NKG2D receptor in mice. Immunity *12*, 721–727.

Cheng, T.P., French, A.R., Plougastel, B.F., Pingel, J.T., Orihuela, M.M., Buller, M.L., and Yokoyama, W.M. (2008). Ly49h is necessary for genetic resistance to murine cytomegalovirus. Immunogenetics *60*, 565–573.

Chiu, Y.H., Macmillan, J.B., and Chen, Z.J. (2009). RNA polymerase III detects cytosolic DNA and induces type I interferons through the RIG-I pathway. Cell *138*, 576–591.

Coles, M.C., McMahon, C.W., Takizawa, H., and Raulet, D.H. (2000). Memory CD8 T-lymphocytes express inhibitory MHC-specific Ly49 receptors. Eur. J. Immunol. *30*, 236–244.

Collins, T.M., Quirk, M.R., and Jordan, M.C. (1994). Biphasic viremia and viral gene expression in leukocytes during acute cytomegalovirus infection of mice. J. Virol. *68*, 6305–6311.

Cooper, M.A., Fehniger, T.A., Fuchs, A., Colonna, M., and Caligiuri, M.A. (2004). NK cell and DC interactions. Trends Immunol. *25*, 47–52.

Corbett, A.J., Forbes, C.A., Moro, D., and Scalzo, A.A. (2007). Extensive sequence variation exists among isolates of murine cytomegalovirus within members of the m02 family of genes. J. Gen. Virol. *88*, 758–769.

Corbett, A.J., Coudert, J.D., Forbes, C.A., and Scalzo, A.A. (2010). Functional consequences of natural sequence variation of murine cytomegalovirus m157 for Ly49 receptor specificity and NK cell activation. J. Immunol. *186*, 1713–1722.

Cretney, E., Degli-Esposti, M.A., Densley, E.H., Farrell, H.E., Davis-Poynter, N.J., and Smyth, M.J. (1999). m144, a murine cytomegalovirus (MCMV)-encoded major histocompatibility complex class I homologue, confers tumor resistance to natural killer cell-mediated rejection. J. Exp. Med. *190*, 435–444.

Dalod, M., Hamilton, T., Salomon, R., Salazar-Mather, T.P., Henry, S.C., Hamilton, J.D., and Biron, C.A. (2003). Dendritic cell responses to early murine cytomegalovirus infection: subset functional specialization and differential regulation by interferon alpha/beta. J. Exp. Med. *197*, 885–898.

Daniels, K.A., Devora, G., Lai, W.C., O'Donnell, C.L., Bennett, M., and Welsh, R.M. (2001). Murine cytomegalovirus is regulated by a discrete subset of natural killer cells reactive with monoclonal antibody to Ly49H. J. Exp. Med. *194*, 29–44.

Davis, A.H., Guseva, N.V., Ball, B.L., and Heusel, J.W. (2008). Characterization of murine cytomegalovirus m157 from infected cells and identification of critical residues mediating recognition by the NK cell receptor Ly49H. J. Immunol. *181*, 265–275.

Desrosiers, M.P., Kielczewska, A., Loredo-Osti, J.C., Adam, S.G., Makrigiannis, A.P., Lemieux, S., Pham, T., Lodoen, M.B., Morgan, K., Lanier, L.L., *et al.* (2005). Epistasis between mouse Klra and major histocompatibility complex class I loci is associated with a new mechanism of natural killer cell-mediated innate resistance to cytomegalovirus infection. Nat. Genet. *37*, 593–599.

Diefenbach, A., Jensen, E.R., Jamieson, A.M., and Raulet, D.H. (2001). Rae1 and H60 ligands of the NKG2D receptor stimulate tumour immunity. Nature *413*, 165–171.

Dietrich, N., Lienenklaus, S., Weiss, S., and Gekara, N.O. (2010). Murine toll-like receptor 2 activation induces type I interferon responses from endolysosomal compartments. PLoS One *5*, e10250.

Dighe, A., Rodriguez, M., Sabastian, P., Xie, X., McVoy, M., and Brown, M.G. (2005). Requisite H2k role in NK cell-mediated resistance in acute murine cytomegalovirus-infected MA/My mice. J. Immunol. *175*, 6820–6828.

Dimasi, N., Sawicki, M.W., Reineck, L.A., Li, Y., Natarajan, K., Margulies, D.H., and Mariuzza, R.A. (2002). Crystal structure of the Ly49I natural killer cell receptor reveals variability in dimerization mode within the Ly49 family. J. Mol. Biol. *320*, 573–585.

Dokun, A.O., Kim, S., Smith, H.R., Kang, H.S., Chu, D.T., and Yokoyama, W.M. (2001). Specific and nonspecific NK cell activation during virus infection. Nat. Immunol. *2*, 951–956.

Ewald, S.E., Engel, A., Lee, J., Wang, M., Bogyo, M., and Barton, G.M. (2011). Nucleic acid recognition by Toll-like receptors is coupled to stepwise processing by cathepsins and asparagine endopeptidase. J. Exp. Med. *208*, 643–651.

Farrell, H.E., Vally, H., Lynch, D.M., Fleming, P., Shellam, G.R., Scalzo, A.A., and Davis-Poynter, N.J. (1997). Inhibition of natural killer cells by a cytomegalovirus MHC class I homologue *in vivo*. Nature *386*, 510–514.

Fernandes-Alnemri, T., Yu, J.W., Datta, P., Wu, J., and Alnemri, E.S. (2009). AIM2 activates the inflammasome and cell death in response to cytoplasmic DNA. Nature *458*, 509–513.

Fodil-Cornu, N., Lee, S.H., Belanger, S., Makrigiannis, A.P., Biron, C.A., Buller, R.M., and Vidal, S.M. (2008). Ly49h-deficient C57BL/6 mice: a new mouse cytomegalovirus-susceptible model remains resistant to unrelated pathogens controlled by the NK gene complex. J. Immunol. *181*, 6394–6405.

Fodil-Cornu, N., Loredo-Osti, J.C., and Vidal, S.M. (2011). NK cell receptor/H2-Dk-dependent host resistance to viral infection is quantitatively modulated by H2q inhibitory signals. PLoS Genet. *7*, e1001368.

French, A.R., Pingel, J.T., Wagner, M., Bubic, I., Yang, L., Kim, S., Koszinowski, U., Jonjic, S., and Yokoyama, W.M. (2004). Escape of mutant double-stranded DNA virus from innate immune control. Immunity *20*, 747–756.

Gordon, S. (1998). The role of the macrophage in immune regulation. Res. Immunol. *149*, 685–688.

Grundy, J.E., Mackenzie, J.S., and Stanley, N.F. (1981). Influence of H-2 and non-H-2 genes on resistance to murine cytomegalovirus infection. Infect. Immun. *32*, 277–286.

Guseva, N.V., Fullenkamp, C.A., Naumann, P.W., Shey, M.R., Ballas, Z.K., Houtman, J.C., Forbes, C.A., Scalzo, A.A., and Heusel, J.W. (2010). Glycosylation contributes to variability in expression of murine cytomegalovirus m157 and enhances stability of interaction with the NK-cell receptor Ly49H. Eur. J. Immunol. *40*, 2618–2631.

den Haan, J.M., Lehar, S.M., and Bevan, M.J. (2000). CD8(+) but not CD8(-) dendritic cells cross-prime cytotoxic T-cells *in vivo*. J. Exp. Med. *192*, 1685–1696.

Hanson, L.K., Slater, J.S., Karabekian, Z., Virgin, H.W.t., Biron, C.A., Ruzek, M.C., van Rooijen, N., Ciavarra, R.P., Stenberg, R.M., and Campbell, A.E. (1999). Replication of murine cytomegalovirus in differentiated macrophages as a determinant of viral pathogenesis. J. Virol. *73*, 5970–5980.

Hasan, M., Krmpotic, A., Ruzsics, Z., Bubic, I., Lenac, T., Halenius, A., Loewendorf, A., Messerle, M., Hengel, H., Jonjic, S., *et al.* (2005). Selective down-regulation of the

NKG2D ligand H60 by mouse cytomegalovirus m155 glycoprotein. J. Virol. *79*, 2920–2930.

Hengel, H., Reusch, U., Gutermann, A., Ziegler, H., Jonjic, S., Lucin, P., and Koszinowski, U.H. (1999). Cytomegaloviral control of MHC class I function in the mouse. Immunol. Rev. *168*, 167–176.

Hengel, H., Reusch, U., Geginat, G., Holtappels, R., Ruppert, T., Hellebrand, E., and Koszinowski, U.H. (2000). Macrophages escape inhibition of major histocompatibility complex class I-dependent antigen presentation by cytomegalovirus. J. Virol. *74*, 7861–7868.

Hochrein, H., Shortman, K., Vremec, D., Scott, B., Hertzog, P., and O'Keeffe, M. (2001). Differential production of IL-12, IFN-alpha, and IFN-gamma by mouse dendritic cell subsets. J. Immunol. *166*, 5448–5455.

Hoebe, K., Du, X., Georgel, P., Janssen, E., Tabeta, K., Kim, S.O., Goode, J., Lin, P., Mann, N., Mudd, S., *et al.* (2003). Identification of Lps2 as a key transducer of MyD88-independent TIR signalling. Nature *424*, 743–748.

Hoglund, P., and Brodin, P. (2010). Current perspectives of natural killer cell education by MHC class I molecules. Nat. Rev. Immunol. *10*, 724–734.

Holtappels, R., Gillert-Marien, D., Thomas, D., Podlech, J., Deegen, P., Herter, S., Oehrlein-Karpi, S.A., Strand, D., Wagner, M., and Reddehase, M.J. (2006). Cytomegalovirus encodes a positive regulator of antigen presentation. J. Virol. *80*, 7613–7624.

Hornung, V., Ablasser, A., Charrel-Dennis, M., Bauernfeind, F., Horvath, G., Caffrey, D.R., Latz, E., and Fitzgerald, K.A. (2009). AIM2 recognizes cytosolic dsDNA and forms a caspase-1-activating inflammasome with ASC. Nature *458*, 514–518.

Iizuka, K., Naidenko, O.V., Plougastel, B.F., Fremont, D.H., and Yokoyama, W.M. (2003). Genetically linked C-type lectin-related ligands for the NKRP1 family of natural killer cell receptors. Nat. Immunol. *4*, 801–807.

Ishii, K.J., Kawagoe, T., Koyama, S., Matsui, K., Kumar, H., Kawai, T., Uematsu, S., Takeuchi, O., Takeshita, F., Coban, C., *et al.* (2008). TANK-binding kinase-1 delineates innate and adaptive immune responses to DNA vaccines. Nature *451*, 725–729.

Ishikawa, H., Ma, Z., and Barber, G.N. (2009). STING regulates intracellular DNA-mediated, type I interferon-dependent innate immunity. Nature *461*, 788–792.

Iyoda, T., Shimoyama, S., Liu, K., Omatsu, Y., Akiyama, Y., Maeda, Y., Takahara, K., Steinman, R.M., and Inaba, K. (2002). The CD8+ dendritic cell subset selectively endocytoses dying cells in culture and *in vivo*. J. Exp. Med. *195*, 1289–1302.

Jonjic, S., Babic, M., Polic, B., and Krmpotic, A. (2008). Immune evasion of natural killer cells by viruses. Curr. Opin. Immunol. *20*, 30–38.

Karlhofer, F.M., Ribaudo, R.K., and Yokoyama, W.M. (1992). The interaction of Ly-49 with H-2Dd globally inactivates natural killer cell cytolytic activity. Trans. Assoc. Am. Physicians *105*, 72–85.

Kavanagh, D.G., Gold, M.C., Wagner, M., Koszinowski, U.H., and Hill, A.B. (2001). The multiple immune-evasion genes of murine cytomegalovirus are not redundant: m4 and m152 inhibit antigen presentation in a complementary and cooperative fashion. J. Exp. Med. *194*, 967–978.

Kielczewska, A., Kim, H.S., Lanier, L.L., Dimasi, N., and Vidal, S.M. (2007). Critical residues at the Ly49 natural killer receptor's homodimer interface determine functional recognition of m157, a mouse cytomegalovirus MHC class I-like protein. J. Immunol. *178*, 369–377.

Kielczewska, A., Pyzik, M., Sun, T., Krmpotic, A., Lodoen, M.B., Munks, M.W., Babic, M., Hill, A.B., Koszinowski, U.H., Jonjic, S., *et al.* (2009). Ly49P recognition of cytomegalovirus-infected cells expressing H2-Dk and CMV-encoded m04 correlates with the NK cell antiviral response. J. Exp. Med. *206*, 515–523.

Kiessling, R., Klein, E., and Wigzell, H. (1975). 'Natural' killer cells in the mouse. I. Cytotoxic cells with specificity for mouse Moloney leukemia cells. Specificity and distribution according to genotype. Eur. J. Immunol. *5*, 112–117.

Kim, Y.M., Brinkmann, M.M., Paquet, M.E., and Ploegh, H.L. (2008). UNC93B1 delivers nucleotide-sensing toll-like receptors to endolysosomes. Nature *452*, 234–238.

Kleijnen, M.F., Huppa, J.B., Lucin, P., Mukherjee, S., Farrell, H., Campbell, A.E., Koszinowski, U.H., Hill, A.B., and Ploegh, H.L. (1997). A mouse cytomegalovirus glycoprotein, gp34, forms a complex with folded class I MHC molecules in the ER which is not retained but is transported to the cell surface. EMBO J. *16*, 685–694.

Kloover, J.S., Grauls, G.E., Blok, M.J., Vink, C., and Bruggeman, C.A. (2002). A rat cytomegalovirus strain with a disruption of the r144 MHC class I-like gene is attenuated in the acute phase of infection in neonatal rats. Arch. Virol. *147*, 813–824.

Krmpotic, A., Hasan, M., Loewendorf, A., Saulig, T., Halenius, A., Lenac, T., Polic, B., Bubic, I., Kriegeskorte, A., Pernjak-Pugel, E., *et al.* (2005). NK cell activation through the NKG2D ligand MULT-1 is selectively prevented by the glycoprotein encoded by mouse cytomegalovirus gene m145. J. Exp. Med. *201*, 211–220.

Krug, A., Uppaluri, R., Facchetti, F., Dorner, B.G., Sheehan, K.C., Schreiber, R.D., Cella, M., and Colonna, M. (2002). IFN-producing cells respond to CXCR3 ligands in the presence of CXCL12 and secrete inflammatory chemokines upon activation. J. Immunol. *169*, 6079–6083.

Krug, A., French, A.R., Barchet, W., Fischer, J.A., Dzionek, A., Pingel, J.T., Orihuela, M.M., Akira, S., Yokoyama, W.M., and Colonna, M. (2004). TLR9-dependent recognition of MCMV by IPC and DC generates coordinated cytokine responses that activate antiviral NK cell function. Immunity *21*, 107–119.

Kubota, A., Kubota, S., Farrell, H.E., Davis-Poynter, N., and Takei, F. (1999). Inhibition of NK cells by murine CMV-encoded class I MHC homologue m144. Cell. Immunol. *191*, 145–151.

Lanier, L.L. (2008). Up on the tightrope: natural killer cell activation and inhibition. Nat. Immunol. *9*, 495–502.

Lee, S.H., Webb, J.R., and Vidal, S.M. (2002). Innate immunity to cytomegalovirus: the Cmv1 locus and its role in natural killer cell function. Microbes Infect. *4*, 1491–1503.

Lee, S.H., Zafer, A., de Repentigny, Y., Kothary, R., Tremblay, M.L., Gros, P., Duplay, P., Webb, J.R., and Vidal, S.M. (2003). Transgenic expression of the activating natural killer receptor Ly49H confers resistance to cytomegalovirus in genetically susceptible mice. J. Exp. Med. *197*, 515–526.

Lemmermann, N.A., Gergely, K., Bohm, V., Deegen, P., Daubner, T., and Reddehase, M.J. (2010). Immune evasion proteins of murine cytomegalovirus preferentially affect cell surface display of recently generated peptide presentation complexes. J. Virol. *84*, 1221–1236.

Lenac, T., Budt, M., Arapovic, J., Hasan, M., Zimmermann, A., Simic, H., Krmpotic, A., Messerle, M., Ruzsics, Z., Koszinowski, U.H., *et al.* (2006). The herpesviral Fc receptor fcr-1 down-regulates the NKG2D ligands MULT-1 and H60. J. Exp. Med. *203*, 1843–1850.

Lisnic, V.J., Krmpotic, A., and Jonjic, S. (2010). Modulation of natural killer cell activity by viruses. Curr. Opin. Microbiol. *13*, 530–539.

Lodoen, M., Ogasawara, K., Hamerman, J.A., Arase, H., Houchins, J.P., Mocarski, E.S., and Lanier, L.L. (2003). NKG2D-mediated natural killer cell protection against cytomegalovirus is impaired by viral gp40 modulation of retinoic acid early inducible 1 gene molecules. J. Exp. Med. *197*, 1245–1253.

Loewendorf, A.I., Steinbrueck, L., Peter, C., Busche, A., Benedict, C.A., and Kay-Jackson, P.C. (2011). The mouse cytomegalovirus glycoprotein m155 inhibits CD40 expression and restricts CD4 T-cell responses. J. Virol. *85*, 5208–5212.

Loo, Y.M., and Gale, M., Jr. (2011). Immune signaling by RIG-I-like receptors. Immunity *34*, 680–692.

Lu, X., Kavanagh, D.G., and Hill, A.B. (2006). Cellular and molecular requirements for association of the murine cytomegalovirus protein m4/gp34 with major histocompatibility complex class I molecules. J. Virol. *80*, 6048–6055.

Mans, J., Natarajan, K., Balbo, A., Schuck, P., Eikel, D., Hess, S., Robinson, H., Simic, H., Jonjic, S., Tiemessen, C.T., *et al.* (2007). Cellular expression and crystal structure of the murine cytomegalovirus major histocompatibility complex class I-like glycoprotein, m153. J. Biol. Chem. *282*, 35247–35258.

Mans, J., Zhi, L., Revilleza, M.J., Smith, L., Redwood, A., Natarajan, K., and Margulies, D.H. (2009). Structure and function of murine cytomegalovirus MHC-I-like molecules: how the virus turned the host defense to its advantage. Immunol. Res. *43*, 264–279.

Mesci, A., Ljutic, B., Makrigiannis, A.P., and Carlyle, J.R. (2006). NKR-P1 biology: from prototype to missing self. Immunol. Res. *35*, 13–26.

Mintern, J.D., Klemm, E.J., Wagner, M., Paquet, M.E., Napier, M.D., Kim, Y.M., Koszinowski, U.H., and Ploegh, H.L. (2006). Viral interference with B7-1 costimulation: a new role for murine cytomegalovirus fc receptor-1. J. Immunol. *177*, 8422–8431.

Noda, S., Aguirre, S.A., Bitmansour, A., Brown, J.M., Sparer, T.E., Huang, J., and Mocarski, E.S. (2006). Cytomegalovirus MCK-2 controls mobilization and recruitment of myeloid progenitor cells to facilitate dissemination. Blood *107*, 30–38.

Ogasawara, K., Benjamin, J., Takaki, R., Phillips, J.H., and Lanier, L.L. (2005). Function of NKG2D in natural killer cell-mediated rejection of mouse bone marrow grafts. Nat. Immunol. *6*, 938–945.

Orange, J.S., and Biron, C.A. (1996). Characterization of early IL-12, IFN-alphabeta, and TNF effects on antiviral state and NK cell responses during murine cytomegalovirus infection. J. Immunol. *156*, 4746–4756.

Orr, M.T., Murphy, W.J., and Lanier, L.L. (2010). 'Unlicensed' natural killer cells dominate the response to cytomegalovirus infection. Nat. Immunol. *11*, 321–327.

Ortaldo, J.R., Mason, A.T., Winkler-Pickett, R., Raziuddin, A., Murphy, W.J., and Mason, L.H. (1999). Ly-49 receptor expression and functional analysis in multiple mouse strains. J. Leukoc. Biol. *66*, 512–520.

Park, B., Brinkmann, M.M., Spooner, E., Lee, C.C., Kim, Y.M., and Ploegh, H.L. (2008). Proteolytic cleavage in an endolysosomal compartment is required for activation of Toll-like receptor 9. Nat. Immunol. *9*, 1407–1414.

Patel, R., Belanger, S., Tai, L.H., Troke, A.D., and Makrigiannis, A.P. (2010). Effect of Ly49 haplotype variance on NK cell function and education. J. Immunol. *185*, 4783–4792.

Pichlmair, A., Schulz, O., Tan, C.P., Naslund, T.I., Liljestrom, P., Weber, F., and Reis e Sousa, C. (2006). RIG-I-mediated antiviral responses to single-stranded RNA bearing 5'-phosphates. Science *314*, 997–1001.

Pooley, J.L., Heath, W.R., and Shortman, K. (2001). Cutting edge: intravenous soluble antigen is presented to CD4 T-cells by CD8- dendritic cells, but cross-presented to CD8 T-cells by CD8+ dendritic cells. J. Immunol. *166*, 5327–5330.

Pyzik, M., Charbonneau, B., Gendron-Pontbriand, E.M., Babic, M., Krmpotic, A., Jonjic, S., and Vidal, S.M. (2011a). Distinct MHC class I-dependent NK cell-activating receptors control cytomegalovirus infection in different mouse strains. J. Exp. Med. *208*, 1105–1117.

Pyzik, M., Gendron-Pontbriand, E.M., and Vidal, S.M. (2011b). The impact of Ly49-NK cell-dependent recognition of MCMV infection on innate and adaptive immune responses. J. Biomed. Biotechnol. *2011*, 641702.

Raftery, M.J., Schwab, M., Eibert, S.M., Samstag, Y., Walczak, H., and Schonrich, G. (2001). Targeting the function of mature dendritic cells by human cytomegalovirus: a multilayered viral defense strategy. Immunity *15*, 997–1009.

Rathinam, V.A., Jiang, Z., Waggoner, S.N., Sharma, S., Cole, L.E., Waggoner, L., Vanaja, S.K., Monks, B.G., Ganesan, S., Latz, E., *et al.* (2010). The AIM2 inflammasome is essential for host defense against cytosolic bacteria and DNA viruses. Nat. Immunol. *11*, 395–402.

Rebsamen, M., Heinz, L.X., Meylan, E., Michallet, M.C., Schroder, K., Hofmann, K., Vazquez, J., Benedict, C.A., and Tschopp, J. (2009). DAI/ZBP1 recruits RIP1 and RIP3 through RIP homotypic interaction motifs to activate NF-kappaB. EMBO Rep. *10*, 916–922.

Redpath, S., Angulo, A., Gascoigne, N.R., and Ghazal, P. (1999). Murine cytomegalovirus infection down-regulates MHC class II expression on macrophages by induction of IL-10. J. Immunol. *162*, 6701–6707.

Riegler, S., Hebart, H., Einsele, H., Brossart, P., Jahn, G., and Sinzger, C. (2000). Monocyte-derived dendritic cells are permissive to the complete replicative cycle of human cytomegalovirus. J. Gen. Virol. *81*, 393–399.

Robbins, S.H., Bessou, G., Cornillon, A., Zucchini, N., Rupp, B., Ruzsics, Z., Sacher, T., Tomasello, E., Vivier, E., Koszinowski, U.H., *et al.* (2007). Natural killer cells promote early CD8 T-cell responses against cytomegalovirus. PLoS Pathog. *3*, e123.

Roberts, T.L., Idris, A., Dunn, J.A., Kelly, G.M., Burnton, C.M., Hodgson, S., Hardy, L.L., Garceau, V., Sweet, M.J., Ross, I.L., *et al.* (2009). HIN-200 proteins regulate caspase activation in response to foreign cytoplasmic DNA. Science *323*, 1057–1060.

Rodriguez, M., Sabastian, P., Clark, P., and Brown, M.G. (2004). Cmv1-independent antiviral role of NK cells revealed in murine cytomegalovirus-infected New Zealand White mice. J. Immunol. *173*, 6312–6318.

Sauer, J.D., Sotelo-Troha, K., von Moltke, J., Monroe, K.M., Rae, C.S., Brubaker, S.W., Hyodo, M., Hayakawa, Y., Woodward, J.J., Portnoy, D.A., *et al.* (2011). The

N-ethyl-N-nitrosourea-induced Goldenticket mouse mutant reveals an essential function of Sting in the *in vivo* interferon response to Listeria monocytogenes and cyclic dinucleotides. Infect. Immun. *79*, 688–694.

Schneider, K., Loewendorf, A., De Trez, C., Fulton, J., Rhode, A., Shumway, H., Ha, S., Patterson, G., Pfeffer, K., Nedospasov, S.A., *et al.* (2008). Lymphotoxin-mediated cross-talk between B cells and splenic stroma promotes the initial type I interferon response to cytomegalovirus. Cell Host Microbe *3*, 67–76.

Shanley, J.D. (1990). *In vivo* administration of monoclonal antibody to the NK 1.1 antigen of natural killer cells: effect on acute murine cytomegalovirus infection. J. Med. Virol. *30*, 58–60.

Sharma, S., and Fitzgerald, K.A. (2011). Innate immune sensing of DNA. PLoS Pathog. *7*, e1001310.

Shellam, G.R., Allan, J.E., Papadimitriou, J.M., and Bancroft, G.J. (1981). Increased susceptibility to cytomegalovirus infection in beige mutant mice. Proc. Natl. Acad. Sci. U.S.A. *78*, 5104–5108.

Shortman, K., and Liu, Y.J. (2002). Mouse and human dendritic cell subtypes. Nat. Rev. Immunol. *2*, 151–161.

Silver, E.T., Lavender, K.J., Gong, D.E., Hazes, B., and Kane, K.P. (2002). Allelic variation in the ectodomain of the inhibitory Ly-49G2 receptor alters its specificity for allogeneic and xenogeneic ligands. J. Immunol. *169*, 4752–4760.

Sjolin, H., Tomasello, E., Mousavi-Jazi, M., Bartolazzi, A., Karre, K., Vivier, E., and Cerboni, C. (2002). Pivotal role of KARAP/DAP12 adaptor molecule in the natural killer cell-mediated resistance to murine cytomegalovirus infection. J. Exp. Med. *195*, 825–834.

Smith, H.R., Heusel, J.W., Mehta, I.K., Kim, S., Dorner, B.G., Naidenko, O.V., Iizuka, K., Furukawa, H., Beckman, D.L., Pingel, J.T., *et al.* (2002). Recognition of a virus-encoded ligand by a natural killer cell activation receptor. Proc. Natl. Acad. Sci. U.S.A. *99*, 8826–8831.

Smith, L.M., Shellam, G.R., and Redwood, A.J. (2006). Genes of murine cytomegalovirus exist as a number of distinct genotypes. Virology *352*, 450–465.

Spits, H., and Di Santo, J.P. (2011). The expanding family of innate lymphoid cells: regulators and effectors of immunity and tissue remodeling. Nat. Immunol. *12*, 21–27.

Steinman, R.M. (1991). The dendritic cell system and its role in immunogenicity. Annu. Rev. Immunol. *9*, 271–296.

Stoddart, C.A., Cardin, R.D., Boname, J.M., Manning, W.C., Abenes, G.B., and Mocarski, E.S. (1994). Peripheral blood mononuclear phagocytes mediate dissemination of murine cytomegalovirus. J. Virol. *68*, 6243–6253.

Swiecki, M., Gilfillan, S., Vermi, W., Wang, Y., and Colonna, M. (2010). Plasmacytoid dendritic cell ablation impacts early interferon responses and antiviral NK and CD8(+) T-cell accrual. Immunity *33*, 955–966.

Szomolanyi-Tsuda, E., Liang, X., Welsh, R.M., Kurt-Jones, E.A., and Finberg, R.W. (2006). Role for TLR2 in NK cell-mediated control of murine cytomegalovirus *in vivo*. J. Virol. *80*, 4286–4291.

Tabeta, K., Georgel, P., Janssen, E., Du, X., Hoebe, K., Crozat, K., Mudd, S., Shamel, L., Sovath, S., Goode, J., *et al.* (2004). Toll-like receptors 9 and 3 as essential components of innate immune defense against mouse cytomegalovirus infection. Proc. Natl. Acad. Sci. U.S.A. *101*, 3516–3521.

Takaoka, A., Wang, Z., Choi, M.K., Yanai, H., Negishi, H., Ban, T., Lu, Y., Miyagishi, M., Kodama, T., Honda, K., *et al.* (2007). DAI (DLM-1/ZBP1) is a cytosolic DNA sensor and an activator of innate immune response. Nature *448*, 501–505.

Tormo, J., Natarajan, K., Margulies, D.H., and Mariuzza, R.A. (1999). Crystal structure of a lectin-like natural killer cell receptor bound to its MHC class I ligand. Nature *402*, 623–631.

Tripathy, S.K., Keyel, P.A., Yang, L., Pingel, J.T., Cheng, T.P., Schneeberger, A., and Yokoyama, W.M. (2008). Continuous engagement of a self-specific activation receptor induces NK cell tolerance. J. Exp. Med. *205*, 1829–1841.

Upton, J.W., Kaiser, W.J., and Mocarski, E.S. (2010). Virus inhibition of RIP3-dependent necrosis. Cell Host Microbe *7*, 302–313.

Vance, R.E., Kraft, J.R., Altman, J.D., Jensen, P.E., and Raulet, D.H. (1998). Mouse CD94/NKG2A is a natural killer cell receptor for the nonclassical major histocompatibility complex (MHC) class I molecule Qa-1(b). J. Exp. Med. *188*, 1841–1848.

Vidal, S.M., Malo, D., Marquis, J.F., and Gros, P. (2008). Forward genetic dissection of immunity to infection in the mouse. Annu. Rev. Immunol. *26*, 81–132.

Vivier, E., Tomasello, E., and Paul, P. (2002). Lymphocyte activation via NKG2D: towards a new paradigm in immune recognition? Curr. Opin. Immunol. *14*, 306–311.

Vivier, E., Raulet, D.H., Moretta, A., Caligiuri, M.A., Zitvogel, L., Lanier, L.L., Yokoyama, W.M., and Ugolini, S. (2011). Innate or adaptive immunity? The example of natural killer cells. Science *331*, 44–49.

Vogler, I., and Steinle, A. (2011). Vis-a-vis in the NKC: genetically linked natural killer cell receptor/ligand pairs in the natural killer gene complex (NKC). J. Innate Immun. *3*, 227–235.

Voigt, S., Mesci, A., Ettinger, J., Fine, J.H., Chen, P., Chou, W., and Carlyle, J.R. (2007). Cytomegalovirus evasion of innate immunity by subversion of the NKR-P1B:Clr-b missing-self axis. Immunity *26*, 617–627.

Voigt, V., Forbes, C.A., Tonkin, J.N., Degli-Esposti, M.A., Smith, H.R., Yokoyama, W.M., and Scalzo, A.A. (2003). Murine cytomegalovirus m157 mutation and variation leads to immune evasion of natural killer cells. Proc. Natl. Acad. Sci. U.S.A. *100*, 13483–13488.

Wagner, M., Gutermann, A., Podlech, J., Reddehase, M.J., and Koszinowski, U.H. (2002). Major histocompatibility complex class I allele-specific cooperative and competitive interactions between immune evasion proteins of cytomegalovirus. J. Exp. Med. *196*, 805–816.

Walzer, T., Dalod, M., Vivier, E., and Zitvogel, L. (2005). Natural killer cell-dendritic cell cross-talk in the initiation of immune responses. Expert Opin. Biol. Ther. *5(Suppl. 1)*, S49–59.

Wang, Z., Choi, M.K., Ban, T., Yanai, H., Negishi, H., Lu, Y., Tamura, T., Takaoka, A., Nishikura, K., and Taniguchi, T. (2008). Regulation of innate immune responses by DAI (DLM-1/ZBP1) and other DNA-sensing molecules. Proc. Natl. Acad. Sci. U.S.A. *105*, 5477–5482.

Ware, C.F. (2005). Network communications: lymphotoxins, LIGHT, and TNF. Annu. Rev. Immunol. *23*, 787–819.

Welsh, R.M., Brubaker, J.O., Vargas-Cortes, M., and O'Donnell, C.L. (1991). Natural killer (NK) cell response to virus infections in mice with severe combined immunodeficiency. The stimulation of NK cells and the NK cell-dependent control of virus infections occur independently of T and B cell function. J. Exp. Med. *173*, 1053–1063.

Xie, X., Dighe, A., Clark, P., Sabastian, P., Buss, S., and Brown, M.G. (2007). Deficient major histocompatibility complex-linked innate murine cytomegalovirus immunity in MA/My.L-H2b mice and viral down-regulation of H-2k class I proteins. J. Virol. *81*, 229–236.

Xie, X., Stadnisky, M.D., and Brown, M.G. (2009). MHC class I Dk locus and Ly49G2+ NK cells confer H-2k resistance to murine cytomegalovirus. J. Immunol. *182*, 7163–7171.

Xie, X., Stadnisky, M.D., Coats, E.R., Ahmed Rahim, M.M., Lundgren, A., Xu, W., Makrigiannis, A.P., and Brown, M.G. (2010). MHC class I D(k) expression in hematopoietic and nonhematopoietic cells confers natural killer cell resistance to murine cytomegalovirus. Proc. Natl. Acad. Sci. U.S.A. *107*, 8754–8759.

Yokoyama, W.M., and Plougastel, B.F. (2003). Immune functions encoded by the natural killer gene complex. Nat. Rev. Immunol. *3*, 304–316.

Zhang, Z., Yuan, B., Bao, M., Lu, N., Kim, T., and Liu, Y.J. (2011). The helicase DDX41 senses intracellular DNA mediated by the adaptor STING in dendritic cells. Nat. Immunol. *12*, 959–965.

Zhi, L., Mans, J., Paskow, M.J., Brown, P.H., Schuck, P., Jonjic, S., Natarajan, K., and Margulies, D.H. (2010). Direct interaction of the mouse cytomegalovirus m152/gp40 immunoevasin with RAE-1 isoforms. Biochemistry *49*, 2443–2453.

Zitvogel, L. (2002). Dendritic and natural killer cells cooperate in the control/switch of innate immunity. J. Exp. Med. *195*, F9–14.

Protective Humoral Immunity

II.10

Michael Mach, Anna-Katharina Wiegers, Nadja Spindler and Thomas H. Winkler

Abstract

The generation of antibodies represents a powerful tool of the adaptive immune system in the battle against viral infections. Targets for antibodies with potential antiviral activity are glycoproteins in the viral envelope and/or on the surface of infected cells. In recent years, considerable progress has been made in our understanding of the protective antibody response against cytomegaloviruses. Animal studies have unambiguously demonstrated the protective capacity of antibodies both in prophylaxis as well as in therapy of existing primary infection or reactivation. A number of human monoclonal antibodies have been isolated which show potent virus-neutralizing capacity and new antibody targets have been identified. However, we still need to expand our knowledge on the mechanisms of virus neutralization by antibodies and the mode of action of protective antibodies *in vivo*. Increasing this knowledge will help us to rationally design strategies to limit the consequences of infections in populations at risk for CMV disease.

Introduction

Over eons, CMVs and their hosts have co-evolved. During this long co-evolution the virus adapted perfectly to the respective host defence systems and vice versa. As a result, infections in immunocompetent hosts are generally asymptomatic and a lifelong persistent/latent infection is established. Development of symptoms or disease is prevented by a multilayered, in large parts redundant, innate as well as adaptive immune response. Yet, none of the immune effector functions is capable of eliminating the virus from the host and, at least when entire populations are considered, preventing horizontal or vertical virus transmission. Transmission is thought to occur via cell free virus, indicating that CMVs have evolved mechanisms to evade complete immune control and allow production of infectious virus. It is unknown whether every infected host permits virus transmission or if 'complete controllers' exist and which antiviral functions would be associated with complete control. To device treatment strategies that keep the virus under control and prevent transmission we need to learn more about the individual immune effector functions that control virus latency, reactivation and transmission.

Antibodies represent an extremely powerful tool of the adaptive immune system to combat pathogens. The tremendous success of vaccines for prevention of viral infections bears witness to that since the correlate of protection following vaccination is virus-neutralizing antibodies (Plotkin, 2008; see also Chapter II.20). In addition, there is a vast number of examples which show protection from viral infection by the passive transfer of polyclonal or monoclonal antibodies. However, for herpesviruses in general and CMV in particular, protective capacity of antibodies is viewed with scepticism or even denied. The question arises whether CMV is truly the exception from the rule or whether this is a misconception and the importance of antibodies for the control of CMV has been underestimated or inadequately analysed so far.

In this chapter we will attempt to summarize what is known about the antiviral activity of CMV-specific antibodies and try to formulate some open questions that need to be addressed in order to enhance our understanding of the antiviral potential of CMV-specific antibodies and use it to improve strategies to prevent/treat the infection.

Effector functions of antibodies

The primary roles of antibodies are to prevent infection and/or reduce the overall viral burden in the host to an extent that infection is contained by the ensuing immune response. This task can be achieved by a variety of different antibody effector functions and before we go into detail with respect to anti-CMV antibodies we would like to briefly summarize some known antiviral functions of immunoglobulins.

In principle, antibodies can act directly on free virus and indirectly on infected cells (Fig. II.10.1). Binding to free virus can result in virus neutralization. By definition, neutralization is the ability of an antibody to bind to free virus thereby inactivating infectivity under defined conditions *in vitro*. Most neutralizing antibodies, but not all, also are able to protect animals *in vivo*. However, whether this protective capacity *in vivo* is based on direct neutralization of free virus or additional antibody-mediated effects is far from clear. *In vivo*, the situation is much more complex since antibodies interact with cells and molecules of the innate immune system and numerous studies have shown that non-neutralizing antibodies can have a protective effect *in vivo*.

Antibodies binding to virions

Despite its importance and decades of research, the mechanism(s) of virus neutralization is still controversially discussed. A number of different mechanisms have been postulated such as

- blocking receptor binding;
- inhibition of viral attachment to or entry into target cells;
- post-entry mechanisms such as prevention of uncoating;
- destabilization of virion structure;
- induction of conformational changes in envelope or capsid proteins following antibody binding; and
- aggregation of virions (Dimmock, 1993).

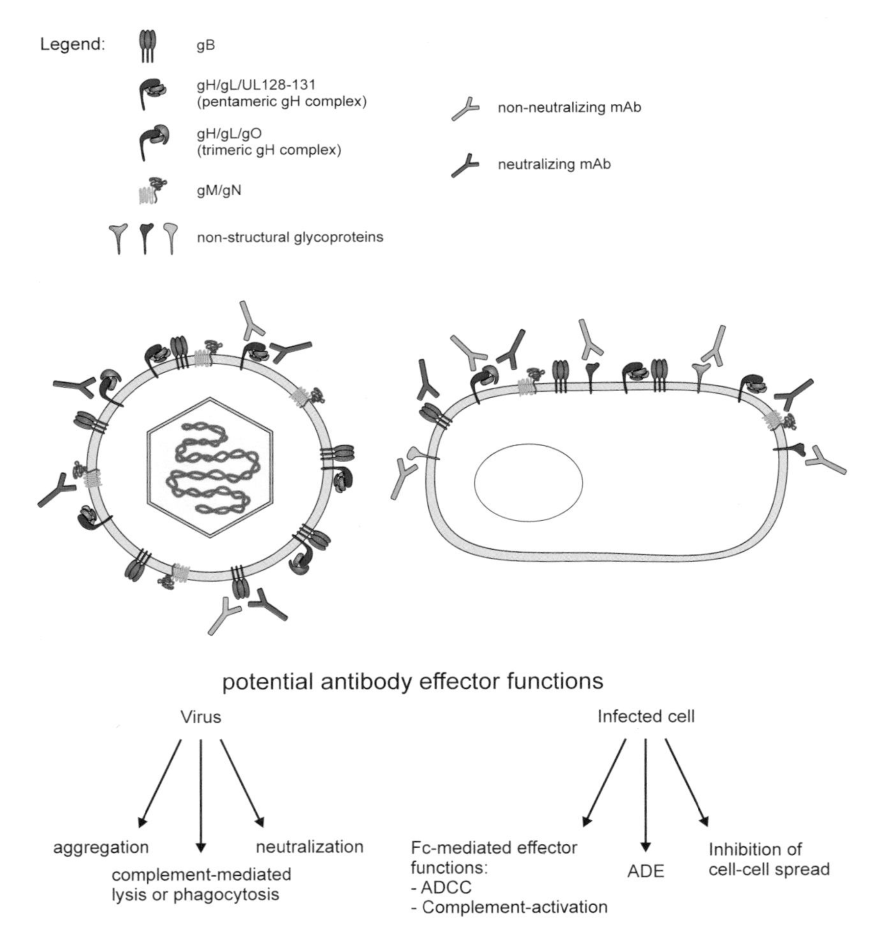

Figure II.10.1 The antiviral activities of antibodies. Schematic representation of potential targets of neutralizing or non-neutralizing antibodies on the surface of HCMV-virions (left panel) or HCMV-infected cells (right panel). Effector functions of antibodies on respective targets are indicated. ADE; antibody-dependent enhancement of infection, ADCC; antibody-dependent cellular cytotoxicity.

Recently, a more simple model was put forward which argues that neutralization is simply a function of critical density of antibody binding targets on the virus and occupancy of these sites by antibody which may be mediated mostly by the bulk of the antibody molecule (Burton *et al.*, 2001).

The mechanism of neutralization is determined by properties of the virion epitope(s) as well as the antibody-isotype that binds to it. It is, for example, easy to conceive that a bulky immunoglobulin (Ig) M antibody binding to the same epitope as an IgG molecule may be much more efficient in sterical obstruction of virus binding to a target molecule. However, more subtle differences such as different IgG subclass may also have an impact on neutralization (Rodrigo *et al.*, 2009).

Binding of complement

Mediated by the Fc-part of the Ig-molecule, complement components can be deposited and this can lead to an increased coating of protein molecules on the surface of virions. According to the occupancy model this will result in an increase in neutralization since it may prevent virion binding to target cells. Another possibility is activation of complement following binding to antibody coated virions. This may result in destruction of extracellular virions.

Inhibition of cell–cell spread

Many enveloped viruses, including herpesviruses, can spread directly from infected to uninfected cells and thereby avoid obstacles to infection such as neutralization by antibodies. Cell–cell spread is well established for CMVs although the mechanism(s) is unclear (see Chapter I.17). Depending on the mode of transport of virions from cell to cell (e.g. fusion at the plasma membrane or fusion within endosomal membranes) it can be expected that antibodies may have differential effects on these processes.

Last but not least, neutralization *in vitro* may also be crucially dependent on variables such as target cell type and source of virus (see below).

Antibodies binding to infected cells

Antibodies specific for viral envelope proteins can also bind to the surface of infected cells, provided the respective proteins are exposed (Fig. II.10.1). Binding of antibody to infected cells can trigger several antiviral activities. Fc-receptor activated effector systems can lead to destruction of the infected cells by antibody-dependent cellular cytotoxicity (ADCC). Given the fact that the individual IgG subclasses are bound by different IgG Fc-receptors with variable efficiency plus the differential expression of the Fc-receptors on different effector cells such as NK cells and monocytes the situation becomes very complex (Nimmerjahn and Ravetch, 2006). Binding of complement to the Fc-part of Ig can lead to destruction of infected cells or to removal of immune complexes via phagocytic cells that carry complement receptors.

Antibodies recognizing viral proteins on the surface of infected cells do not need to have similar specificity as those that act on free virus. Herpesviruses and CMVs in particular expose a large set of glycoproteins on the surface of infected cells of which a considerable fraction is not incorporated into the virion. Moreover, those proteins that are found on both infected cells and free virus may expose structurally different epitopes on either entity and thus be bound by different sets of antibodies. Thus, antibodies which do not recognize extracellular virus can nevertheless function in antiviral activity and thereby broaden the spectrum of antiviral activity of antibodies.

The relative importance of the individual antibody effector functions is dependent on a number of variables such as the cytopathogenicity of the virus or the mode of spread within the host. It seems logical that in infections which produce large numbers of free virus neutralizing antibodies will represent a major defence mechanism, whereas in those cases where virus spreads mainly cell-associated, Fc-mediated effector function will be important.

However, antiviral antibodies can also be detrimental for the host. Antibody dependent enhancement of infection (ADE) has been reported for a number of viruses, mostly RNA viruses but also herpesviruses, mostly in *in vitro* studies (Takada and Kawaoka, 2003). ADE describes the phenomenon, that in some circumstances antibodies may enhance viral infections. One major mechanism through which ADE may work is enhanced infection of Fc-receptor bearing cells such as monocytes and DCs by antibody-coated virus. With the exception of some RNA viruses, the importance of ADE for the *in vivo* situation is unclear.

In most cases, in which protection by antibodies *in vivo* has been shown, the mechanism(s) have not been studied in detail and much has to be learned about antibody effector functions for protection *in vivo*. Increasing this knowledge will be to the benefit of antibody therapy and vaccine development.

CMV and antibodies

Mode of infection and targets of protective antibody

Mammalian CMVs code for >50 glycoproteins (see Chapter I.1). The exact number of structural glycoproteins for the individual viruses is unknown. Proteome analyses (see also Chapter I.6) have identified approximately 20 in HCMV and just four in MCMV (Kattenhorn *et al.*, 2004; Varnum *et al.*, 2004). The major targets for induction of virus-neutralizing antibodies during natural infection have probably been identified. They represent constituents of the protein complexes that are conserved throughout the herpesvirus family, namely glycoprotein (g) B, the different gH/gL complexes and the gM/gN complex. While the function of gB and the different gH/gL complexes is mainly associated with membrane fusion, the function of the gM/gN complex for viral replication is unknown. Neutralizing antibodies have been identified for all of these glycoprotein complexes.

Before we discuss the antiviral potential of antibodies against the individual glycoprotein complexes we briefly need to touch on the entry routes of CMVs, in particular HCMV, since this is relevant for antibody neutralization in some situations. HCMV can infect different target cell types utilizing different entry routes such as fusion at the plasma membrane and/or endocytosis (see Chapter I.17). Although not analysed in detail, it seems plausible that, due to the different environments, fusion at the plasma membrane or endosomes does not follow identical mechanisms.

As shown for HSV, gB and the gH/gL heterodimer constitute the core fusion machinery with gB being the fusion-active molecule (Connolly *et al.*, 2011). It is thought that for CMVs the situation is similar. Fusogenic activity of HCMV gB has also been reported but the activity seems lower than for HSV gB (Gicklhorn *et al.*, 2003; Kinzler and Compton, 2005; Vanarsdall *et al.*, 2008). A notable difference between HSV and CMV is the fact that for CMV the gH/gL complex most likely does not exist in its isolated form but instead recruits additional molecules to form higher order structures; on human CMVs the gH/gL/gO complex (trimeric gH complex) and the gH/gL/pUL128/pUL130/pUL131a complex (pentameric gH complex) (Li *et al.*, 1997; Huber and Compton, 1998). For human CMV, the additional components of the gH/gL complex determine cell tropism to a large extent. While the trimeric gH complex is sufficient for infection of fibroblasts, the pentameric gH complex is required for infection of additional cell types such as endothelial and epithelial cells (Wang and Shenk, 2005; Ryckman *et al.*, 2008). With respect to antibody recognition and neutralization, the different gH complexes become particularly important (see below).

Given the complex entry mechanism of CMVs into target cells one can envisage several potential mechanisms by which antibodies neutralize infection:

- prevention of virus attachment/receptor binding;
- inhibition of conformational changes in gB or gH/gL which are required to execute fusion;
- prevention of gB and gH/gL interaction and thus activation of the fusion complex; and
- blocking the uncoating of virus/capsid release.

For CMV-specific antibodies neutralization mechanisms are largely unknown. However, data from the related herpesviruses can be used to postulate mechanisms which can then be tested for CMV.

Glycoprotein B

Glycoprotein B is the most highly conserved component of the herpesviral core fusion machinery. Studies on deletion mutants have demonstrated that gB is an essential protein for virus replication (Hobom *et al.*, 2000), especially entry, and gB is involved in the membrane fusion process and in cell–cell spread, but is not required for attachment, assembly or egress (Navarro *et al.*, 1993; Tugizov *et al.*, 1994; Bold *et al.*, 1996; Isaacson and Compton, 2009). It has been shown to bind to a variety of cell surface molecules such as heparan sulfate proteoglycans, integrin heterodimers and platelet-derived growth factor-α receptor (Compton *et al.*, 1993; Feire *et al.*, 2004; Soroceanu *et al.*, 2008; see also Chapter I.8). This type I membrane protein evolves from a 906 aa polypeptide (strain AD169), and is posttranslationally glycosylated and cleaved by the cellular endoprotease furin into subunits (Britt and Vugler, 1989; Spaete *et al.*, 1990) that are covalently linked via disulfide bonds. Crystal structure analysis of HSV-1 and EBV gB (Heldwein *et al.*, 2006; Backovic *et al.*, 2009) revealed a multidomain trimer and each protomer consists of five distinct domains. The crystal structures indicate that gB has the characteristics of a fusion protein and is a member of the class III group of viral fusion proteins (Rey, 2006). Since gB is the most conserved glycoprotein among the herpesviruses, the established crystal structure of HSV-1 gB has been used as template for the generation of a molecular model of HCMV gB and to map antigenic binding sites on it (Fig. II.10.2) (Pötzsch *et al.*, 2011).

The serological response to gB as a result of HCMV infection is universal. Assays using recombinant gB have detected gB-specific antibodies in sera from naturally infected individuals without exception, and a large

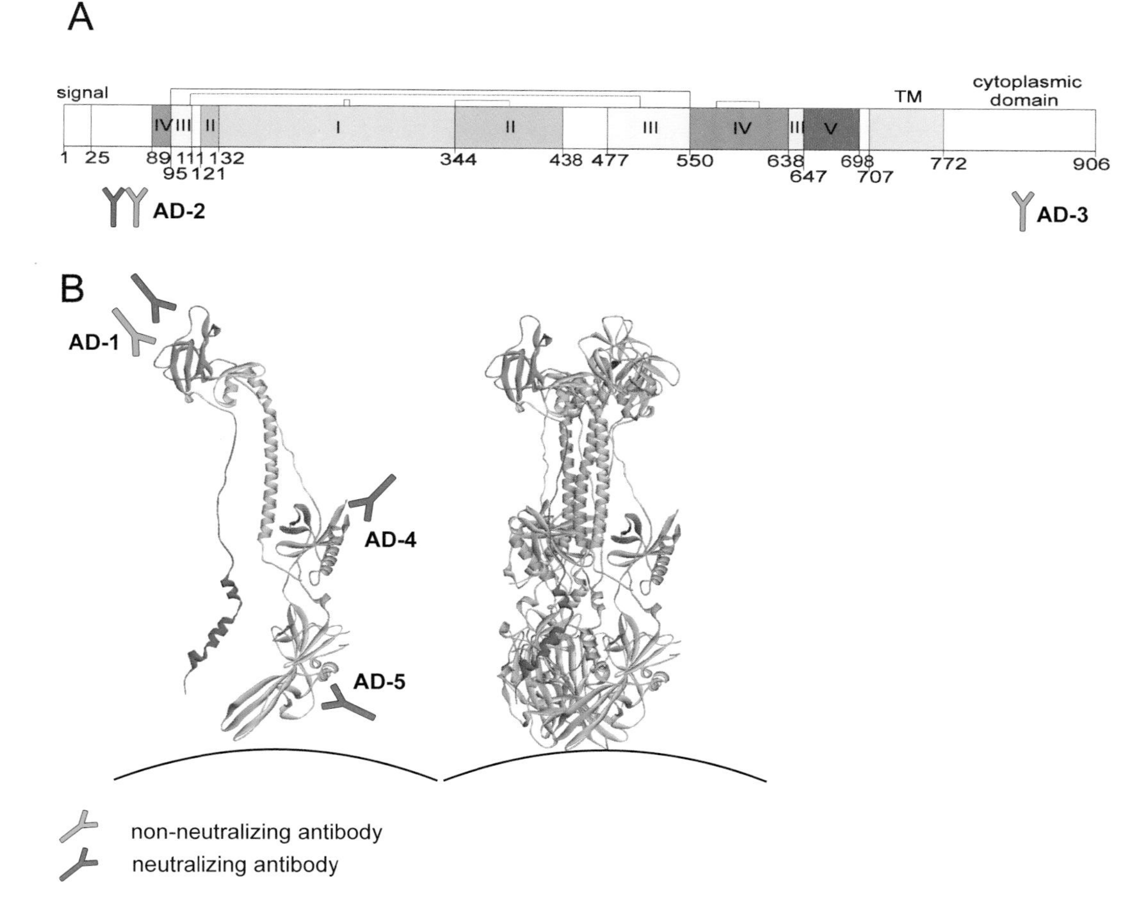

Figure II.10.2 Model of the domain architecture of HCMV gB and antibody binding sites on gB. Regions representing individual domains are displayed in different colours in analogy to the HSV-1 gB structure by Heldwein *et al.* (2006). (A) Linear cartoon of HCMV gB. Numbers of the starting residues are given and brackets indicate disulfide bonds; Signal: signal sequence; TM: transmembrane anchor. (B) Hypothetical domain architecture of HCMV gB. Ribbon diagram of a gB monomer (left) and trimer with two protomers shown in grey (right). Curves indicate virus membrane. Antibodies binding to gB are depicted in (A) and (B); the AD-2 corresponding sequence was not resolved in the original HSV-1 gB structure.

fraction of HCMV-neutralizing antibodies is directed against gB (Britt *et al.*, 1990; Marshall *et al.*, 1992). Five antigenic domains (AD) on HCMV gB have been described so far: AD-1 to AD-5.

AD-1 consists of approximately 80 aa between positions 552–635 of gB of HCMV strain AD169 (Britt and Mach, 1996). The seropositivity rate of AD-1 in sera from HCMV-infected individuals was found to be 100% (Schoppel *et al.*, 1997). However, AD-1 induces antibodies not only with potent neutralizing capacity but also those which do not reduce infectivity when tested *in vitro*. Polyclonal AD-1-specific antibodies, purified from human serum, are incapable of completely neutralizing HCMV even at high concentration, indicating that AD-1 is bound by antibodies with widely differing neutralizing activity which bind competitively (Speckner *et al.*, 1999).

AD-2 is located between aa 50–70 and consists of at least two sites. Site I is common to all strains and induces neutralizing antibodies, whereas site II differs between strains and is recognized by strain-specific antibodies which are non-neutralizing (Meyer *et al.*, 1990, 1992). Antigenicity of AD-2 is considerably lower than that of AD-1 since approximately 50% of sera from HCMV-infected individuals recognize this epitope (Schoppel *et al.*, 1997). Very little is known about the serological response against AD-3 (located between aa 783 and 906) except that in sera from a cohort of healthy blood donors a 100% reactivity was observed (Kniess *et al.*, 1991; Silvestri *et al.*, 1991).

AD-3 is located at the cytosolic/intraviral part of gB and antibodies binding here are non-neutralizing as can be expected from the localization of AD-3. Following isolation of human monoclonal antibodies (hu-mAbs) from healthy seropositive donors, we recently identified two additional antigenic binding sites on HCMV gB: AD-4 and AD-5 (Pötzsch *et al.*, 2011). The molecular model of HCMV gB, as shown in Fig. II.10.2, was a helpful tool to identify these conformation dependent epitopes. AD-4 is a discontinuous sequence and corresponds to protein domain II of HCMV gB (aa 121–132 and aa 344–438). Seropositivity rates in samples from HCMV-positive individuals were found to be > 90%. In contrast to AD-1, all AD-4-specific hu-mAbs were potently neutralizing. In addition, neutralization activity by human polyclonal affinity-purified AD-4-specific IgGs was in the same concentration range as AD-4 specific hu-mAbs indicating that (i) AD-4-specific antibodies neutralize strain-independent, and (ii) the spectrum of anti-AD-4 antibodies does not contain significant concentrations of non-neutralizing antibodies that compete for binding (Pötzsch *et al.*, 2011). AD-5 is formed by approximately 200 aa between positions 133 and 343 and corresponds to protein domain I of HCMV gB. AD-5 is recognized by > 50% of sera from HCMV-infected individuals. AD-5 specific hu-mAbs are also potent neutralizers and, as for AD-4, no non-neutralizing antibodies to this domain have been found so far (Pötzsch *et al.*, 2011).

The gH/gL complexes

Much less information is available on antibody target structures on the gH/gL complexes.

Following natural infection with HCMV, gH-specific antibodies have been detected with frequencies between 95% and 100% (Boppana *et al.*, 1995; Urban *et al.*, 1996). In some human sera, anti-gH antibodies constitute the majority of the neutralizing activity (Urban *et al.*, 1996). The protein regions relevant to induction of antibodies are not well defined (Simpson *et al.*, 1993). Epitope mapping studies of hu-mAbs will prove difficult since coexpression of all components of the respective gH/gL complexes is required for correct folding of the individual proteins.

With respect to the assay systems for evaluation of neutralizing activity of antibodies, the target cells become critical when looking at antibodies specific for the various gH/gL-based complexes. For decades neutralizing antibodies have been evaluated in fibroblast based-neutralization assays using viruses expressing only the trimeric gH complex such as strain AD169. Recent results from Gerna *et al.* (2008) demonstrate that by using viruses that express both types of gH/gL complexes and fibroblasts as target cells, a substantial portion of serum-neutralizing capacity might have been missed. They observed a striking difference in neutralizing activity of human sera when tested on endothelial or epithelial cells compared to fibroblasts, exhibiting a more than 100-fold higher neutralizing titre on epithelial or endothelial cells. This neutralizing antibody response during primary infection measured with endothelial cells is potent and occurs very early, whereas neutralizing antibodies measured with fibroblasts appear late and are less potent. Similarly, HCMV hyperimmunoglobulin preparations showed on average a 48-fold higher neutralizing activity against epithelial cell entry than against fibroblast entry (Cui *et al.*, 2008). Whether similar differences apply for the neutralization of additional cell types needs to be seen.

The basis for the enhanced neutralization capacity of polyspecific antibodies for viruses that require the pentameric gH complex for infection of epithelial and endothelial cells is not clear. However, it is tempting to speculate that the pUL128/pUL130/pUL131a proteins are involved. Support for this assumption comes from recent work on hu-mAbs. The group of Lanzavecchia (Macagno *et al.*, 2010) was able to isolate a panel of hu-mAbs directed against the pentameric gH complex induced during natural infection. These antibodies exhibit an extraordinarily high potency in neutralizing HCMV infection in endothelial, epithelial and myeloid cells with a 90% inhibitory concentration in the picomolar range. However, these antibodies completely fail to neutralize infection of fibroblasts. With the exception of one antibody recognizing individually expressed UL128, the antibodies required coexpression of several components of the pentameric gH complex. In contrast to hu-mAbs, antibodies raised against recombinant proteins or peptides in animals are far less potent, again suggesting that the correct folding and assembly of all protein components of the gH/gL complex are required for optimal antibody induction (Saccoccio *et al.*, 2011).

The gM/gN complex

The gM/gN complex is the third major constituent of the viral envelope (Mach *et al.*, 2000). In fact, gM is the most abundant protein in the viral envelope (Varnum *et al.*, 2004). The presence of both proteins is required for proper modification and transport of the constituents to the virus assembly compartment within infected cells. Using 293T-cells transiently expressing the gM/gN complex, seropositivity rates of 62% have been reported (Mach *et al.*, 2000). The fact that a gN-specific murine mAb neutralizes infectious virus very efficiently in the presence of complement suggests that gM/gN-specific antibodies present in human sera could also have virus-neutralizing activity (Britt and Auger, 1985). Indeed, anti-gM/gN antibodies affinity purified from pooled human CMV immune serum have been shown to neutralize HCMV at concentrations of approximately 0.1 µg/ml, levels similar to the neutralizing activity of anti-gB antibodies found in human immune serum (Shimamura *et al.*, 2006). The target for the neutralizing activity seems to be gN since most of gM is buried in the membrane.

Mechanism(s) of neutralization

Experimental data on potential mechanisms by which the individual antibodies work are scarce. None of the gB-specific hu-mAbs described to date inhibits virion attachment to the cell surface but all anti-gB mAbs function at a post-attachment step (Ohizumi *et al.*, 1992; Pötzsch *et al.*, 2011). As gB is part of the core fusion machinery, anti-gB mAbs most likely work by inhibiting the fusion process directly, either by stabilizing gB in its conformational state/preventing conformational changes or by inhibiting gB–gH-complex interaction. Recent data from HSV suggest that gB-specific murine mAbs which bind to regions corresponding to AD-1, AD-4 or AD-5 of HCMV gB neutralize infection by different mechanisms. Antibodies that bind to the AD-1 corresponding region block fusion but do not interfere with interaction of gB with gH/gL, while those to AD-4 or AD-5 block interaction of gB with gH/gL and thereby inhibit fusion (Atanasiu *et al.*, 2010a,b). Whether the situation is similar for hu-mAbs directed against HCMV gB remains to be determined. For the other antibodies mechanistic information is lacking.

A puzzling fact with respect to neutralization by human IgGs directed against the different glycoproteins is their potency. They seem to separate clearly into two classes. Antibodies against gB, gH and the gM/gN complex, either hu-mAbs or affinity purified from human sera, show 50% neutralization (IC50) at concentrations between 0.2 to 1 µg/ml i.e. in the low nanomolar range (Masuho *et al.*, 1987; Macagno *et al.*, 2010; Pötzsch *et al.*, 2011). In stark contrast, hu-mAbs directed against the pUL128/pUL130/pUL131 subunits of the pentameric gH complex have IC50 values between 0.0007 and 0.02 µg/ml i.e. one to two orders of magnitude lower (Macagno *et al.*, 2010).

At present we can only speculate about this difference:

- The isolated hu-mAbs may represent, by chance, the extremes of a wide spectrum. In viral systems, for which more information is available, hu-mAbs show a wide range of activity and the separation of anti-HCMV hu-mAbs in two classes may not hold true if more antibodies become available.
- The binding affinity of the gH complex antibodies may significantly differ between the two classes of hu-mAbs. This, however, seems unlikely since the K_D of anti-gB hu-mAbs is already in the range of 4×10^{-11} M (Pötzsch *et al.*, 2011) and thus may not be much higher for the gH-complex antibodies. Yet, avidity of the two types of antibody may differ.
- The abundance of infection-relevant and infection-irrelevant molecules on the virion may differ for the individual protein–protein complexes. Hypothetically, envelopes of complex viruses may contain infection-relevant and infection-irrelevant molecules of the same species. For example, not every gH molecule in the virion may be relevant for fusion, but will still bind antibody. Assuming that the pentameric gH complexes are less abundant than the additional gH complexes and that the majority of the pentameric complex is infection-relevant, it is conceivable that lower concentrations of the anti-pUL128pUL130pUL131a antibodies would suffice for neutralization. Moreover, accessibility of the gB and gH complexes for antibody may differ on a single virion. Cryo-electron tomography of HSV virions revealed that the density of envelope proteins is not equally distributed over the surface of the virion which may result in sterical obstruction of binding sites for some proteins (Grunewald *et al.*, 2003).

Antibodies in CMV protection *in vivo*

In the immunocompetent host we can expect a multi-layered response to CMV infection by the innate and adaptive immune system, which is also in large parts redundant. The result is control of virus replication in most organs. In humans it is impossible to dissect the contribution of the different Ig-effector functions to protection. So we have to rely on animal model systems where manipulation of the immune system is possible to a certain extent. One has to keep in mind, however, that animal CMVs have comparable yet not identical biology and pathology. MCMV is the prime model and important insights in immune defence against CMVs originated from this model. However, additional systems like the GPCMV and the primate model RhCMV have also considerably contributed to our understanding of the protective capacity of the humoral immune response to CMVs (see also Chapters II.5 and II.22).

When considering antiviral properties of antibodies it is helpful to differentiate between antibody that is present before infection and effects once infection is established e.g. role of antibody during viral reactivation and antibodies as therapeutic agents.

Pre-existing antibody following adoptive transfer

A number of early studies with MCMV (Araullo-Cruz *et al.*, 1978; Shanley *et al.*, 1981; Lawson *et al.*, 1988) using adoptive transfer of serum from infected or immunized mice showed marked reduction in organ titres following viral challenge. Later on, transfer of mAbs has been used to explore protection from subsequent infection. Farrell and Shellam (1990) used several mAbs with highly divergent *in vitro* neutralizing activity to demonstrate protection from a lethal challenge dose. Interestingly, no correlation was seen between the level of protection and the neutralization titres *in vitro*. Also, they observed an organ-specific effect of antibody.

The benefit of pre-existing antibody in humans for the course of a primary infection is less clear. A recent meta-analysis came to the conclusion that in the context of bone marrow transplantation the prophylactic intravenous administration of immunoglobulins (IVIG) does not have an effect on survival or other outcomes (Raanani *et al.*, 2008; see also Chapter II.16). One has to keep in mind, however, that different preparations of IVIG were included in the meta-analysis. It is well known, that IVIG shows lot to lot variation between batches with respect to virus neutralizing antibodies. For overviews on passive immunization against congenital CMV, see Chapters II.3 and II.5.

Pre-existing antibody induced by vaccination

If antibodies are induced following vaccination (see also Chapter II.20), the situation is not as clear cut as after transfer of serum or monoclonal antibodies since there is always a concurrent T-cell response that may contribute to protection – so, results have to be interpreted with caution. In addition, vaccination may induce a different set of antibodies compared to natural infection with respect to epitope recognition (Cheng *et al.*, 2010). Nevertheless, studies in animal models (MCMV, RhCMV) as well as in humans indicate that antibodies induced by immunization (especially to gB) do have a protective effect, even if subsequent infection cannot be prevented.

For MCMV, Shanley and Wu (2003) reported protection from intranasal (i.n.) challenge (10^5 PFU) following vaccination with replication-deficient adenovirus expressing gB. Notably, a massive virus titre reduction in salivary glands was achieved following challenge infection. However, even for a low dose challenge (10^3 PFU) no sterilizing immunity was achieved. The vaccine gave a significantly better immune response when given i.n. compared to intraperitoneal (i.p.) immunization. In contrast, DNA immunization with a gB-expressing plasmid did not result in production of significant neutralizing antibody titres and resulted in no protection (Morello *et al.*, 2005). When a prime-boost strategy was exerted, in which the boost consisted of inactivated MCMV virions, neutralizing titres similar to natural infections were achieved and nearly complete protection after systemic or mucosal MCMV challenge was seen.

Vaccine-induced reduction in plasma titres was observed following immunization of rhesus monkeys with a gB-based vaccine (Abel *et al.*, 2011; see also Chapter II.22). Finally, recent phase II clinical trials using recombinant gB as vaccine antigen have shown partial protection from maternal and congenital infection and a reduction of the duration of viraemia in transplant recipients (Pass *et al.*, 2009; Griffiths *et al.*, 2011; see also Chapters II.3, II.14 and II.20).

In the experimental passive serum transfer as well as vaccination studies infection was not prevented in most cases despite the presence of antibody at the time of infection; so, no sterilizing immunity was achieved. However, the challenge infection typically involved large doses of virus (typically 10^4 to 10^5 PFU of virus given i.p., a dose that probably is not physiological). Future studies should aim at challenging the protection by infection under more physiological conditions (less virus, mucosal infection). One of the problems is that the route and amount of virus required for horizontal transmission is unknown. However, it is likely to be substantially less than the amount used in most challenge studies. Interestingly, the recent clinical trial with HCMV-seronegative women suggested protection from infection in a fraction of the vaccinees (Pass *et al.*, 2009).

Antibody in existing primary infection (therapy)

We have recently published data strongly supporting the therapeutic potential of antibodies (Klenovsek *et al.*, 2007). We used a recombinant MCMV, expressing the firefly luciferase gene under the control of the immediate early promoter/enhancer of HCMV, allowing for the *in vivo* detection of infected cells (Fig. II.10.3). RAG-1$^{-/-}$ mice, which do not have functional T and B-cells were infected with this virus and tested in serotherapy experiments. Administration of immune serum from C57BL/6 mice 3 days after infection significantly reduced the bioluminescence signal in the animals, indicating reduction of the

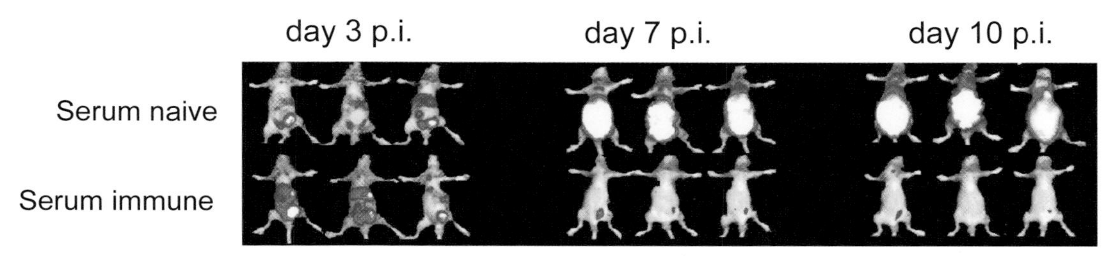

Figure II.10.3 Serotherapy of primary MCMV infection. RAG$^{-/-}$ mice were infected with 10^5 PFU of a luciferase expressing recombinant MCMV (MCMV157luc). Three days after infection sera from naïve mice, or sera from immune mice (250 µL each) were transferred and virus load was detected by *in vivo* imaging. The data were originally published by Klenovsek *et al*. (2007) 'Protection from CMV infection in immunodeficient hosts by adoptive transfer of memory B-cells'. Blood *110*, 3472–3479. © The American Society of Hematology.

number of infected cells to below the detection limit. Determination of the viral titres in several organs, including the salivary glands, confirmed the results of the *in vivo* imaging. The animals were completely protected from the lethal course of the infection that is otherwise invariably seen in these animals. Interestingly, adoptive transfer of memory B-cells from infected, immunocompetent C57BL/6 mice into the RAG-1$^{-/-}$ animals showed also long-term protection. Cekinović *et al.* (2008) used a recently established model of MCMV infection in newborn mice to analyse the contribution of antibody to virus clearance from the brain (see also Chapter II.6). In brains of MCMV-infected newborn mice treated with immune serum 5 to 9 days after infection, the titre of infectious virus was reduced to below detection limit. Treatment with a gB-specific monoclonal antibody also resulted in reduction of virus titre in the brain but was less efficient. Finally, Wirtz *et al.* (2008) could demonstrate that serotherapy in Ig deficient mice was capable of protection from the focal cell–cell spread of virus after tissue infection is established.

Results on protection by passive antibody treatment in humans are less clear. IVIG preparations have been extensively used in transplant patients at risk for CMV disease. However, even after almost 20 years of using this treatment either prophylactically or therapeutically uncertainty about benefits of immunoprophylaxis for the prevention of CMV infection and disease is evident in the literature (Sokos *et al.*, 2002; Bonaros *et al.*, 2008). Noteworthy, a recent study in pregnant women with primary CMV infection showed that IVIG therapy was associated with a significantly lower risk of congenital disease in the babies (Nigro *et al.*, 2005; see also Chapter II.3). MSL-109, a hu-mAb specific for gH, has also been used for CMV therapy/prophylaxis in different patient populations. No statistically significant difference in CMV antigenaemia or viraemia and overall survival rates was seen in transplant recipients or AIDS patients (Boeckh *et al.*, 2001; Borucki *et al.*, 2004). However, as discussed below, HCMV has evolved a highly unusual evasion mechanism for the antiviral activity of MSL-109 and thus, the result of the clinical trial may not be representative of hu-mAb therapy in general (Manley *et al.*, 2011).

Antibody in reactivation

Even in the face of a fully functional immune system, HCMV is thought to persist as a very low level continuous or frequently intermittent infection rather than a predominantly or exclusively quiescent, latent infection demonstrating that immunological control is not complete. Also, as outlined above, pre-existing antibody cannot prevent subsequent infection indicating that induction of sterilizing immunity seems not to be possible. So, one would expect that antibodies also fail to prevent virus recurrence. However, this seems not to be the case. In the MCMV system, combined depletion of CD8$^+$ T-cells, CD4$^+$ T-cells and NK cells does not result in notable virus recurrence (Polic *et al.*, 1998). Only if these cells are depleted in Ig-deficient mice virus recurrence is seen. Thus, in situations where infectious virus is endogenously produced following reactivation, antibodies can control it.

Antibodies in prevention of vertical transmission

Serum prevalence of animal CMVs in wild populations is close to 100% (Smith *et al.*, 1993; Jones-Engel *et al.*, 2006). Likewise the prevalence of HCMV is very close to 100% in regions with low socioeconomic levels in Africa, Asia and South-America (Cannon *et al.*, 2010; see also Chapter II.2). Vertical transmission from mother to the fetus is most common in human as well as animal CMVs. There is strong evidence that transmission of free HCMV as well as infected cells by breast milk and colostrum represent the most important route of transmission from mother to child (Dworsky *et al.*, 1983; Kaye *et al.*, 2008). Evidently, neither HCMV-specific IgG transferred from the mother via the placenta nor IgA in the colostrum and milk is sufficient to prevent vertical transmission of HCMV. It is remarkable, however, that HCMV infection becomes detectable in the child only 6–8 weeks after the first exposure to the virus in milk (Hamprecht *et al.*, 2001; Kaye *et al.*, 2008), indicating that antibodies from the mother could have some role in prevention of transmission and/or virus spread in the newborn. In addition, newborn infants are protected from transfusion-acquired infectious complications when mothers were seropositive, whereas severe complications were frequent after transfusions into newborns from seronegative mothers (Yeager *et al.*, 1981). In an experimental setting using neonatal MCMV infection, serum antibody from the latently infected mothers was able to protect from infection in the neonates (Slavuljica *et al.*, 2010). It is entirely unclear at the moment, why serum antibodies and cellular immunity in the seroimmune populations fail to prevent congenital infection. One reason might be that there is substantial strain diversity and intra-individual diversity (Renzette *et al.*, 2011; Ross *et al.*, 2011) (see also below).

In animal models of congenital infection antibodies did protect the fetus through placental transfer. In the GPCMV model, in which intrauterine transmission occurs like in HCMV, passive antibody transfer of immune serum against whole virus or glycoprotein B was able to significantly reduce placental infection and pregnancy loss (Bratcher *et al.*, 1995; Chatterjee *et al.*, 2001). It is noteworthy that the antibody transfer was not able to prevent infection of the fetus, however (see also Chapter II.5). Importantly, a clinical study with pregnancies under risk for congenital HCMV transmission showed efficacy of HCMV-hyperimmune globulin applied to mothers with confirmed primary HCMV infection during pregnancy (Nigro *et al.*, 2005; see also Chapter II.3).

Evasion from antibody

CMVs have developed a plethora of mechanisms to avoid elimination from the host by the immune response. For the innate and cellular immune response this has been well documented (see Chapters II.7, II.8, II.9, II.17 and II.22). It would be more than naïve to assume that the humoral immune response is an exception.

Several mechanisms could contribute:

Induction of large quantities of non-neutralizing antibodies

The repertoire analysis of anti-gB antibodies during natural infection in healthy seropositive donors has revealed that more than 90% of anti-gB antibodies are non-neutralizing *in vitro* (Pötzsch *et al.*, 2011). Although it has not been analysed, it is probably fair to assume that the situation is similar for the other neutralization relevant antigens in the CMV envelope. At this time we can only speculate on a potential role of the non-neutralizing antibodies for the infection *in vivo*. Besides being irrelevant for the course of the infection, non-neutralizing antibodies might have positive or negative effects. Effector functions mediated via the Fc-portion of the antibodies such as ADCC and/or complement fixation may contribute to elimination of infected cells and thus could lead to enhanced clearance of the infection. That such mechanisms are operative *in vivo* has been shown in different viral systems, including herpesviruses, but not yet for HCMV (Beresford *et al.*, 1988; Chung *et al.*, 2007; Hessell *et al.*, 2007; Chu *et al.*, 2008). On the other hand, infection-enhancing effects could be contributed by mechanisms such as competition of non-neutralizing antibodies for binding to neutralization-relevant protein domains or enhanced infection of Fc-receptor bearing cells such as monocytes, which is one of the major target cell population for HCMV infection *in vivo* (Sinzger *et al.*, 2008).

Additionally, the disproportion of neutralizing versus non-neutralizing antibodies may represent *per se* a mechanism to evade efficient virus control. If we assume that the total concentration of anti-CMV antibodies is more or less constant in an infected host then the fraction of antibodies that have direct antiviral activity may become stoichiometrically critical. In some situations the level of neutralizing antibodies may be below a threshold of biological activity. The low overall neutralizing titres of anti-CMV antibodies compared to other viral systems might be an indication of that.

The MSL-109 phenomenon

Generation of antibody resistant CMVs through mutations in epitopes within a host is probably a rare event. *In vivo*, virus neutralization rests on multiple epitopes on different proteins, and mutations in several epitopes would be required to render a virus resistant to neutralization by the polyspecific antibodies developed during natural infection.

However, regulating the concentration of neutralization relevant proteins in the viral envelope might provide the virus with an escape mechanism at least as far as monoclonal antibodies are concerned. Data reported for the naturally occurring hu-mAb MSL-109 indicates that such a mechanism is indeed operative for CMV. Li *et al.* (1995) analysed a neutralization escape mutant for MSL-109 that was isolated from an antibody-treated patient. The escape variant had developed resistance to several anti-gH mAbs and polyclonal anti-gH antibodies but not anti-gB mAbs or sera. No genetic mutation in gH was found but rather the resistant virus displayed a marked decrease in the amount of gH expressed on the viral envelope. Further evidence that the resistance phenotype was not associated with genetic mutation came from the observation that an anti-gH sensitive revertant virus was generated within a single cell culture passage in the absence of MSL-109. More recently, Manley *et al.* (2011) reported on a MSL-109 resistant virus that was generated *in vitro*. Looking for the resistance mechanism in detail, it was found that MSL-109 does not neutralize CMV by binding to free virus. Instead, neutralization seems to occur by uptake of the antibody into the infected cells by cell surface-exposed gH and subsequent incorporation of this antibody bound to gH complexes into the newly formed viruses at the site of virus assembly. The thus formed resistant virus then uses the Fc-domain of the envelope-incorporated antibody to infect non-immune cells. The mechanism seems to be unique for MSL-109 since other neutralizing gH- and gB-specific hu-mAbs did not show a similar resistance mechanism. Whether the effect is specific for the virus strain that was used in this study remains to be seen. Overall, the studies on resistance to gH-specific antibodies clearly demonstrate that evasion from neutralization is possible, if it rests on a single specificity.

Complement regulation

The fact that many herpesviruses encode proteins that interfere with the complement cascade indicates that evasion of the system seems to be beneficial for the viruses. HCMV has been shown to regulate the complement activity by incorporation of host-cell complement regulatory components such as CD55, CD59 and CD46 (Spear *et al.*, 1995; Spiller *et al.*, 1997). The contribution of the complement system to viral clearance *in vivo* is unclear. MCMV-specific mAbs that required complement *in vitro* for efficient neutralization were as effective *in vivo* in C5-deficient mice as in normal mice (Farrell and Shellam, 1991).

Virus-encoded Fc-receptors

CMVs express proteins on the surface of infected cells that recognize the Fc region of Ig molecules. It is assumed that the viral Fc-γ receptors protect the virus and/or infected cells from the IgG-mediated immune response. For the corresponding proteins of HSV-1 a phenomenon referred to as bipolar bridging has been reported, whereby simultaneous binding of human anti-HSV IgG to both an HSV antigen with its Fab arm and to the Fc-receptor through the Fc-region eliminates potential antiviral effects of IgG (Sprague *et al.*, 2006). HCMV encodes Fc-γ receptors with binding capacity for all four IgG subclasses but not IgA and IgM (Antonsson and Johansson, 2001; Atalay *et al.*, 2002). The relevance for the *in vivo* situation is not known. In MCMV, deletion of the fcr-1 gene, which codes for a Fc-γ receptor, had little impact on the infection kinetics in immunocompetent mice and mutant mice lacking antibodies (Crnkovic-Mertens *et al.*, 1998).

Antigenic variation

It becomes more and more evident that variability of CMV strains is extensive (see Chapter I.1). The differences at the genome level translate into antigenic variability. Antigenic variation is a long known phenomenon in CMV infections. In work already published in 1978, Waner and Weller noted antigenic diversity among HCMV isolates. They used sera produced in rabbits against lysates from HCMV-infected cells and noted significant differences in the complement-dependent neutralization capacity. The findings were later confirmed by Klein *et al.* (1999) using human sera. They observed differences, independent of the overall neutralization capacity, of the sera against individual strains from 6-fold to more than 60-fold. For one isolate complete resistance to neutralization by two human sera was seen. Thus, antigenic variation can render an existing pool of antibodies ineffective for neutralization of heterologous virus isolates.

The antigens that are responsible for these differences have not been characterized in detail. However, if the reduced capacity of human polyclonal sera to neutralize heterologous HCMV-isolates *in vitro* represents a mechanism to escape efficient neutralization *in vivo*, gB, gH and gN would be prime candidates since they contribute significantly to the overall neutralization capacity in human sera (see above). Whether this is indeed the case has so far only been tested for gN. We constructed four recombinant viruses on the genetic background of HCMV strain AD169 which differed only in the expression of the gN genotype. Exchange of gN genotypes had no detectable influence on viral replication, gN expression and gN/gM complex formation. Randomly selected human sera were analysed for neutralizing capacity of the recombinant viruses. Of these sera, 30% showed strain-specific neutralization with more than 7-fold difference in the 50% neutralization titres (Burkhardt *et al.*, 2009).

An indirect piece of evidence that gH and gB could also be involved in the induction of strain-specific antibodies comes from studies of monoclonal antibodies. A number of human or murine mAbs reacting with gB or gH show strain specific neutralization capacity (Rasmussen *et al.*, 1985; Baboonian *et al.*, 1989; Simpson *et al.*, 1993). The differences can be dramatic, resulting in lack of binding to some HCMV strains. Exceptions seem to be antibodies recognizing AD-2, AD-4 and AD-5 of gB.

Thus, the available data justify the conclusion that the antibody response to the major glycoproteins that are targets of neutralizing antibodies could contribute to the observed strain-specific effects. Of note, the antibody response to HCMV gO, another highly polymorphic protein which may be involved in cell–cell spread has not been analysed so far.

A strain-specific antibody response may have a significant impact on protection from reinfection with antigenically different viruses. It seems difficult to superinfect an immune host with a homologous virus. Reddehase and colleagues noted *'it is essentially impossible to superinfect a CMV-immune mouse with the homologous virus, at least when the system is not overwhelmed by unreasonably high virus doses'* even in the absence of T-cells and NK cells (Wirtz *et al.*, 2008). On the other hand, reinfection with antigenically different viruses can easily be demonstrated in humans as well as in animal systems (Boppana *et al.*, 2001; Gorman *et al.*, 2006).

It can be expected that in the immunocompetent host, reinfection with a genetically different virus rarely, if ever, occurs as clinically symptomatic infection, much like the primary infection. However, in immunocompromised hosts reinfection may represent a clinically important problem since a *de novo* adaptive immune response against the reinfecting strain is impaired or in some cases absent, depending on the extent of immunosuppression. Clinical studies in transplant patients have supported the importance of HCMV reinfection in these patients (Grundy *et al.*, 1988; Coaquette *et al.*, 2004; Ishibashi *et al.*, 2007). Moreover, in women who are seropositive for HCMV, reinfection with a different strain can lead to intrauterine transmission and symptomatic congenital infection (Boppana *et al.*, 2001).

Unclear is whether reinfection is inevitable upon contact with a different virus strain or whether in certain cases a cross-protecting immune response can develop which protects from reinfection. Recent data from other viral systems like HIV or influenza have indicated that in a fraction of infected individuals broadly neutralizing antibodies can develop (Walker *et al.*, 2009; Krause *et al.*, 2011).

Conclusion

Considerable progress has been made in recent years in our understanding of the humoral immune response against this family of complex viruses. New antigens in the viral envelope have been identified and human monoclonal antibodies have been isolated which bind to previously unknown targets or extend the antigenic map of known antigens such as gB. In animal systems the antiviral capacity of antibodies in prophylaxis and therapy has clearly been shown, but the road from findings in basic science to successful translation into clinical use may still be quite long and winding.

In our personal view a number of projects should be addressed in the future, as follows.

Isolation of larger panels of hu-mAbs

The advent of highly efficient techniques to isolate these molecules in large numbers will allow for a detailed characterization of the human antibody repertoire against different HCMV strains. The information derived from the few hu-mAbs that have recently been isolated are encouraging since they have already identified new target proteins for potently neutralizing antibodies (the pentameric gH complex) and extended our knowledge of antigenic domains on already previously characterized antigens (gB).

We need to find broadly neutralizing antibodies. This will be extremely important if hu-mAbs are to be developed for use in humans. The remarkable recent success in the identification of broadly neutralizing hu-mAbs against other viruses such as HIV or influenza clearly warrants further systematic approaches for isolation of hu-mAbs against HCMV. It seems that this type of antibody only develops after repeated antigenic stimulation and multiple rounds of affinity maturation. So, the donor population for isolation of anti-HCMV hu-mAbs should be chosen with care. Eventually we may be able to identify antibodies which effectively neutralize a broad range of HCMV isolates and maybe even provide sterilizing immunity if present in high enough concentrations. High-throughput DNA sequencing should help to characterize the strain diversity of HCMVs. A most thorough characterization of the antibody response during HCMV infection will also be very helpful in appraising the immune response following vaccination.

Definition of neutralization mechanisms

It is disappointing that we still have no mechanistic information on how neutralizing antibodies work. We need to define these mechanisms in order to find antibodies which truly attack the 'Achilles heel' of this virus and which cannot be evaded by the virus. Antibodies which block receptor binding would be ideal. However, these may not be easy to find since multiple receptors and different entry routes may be used by CMVs – much like in HSV. The case of MSL-109 also shows that we need to be prepared for surprises.

Definition of effector mechanisms of antibody protection *in vivo*

Antibodies protect from CMV-disease *in vivo*. The published studies in the animal models suggest that polyspecific sera, usually derived from infected donors, have higher protective capacity than monospecific sera or mAbs. If this holds true in further studies using antibodies against different viral antigens, we need to define the underlying mechanism(s). Is neutralization of free virus more efficient when multiple antigens on the virion are bound by antibodies, or is the fraction of antibodies that bind to the infected cell significantly influencing protection *in vivo*? Finally, the contribution of the Fc-part of the antibodies for *in vivo* protection should be defined.

In summary, results from human and animal studies clearly demonstrate that antibodies do what they are supposed to do: they neutralize virus, limit dissemination, and ultimately prevent disease.

Acknowledgement

We would like to thank members of our laboratories, past and present, whose work has contributed to this chapter. We also apologize to colleagues in the field whose work could not be cited due to space limitations. Work in our laboratories was supported by grants from the Deutsche Forschungsgemeinschaft, the Bundesministerium für Bildung und Forschung, the Wilhelm-Sander Stiftung and BayImmuNet.

References

Abel, K., Martinez, J., Yue, Y., Lacey, S.F., Wang, Z., Strelow, L., Dasgupta, A., Li, Z., Schmidt, K.A., Oxford, K.L., *et al.* (2011). Vaccine-induced control of viral shedding following rhesus cytomegalovirus challenge in rhesus macaques. J. Virol. *85*, 2878–2890.

Antonsson, A., and Johansson, P.J. (2001). Binding of human and animal immunoglobulins to the IgG Fc receptor induced by human cytomegalovirus. J. Gen. Virol. *82*, 1137–1145.

Araullo-Cruz, T.P., Ho, M., and Armstrong, J.A. (1978). Protective effect of early serum from mice after cytomegalovirus infection. Infect. Immun. *21*, 840–842.

Atalay, R., Zimmermann, A., Wagner, M., Borst, E., Benz, C., Messerle, M., and Hengel, H. (2002). Identification and expression of human cytomegalovirus transcription units coding for two distinct Fcgamma receptor homologs. J. Virol. *76*, 8596–8608.

Atanasiu, D., Saw, W.T., Cohen, G.H., and Eisenberg, R.J. (2010a). Cascade of events governing cell–cell fusion induced by herpes simplex virus glycoproteins gD, gH/gL, and gB. J. Virol. *84*, 12292–12299.

Atanasiu, D., Whitbeck, J.C., de Leon, M.P., Lou, H., Hannah, B.P., Cohen, G.H., and Eisenberg, R.J. (2010b). Bimolecular complementation defines functional regions of Herpes simplex virus gB that are involved with gH/gL as a necessary step leading to cell fusion. J. Virol. *84*, 3825–3834.

Baboonian, C., Blake, K., Booth, J.C., and Wiblin, C.N. (1989). Complement-independent neutralising monoclonal antibody with differential reactivity for strains of human cytomegalovirus. J. Med. Virol. *29*, 139–145.

Backovic, M., Longnecker, R., and Jardetzky, T.S. (2009). Structure of a trimeric variant of the Epstein–Barr virus glycoprotein B. Proc. Natl. Acad. Sci. U.S.A. *106*, 2880–2885.

Beresford, C.H., Storey, G., Faed, J.M., and Milligan, L.M. (1988). Cytomegalovirus infection and blood transfusion. N. Z. Med. J. *101*, 2–4.

Boeckh, M., Bowden, R.A., Storer, B., Chao, N.J., Spielberger, R., Tierney, D.K., Gallez-Hawkins, G., Cunningham, T., Blume, K.G., Levitt, D., *et al.* (2001). Randomized, placebo-controlled, double-blind study of a cytomegalovirus-specific monoclonal antibody (MSL-109) for prevention of cytomegalovirus infection after allogeneic hematopoietic stem cell transplantation. Biol. Blood Marrow Transplant. *7*, 343–351.

Bold, S., Ohlin, M., Garten, W., and Radsak, K. (1996). Structural domains involved in human cytomegalovirus glycoprotein B-mediated cell–cell fusion. J. Gen. Virol. *77*, 2297–2302.

Bonaros, N., Mayer, B., Schachner, T., Laufer, G., and Kocher, A. (2008). CMV-hyperimmune globulin for preventing cytomegalovirus infection and disease in solid organ transplant recipients: a meta-analysis. Clin. Transplant. *22*, 89–97.

Boppana, S.B., Polis, M.A., Kramer, A.A., Britt, W.J., and Koenig, S. (1995). Virus-specific antibody responses to human cytomegalovirus (HCMV) in human immunodeficiency virus type 1-infected persons with HCMV retinitis. J. Infect. Dis. *171*, 182–185.

Boppana, S.B., Rivera, L.B., Fowler, K.B., Mach, M., and Britt, W.J. (2001). Intrauterine transmission of cytomegalovirus to infants of women with preconceptional immunity. N. Engl. J. Med. *344*, 1366–1371.

Borucki, M.J., Spritzler, J., Asmuth, D.M., Gnann, J., Hirsch, M.S., Nokta, M., Aweeka, F., Nadler, P.I., Sattler, F., Alston, B., *et al.* (2004). A phase II, double-masked, randomized, placebo-controlled evaluation of a human monoclonal anti-Cytomegalovirus antibody (MSL-109) in combination with standard therapy versus standard therapy alone in the treatment of AIDS patients with Cytomegalovirus retinitis. Antiviral Res. *64*, 103–111.

Bratcher, D.F., Bourne, N., Bravo, F.J., Schleiss, M.R., Slaoui, M., Myers, M.G., and Bernstein, D.I. (1995). Effect of passive antibody on congenital cytomegalovirus infection in guinea pigs. J. Infect. Dis. *172*, 944–950.

Britt, W.J., and Auger, D. (1985). Identification of a 65 000 dalton virion envelope protein of human cytomegalovirus. Virus Res. *4*, 31–36.

Britt, W.J., and Mach, M. (1996). Human cytomegalovirus glycoproteins. Intervirology *39*, 401–412.

Britt, W.J., and Vugler, L.G. (1989). Processing of the gp55–116 envelope glycoprotein complex (gB) of human cytomegalovirus. J. Virol. *63*, 403–410.

Britt, W.J., Vugler, L., Butfiloski, E.J., and Stephens, E.B. (1990). Cell surface expression of human cytomegalovirus (HCMV) gp55–116 (gB): use of HCMV-recombinant vaccinia virus-infected cells in analysis of the human neutralizing antibody response. J. Virol. *64*, 1079–1085.

Burkhardt, C., Himmelein, S., Britt, W., Winkler, T., and Mach, M. (2009). Glycoprotein N subtypes of human cytomegalovirus induce a strain-specific antibody response during natural infection. J. Gen. Virol. *90*, 1951–1961.

Burton, D.R., Saphire, E.O., and Parren, P.W. (2001). A model for neutralization of viruses based on antibody coating of the virion surface. Curr. Top. Microbiol. Immunol. *260*, 109–143.

Cannon, M.J., Schmid, D.S., and Hyde, T.B. (2010). Review of cytomegalovirus seroprevalence and demographic characteristics associated with infection. Rev. Med. Virol. *20*, 202–213.

Cekinovic, D., Golemac, M., Pugel, E.P., Tomac, J., Cicin-Sain, L., Slavuljica, I., Bradford, R., Misch, S., Winkler, T.H., Mach, M., *et al.* (2008). Passive immunization reduces murine cytomegalovirus-induced brain pathology in newborn mice. J. Virol. *82*, 12172–12180.

Chatterjee, A., Harrison, C.J., Britt, W.J., and Bewtra, C. (2001). Modification of maternal and congenital cytomegalovirus infection by anti-glycoprotein b antibody transfer in guinea pigs. J. Infect. Dis. *183*, 1547–1553.

Cheng, C., Gall, J.G., Nason, M., King, C.R., Koup, R.A., Roederer, M., McElrath, M.J., Morgan, C.A., Churchyard, G., Baden, L.R., *et al.* (2010). Differential specificity and immunogenicity of adenovirus type 5 neutralizing antibodies elicited by natural infection or immunization. J. Virol. *84*, 630–638.

Chu, C.F., Meador, M.G., Young, C.G., Strasser, J.E., Bourne, N., and Milligan, G.N. (2008). Antibody-mediated protection against genital herpes simplex virus type 2 disease in mice by Fc gamma receptor-dependent and -independent mechanisms. J. Reprod. Immunol. *78*, 58–67.

Chung, K.M., Thompson, B.S., Fremont, D.H., and Diamond, M.S. (2007). Antibody recognition of cell surface-associated NS1 triggers Fc-{gamma} receptor-mediated phagocytosis and clearance of west nile virus-infected cells. J. Virol. *81*, 9551–9555.

Coaquette, A., Bourgeois, A., Dirand, C., Varin, A., Chen, W., and Herbein, G. (2004). Mixed cytomegalovirus glycoprotein B genotypes in immunocompromised patients. Clin. Infect. Dis. *39*, 155–161.

Compton, T., Nowlin, D.M., and Cooper, N.R. (1993). Initiation of human cytomegalovirus infection requires initial interaction with cell surface heparan sulfate. Virology *193*, 834–841.

Connolly, S.A., Jackson, J.O., Jardetzky, T.S., and Longnecker, R. (2011). Fusing structure and function: a structural view of the herpesvirus entry machinery. Nat. Rev. Microbiol. *9*, 369–381.

Crnkovic-Mertens, I., Messerle, M., Milotic, I., Szepan, U., Kucic, N., Krmpotic, A., Jonjic, S., and Koszinowski, U.H. (1998). Virus attenuation after deletion of the cytomegalovirus Fc receptor gene is not due to antibody control. J. Virol. *72*, 1377–1382.

Cui, X., Meza, B.P., Adler, S.P., and McVoy, M.A. (2008). Cytomegalovirus vaccines fail to induce epithelial entry neutralizing antibodies comparable to natural infection. Vaccine *26*, 5760–5766.

Dimmock, N.J. (1993). Neutralization of animal viruses. Curr. Top. Microbiol. Immunol. *183*, 1–149.

Dworsky, M., Yow, M., Stagno, S., Pass, R.F., and Alford, C. (1983). Cytomegalovirus infection of breast milk and transmission in infancy. Pediatrics *72*, 295–299.

Farrell, H.E., and Shellam, G.R. (1990). Characterization of neutralizing monoclonal antibodies to murine cytomegalovirus. J. Gen. Virol. *71*, 655–664.

Farrell, H.E., and Shellam, G.R. (1991). Protection against murine cytomegalovirus infection by passive transfer of neutralizing and non-neutralizing monoclonal antibodies. J. Gen. Virol. *72*, 149–156.

Feire, A.L., Koss, H., and Compton, T. (2004). Cellular integrins function as entry receptors for human cytomegalovirus via a highly conserved disintegrin-like domain. Proc. Natl. Acad. Sci. U.S.A. *101*, 15470–15475.

Gerna, G., Sarasini, A., Patrone, M., Percivalle, E., Fiorina, L., Campanini, G., Gallina, A., Baldanti, F., and Revello, M.G. (2008). Human cytomegalovirus serum neutralizing antibodies block virus infection of endothelial/epithelial cells, but not fibroblasts, early during primary infection. J. Gen. Virol. *89*, 853–865.

Gicklhorn, D., Eickmann, M., Meyer, G., Ohlin, M., and Radsak, K. (2003). Differential effects of glycoprotein B epitope-specific antibodies on human cytomegalovirus-induced cell–cell fusion. J. Gen. Virol. *84*, 1859–1862.

Gorman, S., Harvey, N.L., Moro, D., Lloyd, M.L., Voigt, V., Smith, L.M., Lawson, M.A., and Shellam, G.R. (2006). Mixed infection with multiple strains of murine cytomegalovirus occurs following simultaneous or sequential infection of immunocompetent mice. J. Gen. Virol. *87*, 1123–1132.

Griffiths, P.D., Stanton, A., McCarrell, E., Smith, C., Osman, M., Harber, M., Davenport, A., Jones, G., Wheeler, D.C., O'Beirne, J., *et al.* (2011). Cytomegalovirus glycoprotein-B vaccine with MF59 adjuvant in transplant recipients: a phase 2 randomised placebo-controlled trial. Lancet *377*, 1256–1263.

Grundy, J.E., Lui, S.F., Super, M., Berry, N.J., Sweny, P., Fernando, O.N., Moorhead, J., and Griffiths, P.D. (1988). Symptomatic cytomegalovirus infection in seropositive kidney recipients: reinfection with donor virus rather than reactivation of recipient virus. Lancet *2*, 132–135.

Grunewald, K., Desai, P., Winkler, D.C., Heymann, J.B., Belnap, D.M., Baumeister, W., and Steven, A.C. (2003). Three-dimensional structure of herpes simplex virus from cryo-electron tomography. Science *302*, 1396–1398.

Hamprecht, K., Maschmann, J., Vochem, M., Dietz, K., Speer, C.P., and Jahn, G. (2001). Epidemiology of transmission of cytomegalovirus from mother to preterm infant by breastfeeding. Lancet *357*, 513–518.

Heldwein, E.E., Lou, H., Bender, F.C., Cohen, G.H., Eisenberg, R.J., and Harrison, S.C. (2006). Crystal structure of glycoprotein B from herpes simplex virus 1. Science *313*, 217–220.

Hessell, A.J., Hangartner, L., Hunter, M., Havenith, C.E., Beurskens, F.J., Bakker, J.M., Lanigan, C.M., Landucci, G., Forthal, D.N., Parren, P.W., *et al.* (2007). Fc receptor but not complement binding is important in antibody protection against HIV. Nature *449*, 101–104.

Hobom, U., Brune, W., Messerle, M., Hahn, G., and Koszinowski, U. (2000). Fast screening procedures for random transposon libraries of cloned herpesvirus genomes: Mutational analysis of human cytomegalovirus envelope glycoprotein genes. J. Virol. *74*, 7720–7729.

Huber, M.T., and Compton, T. (1998). The human cytomegalovirus UL74 gene encodes the third component of the glycoprotein H-glycoprotein L-containing envelope complex. J. Virol. *72*, 8191–8197.

Isaacson, M.K., and Compton, T. (2009). Human cytomegalovirus glycoprotein B is required for virus entry and cell-to-cell spread but not for virion attachment, assembly, or egress. J. Virol. *83*, 3891–3903.

Ishibashi, K., Tokumoto, T., Tanabe, K., Shirakawa, H., Hashimoto, K., Kushida, N., Yanagida, T., Inoue, N., Yamaguchi, O., Toma, H., *et al.* (2007). Association of the outcome of renal transplantation with antibody response to cytomegalovirus strain-specific glycoprotein H epitopes. Clin. Infect. Dis. *45*, 60–67.

Jones-Engel, L., Engel, G.A., Heidrich, J., Chalise, M., Poudel, N., Viscidi, R., Barry, P.A., Allan, J.S., Grant, R., and Kyes, R. (2006). Temple monkeys and health implications of commensalism, Kathmandu, Nepal. Emerg. Infect. Dis. *12*, 900–906.

Kattenhorn, L.M., Mills, R., Wagner, M., Lomsadze, A., Makeev, V., Borodovsky, M., Ploegh, H.L., and Kessler, B.M. (2004). Identification of proteins associated with murine cytomegalovirus virions. J. Virol. *78*, 11187–11197.

Kaye, S., Miles, D., Antoine, P., Burny, W., Ojuola, B., Kaye, P., Rowland Jones, S., Whittle, H., van der Sande, M., and Marchant, A. (2008). Virological and immunological correlates of mother-to-child transmission of cytomegalovirus in The Gambia. J. Infect. Dis. *197*, 1307–1314.

Kinzler, E.R., and Compton, T. (2005). Characterization of human cytomegalovirus glycoprotein-induced cell–cell fusion. J. Virol. *79*, 7827–7837.

Klein, M., Schoppel, K., Amvrossiadis, N., and Mach, M. (1999). Strain-specific neutralization of human cytomegalovirus isolates by human sera. J. Virol. *73*, 878–886.

Klenovsek, K., Weisel, F., Schneider, A., Appelt, U., Jonjic, S., Messerle, M., Bradel-Tretheway, B., Winkler, T.H., and Mach, M. (2007). Protection from CMV infection in immunodeficient hosts by adoptive transfer of memory B cells. Blood *110*, 3472–3479.

Kniess, N., Mach, M., Fay, J., and Britt, W.J. (1991). Distribution of linear antigenic sites on glycoprotein gp55 of human cytomegalovirus. J. Virol. *65*, 138–146.

Krause, J.C., Tsibane, T., Tumpey, T.M., Huffman, C.J., Basler, C.F., and Crowe, J.E., Jr. (2011). A broadly neutralizing human monoclonal antibody that recognizes a conserved, novel epitope on the globular head of the influenza H1N1 virus hemagglutinin. J. Virol. *85*, 10905–10908.

Lawson, C.M., Grundy, J.E., and Shellam, G.R. (1988). Antibody responses to murine cytomegalovirus in genetically resistant and susceptible strains of mice. J. Gen. Virol. *69*, 1987–1998.

Li, L., Coelingh, K.L., and Britt, W.J. (1995). Human cytomegalovirus neutralizing antibody-resistant phenotype is associated with reduced expression of glycoprotein H. J. Virol. *69*, 6047–6053.

Li, L., Nelson, J.A., and Britt, W.J. (1997). Glycoprotein H-related complexes of human cytomegalovirus: identification of a third protein in the gCIII complex. J. Virol. *71*, 3090–3097.

Macagno, A., Bernasconi, N.L., Vanzetta, F., Dander, E., Sarasini, A., Revello, M.G., Gerna, G., Sallusto, F., and Lanzavecchia, A. (2010). Isolation of human monoclonal antibodies that potently neutralize human cytomegalovirus infection by targeting different epitopes on the gH/gL/UL128-131A complex. J. Virol. *84*, 1005–1013.

Mach, M., Kropff, B., Dal-Monte, P., and Britt, W.J. (2000). Complex formation of human cytomegalovirus gylcoprotein M (gpUL100) and glycoprotein N (gpUl73). J. Virol. *74*, 11881–11892.

Manley, K., Anderson, J., Yang, F., Szustakowski, J., Oakeley, E.J., Compton, T., and Feire, A.L. (2011). Human cytomegalovirus escapes a naturally occurring neutralizing antibody by incorporating it into assembling virions. Cell Host Microbe *10*, 197–209.

Marshall, G.S., Rabalais, G.P., Stout, G.G., and Waldeyer, S.L. (1992). Antibodies to recombinant-derived glycoprotein B after natural human cytomegalovirus infection correlate with neutralizing activity. J. Infect. Dis. *165*, 381–384.

Masuho, Y., Matsumoto, Y., Sugano, T., Fujinaga, S., and Minamishima, Y. (1987). Human monoclonal antibodies neutralizing human cytomegalovirus. J. Gen. Virol. *68*, 1457–1461.

Meyer, H., Masuho, Y., and Mach, M. (1990). The gp116 of the gp58/116 complex of human cytomegalovirus represents the amino-terminal part of the precursor molecule and contains a neutralizing epitope. J. Gen. Virol. *71*, 2443–2450.

Meyer, H., Sundqvist, V.A., Pereira, L., and Mach, M. (1992). Glycoprotein gp116 of human cytomegalovirus contains epitopes for strain-common and strain-specific antibodies. J. Gen. Virol. *73*, 2375–2383.

Morello, C.S., Ye, M., Hung, S., Kelley, L.A., and Spector, D.H. (2005). Systemic priming-boosting immunization with a trivalent plasmid DNA and inactivated murine cytomegalovirus (MCMV) vaccine provides long-term protection against viral replication following systemic or mucosal MCMV challenge. J. Virol. *79*, 159–175.

Navarro, D., Paz, P., Tugizov, S., Topp, K., La Vail, J., and Pereira, L. (1993). Glycoprotein B of human cytomegalovirus promotes virion penetration into cells, transmission of infection from cell to cell, and fusion of infected cells. Virology *197*, 143–158.

Nigro, G., Adler, S.P., La, T.R., and Best, A.M. (2005). Passive immunization during pregnancy for congenital cytomegalovirus infection. N. Engl. J. Med. *353*, 1350–1362.

Nimmerjahn, F., and Ravetch, J.V. (2006). Fcgamma receptors: old friends and new family members. Immunity. *24*, 19–28.

Ohizumi, Y., Suzuki, H., Matsumoto, Y., Masuho, Y., and Numazaki, Y. (1992). Neutralizing mechanisms of two human monoclonal antibodies against human cytomegalovirus glycoprotein 130/55. J. Gen. Virol. *73*, 2705–2707.

Pass, R.F., Zhang, C., Evans, A., Simpson, T., Andrews, W., Huang, M.L., Corey, L., Hill, J., Davis, E., Flanigan, C., *et al.* (2009). Vaccine prevention of maternal cytomegalovirus infection. N. Engl. J. Med. *360*, 1191–1199.

Plotkin, S.A. (2008). Vaccines: correlates of vaccine-induced immunity. Clin. Infect. Dis. *47*, 401–409.

Polic, B., Hengel, H., Krmpotic, A., Trgovcich, J., Pavic, I., Luccaronin, P., Jonjic, S., and Koszinowski, U.H. (1998). Hierarchical and redundant lymphocyte subset control precludes cytomegalovirus replication during latent infection. J. Exp. Med. *188*, 1047–1054.

Pötzsch, S., Spindler, N., Wiegers, A.K., Fisch, T., Rucker, P., Sticht, H., Grieb, N., Baroti, T., Weisel, F., Stamminger, T., *et al.* (2011). B cell repertoire analysis identifies new antigenic domains on glycoprotein B of human cytomegalovirus which are target of neutralizing antibodies. PLoS. Pathog. *7*, e1002172.

Raanani, P., Gafter-Gvili, A., Paul, M., Ben-Bassat, I., Leibovici, L., and Shpilberg, O. (2008). Immunoglobulin prophylaxis in hematological malignancies and hematopoietic stem cell transplantation. Cochrane. Database. Syst. Rev. CD006501.

Rapp, M., Messerle, M., Lucin, P., and Koszinowski, U.H. (1993). *In vivo* protection studies with MCMV glycoproteins gB and gH expressed by vaccinia virus. In Multidisciplinary Approach to Understanding Cytomegalovirus Disease., Michelson, S., and Plotkin, S.A., eds. (Excerpta Medica, Amsterdam), pp. 327–332.

Rasmussen, L., Mullenax, J., Nelson, M., and Merigan, T.C. (1985). Human cytomegalovirus polypeptides stimulate neutralizing antibody *in vivo*. Virology *145*, 186–190.

Renzette, N., Bhattacharjee, B., Jensen, J.D., Gibson, L., and Kowalik, T.F. (2011). Extensive genome-wide variability of human cytomegalovirus in congenitally infected infants. PLoS Pathog. *7*, e1001344.

Rey, F.A. (2006). Molecular gymnastics at the herpesvirus surface. EMBO Rep. *7*, 1000–1005.

Rodrigo, W.W., Block, O.K.T., Lane, C., Sukupolvi-Petty, S., Goncalvez, A.P., Johnson, S., Diamond, M.S., Lai, C.-J., Rose, R.C., Jin, X., *et al.* (2009). Denguevirus neutralization is modulated by IgG antibody subclass and Fcγ receptor subtype. Virology *394*, 175–182.

Ross, S.A., Novak, Z., Pati, S., Patro, R.K., Blumenthal, J., Danthuluri, V.R., Ahmed, A., Michaels, M.G., Sanchez, P.J., Bernstein, D.I., *et al.* (2011). Mixed infection and strain diversity in congenital cytomegalovirus infection. J. Infect. Dis. *204*, 1003–1007.

Ryckman, B.J., Rainish, B.L., Chase, M.C., Borton, J.A., Nelson, J.A., Jarvis, M.A., and Johnson, D.C. (2008). Characterization of the human cytomegalovirus gH/gL/UL128–131 complex that mediates entry into epithelial and endothelial cells. J. Virol. *82*, 60–70.

Saccoccio, F.M., Sauer, A.L., Cui, X., Armstrong, A.E., Habib, e., Johnson, D.C., Ryckman, B.J., Klingelhutz, A.J., Adler, S.P., and McVoy, M.A. (2011). Peptides from cytomegalovirus UL130 and UL131 proteins induce high titer antibodies that block viral entry into mucosal epithelial cells. Vaccine *29*, 2705–2711.

Schoppel, K., Kropff, B., Schmidt, C., Vornhagen, R., and Mach, M. (1997). The humoral immune response against human cytomegalovirus is characterized by a delayed synthesis of glycoprotein-specific antibodies. J. Infect. Dis. *175*, 533–544.

Shanley, J.D., and Wu, C.A. (2003). Mucosal immunization with a replication-deficient adenovirus vector expressing murine cytomegalovirus glycoprotein B induces mucosal and systemic immunity. Vaccine *21*, 2632–2642.

Shanley, J.D., Jordan, M.C., and Stevens, J.G. (1981). Modification by adoptive humoral immunity of murine cytomegalovirus infection. J. Infect. Dis. *143*, 231–237.

Shimamura, M., Mach, M., and Britt, W.J. (2006). Human cytomegalovirus infection elicits a glycoprotein M (gM)/gN-specific virus-neutralizing antibody response. J. Virol. *80*, 4591–4600.

Silvestri, M., Sundqvist, V.A., Ruden, U., and Wahren, B. (1991). Characterization of a major antigenic region on gp55 of human cytomegalovirus. J. Gen. Virol. *72*, 3017–3023.

Simpson, J.A., Chow, J.C., Baker, J., Avdalovic, N., Yuan, S., Au, D., Co, M.S., Vasquez, M., Britt, W.J., and Coelingh, K.L. (1993). Neutralizing monoclonal antibodies that distinguish three antigenic sites on human cytomegalovirus glycoprotein H have conformationally distinct binding sites. J. Virol. *67*, 489–496.

Sinzger, C., Digel, M., and Jahn, G. (2008). Cytomegalovirus cell tropism. Curr. Top. Microbiol. Immunol. *325*, 63–83.

Slavuljica I., Busche, A., Babić, M., Mitrović, M., Gašparović, I., Cekinović, D., Markova Car, E., Pernjak Pugel, E., Ciković, A., Lisnić, V.J., *et al.* (2010). Recombinant mouse cytomegalovirus expressing a ligand for the NKG2D receptor is attenuated and has improved vaccine properties. J. Clin. Invest. *120*, 4532–4545.

Smith, A.L., Singleton, G.R., Hansen, G.M., and Shellam, G. (1993). A serologic survey for viruses and Mycoplasma pulmonis among wild house mice (Mus domesticus) in southeastern Australia. J. Wildl. Dis. *29*, 219–229.

Sokos, D.R., Berger, M., and Lazarus, H.M. (2002). Intravenous immunoglobulin: appropriate indications and uses in hematopoietic stem cell transplantation. Biol. Blood Marrow Transplant. *8*, 117–130.

Soroceanu, L., Akhavan, A., and Cobbs, C.S. (2008). Platelet-derived growth factor-alpha receptor activation is required for human cytomegalovirus infection. Nature *455*, 391–395.

Spaete, R.R., Saxena, A., Scott, P.I., Song, G.J., Probert, W.S., Britt, W.J., Gibson, W., Rasmussen, L., and Pachl, C. (1990). Sequence requirements for proteolytic processing of glycoprotein B of human cytomegalovirus strain Towne. J. Virol. *64*, 2922–2931.

Spear, G.T., Lurain, N.S., Parker, C.J., Ghassemi, M., Payne, G.H., and Saifuddin, M. (1995). Host-cell-derived complement control proteins CD55 and CD59 are incorporated into the virions of two unrelated enveloped viruses. Human T-cell leukemia/lymphoma virus type I (HTLV-I) and human cytomegalovirus (HCMV). J. Immunol. *155*, 4376–4381.

Speckner, A., Glykofrydes, D., Ohlin, M., and Mach, M. (1999). Antigenic domain 1 of human cytomegalovirus glycoprotein B induces a multitude of different antibodies which, when combined, results in incomplete virus neutralization. J. Gen. Virol. *80*, 2183–2191.

Spiller, O.B., Hanna, S.M., Devine, D.V., and Tufaro, F. (1997). Neutralization of cytomegalovirus virions: the role of complement. J. Infect. Dis. *176*, 339–347.

Sprague, E.R., Wang, C., Baker, D., and Bjorkman, P.J. (2006). Crystal structure of the HSV-1 Fc receptor bound to Fc reveals a mechanism for antibody bipolar bridging. PLoS. Biol. *4*, e148.

Takada, A., and Kawaoka, Y. (2003). Antibody-dependent enhancement of viral infection: molecular mechanisms and *in vivo* implications. Rev. Med. Virol. *13*, 387–398.

Tugizov, S., Navarro, D., Paz, P., Wang, Y., Qadri, I., and Pereira, L. (1994). Function of human cytomegalovirus glycoprotein B: syncytium formation in cells constitutively expressing gB is blocked by virus-neutralizing antibodies. Virology *201*, 263–276.

Urban, M., Klein, M., Britt, W.J., Hassfurther, E., and Mach, M. (1996). Glycoprotein H of human cytomegalovirus is a major antigen for the neutralizing humoral immune response. J. Gen. Virol. *77*, 1537–1547.

Vanarsdall, A.L., Ryckman, B.J., Chase, M.C., and Johnson, D.C. (2008). Human cytomegalovirus glycoproteins gB and gH/gL mediate epithelial cell–cell fusion when expressed either in cis or in trans. J. Virol. *82*, 11837–11850.

Varnum, S.M., Streblow, D.N., Monroe, M.E., Smith, P., Auberry, K.J., Pasa-Tolic, L., Wang, D., Camp, D.G., Rodland, K., Wiley, S., *et al.* (2004). Identification of proteins in human cytomegalovirus (HCMV) particles: the HCMV proteome. J. Virol. *78*, 10960–10966.

Walker, L.M., Phogat, S.K., Chan-Hui, P.Y., Wagner, D., Phung, P., Goss, J.L., Wrin, T., Simek, M.D., Fling, S., Mitcham, J.L., *et al.* (2009). Broad and potent neutralizing antibodies from an African donor reveal a new HIV-1 vaccine target. Science *326*, 285–289.

Waner, J.L., and Weller, T.H. (1978). Analysis of antigenic diversity among human cytomegaloviruses by kinetic neutralization tests with high-titered rabbit antisera. Infect. Immun. *21*, 151–157.

Wang, D., and Shenk, T. (2005). Human cytomegalovirus virion protein complex required for epithelial and endothelial cell tropism. Proc. Natl. Acad. Sci. U.S.A. *102*, 18153–18158.

Wirtz, N., Schader, S.I., Holtappels, R., Simon, C.O., Lemmermann, N.A., Reddehase, M.J., and Podlech, J. (2008). Polyclonal cytomegalovirus-specific antibodies not only prevent virus dissemination from the portal of entry but also inhibit focal virus spread within target tissues. Med. Microbiol. Immunol. *197*, 151–158.

Yeager, A.S., Grumet, F.C., Hafleigh, E.B., Arvin, A.M., Bradley, J.S., and Prober, C.G. (1981). Prevention of transfusion-acquired cytomegalovirus infections in newborn infants. J. Pediatr. *98*, 281–287.

Immunoregulatory Cytokine Networks Discovered and Characterized during Murine Cytomegalovirus Infections

II.11

Marc Dalod and Christine A. Biron

Abstract

In addition to mediating early defence during primary infection, innate immunity delivers immunoregulatory functions to shape innate and adaptive immunity. Basic knowledge on the mechanisms inducing innate cytokine responses and the consequences for orchestrating downstream immunity is being advanced at a dramatic rate. Studies of murine cytomegalovirus (MCMV) infections in mice have contributed many of the breakthrough discoveries in these areas. Although intrinsic differences in viruses, hosts, and infection sites result in unique host–microbe relationships, there are overlapping effects mediated by host cell subsets and molecules. Thus, the information resulting from the leading-edge characterization of responses to MCMV has set the framework for understanding responses to a variety of infections in humans and mice and provided insights on approaches for enhancing resistance to virus-induced diseases. Key areas of progress include sensing of infection, co-ordination of cellular and innate cytokine cascades, consequences for delivery of innate and adaptive immunity to mediate defence, and regulation to protect from immune-mediated damage. These are reviewed here with a focus on the early cytokine/chemokine networks as infection spreads through a host. The surprisingly elegant picture emerging is one of profound flexibility for orchestrating optimal subset innate responses to protect against diverse and complex infectious organisms.

Introduction

From the earliest studies of host–pathogen interactions, it has been clear that there are multiple defences against infection. The first characterization of cellular contributions, defined under conditions examining effects mediated by cells of the innate immune system, evolved to be called innate immunity. The first characterization of defence delivered by soluble factors, mediated by antibodies under the conditions of examination, evolved to be known as adaptive immunity. It is now clear, however, that there are both cellular and soluble mediators of innate and adaptive immunity, and that these two arms of the immune system are communicating with each other to shape overall defence against infection. The cellular constitutes of innate immunity, including polymorphonuclear leucocytes (PMNs), monocyte/macrophage lineage cells (mono/MΦ), and natural killer (NK) cell lymphocytes, play important direct antimicrobial roles with the relative contributions made by each depending on the infectious agent. The list of innate cytokines is extensive and includes products of non-immune as well as innate immune cells. The T- and B-lymphocytes, expressing on their cell surfaces products of rearranging genes as receptors for antigens, are the cellular staples of adaptive immunity with soluble antibodies being products of B-lymphocytes and soluble cytokines being products of T-cells. The composition of subset innate responses, under particular conditions of infection, directs the characteristics of stimulated adaptive responses.

Despite the characterization of the type 1 interferons (IFNs) as early mediators of antiviral defence over fifty years ago (Borden *et al.*, 2007), the remarkable story of the importance of soluble factors in innate immunity has been slow to crystallize. This is, in part, a result of the often pleiotropic effects mediated by various cytokines, with consequences of exposure varying depending on the cell type or condition examined, and of the difficulty in defining the factors as innate mediators of resistance to infection. Soluble factors, first appreciated for their roles in inflammation, are now understood to be part of innate immunity in antimicrobial defence, and the new characterization of pathways regulating primary or alternative signalling from different innate cytokines is helping to explain how the same cytokine can be used to mediate different effects as needed. In addition, subsets of dendritic cells (DCs), originally appreciated for their roles in antigen presentation and stimulation of adaptive immunity, are now known to be important in producing innate cytokines. Finally, recent advances in the characterization of the sensors activating innate immunity, as products of germ-line gene families detecting structures not present or not seen in an uninfected host, have provided the mechanisms used for initiating immune responses and helped clarify the definition of the word 'innate' by demonstrating that although certain elements are basally expressed, much of innate immunity is induced or elevated at early times following infection. As a result of this progress, it is possible to synthesize a picture of the steps from sensing infections to inducing innate responses mediating antiviral and immunoregulatory effects, to the consequences for the development of adaptive immunity.

In addition to the type 1 IFNs, the soluble constituents of innate immunity include interleukin 1 (IL-1), IL-18, IL-6, tumour necrosis factor (TNF), IL-12, and IFN-γ when it is produced by classical, non-T, NK cells, as well as a variety of chemokines. Some of these factors directly induce and deliver antimicrobial effects. They all have immunoregulatory functions. In some cases, their production is linked to a common pathway(s) for stimulation. In others, the responses are elicited in cascades with the production of one factor by a particular cell mediating endocrine or autocrine effects to stimulate cellular production of a subsequent factor. The cytokine effects require not only the expression of receptors for the factors on, and particular signalling pathways in, responding cells but also the absence of viral products blocking the delivery of functions depending on exposure or stimulated 'downstream'. The importance for individual cells and factors to the host can be demonstrated by experimental deletion or inhibition of function, and the potential for inhibiting viral replication is often supported by the demonstration of specific viral avoidance mechanisms expressed in infected cells. Regulating innate responses is critical to protect from damage mediated by innate immune system, and innate responses can act to protect from immunopathology mediated by adaptive immune responses. Studies of murine cytomegalovirus (MCMV) infections of mice have played a major role in elucidating the induction and functions of many of these soluble and cellular mediators of innate immunity. The key discoveries are the focus of this chapter.

As the work is reviewed, it is important to note that MCMV was originally discovered through efforts to find an agent inducing changes comparable to those observed during cytomegalic inclusion disease (CID) in humans. As a consequence, the system is of great usefulness in providing insights to the human condition. To establish the disease in mice, salivary gland extracts from mice were injected into mice to induce a range of disease parameters, including splenomegaly and hepatitis as well as particular cellular morphological changes, i.e. cytomegalic inclusion bodies (McCordock and Smith, 1936). A number of groups have since worked to molecularly characterize different strains of the virus and a wide range of immune responses elicited in infected mice. Although several routes of infection, including intraperitoneal (i.p.), were originally examined, characterization of immune responses to MCMV has been largely done following i.p. infection with moderate to relatively high challenge doses of the virus. Because materials delivered to the peritoneum are rapidly taken up by the lymphatics and dumped into the circulation, the approach has often elucidated details on responses in the context of blood exposure. Under these conditions in immunocompetent mice, the virus is found replicating in spleen and liver at early times after infection and eventually in salivary glands. Many of the immune responses are initiated in the spleen with its marginal zone regions and resident MΦ filtering virus from the blood and initiating communication with the immune system. It is important to remember these aspects of the virus–host relationship as the specifics and generalities of the discoveries resulting from defining immune responses to MCMV are considered.

Cytokine responses and the systemic events and consequences

Because of technological advances, the characterization of cytokine expression during infection has often been done at the level of mRNA expression. This approach can facilitate the acquisition of information on a large number of target genes and on the pathways to induction of gene expression. Because of the reagents available at the time, however, many of the early studies were done at the level of protein expression. Whether the induction is first characterized at the RNA or protein level, the biological importance for any factor is dependent on protein expression, and documentation of individual or combined functions is dependent on blocking their actions in the context of complex immune responses induced during particular infections. This is especially true in assigning required functions to individual cytokines because these factors can have diverse and redundant functions. As a result, although information on the expression of individual cytokines is growing, the biological roles are not yet completely understood for all of the responses.

Composite innate cytokine responses

Many studies combined have demonstrated that MCMV infections induce multiple waves of innate cytokine and chemokine responses. The type 1 IFNs, first characterized as produced by an infected cell to induce antiviral states in neighbouring cells and known to be induced during other viral infections, were shown to be produced during MCMV infections of mice a number of years ago. Studies neutralizing type 1 IFN functions by antibody treatment demonstrated that these cytokines are important in protecting against MCMV replication *in vivo* (Grundy *et al.*, 1982; Chong *et al.*, 1983; Orange and Biron, 1996b). Likewise, efforts to define antiviral functions for TNF led to a few reports suggesting that the factor had antiviral effects when added to culture systems and upon *in vivo* neutralization in several viral systems including MCMV (Pavic *et al.*, 1993; Lucin *et al.*, 1994; Heise and Virgin, 1995; Orange and Biron, 1996b).

The coordination of innate cytokine responses during a viral infection with characteristics overlapping those of pro-inflammatory cytokine cascades, however, came first through studies of the MCMV system (Orange and Biron, 1996a,b; Ruzek *et al.*, 1997, 1999; Biron *et al.*, 1999). At the time, the thinking was that such complex and systemic responses were characteristic of, and limited to, septic conditions induced by Gram-negative bacteria. The breakthrough resulted from efforts to determine whether or not the previously defined role for NK cells in defence against MCMV (Bancroft *et al.*, 1981; Bukowski *et al.*, 1983) included antiviral functions delivered through IFN-γ production as well as elevated lysis of infected target cells. The collective work included the first demonstrated induction and function for IL-12 (Orange and Biron, 1996a) and for NK cell IFN-γ production (Orange *et al.*, 1995; Orange and Biron, 1996a,b) during a viral infection (discussed in more detail below). Control experiments with inactivated virus demonstrated that infection was required for these innate responses. Once the parameters of the responses were extended, it became clear that there is peak systemic cytokine production easily detected in the serum at a time centred around a 36- to 40-hour period after i.p. infections (Orange and Biron, 1996a,b; Ruzek *et al.*, 1997, 1999; Pien *et al.*, 2000; Delale *et al.*, 2005) (schematically compiled in Fig. II.11.1A). Hence, the response coincides with the appearance of virus replication in the spleen but not the extended viral replication generally observed in the liver or the delayed replication in salivary glands. Components include sharp peaks of production for type 1 IFN and extremely high levels of IL-6 as well as more moderate levels of biologically active IL-12 and IFN-γ. There is also a lower but more sustained release of TNF. Surprisingly, although both IL-1α and IL-1β are induced at the mRNA level in infected spleens (Ruzek *et al.*, 1999), it is difficult to detect circulating IL-1. In contrast, IL-18 induction is readily detected in the serum and sustained for longer periods of time (Pien *et al.*, 2000). The peak innate cytokine responses are accompanied by increases in circulating hormones with known immunoregulatory functions, glucocorticoids or specifically in the mouse, corticosterone (Ruzek *et al.*, 1997), but only low to undetectable levels of IL-10, a known immunoregulatory cytokine (Lee *et al.*, 2009). Thus, under the conditions of a spreading viral infection, an innate cytokine storm can be induced with many characteristics overlapping with, and a few distinguishing from, those of systemic infections with other agents.

Cytokines in virus-induced disease

The discovery of the systemic innate cytokine storm stirred efforts to evaluate a role for the cytokines in viral pathogenesis. Here again, work in the MCMV system led to the understanding of the pathways for immune-mediated disease during viral infections. Prior to this time, the critical conditions resulting from systemic cytokine production had been characterized in the context of bacterial infections, but the focus in the context of viral infections had been on immune pathology resulting from adaptive immune responses. There had been reports, however, of TNF contributing to liver disease induced through non-viral pathways and of liver disease in a number of different viral infections also inducing TNF. Examination of TNF-dependent effects during MCMV infection demonstrated that the cytokine can be a major contributor to early liver disease during viral infection (Orange *et al.*, 1997). When TNF is neutralized, the liver pathology characterized as grossly and microscopically visible liver necrotic foci with the appearance of the circulating liver enzymes aspartate aminotransferase (AST) and alanine aminotransferase (ALT) at early times after infection, i.e. days 2 and 3, are significantly ablated. The results are particularly compelling because the inhibition of these parameters of pathology occurs with accompanying increases in viral replication, hepatic cytomegalic inclusion bodies, and inflammatory foci at sites of viral replication in the liver. Thus, viral burden and hepatitis can be separated from parameters of significant liver disease, and there is a cytokine component to virus-induced liver pathology.

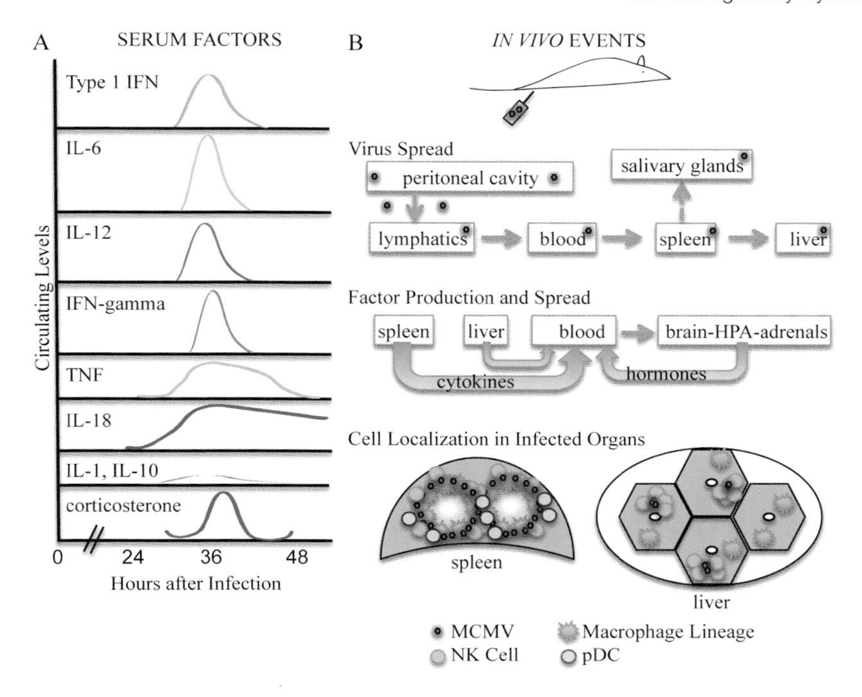

Figure II.11.1 Systemic and regional innate cytokine responses to MCMV. Many of the early responses to MCMV infection have been characterized after moderate to high dose i.p. infection. (A) There are waves of systemic innate cytokine responses, with peak levels in the serum at 36 to 40 hours after infection, including sharp high peaks for type 1 IFN and IL-6, moderate levels of biologically active IL-12 and IFN-γ, a lower but more sustained release of TNF, and IL-18 detected for longer periods. Although IL-1α and IL-1β are induced at the mRNA level in infected spleens, it is difficult to detect circulating IL-1. The cytokine responses are accompanied by increases in circulating glucocorticoids with immunoregulatory functions, but only low to undetectable levels of the immunoregulatory IL-10 cytokine. (B) Overview of viral spread, cell trafficking, cytokine responses, and communication with the neuroendocrine system is presented. Following i.p. injection, virus is rapidly taken up by the lymphatic system and delivered to the blood. The spleen, with MΦ and the unique marginal zone structure, acts as an early filtration and infection site. The liver is also an early site of viral infection with infected hepatocytes having characteristic cytomegalic inclusion bodies. Evidence of viral infection is detected along splenic marginal zones and at focal sites in liver, and NK cells expressing IFN-γ accumulate at these sites from early time-points after infection onwards. There is co-localized expression of type 1 IFNs and IL-12 as well as pDCs in the vicinity of infected marginal zones in the spleen. These conditions are associated with the systemic cytokine responses. Specific functions are known for IL-6 in induction of glucocorticoids through the HPA axis and for TNF in mediating liver pathology. The localization of type 1 IFN is dispersed in the liver. (Representations are compiled from studies cited in the text.)

Characterization of the roles for innate cytokine responses in virus-induced disease was extended through studies examining the benefit, to a competent host, of the induced endogenous glucocorticoid response. Previous work, largely carried out evaluating responses to administered cytokines or to cytokines induced in response to administered lipopolysaccharide from Gram-negative bacteria, had demonstrated cytokine communication with the neuroendocrine system for activation of the hypothalamic-pituitary-adrenal (HPA) axis. The pathway had been defined as delivered through hypothalamus stimulation for production of corticotropin-releasing hormone (CRH), to induce pituitary release of adrenocorticotropin hormone (ACTH) and result in downstream induction of adrenal gland glucocorticoid production. There had, however, never been a demonstration of the pathway in the context of infection, and the reported roles for individual cytokines were inconsistent. The MCMV system was the first to provide definitive evidence for, and to assign a major non-redundant role to a particular cytokine in activating, the circuit during an infection (Ruzek *et al.*, 1997, 1999) (Fig. II.11.1B). In addition to demonstrating activation of the HPA axis and a resulting induction of glucocorticoids, these studies showed that IL-6, rather than IL-1, is largely responsible for the immune-neuroendocrine loop. The dependence on IL-6 here may be a result of the low levels of circulating IL-1 induced during the MCMV infection.

More recent studies indicate that IL-6 might also act through a different pathway to induce glucocorticoids (Silverman *et al.*, 2004), but the major pathway in a competent host is likely to be through the HPA axis.

Because this neuroendocrine response can act as a negative feedback loop to prevent excessive production of pro-inflammatory cytokines, the consequences of blocking the HPA axis were also evaluated in an infection condition for the first time (Ruzek *et al.*, 1999). Removal of the adrenal glands to block the endogenous glucocorticoid response results in increases in IL-12, IFN-γ, TNF, and IL-6 production, as well as in mRNA expression for a wider range of cytokines, including IL-1α and IL-1β, during MCMV infection. The conditions are extremely detrimental to the host and can lead to mortality. Protection is achieved by neutralizing TNF but not IFN-γ. Because TNF and IFN-γ often act synergistically, the lack of requirement for IFN-γ is perhaps surprising but may result from the high type 1 IFN levels present and their overlapping effects. Here again, the regulation of viral burden is separated from pathology; in the absence of the endogenous glucocorticoids, decreases in virus replication accompany increases in TNF and death. Basal states are restored by adding back hormone. Thus, the importance of induced endogenous glucocorticoids in protection against life-threatening disease resulting from infection-elicited cytokine responses is clearly demonstrated during an infection. This effect may be critical for protecting vital visceral organs, such as the liver with its sensitivity to TNF-mediated toxic effects during MCMV infection (Orange *et al.*, 1997) and its high expression of glucocorticoid receptors (Turner *et al.*, 2006). Since the original reports, a role of HPA axis activation in preventing excessive immunopathology upon co-infections by viruses and bacteria has also been demonstrated in mice infected intranasally by influenza virus and challenged intravenously by *Listeria monocytogenes*, with the induced glucocorticoid response necessary to limit pro-inflammatory cytokine production, overall inflammation and ensuing tissue destruction in the lung and liver (Jamieson *et al.*, 2010). Taken together, the studies of systemic responses to MCMV infection have led the way in characterizing complex endogenous innate cytokine responses during viral infections, innate cytokine contributions to virus-induced disease, and the role for these mediators in communication between the immune and neuroendocrine systems to limit immune-mediated pathology during viral infections. All of these discoveries are significant to the broad understanding of virus–host relationships and the balance between immune responses for mediating antimicrobial defence but protecting from immune-mediated disease.

Localization of responses

Since the original characterization of the complex systemic pro-inflammatory cytokine responses to MCMV infection, studies of the induction and function of cytokine and chemokine responses have been broadly expanded and significantly advanced. Much of this work has resulted from evaluating production or expression in cell subsets isolated from mixed populations and/or from mice depleted of particular cytokines, cells, or stimulatory pathways. Before considering this information, however, it is important to keep in mind what is known about the *in situ* localization of cells and responses as revealed through analysis of tissues prepared from MCMV-infected mice. Taken together, these studies provide a framework for a better understanding of the complex virus–host interactions at early times after infections as well as a foundation for interpreting much of the work done evaluating the sensing of infection to stimulate different cellular sources of cytokines and chemokines.

Characterization in tissues

Under basal conditions and after i.p. challenges with many viruses, NK cells are primarily located in the blood, spleen, liver and lungs. The spleen is an early site for filtering blood-borne virus and initiating immune responses, and there is viral replication in the organ. The liver is a key site of viral infection, and viral replication at this site can be observed at times after its decline in the spleen. Type 1 IFNs, stimulated by exposure to either the chemical inducer of type 1 IFN, polyinosinic-polycytidylic acid (PIC) (Ishikawa and Biron, 1993) or viral infections, including MCMV (Salazar-Mather *et al.*, 1996), elicit a dramatic redistribution of splenic lymphocytes with increases in white pulp but decreases in red pulp areas without increases in total yields of splenic leucocytes. Cell trafficking studies of bone marrow populations into spleens of recipient mice isolated at early times following MCMV infection have shown that there is an accumulation of non-T/non-B-cells along the borders of splenic red and white pulp regions, i.e. in proximity to areas identified as marginal zones, with the NK cell phenotype, i.e. $NK1.1^{+}AGMl^{+}F4/80^{-}$ (Salazar-Mather *et al.*, 1996). Interestingly, *in situ* hybridization of tissue sections from mice on day 2 of MCMV infection has demonstrated that proportions of these cells express IFN-γ mRNA (Salazar-Mather *et al.*, 1996). Other immunofluorescence and immunohistochemical studies have extended these observations using a broader range of markers expressed on different cell types including marginal zone or metallophilic MΦ and plasmacytoid DCs (pDCs) as well as local expression of particular cytokines (Andrews *et al.*, 2001; Dokun *et al.*, 2001a; Dalod *et al.*, 2002; Courgette *et al.*, 2008a). MCMV can infect MΦ (Hanson *et al.*, 1999, 2001), and studies have demonstrated MCMV infection in the marginal region borders, with type 1 IFN and IL-12 expression primarily detectable in proximity to these changes (Fig. II.11.1B). In total, the observations suggest an exciting model for induction of NK cell trafficking from bone marrow to secondary compartments to localize the populations in the vicinity of activated mono/MΦ/DCs to receive particular stimulation signals.

An additional or different possibility is that the NK cells are moving to sites of viral infection to mediate defence. Other studies examining the localization of MΦ and NK cells in infected livers, carried out first in the MCMV system, have demonstrated that NK cells are accumulating in proximity to MCMV-infected cells at early times after infection (Orange *et al.*, 1997; Salazar-Mather *et al.*, 1998; Dokun *et al.*, 2001a), and that this response is dependent on the chemokine CCL3, first identified as macrophage inflammatory protein 1α (MIP-1α) (Salazar-Mather *et al.*, 1998). The migration of NK cells into livers results in their delivery of IFN-γ for the local induction of a downstream chemokine CXCL9, first named the monokine induced by IFN-γ (Mig) (Salazar-Mather *et al.*, 2000). Because CXCL9 can promote the recruitment of activated T-cells, this cascade of events not only results in the delivery of NK cell IFN-γ with the potential to activate direct antiviral defence mechanisms but also to mediate immunoregulatory functions. The expression of type 1 IFN in infected livers is dispersed (Salazar-Mather *et al.*, 2002), and the events coordinating NK cell migration into livers are facilitated by type 1 IFN-dependent induction of the chemokine, CCL2, first known as the monocyte chemoattractant protein-1 (MCP-1) (Hokeness *et al.*, 2005), to promote migration of CCL3-producing MΦ into the organ (Salazar-Mather *et al.*, 2002; Hokeness *et al.*, 2005). Systemic type 1 IFN production may also induce local bone marrow monocyte/macrophage production of CCL2, CCL7, and CCL12 to facilitate cellular egress into the blood (Crane *et al.*, 2009). Taken together, the studies of MCMV infection in spleen and liver tissues demonstrate the dramatic trafficking of cells in response to infection and infection-induced cytokines and chemokines. They identify chemokine-to-cytokine-to-chemokine cascades to orchestrate the delivery of antiviral defence, mediated by both innate and adaptive components of immune responses, into infected tissues (Fig. II.11.1B).

Cell sources of innate cytokines

The current list of innate cytokines and chemokines stimulated during early times after MCMV infections is extensive (Table II.11.1). In addition to the type 1 IFNs, IL-12, IL-6, NK cell-produced IFN-γ, and IL-18 responses detected at the protein level in serum, the IL-1 observed in spleens at the level of mRNA expression, and the chemokines noted above, they include lymphotoxin α (LTα), a cytokine important for development of splenic and lymph node architecture, IL-10, a cytokine with pleiotropic anti-inflammatory functions, granulocyte/macrophage colony stimulating factor (GM-CSF), a cytokine promoting differentiation of cells from bone marrow cell precursors, and IL-15, a growth factor important for proliferation of NK cells. These factors have been best studied in the period of the pro-inflammatory cytokine cascade induced at times of expanding viral replication, i.e. 36–40 hours after infection. Most cytokines or chemokines can be produced by several cell types, depending on the signals that induce them. As examples, either innate NK cells or adaptive T-cells can produce IFN-γ, and virtually any nucleated cell type can contribute to type 1 IFN production if appropriately stimulated. It is becoming clear, however, through key discovery made in the MCMV system, that particular cells are in place to have specialized functions for delivering particular cytokine responses under different conditions of challenge. The details of the studies on specialized and more generalized expression of cytokines by cell subsets, as evaluated at the protein and mRNA levels, are reviewed here.

Cytokines/chemokines mainly produced by pDC

A collection of experiments, performed by several independent teams using different experimental approaches, have clearly demonstrated that type 1 IFNs are mainly produced by pDCs around 36 hours after infection (Fig. II.11.2A). Collectively, these studies were the first to show a requirement for a specialized cell in producing type 1 IFN in response to a viral infection *in vivo*. The initial characterization was based on evaluating the production of circulating factors in mice with or without pDC populations, and the secretion of the type 1 IFNs by purified pDC populations isolated from spleens after MCMV infection (Asselin-Paturel *et al.*, 2001; Dalod *et al.*, 2002, 2003). The responses can be observed in both C57BL/6 or 129Sv mice but are more elevated in 129Sv mice with their higher proportions and reactivity of pDCs. The first selective depletions of pDC *in vivo* by administration of antibodies led to a dramatic decrease in serum levels of type 1 IFNs at

Table II.11.1 List of cytokines/chemokines/growth factors induced at 36–40 hours post infection with their cellular sources and documented effects

Factors	Time points	Main cellular sources during MCMV infection[a]	❶Receptor ❷Simplified main signalling pathway ❸Main target cells[c]	General knowledge on biological effects relevant to viral infections	Biological effects during MCMV infection	References
Type I IFN	8–12 h 36 h 44 h	Stromal cells[b] pDC (in liver, mo/MΦ[b]?) ?	❶IFNAR1/IFNAR2 ❷JAK1/TYK2→STAT1/STAT2/IRF9 Other STATs can be involved ❸Ubiquitous	Innate immune response against viral infections Induction of cell-intrinsic antiviral defence mechanisms Activation of immune cells Anti-proliferative and pro-apoptotic functions (often paradoxical effects)	Critical for survival Direct inhibition of viral replication in part through down-modulation of cholesterol metabolism Promotion of DC maturation Inhibition of DC IL-12 production Initiation of a cytokine/chemokine cascade for immune cell recruitment to infected liver Induction of IL-15, activation of NK cell cytotoxic activity and entry into cell cycle	Presti *et al.* (1998), Biron (2001), Dalod *et al.* (2002), Nguyen *et al.* (2002), Salazar-Mather *et al.* (2002), Dalod *et al.* (2003), Krug *et al.* (2004), Strobl *et al.* (2005), Scheu *et al.* (2008), Schneider *et al.* (2008), Courgette *et al.* (2008a,b), Swiecki *et al.* (2010), Blanc *et al.* (2011)
IL-28 (IFN-λ)	36 h	*pDC and CD8α[+] DCs*	❶IL-28RA/IL10RB ❷JAK1→STAT1/STAT2/IRF9; p38→JNK ❸Neutrophils, epithelial cells	Induced by viral infection Innate immune response against viral infections Induction of cell-intrinsic antiviral defence mechanisms	ND	Dalod *et al.* (unpublished)
IL-1α IL-1β	NI 36 h	NT *All DC subsets but not NK, B or CD8 T-cells*	❶IL-1R1/ILRAP ❷IL1R2 is a decoy receptor TOLLIP/MYD88→IRAK4→NF-kB ❸Neutrophils, DCs, T-cells	Proteolytically processed and released by caspase 1 Primarily responsible for inflammation, fever and sepsis Many other metabolic and haematopoietic activities	ND Not needed for glucocorticoid responses	Ruzek *et al.* (1997, 1999)[e]
IL-6	36 h	*All DC subsets and NK cells; not B and CD8 T-cells*	❶IL-6R/gp130 ❷JAK1/TYK2→STAT3; RAS→MAPK→ERK1/2 ❸Myeloid cells, T-cells	Many pro- and anti-inflammatory functions Potent inducer of the acute phase response Crucial for plasma cell differentiation Role in lymphocyte and monocyte differentiation Inhibition of TNF and IL-1 Activation of IL-1RA and IL-10	Prevents immunopathology (cytokine shock) by promoting optimal release of glucocorticoids at 36 h post infection	Ruzek *et al.* (1997, 1999)

IL-10	36 h	*NK cells (low)*	❶IL-10RA/IL10RB ❷JAK1/TYK2→STAT3→SOCS3 ❸All immune cells	Pleiotropic anti-inflammatory effects Blocks NF-kB activity Regulates JAK-STAT signalling Inhibits synthesis of many cytokines Down-regulates MHC-II and co-stimulatory molecules Enhances B-cell survival, proliferation and Ab production	Promotes NK cell survival Prevents CD8 T-cell-mediated immunopathology	Lee *et al.* (2009), Stacey *et al.* (2011)
	4 d	NK cells				
IL-12	36 h	pDC and CD8α[+] DCs, (depends on mouse strain, viral inoculums and type I IFN effects)	❶IL-12RB1/IL12RB2 ❷Tyk2/Jak2→STAT4 ❸NK cells; stimulated T-cells	Essential for induction/ maintenance of Th1 responses Crucial for protection to intracellular pathogens Stimulates NK and T-cell IFN-γ ± cytotoxicity	critical for survival promotes NK cell IFN-γ production	Orange and Biron (1996a,b), Ruzek *et al.* (1997), Nguyen *et al.* (2000), Pien *et al.* (2000), Dalod *et al.* (2002), Krug *et al.* (2004), Courgette *et al.* (2008a)
IL-15	36 h	*High in CD8α[+] and CD11b[+] DCs; low in pDC and B-cells*	❶IL-2Rβ/γ trans-presentation by IL-15Rα ❷JAK1/JAK3→STAT5 ❸NK cells; T-cells; +/-mono	Regulates T and NK cell activation and proliferation Delivers proliferative and anti-apoptotic signals required for differentiation and maintenance of NK and memory CD8 T-cells	Promotes NK cell cytotoxicity, proliferation and survival Promotes granzyme B and perforin mRNA translation in NK cells	Nguyen *et al.* (2000), Fehniger *et al.* (2007)
IL-18	36 h	*High in CD8α[+] and CD11b[+] DCs; low in pDC and B-cells*	❶IL-18R ❷MYD88→IRAK4→NF-kB ❸NK cells, T-cells, neutrophils	Proinflammatory cytokine with IL-12, promotes NK and T-cell IFN-γ can induce severe inflammation	Dispensable for survival to infection Required for NK cell IFN-γ production in spleen	Pien *et al.* (2000), French *et al.* (2006)
LTα	36 h	*High in pDC*	❶Homotrimers→TNFRSF1A/1B/14; ❷TNFRSF1A/14→TRAF2→NF-kB; TNFRSF1A→FADD→Caspase8; TNFRSF1B→TRAF1/2→BIRC2/3; ❸TNFRSF1A: high on stromal cells and mo/MΦ, in all immune cells but B-cells; TNFRSF1B: high on mo/MΦ, medium on stromal cells, T and NK cells; TNFRSF14: all immune cells	Promotes formation of secondary lymphoid organs Inflammatory, immunostimulatory and antiviral functions	Required for survival to infection Promotes type I IFN production Promotes survival of B- and T-lymphocytes	Benedict *et al.* (2001), Banks *et al.* (2005), Schneider *et al.* (2008)

Table II.11.1 (Continued)

Factors	Time points	Main cellular sources during MCMV infection[a]	❶Receptor ❷Simplified main signalling pathway ❸Main target cells[c]	General knowledge on biological effects relevant to viral infections	Biological effects during MCMV infection	References
LTβ	Const.	*B and CD8 T-cells*	❶Heterotrimers with LTα→LTBR ❷TRAF2→NIK→IKKα→NF-kB ❸Stromal cells, mo/MΦ, HSC, DCs			
TNF	36 h	*High in pDC, medium in CD8α⁺ and CD11b⁺ DCs*	❶TNFRSF1A/1B ❷❸see LTα	Pro-inflammatory cytokine regulating many biological processes including cell proliferation, differentiation, apoptosis, lipid metabolism and coagulation Can inhibit viral replication Promotes fever directly or by IL-1 secretion stimulation	Contributes to NK cell IFN-γ production Involved in MCMV-induced liver pathology	Orange and Biron (1996b), Orange *et al.* (1997), Ruzek *et al.* (1997), van Dommelen *et al.* (2006), Courgette *et al.* (2008a)
CCL3 (MIP1α)	36–48 h	*All DC subsets and NK cells* (in liver, F4/80⁺ inflammatory mo/MΦ)	❶CCR1, (CCR4, CCR5) ❷GPCR signalling[d] ❸CCR1 is high on neutrophils, mo/MΦ	Pro-inflammatory and chemokinetic properties Recruitment and activation of polymorphonuclear leucocytes	Required for survival to infection; Contributes to NK cell recruitment and activation in infected liver	Salazar-Mather *et al.* (1998, 2002), Dalod *et al.* (2003)
	5–7 d	CD8 T-cells				
CCL4 (MIP1β)	36 h	*All DC subsets and NK cells*	❶CCR5 ❷GPCR signalling ❸pDC, NK cells, activated T-cells	Pro-inflammatory and chemokinetic properties Chemo-attractant for immune cells including NK cells and monocytes	ND	Dalod *et al.* (unpublished)
	5–7 d	CD8 T-cells				
CXCL9 (Mig)	36 h	*All DC subsets*	❶CXCR3 ❷GPCR signalling ❸pDC, NK cells, activated T-cells, CD8α⁺ DCs	Induced by IFN-γ Chemotactic for activated T-cells	CXCL9, CXCL10 and CXCR3 contribute to control viral replication and damage in liver by promoting infiltration and activation of antiviral CD8 T-cells Limits morbidity but dispensable for survival CXCL11 may also contribute to this function	Salazar-Mather *et al.* (2000), Hokeness *et al.* (2007)
	4–5 d	NT				
CXCL10 (IP-10)	36 h–3 d	All DC subsets; NK, B and CD8 T-cells		Induced by type I IFN and IFN-γ Chemotactic for activated mo/MΦ, DCs, T and NK cells Promotion of T-cell adhesion to endothelial cells		
	7 d	NT				

IFN-γ	36 h 5–7 d	NK and NK T-cells (in liver also) CD8 T-cells (in liver also)	❶IFNGR ❷JAK1/2→STAT1 ❸high on stromal cells, mo/MΦ, DCs, NK cells, T-cells	Antiviral activity and immunoregulatory functions Potent activator of macrophages Can potentiate the effects of type I IFN Critical for innate and adaptive immunity against intracellular pathogens	Critical for survival Direct antiviral effects Induces CXCL9 to promote downstream recruitment of antiviral CD8 T-cells	Orange *et al.* (1995), Orange and Biron (1996a,b), Ruzek *et al.* (1997), Tay and Welsh (1997), Presti *et al.* (1998), Salazar-Mather *et al.* (2000), Nguyen *et al.* (2002), Hokeness *et al.* (2007), Wesley *et al.* (2008)
XCL1	36 h	NK cells (in liver also)	❶XCR1 ❷GPCR signalling ❸CD8α⁺ DCs	Produced by activated NK, NK T and CD8 T-cells Acts specifically on CD8α⁺-type DCs	ND	Dorner *et al.* (2004), Crozat *et al.* (2010; 2011)
CSF2 (GM-CSF)	36 h	*NK cells*	❶CSF2-R ❷JAK2→STAT5 ❸DCs, B1 B-cells, mo/MΦ, neutrophils	Controls the production, differentiation and function of mo/MΦ, DCs and granulocytes	ND	Dalod *et al.* (unpublished)
FLT3-L	Const. 36–48 h	?	❶FLT3 ❷SRC→RAF1→MAP2K1/2→ERK→STAT3→IRF8 ❸DCs, pDC, monocytes, haematopoietic stem cells	Induced during inflammation Synergizes with other growth factors and cytokines Stimulates the proliferation of early haematopoietic cells Controls the development of pDC and CD8α⁺ DCs Controls the differentiation of B and NK cells	Critical for survival; promotes NK cell activation by DCs	Eidenschenk *et al.* (2010)

h = hours: NI = not induced strongly in serum; d = days; Const. = constitutive; NT = not tested; ND = not documented.
Italics are mRNA data only based on gene expression profiling of spleen pDC, CD8α⁺ DCs, CD11b⁺ DCs, NK cells, CD8 T-cells and B-lymphocytes at 36 hours post-MCMV infection (see Fig. II.11.2B).
[a]In spleen unless specified otherwise.
[b]Unless specified, the contribution of monocytes/macrophages (mo/MΦ), of stromal cells and other non-haematopoietic cells is generally unknown.
[c]Based on literature mining and/or analysis of mRNA expression of receptors in the ImmGen compendium (http://www.immgen.org).
[d]GPCR signalling = G protein coupled receptor signalling, not detailed due to the complexity of these signalling pathways and to the lack of information on the specific GPCR pathways triggered downstream of chemokine receptors; for a review on GPCR signalling, see for example: Cotton, M., and Claing, A. (2009). G protein-coupled receptors stimulation and the control of cell migration. Cell. Signal. *21*, 1045–1053.
[e]Total mRNA levels of IL-1α and IL-1β are documented in Ruzek *et al.* (1999).

36 hours after infection (Asselin-Paturel *et al.*, 2001; Dalod *et al.*, 2002; Krug *et al.*, 2004). A possible limitation to the studies is that the two types of antibodies used had the potential to deplete cell types in addition to pDCs; the anti-Ly6G/C antibody can also detect neutrophils and inflammatory monocytes, and the anti-PDCA1/120G8/Bst2 antibody depletes a subset of plasma cells. However, neither neutrophils/monocytes nor plasma cells express type 1 IFN at 36 hours after MCMV infection, and both antibodies abrogate type 1 IFN production with their only common target being pDCs. Other conditions of deficiencies in pDCs have now provided independent evidence of their specific role in the response. Mice constitutively lacking pDC due to a hypomorphic mutation in the Ikaros transcription factor (Allman *et al.*, 2006), and transgenic mice expressing the diphtheria toxin receptor (DTR) specifically in pDC under the control of the human CLEC4C lectin promoter, allowing efficient and specific pDC depletion after treatment with DT *in vivo* (Swiecki *et al.*, 2010), have unambiguously established that the systemic type 1 IFN response at the critical time 36 hours after MCMV infection originates from pDCs.

Type 1 IFNs are products of complex gene families with one IFN-β and approximately 12 IFN-α genes. Extended evaluation of type 1 IFN expression at the level of mRNA expression has demonstrated that pDCs are the only cell type expressing high levels of these cytokines among splenic leucocytes isolated at 36 hours after MCMV infection (Dalod *et al.*, 2003; Courgette *et al.*, 2008a) (see also Fig. II.11.2B). Moreover, flow cytometric single cell analyses of type 1 IFN expression using antibodies for intracellular staining in splenic leucocytes isolated from C57BL/6, 129Sv and BALB/c mice at 36 hours after MCMV infection (Courgette *et al.*, 2008a) or from mutant mice expressing the YFP fluorescent reporter under the control of the IFN-β promoter at 12 hours after infection (IFN-β/YFP mice; Scheu *et al.*, 2008) also show that all positive cells express a combination of markers specific for pDC ($CD11c^{int}120G8^{high}SiglecH^{+}$ or $CD11c^{int}Ly6G/C^{high}B220^{+}CD8\alpha^{+}CD11b^{-}$). Finally, *in situ* identification of type 1 IFN-producing cells in spleen sections by immunohistofluorescence using antibodies against the cytokines or IFN-β/YFP mice demonstrate that they express a combination of markers specific for pDC (Scheu *et al.*, 2008; Courgette *et al.*, 2008a). Of note, at different time points after challenge, other cell types can contribute to type 1 IFN production. In particular, stromal cells are reported to be an initial source at 6–8 hours (Schneider *et al.*, 2008) and may facilitate later production of the cytokines by pDC through a type 1 IFN auto-amplification loop (Dalod *et al.*, 2002; Schneider *et al.*, 2008). Conventional DCs (cDC), identified as being $CD19^{-}CD3^{-}NKp46^{-}CD11c^{high}SiglecH^{-}$, may make a minor contribution around two days after infection (Andoniou *et al.*, 2005; Delale *et al.*, 2005). Lastly, transcriptomic data suggest that pDC are also the major source of LTα at 36 hours after infection (Fig. II.11.2B).

Cytokines/chemokines mainly produced by NK cells

Most cell types express detectable levels of *Ifng* mRNA at steady state (Fig II.11.2B), but these levels are higher in NK cells and induced to much higher levels after infections. NK cells (Fig. II.11.2A) and to a lesser extent NKT-cells can be demonstrated to be sources of IFN-γ production *in vivo* in C57BL/6 mice at 36 hours after infection, through experiments analysing *ex vivo* cytokine production by enriched cell subsets, single cell expression by intracellular staining, and the impact of antibody-mediated NK cell depletion on systemic production (Orange *et al.*, 1995; Nguyen *et al.*, 2002a; Tabeta *et al.*, 2004; Wesley *et al.*, 2008). It is important to note, however, that the first characterization of the NK cell IFN-γ response to MCMV included studies also done in T-cell-deficient mice (Orange et al., 1995). These conditions clearly establish the importance and major contribution made by classical non-T, NK cells. The reciprocal experiment, namely evaluating the contribution made by NKT-cells in the absence of NK cells, has not been done. Taken together, the mRNA and protein production observations emphasize the fact that although the correlation between mRNA and protein expression can be made, it is not always the case. Hence, if the goal is to understand the biologically relevant responses, transcriptomics data must be confirmed by analyses at the protein level. In the case of NK cell IFN-γ production, this has been done conclusively.

Nevertheless, expression of mRNA does indicate a cell's potential to produce a factor. Gene expression profiling data show that NK cells but not B-lymphocytes, CD8 T-lymphocytes, or DC subsets also express high levels of *Xcl1* and *Csf2* mRNA and detectable levels of *Il10* mRNA at 36 hours after MCMV infection (Fig. II.11.2B). Because *Ccl5* shows the same gene expression pattern as *Ifng* in splenic leucocyte subsets isolated from MCMV-infected mice, and because these two cytokines have often been found to be co-expressed under other conditions of immune stimulation, it is possible that NK cells also constitute a major source of CCL5. The specific expression of XCL1 by NK cells in MCMV-infected animals has been established, at the protein level, using intracellular staining (Dorner *et al.*, 2004). Of note, although there had been indications of the potential for NK cells to express IL-10

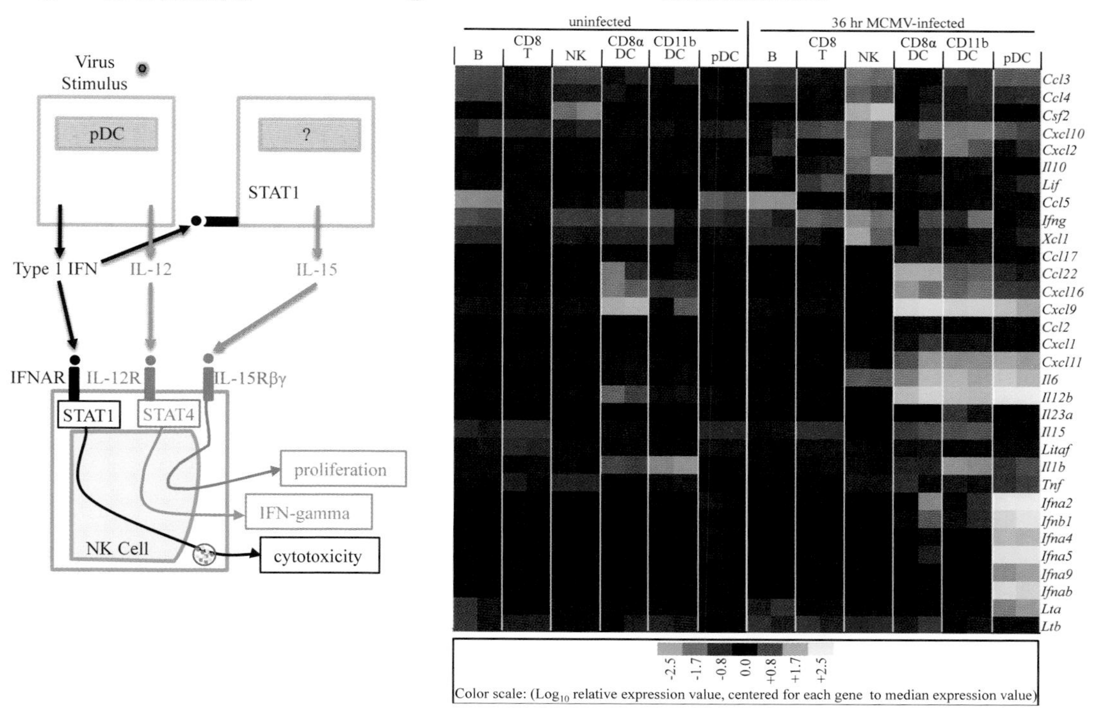

Figure II.11.2 Cellular sources of cytokines. Different cells may be positioned to express cytokines. There are, however, some with specialized functions under different conditions of infection. (A) Evaluation of cellular sources for particular cytokines *in vivo* and/or after isolation have shown that at the protein level, type 1 IFNs and IL-12 are largely produced by pDCs under immunocompetent conditions. In contrast, NK cells are the major producers of IFN-γ, and require IL-12 to generate high levels of this cytokine during MCMV infection. They also use type 1 IFN-induced IL-15 expression to help support their proliferation, but the critical sources of this cytokine remain to be defined *in vivo*. (B) Following MCMV infection, gene expression can be detected at the mRNA level in a variety of cell types, but there are subsets with higher levels of expression. A heat map shows relative mRNA levels for cytokines and chemokines harbouring significant changes in their expression at 36 hours after MCMV infection in at least one of the cell types studied. The six cell types examined are indicated above the heat map: B = B-lymphocytes; CD8 T = CD8 T-cells; NK = NK cells; CD8a DC; CD11b DC; pDC (Courgette *et al.*, 2008a; Baranek *et al.*, 2009; and unpublished data). Genes were regrouped according to the similarity of their expression pattern across all conditions examined, using the Cluster software. Black corresponds to levels around the median expression across all conditions, blue indicates lower level of expression compared with the median, and yellow corresponds to higher expression. (See text for other references.)

under particular conditions of stimulation, the first demonstration of NK cell production of this factor during an infection was in the MCMV system (Lee *et al.*, 2009). Here, NK cells are the major source of the high systemic IL-10 production at day 4 after challenge of perforin-deficient mice failing to clear the virus. Taken together with the expression observed at the transcriptomic level (Fig. II.11.2B) and with the observation that challenges with extremely high doses of the virus induce low levels of IL-10 and IL-10-dependent protection against disease (Oakley *et al.*, 2008), the NK cell IL-10 production seen in the perforin-deficient mice supports the importance of this cytokine source at earlier times during infections of immunocompetent mice. Further experiments are warranted, however, to rigorously analyse the contribution and function of NK cells to IL-10, as well as CCL5 and GM-CSF production during MCMV infections.

Flexibility for multiple cell types to contribute to cytokine and/or chemokine production

The current indications are that different and/or multiple cell types have the potential to contribute to the production of IL-12, IL-6, TNF-α, IL-1β, IL-18, CXCL9 and CXCL11 during MCMV infection. The magnitude of and flexibility for different cellular contributions to responses was first appreciated in the context of examining IL-12 production during MCMV infections of immunocompetent mice or mice deficient in responsiveness to type 1 IFN. These studies demonstrated that in the absence of the peak type 1 IFN production from pDC, higher levels of IL-12 production are contributed by alternative DC populations (Dalod *et al.*, 2002, 2003). Gene expression profiling data on isolated cell subsets from immunocompetent mice show that *Il12*, *Il6*, *Tnf*, *Il1b*, *Cxcl9* and *Cxcl11* mRNA are induced in all three spleen DC subsets, i.e. pDCs, CD8α$^+$ cDC, and CD11b$^+$ cDC, but not NK cells, CD8 T-cells or B-lymphocytes (Fig. II.11.2B). Specific IL-12 and TNF-α expression in DC subsets, but not in lymphocytes, has been confirmed at the protein level by ELISA in *ex vivo* cultures of purified cell subsets or by intracellular staining in splenic leucocytes (Dalod *et al.*, 2002, 2003; Courgette *et al.*, 2008a). Depending on the mouse strains studied and on the dose of viral inoculum used, the respective contribution to IL-12 and TNF-α production of the different DC subsets and of other cell types, including inflammatory monocytes, varies (Courgette *et al.*, 2008a). In particular, the contribution of pDCs to the production of these cytokines is major in 129Sv and BALB/c mice but less so in C57BL/6 mice. Importantly, pDCs are activated specifically in the spleen for the production of cytokines, but not in other organs also supporting viral replication such as the liver or the lung (Courgette *et al.*, 2008a). As noted above, bone marrow mono/MΦ have been proposed to bear a major contribution to systemic production of some chemokines in response to type 1 IFN stimulation at 36 hours after MCMV infection (Crane *et al.*, 2009). In the liver, infiltrating mono/MΦ appear to be the major source of pro-inflammatory cytokines/chemokines, including CCL2 and CCL3 between 24 and 48 hours of infection (Salazar-Mather *et al.*, 2002; Hokeness *et al.*, 2005) as well as possibly CXCL9 and CXCL11 at later time points (Hokeness *et al.*, 2007). Further experiments are warranted to rigorously analyse the requirement for, and levels of contribution of, different cell subsets to IL-6, IL-1β, IL-18, CXCL9, and CXCL11 production during the course of MCMV infection.

Another interesting innate cytokine is IL-15. A cell producing this cytokine uses the chain required for a high affinity IL-15 receptor to trans-present the factor to a responding cell expressing the signalling constituents of the receptor on its cell surface. It is clear that the molecule is induced at the mRNA level in the spleens of MCMV-infected mice (Nguyen *et al.*, 2002a; Dalod *et al.*, 2003). Although it is difficult to detect the protein, studies neutralizing the biological function of IL-15 *in vivo* have definitively established that it is produced to facilitate NK cell proliferation under these conditions of infection (Nguyen *et al.*, 2002a). Gene expression profiling data using multiple isolated cell subsets show that *Il15* mRNA is induced in spleen CD8α$^+$ cDC and CD11b$^+$ cDC, only weakly in B-lymphocytes and not in pDC, NK cells, or CD8 T-cells (Fig. II.11.2B) (Courgette *et al.*, 2008a; and unpublished data). However, mono/MΦ and non-haematopoietic cells are known to be important sources of biologically active IL-15 in steady-state conditions as well as under other types of stimulation. Hence, further experiments are needed to determine the contribution of different cell subsets to IL-15 production as well as its trans-presentation on the IL-15Rα chain to NK cells after infection. Finally, the gene profiling data suggest that CXCL10 can be produced by multiple cell types because the *Cxcl10* mRNA is induced after MCMV infection in all the immune cell types examined (Fig. II.11.2B). This is consistent with the known stimulation of the chemokine by type 1 IFN, but further experiments are warranted to assess the contribution of different cell subsets to overall CXCL10 production.

Positive regulation of cytokine/chemokine cascades and cellular responses

The characterization of specialized cells with unique innate functions, including the expression of particular cytokine profiles, indicates that there is extensive cross-talk for delivery of optimal responses to infection. A focus on establishing links between innate cytokines once induced and downstream cellular and cytokine/chemokine responses provided the first information on the pathways regulating innate cascades and their downstream functions. The question of how innate responses are initiated awaited identification of the sensors used by a host to detect the threat of an infection and stimulate elevated levels of existing and/or new expression of innate system components. Complex products of germ-line gene families are now known to be in place for detecting changes either not expressed or not seen in an uninfected host, and the characterization of the roles for these molecules in stimulating innate immunity during viral infection is advancing. There are several classes of sensors. Those characterized as pattern recognition receptors (PRRs) binding pathogen associated molecular patterns (PAMPs) include members of the toll-like receptors (TLRs), which are expressed on cell surfaces or in endosomal compartments (Arpaia and Barton, 2011). There are also two different classes of cytosolic receptors recognizing viral RNA or DNA in infected cells. These include, among others, an expanding family of sensors inducing type 1 IFN expression, and inflammasomes to activate caspases for processing of biologically active IL-1 and IL-18 (Rathinam and Fitzgerald, 2011). The complex families of germ-line genes encoding receptors for NK cells is an independent class of receptors with the potential to recognize changes induced on cell surfaces as a result of viral infection (Vidal *et al.*, 2011; see also Chapter II.9). They can play a role in inducing or enhancing innate responses. The understanding of the stimulation of these pathways in regulating innate cytokine responses to MCMV is synthesized here (schematically represented in Fig. II.11.3A).

Cytokine to cytokine/chemokine and cellular response pathways

In addition to stimulating direct antiviral functions, type 1 IFNs have a wide range of immunoregulatory effects (Garcia-Sastre and Biron, 2006). The importance of the direct antiviral functions in controlling MCMV infection is supported by the fact that the virus has evolved mechanisms to inhibit the induction of type 1 IFN gene expression (Le *et al.*, 2008) as well as the signalling through the type 1 IFN receptor depending on tyrosine kinase 2 (Tyk2) (Strobl *et al.*, 2005) and the signal transducer and activator of transcription 2 (STAT2) (Zimmermann *et al.*, 2005; see also Chapter I.16). Thus, pathways for stimulating type 1 IFN production by uninfected cells and for using the cytokines to enhance antiviral effects mediated by uninfected immune cells are important to the host. In the context of MCMV infection, type 1 IFNs play an expected role in activating NK cell-mediated lysis by increasing intracellular expression of the perforin molecule that is required for killing functions (Orange and Biron, 1996b; Nguyen *et al.*, 2002a), and a previously identified role in promoting NK cell proliferation (Orange and Biron, 1996b; Nguyen *et al.*, 2002a) (see Fig. II.11.3A). In addition, the type 1 IFNs induce the expression of IL-15 (Nguyen *et al.*, 2002a) along with a variety of chemokines (described above). The receptors for type 1 IFN are broadly expressed on all cell types (Table II.11.1), and binding to these receptors stimulates signalling through multiple STAT molecules for the activation of target genes. Although type 1 IFNs have some promiscuity in using different STAT molecules, the classic signalling pathway used to induce antiviral effects depends on STAT1 and STAT2, and their induction of NK cell-mediated lysis as well as IL-15 expression are both dependent on STAT1 (Nguyen *et al.*, 2002a). Although there are conditions under which type 1 IFNs can induce IFN-γ (Pien *et al.*, 2002; Miyagi *et al.*, 2007; Mack *et al.*, 2011), the systemic, spleen and liver IFN-γ responses to MCMV infection depend on the induction of IL-12, signalling through STAT4 (Orange and Biron, 1996a,b; Nguyen *et al.*, 2002a). The elicited IL-15 promotes NK cell proliferation (Nguyen *et al.*, 2002a). It should be noted, however, that there may be other independent or synergizing pathways for NK cell proliferation. Although IL-18 does not stimulate proliferation of NK cells on its own, it can synergize with IL-15 to do so in culture (French *et al.*, 2006), but this pathway remains to be tested *in vivo*. Also, studies in immunodeficient mice lacking the common gamma chain receptor needed to respond to several proliferative cytokines including IL-15, have an IL-12-dependent proliferation of the Ly49H subset of NK cells recognizing the MCMV product m157, an activating ligand (Sun *et al.*, 2009), but this pathway is only demonstrable under profoundly altered conditions.

Given the pleiotropic and sometimes overlapping effects of these cytokines, it is somewhat surprising that during MCMV infection, individual NK cell responses can be precisely assigned to each, with type 1 IFN inducing killing function, IL-12 stimulating IFN-γ production, and IL-15 promoting proliferation. Studies in other systems are shedding light on the mechanisms likely to be in place to restrict the responses NK cells can have to individual cytokines, particularly type 1 IFNs, during infection (Nguyen *et al.*, 2002b; Miyagi *et al.*, 2007; Mack *et al.*, 2011). This work has demonstrated that NK cells basally express high levels of STAT4 and as a result, have a window of opportunity to respond to type 1 IFN with STAT4 activation and IFN-γ production. Exposure to the type 1 IFNs, however, also activates STAT1 to induce elevated levels of STAT1 protein. Because the type 1 IFN receptor preferentially interacts with STAT1 over STAT4, the elevated STAT1 blocks access to STAT4. In the absence of this pathway, immune responses can be deregulated (Nguyen *et al.*, 2000; Miyagi *et al.*, 2007) to the detriment of the host. The expectation, therefore, is that the strict requirement for IL-12 in inducing NK cell IFN-γ results from a 'conditioning' of the cells to limit the effects mediated by type 1 IFNs and force the requirement for stimulation through a receptor having a preference for STAT4 usage, i.e. IL-12R (Vidal *et al.*, 2011). The production of IL-12 over a narrow window of time is likely to also play a role in protecting against IFN--dependent pathology. These observations provide an exciting model for shaping how cells experience exposure to particular cytokines whereby they are conditioned to respond in particular ways based on the concentrations of individual STAT molecules (Garcia-Sastre and Biron, 2006).

Thus, the production of type 1 IFN, IL-12 and IL-15 are all required to promote the NK cell responses

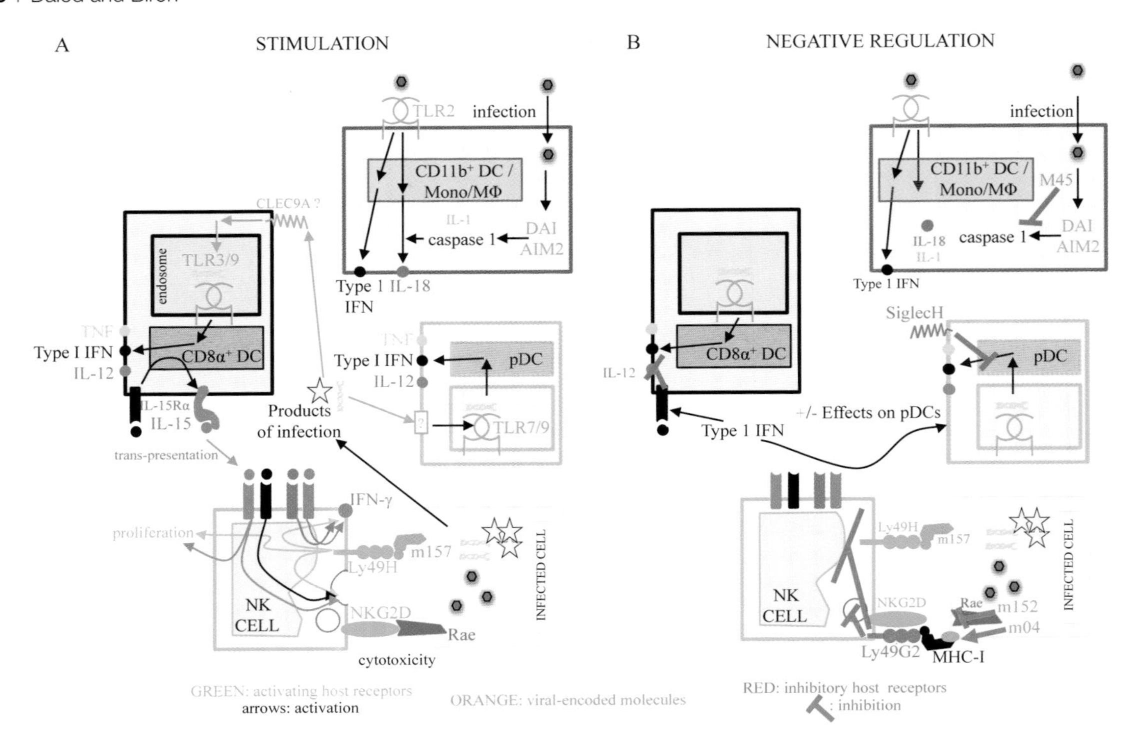

Figure II.11.3 Scheme of the different signals regulating innate cytokine/chemokine production by DCs and NK cells during MCMV infection. (A) The details of the activating pathways are growing. At 36 hours after MCMV infection, pDC are not infected but are stimulated through the endosomal PRRs, largely TLR9 and TLR7 sensing MCMV hypomethylated CpG DNA and dsRNA, respectively, to produce high levels of type 1 IFNs. The pDC are also a major source of IL-12 and TNF depending on the viral inoculum and the strains of mice used and can possibly be stimulated for IL-15 and IL-15Rα expression. The cytokines have non-redundant functions for activating NK cells (see Fig. II.11.2A). The CD11b+ DCs and mono/MΦ express cell surface TLR2, which recognizes a viral glycoprotein and participates in the transcription of certain pro-inflammatory cytokines. The CD11b+ DCs and mono/MΦ express high constitutive levels of the cytosolic helicase, DAI, which recognizes viral DNA, and the inflammasome component, AIM2. DAI promotes the transcription of the type 1 IFN genes. AIM2 leads to the activation of caspase 1 which allows the cleavage of pro-IL-18 and pro-IL-1 proteins to bioactive mature cytokines. NK cells can sense infected cells through a variety of cell–cell interactions. In C57BL/6 mice, the NK cell activating receptor Ly49H allows the recognition of infected cells through binding to the viral molecule, m157, which has structural homology to MHC-I molecules, and is expressed on the membrane of infected cells. In addition to promoting the killing of infected cells by triggering the release of cytotoxic granules, Ly49H enhances NK cell cytokine/chemokine production and proliferation. The NKG2D receptor can also promote NK cell activation and cytokine production through binding to stress self ligands, including Rae induced by infection in host cells. (B) There are, however, a variety of pathways negatively regulating innate responses including those provided by the host to protect from immune-mediated disease and others provided by the virus as immune avoidance mechanisms. The composite of DAP12-associated cell surface receptors inhibit pDC cytokine production during MCMV infections, and the C-type lectin SiglecH is likely to be one of the key molecules. The CD8α+ DCs, CD11b+ DCs, and mono/MΦ can be infected by MCMV *in vivo*. The frequencies of infected cells are low, but the viral M45 molecule interferes with the DAI pathway. Viral genes products interfere with the pathway for presenting stress molecules on the cell surface to be recognized by NKG2D. (See text for references and detailed discussion.)

characterized during MCMV infection, and the induction of IL-15 is downstream of type 1 IFN production. All NK cells either basally express and/or are induced to express receptors for type 1 IFNs, IL-12 and IL-15 (Table II.11.1), and it appears that significant proportions of NK cells are receiving signals from multiple cytokines during MCMV infection. This is evidenced by intracellular combinations of responses, i.e. incorporation of the DNA analogue, BrdU, indicative of proliferation, along with IFN-γ expression, or elevated perforin and IFN-γ expression, with the frequencies of responding cells being such that a subset must be receiving signals from all three cytokines (Nguyen *et al.*, 2002a). The effects of type 1 IFNs, IL-12 and IL-15 can be observed in both the liver and the spleen. Interestingly, the NK cell IFN-γ response to MCMV, in the serum and spleen but not in the liver, is also dependent on IL-18 (Pien *et al.*, 2000). The receptor for IL-18 (IL-18R) signals through a pathway using the IL-1 receptor–associated kinase (IRAK), and the requirement for IL-18 is reflected in a deficiency in the IFN-γ but not the elevated cytotoxic responses of NK cells to MCMV infection in mice rendered IRAK deficient by genetic mutation (Kanakaraj *et al.*, 1999). A more recently described role for IRAK in TLR signalling suggest, however, that the impact on IFN-γ production during MCMV infection may well be a composite of effects resulting from a defect in IL-12 production in addition to one resulting from a defect in IL-18R signalling. There is also evidence that the IκB-like nuclear protein, IκBζ, is expressed in NK cells and essential for the NK cell IFN-γ expression in response to IL-12 and IL-18 (Miyake *et al.*, 2010). A few other cytokines have been evaluated for effects on NK cell responses to MCMV infections. Endogenously produced TNF is not required but does facilitate IFN-γ while inhibiting the proliferative responses of NK cells (Orange and Biron, 1996b). Although the kinetics of the effects are delayed in comparison to the other responses reviewed here, IL-10 receptor blockade experiments support a role for this factor in protecting NK cells from activation-induced death, without altering their ability to provide an IFN-γ response, during MCMV infection (Stacey *et al.*, 2011).

Positive regulation by TLR- and AIM2-dependent signals

Studies in the MCMV system have played a key role in advancing the basic understanding of how the TLR and cytosolic receptors for infectious organisms function during viral infections (Paludan *et al.*, 2011). There have been reports of a role for TLR2 in sensing MCMV (Szomolanyi-Tsuda *et al.*, 2006; Barbalat *et al.*, 2009). The analysis of candidate genes inactivated in targeted knock-out mice as well as the identification of the mutations causing enhanced susceptibility to MCMV infection in mice bearing random mutations induced by *N*-ethyl-*N*-nitrosourea (ENU) treatment clearly demonstrate, however, that the peak production of innate cytokines at 36 hours of MCMV infection first and foremost depends on DC sensing of nucleic acids derived from infected cells through the triggering of endosomal TLRs, most prominently TLR9 but also TLR7 and possibly TLR3 (Krug *et al.*, 2004; Tabeta *et al.*, 2004; Delale *et al.*, 2005; Courgette *et al.*, 2008a,b; see also Chapter II.12). Interestingly, CD8α^{+} cDCs (Dalod *et al.*, 2003), CD11b^{+} cDCs (Andrews *et al.*, 2001), and mono/MΦ (Hanson *et al.*, 1999, 2001; Banks *et al.*, 2005) can be infected by MCMV *in vivo*, but the pDCs responding to MCMV infection are not, themselves, infected (Dalod *et al.*, 2003). A remarkable aspect of these and other studies, however, is that although pDCs are the major producers of both type 1 IFNs and IL-12 during infections of immunocompetent mice (Dalod *et al.*, 2002), IL-12 production is passed over to other populations upon strong downmodulation of type 1 IFN responses as achieved by pDC depletion or by type 1 IFN receptor mutation and both are blocked in the absence of the relevant TLR sensors. These results begin to reveal how sensors on different cell types might elicit different cytokine responses depending on the basal and conditioned states of the cells. The pDCs are very good producers of type 1 IFNs in part because they have high steady state levels of the transcription factors driving their production, such as IRF-7. Other DC subsets are capable of providing an IL-12, but not a type 1 IFN, response but do so only in the absence of type 1 IFNs. Thus, the DC subsets are differentially conditioned in their responses, but the different cell types are being stimulated through overlapping receptors.

Signalling through the TLRs is not the only pathway to type 1 IFN production. Mice deficient for MyD88, a signalling molecule shared by all endosomal TLRs, harbour residual and delayed production of type 1 IFN, at 44 hours after challenge, which suggests the existence of other TLR-independent signals for MCMV sensing *in vivo* during periods of viral replication (Delale *et al.*, 2005; Courgette *et al.*, 2008b). Also, the very early type 1 IFN response to MCMV infection, i.e. at 8 hours after i.p. challenge, is elicited through a TLR-independent pathway in splenic stromal cells depending on lymphotoxin (LT) as well as exposure to virus (Schneider *et al.*, 2008; see also Chapter II.12). This response might be induced by the inoculum dose itself, because its kinetics precedes documented viral replication and is closer in time to those observed following *in vivo* treatments with non-replicating ligands for a variety of sensors. It will be interesting to determine whether or not other cytosolic receptors, expressed more broadly or in most nucleated cells and inducing type 1 IFN expression, are sensing and stimulating MCMV responses *in vivo*. Given the importance of LT for normal splenic architecture and the demonstrated role for marginal zone MΦ in type 1 IFN responses to other viruses (Louten *et al.*, 2006), a prediction is that splenic MΦ, filtering blood, play an important role in the very early response. The demonstration of this TLR-independent response might be indication of recognition by cytosolic sensors with signalling pathways sensitive to inhibition by viral products synthesized during replication. One candidate sensor molecule is the cytosolic DNA-dependent activator of interferon-regulatory factors (DAI). Sequences from the MCMV protein product, M45, can interfere with activation by DAI (Rebsamen *et al.*, 2009), and the viral protein inhibits the activation of the receptor-interacting protein complex (Upton *et al.*, 2010). Thus, DAI may sense MCMV genomic material in infected cells, but co-localized expression of viral proteins presumably shut down the pathway such that it is only accessible before extensive viral replication. Such conditions would explain the difficulty in detecting type 1 IFN production in MCMV-infected macrophage cultures and the *in vivo* requirement for DC sensing through TLRs. Because the TLRs look out from cell surface or endosomal membranes, they provide mechanisms for recognition and signalling prior to intracellular viral infection.

Other members of the class of cytosolic receptors belong to the inflammasome complexes (Rathinam and Fitzgerald, 2011). These receptors catalyse a series of events to result in the activation of caspases for the maturation of pro-IL1 and pro-IL-18 into mature biologically active cytokines (Rathinam *et al.*, 2010). There are many receptors and components, including the 'absent in melanoma 2' molecule (AIM2) sensor and the 'apoptosis-associated speck-like protein containing a caspase activation and recruitment domain' (ASC) adaptor. Here, MCMV was one of several systems used in the original reports to demonstrate a role for the pathway in sensing viral infections. Both AIM2 and ASC are important for detecting MCMV infections because there are drastic reductions of serum IL-18 production at 36 hours after MCMV infection in mice with genetic deficiencies in either gene (Rathinam *et al.*, 2010). The drop is associated with a significant reduction in the per cent of spleen NK cells producing IFN-γ and a significant increase in viral replication in the spleen at 36 hours after infection, which is more pronounced in the *Aim2*-deficient mice on a mixed 129Sv × C57BL/6 genetic background and infected with 10^5 PFU than in the *Asc*-deficient mice on a pure C57BL/6 genetic background and infected with 10^6 PFU. Hence, at 36 hours after MCMV infection, AIM2 plays a critical role in caspase 1 activation and maturation of IL-18, which contributes in the spleen to IFN-γ induction in NK cells and to early control of viral replication. As the AIM2 pathway is required for consecutive IL-1β production from thioglycollate-elicited MΦ or bone marrow-derived GM-CSF DCs upon MCMV infection *in vitro*, it is tempting to speculate that this molecule also activates the inflammasome *in vivo* through the sensing of MCMV DNA. However, the mechanisms leading to exposure of MCMV DNA to AIM2 in the cytosol of infected cells remain to be elucidated. Given the common mechanism for processing IL-1β and IL-18 activated by this pathway, an interesting prediction from these studies is that detection of IL-1 mRNA but not of caspase-processed, secreted protein during MCMV infection (Ruzek *et al.*, 1997, 1999) is indicative of a viral mechanism for blocking release of IL-1. The cellular sources of IL-18 remain to be defined, and the overall cellular distribution of inflammasomes is still under characterization. However, mDC and mono/MΦ express high constitutive levels of the AIM2 component (Crozat *et al.*, 2009; Luber *et al.*, 2010).

Positive regulation by Ly49H and effects of other NK cell receptors

There are many NK cell-activating receptors (see also Chapter II.9). Because the list of virus-induced changes that can be recognized by different NK cell-activating receptors is growing, these molecules can be likened to PRRs. They are products of highly conserved gene families as well as products of highly polygenic and polymorphic gene families such that different mouse strains carry different individual genes as well as different numbers of genes (Vidal *et al.*, 2011). The best characterized is Ly49H. This NK cell activating receptor is a product of the latter group, and different strains of mice can be Ly49H positive or negative. It was first defined as a gene for resistance to MCMV, *cmv-1* (Scalzo *et al.*, 1990). To efficiently contribute to the control of MCMV infection, NK cells must specifically recognize and kill infected cells. In C57BL/6 mice, this is achieved through recognition by the NK cell activation receptor Ly49H of the viral ligand m157 expressed at the surface of infected cells and folded like the class I molecules of the major histocompatibility complex (MHC-I) (Brown *et al.*, 2001; Lee *et al.*, 2001; Arase *et al.*, 2002; Smith *et al.*, 2002; Vidal *et al.*, 2011). However, in addition to allowing the recognition of infected cells and triggering of the degranulation of cytotoxic effectors, Ly49H can directly promote the production of IFN-γ and XCL1 by NK cells as well as NK cell proliferation (Arase *et al.*, 2002; Smith *et al.*, 2002; Dorner *et al.*, 2004; Lee *et al.*, 2009). This pathway is not required for many of the NK cell responses at 36 hours after infection in immunocompetent animals, where innate cytokines are the major factors driving NK cell activation for induction of elevated cytotoxic effector function, cytokine production, and proliferation (Orange and Biron, 1996b; Dokun *et al.*, 2001b; Nguyen *et al.*, 2002a). An activating receptor is, however, required for lysis of a target cell. The Ly49H engagement may contribute to sustained NK cell IFN-γ production at later time points during infection.

This could occur either through direct IFN-γ induction by Ly49H triggering, or because Ly49H triggering delivers survival/proliferation signals that are critical to maintaining NK cells in the face of persistent viral replication and as a result, provides continuous access to their ability to produce cytokine/chemokines in response to these and/or other signals (Dokun *et al.*, 2001b; Lee *et al.*, 2009).

Although the importance of Ly49H in protection against MCMV infection has been conclusively demonstrated, it is not yet clear to what extent Ly49H engagement could compensate to maintain IFN-γ production by NK cells under conditions of dramatically decreased IL-12 production, such as in TLR9-deficient mice, or its role in viral control under conditions of infections with moderate doses of MCMV. Also, many other activating receptors have been shown to be important during different viral infections (Vidal *et al.*, 2011). For the most part, their relative contributions to NK cell activation and function remain to be thoroughly evaluated. There are, however, a few exceptions. There is a growing appreciation for the role of other highly diverse Ly49 activating receptors in defence against MCMV infections on different but specific genetic backgrounds (Kielczewska *et al.*, 2009; Pyzik *et al.*, 2011). Moreover, another conserved activating receptor, NKG2D, broadly expressed on all NK cells, recognizes multiple host stress ligands, including Raeα-ε, H60a-c, and MULT1. The importance of these conserved receptor–ligand pairs is supported by the demonstration that MCMV infection induces the expression of the NKG2D ligands in infected cells, but the virus has evolved mechanisms for blocking their egress to the cell surface to escape NK cell detection (Lenac *et al.*, 2008; Lisnic *et al.*, 2010). Certainly, there is much remaining to be learned about the role for NK cell activating receptors in total and individually in protection against MCMV. It is clear, however, that the understanding of the recognition of virus-infected cells by NK cells is another example of how the MCMV system has paved the way in advancing immunology.

Negative (non-hormone) regulation of cytokine/chemokine cascades

As activated immune responses can mediate host damage, mechanisms are in place to regulate their composition and magnitude. Negative regulation is not required if there is no positive stimulation, and the most effective pathway to control potential damage resulting from immune activation is elimination of the stimulant, i.e. the virus. There are, however, many other pathways in place to help out, such as the induction of endogenous glucocorticoids described above. Others include cytokine-mediated negative regulation of cytokine and cellular responses and effects delivered to particular cell types through regulatory receptors on their cell surfaces (schematically represented in Fig. II.11.3B).

Controlling the infection

For the most part, the intensity of the early production of innate cytokines and chemokines during MCMV infection is commensurate with the level of viral replication in infected animals. UV-inactivated MCMV is a poor inducer of early innate cytokines *in vivo*. In Ly49H$^-$ mice, the production of type 1 IFN and IL-12 by pDC, and IFN-γ by NK cells decreases upon pharmacological inhibition of viral replication *in vivo* (Robbins *et al.*, 2007). In Ly49H$^+$ mice, the production of these cytokines increases with the dose of viral inoculum in parallel with the increase in early viral titres in the spleen, and NK cell depletion at low doses of viral inoculum results in enhanced viral replication and increased systemic production of innate cytokines by other cell types at 36 hours post infection. Hence, by controlling viral replication early after infection, Ly49H$^+$ NK cells limit the ligands available for stimulating pattern receptors *in vivo* to limit activation of other immune cells including DCs and prevent a cytokine storm with its associated tissue damage (Robbins *et al.*, 2007; Bekiaris *et al.*, 2008; Fodil-Cornu *et al.*, 2008). Innate immune cells may also sense information on the viability of the invading pathogens, to tune their production of cytokines accordingly to the infectious risk and avoid unnecessary immunopathological effects. This has been suggested by the demonstration that the immune system detects bacterial mRNA content, the integrity of which is only preserved in live bugs (Sander *et al.*, 2011).

Negative regulation delivered by type 1 IFNs

The effects mediated by type 1 IFN exposure can be contradictory with enhancing and inhibiting consequences on a variety of responses. Examples are that the type 1 IFNs can have both pro- and anti-apoptotic as well as pro- and anti-proliferative functions (Garcia-Sastre and Biron, 2006), and they may deliver these contrasting effects to balance or shape the presence and ratios of DC subsets (Dalod *et al.*, 2003; Hasan *et al.*, 2007; Mattei *et al.*, 2009; Yen and Ganea, 2009; Fuertes Marraco *et al.*, 2011; Swiecki *et al.*, 2011). The first characterization of such effects on DC subsets *in vivo* was the loss of an induced expansion of the proportions of pDCs during

MCMV infections of mice blocked in type 1 IFN induction or function (Dalod *et al.*, 2003), but new studies suggest that the cytokines can also contribute to the natural rapid decline of pDCs during MCMV infection (Swiecki *et al.*, 2011). An independent pathway by which type 1 IFNs may influence DCs is through the induction of ligands for their activating receptors. Curiously, the activation of pDC for the production of innate cytokines during MCMV infection is negatively regulated in a cell-intrinsic manner by an adaptor molecule with immunoreceptor tyrosine-based activation motifs (ITAM), DAP12 (Sjolin *et al.*, 2006). The C-type lectin endocytic receptor SiglecH signals through DAP12 (Blasius *et al.*, 2006). Although the ligand of SiglecH in the mouse is not yet known, in humans, another endocytic receptor specifically expressed in pDCs, ILT7, signals through the ITAM-bearing adaptor, FcεRIγ, and recognizes the host ligand induced by type 1 IFN, BST2 (Cao *et al.*, 2009). Hence, it is possible that SiglecH functions in a similar manner by recognizing an autologous type 1 IFN-induced ligand and stopping pDC cytokine production in a negative feedback loop.

Another example of the contrasting effects mediated through type 1 IFNs to shape DC responses is that although type 1 IFNs can enhance IL-12 and IFN-γ at low concentrations, they inhibit IL-12 induction in response to a variety of stimuli (Cousens *et al.*, 1997) as well as IL-12 induction of IFN-γ (Nguyen *et al.*, 2000) at high but physiologically relevant levels. The effects of inhibiting IL-12 expression have been shown to be specifically delivered during MCMV infection to CD8α$^+$ and CD11b$^+$ DCs, but not to pDC (Dalod *et al.*, 2002; Krug *et al.*, 2004). Taken together, these studies indicate that the type 1 IFNs are mediating complex positive and negative effects on DC subsets to control innate immunity to MCMV. The mechanisms for the differential effects on DCs remain to be defined, but based on studies with human DCs derived in culture, alterations in access to different STAT signalling pathways may be a contributing factor (Longman *et al.*, 2007).

Negative regulation delivered to and by NK cells

The stimulation of NK cells requires both the absence of negative as well as the presence of positive signals (see also Chapter II.9). In addition to the wide range of NK cell activating receptors, there are NK cell inhibitory receptors. Under basal conditions, these receptors receive signals from MHC-I molecules to turn off NK cell stimulation through positive receptors. Viral infection leads to the down-modulation of MHC-I at the surface of infected cells. The effect is in place to avoid presentation of viral peptides to CD8 T-cells and concurrently protect infected cells from adaptive immune defence mechanisms. It does, however, also dampen the signals delivered to NK cells through inhibitory receptors, such as Ly49G2, to increase sensitivity to NK cell-mediated defence mechanisms (Stadnisky *et al.*, 2011). To counter the effect in infected cells, the product of the MCMV gene *m04* forms a stable physical complex with MHC-I thereby rescuing their expression at the cell surface. By stimulating NK cell-inhibitory receptors, these m04-MHC-I complexes efficiently interfere with NK cell-activating receptor signalling in most mouse strains and do so without increasing sensitivity to CD8 T-cell killing (Babic *et al.*, 2010).

Although NK cells have been mostly appreciated for their positive immunoregulatory effects, mediated largely through IFN-γ, as noted above, they also can be induced to produce IL-10. In the case of perforin-deficient mice failing to control MCMV infection, the unregulated replication of the virus stimulated extensive NK cell proliferation and resulted in NK cell maintenance for the IL-10 production. The response is seen around day 4 and is critical to prevent excessive activation of antiviral CD8 T-cells for cytokine production and ensuing TNF- and IFN-γ-dependent immunopathology and death (Lee *et al.*, 2009). Thus, NK cells can be accessed to mediate both positive and negative immunoregulation.

Downstream immune effects

The focus of the work reviewed here has been on the cytokine networks as they interact during the development of innate immunity, but there are also consequences for the adaptive immune responses (Fig. II.11.4). The timing and complexity of adaptive events, as they unfold during the progression of infection and development of immunity, makes the precise definition of direct compared with indirect effects more difficult. Other studies have suggested mechanisms by which innate cytokines induced during infections can have effects on DCs and adaptive lymphocytes, including CD8 T-cells (Cousens *et al.*, 1997; Bahl *et al.*, 2006; Garcia-Sastre and Biron, 2006; Gil *et al.*, 2006; Thompson *et al.*, 2006; Longman *et al.*, 2007). Moreover, once NK cells are activated by innate cytokines, they can modulate the numbers and functions of DCs and T-lymphocytes during MCMV infection. The pathways for NK cell-mediated effects on adaptive immunity include the production of cytokines and the killing of infected cells. Depending on time and context, these can ultimately tune CD8 T-cell responses through

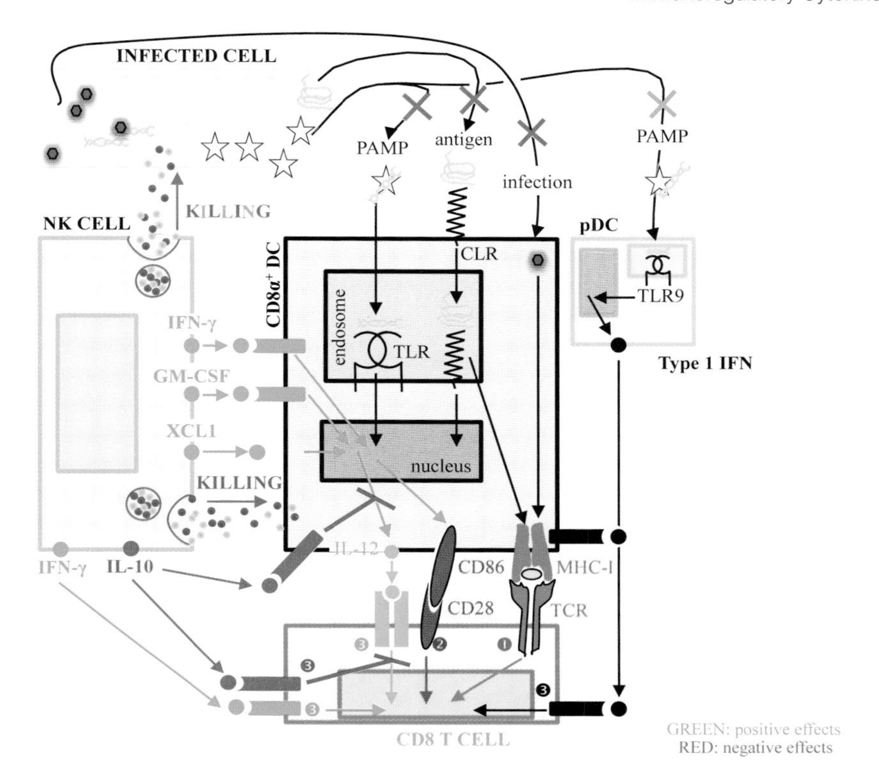

Figure II.11.4 Role of NK cells in modulation of CD8 T-cell responses to MCMV. During MCMV infection, the induction of antiviral CD8 T-cell responses depends on CD8α⁺ DCs and likely on their ability to cross-present viral antigens. However, direct presentation of viral antigens by infected DCs can also contribute, in particular when viral immune avoidance functions are overruled by the enhancing effects of IFN-γ on antigen processing and presentation. The CD8α⁺ DCs capture apoptotic bodies derived from infected cells, through C-type lectin receptors (CLR) such as Clec9A which delivers antigens and PAMPs to endosomes. The PAMPS are sensed by PRRs such as TLR3 which binds viral dsRNA or TLR9 which binds viral hypomethylated CpG DNA. The engagement of PRRs and CLRs determines the ability of DCs to process antigens and present them in association with MHC-I molecules for activation of naïve CD8 T-cells, a process referred to as 'priming'. The specific engagement of the T-cell receptor (TCR) by these MHC–I+peptide complexes constitutes signal 1 (❶) to naïve CD8 T-cells. Signal 2 (❷) consists of the engagement of activating (CD28) co-receptors by costimulatory molecules (e.g. CD86). Signal 3 (❸) consists of cytokines or chemokines. The mixture of signals 1, 2 and 3 that the DCs deliver to the CD8 T-cells will shape the kinetics, intensity and effectiveness of the antiviral CD8 T-cell responses (see also Chapter II.12), and is determined by the nature of the signals received and integrated by the DCs, not only the engagement of TLRs and CLRs but also the cytokines or chemokines produced by other immune cells. NK cells can affect the priming of antiviral CD8 T-cells in both positive (green) and negative (red) ways, through modulating the function of the DCs or through direct effects on the CD8 T-cells themselves, as described in the main text. Killing of infected cells by Ly49H⁺ NK cells has ambivalent effects on CD8 T-cell activation ('KILLING' written in a mixed green/red font).

either positive or negative effects (Fig. II.11.4). Specifically, NK cell might enhance the antigen presentation capacity of DCs, their expression of co-stimulatory molecules, and their production of IL-12 by delivering IFN-γ, GM-CSF, and/or XCL1. NK cell-derived IFN-γ might also directly act on the CD8 T-cells to promote their activation. In contrast, under specific conditions of uncontrolled high viral replication, such as with very high viral inoculums or in perforin-deficient animals, NK cell production of IL-10 can directly act on CD8 T-cells to reduce their stimulation, and this response might also reduce DC immunogenicity (Lee *et al.*, 2009).

NK cell cytotoxic activity can contribute to immune regulation by allowing early delivery of viral antigens from infected cells to DCs for cross-presentation. Killing of infected cells by Ly49H⁺ NK cells has ambivalent effects on CD8 T-cell responses. It prevents DC attrition (Robbins *et al.*, 2007) and promotes early induction of antiviral CD8 T-cell responses (Robbins *et al.*, 2007), likely by preventing detrimental effects on DCs and CD8 T-cells resulting from a cytokine storm around 36 hours after infection (Robbins *et al.*, 2007). On the other hand, NK cell killing of infected cells can also reduce the duration and intensity of antigen presentation, leading to antiviral effector CD8 T-cell responses of shorter duration and lower intensity with consequences for MCMV persistence (Andrews *et al.*, 2010; Mitrovic *et al.*, 2012). However, depending on experimental conditions or mouse strains, the increase in CD8 T-cell responses associated with NK cell control of viral replication can be sustained over long times (Slavuljica *et al.*, 2010; Stadnisky *et al.*, 2011). Clearly, more work is required to better understand the interplay between cytokines, NK cells, DCs and CD8 T-cells during MCMV infection.

Conclusions/perspectives

In summary, careful study of MCMV infections in mice, by numerous investigators and laboratories, has led to a number of key scientific breakthroughs. Many of these discoveries are of major significance to the broad understanding of basic innate immunity. Others provide important insights into the roles of cytokine and cellular networks in shaping endogenous responses to viruses in general and into unique aspects of the virus–host relations in the context of MCMV in particular. For the most part, the work has been advanced by following responses after i.p. infection with moderate to high doses of the virus. This route of infection results in rapid uptake of the virus with delivery to blood and allows the careful examination of responses in serum, spleen, and liver. The approach has made possible the discovery of NK cell contributions to antiviral defence and of innate cytokine cascade responses to viruses and has led to the documentation of specialized pDC cell functions for type 1 IFN production *in vivo*. Because the cytokine responses are elicited at systemic levels, it has also allowed the characterization of a role for the factors in virus-induced disease and for cytokine–neuroendocrine networks. The results of these studies set the stage for much of the initial work characterizing host sensors of viral infections at the level of TLR and cytosolic as well as NK cell receptors. When the discoveries to date are considered in total, it is fair to say that much of what is known about early responses to viral infections is a result of examining immunoregulatory cytokine networks and NK cell contributions to these in the MCMV system. The nature of the basic discoveries, along with the similarities between mouse and human genes and immune responses as well as the opportunities to compare results to those from studies of human CMV, has helped set the framework for understanding immunity to a variety of infections in both species. The rich data are providing unique insights for novel therapeutic interventions to enhance resistance to viral infections and virus-induced diseases.

Despite the progress, there are many unanswered questions remaining, particularly in regard to the biological advantages to the host or virus of particular mechanisms for shaping the cytokine networks as they are induced. As the very earliest cytokine responses are considered, why use common sensors of infection expressed in the same cell, compared with separate sensors in the same or different cells, to induce responses overlapping in time, i.e. type 1 IFNs and IL-12 induced by TLR stimulation versus IL-18 processing induced by activation of inflammasomes? Likewise, why have induction of one innate cytokine dependent on another instead of induced by the same pathway, i.e. IL-12 production by DCs inducing NK cell IFN-γ? Also, with the exception of the delivery of killing to target cells, the conditions requiring NK activating receptors compared with innate cytokines in stimulating NK cell responses remain to be defined. Understanding these general questions will require further developments in the field, but there is a good possibility that the complexities of stimulation provide the most flexibility for sensing infections and/or for getting to a common response by different pathways. The latter would be advantageous to the host in overcoming immune evasion strategies delivered by a virus.

Other issues concern the new insight extended by the MCMV studies that immune cells can be intrinsically conditioned in their responses to particular stimuli during infection. Traits imposed are such that responses can be more restricted than expected. The full understanding, however, of the advantages and consequences of these effects as well as the pathways eliciting them remain elusive. In the case of NK cells, there is a specific requirement for particular innate cytokines in inducing particular NK cell responses to MCMV infection, with type 1 IFNs shaping the cell subset responses. In the case of DCs, there is a requirement for a particular subset in initiating a particular response, but in the absence of this subset, alternative cells can be stimulated through parallel pathways to provide subset responses. Modifying STAT levels is a mechanism for changing responses to the same stimuli. Because a large number of receptors in addition to those for type 1 IFNs use different assortments of STAT molecules to signal, changes in their relative concentrations are likely to shape responses to a variety of factors. Thus, conditions influencing STAT levels may be important in the selection of subset immune responses as needed, and this mechanism could have broad ramifications that have not yet been imagined. New technologies based on gene expression profiling or phosphoproteomics of purified cell types stimulated with individual cytokines, as well as *in vivo* studies examining the impact of cell type-specific inactivation of cytokine receptors or use of reporter mice to track signalling through cytokine receptors at the single cell level, are certain to advance the work. Systems biology approaches (see also Chapter I.7) are also likely to facilitate the understanding of the regulatory circuits controlling innate cytokine production and effects *in vivo* during MCMV infection.

With the pDCs, it is interesting to note that these cells are equipped to sense viruses in a host prior to succumbing to infection themselves because they have the appropriate assortment of TLRs expressed on membranes. They also have high levels of the transcription factors used to induce expression of type 1 IFN genes. Thus, they are excellent producers of the cytokines once stimulated. They deliver this function in the spleen, but not liver or lung, and there are accompanying high levels of cytokines detected in the serum. Once systemic type 1 IFN is induced, it can distribute throughout the host to stimulate antiviral states in uninfected cells. The responses to type 1 IFNs include elevated expression of genes directly mediating antiviral effects as well as certain sensors for infection. Hence, exposure to type 1 IFNs renders uninfected cells better able to detect a virus and block its replication. The understanding, however, of how innate cytokine responses are regulated in different cells and locations, as well as the pathways used by the virus to control cytokine responses, is still superficial. It appears to be more difficult to detect type 1 IFN production by cells other than pDCs, but the studies of liver sections and samples from MCMV-infected mice indicate that the cytokines are being expressed by non-pDCs dispersed throughout this organ but distally separated from virus-infected hepatocytes. Thus, although there are candidate cytosolic sensors for MCMV nucleic material, the tissue observations suggest that even in the liver, the factor producers are not infected cells. The characterization of other cellular sources of innate cytokines in this system is likely to have important implications for the understanding of how the host works to detect and respond at different locations and to sense infections initiating at other sites. This work will have to be advanced by extending characterization of responses to those elicited following different conditions of infections.

Related issues are the questions of cellular origins of mature IL-18 and of how it is induced during *in vivo* MCMV infection in the absence of detectable IL-1. Activation of this pathway by other mechanisms results in processing of both IL-1 and IL-18 immature to mature forms. If the IL-18 response is critical to host efforts at clearing infection, why would not the virus evolve a mechanism for blocking it? Conversely, is IL-1 production sufficiently offensive to the virus that it has evolved a mechanism to block the release of the protein? If so, why? Efforts to address these questions will result in more insights on the molecular interactions been MCMV and its host. As the understanding of the immune avoidance mechanisms for this and other pathways is developed, specific details on the virus–host relationship will be revealed to result in a better understanding of how they have shaped each other's evolution and learned to live together. These studies will require elegant molecular dissection of the role for different MCMV gene products, and further studies of the progress of infection *in vivo*.

In conclusion, work in the mouse with MCMV has had a major impact on immunology, but there are still many unresolved questions. The detailed knowledge that has resulted to date uniquely positions the MCMV system to be used for additional breakthrough discoveries advancing basic immunological knowledge and the understanding of virus–host interactions.

Acknowledgements

The work in the authors' laboratories is supported by the National Institutes of Health, USA (to C.A.B), and by funds from the Centre National de la Recherche Scientifique (CNRS) and from the Agence Nationale de la Recherche (ANR) (to M.D.). The authors thank the members of their laboratories for their contributions to the studies and concepts described in this review and Lara Kallal for reading the manuscript, and apologize for the work of some colleagues that might not have been included due to space constraints.

References

Allman, D., Dalod, M., Asselin-Paturel, C., Delale, T., Robbins, S.H., Trinchieri, G., Biron, C.A., Kastner, P., and Chan, S. (2006). Ikaros is required for plasmacytoid dendritic cell differentiation. Blood *108*, 4025–4034.

Andoniou, C.E., van Dommelen, S.L., Voigt, V., Andrews, D.M., Brizard, G., Asselin-Paturel, C., Delale, T., Stacey, K.J., Trinchieri, G., and Degli-Esposti, M.A. (2005). Interaction between conventional dendritic cells and natural killer cells is integral to the activation of effective antiviral immunity. Nat. Immunol. *6*, 1011–1019.

Andrews, D.M., Farrell, H.E., Densley, E.H., Scalzo, A.A., Shellam, G.R., and Degli-Esposti, M.A. (2001). NK1.1$^+$ cells and murine cytomegalovirus infection: what happens *in situ*? J. Immunol. *166*, 1796–1802.

Andrews, D.M., Scalzo, A.A., Yokoyama, W.M., Smyth, M.J., and Degli-Esposti, M.A. (2003). Functional interactions between dendritic cells and NK cells during viral infection. Nat. Immunol. *4*, 175–181.

Andrews, D.M., Estcourt, M.J., Andoniou, C.E., Wikstrom, M.E., Khong, A., Voigt, V., Fleming, P., Tabarias, H., Hill, G.R., van der Most, R.G., *et al.* (2010). Innate immunity defines the capacity of antiviral T-cells to limit persistent infection. J. Exp. Med. *207*, 1333–1343.

Arase, H., Mocarski, E.S., Campbell, A.E., Hill, A.B., and Lanier, L.L. (2002). Direct recognition of cytomegalovirus by activating and inhibitory NK cell receptors. Science *296*, 1323–1326.

Arpaia, N., and Barton, G.M. (2011). Toll-like receptors: key players in antiviral immunity. Curr. Opin. Virol. *1*, 447–454.

Asselin-Paturel, C., Boonstra, A., Dalod, M., Durand, I., Yessaad, N., Dezutter-Dambuyant, C., Vicari, A., O'Garra, A., Biron, C., Briere, F., *et al.* (2001). Mouse type I IFN-producing cells are immature APCs with plasmacytoid morphology. Nat. Immunol. *2*, 1144–1150.

Babic, M., Pyzik, M., Zafirova, B., Mitrovic, M., Butorac, V., Lanier, L.L., Krmpotic, A., Vidal, S.M., and Jonjic, S. (2010). Cytomegalovirus immunoevasin reveals the physiological role of 'missing self' recognition in natural killer cell dependent virus control *in vivo*. J. Exp. Med. *207*, 2663–2673.

Bahl, K., Kim, S.K., Calcagno, C., Ghersi, D., Puzone, R., Celada, F., Selin, L.K., and Welsh, R.M. (2006). IFN-induced attrition of CD8 T-cells in the presence or absence of cognate antigen during the early stages of viral infections. J. Immunol. *176*, 4284–4295.

Bancroft, G.J., Shellam, G.R., and Chalmer, J.E. (1981). Genetic influences on the augmentation of natural killer (NK) cells during murine cytomegalovirus infection: correlation with patterns of resistance. J. Immunol. *126*, 988–994.

Banks, T.A., Rickert, S., Benedict, C.A., Ma, L., Ko, M., Meier, J., Ha, W., Schneider, K., Granger, S.W., Turovskaya, O., *et al.* (2005). A lymphotoxin-IFN-beta axis essential for lymphocyte survival revealed during cytomegalovirus infection. J. Immunol. *174*, 7217–7225.

Baranek, T., Zucchini, N., and Dalod, M. (2009). Plasmacytoid dendritic cells and the control of herpesvirus infections. Viruses *1*, 383–419,

Barbalat, R., Lau, L., Locksley, R.M., and Barton, G.M. (2009). Toll-like receptor 2 on inflammatory monocytes induces type I interferon in response to viral but not bacterial ligands. Nat. Immunol. *10*, 1200–1207.

Bekiaris, V., Timoshenko, O., Hou, T.Z., Toellner, K., Shakib, S., Gaspal, F., McConnell, F.M., Parnell, S.M., Withers, D., Buckley, C.D., *et al.* (2008). Ly49H$^+$ NK cells migrate to and protect splenic white pulp stroma from murine cytomegalovirus infection. J. Immunol. *180*, 6768–6776.

Benedict, C.A., Banks, T.A., Senderowicz, L., Ko, M., Britt, W.J., Angulo, A., Ghazal, P., and Ware, C.F. (2001). Lymphotoxins and cytomegalovirus cooperatively induce interferon-β, establishing host-virus detente. Immunity *15*, 617–626.

Biron, C.A. (2001). Interferons alpha and beta as immune regulators – a new look. Immunity *14*, 661–664.

Biron, C.A., Nguyen, K.B., Pien, G.C., Cousens, L.P., and Salazar-Mather, T.P. (1999). Natural killer cells in antiviral defense: function and regulation by innate cytokines. Annu. Rev. Immunol. *17*, 189–220.

Blanc, M., Hsieh, W.Y., Robertson, K.A., Watterson, S., Shui, G., Lacaze, P., Khondoker, M., Dickinson, P., Sing, G., Rodriguez-Martin, S., *et al.* (2011). Host defense against viral infection involves interferon mediated down-regulation of sterol biosynthesis. PLoS Biol. *9*, e1000598.

Blasius, A.L., Cella, M., Maldonado, J., Takai, T., and Colonna, M. (2006). Siglec-H is an IPC-specific receptor that modulates type I IFN secretion through DAP12. Blood *107*, 2474–2476.

Borden, E.C., Sen, G.C., Uze, G., Silverman, R.H., Ransohoff, R.M., Foster, G.R., and Stark, G.R. (2007). Interferons at age 50: past, current and future impact on biomedicine. Nat. Rev. Drug Discov. *6*, 975–990.

Brown, M.G., Dokun, A.O., Heusel, J.W., Smith, H.R., Beckman, D.L., Blattenberger, E.A., Dubbelde, C.E., Stone, L.R., Scalzo, A.A., and Yokoyama, W.M. (2001). Vital involvement of a natural killer cell activation receptor in resistance to viral infection. Science *292*, 934–937.

Bukowski, J.F., Woda, B.A., Habu, S., Okumura, K., and Welsh, R.M. (1983). Natural killer cell depletion enhances virus synthesis and virus-induced hepatitis *in vivo*. J. Immunol. *131*, 1531–1538.

Cao, W., Bover, L., Cho, M., Wen, X., Hanabuchi, S., Bao, M., Rosen, D.B., Wang, Y.H., Shaw, J.L., Du, Q., *et al.* (2009). Regulation of TLR7/9 responses in plasmacytoid dendritic cells by BST2 and ILT7 receptor interaction. J. Exp. Med. *206*, 1603–1614.

Chong, K.T., Gresser, I., and Mims, C.A. (1983). Interferon as a defence mechanism in mouse cytomegalovirus infection. J. Gen. Virol. *64*, 461–464.

Cousens, L.P., Orange, J.S., Su, H.C., and Biron, C.A. (1997). Interferon-alpha/beta inhibition of interleukin-12 and interferon-gamma production *in vitro* and endogenously during viral infection. Proc. Natl. Acad. Sci. U.S.A. *94*, 634–639.

Crane, M.J., Hokeness-Antonelli, K.L., and Salazar-Mather, T.P. (2009). Regulation of inflammatory monocyte/macrophage recruitment from the bone marrow during murine cytomegalovirus infection: role for type I interferons in localized induction of CCR2 ligands. J. Immunol. *183*, 2810–2817.

Crozat, K., Vivier, E., and Dalod, M. (2009). Crosstalk between components of the innate immune system: promoting anti-microbial defenses and avoiding immunopathologies. Immunol. Rev. *227*, 129–149.

Crozat, K., Guiton, R., Contreras, V., Feuillet, V., Dutertre, C.A., Ventre, E., Vu Manh, T.P., Baranek, T., Storset, A.K., Marvel, J., *et al.* (2010). The XC chemokine receptor 1 is a conserved selective marker of mammalian cells homologous to mouse CD8α$^+$ dendritic cells. J. Exp. Med. *207*, 1283–1292.

Crozat, K., Tamoutounour, S., Vu Manh, T.P., Fossum, E., Luche, H., Ardouin, L., Guilliams, M., Azukizawa, H., Bogen, B., Malissen, B., *et al.* (2011). Cutting edge: Expression of XCR1 defines mouse lymphoid-tissue resident and migratory dendritic cells of the CD8α$^+$ type. J. Immunol. *187*, 4411–4415.

Dalod, M., Salazar-Mather, T.P., Malmgaard, L., Lewis, C., Asselin-Paturel, C., Briere, F., Trinchieri, G., and Biron, C.A. (2002). Interferon-α/β and interleukin-12 responses to viral infections: pathways regulating dendritic cell cytokine expression *in vivo*. J. Exp. Med. *195*, 517–528.

Dalod, M., Hamilton, T., Salomon, R., Salazar-Mather, T.P., Henry, S.C., Hamilton, J.D., and Biron, C.A. (2003). Dendritic cell responses to early murine cytomegalovirus infection: subset functional specialization and differential regulation by interferon-α/β. J. Exp. Med. *197*, 885–898.

Delale, T., Paquin, A., Asselin-Paturel, C., Dalod, M., Brizard, G., Bates, E.E., Kastner, P., Chan, S., Akira, S., Vicari, A., *et al.* (2005). MyD88-dependent and -independent murine cytomegalovirus sensing for IFN-α release and initiation of immune responses *in vivo*. J. Immunol. *175*, 6723–6732.

Dokun, A.O., Chu, D.T., Yang, L., Bendelac, A.S., and Yokoyama, W.M. (2001a). Analysis of *in situ* NK cell responses during viral infection. J. Immunol. *167*, 5286–5293.

Dokun, A.O., Kim, S., Smith, H.R., Kang, H.S., Chu, D.T., and Yokoyama, W.M. (2001b). Specific and nonspecific NK cell activation during virus infection. Nat. Immunol. *2*, 951–956.

Dorner, B.G., Smith, H.R., French, A.R., Kim, S., Poursine-Laurent, J., Beckman, D.L., Pingel, J.T., Kroczek, R.A., and Yokoyama, W.M. (2004). Coordinate expression of cytokines and chemokines by NK cells during murine cytomegalovirus infection. J. Immunol. *172*, 3119–3131.

Eidenschenk, C., Crozat, K., Krebs, P., Arens, R., Popkin, D., Arnold, C.N., Blasius, A.L., Benedict, C.A., Moresco, E.M., Xia, Y., *et al.* (2010). Flt3 permits survival during infection by rendering dendritic cells competent to activate NK cells. Proc. Natl. Acad. Sci. U.S.A. *107*, 9759–9764.

Fehniger, T.A., Cai, S.F., Cao, X., Bredemeyer, A.J., Presti, R.M., French, A.R., and Ley, T.J. (2007). Acquisition of murine NK cell cytotoxicity requires the translation of a pre-existing pool of granzyme B and perforin mRNAs. Immunity *26*, 798–811.

Fodil-Cornu, N., Lee, S.H., Belanger, S., Makrigiannis, A.P., Biron, C.A., Buller, R.M., and Vidal, S.M. (2008). Ly49h-deficient C57BL/6 mice: a new mouse cytomegalovirus-susceptible model remains resistant to unrelated pathogens controlled by the NK gene complex. J. Immunol. *181*, 6394–6405.

French, A.R., Holroyd, E.B., Yang, L., Kim, S., and Yokoyama, W.M. (2006). IL-18 acts synergistically with IL-15 in stimulating natural killer cell proliferation. Cytokine *35*, 229–234.

Fuertes Marraco, S.A., Scott, C.L., Bouillet, P., Ives, A., Masina, S., Vremec, D., Jansen, E.S., O'Reilly, L.A., Schneider, P., Fasel, N., *et al.* (2011). Type I interferon drives dendritic cell apoptosis via multiple BH3-only proteins following activation by PolyIC *in vivo*. PLoS One *6*, e20189.

Garcia-Sastre, A., and Biron, C.A. (2006). Type 1 interferons and the virus–host relationship: a lesson in detente. Science *312*, 879–882.

Gil, M.P., Salomon, R., Louten, J., and Biron, C.A. (2006). Modulation of STAT1 protein levels: a mechanism shaping CD8 T-cell responses *in vivo*. Blood *107*, 987–993.

Grundy, J.E., Trapman, J., Allan, J.E., Shellam, G.R., and Melief, C.J. (1982). Evidence for a protective role of interferon in resistance to murine cytomegalovirus and its control by non-H-2-linked genes. Infect. Immun. *37*, 143–150.

Hanson, L.K., Slater, J.S., Karabekian, Z., Virgin, H.W., Biron, C.A., Ruzek, M.C., van Rooijen, N., Ciavarra, R.P., Stenberg, R.M., and Campbell, A.E. (1999). Replication of murine cytomegalovirus in differentiated macrophages as a determinant of viral pathogenesis. J. Virol. *73*, 5970–5980.

Hanson, L.K., Slater, J.S., Karabekian, Z., Ciocco-Schmitt, G., and Campbell, A.E. (2001). Products of US22 genes M140 and M141 confer efficient replication of murine cytomegalovirus in macrophages and spleen. J. Virol. *75*, 6292–6302.

Hasan, U.A., Caux, C., Perrot, I., Doffin, A.C., Menetrier-Caux, C., Trinchieri, G., Tommasino, M., and Vlach, J. (2007). Cell proliferation and survival induced by Toll-like receptors is antagonized by type I IFNs. Proc. Natl. Acad. Sci. U.S.A. *104*, 8047–8052.

Heise, M.T., and Virgin, H.W. (1995). The T-cell-independent role of gamma interferon and tumor necrosis factor alpha in macrophage activation during murine cytomegalovirus and herpes simplex virus infections. J. Virol. *69*, 904–909.

Hokeness, K.L., Kuziel, W.A., Biron, C.A., and Salazar-Mather, T.P. (2005). Monocyte chemoattractant protein-1 and CCR2 interactions are required for IFN-alpha/beta-induced inflammatory responses and antiviral defense in liver. J Immunol *174*, 1549–1556.

Hokeness, K.L., Deweerd, E.S., Munks, M.W., Lewis, C.A., Gladue, R.P., and Salazar-Mather, T.P. (2007). CXCR3-dependent recruitment of antigen-specific T-lymphocytes to the liver during murine cytomegalovirus infection. J. Virol. *81*, 1241–1250.

Ishikawa, R., and Biron, C.A. (1993). IFN induction and associated changes in splenic leukocyte distribution. J. Immunol. *150*, 3713–3727.

Jamieson, A.M., Yu, S., Annicelli, C.H., and Medzhitov, R. (2010). Influenza virus-induced glucocorticoids compromise innate host defense against a secondary bacterial infection. Cell Host Microbe *7*, 103–114.

Kanakaraj, P., Ngo, K., Wu, Y., Angulo, A., Ghazal, P., Harris, C.A., Siekierka, J.J., Peterson, P.A., and Fung-Leung, W.P. (1999). Defective interleukin (IL)-18-mediated natural killer and T helper cell type 1 responses in IL-1 receptor-associated kinase (IRAK)-deficient mice. J. Exp. Med. *189*, 1129–1138.

Kielczewska, A., Pyzik, M., Sun, T., Krmpotic, A., Lodoen, M.B., Munks, M.W., Babic, M., Hill, A.B., Koszinowski, U.H., Jonjic, S., *et al.* (2009). Ly49P recognition of cytomegalovirus-infected cells expressing H2-Dk and CMV-encoded m04 correlates with the NK cell antiviral response. J. Exp. Med. *206*, 515–523.

Krug, A., French, A.R., Barchet, W., Fischer, J.A., Dzionek, A., Pingel, J.T., Orihuela, M.M., Akira, S., Yokoyama, W.M., and Colonna, M. (2004). TLR9-dependent recognition of MCMV by IPC and DC generates coordinated cytokine responses that activate antiviral NK cell function. Immunity *21*, 107–119.

Le, V.T., Trilling, M., Zimmermann, A., and Hengel, H. (2008). Mouse cytomegalovirus inhibits beta interferon (IFN-beta) gene expression and controls activation pathways of the IFN-beta enhanceosome. J. Gen. Virol. *89*, 1131–1141.

Lee, S.H., Girard, S., Macina, D., Busa, M., Zafer, A., Belouchi, A., Gros, P., and Vidal, S.M. (2001). Susceptibility to mouse cytomegalovirus is associated with deletion of an activating natural killer cell receptor of the C-type lectin superfamily. Nat. Genet. *28*, 42–45.

Lee, S.H., Kim, K.S., Fodil-Cornu, N., Vidal, S.M., and Biron, C.A. (2009). Activating receptors promote NK cell expansion for maintenance, IL-10 production, and CD8 T-cell regulation during viral infection. J. Exp. Med. *206*, 2235–2251.

Lenac, T., Arapovic, J., Traven, L., Krmpotic, A., and Jonjic, S. (2008). Murine cytomegalovirus regulation of NKG2D ligands. Med. Microbiol. Immunol. *197*, 159–166.

Lisnic, V.J., Krmpotic, A., and Jonjic, S. (2010). Modulation of natural killer cell activity by viruses. Curr. Opin. Microbiol. *13*, 530–539.

Longman, R.S., Braun, D., Pellegrini, S., Rice, C.M., Darnell, R.B., and Albert, M.L. (2007). Dendritic-cell maturation alters intracellular signaling networks, enabling differential effects of IFN-alpha/beta on antigen cross-presentation. Blood *109*, 1113–1122.

Louten, J., van Rooijen, N., and Biron, C.A. (2006). Type 1 IFN deficiency in the absence of normal splenic architecture during lymphocytic choriomeningitis virus infection. J. Immunol. *177*, 3266–3272.

Luber, C.A., Cox, J., Lauterbach, H., Fancke, B., Selbach, M., Tschopp, J., Akira, S., Wiegand, M., Hochrein, H., O'Keeffe, M., *et al.* (2010). Quantitative proteomics reveals subset-specific viral recognition in dendritic cells. Immunity *32*, 279–289.

Lucin, P., Jonjic, S., Messerle, M., Polic, B., Hengel, H., and Koszinowski, U.H. (1994). Late phase inhibition of murine cytomegalovirus replication by synergistic action of interferon-gamma and tumour necrosis factor. J. Gen. Virol. *75*, 101–110.

McCordock, H.A., and Smith, M.G. (1936). The visceral lesions produced in mice by the salivary gland virus of mice. J. Exp. Med. *63*, 303–310.

Mack, E.A., Kallal, L.E., Demers, D.A., and Biron, C.A. (2011). Type 1 interferon induction of natural killer cell gamma interferon production for defense during lymphocytic choriomeningitis virus infection. MBio *2*, mBio.00169–11.

Mattei, F., Bracci, L., Tough, D.F., Belardelli, F., and Schiavoni, G. (2009). Type I IFN regulate DC turnover *in vivo*. Eur. J. Immunol. *39*, 1807–1818.

Mitrovic, M., Arapovic, J., Jordan, S., Fodil-Cornu, N., Ebert, S., Vidal, S.M., Krmpotic, A., Reddehase, M.J., and Jonjic, S. (2012). The NK-cell response to mouse cytomegalovirus infection affects the level and kinetics of the early CD8+ T-cell response. J. Virol. *86*, 2165–2175.

Miyagi, T., Gil, M.P., Wang, X., Louten, J., Chu, W.M., and Biron, C.A. (2007). High basal STAT4 balanced by STAT1 induction to control type 1 interferon effects in natural killer cells. J. Exp. Med. *204*, 2383–2396.

Miyake, T., Satoh, T., Kato, H., Matsushita, K., Kumagai, Y., Vandenbon, A., Tani, T., Muta, T., Akira, S., and Takeuchi, O. (2010). IkappaBzeta is essential for natural killer cell activation in response to IL-12 and IL-18. Proc. Natl. Acad. Sci. U.S.A. *107*, 17680–17685.

Nguyen, K.B., Cousens, L.P., Doughty, L.A., Pien, G.C., Durbin, J.E., and Biron, C.A. (2000). Interferon alpha/beta-mediated inhibition and promotion of interferon gamma: STAT1 resolves a paradox. Nat. Immunol. *1*, 70–76.

Nguyen, K.B., Salazar-Mather, T.P., Dalod, M.Y., Van Deusen, J.B., Wei, X.Q., Liew, F.Y., Caligiuri, M.A., Durbin, J.E., and Biron, C.A. (2002a). Coordinated and distinct roles for IFN-alpha beta, IL-12, and IL-15 regulation of NK cell responses to viral infection. J. Immunol. *169*, 4279–4287.

Nguyen, K.B., Watford, W.T., Salomon, R., Hofmann, S.R., Pien, G.C., Morinobu, A., Gadina, M., O'Shea, J.J., and Biron, C.A. (2002b). Critical role for STAT4 activation by type 1 interferons in the interferon-gamma response to viral infection. Science *297*, 2063–2066.

Oakley, O.R., Garvy, B.A., Humphreys, S., Qureshi, M.H., and Pomeroy, C. (2008). Increased weight loss with reduced viral replication in interleukin-10 knock-out mice infected with murine cytomegalovirus. Clin. Exp. Immunol. *151*, 155–164.

Orange, J.S., and Biron, C.A. (1996a). An absolute and restricted requirement for IL-12 in natural killer cell IFN-gamma production and antiviral defense. Studies of natural killer and T-cell responses in contrasting viral infections. J. Immunol. *156*, 1138–1142.

Orange, J.S., and Biron, C.A. (1996b). Characterization of early IL-12, IFN-alpha/beta, and TNF effects on antiviral state and NK cell responses during murine cytomegalovirus infection. J. Immunol. *156*, 4746–4756.

Orange, J.S., Wang, B., Terhorst, C., and Biron, C.A. (1995). Requirement for natural killer cell-produced interferon gamma in defense against murine cytomegalovirus infection and enhancement of this defense pathway by interleukin-12 administration. J. Exp. Med. *182*, 1045–1056.

Orange, J.S., Salazar-Mather, T.P., Opal, S.M., and Biron, C.A. (1997). Mechanisms for virus-induced liver disease: tumor necrosis factor-mediated pathology independent of natural killer and T-cells during murine cytomegalovirus infection. J. Virol. *71*, 9248–9258.

Paludan, S.R., Bowie, A.G., Horan, K.A., and Fitzgerald, K.A. (2011). Recognition of herpesviruses by the innate immune system. Nat. Rev. Immunol. *11*, 143–154.

Pavic, I., Polic, B., Crnkovic, I., Lucin, P., Jonjic, S., and Koszinowski, U.H. (1993). Participation of endogenous tumour necrosis factor alpha in host resistance to cytomegalovirus infection. J. Gen. Virol. *74*, 2215–2223.

Pien, G.C., Satoskar, A.R., Takeda, K., Akira, S., and Biron, C.A. (2000). Cutting edge: selective IL-18 requirements for induction of compartmental IFN-γ responses during viral infection. J. Immunol. *165*, 4787–4791.

Pien, G.C., Nguyen, K.B., Malmgaard, L., Satoskar, A.R., and Biron, C.A. (2002). A unique mechanism for innate cytokine promotion of T-cell responses to viral infections. J. Immunol. *169*, 5827–5837.

Presti, R.M., Pollock, J.L., Dal Canto, A.J., O'Guin, A.K., and Virgin, H.W. (1998). Interferon- γ regulates acute and latent murine cytomegalovirus infection and chronic disease of the great vessels. J. Exp. Med. *188*, 577–588.

Pyzik, M., Charbonneau, B., Gendron-Pontbriand, E.M., Babic, M., Krmpotic, A., Jonjic, S., and Vidal, S.M. (2011). Distinct MHC class I-dependent NK cell-activating receptors control cytomegalovirus infection in different mouse strains. J. Exp. Med. *208*, 1105–1117.

Rathinam, V.A., and Fitzgerald, K.A. (2011). Innate immune sensing of DNA viruses. Virology *411*, 153–162.

Rathinam, V.A., Jiang, Z., Waggoner, S.N., Sharma, S., Cole, L.E., Waggoner, L., Vanaja, S.K., Monks, B.G., Ganesan, S., Latz, E., *et al.* (2010). The AIM2 inflammasome is essential for host defense against cytosolic bacteria and DNA viruses. Nat. Immunol. *11*, 395–402.

Rebsamen, M., Heinz, L.X., Meylan, E., Michallet, M.C., Schroder, K., Hofmann, K., Vazquez, J., Benedict, C.A., and Tschopp, J. (2009). DAI/ZBP1 recruits RIP1 and RIP3 through RIP homotypic interaction motifs to activate NF-kappaB. EMBO Rep. *10*, 916–922.

Robbins, S.H., Bessou, G., Cornillon, A., Zucchini, N., Rupp, B., Ruzsics, Z., Sacher, T., Tomasello, E., Vivier, E., Koszinowski, U.H., *et al.* (2007). Natural killer cells promote early CD8 T-cell responses against cytomegalovirus. PLoS Pathog. *3*, e123.

Ruzek, M.C., Miller, A.H., Opal, S.M., Pearce, B.D., and Biron, C.A. (1997). Characterization of early cytokine responses and an interleukin (IL)–6-dependent pathway of endogenous glucocorticoid induction during murine cytomegalovirus infection. J. Exp. Med. *185*, 1185–1192.

Ruzek, M.C., Pearce, B.D., Miller, A.H., and Biron, C.A. (1999). Endogenous glucocorticoids protect against cytokine-mediated lethality during viral infection. J. Immunol. *162*, 3527–3533.

Salazar-Mather, T.P., Ishikawa, R., and Biron, C.A. (1996). NK cell trafficking and cytokine expression in splenic compartments after IFN induction and viral infection. J. Immunol. *157*, 3054–3064.

Salazar-Mather, T.P., Orange, J.S., and Biron, C.A. (1998). Early murine cytomegalovirus (MCMV) infection induces liver natural killer (NK) cell inflammation and protection through macrophage inflammatory protein 1alpha (MIP-1alpha)-dependent pathways. J. Exp. Med. *187*, 1–14.

Salazar-Mather, T.P., Hamilton, T.A., and Biron, C.A. (2000). A chemokine-to-cytokine-to-chemokine cascade critical in antiviral defense. J. Clin. Invest. *105*, 985–993.

Salazar-Mather, T.P., Lewis, C.A., and Biron, C.A. (2002). Type I interferons regulate inflammatory cell trafficking and macrophage inflammatory protein 1alpha delivery to the liver. J. Clin. Invest. *110*, 321–330.

Sander, L.E., Davis, M.J., Boekschoten, M.V., Amsen, D., Dascher, C.C., Ryffel, B., Swanson, J.A., Muller, M., and Blander, J.M. (2011). Detection of prokaryotic mRNA signifies microbial viability and promotes immunity. Nature *474*, 385–389.

Scalzo, A.A., Fitzgerald, N.A., Simmons, A., La Vista, A.B., and Shellam, G.R. (1990). Cmv-1, a genetic locus that controls murine cytomegalovirus replication in the spleen. J. Exp. Med. *171*, 1469–1483.

Scheu, S., Dresing, P., and Locksley, R.M. (2008). Visualization of IFN-β production by plasmacytoid versus conventional dendritic cells under specific stimulation conditions *in vivo*. Proc. Natl. Acad. Sci. U.S.A. *105*, 20416–20421.

Schneider, K., Loewendorf, A., De Trez, C., Fulton, J., Rhode, A., Shumway, H., Ha, S., Patterson, G., Pfeffer, K., Nedospasov, S.A., *et al.* (2008). Lymphotoxin-mediated cross-talk between B cells and splenic stroma promotes the initial type I interferon response to cytomegalovirus. Cell Host Microbe *3*, 67–76.

Silverman, M.N., Miller, A.H., Biron, C.A., and Pearce, B.D. (2004). Characterization of an interleukin-6- and adrenocorticotropin-dependent, immune-to-adrenal pathway during viral infection. Endocrinology *145*, 3580–3589.

Sjolin, H., Robbins, S.H., Bessou, G., Hidmark, A., Tomasello, E., Johansson, M., Hall, H., Charifi, F., Karlsson Hedestam, G.B., Biron, C.A., *et al.* (2006). DAP12 signaling regulates plasmacytoid dendritic cell homeostasis and down-modulates their function during viral infection. J. Immunol. *177*, 2908–2916.

Slavuljica, I., Busche, A., Babic, M., Mitrovic, M., Gasparovic, I., Cekinovic, D., Markova Car, E., Pernjak Pugel, E., Cikovic, A., Lisnic, V.J., *et al.* (2010). Recombinant mouse cytomegalovirus expressing a ligand for the NKG2D receptor is attenuated and has improved vaccine properties. J. Clin. Invest. *120*, 4532–4545.

Smith, H.R., Heusel, J.W., Mehta, I.K., Kim, S., Dorner, B.G., Naidenko, O.V., Iizuka, K., Furukawa, H., Beckman, D.L., Pingel, J.T., *et al.* (2002). Recognition of a virus-encoded ligand by a natural killer cell activation receptor. Proc. Natl. Acad. Sci. U.S.A. *99*, 8826–8831.

Stacey, M.A., Marsden, M., Wang, E.C., Wilkinson, G.W., and Humphreys, I.R. (2011). IL-10 restricts activation-induced death of NK cells during acute murine cytomegalovirus infection. J. Immunol. *187*, 2944–2952.

Stadnisky, M.D., Xie, X., Coats, E.R., Bullock, T.N., and Brown, M.G. (2011). Self MHC class I-licensed NK cells enhance adaptive CD8 T-cell viral immunity. Blood *117*, 5133–5141.

Strobl, B., Bubic, I., Bruns, U., Steinborn, R., Lajko, R., Kolbe, T., Karaghiosoff, M., Kalinke, U., Jonjic, S., and Muller, M. (2005). Novel functions of tyrosine kinase 2 in the antiviral defense against murine cytomegalovirus. J. Immunol. *175*, 4000–4008.

Sun, J.C., Ma, A., and Lanier, L.L. (2009). Cutting edge: IL-15-independent NK cell response to mouse cytomegalovirus infection. J. Immunol. *183*, 2911–2914.

Swiecki, M., and Colonna, M. (2011). Type I interferons: diversity of sources, production pathways and effects on immune responses. Curr. Opin. Virol. *1*, 463–475.

Swiecki, M., Gilfillan, S., Vermi, W., Wang, Y., and Colonna, M. (2010). Plasmacytoid dendritic cell ablation impacts early interferon responses and antiviral NK and CD8$^+$ T-cell accrual. Immunity *33*, 955–966.

Swiecki, M., Wang, Y., Vermi, W., Gilfillan, S., Schreiber, R.D., and Colonna, M. (2011). Type I interferon negatively controls plasmacytoid dendritic cell numbers *in vivo*. J. Exp. Med. *208*, 2367–2374.

Szomolanyi-Tsuda, E., Liang, X., Welsh, R.M., Kurt-Jones, E.A., and Finberg, R.W. (2006). Role for TLR2 in NK cell-mediated control of murine cytomegalovirus *in vivo*. J. Virol. *80*, 4286–4291.

Tabeta, K., Georgel, P., Janssen, E., Du, X., Hoebe, K., Crozat, K., Mudd, S., Shamel, L., Sovath, S., Goode, J., *et al.* (2004). Toll-like receptors 9 and 3 as essential components of innate immune defense against mouse cytomegalovirus infection. Proc. Natl. Acad. Sci. U.S.A. *101*, 3516–3521.

Tay, C.H., and Welsh, R.M. (1997). Distinct organ-dependent mechanisms for the control of murine cytomegalovirus infection by natural killer cells. J. Virol. *71*, 267–275.

Thompson, L.J., Kolumam, G.A., Thomas, S., and Murali-Krishna, K. (2006). Innate inflammatory signals induced by various pathogens differentially dictate the IFN-I dependence of CD8 T-cells for clonal expansion and memory formation. J. Immunol. *177*, 1746–1754.

Turner, J.D., Schote, A.B., Macedo, J.A., Pelascini, L.P., and Muller, C.P. (2006). Tissue specific glucocorticoid receptor expression, a role for alternative first exon usage? Biochem. Pharmacol. *72*, 1529–1537.

Upton, J.W., Kaiser, W.J., and Mocarski, E.S. (2010). Virus inhibition of RIP3-dependent necrosis. Cell Host Microbe *7*, 302–313.

Vidal, S.M., Khakoo, S.I., and Biron, C.A. (2011). Natural killer cell responses during viral infections: flexibility and conditioning of innate immunity by experience. Curr. Opin. Virol. *1*, 497–512.

Wesley, J.D., Tessmer, M.S., Chaukos, D., and Brossay, L. (2008). NK cell-like behavior of Valpha14i NK T-cells during MCMV infection. PLoS Pathog. *4*, e1000106.

Yen, J.H., and Ganea, D. (2009). Interferon beta induces mature dendritic cell apoptosis through caspase-11/caspase-3 activation. Blood *114*, 1344–1354.

Zimmermann, A., Trilling, M., Wagner, M., Wilborn, M., Bubic, I., Jonjic, S., Koszinowski, U., and Hengel, H. (2005). A cytomegaloviral protein reveals a dual role for

STAT2 in IFN-γ signaling and antiviral responses. J. Exp. Med. *201*, 1543–1553.

Zucchini, N., Bessou, G., Robbins, S.H., Chasson, L., Raper, A., Crocker, P.R., and Dalod, M. (2008a). Individual plasmacytoid dendritic cells are major contributors to the production of multiple innate cytokines in an organ-specific manner during viral infection. Int. Immunol. *20*, 45–56.

Zucchini, N., Bessou, G., Traub, S., Robbins, S.H., Uematsu, S., Akira, S., Alexopoulou, L., and Dalod, M. (2008b). Cutting edge: Overlapping functions of TLR7 and TLR9 for innate defense against a herpesvirus infection. J. Immunol. *180*, 5799–5803.

Host Genetic Models in Cytomegalovirus Immunology

II.12

Chris A. Benedict, Karine Crozat, Mariapia Degli-Esposti and Marc Dalod

Abstract

Mouse cytomegalovirus (MCMV) was first isolated more than half a century ago (Smith, 1954). Subsequent studies of MCMV in its natural host have yielded enormous information regarding the cellular and molecular immune mechanisms that regulate the various phases of this lifelong β-herpesvirus infection. As the techniques and tools for studying mechanisms of immune defence in mice have advanced, so has our understanding of the specific host pathways that operate to control this complex host–pathogen relationship. In this chapter we will review how various mouse genetic models have defined an initial blueprint for how immune control of MCMV is achieved, both at the level of innate and adaptive immunity, and where we foresee the advancements in this model of CMV infection will come from in the future.

Introduction

The course of MCMV infection is quite complex, as are the immune mechanisms required to control this virus, arising from millions of years of coevolution with its mammalian host (McGeoch *et al.*, 2000). In naïve immunocompetent animals, experimental infection initiates with replication in most peripheral organs, contained in large part by components of the innate immune system. Systemic MCMV replication is mostly controlled during the first week, commensurate with the development of adaptive immunity, but persistent replication continues for weeks to months at select sites (Fig. II.12.1). Then, lifelong latency is established in several tissues and cell types, with both innate and adaptive immunity play roles in containing MCMV re-emergence. Several chapters in this book are dedicated to describing the immune mechanisms and viral counterstrategies that operate in these three distinct phases of MCMV infection. Our task is to summarize the invaluable role that different mouse genetic models have played in elucidating the nature of immune defence to this β-herpesvirus.

MCMV has been studied by taking both forward and reverse genetic approaches (Table II.12.1) (see also Chapter II.9). Genome saturating forward genetic approaches using *N*-ethyl-*N*-nitrosourea (ENU) have revealed that > 3% of mouse genome contributes to the MCMV 'resistome' (Beutler *et al.*, 2005; Crozat *et al.*, 2006). In turn, MCMV infection of mice that are deficient for a specific gene(s) hypothesized to participate in immune control of this virus (i.e. 'knockout' mice), either in all cells or in specific cellular compartments, has also been a very productive approach. Before either of these approaches was readily achievable, comparing the sensitivity and MCMV-specific immune response of various inbred and congenic mouse strains laid the foundation for many important discoveries, and this approach is still used to dissect and refine mechanistic details, particularly in the area of natural killer (NK) cell-dependent control of MCMV infection (Pyzik *et al.*, 2011). Finally, and more recently, mice with fluorescent reporter genes 'knocked-in' at a specific genomic locus have been generated, thus allowing for direct visualization of cells that express specific immune regulating genes *in vivo* during infection.

Given the focus of this chapter, we will discuss viral proteins/strategies that mould MCMV-specific immune defences only briefly, when required to explain the results obtained in mouse genetic models, even though more than half of the viral genome appears to be dedicated to this task (Brune *et al.*, 2000). It is important to remember that advances in our understanding of MCMV immune control sometimes only become clear when studying viral mutants lacking specific viral components. This is because if a viral protein(s) effectively neutralizes a host defence strategy in wild-type mice, studying infection in mice deficient for this pathway can result in no observable phenotype. However, this will not always be the case, as many MCMV strategies are restricted to infected cells only. Consequently, in the MCMV model, studying infection both from the host and the virus perspective has revealed mechanisms of immunity that were previously unappreciated. Here, we

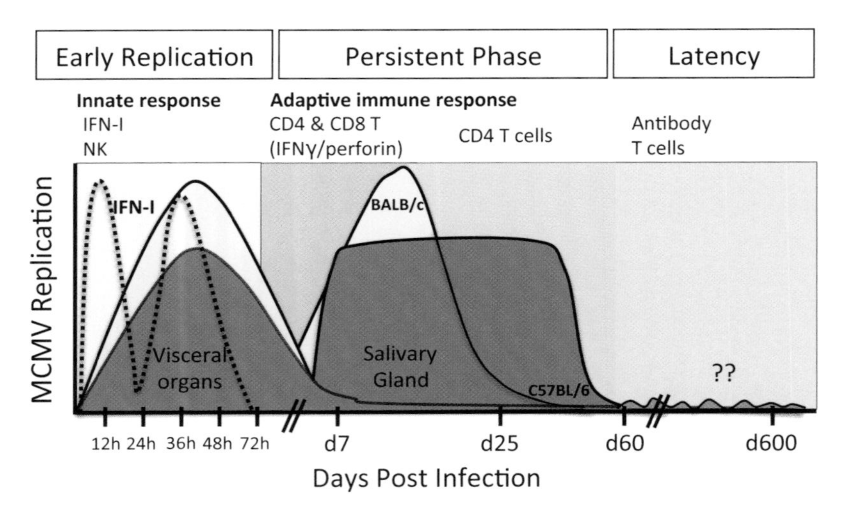

Figure II.12.1 The course of MCMV infection and immune control. Shown are the approximate replication times and relative levels for early and persistent infection in immune competent C57BL/6 (blue curves) and BALB/c (yellow curves) mice. Key mechanisms for innate and adaptive immunity are listed in blue above the infection phases. Dotted line depicts the two 'waves' of IFN-I production induced during the first two days of infection via the i.p. or i.v. route. Orange bumpy line during the latent phase indicates attempted/aborted reactivation events thought to contribute to CD8 T-cell inflation. IFN-I, type I interferon; NK, natural killer cells.

Table II.12.1 Strengths and limitations of genetic mouse models

Methods	Distinctions	Technical approach	Strengths	Limitations
Reverse genetics Gene → phenotype	Hypothesis-driven Selection of a gene(s) of interest	Induced mutagenesis (e.g. knock-out, knock-in or gene trapping)	Leads to targeted modifications The genetic modification is controllable (point mutations, deletions, insertion, ...) Can be ubiquitous or tissue-specific Can be conditional Expedient generation of mutants	May confer embryonic lethality May not result in a phenotype because of functional redundancy May affect the epigenetic regulation of flanking genes Mutants created mixed background Unknown copy number insertions for transgenesis May not be transmitted to progenies
Forward genetics Phenotype → gene	Phenotype-driven Selection based on a biological process	Spontaneous mutations (ex: QTLs) Induced mutations (ex: gene trapping using transposons, use of chemicals such as ENU, or physical methods (e.g. gamma-irradiation))	Random mutations generated Non-redundant mutations selected for large mutation collections generated (dominant, recessive, deletion, rearrangement, missense, nonsense or splicing errors) Generated on pure genetic backgrounds Allows identification of all functional genes for a specific biological process—unbiased May lead to neomorphic or hypomorphic mutations in critical/vital genes Additive mutation effects may enhance a phenotype Inherited through germline transmission	Requires a clear-cut assay to isolate abnormal phenotypes Laborious screening Polygenic forms of phenovariance may result in difficult positional mapping Complex and difficult to characterize genetic rearrangements may arise (ex: gamma-rays) Requires positional cloning to identify the mutation

QTL, quantitative trait loci.

will initially discuss aspects of innate immune defence that are triggered in the first hours/days, and then progress to discuss the responses that operate largely in the persistent/latent phase of infection. Importantly, we will also discuss how these two aspects of immunity to MCMV are not mutually exclusive.

Identifying the cellular sources of type I interferons in response to MCMV infection

The production of type I interferon (IFN-I) is one of the very first responses elicited by the recognition of MCMV by the immune system, as is true for most viruses (see also Chapter I.16). Mice lacking IFN-I signalling show increased sensitivity to MCMV infection, resulting from alterations in various arms of the innate immune response. Here, we will discuss the genetic models that have elucidated the specific cell types and signalling pathways that result in IFN-I production; these studies have provided a critical mechanistic understanding of how these sentinel cytokines shape immune defence to MCMV. In mice infected with MCMV, IFN-I induction occurs in successive, distinct waves over the initial days of infection, these phases being coincident with changing cellular sources producing the IFN-I (Loewendorf and Benedict, 2010). IFN-I is detected very rapidly in C57BL/6 (B6) mice infected intraperitoneally with moderate to high doses (1×10^4–2×10^5 PFU) of MCMV, with both systemic and spleen/liver levels peaking between 6 hours and 12 hours (Grundy *et al.*, 1982; Schneider *et al.*, 2008). IFN-I production then wanes, returning to undetectable levels between 24 hours and 30 hours (Schneider *et al.*, 2008; Courgette *et al.*, 2008a). A second major wave then occurs at around 36 hours, commensurate with the first round of MCMV replication and spread *in vivo*. This burst of IFN-I production coinciding with newly produced virions sharply declines by 48 hours, although ~ 10–20% of peak levels can still be detected at this time (Delale *et al.*, 2005; Courgette *et al.*, 2008a), and becomes undetectable by ~ 72 hours (Delale *et al.*, 2005; Schneider *et al.*, 2008).

Several approaches in mouse genetic models have been taken to delineate the cell types and molecular pathways responsible for regulating IFN-I production over the first days of MCMV infection. Mice genetically deficient for, or antibody-depleted of, specific haematopoietic cell populations have been utilized to address which cell(s) contribute to regulating IFN-I. In turn, both forward and reverse genetic approaches in mice have been fruitful in delineating the innate immune receptors and pathways controlling cell-intrinsic recognition of MCMV.

Lymphotoxin-deficient mice reveal stromal cells as initial sensors of MCMV infection and production of IFN-I

Several studies have revealed a role for lymphotoxins (LTα and LTβ, TNF-related cytokines) in the induction of IFN-I and downstream innate control of MCMV replication (Benedict *et al.*, 2001, 2006; Banks *et al.*, 2005; Schneider *et al.*, 2008). Time-course analyses using mice genetically disrupted for expression of LTα, LTβ and/or LIGHT (the three known ligands for the LTβ-receptor), revealed that the initial peak of systemic and splenic IFN-I, occurring at ~ 8 hours, is dependent upon signalling by the lymphotoxins, while IFN-I production in the liver is unaffected in lymphotoxin-knockout mice. Studies of mice lacking *ltβ* expression only in B-cells revealed these lymphocytes to be the critical source of LTα1β2 (Schneider *et al.*, 2008), the heterotrimeric ligand that is expressed solely on the cell surface. B-cell-dependent IFN-I production was confirmed in mice lacking B- and T-cells (RAG KO), as well as in those lacking only B-cells (IgH–$6^{-/-}$). The LTβR (which binds LTα1β2) is expressed on most stromal and myeloid cell lineages, and the use of WT/LTβR$^{-/-}$ bone marrow (BM) chimeras revealed that receptor expression in splenic stromal cells was required for interactions with LTα1β2-expressing B-cells. Alymphoplasia mice (*aly/aly*, which contain a naturally occurring functional mutation in the NFκB inducing kinase (NIK) (Xiao *et al.*, 2001) also show a defect in the initial wave of stroma-derived IFN-I induced in response to MCMV, suggesting that LTβR activation of the non-canonical NFκB pathway in the stroma is required. Physical separation of spleen components into stromal and haematopoietic cells revealed that the vast majority of IFN-I and MCMV gene expression was restricted to the stroma, confirming the BM chimera results. In total, these results indicate that the initial sensing of MCMV infection in the spleen occurs in stromal cells and is dependent upon their direct interaction with, and/or 'conditioning' by, B-cells (Schneider *et al.*, 2008). It is quite likely that the relevant stromal cells reside in the splenic marginal zone, as these spleen cells are amongst the first to be infected by MCMV (Mercer *et al.*, 1988; Hsu *et al.*, 2009; Benedict, C.A., Fukuyama, S., Verma, S., and Ware, C.F., unpublished observations), and they also may harbour latent viral genome. An early study from Grundy and colleagues showed that various inbred mouse strains produce substantially different levels of systemic IFN-I at 6 hours in response to the same dose of MCMV (Grundy *et al.*, 1982). Importantly, using congenic mice, these studies found that these differences in IFN-I production are not H-2 linked, suggesting that genetic factors regulating IFN-I production by marginal zone stromal cells, including LT, may differ in individual inbred mouse strains.

The second wave of MCMV-induced IFN-I production comes from plasmacytoid dendritic cells

In contrast to the initial IFN-I that is derived from the splenic stroma, systemic and splenic production of IFN-I at 36 hours after MCMV infection is LTαβ-independent, is not abrogated in RAG-KO mice and segregates with CD11c+ dendritic cells (DCs) (Dalod *et al.*, 2002). Mouse genetic models show that this 36 hours IFN-I production is reduced by > 90% in animals that lack plasmacytoid DCs (pDC) either constitutively due to a hypomorphic mutation in the Ikaros transcription factor (Allman *et al.*, 2006) or conditionally upon diphtheria toxin (DT) administration to mice transgenic for ectopic expression of DT receptor (DTR) specifically in pDC (Swiecki *et al.*, 2010). Conventional DCs (cDC) subsets may contribute the proportion of IFN-I that is not pDC-derived at this time (likely < 10%) (Krug *et al.*, 2004; Andoniou *et al.*, 2005; Delale *et al.*, 2005). Hence, pDC are the major producers of splenic and systemic IFN-I at 36 hours after MCMV infection. This has been confirmed by intracellular cytokine staining in total splenic leucocyte suspensions analysed directly *ex vivo* (Courgette *et al.*, 2008a), by immunohistochemical (IHC) studies in wild-type animals (Courgette *et al.*, 2008a) and in IFN-β reporter mice (Scheu *et al.*, 2008), and by antibody depletion of pDC (Dalod *et al.*, 2002; Krug *et al.*, 2004). cDC likely contribute to IFN-I production mainly between 36 and 48 hours (Andoniou *et al.*, 2005; Delale *et al.*, 2005), which is discussed below.

Toll-like receptor (TLR) recognition of MCMV by pDC

What are the cell-intrinsic mechanisms for recognition of MCMV by pDC which lead to IFN-I production? As is the case for pDC-mediated recognition of most pathogens, many studies indicate it is the Toll-like receptors (TLR) and their downstream adaptors that are the orchestrators. The use of mice disrupted for genes required for TLR-signalling, as well as analysis of ENU mutant mice showing increased sensitivity/susceptibility to MCMV, has led to these conclusions. TLR9 signalling is likely to be somewhat redundant for pDC-derived IFN-I during MCMV infection, as variable reductions in 36 hour IFN-I levels have been observed in TLR9$^{-/-}$ animals (Krug *et al.*, 2004; Tabeta *et al.*, 2004; Andoniou *et al.*, 2005; Delale *et al.*, 2005), while B6 mice deficient for both TLR9 and TLR7 show a complete abrogation of this response (Courgette *et al.*, 2008b). Accordingly, mice deficient for MyD88, a cytoplasmic adaptor essential for both TLR7 and 9 signalling, show a dramatic reduction in IFN-I production at 36 hours (Krug *et al.*, 2004; Tabeta *et al.*, 2004; Delale *et al.*, 2005; Courgette *et al.*, 2008b). B6 mice deficient for Pin-1, an isomerase that interacts with IRAK1 and promotes downstream activation of IRF7, show a similar IFN-I production deficit at 36 hours (Tun-Kyi *et al.*, 2011). TLR9 and MyD88 are also required for all spleen DC subsets to produce IL-12 after MCMV has completed its first round of replication *in vivo* (Krug *et al.*, 2004; Tabeta *et al.*, 2004; Delale *et al.*, 2005; Courgette *et al.*, 2008b). As one would therefore expect, TLR9-KO mice show enhanced susceptibility to MCMV-induced morbidity, but to a lesser extent than MyD88-KO or TLR7/9-KO animals (Courgette *et al.*, 2008b).

Most of the published studies have been conducted in mice on a B6 background. Although the reported requirements may hold true in other mouse strains, it is worth noting that the magnitude and kinetics of the IFN-I response induced by MCMV infection appear to be mouse strain specific. B6 mice produce more IFN-I than BALB/c mice at 8 hours after MCMV infection (Schneider *et al.*, 2008), but this is reversed from 30 hours onward, being significantly higher in BALB/c, 129Sv and 129S2 mice (Dalod *et al.*, 2002; Krug *et al.*, 2004; Schneider *et al.*, 2008; Courgette *et al.*, 2008b). This absolute increase in IFN-I levels in BALB/c in the second phase is consistent with the 100- to 1000-fold higher levels of MCMV produced in the spleen in this strain compared to B6, but if normalized to the levels of MCMV *ie1* gene expression, IFN-I levels are still lower than in B6 on a 'per viral transcript' level at these later times (Schneider *et al.*, 2008). Indeed, pharmacological inhibition of MCMV replication in BALB/c mice dramatically decreases pDC production of IFN-I to the levels observed in mice able to naturally control acute infection, like B6 animals (Robbins *et al.*, 2007). This point highlights the multiple factors that must be considered when analysing innate cytokine responses at various times of early infection in mouse strains showing different susceptibility to MCMV replication. Furthermore, although pDCs are the main source of IFN-I in B6, BALB/c and 129S2 mice at 36h p.i. (Krug *et al.*, 2004; Courgette *et al.*, 2008a), more work is required to determine whether the signalling pathways involved are the same as those described for B6 mice both at this time point and earlier. Finally, the likelihood that cell-intrinsic antiviral responses to IFN-I signalling will result in differential outcomes between mouse strains must be considered (Mashimo *et al.*,

2002), one example being mouse strain-specific action of NK cells (see also Chapter II.9).

Screening ENU mutant mice displaying enhanced susceptibility to MCMV infection associated with impaired IFN-I production identified additional molecules involved in regulating the function of nucleic acid-sensing TLRs in most cells (Unc93b1) (Tabeta *et al.*, 2006), or in pDCs specifically (Slc15a4, AP-3, and BLOC-1/2 Hermansky-Pudlak syndrome proteins) (Blasius *et al.*, 2010). MCMV-susceptible ENU mutant mice also suggested a role for TLR3 and its downstream adaptor TRIF in IFN-I, IL12p40 and NK cell IFNγ-production at 36 hours, as well as in control of MCMV replication in the spleen at days 4–5 (Hoebe *et al.*, 2003; Tabeta *et al.*, 2004). However, the importance of this pathway for defence to MCMV is controversial, as there is variability in the results from different studies with regards to the MCMV susceptibility and 36 hours IFN-I production observed in $TLR3^{-/-}$ mice. It is possible that the observed differences are attributable to the use of different mouse backgrounds, different viral strains or different infectious doses of MCMV (Edelmann *et al.*, 2004; Delale *et al.*, 2005), and 'technical' differences in experimental parameters between labs must also be considered.

Production of IFN-I by non-pDCs in the second phase of the innate response

In genetically deficient/mutagenized mice where pDCs are unable to sense MCMV infection, the systemic IFN-I levels observed at 44–48 hours are roughly equivalent to those observed in wild-type mice, albeit the absolute levels are ~ 10–20% those of the 36 hour peak (Delale *et al.*, 2005). The cellular source of this IFN-I remains to be identified, but stromal cells, cDCs or monocytes are potential contributors (Andoniou *et al.*, 2005; Courgette *et al.*, 2008a). Interestingly, the sustained production of IFN-I is independent of TRIF and TLR3 signals and largely independent of MyD88 (Delale *et al.*, 2005), suggesting that a currently unidentified molecular sensor, perhaps a cytosolic DNA sensor (Sharma and Fitzgerald, 2011), may contribute to MCMV recognition in myeloid or stromal lineage cells, as has been proposed for fibroblasts infected with human CMV (DeFilippis *et al.*, 2010). Alternatively, it is possible that multiple TLRs are involved in a semi-redundant manner, which would explain the apparent independence of MyD88 and TRIF signalling observed when these pathways were examined individually (Delale *et al.*, 2005). It is worth noting that currently it is unclear how the various sources and waves of IFN-I production function to regulate innate defences to MCMV infection, although the specific depletion of pDCs leads to a significant, albeit transient increase in viral loads in both 129Sv (Dalod *et al.*, 2002) and B6 (Swiecki *et al.*, 2010) mice. Is it the first wave of IFN-I production from stromal cells that is critical to control replication and activate innate effector functions, the pDC-derived burst at 36 hours and/or the sustained production at 44–48 hours? We would postulate that all are important, and likely function in organ-specific contexts. Emphasizing this point, very little is currently understood about what cell type(s) produce the initial and subsequent waves of IFN-I in the liver of MCMV-infected mice, although it appears that pDCs and LTαβ-dependent cells are not the source (Schneider *et al.*, 2008; Courgette *et al.*, 2008a).

Control of MCMV replication by IFN-I

To this point we have described the cells and molecular mechanisms regulating IFN-I production, but once produced this potent antiviral cytokine can mediate innate control via multiple mechanisms. The role that the first wave of IFN-I produced by stromal cells has in controlling initial MCMV replication and/or spread is currently unclear, although preliminary results suggest it is important (Verma, S., and Benedict, C.A., unpublished). Studying mice genetically deficient for the IFN-I receptor or for critical downstream adaptors (e.g. Tyk 2 and STAT1) has clearly established that IFN-I are critical for limiting MCMV replication in both B6 and 129Sv mouse strains over the first few days of infection (Presti *et al.*, 1998; Strobl *et al.*, 2005; Crozat *et al.*, 2006). These data confirmed the findings from early experiments performed in B6 mice, where increased MCMV replication in spleen and liver was observed using blocking IFN-I antibodies (Orange and Biron, 1996a). IFN-I signalling also promotes splenocyte survival during early MCMV infection (Banks *et al.*, 2005), with massive apoptosis of both lymphoid and myeloid lineage cells seen in B6 $IFN\text{-}IR^{-/-}$ mice, a likely consequence of the extremely high splenic viral loads seen in these mice. Notably, productive viral replication in these spleen cell subsets is not a strict requirement for the observed cell death, as MCMV shows little to no tropism for lymphoid cells.

IFN-I can exert both direct antiviral effects on infected cells, as well as induce immunoregulatory functions through a multitude of mechanisms, with activation of NK cells being one critical component (discussed below: see also Chapter II.9). In general, untangling the NK-dependent and -independent roles for IFN-I action over the initial days of infection is a task which requires more attention. Additionally, the relative importance of IFN-I in dampening MCMV production from infected cells, as opposed to acting on neighbouring, uninfected cells to inhibit spread, remains unclear. The cell-intrinsic antiviral defences instigated by IFN-I signalling during influenza and HIV infection are now documented (Schoggins *et al.*, 2011), and it is almost certain that the interferon stimulated genes (ISGs) identified in these studies will play a role in MCMV defence as well (see also Chapters I.8 and I.16). IFN-I is also an important regulator of cellular metabolism, and recent results show that its restriction of sterol biosynthesis during MCMV infection is a key antiviral function (Blanc *et al.*, 2011). Additional evidence for links between IFN-I and metabolic regulation of MCMV infection comes from mice genetically deficient in methylene-tetrahydrofolate reductase ($MTHFR^{-/-}$), which show enhanced control of MCMV splenic replication at day 3 and reduced serum IFN-I levels at 36 hours (Fodil-Cornu *et al.*, 2009). As for most key innate defence mechanisms, MCMV encodes proteins to inhibit/dampen IFN-I signalling (e.g. M27 inhibition of STAT2), and these function to promote early viral replication, as well as potentially playing a role in viral persistence and/or latency (Zimmermann *et al.*, 2005; see also Chapter I.16). In summary, although much remains to be learned regarding how IFN-I coordinates resistance to MCMV, and how the virus shapes its signalling, it is clear that IFN-I plays a critical role in restricting early replication through a variety of mechanisms.

NK cell-mediated innate defence to MCMV revealed by mouse genetic models

IFN-I is only one key player in coordinating the complex innate defence to MCMV infection. An equally important component is natural killer (NK) cells, which mediate their effector functions through both cytolytic and non-cytolytic mechanisms over the initial days of infection (discussed in greater detail in Chapters II.9 and II.11) (Fig. II.12.2). Genetically mutant mouse models have provided a better understanding of the factors that regulate NK cell activation and downstream control of early MCMV replication, with knockout, ENU-mutagenized and congenic mouse strains all providing extremely valuable information (Table II.12.2). In particular, forward genetics approaches using congenic mice provided the first insight into the importance of NK cell-mediated immune control of MCMV, with the first studies being published more than three decades ago (Bancroft *et al.*, 1981; Shellam *et al.*, 1981; Scalzo *et al.*, 1990).

Cytokine-mediated activation of NK cells

NK cells are activated during MCMV infection to produce IFNγ and acquire cytotoxic function (Bancroft *et al.*, 1981; Orange *et al.*, 1995). Studying beige mice of different genetic backgrounds, which are genetically impaired for NK cell cytotoxic activity, demonstrated that NK cells are required for the high resistance of certain inbred mouse strains to MCMV infection (Shellam *et al.*, 1981; Bukowski *et al.*, 1983). The use of knockout mice lacking specific cytokines, their receptors or downstream STAT adaptors, showed that IFN-I, IL-12, IL-18 and IL-15 all contribute at relative degrees to NK cell activation and proliferation after MCMV infection (Pien *et al.*, 2000; Nguyen *et al.*, 2002; Andrews *et al.*, 2003; Geurs *et al.*, 2009). STATs ultimately lead to downstream transcription of effector cytokines, such as IFNγ, whose induction in NK cells during MCMV infection depends on ATF3 and IkBζ, as recently demonstrated using genetic mouse mutants (Rosenberger *et al.*, 2008; Miyake *et al.*, 2010). This cytokine-mediated activation is independent of any direct recognition of MCMV-infected cells, and direct detection is also dispensable for the acquisition of NK cell effector functions. However, direct sensing is critical for efficient killing of MCMV-infected cells (see Chapter II.9). Innate cytokines, such as IFN-I and IL-12, promote NK cells to load their cytotoxic granules with granzymes and perforin (analogous to loading a quiver with arrows). Then, the direct sensing of infected cells by NK cell activation receptors is required for the lethal hit by triggering the polarized exocytosis of cytotoxic granules towards the target cell, reorganization of the cytoskeleton and activation of molecular motors (analogous to the eye and brain instructing the finger to pull the bow and send an arrow once the target has been formally recognized).

MCMV-induced NK cell activation via specific receptor–ligand interactions

Interbreeding and careful analysis of congenic mouse strains deciphered the molecular basis for the 'resistance' of B6 mice to MCMV infection when compared to strains such as BALB/c. B6 resistance was linked to the *Cmv1* genetic locus (Scalzo *et al.*, 1990), which was later shown to encode the Ly49H NK cell activating receptor (Brown *et al.*, 2001; Lee *et al.*, 2001). The Ly49 family of NK cell receptors in mice (see also Chapter II.9) associate principally with the immunoreceptor tyrosine-based activation motif (ITAM) containing adaptor DAP12 (also known as KARAP/TYROBP) to transduce activation signals in NK cells, but can also associate with the DAP10 adaptor. Shortly after

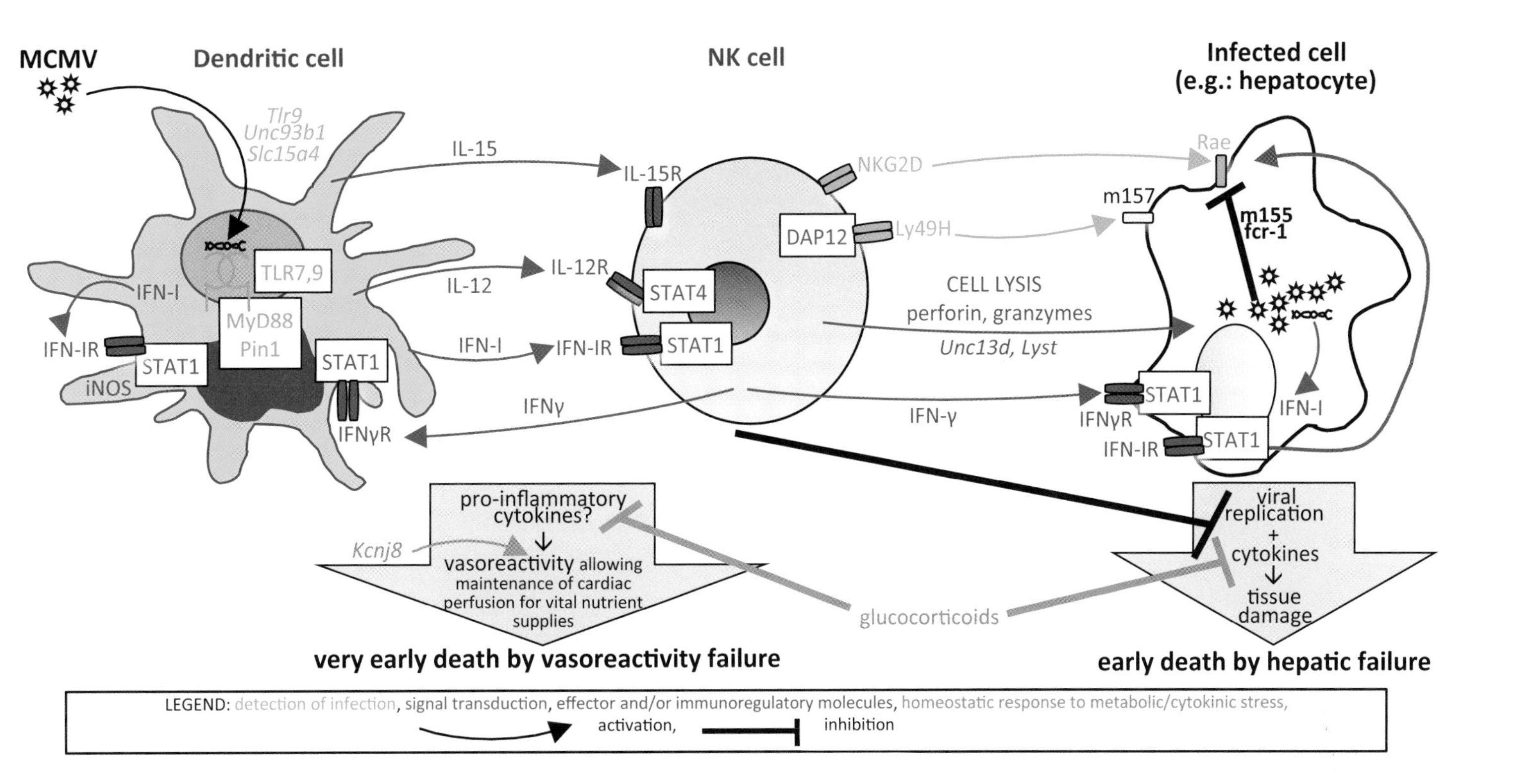

Figure II.12.2 Cellular and molecular interactions that direct innate defences and the outcome of early MCMV infection. Shown are the sensors which detect MCMV or infected cells, cytokines and their receptors, signalling adaptors mediating cellular cross-talk and proteins that regulate the cytolytic and secretory pathways facilitating viral clearance. For example, uninfected DCs sense RNA or hypomethylated CpG DNA sequences from engulfed MCMV particles or infected cells through TLR7 or 9, inducing cytokine production including type I interferons (IFN-I), interleukin 12 (IL-12) and interleukin-15 (IL-15). These cytokines then promote NK cell activation, cytotoxicity and cytokine production (e.g. IFNγ). NK cell IFNγ and cytotoxicity control MCMV replication and dampen innate cytokine production by DCs and other cells. Both the antiviral and immunoregulatory functions of NK cells are crucial to limit tissue damage and prevent hepatic failure. These are all non-redundant factors regulating innate resistance to MCMV. Also depicted are molecules that counteract the detrimental side effects of severe inflammation (glucocorticoids), or that tolerize mice to inflammation-induced metabolic stress (e.g. Kcnj8 regulates vasoconstriction), composing the 'immune homeostatic response' to infection. Many of the depicted factors have been identified through phenotypic screening of mice subjected to ENU mutagenesis. TLR, Toll-like receptor; MyD88, myeloid differentiation primary response gene (88); Pin1, peptidylprolyl cis/trans isomerase, NIMA-interacting 1; Unc93b1, unc-93 homologue B1 (*C. elegans*); Slc15a4, solute carrier family 15, member 4; IFN-IR, type I interferon receptor; IFNγR, receptor for interferon-gamma; STAT, signal transducer and activator of transcription; iNOS: inducible nitric oxide synthase; DAP12: DNAX-activation protein 12, also called TYROBP (TYRO protein tyrosine kinase-binding protein); NKG2D, NK cell receptor D; Rae, retinoic acid early transcript; Unc13d, unc-13 homologue D (*C. elegans*); Lyst, lysosomal trafficking regulator; fcr-1: Fc receptor glycoprotein (Murid herpesvirus 1).

Table II.12.2 Factors required for C57BL/6J mouse to control MCMV infection

Factors	Functions	Type of mutation	Outcome of mutant mice	References
Viral sensing				
Ly49H	Recognition of MCMV-infected, m157-expressing cells by NK cells	QTLs, KO	Death	Brown *et al.* (2001), Lee *et al.* (2001), Cheng *et al.* (2008), Fodil-Cornu *et al.* (2008)
DAP12	Adaptor in Ly49H signalling pathway in NK cells	KO	Sickness	Sjolin *et al.* (2002)
NKG2D	Recognition of 'induced self' ligands on infected cells by NK cells	KO	Resistance	Zafirova *et al.* (2009)
TLR9	Sensing of DNA-derived CpG by DC subsets	KO	Death	Krug *et al.* (2004), Delale *et al.* (2005), Courgette *et al.* (2008a)
Unc93b1	Endosomal trafficking of TLR3,7 and 9 in DCs	ENU	Death	Tabeta *et al.* (2007)
MyD88	TLR7 and TLR9 adaptor in DC subsets	KO, ENU	Death	Krug *et al.* (2004), Delale *et al.* (2005), Courgette *et al.* (2008b)
TLR3	Sensing of infected cell-derived dsRNA by cDCs	KO,ENU	Sickness	Tabeta *et al.* (2004), Edelmann *et al.* (2004)
TRIF/ TICAM-1	Adaptor in TLR3 signalling	KO, ENU	Sickness	Hoebe *et al.* (2003)
Pin1	Required for type I IFN production upon TLR stimulation in DCs	KO	Death	Tun-Kyi *et al.* (2011)
TRAIL-R	Death receptor of the TNFR family	KO	Decreased viral replication	Diehl *et al.* (2004)
Interaction between immune cells				
LTαβ/ LTβR	Regulates lymphoid tissue architecture and homeostasis during infection. Promotes initial IFNα/β production by splenic stromal cells	Tg, KO	Death	Benedict *et al.* (2001, 2006), Banks *et al.* (2005), Schneider *et al.* (2008)
IFN-I and IFN-IR	Required for cell intrinsic antiviral defence Ubiquitous effects	KO	Death	Salazar-Mather *et al.* (2002)
STAT1	Involved in IFN-IR and IFNγR signalling	KO, ENU	Death	Durbin *et al.* (1996) Crozat *et al.* (2006)
TYK2	Involved in IFN-IR signalling pathway	KO	Sickness	Strobl *et al.* (2005)
IL-12 and IL-12R	Required for IFNγ production by NK cells	KO	Death	Orange *et al.* (1996a,b), Pien *et al.* (2000)
ATF3	Transcription factor regulating IFNγ gene expression in NK cells	KO	Resistance	Rosenberger *et al.* (2008)
IkBζ	Induces IFNγ gene expression in response to IL12/IL18 in NK cells	KO	Death	Miyake *et al.* (2010)
IFNγ and IFNγR	Required for MHC up-regulation on APCs, and subsequent maturation	KO, blocking Ab	Death	Orange *et al.* (1995, 1996a,b), Fernandez *et al.* (2000), Salazar-Mather *et al.* (2003), Loh *et al.* (2005)
MIP1α	Produced by macrophages in a IFN-I-dependent manner in the liver, required for NK cell cytokine response	KO	Death	Salazar-Mather *et al.* (2000)
Mig/ CXCL9	Induced by IFNγ in DCs, and required for T-cell recruitment	Blocking Ab	Death	Salazar-Mather *et al.* (2000)
CD30	Required for NK cell survival and cross-talk with LTi	KO	Sickness	Bekiaris *et al.* (2009)
CD1d	Vα14i NKT-cells may promote NK cell activation through CD1d	KO	Sickness	Wesley *et al.* (2008)
CD3e	Activation of CD8 T-cells by DCs	KO	Sickness	Du X., Crozat K et Beutler B. (unpublished observation)
Flt3-L/Flt3	Involved in DCs and NK cell development and in NK cell priming by DCs	ENU	Sickness	Eidenschenk *et al.* (2010)

Factors	Functions	Type of mutation	Outcome of mutant mice	References
Role of specific immune cell types				
pDC	IFN-I production, promotion of CD8 T-cell responses	DTR KI	transient increase in viral load	Swiecki *et al.* (2010)
B-cells	Prevent viral reactivation upon immunosuppression. Promote splenic IFN-I response from stroma.	μMT KO IgH KO	Viral reactivation	Jonjic *et al.* (1994), Reddehase *et al.* (1994), Polic *et al.* (1998), Schneider *et al.* (2008)
T and B-cells	Prevent selection of m157 mutants escaping NK cell control	Rag1-KO SCID	death	French *et al.* (2004)
Immune cell development and maintenance				
IRF1	Required for CD8 T and NK cell development, IFN-I production and ISGs upon TLR stimulation	ENU	Sickness	Fernandez *et al.* (2000)
IRF8	Required for CD8α+ DC, pDC, monocytes, B-cell and Th1 development, and the induction of IFNγ-responsive genes	ENU	Death	Blasius *et al.* (2010)
Slfn2	Survival of peripheral CD8 T-cells and monocytes	ENU	Death	Berger *et al.* (2010)
Gimap5	Required for NK, NKT, T-cell survival and B-cell proliferation	ENU	Death	Barnes *et al.* (2010)
Cytolytic effector functions				
DAF	blocks the formation of the complement membrane attack complex	KO	Sickness	Bani-Ahmad *et al.* (2011)
iNos	Nitric oxide production by monocytes, cell-intrinsic antiviral function	KO	Death	Fernandez *et al.* (2000), Noda *et al.* (2001)
CatC	Unknown mechanism, independent of NK and CD8 T-cell cytolytic functions	KO	Sickness	Andoniou *et al.* (2010)
GzmA	Involved in NK and CD8 T-cell cytolytic function	KO	Sickness-Delayed viral control	Van Dommelen *et al.* (2006)
GzmB	Involved in NK and CD8 T-cell cytolytic function	KO	Sickness-Delayed viral control	Van Dommelen *et al.* (2006)
GzmM	Involved in NK and CD8 T-cell cytolytic function	KO	Sickness-Delayed viral control	Pao *et al.* (2005)
Perforin	Involved in NK and CD8 T-cell cytolytic function	KO	Death	Tay *et al.* (1997), Fernandez *et al.* (2000), Loh *et al.* (2005), Van Dommelen *et al.* (2006)
Unc13d	Involved in NK and CD8 T-cell degranulation of lytic vesicles	ENU	Death	Crozat *et al.* (2007)
Lyst	Involved in NK and CD8 T-cell degranulation of lytic vesicles	QTL	Death	Shellam *et al.* (1981), Bukowski *et al.* (1983)
Host metabolism				
Kir6.1	Potassium inwardly rectifying channel involved in coronary vessel vasodilatation in response to inflammation	ENU/QTL	Death but virus is controlled	Croker *et al.* (2007)
MTHFR	Required for folate metabolism	KO	Resistance	Fodil-Cornu *et al.* (2010)

the time when Ly49H was shown to promote MCMV resistance, B6 mice deficient for DAP12 functions were also found to control early MCMV replication poorly in most organs due to NK cell defects (Sjolin *et al.*, 2002), and DAP10$^{-/-}$ mice show increased viral loads in the salivary gland at day 7 after infection (Orr *et al.*, 2009). Hence, both DAP10 and DAP12 are required for optimal NK cell control of MCMV infection in B6 mice, functioning downstream of Ly49H activation. The generation and analysis of Ly49H transgenic BALB/c mice, as well as the complementary approach of using B6 mice genetically deficient for Ly49H, confirmed that this activating receptor is necessary and sufficient to induce resistance to acute primary MCMV infection in immunocompetent animals, primarily in the spleen (Lee *et al.*, 2003; Cheng *et al.*, 2008; Fodil-Cornu *et al.*, 2008).

Ly49H binds an MCMV-encoded protein, m157, which is expressed at the surface of infected cells as a glycoinositol phospholipid (GPI)-linked protein. In B6 mice, m157 promotes activation of Ly49H-expressing NK cells, as well as their preferential proliferation starting at ~ day 5 of infection (Dokun *et al.*, 2001). Interestingly, signalling via Ly49H is sufficient to bypass the need for cytokines in inducing NK cell proliferation, as Ly49H$^+$ NK cells proliferate both earlier and preferentially in MCMV-infected IFN-IR$^{-/-}$ mice (Geurs *et al.*, 2009). m157 displays homology to MHC class I-like molecules, which normally function as inhibitory ligands for Ly49 family proteins (Arase and Lanier, 2002; Smith *et al.*, 2002). Although m157-Ly49H control of early MCMV replication is a dominant mechanism, several *Cmv1*-independent genomic loci regulating NK cell recognition of MCMV-infected cells have been identified in quantitative trait loci (QTL) studies, illustrating the extraordinary complexity that encompasses the NK cell-mediated control of MCMV infection. We have noted many of these studies in Table II.12.3, as opposed to delineating them all in the text, as they are discussed in detail in other chapters of this book. The complex role that NK cells play during MCMV infection has almost certainly evolved from extended host-virus co-evolution, driving the genetic diversity of both players. Indeed,

Table II.12.3 Mouse strain outcomes after MCMV infection

Strain	QTL/ gene	Ligand	MHC-I haplotypes	Molecule affected	Cellular phenotype	Conferred outcome	References
C57BL/6J	*Cmv1*^{R}/ *Ly49h*	m157	*H-2*^{b}	MHCI-recognizing activating receptor Ly49H	Recognition of infected cells by NK cells	Resistance	Scalzo *et al.* (1990), Brown *et al.* (2001), Lee *et al.* (2001), Fodil-Cornu *et al.* (2008), Cheng *et al.* (2008)
BALB/cByJ	*Cmv1*^{s}/ *Ly49h*$^-$	n.a.	*H-2*^{d}	Lack Ly49H expression	NK cells fail in recognizing infected cells	Susceptibility	Brown *et al.* (2001), Lee *et al.* (2001)
129/SvJ	*Cmv1*^{s}/ *Ly49h*$^-$ *Ly49i*	m157	*H-2*^{bc}	Inhibitory receptor Ly49I	NK cells recognize infected cells but fail to activate	Susceptibility	Arase *et al.* (2002), Bubić *et al.* (2004), Adam *et al.* (2006)
NZW/LacJ	*Cmv2*^{s}	n.d.	*H-2*^{z}	Polygenic X-linked H-2- and NK gene complex loci-independent	n.d.	Susceptibility compared to NZB	Rodriguez *et al.* (2004, 2009)
NZB/B1NJ	*Cmv2*^{R}	n.d.	*H-2*^{d}	Polygenic X-linked H-2- and NK gene complex loci-independent	n.d.	Resistance compared to NZW	Rodriguez *et al.* (2004, 2009)
PWK/Pas	*Cmv1*^{s}/ *Ly49h*$^-$ *Cmv4*^{R}	n.d.	*H-2*^{b}	NK cell recognition?	n.d.	Resistance	Adam *et al.* (2006)
MA/My	*Cmv1*^{s}/ *Ly49h*$^-$ *Cmv3*/ *Ly49p,r and u*	H2-D^k/ m04	*H-2*^{k}	MHC-I H-2D^K recognition receptor Ly49P	Recognition of m04-expressing infected cells by NK cells	Resistance	Desrosier *et al.* (2005), Dighe *et al.* (2005), Kielczewska *et al.* (2009), Xie *et al.* (2009)
FVB/N	*Cmv1*^{s}/ *Ly49h*$^-$ *Cmv3*/ *Ly49p,r and u*	n.d.	*H-2*^{q}	Ly49-dependent inhibitory signals	NK cells fail to recognized infected cells	As susceptible as BALB/c	Fodil-cornu *et al.* (2011)
C3H/HeJ	*Cmv1*^{s}/ *Ly49h*$^-$ *Cmv3?*	n.d.	*H-2*^{k}	n.d.	n.d.	Resistance compared to BALB/c and C57BL/6	Chalmer *et al.* (1977), Grundy *et al.* (1981)

n.a., non-applicable; n.d., not determined.
For other mouse strains, see Adam *et al.* (2006).

use of congenic mouse strains in combination with genetically deficient viruses has revealed how MCMV down-regulates specific MHCI molecules, while also maintaining other MHCI molecules on the cell surface to engage NK cell inhibitory receptors, without activating antiviral CD8 T-cells (Babic *et al.*, 2010).

NK cell control of MCMV replication

NK cells and invariant NK T-cells (iNKT) are the only cells which express high levels of perforin and IFNγ at 36–72 hours of MCMV infection (Orange *et al.*, 1995; Nguyen *et al.*, 2002; Tabeta *et al.*, 2004; Tyznik *et al.*, 2008; Wesley *et al.*, 2008). The use of IFNγ$^{-/-}$ and perforin knockout B6 mice has clearly established a major contribution for these effector molecules in control of early MCMV infection. Perforin$^{-/-}$, IFNγ$^{-/-}$ and IFNγR$^{-/-}$ mice show impaired control of viral replication early in infection, before adaptive immunity has developed, and show enhanced susceptibility to MCMV-induced death (Tay and Welsh, 1997; Fernandez *et al.*, 2000; Loh *et al.*, 2005; van Dommelen *et al.*, 2006). The enhanced MCMV replication observed in MyD88$^{-/-}$ and TLR7/9$^{-/-}$ B6 mice has been largely attributed to decreased cytokine-induced NK cell activation by pDCs. However, and quite surprisingly, recent results in B6 mice engineered to express the DTR in pDCs (thus allowing their conditional depletion following toxin administration) showed only minor and transient reduction in NK cell cytotoxicity, and enhanced IFNγ production by NK cells was observed at early times after infection (Swiecki *et al.*, 2010). Consistently, 5- to 10-fold, dose-dependent, but transient increases in MCMV replication were observed at day 3 in both the spleen and liver of pDC-depleted mice. These data again highlight the 'flexibility' of cellular subsets to produce key innate cytokines, such as IFN-I and IL-12, during MCMV infection (Dalod *et al.*, 2002; Krug *et al.*, 2004; Andoniou *et al.*, 2005; Delale *et al.*, 2005), and suggest that our thinking about the connection between NK cell activation and their antiviral function early during MCMV infection in TLR-deficient mice may need to be re-evaluated.

Although both Perforin$^{-/-}$ and IFNγ$^{-/-}$ mice display increased susceptibility to MCMV infection, mice lacking perforin succumb to infection with much lower viral doses than those lacking IFNγ. Initially this was thought to reflect deregulated viral control and more extensive virus dissemination. However, although B6 mice genetically deficient for granzyme A and B, like perforin-deficient mice, also show impaired control of MCMV replication and high viral loads in all target organs, they ultimately control the virus and survive infection (van Dommelen *et al.*, 2006). The difference in the susceptibility of perforin versus granzymeAB knockout mice was explained to result from the specific inability of the former to limit immunopathology caused by activated myeloid cells, in particular monocytes/macrophages. Interestingly, in perforin-deficient mice, the excessive production of pro-inflammatory cytokines (mainly TNFα) after MCMV infection, results in a disease resembling human haemophagocytic lymphohistiocytosis (HLH). Hence, NK cell perforin-dependent cytotoxicity contributes resistance to MCMV infection in B6 mice both by direct antiviral effects and through the regulation of immunopathology. Importantly, both 129Sv and BALB/c mouse strains genetically deficient for IFNγ-signalling show even more severe defects in control of early MCMV replication (Presti *et al.*, 1998; Sumaria *et al.*, 2009), suggesting that in B6 mice NK cell cytotoxicity mediated predominantly by Ly49H^{+} NK cells is the dominant effector mechanism, at least in the spleen. Interestingly, B6 IL-18$^{-/-}$ mice show a dramatic reduction in NK cell-produced IFNγ selectively in the spleen, but not the liver, and the fact these knockout mice control MCMV replication normally supports this idea (Pien *et al.*, 2000). However, IFNγ plays a critical role in controlling virus in the liver, both in B6 and BALB/c mice (Tay and Welsh, 1997; Loh *et al.*, 2005; Sumaria *et al.*, 2009).

The release of perforin and granzymes by NK cells requires the polarization of granules to the plasma membrane. Several cellular factors function non-redundantly to mediate this release, and the mechanism is shared by melanocytes for exporting granules to the skin or the hair shaft. Therefore, mice displaying a hypopigmented phenotype often show impaired degranulation of NK and CD8 T-cells. Granulated cells of the *beige* mouse, in which the *lyst* gene that regulates lysosomal trafficking is mutated, present giant intracytoplasmic lysosomal structures and defects in the biogenesis of cytolytic granules, accounting for their MCMV susceptibility (Shellam *et al.*, 1981; Bukowski *et al.*, 1983). Similarly, a functionally defective GTPase Rab27α in *ashen* mice restricts the docking of vesicles at the plasma membrane of NK and CD8 T-cells (Menasche *et al.*, 2000), also leading to impaired MCMV control (Crozat and Beutler, unpublished data).

Analysis of the ENU mutant *Jinx* mice, which does not present any hypopigmentation, but were impaired for NK cell control of MCMV replication, unravelled the role of Unc13d (mouse homologue of MUNC13–4) in controlling cytotoxic granule exocytosis (Crozat *et al.*, 2007). The genetic mutation and virus-induced disease phenotype of *Jinx* mice recapitulated those occurring in certain patients suffering from type III familial HLH. Hence, unravelling the genetics of the

MCMV 'resistome' has helped to generate a novel preclinical animal model for studying the pathophysiology of human HLH (Krebs *et al.*, 2011).

Role of cathepsin C and granzyme M in innate defence to MCMV

As discussed above, perforin and granzymes A and B play important roles during MCMV infection, by limiting both MCMV replication and pro-inflammatory cytokine production. The activation of granzymes in inflammatory cells, including NK cells and T-cells, is thought to require processing by a specific dipeptidyl peptidase, namely cathepsin (CatC). Thus, in theory, in the absence of CatC, both granzyme activity and perforin-dependent cytotoxicity should be lost. Interestingly, infection of CatC-deficient mice with MCMV resulted in an outcome differing from that observed in mice lacking granzyme AB. Unlike granzyme AB-deficient mice, where increased MCMV replication is observed in all target organs, CatC$^{-/-}$ mice display organ-specific defects for viral control (Andoniou *et al.*, 2011). Thus, in CatC$^{-/-}$ mice, spleen and lung replication is equivalent to that observed in wild-type mice at days 4 and 10 of infection. However, viral titres are increased in the liver and salivary glands of CatC-deficient mice. Furthermore, the increased susceptibility of CatC$^{-/-}$ mice to MCMV infection occurs despite normal cytolytic function in both NK cells and cytotoxic CD8 T-cells. Although CatC is required for the activation of serine proteases other than those expressed in cytolytic effector cells, including those expressed in neutrophils, it is unlikely that this accounts for the differences in MCMV control observed in CatC-/- mice, as neutrophil depletion does not replicate this defect (Andoniou *et al.*, 2011). A possibility worth exploring is that proteolytic CatC-mediated activities may inhibit viral replication by interfering with essential viral processes, such as viral assembly.

The role of proteases in limiting infection by directly interfering with viral replication is becoming clearer, and for MCMV is illustrated by studies investigating the impact of granzyme M on infection and viral replication, both *in vivo* and *in vitro*. Granzyme M is one of 11 mouse granzymes (A–G and K–N), and has been found to play a role in MCMV infection (Pao *et al.*, 2005). Granzyme M-deficient mice (GzmM$^{-/-}$) display significantly higher viral titres both in the spleen and liver during early infection (days 2–6) compared with wild-type mice. However, these defects are transient, and ultimately MCMV infection is controlled in the absence of granzyme M. One of the most notable features of MCMV infection in GzmM$^{-/-}$ mice is revealed by histological analysis of infected organs, mainly the liver, where increased numbers of viral foci and cytomegalic cells are detected (Pao *et al.*, 2005). These initial findings suggested that granzyme M may play a non-cytotoxic role in limiting MCMV infection, perhaps by interfering with viral replication and/or the release of viral progeny. Subsequent studies have shown that human granzyme M likely inhibits HCMV replication in fibroblasts through direct cleavage of the viral phosphoprotein 71 (van Domselaar *et al.*, 2010). Although equivalent biochemical studies with mouse granzyme M are yet to be undertaken, the combined data are suggestive that granzyme M affects viral replication independently of host cell death via direct cleavage of essential CMV proteins, and whether additional granzymes work in a similar fashion should also likely be considered.

Myeloid lineage cells in innate MCMV defence

Although myeloid lineage cells are normally thought of in the context of MCMV dissemination (see also Chapter I.21), as well as promoting adaptive immunity via their professional APC function, there is some evidence for their role in controlling early MCMV infection via direct or indirect regulation of innate defence (Hanson *et al.*, 1999; Salazar-Mather and Hokeness, 2006; see also Chapter II.11). Recruitment of myeloid cells from the bone marrow and/or blood to the liver occurs during the initial days of MCMV infection and is dependent upon IFN-I signalling and CCR2 ligands, as shown using B6 mice genetically deficient in these pathways (Crane *et al.*, 2009). Once in the liver these 'inflammatory monocytes' produce CCL3, promoting NK cell recruitment, which ultimately assists in control of MCMV replication (Salazar-Mather and Hokeness, 2006). These monocytes may also produce IFN-I in the liver through TLR2-dependent mechanisms, although this has not been proven directly *in vivo* for MCMV (Barbalat *et al.*, 2009). Therefore, in this context monocytes/macrophages play more of a 'bridging' role in innate defence through their ability to coordinate other innate cell populations and cytokine production. B6 knockout mice lacking nitric oxide synthase 2 (NOS2$^{-/-}$) show enhanced MCMV replication in the salivary gland and lung for several weeks after infection (Fernandez *et al.*, 2000; Noda *et al.*, 2001). The timing of this phenotype late in infection suggests that these differences cannot be ascribed to innate defence mediated by myeloid cells, but earlier studies using drug inhibition of nitric oxide production also showed higher replication of MCMV in the liver at day 3 (Tay and Welsh, 1997). Therefore, in general, the available evidence suggests that mature myeloid lineage cells and granulocytes may play a minor role in promoting innate defence to early MCMV infection. Notably, salivary gland replication of MCMV in NOS2$^{-/-}$ mice has been reported to be dramatically enhanced, which might indicate a crucial role for myeloid cells in transporting MCMV to this site or in controlling virus in this tissue.

TNF-family cytokines in innate MCMV defence

Ligands and receptors of the tumour necrosis factor (TNF)-superfamily are expressed by both stromal and haematopoietic cells, and play critical roles in regulating innate and adaptive immunity. Several TNFRs mediate apoptosis through their death domains and downstream caspase activation i.e. 'death receptors', the prototypic ones being TNF-R1 and Fas. Fas-FasL signalling is disrupted in the naturally occurring *lpr* and *gld* mouse mutants, with both strains displaying massive lymphoproliferation (Cohen and Eisenberg, 1991). B6 *lpr* mice show little to no increase in early MCMV replication, and this is also true for B6 TNF-R1$^{-/-}$ mice and for *lpr*/TNFR-1 double deficient mice (Fleck *et al.*, 1998; Fernandez *et al.*, 2000). These mice however do show a reduced ability to control MCMV replication at times later than one week, and exhibit increased immunopathology at these later times, but they do appear to have unaltered NK cell-mediated control at earlier times. TRAIL-R is another death receptor, and B6 TRAIL-R$^{-/-}$ mice also show no defect in early defence to MCMV, controlling viral replication even better than wild-type mice in the spleen (Diehl *et al.*, 2004). At first glance, the absence of phenotype in mice lacking death receptors is somewhat surprising, as NK cells express most TNF-family ligands when activated, but perhaps this reflects in large part the ability of MCMV to inhibit the extrinsic apoptotic pathway (see Chapter I.15). These issues could be clarified by examining viral control in death receptor-deficient mice infected with mutant MCMV lacking the genes shown to inhibit the extrinsic apoptotic pathway.

In B6 mice genetically deficient for LTαβ–LTβ receptor signalling, significantly higher levels of MCMV replication are observed at day 3 in the spleen and liver (10- to 100-fold). These mice also display increased sensitivity to virus-induced cell death at day 5–6 p.i. (Benedict *et al.*, 2001; Banks *et al.*, 2005), consistent with the lower levels of IFN-I produced after MCMV infection. Mice lacking LIGHT, a second ligand for the LTβR, also show a modest 5-fold increase in MCMV replication at day 3 post infection (Banks *et al.*, 2005). Interestingly, mice deficient for LTβR signalling are especially sensitive to high doses of MCMV, whereas infection with moderate to low doses results in a much less severe phenotype (Benedict and colleagues, unpublished observations).

Mechanisms regulating host tolerance to innate MCMV defences revealed by forward genetics

Screening ENU mutant mice for MCMV-induced death has revealed never before encountered phenotypes. For example, *mayday* mice died very suddenly after infection, between 36 and 72 hours, with no evidence of uncontrolled viral replication (Croker *et al.*, 2007). The genetic mutation in Mayday mice is localized to the *Kir6.1* gene, encoding a potassium inwardly rectifying channel that regulates smooth muscle cells and coronary vessel vasodilatation in response to inflammation (Kane *et al.*, 2006). Hence, *mayday* mice are unable to cope with metabolic changes induced by inflammatory cytokines produced early after MCMV infection, and succumb to severe cardiac ischaemia. The response observed in *mayday* mice is a good illustration of how the host defence to infection operates at multiple levels. Tolerance, in the broadest sense, includes all mechanisms that dampen self-harm resulting from host-regulated immune responses, but is separable from pathogen resistance *per se* (Schneider and Ayres, 2008). *Mayday* mice illustrate this point, displaying normal resistance to MCMV replication, but decreased tolerance to inflammation-induced death. A second 'homeostatic' host defence mechanism promoting tolerance to MCMV involves cross-talk between the immune and the neuroendocrine systems, which regulates glucocorticoid production and dampens the effects of cytokine-induced shock. Illustrating this, adrenalectomy of mice prior to MCMV infection leads to a dramatic increase in the production of innate cytokines at 36 h p.i. and results in death (Ruzek *et al.*, 1999). The balance between pathogen resistance and immune tolerance often depends upon the infectious pathogen load. At high MCMV doses, perforin knockout mice show a reduction in both resistance (as assessed by dramatically increased viral replication) and tolerance (as assessed by the development of an HLH syndrome caused by dysregulated cytokine production by myeloid cells and T-lymphocytes, itself resulting from the loss of NK cell-mediated immunoregulation) (van Dommelen *et al.*, 2006). At lower MCMV doses perforin-deficient mice can maintain tolerance through increased IL-10 production by Ly49H^{+} NK cells, thus limiting at least the activation of CD8 T-cells and its pathological consequences (Lee *et al.*, 2009).

Adaptive immune defence to MCMV infection

To this point, we have focused solely on how mouse genetic models have helped to clarify the cell types, cytokines and molecular pathways orchestrating early innate defence to MCMV. However, as is the case for most pathogens, innate defences are not enough, and adaptive immunity is required to ultimately control both systemic and persistent MCMV infection (see Table II.12.4). In humans, when immunity is compromised or naïve, it is clear that 'providing' components of HCMV adaptive immunity can reduce the incidence and/or severity of disease. Virus-specific T-cell immunotherapy can both prevent and reduce HCMV-induced disease in bone marrow transplant patients (Riddell *et al.*, 2000; see also Chapter II.16). Additionally, administering pooled human immunoglobulin containing high titres of anti-HCMV antibodies to pregnant mothers with a primary HCMV infection protects against virus-induced congenital disease in their newborns (Nigro *et al.*, 2005; see also Chapter II.3). Studies adoptively transferring MCMV-specific T-cells in bone marrow ablated mice provided proof of principle that T-cell therapy works (Holtappels *et al.*, 2008; see also Chapter II.17), and many studies in immune competent mice have delineated cellular and molecular mechanisms that regulate the generation and function of these cells. Much less work has been done with regards to humoral immune control of MCMV, perhaps because almost two decades ago it was shown that antibodies do not contribute to controlling primary MCMV infection (Jonjic *et al.*, 1994). However, a few studies examining the role of humoral immunity in limiting MCMV infection in the brain of newborn mice (Cekinovic *et al.*, 2008), or protection mediated by memory B-cells in immunodeficient hosts (Klenovsek *et al.*, 2007), have been informative (see Chapter II.10). Recent work with MCMV has also identified specific mechanisms that promote the 'transition' from innate to adaptive immune control of infection (Robbins *et al.*, 2007; Andrews *et al.*, 2010; Stadnisky *et al.*, 2011), revealing that both host and viral genetic diversity impact this phase by regulating NK–DC cross-talk. These topics are discussed below.

The MCMV-specific T-cell response

Several chapters in this book address the T-cell response generated following CMV infection of various hosts, including humans, monkeys, mice and guinea pigs. Multiple aspects appear to be conserved between these genetically distinct CMVs and their hosts, the most relevant for our discussion purposes being that (1) CD8 and CD4 T-cells responding to a broad range of viral protein-derived epitopes appear to arise in all CMV infections and (2) a relatively small proportion of these epitope-specific responses come to dominate the T-cell memory pool as infection progresses through the persistent phase and into latency. For MCMV, an IE1-derived epitope displayed by H-2L^d in BALB/c mice was the first to be identified (Del Val *et al.*, 1988; Reddehase *et al.*, 1989). Several additional MCMV ORF-derived epitopes targeted by CD8 T-cells and presented by H-2^d have now been characterized in BALB/c mice (Reddehase, 2002; Holtappels *et al.*, 2008). In B6 mice, 24 unique epitopes presented by H-2^b that promote CD8 T-cell responses have been identified (Gold *et al.*, 2002; Munks *et al.*, 2006b). In both BALB/c and B6 mice, a few of these epitope-specific responses come to dominate the CD8 T-cell memory pool during persistence/latency (Holtappels *et al.*, 2000, 2002; Munks *et al.*, 2006a; Snyder *et al.*, 2008), a process often referred to as 'memory inflation' (Karrer *et al.*, 2003).

MCMV epitope-specific CD4 T-cell responses have also been defined in B6 mice. Eighteen 15mer peptides derived from various viral proteins are targets of MCMV-specific CD4 T-cells, with most being presented by I-A (Arens *et al.*, 2008; Walton *et al.*, 2008). It is likely that these 18 responses represent ~25–30% of the total MCMV-specific CD4 T-cell response (Arens *et al.*, 2008). Interestingly, as is the case for some MCMV-specific CD8 T-cell responses, epitope-specific CD4 T-cell responses can also show dramatically different expansion and contraction kinetics. To date, only one inflationary response (specific for the m09$_{133-147}$ epitope) has been identified in the MCMV model, peaking at 6–8 weeks after initial infection, as opposed to the normal 8–9 days (Arens *et al.*, 2008). Currently it is unclear whether inflation occurs for additional epitopes, or in other mouse strains, as is the case for MCMV-specific CD8 T-cell responses. The interest in understanding the molecular and cellular mechanisms that drive these differential populations of epitope-specific T-cells that arise during CMV infection has been intense, particularly because such a large proportion of the circulating memory T-cell pool in infected humans is dedicated to HCMV (see Chapter II.7). In recent years, various mouse genetic models have yielded significant information in this regard and it is now clear that both host and viral factors contribute to this process.

The requirement of CD8 and CD4 T-cells for control of MCMV infection

As many of the initial studies of T-cell control of MCMV were performed in BALB/c mice, depletion

Table II.12.4 Factors involved in mounting an efficient CD8 T-cell response

Mechanism	Factors	Mouse models	Effect/functions	MCMV-specific immune function	Viral control	References
DC ontogeny	Batf3	KO	Transcription factor involved in CD8a+ DC differentiation Batf3-/- mice have reduced CD8α+ DCs in lymphoid organs	Needed for CD8 T-cell priming by CD8α+ DCs	Normal	Torti *et al.* (2011)
Antigenic presentation	MHC-I	β2m-/-	Absence of CD8 T-cells; NK cells are not licensed by their self ligands	Normally contribute to immune control, not absolutely required in B6	Normal but persistent viral load in SG, and lethality at high doses	Polic *et al.* (1996)
	MHC-II	H2-Ab1 or H2-Ea KO	Absence of CD4 T-cells	IFNγ-production by CD4 T-cells essential for control of MCMV in the SG CD4 T-cells provide 'help' for memory inflation of CD8 T-cells	Persistence in SG only	Snyder *et al.* (2009), Walton *et al.* (2011a,b)
	LMP7	KO	Immunoproteasome subunit; required for generating the repertoire of MHC-I-loaded peptides Expressed by APCs	promote 'stable' CD8 T-cells, little role for inflation	n.d.	Hutchinson *et al.* (2011)
Costimulation signal provided by APCs	PD-L1	Blocking Ab	Expressed by DCs and engaged PD1 expressed by T-cells	Up-regulated in MCMV-infected DCs. Restrains T-cell responses in absence of positive cosignals	n.d.	Benedict *et al.* (2008)
	B7.1, B7.2 → CD28 signalling	B7.1 KO B7.2 KO CD28 KO	B7 Expressed by APC and engages CD28/CTLA4 on T-cells	Promote initial expansion of CD8T, little role in memory inflation Required for CD4 T-cell expansion and memory	Persistence in SG up to 100 days pi	Cook *et al.* (2009), Arens *et al.* (2011a,b)
	OX40	KO	cosignallng TNFR expressed by many cell types, regulates T-cell memory	Needed for optimal memory inflation of CD8T	Increased viral load in SG at 14 weeks p.i.	Humphreys *et al.* (2007), Snyder *et al.* (2009)
	CD40L	KO	TNF-family costimulatory ligand expressed largely by professional APC	Promotes inflation of IE3-specific CD8T	Normal in SG at 24 weeks p.i.	Snyder *et al.* (2009)
	4–1BB	KO	cosignallng TNFR expressed by many cell types, regulates T-cell memory	Limits CD8 T-cell expansion, but promotes memory inflation	Normal in SG at 30 days p.i.	Humphreys *et al.* (2010)
	CD4	KO and Antibody depletion	Helper functions, cytokine production, potential direct cytolysis	Production of IFNγ by CD4 T-cells in the salivary gland is essential to control MCMV persistence. Can control systemic infection in the absence of CD8T (acquire better effector function)	Persistence in SG only	Jonjic *et al.* (1989), Lucin *et al.* (1992), Polic *et al.* (1996), Lathbury *et al.* (1997), Snyder *et al.* (2009), Walton *et al.* (2011)

Table II.12.4 (Continued)

Mechanism	Factors	Mouse models	Effect/functions	MCMV-specific immune function	Viral control	References
Costimulation signal provided by APCs	CD8	KO and Antibody depletion	Cytolytic and cytokine effector function	CD8 T-cells provide rapid control in immune ablated mice upon transfer Utilize IFNγ and cytolytic effector activities to control infection	Absolutely required for control in immune ablated mice. Modest increase in their absence in immune competent mice	Reddehase *et al.* (1985; 1987), Jonjic *et al.* (1990), Lathbury *et al.* (1997), Podlech *et al.* (1998), Polic *et al.* (1998), Salem *et al.* (2000)
	Cxcr3	KO	Receptor for monokine induced by IFNγ (Mig) and IFNγ-induced protein 10 (IP-10) expressed by T-cells	Promotes CD8 T-cell infiltration in the liver, and promotes their subsequent activation	MCMV controlled by day 7 in both liver and spleen	Hokeness *et al.* (2007)

and/or adoptive transfer of T-cell subsets was often performed to assess their relative importance in immune control, as knockout mice are not readily available for this strain. It is important to keep in mind the absence of Ly49H$^+$ NK cells in BALB/c mice compared to B6, as this can impact the development of T-cell-mediated immunity (Robbins *et al.*, 2007; Andrews *et al.*, 2010; Stadnisky *et al.*, 2011). It is also clear that there is strong selective pressure to maintain T-cell-mediated control of MCMV, as infection of B6 RAG$^{-/-}$ or B6-SCID (genetically deficient for B and T-cells) results in the rapid emergence of MCMV 'escape mutants' *in vivo* (French *et al.*, 2004). In this case, NK cells maintain control of wild-type MCMV for ~20 days, before m157 mutants unable to interact with Ly49H emerge, replicate uncontrolled, and kill these mice. T-cell control of MCMV in the BALB/c model, often in the context of immune suppression, has been reviewed (Reddehase, 2002), and is also detailed in Chapter II.17.

Gene deficient mice reveal distinct mechanisms regulating the 'stable' and 'inflationary' populations of MCMV-specific CD8 T-cells

The identification of peptide epitopes for tracking 'stable' and 'inflationary' populations of MCMV-specific CD8 T-cells in B6 mice (Munks *et al.*, 2006b) has provided the opportunity to assess the role of a variety of molecules in promoting these responses using knockout mice. One aspect that has been relatively well studied over the last 3–4 years is the role of cosignalling pathways. Using B7.1$^{-/-}$, B7.2$^{-/-}$, B7.1$^{-/-}$/B7.2$^{-/-}$ and CD28$^{-/-}$ mice, it has been shown this prototypic costimulatory pathway is required for the initial expansion of MCMV-specific CD8 T-cells detectable by day 8 (Cook *et al.*, 2009; Arens *et al.*, 2011b). However, as the inflationary responses start to dominate the memory CD8 T-cell pool after ~day 15, little to no defect in the numbers of these populations is seen in mice lacking these costimulatory molecules, while CD8 T-cells which establish stable memory pools remain low in numbers compared to wild-type mice (Arens *et al.*, 2011b). Interestingly, the inhibitory cosignalling receptor PD-1 markedly restricts the initial expansion of MCMV-specific CD8 T-cells if B7-CD28 signalling is absent, but plays a minor role if this costimulatory signal is intact, and high levels of PD-L1 expression maintained in MCMV-infected DCs are a likely important source of ligand (Benedict *et al.*, 2008).

OX40 and 4–1BB are cosignalling receptors of the TNFR family. In contrast to B7-CD28 deficient mice, OX40$^{-/-}$ and 4–1BB$^{-/-}$ B6 mice showed lower levels of inflationary CD8 T-cell populations at later times of infection, with stable CD8 T-cells showing no defect (Humphreys *et al.*, 2007b; 2010). Interestingly, 4–1BB$^{-/-}$ mice actually showed enhanced expansion of all CD8 T-cells at day 8 after MCMV infection in comparison with wild-type mice, consistent with the dichotomous role this TNFR can play as both a positive and negative regulator of T-cells in other mouse models (Croft, 2003). When B6 mice lacking another TNF-family ligand, CD40L, were examined, lower levels of inflationary, IE3-specific CD8 T-cells were seen at ~6 months after infection (Snyder *et al.*, 2009). The important role of TNFR cosignalling ligands and receptors in regulating memory inflation of CD8 T-cells is probably due in large part to their promotion of CD4 T-cell 'help' to these cells (Humphreys *et al.*, 2007b). Consistently, MHCII$^{-/-}$ and

$CD4^{-/-}$ mice show lower levels of inflationary T-cells at late times of infection (Snyder *et al.*, 2009; Arens *et al.*, 2011a; Walton *et al.*, 2011b). Adoptive transfer of IL-2 receptor knockout cells into MCMV-infected B6 mice indicated that IL-2 preferentially promotes inflation of some CD8 T-cells, but not those which remain stable (Bachmann *et al.*, 2007). Recently, infection of B6 mice lacking components of the inducible immunoproteasome ($LMP7^{-/-}$) has shown that inflationary CD8 T-cells are less dependent upon this mechanism of antigen presentation than stable responses (Hutchinson *et al.*, 2011). Finally, infection of $IL\text{-}10^{-/-}$ B6 mice revealed that this immunosuppressive cytokine significantly restricts the expansion of inflationary CD8 T-cells (Jones *et al.*, 2010).

One key unanswered point is the 'source' of MCMV antigen that drives memory inflation of CD8 T-cells (for discussion and a current hypothesis, see Chapter I.22). Recently, a bone marrow chimeric approach, where irradiated BALB/c mice were reconstituted with bone marrow from mice lacking $H2\text{-}L^d$ to track the requirements for inflation of IE1-specific CD8 T-cells, showed that a radio-resistant non-haematopoietic tissue cell is most likely to be the source of persistent MCMV antigen (Seckert *et al.*, 2011). In contrast, priming of MCMV-specific CD8 T-cells likely occurs via uptake of MCMV antigens by uninfected cells, followed by cross-presentation via MHCI. This hypothesis is based on results obtained using MCMV mutants unable to restrict MHCI expression (Munks *et al.*, 2007), or unable to spread from initially infected cells (Mohr *et al.*, 2010; Snyder *et al.*, 2010). In turn, *in vitro* infection of DCs derived from K^{bm1} mutant mice followed by adoptive transfer into WT mice also suggested the importance of cross-presentation (K^{bm1} cannot present the SIINFEKL peptide from ovalbumin, but the recipient mice can) (Benedict *et al.*, 2008). A similar approach infecting H-2 K^b/D^b double deficient cells, followed by transfer into wild-type mice, was in line with this model (Benedict *et al.*, 2008; Snyder *et al.*, 2010). More recently, results from MCMV infection of $Batf3^{-/-}$ B6 mice, which lack $CD103^+$ and $CD8\alpha^+$ DCs (the $CD8\alpha^+$ DC population thought to be the one that primarily promotes cross-presentation *in vivo*), indicate that these APCs facilitate priming of MCMV-specific CD8 T-cells, but do not affect the generation of inflationary memory responses (Torti *et al.*, 2011). Importantly, it should be noted that many of the above conclusions are based on studies undertaken in B6 mice, where the productive infection of cDC is of limited duration due to Ly49H-mediated NK cell responses (Andrews *et al.*, 2010). Hence whether the same processes apply in BALB/c mice remains to be determined.

MCMV-specific CD4 T-cell responses

Studies in $OX40^{-/-}$ B6 mice showed reduced memory levels of MCMV-specific CD4 T-cells, while expansion at day 7 was unaffected, although in this study only the total CD4 T-cell response was analysed after polyclonal restimulation, as MHCII-presented epitopes had not yet been identified (Humphreys *et al.*, 2007b). Further studies in $4\text{-}1BB^{-/-}$ mice showed increased levels of MCMV epitope-specific CD4 T-cells, both at days 7 and 30 after infection (Humphreys *et al.*, 2010). B6 $IL\text{-}10^{-/-}$ mice show significantly enhanced numbers of MCMV epitope-specific CD4 T-cells at 90 days after infection, and antibody neutralization with an anti-IL-10R at the time of infection suggests expansion of these cells in the first week is also restricted by this mechanism (Jones *et al.*, 2010). The expansion of MCMV epitope-specific CD4 T-cells is dramatically reduced in $B7.1/2^{-/-}$ and $CD28^{-/-}$ B6 mice, including m09-specific CD4 T-cells which do not peak until ~ day 40, and enhanced persistence of MCMV in the salivary glands is seen in these mice (Arens *et al.*, 2011a). Notably, deletion of the two MCMV ORF that restrict B7.1 and B7.2 expression also promoted enhanced CD4 T-cell responses and reduced viral persistence (m138 and m147.5), but had no effect on MCMV-specific CD8 T-cells (Arens *et al.*, 2011a). This link between the levels of MCMV-specific CD4 T-cells and the duration and/or magnitude of MCMV replication in the salivary glands holds true in all cases we are aware of in B6 mice studied to date, as suggested by the initial studies in BALB/c mice (Jonjic *et al.*, 1989). IL-10 production by these CD4 T-cells likely plays a critical role in sustaining persistence in this mucosal organ as well (Humphreys *et al.*, 2007a). Additionally, very recent work using various gene-deficient B6 mice and bone marrow chimera approaches has shown that salivary gland-resident professional APC likely activate CD4 T-cells to produce IFNγ, which then controls MCMV replication in acinar duct epithelial cells (Walton *et al.*, 2011a).

Humoral immune control of MCMV

As mentioned, the use of B6 mice genetically deficient for the transmembrane domain of the Ig μ chain ($\mu MT^{-/-}$ mice, lacking all B-cells) formally showed that antibody responses are not required for immune control of primary MCMV infection in immune competent mice, but contribute substantially to limiting the levels of viral replication after severe immune suppression (Jonjic *et al.*, 1994; Reddehase *et al.*, 1994). $\mu MT^{-/-}$ mice were also subsequently used to show that in the absence of antibodies, cell-mediated immune control of reactivated virus was hierarchical and involved CD8 and CD4 T-cells, as well as NK cells, and IFNγ production (Polic *et al.*, 1998). Adoptive transfer of MCMV immune serum from WT BALB/c mice or μMT$^{+/-}$ B6 mice into newly infected mice restricts the spread of MCMV within the infected liver (Wirtz *et al.*, 2008). Likewise, immune serum transfer into intracranially infected newborn mice protects against MCMV-induced death (Cekinovic *et al.*, 2008; see also Chapter II.6). The adoptive transfer of memory B-cells from MCMV-infected mice into naïve mice lacking T and B-cells (RAG-1$^{-/-}$), significantly reduces MCMV replication and protects against virus-induced death (Klenovsek *et al.*, 2007; see also Chapter II.10). The requirements for activation of memory B-cells upon MCMV infection was further dissected using B6 mice genetically deficient for both TNF and LTα, as well as mice depleted of CD11c^{+} dendritic cells through transgenic expression of the diphtheria toxin receptor (Weisel *et al.*, 2010). This recent progress in developing new approaches to address the contribution of antibody and memory B-cells in various mouse models of MCMV infection has great potential to provide clinically valuable information for developing improved therapeutic strategies for combating HCMV-induced disease.

The transition from innate to adaptive immunity

The previous sections established that different innate and adaptive immune cells are activated during MCMV infection. This raises the question of how cross-talk between these different cell types orchestrates antiviral defence, and, in particular, the transition between innate and adaptive immunity. DCs are likely to play a central role in this process, because (i) they act as innate sentinels to detect viral products early during infection, (ii) they sense a variety of additional non-viral danger signals and (iii) they integrate this complex array of 'inputs' to mount the specific 'outputs' required to activate and functionally polarize T-cells (see Fig. II.12.3). We will focus in particular on how NK–DC

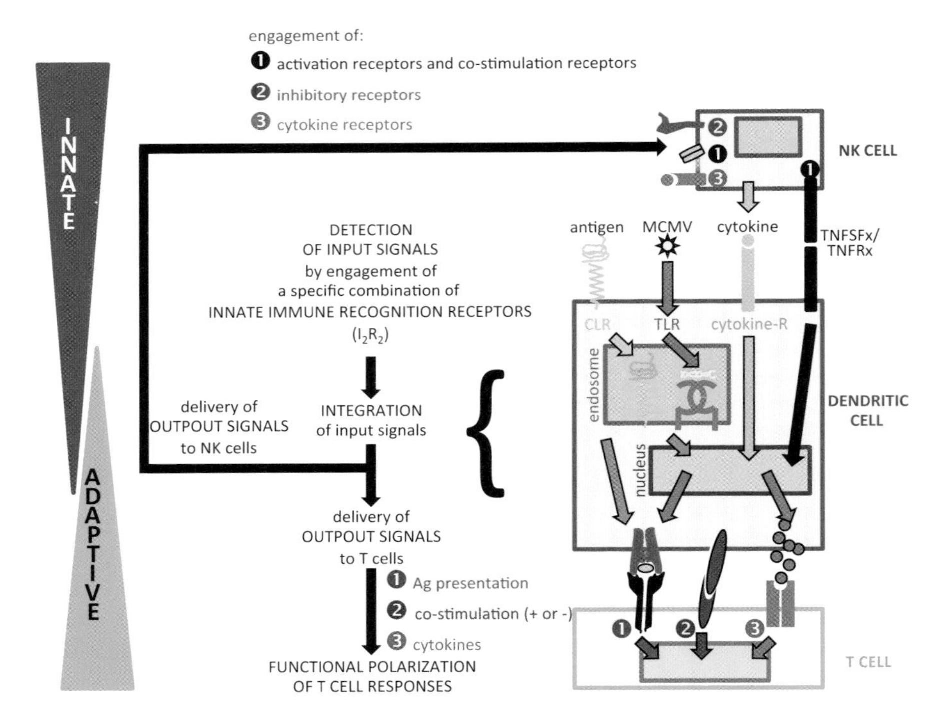

Figure II.12.3 NK–DC cross-talk promotes adaptive immunity to MCMV. A key job of dendritic cells (DCs) is to link innate and adaptive immune responses. This is accomplished by sensing microbial or other danger signals and integrating them through intracellular signalling networks present in cytoplasmic and nuclear compartments, ultimately delivering three distinct but complementary output signals to 'programme' both NK and T-cells to perform specific functions. In turn, DCs receive signals from activated NK cells that can be mediated by cytokines or members of the TNF-family of cytokines and receptors (TNFSFx and TNFRSFx), which are integrated with other input signals to shape T-cell responses. CLR, C-type lectin receptor; cytokine-R, receptor for cytokine; TLR, Toll-like receptor; Ag, antigen.

cross-talk shapes the MCMV-specific T-cell response. As already discussed, innate cytokine production by pDC at 36 hours of infection activates NK-cell effector functions, although in the absence of these cells redundant and/or alternative mechanisms are operable at this time (Swiecki and Colonna, 2010). Along these lines, ENU generated B6 mice lacking a functional Fms-like tyrosine kinase 3 (Flt3, '*warmflash*' mice) succumb to MCMV infection and show markedly higher replication levels in the spleen at day 5 (100- to 1000-fold) (Eidenschenk *et al.*, 2010). As in Flt3 knockout mice, reductions in both pDC and cDC subsets were observed in warmflash mice, as well as decreased total NK-cell numbers. NK-cell activation, based on CD69 cell surface expression and IFNγ production, was also reduced in *warmflash* mice, with more severe defects seen at 24 hour or 48 hour time points, as opposed to 36 hours, suggesting that cDC or stromal production of innate cytokines was hampered in this model. Both pDC and cDC were activated to a lesser degree in response to TLR ligands, and both subsets also showed a reduced ability to facilitate NK-cell activation when co-cultured *in vitro* (Eidenschenk *et al.*, 2010). Importantly, this study illustrates that Flt3 ligand 'conditioning' of both pDC and cDC regulates their activation threshold, and the subsequent NK–DC interactions at early times of MCMV infection which set the stage for transitioning to adaptive immunity.

During MCMV infection, interactions between DCs and NK cells are extensive and shape innate antiviral immunity, but also impact the generation and quality of adaptive immune responses. Reciprocal interactions between DCs and NK cells during MCMV infection were first demonstrated in B6 mice. These studies showed that splenic CD8α$^+$ DCs are lost after MCMV infection if NK control is inefficient or absent, as is the case for BALB/c mice, if B6 mice are depleted of NK cells or treated with a Ly49H blocking antibody. Interestingly these studies also showed that in B6 mice, antibody-mediated ablation of CD8α$^+$ DC compromises the proliferation of Ly49H$^+$ NK cells (Andrews *et al.*, 2003). Although the relevance of the late proliferation of Ly49H$^+$ NK cells to the long-term outcome of MCMV infection is not yet fully understood, these cells have been shown to give rise to NK cells exhibiting aspects of immunological memory. Further, the adoptive transfer of CD11b$^+$ DCs infected with MCMV *in vitro* promotes activation of NK cells and leads to improved control of subsequent MCMV infection (Andoniou *et al.*, 2005). Hence, in addition to pDC, both CD8α$^+$ cDC and CD11b$^+$ cDC can affect NK cell response during MCMV infection.

More recent studies have shown that the extent of DC-NK cell interactions early in infection can impact the generation of adaptive antiviral immunity. The generation of MCMV-specific CD8 T-cells is hastened by NK cell-regulation of IFN-I production (Robbins *et al.*, 2007). Thus, in the presence of Ly49H$^+$ NK cells, effective early control of MCMV reduces the amount of IFN-I produced by pDC in the first days of infection, promoting the generation of CD8 T-cells ~ 1 day earlier than in mice lacking these NK cells. The impact that this more rapid CD8 T-cell response has on MCMV replication still needs to be further defined, but these studies clearly show that NK cells can modulate CD8 T-cell responses through regulation of pDC. Further studies have shown that NK cells can impact the maintenance and function of both CD4 and CD8 MCMV-specific T-cells, impacting their control of persistent infection in the salivary glands (Andrews *et al.*, 2010). The improved T-cell responses generated in the absence of robust NK cell responses (i.e. in Ly49H$^-$ mouse strains) result from increased maintenance of infected APC, including cDC, leading to more priming and sustained recruitment of naïve T-cells to the effector pool. Although MCMV infection of DCs eventually leads to functional impairment, this data suggests that this is preceded by a phase of DC activation during which MCMV-infected DCs can efficiently prime autologous naïve T-cells (Andrews *et al.*, 2001; Mathys *et al.*, 2003; Benedict *et al.*, 2008). Furthermore, immature DCs are a principal target of MCMV infection and viral production (Andrews *et al.*, 2001). In mice lacking effective NK cell responses (Ly49H$^-$), MCMV-infected DCs are not efficiently eliminated and remain as a source of infectious viral progeny. By contrast, the prompt elimination of infected DCs that occurs in Ly49H$^+$ mice removes this source of infectious virus soon after infection and limits the duration of antiviral T-cell priming.

Enhanced antiviral CD8 T-cell responses have recently been reported in a model of MCMV infection where the Ly49G2 NK-cell inhibitory receptor mediates virus resistance (Stadnisky *et al.*, 2011). Improved CD8 T-cell responses were observed at day 6 after MCMV infection in Ly49G2$^+$ mice, but this difference was lost by day 8 when the *in vivo* cytotoxic potential of CD8 T-cells was assessed. These findings are therefore similar to the transiently improved CD8$^+$ T-cell response reported in Ly49H$^+$ mice by Robbins and colleagues, and might correlate with reduced IFN-I responses. Alternatively, it is possible that the impact of NK cell responses on adaptive immunity differs depending on whether stimulatory (Ly49H) versus inhibitory (Ly49G2) NK cell receptors mediate MCMV control. Again, it will be very informative to determine how temporary differences in the development of adaptive immunity ultimately impact control of MCMV infection.

Concluding remarks

We have attempted to recount over 30 years of immune-related research in the mouse model of CMV infection, and in doing so have had to emphasize specific results we believe to have driven the field to where it stands today. As the tools have advanced (e.g. BAC cloned MCMV genomes and genetically engineered mice), so has our understanding of the complex equilibrium that is required to maintain homeostatic immunity in mice faced with this natural pathogen. It is worth noting that two MCMV 'strains' are predominantly used in the field, Smith and K181 (see also Chapter I.2), and since K181 was selected as a more virulent 'variant' of Smith after *in vivo* passage it stands to reason that results obtained with these two strains may differ in some mouse genetic models (Smith, 1954; Misra and Hudson, 1980). In addition, different viral inocula can be used experimentally, the most common being 'tissue culture derived' and 'salivary gland derived' viral preparations. Since MCMV derived from fibroblasts has high levels of multicapsid virions that are not present in preparations from salivary gland acinar epithelial cells (can reach levels of ~90% of total virions, but likely varies between individual labs and with preparation technique) (Chong and Mims, 1981; Kurz *et al.*, 1997), this may also be an important issue to consider. As a specific example, our published experiments have utilized SG-derived virus to characterize the first-burst of IFN-I from splenic stromal cells. We have found that one passage of SG-derived virus in fibroblasts results in 10-fold lower levels of IFN-I being produced per PFU of virus following intraperitoneal injection, and we hypothesize this may be due to the inability of multicapsid particles to traverse the diaphragm after injection (Hsu *et al.*, 2009; Benedict *et al.* unpublished observations). Purification methods for MCMV preparations may also affect experimental results, but we postulate that this will principally impact very early innate responses. Importantly, it is worth noting that passaging viruses *in vitro* might generate progeny with differing infecting and/or spread capacities. This is clearly exemplified by early studies of HCMV, showing that the ability to infect DCs varies considerably when HCMV strains that were propagated in endothelial cell culture were compared with fibroblast-adapted HCMV strains (Riegler *et al.*, 2000). Furthermore, recent studies have shown that propagating MCMV in cell culture can select for viruses carrying inactivating mutations in MCK-2, a virally encoded chemokine protein required for efficient dissemination to in salivary glands (Jordan *et al.*, 2011). These differences should be considered when selecting viral preparations for experimentation and when interpreting past and future results (for further discussion, see also Chapter II.17).

As mentioned earlier, if MCMV effectively neutralizes a particular innate or adaptive immune response, it stands to reason that no phenotype may be observed when infecting mice that are genetically deficient for this pathway with WT MCMV. A specific example of this has come recently from studies using MCMV mutants in the M45 protein, a viral protein with homology to a ribonucleotide reductase that is essential for replication *in vivo* (Brune *et al.*, 2001). Upon analysis of M45, it was found that this protein displayed no reductase enzymatic activity (Lembo *et al.*, 2004), but instead bound to RIP family proteins (Mack *et al.*, 2008; Upton *et al.*, 2008; Rebsamen *et al.*, 2009). In turn, it was demonstrated that M45 binds RIP3 directly and inhibits RIP3-dependent 'necroptosis' in infected cells, and replication of a M45-deficient MCMV mutant is restored to WT levels in $RIP3^{-/-}$ mice. Finally, the identification of the M45–RIP3 interaction led to the elucidation of how RIP proteins and caspase-8 counterbalance each other to protect cells against necroptosis (Kaiser *et al.*, 2011; Oberst *et al.*, 2011; see also Chapter I.15). Taken together, this series of studies elegantly demonstrates how the combination of studying specific MCMV mutants in genetically deficient mice can reveal novel pathways regulating cell-intrinsic immunity.

As we move into the future, we feel that mouse models where fluorescent reporter genes are 'knocked-in' at a genomic locus to allow direct visualization of cells that express specific genes *in vivo* will be used more regularly. This is normally accomplished by inserting the reporter gene (e.g. GFP) in the 3′ untranslated region of a gene of interest through the use of an internal ribosome entry site (IRES), hoping not to disrupt the normal transcriptional regulation of the locus. While to date IFNβ and IL-10 reporter mice are the only ones that have been analysed in the context of MCMV infection (Scheu *et al.*, 2008; Madan *et al.*, 2009), other mice engineered to report expression of various cytokines, chemokines or antimicrobial molecules are becoming available and will undoubtedly be of great utility to identify the cell types involved in immune defence to MCMV. Most importantly, when engineered to express various fluorescent reporters in specific cell types, these mice will offer the opportunity to study the dynamics of antiviral MCMV immunity through *in vivo* real time imaging, as was recently performed in other infections (Kang *et al.*, 2008; Price *et al.*, 2010; Sullivan *et al.*, 2011). This approach has already been taken for the MCMV genome itself, where viral transcription of GFP is induced upon replicating in cells specifically expressing Cre (Sacher *et al.*, 2008). In addition, as 'conditional knockout' mice become more commonly available where floxed genes can be deleted in cell-type-specific Cre mice, as has been done for LTβ deleted specifically in B-cells, more information will be gained with regards to the key cells that provide important effector molecules. We feel that depletion of the IFN-I receptor, or downstream adaptors (e.g. IRF3), in specific cell types will provide very valuable information regarding the mechanism(s) regulated by this critical innate cytokine during MCMV infection.

Finally, we have tried to emphasize throughout this chapter the importance of considering mouse strain-specific issues that regulate innate and adaptive immunity. One relevant example might be recent studies that have shown invariant NK T-cells (iNKT) are robustly activated in response to MCMV infection (Tyznik *et al.*, 2008; Wesley *et al.*, 2008), but that iNKT appear to contribute minimally to controlling replication of MCMV, as was learned from genetically deficient B6 mice lacking these cells ($CD1d^{-/-}$ or $J\alpha18^{-/-}$ mice) (van Dommelen *et al.*, 2003; Wesley *et al.*, 2008). As discussed in details throughout, perhaps these cells would contribute to a greater extent in controlling early replication in immune competent mice if examined in the absence of Ly49H–m157 interactions. Taken together, the mouse has provided, and will continue to provide, key insight into host control of CMV infection. Given that this model is by far the most tractable for studying mechanisms of anti-CMV immunity, we feel it will provide many more key insights into immune control of this virus in the coming decades.

Acknowledgements

We would like to thank Shilpi Verma for figure assistance. Although we have attempted to discuss and reference as many studies as possible due to the breath of the topic covered the work of some colleagues might have not been included and for this we extend our apologies.

References

Adam, S.G., Caraux, A., Fodil-Cornu, N., Loredo-Osti, J.C., Lesjean-Pottier, S., Jaubert, J., Bubic, I., Jonjic, S., Guenet, J.L., Vidal, S.M., *et al.* (2006). Cmv4, a new locus linked to the NK cell gene complex, controls innate resistance to cytomegalovirus in wild-derived mice. J. Immunol. *176*, 5478–5485.

Allman, D., Dalod, M., Asselin-Paturel, C., Delale, T., Robbins, S.H., Trinchieri, G., Biron, C.A., Kastner, P., and Chan, S. (2006). Ikaros is required for plasmacytoid dendritic cell differentiation. Blood *108*, 4025–4034.

Andoniou, C.E., van Dommelen, S.L., Voigt, V., Andrews, D.M., Brizard, G., Asselin-Paturel, C., Delale, T., Stacey, K.J., Trinchieri, G., and Degli-Esposti, M.A. (2005). Interaction between conventional dendritic cells and natural killer cells is integral to the activation of effective antiviral immunity. Nat. Immunol. *6*, 1011–1019.

Andoniou, C.E., Fleming, P., Sutton, V.R., Trapani, J.A., and Degli-Esposti, M.A. (2011). Cathepsin C limits acute viral infection independently of NK cell and CD8+ T-cell cytolytic function. Immunol. Cell Biol. *89*, 540–548.

Andrews, D.M., Andoniou, C.E., Granucci, F., Ricciardi-Castagnoli, P., and Degli-Esposti, M.A. (2001). Infection of dendritic cells by murine cytomegalovirus induces functional paralysis. Nat. Immunol. *2*, 1077–1084.

Andrews, D.M., Scalzo, A.A., Yokoyama, W.M., Smyth, M.J., and Degli-Esposti, M.A. (2003). Functional interactions between dendritic cells and NK cells during viral infection. Nat. Immunol. *4*, 175–181.

Andrews, D.M., Estcourt, M.J., Andoniou, C.E., Wikstrom, M.E., Khong, A., Voigt, V., Fleming, P., Tabarias, H., Hill, G.R., van der Most, R.G., *et al.* (2010). Innate immunity defines the capacity of antiviral T-cells to limit persistent infection. J. Exp. Med. *207*, 1333–1343.

Arase, H., and Lanier, L.L. (2002). Virus-driven evolution of natural killer cell receptors. Microbes Infect. *4*, 1505–1512.

Arens, R., Loewendorf, A., Her, M.J., Schneider-Ohrum, K., Shellam, G.R., Janssen, E., Ware, C.F., Schoenberger, S.P., and Benedict, C.A. (2011a). B7-mediated costimulation of CD4 T-cells constrains cytomegalovirus persistence. J. Virol. *85*, 390–396.

Arens, R., Loewendorf, A., Redeker, A., Sierro, S., Boon, L., Klenerman, P., Benedict, C.A., and Schoenberger, S.P. (2011b). Differential B7-CD28 costimulatory requirements for stable and inflationary mouse cytomegalovirus-specific memory CD8 T-cell populations. J. Immunol. *186*, 3874–3881.

Arens, R., Wang, P., Sidney, J., Loewendorf, A., Sette, A., Schoenberger, S.P., Peters, B., and Benedict, C.A. (2008). Cutting edge: murine cytomegalovirus induces a polyfunctional CD4 T-cell response. J. Immunol. *180*, 6472–6476.

Babic, M., Pyzik, M., Zafirova, B., Mitrovic, M., Butorac, V., Lanier, L.L., Krmpotic, A., Vidal, S.M., and Jonjic, S. (2010). Cytomegalovirus immunoevasin reveals the physiological role of 'missing self' recognition in natural killer cell dependent virus control *in vivo*. J. Exp. Med. *207*, 2663–2673.

Bachmann, M.F., Wolint, P., Walton, S., Schwarz, K., and Oxenius, A. (2007). Differential role of IL-2R signaling for CD8+ T-cell responses in acute and chronic viral infections. Eur. J. Immunol. *37*, 1502–1512.

Bancroft, G.J., Shellam, G.R., and Chalmer, J.E. (1981). Genetic influences on the augmentation of natural killer (NK) cells during murine cytomegalovirus infection: correlation with patterns of resistance. J. Immunol. *126*, 988–994.

Bani-Ahmad, M., El-Amouri, I.S., Ko, C.M., Lin, F., Tang-Feldman, Y., and Oakley, O.R. (2011). The role of decay accelerating factor in the immunopathogenesis of cytomegalovirus infection. Clin. Exp. Immunol. *163*, 199–206.

Banks, T.A., Rickert, S., Benedict, C.A., Ma, L., Ko, M., Meier, J., Ha, W., Schneider, K., Granger, S.W., Turovskaya, O., *et al.* (2005). A lymphotoxin-IFN-beta axis essential for lymphocyte survival revealed during cytomegalovirus infection. J. Immunol. *174*, 7217–7225.

Barbalat, R., Lau, L., Locksley, R.M., and Barton, G.M. (2009). Toll-like receptor 2 on inflammatory monocytes induces type I interferon in response to viral but not bacterial ligands. Nat. Immunol. *10*, 1200–1207.

Barnes, M.J., Aksoylar, H., Krebs, P., Bourdeau, T., Arnold, C.N., Xia, Y., Khovananth, K., Engel, I., Sovath, S., Lampe, K., *et al.* (2010). Loss of T-cell and B cell quiescence precedes the onset of microbial flora-dependent wasting disease and intestinal inflammation in Gimap5-deficient mice. J. Immunol. *184*, 3743–3754.

Benedict, C.A., Banks, T.A., Senderowicz, L., Ko, M., Britt, W.J., Angulo, A., Ghazal, P., and Ware, C.F. (2001). Lymphotoxins and cytomegalovirus cooperatively induce interferon-beta, establishing host-virus detente. Immunity *15*, 617–626.

Bekiaris, V., Gaspal, F., McConnell, F.M., Kim, M.Y., Withers, D.R., Sweet, C., Anderson, G., and Lane, P.J. (2009). NK cells protect secondary lymphoid tissue from cytomegalovirus via a CD30-dependent mechanism. Eur. J. Immunol. *39*, 2800–2808.

Benedict, C.A., De Trez, C., Schneider, K., Ha, S., Patterson, G., and Ware, C.F. (2006). Specific remodeling of splenic architecture by cytomegalovirus. PLoS Pathog. *2*, e16.

Benedict, C.A., Loewendorf, A., Garcia, Z., Blazar, B.R., and Janssen, E.M. (2008). Dendritic cell programming by cytomegalovirus stunts naïve T-cell responses via the PD-L1/PD-1 pathway. J. Immunol. *180*, 4836–4847.

Berger, M., Krebs, P., Crozat, K., Li, X., Croker, B.A., Siggs, O.M., Popkin, D., Du, X., Lawson, B.R., Theofilopoulos, A.N., *et al.* (2010). An Slfn2 mutation causes lymphoid and myeloid immunodeficiency due to loss of immune cell quiescence. Nat. Immunol. *11*, 335–343.

Beutler, B., Georgel, P., Rutschmann, S., Jiang, Z., Croker, B., and Crozat, K. (2005). Genetic analysis of innate resistance to mouse cytomegalovirus (MCMV). Brief Funct. Ggenomic. Proteomic. *4*, 203–213.

Blanc, M., Hsieh, W.Y., Robertson, K.A., Watterson, S., Shui, G., Lacaze, P., Khondoker, M., Dickinson, P., Sing, G., Rodriguez-Martin, S., *et al.* (2011). Host defense against viral infection involves interferon mediated down-regulation of sterol biosynthesis. PLoS Biol. *9*, e1000598.

Blasius, A.L., Arnold, C.N., Georgel, P., Rutschmann, S., Xia, Y., Lin, P., Ross, C., Li, X., Smart, N.G., and Beutler, B. (2010). Slc15a4, AP-3, and Hermansky-Pudlak syndrome proteins are required for Toll-like receptor signaling in plasmacytoid dendritic cells. Proc. Natl. Acad. Sci. U.S.A. *107*, 19973–19978.

Brown, M.G., Dokum, A.O., Heusel, J.W., Smith, H.R., Beckman, D.L., Blattenberger, E.A., Dubbelde, C.E., Stone, L.R., Sclazo, A.A., and Yokoyama, W.M. (2001). Vital involvement of a natural killer cell activation receptor in resistance to viral infection. Science *292*, 934–937.

Brune, W., Messerle, M., and Koszinowski, U.H. (2000). Forward with BACs: new tools for herpesvirus genomics. Trends Genet. *16*, 254–259.

Brune, W., Menard, C., Heesemann, J., and Koszinowski, U.H. (2001). A ribonucleotide reductase homolog of cytomegalovirus and endothelial cell tropism. Science *291*, 303–305.

Bubic, I., Wagner, M., Krmpotic, A., Saulig, T., Kim, S., Yokoyama, W.M., Jonjic, S., and Koszinowski, U.H. (2004). Gain of virulence caused by loss of a gene in murine cytomegalovirus. J. Virol. *78*, 7536–7544.

Bukowski, J.F., Woda, B.A., Habu, S., Okumura, K., and Welsh, R.M. (1983). Natural killer cell depletion enhances virus synthesis and virus-induced hepatitis *in vivo*. J. Immunol. *131*, 1531–1538.

Cekinovic, D., Golemac, M., Pugel, E.P., Tomac, J., Cicin-Sain, L., Slavuljica, I., Bradford, R., Misch, S., Winkler, T.H., Mach, M., *et al.* (2008). Passive immunization reduces murine cytomegalovirus-induced brain pathology in newborn mice. J. Virol. *82*, 12172–12180.

Chalmer, J.E., Mackenzie, J.S., and Stanley, N.F. (1977). Resistance to murine cytomegalovirus linked to the major histocompatibility complex of the mouse. J. Gen. Virol. *37*, 107–114.

Cheng, T.P., French, A.R., Plougastel, B.F., Pingel, J.T., Orihuela, M.M., Buller, M.L., and Yokoyama, W.M. (2008). Ly49h is necessary for genetic resistance to murine cytomegalovirus. Immunogenetics *60*, 565–573.

Chong, K.T., and Mims, C.A. (1981). Murine cytomegalovirus particle types in relation to sources of virus and pathogenicity. J. Gen. Virol. *57*, 415–419.

Cohen, P.L., and Eisenberg, R.A. (1991). Lpr and gld: single gene models of systemic autoimmunity and lymphoproliferative disease. Annu. Rev. Immunol. *9*, 243–269.

Cook, C.H., Chen, L., Wen, J., Zimmerman, P., Zhang, Y., Trgovcich, J., Liu, Y., and Gao, J.X. (2009). CD28/B7-mediated co-stimulation is critical for early control of murine cytomegalovirus infection. Viral Immunol. *22*, 91–103.

Crane, M.J., Hokeness-Antonelli, K.L., and Salazar-Mather, T.P. (2009). Regulation of inflammatory monocyte/macrophage recruitment from the bone marrow during murine cytomegalovirus infection: role for type I interferons in localized induction of CCR2 ligands. J. Immunol. *183*, 2810–2817.

Croft, M. (2003). Costimulation of T-cells by OX40, 4-1BB, and CD27. Cytokine Growth Factor Rev. *14*, 265–273.

Croker, B., Crozat, K., Berger, M., Xia, Y., Sovath, S., Schaffer, L., Eleftherianos, I., Imler, J.L., and Beutler, B. (2007). ATP-sensitive potassium channels mediate survival during infection in mammals and insects. Nat. Genet. *39*, 1453–1460.

Crozat, K., Georgel, P., Rutschmann, S., Mann, N., Du, X., Hoebe, K., and Beutler, B. (2006). Analysis of the MCMV resistome by ENU mutagenesis. Mamm. Genome. *17*, 398–406.

Crozat, K., Hoebe, K., Ugolini, S., Hong, N.A., Janssen, E., Rutschmann, S., Mudd, S., Sovath, S., Vivier, E., and Beutler, B. (2007). Jinx, an MCMV susceptibility phenotype caused by disruption of Unc13d: a mouse model of type 3 familial hemophagocytic lymphohistiocytosis. J. Exp. Med. *204*, 853–863.

Dalod, M., Salazar-Mather, T.P., Malmgaard, L., Lewis, C., Asselin-Paturel, C., Briere, F., Trinchieri, G., and Biron, C.A. (2002). Interferon alpha/beta and interleukin 12 responses to viral infections: pathways regulating dendritic cell cytokine expression *in vivo*. J. Exp. Med. *195*, 517–528.

DeFilippis, V.R., Alvarado, D., Sali, T., Rothenburg, S., and Fruh, K. (2010). Human cytomegalovirus induces the interferon response via the DNA sensor ZBP1. J. Virol. *84*, 585–598.

Del Val, M., Volkmer, H., Rothbard, J.B., Jonjic, S., Messerle, M., Schickedanz, J., Reddehase, M.J., and Koszinowski, U.H. (1988). Molecular basis for cytolytic T-lymphocyte recognition of the murine cytomegalovirus immediate-early protein pp89. J. Virol. *62*, 3965–3972.

Delale, T., Paquin, A., Asselin-Paturel, C., Dalod, M., Brizard, G., Bates, E.E., Kastner, P., Chan, S., Akira, S., Vicari, A., *et al.* (2005). MyD88-dependent and -independent murine cytomegalovirus sensing for IFN-alpha release and initiation of immune responses *in vivo*. J. Immunol. *175*, 6723–6732.

Desrosiers, M.P., Kielczewska, A., Loredo-Osti, J.C., Adam, S.G., Makrigiannis, A.P., Lemieux, S., Pham, T., Lodoen, M.B., Morgan, K., Lanier, L.L., *et al.* (2005). Epistasis between mouse Klra and major histocompatibility complex class I loci is associated with a new mechanism of natural killer cell-mediated innate resistance to cytomegalovirus infection. Nat. Genet. *37*, 593–599.

Diehl, G.E., Yue, H.H., Hsieh, K., Kuang, A.A., Ho, M., Morici, L.A., Lenz, L.L., Cado, D., Riley, L.W., and Winoto, A. (2004). TRAIL-R as a negative regulator of innate immune cell responses. Immunity *21*, 877–889.

Dighe, A., Rodriguez, M., Sabastian, P., Xie, X., McVoy, M., and Brown, M.G. (2005). Requisite H2k role in NK cell-mediated resistance in acute murine cytomegalovirus-infected MA/My mice. J. Immunol. *175*, 6820–6828.

Dokun, A.O., Kim, S., Smith, H.R., Kang, H.S., Chu, D.T., and Yokoyama, W.M. (2001). Specific and nonspecific NK cell activation during virus infection. Nat. Immunol. *2*, 951–956.

van Dommelen, S.L., Tabarias, H.A., Smyth, M.J., and Degli-Esposti, M.A. (2003). Activation of natural killer (NK) T-cells during murine cytomegalovirus infection enhances the antiviral response mediated by NK cells. J. Virol. *77*, 1877–1884.

van Dommelen, S.L., Sumaria, N., Schreiber, R.D., Scalzo, A.A., Smyth, M.J., and Degli-Esposti, M.A. (2006). Perforin and granzymes have distinct roles in defensive immunity and immunopathology. Immunity *25*, 835–848.

Durbin, J.E., Hackenmiller, R., Simon, M.C., and Levy, D.E. (1996). Targeted disruption of the mouse Stat1 gene results in compromised innate immunity to viral disease. Cell *84*, 443–450.

Edelmann, K.H., Richardson-Burns, S., Alexopoulou, L., Tyler, K.L., Flavell, R.A., and Oldstone, M.B. (2004). Does Toll-like receptor 3 play a biological role in virus infections? Virology *322*, 231–238.

Eidenschenk, C., Crozat, K., Krebs, P., Arens, R., Popkin, D., Arnold, C.N., Blasius, A.L., Benedict, C.A., Moresco, E.M., Xia, Y., *et al.* (2010). Flt3 permits survival during infection by rendering dendritic cells competent to activate NK cells. Proc. Natl. Acad. Sci. U.S.A. *107*, 9759–9764.

Fernandez, J.A., Rodrigues, E.G., and Tsuji, M. (2000). Multifactorial protective mechanisms to limit viral replication in the lung of mice during primary murine cytomegalovirus infection. Viral Immunol. *13*, 287–295.

Fleck, M., Kern, E.R., Zhou, T., Podlech, J., Wintersberger, W., Edwards, C.K., and Mountz, J.D. (1998). Apoptosis mediated by Fas but not tumor necrosis factor receptor 1 prevents chronic disease in mice infected with murine cytomegalovirus. J. Clin. Invest. *102*, 1431–1443.

Fodil-Cornu, N., Lee, S.H., Belanger, S., Makrigiannis, A.P., Biron, C.A., Buller, R.M., and Vidal, S.M. (2008). Ly49h-deficient C57BL/6 mice: a new mouse cytomegalovirus-susceptible model remains resistant to unrelated pathogens controlled by the NK gene complex. J. Immunol. *181*, 6394–6405.

Fodil-Cornu, N., Kozij, N., Wu, Q., Rozen, R., and Vidal, S.M. (2009). Methylenetetrahydrofolate reductase (MTHFR) deficiency enhances resistance against cytomegalovirus infection. Genes Immun. *10*, 662–666.

Fodil-Cornu, N., Loredo-Osti, J.C., and Vidal, S.M. (2011). NK cell receptor/H2-Dk-dependent host resistance to viral infection is quantitatively modulated by H2q inhibitory signals. PLoS Genetics *7*, e1001368.

French, A.R., Pingel, J.T., Wagner, M., Bubic, I., Yang, L., Kim, S., Koszinowski, U., Jonjic, S., and Yokoyama, W.M. (2004). Escape of mutant double-stranded DNA virus from innate immune control. Immunity *20*, 747–756.

Geurs, T.L., Zhao, Y.M., Hill, E.B., and French, A.R. (2009). Ly49H engagement compensates for the absence of type I interferon signaling in stimulating NK cell proliferation during murine cytomegalovirus infection. J. Immunol. *183*, 5830–5836.

Gold, M.C., Munks, M.W., Wagner, M., Koszinowski, U.H., Hill, A.B., and Fling, S.P. (2002). The murine cytomegalovirus immunomodulatory gene m152 prevents recognition of infected cells by M45-specific CTL but does not alter the immunodominance of the M45-specific CD8 T-cell response *in vivo*. J. Immunol. *169*, 359–365.

Grundy, J.E., Mackenzie, J.S., and Stanley, N.F. (1981). Influence of H-2 and non-H-2 genes on resistance to murine cytomegalovirus infection. Infect. Immun. *32*, 277–286.

Grundy, J.E., Trapman, J., Allan, J.E., Shellam, G.R., and Melief, C.J. (1982). Evidence for a protective role of interferon in resistance to murine cytomegalovirus and its control by non-H-2-linked genes. Infect. Immun. *37*, 143–150.

Hanson, L.K., Slater, J.S., Karabekian, Z., Virgin, H.W.T., Biron, C.A., Ruzek, M.C., van Rooijen, N., Ciavarra, R.P., Stenberg, R.M., and Campbell, A.E. (1999). Replication of murine cytomegalovirus in differentiated macrophages as a determinant of viral pathogenesis. J. Virol. *73*, 5970–5980.

Hoebe, K., Du, X., Georgel, P., Janssen, E., Tabeta, K., Kim, S.O., Goode, J., Lin, P., Mann, N., Mudd, S., *et al.* (2003). Identification of Lps2 as a key transducer of MyD88-independent TIR signalling. Nature *424*, 743–748.

Hokeness, K.L., Deweerd, E.S., Munks, M.W., Lewis, C.A., Gladue, R.P., and Salazar-Mather, T.P. (2007). CXCR3-dependent recruitment of antigen-specific T-lymphocytes to the liver during murine cytomegalovirus infection. J. Virol. *81*, 1241–1250.

Holtappels, R., Pahl-Seibert, M.F., Thomas, D., and Reddehase, M.J. (2000). Enrichment of immediate-early 1 (m123/pp89) peptide-specific CD8 T-cells in a pulmonary CD62L(lo) memory-effector cell pool during latent murine cytomegalovirus infection of the lungs. J. Virol. *74*, 11495–11503.

Holtappels, R., Thomas, D., Podlech, J., and Reddehase, M.J. (2002). Two antigenic peptides from genes m123 and m164 of murine cytomegalovirus quantitatively dominate CD8 T-cell memory in the H-2d haplotype. J. Virol. *76*, 151–164.

Holtappels, R., Bohm, V., Podlech, J., and Reddehase, M.J. (2008). CD8 T-cell-based immunotherapy of cytomegalovirus infection: 'proof of concept' provided by the murine model. Med. Microbiol. Immunol. *197*, 125–134.

Hsu, K.M., Pratt, J.R., Akers, W.J., Achilefu, S.I., and Yokoyama, W.M. (2009). Murine cytomegalovirus displays selective infection of cells within hours after systemic administration. J. Gen. Virol. *90*, 33–43.

Humphreys, I.R., de Trez, C., Kinkade, A., Benedict, C.A., Croft, M., and Ware, C.F. (2007a). Cytomegalovirus exploits IL-10-mediated immune regulation in the salivary glands. J. Exp. Med. *204*, 1217–1225.

Humphreys, I.R., Loewendorf, A., de Trez, C., Schneider, K., Benedict, C.A., Munks, M.W., Ware, C.F., and Croft, M. (2007b). OX40 costimulation promotes persistence of cytomegalovirus-specific CD8 T-cells: A CD4-dependent mechanism. J. Immunol. *179*, 2195–2202.

Humphreys, I.R., Lee, S.W., Jones, M., Loewendorf, A., Gostick, E., Price, D.A., Benedict, C.A., Ware, C.F., and Croft, M. (2010). Biphasic role of 4–1BB in the regulation of mouse cytomegalovirus-specific CD8(+) T-cells. Eur. J. Immunol. *40*, 2762–2768.

Hutchinson, S., Sims, S., O'Hara, G., Silk, J., Gileadi, U., Cerundolo, V., and Klenerman, P. (2011). A dominant role for the immunoproteasome in CD8+ T-cell responses to murine cytomegalovirus. PLoS One *6*, e14646.

Jones, M., Ladell, K., Wynn, K.K., Stacey, M.A., Quigley, M.F., Gostick, E., Price, D.A., and Humphreys, I.R. (2010). IL-10 restricts memory T-cell inflation during cytomegalovirus infection. J. Immunol. *185*, 3583–3592.

Jonjic, S., Mutter, W., Weiland, F., Reddehase, M.J., and Koszinowski, U.H. (1989). Site-restricted persistent cytomegalovirus infection after selective long-term depletion of CD4+ T-lymphocytes. J. Exp. Med. *169*, 1199–1212.

Jonjic, S., Pavic, I., Lucin, P., Rukavina, D., and Koszinowski, U.H. (1990). Efficacious control of cytomegalovirus infection after long-term depletion of CD8+ T-lymphocytes. J. Virol. *64*, 5457–5464.

Jonjic, S., Pavic, I., Polic, B., Crnkovic, I., Lucin, P., and Koszinowski, U.H. (1994). Antibodies are not essential for the resolution of primary cytomegalovirus infection but limit dissemination of recurrent virus. J. Exp. Med. *179*, 1713–1717.

Jordan, S., Krause, J., Prager, A., Mitrovic, M., Jonjic, S., Koszinowski, U.H., and Adler, B. (2011). Virus progeny of murine cytomegalovirus bacterial artificial chromosome pSM3fr show reduced growth in salivary Glands due to a fixed mutation of MCK-2. J. Virol. *85*, 10346–10353.

Kaiser, W.J., Upton, J.W., Long, A.B., Livingston-Rosanoff, D., Daley-Bauer, L.P., Hakem, R., Caspary, T., and Mocarski, E.S. (2011). RIP3 mediates the embryonic lethality of caspase-8-deficient mice. Nature *471*, 368–372.

Kane, G.C., Lam, C.F., O'Cochlain, F., Hodgson, D.M., Reyes, S., Liu, X.K., Miki, T., Seino, S., Katusic, Z.S., and Terzic, A. (2006). Gene knockout of the KCNJ8-encoded Kir6.1 K(ATP) channel imparts fatal susceptibility to endotoxemia. FASEB J. *20*, 2271–2280.

Kang, S.J., Liang, H.E., Reizis, B., and Locksley, R.M. (2008). Regulation of hierarchical clustering and activation of innate immune cells by dendritic cells. Immunity *29*, 819–833.

Karrer, U., Sierro, S., Wagner, M., Oxenius, A., Hengel, H., Koszinowski, U.H., Phillips, R.E., and Klenerman, P. (2003). Memory inflation: continuous accumulation of antiviral CD8+ T-cells over time. J. Immunol. *170*, 2022–2029. Published correction in J. Immunol. *171*, 3895.

Kielczewska, A., Pyzik, M., Sun, T., Krmpotic, A., Lodoen, M.B., Munks, M.W., Babic, M., Hill, A.B., Koszinowski, U.H., Jonjic, S., *et al.* (2009). Ly49P recognition of cytomegalovirus-infected cells expressing H2-Dk and CMV-encoded m04 correlates with the NK cell antiviral response. J. Exp. Med. *206*, 515–523.

Klenovsek, K., Weisel, F., Schneider, A., Appelt, U., Jonjic, S., Messerle, M., Bradel-Tretheway, B., Winkler, T.H., and Mach, M. (2007). Protection from CMV infection in immunodeficient hosts by adoptive transfer of memory B cells. Blood *110*, 3472–3479.

Krebs, P., Crozat, K., Popkin, D., Oldstone, M.B., and Beutler, B. (2011). Disruption of MyD88 signaling suppresses hemophagocytic lymphohistiocytosis in mice. Blood *117*, 6582–6588.

Krug, A., French, A.R., Barchet, W., Fischer, J.A., Dzionek, A., Pingel, J.T., Orihuela, M.M., Akira, S., Yokoyama, W.M., and Colonna, M. (2004). TLR9-dependent recognition of MCMV by IPC and DC generates coordinated cytokine responses that activate antiviral NK cell function. Immunity *21*, 107–119.

Kurz, S., Steffens, H.P., Mayer, A., Harris, J.R., and Reddehase, M.J. (1997). Latency versus persistence or intermittent recurrences: evidence for a latent state of murine cytomegalovirus in the lungs. J. Virol. *71*, 2980–2987.

Lathbury, L.J., Allan, J.E., Shellam, G.R., and Scalzo, A.A. (1996). Effect of host genotype in determining the relative roles of natural killer cells and T-cells in mediating protection against murine cytomegalovirus infection. J. Gen. Virol. *77*, 2605–2613.

Lee, S.H., Zafer, A., de Repentigny, Y., Kothary, R., Tremblay, M.L., Gros, P., Duplay, P., Webb, J.R., and Vidal, S.M. (2003). Transgenic expression of the activating natural killer receptor Ly49H confers resistance to cytomegalovirus in genetically susceptible mice. J. Exp. Med. *197*, 515–526.

Lee, S.H., Kim, K.S., Fodil-Cornu, N., Vidal, S.M., and Biron, C.A. (2009). Activating receptors promote NK cell expansion for maintenance, IL-10 production, and CD8 T-cell regulation during viral infection. J. Exp. Med. *206*, 2235–2251.

Lee, S.H., Girard, S., Macina, D., Busa, M., Zafer, A., Belouchi, A., Gros, P., and Vidal, S.M. (2001). Susceptibility to mouse cytomegalovirus is associated with deletion of an activating natural killer cell receptor of the C-type lectin superfamily. Nat. Genet. *28*, 42–45.

Lembo, D., Donalisio, M., Hofer, A., Cornaglia, M., Brune, W., Koszinowski, U., Thelander, L., and Landolfo, S. (2004). The ribonucleotide reductase R1 homolog of murine cytomegalovirus is not a functional enzyme subunit but is required for pathogenesis. J. Virol. *78*, 4278–4288.

Loewendorf, A., and Benedict, C.A. (2010). Modulation of host innate and adaptive immune defenses by cytomegalovirus: Timing is everything. J. Intern. Med. *267*, 483–501.

Loh, J., Chu, D.T., O'Guin, A.K., Yokoyama, W.M., and Virgin, H.W. (2005). Natural killer cells utilize both perforin and gamma interferon to regulate murine cytomegalovirus infection in the spleen and liver. J. Virol. *79*, 661–667.

Lucin, P., Pavic, I., Polic, B., Jonjic, S., and Koszinowski, U.H. (1992). Gamma interferon-dependent clearance of cytomegalovirus infection in salivary glands. J. Virol. *66*, 1977–1984.

McGeoch, D.J., Dolan, A., and Ralph, A.C. (2000). Toward a comprehensive phylogeny for mammalian and avian herpesviruses. J. Virol. *74*, 10401–10406.

Mack, C., Sickmann, A., Lembo, D., and Brune, W. (2008). Inhibition of proinflammatory and innate immune signaling pathways by a cytomegalovirus RIP1-interacting protein. Proc. Natl. Acad. Sci. U.S.A. *105*, 3094–3099.

Madan, R., Demircik, F., Surianarayanan, S., Allen, J.L., Divanovic, S., Trompette, A., Yogev, N., Gu, Y., Khodoun, M., Hildeman, D., *et al.* (2009). Nonredundant roles for B cell-derived IL-10 in immune counter-regulation. J. Immunol. *183*, 2312–2320.

Mashimo, T., Lucas, M., Simon-Chazottes, D., Frenkiel, M.P., Montagutelli, X., Ceccaldi, P.E., Deubel, V., Guenet, J.L., and Despres, P. (2002). A nonsense mutation in the gene encoding 2′–5′-oligoadenylate synthetase/L1 isoform is associated with West Nile virus susceptibility in laboratory mice. Proc. Natl. Acad. Sci. U.S.A. *99*, 11311–11316.

Mathys, S., Schroeder, T., Ellwart, J., Koszinowski, U.H., Messerle, M., and Just, U. (2003). Dendritic cells under influence of mouse cytomegalovirus have a physiologic dual role: to initiate and to restrict T-cell activation. J. Infect. Dis. *187*, 988–999.

Menasche, G., Pastural, E., Feldmann, J., Certain, S., Ersoy, F., Dupuis, S., Wulffraat, N., Bianchi, D., Fischer, A., Le Deist, F., *et al.* (2000). Mutations in RAB27A cause Griscelli syndrome associated with haemophagocytic syndrome. Nat. Genet. *25*, 173–176.

Mercer, J.A., Wiley, C.A., and Spector, D.H. (1988). Pathogenesis of murine cytomegalovirus infection: identification of infected cells in the spleen during acute and latent infections. J. Virol. *62*, 987–997.

Misra, V., and Hudson, J.B. (1980). Minor base sequence differences between the genomes of two strains of murine cytomegalovirus differing in virulence. Arch. Virol. *64*, 1–8.

Miyake, T., Satoh, T., Kato, H., Matsushita, K., Kumagai, Y., Vandenbon, A., Tani, T., Muta, T., Akira, S., and Takeuchi, O. (2010). IkappaBzeta is essential for natural killer cell activation in response to IL-12 and IL-18. Proc. Natl. Acad. Sci. U.S.A. *107*, 17680–17685.

Mohr, C.A., Arapovic, J., Muhlbach, H., Panzer, M., Weyn, A., Dolken, L., Krmpotic, A., Voehringer, D., Ruzsics, Z., Koszinowski, U., *et al.* (2010). A spread-deficient cytomegalovirus for assessment of first-target cells in vaccination. J. Virol. *84*, 7730–7742.

Munks, M.W., Cho, K.S., Pinto, A.K., Sierro, S., Klenerman, P., and Hill, A.B. (2006a). Four distinct patterns of memory CD8 T-cell responses to chronic murine cytomegalovirus infection. J. Immunol. *177*, 450–458.

Munks, M.W., Gold, M.C., Zajac, A.L., Doom, C.M., Morello, C.S., Spector, D.H., and Hill, A.B. (2006b). Genome-wide analysis reveals a highly diverse CD8 T-cell response to murine cytomegalovirus. J. Immunol. *176*, 3760–3766.

Munks, M.W., Pinto, A.K., Doom, C.M., and Hill, A.B. (2007). Viral interference with antigen presentation does not alter acute or chronic CD8 T-cell immunodominance in murine cytomegalovirus infection. J. Immunol. *178*, 7235–7241.

Nguyen, K.B., Salazar-Mather, T.P., Dalod, M.Y., Van Deusen, J.B., Wei, X.Q., Liew, F.Y., Caligiuri, M.A., Durbin, J.E., and Biron, C.A. (2002). Coordinated and distinct roles for IFN-alpha beta, IL-12, and IL-15 regulation of NK cell responses to viral infection. J. Immunol. *169*, 4279–4287.

Nigro, G., Adler, S.P., La Torre, R., and Best, A.M. (2005). Passive immunization during pregnancy for congenital cytomegalovirus infection. New Engl. J. Med. *353*, 1350–1362.

Noda, S., Tanaka, K., Sawamura, S., Sasaki, M., Matsumoto, T., Mikami, K., Aiba, Y., Hasegawa, H., Kawabe, N., and Koga, Y. (2001). Role of nitric oxide synthase type 2 in acute infection with murine cytomegalovirus. J. Immunol. *166*, 3533–3541.

Oberst, A., Dillon, C.P., Weinlich, R., McCormick, L.L., Fitzgerald, P., Pop, C., Hakem, R., Salvesen, G.S., and Green, D.R. (2011). Catalytic activity of the caspase-8-FLIP(L) complex inhibits RIPK3-dependent necrosis. Nature *471*, 363–367.

Orange, J.S., and Biron, C.A. (1996a). Characterization of early IL-12, IFN-alpha beta, and TNF effects on antiviral state and NK cell responses during murine cytomegalovirus infection. J. Immunol. *156*, 4746–4756.

Orange, J.S., and Biron, C.A. (1996b). An absolute and restricted requirement for IL-12 in natural killer cell IFN-gamma production and antiviral defense. Studies of natural killer and T-cell responses in contrasting viral infections. J. Immunol. *156*, 1138–1142.

Orange, J.S., Wang, B., Terhorst, C., and Biron, C.A. (1995). Requirement for natural killer cell-produced interferon gamma in defense against murine cytomegalovirus infection and enhancement of this defense pathway by interleukin 12 administration. J. Exp. Med. *182*, 1045–1056.

Orr, M.T., Sun, J.C., Hesslein, D.G., Arase, H., Phillips, J.H., Takai, T., and Lanier, L.L. (2009). Ly49H signaling through DAP10 is essential for optimal natural killer cell responses to mouse cytomegalovirus infection. J. Exp. Med. *206*, 807–817.

Pien, G.C., Satoskar, A.R., Takeda, K., Akira, S., and Biron, C.A. (2000). Cutting edge: selective IL-18 requirements for induction of compartmental IFN-gamma responses during viral infection. J. Immunol. *165*, 4787–4791.

Podlech, J., Holtappels, R., Wirtz, N., Steffens, H.P., and Reddehase, M.J. (1998). Reconstitution of CD8 T-cells is essential for the prevention of multiple-organ cytomegalovirus histopathology after bone marrow transplantation. J. Gen. Virol. *79*, 2099–2104.

Polic, B., Jonjic, S., Pavic, I., Crnkovic, I., Zorica, I., Hengel, H., Lucin, P., and Koszinowski, U.H. (1996). Lack of MHC class I complex expression has no effect on spread and control of cytomegalovirus infection *in vivo*. J. Gen. Virol. *77*, 217–225.

Polic, B., Hengel, H., Krmpotic, A., Trgovcich, J., Pavic, I., Luccaronin, P., Jonjic, S., and Koszinowski, U.H. (1998). Hierarchical and redundant lymphocyte subset control precludes cytomegalovirus replication during latent infection. J. Exp. Med. *188*, 1047–1054.

Presti, R.M., Pollock, J.L., Dal Canto, A.J., O'Guin, A.K., and Virgin, H.W.T. (1998). Interferon gamma regulates acute and latent murine cytomegalovirus infection and chronic disease of the great vessels. J. Exp. Med. *188*, 577–588.

Price, A.E., Liang, H.E., Sullivan, B.M., Reinhardt, R.L., Eisley, C.J., Erle, D.J., and Locksley, R.M. (2010). Systemically dispersed innate IL-13-expressing cells in type 2 immunity. Proc. Natl. Acad. Sci. U.S.A. *107*, 11489–11494.

Pyzik, M., Gendron-Pontbriand, E.M., and Vidal, S.M. (2011). The impact of Ly49-NK cell-dependent recognition of MCMV infection on innate and adaptive immune responses. J. Biomed. Biotechnol. *2011*, 641702.

Rebsamen, M., Heinz, L.X., Meylan, E., Michallet, M.C., Schroder, K., Hofmann, K., Vazquez, J., Benedict, C.A., and Tschopp, J. (2009). DAI/ZBP1 recruits RIP1 and RIP3 through RIP homotypic interaction motifs to activate NF-kappaB. EMBO Rep. *10*, 916–922.

Reddehase, M.J. (2002). Antigens and immunoevasins: opponents in cytomegalovirus immune surveillance. Nat. Rev. Immunol. *2*, 831–844.

Reddehase, M.J., Rothbard, J.B., and Koszinowski, U.H. (1989). A pentapeptide as minimal antigenic determinant for MHC class I-restricted T-lymphocytes. Nature *337*, 651–653.

Reddehase, M.J., Weiland, F., Munch, K., Jonjic, S., Luske, A., and Koszinowski, U.H. (1985). Interstitial murine cytomegalovirus pneumonia after irradiation: characterization of cells that limit viral replication during established infection of the lungs. J. Virol. *55*, 264–273.

Reddehase, M.J., Mutter, W., Munch, K., Buhring, H.J., and Koszinowski, U.H. (1987). CD8-positive T-lymphocytes specific for murine cytomegalovirus immediate-early antigens mediate protective immunity. J. Virol. *61*, 3102–3108.

Reddehase, M.J., Balthesen, M., Rapp, M., Jonjic, S., Pavic, I., and Koszinowski, U.H. (1994). The conditions of primary infection define the load of latent viral genome in organs and the risk of recurrent cytomegalovirus disease. J. Exp. Med. *179*, 185–193.

Riddell, S.R., Warren, E.H., Lewinsohn, D., Mutimer, H., Topp, M., Cooper, L., de Fries, R., and Greenberg, P.D. (2000). Application of T-cell immunotherapy for human viral and malignant diseases. In Ernst Schering Res Found Workshop, pp. 53–73.

Riegler, S., Hebart, H., Einsele, H., Brossart, P., Jahn, G., and Sinzger, C. (2000). Monocyte-derived dendritic cells are permissive to the complete replicative cycle of human cytomegalovirus. J. Gen. Virol. *81*, 393–399.

Robbins, S.H., Bessou, G., Cornillon, A., Zucchini, N., Rupp, B., Ruzsics, Z., Sacher, T., Tomasello, E., Vivier, E., Koszinowski, U.H., *et al.* (2007). Natural killer cells promote early CD8 T-cell responses against cytomegalovirus. PLoS Pathog. *3*, e123.

Rodriguez, M., Sabastian, P., Clark, P., and Brown, M.G. (2004). Cmv1-independent antiviral role of NK cells revealed in murine cytomegalovirus-infected New Zealand White mice. J. Immunol. *173*, 6312–6318.

Rodriguez, M.R., Lundgren, A., Sabastian, P., Li, Q., Churchill, G., and Brown, M.G. (2009). A Cmv2 QTL on chromosome X affects MCMV resistance in New Zealand male mice. Mammal. Gen. *20*, 414–423.

Rosenberger, C.M., Clark, A.E., Treuting, P.M., Johnson, C.D., and Aderem, A. (2008). ATF3 regulates MCMV infection in mice by modulating IFN-gamma expression in natural killer cells. Proc. Natl. Acad. Sci. U.S.A. *105*, 2544–2549.

Ruzek, M.C., Pearce, B.D., Miller, A.H., and Biron, C.A. (1999). Endogenous glucocorticoids protect against cytokine-mediated lethality during viral infection. J. Immunol. *162*, 3527–3533.

Sacher, T., Podlech, J., Mohr, C.A., Jordan, S., Ruzsics, Z., Reddehase, M.J., and Koszinowski, U.H. (2008). The major virus-producing cell type during murine cytomegalovirus infection, the hepatocyte, is not the source of virus dissemination in the host. Cell Host Microbe *3*, 263–272.

Salazar-Mather, T.P., and Hokeness, K.L. (2003). Calling in the troops: regulation of inflammatory cell trafficking through innate cytokine/chemokine networks. Viral Immunol. *16*, 291–306.

Salazar-Mather, T.P., and Hokeness, K.L. (2006). Cytokine and chemokine networks: pathways to antiviral defense. Curr. Top. Microbiol. Immunol. *303*, 29–46.

Salazar-Mather, T.P., Hamilton, T.A., and Biron, C.A. (2000). A chemokine-to-cytokine-to-chemokine cascade critical in antiviral defense. J. Clin. Invest. *105*, 985–993.

Salazar-Mather, T.P., Lewis, C.A., and Biron, C.A. (2002). Type I interferons regulate inflammatory cell trafficking and macrophage inflammatory protein 1alpha delivery to the liver. J. Clin. Invest. *110*, 321–330.

Salem, M.L., and Hossain, M.S. (2000). *In vivo* acute depletion of CD8(+) T-cells before murine cytomegalovirus infection up-regulated innate antiviral activity of natural killer cells. Int. J. Immunopharm. *22*, 707–718.

Scalzo, A.A., Fitzgerald, N.A., Simmons, A., La Vista, A.B., and Shellam, G.R. (1990). Cmv-1, a genetic locus that controls murine cytomegalovirus replication in the spleen. J. Exp. Med. *171*, 1469–1483.

Scheu, S., Dresing, P., and Locksley, R.M. (2008). Visualization of IFNbeta production by plasmacytoid versus conventional dendritic cells under specific stimulation conditions *in vivo*. Proc. Natl. Acad. Sci. U.S.A. *105*, 20416–20421.

Schneider, D.S., and Ayres, J.S. (2008). Two ways to survive infection: what resistance and tolerance can teach us about treating infectious diseases. Nat. Rev. Immunol. *8*, 889–895.

Schneider, K., Loewendorf, A., De Trez, C., Fulton, J., Rhode, A., Shumway, H., Ha, S., Patterson, G., Pfeffer, K., Nedospasov, S.A., *et al.* (2008). Lymphotoxin-mediated cross-talk between B cells and splenic stroma promotes the initial type I interferon response to cytomegalovirus. Cell Host Microbe *3*, 67–76.

Schoggins, J.W., Wilson, S.J., Panis, M., Murphy, M.Y., Jones, C.T., Bieniasz, P., and Rice, C.M. (2011). A diverse range of gene products are effectors of the type I interferon antiviral response. Nature *472*, 481–485.

Seckert, C.K., Schader, S.I., Ebert, S., Thomas, D., Freitag, K., Renzaho, A., Podlech, J., Reddehase, M.J., and Holtappels, R. (2011). Antigen-presenting cells of haematopoietic origin prime cytomegalovirus-specific CD8 T-cells but are not sufficient for driving memory inflation during viral latency. J. Gen. Virol. *92*, 1994–2005.

Sharma, S., and Fitzgerald, K.A. (2011). Innate immune sensing of DNA. PLoS Pathog. *7*, e1001310.

Shellam, G.R., Allan, J.E., Papadimitriou, J.M., and Bancroft, G.J. (1981). Increased susceptibility to cytomegalovirus infection in beige mutant mice. Proc. Natl. Acad. Sci. U.S.A. *78*, 5104–5108.

Sjolin, H., Tomasello, E., Mousavi-Jazi, M., Bartolazzi, A., Karre, K., Vivier, E., and Cerboni, C. (2002). Pivotal role of KARAP/DAP12 adaptor molecule in the natural killer cell-mediated resistance to murine cytomegalovirus infection. J. Exp. Med. *195*, 825–834.

Smith, H.R., Heusel, J.W., Mehta, I.K., Kim, S., Dorner, B.G., Naidenko, O.V., Iizuka, K., Furukawa, H., Beckman, D.L., Pingel, J.T., *et al.* (2002). Recognition of a virus-encoded ligand by a natural killer cell activation receptor. Proc. Natl. Acad. Sci. U.S.A. *99*, 8826–8831.

Smith, M.G. (1954). Propagation of salivary gland virus of the mouse in tissue cultures. Proc. Soc. Exp. Biol. Med. *86*, 435–440.

Snyder, C.M., Cho, K.S., Bonnett, E.L., van Dommelen, S., Shellam, G.R., and Hill, A.B. (2008). Memory inflation during chronic viral infection is maintained by continuous production of short-lived, functional T-cells. Immunity *29*, 650–659.

Snyder, C.M., Loewendorf, A., Bonnett, E.L., Croft, M., Benedict, C.A., and Hill, A.B. (2009). CD4+ T-cell help has an epitope-dependent impact on CD8+ T-cell memory inflation during murine cytomegalovirus infection. J. Immunol. *183*, 3932–3941.

Snyder, C.M., Allan, J.E., Bonnett, E.L., Doom, C.M., and Hill, A.B. (2010). Cross-presentation of a spread-defective MCMV is sufficient to prime the majority of virus-specific CD8+ T-cells. PLoS One *5*, e9681.

Stadnisky, M.D., Xie, X., Coats, E.R., Bullock, T.N., and Brown, M.G. (2011). Self MHC class I-licensed NK cells enhance adaptive CD8 T-cell viral immunity. Blood *117*, 5133–5141.

Strobl, B., Bubic, I., Bruns, U., Steinborn, R., Lajko, R., Kolbe, T., Karaghiosoff, M., Kalinke, U., Jonjic, S., and Muller, M. (2005). Novel functions of tyrosine kinase 2 in the antiviral defense against murine cytomegalovirus. J. Immunol. *175*, 4000–4008.

Sullivan, B.M., Liang, H.E., Bando, J.K., Wu, D., Cheng, L.E., McKerrow, J.K., Allen, C.D., and Locksley, R.M. (2011). Genetic analysis of basophil function *in vivo*. Nat. Immunol. *12*, 527–535.

Sumaria, N., van Dommelen, S.L., Andoniou, C.E., Smyth, M.J., Scalzo, A.A., and Degli-Esposti, M.A. (2009). The roles of interferon-gamma and perforin in antiviral immunity in mice that differ in genetically determined NK-cell-mediated antiviral activity. Immunol. Cell Biol. *87*, 559–566.

Swiecki, M., and Colonna, M. (2010). Unraveling the functions of plasmacytoid dendritic cells during viral infections, autoimmunity, and tolerance. Immunol. Rev. *234*, 142–162.

Swiecki, M., Gilfillan, S., Vermi, W., Wang, Y., and Colonna, M. (2010). Plasmacytoid dendritic cell ablation impacts early interferon responses and antiviral NK and CD8(+) T-cell accrual. Immunity *33*, 955–966.

Tabeta, K., Georgel, P., Janssen, E., Du, X., Hoebe, K., Crozat, K., Mudd, S., Shamel, L., Sovath, S., Goode, J., *et al.* (2004). Toll-like receptors 9 and 3 as essential components of innate immune defense against mouse cytomegalovirus infection. Proc. Natl. Acad. Sci. U.S.A. *101*, 3516–3521.

Tabeta, K., Hoebe, K., Janssen, E.M., Du, X., Georgel, P., Crozat, K., Mudd, S., Mann, N., Sovath, S., Goode, J., *et al.* (2006). The Unc93b1 mutation 3d disrupts exogenous antigen presentation and signaling via Toll-like receptors 3, 7 and 9. Nat. Immunol. *7*, 156–164.

Tay, C.H., and Welsh, R.M. (1997). Distinct organ-dependent mechanisms for the control of murine cytomegalovirus infection by natural killer cells. J. Virol. *71*, 267–275.

Torti, N., Walton, S.M., Murphy, K.M., and Oxenius, A. (2011). Batf3 transcription factor-dependent DC subsets in murine CMV infection: Differential impact on T-cell priming and memory inflation. Eur. J. Immunol. *41*, 2612–2618.

Tun-Kyi, A., Finn, G., Greenwood, A., Nowak, M., Lee, T.H., Asara, J.M., Tsokos, G.C., Fitzgerald, K., Israel, E., Li, X., *et al.* (2011). Essential role for the prolyl isomerase Pin1 in Toll-like receptor signaling and type I interferon-mediated immunity. Nat. Immunol. *12*, 733–741.

Tyznik, A.J., Tupin, E., Nagarajan, N.A., Her, M.J., Benedict, C.A., and Kronenberg, M. (2008). Cutting edge: the mechanism of invariant NKT-cell responses to viral danger signals. J Immunol *181*, 4452–4456.

Upton, J.W., Kaiser, W.J., and Mocarski, E.S. (2008). Cytomegalovirus M45 cell death suppression requires receptor-interacting protein (RIP) homotypic interaction motif (RHIM)-dependent interaction with RIP1. J. Biol. Chem. *283*, 16966–16970.

Walton, S.M., Wyrsch, P., Munks, M.W., Zimmermann, A., Hengel, H., Hill, A.B., and Oxenius, A. (2008). The dynamics of mouse cytomegalovirus-specific CD4 T-cell responses during acute and latent infection. J. Immunol. *181*, 1128–1134.

Walton, S.M., Mandaric, S., Torti, N., Zimmermann, A., Hengel, H., and Oxenius, A. (2011a). Absence of cross-presenting cells in the salivary gland and viral immune evasion confine cytomegalovirus immune control to effector CD4 T-cells. PLoS Pathog. *7*, e1002214.

Walton, S.M., Torti, N., Mandaric, S., and Oxenius, A. (2011b). T-cell help permits memory CD8(+) T-cell inflation during cytomegalovirus latency. Eur. J. Immunol. *41*, 2248–2259.

Weisel, F.J., Appelt, U.K., Schneider, A.M., Horlitz, J.U., van Rooijen, N., Korner, H., Mach, M., and Winkler, T.H. (2010). Unique requirements for reactivation of virus-specific memory B-lymphocytes. J. Immunol. *185*, 4011–4021.

Wesley, J.D., Tessmer, M.S., Chaukos, D., and Brossay, L. (2008). NK cell-like behavior of Valpha14i NK T-cells during MCMV infection. PLoS Pathog. *4*, e1000106.

Wirtz, N., Schader, S.I., Holtappels, R., Simon, C.O., Lemmermann, N.A., Reddehase, M.J., and Podlech, J. (2008). Polyclonal cytomegalovirus-specific antibodies not only prevent virus dissemination from the portal of entry but also inhibit focal virus spread within target tissues. Med. Micro. Immunol. *197*, 151–158.

Xiao, G., Harhaj, E.W., and Sun, S.C. (2001). NF-kappaB-inducing kinase regulates the processing of NF-kappaB2 p100. Mol. Cell. *7*, 401–409.

Xie, X., Stadnisky, M.D., and Brown, M.G. (2009). MHC class I Dk locus and Ly49G2+ NK cells confer H-2k resistance to murine cytomegalovirus. J. Immunol. *182*, 7163–7171.

Zafirova, B., Mandaric, S., Antulov, R., Krmpotic, A., Jonsson, H., Yokoyama, W.M., Jonjic, S., and Polic, B. (2009). Altered NK cell development and enhanced NK cell-mediated resistance to mouse cytomegalovirus in NKG2D-deficient mice. Immunity *31*, 270–282.

Zimmermann, A., Trilling, M., Wagner, M., Wilborn, M., Bubic, I., Jonjic, S., Koszinowski, U., and Hengel, H. (2005). A cytomegaloviral protein reveals a dual role for STAT2 in IFN-{gamma} signaling and antiviral responses. J. Exp. Med. *201*, 1543–1553.

Zucchini, N., Bessou, G., Robbins, S.H., Chasson, L., Raper, A., Crocker, P.R., and Dalod, M. (2008a). Individual plasmacytoid dendritic cells are major contributors to the production of multiple innate cytokines in an organ-specific manner during viral infection. Int. Immunol. *20*, 45–56.

Zucchini, N., Bessou, G., Traub, S., Robbins, S.H., Uematsu, S., Akira, S., Alexopoulou, L., and Dalod, M. (2008b). Cutting edge: Overlapping functions of TLR7 and TLR9 for innate defense against a herpesvirus infection. J. Immunol. *180*, 5799–5803.

Clinical Cytomegalovirus Research: Thoracic Organ Transplantation

II.13

Robin K. Avery

Abstract

Heart and lung transplantation can be lifesaving therapies for end-stage organ disease in some patients. Despite advances in antiviral prevention, cytomegalovirus infection is still an important issue in post-transplant management and may contribute to survival-limiting dysfunction of the transplanted organ. This chapter reviews the clinical presentations and risk factors for CMV infection in thoracic transplant recipients, as well as the direct and indirect effects of CMV. Recent studies on CMV-specific immunity and allograft dysfunction have shed further light on the differential benefits of prophylaxis and pre-emptive strategies for CMV prevention, as well as the potential benefits of an extended duration of prophylaxis. Finally, the development of newer anti-CMV agents holds promise for therapeutic management in the future.

Introduction

In the early era of thoracic solid organ transplantation, CMV infection was one of the most feared post-transplant infectious complications, particularly CMV pneumonitis in lung and heart-lung transplant recipients (Dummer *et al.*, 1985; Duncan *et al.*, 1991; Kroshus *et al.*, 1997). While both prophylaxis and pre-emptive therapy have decreased CMV risk considerably, there is still substantial morbidity associated with CMV in this setting (Danziger-Isakov *et al.*, 2009; Humar and Snydman, 2009). Heart and lung transplantation represent two settings in which CMV infection may have adverse consequences for the subsequent function of the transplanted organ (allograft dysfunction) (Grattan *et al.*, 1989; Duncan *et al.*, 1991; Everett *et al.*, 1992; Bando *et al.*, 1995; Kroshus *et al.*, 1997). Hence, prevention of CMV may also help to preserve allograft function for the long term, although this area remains controversial. In the prophylaxis era, late post-transplant CMV episodes and antiviral-resistant CMV pose challenges for the clinician and the patient (Limaye, 2002; Singh, 2005). The availability of valganciclovir, a highly bioavailable oral derivative of ganciclovir, has raised the possibility of longer-term courses of viral suppression (Zamora *et al.*, 2004), although bone marrow and other toxicities may be problematic. A recent randomized trial of longer prophylaxis (3 vs. 12 months) in lung recipients has shown significant reductions in CMV-related outcomes that persist well beyond discontinuation of antiviral medications (Palmer *et al.*, 2010; Finlen Copeland *et al.*, 2011). However, many questions still remain to be answered, and development of newer anti-CMV agents with improved side effect profiles will be important to the transplant community.

Clinical manifestations of CMV in heart and lung transplant recipients

As with other solid organ transplant recipients, clinical manifestations of CMV infection in thoracic organ transplant recipients fall into three broad categories: asymptomatic viraemia, CMV syndrome, and tissue-invasive CMV (Humar and Snydman, 2009). The last two categories constitute symptomatic CMV disease. These three categories generally correspond to the magnitude of the CMV viral load in peripheral blood, although there are exceptions. Asymptomatic viraemia is generally associated with a low CMV viral load in DNA copies/ml (or in number of positive cells, if pp65 antigenaemia testing is being used). At moderately elevated viral loads, patients tend to develop CMV syndrome, a flulike illness which may be characterized by fevers, chills, malaise, myalgias, leucopenia, thrombocytopenia, and mild elevation of liver function tests. At high levels of blood viral load, tissue-invasive CMV disease is more likely to be present; that is, if a biopsy of an end-organ is performed, viral inclusions may be seen on histopathology, or a positive result may be obtained on CMV immunostaining. Common tissue localizations include CMV pneumonitis (see Fig.

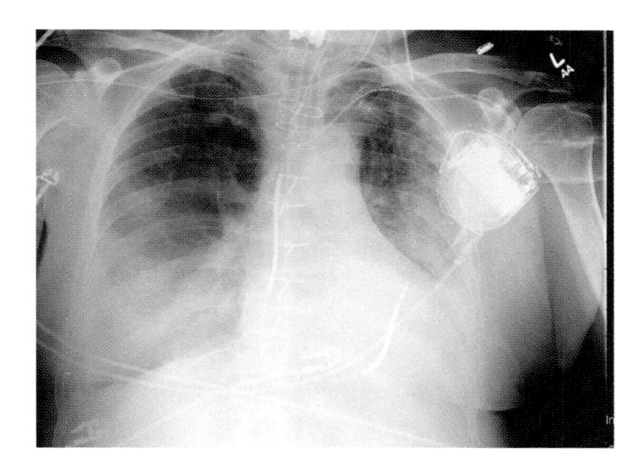

Figure II.13.1 Cytomegalovirus pneumonitis in a single lung transplant recipient: chest radiograph.

II.13.1), esophagitis, gastritis, enteritis, colitis, hepatitis, and less frequently meningoencephalitis. In lung transplant recipients, reactivation in the lung allograft is common, likely because of donor origin of the virus in some cases, or as a result of intragraft inflammatory processes. Tissue-invasive CMV disease can occasionally occur with a low or even undetectable peripheral blood viral load, particularly when the tissue involved is in the gastrointestinal tract. CMV retinitis in solid organ recipients is uncommon by contrast with AIDS patients, but may occur after several recurrences of CMV viraemia.

It is important, when reviewing published studies, to understand which of these outcomes is being assessed by the investigators. For example, the probability of detecting asymptomatic viraemia will increase with the frequency of monitoring for viral load. Studies that do not monitor asymptomatic viraemia frequently may not reflect the true incidence of viraemia. It is also important, when designing studies of CMV prevention or interventions, to adhere to internationally accepted definitions (Ljungman *et al.*, 2002; Humar and Snydman, 2009; Kotton *et al.*, 2010; Snydman *et al.*, 2011).

Direct clinical manifestations of CMV in heart recipients are often similar to those in other organ transplant recipients, but lung transplant recipients appear to be particularly prone to developing CMV pneumonitis, with possible allograft injury as a sequela (Duncan *et al.*, 1991; Kroshus *et al.*, 1997). Lung recipients also appear to have more recurrences of CMV and more ganciclovir-resistant CMV (Isada *et al.*, 2002; Limaye, 2002) than any other solid organ transplant recipients (with the possible exception of intestinal transplant recipients); lung recipients also seem to benefit from longer durations of prophylaxis (see below) (Palmer *et al.*, 2010; Finlen Copeland *et al.*, 2011).

Risk factors for CMV infection in thoracic transplant recipients

As with other solid organ transplant recipients, the strongest risk factor for CMV incidence and severity is the donor seropositive, recipient seronegative (D^+/R^-) status, in which the recipient acquires a CMV primary infection from the donor, without antecedent immunity to limit viral load and morbidity (Krogsgaard *et al.*, 1994; Miller *et al.*, 1994; Grossi *et al.*, 1995; Humar and Snydman, 2009; Delgado *et al.*, 2011). Other risk factors include intensification of immunosuppression (Grossi *et al.*, 1992), particularly the use of antilymphocyte therapy such as thymoglobulin or OKT3 for steroid-refractory acute rejection (Costanzo-Nordin *et al.*, 1992; Miller *et al.*, 1994; Issa and Fishman, 2009). Antilymphocyte therapy further impairs cellular immune defences, particularly virus-specific cytotoxic T-lymphocytes, allowing for more rapid and unchecked CMV replication. The amount of lymphoid tissue and the latent viral load in the donated organ can also add to CMV risk from the donor. In the mouse model, the lung has been identified as an organ in which latent viral load is particularly high (Balthesen *et al.*, 1993). Concurrent or recent bacterial infections, particularly sepsis or other infections where cytokines are up-regulated, can trigger CMV reactivation (Mutimer *et al.*, 1997; Rubin, 2007). Other risk factors, such as genetic polymorphisms of the immune system including toll-like receptors (TLR) (Kijpittayarit *et al.*, 2007) and interferon-gamma production (Mitsani *et al.*, 2011), constitute an active area of research.

The choice of immunosuppressive regimen can also affect the risk for developing overt CMV (Table II.13.1). Most studies have not shown a significant difference between the calcineurin inhibitors cyclosporine and tacrolimus in this regard (Keenan *et al.*, 1995; Grimm *et al.*, 2006; Hachem *et al.*, 2007), although one study showed an increase in viral infections in the cyclosporine group (Sanchez-Lazaro *et al.*, 2011). Regimens containing mycophenolate mofetil (MMF) may be associated with higher rates of CMV infection than those that contain azathioprine, as seen in a large multicentre heart transplant study (Eisen *et al.*, 2005), although a study of MMF versus azathioprine in lung recipients did not show any difference in CMV risk (Palmer *et al.*, 2001). Results concerning the use of IL-2 receptor inhibitors such as basiliximab and daclizumab for induction have been conflicting with regards to CMV risk (Carrier *et al.*, 2007; Mattei *et al.*, 2007; Mullen *et al.*, 2007). Most interestingly, recent work has suggested that the mammalian target-of-rapamycin (mTOR) inhibitors everolimus and sirolimus are associated with lower incidence of CMV (Eisen *et al.*, 2003; Kobashigawa

Table II.13.1 CMV incidence in selected randomized trials of different immunosuppressive regimens in heart and lung transplant recipients

Author (year)	Patients and N	Randomization	CMV/infections	Comments
Macdonald (1993)	41 heart recipients	ATG vs. OKT3(plus pred/ AZA/CYA)	OKT3: more viral infections	Rejection similar
Keenan (1995)	133 lung recipients	TAC vs. CYA	Viral infections similar	CYA more bacterial, less fungal risk
Palmer (2001)	81 lung recipients	MMF vs. AZA (plus pred/ CYA)	CMV-I similar	Rejection similar
Eisen (2003), Hill (2007)	634 heart recipients	Pred/CYA plus everolimus 1.5 mg, 3 mg, or AZA	CMV-I 3-fold less in the everolimus groups	CAV less in the everolimus groups
Kobashigawa (2006)	343 heart recipients	TAC/MMF vs. TAC/SRL vs. CYA/MMF	TAC/SRL fewer viral but more fungal inf.	Less rejection in the TAC groups
Eisen (2005)	650 heart recipients	MMF vs. AZA (plus pred/ CYA)	More CMV-I in MMF group	MMF reduced mortality and graft loss
Grimm (2006)	314 heart recipients	TAC vs. CYA (plus pred/ AZA)	No difference	Less rejection in the TAC group
Mattei (2007)	80 heart recipients	Basiliximab vs. ATG induction	Infectious deaths less with basiliximab	
Mullen (2007)	50 lung recipients	Daclizumab vs. ATG induction	Daclizumab: more CMV-I	Daclizumab had more D^+/R^- pts
Hachem (2007)	90 lung recipients	TAC vs. CYA (plus pred/ AZA)	No difference	Less rejection in the TAC group
Carrier (2007)	35 heart recipients	Basiliximab vs. ATG induction	More CMV-I in the ATG group	
Lehmkuhl (2009)	176 heart recipients	Everolimus/reduced CYA vs. MMF/CYA	CMV-I in 4.4% vs. 16.9%	
Sanchez-Lazaro (2011)	106 heart recipients	TAC vs. CYA (plus steroids, daclizumab/ MMF)	CYA more viral infections	TAC less rejection
Bhorade (2011)	181 lung recipients	TAC/SRL vs. TAC/AZA	CMV-I less in SRL arm	SRL: high rate of adverse events

ATG, antithymocyte globulin; AZA, azathioprine; CAV, cardiac allograft vasculopathy; CMV-I, CMV infection; CYA, cyclosporine; D^+/R^-, donor seropositive, recipient seronegative; Pred, prednisone; OKT3, muromonab anti-CD3; SRL, sirolimus; TAC,tacrolimus.

et al., 2006; Delgado *et al.*, 2011), and (in the case of everolimus) a lower incidence of cardiac allograft vasculopathy when compared with regimens containing azathioprine (Eisen *et al.*, 2003; Hill *et al.*, 2007). A more recent trial of everolimus versus MMF also showed significant reductions in CMV infection and disease in the everolimus group (Vigano *et al.*, 2010). Whether or not those two phenomena (reduced CMV incidence and reduced risk of cardiac allograft vasculopathy) are causally related is a matter of great interest. With the development of any new immunosuppressive medication, it is important to examine not only its efficacy in preventing rejection, but also its effects on infection risk, specifically CMV. Table II.13.1 summarizes selected randomized trials of different immunosuppressive regimens in thoracic transplantation, and their effects on CMV incidence.

Direct and indirect effects of CMV in thoracic transplantation

The direct and indirect effects of CMV will be discussed in detail in Chapter II.14 on abdominal organ transplantation. Direct effects refer to the categories of infectious syndromes described above. Indirect effects include predisposition to fungal and other opportunistic infections in the aftermath of a CMV infection (Rubin, 1989). Indirect effects also may include forms of allograft dysfunction (Grattan *et al.*, 1989; Duncan *et al.*, 1991; Bando *et al.*, 1995; Kroshus *et al.*, 1997;

Rubin, 2007): a controversial area, but one in which intriguing new evidence has emerged in support of a role for CMV, as discussed below (Potena *et al.*, 2003, 2009; Tu *et al.*, 2006).

Allograft dysfunction, formerly termed 'chronic rejection,' is a survival-limiting condition in thoracic transplantation, and is a major focus of current research. In cardiac transplantation, the most significant condition is cardiac allograft vasculopathy (CAV), otherwise known as transplant coronary artery disease. Unlike standard atherosclerosis, which involves focal stenoses and plaques, CAV involves diffuse intimal thickening which is difficult to bypass surgically (Fig. II.13.2). Formerly diagnosed by coronary arteriography, CAV is now most accurately assessed by the use of intravascular ultrasound (IVUS) techniques (Fig. II.13.3A and B), which allow for calculation of the intimal thickness and the residual vessel lumen (Kapadia *et al.*, 2000; Tuzcu *et al.*, 2005). The International Society of Heart and Lung Transplantation (ISHLT) has adopted consensus definitions for nomenclature of the stages of CAV (Table II.13.2) (Mehra *et al.*, 2010). Many factors can trigger CAV: some are immunologic, some infectious, and some are responses to injury of other kinds (Young, 2000).

The impact of CMV on cardiac allograft function has not been uniform across all studies, but the preponderance of evidence suggests a role for CMV in this process (Koskinen *et al.*, 1999; Potena and Valantine, 2007). Early reports suggested that both symptomatic CMV disease and high-risk CMV D^+/R^- status were associated with greater risk for developing CAV over time (Grattan *et al.*, 1989; Koskinen *et al.*, 1993; Fateh-Moghadam *et al.*, 2003). However, some studies have not confirmed this finding (Luckraz *et al.*, 2003; Mahle *et al.*, 2009). A paediatric heart transplant study found an association of the presence of viral genomes in myocardial biopsy tissue with higher risk for adverse outcomes including graft dysfunction (Shirali *et al.*, 2001). However, the most common virus detected in myocardial tissue in this study was adenovirus, far more commonly than CMV (Shirali *et al.*, 2001). It has been suggested that the heterogeneity in results of these past studies might relate to differential effects of different strains of CMV (Hosenpud, 1999; Srivastava *et al.*, 1999).

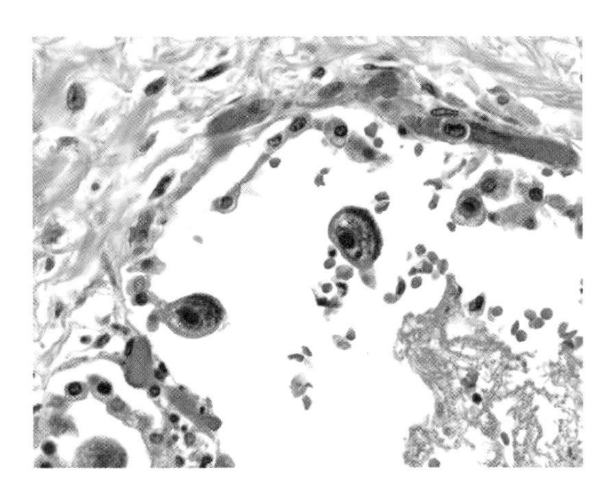

Figure II.13.2 Cytomegalovirus inclusion in type 2 pneumocyte from a patient with CMV pneumonitis. Note the enlarged cells (two to three times larger than adjacent respiratory pneumocytes). Haematoxylin and eosin, 400×. Courtesy of Dr. Carol Farver.

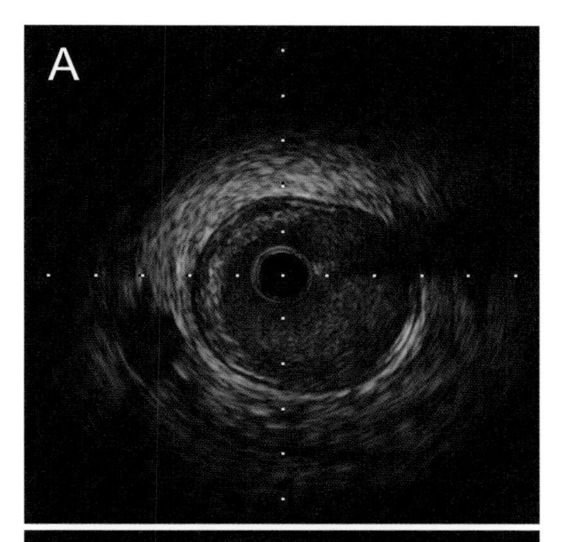

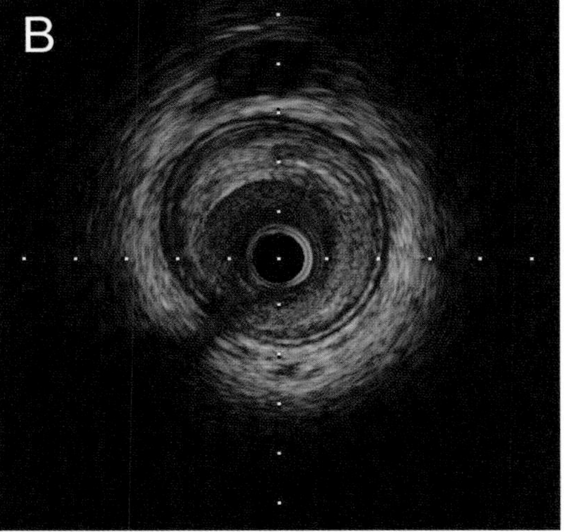

Figure II.13.3 Cardiac allograft vasculopathy visualized by intravascular ultrasound at baseline (A: 4 weeks post-transplant) and at 1 year post-transplant (B) showing progression of a lesion in the identical coronary vessel location. Courtesy of Dr. Randall Starling and Mr. William Magyar.

Table II.13.2 International Society for Heart and Lung Transplantation (ISHLT) recommended nomenclature for cardiac allograft vasculopathy (adapted from Mehra MR *et al.* (2010) J. Heart Lung Transplant *29*, 717–727)

ISHLT CAV_0 (Not significant)	No detectable angiographic lesion
ISHLT CAV_1 (Mild):	Angiographic left main <50%, or primary vessel with maximum lesion of <70%, or any branch stenosis <70% (including diffuse narrowing) without allograft dysfunction
ISHLT CAV_2 (Moderate)	Angiographic left main <50%; a single primary vessel ≥70%, or isolated branch stenosis ≥70% in branches of two systems, without allograft dysfunction
ISHLT CAV_3 (Severe)	Angiographic left main ≥50%, or two or more primary vessels ≥70% stenosis, or isolated branch stenosis ≥70% in all three systems; or ISHLT CAV_1 or CAV_2 with allograft dysfunction (defined as LVEF ≤45% usually in the presence of regional wall motion abnormalities) or evidence of significant restrictive physiology

In addition, some investigators have suggested that it may not be the high-viral-load, highly symptomatic CMV episode that is most deleterious to cardiac function; rather, long-term exposure to low-levels of CMV in a subclinical fashion may have a more profound effect on the allograft (Everett *et al.*, 1992; Arkonac *et al.*, 1997). For example, Everett *et al.* (1992) studied 129 heart recipients and found that neither CMV infection nor primary CMV infection was associated with an increased risk for cardiac allograft vasculopathy, but persistent viraemia for at least 4 months was associated with increased risk for CAV (47% vs. 18%, $P=0.01$). This suggests that it may be the duration, rather than the magnitude, of the CMV exposure which confers increased risk.

It may be subtler measures of allograft dysfunction that are affected by CMV. With regards to symptomatic CMV, a study by Potena *et al.* (2003) found that CMV infection requiring treatment was associated with inadequate coronary artery remodelling and ultimately lumen loss. There are a variety of mechanisms that could contribute to these phenomena. CMV infection is associated with high levels of asymmetric dimethylarginine (ADMA) which impairs the endothelial nitric oxide synthase pathway, thereby increasing oxidative stress in heart recipients (Weis *et al.*, 2004); CMV infection has been shown to induce anti-endothelial cell antibodies (Toyoda *et al.*, 1997); CMV infection in a rat model is associated with increased expression of MHC Class II antigens (Ustinov *et al.*, 1994), and also with increased levels of chemokines that contribute to later development of transplant vasculopathy (Streblow *et al.*, 2003; see also Chapter II.15). A study in paediatric heart recipients found that CMV infection is associated with systemic endothelial dysfunction (Simmonds *et al.*, 2008). Another study reported that CMV episodes are associated with impaired coronary endothelial function, and D^+/R^- heart recipients have distal epithelial endocardial dysfunction and higher rates of adverse allograft outcomes and death (Petrakopoulou *et al.*, 2004).

The most significant form of allograft dysfunction in lung transplantation is termed bronchiolitis obliterans syndrome (BOS) (Bando *et al.*, 1995; Kroshus *et al.*, 1997; Estenne *et al.*, 2002). BOS is a phenomenon of progressive airflow limitation, affecting small airways of the lung, which is survival-limiting. It can have a rapid and fulminant course, or a more indolent course. As in the case of CAV, multiple risk factors have been identified, including acute rejection, respiratory viral infections, anti-donor immunity, and others. Once BOS develops, therapies are generally disappointing, although some inflammatory, neutrophilic forms of BOS appear to respond to azithromycin (Vos *et al.*, 2010). While bronchiolitis obliterans ('BO' or 'OB') is a histopathological definition, the more commonly used term 'BOS' refers to the ratio of the current pulmonary function tests (particularly the FEV_1 or forced expiratory volume in 1 second) compared with the average of the best two post-transplant measures. In other words, BOS represents the decline in pulmonary function over time, compared with the highest levels of function achieved post-transplant. The ISHLT has adopted a consensus definition of the stages of BOS (Table II.13.3) (Estenne *et al.*, 2002).

CMV has been cited as a risk factor for BOS since the earliest days of lung transplantation (Duncan *et al.*, 1991; Bando *et al.*, 1995; Kroshus *et al.*, 1997). Both CMV disease (particularly CMV pneumonitis) and also CMV D^+/R^- status have been linked to BOS risk in previous studies (Duncan *et al.*, 1991; Bando *et al.*, 1995; Kroshus *et al.*, 1997; Thomas *et al.*, 2009). This appears to have been especially true in the early years when CMV prevention strategies were less developed, and immunosuppressive regimens were less refined. It is less clear if CMV still has this degree of impact in the current era; a study of 341 lung recipients by Tamm *et al.* (2004) showed no adverse impact of treated

Table II.13.3 International Society for Heart and Lung Transplantation (ISHLT) classification and staging of bronchiolitis obliterans syndrome (BOS) (adapted from Estenne M *et al*, J Heart Lung Transplant 2002; 21:297–310)

BOS 0	FEV_1 >90% of baseline[a] and FEF_{25-75} >75% of baseline
BOS 0-p	FEV_1 81–90% of baseline and/or FEF_{25-75} ≤75% of baseline
BOS 1	FEV_1 66–80% of baseline
BOS 2	FEV_1 51–65% of baseline
BOS 3	FEV_1 50% or less of baseline

FEV_1, forced expiratory volume in 1 second; FEF_{25-75}, mid-expiratory flow rate.
[a]Baseline refers to the average of the two highest pulmonary function test results achieved post-transplant, obtained at least 3 weeks apart, and without use of an inhaled bronchodilator preceding the study.

CMV pneumonitis on survival, BOS, or time to BOS. However, in contradistinction to the preceding study, Snyder *et al.* (2011a) studied a cohort of 231 lung recipients who had received short-course ganciclovir prophylaxis between 2000 and 2004, 21% of whom had developed CMV pneumonitis. CMV pneumonitis, even when treated, was associated with increased risk for BOS and death, and this result persisted when considering CMV pneumonitis as a time-dependent predictor, and through multivariable analysis (Snyder *et al.*, 2011a). Thus, in the modern era, there are varying results concerning the impact of CMV pneumonitis on the lung allograft and other outcomes, but at least some studies suggest a persistent effect even in the era of prophylaxis and effective antiviral therapy.

Recent work has provided more insights into the pathogenesis of CMV-associated injury to the lung allograft. In a rat tracheal allograft model, CMV infection enhances the development of bronchiolitis obliterans, increases MHC class II expression in lung epithelium, and is associated with a fibroproliferative process involving up-regulation of expression of platelet-derived growth factor (PDGF) (Koskinen *et al.*, 1997), and increased interleukin 2 (IL-2) and tumour necrosis factor-alpha (TNF-alpha) expression (Tikkanen *et al.*, 2001). These effects could be prevented by administration of antiviral prophylaxis with ganciclovir (Tikkanen *et al.*, 2001). Most recently, chemokines CCL2, 3, and 5 were found to be elevated in the bronchoalveolar lavage fluid of lung recipients with CMV pneumonitis (Weigt *et al.*, 2008). These elevations could provide an explanation for later development of fibroproliferation as a response to inflammation and injury.

Immune control of CMV in thoracic transplantation

The ability of the host's immune system to recognize CMV-infected cells and to control CMV replication is crucial to successful suppression of reactivation episodes. Virus-specific T-lymphocytes are particularly important in this regard, and their numbers and function depend on donor and recipient CMV serostatus, among other factors. Sester *et al.* (2005) studied CMV-specific CD4 T-cells in kidney, heart, and lung transplant recipients. Although kidney and heart recipients had CMV-CD4 T-cell frequencies similar to those of controls, lung recipients had significantly lower levels of these cells, and this correlated with risk for CMV infection episodes (Sester *et al.*, 2005). Gerna *et al.* (2006) studied CMV-specific CD4 and CD8 T-lymphocytes in heart, lung, and kidney recipients, and found two patterns of early versus late response in reconstitution of CMV-specific immunity. Most early responders had self-limited CMV episodes while late responders had more symptomatic CMV episodes requiring antiviral therapy (Gerna *et al.*, 2006). Zeevi and colleagues (1992) demonstrated that primed CMV-specific lymphocytes accumulate in the lung allograft during CMV infection and may be detectable for months in bronchoalveolar lavage samples. This group also found that CMV-specific memory responses were highest in the D^+/R^+ group, in a study of 99 lung recipients (Zeevi *et al.*, 1998). Although 16 of 17 D^+/R^- patients eventually developed evidence of CMV-specific helper responses, those that had delayed development of such immunity experienced multiple episodes of recurrent CMV infection (Zeevi *et al.*, 1998). Zeevi *et al.* (1999) also described two types of CMV-specific helper responses in CMV R^+ lung recipients: one that was associated with control of CMV recurrences, and another that was detectable during active CMV disease. More recent work by this group has characterized CMV-specific responses in patients treated with alemtuzumab (anti-CD52 monoclonal antibody), which eliminates most T-cells for a period of 6 months. Despite this profound depletion, 80% of CMV R^+ lung transplant patients had memory responses to CMV, but only two of five CMV

D^+/R^- lung transplant patients developed memory responses (Zeevi *et al.*, 2007). Thus, although CMV-specific memory cells appear to be spared by alemtuzumab, the CMV D^+/R^- lung recipient is still subject to severe CMV complications.

Another line of evidence concerning D^+/R^- lung recipients suggests that there are differential responses between those who seroconvert to CMV and those who do not, and that CMV-specific immune function may vary between compartments. Shlobin *et al.* (2006) found that 13 of 22 D^+/R^- lung recipients seroconverted to CMV IgG positivity, and 90% of these seroconverters developed CMV-specific $CD8^+$, interferon-gamma (IFN-gamma)$^+$ T-cells in peripheral blood. BAL samples included detectable CMV-specific $CD4^+$ and $CD8^+$lymphocytes, including some patients who did not have detectable CMV-specific $CD4^+$ lymphocytes in peripheral blood. Memory responses persisted in both peripheral blood and in the lung allograft after primary infection in those individuals. Seroconversion and acquisition of CMV-specific immunity was associated with freedom from CMV disease and preservation of allograft function (Shlobin *et al.*, 2006). Again, the challenge is that subpopulation of D^+/R^- patients who fail to develop CMV-specific immune function.

The question of whether administration of antiviral prophylaxis delays or prevents development of CMV-specific immune responses was addressed by Snyder *et al.* (2011b). This group studied CMV R^+ lung recipients who were receiving valganciclovir for a median of 6 months, and found that CMV-specific immune responses of peripheral blood $CD4^+$ and $CD8^+$ lymphocytes to HLA-restricted pp65 and IE-1 were comparable in these prophylaxed patients to R^+ transplant recipients not on prophylaxis. Therefore, at least in this recipient seropositive group, extended valganciclovir prophylaxis did not appear to inhibit CMV-specific cellular immune responses (Snyder *et al.*, 2011b).

The connection between CMV-specific immune responses and allograft function has been explored in several studies in cardiac transplant recipients as well. In one recent example, Tu *et al.* (2006) studied 39 CMV R^+ heart transplant patients who received prophylactic ganciclovir, and none of whom developed symptomatic CMV disease. Development of an early CMV-specific $CD4^+$ response was correlated with protection from high mean and peak viral loads, acute rejection, and loss of allograft coronary artery lumen and whole vessel area (Tu *et al.*, 2006). This study has several important implications. First, all of the CMV infection in this study was asymptomatic and subclinical. Second, not only CMV events and severity, but also allograft vasculopathy appeared to be ameliorated in the presence of a CMV-specific immune response (Tu *et al.*, 2006).

Prophylaxis and pre-emptive therapy for CMV in thoracic transplantation

Historically, CMV prevention strategies after solid organ transplantation have fallen into three categories (Table II.13.4): prophylaxis, pre-emptive therapy, or neither (deferred therapy). The last category, in which antiviral therapy is administered based on symptoms alone, has been largely abandoned in thoracic transplantation because of the inability to prevent highly symptomatic CMV disease. Most centres use either prophylaxis, pre-emptive therapy, a combination of both (hybrid strategy), or a risk-stratified approach of prophylaxis in D^+/R^- and pre-emptive therapy in R^+ patients (Zuk *et al.*, 2010). A recent survey of lung transplant programmes showed that 94.9% use prophylaxis for the D^+/R^- group, while 86.4% use prophylaxis and 13.6% use pre-emptive therapy for the R^+ group (Zuk *et al.*, 2010).

Prophylaxis refers to administration of an antiviral agent to all patients in a group, whereas pre-emptive therapy refers to administration of an antiviral agent only to those who develop evidence of CMV infection on a sensitive early detection test, most often the blood CMV PCR or pp65 antigenaemia assay (Gerna *et al.*, 1998; Guiver *et al.*, 2001). Both strategies have been shown to have efficacy in particular patient groups (Kalil *et al.*, 2005; Humar and Snydman, 2009). Advocates of pre-emptive therapy cite the advantages of less cost, toxicity, and resistance; reduction in 'late CMV', and more opportunity to develop CMV-specific immunity (Singh *et al.*, 2008). Potential disadvantages of pre-emptive therapy include some patients who already have highly symptomatic disease on first detection (Casillo *et al.*, 2004; Monforte *et al.*, 2005); complicated logistics (Kunzle *et al.*, 2000) and protocol noncompliance particularly in transplant programmes with many non-local patients; and possibly the loss of prophylactic activity against other herpesviruses such as EBV and human herpesvirus-6 since pre-emptive therapy is solely CMV-triggered. Prophylaxis has a long track record through many studies and its benefits in prevention of both direct and indirect CMV effects, as well as cost-effectiveness, have been well established (Snydman, 2006). Nonetheless, the existence of 'late CMV' after discontinuation of prophylaxis is a matter for concern (Singh, 2005).

A number of studies have demonstrated potential utility of pre-emptive therapy in thoracic recipients, particularly in heart transplantation, using a variety of assays to trigger antiviral therapy (Paniagua *et al.*, 2002; Gerna *et al.*, 2003, 2007; Villa *et al.*, 2003; Casillo *et al.*, 2004; De Santo *et al.*, 2005; Diaz-Pedroche *et al.*, 2005; Monforte *et al.*, 2005). Egan *et al.* (1998) studied pre-emptive therapy with i.v. ganciclovir in 19 heart or lung transplant recipients, compared with historical controls who received deferred therapy based on symptoms, and found a significant decrease in CMV disease in the pre-emptive group (0/19 vs. 5/18; $P=0.019$). Gerna *et al.* (2009) studied pre-emptive therapy based on blood and BAL CMV PCR in lung recipients, and found that monitoring both compartments plus CMV-specific cellular immunity was useful. Senechal *et al.* (2003) used pp65 antigenaemia as a trigger for pre-emptive therapy in 94 R^+ heart recipients, finding that two-thirds of this group did not require antiviral therapy. Although a lower incidence of antiviral resistance has often been cited as a potential advantage of pre-emptive therapy, there are reports of ganciclovir resistance developing in heart recipients (Gilbert *et al.*, 2001; Casillo *et al.*, 2004) and in lung recipients treated with pre-emptive therapy (Limaye *et al.*, 2002).

Prophylaxis for CMV in thoracic transplant recipients has a long history, and due to the large number of studies published, only a few highlights will be cited here (Humar and Snydman, 2009). Merigan *et al.* (1992) performed a randomized trial in heart recipients using a 4-week i.v. ganciclovir regimen. Symptomatic CMV disease was reduced from 46% to 9% in the R^+ group who received ganciclovir compared with the R^+ group that received no prophylaxis, but the D^+/R^- group did not experience a comparable reduction. A post-hoc analysis with longer follow-up (4.7 years) of the original study participants by Valantine *et al.* (1999) showed a reduction in cardiac allograft vasculopathy in those patients treated with ganciclovir in the group who were not receiving calcium blockers. Recent results by Palmer *et al.* (2010) in a randomized trial comparing 3 versus 12 months of valganciclovir prophylaxis in lung recipients have shown that prophylaxis courses longer than three months confer benefit in terms of long-term freedom from CMV events. In this multicentre study of 136 lung recipients, CMV disease was reduced from 32% to 4% and CMV infection from 64% to 10% with the 12-month prophylaxis course. Disease severity as measured by the median CMV viral load was lower in the 12-month group (3200 vs. 100,000 copies/ml) (Palmer *et al.*, 2010). There was a low incidence of 'late CMV' in the 6 months following discontinuation of prophylaxis, contrary to the expectation that prophylaxis might just delay the onset of CMV disease. A subsequent 4-year follow-up study demonstrated that these benefits in freedom from CMV outcomes persist longer-term (Finlen Copeland *et al.*, 2011). Although cost and toxicity may be issues with longer courses of prophylaxis, adverse events including neutropenia were not significantly different in this study between the two groups with different lengths of valganciclovir prophylaxis (Finlen Copeland *et al.*, 2011).

In terms of assessing the differential effects of prophylaxis versus pre-emptive therapy on long-term outcomes and allograft function, there is a paucity of randomized controlled trials that compare these modalities head-to-head. An intriguing recent study by Potena *et al.* (2009) in 40 heart recipients who received pre-emptive therapy (earlier era) or prophylaxis (more recent) showed that prophylaxis was associated with later CMV onset, lower peak CMV viral load, and less symptomatic CMV disease. Allograft vasculopathy, as measured by the 1-year increase in maximal intimal thickening, was significantly less in the prophylaxis group (Potena *et al.*, 2009). If these results could be confirmed in a larger, randomized trial, this would be helpful in progress towards defining the optimal form of CMV prevention.

Newer agents for treatment of refractory CMV

Both prophylactic and pre-emptive regimens can lead to development of antiviral resistance (primarily UL97 mutations and occasionally UL54 mutations; see also Chapter II.19). Ganciclovir resistance is a difficult problem in lung recipients, with their more frequent recurrences and symptomatic CMV disease (Limaye, 2002). The current licensed drugs, foscarnet and cidofovir, both have significant toxicities, particularly nephrotoxicity. A regimen of half-dose i.v. ganciclovir plus escalating dose foscarnet has been reported as having been effective in six patients (Mylonakis *et al.*, 2002). However, there are some patients who become resistant or refractory to all licensed therapies. The rheumatoid arthritis drug leflunomide has novel anti-CMV activity and has been utilized for long-term viral suppression in some patients with multiple failed prior therapies (Avery *et al.*, 2010b). Maribavir, an investigational benzimidazole antiviral agent, has been used successfully in therapy of multiresistant CMV in several solid organ transplant recipients (Avery *et al.*, 2010a), although maribavir resistance in a heart transplant recipient has already been described (Strasfeld *et al.*, 2010). CMX001, an investigational oral lipid conjugate of cidofovir, has been reported to be efficacious in treatment of adenovirus disease (Florescu *et al.*, 2011), and trials in solid organ transplant patients with CMV are under way. Finally, a lung recipient with CMV that was refractory to multiple antiviral agents responded to the investigational viral terminase complex

Table II.13.4 Randomized trials of CMV prevention in heart and lung transplant recipients

Author (year)	Patients and N	Randomization	Main results	Comments
Merigan (1992)	112 heart recipients	i.v. GCV 28 days vs. placebo	CMV-D in 9% vs. 46% of R+ but not in D+/R−	See Valantine (1999), re post-hoc late CAV
Boland (1993)	28 D+/R− heart and kidney recipients	CMVIg to week 7 vs. no CMVIg	CMV-I 50% in both groups	No reduction in severity
Duncan (1994)	25 lung recipients	i.v.GCV 3 wks then ACV, vs. i.v. GCV to day 90	CMV-I 75% vs. 15%; BOS 54% vs. 17% at 1 yr	CMV and BOS cumulative rates similar by 2 years
Aguado (1995)	31 R+ heart recipients	CMVIg vs. i.v. GCV	CMV-D 40% vs. 6%	All patients received OKT3
Macdonald (1995)	56 heart recipients	i.v. GCV 5 mg/kg thrice weekly × 6 weels vs. placebo	CMV-D in 11% vs. 71% of D+/R−but not in R+	In R+, GCV reduced morbidity
Hertz (1998)	72 lung recipients	i.v.GCV 5 mg/kg b.i.d. × 2 weeks then Daily vs. thrice weekly to day 90	No significant difference in CMV-I or CMV-D	Some centres have noted GCV-R with the thrice weekly
Rubin (2000)	155 D+/R− heart, kidney, liver recipients	i.v. GCV 5–10 days then OGCV vs. ACV to 12 weeks	CMV-I 32% vs. 50%	Tissue-invasive disease significantly less with OGCV
Egan (2002)	27 R+ heart recipients	VACV vs. ACV to day 90	VACV: CMV-D and antigenaemia delayed	VACV: fewer opportunistic infections
Kruger (2003)	44 R+ lung recipients	CMVIg vs. no CMVIg	No difference in CMV-I or pneumonitis	No difference in rejection, BOS, survival
Gerna (2003)	82 heart and lung recipients	Pre-emptive: pp67mRNA vs. pp65 antigenaemia	pp67: number of treated/ infected pts higher	No patient developed CMV-D
Vrtovec (2004)	59 R+ heart recipients	37 with + pp65: i.v. GCV vs. CMVIg	No CMV-D in either group	133 controls with no testing; 20% CMV-D
Paya (2004)	364 D+/R− kidney, liver, heart, pancreas	VGCV vs. OGCV to day 100	CMV-I lower during VGCV, but later equal	CMV-D same at 1 year; VGCV no GCV-R
Yamani (2005)	23 heart recipients with hypogammaglobulinaemia	CMVIg vs. placebo	CMV-I in 15% vs. 60%	CMVIg trend to less rejection
Gerna (2007)	200 solid organ recipients	Pre-emptive: DNAemia vs. pp65	CMV-I same; fewer DNAemia pts were treated	4 total patients with CMV-D prior to reaching cut-off
Palmer (2010)	136 lung recipients	3 vs. 12 months' valganciclovir	CMV-D in 32% vs. 4%; CMV-I 64% vs. 10%; median viral load 110,000 vs. 3200 copies/ml	Effects persist out to 4 yreas in the Duke subgroup; low incidence of late CMV

ACV, acyclovir; b.i.d., twice daily; CMV-D, CMV disease (symptomatic); CMV-I, CMV infection; CMVIg, CMV hyperimmune globulin; D+/R−, donor seropositive, recipient seronegative; GCV, ganciclovir; GCV-R, ganciclovir-resistant CMV; Inf, infection; i.v. GCV, intravenous ganciclovir; OGCV, oral ganciclovir; OKT3, muromonab anti-CD3; R+, recipient seropositive; VACV, valacyclovir; VGCV, valganciclovir.

inhibitor AIC246 (Kaul *et al.*, 2011). Given the severity of illness in some thoracic transplant recipients with ganciclovir-resistant CMV, it is to be hoped that the above investigational drugs will become available in the future. These drugs would also be useful in patients with intolerance to ganciclovir derivatives, particularly neutropenia.

Conclusion and directions for future research

There has been considerable progress in CMV prevention and treatment in thoracic organ transplantation over the last two decades. Both prophylaxis and preemptive therapy have been shown to be efficacious in various settings, but prophylaxis is favoured by nearly 95% of lung transplant centres for the highest-risk D^+/R^- group and for the majority of R^+ lung recipients (Zuk *et al.*, 2010). An extended course of 12 months of prophylaxis has been shown to confer greater long-term freedom from CMV events in lung recipients compared with a 3-month course (Palmer *et al.*, 2010; Finlen Copeland *et al.*, 2011). Recent intriguing work from multiple centres has shed further light on the role of CMV-specific immunity and of CMV prevention strategies on the risk for development of symptomatic CMV, and for allograft dysfunction in the form of cardiac allograft vasculopathy (heart recipients) and bronchiolitis obliterans syndrome (lung recipients). Differential effects on CMV of newer immunosuppressive agents such as mTOR inhibitors are of interest. Ideally, future randomized trials should be multicentre and international; should include meticulous monitoring of viral load according to current guidelines (Humar and Michaels, 2006), as well as sequential measurements of CMV-specific immunity; should include long-term follow-up, and should be powered to detect differences in the development of allograft dysfunction over time. Potential creative solutions for late-onset CMV disease and ganciclovir-resistant CMV will be important. Although ganciclovir derivatives have played a major role in the reduction of CMV outcomes over time, the development of new, effective, orally bioavailable anti-CMV drugs with favourable toxicity profiles would be a welcome development in the care of thoracic transplant recipients.

References

Aguado, J.M., Gomez-Sanchez, M.A., Lumbreras, C., Delgado, J., Lizasoain, M., Otero, J.R., Rufilanchas, J.J., and Noriega, A.R. (1995). Prospective randomized trial of efficacy of ganciclovir versus that of anti-cytomegalovirus (CMV) immunoglobulin to prevent CMV disease in CMV-seropositive heart transplant recipients treated with OKT3. Antimicrob. Agents Chemother. *39*, 1643–1645.

Arkonac, B., Mauck, K.A., Chou, S., and Hosenpud, J.D. (1997). Low multiplicity cytomegalovirus infection of human aortic smooth muscle cells increases levels of major histocompatibility complex class I antigens and induces a proinflammatory cytokine milieu in the absence of cytopathology. J. Heart Lung Transplant. *16*, 1035–1045.

Avery, R.K., Marty, F.M., Strasfeld, L., Lee, I., Arrieta, A., Chou, S., Tatarowicz, W., and Villano, S. (2010a). Oral maribavir for treatment of refractory or resistant cytomegalovirus infections in transplant recipients. Transpl. Infect. Dis. *12*, 489–496.

Avery, R.K., Mossad, S.B., Poggio, E., Lard, M., Budev, M., Bolwell, B., Waldman, W.J., Braun, W., Mawhorter, S.D., Fatica, R., *et al.* (2010b). Utility of leflunomide in the treatment of complex cytomegalovirus syndromes. Transplantation *90*, 419–426.

Balthesen, M., Messerle, M., and Reddehase, M.J. (1993). Lungs are a major organ site of cytomegalovirus latency and recurrence. J. Virol. *67*, 5360–5366.

Bando, K., Paradis, I.L., Similo, S., Konishi, H., Komatsu, K., Zullo, T.G., Yousem, S.A., Close, J.M., Zeevi, A., Duquesnoy, R.J., *et al.* (1995). Obliterative bronchiolitis after lung and heart-lung transplantation. An analysis of risk factors and management. J. Thorac. Cardiovasc. Surg. *110*, 4–13; discussion 13–14.

Bhorade, S., Ahya, V.N., Baz, M.A., Valentine, V.G., Arcasoy, S.M., Love, R.B., Seethamraju, H., Alex, C.G., Bag, R., Deoliveira, N.C., *et al.* (2011). Comparison of sirolimus with azathioprine in a tacrolimus-based immunosuppressive regimen in lung transplantation. Am. J. Respir. Crit. Care Med. *183*, 379–387.

Boland, G.J., Ververs, C., Hene, R.J., Jambroes, G., Donckerwolcke, R.A., and de Gast, G.C. (1993). Early detection of primary cytomegalovirus infection after heart and kidney transplantation and the influence of hyperimmune globulin prophylaxis. Transpl. Int. *6*, 34–38.

Carrier, M., Leblanc, M.H., Perrault, L.P., White, M., Doyle, D., Beaudoin, D., and Guertin, M.C. (2007). Basiliximab and rabbit anti-thymocyte globulin for prophylaxis of acute rejection after heart transplantation: a non-inferiority trial. J. Heart Lung Transplant. *26*, 258–263.

Casillo, R., Grimaldi, M., Ragone, E., Maiello, C., Marra, C., De Santo, L., Amarelli, C., Romano, G., Della Corte, A., Portella, G., *et al.* (2004). Efficacy and limitations of preemptive therapy against cytomegalovirus infections in heart transplant patients. Transplant. Proc. *36*, 651–653.

Costanzo-Nordin, M.R., Swinnen, L.J., Fisher, S.G., O'Sullivan, E.J., Pifarre, R., Heroux, A.L., Mullen, G.M., and Johnson, M.R. (1992). Cytomegalovirus infections in heart transplant recipients: relationship to immunosuppression. J. Heart Lung Transplant. *11*, 837–846.

Danziger-Isakov, L.A., Worley, S., Michaels, M.G., Arrigain, S., Aurora, P., Ballmann, M., Boyer, D., Conrad, C., Eichler, I., Elidemir, O., *et al.* (2009). The risk, prevention, and outcome of cytomegalovirus after pediatric lung transplantation. Transplantation *87*, 1541–1548.

De Santo, L.S., Romano, G., Mastroianni, C., Roberta, C., Della Corte, A., Amarelli, C., Maiello, C., Giannolo, B., Marra, C., Ragone, E., *et al.* (2005). Role of immunosuppressive regimen on the incidence and characteristics of cytomegalovirus infection in heart transplantation: a single-center experience with preemptive therapy. Transplant. Proc. *37*, 2684–2687.

Delgado, J.F., Manito, N., Almenar, L., Crespo-Leiro, M., Roig, E., Segovia, J., Vazquez de Prada, J.A., Lage, E., Palomo, J., Camprecios, M., *et al.* (2011). Risk factors associated with cytomegalovirus infection in heart transplant patients: a prospective, epidemiological study. Transpl. Infect. Dis. *13*, 136–144.

Diaz-Pedroche, C., Lumbreras, C., Del Valle, P., San Juan, R., Hernando, S., Folgueira, D., Andres, A., Delgado, J.,

Meneu, J.C., Morales, J.M., *et al.* (2005). Efficacy and safety of valgancyclovir as preemptive therapy for the prevention of cytomegalovirus disease in solid organ transplant recipients. Transplant. Proc. *37*, 3766–3767.

Dummer, J.S., White, L.T., Ho, M., Griffith, B.P., Hardesty, R.L., and Bahnson, H.T. (1985). Morbidity of cytomegalovirus infection in recipients of heart or heart-lung transplants who received cyclosporine. J. Infect. Dis. *152*, 1182–1191.

Duncan, A.J., Dummer, J.S., Paradis, I.L., Dauber, J.H., Yousem, S.A., Zenati, M.A., Kormos, R.L., and Griffith, B.P. (1991). Cytomegalovirus infection and survival in lung transplant recipients. J. Heart Lung Transplant. *10*, 638–644; discussion 645–636.

Duncan, S.R., Grgurich, W.F., Iacono, A.T., Burckart, G.J., Yousem, S.A., Paradis, I.L., Williams, P.A., Johnson, B.A., and Griffith, B.P. (1994). A comparison of ganciclovir and acyclovir to prevent cytomegalovirus after lung transplantation. Am. J. Respir. Crit. Care Med. *150*, 146–152.

Egan, J.J., Lomax, J., Barber, L., Lok, S.S., Martyszczuk, R., Yonan, N., Fox, A., Deiraniya, A.K., Turner, A.J., and Woodcock, A.A. (1998). Preemptive treatment for the prevention of cytomegalovirus disease: in lung and heart transplant recipients. Transplantation *65*, 747–752.

Egan, J.J., Carroll, K.B., Yonan, N., Woodcock, A., and Crisp, A. (2002). Valacyclovir prevention of cytomegalovirus reactivation after heart transplantation: a randomized trial. J. Heart Lung Transplant *21*, 460–466.

Eisen, H.J., Tuzcu, E.M., Dorent, R., Kobashigawa, J., Mancini, D., Valantine-von Kaeppler, H.A., Starling, R.C., Sorensen, K., Hummel, M., Lind, J.M., *et al.* (2003). Everolimus for the prevention of allograft rejection and vasculopathy in cardiac-transplant recipients. N. Engl. J. Med. *349*, 847–858.

Eisen, H.J., Kobashigawa, J., Keogh, A., Bourge, R., Renlund, D., Mentzer, R., Alderman, E., Valantine, H., Dureau, G., Mancini, D., *et al.* (2005). Three-year results of a randomized, double-blind, controlled trial of mycophenolate mofetil versus azathioprine in cardiac transplant recipients. J. Heart Lung Transplant. *24*, 517–525.

Estenne, M., Maurer, J.R., Boehler, A., Egan, J.J., Frost, A., Hertz, M., Mallory, G.B., Snell, G.I., and Yousem, S. (2002). Bronchiolitis obliterans syndrome 2001: an update of the diagnostic criteria. J. Heart Lung Transplant. *21*, 297–310.

Everett, J.P., Hershberger, R.E., Norman, D.J., Chou, S., Ratkovec, R.M., Cobanoglu, A., Ott, G.Y., and Hosenpud, J.D. (1992). Prolonged cytomegalovirus infection with viremia is associated with development of cardiac allograft vasculopathy. J. Heart Lung Transplant. *11*, S133–137.

Fateh-Moghadam, S., Bocksch, W., Wessely, R., Jager, G., Hetzer, R., and Gawaz, M. (2003). Cytomegalovirus infection status predicts progression of heart-transplant vasculopathy. Transplantation *76*, 1470–1474.

Finlen Copeland, C.A., Davis, W.A., Snyder, L.D., Banks, M., Avery, R., Davis, R.D., and Palmer, S.M. (2011). Long-term efficacy and safety of 12 months of valganciclovir prophylaxis compared with 3 months after lung transplantation: a single-center, long-term follow-up analysis from a randomized, controlled cytomegalovirus prevention trial. J. Heart Lung Transplant. *30*, 990–996.

Florescu, D.F., Pergam, S.A., Neely, M.N., Qiu, F., Johnston, C., Way, S.S., Sande, J., Lewinsohn, D.A., Guzman-Cottrill, J.A., Graham, M.L., *et al.* (2011). Safety and efficacy of CMX001 as salvage therapy for severe adenovirus infections in immunocompromised patients. Biol. Blood Marrow Transplant. *18*, 731–738.

Gerna, G., Zavattoni, M., Baldanti, F., Sarasini, A., Chezzi, L., Grossi, P., and Revello, M.G. (1998). Human cytomegalovirus (HCMV) leukodnaemia correlates more closely with clinical symptoms than antigenemia and viremia in heart and heart-lung transplant recipients with primary HCMV infection. Transplantation *65*, 1378–1385.

Gerna, G., Baldanti, F., Lilleri, D., Parea, M., Torsellini, M., Castiglioni, B., Vitulo, P., Pellegrini, C., Vigano, M., Grossi, P., *et al.* (2003). Human cytomegalovirus pp67 mRNAemia versus pp65 antigenemia for guiding preemptive therapy in heart and lung transplant recipients: a prospective, randomized, controlled, open-label trial. Transplantation *75*, 1012–1019.

Gerna, G., Lilleri, D., Fornara, C., Comolli, G., Lozza, L., Campana, C., Pellegrini, C., Meloni, F., and Rampino, T. (2006). Monitoring of human cytomegalovirus-specific CD4 and CD8 T-cell immunity in patients receiving solid organ transplantation. Am. J. Transplant. *6*, 2356–2364.

Gerna, G., Baldanti, F., Torsellini, M., Minoli, L., Vigano, M., Oggionnis, T., Rampino, T., Castiglioni, B., Goglio, A., Colledan, M., *et al.* (2007). Evaluation of cytomegalovirus DNAaemia versus pp65-antigenaemia cutoff for guiding preemptive therapy in transplant recipients: a randomized study. Antivir. Ther. *12*, 63–72.

Gerna, G., Lilleri, D., Rognoni, V., Agozzino, M., Meloni, F., Oggionni, T., Pellegrini, C., Arbustini, E., and D'Armini, A.M. (2009). Preemptive therapy for systemic and pulmonary human cytomegalovirus infection in lung transplant recipients. Am. J. Transplant. *9*, 1142–1150.

Gilbert, C., LeBlanc, M.H., and Boivin, G. (2001). Case study: rapid emergence of a cytomegalovirus UL97 mutant in a heart-transplant recipient on pre-emptive ganciclovir therapy. Herpes *8*, 80–82.

Grattan, M.T., Moreno-Cabral, C.E., Starnes, V.A., Oyer, P.E., Stinson, E.B., and Shumway, N.E. (1989). Cytomegalovirus infection is associated with cardiac allograft rejection and atherosclerosis. JAMA *261*, 3561–3566.

Grimm, M., Rinaldi, M., Yonan, N.A., Arpesella, G., Arizon Del Prado, J.M., Pulpon, L.A., Villemot, J.P., Frigerio, M., Rodriguez Lambert, J.L., Crespo-Leiro, M.G., *et al.* (2006). Superior prevention of acute rejection by tacrolimus vs. cyclosporine in heart transplant recipients--a large European trial. Am. J. Transplant. *6*, 1387–1397.

Grossi, P., De Maria, R., Caroli, A., Zaina, M.S., and Minoli, L. (1992). Infections in heart transplant recipients: the experience of the Italian heart transplantation program. Italian Study Group on Infections in Heart Transplantation. J. Heart Lung Transplant. *11*, 847–866.

Grossi, P., Minoli, L., Percivalle, E., Irish, W., Vigano, M., and Gerna, G. (1995). Clinical and virological monitoring of human cytomegalovirus infection in 294 heart transplant recipients. Transplantation *59*, 847–851.

Guiver, M., Fox, A.J., Mutton, K., Mogulkoc, N., and Egan, J. (2001). Evaluation of CMV viral load using TaqMan CMV quantitative PCR and comparison with CMV antigenemia in heart and lung transplant recipients. Transplantation *71*, 1609–1615.

Hachem, R.R., Yusen, R.D., Chakinala, M.M., Meyers, B.F., Lynch, J.P., Aloush, A.A., Patterson, G.A., and Trulock, E.P. (2007). A randomized controlled trial of tacrolimus versus cyclosporine after lung transplantation. J. Heart Lung Transplant. *26*, 1012–1018.

Hertz, M.I., Jordan, C., Savik, S.K., Fox, J.M., Park, S., Bolman, R.M., and Dosland-Mullan, B.M. (1998) Randomized trial of daily versus three-times-weekly prophylactic ganciclovir after lung and heart-lung transplantation. J. Heart Lung Transplant *17*, 913–920.

Hill, J.A., Hummel, M., Starling, R.C., Kobashigawa, J.A., Perrone, S.V., Arizon, J.M., Simonsen, S., Abeywickrama, K.H., and Bara, C. (2007). A lower incidence of cytomegalovirus infection in de novo heart transplant recipients randomized to everolimus. Transplantation *84*, 1436–1442.

Hosenpud, J.D. (1999). Coronary artery disease after heart transplantation and its relation to cytomegalovirus. Am. Heart J. *138*, S469–472.

Humar, A., and Michaels, M. (2006). American Society of Transplantation recommendations for screening, monitoring and reporting of infectious complications in immunosuppression trials in recipients of organ transplantation. Am. J. Transplant. *6*, 262–274.

Humar, A., and Snydman, D. (2009). Cytomegalovirus in solid organ transplant recipients. Am. J. Transplant. *9 Suppl 4*, S78–86.

Isada, C.M., Yen-Lieberman, B., Lurain, N.S., Schilz, R., Kohn, D., Longworth, D.L., Taege, A.J., Mossad, S.B., Maurer, J., Flechner, S.M., *et al.* (2002). Clinical characteristics of 13 solid organ transplant recipients with ganciclovir-resistant cytomegalovirus infection. Transpl. Infect. Dis. *4*, 189–194.

Issa, N.C., and Fishman, J.A. (2009). Infectious complications of antilymphocyte therapies in solid organ transplantation. Clin. Infect. Dis. *48*, 772–786.

Kalil, A.C., Levitsky, J., Lyden, E., Stoner, J., and Freifeld, A.G. (2005). Meta-analysis: the efficacy of strategies to prevent organ disease by cytomegalovirus in solid organ transplant recipients. Ann. Intern. Med. *143*, 870–880.

Kapadia, S.R., Ziada, K.M., L'Allier, P.L., Crowe, T.D., Rincon, G., Hobbs, R.E., Bott-Silverman, C., Young, J.B., Nissen, S.E., and Tuzcu, E.M. (2000). Intravascular ultrasound imaging after cardiac transplantation: advantage of multi-vessel imaging. J. Heart Lung Transplant. *19*, 167–172.

Kaul, D.R., Stoelben, S., Cober, E., Ojo, T., Sandusky, E., Lischka, P., Zimmermann, H., and Rubsamen-Schaeff, H. (2011). First report of successful treatment of multidrug-resistant cytomegalovirus disease with the novel anti-CMV compound AIC246. Am. J. Transplant. *11*, 1079–1084.

Keenan, R.J., Konishi, H., Kawai, A., Paradis, I.L., Nunley, D.R., Iacono, A.T., Hardesty, R.L., Weyant, R.J., and Griffith, B.P. (1995). Clinical trial of tacrolimus versus cyclosporine in lung transplantation. Ann. Thorac. Surg. *60*, 580–584; discussion 584–585.

Kijpittayarit, S., Eid, A.J., Brown, R.A., Paya, C.V., and Razonable, R.R. (2007). Relationship between Toll-like receptor 2 polymorphism and cytomegalovirus disease after liver transplantation. Clin. Infect. Dis. *44*, 1315–1320.

Kobashigawa, J.A., Miller, L.W., Russell, S.D., Ewald, G.A., Zucker, M.J., Goldberg, L.R., Eisen, H.J., Salm, K., Tolzman, D., Gao, J., *et al.* (2006). Tacrolimus with mycophenolate mofetil (MMF) or sirolimus vs. cyclosporine with MMF in cardiac transplant patients: 1-year report. Am. J. Transplant. *6*, 1377–1386.

Koskinen, P.K., Nieminen, M.S., Krogerus, L.A., Lemstrom, K.B., Mattila, S.P., Hayry, P.J., and Lautenschlager, I.T. (1993). Cytomegalovirus infection accelerates cardiac allograft vasculopathy: correlation between angiographic and endomyocardial biopsy findings in heart transplant patients. Transpl. Int. *6*, 341–347.

Koskinen, P.K., Kallio, E.A., Bruggeman, C.A., and Lemstrom, K.B. (1997). Cytomegalovirus infection enhances experimental obliterative bronchiolitis in rat tracheal allografts. Am. J. Respir. Crit. Care Med. *155*, 2078–2088.

Koskinen, P.K., Kallio, E.A., Tikkanen, J.M., Sihvola, R.K., Hayry, P.J., and Lemstrom, K.B. (1999). Cytomegalovirus infection and cardiac allograft vasculopathy. Transpl. Infect. Dis. *1*, 115–126.

Kotton, C.N., Kumar, D., Caliendo, A.M., Asberg, A., Chou, S., Snydman, D.R., Allen, U., and Humar, A. (2010). International consensus guidelines on the management of cytomegalovirus in solid organ transplantation. Transplantation *89*, 779–795.

Krogsgaard, K., Boesgaard, S., Aldershvile, J., Arendrup, H., Mortensen, S.A., and Petterson, G. (1994). Cytomegalovirus infection rate among heart transplant patients in relation to anti-thymocyte immunoglobulin induction therapy. Copenhagen Heart Transplant Group. Scand. J. Infect. Dis. *26*, 239–247.

Kroshus, T.J., Kshettry, V.R., Savik, K., John, R., Hertz, M.I., and Bolman, R.M., 3rd. (1997). Risk factors for the development of bronchiolitis obliterans syndrome after lung transplantation. J. Thorac. Cardiovasc. Surg. *114*, 195–202.

Kruger, R.M., Paranjothi, S., Storch, G.A., Lynch, J.P., and Trulock, E.P. (2003). Impact of prophylaxis with cytogam alone on the incidence of CMV viremia in CMV-seropositive lung transplant recipients. J. Heart Lung Transplant *22*, 754–763.

Kunzle, N., Petignat, C., Francioli, P., Vogel, G., Seydoux, C., Corpataux, J.M., Sahli, R., and Meylan, P.R. (2000). Preemptive treatment approach to cytomegalovirus (CMV) infection in solid organ transplant patients: relationship between compliance with the guidelines and prevention of CMV morbidity. Transpl. Infect. Dis. *2*, 118–126.

Lehmkuhl, H.B., Arizon, J., Vigano, M., Almenar, L., Gerosa, G., Maccherini, M., Varnous, S., Musumeci, F., Hexham, J.M., Mange, K.C., *et al.* (2009). Everolimus with reduced cyclosporine versus MMF with standard cyclosporine in de novo heart transplant recipients. Transplantation *88*, 115–122.

Limaye, A.P. (2002). Ganciclovir-resistant cytomegalovirus in organ transplant recipients. Clin. Infect. Dis. *35*, 866–872.

Limaye, A.P., Raghu, G., Koelle, D.M., Ferrenberg, J., Huang, M.L., and Boeckh, M. (2002). High incidence of ganciclovir-resistant cytomegalovirus infection among lung transplant recipients receiving preemptive therapy. J. Infect. Dis. *185*, 20–27.

Ljungman, P., Griffiths, P., and Paya, C. (2002). Definitions of cytomegalovirus infection and disease in transplant recipients. Clin. Infect. Dis. *34*, 1094–1097.

Luckraz, H., Charman, S.C., Wreghitt, T., Wallwork, J., Parameshwar, J., and Large, S.R. (2003). Does cytomegalovirus status influence acute and chronic rejection in heart transplantation during the ganciclovir prophylaxis era? J. Heart Lung Transplant. *22*, 1023–1027.

Macdonald, P.S., Mundy, J., Keogh, A.M., Chang, V.P., and Spratt, P.M. (1993). A prospective randomized study of prophylactic OKT3 versus equine antithymocyte globulin after heart transplantation—increased morbidity with OKT3. Transplantation *55*, 110–116.

Macdonald, P.S., Keogh, A.M., Marshman, D., Richens, D., Harvison, A., Kaan, AM., and Spratt, P.M. (1995). A double-blind placebo-controlled trial of low-dose ganciclovir to prevent cytomegalovirus disease after heart transplantation. J. Heart Lung Transplant *14*, 32–38.

Mahle, W.T., Fourshee, M.T., Naftel, D.M., Alejos, J.C., Caldwell, R.L., Uzark, K., Berg, A., and Kanter, K.R. (2009). Does cytomegalovirus serology impact outcome after pediatric heart transplantation? J. Heart Lung Transplant. *28*, 1299–1305.

Mattei, M.F., Redonnet, M., Gandjbakhch, I., Bandini, A.M., Billes, A., Epailly, E., Guillemain, R., Lelong, B., Pol, A., Treilhaud, M., *et al.* (2007). Lower risk of infectious deaths in cardiac transplant patients receiving basiliximab versus anti-thymocyte globulin as induction therapy. J. Heart Lung Transplant. *26*, 693–699.

Mehra, M.R., Crespo-Leiro, M.G., Dipchand, A., Ensminger, S.M., Hiemann, N.E., Kobashigawa, J.A., Madsen, J., Parameshwar, J., Starling, R.C., and Uber, P.A. (2010). International Society for Heart and Lung Transplantation working formulation of a standardized nomenclature for cardiac allograft vasculopathy-2010. J. Heart Lung Transplant. *29*, 717–727.

Merigan, T.C., Renlund, D.G., Keay, S., Bristow, M.R., Starnes, V., O'Connell, J.B., Resta, S., Dunn, D., Gamberg, P., Ratkovec, R.M., *et al.* (1992). A controlled trial of ganciclovir to prevent cytomegalovirus disease after heart transplantation. N. Engl. J. Med. *326*, 1182–1186.

Miller, L.W., Naftel, D.C., Bourge, R.C., Kirklin, J.K., Brozena, S.C., Jarcho, J., Hobbs, R.E., and Mills, R.M. (1994). Infection after heart transplantation: a multiinstitutional study. Cardiac Transplant Research Database Group. J. Heart Lung Transplant. *13*, 381–392; discussion 393.

Mitsani, D., Nguyen, M.H., Girnita, D.M., Spichty, K., Kwak, E.J., Silveira, F.P., Toyoda, Y., Pilewski, J.M., Crespo, M., Bhama, J.K., *et al.* (2011). A polymorphism linked to elevated levels of interferon-gamma is associated with an increased risk of cytomegalovirus disease among Caucasian lung transplant recipients at a single center. J. Heart Lung Transplant. *30*, 523–529.

Monforte, V., Roman, A., Gavalda, J., Bravo, C., Gispert, P., Pahissa, A., and Morell, F. (2005). Preemptive therapy with intravenous ganciclovir for the prevention of cytomegalovirus disease in lung transplant recipients. Transplant. Proc. *37*, 4039–4042.

Mullen, J.C., Oreopoulos, A., Lien, D.C., Bentley, M.J., Modry, D.L., Stewart, K., Winton, T.L., Jackson, K., Doucette, K., Preiksaitis, J., *et al.* (2007). A randomized, controlled trial of daclizumab vs anti-thymocyte globulin induction for lung transplantation. J. Heart Lung Transplant. *26*, 504–510.

Mutimer, D., Mirza, D., Shaw, J., O'Donnell, K., and Elias, E. (1997). Enhanced (cytomegalovirus) viral replication associated with septic bacterial complications in liver transplant recipients. Transplantation *63*, 1411–1415.

Mylonakis, E., Kallas, W.M., and Fishman, J.A. (2002). Combination antiviral therapy for ganciclovir-resistant cytomegalovirus infection in solid-organ transplant recipients. Clin. Infect. Dis. *34*, 1337–1341.

Palmer, S.M., Baz, M.A., Sanders, L., Miralles, A.P., Lawrence, C.M., Rea, J.B., Zander, D.S., Edwards, L.J., Staples, E.D., Tapson, V.F., *et al.* (2001). Results of a randomized, prospective, multicenter trial of mycophenolate mofetil versus azathioprine in the prevention of acute lung allograft rejection. Transplantation *71*, 1772–1776.

Palmer, S.M., Limaye, A.P., Banks, M., Gallup, D., Chapman, J., Lawrence, E.C., Dunitz, J., Milstone, A., Reynolds, J., Yung, G.L., *et al.* (2010). Extended valganciclovir prophylaxis to prevent cytomegalovirus after lung transplantation: a randomized, controlled trial. Ann. Intern. Med. *152*, 761–769.

Paniagua, M.J., Crespo-Leiro, M.G., De la Fuente, L., Tabuyo, T., Mosquera, I., Canizares, A., Naya, C., Farina, P., Juffe, A., and Castro-Beiras, A. (2002). Prevention of cytomegalovirus disease after heart transplantation: preemptive therapy with 7 days' intravenous ganciclovir. Transplant. Proc. *34*, 69–70.

Paya, C., Humar, A., Dominguez, E., Washburn, K., Blumberg, E., Alexander, B., Freeman, R., Heaton, N., and Pescovitz, M.D. (2004) Efficacy and safety of valganciclovir vs. oral ganciclovir for prevention of cytomegalovirus disease in solid organ transplant recipients. Am. J. Transplant *4*, 611–620.

Petrakopoulou, P., Kubrich, M., Pehlivanli, S., Meiser, B., Reichart, B., von Scheidt, W., and Weis, M. (2004). Cytomegalovirus infection in heart transplant recipients is associated with impaired endothelial function. Circulation *110*, II207–212.

Potena, L., and Valantine, H.A. (2007). Cytomegalovirus-associated allograft rejection in heart transplant patients. Curr. Opin. Infect. Dis. *20*, 425–431.

Potena, L., Grigioni, F., Ortolani, P., Magnani, G., Marrozzini, C., Falchetti, E., Barbieri, A., Bacchi-Reggiani, L., Lazzarotto, T., Marzocchi, A., *et al.* (2003). Relevance of cytomegalovirus infection and coronary-artery remodeling in the first year after heart transplantation: a prospective three-dimensional intravascular ultrasound study. Transplantation *75*, 839–843.

Potena, L., Grigioni, F., Magnani, G., Lazzarotto, T., Musuraca, A.C., Ortolani, P., Coccolo, F., Fallani, F., Russo, A., and Branzi, A. (2009). Prophylaxis versus preemptive anti-cytomegalovirus approach for prevention of allograft vasculopathy in heart transplant recipients. J. Heart Lung Transplant. *28*, 461–467.

Rubin, R.H. (1989). The indirect effects of cytomegalovirus infection on the outcome of organ transplantation. JAMA *261*, 3607–3609.

Rubin, R.H. (2007). The pathogenesis and clinical management of cytomegalovirus infection in the organ transplant recipient: the end of the 'silo hypothesis'. Curr. Opin. Infect. Dis. *20*, 399–407.

Rubin, R.H., Kemmerly, S.A., Conti, D., Doran, M., Murray, B.M., Neylan, J.F., Pappas, C., Pitts, D., Avery, R., Pavlakis, M., *et al.* (2000) Prevention of primary cytomegalovirus disease in organ transplant recipients with oral ganciclovir or oral acyclovir prophylaxis. Transpl. Infect. Dis. *2*, 112–117.

Sanchez-Lazaro, I.J., Almenar, L., Martinez-Dolz, L., Buendia-Fuentes, F., Aguero, J., Navarro-Manchon, J., Vicente, J.L., and Salvador, A. (2011). A prospective randomized study comparing cyclosporine versus tacrolimus combined with daclizumab, mycophenolate mofetil, and steroids in heart transplantation. Clin. Transplant. *25*, 606–613.

Senechal, M., Dorent, R., du Montcel, S.T., Fillet, A.M., Ghossoub, J.J., Dubois, M., Pavie, A., and Gandjbakhch, I. (2003). Monitoring of human cytomegalovirus infections in heart transplant recipients by pp65 antigenemia. Clin. Transplant. *17*, 423–427.

Sester, U., Gartner, B.C., Wilkens, H., Schwaab, B., Wossner, R., Kindermann, I., Girndt, M., Meyerhans, A., Mueller-Lantzsch, N., Schafers, H.J., *et al.* (2005). Differences in CMV-specific T-cell levels and long-term susceptibility to CMV infection after kidney, heart and lung transplantation. Am. J. Transplant. *5*, 1483–1489.

Shirali, G.S., Ni, J., Chinnock, R.E., Johnston, J.K., Rosenthal, G.L., Bowles, N.E., and Towbin, J.A. (2001). Association of viral genome with graft loss in children after cardiac transplantation. N. Engl. J. Med. *344*, 1498–1503.

Shlobin, O.A., West, E.E., Lechtzin, N., Miller, S.M., Borja, M., Orens, J.B., Dropulic, L.K., and McDyer, J.F. (2006). Persistent cytomegalovirus-specific memory responses in the lung allograft and blood following primary infection in lung transplant recipients. J. Immunol. *176*, 2625–2634.

Simmonds, J., Fenton, M., Dewar, C., Ellins, E., Storry, C., Cubitt, D., Deanfield, J., Klein, N., Halcox, J., and Burch, M. (2008). Endothelial dysfunction and cytomegalovirus replication in pediatric heart transplantation. Circulation *117*, 2657–2661.

Singh, N. (2005). Late-onset cytomegalovirus disease as a significant complication in solid organ transplant recipients receiving antiviral prophylaxis: a call to heed the mounting evidence. Clin. Infect. Dis. *40*, 704–708.

Singh, N., Wannstedt, C., Keyes, L., Mayher, D., Tickerhoof, L., Akoad, M., Wagener, M.M., and Cacciarelli, T.V. (2008). Valganciclovir as preemptive therapy for cytomegalovirus in cytomegalovirus-seronegative liver transplant recipients of cytomegalovirus-seropositive donor allografts. Liver. Transpl. *14*, 240–244.

Snyder, L.D., Finlen-Copeland, C.A., Turbyfill, W.J., Howell, D., Willner, D.A., and Palmer, S.M. (2011a). Cytomegalovirus pneumonitis is a risk for bronchiolitis obliterans syndrome in lung transplantation. Am. J. Respir. Crit. Care Med. *181*, 1391–1396.

Snyder, L.D., Medinas, R., Chan, C., Sparks, S., Davis, W.A., Palmer, S.M., and Weinhold, K.J. (2011b). Polyfunctional cytomegalovirus-specific immunity in lung transplant recipients receiving valganciclovir prophylaxis. Am. J. Transplant. *11*, 553–560.

Snydman, D.R. (2006). The case for cytomegalovirus prophylaxis in solid organ transplantation. Rev. Med. Virol. *16*, 289–295.

Snydman, D.R., Limaye, A.P., Potena, L., and Zamora, M.R. (2011). Update and review: state-of-the-art management of cytomegalovirus infection and disease following thoracic organ transplantation. Transplant. Proc. *43*, S1–S17.

Srivastava, R., Curtis, M., Hendrickson, S., Burns, W.H., and Hosenpud, J.D. (1999). Strain specific effects of cytomegalovirus on endothelial cells: implications for investigating the relationship between CMV and cardiac allograft vasculopathy. Transplantation *68*, 1568–1573.

Strasfeld, L., Lee, I., Tatarowicz, W., Villano, S., and Chou, S. (2010). Virologic characterization of multidrug-resistant cytomegalovirus infection in 2 transplant recipients treated with maribavir. J. Infect. Dis. *202*, 104–108.

Streblow, D.N., Kreklywich, C., Yin, Q., De La Melena, V.T., Corless, C.L., Smith, P.A., Brakebill, C., Cook, J.W., Vink, C., Bruggeman, C.A., *et al.* (2003). Cytomegalovirus-mediated up-regulation of chemokine expression correlates with the acceleration of chronic rejection in rat heart transplants. J. Virol. *77*, 2182–2194.

Tamm, M., Aboyoun, C.L., Chhajed, P.N., Rainer, S., Malouf, M.A., and Glanville, A.R. (2004). Treated cytomegalovirus pneumonia is not associated with bronchiolitis obliterans syndrome. Am. J. Respir. Crit. Care Med. *170*, 1120–1123.

Thomas, L.D., Milstone, A.P., Miller, G.G., Loyd, J.E., and Dummer, J.S. (2009). Long-term outcomes of cytomegalovirus infection and disease after lung or heart-lung transplantation with a delayed ganciclovir regimen. Clin. Transplant. *23*, 476–483.

Tikkanen, J.M., Kallio, E.A., Bruggeman, C.A., Koskinen, P.K., and Lemstrom, K.B. (2001). Prevention of cytomegalovirus infection-enhanced experimental obliterative bronchiolitis by antiviral prophylaxis or immunosuppression in rat tracheal allografts. Am. J. Respir. Crit. Care Med. *164*, 672–679.

Toyoda, M., Galfayan, K., Galera, O.A., Petrosian, A., Czer, L.S., and Jordan, S.C. (1997). Cytomegalovirus infection induces anti-endothelial cell antibodies in cardiac and renal allograft recipients. Transpl. Immunol. *5*, 104–111.

Tu, W., Potena, L., Stepick-Biek, P., Liu, L., Dionis, K.Y., Luikart, H., Fearon, W.F., Holmes, T.H., Chin, C., Cooke, J.P., *et al.* (2006). T-cell immunity to subclinical cytomegalovirus infection reduces cardiac allograft disease. Circulation *114*, 1608–1615.

Tuzcu, E.M., Kapadia, S.R., Sachar, R., Ziada, K.M., Crowe, T.D., Feng, J., Magyar, W.A., Hobbs, R.E., Starling, R.C., Young, J.B., *et al.* (2005). Intravascular ultrasound evidence of angiographically silent progression in coronary atherosclerosis predicts long-term morbidity and mortality after cardiac transplantation. J. Am. Coll. Cardiol. *45*, 1538–1542.

Ustinov, J.A., Lahtinen, T.T., Bruggeman, C.A., Hayry, P.J., and Lautenschlager, I.T. (1994). Direct induction of class II molecules by cytomegalovirus in rat heart microvascular endothelial cells is inhibited by ganciclovir (DHPG). Transplantation *58*, 1027–1031.

Valantine, H.A., Gao, S.Z., Menon, S.G., Renlund, D.G., Hunt, S.A., Oyer, P., Stinson, E.B., Brown, B.W., Jr., Merigan, T.C., and Schroeder, J.S. (1999). Impact of prophylactic immediate posttransplant ganciclovir on development of transplant atherosclerosis: a post hoc analysis of a randomized, placebo-controlled study. Circulation *100*, 61–66.

Vigano, M., Dengler, T., Mattei, M.F., Poncelet, A., Vanhaecke, J., Vermes, E., Kleinloog, R., Li, Y., Gezahegen, Y., and Delgado, J.F. (2010). Lower incidence of cytomegalovirus infection with everolimus versus mycophenolate mofetil in de novo cardiac transplant recipients: a randomized, multicenter study. Transpl. Infect. Dis. *12*, 23–30.

Villa, M., Lage, E., Ballesteros, S., Canas, E., Sanchez, M., Ordonez, A., Borrego, J.M., Hinojosa, R., and Cisneros, J.M. (2003). Preemptive therapy for the prevention of cytomegalovirus disease following heart transplantation directed by PP65 antigenemia. Transplant. Proc. *35*, 732–734.

Vos, R., Vanaudenaerde, B.M., Ottevaere, A., Verleden, S.E., De Vleeschauwer, S.I., Willems-Widyastuti, A., Wauters, S., Van Raemdonck, D.E., Nawrot, T.S., Dupont, L.J., *et al.* (2010). Long-term azithromycin therapy for bronchiolitis obliterans syndrome: divide and conquer? J. Heart Lung Transplant. *29*, 1358–1368.

Vrtovec, B., Thomas, C.D., Radovancevic, R., Frazier, O.H., and Radovancevic, B. (2004). Comparison of intravenous ganciclovir and cytomegalovirus hyperimmune globulin pre-emptive treatment in cytomegalovirus-positive heart transplant recipients. J. Heart Lung Transplant *23*, 461–465.

Weigt, S.S., Elashoff, R.M., Keane, M.P., Strieter, R.M., Gomperts, B.N., Xue, Y.Y., Ardehali, A., Gregson, A.L., Kubak, B., Fishbein, M.C., *et al.* (2008). Altered levels of CC chemokines during pulmonary CMV predict BOS and mortality post-lung transplantation. Am. J. Transplant. *8*, 1512–1522.

Weis, M., Kledal, T.N., Lin, K.Y., Panchal, S.N., Gao, S.Z., Valantine, H.A., Mocarski, E.S., and Cooke, J.P. (2004). Cytomegalovirus infection impairs the nitric oxide synthase pathway: role of asymmetric dimethylarginine in transplant arteriosclerosis. Circulation *109*, 500–505.

Yamani, M.H., Avery, R., Mawhorter, S.D., McNeill, A., Cook, D., Ratliff, N.B., Pelegrin, D., Colosimo, P., Kiefer, K., Ludrosky, K., *et al.* (2005). The impact of CytoGam on cardiac transplant recipients with moderate hypogammaglobulinemia: a randomized single-center study. J. Heart Lung Transplant *24*, 1766–1769.

Young, J.B. (2000). Perspectives on cardiac allograft vasculopathy. Curr. Atheroscler. Rep. *2*, 259–271.

Zamora, M.R., Nicolls, M.R., Hodges, T.N., Marquesen, J., Astor, T., Grazia, T., and Weill, D. (2004). Following universal prophylaxis with intravenous ganciclovir and cytomegalovirus immune globulin, valganciclovir is safe and effective for prevention of CMV infection following lung transplantation. Am. J. Transplant. *4*, 1635–1642.

Zeevi, A., Uknis, M.E., Spichty, K.J., Tector, M., Keenan, R.J., Rinaldo, C., Yousem, S., Duncan, S., Paradis, I., Dauber, J., *et al.* (1992). Proliferation of cytomegalovirus-primed lymphocytes in bronchoalveolar lavages from lung transplant patients. Transplantation *54*, 635–639.

Zeevi, A., Morel, P., Spichty, K., Dauber, J., Yousem, S., Williams, P., Grgurich, W., Pham, S., Iacono, A., Keenan, R., *et al.* (1998). Clinical significance of CMV-specific T helper responses in lung transplant recipients. Hum. Immunol. *59*, 768–775.

Zeevi, A., Spichty, K., Banas, R., Cai, J., Donnenberg, V.S., Donnenberg, A.D., Ahmed, M., Dauber, J., Iacono, A., Keenan, R., *et al.* (1999). Clinical significance of cytomegalovirus-specific T helper responses and cytokine production in lung transplant recipients. Intervirology *42*, 291–300.

Zeevi, A., Husain, S., Spichty, K.J., Raza, K., Woodcock, J.B., Zaldonis, D., Carruth, L.M., Kowalski, R.J., Britz, J.A., and McCurry, K.R. (2007). Recovery of functional memory T-cells in lung transplant recipients following induction therapy with alemtuzumab. Am. J. Transplant. *7*, 471–475.

Zuk, D.M., Humar, A., Weinkauf, J.G., Lien, D.C., Nador, R.G., and Kumar, D. (2010). An international survey of cytomegalovirus management practices in lung transplantation. Transplantation *90*, 672–676.

Clinical Cytomegalovirus Research: Liver and Kidney Transplantation

II.14

Vincent C. Emery, Richard S.B. Milne and Paul D. Griffiths

Abstract

CMV infection significantly impacts on the success of transplantation of abdominal organs. There are a range of both direct and indirect effects attributable to CMV. In this chapter we survey our current understanding of CMV pathogenesis, the immune control of CMV replication in these transplanted patients, the antiviral chemotherapeutic options available for managing infection/disease and consider the newer options available for drug therapy and for vaccination.

Introduction

Kidney and liver transplantation have become the optimal clinical management strategy for patients with end-organ disease affecting these two sites. While the deceased donor remains an important contributor to the donor pool there is an increasing amount of living donor kidney and partial liver transplantation occurring worldwide and the use of split livers has also impacted on the number of organs available (Florman and Miller, 2006; Shrestha, 2009). The success of these transplant programmes has been partly attributable to the progress in immunosuppressive drugs which have minimized acute graft rejection although there is little evidence that long-term graft survival has improved significantly over the last 10 years (Tantravahi *et al.*, 2007). As a consequence of immunosuppressive therapy, chronic virus infections such as cytomegalovirus (CMV) can exert their full pathogenic potential. In this chapter, we survey the risk factors for CMV infection and disease after kidney and liver transplantation, the important role that the immune system plays despite the immunocompromised state of the host, the effects of CMV on long-term graft function and the therapeutic options used to control CMV including recent advances in CMV vaccines in these patient populations.

Risk Factors for CMV infection and disease after renal transplantation

Risk of CMV infection and disease post transplant is typically stratified based on donor and recipient CMV IgG serostatus. There are four possible donor/recipient combinations: D^+R^-, D^+R^+, D^-R^+ and D^-R^-, with the D^+R^- group at greatest risk for CMV infection and disease because the recipient immune system is naïve to CMV. Natural history studies indicate that the transmission rate in the D^+R^- context is at least 90% and prior to the introduction of antiviral therapy (see later) approximately half of the infected individuals would suffer CMV disease which was associated with high morbidity and mortality (Falagas and Snydman, 1995; Razonable and Paya, 2003; Parsaik *et al.*, 2011). Recipients who are already CMV seropositive are at risk of reactivation of endogenous virus and those who receive an organ from a CMV-positive donor are additionally at risk of reinfection with the donor strain. The risk of active CMV infection in both these populations is relatively high but the incidence of disease is much lower than that observed in the D^+R^- population. The majority of studies ascribe a very low risk for infection and disease to the D^-R^- population. While these serological groupings have been useful for stratifying patients, other changes in clinical management have also impacted on the risk of CMV infection and disease. One of the most significant influences has been the increased deployment of induction immunosuppressive therapy with agents such as anti-CD25 antibodies and anti-thymocyte globulin both of which are widely used in kidney transplant patients and increase the risk for CMV infection in both D^+R^- and R^+ groups (Webster *et al.*, 2010). In the near future we will see other newer immunosuppressants being deployed to minimize organ rejection. Interestingly some of these such as belatacept, have been associated with no impact on the risk of CMV disease compared to cyclosporin A (Vincenti *et al.*, 2010) whereas others, such as the mammalian target of rapamycin (mTOR) inhibitors (sirolimus, everolimus) have been associated with a decrease in CMV disease (Kahan, 2008).

The direct and indirect effects of CMV

The pathological consequences of CMV have been divided into the direct effects and indirect effects (Fig. II.14.1) (Fishman, 2007). The direct effects include pathologies where CMV infection can be demonstrated in individual patients together with organ or system dysfunction. In the absence of prophylactic therapy (see later) direct effects usually manifest in the early phase post kidney or liver transplantation, frequently in the first 60–90 days. Although the incidence of CMV disease differs between the donor-recipient serological groups there is no evidence that the type of disease differs between these risk groups. However, there are data indicating that certain diseases are more prevalent in some groups: for example, liver transplant recipients are more likely to suffer from CMV hepatitis compared to the kidney transplant recipient, probably due to reactivation of virus in the transplanted organ.

The indirect effects of CMV have been relatively controversial because they are observed only in populations and because CMV detection has not been related to the time of the effect being observed. Furthermore, a mechanistic basis for the variety of effects observed has been lacking. Over recent years this situation has changed with the availability of sensitive *in situ* hybridization methods for detecting CMV DNA in target organs and optimized immunohistochemistry detection and a growing appreciation of the multiple ways that the CMV proteome interferes with key aspects of cell biology. Thus, the CMV US28 chemokine receptor leads to increased smooth muscle cell migration and promotes angiogenesis via cyclooxygenase 2 (COX-2) while infection up-regulates a range of cellular adhesion molecules including the leukotriene LTB4, ICAM- and VCAM-1 in transplanted livers and ICAM-1, PDGF and TGF-beta resulting in vascular thickening in transplanted kidneys. In addition, CMV decreases matrix metalloproteinase activity and increases tissue inhibitor of metalloproteinases (TIMP) in infected macrophages. In addition, CMV can directly affect dendritic cells altering their capacity for antigen presentation, pancreatic beta cells and micro- and macrovasculature endothelial cells (Dzabic and Söderberg-Nauclér, 2011). Though, as yet, none of these virus-induced alterations of host systems has been formally associated with CMV pathology *in vivo* they provide a range of mechanisms by which CMV could contribute to the indirect effects summarized in Fig. II.14.1.

A pertinent question has been whether CMV infection in the transplanted organ is a transient phenomenon or whether it precedes the occurrence of CMV viraemia and affects long-term graft function; indeed, the origin and producer cells in CMV viraemia remain unknown. Two groups have provided evidence that the presence of CMV DNA in the graft, as detected by *in situ* hybridization, does affect long-term graft

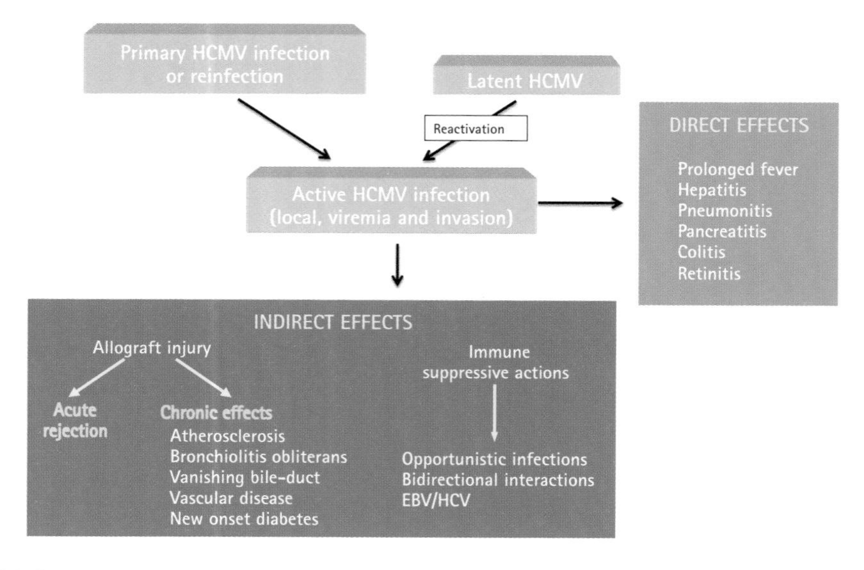

Figure II.14.1 Summary of the drivers for stimulating replication of CMV in the transplanted organ and the direct and indirect effects associated with CMV infection.

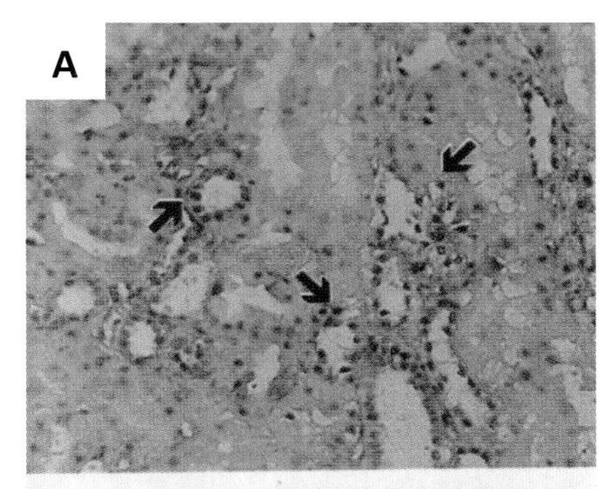

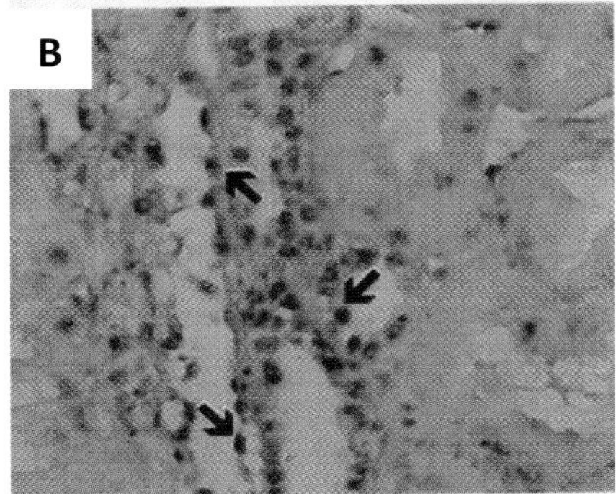

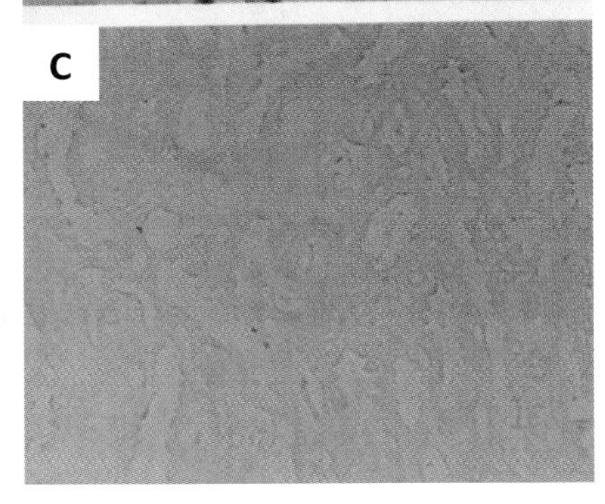

Figure II.14.2 (A,B) Low and high magnification images of *in-situ* staining using a CMV specific riboprobe illustrating the extensive presence of CMV DNA in renal tubule epithelial cells and some endothelial cells (arrowed) in a kidney biopsy taken 7 days after transplantation. (C) Negative control using a riboprobe specific for HHV-7 DNA. (Taken from Li *et al.*, 2010.)

survival. In a study by Li and colleagues (Li *et al.*, 2010), CMV DNA in the kidney was detected early after transplantation, approximately 14 days prior to the detection of CMV viraemia (Fig. II.14.2). The infection was extensive involving renal tubular epithelial cells and some endothelial cells and was apparent in 70% of patients analysed. Although there was no formal association with biopsy proven graft rejection it is worth noting that CMV DNA was detected in biopsies from 50% of patients with clinical signs of rejection but with no pathology indicative of graft rejection. These data argue that CMV infection in the kidney may produce a clinical picture of rejection which is distinct from true acute rejection mediated by a host immune response mounted against the graft. Long term follow up of these patients revealed that early presence of CMV DNA in the kidney was associated with poorer graft function at 1 year and beyond. In a separate study from Finland similar data have been provided showing that persistent kidney infection with CMV was associated with poorer renal function at 1 and 3 years post transplantation and was the dominant risk factor in a multivariable analysis including recognized risk factors such as degree of HLA mismatch between donor and recipient (Helantera *et al.*, 2006).

CMV replication kinetics in the kidney and liver transplant recipient

Work from many groups has shown that viral load levels are highest in the D^+R^- population and are closely associated with the probability of developing CMV disease; indeed the sigmoidal relationship between viral load and risk of disease (Fig. II.14.3) is now well established. In addition, cumulative viral load also plays an important role in pathogenesis (Regoes *et al.*, 2006). CMV replication *in vivo* is a highly dynamic process and viral genome doubling times of between 1–2 days in kidney or liver transplant recipients are not uncommon (Emery *et al.*, 1999). The immune naïve individual has been shown to have a more rapid CMV replication rate compared to the recipient seropositive individuals

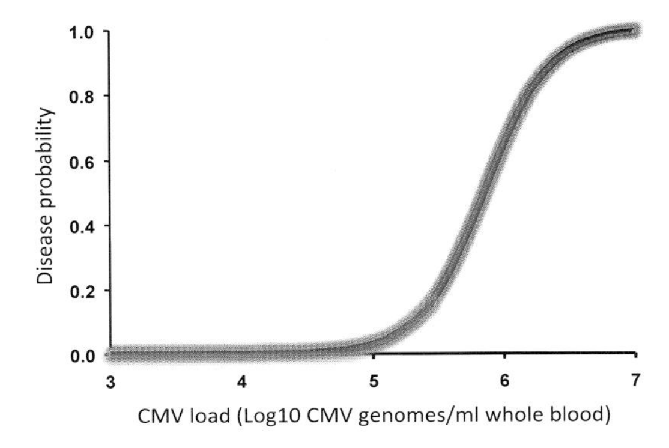

Figure II.14.3 Probability of CMV disease developing accordingly to viral load in blood in liver transplant recipients (taken from Cope *et al.*, 1998). The green shadows indicate the low probability associated with relatively low viral loads whereas at a critical threshold of virus load the probability increases dramatically (indicated by the red shadow).

reflecting a functional immune response against CMV (Emery *et al.*, 2002). These data have been used to provide an estimate of the R_0 (basic reproductive number; the number of newly infected cells arising from one infected cell when target cells are unlimited) for CMV in the organ transplant recipient (Emery *et al.*, 2002). The data suggest that primary infection in a D^+R^- patient population is associated with an R_0 of between 11 and 18 whereas in the immune experienced recipient this number drops to about 2.4. While these data will be influenced by immunosuppressive regimens they provide some support for approaches that can induce, boost, or augment immunity so that control of CMV replication can be achieved (see later). The net result of such an intervention would be the reduced incidence of CMV disease (Fig. II.14.3). The use of viral replication kinetics to identify patients at risk of CMV disease has also been reported although at present has not found its way into the routine diagnostic laboratory as a parameter for risk stratification (Emery *et al.*, 2000b). Nevertheless, routine measurement of CMV viral load using real-time quantitative (qPCR) has become a mainstay of patient management and has allowed the deployment of pre-emptive therapeutic strategies (see below).

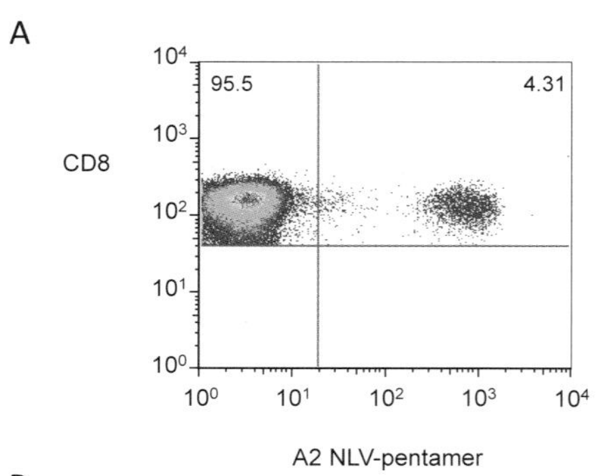

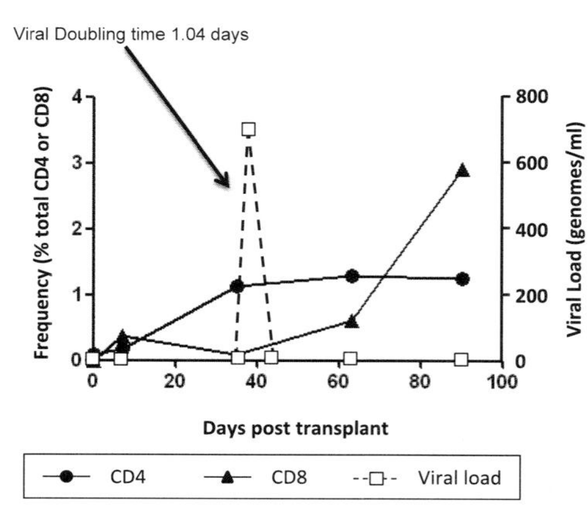

Figure II.14.4 (A) Example of the CMV pp65 NLV epitope specific CD8 T-cell responses in a CMV-seropositive renal transplant recipient experiencing an active CMV infection at day 35 post transplant. In this case, peripheral blood lymphocytes were stained directly with a Class I HLA pentamer refolded with the NLV peptide and an antibody against CD8, illustrating that 4.31% of the $CD8^+$ T-cells recognized this epitope. (B) Course of the cell-mediated immune response in another CMV-seropositive renal transplant recipient. CD4 response to whole CMV lysate (solid line, circles) and CD8 response to pp65 peptide pool (solid line triangles) are shown. Viral load with calculated doubling time is also indicated (dotted line) to illustrate the rapid control of viraemia.

Immune control of CMV in the kidney and liver transplant recipient

It is important to realize that, despite the receipt of immunosuppressive drugs, the immune response against CMV remains an important component of the control of replication. Significant progress has been made in understanding the breadth of the immune response against CMV in both the immunocompetent and immunocompromised host over the last 10 years (Gerna *et al.*, 2006; La Rosa *et al.*, 2007; Egli *et al.*, 2008; Nebbia *et al.*, 2008; Zelini *et al.*, 2010; Fornara *et al.*, 2011; Gerna *et al.*, 2011). This progress has been facilitated by the availability of new reagents such as Class I HLA tetramers and pentamers and also the use of cytokine specific assays such as the ELISPOT and intracellular cytokine profiling (Fig. II.14.4). The current view is that control of CMV replication in the immunocompromised host does not depend solely on the qualitative presence of CD4 and CD8 T-cells recognizing CMV antigens but is more driven by the quality of the response. Thus, a number of studies have shown that polyfunctional T-cells are protective against high-level replication and that presence of a high proportion of CD8 T-cells that produce cytokines such as IFNγ is associated with reduced risk of developing high-level replication and CMV disease (Crough *et al.*, 2007; Mattes *et al.*, 2008). These studies have also allowed specific thresholds to be identified for CD4 and CD8 T-cells that appear to be predictive of CMV replication that may require treatment. However, these assays are not yet routinely deployed in the diagnostic laboratory to aid CMV management. During episodes of CMV replication, the CMV-specific CD4 and CD8 T-cells express high levels of PD-1 indicating that these cells have a limited proliferative potential and may contribute to the reduced effectiveness of the immune system to rapidly control replication (La *et al.*, 2008;

Sester *et al.*, 2008; Krishnan *et al.*, 2010). As well as contributing to a refined understanding of the nature of the protective response, these observations suggest that the novel therapeutic approach of immune receptor blockade should be evaluated.

An area which has been somewhat neglected in recent years has been the humoral immune response to CMV. The titre of IgM antibodies has been associated with the probability of disease after liver transplantation although in multivariable models viral load remained the dominant factor (Emery *et al.*, 2000a). More recently, the identification of novel neutralizing antibody targets and the encouraging results of vaccination studies (summarized below) in which antibodies appeared to play an important role in the control of CMV replication suggest that a re-evaluation of the role of antibody in controlling CMV post transplant is appropriate.

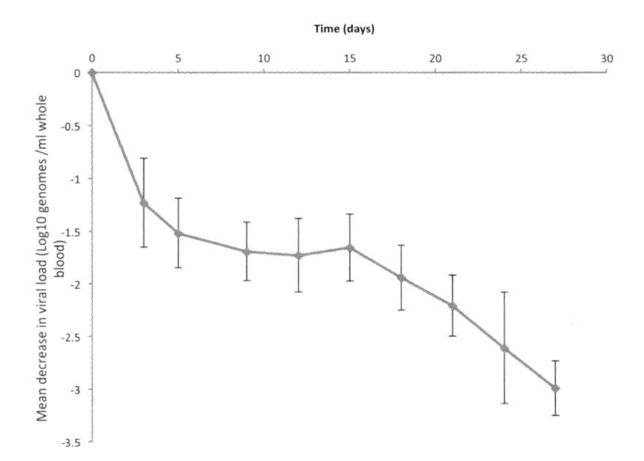

Figure II.14.5 Decline in CMV viral load after therapy. Data are from 45 CMV-seropositive solid organ transplant recipients managed with pre-emptive therapy (ganciclovir or valganciclovir). Day 0 represents the day of peak viral load prior to the onset of decline as a result of antiviral therapy. Error bars indicate standard deviations. Viral load in whole blood was measured using quantitative real time PCR (Emery *et al.*, unpublished data, 2011).

Antiviral approaches to managing CMV after kidney and liver transplantation

Since the introduction of effective anti CMV therapy (Fig. II.14.5) in the late 1980s two approaches to the management of CMV after transplantation have evolved: prophylaxis and pre-emptive therapy (see Table II.14.1) (Hodson *et al.*, 2005, 2008; Khoury *et al.*, 2006). Prophylaxis has generally been targeted at high risk patients, usually D^+R^-, and all the pivotal licensing studies have been performed in this patient group (Paya *et al.*, 2004). Pre-emptive therapy relies on the detection of ongoing viral replication evidenced by the presence of antigenaemia or DNA in whole blood or plasma (Strippoli *et al.*, 2006). In many centres the adoption of prophylaxis has been driven by logistics in so far as patients who reside some distance from the hospital transplant centre cannot be monitored with the frequency required to deliver effective pre-emptive therapy (at least once/week). There have been many trials of prophylaxis with aciclovir, valaciclovir, ganciclovir and valganciclovir in renal and liver transplant recipients (see Table II.14.2). The current consensus recommendations for therapy in high risk patients involve the deployment of valganciclovir prophylaxis at 900mg (reduced if renal dysfunction is present) once daily for 3–6 months (Kotton *et al.*, 2010). This recommendation is based upon randomized studies of valganciclovir versus oral ganciclovir prophylaxis for 3 months in D^+R^- solid organ transplant patients (Paya *et al.*, 2004; Humar *et al.*, 2010a,b) and the more recent randomized study comparing 3 months versus 6 months valganciclovir prophylaxis in D^+R^- renal transplant recipients (Humar *et al.*, 2010a). Of note was the benefit of 6 months prophylaxis on the incidence of late CMV syndrome and disease following the cessation of prophylaxis. In the IMPACT study, late CMV disease was reduced to 16% in the 6-month valganciclovir arm

Table II.14.1 Advantages and disadvantages of prophylaxis versus pre-emptive therapy for CMV

Prophylaxis	Pre-emptive therapy
Targeted at high risk patients	Initiated based on viral load thresholds reducing unnecessary treatment
Reduces the direct and some indirect effects of CMV	Patients may require multiple treatment episodes
Post prophylaxis infection and disease still occurs and may involve resistant strains	Significantly reduced the direct effects of CMV
Costs of extended prophylaxis	Potential impact on the indirect effects of CMV less documented
Predictors for late CMV disease not easily defined	Requires frequent sampling and rapid action

Table II.14.2 Summary of important randomized trials evaluating prophylactic strategies for preventing CMV post liver or kidney transplantation

Organ transplant	Drug/protocol	Effects on CMV infection	Effects on CMV disease	Effects on patient/graft survival	Reference
Kidney	Aciclovir	Reduced	Reduced	None	Balfour *et al.* (1989)
Kidney	Valaciclovir	Reduced	Reduced	Significantly reduced graft rejection	Lowance *et al.* (1999)
Kidney	IFN	Reduced	None	None	Cheesman *et al.* (1979)
Kidney	IFN	Reduced	Reduced	None	Hirsch *et al.* (1983)
Kidney	IFN	Reduced	None	None	Lui *et al.* (1992)
Kidney	IgG	None	None	None	Metselaar *et al.* (1989)
Liver	IgG	None	Reduced	None	Snydman *et al.* (1993)
Liver and kidney[a]	GCV vs. VGCV prophylaxis	None	None[b]	None	Paya *et al.* (2004)
Kidney	VGCV (100d vs. 200d)	Reduced	Reduced	None	Humar *et al.* (2010)
Kidney	Valaciclovir prophylaxis vs. VGCV pre-emptive therapy	Reduced in prophylaxis arm	None	Reduced acute rejection in prophylaxis arm	Reischig *et al.* (2008)
Kidney	GCV prophylaxis vs. VGCV pre-emptive	Reduced in prophylaxis arm	Reduced in pre-emptive arm	Improved in prophylaxis arm	Kliem *et al.* (2009)
Liver	GCV	Reduced	Reduced	None	Gane *et al.* (1998)

IFN, interferon alpha; GCV, oral ganciclovir; VGCV, valganciclovir.
The effects on CMV infection, disease and long-term outcomes such as graft or patient survival.
[a]This trial involved other solid organ transplant patients.
[b]The incidence of CMV disease in liver patients exposed to VGCV was higher than that observed in the oral GCV arm.

compared to 32% in the 3 month arm. These benefits of extending prophylaxis must be considered in the light of total expenditure per patient (Blumberg *et al.*, 2010). As yet the risk factors associated with the occurrence of post-prophylaxis infection and syndrome/disease are poorly recognized, with CD8 T-cell responses at day 100 in D^+R^- patients the only current marker for protection against late CMV disease (Kumar *et al.*, 2009). Recent data indicate that, due to the extended time frame of late CMV disease after prophylaxis and the rapid replication of CMV, a pre-emptive approach during the post-prophylactic period is sub-optimal (Lisboa *et al.*, 2011). Thus, the prevention and control of late-onset disease represent major current challenges in CMV management.

The alternative management strategy to prophylaxis relies on viral replication triggered pre-emptive therapy. In recent years this has been based upon viral load thresholds in order to minimize the unnecessary treatment of individuals who have a low probability to progressing to CMV disease whilst ensuring those at high risk are placed on therapy rapidly to prevent progression to high viral loads and thence disease. Pre-emptive therapy is more frequently deployed in medium risk patients although some centres deploy it for high risk patients with significant success. As stated previously, pre-emptive therapy requires excellent co-ordination between the physician and diagnostic laboratory and requires samples to be obtained from patients on a weekly or better twice-weekly basis.

Comparisons between prophylaxis and pre-emptive therapy approaches have been performed through randomized controlled studies (Khoury *et al.*, 2006; Kliem *et al.*, 2008) and in meta-analysis (see above) of a number of prophylactic and pre-emptive therapy trials concluding that both approaches offer excellent control of CMV disease and are equally cost-effective even after considering the costs of CMV surveillance by real time PCR. However, there is more evidence supporting the benefit of prophylaxis on the indirect effects of CMV with one study showing improved long-term graft survival in kidney patients randomized to receive prophylaxis compared to pre-emptive therapy (Kliem *et al.*, 2008). However, in another study this effect was not observed. (Spinner *et al.*, 2010)

An area of controversy remains with respect to the viral load thresholds that should be used to initiate therapy. Some studies have shown that an antigenaemia level of 50 positive cells/400,000 PBMCs (equivalent to approximately 8600 genomes/ml blood) was appropriate to control CMV disease while Gerna and colleagues have reported a comparison of DNAemia threshold of 300,000 genomes/ml blood to an antigenaemia threshold of 100 positive cells resulted in the treatment of 23/99 patients in the DNAemia arm compared to 42/101 in the antigenaemia arm ($P=0.01$). Of note, four patients (three in the antigenaemia arm, one in the DNAemia arm) went on to get CMV disease in this study indicating the viral load threshold used was not sufficiently robust (Gerna *et al.*, 2007, 2008, 2009). It is noteworthy that standardization of CMV PCR assays is an urgent requirement based on a number of multicentre studies (Paya *et al.*, 2004; Caliendo *et al.*, 2009; Lilleri *et al.*, 2009) and an international WHO standard has recently become available from the National Institute of Biological Standards and Control (NIBSC) UK which should facilitate comparisons between different laboratories and different assays.

The current guidelines suggest that pre-emptive treatment of CMV infection and treatment of CMV syndrome and disease should use intravenous ganciclovir or valganciclovir at 900mg twice daily (Kotton *et al.*, 2010). Observational trials and randomized trials have shown that these two drugs are equally effective with respect to the kinetics of viral load decline. The advantage of valganciclovir is that patients do not require hospitalization although it is imperative that patients continue to take the drug even after the resolution of CMV symptoms since control of replication below the level of detection in current quantitative PCR assays appears to be important for successful resolution of disease. A factor associated with both pre-emptive therapy and treatment of overt CMV syndrome/disease is the recurrence of a further episode of replication requiring treatment. This issue remains important as it may contribute to the development of drug resistance. Although accurate estimates of the frequency of second episodes of replication are not available it is likely that up to 30% of patients will suffer a second episode of replication. Risk factors for the recurrence of replication include failure to reduce viral loads below detectable levels in whole blood or plasma within 21 days, high viral loads and a slow viral load decline rate (Humar *et al.*, 2003).

Currently, drug resistant strains of CMV have not proven to be a major clinical problem in the solid organ transplant setting. There are case reports of resistant strains which often reflect sub-optimal management of CMV infection i.e. multiple treatment episodes without ensuring viral loads are controlled below the level of detection or inappropriate dosing of drug. The incidence of CMV drug resistance in large randomized prophylactic studies supports this notion that drug resistance is uncommon (Boivin *et al.*, 2009). However, in patients with extended periods of therapy and ongoing CMV replication, the dynamics of replication and the error rate of the DNA polymerase combine to increase the risk of drug resistant virus developing (Limaye *et al.*, 2000; Limaye, 2002). Drug resistance against the currently deployed antiviral drugs maps to either the UL97 kinase (for ganciclovir) or the UL54 DNA polymerase (for ganciclovir, foscarnet and cidofovir) (Chou and Marousek, 2006; see also Chapter II.19). Temporally, UL97 kinase mutations develop first, are associated with low to medium level resistance and tend to be focused on key amino acids (positions 460, 520, 594, 595 and 603). Subsequently, UL54 mutations can appear if the patient is maintained on ganciclovir therapy and these can confer cross resistance against foscarnet and cidofovir. Drug resistance should be suspected if patient viral loads remain static or continue to increase over the first 7 days following initiation of antiviral therapy for CMV (Kotton *et al.*, 2010). Paradoxically, a subset of patients treated pre-emptively show an increase in viral load over the first 7 days of therapy which is driven by the natural replication kinetics of CMV i.e. patients where CMV is replicating rapidly before the introduction of antiviral therapy are more likely to exhibit an overshoot in CMV load before it starts to decline (Gerna *et al.*, 2005; Buyck *et al.*, 2010). This pattern is difficult to discern from patients with true drug resistance but in the majority of cases where this is observed in the first therapeutic intervention drug resistance is unlikely to be present.

New approaches to managing CMV after kidney and liver transplantation

As stated above ganciclovir and its pro-drug valganciclovir have become the dominant antiviral agents used to control and manage CMV infection and disease after renal and liver transplantation. However, despite the drug having a high potency against CMV replication (Emery and Griffiths, 2000) and having proven clinical benefit in a variety of studies, the safety profile of ganciclovir is suboptimal. For example, ganciclovir is bone marrow suppressive and so use in neutropenic patients is limited and extended use has been associated with the development of drug resistance although this is not a major problem in the kidney and liver transplant recipient. Clearly, new drugs with similar or better potency and an improved safety profile would be a valuable addition to the management of CMV after transplantation. In addition, drugs that could be used in combination with ganciclovir may also improve the overall control of replication with benefits to both short and long-term graft function and patient well being. However, new drugs for managing CMV have been difficult to develop for a number of reasons, for example, the difficulty of producing compounds against certain attractive targets, such as the CMV protease, with low IC_{50}s and the perception that the market size for a CMV antiviral is relatively small. Notwithstanding these comments, in 2011 a variety of lead antiviral compounds are in development for CMV (reviewed in Emery and Milne, 2011). The drug which has shown most promise for CMV is maribavir, a benzimidazole based compound, which inhibits the CMV UL97 protein kinase (Biron *et al.*, 2002; Shannon-Lowe and Emery, 2010). (As UL97 activates ganciclovir to its monophosphate form, combination therapy is not an option) (Chou and Marousek, 2006). *In vitro*, maribavir is a potent inhibitor of CMV replication although *in vivo* it is highly protein bound which may impact on its performance characteristics. The drug has been evaluated in phase 2 trials in stem cell transplant (SCT) patients (Winston *et al.*, 2008) and an optimized dose obtained from this trial used for phase 3 studies in SCT and solid organ transplant recipients. The results of these studies were disappointing in so far as prophylaxis with maribavir showed no significant reduction in CMV infection or disease in the SCT trial (Marty *et al.*, 2011) and the solid organ transplant trial was stopped early on the advice of the data safety monitoring board due to lack of efficacy. Thus, a compound which showed impressive *in vitro* inhibition of CMV replication was unable to demonstrate efficacy *in vivo* at the doses used. Two other drugs which are currently being evaluated in phase 2 studies include the CMV maturation inhibitor AIC246 (Goldner *et al.*, 2011) which has shown impressive *in vitro* and *in vivo* activity in phase 2a studies in renal transplant recipients (Lischka *et al.*, 2010) and also a hexadecyloxypropyl ester of cidofovir which is approximately 100-fold more active than the parental cidofovir (Williams-Aziz *et al.*, 2005) and is likely to undergo phase 2 studies in solid organ transplant recipients after similar studies in SCT patients.

Vaccines and immunotherapy

The documented effect of prior natural CMV immunity on virus replication, even in the face of immunosuppressive therapy post-transplant (for example Emery *et al.*, 2002), indicates that immunotherapeutic approaches may be appropriate for management of CMV in this setting. Such therapy would be applicable to CMV-positive recipients, in whom responses would be boosted, as well as CMV negative recipients in whom a response would be stimulated *de novo*.

In a recent phase 2 double-blind placebo-controlled trial, a subunit vaccine containing recombinant gB with MF59 adjuvant has been evaluated (Griffiths *et al.*, 2011). Patients on the waiting list for renal or liver transplantation were given three doses of vaccine or placebo. They were then followed up after transplantation to evaluate the effect of the vaccine on parameters of CMV replication. Both CMV-seropositive and seronegative patients were included in the study. The vaccine was expected to induce primarily a B-cell response and all patients who received the vaccine generated an IgG response. There was no evidence of a vaccine specific CD4 or CD8 response. The duration of viraemia and of antiviral therapy after transplantation were reduced in both CMV-seropositive and seronegative vaccine recipients. There was no effect on incidence of viraemia in the seropositive recipients, however, in CMV seronegative vaccine recipients who received a CMV-positive organ, the incidence of CMV infection was significantly reduced. In the vaccine recipients, gB specific IgG levels pre-transplant were inversely correlated with the duration of CMV viraemia post transplant.

This trial demonstrates the potential of a CMV vaccine for preventing primary infection or reducing its severity in the high risk $D^{+}R^{-}$ patient group. It also illustrates the utility of a vaccine delivered in the immunotherapeutic context for controlling CMV replication in the more numerous R^{+} patient group. The study also highlights the somewhat overlooked contribution of antibody to preventing CMV infection and controlling CMV replication. Two tantalizing possibilities arise: 1. Routine immunization of transplant patients. 2. Therapeutic deployment of selected anti-CMV immunoglobulins. Antibodies recognizing gB are clearly candidates for inclusion in such a preparation. Antibodies targeting the recently identified antigens encoded in the UL128–131 region are an obvious second specificity (Macagno *et al.*, 2010). Such preparations can now be produced with ease so a clinical trial of such an intervention could be initiated in the near future.

Of note, two other CMV vaccine candidates, an alphavirus based replicon system producing pp65 and glycoprotein B (Alphavax) and a DNA vaccine expressing the same two antigens (Vical) have recently been licensed for further development to Novartis and Astellas respectively. We envisage that further evaluation of these prototype vaccines will also be performed in solid organ transplant patients over the next 3 years.

Future perspective: individualized management of CMV

With the recent advances in understanding of CMV specific immune responses in the transplant setting as well as of viral replication dynamics and their relationship to risk of disease, we are on the verge of a new era of individualized risk assessment and management. A viable aim would be the generation of an integrated algorithm for predicting risk of CMV, need for therapy, likely duration of therapy and hence on this basis to assign the appropriate management approach. Application of such an algorithm alongside the adoption of universal immunotherapy might be expected to challenge current management approaches and revolutionize the control of CMV post transplant.

References

Biron, K.K., Harvey, R.J., Chamberlain, S.C., Good, S.S., Smith, A.A., III, Davis, M.G., Talarico, C.L., Miller, W.H., Ferris, R., Dornsife, R.E., *et al.* (2002). Potent and selective inhibition of human cytomegalovirus replication by 1263W94, a benzimidazole L-riboside with a unique mode of action. Antimicrob. Agents Chemother. *46*, 2365–2372.

Blumberg, E.A., Hauser, I.A., Stanisic, S., Mueller, E., Berenson, K., Gahlemann, C.G., Humar, A., and Jardine, A.G. (2010). Prolonged prophylaxis with valganciclovir is cost effective in reducing posttransplant cytomegalovirus disease within the USA. Transplantation *90*, 1420–1426.

Boivin, G., Goyette, N., Rollag, H., Jardine, A.G., Pescovitz, M.D., Asberg, A., Ives, J., Hartmann, A., and Humar, A. (2009). Cytomegalovirus resistance in solid organ transplant recipients treated with intravenous ganciclovir or oral valganciclovir. Antivir. Ther. *14*, 697–704.

Buyck, H.C., Griffiths, P.D., and Emery, V.C. (2010). Human cytomegalovirus (HCMV) replication kinetics in stem cell transplant recipients following anti-HCMV therapy. J. Clin. Virol. *49*, 32–36.

Caliendo, A.M., Shahbazian, M.D., Schaper, C., Ingersoll, J., Abdul-Ali, D., Boonyaratanakornkit, J., Pang, X.L., Fox, J., Preiksaitis, J., and Schonbrunner, E.R. (2009). A commutable cytomegalovirus calibrator is required to improve the agreement of viral load values between laboratories. Clin. Chem. *55*, 1701–1710.

Chou, S., and Marousek, G.I. (2006). Maribavir antagonizes the antiviral action of ganciclovir on human cytomegalovirus. Antimicrob. Agents Chemother. *50*, 3470–3472.

Crough, T., Fazou, C., Weiss, J., Campbell, S., Davenport, M.P., Bell, S.C., Galbraith, A., McNeil, K., and Khanna, R. (2007). Symptomatic and asymptomatic viral recrudescence in solid-organ transplant recipients and its relationship with the antigen-specific CD8(+) T-cell response. J. Virol. *81*, 11538–11542.

Dzabic, M., and Söderberg-Nauclér, C. (2011). Indirect effects of cytomegalovirus: a challenge for improved long-term outcome in transplant recipients. Br. J. Transpl. *Suppl.*, 9–13.

Egli, A., Binet, I., Binggeli, S., Jager, C., Dumoulin, A., Schaub, S., Steiger, J., Sester, U., Sester, M., and Hirsch, H.H. (2008). Cytomegalovirus-specific T-cell responses and viral replication in kidney transplant recipients. J. Transl. Med. *6*, 29.

Emery, V.C., and Griffiths, P.D. (2000). Prediction of cytomegalovirus load and resistance patterns after antiviral chemotherapy. Proc. Natl. Acad. Sci. U.S.A. *97*, 8039–8044.

Emery, V.C., Cope, A.V., Bowen, E.F., Gor, D., and Griffiths, P.D. (1999). The dynamics of human cytomegalovirus replication *in vivo*. J. Exp. Med. *190*, 177–182.

Emery, V.C., Cope, A.V., Sabin, C.A., Burroughs, A.K., Rolles, K., Lazzarotto, T., Landini, M.P., Brojanac, S., Wise, J., and Maine, G.T. (2000a). Relationship between IgM antibody to human cytomegalovirus, virus load, donor and recipient serostatus, and administration of methylprednisolone as risk factors for cytomegalovirus disease after liver transplantation. J. Infect. Dis. *182*, 1610–1615.

Emery, V.C., Sabin, C.A., Cope, A.V., Gor, D., Hassan-Walker, A.F., and Griffiths, P.D. (2000b). Application of viral-load kinetics to identify patients who develop cytomegalovirus disease after transplantation. Lancet *355*, 2032–2036.

Emery, V.C., Hassan-Walker, A.F., Burroughs, A.K., and Griffiths, P.D. (2002). Human cytomegalovirus (HCMV) replication dynamics in HCMV-naïve and -experienced immunocompromised hosts. J. Infect. Dis. *185*, 1723–1728.

Falagas, M.E., and Snydman, D.R. (1995). Recurrent cytomegalovirus disease in solid-organ transplant recipients. Transplant. Proc. *27*, 34–37.

Fishman, J.A. (2007). Infection in solid-organ transplant recipients. N. Engl. J. Med. *357*, 2601–2614.

Florman, S., and Miller, C.M. (2006). Live donor liver transplantation. Liver Transpl. *12*, 499–510.

Fornara, C., Lilleri, D., Revello, M.G., Furione, M., Zavattoni, M., Lenta, E., and Gerna, G. (2011). Kinetics of Effector Functions and Phenotype of Virus-Specific and gammadelta T Lymphocytes in Primary Human Cytomegalovirus Infection During Pregnancy. J. Clin. Immunol. *31*, 1054–1064.

Gerna, G., Lilleri, D., Zecca, M., Alessandrino, E.P., Baldanti, F., Revello, M.G., and Locatelli, F. (2005). Rising antigenemia levels may be misleading in pre-emptive therapy of human cytomegalovirus infection in allogeneic hematopoietic stem cell transplant recipients. Haematologica *90*, 526–533.

Gerna, G., Lilleri, D., Fornara, C., Comolli, G., Lozza, L., Campana, C., Pellegrini, C., Meloni, F., and Rampino, T. (2006). Monitoring of human cytomegalovirus-specific CD4 and CD8 T-cell immunity in patients receiving solid organ transplantation. Am. J Transplant. *6*, 2356–2364.

Gerna, G., Baldanti, F., Torsellini, M., Minoli, L., Vigano, M., Oggionnis, T., Rampino, T., Castiglioni, B., Goglio, A., Colledan, M., *et al.* (2007). Evaluation of cytomegalovirus DNAaemia versus pp65-antigenaemia cutoff for guiding preemptive therapy in transplant recipients: a randomized study. Antivir. Ther. *12*, 63–72.

Gerna, G., Lilleri, D., Caldera, D., Furione, M., Zenone, B.L., and Alessandrino, E.P. (2008). Validation of a DNAemia cutoff for preemptive therapy of cytomegalovirus infection in adult hematopoietic stem cell transplant recipients. Bone Marrow Transplant. *41*, 873–879.

Gerna, G., Lilleri, D., Rognoni, V., Agozzino, M., Meloni, F., Oggionni, T., Pellegrini, C., Arbustini, E., and D'Armini, A.M. (2009). Preemptive therapy for systemic and pulmonary human cytomegalovirus infection in lung transplant recipients. Am. J. Transplant. *9*, 1142–1150.

Gerna, G., Lilleri, D., Chiesa, A., Zelini, P., Furione, M., Comolli, G., Pellegrini, C., Sarchi, E., Migotto, C., Bonora, M.R., *et al.* (2011). Virologic and immunologic monitoring of cytomegalovirus to guide preemptive therapy in solid-organ transplantation. Am. J. Transplant. *11*, 2463–2471.

Goldner, T., Hewlett, G., Ettischer, N., Ruebsamen-Schaeff, H., Zimmermann, H., and Lischka, P. (2011). The novel anticytomegalovirus compound AIC246 (Letermovir) inhibits human cytomegalovirus replication through a specific antiviral mechanism that involves the viral terminase. J. Virol. *85*, 10884–10893.

Griffiths, P.D., Stanton, A., McCarrell, E., Smith, C., Osman, M., Harber, M., Davenport, A., Jones, G., Wheeler, D.C., O'Beirne, J., *et al.* (2011). Cytomegalovirus glycoprotein-B vaccine with MF59 adjuvant in transplant recipients: a phase 2 randomised placebo-controlled trial. Lancet *377*, 1256–1263.

Helantera, I., Koskinen, P., Finne, P., Loginov, R., Kyllonen, L., Salmela, K., Gronhagen-Riska, C., and Lautenschlager, I. (2006). Persistent cytomegalovirus infection in kidney allografts is associated with inferior graft function and survival. Transpl. Int. *19*, 893–900.

Hodson, E.M., Craig, J.C., Strippoli, G.F., and Webster, A.C. (2008). Antiviral medications for preventing cytomegalovirus disease in solid organ transplant recipients. Cochrane. Database. Syst. Rev. CD003774.

Hodson, E.M., Jones, C.A., Webster, A.C., Strippoli, G.F., Barclay, P.G., Kable, K., Vimalachandra, D., and Craig, J.C. (2005). Antiviral medications to prevent cytomegalovirus disease and early death in recipients of solid-organ transplants: a systematic review of randomised controlled trials. Lancet *365*, 2105–2115.

Humar, A., Kumar, D., Gilbert, C., and Boivin, G. (2003). Cytomegalovirus (CMV) glycoprotein B genotypes and response to antiviral therapy, in solid-organ-transplant recipients with CMV disease. J. Infect. Dis. *188*, 581–584.

Humar, A., Lebranchu, Y., Vincenti, F., Blumberg, E.A., Punch, J.D., Limaye, A.P., Abramowicz, D., Jardine, A.G., Voulgari, A.T., Ives, J., *et al.* (2010a). The efficacy and safety of 200 days valganciclovir cytomegalovirus prophylaxis in high-risk kidney transplant recipients. Am. J. Transplant. *10*, 1228–1237.

Humar, A., Limaye, A.P., Blumberg, E.A., Hauser, I.A., Vincenti, F., Jardine, A.G., Abramowicz, D., Ives, J.A., Farhan, M., and Peeters, P. (2010b). Extended valganciclovir prophylaxis in D+/R- kidney transplant recipients is associated with long-term reduction in cytomegalovirus disease: two-year results of the IMPACT study. Transplantation *90*, 1427–1431.

Kahan, B.D. (2008). Fifteen years of clinical studies and clinical practice in renal transplantation: reviewing outcomes with de novo use of sirolimus in combination with cyclosporine. Transplant. Proc. *40*, S17–S20.

Khoury, J.A., Storch, G.A., Bohl, D.L., Schuessler, R.M., Torrence, S.M., Lockwood, M., Gaudreault-Keener, M., Koch, M.J., Miller, B.W., Hardinger, K.L., *et al.* (2006). Prophylactic versus preemptive oral valganciclovir for the management of cytomegalovirus infection in adult renal transplant recipients. Am. J. Transplant. *6*, 2134–2143.

Kliem, V., Fricke, L., Wollbrink, T., Burg, M., Radermacher, J., and Rohde, F. (2008). Improvement in long-term renal graft survival due to CMV prophylaxis with oral ganciclovir: results of a randomized clinical trial. Am. J. Transplant. *8*, 975–983.

Kotton, C.N., Kumar, D., Caliendo, A.M., Asberg, A., Chou, S., Snydman, D.R., Allen, U., and Humar, A. (2010). International consensus guidelines on the management of cytomegalovirus in solid organ transplantation. Transplantation *89*, 779–795.

Krishnan, A., Zhou, W., Lacey, S.F., Limaye, A.P., Diamond, D.J., and La, R.C. (2010). Programmed death-1 receptor and interleukin-10 in liver transplant recipients at high risk for late cytomegalovirus disease. Transpl. Infect. Dis. *12*, 363–370.

Kumar, D., Chernenko, S., Moussa, G., Cobos, I., Manuel, O., Preiksaitis, J., Venkataraman, S., and Humar, A. (2009). Cell-mediated immunity to predict cytomegalovirus disease in high-risk solid organ transplant recipients. Am. J. Transplant. *9*, 1214–1222.

La Rosa, C., Limaye, A.P., Krishnan, A., Longmate, J., and Diamond, D.J. (2007). Longitudinal assessment of cytomegalovirus (CMV)-specific immune responses in liver transplant recipients at high risk for late CMV disease. J. Infect. Dis. *195*, 633–644.

La Rosa, C., Krishnan, A., Longmate, J., Martinez, J., Manchanda, P., Lacey, S.F., Limaye, A.P., and Diamond, D.J. (2008). Programmed death-1 expression in liver transplant recipients as a prognostic indicator of cytomegalovirus disease. J. Infect. Dis. *197*, 25–33.

Li, Y.T., Emery, V.C., Surah, S., Jarmulowicz, M., Sweny, P., Kidd, I.M., Griffiths, P.D., and Clark, D.A. (2010). Extensive human cytomegalovirus (HCMV) genomic DNA in the renal tubular epithelium early after renal transplantation: Relationship with HCMV DNAemia and long-term graft function. J. Med. Virol. *82*, 85–93.

Lilleri, D., Gerna, G., Fornara, C., Chiesa, A., Comolli, G., Zecca, M., and Locatelli, F. (2009). Human cytomegalovirus-specific T-cell reconstitution in young patients receiving T-cell-depleted, allogeneic hematopoietic stem cell transplantation. J. Infect. Dis. *199*, 829–836.

Limaye, A.P. (2002). Antiviral resistance in cytomegalovirus: an emerging problem in organ transplant recipients. Semin. Respir. Infect. *17*, 265–273.

Limaye, A.P., Corey, L., Koelle, D.M., Davis, C.L., and Boeckh, M. (2000). Emergence of ganciclovir-resistant cytomegalovirus disease among recipients of solid-organ transplants. Lancet *356*, 645–649.

Lisboa, L.F., Preiksaitis, J.K., Humar, A., and Kumar, D. (2011). Clinical utility of molecular surveillance for cytomegalovirus after antiviral prophylaxis in high-risk solid organ transplant recipients. Transplantation. *92*, 1063–1068.

Lischka, P., Hewlett, G., Wunberg, T., Baumeister, J., Paulsen, D., Goldner, T., Ruebsamen-Schaeff, H., and Zimmermann, H. (2010). *In vitro* and *in vivo* activities of the novel anticytomegalovirus compound AIC246. Antimicrob. Agents Chemother. *54*, 1290–1297.

Macagno, A., Bernasconi, N.L., Vanzetta, F., Dander, E., Sarasini, A., Revello, M.G., Gerna, G., Sallusto, F., and Lanzavecchia, A. (2010). Isolation of human monoclonal antibodies that potently neutralize human cytomegalovirus infection by targeting different epitopes on the gH/gL/UL128–131A complex. J. Virol. *84*, 1005–1013.

Marty, F.M., Ljungman, P., Papanicolaou, G.A., Winston, D.J., Chemaly, R.F., Strasfeld, L., Young, J.A., Rodriguez, T., Maertens, J., Schmitt, M., *et al.* (2011). Maribavir prophylaxis for prevention of cytomegalovirus disease in recipients of allogeneic stem-cell transplants: a phase 3, double-blind, placebo-controlled, randomised trial. Lancet Infect. Dis. *11*, 284–292.

Mattes, F.M., Vargas, A., Kopycinski, J., Hainsworth, E.G., Sweny, P., Nebbia, G., Bazeos, A., Lowdell, M., Klenerman, P., Phillips, R.E., *et al.* (2008). Functional impairment of cytomegalovirus specific cd8 T-cells predicts high-level replication after renal transplantation. Am. J. Transplant. *8*, 990–999.

Nebbia, G., Mattes, F.M., Smith, C., Hainsworth, E., Kopycinski, J., Burroughs, A., Griffiths, P.D., Klenerman, P., and Emery, V.C. (2008). Polyfunctional cytomegalovirus-specific CD4+ and pp65 CD8+ T-cells protect against high-level replication after liver transplantation. Am. J Transplant. *8*, 2590–2599.

Parsaik, A.K., Bhalla, T., Dong, M., Rostambeigi, N., Dierkhising, R.A., Dean, P., Abraham, R., Prieto, M., Kremers, W.K., Razonable, R.R., *et al.* (2011). Epidemiology of cytomegalovirus infection after pancreas transplantation. Transplantation. *92*, 1044–1050.

Paya, C., Humar, A., Dominguez, E., Washburn, K., Blumberg, E., Alexander, B., Freeman, R., Heaton, N., and Pescovitz, M.D. (2004). Efficacy and safety of valganciclovir vs. oral ganciclovir for prevention of cytomegalovirus disease in solid organ transplant recipients. Am. J. Transplant. *4*, 611–620.

Razonable, R.R., and Paya, C.V. (2003). Herpesvirus infections in transplant recipients: current challenges in the clinical management of cytomegalovirus and Epstein–Barr virus infections. Herpes. *10*, 60–65.

Regoes, R.R., Bowen, E.F., Cope, A.V., Gor, D., Hassan-Walker, A.F., Prentice, H.G., Johnson, M.A., Sweny, P., Burroughs, A.K., Griffiths, P.D., *et al.* (2006). Modelling cytomegalovirus replication patterns in the human host: factors important for pathogenesis. Proc. Biol. Sci. *273*, 1961–1967.

Sester, U., Presser, D., Dirks, J., Gartner, B.C., Kohler, H., and Sester, M. (2008). PD-1 expression and IL-2 loss of cytomegalovirus- specific T-cells correlates with viremia and reversible functional anergy. Am. J Transplant. *8*, 1486–1497.

Shannon-Lowe, C.D., and Emery, V.C. (2010). The effects of maribavir on the autophosphorylation of ganciclovir resistant mutants of the cytomegalovirus UL97 protein. Herpesviridae. *1*, 4.

Shrestha, B.M. (2009). Strategies for reducing the renal transplant waiting list: a review. Exp. Clin. Transplant. *7*, 173–179.

Spinner, M.L., Saab, G., Casabar, E., Bowman, L.J., Storch, G.A., and Brennan, D.C. (2010). Impact of prophylactic versus preemptive valganciclovir on long-term renal allograft outcomes. Transplantation *90*, 412–418.

Strippoli, G.F., Hodson, E.M., Jones, C.J., and Craig, J.C. (2006). Pre-emptive treatment for cytomegalovirus viraemia to prevent cytomegalovirus disease in solid organ transplant recipients. Cochrane. Database. Syst. Rev. CD005133.

Tantravahi, J., Womer, K.L., and Kaplan, B. (2007). Why hasn't eliminating acute rejection improved graft survival? Annu. Rev. Med. *58*, 369–385.

Vincenti, F., Blancho, G., Durrbach, A., Friend, P., Grinyo, J., Halloran, P.F., Klempnauer, J., Lang, P., Larsen, C.P., Muhlbacher, F., *et al.* (2010). Five-year safety and efficacy of belatacept in renal transplantation. J. Am. Soc. Nephrol. *21*, 1587–1596.

Webster, A.C., Ruster, L.P., McGee, R., Matheson, S.L., Higgins, G.Y., Willis, N.S., Chapman, J.R., and Craig, J.C. (2010). Interleukin 2 receptor antagonists for kidney transplant recipients. Cochrane. Database. Syst. Rev. CD003897.

Williams-Aziz, S.L., Hartline, C.B., Harden, E.A., Daily, S.L., Prichard, M.N., Kushner, N.L., Beadle, J.R., Wan, W.B., Hostetler, K.Y., and Kern, E.R. (2005). Comparative activities of lipid esters of cidofovir and cyclic cidofovir against replication of herpesviruses *in vitro*. Antimicrob. Agents Chemother. *49*, 3724–3733.

Winston, D.J., Young, J.A., Pullarkat, V., Papanicolaou, G.A., Vij, R., Vance, E., Alangaden, G.J., Chemaly, R.F., Petersen, F., Chao, N., *et al.* (2008). Maribavir prophylaxis for prevention of cytomegalovirus infection in allogeneic stem-cell transplant recipients: a multicenter, randomized, double-blind, placebo-controlled, dose-ranging study. Blood *111*, 5403–5410.

Zelini, P., Lilleri, D., Comolli, G., Rognoni, V., Chiesa, A., Fornara, C., Locatelli, F., Meloni, F., and Gerna, G. (2010). Human cytomegalovirus-specific CD4(+) and CD8(+) T-cell response determination: comparison of short-term (24h) assays vs long-term (7-day) infected dendritic cell assay in the immunocompetent and the immunocompromised host. Clin. Immunol. *136*, 269–281.

The Rat Model of Cytomegalovirus Infection and Vascular Disease

II.15

Sebastian Voigt, Jakob Ettinger and Daniel N. Streblow

Abstract

Infection of rats with RCMV serves as an important model to study human CMV-related diseases. Two prototype RCMV isolates, the English isolate of RCMV (RCMV-E) and the Maastricht isolate (RCMV-M) have been described in the literature that differ significantly from each other by genome size, restriction fragment length polymorphism, and genome content. According to genome size, the RCMV-M genome is similar to that of murine CMV whereas RCMV-E has a substantially smaller genome consisting of only 203 kbp. In addition, RCMV-E encodes several genes that are not part of the other rodent CMV genomes, which makes it a distinct virus. In this chapter, we will first highlight these particular genes and their impact on the immune system as well as on recent advances in the understanding of RCMV biology. In the second part, we will focus on the use of RCMV-M in studying vascular disease.

Introduction – Rat Cytomegalovirus

Publications on RCMV have focused on two viruses, namely the English and the Maastricht isolates (RCMV-E and RCMV-M, respectively), that have become the prototype rat CMVs. Initial publications of RCMV-M have described its role in virus dissemination *in vivo* and general aspects of RCMV-M biology including infected cell types and the interferon response (Bruggeman *et al.*, 1983a,b; Vossen *et al.*, 1996; van der Strate *et al.*, 2003). Thereafter, RCMV-M animal models were established for therapeutic intervention studies that included BMT (Stals *et al.*, 1990, 1993), solid organ transplantation (Steinhoff *et al.*, 1995, 1996; Koskinen *et al.*, 1996; Kloover *et al.*, 2000; Martelius *et al.*, 2001; Orloff *et al.*, 2002) and diabetes (Hillebrands *et al.*, 2003; van der Werf *et al.*, 2003). In the following, RCMV-M has been successfully used to define the role of CMV in vascular disease (Lemström *et al.*, 1993a, 1994a,b; c; Bruning *et al.*, 1994a; Persoons *et al.*, 1997; Streblow *et al.*, 2003; Hillebrands *et al.*, 2005).

Rabson *et al.* (1969) first described the isolation of an RCMV that was found in roof rats (*Rattus rattus*) in Panama (Rabson *et al.*, 1969). However, it took more than a decade until the two main RCMVs RCMV-E and RCMV-M (the latter initially reported as RA-1), were independently isolated in 1982 from the wild brown rat, *Rattus norvegicus* (Bruggeman *et al.*, 1982; 1983a; Priscott and Tyrrell, 1982). As with all cytomegaloviruses, both rat viruses are species specific; a characteristic that holds true in *in vivo* infection as well as in TC since RCMVs replicate in rat but not in mouse embryo fibroblasts. Notably, species specificity of RCMVs is thus stricter than it is for MCMV, which replicates in embryo fibroblasts from both rodent species. So far, species specificity has referred to the organism that becomes infected, e.g. RCMV cannot infect a mouse, but has not addressed the question if a certain rat or mouse species is susceptible to infection with a given MCMV or RCMV isolate (see also Chapter I.18). With regard to RCMV, it is unknown whether or not RCMV-E for example can infect either rat species *norvegicus* or *rattus* or even both since animal infection experiments have exclusively used *norvegicus* as the species. Both RCMV-E and RCMV-M induce an antibody response, have a strong tropism for infecting and persisting long-term in the SG, and they establish latency (Priscott and Tyrrell, 1982; Bruggeman *et al.*, 1983a; Sandford *et al.*, 2010).

Infection of laboratory rats can be lethal in newborns (Priscott and Tyrrell, 1982) but otherwise virus infection does not become symptomatic if the infected animals are immune competent (Bruggeman *et al.*, 1983a). Other RCMVs have been isolated from both *Rattus rattus* and *Rattus norvegicus* in Australia (Smith *et al.*, 2004). Using PCR, DNA sequencing and restriction length polymorphism, Smith and colleagues found that their *Rattus norvegicus* isolates are related to RCMV-E whereas the *Rattus rattus* isolates were distinct viruses but genetically resembled RCMV-M rather than RCMV-E. PCR analyses with gene-specific primers in organs from wild rats trapped in Germany and Thailand revealed that the majority of these animals carry RCMV. Of 64 animals investigated, 44 (nearly 69%) contained RCMV-E. In contrast, only two rats were infected with RCMV-M (3.1%), suggesting a predominance of RCMV-E in this population (S. Voigt, unpublished data). Preliminary sequence data from an RCMV isolate obtained from a wild rat in Berlin reveal that a virus similar to RCMV-E is present in rats and therefore existing in nature (S. Voigt, unpublished data).

Genomic organization of RCMV

To characterize and classify RCMV-M, portions of the genome were sequenced (Beuken *et al.*, 1996; Vink *et al.*, 1996, 1997) until the full genome sequence was reported in 2000 (Vink *et al.*, 2000). In general, CMV genomes have been reported to contain approximately 230–240 kbp of double stranded DNA (Mocarski *et al.*, 2007). In accordance with this, MCMV (Smith isolate) and RCMV-M are 230,278 bp and 230,138 bp in size, respectively, which differs significantly from the size of RCMV-E with only approximately 203 kbp (Burns *et al.*, 1988; S. Voigt, unpublished data). RCMV-E is collinear to the MCMV and RCMV-M genomes (Fig. II.15.1) but the termini contain vastly different genes. The termini do not contain repeats as have been shown for MCMV and RCMV-M, although several repeat regions are located within these rodent CMV genomes.

Based on their genome size, genome content, restriction fragment length polymorphism and characterization of the major IE region, it has been pointed out that RCMV-E and RCMV-M are distinct viruses and not just RCMV strains (Burns *et al.*, 1988; Sandford and Burns, 1996; Beisser *et al.*, 1998; Vink *et al.*, 2000; Smith *et al.*, 2004; Voigt *et al.*, 2005). RCMV-M contains a predicted 166 genes of which 113 are homologous to ORFs present in MCMV or HCMV. In comparison with MCMV and RCMV-M, only 140 ORFs could be determined for RCMV-E (Ettinger and S. Voigt, unpublished data). In addition, RCMV-E encodes 19 novel ORFs neither present in MCMV or RCMV-M.

Whereas MCMV and RCMV-M are equal in size and have similar genes, RCMV-E differs from MCMV and RCMV-M in its genome content, which becomes most obvious at the termini. At the left RCMV-E terminus, the first ORF with homology to MCMV or RCMV-M is e20, i.e. genes such as m04 or m06 that have an important immunomodulatory function in MCMV are not present in RCMV-E. As far as the protein coding sequences are concerned, it could be assumed that the RCMV genomes are closer related to each other than one of the RCMVs with MCMV. However, a comparison of sequence identity on the nucleotide level reveals that RCMV-M is more similar to MCMV than to the English isolate of RCMV. On the protein level, the difference between RCMV-E and MCMV is less pronounced than on the nucleotide level (Table II.15.1; Ettinger and S. Voigt, unpublished data). These data also argue in favour of RCMV-E and RCMV-M as being distinct viruses and have led to the proposal that RCMV-E be separately classified as *murid herpesvirus 8* by the International Committee on Taxonomy of Viruses (ICTV).

RCMV gene expression

RCMV productively infects a number of different types of cultured cells. These include and are not limited to vascular EC, vascular smooth muscle cells (SMCs), epithelial cells, fibroblasts, DCs and macrophages. However, similar to HCMV, virus production and the kinetics of infection can vary between cell types. Under productive infection conditions, CMV gene expression occurs in the three distinct kinetic classes named IE, E and L. However, it was thought that CMV gene expression *in vivo* differs since the non-dividing nature of the cellular environment in the infected animal may not be conducive to a normal productive infection. In addition, productive CMV infections persist, mainly if not exclusively in the salivary glands, for long periods of time even in the presence of robust immune responses. Determining the viral genes that are expressed *in vivo*, especially during persistent and latent infections, is of critical importance to fully understand CMV persistence. Thus, animal models represent an excellent system to identify *in vivo* viral gene profiles and determine differences between *in vitro* and *in vivo* viral gene expression patterns. As described above, the RCMV-M genome is about 230 kbp and encodes approximately 170 predicted ORFs (Vink *et al.*, 2000). Most CMV gene expression studies have focused on individual viral genes or a subset of viral genes. To profile the complete viral transcriptome, CMV microarrays have been generated containing probes for each of the predicted viral ORFs (Chambers *et al.*, 1999; Goodrum *et al.*, 2002; Tang *et al.*, 2006; Streblow *et al.*, 2007). These studies have led to a number of interesting findings about viral gene expression during productive and non-productive infections. To determine whether CMV gene expression occurs in a cell type specific manner RNA from infection-susceptible rat fibroblasts, aortic SMC, aortic EC, alveolar macrophages (NR8383, ATCC), epithelial

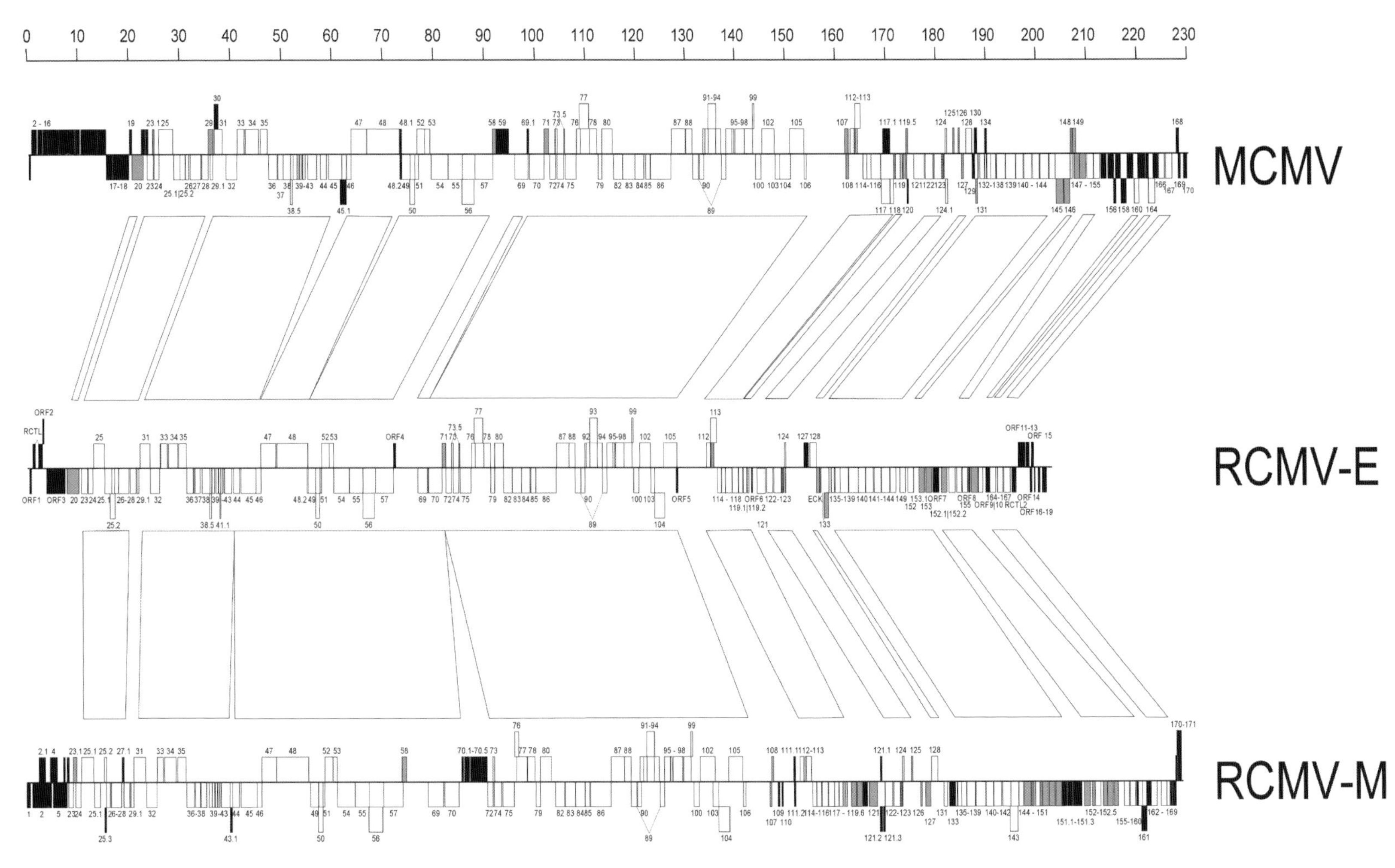

Figure II.15.1 Genome organization of MCMV, RCMV-E, and RCMV-M. Conserved ORFs among all three viruses are marked in white, ORFs conserved in two viruses are coloured in grey, and non-conserved genes are marked as black boxes.

Table II.15.1 Comparison of ORFs in RCMV-E, RCMV-M[a] and MCMV Smith[b]

				Nucleotide identity (%)			Protein identity (%)			
				RCMV-E		RCMV-M	RCMV-E		RCMV-M	
ORF	Length (nt)	Length (aa)	MW (kDa)	RCMV-M	MCMV	MCMV	RCMV-M	MCMV	MCMV	Comments
ORF 1	381	127	13.7							
RCTL	777	181	20.8							
ORF 2	300	100	11.2							
ORF 3	3681	1227	124.0							
e20	2307	769	85.2		45.3			26.1		–
E23	1173	391	43.4	55.2	58.6	62.5	45.3	54.5	48.3	US22 family
E24	966	322	35.7	52.4	58.3	58.6	45.7	48.8	45.2	US22 family
E25	2097	699	78.6	49.3	51.1	53.6	36.7	36.5	32.6	UL25 family
e25.1	1365	455	51.6	47.0	49.4	55.6	35.4	31.4	29.7	US22 family
e25.2	957	319	35.7	38.1	37.8	46.4	14.7	15.2	12.0	US22 family
E26	741	247	27.3	49.4	58.4	57.9	39.1	50.3	41.8	US22 family
E27	1971	657	76.0	51.3	55.2	62.6	41.5	52.4	42.8	–
E28	1149	383	43.4	53.7	53.8	59.3	47.3	45.8	41.6	US22 family
e29.1	411	137	15.6	44.8	48.7	46.1	26.8	29.7	21.9	–
E31	1995	665	75.3	54.2	55.8	62.0	45.2	41.2	39.1	–
E32	1851	617	69.9	44.7	48.8	52.1	31.2	34.3	38.0	Tegument phosphoprotein
E33	1417	445	49.8	61.8	66.1	73.8	57.4	65.9	65.2	GPCR family
E34	1821	607	69.6	51.6	54.3	55.5	38.7	38.5	37.7	Nuclear egress
E35	1518	506	57.4	51.6	51.3	60.4	45.7	39.6	45.8	UL25 family
E36	1502	475	53.6	57.6	58.6	64.0	57.6	58.9	53.5	US22 family
E37	966	322	36.3	50.3	54.7	54.9	39.2	46.8	40.5	–
E38	1542	514	56.8	50.2	52.9	58.6	34.1	36.8	34.2	–
e38.5	435	145	16.1	51.5	51.2	49.1	33.1	31.5	26.1	–
e39	513	171	19.4	42.9	43.0	46.2	27.3	18.8	22.2	–
e40	372	124	14.0	51.7	51.9	54.9	32.1	40.7	38.3	–
e41	384	128	13.7	50.7	52.3	50.6	30.0	36.7	33.3	–
e41.1	168	56	6.0	56.0	51.7	49.4	44.6	46.6	37.9	–
e42	366	122	13.2	49.7	45.0	50.9	30.5	22.7	22.2	–
E43	1482	494	55.8	51.7	54.1	56.1	43.8	43.8	39.1	US22 family
E44	1161	387	43.0	62.1	66.8	75.6	63.6	70.8	68.7	DNA binding-protein
E45	3042	1014	111.7	50.5	51.6	60.9	41.5	42.1	41.8	Ribonucleotide reductase
E46	885	295	33.2	66.1	68.5	77.4	69.5	76.6	73.9	Minor capsid-binding protein
E47	2877	959	110.8	58.5	56.7	65.8	54.9	53.2	55.5	–
E48	6090	2030	231.3	56.0	56.2	64.3	50.1	52.8	50.3	–
e48.2	240	80	8.4	59.3	57.5	63.7	58.6	51.5	50.0	–
E49	1491	497	57.0	63.9	65.5	73.4	70.4	72.4	72.5	–
E50	813	271	30.6	57.5	59.2	63.5	50.5	51.1	51.7	–
E51	327	109	12.2	65.7	72.5	72.1	62.0	71.7	58.9	–
E52	1521	507	58.5	63.6	66.9	75.2	61.8	69.3	67.0	–
E53	843	281	32.7	62.6	64.4	72.2	63.0	58.1	61.7	–
E54	3123	1041	117.7	61.2	65.0	66.4	61.4	67.8	57.7	DNA polymerase
E55	2640	880	98.8	56.6	57.6	60.4	54.5	55.1	52.9	gB
E56	2322	774	86.4	66.3	76.4	70.9	64.3	77.5	66.4	–

Table II.15.1 (Continued)

ORF	Length (nt)	Length (aa)	MW (kDa)	Nucleotide identity (%) RCMV-E		RCMV-M	Protein identity (%) RCMV-E		RCMV-M	Comments
				RCMV-M	MCMV	MCMV	RCMV-M	MCMV	MCMV	
E57	3489	1163	130.1	54.8	65.5	65.0	51.7	68.3	51.3	Major DNA-binding protein
ORF 4	474	158	18.0							–
E69	1914	638	72.9	48.3	53.6	52.3	31.0	39.3	32.9	–
E70	2652	884	102.2	54.5	60.0	63.4	46.3	58.7	49.7	Helicase primase
E71	729	243	27.6		59.7			58.5		–
E72	1026	342	38.0	55.3	53.3	58.6	44.8	36.5	36.9	UTPase
e73.5	1650	78	8.5		52.5			39.8		–
E73	366	122	13.8	57.4	52.0	56.0	50.0	43.6	46.1	–
e74	1086	362	41.5	44.8	47.8	43.2	24.5	26.7	23.1	–
E75	2151	717	81.4	53.6	54.9	60.9	49.7	50.6	45.6	gH
E76	756	252	28.5	58.9	62.1	68.3	51.8	51.8	52.1	–
E77	1743	581	65.1	59.4	60.6	65.9	60.2	62.8	56.3	UL25 family
E78	1314	438	47.6	45.5	46.1	48.7	21.9	24.8	22.1	GPCR family
E79	777	259	29.5	62.9	66.0	75.0	67.2	72.2	71.4	–
E80	1686	562	61.5	51.9	50.4	58.4	43.2	44.9	44.2	–
E82	1707	569	63.1	45.5	44.4	52.4	28.2	23.4	30.8	UL82 family
E83	1770	590	66.9	44.4	43.5	51.4	26.0	19.7	25.0	UL82 family
E84	1656	552	61.9	45.1	48.5	55.4	33.3	35.6	33.3	UL82 family
E85	903	301	33.4	67.0	67.9	74.4	68.0	72.1	68.6	Minor capsid protein
E86	4044	1348	152.1	69.5	69.8	80.5	75.9	77.3	77.2	Major capsid protein
E87	2481	827	93.6	65.0	62.6	73.1	71.9	69.0	71.8	–
E88	1239	413	46.0	55.4	58.9	64.5	53.9	61.5	53.8	–
E89	6329	671	76.9	72.3	74.3	83.6	83.8	86.4	84.6	–
e90	807	269	30.2	42.6	48.4	46.0	21.5	27.7	20.8	–
E91	282	94	10.2	63.4	68.4	61.3	40.8	59.1	41.2	–
E92	687	229	25.5	69.1	72.2	78.2	77.0	80.3	81.5	–
E93	1530	510	58.4	52.7	56.2	63.3	46.2	53.4	50.6	–
E94	1023	341	37.4	52.2	54.6	64.0	48.1	47.1	51.3	–
E95	1155	385	42.3	64.6	65.0	72.0	68.1	67.9	65.0	–
E96	372	124	13.8	55.8	55.9	59.3	50.0	45.9	42.5	–
E97	1710	570	63.5	54.6	59.3	60.5	53.6	56.6	52.8	Phosphotransferase
E98	1599	533	59.4	57.4	57.7	63.3	48.3	50.6	46.8	Exonuclease
E99	291	97	10.3	49.3	53.1	56.9	44.4	43.0	40.3	Tegument phosphoprotein
E100	1056	352	39.7	66.7	71.0	77.5	68.8	76.6	70.2	gM
E102	2187	729	81.9	49.0	52.3	55.9	29.5	43.4	28.8	Helicase-primase subunit
E103	789	263	29.7	59.6	65.3	64.4	45.2	56.9	44.9	Helicase-primase subunit
E104	2028	676	78.1	62.2	64.5	71.6	60.0	68.3	63.5	–
E105	2640	880	99.5	59.1	63.8	66.6	58.1	70.7	57.8	DNA helicase
ORF 5	336	112	12.8							–
E112	1385	311	33.5	53.8	55.5	54.9	41.8	35.4	41.9	Phosphoprotein
E113	1107	369	40.1	48.5	54.9	47.4	30.5	43.1	26.8	–
E114	753	251	28.3	63.2	65.3	67.7	63.2	71.2	68.1	Uracil DNA glycosylase
E115	801	267	30.5	53.1	54.7	54.0	38.0	47.1	39.4	gL
E116	1326	442	47.8	43.2	45.9	43.9	15.3	20.7	14.3	–

ORF	Length (nt)	Length (aa)	MW (kDa)	Nucleotide identity (%) RCMV-E		RCMV-M	Protein identity (%) RCMV-E		RCMV-M	Comments
				RCMV-M	MCMV	MCMV	RCMV-M	MCMV	MCMV	
e117	1230	410	45.5	44.7	45.3	44.3	27.2	24.9	20.7	–
E118	831	265	30.5	48.1	47.8	49.7	30.4	34.8	25.5	–
e119.1	630	210	23.8	51.9	47.0	45.7	30.9	28.4	24.3	–
e119.2	321	107	12.0	51.2	48.0	46.1	34.2	29.8	28.9	–
ORF 6	237	79	9.0							–
e120.1	951	317	43.6		45.0			25.8		–
E121	1737	579	64.4	43.3	42.9	43.3	13.3	15.3	13.2	–
IE2	3370	521	58.3	47.3	46.6	50.0	29.5	31.8	35.4	Immediate early protein 2
IE1	1886	567	63.0	44.7	45.7	44.6	18.5	21.8	20.5	Immediate early protein 1
e124	246	82	8.9	39.7	62.8	38.9	12.9	58.6	12.4	–
e127	855	285	31.8							CD200/OX2 homologue
e128	1239	413	46.6	52.6	50.4	51.3	40.3	42.3	42.7	US22 family
ECK	1001	306	34.4	43.9	43.5	44.1	19.7	18.2	21.9	CC chemokine
e133/ sgg1	1148	296	32.9	45.7	46.1	45.5	29.8	29.1	24.1	–
e135	381	127	14.0	41.9	43.5	48.1	26.8	27.9	33.1	–
e136	591	197	22.9	48.9	50.9	47.2	39.7	37.2	34.1	–
e137	987	329	36.0	50.3	48.2	48.9	34.7	34.3	38.3	–
e138	1572	524	58.5	42.4	43.4	43.8	22.8	20.5	21.9	FCR
e139	1899	633	71.8	52.5	55.9	54.2	44.5	49.3	42.8	US22 family
e140	1392	464	52.7	51.8	54.4	51.8	44.9	48.5	45.1	US22 family
e141	1464	488	56.3	51.8	52.8	53.5	41.5	44.2	44.8	US22 family
e142	1338	446	50.9	57.1	59.1	61.0	51.1	56.7	57.1	US22 family
e143	1581	527	59.5	54.1	56.0	54.9	49.3	47.9	43.2	US22 family
e144	1149	383	43.7	48.2	49.0	46.3	29.6	33.2	31.5	MHC-I homologue
e149	1116	372	42.0	41.4	41.4	39.9	15.1	11.4	10.3	–
e152	1173	391	44.9	42.9	46.0	39.8	15.2	20.9	15.5	m145 family
e153.1	1101	367	43.0		41.0			22.3		m145 family
e153	1218	406	46.5		42.5			20.5		m145 family
ORF 7	1110	370	42.7							
e152.1	1089	363	41.6		42.9			19.5		m145 family
e152.2	1299	433	49.5	44.2	40.7	42.8	16.0	17.4	18.2	m145 family
e155	945	315	35.6	44.7	42.0	40.4	22.7	16.7	16.8	m145 family
e155.1	1074	358	41.3	41.3	42.9	40.4	19.4	21.7	16.8	m145 family
ORF 8	348	116	12.8							
e159	1359	453	48.9		45.1			24.2		–
e160	804	268	30.2	42.7	41.0	43.4	23.6	16.7	19.4	–
ORF 9	339	113	12.7							
ORF 10	399	133	14.8							
e164	1161	387	43.0	42.9	46.8	43.7	18.6	31.1	21.5	–
e166	1101	367	41.6	44.9	45.4	48.6	27.3	25.7	26.7	–
e167	1284	428	47.6	43.9	53.4	44.2	14.0	41.5	13.6	–
RCTL2	822	183	21.5							
ORF 11	744	248	27.6							

Table II.15.1 (Continued)

				Nucleotide identity (%)			Protein identity (%)			
				RCMV-E		RCMV-M	RCMV-E		RCMV-M	
ORF	Length (nt)	Length (aa)	MW (kDa)	RCMV-M	MCMV	MCMV	RCMV-M	MCMV	MCMV	Comments
ORF 12	546	182	20.3							
ORF 13	639	213	22.8							
ORF 14	363	121	13.1							
ORF 15	447	149	16.4							
ORF 16	201	67	7.6							
ORF 17	336	112	12.3							
ORF 18	351	117	12.6							
ORF 19	474	158	18.0							

[a]The Genbank accession number for RCMV-M is AF232689.
[b]The Genbank accession number for MCMV Smith is NC_004065. Maximum sequence identity is coloured in grey.

cells and BM-derived DCs infected with RCMV for 24 hours were analysed by RCMV-specific microarrays. Interestingly, the viral gene profiles and levels of expression vary among the cell types (Fig. II.15.2). Fibroblasts are the most prolific cell type, expressing nearly all of the viral genes by 24 hpi. Macrophages and epithelial cells have the most limited gene expression profiles and the other cell types show intermediate levels of expression. Many RCMV genes are expressed in all TC cell types including R78, r119.2, r127, r128, r149, and r152.4. However, a number of genes are expressed exclusively in one cell type or another demonstrating that CMV gene expression occurs in a cell-type specific manner including r5, R32, R33, R46, r70.1, r111.2, r125, r148, and r152.2. How CMV gene expression is controlled is still unclear but this phenomenon has interesting implications for viral infections *in vivo*.

Studies using the RCMV/rat infection model have also contributed to our understanding of *in vivo* CMV gene expression (Streblow *et al.*, 2007). RCMV transcriptome analysis was performed on RNA isolated from PBMCs, lung, spleen and SG tissues (e.g. submandibular gland; SMG) from a cohort of acutely infected cardiac transplant recipients (Streblow *et al.*, 2005, 2007, 2008b). In general, the RCMV gene expression profile is tissue type specific and does not always reflect the overall level of viral RNA or DNA. For example, 27 viral genes are expressed in PBMCs greater than 2-fold over background, 62 viral genes in spleen and 17 in the lung (Fig. II.15.3 and Table II.15.2). A number of the viral genes expressed in PBMCs were similar to those expressed in other tissues such as spleen, lung or SMG. However, a subset of genes that includes G protein-coupled receptor(s) (GPCR) were either disproportionately regulated (up or down) or their expression was novel to PBMCs. For instance, R78, a viral GPCR (vGPCR), is one of the most highly expressed RCMV genes in PBMCs but its expression is minimal in SMG even though total viral mRNA levels are much higher in the SMG. Other genes like R3, R4, R38, R171, and R171.1 are also expressed in PBMCs but not SMG suggesting that specific gene profiles are required during different infection conditions or environments. There does not appear to be a major effect of the timing of infection on viral gene expression. In the salivary glands, viral gene expression profiles are identical at 14, 21 and 28 dpi (Streblow *et al.*, 2007), although this point has not been fully explored. Interestingly, the number of transcriptionally highly active genes *in vivo* is quite low compared to infected cultured fibroblasts. In fact, most of the highly expressed genes from tissues are clustered on the ends of the RCMV genome, which contain genes hypothesized to be involved in host cell manipulation and/or immune evasion. Expression of the immune modulators occurs at higher levels than previously thought and differentially from those genes involved in virus replication. This altered level of gene expression might allow the virus to persist by turning over small amounts of infectious virus while remaining undetected by the immune system. By studying viral gene expression at the genome level we avoid biasing our findings to one specific gene and have the potential to identify any and all viral genes expressed *in vivo*. Sequencing of the entire viral transcriptome will be used in the future to identify all viral genes and RNAs including those not yet predicted to exist.

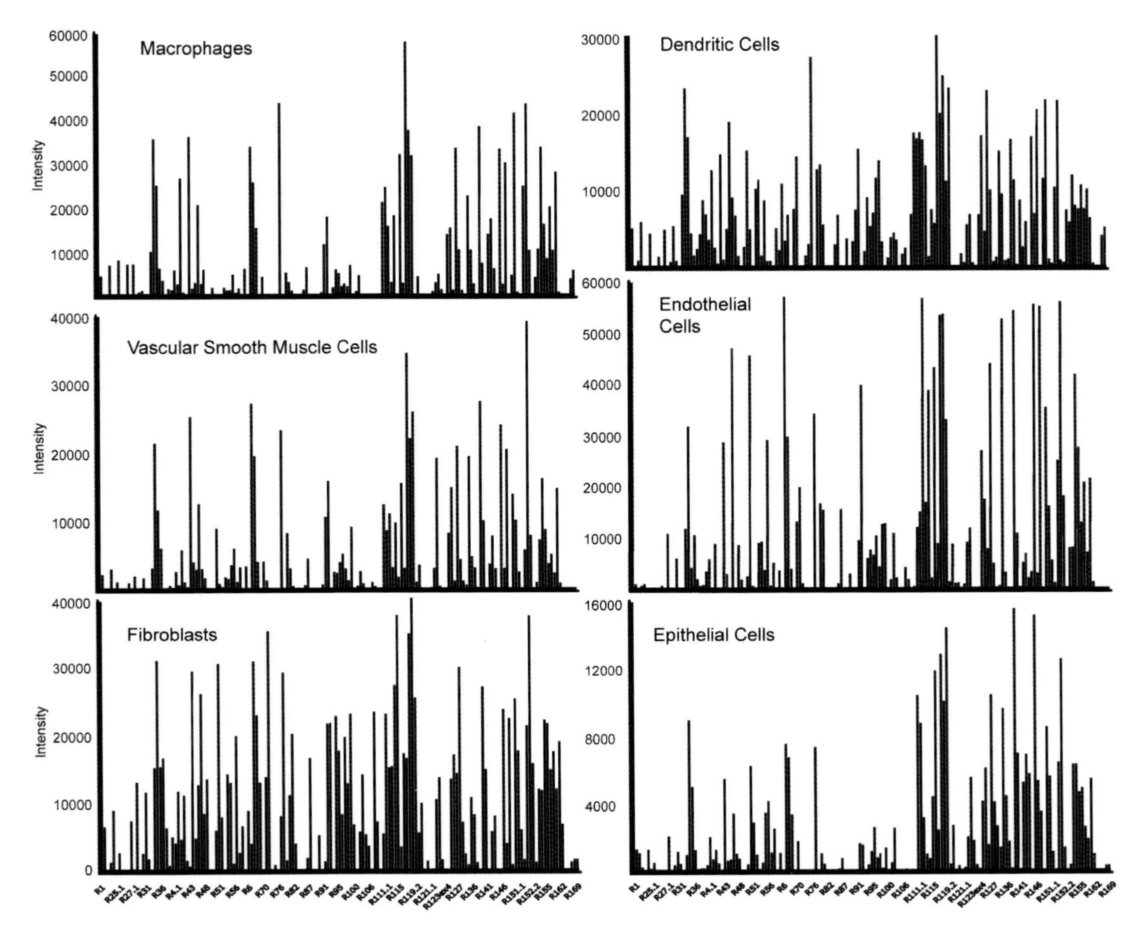

Figure II.15.2 Comparison of RCMV *in vitro* gene expression profiles. Shown are the fluorescence intensities from microarray analysis in RCMV-infected macrophages, DCs, aortic SMC, aortic EC, fibroblasts and epithelial cells. Total RNA was isolated using Trizol and then amplified, labelled and hybridized to RCMV-specific microarrays containing two specific 70-mer oligonucleotide probes for each of the RCMV-M ORFs. The viral ORFs are listed on the *x*-axis of the graphs. (Adapted, Copyright © American Society for Microbiology, Journal of Virology, *81*, 2007, pp. 3816–3826, No. 1, doi:10.1128/JVI.02425-06).

RCMV miRNAs

The discovery of miRNAs is a significant recent advance in biology. miRNAs are typically non-coding RNAs involved in posttranscriptional regulation through RNA interference. Over 140 novel miRNAs have been identified from DNA viruses. The majority of these viral miRNAs have been identified in the herpesvirus family, although SV40 and adenoviruses also encode them. While the CMV encoded miRNAs are discussed in great detail by Hancock and colleagues in Chapter I.5, we highlight here the salient features of those encoded by RCMV. The RCMV-M miRNAs were discovered utilizing a cloning/deep sequencing approach from total RNA isolated from RCMV-infected TC fibroblasts and separately from persistently infected SG at 21 dpi. Comparison of these two data sets indicates that RCMV encodes 24 miRNAs that map to hairpin structures present within the viral genome (Meyer *et al.*, 2011). The RCMV miRNAs are encoded in clusters distributed across the viral genome (Fig. II.15.4), which is different than the observed clustering of the alpha and gamma herpesvirus miRNAs that are limited to the latency-associated regions. The largest cluster in RCMV includes nine miRNAs in a small 700 bp region of the genome located near RCMV ORFs r111.1 and r111.2. Another cluster near RCMV ORF r1 encodes four miRNAs. Both of these miRNA clusters are conserved at the positional level between RCMV and MCMV albeit without any major direct sequence homology (Buck *et al.*, 2007; Meyer *et al.*, 2011). Interestingly, both RCMV and MCMV contain miRNAs within the viral origin of replication (OriLyt), however,

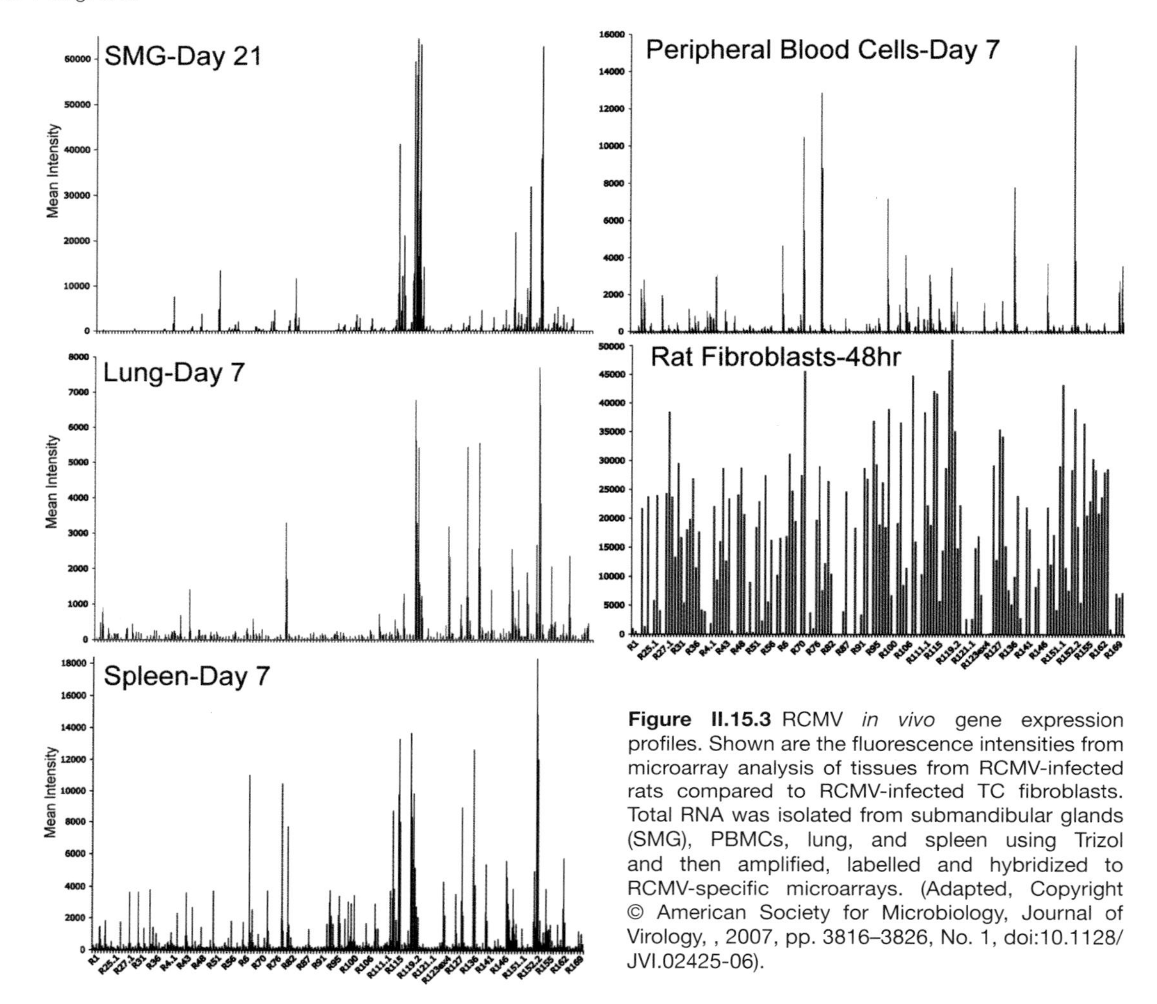

Figure II.15.3 RCMV *in vivo* gene expression profiles. Shown are the fluorescence intensities from microarray analysis of tissues from RCMV-infected rats compared to RCMV-infected TC fibroblasts. Total RNA was isolated from submandibular glands (SMG), PBMCs, lung, and spleen using Trizol and then amplified, labelled and hybridized to RCMV-specific microarrays. (Adapted, Copyright © American Society for Microbiology, Journal of Virology, , 2007, pp. 3816–3826, No. 1, doi:10.1128/JVI.02425-06).

the implications of this finding during active replication, latency or reactivation are yet unknown.

There are at least two highly relevant questions with regards to the CMV-encoded miRNAs including (1) What are the functional targets and roles of the viral miRNAs?; and (2) How is miRNA expression controlled during natural infection scenarios? It appears, at least with RCMV, but it is probably true for all CMVs that miRNA expression is tissue specific and linked to the different stages of replication. Of the 24 RCMV-miRNAs, 22 were expressed in the fibroblasts while four of them (miR-r1-4, miR-r43.1-2, miR-R90-1, and miR-r111.2-1) were unique to fibroblasts and not expressed in persistently infected SG (Meyer *et al.*, 2011). Similarly, 20 viral miRNA were detected in the SG; and two of these miRNAs, miR-r95.1-1 and miR-r170-1, were not expressed in fibroblasts. A kinetic Northern blot analysis revealed that most of the RCMV miRNAs are expressed with early kinetics, with only a few showing late viral gene expression kinetics. Quantitative RT-PCR analysis for RCMV miR-R87-1 and miR-r111.1-2 in tissue samples from infected rats indicated that miR-R87-1 is expressed during acute infection conditions (7 dpi) in every tissue tested. However, RCMV miR-r111.1-2 was almost exclusively expressed in the SG and was increased at times of persistence (28 dpi). Heterogeneity at the viral miRNA 3′ ends is common and was observed for the RCMV miRNAs. Whilst much of the variability was conserved between the samples obtained from *in vitro* and *in vivo* infection conditions, there exist a number of specific examples where the percentage of the particular miRNA isoform differed dramatically. For instance, miR-6-1 has four major isoforms but only three of these were dominant in the fibroblasts. The fourth miR-6-1 isoform was the most highly expressed isoform in SMG tissues (23%) but was expressed at very low levels in fibroblasts (3%). Similarly, an isoform of miR-OriLyt-2

Table II.15.2 RCMV gene expression in tissues from infected rats

RCMV ORF	Kinetic class[a]	Spleen	Lung	Liver	Kidney	SMG	Allograft heart	PBMC	Comments
r2.1	L	809	0	4420	1239	0	0	0	
r3	L	0	0	0	0	0	0	1675	
r4	E	1009	0	1122	0	0	0	1969	
R23	E	594	0	744	0	0	0	1534	US22 Family; tegument
r25.1	E	1462	0	777	0	0	2319	0	US22 Family
R25	E	1158	0	504	0	0	1917	0	UL25 Family; tegument
R27	E	583	0	0	0	0	1024	0	
R28	E	1434	0	632	823	0	1160	761	US22 Family
R29	E	647	0	0	5504	0	949	0	US22 Family
R35	E	0	0	0	0	3058	0	793	UL25 Family; tegument phosphoprotein
R38	E	901	0	982	0	0	942	2888	Virion envelope glycoprotein
r41	E	1650	602	1403	0	0	705	995	
R43	E	976	0	0	0	1135	1568	626	US22 Family, tegument
R49	E	1356	0	0	628	0	905	0	
R55	E	683	0	0	0	754	2929	0	Glycoprotein B
R69	E	861	0	827	0	0	0	3012	Transactivator; tegument
r70.2	E	4172	0	555	0	0	4452	0	m145 gene family
r70.3	E	1175	0	0	0	0	1534	0	m145 gene family
R72	E	0	0	0	0	1034	0	6505	UTPase
R73	L	1779	0	454	658	2286	2143	0	Glycoprotein N
R78	E	4519	1812	7293	777	1274	1819	10194	GPCR homologue
R80	E	3347	0	0	725	6118	4571	0	Assembly protein/proteinase
R82	E	0	0	809	0	1747	1244	0	Upper matrix protein; pp71, transactivator
R93	E	687	0	0	0	0	2272	0	Tegument
R94	E	2938	0	637	918	0	2436	0	Tegument; binds ssDNA
R96	E	2382	0	0	464	969	1514	0	Tegument
R99	E	1276	0	0	474	0	2106	3909	Tegument phosphoprotein; pp28
R100	L	1416	0	0	0	1592	3542	0	Glycoprotein M
R102	E	1526	0	0	0	1405	1955	0	Helicase/primase complex component
r106	L	734	0	607	0	0	0	2475	
r109	E	1576	0	0	523	0	1859	0	
R113	E	1677	0	789	0	1083	0	2667	Transcriptional activation
R114	E	5259	0	736	1396	21637	6256	0	Uracil DNA glycosylase
R115	E	849	0	0	665	4678	1518	0	Glycoprotein L
R116	E	10322	865	1434	3268	10578	6189	1104	
r119.1	E	556	0	0	0	27712	0	0	
r119.2	E	11846	5237	18217	6721	52110	1885	2585	
r119.3	E	6881	3255	9417	1577	38222	976	1022	
r119.4	E	1974	972	2064	691	6389	0	744	
r124	L	2862	2517	2825	0	765	0	1309	
r128	IE	1364	0	2001	0	675	0	0	US22 family
r133	L	4697	2610	7068	567	1748	1362	1314	Homologue of MCMV SGG1
r138	L	7498	3389	9199	856	1681	2048	5855	Fc receptor
r142	E	3164	597	1954	814	1464	3625	0	US22 family
r149	E	4337	1976	4075	922	5653	593	2170	
r150	E	1068	0	1462	0	2042	0	0	m145 gene family

Table II.15.2 (Continued)

RCMV ORF	Kinetic class[a]	Spleen	Lung	Liver	Kidney	SMG	Allograft heart	PBMC	Comments
r151.1	E	958	644	1652	0	1787	0	0	
r151.3	E	599	0	0	0	3071	0	0	
r151	E	2077	1487	1837	570	13367	0	0	m145 gene family
r152.3	E	599	1289	2350	0	729	0	0	m145 gene family
r152.4	E	15024	6077	13520	6759	35768	5592	11314	m145 gene family
r157	E	2044	959	1828	0	1453	0	0	m145 gene family
r158	E	1375	0	631	509	2805	0	0	
r166	E	3333	1312	5271	0	564	792	0	
r171.1	E	461	0	854	0	0	0	2129	
r171	E	780	0	865	0	0	0	2192	

[a]Viral kinetic phase class: immediate early (IE), early (E), late (L). Shown are relative gene expression levels. (Adapted, Copyright © American Society for Microbiology, Journal of Virology, *81*, 2007, pp. 3816–3826, No. 1, doi:10.1128/JVI.02425-06.)

was only detected in the *in vivo* sample (60% of detected sequences for this miRNA) and this isoform was not present at all in infected fibroblasts. These findings suggest that miRNA expression is dynamic on all levels and that further characterization is required to elucidate their role in viral replication and pathogenesis.

RCMV genes involved in immune evasion and manipulation

Both the English and the Maastricht RCMV isolates contain several immunomodulatory genes that can be found in mammalian cells and are believed to have been captured from the host genome during evolution. These hijacked genes have a strong impact on viral pathogenesis with the aim to help the virus circumvent an immunological response to infection. Among these are molecules that imitate MHC-I gene expression, virally encoded chemokines and GPCR that function as receptors for chemokines, a CD200 analogue only present in RCMV-E, and C-type lectin-like gene products that are homologous to ligands of NK cell receptors.

Interference with MHC molecules and MHC imitation

Similar to MCMV and HCMV, RCMV-M interferes with cell surface expression of MHC-I molecules on infected cells (Hassink *et al.*, 2005; Baca Jones *et al.*, 2009). In the study by Hassink *et al.* (2005), RCMV-M infection of rat fibroblasts induced a temporal MHC-I down-regulation that is not due to heavy chain degradation but rather delayed heavy chain maturation. However, RCMV-M infection of rat BM-derived DCs resulted in depletion of both MHC-I and -II (Baca Jones *et al.*, 2009). This depletion was mediated by early viral genes and occurred as early as 24 hpi. Interestingly, MHC-II depletion in RCMV-M infected cells was not mediated by the induction of host IL-10 and RCMV-M is not known to encode a vIL-10, suggesting that other viral genes are involved in this process. Thus far, the viral genes responsible for these effects on MHC-I and -II have not been identified (Hassink *et al.*, 2005; Baca Jones *et al.*, 2009).

As described above, MHC-I molecules are a major target of herpesviruses, and HCMV and MCMV in particular have developed multiple ways to interfere with the presentation of viral peptides to $CD8^+$ T-lymphocytes. Other than blocking this pathway, CMV also mimics MHC-I function since these molecules are down-regulated from the cell surface upon infection (Ahn *et al.*, 1996; Wiertz *et al.*, 1996; Lacayo *et al.*, 2003). As for HCMV and MCMV, both RCMVs encode an MHC-I homologue that might help the virus to prevent recognition by NK cells. A functional analysis of RCMV-M r144, an MHC-I polypeptide heavy chain homologue, revealed that this gene is neither essential *in vitro* nor *in vivo* and replicates as WT RCMV-M in immunocompromised rats (Beisser *et al.*, 2000; Kloover *et al.*, 2002). Nevertheless, WT virus infection induced higher numbers of macrophages and $CD8^+$ T-cells in infected footpads. This was also examined and confirmed three days after infection of neonatal rats. At this time point, but not later, a significant decrease in infiltrating NK cells and macrophages was seen in the spleen of pups infected with the r144 knockout virus. Also, Δr144 replicated to lower titres as

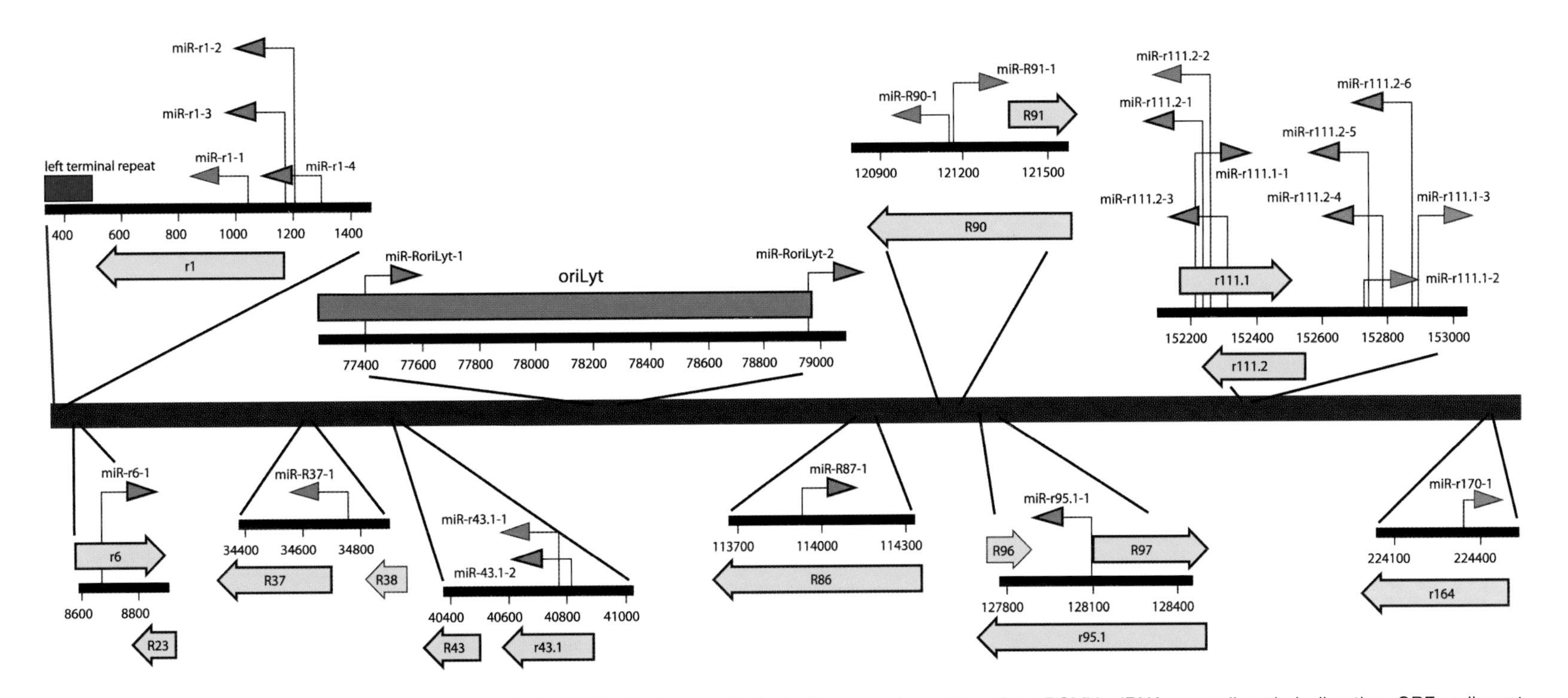

Figure II.15.4 Genomic position of the RCMV-M encoded miRNAs. Red arrows indicate the genomic position of the RCMV miRNAs as well as their direction. ORFs adjacent to the miRNAs are shown as grey-labelled block arrows. The location of the RCMV origin of replication is shown as a blue box. (Adapted, Copyright © American Society for Microbiology, Journal of Virology, *85*, 2011, pp. 378–389, No. 1, doi:10.1128/JVI.01900–10).

compared with WT RCMV-M in the spleen, and it was concluded that r144 might interfere with the immune cells named above but this still needs to be proven (Beisser *et al.*, 2000; Kloover *et al.*, 2002).

Homology to host CD200 molecules

Further, RCMV-M encodes r127 that is translated into a protein with DNA-binding activity and homology to the non-structural or Rep (replication) proteins of parvoviruses. An analogous ORF to r127 with a conserved genomic position and orientation, U94, is present in HHV-6A and HHV-6B (van Cleef *et al.*, 2004). Whereas r127 homologues are not present in CMVs, a positional homologue of r127 exists in RCMV-E, e127. However, e127 encodes an analogue of CD200, a member of the Ig superfamily that is expressed on a variety of cells including B-cells, activated T-cells, EC, neurons, macrophages and DCs. CD200 binds to the similarly structured receptor CD200R. CD200R contains a tyrosine residue in its cytoplasmic tail that is capable of conducting intracellular signalling upon phosphorylation. CD200R was initially reported to be primarily expressed on myeloid cells such as DCs and macrophages but is also present on B and T-cells (Wright *et al.*, 2003; Ahmed and Voigt, unpublished observations). Engagement of the endogenous, cellular CD200 ligand with CD200R inhibits macrophage activation and silences an inflammatory response (Hoek *et al.*, 2000; Wright *et al.*, 2003). Recently, the RCMV-E e127 gene product (vCD200) has been shown to bind to CD200R with identical kinetics as the cellular CD200 protein (Foster-Cuevas *et al.*, 2011).

Interference with natural killer cell activity

In recent years, C-type lectin-like proteins emerged as molecules that are expressed on the surface of host cells and play an important role in guiding NK cell activity. The first C-type lectin-like herpesviral protein was described for RCMV-E and named RCMV C-type lectin (RCTL) (Voigt *et al.*, 2001). Known cellular C-type lectin-like proteins involved in immune recognition included those expressed on NK cells, and since sequence comparisons revealed significant homology, it was thought that the viral homologue could potentially interfere with the immune system. It was hypothesized that the virus had captured the gene from the host; however, the rat genome sequence was then not yet available. An initial characterization of the *rctl* gene showed that it consists of five exons. This is an uncommon feature in CMVs and herpesviruses in general since splicing is restricted to a few genes, the most prominent being the IE genes 1 and 2 (i.e. *ie3* in MCMV). Another unusual finding was a GATA instead of a TATA box in the core promoter region. Moreover, *rctl* exon 1 is not translated and translation beginning in exon 2 is not initiated by an AUG (absent in the entire ORF) but rather a GCC as determined by Edman degradation. At that time, the genome sequence of RCMV-M was published but a similar gene was not identified in that isolate (Vink *et al.*, 2000). A computational re-annotation of the RCMV-M genome, however, revealed a spliced ORF r153 that consists of only four exons and is located at the opposite (right) terminus of the RCMV-M genome (Brocchieri *et al.*, 2005). RCTL and r153 share 48% identity at the aa level.

After the rat genome had been sequenced and published, a sequence comparison with RCTL indicated that a gene named *clr-b* (also *clec2d11*) is likely to be the gene that had been copied from the host into the viral genome. Both *rctl* and *clr-b* genes share the same intron/exon structure and the corresponding proteins contain a characteristic C-type lectin-like domain encoded by the respective third exon. Since the clr-b gene product Clr-b is a ligand for the inhibitory NK receptor NKR-P1B, it was presumed that RCTL might interfere with this interaction. RT-PCR analysis of clr-b transcripts in mock-infected and infected cells showed that both clr-b mRNA and protein on the cell surface are down-regulated after virus infection. However, this was not due to RCTL because infection with an RCTL mutant virus induced the same response (Voigt *et al.*, 2007). It appears that RCMV-E has evolved a decoy ligand to counteract the impact caused by the loss of the endogenous Clr-b ligand on the infected cell surface. It is possible that the cell responded to the infection by a 'danger' mechanism that resulted in the down-regulation of Clr-b. This down-regulation of an inhibitory or 'self' ligand would cause the infected cell to be 'missing self'. NK cell activity is balanced by the interplay between activating and inhibitory receptors that engage their cognate ligands. In this case, the normally inhibitory axis is interrupted and this would lead to dominance of the activating signal and therefore the NK cell would kill the infected cell. However, RCTL replaces Clr-b on the cell surface upon infection and thus re-establishes the 'self' condition, resulting in NK inhibition and protection of the infected cell (Fig. II.15.5). Moreover, the NKR-P1 receptors have similarly changed and developed polymorphisms to evade the viral decoy strategy on the one hand but also retain 'self' recognition with endogenous ligands, likely to avoid autoimmunity. This became obvious when different rat strains were infected with WT and ΔRCTL RCMV-E and NK cells were depleted. Whereas WT RCMV-E and a revertant virus grew to normal titres

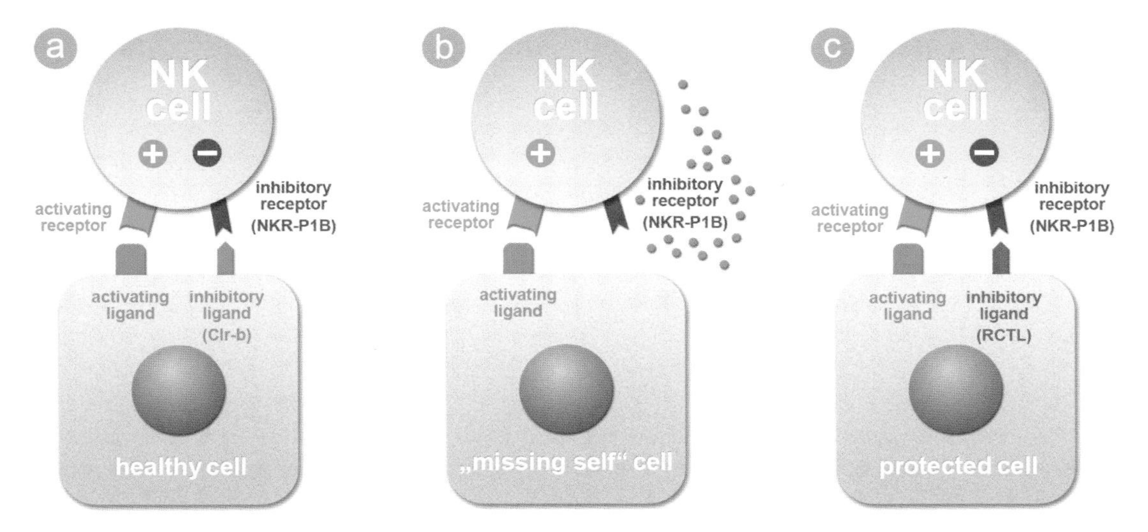

Figure II.15.5 RCTL restores 'self' protection. (a) Normal healthy cells express activating and inhibitory ligands on their surface to prevent lysis by NK cells. (b) Upon infection, inhibitory ligands such as Clr-b (blue) are lost and the infected cell is killed by the NK cell. (c) RCTL functions as a decoy ligand (purple) for the inhibitory NKR-P1B receptor (red) on NK cells and protects the RCMV-infected cell from being lysed by the NK cell.

irrespective of NK cell depletion in Wistar Albino Glaxo (WAG) rats, ΔRCTL was only detectable in spleen or liver when NK cells had been depleted (Fig. II.15.6). However, in another rat strain (Sprague–Dawley), a strain-dependent allelic divergence of the NKR-P1B receptor sequence modulates the interaction between the host and RCMV because a polymorphism in NKR-P1B has led to the loss of RCTL binding. After infection, WT, revertant and ΔRCTL viruses grew to similar titres (albeit consistently lower compared with WAG rats in the absence of NK depletion since the loss of Clr-b could not be compensated by RCTL) in the liver and the spleen. In summary, the NKR-P1B–Clr-b interaction system could provide the organism with a mechanism to distinguish self from non-self. A novelty of this description is that RCTL is a non-MHC decoy ligand involved in viral evasion of NK cell recognition whereas the 'missing-self' hypothesis so far only referred to MHC molecules. The evolvement of RCTL as a viral NK control evasion strategy might suggest that this system represents an attractive target for infectious agents and is therefore of broader importance. This is underscored by the fact that recognition by NKR-P1 is conserved in humans, which might open potential avenues for new antiviral strategies.

RCMV-encoded chemokines and chemokine receptors

Another important class of immunomodulatory genes encoded by RCMV include chemokines, a family of molecules that take part in the trafficking of leucocytes where they can act as chemoattractants to initiate an inflammatory response. Chemokines are widely conserved among the CMVs, which suggest an important role for these molecules in viral replication and survival. They are divided into four classes based on a conserved cysteine motif at the N-terminus. Both RCMV-M and -E possess genes coding for beta (CC-) chemokines. In RCMV-M, two beta chemokines, r129 and r131 exist that might have originated from duplication of a common ancestor gene (Kaptein *et al.*, 2004; Voigt *et al.*, 2005). These two genes are similar to HCMV UL128 and UL130, respectively. In contrast, RCMV-E contains a spliced version, e129/131 that is similarly structured as the murine homologue m129/131 (MacDonald *et al.*, 1999; Voigt *et al.*, 2005). For more detail about MCMV m129/131 please see Chapters I.21 and II.17. The function of RCMV-E e129/131 is unknown, however, RCMV-M r131 has been eliminated from the viral genome and shown to be involved in viral tropism and replication. Acute infection of rats with an r131 deletion mutant resulted in decreased viral titres in SG as well as in spleen and liver tissues. In addition, a significantly lower infiltration rate of leucocytes in infected footpads was observed, thereby indicating a role for r131 in leucocyte attraction (Kaptein *et al.*, 2004). In Boyden migration TC-based assays, RCMV-M r129 protein causes migration of peripheral blood lymphocytes, BM-derived lymphocytes, splenocytes as well as a macrophage cell line (NR8383, ATCC). Migration assays were performed on specific immune cell subsets

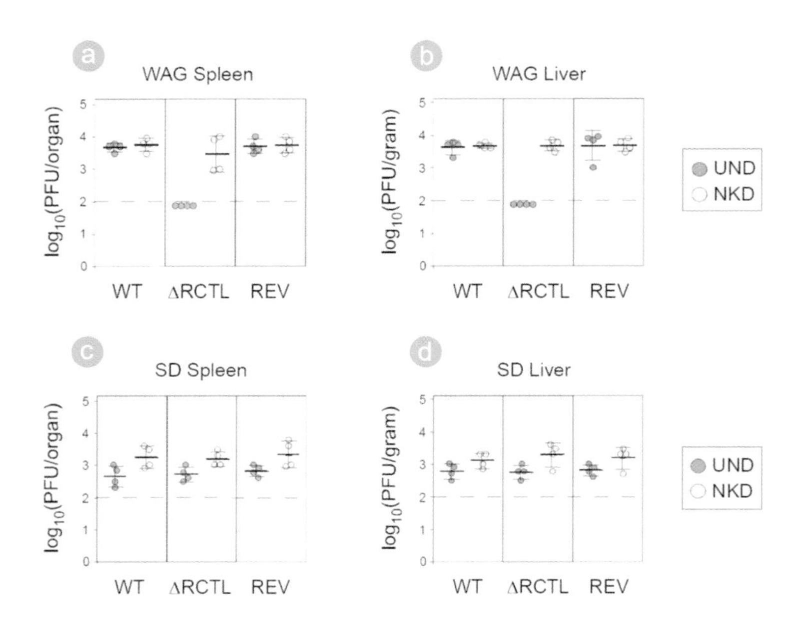

Figure II.15.6 RCTL increases RCMV virulence *in vivo*. In Wistar Albino Glaxo (WAG) rats, RCMV WT and revertant virus (REV) replicate either in the presence or absence of NK cells. However, mutant virus (ΔRCTL) is not able to establish a viral infection in the presence of NK cells (a,b). This severe loss of viral growth can be prevented by the depletion of NK cells. In contrast to WAG rats all three viruses replicate to almost the same levels in Sprague Dawley (SD) rats (c and d). In NK-undepleted (UND) animals, the viral loads remain about 1 LOG below those detected in WAG rats. In NK-depleted (NKD) animals, viral loads rise to similar levels as in WAG rats. (Reprinted and adapted from Immunity, Vol. 26 No. 5, Voigt *et al.*, Cytomegalovirus evasion from innate immunity by subversion of the NKR-P1B:Clr-b missing-self axis, pp. 617–627, Copyright © 2007, with permission from Elsevier).

isolated from splenocytes using antibody-mediated cell sorting, and naïve/central memory CD4$^+$ T-cells were shown to be a major target of r129 (J. Vomaske and D.N. Streblow, unpublished data). By this means, the rodent CMVs might recruit leucocytes to the site of infection and subsequently use them as vehicles to foster virus dissemination or misdirect the adaptive immune response (for related findings in the MCMV model, see Chapter I.21).

To complement this mode of action, the RCMVs also express GPCR, a class of seven transmembrane domain-containing receptors that are present on various cell types. Viral GPCR (vGPCR) bind to host chemokines and are involved in cell signalling and migration. To examine their putative role in intracellular signalling pathways, vGPCR have been tested for their ability to mobilize calcium, increase inositol phosphate, activate gene transcription by NFκB, and promote cellular migration. RCMV-M encodes two vGPCR, the UL33- and UL78-like genes, but the virus does not contain any US27 or US28 homologues (Gruijthuijsen *et al.*, 2002; Vink *et al.*, 2000). The R33 vGPCR did not bind RANTES but was constitutively activating phospholipase C with the consequences being unknown (Gruijthuijsen *et al.*, 2002). R33 is also implicated in cellular migration because RCMV-ΔR33 infected smooth muscle cells failed to migrate when compared to cells infected with WT RCMV-M (Streblow *et al.*, 2005). The precise function of R78 remains unknown. While RCMV-R33 deletion mutant is dispensable for viral replication in TC, the R78 knockout virus or a virus encoding a truncated version of R78 replicated to lower titres after *in vitro* infection (Beisser *et al.*, 1998, 1999). Both the R33 and the R78 deletion mutants resulted in severe attenuation *in vivo*. RCMV-ΔR33 failed to reach and/or replicate in the submandibular glands of infected rats but replicated to WT levels in all other tissues. To the contrary, the RCMV-ΔR78 virus could not reach or did not replicate efficiently in the spleen; RCMV-M WT infection *in vivo* was associated with a higher mortality rate compared with ΔR78 virus infection (Beisser *et al.*, 1998, 1999; Gruijthuijsen *et al.*, 2002; Kaptein *et al.*, 2003). In addition, RCMV-ΔR33 infected rat heart allograft recipients demonstrated an increase in time to chronic rejection and a reduced severity of transplant vascular sclerosis (TVS) compared to WT infected rats, which suggests that R33 is important for CMV disease as well (Streblow *et al.*, 2005).

RCMV infection as a model for human disease

RCMV *in vivo* infections are described as having three phases of replication: (1) acute viremic phase; (2) persistence phase; and (3) clinical latency. Acute infection results in widespread infection in most tissues throughout the rat. The route of infection, virus dosage, and age/immune status of the infected rat can have dramatic effects on RCMV pathogenesis. For example, infection of newborn rats can lead to severe disease and death; whereas, infections of immune competent adult rats usually results in asymptomatic infections. In order to mimic clinically relevant conditions, immunosuppressive agents such as cyclosporine A (CsA) or gamma-irradiation are often used to induce transient immune deficiency, which allows the virus to establish infection, spread to most organs of the infected rat, and produce high titres of virus. Intraperitoneal and intravascular infections spread rapidly throughout the rat because they quickly gain access to the blood stream. However, widespread viraemia from s.c. infections can take many days longer to develop because virus needs to replicate at the initial site of infection and then spread systemically from the local environment. Typically the RCMV acute infection phase is characterized by widespread tissue viraemia that peaks around 3 to 7 dpi. At these early times of infection, virus titres are highest in the spleen and liver although viral load can also be detected in heart, lung, kidneys and SG. Virus infected cells are also readily detectable in PBMCs and BM-derived lymphocytes (D.N. Streblow, unpublished results) during these early days following inoculation. The acute phase ends between 7 and 10 dpi when viral loads begin to drop. By day 14, consistent viral loads are limited to the SG. RCMV has a high propensity to persist in the SG tissues of infected rats. In fact, high titre virus stocks are routinely produced at 28 dpi from the SG of infected rats and this stock is superior to TC generated stocks at infecting rats. Striated duct cells of the submandibular glands are the main target of RCMV infection and an important site of viral persistence. These cells contain large vesicles filled with particles. Viral gene expression is shifted in the SG towards a profile that is more than likely involved in persistence. There is a low-level chronic expression of viral genes involved in virus replication but also a high expression level of viral genes involved in immune evasion and host manipulation (Streblow *et al.*, 2007). As such, low amounts of virus are produced from a large number of cells but there is a very limited number of inflammatory cells that infiltrate the submandibular glands. The result is long-term persistence for over 6–12 months following the initial infection. After this time the virus remains in a latent state existing in monocytic cells of the blood and in bone marrow cells (Bruggeman *et al.*, 1985). The virus may also remain latent in vascular wall cells including EC and SMC. Latent virus can be reactivated from these cells when the rats experience conditions that stress their immune system including immunosuppression and transplantation (Yagyu *et al.*, 1992; Bruggeman, 1993; Orloff *et al.*, 2011). The precise mechanisms that govern reactivation in latently infected rats remain unknown.

Fetal transmission model

Since HCMV can be transferred from mother to child, many efforts have been undertaken to study this process in fetal transmission models. The only functional *in vivo* model of CMV *in utero* transmission is that of the guinea pig, which is described in Chapter II.5. Murine models with MCMV were not successful and could not prove placental transmission (Medearis, 1964; Johnson, 1969). Nevertheless, a mouse newborn infection model (see also Chapter II.6) has shown that virus-specific antibodies protect the development of the central nervous system from MCMV infection (Cekinovic *et al.*, 2008). In their initial description, Priscott and Tyrrell attempted to grow virus from one wild rat that coincidentally happened to be pregnant but they were unable to detect cytopathic effects when they inoculated rat embryo fibroblasts with embryo homogenates (Priscott and Tyrell, 1982). Following this path, a group in Malaysia investigated placental transmission with a newly isolated RCMV called ALL-3 that was reported to be capable of causing fetal transmission (Loh *et al.*, 2006). However, to date, experimental infection of pregnant rats with RCMV-E has shown no successful fetal transmission (S. Voigt, unpublished data). Given the fact that rats and mice share a similar placental structure, RCMV, as MCMV, might not be a useful tool to study fetal transmission, but certain RCMV isolates such as ALL-3 could exist that leave this possibility open.

CMV infection accelerates vascular disease

HCMV infections have been associated with accelerated vascular disease including atherosclerosis, restenosis following angioplasty as well as transplant vascular sclerosis (TVS) associated with chronic solid organ allograft rejection. Mechanistic studies describing the role of HCMV in vascular disease have been difficult to perform and interpret because of the high prevalence of CMV in the population; the nature of the virus infection (lytic vs. latent); and the timing of initial infection. While epidemiological studies have provided invaluable information about the link between HCMV and vascular disease development, animal models have provided important information regarding the mechanisms involved in the acceleration of the disease process. The rat/RCMV model has provided a number of key observations defining the role of CMV in these vascular diseases.

Atherosclerosis

Cardiovascular disease involving the heart and/or blood vessels is the most common cause of death in the world. There are a number of risk factors associated with atherosclerosis including hypertension, hyperlipidaemia, diabetes, cigarette smoking as well as infectious agents such as bacteria and viruses including CMV. Atherosclerosis is a disease that affects the large and medium blood vessels forming in or around areas of turbulent blood flow such as vessel bifurcations. Atherosclerosis is initiated by an injurious event due to trauma, diet, chemicals or pathogens. The damaged vessel endothelium becomes activated producing chemokines, cytokines, growth factors as well as surface adhesion molecules that attract and activate platelets and monocytes. Low-density lipoproteins (LDLs) accumulate in the endothelium and are converted by enzymes and free oxygen radicals into oxidized LDLs. Vascular diseases, in general, are chronic inflammatory diseases. Atherosclerosis is no exception, as early lesions contain high levels of activated macrophages and T-cells. The activated macrophages have increased levels of LDL receptors on their surface and thus accumulate large deposits of fat, converting them into foam cells enhancing the inflammatory process. Smooth muscle cells present in the vessel wall migrate into the damaged areas where they are stimulated to proliferate. Atherosclerosis causes the thickening of the arterial wall, which restricts blood flow. In addition, the fibrous cap present on late stage atherosclerotic lesions are unstable making them prone to rupturing, which will then reform due to the presentation of the underlying exposed matrix. However, if the dislodged fragment is large enough it can result in myocardial infarction or stroke depending upon the location of the blocked vessel.

CMV and atherosclerosis

A number of epidemiological studies have indicated that HCMV infection correlates with an increased risk of atherosclerosis. In addition, a number of studies have also shown that CMV does not correlate with increased prevalence of atherosclerosis. In a study by Muhlestein and colleagues, it was determined that HCMV-seropositive patients have approximately a 3-fold increase in coronary artery disease (Muhlestein *et al.*, 2000). This was especially evident in patients with elevated C-reactive protein (CRP), which is a general marker of inflammation. In addition, high levels of CMV antibody have been shown to correlate with atherosclerotic disease (Nieto *et al.*, 1996). Similarly, HCMV DNA and antigens were shown to be present in the arterial wall of 76% of patients with ischaemic heart disease (Horvath *et al.*, 2000). Another study by Hendrix showed that about 90% of grade III atherosclerotic lesions versus 50% of grade I lesions were positive for HCMV DNA by PCR (Hendrix *et al.*, 1990). Samples acquired from patients that suddenly died revealed that HCMV DNA and/or antigens are present in fatty streaks and normal appearing portions of vessels that were close in proximity to atherosclerotic lesions, but were not present in late atherosclerotic lesions (Pampou *et al.*, 2000). EC and smooth muscle cells appear to be important sites for the virus and may potentiate atherosclerosis through direct and indirect mechanisms. Direct infection of these vascular cells increases expression of endothelial adhesion molecules (ICAM-1 and VCAM-1), which promote leucocyte adhesion and the inflammatory response (Steinhoff *et al.*, 1995). HCMV also increases chemokine and cytokine expression associated with immune cell migration and activation, likewise promoting the inflammatory response (Michelson *et al.*, 1997; Streblow *et al.*, 1999). CMV also alters macrophage lipid metabolism by increasing expression of the scavenger receptor, which induces LDL uptake and deposition (Zhou *et al.*, 1996a). This process would have the added effect of accelerating fatty streak formation in early lesions and promoting atherosclerosis.

RCMV/rat models have not typically been used in strict atherosclerosis studies. However, a number of MCMV/mouse studies have shown that MCMV infection accelerates atherosclerosis even in the absence of an excess of diet fat (Burnett *et al.*, 2001; Streblow *et al.*, 2001). These studies have used atherosclerosis-prone mice lacking either ApoE or LDL-R. MCMV increases T-cell and macrophage infiltration and expression of MCP-1 and TNF-α. These studies have suggested that MCMV induces atherosclerosis through the activation of both local and systemic pro-inflammatory responses.

Arterial restenosis

Angioplasty and stent placement is used to increase blood flow through an artery that is undergoing vascular disease. Since the process of placing the stent requires a balloon catheter to compress the affected area of the vessel wall, which usually is occluded by an atherosclerotic plaque, the vessel wall becomes damaged. Restenosis is the formation of new lesions that occur during the repair process healing the damaged vessel wall following angioplasty. Two types of restenosis can occur: (1) immediate thrombosis due to blood clotting at the site of injury; and (2) eventual occlusion due to SMC migration and proliferation. This second form is similar to atherosclerosis in that both vascular diseases are driven by immune and repair processes. This vascular lesion process usually occurs within 6 months resulting in vessel blockage requiring angioplasty and stent placement to reopen the vessel.

CMV and arterial restenosis

The initial link between prior HCMV infection and the risk of restenosis came from a study by Zhou and Epstein (Zhou *et al.*, 1996b). They demonstrated that 43% of HCMV-positive patients developed restenosis compared to only 8% of CMV-negative control patients. High anti-HCMV antibody titres correlated with an increase in coronary artery disease in a separate study by Blum and colleagues as well (Blum *et al.*, 1998). A more recent study by Mueller and colleagues showed that previous HCMV infection was associated with increased restenosis risk and reduced blood vessel diameter following aggressive angioplasty with provisional stenting (Mueller *et al.*, 2003). Speir and colleagues demonstrated that the HCMV immediate early protein IE86 interacts with and inactivates the cellular tumour suppressor protein p53 providing a potential mechanistic link for HCMV in promoting SMC proliferation during coronary restenosis (Speir *et al.*, 1994, 1995). In their study nearly 80% of the CMV-positive samples of SMC isolated from restenotic vessels had very low levels of p53 activity, suggesting that the virus inactivates p53 *in vivo*, which could cause unregulated cellular growth. In addition, CMV infection may enhance endothelial damage by inducing procoagulant activity and the expression of endothelial adhesion molecules. Additionally, the virus may also participate and enhance the repair process associated with restenosis by increasing expression and release of a number of growth factors that promote EC wound healing and angiogenesis (Dumortier *et al.*, 2008; Streblow *et al.*, 2008b; Botto *et al.*, 2010), which is discussed further in Chapter I.6.

Rat models of carotid artery balloon angioplasty have been used to study the role of CMV in arterial restenosis. As described above, RCMV exists in a lifelong latent state in an infected rat, and angioplasty-induced injury to the vessel wall promotes reactivation of this latent CMV. This combined injury and reactivation induce a sequence of events that causes excessive accumulation of SMC in the developing neointima contributing to restenosis. A study by Epstein and colleagues using a rat carotid-injury model demonstrated a 40% greater neointimal formation in RCMV-infected animals when compared to controls (Zhou *et al.*, 1995). RCMV infection stimulates SMC migration and proliferation, which are two critical components of neointimal hyperplasia during restenosis (Kloppenburg *et al.*, 2005; Streblow *et al.*, 2005).

Transplant vascular sclerosis: chronic allograft rejection of solid organ transplants

Solid organ transplantation remains the only viable cure for patients with end-stage organ failure. While significant advances have been made in solid organ transplantation techniques and post-surgical care, chronic allograft rejection (CR) remains a very important cause of morbidity and mortality in transplant recipients. The vascular lesion that is associated with CR is TVS, which occurs in 10–20% of transplant patients per year with a 5-year prevalence rate of 60% (Hosenpud *et al.*, 1992). The primary TVS lesion shows the accumulation of macrophages, T-cells, NK cells and B-cells in the vessel intima. TVS progresses to a diffuse concentric intimal proliferation of SMC accompanied with the accumulation of macrophages. TVS is associated with a low-grade inflammatory process occurring in the sub-endothelial layer of allograft vessels. There are several risk factors for CR including: donor age, bouts of acute rejection, hypertension, hypercholesterolemia and CMV. The only effective treatment for CR is re-transplantation making the determination of mechanisms involved in TVS and CR important for preventing this disease. With the advent of more potent immunosuppressive agents which target T-lymphocyte dependent immune responses, the risk of CMV infection is ever increasing in the solid organ and bone marrow transplant populations, which underscores the importance of focusing investigation on the sequelae of CMV infection in these patients.

CMV accelerates TVS and chronic allograft rejection

CMV infection is of clinical significance in solid organ and bone marrow transplant recipients because 60–80% of the donor/recipient population is latently infected with HCMV (Britt, 2008; see also Chapters II.13, II.14, and II.16). A number of studies have examined the association of HCMV infection with increased allograft rejection. Overall the incidence of post-transplant disease attributable to HCMV ranges from between 20% and 60% (Britt and Alford, 1996).

HCMV infection doubles the rate of liver graft failure at 3 years post transplantation (de Otero *et al.*, 1998; Rubin, 1999) and doubles the rate of rejection of cardiac allografts at 5 years post transplantation (Grattan *et al.*, 1989), which is linked to an increase in graft TVS. Accordingly, there is an increased incidence of CMV DNA detected in SMC, macrophages and EC of the coronary arteries of heart transplant patients with accelerated atherosclerotic pathology (Wu *et al.*, 1992). The HCMV serological status of the donor–recipient pair is one of the most important factors determining severe HCMV disease, and the effect on the transplanted graft (Fitzgerald *et al.*, 2004). Transplant recipients at highest risk are negative for CMV and receiving a CMV-positive donor graft. CMV-seropositive patients with intermediate risk are those receiving allografts from either seronegative or seropositive donors. CMV negative recipients of grafts from negative donors have the lowest risk of developing CMV disease. Importantly, studies using the CMV chemotherapeutic agent ganciclovir in transplant patients have provided significant evidence that CMV replication has an effect on the development of TVS. CMV-positive heart transplant recipients or patients receiving CMV-positive organs treated with ganciclovir for 28 days post-transplant and followed for an additional 120 days had significantly decreased incidence of primary CMV disease (Merigan *et al.*, 1992). These same patients were followed for 4–6 years post transplant and the ganciclovir-treated group had reduced incidence of TVS compared with patients who did not receive anti-CMV chemotherapy (Valantine *et al.*, 1999). While there exists a strong epidemiological link between CMV infection and accelerated graft rejection, determining the mechanisms involved in HCMV-associated TVS has been difficult because the high prevalence of HCMV results in a lack of negative controls. Furthermore, a clear mechanistic link between HCMV infection and TVS has been clouded by the multifactorial aetiology of TVS, especially when coupled with the lifelong nature of HCMV infection. Therefore, tractable animal models of solid organ transplantation may provide a key avenue to dissecting the mechanisms that link the virus to TVS/CR.

Rat transplantation models have been important for determining the role of CMV infection in the acceleration of TVS and CR. In several rat models of transplantation, RCMV-M accelerates graft failure via a TVS process with similar histology to that seen in human transplants (Lemström *et al.*, 1993b, 1995; Bruning *et al.*, 1994b; Orloff, 1999, 2002). Acute RCMV-M infection significantly increases the severity of TVS in graft vessels and decreases the time to develop TVS and graft failure in a rat model of heart transplantation (Orloff *et al.*, 1999, 2000). Importantly, the effects of CMV on TVS acceleration are not organ specific but occur in a broad range of solid organ transplants including heart, kidney, lung, and small bowel (Orloff, 1999; Orloff *et al.*, 1999, 2000; Soule *et al.*, 2006). Syngeneic transplants and transplants in which the recipient has been tolerized to the graft organ via bone marrow transplantation do not develop TVS or accelerated CR, suggesting that the progression of TVS requires an interplay between CMV replication and the alloreactive response of the recipient's immune system to the host organ (Orloff, 1999; Orloff *et al.*, 2002). Additionally, in RCMV-infected recipients of heart transplants a dramatic increase in expression of adhesion molecules, growth factors, cytokines and chemokines is temporally associated with the increased infiltration of T-cells and macrophages in the vessels of the allograft compared to uninfected allograft recipients (Streblow *et al.*, 2003, 2008a,b). A similar association of RCMV infection with increased chemokine expression and chronic allograft nephropathy was observed in a rat model of kidney transplantation (Soule *et al.*, 2006). These data suggest that RCMV infection requires an allo-immune environment for the acceleration of TVS.

Viral genes such as GPCRs that can be expressed in non-productively infected cells have been implicated as potential mediators in long-term diseases such as vascular disease (Speir *et al.*, 1994). Several CMV genes have been implicated in the development of vascular disease, including IE and E phase genes. These genes are attractive candidates for the acceleration of vascular pathologies since latently infected cells can exhibit partial viral gene expression without producing virus (see the related mouse model for partial viral gene expression during CMV latency, Chapter I.22). The RCMV-encoded chemokine receptor R33 is necessary for RCMV-mediated migration of SMC *in vitro*. Significantly, infection of rat heart transplant recipients with RCMV lacking R33 resulted in increased graft survival and decreased TVS disease compared to wild-type RCMV-infected transplant recipients (Streblow *et al.*, 2005). These data provide strong evidence that cellular chemotaxis resulting from CMV-encoded chemokine receptors plays a role in the inflammatory processes that lead to TVS. Similarly, the HCMV-encoded chemokine receptor US28 is necessary and sufficient for HCMV-mediated SMC and macrophage migration *in vitro* (Streblow *et al.*, 1999; Vomaske *et al.*, 2009).

As mentioned above, in human transplant recipients the source of CMV infection (donor vs. recipient) and the status of viral infection (active vs. latent) can be used to stratify the risk of developing viral disease and associated acceleration of TVS and CR. The effect of the infection status and source on the development of CMV-accelerated TVS was evaluated using the rat heart transplant model. Acutely infected recipients or donors were compared to naïve recipients receiving latently infected donor hearts as well as latently infected recipients of uninfected grafts (Orloff *et al.*, 2011). Acutely infected recipients rejected their naïve donor grafts earlier (POD 43 vs. 53, $P=0.01$; POD = postoperative day) and with increased TVS or neointimal formation (NI = 74 vs. 50, $P=0.01$; NI = neointimal index) compared with naïve recipients of grafts from acutely infected donor hearts, whereas uninfected controls rejected at approximately day 90. Importantly, similar to the human scenario, naïve recipients of latently infected donor hearts rejected their grafts significantly earlier than did the latently infected recipients (POD 54 vs. 63, $P=0.01$; with more severe TVS at the earlier rejection time NI = 63 vs. 51, $P=0.01$). These data demonstrate that the status of CMV infection (acute vs. latent) and the source (donor vs. recipient infection) affect the degree of CMV-acceleration of TVS and CR. We propose that latent donor infection results in an alteration in the graft prior to transplantation, which subsequently promotes a susceptibility to accelerated rejection of the graft upon transplantation into the recipient.

To determine the role of viral replication in acute and latent RCMV-accelerated rejection, allograft recipients (acutely infected recipients of naïve donor hearts and naïve recipients of latently infected donor hearts) were treated with ganciclovir (20 mg/kg/day i.p. for 30 days). The antiviral drug treatment reduced RCMV-accelerated allograft TVS and CR in acutely infected rats, suggesting that viral replication is critical in RCMV-accelerated allograft disease processes. However, the CMV-accelerated TVS and CR were not affected by treatment with ganciclovir in recipients of latently infected allografts. There are two possible reasons for this finding: (1) virus replication in the donor allograft prior to transplantation promotes a predilection to rejection due to increased immune activation in the organ. (2) IE/E viral genes, expressed by latently infected cells, are sufficient to promote the acceleration of TVS, and attempts at inhibition of viral replication are targeting only the downstream process; or a combination of both scenarios may be involved in graft TVS and CR. Immunostaining with antibodies directed against T-cells and macrophages of thin sections from latently infected (180 dpi) non-transplanted hearts as well as uninfected control hearts showed the presence of large clusters of resident macrophages and $CD8\alpha^+$ T-cells within the hearts of the latently infected animals prior to transplantation. These clusters of immune cells were not present in uninfected controls. Flow cytometric analysis of hearts from infected rats at 21 and 180 dpi indicated that they had an increased percentage of both $CD4^+$ and $CD8^+$ T-cells during latent infection (180 dpi) when compared to both uninfected controls and tissues from 21-day infected rats. In addition, the T-cells of the latently infected rat hearts have an effector-memory phenotype (93% $CD4^+CD62L^{low}$) and were dividing. Notably, this situation is highly reminiscent of the accumulation of effector-memory $CD8^+$ T-cells in lungs latently infected with MCMV (Holtappels *et al.*, 2000), a phenomenon linked to 'memory inflation' (see Chapter I.22). These findings suggest that the RCMV latently infected donor hearts contain a substantial number of activated, proliferating T-cells prior to transplantation and these cells may promote rejection by enhancing the inflammatory response.

Thus, it appears that CMV promotes rejection on many different levels including even prior to transplantation of the graft itself (Fig. II.15.7). It will be interesting to determine the viral genes involved in attracting the cells into tissues and whether this is an antiviral response or whether the persistent virus infection induces the formation of these lymphoid structures. Might the viral chemokines be involved in this process? Maybe they drive continued recruitment of inflammatory cells all the while the infected cell remains masked by viral genes preventing recognition and maintaining 'self protection'. The rat/RCMV model will undoubtedly be important for determining the mechanisms involved in these processes and for identifying the role of CMV in the development of vascular disease.

Conclusions

The rat/RCMV model has provided a number of key observations that have helped to elucidate the role of CMV in vascular disease and solid organ transplant rejection. However, the murine model of CMV infection clearly has an advantage over the rat model with regard to the availability of genetic knockout animals. Nevertheless, a new strategy has emerged that employs transcriptase-activated like effectors (TALEs) that were initially identified in the phytopathogen *Xanthomonas*. When combined with a nuclease, these TALEs can be efficiently used to excise a specific gene, resulting in a knockout animal as has successfully been shown in the rat recently (Tesson *et al.*, 2011). This new method will have to prove its effectiveness in the future, but it might represent a substantial breakthrough in the generation of knockout rats and could provide researchers with valuable tools to study human disease. Significant differences between the mouse and the human immune system exist, yet it is neither clear nor suggested that the rat might be a superior model. Apart from the atherosclerosis and vascular disease model discussed here,

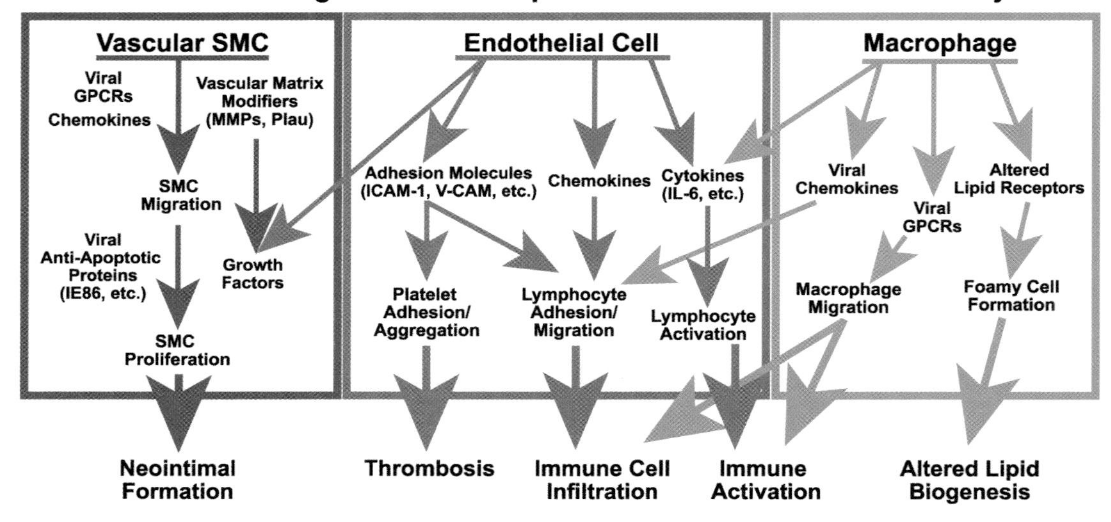

Figure II.15.7 CMV accelerates vascular disease and allograft rejection through promotion of a pathogenic wound repair mechanism. CMV infection of SMC is thought to promote vessel neointimal formation during vascular disease. SMC infected with CMV express viral GPCR that promote cellular migration in response to the enhanced chemokine secretion during infection and vessel injury. At the same time infection induces the production of bioactive enzymes and matrix modifiers such as MMPs and urokinase (Plau) that mediate the dissolution of the vessel matrix separating the endothelial layer from the lower SMC layers and also release latent growth factors that promote SMC migration and proliferation. HCMV IE86 inactivates tumour suppressor protein p53, which increases SMC proliferation, while other viral proteins have anti-apoptotic functions that would prevent cellular death. Infected and/ or injured EC up-regulate adhesion molecules and chemokines/cytokines resulting in increased platelet adhesion and aggregation as well as lymphocyte migration and activation. Infected macrophages have altered lipid receptor expression resulting in the formation of foamy cells. In general, CMV infection promotes a pathogenic wound healing response through the alteration of lipid biogenesis, thrombus formation, as well as increased lymphocyte and SMC migration and activation.

another application for rats is in BMT, an already well-established model in the mouse (see Chapter II.17). This model is similarly useful to study GvHD and can be carried out in rats due to their large body size and their robustness that facilitates follow-up observations. In addition, the rat will continue to have an important role in the elucidation of CMV's role in vascular disease.

Acknowledgements

We would like to thank Michael Wenda for graphics assistance.

References

Ahn, K., Angulo, A., Ghazal, P., Peterson, P.A., Yang, Y., and Früh, K. (1996). Human cytomegalovirus inhibits antigen presentation by a sequential multistep process. Proc. Natl. Acad. Sci. U.S.A. *93*, 10990–10995.

Baca Jones, C.C., Kreklywich, C.N., Messaoudi, I., Vomaske, J., McCartney, E., Orloff, S.L., Nelson, J.A., and Streblow, D.N. (2009). Rat cytomegalovirus infection depletes MHC II in bone marrow derived dendritic cells. Virology *388*, 78–90.

Beisser, P.S., Kaptein, S.J., Beuken, E., Bruggeman, C.A., and Vink, C. (1998). The Maastricht strain and England strain of rat cytomegalovirus represent different betaherpesvirus species rather than strains. Virology *246*, 341–351.

Beisser, P.S., Grauls, G., Bruggeman, C.A., and Vink, C. (1999). Deletion of the R78 G protein-coupled receptor gene from rat cytomegalovirus results in an attenuated, syncytium-inducing mutant strain. J. Virol. *73*, 7218–7230.

Beisser, P.S., Kloover, J.S., Grauls, G.E., Blok, M.J., Bruggeman, C.A., and Vink, C. (2000). The r144 major histocompatibility complex class I-like gene of rat cytomegalovirus is dispensable for both acute and long-term infection in the immunocompromised host. J. Virol. *74*, 1045–1050.

Beuken, E., Slobbe, R., Bruggeman, C.A., and Vink, C. (1996). Cloning and sequence analysis of the genes encoding DNA polymerase, glycoprotein B, ICP18.5 and major DNA-binding protein of rat cytomegalovirus. J. Gen. Virol. *77*, 1559–1562.

Blum, A., Giladi, M., Weinberg, M., Kaplan, G., Pasternack, H., Laniado, S., and Miller, H. (1998). High anti-cytomegalovirus (CMV) IgG antibody titer is associated with coronary artery disease and may predict post-coronary balloon angioplasty restenosis. Am. J. Cardiol. *81*, 866–868.

Botto, S., Streblow, D.N., DeFilippis, V., White, L., Kreklywich, C.N., Smith, P.P., and Caposio, P. (2010). IL-6 in human cytomegalovirus secretome promotes angiogenesis and survival of endothelial cells through the stimulation of survivin. Blood *117*, 352–361.

Britt, W. (2008). Manifestations of human cytomegalovirus infection: proposed mechanisms of acute and chronic disease. Curr. Top. Microbiol. Immunol. *325*, 417–470.

Britt, W.J., and Alford, C.A. (1996). Cytomegalovirus. In Fields Virology, Fields, B.N., Knipe, D.M., and Howley, P.M., eds. (Lippincott-Raven Publishers, Philadelphia, PA), pp. 2493–2523.

Brocchieri, L., Kledal, T.N., Karlin, S., and Mocarski, E.S. (2005). Predicting coding potential from genome sequence: application to betaherpesviruses infecting rats and mice. J. Virol. *79*, 7570–7596.

Bruggeman, C.A. (1993). Cytomegalovirus and latency: an overview. Virchows Arch. B Cell. Pathol. Incl. Mol. Pathol. *64*, 325–333.

Bruggeman, C.A., Meijer, H., Dormans, P.H., Debie, W.M., Grauls, G.E., and van Boven, C.P. (1982). Isolation of a cytomegalovirus-like agent from wild rats. Arch. Virol. *73*, 231–241.

Bruggeman, C.A., Debie, W.M., Grauls, G., Majoor, G., and van Boven, C.P. (1983a). Infection of laboratory rats with a new cytomegalo-like virus. Arch. Virol. *76*, 189–199.

Bruggeman, C.A., Schellekens, H., Grauls, G., Debie, W.M., and van Boven, C.P. (1983b). Rat cytomegalovirus: induction of and sensitivity to interferon. Antiviral. Res. *3*, 315–324.

Bruggeman, C.A., Meijer, H., Bosman, F., and van Boven, C.P. (1985). Biology of rat cytomegalovirus infection. Intervirology *24*, 1–9.

Bruning, J.H., Persoons, M., Lemström, K., Stals, F.S., De Clercq, E., and Bruggeman, C.A. (1994a). Enhancement of transplantation-associated atherosclerosis by CMV, which can be prevented by antiviral therapy in the form of HPMPC. Transpl. Int. *7*, S365–370.

Bruning, J.H., Persoons, M.C.J., Lemström, K.B., Stals, F.S., De Clereq, E., and Bruggeman, C.A. (1994b). Enhancement of transplantation associated atherosclerosis by CMV, which can be prevented by antiviral therapy in the form of HPMPC. Transplant. Int. *7*, 365–370.

Buck, A.H., Santoyo-Lopez, J., Robertson, K.A., Kumar, D.S., Reczko, M., and Ghazal, P. (2007). Discrete clusters of virus-encoded micrornas are associated with complementary strands of the genome and the 7.2-kilobase stable intron in murine cytomegalovirus. J. Virol. *81*, 13761–13770.

Burnett, M.S., Gaydos, C.A., Madico, G.E., Glad, S.M., Paigen, B., Quinn, T.C., and Epstein, S.E. (2001). Atherosclerosis in apoE knockout mice infected with multiple pathogens. J. Infect. Dis. *183*, 226–231.

Burns, W.H., Barbour, G.M., and Sandford, G.R. (1988). Molecular cloning and mapping of rat cytomegalovirus DNA. Virology *166*, 140–148.

Cekinovic, D., Golemac, M., Pugel, E.P., Tomac, J., Cicin-Sain, L., Slavuljica, I., Bradford, R., Misch, S., Winkler, T.H., Mach, M., *et al.* (2008). Passive immunization reduces murine cytomegalovirus-induced brain pathology in newborn mice. J. Virol. *82*, 12172–12180.

Chambers, J., Angulo, A., Amaratunga, D., Guo, H., Jiang, Y., Wan, J.S., Bittner, A., Früh, K., Jackson, M.R., Peterson, P.A., *et al.* (1999). DNA microarrays of the complex human cytomegalovirus genome: profiling kinetic class with drug sensitivity of viral gene expression. J. Virol. *73*, 5757–5766.

van Cleef, K.W., Scaf, W.M., Maes, K., Kaptein, S.J., Beuken, E., Beisser, P.S., Stassen, F.R., Grauls, G.E., Bruggeman, C.A., and Vink, C. (2004). The rat cytomegalovirus homologue of parvoviral rep genes, r127, encodes a nuclear protein with single- and double-stranded DNA-binding activity that is dispensable for virus replication. J. Gen. Virol. *85*, 2001–2013.

Dumortier, J., Streblow, D.N., Moses, A.V., Jacobs, J.M., Kreklywich, C.N., Camp, D., Smith, R.D., Orloff, S.L., and Nelson, J.A. (2008). Human cytomegalovirus secretome contains factors that induce angiogenesis and wound healing. J. Virol. *82*, 6524–6535.

Fitzgerald, J.T., Gallay, B., Taranto, S.E., McVicar, J.P., Troppmann, C., Chen, X., McIntosh, M.J., and Perez, R.V. (2004). Pretransplant recipient cytomegalovirus seropositivity and hemodialysis are associated with decreased renal allograft and patient survival. Transplantation *77*, 1405–1411.

Foster-Cuevas, M., Westerholt, T., Ahmed, M., Brown, M.H., Barclay, A.N., and Voigt, S. (2011). Cytomegalovirus e127 protein interacts with the inhibitory CD200 receptor. J. Virol. *85*, 6055–6059.

Goodrum, F.D., Jordan, C.T., High, K., and Shenk, T. (2002). Human cytomegalovirus gene expression during infection of primary hematopoietic progenitor cells: A model for latency. Proc. Natl. Acad. Sci. U.S.A. *99*, 16255–16260.

Grattan, M.T., Moreno-Cabral, C.E., Starnes, V.A., Oyer, P.E., Stinson, E.B., and Shumway, N.E. (1989). Cytomegalovirus infection is associated with cardiac allograft rejection and atherosclerosis. JAMA *261*, 3561–3566.

Gruijthuijsen, Y.K., Casarosa, P., Kaptein, S.J., Broers, J.L., Leurs, R., Bruggeman, C.A., Smit, M.J., and Vink, C. (2002). The rat cytomegalovirus R33-encoded G protein-coupled receptor signals in a constitutive fashion. J. Virol. *76*, 1328–1338.

Hassink, G.C., Duijvestijn-van Dam, J.G., Koppers-Lalic, D., van Gaans-van den Brink, J., van Leeuwen, D., Vink, C., Bruggeman, C.A., and Wiertz, E.J. (2005). Rat cytomegalovirus induces a temporal down-regulation of major histocompatibility complex class I cell surface expression. Viral Immunol. *18*, 607–615.

Hendrix, M.G., Salimans, M.M., van Boven, C.P., and Bruggeman, C.A. (1990). High prevalence of latently present cytomegalovirus in arterial walls of patients suffering from grade III atherosclerosis. Am. J. Pathol. *136*, 23–28.

Hillebrands, J.L., van der Werf, N., Klatter, F.A., Bruggeman, C.A., and Rozing, J. (2003). Role of peritoneal macrophages in cytomegalovirus-induced acceleration of autoimmune diabetes in BB-rats. Clin. Dev Immunol. *10*, 133–139.

Hillebrands, J.L., van Dam, J.G., Onuta, G., Klatter, F.A., Grauls, G., Bruggeman, C.A., and Rozing, J. (2005). Cytomegalovirus-enhanced development of transplant arteriosclerosis in the rat; effect of timing of infection and recipient responsiveness. Transpl. Int. *18*, 735–742.

Hoek, R.M., Ruuls, S.R., Murphy, C.A., Wright, G.J., Goddard, R., Zurawski, S.M., Blom, B., Homola, M.E., Streit, W.J., Brown, M.H., *et al.* (2000). Down-regulation of the macrophage lineage through interaction with OX2 (CD200). Science *290*, 1768–1771.

Holtappels, R., Pahl-Seibert, M.F., Thomas, D., and Reddehase, M.J. (2000). Enrichment of immediate-early 1 (m123/pp89) peptide-specific CD8 T-cells in a pulmonary CD62L(lo) memory-effector cell pool during latent murine cytomegalovirus infection of the lungs. J. Virol. *74*, 11495–11503.

Horvath, R., Cerny, J., Benedik, J., Hokl, J., and Jelinkova, I. (2000). The possible role of human cytomegalovirus (HCMV) in the origin of atherosclerosis. J. Clin. Virol. *16*, 17–24.

Hosenpud, J.D., Shipley, G.D., and Wagner, C.R. (1992). Cardiac allograft vasculopathy: current concepts, recent developments and future directions. J. Heart Lung Transplant. *11*, 9–23.

Johnson, K.P. (1969). Mouse cytomegalovirus: placental infection. J. Infect. Dis. *120*, 445–450.

Kaptein, S.J., Beisser, P.S., Gruijthuijsen, Y.K., Savelkouls, K.G., van Cleef, K.W., Beuken, E., Grauls, G.E., Bruggeman, C.A., and Vink, C. (2003). The rat cytomegalovirus R78 G protein-coupled receptor gene is required for production of infectious virus in the spleen. J. Gen. Virol. *84*, 2517–2530.

Kaptein, S.J., van Cleef, K.W., Gruijthuijsen, Y.K., Beuken, E.V., Van Buggenhout, L., Beisser, P.S., Stassen, F.R., Bruggeman, C.A., and Vink, C. (2004). The r131 gene of rat cytomegalovirus encodes a proinflammatory CC chemokine homolog which is essential for the production of infectious virus in the salivary glands. Virus Genes *29*, 43–61.

Kloover, J.S., Soots, A.P., Krogerus, L.A., Kauppinen, H.O., Loginov, R.J., Holma, K.L., Bruggeman, C.A., Ahonen, P.J., and Lautenschlager, I.T. (2000). Rat cytomegalovirus infection in kidney allograft recipients is associated with increased expression of intracellular adhesion molecule- 1 vascular adhesion molecule-1, and their ligands leukocyte function antigen-1 and very late antigen-4 in the graft. Transplantation *69*, 2641–2647.

Kloover, J.S., Grauls, G.E., Blok, M.J., Vink, C., and Bruggeman, C.A. (2002). A rat cytomegalovirus strain with a disruption of the r144 MHC class I-like gene is attenuated in the acute phase of infection in neonatal rats. Arch. Virol. *147*, 813–824.

Kloppenburg, G., de Graaf, R., Herngreen, S., Grauls, G., Bruggeman, C., and Stassen, F. (2005). Cytomegalovirus aggravates intimal hyperplasia in rats by stimulating smooth muscle cell proliferation. Microbes Infect. *7*, 164–170.

Koskinen, P., Lemström, K., Mattila, S., Häyry, P., and Nieminen, M.S. (1996). Cytomegalovirus infection associated accelerated heart allograft arteriosclerosis may impair the late function of the graft. Clin. Transplant. *10*, 487–493.

Lacayo, J., Sato, H., Kamiya, H., and McVoy, M.A. (2003). Down-regulation of surface major histocompatibility complex class I by guinea pig cytomegalovirus. J. Gen. Virol. *84*, 75–81.

Lemström, K., Persoons, M., Bruggeman, C., Ustinov, J., Lautenschlager, I., and Häyry, P. (1993a). Cytomegalovirus infection enhances allograft arteriosclerosis in the rat. Transplant. Proc. *25*, 1406–1407.

Lemström, K.B., Bruning, J.H., Bruggeman, C.A., Lautenschlager, I.T., and Häyry, P.J. (1993b). Cytomegalovirus infection enhances smooth muscle cell proliferation and intimal thickening of rat aortic allografts. J. Clin. Invest. *92*, 549–558.

Lemström, K.B., Aho, P.T., Bruggeman, C.A., and Häyry, P.J. (1994a). Cytomegalovirus infection enhances mRNA expression of platelet-derived growth factor-BB and transforming growth factor-beta 1 in rat aortic allografts. Possible mechanism for cytomegalovirus-enhanced graft arteriosclerosis. Arterioscler. Thromb. *14*, 2043–2052.

Lemström, K.B., Bruning, J.H., Bruggeman, C.A., Koskinen, P.K., Aho, P.T., Yilmaz, S., Lautenschlager, I.T., and Häyry, P.J. (1994b). Cytomegalovirus infection-enhanced allograft arteriosclerosis is prevented by DHPG prophylaxis in the rat. Circulation *90*, 1969–1978.

Lemström, K.B., Bruning, J.H., Bruggeman, C.A., Lautenschlager, I.T., and Häyry, P.J. (1994c). Triple drug immunosuppression significantly reduces immune activation and allograft arteriosclerosis in cytomegalovirus-infected rat aortic allografts and induces early latency of viral infection. Am. J. Pathol. *144*, 1334–1347.

Lemström, K., Koskinen, P., Krogerus, L., Daemen, M., Bruggeman, C.A., and Häyry, P.J. (1995). Cytomegalovirus antigen expression, endothelial cell proliferation, and intimal thickening in rat cardiac allografts after cytomegalovirus infection. Circulation *92*, 2594–2604.

Loh, H.S., Mohd-Lila, M.A., Abdul-Rahman, S.O., and Kiew, L.J. (2006). Pathogenesis and vertical transmission of a transplacental rat cytomegalovirus. Virol. J. *3*, 42.

MacDonald, M.R., Burney, M.W., Resnick, S.B., and Virgin, H.W. (1999). Spliced mRNA encoding the murine cytomegalovirus chemokine homolog predicts a beta chemokine of novel structure. J. Virol. *73*, 3682–3691.

Martelius, T.J., Blok, M.J., Inkinen, K.A., Loginov, R.J., Hockerstedt, K.A., Bruggeman, C.A., and Lautenschlager, I.T. (2001). Cytomegalovirus infection, viral DNA, and immediate early-1 gene expression in rejecting rat liver allografts. Transplantation *71*, 1257–1261.

Medearis, D.N., Jr. (1964). Mouse cytomegalovirus infection. 3. attempts to produce intrauterine infections. Am. J. Hyg. *80*, 113–120.

Merigan, T.C., Renlund, D.G., Keay, S., Bristow, M.R., Starnes, V., O'Connell, J.B., Resta, S., Dunn, D., Gamberg, P., Ratkovec, R.M., *et al.* (1992). A controlled trial of ganciclovir to prevent cytomegalovirus disease after heart transplantation. N. Engl. J. Med. *326*, 1182–1186.

Meyer, C., Grey, F., Kreklywich, C.N., Andoh, T.F., Tirabassi, R.S., Orloff, S.L., and Streblow, D.N. (2011). Cytomegalovirus microRNA expression is tissue specific and is associated with persistence. J. Virol. *85*, 378–389.

Michelson, S., Dal Monte, P., Zipeto, D., Bodaghi, B., Laurent, L., Oberlin, E., Arenzana-Seisdedos, F., Virelizier, J.L., and Landini, M.P. (1997). Modulation of RANTES production by human cytomegalovirus infection of fibroblasts. J. Virol. *71*, 6495–6500.

Mocarski, E.S., Shenk, T., and Pass, R.F. (2007). Cytomegaloviruses. In Fields Virology, Knipe, D.M.,Howley, P.M., Griffin, D.E., Lamb, R.A., Martin, M.A., Roizman, B., and Straus, S.E., eds. (Lippincott Williams & Wilkins, Philadelphia, PA), pp. 2701–2772.

Mueller, C., Hodgson, J.M., Bestehorn, H.P., Brutsche, M., Perruchoud, A.P., Marsch, S., Roskamm, H., and Buettner, H.J. (2003). Previous cytomegalovirus infection and restenosis after aggressive angioplasty with provisional stenting. J. Interv. Cardiol. *16*, 307–313.

Muhlestein, J.B., Horne, B.D., Carlquist, J.F., Madsen, T.E., Bair, T.L., Pearson, R.R., and Anderson, J.L. (2000).

Cytomegalovirus seropositivity and C-reactive protein have independent and combined predictive value for mortality in patients with angiographically demonstrated coronary artery disease. Circulation *102*, 1917–1923.

Nieto, F.J., Adam, E., Sorlie, P., Farzadegan, H., Melnick, J.L., Comstock, G.W., and Szklo, M. (1996). Cohort study of cytomegalovirus infection as a risk factor for carotid intimal-medial thickening, a measure of subclinical atherosclerosis. Circulation *94*, 922–927.

Orloff, S.L. (1999). Elimination of donor-specific alloreactivity by bone marrow chimerism prevents cytomegalovirus accelerated transplant vascular sclerosis in rat small bowel transplants. J. Clin. Virol. *12*, 142.

Orloff, S.L., Yin, Q., Coreless, C.L., Loomis, C.B., Rabkin, J.M., and Wagner, C.R. (1999). A rat small bowel transplant model of chronic rejection: histopathologic characteristics. Transplantation *68*, 766–779.

Orloff, S.L., Yin, Q., Corless, C.L., Orloff, M.S., Rabkin, J.M., and Wagner, C.R. (2000). Tolerance induced by bone marrow chimerism prevents transplant vascular sclerosis in a rat model of small bowel transplant chronic rejection. Transplantation *69*, 1295–1303.

Orloff, S.L., Streblow, D.N., Soderberg-Naucler, C., Yin, Q., Kreklywich, C., Corless, C.L., Smith, P.A., Loomis, C.B., Mills, L.K., Cook, J.W., *et al.* (2002). Elimination of donor-specific alloreactivity prevents cytomegalovirus-accelerated chronic rejection in rat small bowel and heart transplants. Transplantation *73*, 679–688.

Orloff, S.L., Hwee, Y.K., Kreklywich, C., Andoh, T.F., Hart, E., Smith, P.A., Messaoudi, I., and Streblow, D.N. (2011). Cytomegalovirus latency promotes cardiac lymphoid neogenesis and accelerated allograft rejection in CMV naïve recipients. Am. J. Transplant. *11*, 45–55.

de Otero, J., Gavalda, J., Murio, E., Vargas, V., Rosello, J., Calico, I., Margarit, C., and Pahissa, A. (1998). Cytomegalovirus disease as a risk factor for graft loss and death after orthotopic liver transplantation. Clin. Invest. Dis. *26*, 865–870.

Pampou, S., Gnedoy, S.N., Bystrevskaya, V.B., Smirnov, V.N., Chazov, E.I., Melnick, J.L., and DeBakey, M.E. (2000). Cytomegalovirus genome and the immediate-early antigen in cells of different layers of human aorta. Virchows Arch. *436*, 539–552.

Persoons, M.C., Daemen, M.J., van Kleef, E.M., Grauls, G.E., Wijers, E., and Bruggeman, C.A. (1997). Neointimal smooth muscle cell phenotype is important in its susceptibility to cytomegalovirus (CMV) infection: a study in rat. Cardiovasc. Res. *36*, 282–288.

Priscott, P.K., and Tyrrell, D.A. (1982). The isolation and partial characterisation of a cytomegalovirus from the brown rat, Rattus norvegicus. Arch. Virol. *73*, 145–160.

Rabson, A.S., Edgcomb, J.H., Legallais, F.Y., and Tyrrell, S.A. (1969). Isolation and growth of rat cytomegalovirus *in vitro*. Proc. Soc. Exp. Biol. Med. *131*, 923–927.

Rubin, R.H. (1999). Importance of CMV in the transplant population. Transplant. Infect. Dis. *1*, 3–7.

Sandford, G.R., and Burns, W.H. (1996). Rat cytomegalovirus has a unique immediate early gene enhancer. Virology *222*, 310–317.

Sandford, G.R., Schumacher, U., Ettinger, J., Brune, W., Hayward, G.S., Burns, W.H., and Voigt, S. (2010). Deletion of the rat cytomegalovirus immediate-early 1 gene results in a virus capable of establishing latency, but with lower levels of acute virus replication and latency that compromise reactivation efficiency. J. Gen. Virol. *91*, 616–621.

Smith, L.M., Tonkin, J.N., Lawson, M.A., and Shellam, G.R. (2004). Isolates of cytomegalovirus (CMV) from the black rat *Rattus rattus* form a distinct group of rat CMV. J. Gen. Virol. *85*, 1313–1317.

Soule, J.L., Streblow, D.N., Andoh, T.F., Kreklywich, C.N., and Orloff, S.L. (2006). Cytomegalovirus accelerates chronic allograft nephropathy in a rat renal transplant model with associated provocative chemokine profiles. Transplant. Proc. *38*, 3214–3220.

Speir, E., Modali, R., Huang, E.S., Leon, M.B., Shawl, F., Finkel, T., and Epstein, S.E. (1994). Potential role of human cytomegalovirus and p53 interaction in coronary restenosis. Science *265*, 391–394.

Speir, E., Huang, E.S., Modali, R., Leon, M.B., Shawl, F., Finkel, T., and Epstein, S.E. (1995). Interaction of human cytomegalovirus with p53: possible role in coronary restenosis. Scand. J. Infect. Dis. Suppl. *99*, 78–81.

Stals, F.S., Bosman, F., van Boven, C.P., and Bruggeman, C.A. (1990). An animal model for therapeutic intervention studies of CMV infection in the immunocompromised host. Arch. Virol. *114*, 91–107.

Stals, F.S., Zeytinoglu, A., Havenith, M., de Clercq, E., and Bruggeman, C.A. (1993). Rat cytomegalovirus-induced pneumonitis after allogeneic bone marrow transplantation: effective treatment with (S)-1-(3-hydroxy-2-phosphonyl-methoxypropyl)cytosine. Antimicrob. Agents Chemother. *37*, 218–223.

Steinhoff, G., You, X.M., Steinmuller, C., Boeke, K., Stals, F.S., Bruggeman, C.A., and Haverich, A. (1995). Induction of endothelial adhesion molecules by rat cytomegalovirus in allogeneic lung transplantation in the rat. Scand. J. Infect. Dis. Suppl. *99*, 58–60.

Steinhoff, G., You, X.M., Steinmuller, C., Bauer, D., Lohmann-Matthes, M.L., Bruggeman, C.A., and Haverich, A. (1996). Enhancement of cytomegalovirus infection and acute rejection after allogeneic lung transplantation in the rat. Transplantation *61*, 1250–1260.

van der Strate, B.W., Hillebrands, J.L., Lycklama a Nijeholt, S.S., Beljaars, L., Bruggeman, C.A., Van Luyn, M.J., Rozing, J., The, T.H., Meijer, D.K., Molema, G., *et al.* (2003). Dissemination of rat cytomegalovirus through infected granulocytes and monocytes *in vitro* and *in vivo*. J. Virol. *77*, 11274–11278.

Streblow, D.N., Söderberg-Nauclér, C., Vieira, J., Smith, P., Wakabayashi, E., Rutchi, F., Mattison, K., Altschuler, Y., and Nelson, J.A. (1999). The human cytomegalovirus chemokine receptor US28 mediates vascular smooth muscle cell migration. Cell 99, 511–520.

Streblow, D.N., Orloff, S.L., and Nelson, J.A. (2001). Do pathogens accelerate atherosclerosis? J. Nutr. *131*, 2798S-2804S.

Streblow, D.N., Kreklywich, C., Yin, Q., De La Melena, V.T., Corless, C.L., Smith, P.A., Brakebill, C., Cook, J.W., Vink, C., Bruggeman, C.A., *et al.* (2003). Cytomegalovirus-mediated up-regulation of chemokine expression correlates with the acceleration of chronic rejection in rat heart transplants. J. Virol. *77*, 2182–2194.

Streblow, D.N., Kreklywich, C.N., Smith, P., Soule, J.L., Meyer, C., Yin, M., Beisser, P., Vink, C., Nelson, J.A., and Orloff, S.L. (2005). Rat cytomegalovirus-accelerated transplant vascular sclerosis is reduced with mutation of the chemokine-receptor R33. Am. J. Transplant. *5*, 436–442.

Streblow, D.N., van Cleef, K.W., Kreklywich, C.N., Meyer, C., Smith, P., Defilippis, V., Grey, F., Früh, K., Searles, R., Bruggeman, C., *et al.* (2007). Rat cytomegalovirus gene expression in cardiac allograft recipients is tissue specific and does not parallel the profiles detected *in vitro*. J. Virol. *81*, 3816–3826.

Streblow, D.N., Dumortier, J., Moses, A.V., Orloff, S.L., and Nelson, J.A. (2008a). Mechanisms of cytomegalovirus-accelerated vascular disease: induction of paracrine factors that promote angiogenesis and wound healing. Curr. Top. Microbiol. Immunol. *325*, 397–415.

Streblow, D.N., Kreklywich, C.N., Andoh, T., Moses, A.V., Dumortier, J., Smith, P.P., Defilippis, V., Früh, K., Nelson, J.A., and Orloff, S.L. (2008b). The role of angiogenic and wound repair factors during CMV-accelerated transplant vascular sclerosis in rat cardiac transplants. Am. J. Transplant. *8*, 277–287.

Tang, Q., Murphy, E.A., and Maul, G.G. (2006). Experimental confirmation of global murine cytomegalovirus open reading frames by transcriptional detection and partial characterization of newly described gene products. J. Virol. *80*, 6873–6882.

Tesson, L., Usal, C., Menoret, S., Leung, E., Niles, B.J., Remy, S., Santiago, Y., Vincent, A.I., Meng, X., Zhang, L., *et al.* (2011). Knockout rats generated by embryo microinjection of TALENs. Nat. Biotechnol. *29*, 695–696.

Valantine, H.A., Gao, S.Z., Menon, S.G., Renlund, D.G., Hunt, S.A., Oyer, P., Stinson, E.B., Brown, B.W., Merigan, T.C., and Schroeder, J.S. (1999). Impact of prophylactic immediate post-transplant ganciclovir on development of transplant atherosclerosis. A post-hoc analysis of a randomized, placebo-controlled study. Circulation *100*, 61–66.

Vink, C., Beuken, E., and Bruggeman, C.A. (1996). Structure of the rat cytomegalovirus genome termini. J. Virol. *70*, 5221–5229.

Vink, C., Beuken, E., and Bruggeman, C.A. (1997). Cloning and functional characterization of the origin of lytic-phase DNA replication of rat cytomegalovirus. J. Gen. Virol. *78*, 2963–2973.

Vink, C., Beuken, E., and Bruggeman, C.A. (2000). Complete DNA sequence of the rat cytomegalovirus genome. J. Virol. *74*, 7656–7665.

Voigt, S., Sandford, G.R., Ding, L., and Burns, W.H. (2001). Identification and characterization of a spliced C-type lectin-like gene encoded by rat cytomegalovirus. J. Virol. *75*, 603–611.

Voigt, S., Sandford, G.R., Hayward, G.S., and Burns, W.H. (2005). The English strain of rat cytomegalovirus (CMV) contains a novel captured CD200 (vOX2) gene and a spliced CC chemokine upstream from the major immediate-early region: further evidence for a separate evolutionary lineage from that of rat CMV Maastricht. J. Gen. Virol. *86*, 263–274.

Voigt, S., Mesci, A., Ettinger, J., Fine, J.H., Chen, P., Chou, W., and Carlyle, J.R. (2007). Cytomegalovirus evasion of innate immunity by subversion of the NKR-P1B:Clr-b missing-self axis. Immunity *26*, 617–627.

Vomaske, J., Melnychuk, R.M., Smith, P.P., Powell, J., Hall, L., DeFilippis, V., Früh, K., Smit, M., Schlaepfer, D.D., Nelson, J.A., *et al.* (2009). Differential ligand binding to a human cytomegalovirus chemokine receptor determines cell type-specific motility. PLoS Pathog. *5*, e1000304.

Vossen, R.C., van Dam-Mieras, M.C., and Bruggeman, C.A. (1996). Cytomegalovirus infection and vessel wall pathology. Intervirology *39*, 213–221.

van der Werf, N., Hillebrands, J.L., Klatter, F.A., Bos, I., Bruggeman, C.A., and Rozing, J. (2003). Cytomegalovirus infection modulates cellular immunity in an experimental model for autoimmune diabetes. Clin. Dev. Immunol. *10*, 153–160.

Wiertz, E.J., Jones, T.R., Sun, L., Bogyo, M., Geuze, H.J., and Ploegh, H.L. (1996). The human cytomegalovirus US11 gene product dislocates MHC class I heavy chains from the endoplasmic reticulum to the cytosol. Cell *84*, 769–779.

Wright, G.J., Cherwinski, H., Foster-Cuevas, M., Brooke, G., Puklavec, M.J., Bigler, M., Song, Y., Jenmalm, M., Gorman, D., McClanahan, T., *et al.* (2003). Characterization of the CD200 receptor family in mice and humans and their interactions with CD200. J. Immunol. *171*, 3034–3046.

Wu, T.C., Hruban, R.H., Ambinder, R.F., Pizzorno, M., Cameron, D.E., Baumgartner, W.A., Reitz, B.A., Hayward, G.S., and Hutchins, G.M. (1992). Demonstration of cytomegalovirus nucleic acids in the coronary arteries of transplanted hearts. Am. J. Pathol. *140*, 739–747.

Yagyu, K., Steinhoff, G., Duijvestijn, A.M., Bruggeman, C.A., Matsumoto, H., and van Breda Vriesman, P.J. (1992). Reactivation of rat cytomegalovirus in lung allografts: an experimental and immunohistochemical study in rats. J. Heart Lung Transplant. *11*, 1031–1040.

Zhou, Y.F., Shou, M., Guzman, R., Guetta, E., Finkel, T., and Epstein, S.E. (1995). Cytomegalovirus infection increases neointimal fromation in the rat model of balloon injury. J. Am. Coll. Cardiol. *25*, 242a.

Zhou, Y.F., Guetta, E., Yu, Z.X., Finkel, T., and Epstein, S.E. (1996a). Human cytomegalovirus increases modified low density lipoprotein uptake and scavenger receptor mRNA expression in vascular smooth muscle cells. J. Clin. Invest. *98*, 2129–2138.

Zhou, Y.F., Leon, M.B., Waclawiw, M.A., Popma, J.J., Yu, Z.X., Finkel, T., and Epstein, S.E. (1996b). Association between prior cytomegalovirus infection and the risk of restenosis after coronary atherectomy. N. Engl. J. Med. *335*, 624–630.

Clinical Cytomegalovirus Research: Haematopoietic Cell Transplantation

II.16

Sachiko Seo and Michael Boeckh

Abstract

HCMV infection remains an important complication after haematopoietic cell transplantation (HCT), although significant progress in the management of HCMV infection and disease has been made in the last two decades. A major achievement has been the optimization of pre-emptive therapy strategies based on surveillance by the pp65 antigenaemia assay or HCMV DNA or RNA detection. However, current strategies are limited by the toxicity of antiviral agents and breakthrough disease in highly immunosuppressed individuals. Antiviral resistance occurs infrequently after HCT, but can be a considerable management challenge. This chapter outlines the current topics in host immunity, diagnosis, prevention, treatment of HCMV disease in HCT recipients, and future directions of HCMV management.

Introduction

HCMV continues to be an important complication after HCT. Major progress has been made over the past 15 years in HCMV management, mainly in preventative strategies. The most commonly used strategies are pre-emptive therapy based on virological monitoring and upfront prophylaxis where an antiviral drug is given based on HCMV serostatus. The recent development of new antiviral drugs and vaccines is likely to provide a wider spectrum of options for prevention and treatment than is presently available. This chapter reviews host immunity, clinical manifestation, diagnosis, risk factors, management of HCMV infection or disease, and new therapeutic options in HCT recipients.

HCMV and host immunity

Both adaptive and innate immunity to HCMV are important in disease control after HCT. General aspects of HCMV-specific immunity are reviewed in Chapters II.7, II.8 and II.10.

Adaptive immunity

T-cell-mediated cellular immunity plays an important role in control of HCMV replication in HCT recipients. Lack of HCMV-specific $CD8^+$ and $CD4^+$ T-cells after transplantation has been associated with progressive HCMV infection and disease (Fuji *et al.*, 2011).

HCMV-specific $CD4^+$ T-cells are important for maintaining HCMV-specific $CD8^+$ T-cell response (Fuji *et al.*, 2011). Previous studies showed that lack of HCMV specific-$CD4^+$ T-cells is related to late HCMV disease and poor outcome (Boeckh *et al.*, 2003a). More recent studies have examined the role of epitope (viral peptide)-specific, HLA-peptide multimer-stained cells and polyfunctional CMV-specific T-cells for defining protective immunity (Lilleri *et al.*, 2008; Gratama *et al.*, 2010; Guerrero *et al.*, 2012). Initial results suggest that these cells are potentially useful for identifying patients at high risk for subsequent HCMV disease.

Little is known about the role of humoral immunity in the control of HCMV infection after HCT. After natural infection, antibodies to multiple different HCMV proteins including gB and gH are produced (Rasmussen *et al.*, 1991). The antibodies can neutralize the virus *in vitro*, but show no inhibition of primary infection and mild effect to prevent severe disease (Schoppel *et al.*, 1998). No study examined the relative role of humoral and T-cell responses in the control of HCMV after HCT.

Innate immunity

Innate immunity is also involved in the regulation of HCMV infection. An earlier report showed that NK cell deficiency is associated with severe HCMV disease (Biron *et al.*, 1989). In an animal study, NK cells impeded murine CMV replication (Brown *et al.*, 2001). More recently, KIR genotype has been shown to affect the development of HCMV infection after HCT, although the exact mechanism remains unclear (Gallez-Hawkins *et al.*, 2011).

During binding to a target cell, HCMV triggers inflammatory cytokine production. One of the mechanisms is due to interaction of gB and gH with TLR2 (Boehme *et al.*, 2006; see Chapter I.8). Among liver transplant recipients, polymorphisms in TLR2 correlated with HCMV infection (Kijpittayarit *et al.*, 2007). Other TLRs, including TLR3 and TLR9, have been reported to be important to prevent murine CMV replication. Polymorphisms in CCR5 and IL-10 are associated with HCMV disease after HCT, whereas polymorphisms in monocyte chemoattractant protein-1 are associated with HCMV reactivation (Loeffler *et al.*, 2006).

Diagnostic methods

Conventional tissue culture takes several weeks to detect HCMV (Boeckh and Boivin, 1998), thereby limiting its clinical utility except for definitive diagnosis and for providing an isolate.

The shell-vial technique is a combination of viral culture and detection assay by monoclonal antibodies. The sample is inoculated onto human fibroblasts, followed by centrifugation to increase infectivity and incubation for 18 to 24 hours. Subsequently, identification by monoclonal antibodies directed at HCMV antigens is performed (Boeckh and Boivin, 1998). Although this assay has shortened the duration of diagnosis (1–2 days), its sensitivity is insufficient for optimized viral surveillance in blood, but it remains an invaluable diagnostic tool for optimized diagnosis in tissue and bronchoalveolar lavage (BAL) samples (Crawford *et al.*, 1988; Hackman *et al.*, 1994; Ljungman *et al.*, 2002b).

Histological detection of nuclear inclusions called 'owl's eye' in tissue specimens is a highly specific but insensitive way to diagnose invasive HCMV disease. As the inclusion bodies are also detected in other herpesvirus infections, immunohistochemical technique or *in situ* hybridization to identify HCMV antigens is commonly used together. Detection of HCMV antigen in peripheral leucocytes (antigenaemia) is a popular method for detection of HCMV in blood because of its rapidity and high sensitivity. The antigenaemia assay is based on direct detection of structural viral matrix protein pp65 (Boeckh and Boivin, 1998). The assay is predictive for HCMV disease and can therefore be used for HCMV surveillance and initiation of pre-emptive therapy (Nichols *et al.*, 2001; Pollack *et al.*, 2011). Limitations of the assay include its dependency on circulating neutrophils (at least 200/mm^3), its high variability of quantitative results (Boeckh and Boivin, 1998), and its low sensitivity for HCMV gastrointestinal disease and retinitis (Boeckh *et al.*, 1996a; Jang *et al.*, 2009).

qPCR is the most sensitive method for detecting HCMV and coincidently maintains high specificity when used in a quantitative fashion (Boeckh *et al.*, 2004). In addition, results can be usually available within 24 hours. qPCR testing has become the standard method for HCMV monitoring in blood (either whole blood or plasma) at many centres (Pollack *et al.*, 2011). Compared to the pp65 antigenaemia assay, qPCR testing allows detection of HCMV during leucopenia, has a high dynamic range and accuracy of quantitation of HCMV viral load, and provides a more accurate correlation with response to treatment after antiviral therapy (Boeckh and Boivin, 1998; Einsele *et al.*, 2000; Boeckh *et al.*, 2003a; Ljungman *et al.*, 2006). The recently described international standard will be useful to compare quantitative results between different in-house assays (Madej *et al.*, 2010). The role of qPCR in the diagnosis of HCMV disease is poorly defined with the exception of HCMV encephalitis (Ljungman *et al.*, 2002b). Owing to its high sensitivity and negative predictive value, it is useful to rule out CMV disease, but correlations of relevant viral load levels in BAL and tissue have not been established. Studies are needed to determine these thresholds because asymptomatic shedding in the lung and other organs are common after HCT (Schmidt *et al.*, 1991).

HCMV mRNA detection by nucleic acid sequence-based amplification (NASBA) is not quantitative and appears to be slightly less sensitive than DNA detection (Lehto *et al.*, 2005). However, since it is a marker of active viral replication, mRNA detection may be more predictive for HCMV disease. Pre-emptive therapy based on detection of HCMV mRNA by NASBA in blood samples has shown similar results to DNA PCR- or pp65 antigenaemia-based pre-emptive therapy strategies after HCT in two small studies (Gerna *et al.*, 2003; Hebart *et al.*, 2011). However, sufficiently powered equivalency or non-inferiority studies have not been reported. For a detailed discussion of HCMV diagnostic laboratory testing, see the specific Chapter II.18.

Clinical manifestations and diagnosis

HCMV 'infection' and 'disease' should be distinguished. HCMV infection refers to the detection of HCMV in samples like plasma, whole blood, urine, or throat swab in the absence of symptoms, whereas HCMV disease indicates the presence of clinical signs and symptoms along with the detection of HCMV in the affected organ. While detection of the virus in the affected organ is required for the diagnosis of HCMV disease, one notable exception is HCMV retinitis, which can be diagnosed with typical ophthalmological signs alone (Ljungman *et al.*, 2002b). HCMV detection by culture and/or histology in a tissue sample is required for diagnosis of HCMV diseases except for encephalitis.

Primary infection occurs in HCMV seronegative patients, typically in childhood via breast milk, saliva, or urine (Ljungman *et al.*, 2002b; Mocarski *et al.*, 2007). HCMV can also be acquired via blood products or genital secretions (Mocarski *et al.*, 2007). Recurrent infection is categorized into two groups, i.e. reinfection and reactivation. Reinfection is present when a different viral strain is detected in a seropositive patient, while reactivation is the reactivation of latent viruses. In clinical practice, the two cannot be easily distinguished with standard diagnostic tests. Epidemiological data suggest that the dominant mechanism of HCMV active infection in seropositive recipients is due to reactivation (Nichols *et al.*, 2002).

International definitions of HCMV diseases have been published in 2002 (Ljungman *et al.*, 2002b). Almost any organ can be involved in HCMV disease and therefore both general symptoms and organ-specific clinical manifestations are important for diagnosis (Ljungman *et al.*, 2002b). Systemic use of corticosteroid after transplantation can obscure clinical symptoms such as fever. Most HCMV disease cases occur between 3 and 12 weeks after HCT. However, 4–15% of seropositive recipients and of seronegative recipients who acquired HCMV from a seropositive donor during the first 3 months develop late HCMV disease. The outcome of late HCMV disease is similar to early HCMV disease (Boeckh *et al.*, 2003b).

Pneumonia

HCMV pneumonia is the most important HCMV disease manifestation in HCT recipients due to its high mortality rate of 50–60% (Erard *et al.*, 2007). Despite advances in prevention of HCMV disease, the outcome of HCMV pneumonia has not changed significantly in the past twenty years (Ljungman *et al.*, 1992; Erard *et al.*, 2007). HCMV pneumonia often originates with fever, non-productive cough, hypoxia, and interstitial infiltrates on radiography. Rarely, nodules may be observed on radiography. The onset of symptoms can occur over 1–2 weeks, but rapid progression to respiratory failure is often seen. The diagnosis of HCMV pneumonia requires detection of HCMV by rapid culture, direct fluorescent antibody testing, or cytology in BAL, or by rapid culture, histology, or immunohistochemistry in lung biopsy specimens. The role of PCR testing on BAL fluid remains controversial for definitive diagnosis of HCMV pneumonia (Ljungman *et al.*, 2002b). The high sensitivity of the PCR assay will detect asymptomatic pulmonary shedding, which is present in approximately one third of seropositive HCT recipients (Schmidt *et al.*, 1991). To date, no study has established viral load thresholds in BAL that distinguish between asymptomatic shedding and HCMV pneumonia. However, the high negative predictive values of PCR assays can be helpful to rule out HCMV pneumonia.

Gastrointestinal disease

HCMV can affect any part of the gastrointestinal (GI) tract from oesophagus to rectum. Oesophagitis typically presents with odynophagia, gastritis with nausea and/or epigastric pain, and colitis with abdominal pain, diarrhoea, and haematochezia. Microscopic findings range from patchy erythema and microerosions to deep ulcers into the submucosal layers. Visual differentiation of these lesions from other diseases like GvHD is often difficult. The diagnosis of GI disease depends on detection of HCMV in biopsy specimens by rapid culture, histology, or immunohistochemistry. A combination of two or more diagnostic tests increases the diagnostic yield (Hackman *et al.*, 1994). Similar to BAL, the role of PCR in the diagnosis of HCMV GI disease remains poorly defined due the possibility of asymptomatic low-level tissue reactivation and the lack of established thresholds that correlate with invasive disease. HCMV GI disease can occur without preceding or concurrent pp65 antigenaemia or HCMV DNA detection in blood (Jang *et al.*, 2009).

Retinitis

HCMV retinitis is relatively uncommon in HCT recipients (Crippa *et al.*, 2001). HCMV retinitis generally occurs later than day 100 after HCT. Prior HCMV reactivation, delayed lymphocyte engraftment, and chronic GvHD are raised as risk factors of HCMV retinitis (Crippa *et al.*, 2001). Decreased visual acuity or blurred vision is a typical presenting symptom, and approximately 60% of patients have involvement of both eyes (Crippa *et al.*, 2001). To avoid blindness, diagnosis should be quickly performed by an ophthalmologist, confirming a characteristic ophthalmoscopic picture of necrotizing retinitis with or without haemorrhage (Carmichael, 2012). Detection of HCMV by PCR in a sample from aqueous humour can support the diagnosis (Ando *et al.*, 2002).

Encephalitis

HCMV may occasionally cause encephalitis. Encephalitis can be diagnosed by PCR using a sample from cerebrospinal fluid, in addition to general symptoms of encephalitis (Fink *et al.*, 2010). HCMV encephalitis is generally a late-onset disease following repeated CMV reactivation in the absence of other HCMV diseases (Reddy *et al.*, 2010). Most cases show ganciclovir-resistant mutation and consequently high mortality. Low T-cell count or T-cell depletion regimen are important risk factors.

Risk factors

HCMV serostatus

In allogeneic HCT recipients, the most important risk factor for HCMV disease is the serological status of the donor (D) and recipient (R). In the current era of leucoreduction, the risk of HCMV disease is very low (1–3%) in seronegative recipients of a seronegative donor product (D^-/R^-). In D^-/R^- patients, HCMV can be acquired by transfused blood products or contact with other seropositive individuals that shed HCMV (Nichols *et al.*, 2002; Mocarski *et al.*, 2007). The efficacy of pre-emptive therapy in D^-/R^- patients has not been evaluated in randomized trials, but monitoring of HCMV infection and pre-emptive therapy has been shown to be beneficial in a large cohort study (Nichols *et al.*, 2003).

In seronegative recipients transplanted from seropositive donor (D^+/R^-), 20–30% of the patients develop primary HCMV infection due to transmission of latent HCMV by stem cell or marrow product; the total nucleated cell count of the stem cell product is an important risk factor for transmission, while recipient factors were not significantly associated with transmission (Nichols *et al.*, 2002; Pergam *et al.*, 2012).

Without prophylaxis, approximately 80% of HCMV-seropositive patients (D^-/R^+ plus D^+/R^+) experience active HCMV infection after allogeneic HCT. Even with use of pre-emptive antiviral agents, HCMV disease occurs in approximately 3–5% of seropositive patients during the first 3 months, and in 7–10% of seropositive recipients during the first year after HCT (Nakamae *et al.*, 2009; Pollack *et al.*, 2012). This represents a major decline compared to the era before pre-emptive therapy during which the incidence was approximately 25%. An HCMV-seropositive recipient is generally at higher risk for non-relapse mortality (NRM) than a D^-/R^- patient because of direct and indirect effects of HCMV, occasional drug resistant disease and adverse effects of the antivirals used to treat HCMV infection and disease (i.e. ganciclovir, valganciclovir, or foscarnet) (Ljungman *et al.*, 1998; Boeckh and Nichols, 2004; Nakamae *et al.*, 2011). In seropositive recipients, use of an HCMV-seropositive donor is associated with a lower proportion of high viral load in the recipient and lower NRM in some studies. However, these findings have not been consistently reported, thus it remains controversial whether a seropositive donor should be selected if several otherwise equally suitable donors are available (Boeckh *et al.*, 2004; Zhou *et al.*, 2009).

Immunosuppressive agents

The widespread use of immunosuppressive drugs for prophylaxis and treatment of GvHD after allogeneic HCT leads to an increased risk of HCMV disease. The use of corticosteroids of more than 1 mg/kg/day is considered a risk factor for HCMV reactivation and impaired response to antiviral therapy (Nichols *et al.*, 2001). Alemutuzumab is an anti-CD52 monoclonal antibody and can be used as a part of conditioning regimen or GvHD prophylaxis. Alemutuzumab induces both $CD4^+$ and $CD8^+$ lymphopenia, which lasts for up to 9 months. HCMV infection typically occurs during 3–6 weeks after the use of alemutuzumab (O'Brien *et al.*, 2006). HCT recipients who receive alemutuzumab have a higher rate of HCMV infection compared with the control (Delgado *et al.*, 2008). A small study reported that the use of mycophenolate mofetile after HCT also increases the risk of HCMV infection (Hambach *et al.*, 2002). Antithymocyte globlin (ATG) is a common agent in conditioning regimen to prevent GvHD by *in vivo* T-cell depletion. ATG affects HCMV reactivation in a dose-dependent fashion (Hamadani *et al.*, 2009), however, the use of ATG did not make a difference in the incidence of CMV infection after the adjustment for cytomegalovirus status at baseline (Finke *et al.*, 2009).

Sirolimus/tacrolimus regimens show decreased HCMV reactivation compared with tacrolimus only (Marty *et al.*, 2007). Although the mechanism of decreased HCMV infection using sirolimus has not been completely described, rictor (rapamycin-insensitive companion of mTOR)-mediated blockade, histone modification, chromatin remodelling, or HCMV-specific lymphocyte immune recovery have been raised as possible mechanisms (Kudchodkar *et al.*, 2006).

Conditioning regimen

Myeloablative conditioning regimens are related to higher HCMV viral loads compared to non-myeloablative regimens (Nakamae *et al.*, 2009). This phenomenon may be explained by maintained host immunity against HCMV after non-myeloablative regimens (Maris *et al.*, 2003). There was no difference in occurrence of HCMV disease between myeloablative and non-myeloablative conditionings (Ljungman *et al.*, 2006; Nakamae *et al.*, 2009). Also, the intensity of the conditioning regimen did not affect the outcome of HCMV disease (Nakamae *et al.*, 2009).

Rituximab is a monoclonal antibody directed against $CD20^{+}$ cells and widely used in a part of chemotherapy or conditioning regimens for lymphomas. One relatively small study indicated that the use of rituximab increased the incidence of HCMV infection (Lee *et al.*, 2008). Additional studies are needed to define the effect of rituximab on HCMV risk.

Donor source

In autologous transplantation, approximately 40% of HCMV-seropositive patients experience HCMV infection, although the development to HCMV disease is rare (Boeckh *et al.*, 1996b). Once HCMV pneumonia occurs, the outcome is as poor as after allogeneic HCT (Reusser *et al.*, 1990; Boeckh *et al.*, 1996b). $CD34^{+}$ cell-selected grafts and the use of total body irradiation, fludarabine, rituximab, or high-dose corticosteroids are risk factors for HCMV disease after autologous HCT (Holmberg *et al.*, 1999; Lee *et al.*, 2008).

In allogeneic HCT, 70–80% of HCMV-seropositive patients reactivate HCMV. HLA mismatched family donor status or HLA matched or mismatched unrelated donor status is considered a strong risk factor for higher viral load and HCMV disease (Einsele *et al.*, 2000; Marty *et al.*, 2007; Nakamae *et al.*, 2009). Recurrence of HCMV infection in late phase occurs more frequently in an unrelated donor than in a related donor setting.

Stem cell source

Although earlier studies suggested a lower risk of HCMV infection in peripheral blood stem cell compared to bone marrow transplants (Ljungman *et al.*, 2006; Walker *et al.*, 2007), a recent analysis from a randomized trial showed peripheral blood stem cell transplantation was associated with a higher rate of HCMV infection and disease during the first 100 days, which was temporally associated with a lower number of functional HCMV-specific T-cells (Guerrero *et al.*, 2012). Beyond 3 months, however, HCMV-specific immune responses recovered equally in both transplantation forms, thereby leading to similar rates of infection and disease during the late period.

Umbilical cord blood cells, an important alternative cell source, are typically HCMV negative. The cord blood graft contains immature T-cells and leads to delayed immune reconstitution, which results in a high incidence of HCMV infection (Komanduri *et al.*, 2007). Incidence rates of 21–80% for HCMV infection and 6–21% for HCMV disease have been reported with different HCMV prophylaxis regimens (Matsumura *et al.*, 2007; Walker *et al.*, 2007). HCMV reactivation tends to be more common in cord blood recipients than in bone marrow or peripheral blood stem cell recipients (Ljungman *et al.*, 2006; Walker *et al.*, 2007; Mikulska *et al.*, 2012). Intensive prophylaxis and early pre-emptive therapy for HCMV after cord blood transplantation results in a decrease of the incidence for HCMV disease, which suggests cord blood transplant recipients require a more aggressive approach towards HCMV prevention (Milano *et al.*, 2011).

T-cell depletion in allogeneic HCT is also a risk factor as well as in autologous HCT. T-cell depleted graft increased HCMV infection, but did not change the incidence of HCMV disease and survival (van Burik *et al.*, 2007).

Graft-versus-host disease

GvHD is the most important post-transplant risk factor in HCMV infection and disease (Ljungman *et al.*, 2006; Matsumura *et al.*, 2007; Walker *et al.*, 2007). Acute GvHD, especially grades II and higher, is significantly related to increased HCMV reactivation, viral load increases after start of pre-emptive therapy, and HCMV disease (Nichols *et al.*, 2001; Ljungman *et al.*, 2006; Marty *et al.*, 2007; Matsumura *et al.*, 2007; Nakamae *et al.*, 2009). Chronic GvHD is a risk factor of late HCMV disease (Einsele *et al.*, 2000; Boeckh *et al.*, 2003a; Nakamae *et al.*, 2009). GvHD can cause lymphopenia and impaired T-cell responses, and most drugs used to treat GvHD further increase the suppressive effect on T-cell function.

Viral load

In the pre-emptive therapy era, only a limited number of patients with HCMV reactivation progress to HCMV disease. However, both initial viral load and viral load increases are highly significant risk factors for development of HCMV disease, in the absence of antivirals (Emery *et al.*, 2000) and in more recent cohorts receiving pre-emptive therapy (Nakamae *et al.*, 2009).

Management of HCMV infection and disease

Antiviral agents

Characteristics of antiviral agents with activity against HCMV are shown in Table II.16.1. An overview of their mechanisms of action is provided in Chapter II.19. High-dose acyclovir has been shown to improve survival in the pre-pre-emptive therapy era (Prentice *et al.*, 1994). Valacyclovir is a valin-ester prodrug of acyclovir with a higher oral bioavailability. A randomized trial performed in the pre-emptive therapy era showed that high-dose valacyclovir was more effective than acyclovir in reducing HCMV infection, but no effect on HCMV disease and survival was shown (Ljungman *et al.*, 2002a).

The most important drug for management of HCMV disease is ganciclovir. Ganciclovir has been demonstrated to reduce the risk of HCMV infection and disease compared with placebo (Goodrich *et al.*, 1993; Boeckh *et al.*, 1996a; Winston *et al.*, 2003). Neutropenia is the most important adverse effect of ganciclovir and occurs in approximately 30% of HCT recipients (Nakamae *et al.*, 2011). Neutropenia is generally reversible by dose reduction, drug discontinuation and/or support of granulocyte-colony stimulating factor. Measurement of ganciclovir concentration may be helpful to reduce the risk of toxicity, especially in the situation of pre-existing renal impairment, however, there are no clearly established levels that correlate with efficacy and toxicity.

Valganciclovir is an orally available prodrug of ganciclovir (Einsele *et al.*, 2006; Winston *et al.*, 2006) and serum concentrations of valganciclovir (administered at 900 mg/day) are similar to those of intravenous ganciclovir (administered at 5 mg/kg/day). However, the fixed dose of valganciclovir may lead to higher serum concentrations of ganciclovir in patients with low body weight (Einsele *et al.*, 2006). Although no statistically powered efficacy trials have been performed, valganciclovir appears to be associated with similar efficacy and toxicity rates as intravenous ganciclovir (Ayala *et al.*, 2006; Einsele *et al.*, 2006).

Foscarnet is a pyrophosphate analogue that binds and competitively inhibits the HCMV DNA polymerase. Foscarnet shows similar efficacy to ganciclovir in pre-emptive therapy (Reusser *et al.*, 2002). Common adverse effects are nephrotoxicity and electrolyte abnormalities, and occasionally ulcers in the urinary tract. The drug is only used as a second-line agent at most centres, most likely due to its toxicity profile, its cost and the requirement for intravenous administration and prehydration (Pollack *et al.*, 2011).

Cidofovir is a cytosine nucleotide analogue that does not require the phosphorylation by viral enzymes for antiviral activity. The long half-life of cidofovir allows once-per-week dosing schedule (Ljungman *et al.*, 2001). A main side effect of cidofovir is severe nephrotoxicity, requiring prehydration and medication with probenecid.

Leflunomide is known as a drug for rheumatoid arthritis. The antiviral mechanism is inhibition of virion assembly by a metabolite of leflunomide, therefore cross-resistance with other agents is rare. The drug has not been evaluated in randomized trials but several single patient reports or small case series suggested possible efficacy of leflunomide in refractory HCMV disease (Avery *et al.*, 2010). Adverse effects have been reported to be diarrhoea, anaemia and hepatotoxicity; fatal liver toxicity has been reported (Avery *et al.*, 2010). The drug is teratogenic in animal studies.

Fomivirsen is an antisense RNA molecule to inhibit the viral replication that was approved for intravitreal treatment of HCMV retinitis in patients with AIDS (Jabs and Griffiths, 2002); however, it is no longer marketed owing to the decline in HCMV retinitis.

Prevention in HCMV seronegative patients

For seronegative recipients, the main issue is to prevent primary HCMV infection. HCMV seronegative donors are preferentially selected for HCMV seronegative recipients if multiple donors are available (Boeckh and Ljungman, 2009), although a recent analysis has shown little impact on the overall outcome if a seropositive donor has been chosen (Pollack *et al.*, 2012). No study has examined the relative importance of HLA match versus HCMV serology and other factors such as age, gender, and donor blood group, however, optimized HLA matching is probably the more important variable (Boeckh and Ljungman, 2009).

To reduce the risk of HCMV transmission, blood products from HCMV seronegative donors are widely used in the setting of D^-/R^- (Pollack *et al.*, 2011). However, the adequate supply of HCMV seronegative products is increasingly difficult. Therefore, a widely used alternative option is to use blood products after filtration or apheresis to reduce the number of leucocytes (Nichols *et al.*, 2003).

Prophylaxis by antiviral agents

A previous randomized study using high-dose acyclovir (500 mg/m^2, three times daily) showed reduced HCMV infection and increased overall survival (Prentice *et al.*, 1994). Recently, high-dose valacyclovir (2 g, 3–4 times daily) prophylaxis following high-dose

Table II.16.1 Antiviral agents; characteristics and indication (adult dosing)

Drug name	Administration route	Mechanism	Major indication	Prophylaxis	Pre-emptive therapy	Treatment	Major toxicity
Acyclovir	i.v., orally	Inhibition of DNA polymerase	Prophylaxis	i.v.: 500 mg/m^2 every 8 h; orally: 800 mg (or 600 mg for BW <40 kg), four times daily	Not recommended	Not recommended	Nephrotoxicity, headache, nausea
Valacyclovir	Orally	Inhibition of DNA polymerase	Prophylaxis	2 g, three or four times daily	Not recommended	Not recommended	Gastrointestinal discomfort, neutropenia
Ganciclovir	i.v., intravitreally	Inhibition of DNA polymerase	All	5 mg/kg once daily	Induction: 5 mg/kg every 12 h for 7–14 days Maintenance: 5 mg/kg once daily	Longer induction dosing, otherwise the same	Myelosuppression, nephrotoxicity
Valganciclovir	Orally	Inhibition of DNA polymerase	Pre-emptive therapy	900 mg once daily (BW ≥40 kg)	Induction: 900 mg twice daily (BW ≥40 kg) for 7–14 days; maintenance: 900 mg once daily	Longer induction dosing, otherwise the same	Myelosuppression
Foscarnet	i.v.	Inhibition of DNA polymerase	Pre-emptive therapy, treatment	Not established	Induction: 60–90 mg/kg every 12 h for 7–14 days Maintenance: 90 mg/kg once daily	Induction: 90 mg/kg every 12 h, 14–21 days; maintenance: 90 mg/kg/day	Nephrotoxicity, electrolyte imbalance
Cidofovir	i.v.	Inhibition of DNA polymerase	Salvage treatment	Not established	Same as treatment	Induction: 5 mg/kg/week for 2–3 doses, Maintenance: 5 mg/kg every 2 weeks	Nephrotoxicity
Leflunomide	Orally	Inhibition of viral capsid assembly by down-modulated cell signalling	Salvage treatment	Not established	Not established	Induction: 100 mg once daily, Maintenance: 40 mg once daily	Hepatotoxicity, diarrhoea
Fomivirsen[a]	Intravitreally	Antisense of IE2 mRNA	Salvage treatment	Not indicated	Not indicated	Induction: 330μg every 2 weeks, Maintenance: 330μg once monthly	High intraocular pressure, inflammation

[a]Fomivirsen is no longer marketed.
BW, body weight.

intravenous acyclovir has shown to reduce the risk for HCMV infection with no survival benefit compared with high-dose oral acyclovir (800 mg, four times daily) (Ljungman *et al.*, 2002a).

Ganciclovir prophylaxis has also been studied in several randomized trials (Goodrich *et al.*, 1993; Boeckh *et al.*, 1996a). Most studies showed a reduction in the risk of HCMV infection and disease compared to controls, but no survival advantage was demonstrated. The lack of a survival benefit in these studies is likely due to adverse effects of ganciclovir. Also, these studies had relatively small sample sizes and were primarily powered to detect differences in HCMV infection and disease. No randomized clinical trials on prophylaxis with valganciclovir or foscarnet in HCT recipients have been reported.

In the current pre-emptive therapy era, antiviral prophylaxis is sometimes used in high risk patients with a high degree of immunosuppressive treatment (O'Brien *et al.*, 2006; Pollack *et al.*, 2011). According to the international survey, approximately 30% of the centres use high-dose acyclovir or valacyclovir as prophylaxis for high-risk patients, especially in the paediatric setting (Pollack *et al.*, 2011). Only 30% of the centres still use the ganciclovir or foscarnet prophylaxis. Cord blood transplant recipients may require more intense strategies, consisting of high-dose acyclovir/valacyclovir prophylaxis combined with early pre-emptive therapy initiated already at very low viral load levels (Milano *et al.*, 2011).

Prophylaxis by IVIG

Pooled or CMV-specific intravenous immunoglobulin (IVIG) for the prophylaxis of primary HCMV infection is only moderately effective (Bowden *et al.*, 1991; Ruutu *et al.*, 1997). Likewise, the effect of immunoglobulin on reducing HCMV infection in seropositive patients is modest, and no survival benefit among the patients receiving immunoglobulin has been reported in any study or a recent meta-analysis (Zikos *et al.*, 1998; Raanani *et al.*, 2009). HCMV prophylaxis using an HCMV gH-specific monoclonal antibody also showed negative results (Boeckh *et al.*, 2001). Recent studies have reported the increase of veno-occlusive disease after IVIG (Raanani *et al.*, 2009). Therefore, the prophylactic use of immunoglobulin is currently not recommended.

Pre-emptive therapy

Pre-emptive therapy is a strategy to start antiviral agents based on the monitoring of HCMV viral load. A recent international survey revealed that pre-emptive therapy is the most commonly used strategy (Pollack *et al.*, 2011). The success of pre-emptive therapy is largely dependent on the early detection of HCMV. Pre-emptive therapy is generally initiated after a positive result of HCMV pp65 antigenaemia, HCMV DNA, or mRNA in blood. The measurement of HCMV DNA load has become popular, but there is presently no validated or standardized threshold. Although several comparative studies of pp65 antigenaemia, HCMV DNA, or HCMV mRNA surveillance have been reported (Gerna *et al.*, 2003; Lehto *et al.*, 2005; Pollack *et al.*, 2012), there is no consensus whether any of these strategies is superior except for that HCMV DNA or mRNA is effective for detection of HCMV infection before engraftment. Most centres use weekly surveillance during the first 3 months and longer periods (up to 1 year) in patients at high risk for late HCMV disease (Boeckh *et al.*, 2003a; Pollack *et al.*, 2011).

Timing of pre-emptive therapy also varies. Individualized surveillance based on patients' risk factors has also been reported and appeared to lead to less use of ganciclovir at similar occurrence rates of HCMV disease (Kanda *et al.*, 2001). Because pre-emptive therapy based on surveillance by pp65 antigenaemia, HCMV DNA or mRNA is highly effective, the incidence of HCMV pneumonia has decreased significantly (Boeckh *et al.*, 2003b; Pollack *et al.*, 2012). However, reduction of HCMV GI disease has been less dramatic, likely due to occurrence of HCMV GI disease in the absence of systemic viral load (Jang *et al.*, 2009).

Ganciclovir is the most frequently used agent for pre-emptive therapy, with foscarnet serving as the second-line agent in neutropenic patients (Pollack *et al.*, 2011). A randomized trial showed similar results of ganciclovir and foscarnet when used for pre-emptive therapy, however, the toxicity profiles differed (Reusser *et al.*, 2002). Several small studies reported the efficacy of valganciclovir in pre-emptive therapy, consistent with the results that valganciclovir provides adequate systemic exposure in HCT recipients (Ayala *et al.*, 2006; Einsele *et al.*, 2006; Chawla *et al.*, 2011); however, adequately powered randomized efficacy trials have not been reported. Owing to its lack of marrow toxicity, foscarnet is an attractive choice for patients before engraftment or during neutropenia. Limited published data exist for cidofovir in HCT recipients (Ljungman *et al.*, 2001). It is considered an alternative for secondary treatment in the setting of drug resistance, foscarnet toxicity, or co-infection with adenovirus. Pre-emptive therapy generally stops when consecutive negative results of pp65 HCMV antigenaemia or one time negative result of PCR are detected (Pollack *et al.*, 2011). Fig. II.16.1 shows a schema of the principle of pre-emptive therapy.

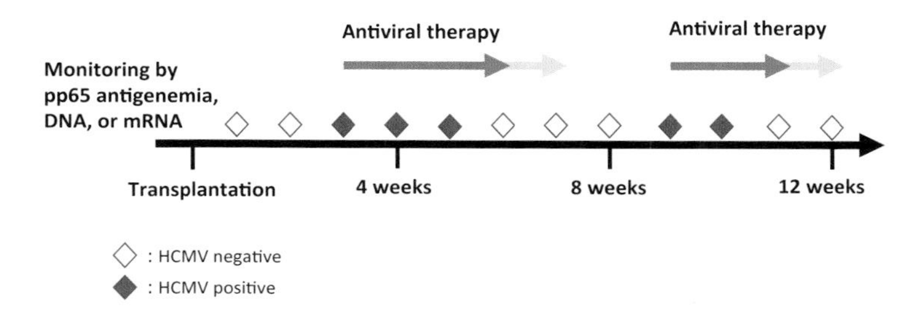

Figure II.16.1 The principle of pre-emptive therapy. HCMV infection is monitored weekly by pp65 antigenaemia, HCMV DNA, or HCMV mRNA. When the test becomes positive, an antiviral drug is generally initiated. Quantitative thresholds may vary based on the test used and the level of immunosuppression of the patient. The therapy is discontinued when one negative result of DNA or mRNA (blue arrow) or consecutive negative results of pp65 antigenaemia (light blue arrow) are obtained. Surveillance and pre-emptive therapy generally continues for 3 months after transplantation but may be continued in patients at risk for late HCMV disease.

Treatment

HCMV disease should be treated with antiviral agents such as ganciclovir or foscarnet. Induction therapy is generally continued for at least 2 weeks (preferably 3 weeks, if tolerated), with subsequent maintenance therapy for another 3–4 weeks. Treatment should be continued until resolution of clinical manifestations and clearance of HCMV in blood (if it was detectable). Follow-up BALs or endoscopies are usually not done in routine clinical practice in patients who respond clinically and virologically.

Pneumonitis

Current standard treatment of HCMV pneumonia is ganciclovir (or foscarnet) in combination with IVIG because several earlier non-randomized studies showed the combination improved the survival rate in comparison with only antiviral therapy (Ljungman *et al.*, 1992). However, recent studies found more limited effect of concomitant immunoglobulin (Ljungman *et al.*, 1992; Machado *et al.*, 2000; Erard *et al.*, 2007). Pooled and HCMV-specific immunoglobulin appeared to show similar efficacies in one study (Ljungman *et al.*, 1992).

Gastrointestinal disease

Standard care of HCMV Gl disease is intravenous ganciclovir (or foscarnet if neutropenia is present). Because mucosal ulcers are often extensive and recurrence of GI disease is high, longer treatment courses (i.e. 3–4 weeks for induction, followed by maintenance dosing for several weeks) is recommended (Reed *et al.*, 1990). Owing to lack of systemic viral load in many cases, virological monitoring of the response to treatment is often not possible. There are no data to support IVIG use for GI disease (Ljungman *et al.*, 1998). The potential of valganciclovir in the treatment of HCMV GI disease has not been well studied. However, a small pre-emptive treatment study in patients with grade II intestinal acute GvHD showed that valganciclovir had similar effects on viral load reductions compared to intravenous ganciclovir (Einsele *et al.*, 2006). Therefore it seems feasible to use valganciclovir for the treatment of HCMV GI disease during the maintenance phase in patients without serious GI GvHD, provided that there is adequate oral intake. Whether mild cases can be treated initially is unclear.

Retinitis

HCMV retinitis should be treated with systemic antiviral agents (Okamoto *et al.*, 1997). Frequent monitoring by an experienced ophthalmologist is critical. The indication for local therapy like intravitreal ganciclovir injections or an implant is determined by severity of retinitis (proximity to zone 1) or efficacy of or ability to provide systemic treatment (Chang and Dunn, 2005).

Antiviral drug resistance

Drug resistance remains rare after HCT, but can occur with all drugs. Most drug resistance has been reported with ganciclovir and valganciclovir because they are most commonly used. Risk factors for drug resistance are prolonged antiviral therapy, drug use in suboptimal levels, and lack of host immunity to HCMV (Chou, 2001). When a drug-naïve patient is treated with an antiviral drug, a transient increase of viral load occurs in approximately one-third of patients, mainly in patients treated with higher doses of immunosuppressive agents such as corticosteroid (Nichols *et al.*, 2001). In most of these cases, the increase of viral load is not due to drug resistance, although early drug resistance has been reported in children (Wolf *et al.*, 1998). True drug resistance should be suspected in patients who had significant prior drug exposure and had increasing viral loads for more than 2 weeks during antiviral therapy.

For diagnosis of drug resistance, genotypic testing should be performed (Fig. II.16.2) (see also Chapter II.19). In patients treated with ganciclovir or valganciclovir, resistance is typically due to mutations in the UL97 gene, followed by mutations in the UL54-encoded DNA polymerase. Interestingly, different UL97 mutations show various levels of ganciclovir resistance. Therefore, in some cases, ganciclovir can be continued even after the detection of UL97 mutation (Iwasenko *et al.*, 2009).

Treatment for drug-resistant HCMV is to switch to an alternative drug. Since genotypic testing may take several days, the change to an alternative drug before obtaining the result is recommended. Ganciclovir is generally changed to foscarnet because its activity is independent of phosphorylation by the UL97 gene product. For ganciclovir-resistant HCMV, the efficacy of a combination therapy of foscarnet and ganciclovir remains unknown, although a combination of both drugs at full dose has been used in selected cases (Drew, 2006; Boeckh, 2011). When mutation of UL54 that leads to foscarnet resistance is detected, continuation of ganciclovir is recommended. No report has been published about cross-resistance between foscarnet and ganciclovir (Chou *et al.*, 2003). Cidofovir can be used as an alternative drug for ganciclovir-resistant HCMV, but cross-resistance exists in cases with mutation of UL54 (Chou *et al.*, 2003; Schreiber *et al.*, 2009). Therefore, before switching ganciclovir to cidofovir, mutation of UL54 should be evaluated.

Response to treatment is monitored by viral load in blood. In patients who do not respond to a new drug, few options remain except for new drugs based on experimental data (Boeckh, 2011).

New therapeutic options

New antiviral agents

Several novel agents for HCMV disease are under development (Table II.16.2) (see also Chapter II.19).

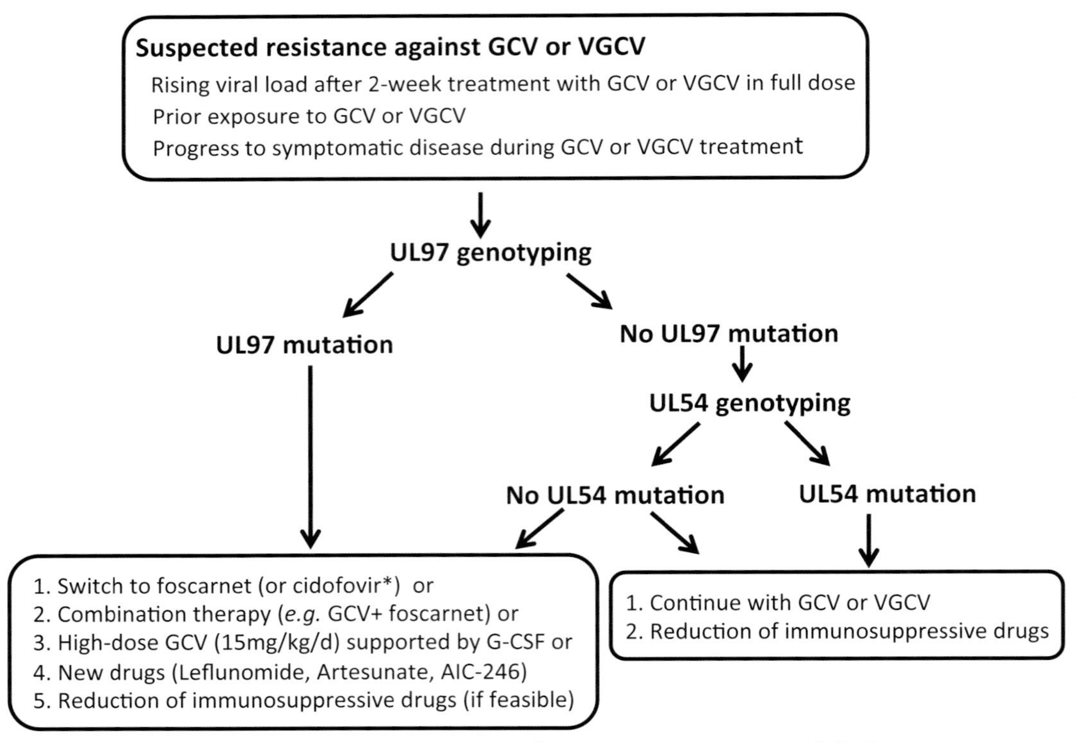

Figure II.16.2 Management of antiviral drug resistance. Abbreviations: GCV, ganciclovir; VGCV, valganciclovir; G-CSF, granulocyte-colony stimulating factor.

Table II.16.2 New antiviral agents; characteristics and progress

Drug name	Administration route	Mechanism	Clinical trial status	Toxicity
Maribavir	Orally	Inhibition of UL97 kinase	Completed phase 3 study for low-dose prophylaxis, negative results; higher dose treatment trials ongoing	Taste disturbance
Artesunate	i.v., orally	Inhibition of viral replication by down-modulation of cell signalling	Case reports	Myelosuppression, neurotoxicity
CMX-001	Orally	Inhibition of DNA polymerase	Phase 2 prophylaxis study completed	Diarrhoea
AIC246 (Letermovir)	Orally	Inhibition of genome packaging by small molecule compound	Phase 2 prophylaxis study completed	Unknown

Maribavir, a UL97 protein kinase inhibitor, is one of several promising drugs because of oral availability and mild toxicity such as taste disturbance (Dropulic and Cohen, 2010). A phase II dose-ranging study showed a reduction of HCMV infection even with the lowest dose (100 mg twice daily) (Winston *et al.*, 2008). However, use of the same dose in a phase III study failed to prevent HCMV disease (Marty *et al.*, 2011). Higher doses of maribavir are being evaluated for treatment of refractory or resistant HCMV disease (Marty and Boeckh, 2011). Maribavir has activity against ganciclovir- or cidofovir-resistant HCMV (Drew *et al.*, 2006).

Artesunate is an anti-malaria drug that shows broad antiviral activity against herpes viruses, hepatitis viruses, and HIV because it functions as a down-modulator of NF-kB or Sp1 pathways (Wolf *et al.*, 2011). Effectiveness of artesunate in a multidrug resistant HCMV strain has been reported (Shapira *et al.*, 2008). No systematic evaluation of this drug for HCMV treatment has been conducted.

CMX-001 is a lipid-conjugated nucleotide analogue of cidofovir that can be administered orally. Its activity against HCMV is approximately 300- to 400-fold higher than that of cidofovir (Dropulic and Cohen, 2010). As CMX001 does not accumulate in kidney, nephrotoxicity has not been observed. Preliminary results of a recent phase II dose-escalation study in HCT recipients showed a reduction of HCMV infection in patients receiving CMX001 for prophylaxis started at engraftment (Marty *et al.*, 2012). The most common side effect was diarrhoea. No haematotoxic or nephrotoxic effects have been reported.

AIC-246 (letermovir) is an inhibitor of HCMV replication (Lischka *et al.*, 2010; Price and Prichard, 2011). The drug is given orally. A recent case report suggested that AIC-246 was effective for multidrug-resistant HCMV disease (Kaul *et al.*, 2011). Preliminary results of a phase II dose-escalation study in HCT recipients showed reduction of HCMV infection in HCMV-seropositive patients receiving the drug at engraftment; the adverse event rates in the letermovir groups were similar to those observed in placebo patients (Ehninger and Zimmermann, 2012).

Vaccination

An HCMV vaccine is one of potential ways to prevent primary HCMV infection in seronegative recipients, or reactivation or progressive HCMV disease in seropositive recipients (see also Chapter II.20 on HCMV vaccines). A phase I study using a compound of two plasmids which encode pp65 and gB showed immunogenicity without severe toxicity in HCMV seronegative healthy adults (Wloch *et al.*, 2008). Recently, a phase II study of this vaccine showed increased HCMV-specific T-cell responses and a reduction of HCMV infection in patients after vaccination (Kharfan-Dabaja *et al.*, 2012). Another group conducted a randomized placebo-controlled study of HCMV gB vaccine in organ transplant recipients (Griffiths *et al.*, 2011). Significant vaccine effects on antibody titres and the duration of HCMV infection were observed. Larger phase III studies are planned. Other vaccine candidates are being developed, e.g. an HCMV peptide vaccine using the HLA A*0201 pp65 cytotoxic $CD8^+$ T-cell epitope (La Rosa *et al.*, 2012), an alphavirus replicon particle vaccine expressing gB or a pp65/IE1 fusion protein (Bernstein *et al.*, 2009), and a trivalent DNA vaccine consisting of gB, pp65, and IE1 (VCL-CT02) followed by administration of live-attenuated HCMV (Towne strain) (Jacobson *et al.*, 2009).

Immunotherapy

HCMV-specific T-cells are known to function for prevention of HCMV disease after HCT (Crough and Khanna, 2009). The first study of adoptive T-cell therapy for HCMV disease was performed using HCMV-specific CD8+ T-cells isolated from peripheral blood cells of transplant donors (Walter *et al.*, 1995). Subsequently, many strategies to expand HCMV-specific T-cells by stimulation of PBMCs or DCs with CMV lysate or peptide have been reported (Mui *et al.*, 2010). However, none of them is standardized for routine use because of complicated procedures and regulation of good manufacturing practices (Fuji *et al.*, 2011). Most studies use stimulated T-cells, while one study using natural donor-derived HCMV-specific CD8+ T-cells selected by HLA-peptide tetramers has been reported (Cobbold *et al.*, 2005). Recent adoptive transfer of donor-derived HCMV-specific T-cells demonstrated antiviral effects without severe side effects in several different cohort studies (Feuchtinger *et al.*, 2010; Peggs *et al.*, 2011; Schmitt *et al.*, 2011). Further investigation to standardize the procedures will be needed.

Conclusions and future research needs

The strategy of pre-emptive therapy substantially reduced the incidence of HCMV disease after HCT in the past 20 years. However, once HCMV disease occurs, the outcome remains poor, especially in HCMV pneumonia. The recent phase III maribavir trial, which used a comparator of ganciclovir-based pre-emptive therapy, documented how effective this therapy is today, with very low HCMV disease rates during the first 3 months after HCT. Since pre-emptive therapy is the current standard comparator in the evaluation of novel HCMV therapeutics, future clinical trials will have to use clinically meaningful viral load levels in blood as part of the efficacy study endpoint. Although pre-emptive therapy is highly effective in most settings, intensified strategies are required for patients with the highest level of immunosuppression, such as cord blood transplant recipients. New treatment options for HCMV are on the horizon, including small molecules such as CMX-001, letermovir, and maribavir at higher doses, as well as immune augmentation strategies, including vaccination and adoptive T-cell transfer, and nonspecific strategies involving keratinocyte growth factor and IL-7 (Holland and van den Brink, 2009). Finally, the quantitative assessment of HCMV-specific T-cell immunity is now feasible and initial data on predictive thresholds for protective immunity are forthcoming (Boeckh, 2011). This strategy could be used to withhold therapy in patients more than 100 days after transplantation; however, a systematic evaluation in a randomized trial is needed to determine whether such strategy can safely be used.

Acknowledgements

Michael Boeckh was supported by the National Institutes of Health (CA 18029, HL093294).

References

Ando, Y., Terao, K., Narita, M., Oguchi, Y., Sata, T., and Iwasaki, T. (2002). Quantitative analyses of cytomegalovirus genome in aqueous humor of patients with cytomegalovirus retinitis. Jpn. J. Ophthalmol. *46*, 254–260.

Avery, R.K., Mossad, S.B., Poggio, E., Lard, M., Budev, M., Bolwell, B., Waldman, W.J., Braun, W., Mawhorter, S.D., Fatica, R., *et al.* (2010). Utility of leflunomide in the treatment of complex cytomegalovirus syndromes. Transplantation *90*, 419–426.

Ayala, E., Greene, J., Sandin, R., Perkins, J., Field, T., Tate, C., Fields, K.K., and Goldstein, S. (2006). Valganciclovir is safe and effective as pre-emptive therapy for CMV infection in allogeneic hematopoietic stem cell transplantation. Bone Marrow Transplant. *37*, 851–856.

Bernstein, D.I., Reap, E.A., Katen, K., Watson, A., Smith, K., Norberg, P., Olmsted, R.A., Hoeper, A., Morris, J., Negri, S., *et al.* (2009). Randomized, double-blind, Phase 1 trial of an alphavirus replicon vaccine for cytomegalovirus in CMV seronegative adult volunteers. Vaccine *28*, 484–493.

Biron, C.A., Byron, K.S., and Sullivan, J.L. (1989). Severe herpesvirus infections in an adolescent without natural killer cells. N. Engl. J. Med. *320*, 1731–1735.

Boeckh, M. (2011). Complications, diagnosis, management, and prevention of CMV infections: current and future. Hematology Am. Soc. Hematol. Educ. Program *2011*, 305–309.

Boeckh, M., and Boivin, G. (1998). Quantitation of cytomegalovirus: methodologic aspects and clinical applications. Clin. Microbiol. Rev. *11*, 533–554.

Boeckh, M., and Ljungman, P. (2009). How we treat cytomegalovirus in hematopoietic cell transplant recipients. Blood *113*, 5711–5719.

Boeckh, M., and Nichols, W.G. (2004). The impact of cytomegalovirus serostatus of donor and recipient before hematopoietic stem cell transplantation in the era of antiviral prophylaxis and preemptive therapy. Blood *103*, 2003–2008.

Boeckh, M., Gooley, T.A., Myerson, D., Cunningham, T., Schoch, G., and Bowden, R.A. (1996a). Cytomegalovirus pp65 antigenemia-guided early treatment with ganciclovir versus ganciclovir at engraftment after allogeneic marrow transplantation: a randomized double-blind study. Blood *88*, 4063–4071.

Boeckh, M., Stevens-Ayers, T., and Bowden, R.A. (1996b). Cytomegalovirus pp65 antigenemia after autologous marrow and peripheral blood stem cell transplantation. J. Infect. Dis. *174*, 907–912.

Boeckh, M., Bowden, R.A., Storer, B., Chao, N.J., Spielberger, R., Tierney, D.K., Gallez-Hawkins, G.,

Cunningham, T., Blume, K.G., Levitt, D., *et al.* (2001). Randomized, placebo-controlled, double-blind study of a cytomegalovirus-specific monoclonal antibody (MSL-109) for prevention of cytomegalovirus infection after allogeneic hematopoietic stem cell transplantation. Biol. Blood Marrow Transplant. *7*, 343–351.

Boeckh, M., Leisenring, W., Riddell, S.R., Bowden, R.A., Huang, M.L., Myerson, D., Stevens-Ayers, T., Flowers, M.E., Cunningham, T., and Corey, L. (2003a). Late cytomegalovirus disease and mortality in recipients of allogeneic hematopoietic stem cell transplants: importance of viral load and T-cell immunity. Blood *101*, 407–414.

Boeckh, M., Nichols, W.G., Papanicolaou, G., Rubin, R., Wingard, J.R., and Zaia, J. (2003b). Cytomegalovirus in hematopoietic stem cell transplant recipients: Current status, known challenges, and future strategies. Biol. Blood Marrow Transplant. *9*, 543–558.

Boeckh, M., Huang, M., Ferrenberg, J., Stevens-Ayers, T., Stensland, L., Nichols, W.G., and Corey, L. (2004). Optimization of quantitative detection of cytomegalovirus DNA in plasma by real-time PCR. J. Clin. Microbiol. *42*, 1142–1148.

Boehme, K.W., Guerrero, M., and Compton, T. (2006). Human cytomegalovirus envelope glycoproteins B and H are necessary for TLR2 activation in permissive cells. J. Immunol. *177*, 7094–7102.

Bowden, R.A., Fisher, L.D., Rogers, K., Cays, M., and Meyers, J.D. (1991). Cytomegalovirus (CMV)-specific intravenous immunoglobulin for the prevention of primary CMV infection and disease after marrow transplant. J. Infect. Dis. *164*, 483–487.

Brown, M.G., Dokun, A.O., Heusel, J.W., Smith, H.R., Beckman, D.L., Blattenberger, E.A., Dubbelde, C.E., Stone, L.R., Scalzo, A.A., and Yokoyama, W.M. (2001). Vital involvement of a natural killer cell activation receptor in resistance to viral infection. Science *292*, 934–937.

van Burik, J.A., Carter, S.L., Freifeld, A.G., High, K.P., Godder, K.T., Papanicolaou, G.A., Mendizabal, A.M., Wagner, J.E., Yanovich, S., and Kernan, N.A. (2007). Higher risk of cytomegalovirus and aspergillus infections in recipients of T-cell-depleted unrelated bone marrow: analysis of infectious complications in patients treated with T-cell depletion versus immunosuppressive therapy to prevent graft-versus-host disease. Biol. Blood Marrow Transplant. *13*, 1487–1498.

Carmichael, A. (2012). Cytomegalovirus and the eye. Eye (Lond) *26*, 237–240.

Chang, M., and Dunn, J.P. (2005). Ganciclovir implant in the treatment of cytomegalovirus retinitis. Expert. Rev. Med. Devices *2*, 421–427.

Chawla, J.S., Ghobadi, A., Mosley, J., 3rd, Verkruyse, L., Trinkaus, K., Abboud, C.N., Cashen, A.F., Stockerl-Goldstein, K.E., Uy, G.L., Westervelt, P., *et al.* (2011). Oral valganciclovir versus ganciclovir as delayed pre-emptive therapy for patients after allogeneic hematopoietic stem cell transplant: a pilot trial (04–0274) and review of the literature. Transpl. Infect. Dis. *14*, 259–267.

Chou, S., Lurain, N.S., Thompson, K.D., Miner, R.C., and Drew, W.L. (2003). Viral DNA polymerase mutations associated with drug resistance in human cytomegalovirus. J. Infect. Dis. *188*, 32–39.

Chou, S.W. (2001). Cytomegalovirus drug resistance and clinical implications. Transpl. Infect. Dis. *3(Suppl. 2)*, 20–24.

Cobbold, M., Khan, N., Pourgheysari, B., Tauro, S., McDonald, D., Osman, H., Assenmacher, M., Billingham, L., Steward, C., Crawley, C., *et al.* (2005). Adoptive transfer of cytomegalovirus-specific CTL to stem cell transplant patients after selection by HLA-peptide tetramers. J. Exp. Med. *202*, 379–386.

Crawford, S.W., Bowden, R.A., Hackman, R.C., Gleaves, C.A., Meyers, J.D., and Clark, J.G. (1988). Rapid detection of cytomegalovirus pulmonary infection by bronchoalveolar lavage and centrifugation culture. Ann. Intern. Med. *108*, 180–185.

Crippa, F., Corey, L., Chuang, E.L., Sale, G., and Boeckh, M. (2001). Virological, clinical, and ophthalmologic features of cytomegalovirus retinitis after hematopoietic stem cell transplantation. Clin. Infect. Dis. *32*, 214–219.

Crough, T., and Khanna, R. (2009). Immunobiology of human cytomegalovirus: from bench to bedside. Clin. Microbiol. Rev. *22*, 76–98.

Delgado, J., Pillai, S., Benjamin, R., Caballero, D., Martino, R., Nathwani, A., Lovell, R., Thomson, K., Perez-Simon, J.A., Sureda, A., *et al.* (2008). The effect of *in vivo* T-cell depletion with alemtuzumab on reduced-intensity allogeneic hematopoietic cell transplantation for chronic lymphocytic leukemia. Biol. Blood Marrow Transplant. *14*, 1288–1297.

Drew, W.L. (2006). Is combination antiviral therapy for CMV superior to monotherapy? J. Clin. Virol. *35*, 485–488.

Drew, W.L., Miner, R.C., Marousek, G.I., and Chou, S. (2006). Maribavir sensitivity of cytomegalovirus isolates resistant to ganciclovir, cidofovir or foscarnet. J. Clin. Virol. *37*, 124–127.

Dropulic, L.K., and Cohen, J.I. (2010). Update on new antivirals under development for the treatment of double-stranded DNA virus infections. Clin. Pharmacol. Ther. *88*, 610–619.

Ehninger, G., and Zimmermann, H. (2012). Latermovir for prevention of HCMV infection after SCT: results of a randomized, double-blind, placebo-controlled trial. In 38th EBMT Annual Meeting (Geneva, Switzerland).

Einsele, H., Hebart, H., Kauffmann-Schneider, C., Sinzger, C., Jahn, G., Bader, P., Klingebiel, T., Dietz, K., Loffler, J., Bokemeyer, C., *et al.* (2000). Risk factors for treatment failures in patients receiving PCR-based preemptive therapy for CMV infection. Bone Marrow Transplant. *25*, 757–763.

Einsele, H., Reusser, P., Bornhauser, M., Kalhs, P., Ehninger, G., Hebart, H., Chalandon, Y., Kroger, N., Hertenstein, B., and Rohde, F. (2006). Oral valganciclovir leads to higher exposure to ganciclovir than intravenous ganciclovir in patients following allogeneic stem cell transplantation. Blood *107*, 3002–3008.

Emery, V.C., Sabin, C.A., Cope, A.V., Gor, D., Hassan-Walker, A.F., and Griffiths, P.D. (2000). Application of viral-load kinetics to identify patients who develop cytomegalovirus disease after transplantation. Lancet *355*, 2032–2036.

Erard, V., Gutherie, K.A., Smith, J., Chien, J., Corey, L., and Boeckh, M. (2007). Cytomegalovirus pneumonia (CMV-IP) after hematopoietic cell transplantation (HCT); outcomes and factors associated with mortality. In 47th The Interscience Conference on Antimicrobial Agents and Chemotherapy (Chicago, IL).

Feuchtinger, T., Opherk, K., Bethge, W.A., Topp, M.S., Schuster, F.R., Weissinger, E.M., Mohty, M., Or, R., Maschan, M., Schumm, M., *et al.* (2010). Adoptive transfer of pp65-specific T-cells for the treatment of chemorefractory cytomegalovirus disease or reactivation after haploidentical and matched unrelated stem cell transplantation. Blood *116*, 4360–4367.

Fink, K.R., Thapa, M.M., Ishak, G.E., and Pruthi, S. (2010). Neuroimaging of pediatric central nervous system cytomegalovirus infection. Radiographics *30*, 1779–1796.

Finke, J., Bethge, W.A., Schmoor, C., Ottinger, H.D., Stelljes, M., Zander, A.R., Volin, L., Ruutu, T., Heim, D.A., Schwerdtfeger, R., *et al.* (2009). Standard graft-versus-host disease prophylaxis with or without anti-T-cell globulin in haematopoietic cell transplantation from matched unrelated donors: a randomised, open-label, multicentre phase 3 trial. Lancet Oncol. *10*, 855–864.

Fuji, S., Kapp, M., Grigoleit, G.U., and Einsele, H. (2011). Adoptive immunotherapy with virus-specific T-cells. Best Pract. Res. Clin. Haematol. *24*, 413–419.

Gallez-Hawkins, G.M., Franck, A.E., Li, X., Thao, L., Oki, A., Gendzekhadze, K., Dagis, A., Palmer, J., Nakamura, R., Forman, S.J., *et al.* (2011). Expression of activating KIR2DS2 and KIR2DS4 genes after hematopoietic cell transplantation: relevance to cytomegalovirus infection. Biol. Blood Marrow Transplant. *17*, 1662–1672.

Gerna, G., Lilleri, D., Baldanti, F., Torsellini, M., Giorgiani, G., Zecca, M., De Stefano, P., Middeldorp, J., Locatelli, F., and Revello, M.G. (2003). Human cytomegalovirus immediate-early mRNAemia versus pp65 antigenemia for guiding pre-emptive therapy in children and young adults undergoing hematopoietic stem cell transplantation: a prospective, randomized, open-label trial. Blood *101*, 5053–5060.

Goodrich, J.M., Bowden, R.A., Fisher, L., Keller, C., Schoch, G., and Meyers, J.D. (1993). Ganciclovir prophylaxis to prevent cytomegalovirus disease after allogeneic marrow transplant. Ann. Intern. Med. *118*, 173–178.

Gratama, J.W., Boeckh, M., Nakamura, R., Cornelissen, J.J., Brooimans, R.A., Zaia, J.A., Forman, S.J., Gaal, K., Bray, K.R., Gasior, G.H., *et al.* (2010). Immune monitoring with iTAg MHC Tetramers for prediction of recurrent or persistent cytomegalovirus infection or disease in allogeneic hematopoietic stem cell transplant recipients: a prospective multicenter study. Blood *116*, 1655–1662.

Green, M.L., Leisenring, W., Stachel, D., Pergam, S.A., Sandmaier, B., Wald, A., Corey, L., and Boeckh, M. (2012). Efficacy of a viral load-based, risk-adapted, preemptive treatment strategy for prevention of Cytomegalovirus disease after hematopoietic cell transplantation. Biol. Blood Marrow Transplant. *18*, 1687–1699.

Griffiths, P.D., Stanton, A., McCarrell, E., Smith, C., Osman, M., Harber, M., Davenport, A., Jones, G., Wheeler, D.C., O'Beirne, J., *et al.* (2011). Cytomegalovirus glycoprotein-B vaccine with MF59 adjuvant in transplant recipients: a phase 2 randomised placebo-controlled trial. Lancet *377*, 1256–1263.

Guerrero, A., Riddell, S.R., Storek, J., Stevens-Ayers, T., Storer, B., Zaia, J.A., Forman, S., Negrin, R.S., Chauncey, T., Bensinger, W., *et al.* (2012). Cytomegalovirus viral load and virus-specific immune reconstitution after peripheral blood stem cell versus bone marrow transplantation. Biol. Blood Marrow Transplant. *18*, 66–75.

Hackman, R.C., Wolford, J.L., Gleaves, C.A., Myerson, D., Beauchamp, M.D., Meyers, J.D., and McDonald, G.B. (1994). Recognition and rapid diagnosis of upper gastrointestinal cytomegalovirus infection in marrow transplant recipients. A comparison of seven virologic methods. Transplantation *57*, 231–237.

Hamadani, M., Blum, W., Phillips, G., Elder, P., Andritsos, L., Hofmeister, C., O'Donnell, L., Klisovic, R., Penza, S., Garzon, R., *et al.* (2009). Improved nonrelapse mortality and infection rate with lower dose of antithymocyte globulin in patients undergoing reduced-intensity conditioning allogeneic transplantation for hematologic malignancies. Biol. Blood Marrow Transplant. *15*, 1422–1430.

Hambach, L., Stadler, M., Dammann, E., Ganser, A., and Hertenstein, B. (2002). Increased risk of complicated CMV infection with the use of mycophenolate mofetil in allogeneic stem cell transplantation. Bone Marrow Transplant. *29*, 903–906.

Hebart, H., Lengerke, C., Ljungman, P., Paya, C.V., Klingebiel, T., Loeffler, J., Pfaffenrath, S., Lewensohn-Fuchs, I., Barkholt, L., Tomiuk, J., *et al.* (2011). Prospective comparison of PCR-based vs late mRNA-based preemptive antiviral therapy for HCMV infection in patients after allo-SCT. Bone Marrow Transplant. *46*, 408–415.

Holland, A.M., and van den Brink, M.R. (2009). Rejuvenation of the aging T-cell compartment. Curr. Opin. Immunol. *21*, 454–459.

Holmberg, L.A., Boeckh, M., Hooper, H., Leisenring, W., Rowley, S., Heimfeld, S., Press, O., Maloney, D.G., McSweeney, P., Corey, L., *et al.* (1999). Increased incidence of cytomegalovirus disease after autologous CD34-selected peripheral blood stem cell transplantation. Blood *94*, 4029–4035.

Iwasenko, J.M., Scott, G.M., Rawlinson, W.D., Keogh, A., Mitchell, D., and Chou, S. (2009). Successful valganciclovir treatment of post-transplant cytomegalovirus infection in the presence of UL97 mutation N597D. J. Med. Virol. *81*, 507–510.

Jabs, D.A., and Griffiths, P.D. (2002). Fomivirsen for the treatment of cytomegalovirus retinitis. Am. J. Ophthalmol. *133*, 552–556.

Jacobson, M.A., Adler, S.P., Sinclair, E., Black, D., Smith, A., Chu, A., Moss, R.B., and Wloch, M.K. (2009). A CMV DNA vaccine primes for memory immune responses to live-attenuated CMV (Towne strain). Vaccine *27*, 1540–1548.

Jang, E.Y., Park, S.Y., Lee, E.J., Song, E.H., Chong, Y.P., Lee, S.O., Choi, S.H., Woo, J.H., Kim, Y.S., and Kim, S.H. (2009). Diagnostic performance of the cytomegalovirus (CMV) antigenemia assay in patients with CMV gastrointestinal disease. Clin. Infect. Dis. *48*, e121–124.

Kanda, Y., Mineishi, S., Saito, T., Seo, S., Saito, A., Suenaga, K., Ohnishi, M., Niiya, H., Nakai, K., Takeuchi, T., *et al.* (2001). Pre-emptive therapy against cytomegalovirus (CMV) disease guided by CMV antigenemia assay after allogeneic hematopoietic stem cell transplantation: a single-center experience in Japan. Bone Marrow Transplant. *27*, 437–444.

Kaul, D.R., Stoelben, S., Cober, E., Ojo, T., Sandusky, E., Lischka, P., Zimmermann, H., and Rubsamen-Schaeff, H. (2011). First report of successful treatment of multidrug-resistant cytomegalovirus disease with the novel anti-CMV compound AIC246. Am. J. Transplant. *11*, 1079–1084.

Kharfan-Dabaja, M.A., Boeckh, M., Wilck, M.B., Langston, A.A., Chu, A.H., Wloch, M.K., Guterwill, D.F., Smith, L.R., Rolland, A.P., and Kenney, R.T. (2012). A novel therapeutic cytomegalovirus DNA vaccine in allogeneic haemopoietic stem-cell transplantation: a randomised, double-blind, placebo-controlled, phase 2 trial. Lancet Infect. Dis. *1*, 555–562.

Kijpittayarit, S., Eid, A.J., Brown, R.A., Paya, C.V., and Razonable, R.R. (2007). Relationship between Toll-like receptor 2 polymorphism and cytomegalovirus disease after liver transplantation. Clin. Infect. Dis. *44*, 1315–1320.

Komanduri, K.V., St John, L.S., de Lima, M., McMannis, J., Rosinski, S., McNiece, I., Bryan, S.G., Kaur, I., Martin, S., Wieder, E.D., *et al.* (2007). Delayed immune reconstitution after cord blood transplantation is characterized by impaired thymopoiesis and late memory T-cell skewing. Blood *110*, 4543–4551.

Kudchodkar, S.B., Yu, Y., Maguire, T.G., and Alwine, J.C. (2006). Human cytomegalovirus infection alters the substrate specificities and rapamycin sensitivities of raptor- and rictor-containing complexes. Proc. Natl. Acad. Sci. U.S.A. *103*, 14182–14187.

La Rosa, C., Longmate, J., Lacey, S.F., Kaltcheva, T., Sharan, R., Marsano, D., Kwon, P., Drake, J., Williams, B., Denison, S., *et al.* (2012). Clinical evaluation of safety and immunogenicity of PADRE-cytomegalovirus (CMV) and tetanus–CMV fusion peptide vaccines with or without PF03512676 adjuvant. J. Infect. Dis. *205*, 1294–1304.

Lee, M.Y., Chiou, T.J., Hsiao, L.T., Yang, M.H., Lin, P.C., Poh, S.B., Yen, C.C., Liu, J.H., Teng, H.W., Chao, T.C., *et al.* (2008). Rituximab therapy increased post-transplant cytomegalovirus complications in Non-Hodgkin's lymphoma patients receiving autologous hematopoietic stem cell transplantation. Ann. Hematol. *87*, 285–289.

Lehto, J.T., Lemstrom, K., Halme, M., Lappalainen, M., Lommi, J., Sipponen, J., Harjula, A., Tukiainen, P., and Koskinen, P.K. (2005). A prospective study comparing cytomegalovirus antigenemia, DNAemia and RNAemia tests in guiding pre-emptive therapy in thoracic organ transplant recipients. Transpl. Int. *18*, 1318–1327.

Lilleri, D., Fornara, C., Chiesa, A., Caldera, D., Alessandrino, E.P., and Gerna, G. (2008). Human cytomegalovirus-specific CD4+ and CD8+ T-cell reconstitution in adult allogeneic hematopoietic stem cell transplant recipients and immune control of viral infection. Haematologica *93*, 248–256.

Lischka, P., Hewlett, G., Wunberg, T., Baumeister, J., Paulsen, D., Goldner, T., Ruebsamen-Schaeff, H., and Zimmermann, H. (2010). *In vitro* and *in vivo* activities of the novel anticytomegalovirus compound AIC246. Antimicrob. Agents Chemother. *54*, 1290–1297.

Ljungman, P., Engelhard, D., Link, H., Biron, P., Brandt, L., Brunet, S., Cordonnier, C., Debusscher, L., de Laurenzi, A., Kolb, H.J., *et al.* (1992). Treatment of interstitial pneumonitis due to cytomegalovirus with ganciclovir and intravenous immune globulin: experience of European Bone Marrow Transplant Group. Clin. Infect. Dis. *14*, 831–835.

Ljungman, P., Aschan, J., Lewensohn-Fuchs, I., Carlens, S., Larsson, K., Lonnqvist, B., Mattsson, J., Sparrelid, E., Winiarski, J., and Ringden, O. (1998). Results of different strategies for reducing cytomegalovirus-associated mortality in allogeneic stem cell transplant recipients. Transplantation *66*, 1330–1334.

Ljungman, P., Deliliers, G.L., Platzbecker, U., Matthes-Martin, S., Bacigalupo, A., Einsele, H., Ullmann, J., Musso, M., Trenschel, R., Ribaud, P., *et al.* (2001). Cidofovir for cytomegalovirus infection and disease in allogeneic stem cell transplant recipients. The Infectious Diseases Working Party of the European Group for Blood and Marrow Transplantation. Blood *97*, 388–392.

Ljungman, P., de La Camara, R., Milpied, N., Volin, L., Russell, C.A., Crisp, A., and Webster, A. (2002a). Randomized study of valacyclovir as prophylaxis against cytomegalovirus reactivation in recipients of allogeneic bone marrow transplants. Blood *99*, 3050–3056.

Ljungman, P., Griffiths, P., and Paya, C. (2002b). Definitions of cytomegalovirus infection and disease in transplant recipients. Clin. Infect. Dis. *34*, 1094–1097.

Ljungman, P., Perez-Bercoff, L., Jonsson, J., Avetisyan, G., Sparrelid, E., Aschan, J., Barkholt, L., Larsson, K., Winiarski, J., Yun, Z., *et al.* (2006). Risk factors for the development of cytomegalovirus disease after allogeneic stem cell transplantation. Haematologica *91*, 78–83.

Loeffler, J., Steffens, M., Arlt, E.M., Toliat, M.R., Mezger, M., Suk, A., Wienker, T.F., Hebart, H., Nurnberg, P., Boeckh, M., *et al.* (2006). Polymorphisms in the genes encoding chemokine receptor 5, interleukin-10, and monocyte chemoattractant protein 1 contribute to cytomegalovirus reactivation and disease after allogeneic stem cell transplantation. J. Clin. Microbiol. *44*, 1847–1850.

Machado, C.M., Dulley, F.L., Boas, L.S., Castelli, J.B., Macedo, M.C., Silva, R.L., Pallota, R., Saboya, R.S., and Pannuti, C.S. (2000). CMV pneumonia in allogeneic BMT recipients undergoing early treatment of pre-emptive ganciclovir therapy. Bone Marrow Transplant. *26*, 413–417.

Madej, R.M., Davis, J., Holden, M.J., Kwang, S., Labourier, E., and Schneider, G.J. (2010). International standards and reference materials for quantitative molecular infectious disease testing. J. Mol. Diagn. *12*, 133–143.

Maris, M., Boeckh, M., Storer, B., Dawson, M., White, K., Keng, M., Sandmaier, B., Maloney, D., Storb, R., and Storek, J. (2003). Immunologic recovery after hematopoietic cell transplantation with nonmyeloablative conditioning. Exp. Hematol. *31*, 941–952.

Marty, F.M., and Boeckh, M. (2011). Maribavir and human cytomegalovirus-what happened in the clinical trials and why might the drug have failed? Curr. Opin. Virol. *1*, 555–562.

Marty, F.M., Bryar, J., Browne, S.K., Schwarzberg, T., Ho, V.T., Bassett, I.V., Koreth, J., Alyea, E.P., Soiffer, R.J., Cutler, C.S., *et al.* (2007). Sirolimus-based graft-versus-host disease prophylaxis protects against cytomegalovirus reactivation after allogeneic hematopoietic stem cell transplantation: a cohort analysis. Blood *110*, 490–500.

Marty, F.M., Ljungman, P., Papanicolaou, G.A., Winston, D.J., Chemaly, R.F., Strasfeld, L., Young, J.A., Rodriguez, T., Maertens, J., Schmitt, M., *et al.* (2011). Maribavir prophylaxis for prevention of cytomegalovirus disease in recipients of allogeneic stem-cell transplants: a phase 3, double-blind, placebo-controlled, randomised trial. Lancet Infect. Dis. *11*, 284–292.

Marty, F.M., Winston, D., Rowley, S.D., Boeckh, M., Vance, E., Papanicolaou, G., Robertson, A., Godkin, S., and Painter, W. (2012). CMX001 for prevention and control of CMV infection in CMV-seropositive allogeneic stem-cell transplant recipients: A phase 2 randomized, double-blind, placebo-controlled, dose-escalation trial of safety, tolerability and antiviral activity. In BMT tandem meeting (San Diego, CA).

Matsumura, T., Narimatsu, H., Kami, M., Yuji, K., Kusumi, E., Hori, A., Murashige, N., Tanaka, Y., Masuoka, K., Wake, A., *et al.* (2007). Cytomegalovirus infections following umbilical cord blood transplantation using reduced intensity conditioning regimens for adult patients. Biol. Blood Marrow Transplant. *13*, 577–583.

Mikulska, M., Raiola, A.M., Bruzzi, P., Varaldo, R., Annunziata, S., Lamparelli, T., Frassoni, F., Tedone, E., Galano, B., Bacigalupo, A., *et al.* (2012). CMV infection after transplant from cord blood compared to other alternative donors: the importance of donor-negative CMV serostatus. Biol. Blood Marrow Transplant. *18*, 92–99.

Milano, F., Pergam, S.A., Xie, H., Leisenring, W.M., Gutman, J.A., Riffkin, I., Chow, V., Boeckh, M.J., and Delaney, C. (2011). Intensive strategy to prevent CMV disease in seropositive umbilical cord blood transplant recipients. Blood *118*, 5689–5696.

Mocarski, E.J., Shenk, T., and Pass, R. (2007). Cytomegaloviruses. In Fields Virology (Lippincott Williams & Wilkins, Philadelphia, PA), pp. 2702–2772.

Mui, T.S., Kapp, M., Einsele, H., and Grigoleit, G.U. (2010). T-cell therapy for cytomegalovirus infection. Curr. Opin. Organ Transplant. *15*, 744–750.

Nakamae, H., Kirby, K.A., Sandmaier, B.M., Norasetthada, L., Maloney, D.G., Maris, M.B., Davis, C., Corey, L., Storb, R., and Boeckh, M. (2009). Effect of conditioning regimen intensity on CMV infection in allogeneic hematopoietic cell transplantation. Biol. Blood Marrow Transplant. *15*, 694–703.

Nakamae, H., Storer, B., Sandmaier, B.M., Maloney, D.G., Davis, C., Corey, L., Storb, R., and Boeckh, M. (2011). Cytopenias after day 28 in allogeneic hematopoietic cell transplantation: impact of recipient/donor factors, transplant conditions and myelotoxic drugs. Haematologica *96*, 1838–1845.

Nichols, W.G., Corey, L., Gooley, T., Drew, W.L., Miner, R., Huang, M., Davis, C., and Boeckh, M. (2001). Rising pp65 antigenemia during preemptive anticytomegalovirus therapy after allogeneic hematopoietic stem cell transplantation: risk factors, correlation with DNA load, and outcomes. Blood *97*, 867–874.

Nichols, W.G., Corey, L., Gooley, T., Davis, C., and Boeckh, M. (2002). High risk of death due to bacterial and fungal infection among cytomegalovirus (CMV)-seronegative recipients of stem cell transplants from seropositive donors: evidence for indirect effects of primary CMV infection. J. Infect. Dis. *185*, 273–282.

Nichols, W.G., Price, T.H., Gooley, T., Corey, L., and Boeckh, M. (2003). Transfusion-transmitted cytomegalovirus infection after receipt of leukoreduced blood products. Blood *101*, 4195–4200.

O'Brien, S.M., Keating, M.J., and Mocarski, E.S. (2006). Updated guidelines on the management of cytomegalovirus reactivation in patients with chronic lymphocytic leukemia treated with alemtuzumab. Clin. Lymphoma Myeloma *7*, 125–130.

Okamoto, T., Okada, M., Mori, A., Saheki, K., Takatsuka, H., Wada, H., Tamura, A., Fujimori, Y., Takemoto, Y., Kanamaru, A., *et al.* (1997). Successful treatment of severe cytomegalovirus retinitis with foscarnet and intraocular injection of ganciclovir in a myelosuppressed unrelated bone marrow transplant patient. Bone Marrow Transplant. *20*, 801–803.

Peggs, K.S., Thomson, K., Samuel, E., Dyer, G., Armoogum, J., Chakraverty, R., Pang, K., Mackinnon, S., and Lowdell, M.W. (2011). Directly selected cytomegalovirus-reactive donor T-cells confer rapid and safe systemic reconstitution of virus-specific immunity following stem cell transplantation. Clin. Infect. Dis. *52*, 49–57.

Pergam, S.A., Xie, H., Sandhu, R., Pollack, M., Smith, J., Stevens-Ayers, T., Ilieva, V., Kimball, L.E., Huang, M.L., Hayes, T.S., *et al.* (2012). Efficiency and risk factors for CMV transmission in seronegative hematopoietic stem cell recipients. Biol. Blood Marrow Transplant. *18*, 1391–1400.

Pollack, M., Heugel, J., Xie, H., Leisenring, W., Storek, J., Young, J.A., Kukreja, M., Gress, R., Tomblyn, M., and Boeckh, M. (2011). An international comparison of current strategies to prevent herpesvirus and fungal infections in hematopoietic cell transplant recipients. Biol. Blood Marrow Transplant. *17*, 664–673.

Prentice, H.G., Gluckman, E., Powles, R.L., Ljungman, P., Milpied, N., Fernandez Ranada, J.M., Mandelli, F., Kho, P., Kennedy, L., and Bell, A.R. (1994). Impact of long-term acyclovir on cytomegalovirus infection and survival after allogeneic bone marrow transplantation. European Acyclovir for CMV Prophylaxis Study Group. Lancet *343*, 749–753.

Price, N.B., and Prichard, M.N. (2011). Progress in the development of new therapies for herpesvirus infections. Curr. Opin. Virol. *1*, 548–554.

Raanani, P., Gafter-Gvili, A., Paul, M., Ben-Bassat, I., Leibovici, L., and Shpilberg, O. (2009). Immunoglobulin prophylaxis in hematopoietic stem cell transplantation: systematic review and meta-analysis. J. Clin. Oncol. *27*, 770–781.

Rasmussen, L., Matkin, C., Spaete, R., Pachl, C., and Merigan, T.C. (1991). Antibody response to human cytomegalovirus glycoproteins gB and gH after natural infection in humans. J. Infect. Dis. *164*, 835–842.

Reddy, S.M., Winston, D.J., Territo, M.C., and Schiller, G.J. (2010). CMV central nervous system disease in stem-cell transplant recipients: an increasing complication of drug-resistant CMV infection and protracted immunodeficiency. Bone Marrow Transplant. *45*, 979–984.

Reed, E.C., Wolford, J.L., Kopecky, K.J., Lilleby, K.E., Dandliker, P.S., Todaro, J.L., McDonald, G.B., and Meyers, J.D. (1990). Ganciclovir for the treatment of cytomegalovirus gastroenteritis in bone marrow transplant patients. A randomized, placebo-controlled trial. Ann. Intern. Med. *112*, 505–510.

Reusser, P., Fisher, L.D., Buckner, C.D., Thomas, E.D., and Meyers, J.D. (1990). Cytomegalovirus infection after autologous bone marrow transplantation: occurrence of cytomegalovirus disease and effect on engraftment. Blood *75*, 1888–1894.

Reusser, P., Einsele, H., Lee, J., Volin, L., Rovira, M., Engelhard, D., Finke, J., Cordonnier, C., Link, H., and Ljungman, P. (2002). Randomized multicenter trial of foscarnet versus ganciclovir for preemptive therapy of cytomegalovirus infection after allogeneic stem cell transplantation. Blood *99*, 1159–1164.

Ruutu, T., Ljungman, P., Brinch, L., Lenhoff, S., Lonnqvist, B., Ringden, O., Ruutu, P., Volin, L., Albrechtsen, D., Sallerfors, B., *et al.* (1997). No prevention of cytomegalovirus infection by anti-cytomegalovirus hyperimmune globulin in seronegative bone marrow transplant recipients. The Nordic BMT Group. Bone Marrow Transplant. *19*, 233–236.

Schmidt, G.M., Horak, D.A., Niland, J.C., Duncan, S.R., Forman, S.J., and Zaia, J.A. (1991). A randomized, controlled trial of prophylactic ganciclovir for cytomegalovirus pulmonary infection in recipients of allogeneic bone marrow transplants; The City of Hope-Stanford-Syntex CMV Study Group. N. Engl. J. Med. *324*, 1005–1011.

Schmitt, A., Tonn, T., Busch, D.H., Grigoleit, G.U., Einsele, H., Odendahl, M., Germeroth, L., Ringhoffer, M., Ringhoffer, S., Wiesneth, M., *et al.* (2011). Adoptive transfer and selective reconstitution of streptamer-selected cytomegalovirus-specific CD8+ T-cells leads to virus clearance in patients after allogeneic peripheral blood stem cell transplantation. Transfusion *51*, 591–599.

Schoppel, K., Schmidt, C., Einsele, H., Hebart, H., and Mach, M. (1998). Kinetics of the antibody response against human cytomegalovirus-specific proteins in allogeneic bone marrow transplant recipients. J. Infect. Dis. *178*, 1233–1243.

Schreiber, A., Harter, G., Schubert, A., Bunjes, D., Mertens, T., and Michel, D. (2009). Antiviral treatment of cytomegalovirus infection and resistant strains. Expert Opin. Pharmacother. *10*, 191–209.

Shapira, M.Y., Resnick, I.B., Chou, S., Neumann, A.U., Lurain, N.S., Stamminger, T., Caplan, O., Saleh, N., Efferth, T., Marschall, M., *et al.* (2008). Artesunate as a potent antiviral agent in a patient with late drug-resistant cytomegalovirus infection after hematopoietic stem cell transplantation. Clin. Infect. Dis. *46*, 1455–1457.

Walker, C.M., van Burik, J.A., De For, T.E., and Weisdorf, D.J. (2007). Cytomegalovirus infection after allogeneic transplantation: comparison of cord blood with peripheral blood and marrow graft sources. Biol. Blood Marrow Transplant. *13*, 1106–1115.

Walter, E.A., Greenberg, P.D., Gilbert, M.J., Finch, R.J., Watanabe, K.S., Thomas, E.D., and Riddell, S.R. (1995). Reconstitution of cellular immunity against cytomegalovirus in recipients of allogeneic bone marrow by transfer of T-cell clones from the donor. N. Engl. J. Med. *333*, 1038–1044.

Winston, D.J., Yeager, A.M., Chandrasekar, P.H., Snydman, D.R., Petersen, F.B., and Territo, M.C. (2003). Randomized comparison of oral valacyclovir and intravenous ganciclovir for prevention of cytomegalovirus disease after allogeneic bone marrow transplantation. Clin. Infect. Dis. *36*, 749–758.

Winston, D.J., Baden, L.R., Gabriel, D.A., Emmanouilides, C., Shaw, L.M., Lange, W.R., and Ratanatharathorn, V. (2006). Pharmacokinetics of ganciclovir after oral valganciclovir versus intravenous ganciclovir in allogeneic stem cell transplant patients with graft-versus-host disease of the gastrointestinal tract. Biol. Blood Marrow Transplant. *12*, 635–640.

Winston, D.J., Young, J.A., Pullarkat, V., Papanicolaou, G.A., Vij, R., Vance, E., Alangaden, G.J., Chemaly, R.F., Petersen, F., Chao, N., *et al.* (2008). Maribavir prophylaxis for prevention of cytomegalovirus infection in allogeneic stem cell transplant recipients: a multicenter, randomized, double-blind, placebo-controlled, dose-ranging study. Blood *111*, 5403–5410.

Wloch, M.K., Smith, L.R., Boutsaboualoy, S., Reyes, L., Han, C., Kehler, J., Smith, H.D., Selk, L., Nakamura, R., Brown, J.M., *et al.* (2008). Safety and immunogenicity of a bivalent cytomegalovirus DNA vaccine in healthy adult subjects. J. Infect. Dis. *197*, 1634–1642.

Wolf, D.G., Yaniv, I., Honigman, A., Kassis, I., Schonfeld, T., and Ashkenazi, S. (1998). Early emergence of ganciclovir-resistant human cytomegalovirus strains in children with primary combined immunodeficiency. J. Infect. Dis. *178*, 535–538.

Wolf, D.G., Shimoni, A., Resnick, I.B., Stamminger, T., Neumann, A.U., Chou, S., Efferth, T., Caplan, O., Rose, J., Nagler, A., *et al.* (2011). Human cytomegalovirus kinetics following institution of artesunate after hematopoietic stem cell transplantation. Antiviral. Res. *90*, 183–186.

Zhou, W., Longmate, J., Lacey, S.F., Palmer, J.M., Gallez-Hawkins, G., Thao, L., Spielberger, R., Nakamura, R., Forman, S.J., Zaia, J.A., *et al.* (2009). Impact of donor CMV status on viral infection and reconstitution of multifunction CMV-specific T-cells in CMV-positive transplant recipients. Blood *113*, 6465–6476.

Zikos, P., Van Lint, M.T., Lamparelli, T., Gualandi, F., Occhini, D., Mordini, N., Berisso, G., Bregante, S., and Bacigalupo, A. (1998). A randomized trial of high dose polyvalent intravenous immunoglobulin (HDIgG) vs. Cytomegalovirus (CMV) hyperimmune IgG in allogeneic hemopoietic stem cell transplants (HSCT). Haematologica *83*, 132–137.

Murine Model for Cytoimmunotherapy of CMV Disease after Haematopoietic Cell Transplantation

II.17

Rafaela Holtappels, Stefan Ebert, Jürgen Podlech, Annette Fink, Verena Böhm, Niels A.W. Lemmermann, Kirsten Freitag, Angélique Renzaho, Doris Thomas and Matthias J. Reddehase

Abstract

Cytomegalovirus (CMV) disease is a clinically relevant complication in haematopoietic (stem) cell transplantation (HCT). The murine model of CMV infection in the phase of haematopoietic reconstitution after experimental HCT has been pivotal in defining efficient endogenous reconstitution of donor (D)-derived antiviral $CD8^+$ T-cells as the decisive immune parameter for the control of lytic CMV replication and the prevention of acute manifestations of end-organ disease in HCT recipients (R). Endogenous reconstitution of protective immunity occurs with some delay, since haematopoietic stem cells and T-cell lineage lymphopoietic progenitors need to engraft in bone marrow (BM) stroma, migrate to the thymus for thymic self/non-self T-cell receptor specificity selection, and emigrate as mature but naïve $CD44^{low}$ T-cells to the periphery where they encounter viral antigen to become 'primed', expand clonally, and differentiate into antiviral effector and memory cell subsets. These processes take time, and depending on when CMV reactivates in HCT recipients, endogenous reconstitution may come too late. In addition, the murine model has shown that CMV directly interferes with endogenous reconstitution by infecting BM stromal cells and inhibiting the expression of stromal cell-derived haematopoietins, specifically of stem cell factor (SCF). Depending on the precise conditions, BM pathogenesis of CMV, often referred to as 'myelosuppression', can range from reduced engraftment to a complete graft failure resulting in BM aplasia. Substituting HCT with already primed, 'ready-to-go' CMV-specific $CD8^+$ effector and/or memory T-cells, also known as 'adoptive T-cell transfer' or 'pre-emptive cytoimmunotherapy', can provide immediate protection. Key parameters of protection by $CD8^+$ T-cells were revealed by the murine model: (i) memory cell subsets isolated *ex vivo* from immune donors are ~100-fold more efficient than cell culture-propagated short-term cytolytic effector cell lines of identical specificity; (ii) the magnitude of the primary immune response to a viral epitope, its 'immunodominance', does not correlate with protection mediated by cognate T-cells; (iii) viral immune evasion proteins determine if an epitope is 'protective'; and (iv) genetic deletion of dominant protective epitopes in either the donor's or the recipient's virus has little impact on protection by a broadly specific T-cell population, predicting 'robustness' of T-cell immunotherapy towards antigenic variation in virus strains or clinical isolates.

Introduction

Haematopoietic (stem) cell transplantation (HCT), in its diverse modifications (see the preceding Chapter II.16), is the ultimate therapy option for those types of haematopoietic cell malignancies that resist standard protocols of radiotherapy and chemotherapy. Though powerful for therapy of some types of lymphomas and leukaemias, graft-versus-host disease (GvHD) in allo-HCT (usually HLA-matched but minor histocompatibility loci mismatched), tumour relapse from 'minimal residual disease/leukaemia (MRD/L)', and infections limit therapy success. Despite prophylactic and pre-emptive therapy with antivirals (see Chapters II.16 and II.19), HCMV infection, more than any other infection, remains a health risk to the patient and a medical challenge to haematologists and their colleagues in medical virology. Of multiple organ manifestations, interstitial pneumonia is the most redoubtable (Chapter II.16). Peritransplant acute HCMV infection is a rare, since avoidable, accident of transmission of virus from a virus shedding contact person. Risk arises from reactivation of latent virus (see Chapters I.19, I.20, II.7 and II.16) in the transplant from a CMV-seropositive (anamnestic antibody indicating latent infection) donor (constellation D^+R^-), in diverse tissues of a seropositive recipient (constellation D^-R^+) or in both (constellation D^+R^+). Whilst risk associated with the donor can, in principle, be avoided by donor selection provided that a choice exists between seropositive and seronegative HLA-compatible donors, which usually isn't the case, the risk associated with the recipient is fate.

Here we will review our experience with a murine model of syngeneic experimental HCT and infection with murine CMV (mCMV) that, compared to any clinical situation, is greatly reduced in variables, but is nonetheless still highly complex. Reductionism includes (i) absence of underlying malignant disease and thus of relapse risk; (ii) associated with this, also absence of pre-existing defects from chemotherapy and medication; (iii) complete genetic matching avoiding complications by GvHD or host-versus-graft reaction (HvGR); (iv) specified pathogen-free (SPF) conditions avoiding interferences between multiple pathogens; (v) defined virus strains and usually infection with single strains to avoid virus recombinations, transactivation, and competition; (vi) absence of antivirals that could also be marrow-toxic; and (vii) defined route, dose, and time of infection. If all these parameters were allowed to be variable, from our experience we would predict that only the unpredictability of the outcome in an individual HCT recipient will be predictable.

Nonetheless, although the murine model has no one-to-one clinical correlate, it has proven valid in unveiling very basic principles of some medical relevance. Specifically, the dominance of the major immediate-early (MIE) gene locus in the immune response to CMVs has been discovered for mCMV (Reddehase and Koszinowski, 1984; Reddehase *et al.*, 1984; 1989; reviewed in Reddehase, 2000, 2002), 15 years before it was appreciated for HCMV (Kern *et al.*, 1999; Bunde *et al.*, 2005; Sylwester *et al.*, 2005), and pre-emptive immunotherapy of lethal CMV disease by adoptive transfer of CMV-specific $CD8^+$ T-cells as a preclinical proof-of-concept for protection from CMV (Reddehase *et al.*, 1985; reviewed in Reddehase, 2002; Holtappels *et al.*, 2008a) preceded successful clinical trials (Greenberg *et al.*, 1991; Riddell *et al.*, 1992; Walter *et al.*, 1995; Peggs *et al.*, 2003; Cobbold *et al.*, 2005; Feuchtinger *et al.*, 2010; Schmitt *et al.*, 2011) (see also Chapter II.16). Likewise, the murine model was instrumental in identifying principles of CMV evasion from innate and adaptive immunity (see below, as well as Chapters II.7 to II.9) and in the discovery of 'memory inflation' (Holtappels *et al.*, 2000a), its link to CMV latency (Chapter I.22), and its application in new-generation 'memory cell vaccines' (Chapter II.21).

Key parameters of the murine model of HCT and CMV infection

Three fundamental conditions must be fulfilled in the murine model to match the cornerstones of clinical CMV disease after HCT: (i) HCT, in absence of infection, should lead to full lympho-haematopoietic reconstitution with a 100% survival rate; (ii) CMV infection should be controlled and allow a 100% survival rate in the immunocompetent or fully reconstituted host but lead to disease with rates of morbidity and mortality in immunocompromised HCT recipients depending on the efficacy of immune reconstitution; and (iii) model CMV disease should mimic clinical CMV disease in its wide range of organ manifestations based on virus tropism for many cell types. In particular, the lung must be a target of infection leading to interstitial pneumonia, a main manifestation of CMV disease in clinical HCT.

Haematopoietic cell transplantation

The basic modul of the model is syngeneic HCT with CMV-naïve BALB/c (*H-2^d^*) mice as both haematopoietic cell (HC) donors (D) and recipients (R), a constellation found in clinical HCT only between identical twins with no CMV infection history (D^-R^-). The HCs are usually BM cells (BMCs), which comprise pluripotent haematopoietic stem cells as well as lineage-committed progenitor cells and haematopoietic lineage cells in more advanced stages of differentiation. Recipients are immunocompromised by haematoablative total-body γ-irradiation on the day before HCT is performed by i.v. infusion of BMCs. Within 2 days, the ablative conditioning leads to an 'empty' BM (BM aplasia), leaving intact only the network of reticular stromal cells (RSCs) (Mutter *et al.*, 1988; Mayer *et al.*, 1997; Steffens *et al.*, 1998b). Associated with BM aplasia is a pancytopenia in blood, which also includes thrombocytopenia. HCT (for conditions, see Fig. II.17.3, green shaded square in the left panel) leads to repopulation of the empty BM stroma by engraftment of haematopoietic stem and progenitor cells. Repopulation visible as colonies early after HCT originates from myelomonocytic and erythroid progenitor cells (Fig. II.17.1A). As shown for chimeric mice generated by HCT with female (*sry*$^-$) donors and male (*sry*$^+$) recipients, BMCs include recipient-derived (here *sry*$^+$) RSCs, but in a serial HCT these cells fail to engraft in the BM stroma of female (*sry*$^-$) recipients (Mayer *et al.*, 1997). Thus, recipients of HCT become BM chimeras

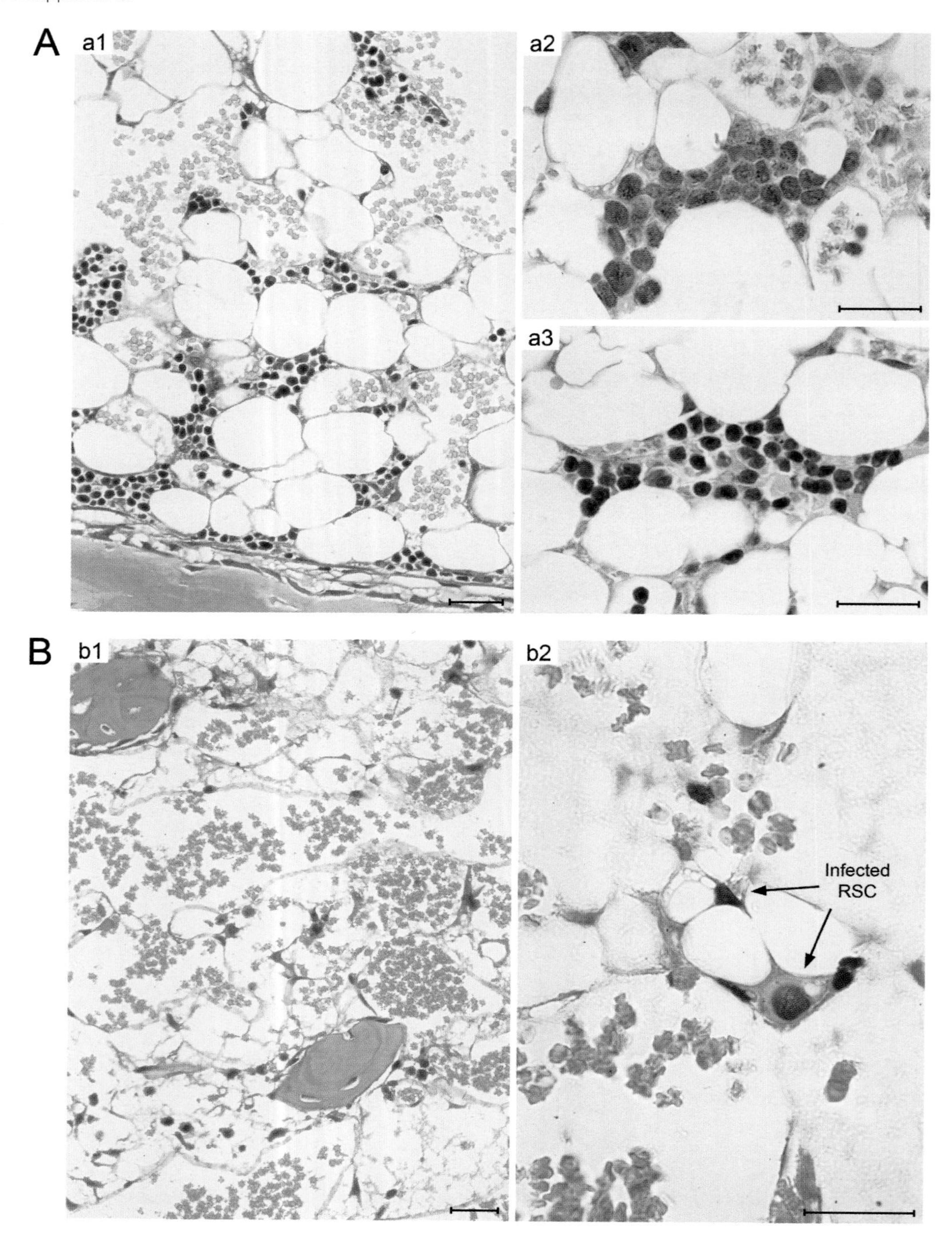

Figure II.17.1 Inhibition of haematopoietic cell engraftment by CMV. (A) Repopulation of BM stroma on day 14 after haematoablative treatment and syngeneic HCT performed with 10,000 BMCs. (a1) Section of a femoral diaphysis stained with haematoxylin–eosin. Overview, showing haematopoietic colonies. (a2) Myelomonocytic colony positive for specific esterase. (a3) Erythroblastic colony negative for specific esterase. (B) BM aplasia due to graft failure caused by mCMV infection. (b1) Overview, showing empty stromal network in a femoral epiphysis. (b2) Infected reticular stromal cells (RSC), identified by brown IHC staining of intranuclear IE1 protein. Bar markers: 25 μm. Reproduced in rearranged form from the Journal of Virology (1998) *72*, 5010 (a1–b1; Steffens *et al*., 1998b) and the Journal of Virology (1997) *71*, 4592 (b2; Mayer *et al*., 1997) with permission by the American Society for Microbiology.

with donor-genotype haematopoietic cells and recipient-genotype stromal cells.

Choice of virus for the HCT model

Since CMVs are host-species specific (Chapter I.18), animal model systems are required for experimental studies. In building up a model, the choice of the virus can be decisive for all results and conclusions; so, this is a critical issue. Based on virus–host co-speciation, resulting in an intimate mutual adaptation, no animal CMV in its specific host can be a fully appropriate model for HCMV in humans (see also Chapters I.1, I.2, II.5, II.15, and II.22). Non-human primate models (Chapter II.22), though the respective viruses are genetically closest to HCMV, like their hosts are genetically closest to humans (Chapter I.1), have revealed pathogenetic similarities but also differences. Humanized mouse models (Chapter I.23) are promising, but even xenograft chimeras with human haematopoietic cells and human tissue implants cannot recapitulate all aspects of CMV biology that involve cytokine networks mediating a cross-talk between immune cells and a wide array of stromal and parenchymal host cell types (Chapter II.11) or that require an intact organ system. So, one should always appreciate both strengths and limitations of specific models.

Clinicians might argue that one cannot model the pathogenesis of HCMV with anything else but HCMV itself – but what actually is HCMV? As is becoming increasingly clear, HCMV strains/variants can differ from each other both genetically (Chapter I.1) and phenotypically (Chapter I.17) and, in particular, long-term cell culture-propagated 'laboratory strains' differ fundamentally from recent clinical isolates. Specifically, in the Thy/liv SCID-hu mouse model, the low-passage strain Toledo replicated in the human tissue implants, whereas the high-passage laboratory strains AD169 and Towne failed or replicated to a very low level only, a difference attributed to tissue tropism determinants present in Toledo and clinical isolates but missing in the laboratory strains (discussed in detail in Chapter I.23). So, AD169 and Towne, which are both widely used in HCMV research, already failed in an exam for an *in vivo* model employing humanized mice. Likewise, it can be assumed that no clinical isolate will exactly represent any other clinical isolate.

Though long-term tissue culture-propagated strains of mCMV also show genetic differences to wild-derived low-passage isolates, and there is also genetic variance between isolates, mCMV, on the whole, appears to be genetically much more stable than HCMV (Chapter I.2). Specifically, loss of broad cell-type tropism during propagation in fibroblasts does not appear to be a problem with mCMV; – if this relates to the fact that mCMV is propagated in fetal fibroblasts has never been considered. Initiated by U.H. Koszinowski in 1980, our models, traditionally, are all based on the Smith strain (ATCC VR-194, meanwhile reaccessioned as VR-1399) of mCMV as a standard, and we use mostly the subcutaneous, intraplantar route of local infection that involves the popliteal lymph node (PLN) as the regional lymph node (RLN) that drains the site of infection. This makes a major difference compared to models using systemic infections that bypass peripheral pathogen recognition mechanisms (see a related discussion in Chapter II.11).

A matter of faith is the source of the virus to be used in experiments: mouse fetal (embryonic) fibroblast (MEF)-propagated, sucrose gradient ultracentrifugation-purified 'tissue culture' virus (TC-mCMV) or more 'virulent' virus passaged *in vivo* in mouse salivary glands (SGs) and used as a diluted, partially purified tissue homogenate (SG-mCMV; in the literature often referred to as SGV). The differences between virus preparations from these sources have been reviewed elsewhere (Lemmermann *et al.*, 2010b; and references therein), and a detailed discussion of *pros* and *cons* is beyond the scope of this chapter (see also Chapter II.12 for discussion). Reservations about SG-mCMV are based on the concern that factors other than virus present in organ homogenate, which are undefined in terms of their composition and concentrations, might impact on early stages of host infection at the viral entry site by modulating intrinsic antiviral defence and innate immune response. Not only could SG homogenate contain host hormones (SGs in male mice are rich in nerve growth factor) and numerous biologically active molecules of the infected cells' secretome (Chapter I.6) but also a plethora of cytokines, including IL-10, secreted by activated leucocytes infiltrating the infected mucosal tissue (Cavanaugh *et al.*, 2003; Humphreys *et al.*, 2007; reviewed in Campbell *et al.*, 2008). The latter factors are not taken into account when 'control' experiments use SG homogenate from uninfected mice. Specifically, IL-10 in the virus source used for infection could indirectly contribute to 'virulence' by dampening the antiviral immune response early on. Particularly suspect are effects of SG-mCMV that do not correlate with virus replication and are not observed with TC-mCMV. Because of all these arguments, and to meet in animal models the GLP-standards that are routine in clinical science, we have decided to build our models on the use of TC-mCMV.

Only very recently it has been recognized that the ATCC-distributed Smith strain represents a mixture of at least two variants (Jordan *et al.*, 2011) differing in the expression of the viral chemokine MCK-2 (encoded by m131-m129), a ligand of chemokine receptor CX3CR1, that plays a dual role in mCMV pathogenesis. On the one hand, by recruiting mCMV-nonpermissive inflammatory monocytes (IMs) from the BM to the entry site of infection, MCK-2 delays viral clearance indirectly by dampening the cytolytic T-lymphocyte (CTL) response. On the other hand, by recruiting mCMV-permissive patrolling monocytes (PMs) that acquire virus, MCK-2 contributes to virus dissemination in the host, in particular dissemination to the SGs, which are sites of virus replication and secretion important for host-to-host transmission (for more details and references, see Chapter I.21). The two variants differ in the MCK-2 coding sequence by a single nucleotide insertion/deletion polymorphism leading to different open reading frames resulting in full-length 280-aa or shortened 144-aa MCK-2, respectively, the latter of which shows reduced dissemination to SGs (Jordan *et al.*, 2011). The current 'Munich batch' of mCMV, the one studied by Jordan and colleagues, is derived from ATCC VR-194/1981, and the mixture was reported to be ~ 1:3 in favour of the short MCK-2 variant, whereas the full-length MCK-2 variant predominates in a recent batch of ATCC VR-1399 (Jordan *et al.*, 2011). Notably, the current 'Munich batch' and the current 'Mainz batch' are both derived from a batch 12/1992 originating from ATCC VR-194/1981, which is still stored in Mainz and shows a mixture of the two variants (Fig. II.17.2C, left panel) similar to the further propagated 'Munich batch'. Importantly, BAC cloning of the mCMV-WT.Smith genome has picked up the short MCK-2 variant, so that the BAC plasmid pSM3fr and corresponding virus MW97.01 (Messerle *et al.*, 1997; Wagner *et al.*, 1999) (here referred to as mCMV-WT.BAC) and numerous mCMV mutants based on pSM3fr (see Chapter I.3) encode the short version of MCK-2. Jordan and colleagues (2011) have reinserted the missing nucleotide in the BAC plasmid to generate mCMV-WT.BAC3.3 (pSM3fr-MCK-2fl clone 3.3) encoding the full-length MCK-2 on the genetic background of pSM3fr.

To evaluate the functional impact of MCK-2 polymorphism on virus growth in the immunocompromised host model in absence of HCT, we determined the log-linear growth curves and doubling times in lungs, spleen, liver, and SGs (Fig. II.17.2) comparing viruses mCMV-WT.Smith 'Mainz batch' 12/1992 (long:short MCK–2 ratio of ~ 1:3), mCMV-WT.BAC (MW97.01; short MCK-2) (Wagner *et al.*, 1999), mCMV-WT.BAC3.3 (long MCK-2) (Jordan *et al.*, 2011) and mCMV-K181.BAC (long MCK-2) (Redwood *et al.*, 2005). In none of the organs, SGs included, were significant differences observed when considering the 95% confidence intervals of the doubling times, and trends to differences were not consistent for all organs and did not correlate with either MCK-2 variant; specifically, variants expressing full-length MCK-2 did not show any replication advantage in SGs (Fig. II.17.2B). These findings in mice after haematoablation show that none of the tested viruses is attenuated with regard to virus replication in different organs composed of different tissues and cell types, and also imply that virus dissemination to SGs does not depend on recruitment of IMs from the BM (see Chapter I.21).

For establishing murine disease models and for interpreting results, the differences in MCK-2 variants must be considered; however, there is no 'wrong' or 'correct' variant. In the immunocompetent or immune reconstituted murine host, full-length MCK-2 virus variants may be expected to disseminate more efficiently via PMs to the SGs for transmission, whereas short MCK-2 variants may be more immunogenic due to failure in recruiting immunosuppressive IMs from the BM. Both components are present in mCMV-WT.Smith. In our experience, a 10-fold increase in intraplantar infection dose from 10,000 to 100,000 PFU translates into an only moderate increase in infection of the RLN in immunocompetent mice (Böhm *et al.*, 2008b); thus, a variant ratio of ~ 1:3 can be assumed to play little role after infection with 100,000 PFU when both variants contribute > 10,000 PFU. In addition, we did not note a significant attenuation of mCMV-WT.BAC compared with mCMV-WT.Smith in immunocompetent mice or after HCT. A direct comparison between mCMV-WT.BAC (MW97.01) and mCMV-WT.BAC3.3 in these systems remains to be performed.

As will be shown below for selected organs, mCMV-WT.Smith can cause a lethal, full-blown end-organ CMV disease in the immunocompromised host, involving many organs with replication in diverse cell types, such as fibroblasts, endothelial cells (ECs), smooth muscle cells (SMCs), cardiac myocytes, macrophages, dendritic cells (DCs) and a wide array of epithelial cell types, including hepatocytes, enterocytes, pneumocytes, and glandular epithelial cells (Reddehase *et al.*, 1985; Podlech *et al.*, 1998a). Thus, mCMV-WT.Smith in the immunocompromised murine host recapitulates key features of human clinical strain CMV disease.

CMV infection of HCT recipients

CMV infection enhances mortality risk after HCT

For reducing the variables in the model, timing, route, and dose of infection of experimental HCT recipients were kept constant; specifically, CMV-naïve 8- to 10-week-old BALB/c mice were experimentally

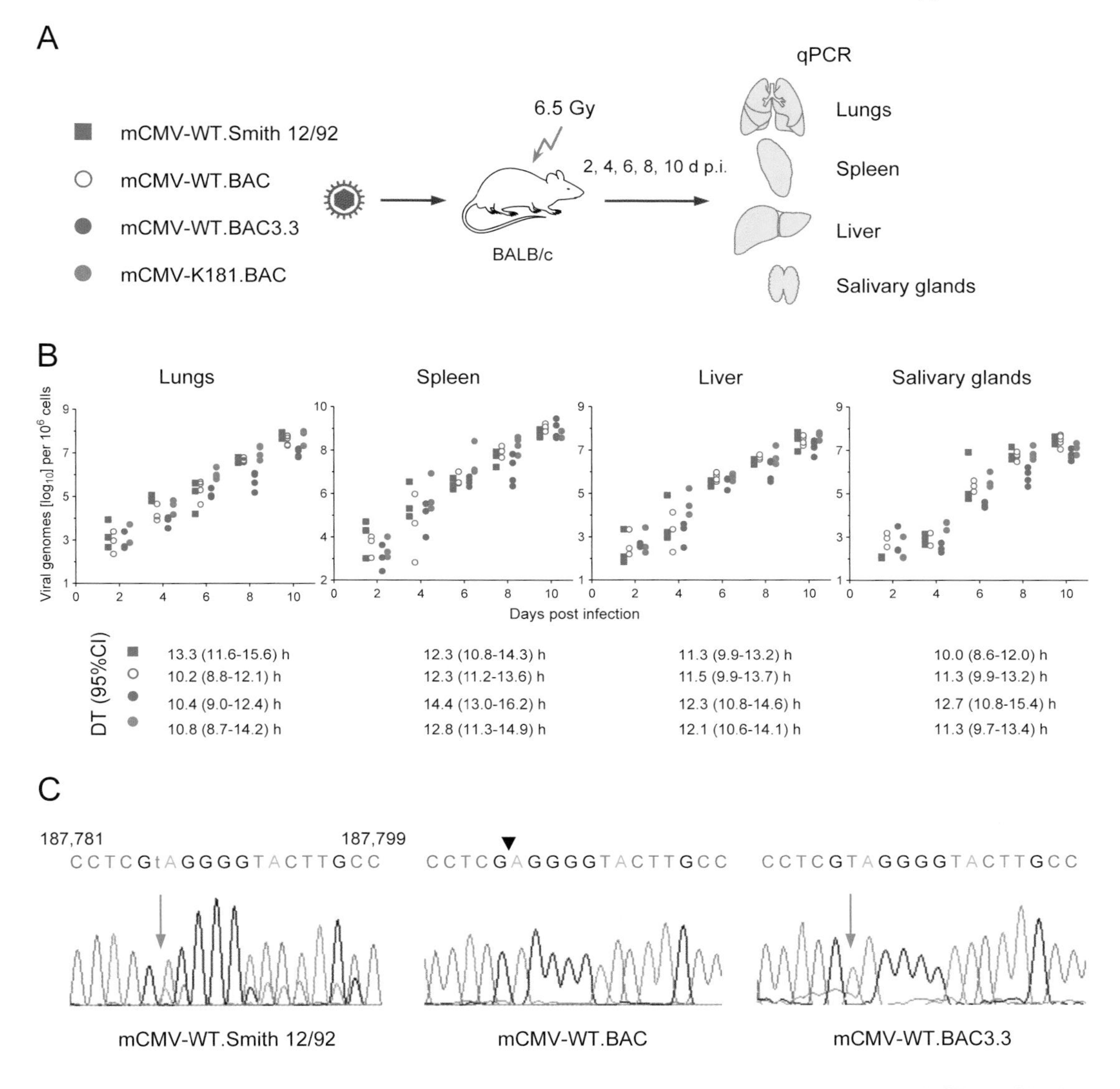

Figure II.17.2 Virus growth in organs of BALB/c mice immunocompromised by haematoablative HCT conditioning. (A) Virus strains and experimental protocol. (B) Log-linear virus replication determined by *M55/gB* gene-specific qPCR. DT (95% CI), doubling times and corresponding 95% confidence intervals determined by linear regression analysis. (C) Sequence chromatograms for the region of interest in the MCK-2 viral chemokine coding gene. The arrowhead points to the position of the missing nucleotide causing a frameshift. Red arrows point to the nucleotide in virus genomes coding for full-length MCK-2. (Left) Sequenced from a sucrose gradient-purified stock preparation. (Centre and right) Sequenced from virus recovered from the salivary glands on day 8.

infected by the intraplantar route with a dose of 100,000 PFU of TC-derived and purified mCMV-WT. Smith 2 hours after HCT. This protocol models the worst case scenario of a CMV reactivation in either donor BMC (D^+R^-) or recipient's tissues (D^-R^+) or both (D^+R^+) occurring without delay after HCT, which is a rare clinical situation, but can exist in unfortunate cases. On this highly reductionistic basis, we have made a two-dimensional evaluation of the impact of two HCT variables on the survival of uninfected versus infected HCT recipients: (variable 1) completeness of recipients' haematoablation prior to HCT by varying the dose of γ-irradiation, and (variable 2) the power of reconstitution by varying the number of transplanted donor BMCs (Fig. II.17.3). The data impressively show that CMV infection can drastically reduce the overall survival chance after HCT, but also show that based on the contributions of transplanted donor

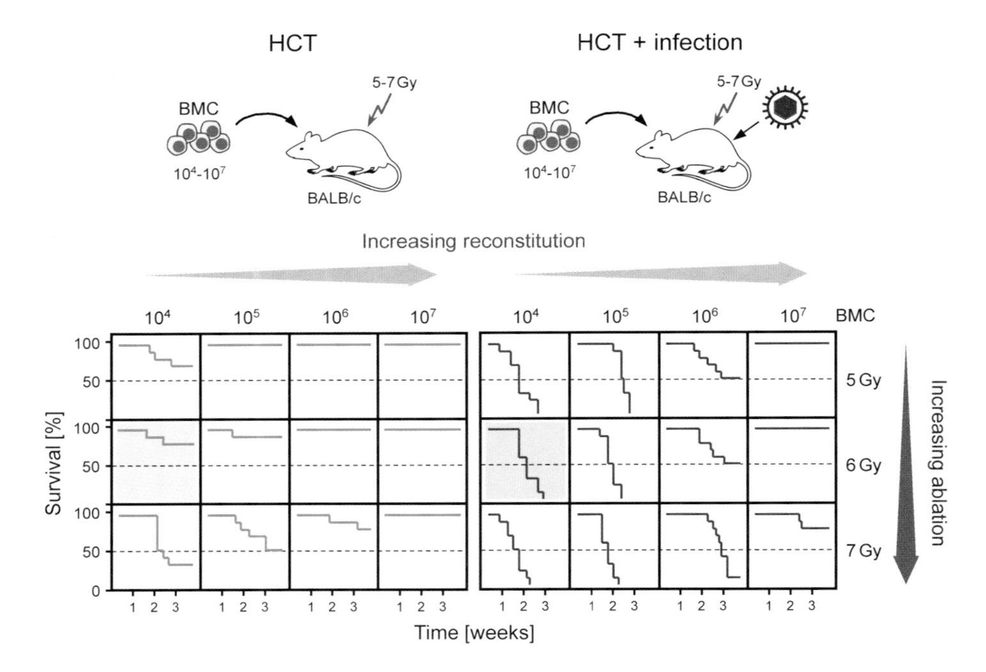

Figure II.17.3 Lethality of CMV infection after experimental HCT. (Left panel) Kaplan–Meier survival curves depending on bone marrow cell (BMC) dose and intensity of the preceding haematoablative conditioning. (Right panel) Impact of infection. Green and red shaded squares indicate the HCT conditions corresponding to Fig. II.17.1A and B, respectively.

haematopoiesis (increasing with increasing numbers of BMC) and recipients' residual haematopoiesis (decreasing with increasing ablation) clinical outcomes can range from 0% to 100% survival. One can only intuit the difficulty in predicting outcome when more variables add further dimensions of complexity to the system, as it is the case in clinical settings.

Importantly, as one can learn from Chapter II.16 on clinical HCT, most HCMV disease cases are diagnosed between 3 and 12 weeks after HCT. So, with CMV-associated mortality occurring between 2 and 3 weeks at an intermediate dose of 10^6 transplanted donor BMCs, the time course in the mouse model of experimental infection soon after HCT is somewhat faster but not far from the time range reported for clinical CMV disease after an unpredictable, variable onset of virus reactivation.

CMV infection inhibits the engraftment of transplanted HCs in recipients' BM stroma

A possible pathomechanism of CMV infection is CMV-mediated HC graft failure, which directly prevents haematopoietic reconstitution, including reconstitution of the T lymphopoietic lineage and thus, in the end, also of antiviral CD8$^+$ T-cells. So, by a fatal feed-back loop, CMV prevents reconstitution of antiviral immunity, and this, in consequence, allows an uncontrolled spread of cytopathogenic infection resulting in lethal end-organ CMV disease. The repopulation of BM stroma by myeloid lineage progenitor cells seen in absence of infection after HCT performed with a low dose of donor BMCs (Fig. II.17.1A; corresponding to the green shaded square in Fig. II.17.3, left panel) fails in presence of infection, resulting in a complete BM aplasia and death (Fig. II.17.1B, image b1, corresponding to the red shaded square in Fig. II.17.3, right panel). IHC specific for the intranuclear IE1 protein of mCMV revealed infection of RSCs (Fig. II.17.1B, image b2) that form the stromal network that is essential for successful HC homing, self-renewal, and differentiation. Sex-mismatched HCT with male (*sry*$^+$) donor BMCs and infected female (*sry*$^-$) recipients followed by quantitation of BM repopulating HCs with *sry* gene-specific qPCR directly revealed an engraftment failure of donor HCs (Steffens *et al.*, 1998b). Though RSC infection *in situ* is not cytolytic and does not destroy the stromal network physically, stromal function was impaired, since engraftment failure was found to be associated with reduced expression of the stromal cell-derived haematopoietins stem cell factor (SCF), granulocyte colony-stimulating factor (G-CSF), and IL-6 (Mayer *et al.*, 1997). Stromal cells enriched from BM of *sry* chimeras by three cell-culture passages were found to be recipient-derived (sry$^-$), express the cell membrane-bound form of SCF and secrete IL-6. These cells proved to be infectable by mCMV, though to a limited extent as indicated by IE1 and E1 protein expression, showed an unusual cell surface marker phenotype (MHC-I$^+$MHC-II$^-$CD4$^+$CD11b$^+$CD44$^+$CD71$^+$) and coexpressed the SMC and EC markers α-SM actin and von Willebrand factor (vWF; in Weibel-Palade bodies), respectively, as detected by two-colour immunofluorescence (authors' unpublished data).

Reconstitution of CD8$^+$ T-cells is essential for surviving CMV infection after HCT

As revealed by the 2-parameter matrices of Kaplan–Meier survival curves (Fig. II.17.3), higher numbers of transplanted BMCs (> 10^6 cells) prevent lethality of CMV. Though high doses of BMCs overcome CMV-associated BM aplasia and allow haematopoietic reconstitution to take place, functional deficiency of the BM stroma after clearance of acute infection manifests as a durably reduced number of pluripotent haematopoietic stem cells, HSCs, as revealed by reduced BM-repopulating capacity of BMCs from infected donors in serially performed HCTs (Reddehase *et al.*, 1992).

Survival from CMV infection is associated with control of virus replication in organs, which, in the lungs, peaks at 2–3 weeks after HCT, followed by a rapid decline that correlates with an immense recruitment of endogenously reconstituted T-cells, largely dominated by the CD8$^+$ subpopulation (Fig. II.17.4A) (Holtappels *et al.*, 1998). Functional relevance of these tissue-infiltrating CD8$^+$ T-cells (mostly KLRG1$^+$CD62L$^-$ short-lived effector cells, SLECs; see Chapter I.22, Fig. I.22.7) in antiviral protection after

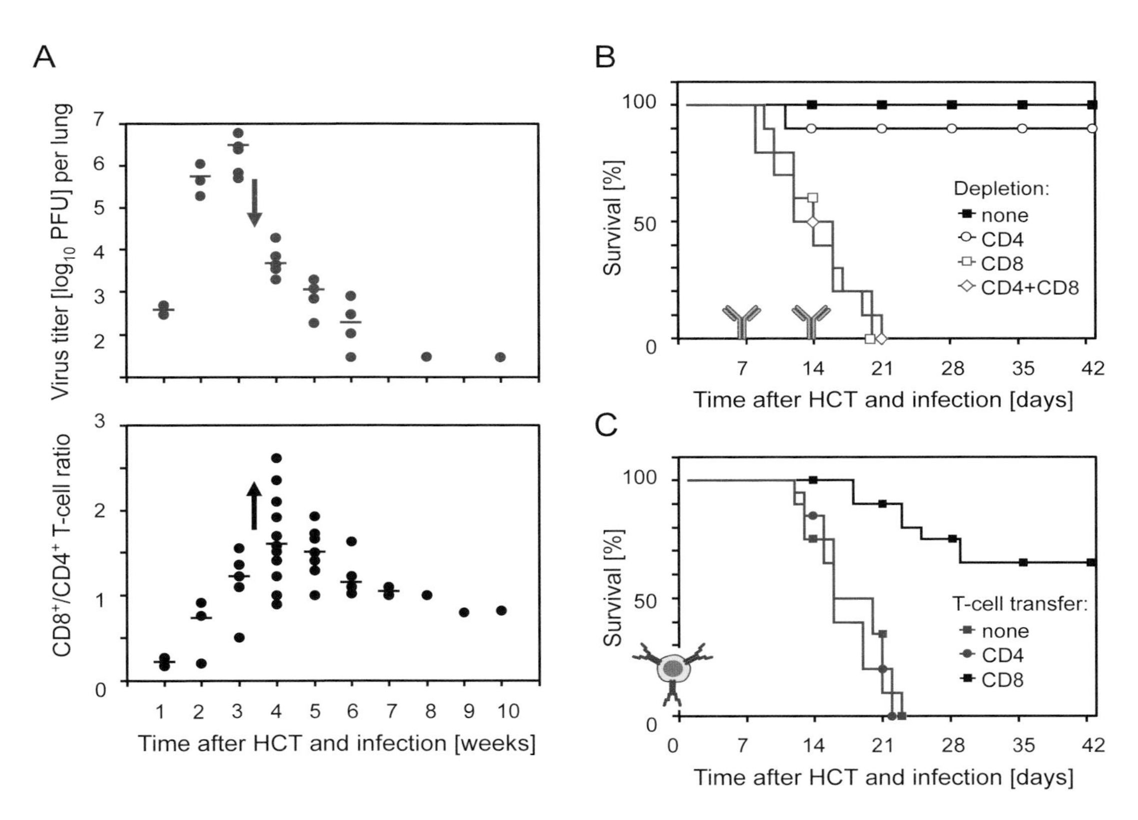

Figure II.17.4 Survival from CMV infection after HCT depends on the reconstitution of CD8$^+$ T-cells. (A) Decline of virus growth in the lungs coincides with recruitment of tissue-infiltrating T-cells, predominantly of the CD8$^+$ subset. Syngeneic HCTs were performed with 10^7 BMCs, a condition under which all recipients survive (see Fig. II.17.3). For details, see Journal of Virology (1998) *72*, 7206 (Holtappels *et al.*, 1998). (B) Depletion of CD8$^+$ T-cells during their reconstitution and recruitment to tissues after HCT is absolutely lethal. Reproduced in modified form from the Journal of Virology (2000) *74*, 7498 (Podlech *et al.*, 2000). (C) Adoptive transfer of virus-specific CD8$^+$ T-cells reduces mortality in recipients of HCT performed with a low dose of BMCs. Reproduced in modified form from the Journal of Virology (1998) *72*, 1799 (Steffens *et al.*, 1998a); permissions by the American Society for Microbiology.

HCT became obvious from depleting CD8⁺ or CD4⁺ T-cells *in vivo* during an ongoing reconstitution after HCT (Fig. II.17.4B) (Podlech *et al.*, 1998a, 2000). Whereas depletion of CD4⁺ cells (mostly but not only T-cells) on day 7 and day 14 after HCT did not end up in significant mortality, depletion of CD8⁺ cells (mostly but not only T-cells), invariably results in 100% mortality. Thus, apparently, survival from infection after HCT critically depends on efficient endogenous reconstitution of CD8⁺ T-cells.

Pre-emptive CD8⁺ T-cell therapy of CMV disease improves the survival rate after HCT

In reverse, under conditions of poor engraftment and insufficient reconstitution (recall Fig. II.17.3), supplementation of transplanted HCs with pre-primed 'ready-to-go' virus-specific CD8⁺ T-cells should reduce the risk of CMV disease and associated mortality. Such an approach is known as 'adoptive (cell) transfer' (AT), 'adoptive cell therapy', or 'donor lymphocyte infusion' (DLI) aimed at providing the HCT recipient with adopted cellular immunity. As shown in Fig. II.17.4C (adapted from Steffens *et al.*, 1998a), AT of polyclonal virus-specific CD8⁺ T-cells, performed as a pre-emptive cell therapy on the day of an undersized HCT (aimed to model a situation of poor graft take) and infection, significantly improved the survival rate, whereas CD4⁺ T-cells from the same primed donor cell source did not. This result is reciprocally identical to the result of T-cell subset depletion (Fig. II.17.4B) and excludes a critical role for CD8⁺ cell types of other provenance, thereby confirming CD8⁺ T-cells as effector cells able on their own to control acute CMV infection.

It is important to emphasize that AT of CMV-specific CD8⁺ T-cells is beneficial also under conditions of HCT that are not associated with mortality from CMV. As shown by Steffens and colleagues, supplementing transplanted HCs with graded numbers of virus-specific CD8⁺ T-cells leads to a dose increase-dependent decrease in peak virus titres in various organs, faster clearance of productive infection, lower load of latent virus genome in organs, and a lower risk of virus reactivation from latency (Steffens *et al.*, 1998a) (for virus latency, see also Chapter I.22).

Tissue-infiltrating antiviral CD8⁺ T-cells limit *in situ* virus spread

That tissue-infiltrating CD8⁺ T-cells mediate protection by controlling tissue-destructing virus spread was safe to assume and is supported by histological evidence (Fig. II.17.5: A, liver; B, lungs). Under conditions of uninfluenced reconstitution (images a1 and b1), infected tissue cells are rare and are confined to foci of T-cell infiltrates. Apparently, the CD3ε⁺ T-cells (black-stained) are attracted by the infected cells (red-stained) and literally beleaguer them. In contrast, depletion of both T-cell subsets, under otherwise identical conditions, results in a fulminant, disseminated tissue infection corresponding to mortality (images a2 and b2). In an approach to identify the T-cell subset that is responsible for the observed protection, depletion of CD4⁺ T-cells did not prevent the formation of infection-confining foci, apparently composed of CD3ε⁺CD8⁺ T-cells (panel a3), whereas depletion of CD8⁺ T-cells prevented the formation of infection-confining foci, resulting in disseminated tissue infection (panel a4). Notably, CD3ε⁺CD4⁺ T-cells infiltrated the tissue, but were found in scattered distribution unable to form foci and prevent virus spread, which may relate to the fact that infected cells in parenchymal tissue, hepatocytes in the specific case of Fig. II.17.5A, are MHC-II⁻. Though we have decided to discuss here the events in liver and lungs as examples, it is important to note that the original publication (Podlech *et al.*, 1998a) shows the same rules apply also to other organs and tissues, including spleen, small intestine, heart, adrenal (suprarenal gland) cortex, kidney glomeruli, and the choroid plexus of brain ventricles. It should be mentioned, however, that a difference between CD4⁺ and CD8⁺ T-cell depletion was less pronounced for SGs, but again with more SG tissue cells being infected after CD8⁺ T-cell depletion, and that infection of adrenal medulla was not controlled by either T-cell subset, which may contribute to hormonal dysregulation and wasting syndrome in CMV-infected HCT recipients.

The result for the SGs may surprise in the light of reports on SG infection being controlled primarily by CD4⁺ T-cells (Jonjic *et al.*, 1989; reviewed in Campbell *et al.*, 2008). These reports, however, apply to the termination of 'persistent' infection of a particular cell type, the glandular epithelial cell that sequesters virions in vacuoles for secretion into the salivary duct and evades recognition by CD8⁺ T-cells (Walton *et al.*, 2011). In immunocompromised mice and after HCT, mCMV replicates also in SG connective tissue, and this component of SG infection is terminated by CD8⁺ T-cells like it is the case in most other tissues (Reddehase *et al.*, 1985; and unpublished TEM studies by F. Weiland, co-author of the referenced work). The exceptional situation in adrenal medulla after HCT remained unexplored.

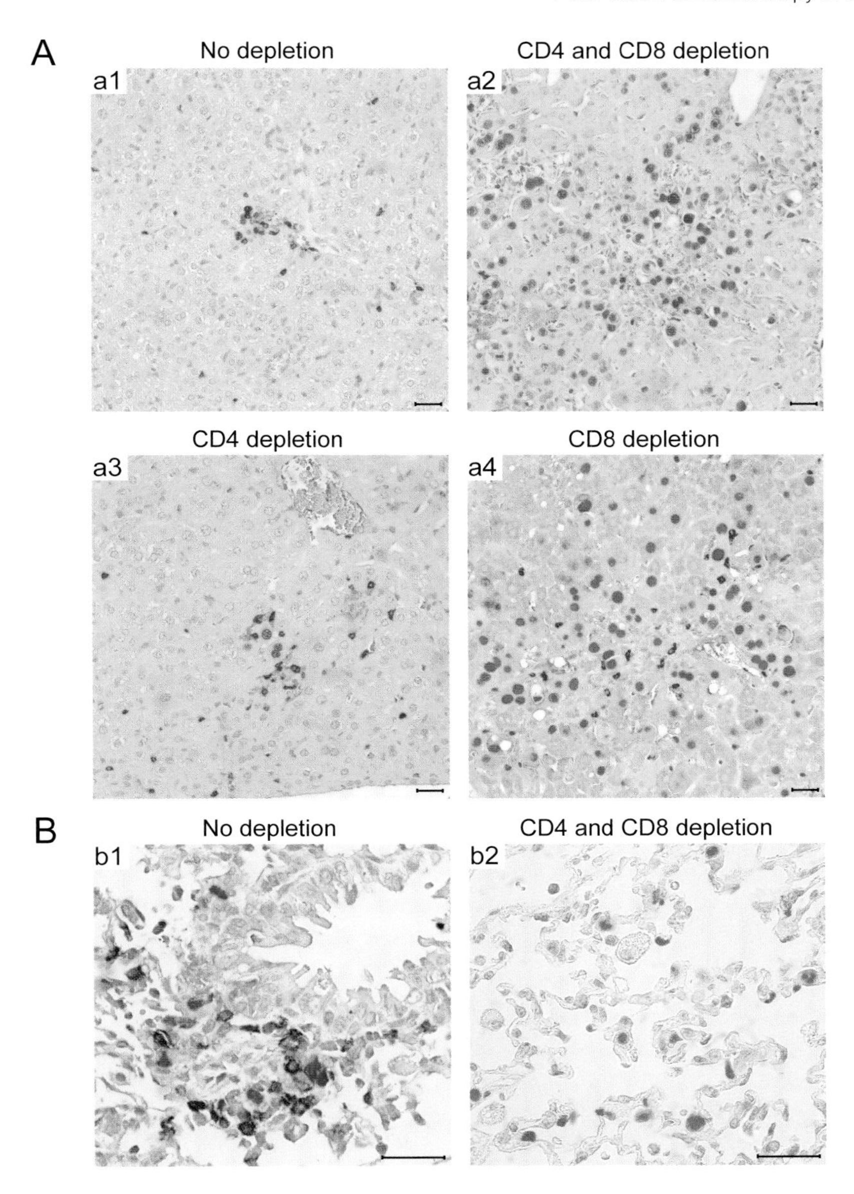

Figure II.17.5 Focal T-cell infiltrates limit virus spread in tissues. Depletion of CD8+ T-cells during reconstitution after HCT leads to disseminated infection and viral histopathology. (A) Liver tissue sections. (B) Lung tissue sections. Two-colour IHC with black staining of CD3ε, identifying T-cells, and red staining of the intranuclear viral protein IE1, identifying infected cells. In the liver, infected cells are mostly hepatocytes, but also some liver sinusoidal endothelial cells (LSEC) and Kupffer cells. Counterstaining with haematoxylin. Bar markers represent 25 µm. The images were taken from tissue sections of mice corresponding to Fig. II.17.4B. The peribronchiolar focus in image B, b1 has been shown previously as the cover photograph and b2 in the corresponding article (Podlech *et al.*, 2000) of the Journal of Virology (August 2000), volume *74*, number 16. Images reproduced with permission by the American Society for Microbiology.

No role for CD4+ T-cells?

From all the aforementioned compiled evidence, we conclude that prevention of acute CMV disease after HCT is primarily a function of CD8+ effector T-cells, whereas CD4+ T-cells do not show an immediate, protective effector function and also appear not to be required as helper cells for the priming, the recruitment to non-lymphoid tissue sites, and the effector function of CD8+ T-cells. This, however, should not be mistaken for an argument against a role of CD4+

T-cells, directly or indirectly, in the long-term surveillance of CMV (see Chapter II.7 for HCMV and Chapter II.12 for mCMV). As one can learn from Chapter II.16 on clinical HCT, CD4+ T-cells are important for maintaining the HCMV-specific CD8+ T-cell response (Fuji *et al.*, 2011) and lack of CD4+ T-cells is related to late HCMV disease and poor outcome (Boeckh *et al.*, 2003). The need of CD4+ T-cells for an efficient antibody response is undisputed (Chapter II.10).

The murine model meets the criteria

In summary, the murine model of CMV infection in an HCT setting matches the cornerstones of clinical CMV disease after HCT: (i) HCT, in absence of infection, leads to full lympho-haematopoietic reconstitution with a 100% survival rate; (ii) CMV infection is controlled by CD8+ T-cells and allows a 100% survival rate in the fully reconstituted HCT recipient but leads to disease with rates of morbidity and mortality that are correlated with decreasing efficacy of immune reconstitution; and (iii) model CMV disease mimics clinical CMV disease in its wide range of organ manifestations based on virus tropism for many cell types. In particular, the lung is a target of infection leading to interstitial pneumonia, a main manifestation of CMV disease in clinical HCT.

Compendium of the CD8+ T-cell response to CMV in the BALB/c mouse model

Prompted by the promising results discussed above, the identification of MHC-I restricted mCMV epitopes, i.e. antigenic peptides presented to cognate CD8+ T-cells by MHC-I molecules, became a focus of research aimed at developing preclinical models of specific CD8+ T-cell immunotherapy of CMV disease in HCT recipients, as well as model vaccines for priming and expanding antiviral CD8+ T-cells with the idea to accelerate and intensify virus control. As work on the CD8+ T-cell response to CMV in murine models has been reviewed and updated over the years (Reddehase, 2000, 2002; Pinto and Hill, 2005; Holtappels *et al.*, 2006b, 2008a; Doom and Hill, 2008; Lemmermann *et al.*, 2010b; 2011a,b) and is also covered under a wide range of aspects in other chapters of this book (Chapters I.22, II.11, and II.12) we refrain from comprehensive referencing. Instead, we focus here on some aspects that we consider instructive for approaches to CD8+ T-cell-based immunotherapy.

Antigen specificity profiling

Identification of antigenic peptides

The first antigenic peptide to be identified for a CMV was the mCMV IE1 protein-derived peptide *YPHFMPTNL* presented by the MHC-I molecule L^d (Reddehase *et al.*, 1989). The way to it was a mixture of 'intuition', 'forward immunology' and, in the final step, a kind of 'reverse immunology' using a prediction algorithm that was based on amphipathic α-helix motifs. The algorithm eventually did not survive the test of time but served its purpose for discovering the IE1 epitope. In the pre-Townsend era of immunology (Townsend *et al.*, 1986), when antigenic peptides were unknown and when viral antigens recognized by T-cells were thought to be glycoproteins localizing to the cell membrane of the infected cell, it was found that cells arrested by metabolic inhibitors in the IE phase of the lytic viral cycle to selectively express the intranuclear, transactivatory IE proteins (see Chapter I.10), can be recognized, contrary to the then prevailing doctrine, by cytolytic CD8+ T-lymphocytes, CTLs (Reddehase and Koszinowski, 1984; Reddehase *et al.*, 1984). Expression of the ORF of the most abundant IE protein (IE1) in vaccinia virus, followed by nested intervals of sequence truncations, narrowed down the antigenic sequence to 96 aa, and screening of N-terminally and C-terminally shortened length variants of a 19-aa peptide containing a predicted amphipathic 'Rothbard' motif eventually ended up in the 9-aa IE1 peptide (reviewed in Reddehase, 2002). It should take another 11 years until the second antigenic peptide of mCMV was identified in the E phase protein m04/gp34 (Holtappels *et al.*, 2000b).

Today, epitope identification by serendipity is replaced with more systematic approaches using genome-wide libraries of overlapping synthetic peptides (Sylwester *et al.*, 2005) or ORF libraries of transfection plasmids (Fig. II.17.6) (Munks *et al.*, 2006b) combined with 'reverse immunology' prediction of antigenic sequences with bioinformatic algorithms based on MHC-I anchor residues of peptides and proteasomal cleavage sites (Rammensee *et al.*, 1997; Tenzer *et al.*, 2005). Once an annotated ORF is identified to code for an antigenic peptide that is presented by the transfected cells and stimulates IFN-γ expression in CMV-experienced (acutely primed or memory) cells expressing the cognate T-cell receptor (TCR) in a polyclonal population of CD8+ T-cells (Fig. II.17.6), it is essential first to determine the presenting MHC-I molecule to narrow down the number of candidate peptides predicted by the algorithms.

It must be noted, however, that the ORF library approach, though undoubtedly representing a great

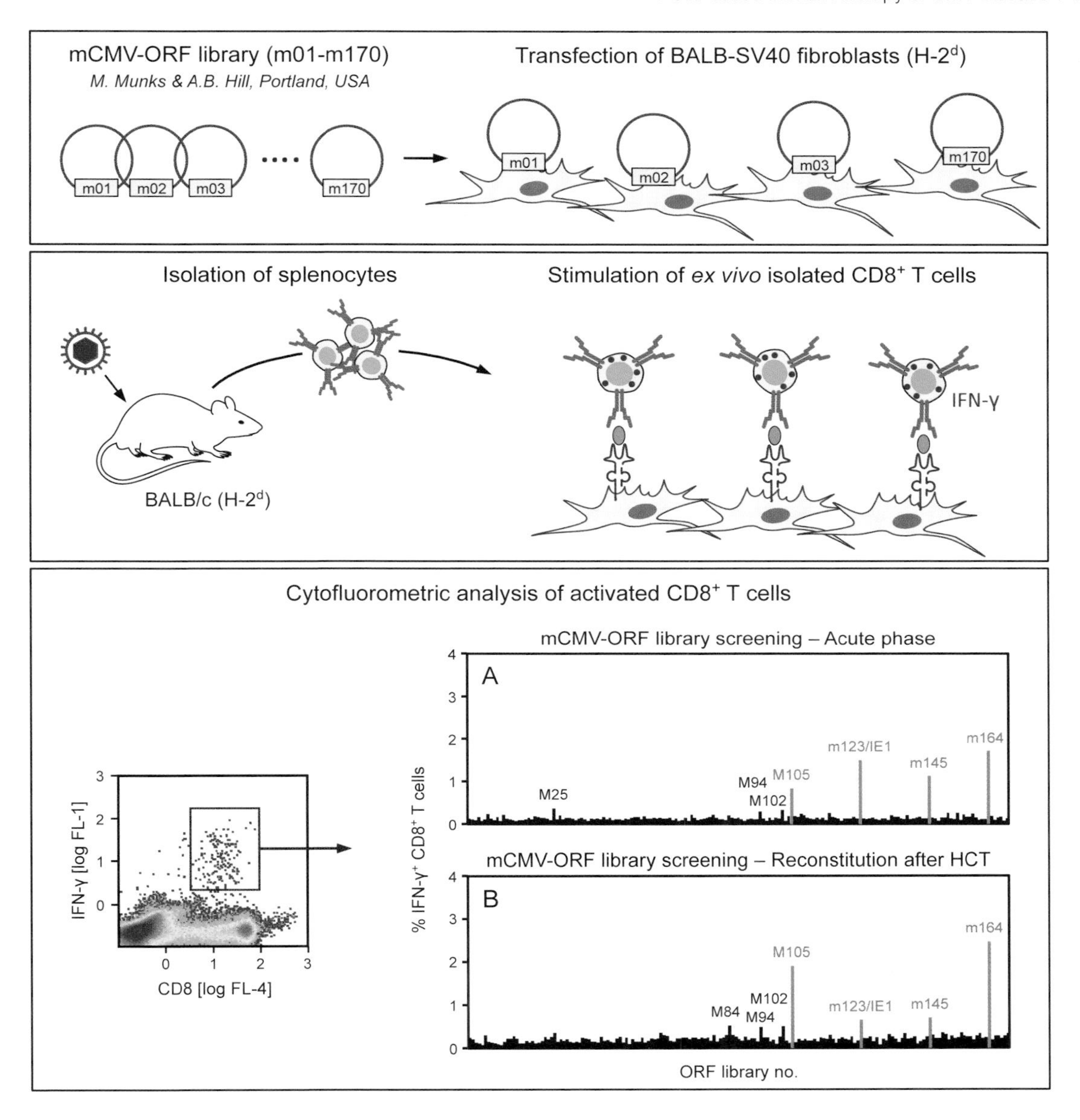

Figure II.17.6 ORF-library of transfection plasmids for viral genome-wide antigenicity screening. Transfection plasmids serve to express antigenic sequences from annotated ORFs in cells used for the stimulation of lymphocyte populations. Responding CD8+ T-cells, indicating that the corresponding ORF encodes antigenic peptide(s), are identified by cytofluorometric detection of intracellular IFN-γ. FL, fluorescence channel and log intensity. Dots, representing cells, are displayed as colour-coded density plots (with red and blue representing highest and lowest cell numbers, respectively). (A) Library screening for spleen cells on day 7 of the acute immune response after intraplantar infection of immunocompetent BALB/c mice. (B) Library screening for tissue-infiltrating, interstitial lymphocytes isolated during the acute response from the lungs at 4 weeks after syngeneic HCT and infection. Response magnitudes to ORFs for which antigenic peptides are already known in their amino acid sequence (see Table II.17.1) are shown as green bars.

advance, may miss antigenic peptides for which the frequency of cognate CD8+ T-cells in the specificity repertoire is low. Gaps can also be caused by insufficient transfection efficacy and by the fact that the library is based on the original annotation (Rawlinson *et al.*, 1996) that does not take into account all possible splice products and more recent identification of new ORFs as well as dismissing of discredited

ORFs (Chapter I.2). This, as well as the possibility that a transfection plasmid can code for more than one antigenic peptide, is also the reason why the frequencies of responding, IFN-γ expressing cells can be biased. Once the antigenic peptides are identified, the actual frequencies of epitope-specific CD8+ T-cells are more reliably revealed by TCR-specific MHC-I-peptide multimers or by stimulation of CD8+ T-cells with an optimized concentration of synthetic peptide of interest followed by quantitating IFN-γ (or other cytokine)-secreting cells in an enzyme-linked immunospot (ELISpot) assay, in which each spot on a filter represents a single epitope-specific CD8+ T-cell (for an instructive, more recent example, see Böhm *et al.*, 2008b).

Table II.17.1 lists all currently known antigenic peptides of mCMV presented by MHC-I molecules of mouse haplotype *H-2^d^* (as for the BALB/c strain). Antigenic mCMV peptides presented by MHC-I molecules of mouse haplotype *H-2^b^* (as for the C57BL/6 strain) have been reported from A.B. Hill's group (OHSU, Portland, Oregon, USA) by Munks and colleagues (Munks *et al.*, 2006b). It should be noted that some of the peptides identified for the BALB/c model, namely those classified as subdominant with respect to the CD8+ T-cell frequencies elicited by them, are not constantly or, in some cases, not at all revealed by the ORF library. On the other hand, the ORF library also predicts the existence of currently still unidentified antigenic peptides, for instance peptides from ORFs M25, M94, and M102.

Use of the ORF library for response monitoring

Notwithstanding the discussed limitations of the ORF library, monitoring response profiles with the ORF library are instrumental to qualitative comparisons of CD8+ T-cell responses (i) in different mouse strains; (ii) at different stages of infection; (iii) to different virus variants; (iv) in different experimental settings which model clinical settings, (v) at different organ sites (e.g. lymphoid versus non-lymphoid); (vi) after infection at different age (e.g. adult versus neonatal); and, finally, (vii) under any other conditions of interest.

Table II.17.1 Antigenic peptides of mCMV in haplotype *H-2^d^*

Annotated ORF	Peptide sequence	Presenting MHC-I	Acute response dominance	Memory inflation	Processing efficiency*	Immunotherapy
m04	YGPSLYRRF(F)	D^d^	Subdominant	No	**	Protective[a]
m18	SGPSRGRII	D^d^	Subdominant/intermediate	No	700	Protective[b]
M45	VGPALGRGL	D^d^	Subdominant	No	3300	Protective[c]
M83	YPSKEPFNF	L^d^	Subdominant	No	100	Protective[e]
M84	AYAGLFTPL	K^d^	Subdominant	No	**	Protective[e]
M105	TYWPVVSDI	K^d^	Dominant	Variable	400	Partly protective[f]
m123	YPHFMPTNL	L^d^	Dominant	Yes	12,000	Protective[a,g,h]
m145	CYYASRTKL	K^d^	Dominant	Variable	6000	Protective[f]
m164	AGPPRYSRI	D^d^	Dominant	Yes	13,000	Protective[g,h]

*Data show numbers of peptide molecules per infected cell in the L phase (fibroblasts/MEFs expressing the constitutive proteasome), which actually reflects the number of pMHC-I complexes per cell, as peptides that do not bind to MHC-I molecules are rapidly degraded. Thus, 'processing' is here meant as the sum of processes that determine the steady-state amount of pMHC-I complexes, which include proteasomal processing of the antigenic protein to precursor peptides, transport of those into the ER, their N-terminal trimming, loading of the finalized antigenic peptide onto MHC-I molecules (Chapter I.22; Wearsch and Cresswell, 2008), turnover, and immune evasion protein functions such as pMHC-I retention by m152 and degradation by m06 (Reddehase, 2002). All listed peptides are generated by proteasomal processing as revealed with proteasome inhibitors (R.H., unpublished information).
**Not determined.
[a]Holtappels *et al.* (2000b).
[b]Holtappels *et al.* (2006b).
[c]Holtappels *et al.* (2009).
[e]Holtappels *et al.* (2001).
[f]Ebert *et al.* (2012).
[g]Holtappels *et al.* (2002).
[h]Holtappels *et al.* (2008b).

Application examples already exist for options (i–v):

i Comparing the response profiles between BALB/c mice and the L^d gene deletion mutant BALB/c-H-2^{dm2} verified absence of response to ORF m123/IE1 in the mutant mouse strain but did not reveal global changes (Seckert *et al.*, 2011).

ii Comparing the acute response and the memory response in the spleen confirmed the focusing of memory on two antigenic ORFs, m123/IE1 and m164, known to code for antigenic peptides that drive 'memory inflation' (Holtappels *et al.*, 2002; Seckert *et al.*, 2011) (see Chapter I.22). Before this result, the existence of further ORFs coding for 'memory inflation'-inducing epitopes had to be considered.

iii Deletion of immunodominant epitopes IE1 and m164 by respective codon mutations in a recombinant virus was found to result in an increased memory response to the ORF m145 epitope but with no global alterations (Holtappels *et al.*, 2008c) (see below), whereas deletion of genes coding for immune evasion proteins of the MHC-I antigen presentation pathway (see below) did not qualitatively alter the acute response specificity profile (Lemmermann *et al.*, 2011a, 2012).

iv/v The specificity repertoire constituting the acute day 7 response to infection in the spleen of immunocompetent BALB/c mice (Fig. II.17.6A) did not substantially differ from the composition of pulmonary CD8$^+$ T-cell-infiltrates at 4 weeks after syngeneic HCT with BALB/c mice as donors and recipients (Fig. II.17.6B). Although a closer look suggests some fluctuation in epitope-specific response magnitudes that alters proportions, for instance a particularly high response to the M105 epitope in lung infiltrates after HCT (Fig. II.17.6B), we need more experience on variance before interpretation is warranted.

The overall impression from ORF library response profiles is that the specificity armoury recruitable to fight the virus is remarkably constant and limited, suggesting that it is genetically fixed at either the TCR repertoire and affinity level or at the MHC-I haplotype and peptide presentation level or all combined. It is worth recalling that the CD8$^+$ T-cell response to HCMV in an individual is also directed against a limited set of epitopes, namely to a median value of 8 with a range between 1 and 32 – not to be mixed up with a 'broad response' on the population level covering HLA polymorphism (Sylwester *et al.*, 2005) (see also Chapter II.7).

The primary response in draining regional lymph nodes

The host's response to a local CMV exposure

After local infection, the adaptive immune response is initiated in the draining regional lymph node (RLN), which is the popliteal lymph node (PLN) in the specific case of intraplantar (footpad) infection (Fig. II.17.7). Local virus replication at the entry site does not appear to contribute much to the initiation of T-cell priming after infection with free virions; rather, virus reaches the RLN via the lymphatics, probably within minutes, and replicates in professional antigen-presenting cells (profAPCs), including macrophages and dendritic cells (DCs), that localize to the RLN cortex in a demarcated zone beneath the subcapsular sinus (SCS), the peri-SCS. Notably, though originally described in elegant intravital microscopy studies (Hickman *et al.*, 2008) for taxonomically and biologically distant viruses unrelated to CMV, namely a DNA virus, vaccinia virus (VV), and a negative-strand RNA virus, vesicular stomatitis virus (VSV), mCMV-infected gB-expressing cells in the PLN conspicuously also sharply localize to the peri-SCS. Infected cells appear to include DCs with gB-stained membrane protrusions and an intranuclear inclusion body that is characteristic of a late stage in the viral replication cycle (Fig. II.17.7A) (Böhm *et al.*, 2008b). Infection can be visualized only in PLNs of an immunocompromised host in which virus gene expression levels (day 3–7) are high enough for IHC detection of gB. In RLN of an immunocompetent host (Fig. II.17.7B, b1), 100-fold lower levels are reached by day 2 and decline thereafter due to innate and early adaptive immune response (Böhm *et al.*, 2008b). As a consequence, infection does not become visible by IHC at any time, but IE1 gene expression can be quantitated by RT-qPCR at the entry site, the footpad, and in the PLN (Fig. II.17.7C).

Cytofluorometric analysis of the CD8$^+$ T-cells in a day-7 PLN (Fig. II.17.7B, b3; b4, CD44 not shown) reveals five major populations, CD62L^{high}KLRG1^{low} cells, comprising CD44^{low} naïve cells (~20% of all CD8$^+$ T-cells) and CD44^{high} central memory T-cells (T_{CM}; accounting for ~70% of all CD8$^+$ T-cells), CD44^{high}CD62L^{low}KLRG1^{low} (~7% of all CD8$^+$ T-cells; comprising early effector cells, EEC, and effector-memory T-cells, T_{EM}; which are not distinguished here), and CD44^{high}CD62L^{low}KLRG1^{high} short-lived effector cells, SLECs (~3% of all CD8$^+$ T-cells) (for differentiation and activation stages of CD8$^+$ T-cells, see the reviews by Lefrançois and Obar, 2010; Obar and Lefrançois, 2010a,b). At first glance

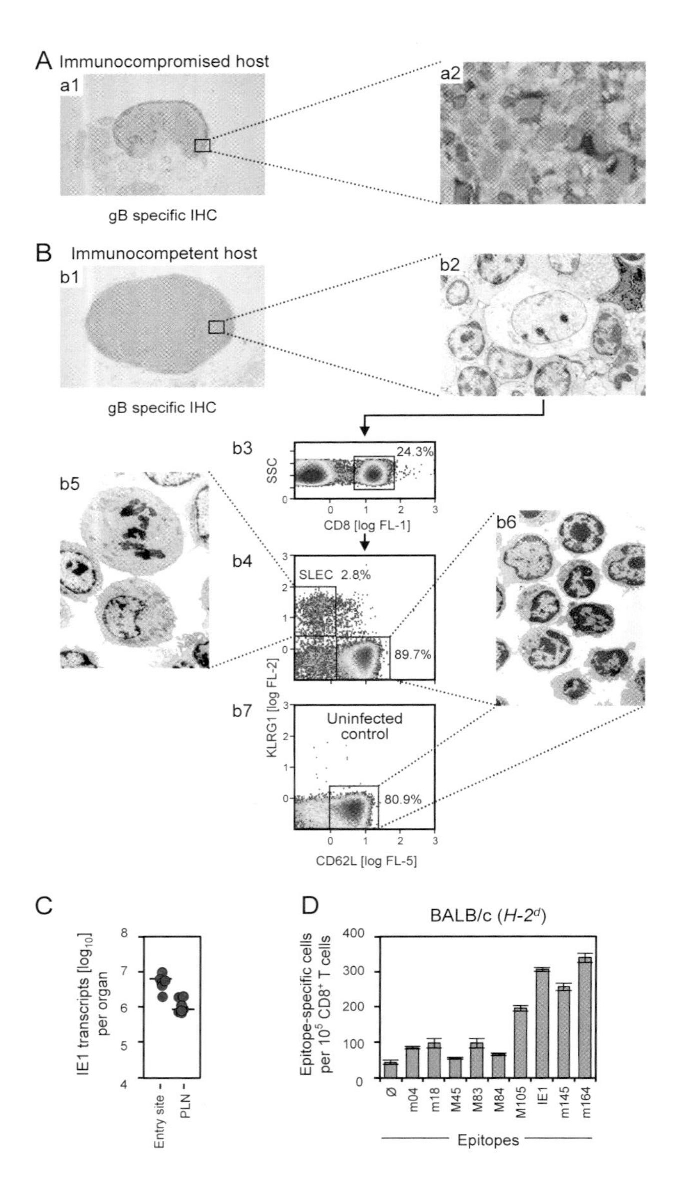

Figure II.17.7 A look into the PLN after acute intraplantar infection. (A) Infection localizing to a highly demarcated zone underneath the subcapsular sinus in the cortex of the PLN of an immunocompromised BALB/c mouse on day 4 after haematoablative treatment. Infected cells are identified by IHC specific for M55/gB (red staining). (a1) Whole organ section overview. (a2) Indicated area resolved to greater detail. (B) PLN of an immunocompetent BALB/c mouse during immunological 'priming' after intraplantar infection. (b1) Whole organ section overview on day 2. (b2) TEM image of PLN lymphocytes *in situ*. (b3, b4) Cytofluorometric phenotyping of PLN lymphocytes on day 7 after intraplantar infection. (b7) Cytofluorometric phenotyping of lymphocytes from pooled axillary and mandibular lymph nodes of age-matched uninfected BALB/c mice. SSC, side scatter in the lymphocyte gate. FL, fluorescence channel and log intensity. Dots, representing cells, are displayed as colour-coded density plots (with red and blue representing highest and lowest cell numbers, respectively). Percentages of main interest are indicated for gated areas. SLEC, short-lived effector cells. The $CD62L^{high}$ $KLRG1^{low}$ population is comprised of $CD44^{low}$ naïve cells and $CD44^{high}$ T_{CM}. (b5, b6) TEM images showing lymphoblasts (one in mitosis) and small lymphocytes, respectively. (C) Quantitation of *IE1* transcripts by RT-qPCR at the entry site, the footpad, and in the corresponding PLN on day 2. (D) Quantitation of epitope-specific $CD8^{+}$ T-cells by ELISpot assay in the draining PLN on day 7 after intraplantar infection. For the sequences of the indicated antigenic peptides, see Table II.17.1. Images a1, a2, and b1 are reproduced from the Journal of Virology (2008) *82*, 11641 (Böhm *et al.*, 2008b); with permission by the American Society for Microbiology. Images b2, b5, and b6 are from M.J. Reddehase, 1984, PhD thesis; courtesy of F. Weiland (then at the Federal Research Center for Virus Diseases of Animals, Tuebingen, Germany).

one may be surprised to find the majority of $CD8^+$ T-cells in the T_{CM} population at the maximum of the antiviral immune response in a draining RLN, but a very similar subset composition of $CD8^+$ T-cells from a pool of axillary and mandibular LNs of uninfected BALB/c mice (Fig. II.17.7B, b7) made it evident that most T_{CM} in RLNs represent memory of preceding antigen encounters unrelated to CMV. The only obvious difference between infected and control RLNs is the SLEC population found exclusively in the PLN that drains the entry site of infection. Morphologically, these cells are lymphoblasts, whereas the naïve cells and T_{CM} are small lymphocytes (Fig. II.17.7B, b5, b6). Whilst it is apparent that CMV epitope-specific cells are in the SLEC population, it is an open question if all arise from the priming of naïve, recent thymic emigrants or if, possibly, restimulation of cross-reactive T_{CM} contributes.

As has been shown by Böhm and colleagues, a viral epitope-specific $CD8^+$ T-cell response in the PLN can be elicited with an infectious dose as low as 10 PFU, and primed 'novice' $CD8^+$ T-cells (likely early SLECs), being already capable of secreting IFN-γ and participating in the control of infection, are detectable from day 3 onward (Böhm *et al.*, 2008b).

Epitope hierarchy in the acute response magnitude does not correspond to antigen processing in infected cells

Screening for CMV epitope-specificities with an ELISpot assay (Fig. II.17.7D) by stimulating $CD8^+$ T-cells with the respective synthetic antigenic peptides listed in Table II.17.1 revealed acute response frequency hierarchies in draining PLNs (measured on day 7 after intraplantar virus exposure) on which the classification of epitopes as 'dominant' or 'subdominant' is based (Table II.17.1, column 4; m18 is classified as being subdominant or of intermediate immunodominance, because it occasionally induces a higher response). The immunodominance classification by acute response magnitude apparently does not correlate with the number of peptides, equalling peptide-loaded MHC-I (pMHC-I) complexes, generated by the processing of the respective antigenic proteins in infected cells (Table II.17.1, columns 4 and 6). This is particularly obvious for the M45 epitope, for which pMHC-I complexes are efficiently generated (Holtappels *et al.*, 2009) and which nevertheless induces a $CD8^+$ T-cell response close to the limit of detection, whereas m145 and M105 are immunodominant, although processing in infected cells generates these peptides in similar and even lower amounts, respectively (Table II.17.1, column 6). As will be discussed below, current evidence suggests that priming is not predominantly accomplished by epitope presentation on infected cells.

The shaping of memory

In BALB/c mice, the two most dominant specificities of the acute response, m123/IE1 and m164, also dominate the memory response (Holtappels *et al.*, 2002; Seckert *et al.*, 2011) that is shaped by 'memory inflation' (Table II.17.1, column 5) (see Chapter I.22). Such a congruity between acute response and memory, however, is not the rule. As revealed by examples in the C57BL/6 model, subdominant acute responses to certain epitopes can become dominant in the memory phase, and dominant acute responses to certain other epitopes vanish in the memory phase (Munks *et al.*, 2006a).

It is becoming increasingly clear that CMV-specific $CD8^+$ T-cell priming is accomplished primarily by uninfected DCs that take up and process viral proteins for presentation, a priming mode known as 'cross-presentation' (Gold *et al.*, 2002; Böhm *et al.*, 2008b; Snyder *et al.*, 2010; Torti *et al.*, 2011b; and references therein). In contrast, shaping of memory by 'memory inflation' is brought about primarily by 'direct antigen presentation' on the surface of infected host tissue cells, which are mostly non-haematopoietic cell types (constitutively expressing the 'constitutive proteasome') distinct from haematopoietic profAPCs (constitutively expressing the immunoproteasome) (Hutchinson *et al.*, 2011; Seckert *et al.*, 2011; Torti *et al.*, 2011a; and references therein). Accordingly, 'memory inflation', rather than priming in the acute response, might correlate with the efficacy of antigen processing in an infected cell of a non-haematopoietic cell type that expresses the constitutive proteasome (Table II.17.1, columns 5 and 6). It is plausible to predict that generation of high numbers of pMHC-I complexes increases the chance for presentation at the cell surface where recognition by TCRs of cognate $CD8^+$ T-cells takes place. Whilst, in accordance with the prediction, 'memory inflation'-inducing epitopes m123/IE1 and m164 are indeed generated in high numbers in productively infected fibroblasts/MEFs, many other conditions need to be fulfilled for an epitope to induce 'memory inflation' (for a detailed discussion, see Chapter I.22). We currently believe that the driver of 'memory inflation', which determines the specificity repertoire of the memory $CD8^+$ T-cell population, is a stochastic transcriptional activity during viral latency and/or viral gene expression resulting from uncompleted reactivation (see Chapter I.22).

Experimental modelling of cytoimmunotherapy

Retrospect

In retrospect, lymphocyte transfer experiments in the mouse model revealing a protective antiviral function of T-cells (Starr and Allison, 1977), specifically of H-2-restricted cytotoxic T-cells (Ho, 1980), can be viewed as the first experimental approaches to a T-cell therapy of CMV disease. The idea of CMV-specific CD8$^+$ T-cell therapy in model HCT recipients after haematoablative conditioning was largely promoted by demonstrating that specifically primed CD8$^+$CD4$^-$ (at that time Lyt-2$^+$L3T4$^-$) T-lymphocytes protect against interstitial pneumonia and other organ manifestations of lethal CMV disease not only when AT was performed at the same time as experimental infection was initiated (then named 'prophylactic AT', which we would today call a 'pre-emptive immunotherapy') but also when AT was performed with delay at a time when the infection was already established in the end-organs of CMV disease (then named 'therapeutic AT', which we would today call an 'immunotherapy') (Reddehase *et al.*, 1985; 1987a) (reviewed in Reddehase, 2002; Holtappels *et al.*, 2008a). These experiments were actually performed with acutely primed CD8$^+$ T-cells derived from draining PLNs, of which we know today that they contained polyclonal and broadly specific CD44^{high}CD62L^{low}KLRG1^{high} SLECs (Fig. II.17.7B) with an mCMV epitope specificity distribution most likely very similar to that shown in Fig. II.17.7D. Soon afterwards, protection was also demonstrated with an IL-2 driven, polyclonal, short-term line of cytolytic T-lymphocytes (CTLL) generated from primed draining PLN lymphocytes (Reddehase *et al.*, 1987b) as well as with CD8$^+$ memory T-cells from the spleen of latently infected mice (Reddehase *et al.*, 1988), of which we know in retrospect that they consisted of T_{CM} and T_{EM} ($T_{CM} > T_{EM}$), with predominance of cells specific for the epitopes m123/IE1 and m164 (Holtappels *et al.*, 2002). Importantly, this work (Reddehase *et al.*, 1988) showed that co-transfer of CD4$^+$ memory T-cells did not improve the protective effect of CD8$^+$ T-cells against acute infection, and this applied even to an experimental co-transfer trial with a 4:1 excess of the potential helper cells. As discussed earlier in this chapter, though CD4$^+$ T-cells are apparently not contributing much to fighting acute infection, these cells support long-term virus control by maintaining the protective CD8$^+$ memory T-cell pool.

Selected special lectures held by murine cytomegalovirus on current topics relevant to cytoimmunotherapy of clinical CMV disease

Lecture 1: viral epitope specificity of protection

For a long time, antigen specificity of the protection by CMV-primed CD8$^+$ T-cells was concluded, reasonably, from the finding that unprimed CD8$^+$ T-cells present in control RLNs of uninfected mice were not protective when transferred in comparable numbers (Reddehase *et al.*, 1985), though transfer of high numbers gives protection, probably by reconstituting the recipients with a sufficient number of viral epitope-specific naïve cells or cross-reactive T_{CM} primed or restimulated, respectively, by antigens expressed in the AT recipient (Holtappels *et al.*, 2008c). However, unprimed control RLNs do not contain any SLECs (Fig. II.17.7, b7), and so one might have argued that highly activated CD62L^{low}KLRG1^{high} cells, as well as CTLL known to express the activatory NK cell receptor NKG2D as a costimulatory receptor, might exert an antiviral effector function by interacting with ligands on infected cells unrelated to the interaction between pMHC–I complexes and a specific TCR. This issue was settled by showing that AT of IE1-epitope specific *ex vivo* sort-purified CD8$^+$ memory T-cells (T_{CM} and T_{EM}) (Fig. II.17.8A), as well as cells of a polyclonal but IE1-monospecific CTLL, protected recipients against infection with viruses expressing the IE1 epitope (WT and IE1.A176L revertant), but failed to protect against a virus mutant in which the IE1 epitope was annulled by a point mutation L176A of the C-terminal residue that is critical for both, proteasomal cleavage and, as an anchor, for binding to the presenting MHC-I molecule, which is L^d in the specific example (Fig. II.17.8B) (Böhm *et al.*, 2008a). Interestingly, missing presentation of the IE1 peptide in tissues infected with the L176A mutant also abolished the recruitment of IE1-specific CD8$^+$ T-cells and thus the formation of focal infiltrates that prevent spread of the infection (Böhm *et al.*, 2008a).

Lecture 2: loss of antiviral potency in CTLL

As CMV-specific cells are present in the donor lymphocyte population of healthy (latently infected) CMV carriers in only low absolute numbers, early clinical trials of CMV-specific CD8$^+$ T-cell pre-emptive therapy used amplification by expansion to CTLL in cell culture to reach therapeutically efficient cell numbers (see Chapters II.7 and II.16). This is a logistic problem that has prevented CMV cytoimmunotherapy from becoming clinical routine. The murine model was

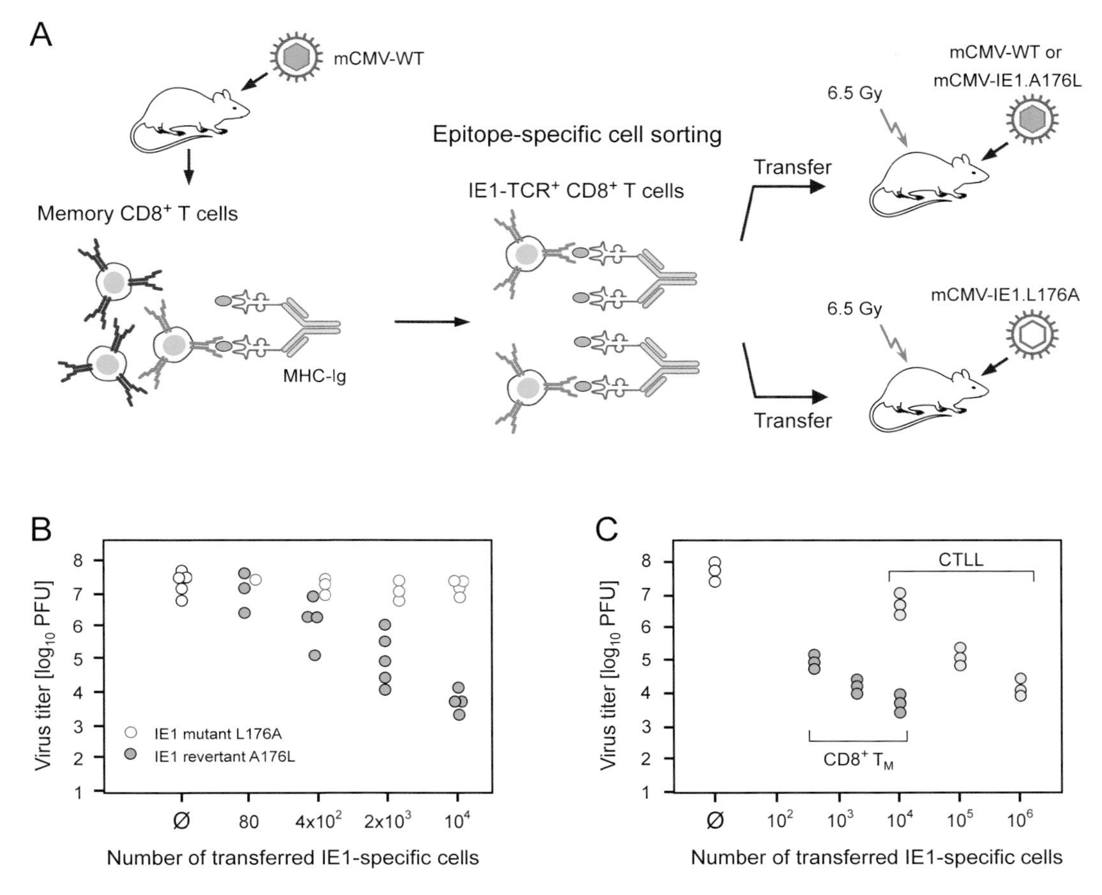

Figure II.17.8 IE1 epitope-specific pre-emptive cytoimmunotherapy of CMV infection in the murine model. (A) Process flow diagram for the cytofluorometric sort-purification of memory $CD8^+$ T-cells expressing IE1-TCRs specific for the pMHC–I complex IE1-L^d, identified by using fluorochrome-conjugated pMHC-I multimers (originally, IE1 peptide-loaded MHC-I [L^d]-Ig dimers). The sorted cells were then transferred into immunocompromised BALB/c mice infected with virus expressing the IE1 epitope (mCMV-WT.BAC or revertant virus mCMV-IE1.A176L) or with virus not expressing the IE1 epitope because of a codon mutation replacing leucine with alanine in position 9 of the IE1 peptide (mCMV-IE1.L176A) (Simon *et al.*, 2006; Böhm *et al.*, 2008a, Lemmermann *et al.*, 2011b). Virus titres were determined in the spleen on day 12 after cell transfer. (B) Epitope specificity of antiviral activity. (C) Different antiviral efficacies of IE1 epitope-specific *ex vivo* sorted $CD8^+$ memory T-cells ($CD8^+$-T_M) and an IE1 epitope-specific short-term CTLL. Based on data from Pahl-Seibert and colleagues (Pahl-Seibert *et al.*, 2005; and M.F. Pahl-Seibert, 2003, PhD thesis). Reproduced in modified form from this book's first edition (Holtappels *et al.*, 2006b), where one can find also immunohistological images that illustrate the control of virus spread in adrenal (suprarenal) gland and liver.

vanguard in documenting that the advantage hoped for by the expansion in cell numbers is almost annulled by reduction in functional efficacy. Specifically, though care was taken to limit cell culture of memory $CD8^+$ T-cells to the minimum of three rounds of stimulation with antigenic IE1 peptide required to reach mono-specificity while retaining polyclonality (as evidenced by broad TCR Vβ usage) and expression of the CD8 co-receptor for increasing avidity, the resulting short-term IE1-CTLL proved to be ~ 100-fold less efficient in controlling CMV infection after AT compared with *ex vivo* pMHC-multimer sort-purified and directly transferred IE1-specific $CD8^+$ memory T-cells (Pahl-Seibert *et al.*, 2005) (Fig. II.17.8C). At about the same time, clinical research has confirmed a high antiviral potency of *ex vivo* isolated CMV-specific memory $CD8^+$ T-cells in AT recipients (Cobbold *et al.*, 2005), but for understandable reasons, a direct efficacy comparison between a short-term CTLL and *ex vivo* sorted memory cells specific for the very same viral epitope remained a unique contribution made by the mouse model.

The difference in antiviral efficacy is clearly not explainable by functional avidity of target cell recognition, as the functional avidity of IE1 pMHC-I complex recognition by the CTLL was a $\log_{10}$-step higher (a proportion of the CTLL cells recognized target cells exogenously loaded with 10^{-12} M IE1 peptide) than by the IE1-specific *ex vivo* memory cells among which there were none that recognized such low concentration (Pahl-Seibert *et al.*, 2005). Whilst the reason for the difference was not explored further by Pahl-Seibert and colleagues, with the advanced immunological knowledge of today it is reasonable to postulate that the known longevity and high proliferative potential of T_{CM}, generating new SLECs exponentially with every antigen encounter, favour antiviral efficacy as compared with cells administered to the AT recipient already in the stage of terminally differentiated SLECs (Weninger *et al.*, 2002; Welsh *et al.*, 2004; Boyman *et al.*, 2009; Lefrançois and Obar, 2010; Obar and Lefrançois, 2010a,b).

Lecture 3: protection by immunotherapy does not care about immunodominance

The identification of viral epitopes and of their hierarchy in terms of magnitude of the CD8+ T-cell response that they elicit in the host during natural (for HCMV) or experimental (in animal models) infection, i.e. their inherent 'immunodominance', was long spurred – and it is often still so – by the hope that quantity would indicate functional relevance in controlling infection. Accordingly, immunodominant epitopes are being considered as promising candidates for use in cytoimmunotherapy and in vaccines. In retrospect, one may wonder why it was rarely discussed that high numbers might result from many rounds of proliferation induced by frequent restimulation with persistent antigen – and thus reflect a limited per-cell capacity to clear the antigen source. Moreover, high amounts of antigen may favour the induction and expansion of CD8+ T-cells with low TCR affinity and low overall functional avidity.

The murine model of AT allowed an assessment of a possible correlation between 'immunodominance' of an epitope in the acute or in the memory response (Table II.17.1, columns 4 and 5) and the protective capacity of a CTLL specific for that epitope (column 7). The data gave a clear message: CTLL specific for 'subdominant' epitopes, including epitopes that elicit acute responses with CD8+ T-cell frequencies close to the detection limit, e.g. m04, M45 and M84 (Fig. II.17.7D), proved to be at least as efficient in clearing acute infection upon AT than were CTLL specific for 'immunodominant' epitopes such as m123/IE1 and m164 (see Table II.17.1 for references). So, the data revealed the usefulness of 'subdominant' epitopes, but a speculation that 'immunodominant' epitopes, in general, were associated with low protective capacity was also refuted by the model; – the conclusion rather is that the immunodominance status of an epitope does not matter at all. It seemed reasonable to assume that a high efficacy of antigen processing, resulting in a high number of pMHC-I complexes within the cell, would also increase the chance of cell surface display for the recognition by protective CD8+ effector cells. The data on processing (Table II.17.1, column 6), however, did not allow a direct prediction of protection, as even 100-fold differences in the amount of generated pMHC-I complexes (e.g. between M83 on the one hand and m123/IE1 or m164 on the other hand) did not nearly entail similar differences in the protection achievable with the corresponding CTLLs. As epitope specificity of protection (see Lecture 1) implies the generation and presentation of at least some pMHC-I complexes at the cell surface, we believe in a threshold that likely differs between epitopes based on the avidity of TCR–pMHC-I interaction.

Notable cases of protection failure are epitopes M105 (Ebert *et al.*, 2012) and an epitope from ORF M45 that is presented by the MHC-I molecule D^b in C57BL/6 mice (Gold *et al.*, 2002; Holtappels *et al.*, 2004), both of which are classified as being 'immunodominant' in the acute response. Though processing of the D^b-presented M45 peptide is reported to be vanishingly low, namely just ~ 10 peptide molecules per cell (Holtappels *et al.*, 2009), this alone cannot provide a comprehensive explanation since M105 largely fails in protection despite ~400 peptide molecules per cell, whereas for M83 ~100 peptide molecules per cell suffice for protection (Table II.17.1, column 6). Besides the affinity/avidity of TCR–pMHC-I interaction trafficking of pMHC-I complexes from the loading compartment, the ER, to the cell surface is another potential variable that needs to be considered (see Lecture 5).

It may be instructive to note that pre-treatment of cells with IFN-γ can increase processing and presentation by induction of the immunoproteasome and up-regulation of MHC-I synthesis (Reddehase, 2002; Fink *et al.*, 2012), but it is our experience with M45-D^b that the failure in protection *in vivo* better corresponds to the low amount of peptide generated by the constitutive proteasome (R.H., unpublished data).

In parallel to our studies on antiviral intervention by AT, the conclusion that 'subdominant' epitopes qualify for inducing protective CD8+ T-cell responses to CMV was found also in the murine model of viral DNA vaccination by the group of D.H. Spector (referenced in Holtappels *et al.*, 2008a) (see also Chapter II.20 on vaccination against CMV infections).

Lecture 4: escape variants of CD8⁺ T-cell epitopes are no issue of serious concern in CMV immunotherapy

Although the mutation rate is low in genomes of CMVs, many virus variants have evolved and circulate in host populations. Co-infections as well as consecutive infections appear to be more frequent than was previously thought, and are increasingly recognized as being pathogenetically and medically relevant (see Chapters I.1, I.2, II.22).

We are not aware of studies on the variance of $CD8^+$ T-cell epitopes in clinical strains and isolates of HCMV, but it is evident that the latent virus variant harboured by a positive HCT/AT HLA-matched donor will unlikely be the same as the one that may reactivate in tissues of a latently infected recipient or the one transmitted accidentally to an uninfected recipient. As a consequence, the memory specificity repertoire transferred with donor T-cells might not completely match the epitope repertoire presented on infected tissue cells of the recipient. So, more precisely, an HCT/AT constellation D^+R^+ should rather be written $D^{var1}R^{var2}$, although, fortunately, the genetic variance may only rarely concern T-cell epitopes. For mCMV, however, natural sequence variation in the L^d-presented m123/IE1 epitope has been described for wild-derived virus isolates (Lyons *et al.*, 1996), and, as mCMV appears to be genetically more stable than HCMV (Chapters I.1 and I.2), there exists no rationale to deny the possibility of epitope sequence variations existing in HCMV as well. Theoretically, such an incomplete fit could limit the efficacy of $CD8^+$ T-cell-based immunotherapy or, in the worst case, even result in therapy failure.

This question was approached in the murine model by AT with BALB/c donors and recipients experimentally infected with viruses that differ in the expression of 'immunodominant epitopes' (IDEs) (Holtappels *et al.*, 2008c) (Fig. II.17.9). As said earlier in this chapter (see also Chapter I.22), the 'memory inflation'-shaped memory $CD8^+$ T-cell repertoire is focused in the BALB/c mouse strain on just two epitopes, memory IDEs, namely m123/IE1 and m164 (Table II.17.1). It may be worth noting that memory cells were deliberately chosen for AT in the mouse model, because obviously, human AT donors are always in the memory state. Using BAC mutagenesis (see Chapter I.3 and Lemmermann *et al.*, 2011b) we generated a recombinant mCMV (ΔIDE) for which both these antigenic peptides are deleted by codon mutations replacing the C-terminal peptide residues with alanines, L176A and I175A, respectively. Accordingly, virus ΔIDE-rev was generated by backmutations A176L and A175I (Holtappels *et al.*, 2008c). So, infection of the donor with the mutant virus removes both specificities from the donor memory $CD8^+$ T-cell repertoire and infection of the recipient annuls presentation of these two epitopes on infected tissue cells.

Fig. II.17.9 shows a 2 × 2 criss-cross AT performed for evaluating the outcomes of all four possible combinations (Holtappels *et al.*, 2008c). A lower control of infection in the AT recipients was seen, in particular in lungs and spleen, only when the recipients expressed IDEs for which the $CD8^+$ T-cells were missing in the donor population, suggesting that IDEs inhibit the presentation of subdominant epitopes on the target cell level. Strikingly, in cases of epitope match, presence or absence of IDEs did not matter at all. By high donor cell numbers, however, protection was secured for all tested combinations. So, despite some sophisticated difference in efficacies, the take home message is that success of AT is remarkably robust against even major antigenic differences between the virus variants harboured by donor and recipient. The findings also imply that in absence of IDEs subdominant viral epitopes assume responsibility.

In conclusion, it appears that in the genetically complex CMVs, antigenic drift or antigenicity loss mutations in certain antigenic proteins are buffered by redundancy.

Lecture 5: viral immune evasion proteins limit AT efficacy but do not prevent protection

In the US2–11 region of its genome, HCMV codes for proteins that specifically interfere with the presentation of antigenic peptides in the MHC-I pathway, a phenomenon best known under the term 'immune evasion' (see Chapter II.7). Apparently, however, inhibition of pMHC-I presentation is not complete and the expression of 'immune evasion' genes must be regulated in a highly balanced way, as otherwise $CD8^+$ T-cells would not be involved at all in the control of infection. Specifically, if an 'immune evasion' applies in the literal sense of the term, attempts to develop a $CD8^+$ T-cell-based immunotherapy of CMV disease would have no prospects of success.

The murine model has been pivotal in the discovery of MHC-I pathway immune evasion of CMVs (for reviews, see Reddehase, 2002; Powers *et al.*, 2008; Hansen and Bouvier, 2009), and it is instrumental in defining the role these viral immune counter measures may play *in vivo*. Rather than reviewing the field

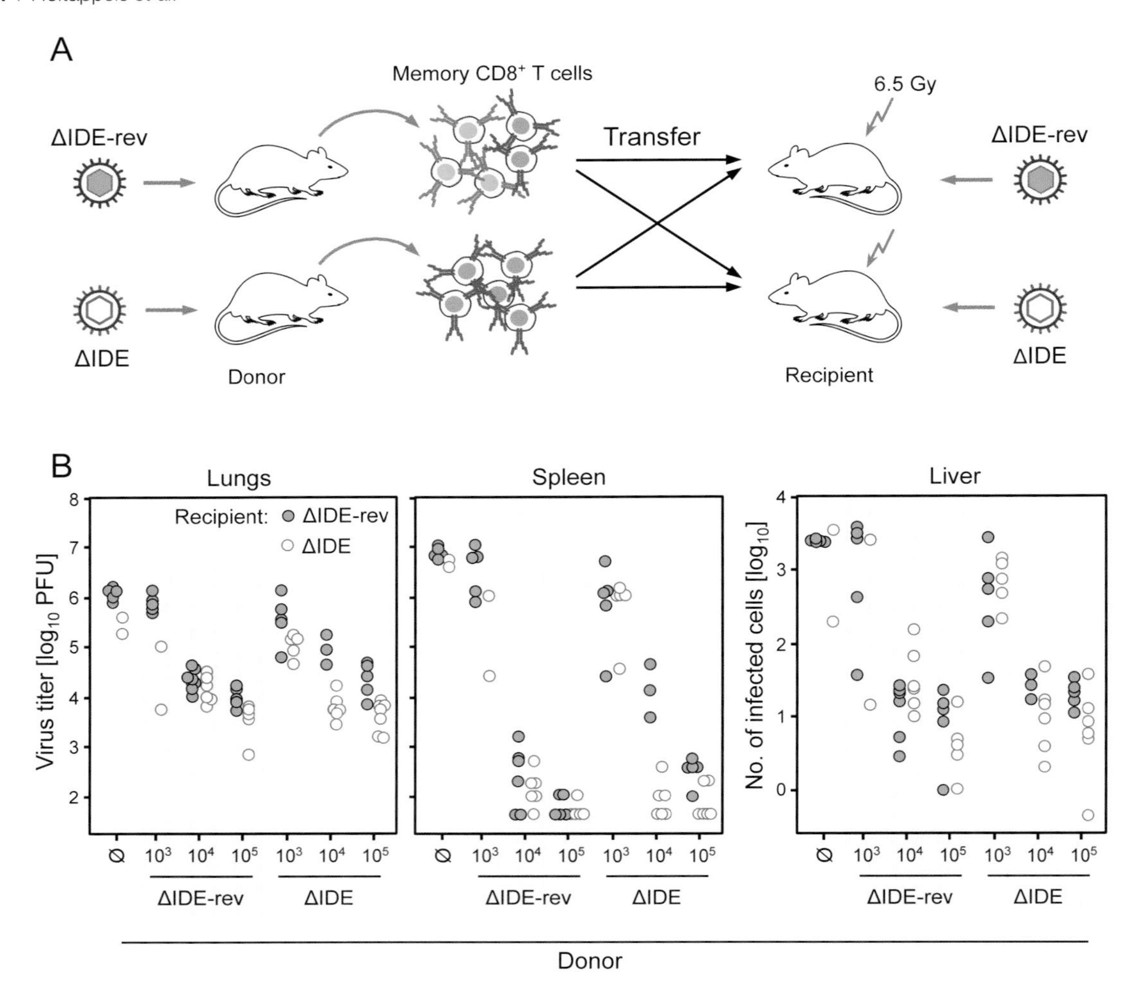

Figure II.17.9 Immunotherapy with polyclonal virus-primed memory CD8+ T-cells is effective even when donor and recipient virus variants do not express immunodominant epitopes (IDEs). (A) Scheme of a 2 × 2 criss-cross AT approach, with donors and recipients infected with virus either expressing IDEs (in the specific case, IE1 and m164) or not expressing IDEs, i.e. viruses mCMV-ΔIDE-rev and mCMV-ΔIDE, respectively. (B) Results from all four possible AT combinations, based on data from Holtappels and colleagues (Holtappels *et al.*, 2008c).

of mCMV immune evasion, which has been done extensively in the past and also more recently (Doom and Hill, 2008; Lemmermann *et al.*, 2011a; 2012), we will here focus on the question to what extent the efficacy of AT is affected by inhibition of pMHC-I presentation.

Just briefly, as background knowledge, mCMV encodes two proteins that undisputedly function as negative 'viral regulators of direct antigen presentation' (here referred to as vRAPs) in the MHC-I pathway, namely m06-gp48 and m152-gp40. A third protein, m04-gp34, may play a subtle role, but has recently been recognized as a negative regulator of NK cell activation (see Chapter II.9). Both mCMV vRAPs interfere with the trafficking of recently folded pMHC-I complexes from the peptide loading compartment in the ER to the cell surface (Lemmermann *et al.*, 2010a). In the terminology introduced by Hansen and Bouvier (2009), m152-gp40 functions as a 'retainer' by interacting transiently with pMHC-I complexes, retaining them in a cis-Golgi, ER-Golgi intermediate compartment (ERGIC), whereas m06-gp48 functions as a 'sorter' by binding stably to pMHC-I complexes and sorting them to lysosomes for degradation. Presentation studies in cells infected with mCMV mutants expressing either of these proteins selectively (Wagner *et al.*, 2002) have shown that m152-gp40 is much more potent than m06-gp48 in inhibiting pMHC-I cell surface presentation (Holtappels *et al.*, 2006a; Pinto *et al.*, 2006; Lemmermann *et al.*, 2012).

For evaluating the impact of vRAPs on the efficacy of AT in the murine model of cytoimmunotherapy, *ex vivo* isolated memory CD8⁺ T-cells from donors primed by infection with either mCMV-WT.BAC expressing vRAPs (briefly WT) or mutant virus mCMV-ΔvRAP not expressing vRAPs (briefly ΔvRAP) (Wagner *et al.*, 2002) were transferred in a 2×2 criss-cross approach into recipients also infected with either of these viruses (Fig. II.17.10; and V. Böhm, 2009; PhD thesis). At a glance, whereas the provenance of the donors' memory cells did not make any difference in the protective capacity, which is in good agreement with the identical specificity composition (Holtappels *et al.*, 2006b; see also Chapter I.22 for the vRAP independence of 'memory inflation'), viral loads were 1–2 $\log_{10}$ grades higher after infection of recipients with virus expressing vRAPs. So, vRAPs, as predicted, reduce the susceptibility of infected host tissue cells to be controlled by CD8⁺ T-cells, but the inhibition is far from being complete for a polyclonal CD8⁺ donor T-cell population.

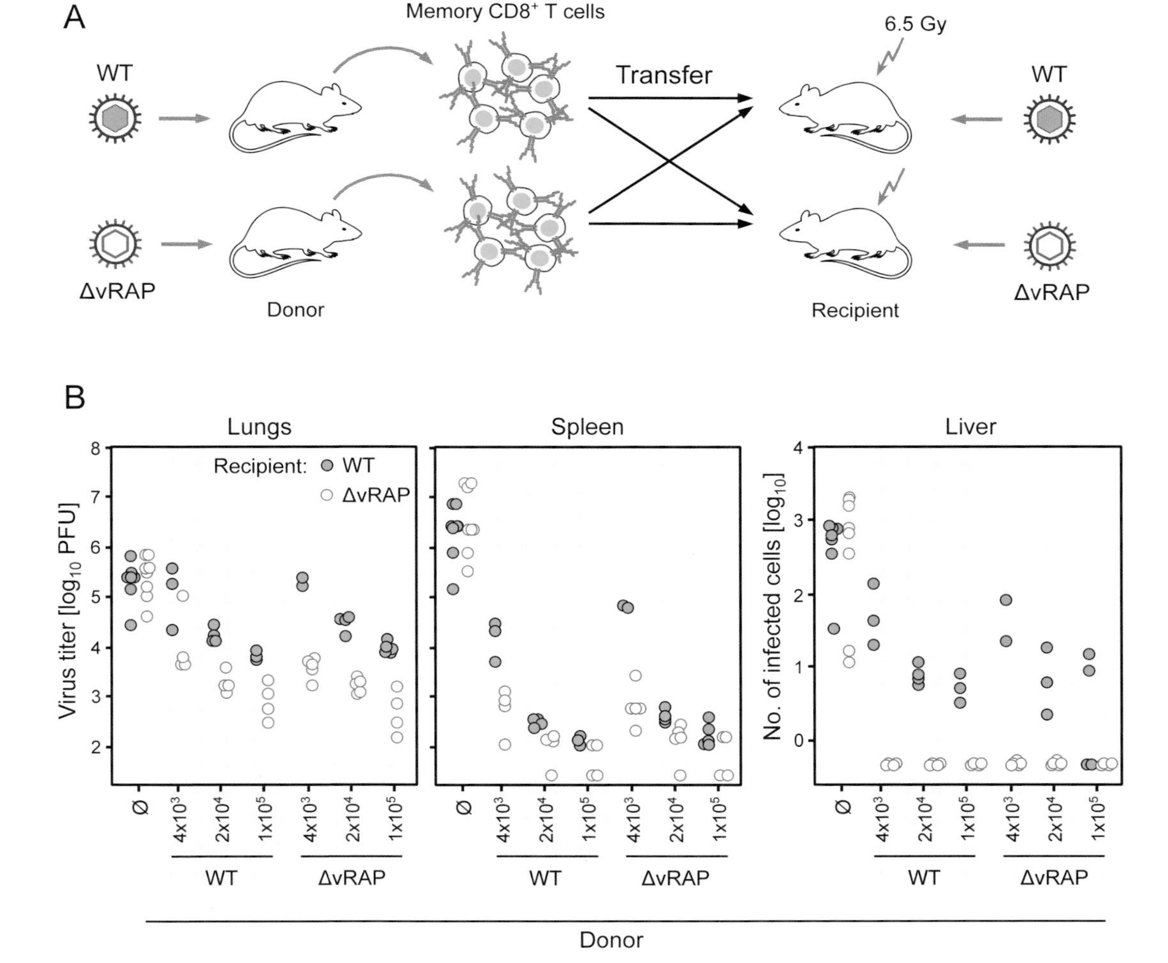

Figure II.17.10 Role for 'immune evasion' proteins inhibiting the MHC class-I pathway of antigen presentation. The efficacy of immunotherapy with polyclonal virus-primed memory CD8⁺ T-cells is independent of 'viral regulators of (direct) antigen presentation' (vRAPs) encoded by the donor virus variant but is reduced when vRAPs inhibit pMHC-I presentation in tissues of AT recipients. (A) Scheme of a 2×2 criss-cross AT approach, with donors and recipients infected with virus either expressing vRAPs or not expressing vRAPs, i.e. viruses mCMV-WT.BAC and mCMV-ΔvRAP, respectively. (B) Results from all four possible AT combinations, based on data from V. Böhm (2009), PhD thesis.

It should be noted, however, that the presentation of particular epitopes can be prevented completely, resulting in loss of protection (Holtappels *et al.*, 2004), especially when the corresponding peptide is generated in the infected cell in low amounts. The prototype example of a 'non-protective epitope' is the D^b-presented M45 peptide *HGIRNASFI* (Gold *et al.*, 2002; Holtappels *et al.*, 2004) of which only ~ 10 molecules can be isolated per cell infected with WT virus (Holtappels *et al.*, 2009). Interestingly, this low number is even an overestimate of processing efficiency, as it represents the number of accumulated pMHC-I molecules trapped in the ERGIC by the action of m152-gp40, whereas the peptide yield from cells infected with the vRAP deletion mutant is just 1–2 molecules per cell (Holtappels *et al.*, 2009). This vanishingly low number makes it easy for the vRAPs to catch really all M45–D^b complexes on their way to the cell surface. In C57BL/6 AT recipients infected with the vRAP deletion mutant, the same M45-D^b-specific $CD8^+$ T-cells (CTLL as well as *ex vivo* sorted memory cells) that failed to protect against WT virus were found to protect. This result proved that failure in protection against WT virus was not caused by an intrinsically inefficient effector function of the $CD8^+$ T-cells but by vRAP-mediated complete inhibition of M45-D^b presentation in the infected tissues of the AT recipients (Holtappels *et al.*, 2004). With this insight into the mechanism, 'non-protective epitopes' are not non-protective *per se* but only in the context of vRAP expression. This applies also to the M105-K^d epitope (Ebert *et al.*, 2012).

Fig. II.17.11 illustrates the current view of how to explain the apparent dichotomy between epitope-specific response magnitude and protection against infection. Key to the understanding is that $CD8^+$ T-cell priming occurs primarily through antigen cross-presentation by uninfected profAPCs, DCs in particular, that have taken up and processed antigens derived from infected cells in the course of cell death, whereas delivery of antiviral effector functions to protect from CMV spread and disease depends on vRAP-regulated direct antigen presentation by infected host tissue cells.

Strikingly, both currently known 'non-protective' epitopes are classified as being 'immunodominant' by virtue of acute response magnitude. Apparently, the host-virus balance evolved during co-speciation (see Chapter I.1 and I.2), which occurs on an MHC-polymorphic population level, tolerated the luxury that individuals (with a mouse inbred strain being an 'individual' too) mount high primary $CD8^+$ T-cell responses to viral epitopes that do not contribute to antiviral protection.

Lecture forecast

What will be the next lecture delivered by murine cytomegalovirus? As we have seen in this chapter, it was a long and winded road to go, before we even began to understand CMV disease and principles of immune control in a highly reductionistic model of HCT and $CD8^+$ T-cell-based immunotherapy. An obvious way to go in order to bring this model a further step closer to clinical HCT is switching from syngeneic HCT, which has modelled the clinically rare case of identical twin donor and recipient, to allo-HCT. Our experience with a single-MHC-I molecule-mismatched, minor-H-matched model using BALB/c mice and the MHC-I L^d gene deletion mutant BALB/c-H-2^{dm2} electively as HCT donors or recipients (Alterio de Goss *et al.*, 1998; Podlech *et al.*, 1998b; Seckert *et al.*, 2011) has already revealed a major impact of just a single MHC-I molecule disparity. Whereas, in absence of infection, HCT performed in a genetic graft-versus-host (GvH) direction, that is L^{d-} donor HCs transplanted to L^{d+} recipients, resulted in successful reconstitution with no evidence for GvH disease, mCMV infection of recipients was 100% lethal within 3 weeks with fulminant virus spread in all typical end-organs of CMV disease but no overt signs or functional evidence of an immunological GvH reaction. It rather appears that the single MHC-I molecule disparity interferes critically with the $CD8^+$ T-cell control of infection. The immediate aim is to find out if the failure is on the side of $CD8^+$ T-cell reconstitution and/or function, or on the side of antigen presentation by host tissue cells. In the first case, AT could be curative, whereas in the second case AT is likely to fail.

Since, in clinical allo-HCT, donors and recipients are usually HLA-matched but minor-H loci mismatched, our model for the future will be HCT and AT performed between BALB.B and C57BL/6 mice in either direction.

Acknowledgements

We would like to thank members of our laboratory past and present whose work over the years has contributed to this chapter. We would also like to apologize to colleagues in the field whose work may have escaped our attention or did not fit to the chapter's focus and is hopefully appreciated in cross-referenced other chapters of this book.

Work of the authors that led to data reviewed in this chapter was funded by the Deutsche Forschungsgemeinschaft, Collaborative Research Grant (Sonderforschungsbereich) 490, individual projects E3 'Persistence of murine cytomegalovirus after modulation of the CD8 T-cell immunome' (R.H. and S.E), and

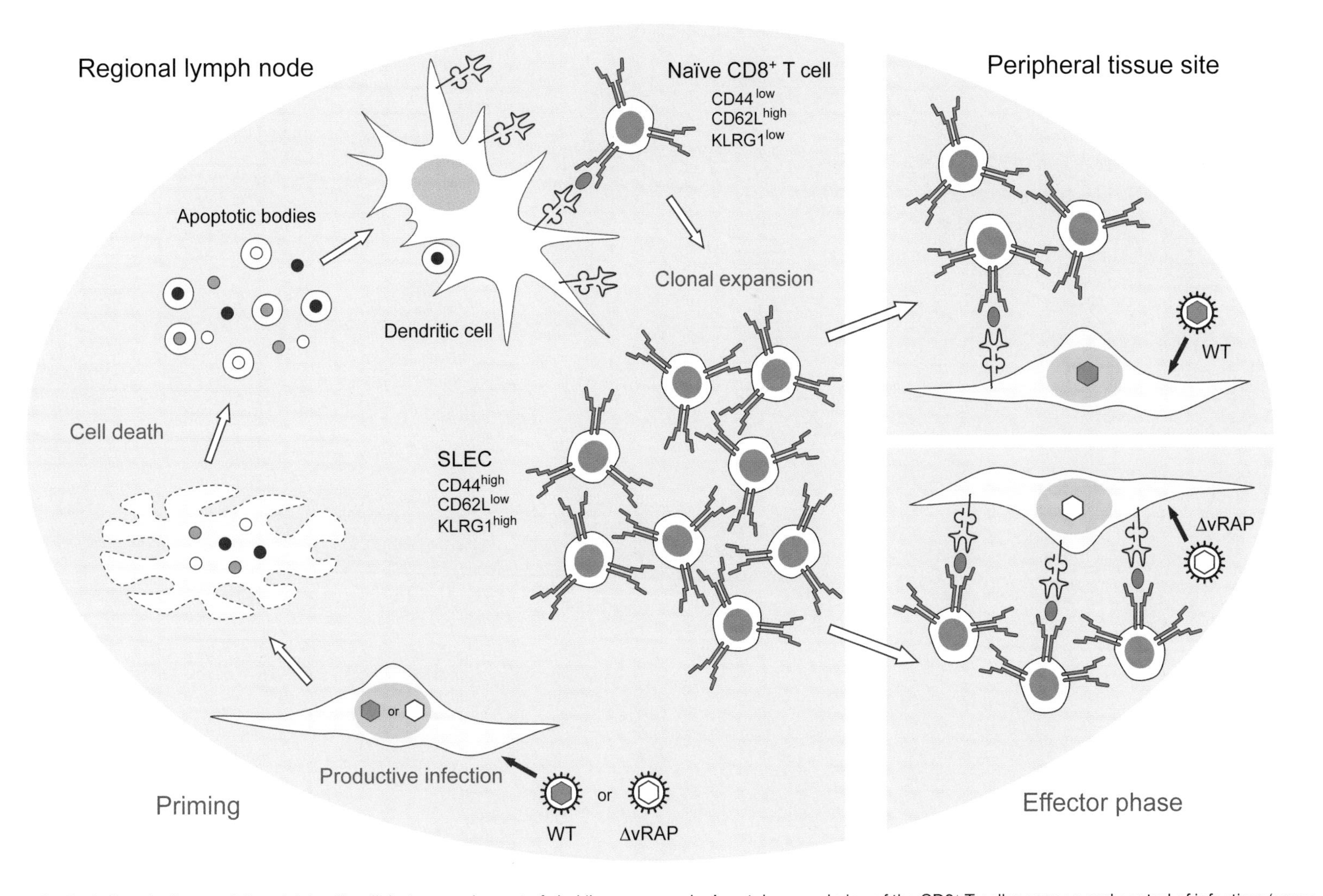

Figure II.17.11 Concluding model explaining the dichotomous impact of viral 'immune evasion' proteins on priming of the CD8+ T-cell response and control of infection: 'cross-presentation' governs priming and 'direct presentation' governs effector function.

E4 'Antigen presentation under the influence of murine cytomegalovirus immune evasion proteins' (A.F., V.B., M.J.R.,) as well as Clinical Research Group (KFO) 183, individual project 'Establishment of a challenge model for optimizing the immunotherapy of cytomegalovirus disease' (N.A.W.L. and M.J.R.).

References

Alterio de Goss, M., Holtappels, R., Steffens, H.P., Podlech, J., Angele, P., Dreher, L., Thomas, D., and Reddehase, M.J. (1998). Control of cytomegalovirus in bone marrow transplantation chimeras lacking the prevailing antigen-presenting molecule in recipient tissues rests primarily on recipient-derived CD8 T-cells. J. Virol. *72*, 7733–7744.

Boeckh, M., Leisenring, W., Riddell, S.R., Bowden, R.A., Huang, M.L., Myerson, D., Stevens-Ayers, T., Flowers, M.E., Cunningham, T., and Corey, L. (2003). Late cytomegalovirus disease and mortality in recipients of allogeneic hematopoietic stem cell transplants: importance of viral load and T-cell immunity. Blood *101*, 407–414.

Böhm, V., Podlech, J., Thomas, D., Deegen, P., Pahl-Seibert, M.F., Lemmermann, N.A., Grzimek, N.K., Oehrlein-Karpi, S.A., Reddehase, M.J., and Holtappels, R. (2008a). Epitope-specific *in vivo* protection against cytomegalovirus disease by CD8 T-cells in the murine model of preemptive immunotherapy. Med. Microbiol. Immunol. *197*, 135–144.

Böhm, V., Simon, C.O., Podlech, J., Seckert, C.K., Gendig, D., Deegen, P., Gillert-Marien, D., Lemmermann, N.A., Holtappels, R., and Reddehase, M.J. (2008b). The immune evasion paradox: immunoevasins of murine cytomegalovirus enhance priming of CD8 T-cells by preventing negative feedback regulation. J. Virol. *82*, 11637–11650.

Boyman, O., Létourneau, S., Krieg, C., and Sprent, J. (2009). Homeostatic proliferation and survival of naïve and memory T-cells. Eur. J. Immunol. *39*, 2088–2094.

Bunde, T., Kirchner, A., Hoffmeister, B., Habedank, D., Hetzer, R., Cherepnev, G., Proesch, S., Reinke, P., Volk, H.D., Lehmkuhl, H., *et al.* (2005). Protection from cytomegalovirus after transplantation is correlated with immediate early 1-specific CD8 T-cells. J. Exp. Med. *201*, 1031–1036.

Campbell, A.E., Cavanaugh, V.J., and Slater, J.S. (2008). The salivary glands as a privileged site of cytomegalovirus immune evasion and persistence. Med. Microbiol. Immunol. *197*, 205–213.

Cavanaugh, V.J., Deng, Y., Birkenbach, M.P., Slater, J.S., and Campbell, A.E. (2003). Vigorous innate and virus-specific cytotoxic T-lymphocyte responses to murine cytomegalovirus in the submaxillary salivary gland. J. Virol. *77*, 1703–1717.

Cobbold, M., Khan, N., Pourgheysari, B., Tauro, S., McDonald, D., Osman, H., Assenmacher, M., Billingham, L., Steward, C., Crawley, C., *et al.* (2005). Adoptive transfer of cytomegalovirus-specific CTL to stem cell transplant patients after selection by HLA-peptide tetramers. J. Exp. Med. *202*, 379–386.

Doom, C.M., and Hill, A.B. (2008). MHC class I immune evasion in MCMV infection. Med. Microbiol. Immunol. *197*, 191–204.

Ebert, S., Podlech, J., Gillert-Marien, D., Gergely, K.M., Büttner, J.K., Fink, A., Freitag, K., Thomas, D., Reddehase, M.J., and Holtappels, R. (2012). Parameters determining the efficacy of adoptive CD8 T-cell therapy of cytomegalovirus infection. Med. Microbiol. Immunol. *201*, 527–539.

Feuchtinger, T., Opherk, K., Bethge, W.A., Topp, M.S., Schuster, F.R., Weissinger, E.M., Mohty, M., Or, R., Maschan, M., Schumm, M., *et al.* (2010). Adoptive transfer of pp65-specific T-cells for the treatment of chemorefractory cytomegalovirus disease or reactivation after haploidentical and matched unrelated stem cell transplantation. Blood *116*, 4360–4367.

Fink, A., Lemmermann, N.A., Gillert-Marien, D., Thomas, D., Freitag, K., Böhm, V., Wilhelmi, V., Reifenberg, K., Reddehase, M.J., and Holtappels, R. (2012). Antigen presentation under the influence of 'immune evasion' proteins and its modulation by interferon-gamma: implications for immunotherapy of cytomegalovirus infection with antiviral CD8 T cells. Med. Microbiol. Immunol. *201*, 513–525.

Fuji, S., Kapp, M., Grigoleit, G.U., and Einsele, H. (2011). Adoptive immunotherapy with virus-specific T-cells. Best Pract. Res. Clin. Haematol. *24*, 413–419.

Gold, M.C., Munks, M.W., Wagner, M., Koszinowski, U.H., Hill, A.B., and Fling, S.P. (2002). The murine cytomegalovirus immunomodulatory gene m152 prevents recognition of infected cells by M45-specific CTL but does not alter the immunodominance of the M45-specific CD8 T-cell response *in vivo*. J. Immunol. *169*, 359–365.

Greenberg, P.D., Reusser, P., Goodrich, J.M., and Riddell, S.R. (1991). Development of a treatment regimen for human cytomegalovirus (CMV) infection in bone marrow transplantation recipients by adoptive transfer of donor-derived CMV-specific T-cell clones expanded *in vitro*. Ann. N. Y. Acad. Sci. *636*, 184–195.

Hansen, T.H., and Bouvier, M. (2009). MHC class I antigen presentation: learning from viral evasion strategies. Nat. Rev. Immunol. *9*, 503–513.

Hickman, H.D., Takeda, K., Skon, C.N., Murray, F.R., Hensley, S.E., Loomis, J., Barber, G.N., Bennink, J.R., and Yewdell, J.W. (2008). Direct priming of antiviral CD8+ T-cells in the peripheral interfollicular region of lymph nodes. Nat. Immunol. *9*, 155–165.

Ho, M. (1980). Role of specific cytotoxic lymphocytes in cellular immunity against murine cytomegalovirus. Infect. Immun. *27*, 767–776.

Holtappels, R., Podlech, J., Geginat, G., Steffens, H.P., Thomas, D., and Reddehase, M.J. (1998). Control of murine cytomegalovirus in the lungs: relative but not absolute immunodominance of the immediate-early 1 nonapeptide during the antiviral cytolytic T-lymphocyte response in pulmonary infiltrates. J. Virol. *72*, 7201–7012.

Holtappels, R., Pahl-Seibert, M.F., Thomas, D., and Reddehase, M.J. (2000a). Enrichment of immediate-early 1 (m123/pp89) peptide-specific CD8 T-cells in a pulmonary CD62L(lo) memory-effector cell pool during latent murine cytomegalovirus infection of the lungs. J. Virol. *74*, 11495–11503.

Holtappels, R., Thomas, D., Podlech, J., Geginat, G., Steffens, H.P., and Reddehase, M.J. (2000b). The putative natural killer decoy early gene m04 (gp34) of murine cytomegalovirus encodes an antigenic peptide recognized by protective antiviral CD8 T-cells. J. Virol. *74*, 1871–1884.

Holtappels, R., Podlech, J., Grzimek, N.K., Thomas, D., Pahl-Seibert, M.F., and Reddehase, M.J. (2001). Experimental preemptive immunotherapy of murine cytomegalovirus disease with CD8 T-cell lines specific for ppM83 and pM84, the two homologs of human cytomegalovirus tegument protein ppUL83 (pp65). J. Virol. *75*, 6584–6600.

Holtappels, R., Thomas, D., and Reddehase, M.J. (2002). Two antigenic peptides from genes m123 and m164 of murine cytomegalovirus quantitatively dominate CD8 T-cell memory in the H-2d haplotype. J. Virol. *76*, 151–164.

Holtappels, R., Podlech, J., Pahl-Seibert, M.F., Jülch, M., Thomas, D., Simon, C.O., Wagner, M., and Reddehase, M.J. (2004). Cytomegalovirus misleads its host by priming of CD8 T-cells specific for an epitope not presented in infected tissues. J. Exp. Med. *199*, 131–136.

Holtappels, R., Gillert-Marien, D., Thomas, D., Podlech, J., Deegen, P., Herter, S., Oehrlein-Karpi, S.A., Strand, D., Wagner, M., and Reddehase, M.J. (2006a). Cytomegalovirus encodes a positive regulator of antigen presentation. J. Virol. *80*, 7613–7624.

Holtappels, R., Munks, M.W., Podlech, J., and Reddehase, M.J. (2006b). CD8 T-cell-based immunotherapy of cytomegalovirus disease in the mouse model of the immunocompromised bone marrow transplantation recipient. In Cytomegaloviruses: Molecular Biology and Immunology, Reddehase, M.J., ed. (Caister Academic Press, Norfolk, UK), pp. 383–418.

Holtappels, R., Böhm, V., Podlech, J., and Reddehase, M.J. (2008a). CD8 T-cell-based immunotherapy of cytomegalovirus infection: 'proof of concept' provided by the murine model. Med. Microbiol. Immunol. *197*, 125–134.

Holtappels, R., Janda, J., Thomas, D., Schenk, S., Reddehase, M.J., and Geginat, G. (2008b). Adoptive CD8 T-cell control of pathogens cannot be improved by combining protective epitope specificities. J. Infect. Dis. *197*, 622–629.

Holtappels, R., Simon, C.O., Munks, M.W., Thomas, D., Deegen, P., Kühnapfel, B., Däubner, T., Emde, S.F., Podlech, J., Grzimek, N.K., *et al.* (2008c). Subdominant CD8 T-cell epitopes account for protection against cytomegalovirus independent of immunodomination. J. Virol. *82*, 5781–5796.

Holtappels, R., Thomas, D., and Reddehase, M.J. (2009). The efficacy of antigen processing is critical for protection against cytomegalovirus disease in the presence of viral immune evasion proteins. J. Virol. *83*, 9611–9615.

Humphreys, I.R., de Trez, C., Kinkade, A., Benedict, C.A., Croft, M., and Ware, C.F. (2007). Cytomegalovirus exploits IL-10-mediated immune regulation in the salivary glands. J. Exp. Med. *204*, 1217–1225.

Hutchinson, S., Sims, S., O'Hara, G., Silk, J., Gileadi, U., Cerundolo, V., and Klenerman, P. (2011). A dominant role for the immunoproteasome in CD8+ T-cell responses to murine cytomegalovirus. PLoS One *6*, e14646.

Jonjic, S., Mutter, W., Weiland, F., Reddehase, M.J., and Koszinowski, U.H. (1989). Site-restricted persistent cytomegalovirus infection after selective long-term depletion of CD4+ T-lymphocytes. J. Exp. Med. *169*, 1199–1212.

Jordan, S., Krause, J., Prager, A., Mitrovic, M., Jonjic, S., Koszinowski, U.H., and Adler, B. (2011). Virus progeny of murine cytomegalovirus bacterial artificial chromosome pSM3fr show reduced growth in salivary glands due to a fixed mutation of MCK-2. J. Virol. *85*, 10346–10353.

Kern, F., Surel, I.P., Faulhaber, N., Frömmel, C., Schneider-Mergener, J., Schönemann, C., Reinke, P., and Volk, H.D. (1999). Target structures of the CD8(+)-T-cell response to human cytomegalovirus: the 72-kilodalton major immediate-early protein revisited. J. Virol. *73*, 8179–8184.

Lefrançois, L., and Obar, J.J. (2010). Once a killer, always a killer: from cytotoxic T-cell to memory cell. Immunol. Rev. *235*, 206–218.

Lemmermann, N.A., Gergely, K., Böhm, V., Deegen, P., Däubner, T., and Reddehase, M.J. (2010a). Immune evasion proteins of murine cytomegalovirus preferentially affect cell surface display of recently generated peptide presentation complexes. J. Virol. *84*, 1221–1236.

Lemmermann, N.A., Podlech, J., Seckert, C.K., Kropp, K.A., Grzimek, N.K., Reddehase, M.J., and Holtappels, R. (2010b). CD8 T-cell immunotherapy of cytomegalovirus disease in the murine model. In Methods in Microbiology: Immunology of Infection, Kabelitz, D., and Kaufmann, S.H.E., eds. (Academic Press, London, UK), pp. 369–420.

Lemmermann, N.A., Böhm, V., Holtappels, R., and Reddehase, M.J. (2011a). *In vivo* impact of cytomegalovirus evasion of CD8 T-cell immunity: facts and thoughts based on murine models. Virus Res. *157*, 161–174.

Lemmermann, N.A., Kropp, K.A., Seckert, C.K., Grzimek, N.K.A., and Reddehase, M.J. (2011b). Reverse genetics modification of cytomegalovirus antigenicity and immunogenicity by CD8 T-cell epitope deletion and insertion. J. Biomed. Biotechnol. 2011, 812742.

Lemmermann, N.A., Fink, A., Podlech, J., Ebert, S., Wilhelmi, V., Böhm, V., Holtappels, R., and Reddehase, M.J. (2012). Murine cytomegalovirus immune evasion proteins operative in the MHC class I pathway of antigen processing and presentation: state of knowledge, revisions, and questions. Med. Microbiol. Immunol. *201*, 497–512.

Lyons, P.A., Allan, J.E., Carrello, C., Shellam, G.R., and Scalzo, A.A. (1996). Effect of natural sequence variation at the H-2Ld-restricted CD8+ T-cell epitope of the murine cytomegalovirus ie1-encoded pp89 on T-cell recognition. J. Gen. Virol. *77*, 2615–2623.

Mayer, A., Podlech, J., Kurz, S., Steffens, H.P., Maiberger, S., Thalmeier, K., Angele, P., Dreher, L., and Reddehase, M.J. (1997). Bone marrow failure by cytomegalovirus is associated with an *in vivo* deficiency in the expression of essential stromal hemopoietin genes. J. Virol. *71*, 4589–4598.

Messerle, M., Crnkovic, I., Hammerschmidt, W., Ziegler, H., and Koszinowski, U.H. (1997). Cloning and mutagenesis of a herpesvirus genome as an infectious bacterial artificial chromosome. Proc. Natl. Acad. Sci. U.S.A. *94*, 14759–14763.

Munks, M.W., Cho, K.S., Pinto, A.K., Sierro, S., Klenerman, P., and Hill, A.B. (2006a). Four distinct patterns of memory CD8 T-cell responses to chronic murine cytomegalovirus infection. J. Immunol. *177*, 450–458.

Munks, M.W., Gold, M.C., Zajac, A.L., Doom, C.M., Morello, C.S., Spector, D.H., and Hill, A.B. (2006b). Genome-wide analysis reveals a highly diverse CD8 T-cell response to murine cytomegalovirus. J. Immunol. *176*, 3760–3766.

Mutter, W., Reddehase, M.J., Busch, F.W., Bühring, H.J., and Koszinowski, U.H. (1988). Failure in generating hemopoietic stem cells is the primary cause of death from

cytomegalovirus disease in the immunocompromised host. J. Exp. Med. *167*, 1645–1658.
Obar, J.J., and Lefrançois, L. (2010a). Early events governing memory CD8+ T-cell differentiation. Int. Immunol. *22*, 619–625.
Obar, J.J., and Lefrançois, L. (2010b). Memory CD8+ T-cell differentiation. Ann. N. Y. Acad. Sci. *1183*, 251–266.
Pahl-Seibert, M.F., Jülch, M., Podlech, J., Thomas, D., Deegen, P., Reddehase, M.J., and Holtappels, R. (2005). Highly-protective *in vivo* function of cytomegalovirus IE1 epitope-specific memory CD8 T-cells purified by T-cell receptor-based cell sorting. J. Virol. *79*, 5400–5413.
Peggs, K.S., Verfuerth, S., Pizzey, A., Khan, N., Guiver, M., Moss, P.A., and Mackinnon, S. (2003). Adoptive cellular therapy for early cytomegalovirus infection after allogeneic stem-cell transplantation with virus-specific T-cell lines. Lancet *362*, 1375–1377.
Pinto, A.K., and Hill, A.B. (2005). Viral interference with antigen presentation to CD8+ T-cells: lessons from cytomegalovirus. Viral Immunol. *18*, 434–444.
Pinto, A.K., Munks, M.W., Koszinowski, U.H., and Hill, A.B. (2006). Coordinated function of murine cytomegalovirus genes completely inhibits CTL lysis. J. Immunol. *177*, 3225–3234.
Podlech, J., Holtappels, R., Wirtz, N., Steffens, H.P., and Reddehase, M.J. (1998a). Reconstitution of CD8 T-cells is essential for the prevention of multiple-organ cytomegalovirus histopathology after bone marrow transplantation. J. Gen. Virol. *79*, 2099–2104.
Podlech, J., Steffens, H.P., Holtappels, R., Mayer, A., Alterio de Goss, M., Oettel, O., Wirtz, N., Maiberger, S., and Reddehase, M.J. (1998b). Cytomegalovirus pathogenesis after experimental bone marrow transplantation. Monogr. Virol. *21*, 119–128.
Podlech, J., Holtappels, R., Pahl-Seibert, M.F., Steffens, H.P., and Reddehase, M.J. (2000). Murine model of interstitial cytomegalovirus pneumonia in syngeneic bone marrow transplantation: persistence of protective pulmonary CD8 T-cell infiltrates after clearance of acute infection. J. Virol. *74*, 7496–7507.
Powers, C., DeFilippis, V., Malouli, D., and Früh, K. (2008). Cytomegalovirus immune evasion. Curr. Top. Microbiol. Immunol. *325*, 333–359.
Rammensee, H.-G., Bachmann, J., and Stevanovic, S. (1997). MHC Ligands and Peptide Motifs (Molecular Biology Intelligence Unit, Landes Bioscience, Austin, TX).
Rawlinson, W.D., Farrell, H.E., and Barrell, B.G. (1996). Analysis of the complete DNA sequence of murine cytomegalovirus. J. Virol. *70*, 8833–8849.
Reddehase, M.J. (2000). The immunogenicity of human and murine cytomegaloviruses. Curr. Opin. Immunol. *12*, 390–396, 738.
Reddehase, M.J. (2002). Antigens and immunoevasins: opponents in cytomegalovirus immune surveillance. Nat. Rev. Immunol. *2*, 831–844.
Reddehase, M.J., and Koszinowski, U.H. (1984). Significance of herpesvirus immediate early gene expression in cellular immunity to cytomegalovirus infection. Nature *312*, 369–371.
Reddehase, M.J., Keil, G.M., and Koszinowski, U.H. (1984). The cytolytic T-lymphocyte response to the murine cytomegalovirus. II. Detection of virus replication stage-specific antigens by separate populations of *in vivo* active cytolytic T-lymphocyte precursors. Eur. J. Immunol. *14*, 56–61.
Reddehase, M.J., Weiland, F., Münch, K., Jonjic, S., Lüske, A., and Koszinowski, U.H. (1985). Interstitial murine cytomegalovirus pneumonia after irradiation: characterization of cells that limit viral replication during established infection of the lungs. J. Virol. *55*, 264–273.
Reddehase, M.J., Mutter, W., and Koszinowski, U.H. (1987a). *In vivo* application of recombinant interleukin 2 in the immunotherapy of established cytomegalovirus infection. J. Exp. Med. *165*, 650–656.
Reddehase, M.J., Mutter, W., Münch, K., Bühring, H.J., and Koszinowski, U.H. (1987b). CD8-positive T-lymphocytes specific for murine cytomegalovirus immediate-early antigens mediate protective immunity. J. Virol. *61*, 3102–3108.
Reddehase, M.J., Jonjic, S., Weiland, F., Mutter, W., and Koszinowski, U.H. (1988). Adoptive immunotherapy of murine cytomegalovirus adrenalitis in the immunocompromised host: CD4-helper-independent antiviral function of CD8-positive memory T-lymphocytes derived from latently infected donors. J. Virol. *62*, 1061–1065.
Reddehase, M.J., Rothbard, J.B., and Koszinowski, U.H. (1989). A pentapeptide as minimal antigenic determinant for MHC class I-restricted T-lymphocytes. Nature *337*, 651–653.
Reddehase, M.J., Dreher-Stumpp, L., Angele, P., Balthesen, M., and Susa, M. (1992). Hematopoietic stem cell deficiency resulting from cytomegalovirus infection of bone marrow stroma. Ann. Hematol. *64*, 125–127.
Redwood, A.J., Messerle, M., Harvey, N.L., Hardy, C.M., Koszinowski, U.H., Lawson, M.A., and Shellam, G.R. (2005). Use of a murine cytomegalovirus K181-derived bacterial artificial chromosome as a vaccine vector for immunocontraception. J. Virol. *79*, 2998–3008.
Riddell, S.R., Watanabe, K.S., Goodrich, J.M., Li, C.R., Agha, M.E., and Greenberg, P.D. (1992). Restoration of viral immunity in immunodeficient humans by the adoptive transfer of T-cell clones. Science *257*, 238–241.
Schmitt, A., Tonn, T., Busch, D.H., Grigoleit, G.U., Einsele, H., Odendahl, M., Germeroth, L., Ringhoffer, M., Ringhoffer, S., Wiesneth, M., *et al.* (2011). Adoptive transfer and selective reconstitution of streptamer-selected cytomegalovirus-specific CD8+ T-cells leads to virus clearance in patients after allogeneic peripheral blood stem cell transplantation. Transfusion *51*, 591–599.
Seckert, C.K., Schader, S.I., Ebert, S., Thomas, D., Freitag, K., Renzaho, A., Podlech, J., Reddehase, M.J., and Holtappels, R. (2011). Antigen-presenting cells of haematopoietic origin prime cytomegalovirus-specific CD8 T-cells but are not sufficient for driving memory inflation during viral latency. J. Gen. Virol. *92*, 1994–2005.
Simon, C.O., Holtappels, R., Tervo, H.M., Böhm, V., Däubner, T., Oehrlein-Karpi, S.A., Kühnapfel, B., Renzaho, A., Strand, D., Podlech, J., *et al.* (2006). CD8 T-cells control cytomegalovirus latency by epitope-specific sensing of transcriptional reactivation. J. Virol. *80*, 10436–10456.
Snyder, C.M., Allan, J.E., Bonnett, E.L., Doom, C.M., and Hill, A.B. (2010). Cross-presentation of a spread-defective MCMV is sufficient to prime the majority of virus-specific CD8+ T-cells. PLoS One *5*, e9681.
Starr, S.E., and Allison, A.C. (1977). Role of T-lymphocytes in recovery from murine cytomegalovirus infection. Infect. Immun. *17*, 458–462.
Steffens, H.P., Kurz, S., Holtappels, R., and Reddehase, M.J. (1998a). Preemptive CD8 T-cell immunotherapy of acute

cytomegalovirus infection prevents lethal disease, limits the burden of latent viral genomes, and reduces the risk of virus recurrence. J. Virol. *72*, 1797–1804.

Steffens, H.P., Podlech, J., Kurz, S., Angele, P., Dreis, D., and Reddehase, M.J. (1998b). Cytomegalovirus inhibits the engraftment of donor bone marrow cells by down-regulation of hemopoietin gene expression in recipient stroma. J. Virol. *72*, 5006–5015.

Sylwester, A.W., Mitchell, B.L., Edgar, J.B., Taormina, C., Pelte, C., Ruchti, F., Sleath, P.R., Grabstein, K.H., Hosken, N.A., Kern, F., *et al.* (2005). Broadly targeted human cytomegalovirus-specific CD4+ and CD8+ T-cells dominate the memory compartments of exposed subjects. J. Exp. Med. *202*, 673–685.

Tenzer, S., Peters, B., Bulik, S., Schoor, O., Lemmel, C., Schatz, M.M., Kloetzel, P.M., Rammensee, H.G., Schild, H., and Holzhütter, H.G. (2005). Modeling the MHC class I pathway by combining predictions of proteasomal cleavage, TAP transport and MHC class I binding. Cell. Mol. Life Sci. *62*, 1025–1037.

Torti, N., Walton, S.M., Brocker, T., Rulicke, T., and Oxenius, A. (2011a). Non-hematopoietic cells in lymph nodes drive memory CD8 T-cell inflation during murine cytomegalovirus infection. PLoS Pathog. *7*, e1002313.

Torti, N., Walton, S.M., Murphy, K.M., and Oxenius, A. (2011b). Batf3 transcription factor-dependent DC subsets in murine CMV infection: differential impact on T-cell priming and memory inflation. Eur. J. Immunol. *41*, 2612–2618.

Townsend, A.R., Rothbard, J., Gotch, F.M., Bahadur, G., Wraith, D., and McMichael, A.J. (1986). The epitopes of influenza nucleoprotein recognized by cytotoxic T-lymphocytes can be defined with short synthetic peptides. Cell *44*, 959–968.

Wagner, M., Jonjic, S., Koszinowski, U.H., and Messerle, M. (1999). Systematic excision of vector sequences from the BAC-cloned herpesvirus genome during virus reconstitution. J. Virol. *73*, 7056–7060.

Wagner, M., Gutermann, A., Podlech, J., Reddehase, M.J., and Koszinowski, U.H. (2002). Major histocompatibility complex class I allele-specific cooperative and competitive interactions between immune evasion proteins of cytomegalovirus. J. Exp. Med. *196*, 805–816.

Walter, E.A., Greenberg, P.D., Gilbert, M.J., Finch, R.J., Watanabe, K.S., Thomas, E.D., and Riddell, S.R. (1995). Reconstitution of cellular immunity against cytomegalovirus in recipients of allogeneic bone marrow by transfer of T-cell clones from the donor. N. Engl. J. Med. *333*, 1038–1044.

Walton, S.M., Mandaric, S., Torti, N., Zimmermann, A., Hengel, H., and Oxenius, A. (2011). Absence of cross-presenting cells in the salivary gland and viral immune evasion confine cytomegalovirus immune control to effector CD4 T-cells. PLoS Pathog. *7*, e1002214.

Wearsch, P.A., and Cresswell, P. (2008). The quality control of MHC class I peptide loading. Curr. Opin. Cell Biol. *20*, 624–631.

Welsh, R.M., Selin, L.K., and Szomolanyi-Tsuda, E. (2004). Immunological memory to viral infections. Annu. Rev. Immunol. *22*, 711–743.

Weninger, W., Manjunath, N., and von Andrian, U.H. (2002). Migration and differentiation of CD8+ T-cells. Immunol. Rev. *186*, 221–233.

State of the Art and Trends in Cytomegalovirus Diagnostics

II.18

Maria Grazia Revello and Giuseppe Gerna

Abstract

Diagnosis of human cytomegalovirus (HCMV) infection is required in two clinical situations: the immunocompetent HCMV-seronegative pregnant woman, and the immunocompromised patient. In the case of pregnant women, diagnosis is primarily based on serology, but also on detection of viral DNA in blood. Viral DNA in blood may be quantified by the following assays: viraemia (infectious virus in blood), antigenaemia (leucocytes carrying virus or viral products) or DNAemia. In pregnant women diagnosis of primary infection is mandatory, in view of the possible virus transmission to the fetus. In the fetus, virus infection is diagnosed by performing viral assays on amniotic fluid and, when appropriate, on fetal blood. In newborns, diagnosis of congenital infection may be achieved by virus detection at birth. In immunocompromised transplanted patients virus quantification in blood is also mandatory to determine optimal time to intervene by antiviral treatment, unless a prophylaxis approach is adopted. Repeated antiviral drug courses may cause emergence of drug-resistant HCMV strains harbouring mutations in the UL97 or UL54 genes. Control of HCMV infection is guided by the immune response, which is elicited by viral infection at the level of both humoral and T-cell immunity.

Introduction

Human cytomegalovirus (HCMV), first detected in the 1950s, was somewhat difficult to isolate, since it is strictly host species-specific and cell-associated, and was difficult to cultivate in the laboratory. Initially, pathogenicity appeared to be restricted to newborns with congenital infection, while in healthy individuals the only relatively infrequent clinical disease was HCMV infectious mononucleosis. It became evident early on that congenital infection was mostly related to HCMV primary infection occurring in the mother during pregnancy. With the advent of AIDS and the transplantation era, HCMV was recognized as responsible for the death of a number of immunocompromised patients. This prompted the development of a number of diagnostic procedures as well as specific antiviral drugs. In the immunocompromised patient population, the viral load level in blood was soon related to the appearance of clinical symptoms. Antiviral drugs controlled the infection in most cases, however, in the presence of only the humoral immune response, frequent HCMV reactivation episodes occurred, until the HCMV-specific T-cell response developed or was reconstituted. In this chapter, both aspects of the immune response will be briefly discussed in their relationship with diagnostic implications.

Serology

Determination of HCMV-specific immunoglobulin (Ig) of the G (IgG) and M (IgM) class represents the most widely used approach for diagnosis of HCMV infection and/or determination of virus-specific immune status. HCMV-specific antibodies can be detected either by their physical binding to viral components or by their biological function. The former approach, which allows for differential determination of virus-specific IgG and IgM, mainly relies on the immunoassay technology, whereas the assessment of biological functions is more complex and requires *in vitro* neutralization assays. Former assays are mainly used for diagnostic purposes, whereas functional assays are performed mainly for research purposes or for interpretation of the antibody response.

Generalities in diagnostic assays for HCMV-specific antibody determination

The enzyme immunoassay (EIA) and enzyme-linked immunosorbent assay (ELISA) are the most widely used techniques for detection of HCMV-specific antibodies. Traditional ELISA assays use chromogenic reporters and suitable substrates resulting in a colour or fluorescence change (enzyme-linked fluorescent assay, ELFA) in the reaction well. Newer ELISA-like techniques use chemiluminescent reporters (chemilumiscence immunoassay, CLIA). The great majority of commercially available assays are based on the indirect ELISA/CLIA principle.

Briefly, HCMV antigen (represented by cell lysate and/or recombinant proteins) is coated onto a solid phase consisting of polystyrene wells or paramagnetic beads. For IgG determination, test serum is added and incubated for a variable period of time, then a conjugated secondary anti-IgG antibody is added. Results are expressed either qualitatively or quantitatively as an index or in arbitrary units as an international standard is not available for HCMV IgG.

As for IgM determination either indirect or capture immunoassays can be used. With the indirect immunoassay, test sera are pre-treated with antibodies directed against human IgG (anti-huIgG) to prevent binding of IgG-reactive rheumatoid factor IgM that would be detected by secondary conjugated anti-IgM antibody, and reduce competition between IgG and IgM for antigen bound to the solid-phase. In capture assays, IgM are immunologically captured by an anti-human IgM monoclonal or polyclonal antibody bound to the solid phase. Capture assays are less prone to non-specific IgM reactivities and reduce the competition of virus-specific IgG for HCMV antigen. IgM results are usually expressed as an index, consisting of the ratio of test serum and cut-off serum signals.

Generally, a single analyte, namely IgG or IgM to HCMV, is measured in each assay. Recently, multiplex flow immunoassay (MFI) technology has emerged as a novel approach to simultaneously assess the serological response to a number of infectious agents, including HCMV (Binnicker *et al.*, 2010).

A number of manual, semi- or fully automated assays for IgG, IgM, and IgG avidity determination are commercially available. Automated systems are generally preferred (even though they require dedicated instrumentation) because they allow high throughput, and complete automation of HCMV antibody panel determination including reflex testing (i.e. automated avidity determination in case of positive IgM result). Currently available automated assays are listed in Table II.18.1. It should be emphasized that automated assays, in order to reduce incubation time and increase turnaround activity, require an excess of antigen and conjugate as well as lower test sample dilution. This implies that the performance characteristics of automated assays are generally inferior to those provided by manual assays in which immunological reactions are allowed to occur under equilibrium conditions.

HCMV-specific antibody response

IgM antibody

The IgM antibody response following primary HCMV infection slightly precedes IgG development, reaches a plateau in the first month after onset of infection and then slowly declines in the following 3–6 months (Fig. II.18.1). However, the individual variability and sensitivity of the assay employed may greatly influence these kinetics. Indeed, a very brief IgM antibody response may be observed in some individuals, whereas a low-level IgM antibody response has been reported to persist for several months after a primary infection (Revello and Gerna, 2002). Two representative examples of these extremes are shown in Fig. II.18.2A.

The specificity of IgM results is of great concern, particularly when IgM determination is carried on in pregnant women in conjunction with IgG determination in order to establish HCMV-specific immune status and detect recent asymptomatic primary infection. Interference with rheumatoid factors, IgM reactive with cellular antigens present in the antigen preparation, crossreactivity or polyclonal stimulation during other viral infections are all potential causes for non-specific results. However, pre-adsorption of sera with anti-huIgG, the use of capture rather than indirect immunoassays and the use of control antigen have all contributed to reduce non-specific results.

In addition, since the viral proteins pp150 (UL32), pp65 (UL83), pp38 (UL80a), pp52 (UL44), and p130 (UL57) have been shown to be specifically recognized by IgM antibody following primary infection, the use of immunoassays based on the above recombinant proteins was advocated to improve specificity and reduce result variability among different assays. Indeed, the first fully automated IgM assay, developed in 2000, was based on a mixture of pp150, pp65, pp52, and pp38 recombinant proteins (AxSYM, Abbott). Although sensitivity was claimed to be very high (Maine *et al.*, 2000), specificity was rather low (Lazzarotto *et al.*, 2001). Thus, a second generation HCMV IgM assay was developed by Abbott (Architect) which employs a mixture of HCMV lysate and recombinant viral proteins (Table II.18.1). According to available information, all the remaining automated IgM assays, but one, use variably purified and inactivated HCMV lysate as antigen. The only exception is the Elecsys CMV IgM (Roche Diagnostics, Germany) kit in which a cocktail of recombinant proteins is employed (Table II.18.1).

In addition to primary infection, HCMV-specific IgM can be detected during recurrent infection. This is well documented in immunocompromised patients whereas, to the best of our knowledge, no study has ever specifically addressed this issue in immunocompetent

Table II.18.1 Characteristics of automated assays commercially available for determination of HCMV-specific IgG, IgM and IgG avidity

Manufacturer platform	Menu	Antigen	Technology	Incubation time	CE/FDA approval
Abbott					
AxSYM	IgG	AD169 lysate	Indirect MEIA	20 min	CE/FDA
	IgM	pp150, pp52, pp65, pp38 recombinant proteins	Indirect MEIA	20 min	CE
Architect	IgG	AD169 lysate	Indirect CLIA	29 min	CE
	IgM	AD169 lysate + pp150/pp52 fusion protein	Indirect CLIA[a]	29 min	CE
	IgG avidity	AD169 lysate	Indirect CLIA	29 min	CE
Beckman Coulter					
Access	IgG	AD169 lysate	Indirect CLIA	35 min	CE
	IgM	AD169 lysate	Indirect CLIA[a]	55 min	CE
BioMérieux					
VIDAS	IgG	AD169 lysate	Indirect ELFA	40 min	CE/FDA
	IgM	AD169 lysate	Indirect ELFA[a]	60 min	CE/FDA
	IgG avidity	AD169 lysate	Indirect ELFA	40 min	CE
Bio-Rad					
BioPlex 2200	ToRC IgG	AD169 lysate	Indirect MFI	45 min	CE/FDA
	ToRC IgM	AD169 lysate	Indirect MFI[a]	45 min	CE
DiaSorin					
Liaison	IgG	AD169 lysate	Indirect CLIA	35 min	CE/FDA
	IgM	AD169 lysate	Indirect CLIA[a]	45 min	CE/FDA
	IgG avidity	AD169 lysate	Indirect CLIA	45 min	CE
Ortho					
Vitros	IgG	AD169 lysate	Indirect CLIA	35 min	CE/FDA
	IgM	AD169 lysate	Capture CLIA	68 min	CE/FDA
Roche					
Elecsys	IgG	Recombinant proteins	Sandwich ECLIA	18 min	CE
	IgM	Recombinant proteins	Capture ECLIA	18 min	CE
	IgG avidity	Recombinant proteins	Sandwich ECLIA	18 min	CE
Siemens					
Immulite	IgG	AD169 lysate	Indirect CLIA	60 min	CE/FDA[b]
	IgM	AD169 lysate	Indirect CLIA[a]	90 min	CE/FDA[b]

Abbreviations: CE, Conformité Européenne; CLIA, chemiluminescence immunoassay; ECLIA, electro-chemiluminescence immunoassay; ELFA, enzyme-linked fluorescence assay; FDA, Food and Drug Administration; MEIA, microparticle enzyme immunoassay; MIF, multiplex flow immunoassay; NA, not available.
[a]Sera are pre-treated with antibody directed against human IgG.
[b]Not approved by FDA for blood donors screening.

individuals. Rather, in one study we observed that out of 22 HCMV-seropositive breastfeeding women with positive virus isolation from milk, no subject was IgM-positive as determined by two in-house developed capture ELISAs (Revello *et al.*, 1999). Nevertheless, it cannot be excluded that, even in immunocompetent individuals, transient IgM levels might be detected by highly sensitive assays during recurrent infections.

Past and recent studies have reported a wide variability in the performance of HCMV IgM assays. Ultimately, the performance of a given kit in terms of sensitivity and specificity depends on the cut-off selected. In our opinion, the issue of specificity should be a major concern when developing or selecting an assay for HCMV-specific IgM determination in consideration of the following: (i) the target population for HCMV serology is primarily represented by pregnant women, (ii) even though vertical transmission

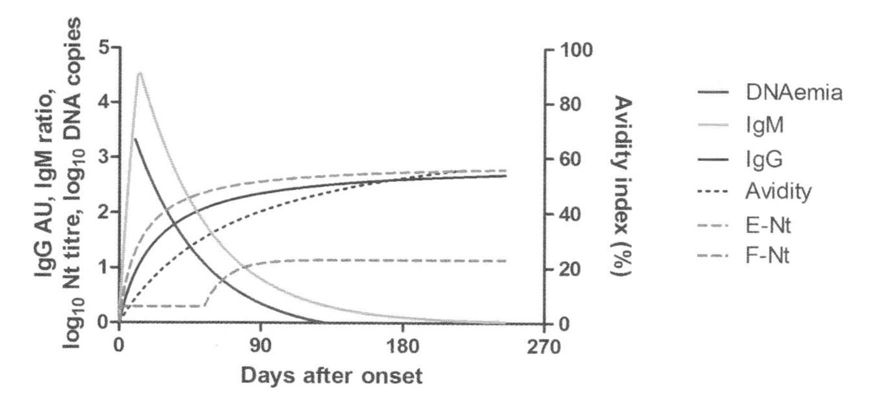

Figure II.18.1 Kinetics (median values) of HCMV-specific antibody and DNAemia as determined in 151 sequential serum samples from 43 pregnant women with primary HCMV infection.

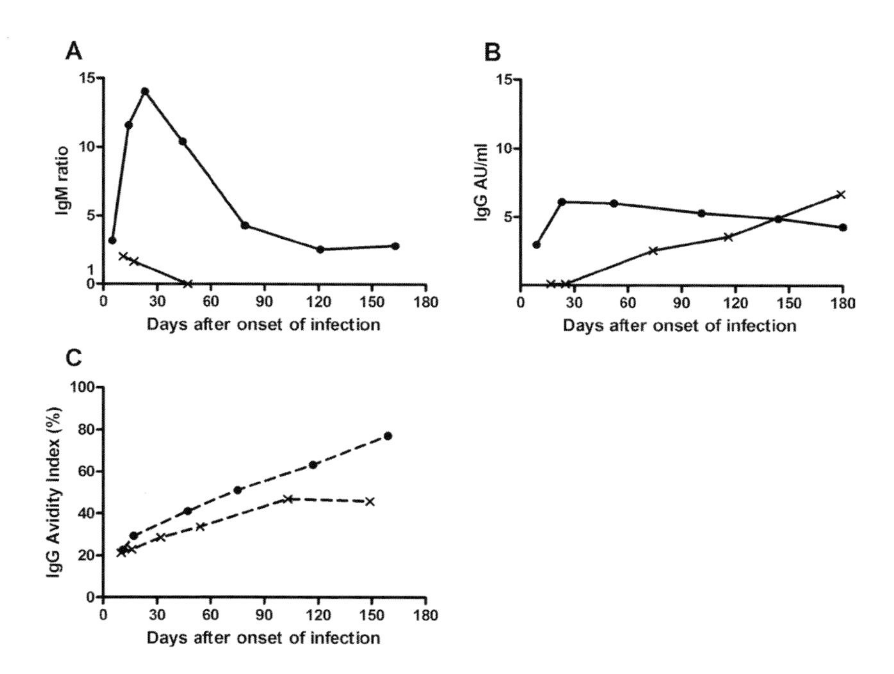

Figure II.18.2 Examples of individual variability in the development of HCMV-specific IgM (A), IgG (B) and IgG avidity maturation (C) responses. Serological follow-up of two different pregnant women with primary HCMV infection are shown in A, B, and C, respectively. Commercially available ELISA assays were used for IgM and IgG antibody determination. An in-house developed assay was used for IgG avidity determination.

following primary infection is substantial (about 40%), the seroconversion rate in pregnancy is rather low (about 2%), (iii) only a minority of HCMV-infected newborns develop symptoms/sequelae, (iv) no proven treatment is available for preventing intrauterine transmission and (v) interpretation of positive IgM results requires a battery of additional tests which, in turn, need to be interpreted (see also Chapter II.3).

IgG antibody

The appearance of virus-specific IgG in a previously seronegative subject (seroconversion) represents the most reliable approach to the diagnosis of primary infection. It is recognized that IgG do follow IgM antibody appearance but, unlike IgM, they persist for life (Fig. II.18.1). Indeed, determination of virus-specific IgG is the quickest assay to discriminate between immunity and susceptibility to HCMV in a given subject at any time. Exceptions to this rule are represented by immunocompromised patients in whom passively acquired antibodies or immunosuppression sometimes make serological determination of HCMV-specific immune status unreliable (see below).

Studies addressing the ontogeny of IgG response following primary HCMV infection have shown that IgG antibody to structural phosphoproteins (such as pp150, pp65, and pp28) and non-structural proteins (such as p52 and p72) are promptly synthesized, whereas the appearance of glycoprotein-specific antibodies is delayed (Schoppel *et al.*, 1997). Moreover, the kinetics of the appearance of IgG to different antigens is not synchronized and individual variations have been reported (Daiminger *et al.*, 1998). However, individual antibody profiles to different antigens seem to remain qualitatively stable over the years, whereas quantitative variations have been observed. A lysate of HCMV-infected cells (which includes structural and non-structural proteins as well as glycoproteins) is generally used as a solid-phase in diagnostic assays for IgG determination (Table II.18.1).

Two representative examples of IgG kinetics observed in two pregnant women are shown in Fig. II.18.2B. Correct determination of HCMV immune status is of critical importance in at least three groups of subjects, i.e. pregnant women (see also Chapter II.3), candidates for solid organ or haematopoietic stem cell donation/transplantation (see also Chapters II.13, II.14 and II.16), and blood donors. In the absence of an International Standard, definition of a cut-off value in a given assay is generally accomplished by comparing its performance with that of a similar commercial assay. The use of neutralization as a reference test to define the best cut-off for optimal discrimination between positive and negative ELISA results has been reported (Gerna *et al.*, 1992a).

IgG avidity

The terms 'affinity' and 'avidity' are sometimes inappropriately used as synonyms. In fact, although both terms refer to the concept of binding strength, affinity is the strength of binding of an antibody combining site to one corresponding antigen epitope, whereas avidity is the total strength of the antigen-binding capacity of the mixture of polyclonal antibodies. It has long been known that antibodies produced early after an infection react weakly and those produced later react more strongly forming more stable aggregates with the relevant antigen. This phenomenon has been exploited in the laboratory by developing assays in which the strength of antibody-antigen binding is measured by using chemical compounds, such as 6 M urea solution, which disrupts weak antibody–antigen complexes, but has little effect on strong antibody-antigen binding. As a result, low avidity antibodies can be easily differentiated from high avidity antibodies.

The principle of the assay is rather simple and is based on the dissociation of previously formed IgG-HCMV antigens immunocomplexes by a dissociating agent followed by calculation of the ratio between reactivity (expressed as optical density, relative light units, or relative fluorescence units) in wells treated with dissociating buffer and reactivity of untreated wells (avidity index). An original (and so far unique) approach to determine avidity of HCMV-specific IgG was developed by Abbott for the Architect platform. In this assay (based on the CLIA technology) test serum is pre-treated either with a suspension of HCMV-infected cell lysate (to remove high avidity IgG) or with buffer (control solution). The IgG avidity index is then measured by calculating the ratio of signals obtained with or without the pre-treatment step.

Since the original report (Blackburn *et al.*, 1991), a number of studies have confirmed the potential usefulness of IgG avidity determination in distinguishing between primary (with low IgG avidity) and non-primary (with high avidity IgG) HCMV infection, particularly in the pregnancy setting (Grangeot-Keros *et al.*, 1997; Bodeus *et al.*, 1998; Lazzarotto *et al.*, 1999). However, in order to correctly interpret IgG avidity results, one must take into consideration that (i) avidity cannot be reliably measured when IgG level is too low; (ii) individual variation in the rate of avidity maturation does exist; and (iii) different assays are subject to variations. The latter issue is of particular concern given the reported high variability in the performance of commercial assays (Revello *et al.*, 2010). Taken together, the above caveats indicate that standardization of avidity assays is lacking, and that IgG avidity determination cannot be used alone for the diagnosis of acute/recent primary HCMV infection.

Finally, little is known about avidity maturation to different viral proteins. The use of infected cell lysate in both IgG and IgG avidity determination should obviate major individual variability. Nevertheless, it is not unusual to observe strikingly different kinetics in avidity maturation as exemplified in Fig. II.18.2C. As reported in Table II.18.1, only one recently developed automated kit employs a cocktail of recombinant proteins for avidity determination (Elecsys CMV avidity, Roche Diagnostics, Germany).

Neutralizing and glycoprotein-specific antibody

Only a minor fraction of antibodies raised against viral proteins have direct antiviral activity *in vitro*. These antibodies are referred to as neutralizing antibodies (Nt-Ab) and possess a relatively high affinity and/or avidity for exposed structures on the virus surface. Nt activity and high avidity have been involved in the explanation of the potential benefit of passive immunization (Law and Hangartner, 2008).

As for Nt-Ab, HCMV could be considered a prototype of viruses that elicit a delayed neutralizing response which favours lifetime persistence in the host and is poorly protective because of its delayed appearance. Indeed, it is well known that (i) recurrent HCMV infections occur in the presence of circulating Nt-Ab; (ii) preconceptional maternal immunity, albeit substantially protective, still does not completely prevent congenital infection; and (iii) the Nt-Ab response until recently has been considered quite delayed following primary infection.

With reference to the latter statement, it is important to stress that *in vitro* determination of Nt activity in human sera, as routinely performed so far, is based on the use of HCMV laboratory strains, such as the AD169 strain, and human fibroblasts. The classical 14-day plaque reduction assay or the more rapid (7- or 3-day) micro-neutralization assay can be used. Nt-Ab titres are expressed as the highest serum dilution causing 50%, 90% or 99% (depending on the end-point used) reduction in the number of plaques or focus forming units (FFU) compared with counts in control wells.

Recently, it has been discovered that the UL131A-128 locus of the HCMV genome is indispensable for the infection of endothelial cells (Hahn *et al.*, 2004) and epithelial cells (Wang and Shenk, 2005), and that this locus is functionally missing in highly cell culture-adapted HCMV laboratory strains which no longer exhibit either endothelial cell or leucocyte tropism (Gerna *et al.*, 2006b). When human sera were tested for NT-Ab on both endothelial and epithelial cells instead of human fibroblasts, and a wild HCMV prototype strain (VR1814) was used, Nt activity was detected much earlier (within 10 days after the onset) and at much higher titres (32–64×) compared to Nt activity determined with fibroblasts (Gerna *et al.*, 2008b).

Thus, two quite distinct Nt-Ab responses can be determined *in vitro*, depending on the HCMV strain and cellular substrate used (Fig. II.18.1). Since the pentameric glycoprotein complex gH/gL/pUL128–131 is present in the HCMV envelope and is required for infection of both endothelial and epithelial cells, whereas the glycoprotein complex gH/gL/gO is required for infection of human fibroblasts (Kinzler *et al.*, 2002; Wang and Shenk, 2005; see also Chapter I.17), it is reasonable to infer that neutralizing activity in endothelial/epithelial cells is preferentially mediated by antibodies directed against the pentameric complex, whereas antibodies to gB and gH are primarily involved in HCMV neutralization in fibroblasts.

The IgG response to the gH/gL/pUL128–131 protein complex during primary and recurrent HCMV infection was recently investigated by both immunofluorescence and ELISA using epithelial cells infected with one or more adenoviral vectors, each expressing one of the five proteins of the complex (Genini *et al.*, 2011). ELISA-IgG seroconversion to the complex was detectable in all cases of primary infection examined (Fig. II.18.3), although slightly later (17–68, median 55.5 days) compared to the neutralizing antibody response as determined with HCMV (VR1814)-infected ARPE-19 epithelial cells, but much earlier than neutralizing activity as determined with HCMV (VR1814)-infected fibroblasts.

The same delay observed *in vitro* for the Nt-Ab response when measured with HCMV laboratory strains has been reported to occur also with glycoprotein (gB, gH)-specific antibodies (Schoppel *et al.*, 1997). From a diagnostic point of view, this delayed appearance of Nt-Ab is considered a valuable aid for the interpretation of HCMV serology in pregnant women, particularly when only a single serum sample is available (Eggers *et al.*, 1998).

Viral detection

Introduction to diagnostic assays

Following its discovery in the second half of the 1950s, HCMV was isolated and identified in HFF cell cultures, where it grew slowly as a cell-associated virus spreading from cell-to-cell and eventually resulting in slow-expanding cell foci (or plaques). Initially, in the presence of low virus amount in the sample examined, it took up to 2–3 weeks for the virus to develop a CPE focus revealing the presence of virus in the sample. In the mid 1970s, IHC, such as immunoperoxidase and immunofluorescence, mostly using convalescent-phase human sera, as well as ISH techniques, allowed for faster HCMV identification. However, it was in the mid 1980s that the development of the hybridoma technology allowed a significant improvement in the diagnosis of HCMV infection. Overall, mAbs recognizing different viral proteins were employed to develop rapid diagnostic assays allowing virus detection within 3–6 to 24

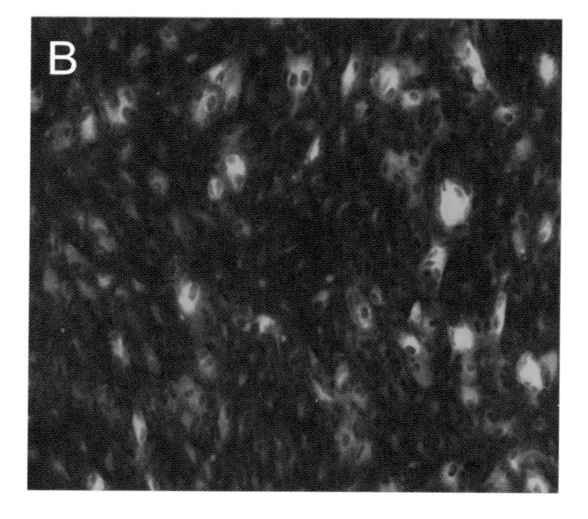

Figure II.18.3 A. Primary HCMV infection: acute-phase serum not reactive with epithelial cells expressing the HCMV gH/gL/pUL128–131 pentamer complex. B. Convalescent-phase serum from the same patient staining epithelial cells expressing the pentamer complex. In both panels, ARPE-19 cells were infected with five adenovirus vectors, each carrying a single gene of the pentamer complex.

hours either in clinical sample cells or in cell cultures following sample inoculation.

Among the rapid assays developed in cell cultures, the most useful was the so-called *shell-vial* assay. This assay was based on the use of the shell vial technique (Gleaves *et al.*, 1984) and allowed detection of HCMV-infected cells prior to CPE appearance. It was generally applied to detection of infectious virus in peripheral blood leucocytes (polymorphonuclear leucocytes and monocytes) of subjects with primary HCMV infection, newborns with congenital HCMV infection, or immunocompromised transplanted patients with disseminated HCMV infection. The shell-vial assay (or viraemia assay) was performed by inoculating an HFF monolayer with a test sample and visualizing infected cell nuclei with an IHC technique 16–24 hours after inoculation (Gerna *et al.*, 1990) using a p72/IE1-specific mAb. Subsequently, a pool of p72 mAbs was used to overcome problems of virus identification due to mutations in the IE region (Gerna *et al.*, 2003a).

A major breakthrough in the diagnosis of disseminated HCMV infection was achieved at the end of 1980s with the introduction of the *antigenaemia* assay, which was developed simultaneously by two groups of researchers in Groningen, The Netherlands, and in Pavia, Italy (van der Bij *et al.*, 1988; Revello *et al.*, 1989). This assay was aimed at detecting the presence of HCMV pp65 protein in nuclei of blood leucocytes by using a pool of pp65-specific mAbs (Gerna *et al.*, 1992b) and an IHC technique.

Antigenaemia offers a greater sensitivity than viraemia and a shorter turnaround time, requiring about 2 hours to complete. The assay is relatively easy to perform, but it may not be possible to perform with a neutrophil count less than 1000 cells/μl blood, and should be completed within 6–8 hours of collection to prevent a decrease in sensitivity.

However, the major achievement of the last two decades has been the development of *PCR* and *RT-PCR* techniques for detection of HCMV genomic DNA and RNA transcripts in clinical samples following amplification of target sequences by a cyclic enzymatic procedure. In addition, methods for detecting HCMV DNA through signal amplification have been developed using either branched DNA probes (Kolberg *et al.*, 1996) or RNA probes (Hebart *et al.*, 1998). A different amplification technique (nucleic acid sequence-based amplification, NASBA) was also developed for detection of HCMV transcripts in blood (Gerna *et al.*, 1999). More recently, automatic extraction procedures and a real-time readout format have improved the standardization of PCR conditions.

HCMV quantification

Viraemia assay

Quantification of viral load in blood was found to possess a high positive predictive value for the development of HCMV disease (Grossi *et al.*, 1995) and to correlate with virus replication. However, the shell vial assay (Fig. II.18.4B), which was based on the finding that a single PBL was able to infect a single HFF cell (i.e. the number of p72-positive HFF nuclei correlated with the number of leucocytes carrying infectious virus) lacked sensitivity (Gerna *et al.*, 2001).

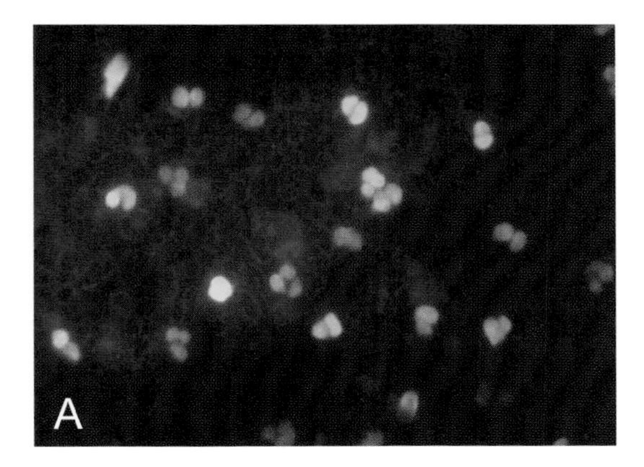

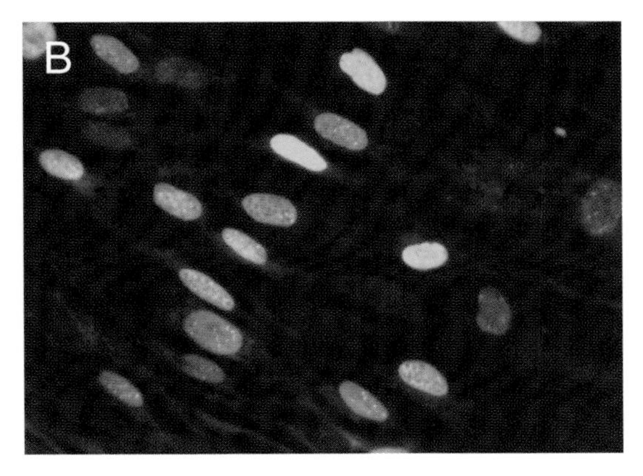

Figure II.18.4 (A) Antigenaemia assay, showing the presence of HCMV pp65 in the nucleus of some peripheral blood leucocytes. Indirect immunofluorescence. (B) Viraemia assay, showing the presence of p72/IE1-positive cells in the fibroblast cell monolayer, following inoculation of leucocytes carrying infectious virus from a patient with disseminated HCMV infection.

Antigenaemia assay

A major advance in the quantification of viral load was the introduction of the antigenaemia assay (Fig. II.18.4A). Following twice a week monitoring of antigenaemia levels in solid-organ transplant recipients (SOTR), threshold values associated with appearance of HCMV disease were identified (Grossi *et al.*, 1995). Thus, antigenaemia levels preceding the appearance of clinical symptoms were identified to start presymptomatic (pre-emptive) therapy in transplanted patients (Locatelli *et al.*, 1994; Grossi *et al.*, 1995; Boeckh, 1999). The quantification of HCMV antigenaemia also allowed monitoring of antiviral treatment, including the emergence of drug-resistant HCMV strains (Baldanti *et al.*, 1998, 2004). In addition, in the late 1990s, the *in vitro* transfer of HCMV pp65 into PBL allowed standardization of the assay (Gerna *et al.*, 1998a; Revello *et al.*, 1998a). However, standardization of the assay could not avoid two major limitations: one, is that it cannot be used in HSC transplant recipients during engraftment (Limaye *et al.*, 1997); and the other, is that a paradoxical rise in antigenaemia levels during GCV treatment may occur independently of viral replication (Gerna *et al.*, 1998b, 2003b, 2005).

PCR-based molecular assays

Pitfalls of the antigenaemia assay were overcome by the introduction of PCR-based molecular assays (Gerna *et al.*, 1991; Humar *et al.*, 1999; Emery *et al.*, 2000; Razonable *et al.*, 2003), which have been rapidly replaced by real-time PCR techniques. These techniques are more precise and rapid, possess a broader linear range and a higher output, and entail a lower risk of carryover contamination (Mengelle *et al.*, 2003). Although viral DNA can be quantified in a number of clinical samples, including different blood fractions (leucocytes or plasma) or whole blood (Gerna *et al.*, 1994), whole blood is now accepted as the specimen of choice for disseminated infections since it allows earlier and higher viral load detection compared to plasma. Leucocytes are no longer used owing to the cumbersome procedure and somewhat unrealistic evaluation of viral load.

The use of real-time PCR does not avoid the risk of cross-contamination between clinical samples. To reduce this risk, suitable automated extraction platforms are available. As for target amplification, oligonucleotide primers and probes for amplification and detection of nucleic acid, respectively, are selected from conserved nucleotide sequences within a viral gene; these products constitute the first level of sensitivity and specificity for quantitative real-time PCR. Together with other components, this assay is subsequently adjusted to permit the polymerase enzyme to function optimally and to produce sensitive and specific signals from labelled probes that are proportional to the amount of the target DNA present in the blood sample (Espy *et al.*, 2006).

In recent years, a variety of commercial and laboratory-developed assays have been utilized for CMV DNA quantification. To assess interlaboratory variability in quantitative HCMV viral load testing, a panel of samples was distributed to 33 laboratories in the USA, Canada and Europe, where testing was performed using commercial reagents or laboratory-developed assays (Pang *et al.*, 2009). Variations observed in reported results for individual samples ranged from $2.0\log_{10}$ (minimum) to $4.3\log_{10}$ (maximum), and variation was greater with low viral load values. The use of commercially available reagents

Table II.18.2 Real time PCR for HCMV detection in blood

Manufacturer	Assay	Target	Specimen	Range (copies/ml)	Automation
Roche	COBAS® Ampliprep/ COBAS® TaqMan CMV	UL54	Plasma	150–10^{7a}	Complete
Abbott	Real-Time CMV	Unknown	Whole blood	40–10^8	Complete
			Plasma	20–10^8	Complete
QIAGEN	*Artus*® CMV qs-RGQ	UL122	Plasma	79–10^8	Complete
Argene-BioMérieux	CMV r-gene™	UL83	Whole blood Plasma	Up to 10^8 Up to 10^8	ASR ASR
Nanogen	Q-CMV Real-Time	UL123	Not specified	20–5×10^5	ASR
Cepheid	Affigene® CMV trender	Unknown	Whole blood Plasma	10^2–10^7 10^2–10^7	ASR ASR

Abbreviation: ASR, analyte-specific reagent.
[a]Conversion to International Units (IU)/ml available, corresponding to 137–9.1×10^6.

and procedures was associated with less variability compared with laboratory-developed assays. Another study evaluated 15 Italian laboratories belonging to different transplantation centres who performed testing using five commercial reagents and two laboratory-developed assays (Lilleri *et al.*, 2009). The variability range was wide (about 2 $\log_{10}$) for sample containing low amounts of HCMV DNA (< 1000 copies/ml), but it decreased with increasing concentrations of HCMV DNA. For HCMV DNA levels ≥ 5000 copies/ml, the different methods provided results within a ± 0.5 $\log_{10}$ variability range, while the 80% range (range in which 80% of the results obtained will fall) was within ± 0.3 $\log_{10}$ or less. If we consider that the precision of quantitative nucleic acid amplification techniques (QNAT) used to test viral load is such that changes in values should be greater than 3-fold (0.5 $\log_{10}$) to represent biologically important changes in viral replication (Caliendo *et al.*, 2001), an acceptable level of variability was reached among different methods for HCMV DNA quantitation in samples containing a clinically significant viral DNA amount. Intralaboratory variance was negligible.

Just a few months ago the first WHO International Standard for Human Cytomegalovirus DNA quantification, (NIBSC code 09/162, version 3.0. dated 30 November 2010) was made available. It is intended to be used in the standardization of NAT-based assays for HCMV. The standard consists of a whole virus preparation of the HCMV Merlin strain (Dolan *et al.*, 2004) formulated in an universal buffer containing Tris-HCl and human serum albumin. The material is lyophilized in 1.0 ml aliquots and stored at −20°C. The standard was evaluated in a worldwide collaborative study involving 32 laboratories performing a range of NAT-based assays for HCMV (Freyer *et al.*, 2010). The standard has been assigned a concentration of 5,000,000 International Units (IU) when reconstituted in 1.0 ml nuclease-free water. Once reconstituted, it should be diluted in the matrix appropriate to the material being calibrated and should be extracted prior to HCMV DNA measurement.

Another limitation to the optimal use of real-time PCR is the genomic variability of HCMV. In this regard, not all primer–probe combinations are equally efficient in amplifying viral DNA. We observed in our system that in 10–20% of samples tested, figures provided by different assays may exceed the 0.5 $\log_{10}$ range. In this respect, sequence analysis of the real-time target region (US8) showed a substitution (from GG to AA) of the 3'nucleotides (nt position 271–272) of the genome sequence recognized by the TaqMan probe (Lilleri *et al.*, 2009).

A variety of commercial molecular assays have been developed (Table II.18.2). One of the more widely used is COBAS Amplicor CMV Monitor test (Roche Diagnostics, Indianapolis, IN), an end-point detection PCR-based assay that amplifies a 365 base pair region of the HCMV polymerase gene. The dynamic range of the assay using a plasma-based standard is reported to be between 400 and 50,000 copies/ml of plasma (Caliendo *et al.*, 2001). The assay has been designed for use with plasma, leucocytes, and whole blood specimens. Another commonly used assay is the Hybrid Capture System CMV DNA test (version 2; Digene Corporation, Gaithersburg, MD), a signal amplification method using an RNA probe that targets 17% of the HCMV genome. The target is detected by employing antibodies that specifically bind RNA:DNA hybrids. The dynamic range is 1400–560,000 copies/

ml (Wattanamano *et al.*, 2000). The assay has been designed for whole blood specimens.

Recently, several analyte-specific reagents (ASRs) for HCMV have become available. One example is Artus CMV ASR (Hamburg, Germany; primers, probes, internal control, standards targeting a 105-base pair region of the major IE coding sequence), which is distributed in one version by Abbott Molecular (Des Plaines, IL), and in another version by QIAGEN (Valencia, CA). In a recent study (Caliendo *et al.*, 2007), the performance characteristics of the Abbott test and the QIAGEN test compared to Amplicor CMV Monitor and Hybrid Capture assays were evaluated. For plasma specimens, the Abbott test had a limit of detection of 2.3 $\log_{10}$ copies/ml and a linear range up to at least 6.0 $\log_{10}$ copies/ml. Viral load obtained from plasma specimens tested by Abbott and QIAGEN tests were in close agreement (mean difference 0.144 $\log_{10}$ copies/ml). These data support the concept that Abbott and QIAGEN reagents provide laboratories with additional tools that can be used reliably for HCMV viral load testing.

Diagnosis of HCMV infection in different clinical settings

Pregnancy

Pregnant women represent the most important and critical target for diagnosis of HCMV infection given the risk of fetal infection associated with primary HCMV infection in pregnancy (see also Chapter II.3). Routine HCMV testing of pregnant women is not recommended in any country. However, determination of HCMV immune status is *de facto* performed in a large proportion of pregnant women at least in some European countries such as Italy, Belgium, Germany and France. This information is based on (i) personal experience relevant to Italy; (ii) available literature data; and (iii) the fact that it is known that several million HCMV IgG/IgM tests are performed yearly in Europe for HCMV antibody testing of pregnant women (personal communication).

Primary infection

Since maternal infections are mostly asymptomatic, it is common practice to test pregnant women for the presence of both IgG and IgM in order to detect subclinical infections. Such an approach, however, carries the potential risk of detecting HCMV-specific IgM in an higher number of women than those actually experiencing an acute/recent primary infection. Indeed, in our experience as well as in the experience of others, only about 20% of positive IgM results are actually due to an acute/recent primary HCMV infection (Revello and Gerna, 2002). A correct and quick interpretation of any IgM positivity is therefore mandatory in order to avoid anxiety and unnecessary pregnancy termination (Guerra *et al.*, 2007).

It is beyond the scope of this chapter to discuss the technicalities involved in performing a correct diagnosis of primary HCMV infection (Revello and Gerna, 2002). However, some general rules may be of help.

Diagnosis of acute/recent primary infection can be reliably achieved only when two or more of the following parameters are concomitant: (i) IgG seroconversion; (ii) presence of virus-specific IgM; (iii) presence of low IgG avidity; (iv) presence of DNAemia; (v) absence of neutralizing antibody (as determined in HFF). In no instance, should a diagnosis of acute/recent primary HCMV infection be based on the presence of only one of the above criteria for the following reasons: (i) passively acquired IgG antibody, specimen mix-up or technical problems in assay performance (particularly when fully automatized systems are used), may result in false seroconversions; (ii) IgM positivity may be due to interfering factors, such as cross-reactivity, polyclonal activation due to other infections or IgM persistence; (iii) low/intermediate IgG avidity may be detected for many months after the onset of infection depending on individual variability and the assay used (Revello *et al.*, 2010); (iv) the diagnostic value of DNAemia in immunocompetent individuals has to be defined in each laboratory since, contrary to our long-standing experience of exclusive detection in immunocompetent subjects undergoing primary infection (Revello *et al.*, 1998b, 2001), circulating DNA has been recently detected also in healthy seropositive women (Arora *et al.*, 2010); (v) Nt-Ab may be absent either in the early phase of primary infection or in seronegative individuals.

From a diagnostic standpoint, determination of IgG avidity, DNAemia and neutralizing antibodies are considered indispensable adjunctive tests for interpretation of positive IgM results. Only reference laboratories with an array of additional commercial and non-commercial assays, and virologists with specific experience and competence in interpreting laboratory data should attempt to interpret correctly a positive IgM result.

Once diagnosis of primary infection is confirmed, defining its presumed onset is important for prognosis, management of pregnancy as well as counselling. Anamnestic data and/or clinical history will help in timing the onset as long as they fit chronologically with serological/virological data (Revello and Gerna, 2002). Gestation timing at the first testing is crucial for defining whether primary infection occurred before or during pregnancy (Revello *et al.*, 2002). More specifically, pregnant women should be tested as early as possible (ideally no later than 12 weeks of gestation). First testing after 20 weeks' gestation should be avoided unless suggested by clinical symptoms. Preconceptional testing represents the best way to assess HCMV immune status of women in child-bearing age.

Recurrent infection

Recurrencies include reactivations of the endogenous HCMV strain(s) as well as reinfection with new HCMV strains. From the diagnostic standpoint, a recurrent infection is diagnosed in an immunocompetent IgG-seropositive individual, whenever infectious HCMV or viral DNA is recovered from bodily fluids (urine, saliva, genital secretions, human milk, etc.) in the absence of serological markers of recent primary infection (presence of IgM antibody and/or low-intermediate IgG avidity).

It must be emphasized that the presence of IgM antibody does not substantiate *per se* diagnosis of recurrent HCMV infection. In fact, as already mentioned above, the issue of whether recurrent HCMV infections elicit an IgM response in the immunocompetent host (in pregnant women, in particular) has never been properly addressed. Similarly, an increase in the IgG antibody level cannot be considered a marker of an ongoing recurrence in the absence of virus detection/isolation. Therefore, serological monitoring of HCMV IgG-seropositive pregnant women has no clinical value.

Recent reports seem to indicate that reinfections rather than reactivations are more likely to be responsible for symptomatic congenital infection (Yamamoto *et al.*, 2010). Reactivations and reinfections cannot be distinguished by using routine serological/virological techniques. The detection of IgG antibodies with new antigenic specificities to the surface glycoproteins H and B has been proposed as a tool for diagnosing reinfections (Novak *et al.*, 2009). However, at the moment, such an approach is restricted to investigational purposes.

Finally, the search for HCMV in bodily fluids in pregnant women has no utility either for diagnosis or management of HCMV infection and is therefore not recommended.

Maternal prognostic markers of in utero transmission

Maternal markers that could reliably identify pregnant women at risk of transmitting the HCMV infection to the fetus have not been found yet. In the past, a limited number of studies have been performed to determine whether specific deficits in the antibody response were associated with intrauterine transmission with variable and sometimes contrasting results (Table II.18.3). A recent study performed by our group failed to show a significant difference in the kinetics of antibodies neutralizing the infection of endothelial cells (HUVEC) between a group of 18 transmitter and a group of 23 non-transmitter mothers (Revello *et al.*, 2010).

On the other hand, major differences between women who transmitted the infection and those who did not were observed when HCMV-specific cell-mediated immunity was investigated (Fig. II.18.5). In fact, a significantly lower and delayed lymphoproliferative response was reported in transmitter compared to

Table II.18.3 HCMV-specific antibody response in transmitter (TR) and non-transmitter (non-TR) mothers

Reference	Parameter	Assay	Finding
Alford *et al.* (1988)	IgG	RIP	More intense in TR
Boppana and Britt (1995)	IgM	RIA	Higher titres in TR
Lazzarotto *et al.* (1998)	IgM	IB	Higher number of reactive bands in TR
Revello *et al.* (2006)	IgM	ELISA	No difference
Boppana and Britt (1995)	IgG to gB	RIA	Increased in TR
Eggers *et al.* (2001)	IgG to gB	IB	Increased in TR
	IgG to gH	IB	No difference
Boppana and Britt (1995)	Nt-Ab	NTA on fibroblasts	Increased in non-TR
Revello and Gerna (2010)	Nt-Ab	NTA on HUVEC	No difference
Boppana and Britt (1995)	Avidity	RIA	Lower in TR
Revello *et al.* (2006)	Avidity	ELISA	No difference

Abbreviations: HUVEC, human umbilical vein endothelial cell; IB, immunoblot; NTA, neutralization assay; RIA, radioimmunoassay; RIP, radioimmunoprecipitation.

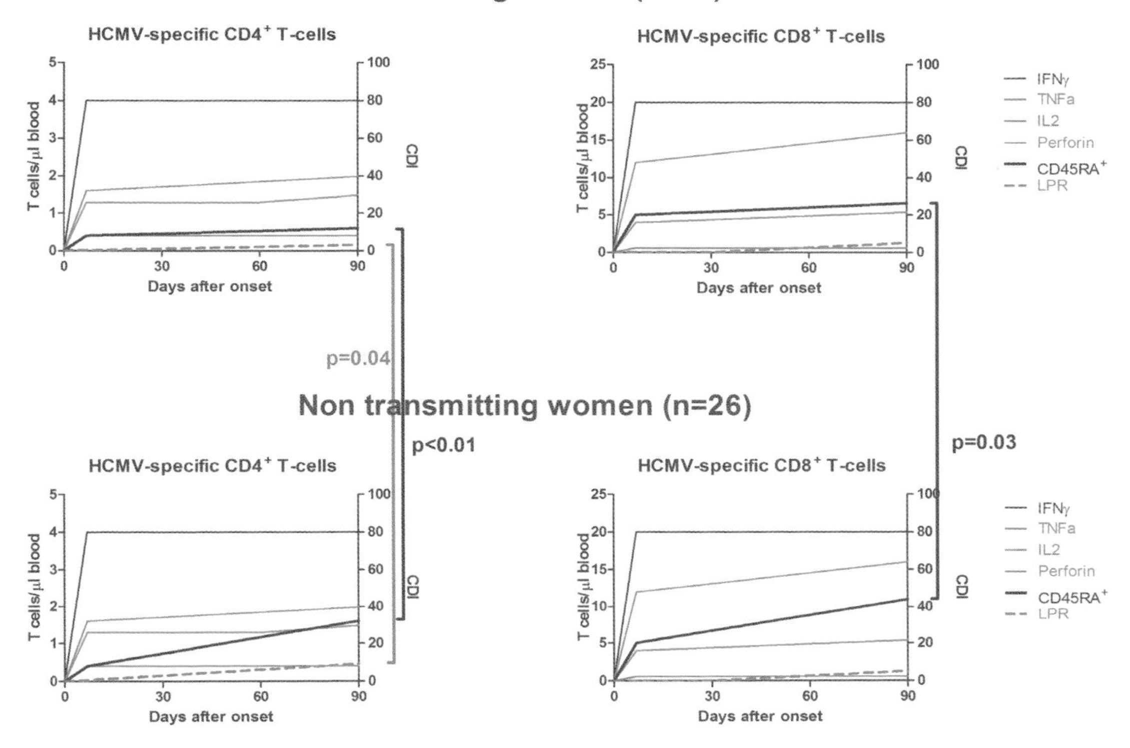

Figure II.18.5 Comparison of different immunological parameters between a group of pregnant women transmitting and a group of women not transmitting the virus to the fetus. Two parameters were significantly different between the two groups: lymphoproliferative response (LPR, dotted purple line) for CD4+ ($P=0.04$) T-cells, and CD45RA+ T-cells (dark blue line) for HCMV-specific CD4+ ($P=0.007$), and CD8+ ($P=0.03$) T-cells. CDI, cell division index.

non-transmitter mothers (Revello *et al.*, 2006; Lilleri *et al.*, 2007a). Furthermore, it was found that a higher percentage of HCMV-specific effector memory T-cells that revert to the RA isoform of CD45 (CD45RA+CCR7+) during the first months after infection, was associated with a lower risk of transmitting HCMV infection to the fetus (Lilleri *et al.*, 2008, Fornara *et al.*, 2011) Thus, the development of the cell-mediated immune response appears to be crucial for the control of vertical transmission of HCMV infection.

Unfortunately, none of the parameters listed above can be used for counselling on an individual basis since results partially overlap between transmitter and non-transmitter mothers.

Fetus

Diagnosis of congenital infection during fetal life may be accomplished only by a search for virus or viral components in amniotic fluid (AF) following amniocentesis or in fetal blood (FB) following cordocentesis.

Amniocentesis

HCMV presence in AF can be detected by conventional, rapid and molecular assays. HCMV isolation from AF has represented for quite some time the only approach to virus detection in AF, and is considered the gold standard for prenatal diagnosis. As mentioned above, by using mAbs to the major IE protein p72 (Gerna *et al.*, 2003a) and the shell vial technique (Gleaves *et al.*, 1984), HCMV can be detected within 24 hours of sample collection. However, virus isolation is still considered a very valuable assay for confirmation of results obtained by molecular techniques.

Molecular techniques (PCR, nested PCR and real-time PCR) have greatly improved the sensitivity of HCMV DNA detection in AF. Studies conducted in large series of pregnancies have reported sensitivities ranging around 80–90% (Liesnard *et al.*, 2000; Enders *et al.*, 2001; Revello and Gerna, 2002). Importantly, specificity has been reported to be very high (98–100%), thus indicating that, provided that these highly sensitive assays are performed properly, even small amounts of viral DNA in AF correlate with congenital infection at birth. The advent of the real-time PCR methodology (which avoids the risk of false-positive results due to carryover products) represented a great improvement in the specificity of prenatal diagnosis results.

However, it must be kept in mind that even when the most sensitive techniques are used and invasive procedures are performed at the most opportune gestation time, a small number of false-negative prenatal diagnosis results cannot be avoided. In our experience, 8% of negative prenatal diagnosis results are not confirmed at birth (Revello *et al.*, 2011). Delayed transmission, i.e. transmission occurring after amniocentesis represents an objective limitation of prenatal diagnosis and, as such, it should be discussed with the woman/couple during counselling sessions in order to avoid recriminations, in case of negative results at amniocentesis which are not confirmed at birth.

Finally, it must again be emphasized that prenatal diagnosis is a very delicate task leading to irrevocable decisions and, thus, it should never be based on the result of a single assay.

Cordocentesis

In general, examination of FB does not improve the sensitivity of prenatal diagnosis of fetal infection with respect to virus detection in AF. In fact, determination of virus-specific IgM in fetal blood has a limited diagnostic value due to its low (25% to 75%) sensitivity (Revello *et al.*, 1999). Similarly, low sensitivities have been reported for detection of virus or viral components in fetal blood. Only determination of DNAemia seems to provide sensitivity levels comparable to those obtained in AF (92% to 100%) (Enders *et al.*, 2001; Benoist *et al.*, 2008), when cordocentesis is performed after 30 weeks' gestation.

However, examination of fetal blood may confirm results obtained with AF, and when tests are performed quantitatively, may provide important prognostic information (see below). In addition, other haematological, biochemical and immunological non-specific parameters may be assessed in fetal blood and may contribute to a more comprehensive prognostic evaluation of fetal infection.

Fetal prognostic markers of congenital disease

Once diagnosis of fetal infection is made, the main issue is whether it is possible to predict which infected fetuses will be symptomatic at birth or later in life. The clinical significance of HCMV load in AF has been repeatedly investigated over the years by a number of researchers. Apart from initial reports from a single group showing that high DNA levels were predictive of poor outcome, other studies have clearly indicated that other variables such as gestational age at time of amniocentesis and the time elapsed since maternal infection could influence viral load irrespective of fetal outcome. In particular, while a low viral load in AF was consistently found to be associated with asymptomatic congenital infection, it was observed that high viral load in AF was associated with either symptomatic or asymptomatic congenital infection.

The delicate issue of predicting fetal outcome has been recently addressed by our group in a retrospective study, in which a cohort of pregnant women who suffered from primary HCMV infection at different times during gestation and who underwent amniocentesis and cordocentesis was examined (Fabbri *et al.*, 2010). A panel of non-viral and viral assays was performed on fetal blood samples, and the results were compared between infected and non-infected, and between symptomatic and asymptomatic fetuses. The best non-viral factor for differentiating symptomatic from asymptomatic congenital infections was β_2-microglobulin followed by platelet count, whereas the best virological markers were IgM antibody and DNAemia. When two virological markers at established cutoffs (3.0 for IgM ratio and 30,000 DNA copies/ml blood for DNAemia level or per 10^7 peripheral blood leucocytes), along with β_2-microglobulin at the established cutoff (11.5 mg/l, and platelet count at the established cutoff of 50,000/μl blood) were analysed in combination (and when at least three out of four were abnormal), the diagnostic efficacy was very acceptable with 100% specificity, 100% positive predictive value, 94% negative predictive value and 86% sensitivity. In another recent study, low platelet count determination was shown to be an independent indicator of poor outcome (Benoist *et al.*, 2008). Therefore, fetal blood sampling appears justified whenever additional information on fetal conditions are requested by the pregnant woman/couple in order to make an informed decision regarding termination/continuation of pregnancy, after diagnosis of fetal infection.

Newborn

Examination of newborns for congenital HCMV infection is required to confirm a prenatal diagnosis as well as for investigating transmission to fetus following primary infection in the mother. Specimens must be collected within the first 2 weeks of life in order to exclude peri-postnatal infection acquired via contacts with infected vaginal secretions or breast milk.

Virus detection

The gold standard diagnostic assay is virus recovery from urine or saliva samples. PCR for HCMV DNA detection in urine has been reported to be 100% sensitive and specific when compared with conventional virus isolation (Demmler *et al.*, 1988). However, urine is not an easy sample to collect for screening purposes. A recent study conducted on 34,989 infants reported that real-time PCR performed on saliva (either liquid or dried) specimens was 100% and 97.4% sensitive, respectively, and 100% specific when compared to rapid culture (Boppana *et al.*, 2011). Some potential problems with the use of saliva for screening for congenital HCMV infection may arise from oral contamination by virus present in the maternal genital tract or breast milk. However, collection of saliva samples on the second day of life should minimize the chance of false-positive results.

HCMV DNA detection in blood

Detection of HCMV DNA in neonatal blood is an additional approach potentially useful for screening purposes since viral DNA has been detected by PCR both in serum and in peripheral blood leucocytes of both symptomatic and asymptomatic HCMV-infected newborns (Nelson *et al.*, 1995; Revello *et al.*, 1999). Moreover, since dried blood spot (DBS) are routinely collected in the first 3–5 days of life in many countries for neonatal screening of metabolic and hereditary diseases, the search for viral HCMV DNA in DBS has attracted considerable interest. Despite the potential advantages of such an approach (DBSs can be stored for years, thus they may provide an important potential tool for retrospective diagnosis), the actual sensitivity and specificity of HCMV DNA detection in DBSs remains to be assessed. In fact, sensitivities varying from 71% to 100% and specificities around 100% have been reported (reviewed in Barbi *et al.*, 2006), whereas a recent study performed prospectively on 20,448 infants reported very low sensitivity (28.3–34.4% when a single or two primers were used, respectively) when compared with saliva rapid culture (Boppana *et al.*, 2010). On the other hand, specificity was excellent (99.9%). Finally, IgM antibody determination at birth is of limited value given its low sensitivity (Revello *et al.*, 1999).

Immunocompromised patients

Diagnostic and therapeutic impact of HCMV DNA quantification

The quantification of HCMV DNA has allowed the identification of threshold levels associated with onset of clinical symptoms and, thus, prevention of onset of HCMV disease by adopting therapeutic interventions prior to symptom onset (pre-emptive therapy). This approach is advantageous as compared with universal prophylaxis (treatment of all patients from the day of transplant for a period of 3–6 months) as a smaller number of patients are treated for a shorter period of time. Other advantages of pre-emptive therapy include savings in terms of drug toxicity and patient management costs (Kusne *et al.*, 1999), and, for some authors, this approach has an efficacy comparable to that of prophylaxis in preventing the indirect effects of HCMV infection, such as graft failure/rejection and fungal/bacterial infections (Singh *et al.*, 2005, 2006).

Today, several transplantation centres (and particularly HSCT centres) still do not use any cut-off for initiation of pre-emptive therapy, thereby starting antiviral treatment upon first virus detection in blood; and cut-offs used for HSCTR and SOTR in different centres vary greatly. Using continuous monitoring of viral load (twice a week), we adopted a cut-off of 300,000 DNA copies/ml whole blood before starting pre-emptive therapy in SOTR, while we use a cut-off of 30,000 DNA copies/ml whole blood in HSCTR. In SOTR, the above reported DNAemia cut-off was eventually selected after a trial in which it was compared with a previously selected antigenaemia cut-off of 100 pp65-positive/$2x10^5$ leucocytes (Gerna *et al.*, 2007). Results obtained on a large number of patients showed that: (i) the selected DNAemia cut-off significantly reduced the number of patients receiving treatment compared with antigenaemia; (ii) was capable of guiding pre-emptive therapy of both primary and reactivated HCMV infections; and (iii) did not significantly alter the overall duration of treatment. Notwithstanding the results of our study, several transplantation centres still use a differential approach to preventative treatment of primary and reactivated infections. Recent guidelines published by The Transplantation Society (TTS) of America suggested prolonging prophylaxis from 3 to 6 months in D^+/R^- SOT patients (Kotton *et al.*, 2010). Since our cut-off was determined on a large number of SOTR, and as we did not encounter problems with HCMV disease, we continue to use our cut-off which is higher than those used by other transplantation centres.

Based on previous observations in HSCTR, where the use of a quantitative, rather than qualitative, DNAemia cut-off appeared suitable to increase the negative predictive value without altering the positive predictive value compared to qualitative antigenaemia (Lilleri *et al.*, 2004), a randomized prospective open-label study aimed at defining the effectiveness of the real-time DNAemia vs. antigenaemia cut-off as the guiding parameter for pre-emptive therapy in paediatric

HSCTR was performed (Lilleri *et al.*, 2007b). Using a cut-off of 10,000 DNA copies/ml whole blood vs. first positive antigenaemia, a significantly lower number of patients required treatment in the DNAemia arm with respect to the antigenaemia arm (Lilleri *et al.*, 2007b). Thus, DNAemia cut-off-guided pre-emptive treatment was shown to be safe and effective in controlling HCMV infection in HSCTR, thereby avoiding unnecessary treatment of a significant proportion of patients. Similar results were obtained in an adult HSCTR population (Gerna *et al.*, 2008a). A similar approach was proposed by another group (Verkruyse *et al.*, 2006).

Diagnosis of antiviral drug resistance

The acquisition of mutations in UL97 phosphotransferase appears to be a crucial step in the selection of GCV-resistant HCMV strains. HCMV strains with mutations in key regions of the viral enzyme have been proven to be highly resistant to GCV (Chou *et al.*, 1995; Lurain *et al.*, 1996; Baldanti *et al.*, 1998). In addition to UL97, mutations in UL54 impact GCV susceptibility (Cihlar *et al.*, 1998). However, mutations in UL54 are less common and have been reported only in patients harbouring GCV-resistant UL97-mutant strains. HCMV strains with mutations in both UL97 and UL54 showed very high levels of resistance to GCV, but retained foscarnet susceptibility. Mutations in UL54 are responsible for foscarnet resistance (Baldanti *et al.*, 1996) and are responsible for simultaneous resistance to cidofovir and GCV (Cihlar *et al.*, 1998).

Methods to determine antiviral drug resistance are either phenotypic or genotypic. Among phenotypic methods the most popular are the plaque reduction assay, *in situ* ELISA, virus yield and DNA reduction assay. Genotypic methods include sequencing, restriction fragment length polymorphism analysis and probe-specific hybridization or primer-specific amplification.

Methods determining the level of viral replication *in vitro* in the presence of the drug provide direct evidence of susceptibility or resistance to antiviral compounds. On the other hand, the use of molecular techniques for detection of drug-resistance associated mutations can greatly reduce the time required for the identification of HCMV drug-resistant strains (Baldanti and Gerna, 2003; see also Chapter II.19).

Diagnosis of HCMV infection by cell-mediated immunity

Adaptive immune responses to HCMV infection consist of both humoral (serological) and T-cell-mediated immune responses. HCMV is a potent immunogen, eliciting a strong immune response from both arms of the immune system (see also Chapters II.7 and II.10). While the protective role of the antibody (including neutralizing antibodies) response remains to be elucidated, based on the observation that HCMV disease usually occurs in immunocompromised patients with profound cellular immune deficiency, it is well accepted that the cell-mediated immune response is the major mechanism of response and protection against HCMV infection/disease. In some cases, namely in young immunocompromised patients, it is impossible to discriminate between seronegative patients and patients with low antibody titres deriving from blood product administration. Under these circumstances, measurement of T-cell immunity may be decisive.

Within the T-cell immune response, HCMV-specific $CD4^+$ and $CD8^+$ T-cells, as well as $\gamma\delta$ T-cells all seem to play an important role in the response and in the control of HCMV infection. This conclusion was reached in both SOT and HSCT recipients, in that $CD4^+$ T-cells were required for expansion of $CD8^+$ CTL (Einsele *et al.*, 2002). The most important targets of HCMV-specific $CD4^+$ T-cells are products of UL55 (gB) and UL83 (pp65) (Crough and Kanna, 2009), while the most immunodominant targets for $CD8^+$ T-cells are pUL123 (IE-1), pUL122 (IE-2) and pUL83 (pp65) (Waldrop *et al.*, 1997; Sylwester *et al.*, 2005; Crough and Kanna, 2009). Finally, the $\gamma\delta$ T-cell subset is clearly involved in the anti-HCMV immune response and their expansion is associated with resolution of HCMV infection (Knight *et al.*, 2010).

Among immunological methods for determination of the T-cell response, the most widely used are peptide-MHC class I tetramers (Gratama *et al.*, 2001; 2010) and, among functional assays, the intracellular cytokine staining (ICS), and the enzyme-linked immunospot (ELISPOT) assay both measuring T-cells producing cytokines such as IFN-γ. Finally, the commercially available quantiFERON-CMV assay measures IFN-γ released into the blood.

In our laboratory, a novel assay measuring the number of $CD4^+$ and $CD8^+$ T-cells producing cytokines after stimulation with HCMV-infected autologous DCs, was developed (Lozza *et al.*, 2005). ICS and flow cytometry analysis allowed quantification of functional HCMV-specific $CD4^+$ and $CD8^+$ T-cells. By using this assay, it was possible not only to determine the immune response to HCMV in primary HCMV infection (Fornara *et al.*, 2011), but also to establish, in the immunocompromised transplanted patients, levels of HCMV-specific $CD4^+$ and $CD8^+$ T-cells conferring protection against HCMV reactivation. These levels were 0.4 $CD4^+$ and $CD8^+$ T-cells each/μl blood in the immunocompetent subject (Lozza *et al.*, 2005), again

0.4 T-cells/μl blood for both $CD4^+$ and $CD8^+$ in SOTR (Gerna *et al.*, 2006a), and 1.0 $CD4^+$ and 3.0 $CD8^+$ T-cells/μl blood in HSCTR (Lilleri *et al.*, 2006).

Conclusions

Major advancements have been achieved in recent years in the diagnosis and management of HCMV infection in different clinical settings. These achievements have been made possible thanks to developments in molecular biology, as well as molecular and clinical immunology. Diagnosis of primary HCMV infection has greatly improved, while virus detection in clinical samples has reached a high level of sensitivity, thus allowing correlation of viral load with clinical symptoms and, thus, knowledge of when to intervene. Antiviral drugs control HCMV infection, but repeated treatment courses must be administered (with the risk of emergence of drug-resistant HCMV strains) to control reactivation episodes in the immunocompromised patient, if the immune system is not reconstituted. The main target of future research activity in this field is the development of a protective vaccine which could prevent HCMV infection in the immunocompetent woman in child-bearing age and control reactivation episodes in the immunocompromised (transplanted) patient.

Acknowledgements

The authors are grateful to Milena Furione for her help in writing viral DNA PCR quantification methods by PCR, and to Daniele Lilleri and Vanina Rognoni for preparing the Figures. We are also indebted to Daniela Sartori for continuous and thoughtful editing assistance, and to Laurene Kelly for revision of the English. This work was supported by grants from Fondazione Carlo Denegri, Torino, and Fondazione CARIPLO, Milano, Italy, and by Ministero della Salute, Ricerca Corrente Fondazione IRCCS Policlinico San Matteo (grant 80206).

References

Alford, C.A., Hayes, K., and Britt, W. (1988). Primary cytomegalovirus infection in pregnancy: Comparison of antibody responses to virus-encoded proteins between women with and without intrauterine infection. J. Infect. Dis. *158*, 917–924.

Arora, N., Novak, Z., Fowler, K.B., Boppana, S.B., and Ross, S.A. (2010). Cytomegalovirus viruria and DNAemia in healthy seropositive women. J. Infect. Dis. *202*, 1800–1803.

Baldanti, F., and Gerna, G. (2003). Human cytomegalovirus resistance to antiviral drugs: diagnosis, monitoring and clinical impact. J. Antimicrob. Chemother. *52*, 324–330.

Baldanti, F., Underwood, M.R., Stanat, S.C., Biron, K.K., Chou, S., Sarasini, A., Silini, E., and Gerna, G. (1996). Single amino acid changes in the DNA polymerase confer foscarnet resistance and slow-growth phenotype, while mutations in the UL97-encoded phosphotransferase confer ganciclovir resistance in three double-resistant human cytomegalovirus strains recovered from patients with AIDS. J. Virol. *70*, 1390–1395.

Baldanti, F., Simoncini, L., Sarasini, A., Zavattoni, M., Grossi, P., Revello, M.G., and Gerna, G. (1998). Ganciclovir resistance as a result of oral ganciclovir in a heart transplant recipient with multiple human cytomegalovirus strains in blood. Transplantation *66*, 324–329.

Baldanti, F., Lilleri, D., Campanini, G., Comolli, G., Ridolfo, A.L., Rusconi, S., and Gerna, G. (2004). Human cytomegalovirus double resistance in a donor-positive/recipient-negative lung transplant patient with an impaired CD4-mediated specific immune response. J. Antimicrob. Chemother. *53*, 536–539.

Barbi, M., Binda, S., and Caroppo, S. (2006). Diagnosis of congenital CMV infection via dried blood spots. Rev. Med. Virol. *16*, 385–392.

Benoist, G., Salomon, U., Jacquemard, F., Daffos, F., and Ville, Y. (2008). The prognostic value of ultrasound abnormalities and biological parameters in blood of fetuses infected with cytomegalovirus. BJOG *115*, 823–829.

Binnicker, M.J., Jespersen, D.J., and Harring, J.A. (2010). Multiplex detection of IgG and IgM class antibodies to *Toxoplasma gondii*, rubella virus, and cytomegalovirus using a novel multiplex flow immunoassay. Clin. Vaccine Immunol. *17*, 1734–1738.

van der Bij, W., Torensma, R., van Son, W.J., Anema, J., Schirm, J., Tegzess, A.M., and The, T.H. (1988). Rapid immunodiagnosis of active cytomegalovirus infection by monoclonal antibody staining of blood leucocytes. J. Med. Virol. *25*, 179–188.

Blackburn, N.K., Besselaart, T.G., Schoub, B.D., and O'Connell, K.F. (1991). Differentiation of primary cytomegalovirus infection from reactivation using the urea denaturation test for measuring antibody avidity. J. Med. Virol. *33*, 6–9.

Bodeus, M., Feyder, S., and Goubau, P. (1998). Avidity of IgG antibodies distinguishes primary from non-primary cytomegalovirus infection in pregnant women. Clin. Diagn. Virol. *9*, 9–16.

Boeckh, M. (1999). Current antiviral strategies for controlling cytomegalovirus in hematopoietic stem cell transplant recipients: prevention and therapy. Transpl. Infect. Dis. *1*, 165–178.

Boppana, S.B., and Britt, W.J. (1995). Antiviral antibody response and intrauterine transmission after primary maternal cytomegalovirus infection. J. Infect. Dis. *177*, 1115–1121.

Boppana, S.B., Ross, S.A., Novak, Z., Shimamura, M., Tolan, R.W. Jr, Palmer, A.L., Ahmed, A., Michaels, M.G., Sánchez, P.J., Bernstein, D.I., *et al.* (2010). Dried blood spot real-time polymerase chain reaction assays to screen newborns for congenital cytomegalovirus infection. JAMA *303*, 1375–1382.

Boppana, S.B., Ross, S.A., Shimamura, M., Palmer, A.L., Ahmed, A., Michaels, M.G., Sànchez, P.J., Bernstein, D.I., Tolan, E.W., Novak, Z., *et al.* (2011). Saliva polymerase-chain-reaction assay for cytomegalovirus screening in newborns. N. Engl. J. Med. *346*, 2111–2118.

Caliendo, A.M., Ingersoll, J., Fox-Canale, A.M., Pargman, S., Bythwood, T., Hayden, M.K., Bremer, J.W., and Lurain, N.S. (2007). Evaluation of real-time PCR laboratory-developed tests using analyte-specific reagents for cytomegalovirus quantification. J. Clin. Microbiol. *45*, 1723–1727.

Caliendo, A.M., Schuurman, R., Yen-Lieberman, B., Spector, S.A., Andersen, J., Manjiry, R., Crumpacker, C., Lurain, N.S., Erice, A., and CMV Working Group of the Complications of HIV Disease RAC, AIDS Clinical Trials Group. (2001). Comparison of quantitative and qualitative PCR assays for cytomegalovirus DNA in plasma. J. Clin. Microbiol. *39*, 1334–1338.

Chou, S., Guentzel, S., Michels, K.R., Miner, R.C., and Drew, W.L. (1995). Frequency of UL97 phosphotransferase mutations related to ganciclovir resistance in clinical cytomegalovirus isolates. J. Infect. Dis. *172*, 239–242.

Cihlar, T., Fuller, M.D., and Cherrington, J.M. (1998). Characterization of drug resistance-associated mutations in the human cytomegalovirus DNA polymerase gene by using recombinant mutant viruses generated from overlapping DNA fragments. J. Virol. *72*, 5927–5936.

Crough, T., and Khanna, R. (2009). Immunobiology of human cytomegalovirus: from bench to bedside. Clin. Microbiol. Rev. *22*, 76–98.

Daiminger, A., Bader, U., and Enders, G. (1998). An enzyme linked immunoassay using recombinant antigens for differentiation of primary from secondary or past CMV infections in pregnancy. J. Clin. Virol. *11*, 93–102.

Demmler, G.J., Buffone, G.J., Schimbor, C.M., and May, R.A.(1988). Detection of cytomegalovirus in urine from newborns by using polymerase chain reaction DNA amplification. J. Infect. Dis. *158*, 1177–1184.

Dolan, A., Cunningham, C., Hector, R.D., Hassan-Walker, A.F., Lee, L., Addison, C., Dargan, D.J., McGeoch, D.J., Gatherer, D., Emery, V.C., *et al.* (2004). Genetic content of wild-type human cytomegalovirus. J. Gen. Virol. *85*, 1301–1312.

Eggers, M., Metzger, C., and Enders, G. (1998). Differentiation between acute primary and recurrent human cytomegalovirus infection in pregnancy, using a microneutralization assay. J. Med. Virol. *56*, 351–358.

Eggers, M., Radsak, K., Enders, G., and Reschke, M. (2001). Use of recombinant glycoprotein antigen gB and gH for diagnosis of primary human cytomegalovirus infection during pregnancy. J. Med. Virol. *63*, 135–142.

Einsele, H., Roosnek, E., Rufer, N., Sinzger, C., Riegler, S., Löffler, J., Grigoleit, U., Moris, A., Rammensee, H.G., Kanz, L., *et al.* (2002). Infusion of cytomegalovirus (CMV)-specific T-cells for the treatment of CMV infection not responding to antiviral chemotherapy. Blood *99*, 3916–3922.

Emery, V.C., Sabin, C.A., Cope, A.V., Gor, D., Hassan-Walker, A.F., and Griffiths, P.D. (2000). Application of viral-load kinetics to identify patients who develop cytomegalovirus disease after transplantation. Lancet *355*, 2032–2036.

Enders, G., Bäder, U., Lindemann, L., Schalasta, G., and Daiminger, A. (2001). Prenatal diagnosis of congenital cytomegalovirus infection in 189 pregnancies with known outcome. Prenat. Diagn. *21*, 362–377.

Espy, M.J., Uhl, J.R., Sloan, L.M., Buckwalter, S.P., Jones, M.F., Vetter, E.A., Yao, J.D., Wengenack, N.L., Rosenblatt, J.E., Cockerill, F.R. 3rd, *et al.* (2006). Real-time PCR in clinical microbiology: applications for routine laboratory testing. Clin. Microbiol. Rev. *19*,165–256. Review. Erratum in: Clin. Microbiol. Rev. (2006) *19*, 595.

Fabbri, E., Revello, M.G., Furione, M., Zavattoni, M., Lilleri, D., Tassis, B., Quarenghi, A., Rustico, M., Nicolini, U., Ferrazzi, E., *et al.* (2011). Prognostic markers of symptomatic congenital human cytomegalovirus infection in fetal blood. BJOG *118*, 448–456.

Fornara, C., Lilleri, D., Revello, M.G., Furione, M., Zavattoni, M., Lenta, E., and Gerna, G. (2011). Kinetics of effector functions and phenotype of virus-specific and $\gamma\delta$ T-lymphocytes in primary human cytomegalovirus infection during pregnancy. J. Clin. Immunol. *31*, 1054–1064.

Freyer, J.F., Heath, A.B., Anderson, R., Minor, P.D., and the Collaborative Study Group. (2010). Collaborative study to evaluate the proposed 1st WHO International Standard for human cytomegalovirus (HCMV) for nucleic acid amplification (NAT)-based assays. WHO/BS/10.2138, pp. 1–40. Geneva, 18–22 Oct. 2010.

Genini, E., Percivalle, E., Sarasini, A., Revello, M.G., Baldanti, F., and Gerna, G. (2011). Serum antibody response to the gH/gL/pUL128–131 five-protein complex of human cytomegalovirus (HCMV) in primary and reactivated HCMV infections. J. Clin. Virol. *52*, 113–118.

Gerna, G., Revello, M.G., Percivalle, E., Zavattoni, M., Parea, M., and Battaglia, M. (1990). Quantification of human cytomegalovirus viremia by using monoclonal antibodies to different viral proteins. J. Clin. Microbiol. *28*, 2681–2688.

Gerna, G., Zipeto, D., Parea, M., Revello, M.G., Silini, E., Percivalle, E., Zavattoni, M., Grossi, P., and Milanesi, G. (1991). Monitoring of human cytomegalovirus infections and ganciclovir treatment in heart transplant recipients by determination of viremia, antigenemia and DNAemia. J. Infect. Dis. *164*, 488–498.

Gerna, G., Revello, M.G., Palla, M., Percivalle, E., and Torsellini, M. (1992a). Microneutralization as a reference method for selection of the cut-off of an enzyme-linked immunosorbent assay for detection of IgG antibody to human cytomegalovirus. Microbiologica *15*, 177–182.

Gerna, G., Revello, M.G., Percivalle, E., and Morini, F. (1992b). Comparison of different immunostaining techniques and monoclonal antibodies to the lower matrix phosphoprotein (pp65) for optimal quantitation of human cytomegalovirus antigenemia. J. Clin. Microbiol. *30*, 1232–1237.

Gerna, G., Furione, M., Baldanti, F., and Sarasini, A. (1994). Comparative quantitation of human cytomegalovirus DNA in blood leukocytes and plasma of transplant and AIDS patients. J. Clin. Microbiol. *32*, 2709–2717.

Gerna, G., Percivalle, E., Torsellini, M., and Revello, M.G. (1998a). Standardization of the human cytomegalovirus antigenemia assay by means of *in vitro*-generated pp65-positive peripheral blood polymorphonuclear leukocytes. J. Clin. Microbiol. *36*, 3585–3589.

Gerna, G., Zavattoni, M., Percivalle, E., Grossi, P., Torsellini, M., and Revello, M.G. (1998b). Rising levels of human cytomegalovirus (HCMV) antigenemia during initial antiviral treatment of solid-organ transplant recipients with primary HCMV infection. J. Clin. Microbiol. *36*, 1113–1116.

Gerna, G., Baldanti, F., Middeldorp, J.M., Furione, M., Zavattoni, M., Lilleri, D., and Revello, M.G. (1999). Clinical significance of expression of human cytomegalovirus pp67 late transcript in heart, lung, and bone marrow transplant recipients as determined by nucleic acid sequence-based amplification. J. Clin. Microbiol. *37*, 902–911.

Gerna, G., Baldanti, F., Grossi, P., Locatelli, F., Colombo, P., Vigaǹo, M., and Revello, M.G. (2001). Diagnosis and monitoring of human cytomegalovirus infection in transplant recipients. Rev. Med. Microbiol. *12*, 155–175.

Gerna, G., Baldanti, F., Percivalle, E., Zavattoni, M., Campanini, G., and Revello, M.G. (2003a). Early identification of human cytomegalovirus strains by the shell vial assay is prevented by a novel amino acid substitution in UL123 IE1 gene product. J. Clin. Microbiol. *41*, 4494–4495.

Gerna, G., Sarasini, A., Lilleri, D., Percivalle, E., Torsellini, M., Baldanti, F., and Revello, M.G. (2003b). In vitro model for the study of the dissociation of increasing antigenemia and decreasing DNAemia and viremia during treatment of human cytomegalovirus infection with ganciclovir in transplant recipients. J. Infect. Dis. *188*, 1639–1647.

Gerna, G., Lilleri, D., Zecca, M., Alessandrino, E.P., Baldanti, F., Revello, M.G., and Locatelli, F. (2005). Rising antigenemia levels may be misleading in pre-emptive therapy of human cytomegalovirus infection in allogeneic hematopoietic stem cell transplant recipients. Haematologica *90*, 526–533.

Gerna, G., Lilleri, D., Fornara, C., Comolli, G., Lozza, L., Campana, C., Pellegrini, C., Meloni, F., and Rampino, T. (2006a). Monitoring of human cytomegalovirus-specific CD4+ and CD8+ T-cell immunity in patients receiving solid organ transplantation. Am. J. Transplant. *6*, 2356–2364.

Gerna, G., Sarasini, A., Genini, E., Percivalle, E., and Revello, M.G. (2006b). Prediction of endothelial cell tropism of human cytomegalovirus strains. J. Clin. Virol. *35*, 470–473.

Gerna, G., Baldanti, F., Torsellini, M., Minoli, L., Vigano', M., Oggionni, T., Rampino, T., Castiglioni, B., Goglio, A., Colledan, M., Mammana, C., Nozza, F., and Lilleri, D. (2007). Evaluation of human cytomegalovirus DNAemia vs pp65-antigenemia cutoff for guiding preemptive therapy in transplant recipients: A randomized study. Antivir. Ther. *12*, 63–72.

Gerna, G., Lilleri, D., Caldera, D., Furione, M., Zenone Bragotti, L., and Alessandrino, E.P. (2008a). Validation of a DNAemia cut-off for pre-emptive therapy of cytomegalovirus infection in adult hematopoietic stem cell transplant recipients. Bone Marrow Transplant. *41*, 873–879.

Gerna, G., Sarasini, A., Patrone, M., Percivalle, E., Fiorina, L., Campanini, G., Gallina, A., Baldanti, F., and Revello, M.G. (2008b). Human cytomegalovirus serum neutralizing antibodies block virus infection of endothelial/epithelial cells, but not fibroblasts, early during primary infection. J. Gen. Virol. *89*, 853–865.

Gleaves, C.A., Smith, T.F., Shuster, E.A., and Pearson, G.R. (1984). Rapid detection of cytomegalovirus in MRC-5 cells inoculated with urine specimens by using low speed centrifugation and monoclonal antibody to an early antigen. J. Clin. Microbiol. *19*, 917–919.

Grangeot-Keros, L., Mayaux, M.J., Lebon, P., Freymuth, F., Eugene, G., Stricker, R., and Dussaix, E. (1997). Value of cytomegalovirus (CMV) IgG avidity index for the diagnosis of primary CMV infection in pregnant women. J. Infect. Dis. *175*, 944–946.

Gratama, J.W., van Esser, J.W., Lamers, C.H., Tournay, C., Löwenberg, B., Bolhuis, R.L., and Cornelissen, J.J. (2001). Tetramer-based quantification of cytomegalovirus (CMV)-specific CD8+ T-lymphocytes in T-cell-depleted stem cell grafts and after transplantation may identify patients at risk for progressive CMV infection. Blood *98*, 1358–1364.

Gratama, J.W., Boeckh, M., Nakamura, R., Cornelissen, J.J., Brooimans, R.A., Zaia, J.A., Forman, S.J., Gaal, K., Bray, K.R., Gasior, G.H., *et al.* (2010). Immune monitoring with iTAg MHC Tetramers for prediction of recurrent or persistent cytomegalovirus infection or disease in allogeneic hematopoietic stem cell transplant recipients: a prospective multicenter study. Blood *116*, 1655–1662.

Grossi, P., Minoli, L., Percivalle, E., Irish, W., Vigaǹo, M., and Gerna, G. (1995). Clinical and virological monitoring of human cytomegalovirus infection in 294 heart transplant recipients. Transplantation *59*, 847–851.

Guerra, B., Simonazzi, G., Banfi, A., Lazzarotto, T., Farina, A., Lanari, M., and Rizzo, N. (2007). Impact of diagnostic and confirmatory tests and prenatal counseling on the rate of pregnancy termination among women with positive cytomegalovirus immunoglobulin M antibody titers. Am. J. Obstet. Gynecol. *196*, 221e1–221e6.

Hahn, G., Revello, M.G., Patrone, M., Percivalle, E., Campanini, G., Sarasini, A., Wagner, M., Gallina, A., Milanesi, G., Koszinowski, U., *et al.* (2004). Human cytomegalovirus UL131–128 genes are indispensable for virus growth in endothelial cells and virus transfer to leukocytes. J. Virol. *78*, 10023–10033. Erratum in: (2009) J. Virol. *83*, 6323.

Hebart, H., Gamer, D., Loeffler, J., Mueller, C., Sinzger, C., Jahn, G., Bader, P., Klingebiel, T., Kanz, L., and Einsele, H. (1998). Evaluation of Murex CMV DNA hybrid capture assay for detection and quantitation of cytomegalovirus infection in patients following allogeneic stem cell transplantation. J. Clin. Microbiol. *36*, 1333–1337.

Humar, A., Gregson, D., Caliendo, A.M., Mcgeer, A., Malkan, G., Krajden, M., Corey, P., Greig, P., Walmsley, S., Levy, G., *et al.* (1999). Clinical utility of quantitative cytomegalovirus viral load determination for predicting cytomegalovirus disease in liver transplant recipients. Transplantation *68*, 1305–1311.

Kinzler, E.R., Theiler, R.N., and Compton, T.(2002). Expression and reconstitution of the gH/gL/gO complex of human cytomegalovirus. J. Clin. Virol. *25*, S87–S95.

Knight, A., Madrigal, A.J., Grace, S., Sivakumaran, J., Kottaridis, P., Mackinnon, S., Travers, P.J., and Lowdell, M.W. (2010). The role of Vδ2-negative γδ T-cells during cytomegalovirus reactivation in recipients of allogeneic stem cell transplantation. Blood *116*, 2164–2172.

Kolberg, J., Miner, R., Hoo, B., Kelso, R., Wiegard, J., Jekic-McMullen, D., and Drew, W.L. (1996). Development of a branched DNA signal amplification assay to monitor patients on anti-CMV therapies. Biologicals *24*, 216.

Kotton, C.N., Kumar, D., Caliendo, A.M., Åsberg, A., Chou, S., Snydman, D.R., Allen, U., and Humar, A. on behalf of the Transplantation Society International CMV Consensus Group. (2010). International Consensus Guideliness on the Management of cytomegalovirus in solid organ transplantation. Transplantation *89*, 779–795.

Kusne, S., Grossi, P., Irish, W., St George, K., Rinaldo, C., Rakela, J., and Fung, J. (1999). Cytomegalovirus pp65 antigenemia monitoring as a guide for preemptive therapy: a cost effective strategy for prevention of cytomegalovirus disease in adult liver transplant recipients. Transplantation *68*, 1125–1131.

Law, M., and Hangartner, L. (2008). Antibodies against viruses: passive and active immunization. Curr. Opin. Immunol. *20*, 486–492.

Lazzarotto, T., Ripalti, A., Bergamini, G., Battista, M.C., Spezzacatena, P., Campanini, F., Pradelli, P., Varani, S., Bagrielli, L., Maine, G.T., *et al.* (1998). Development of a new cytomegaloviurs (CMV) immmunoglobulin M (IgM) immunoblot for detection of CMV-specific IgM. J. Clin. Microbiol. *36*, 3337–3341.

Lazzarotto, T., Spezzacatena, P., Varani, S., Gabrielli, L., Pradelli, P., Guerra, B., and Landini, M.P. (1999). Anticytomegalovirus (anti-CMV) immunoglobulin G avidity in identification of pregnant women at risk of transmitting congenital CMV infection. Clin. Diagn. Lab. Immunol. *6*, 127–129.

Lazzarotto, T., Galli, C., Pulvirenti, R., Rescaldani, R., Vezzo, R., La Gioia, A., Martinelli, C., La Rocca, S., Agresti, G., Grillner, L., *et al.* (2001). Evaluation of the Abbott AxSYM cytomegalovirus (CMV) immunoglobulin M (IgM) assay in conjunction with other CMV IgM tests and a CMV IgG avidity assay. Clin. Diagn. Lab. Immunol. *8*, 196–198.

Liesnard, C., Donner, C., Brancart, F., Gosselin, F., Delforge, L., and Rodesch, F. (2000). Prenatal diagnosis of congenital cytomegalovirus infection: prospective study of 237 pregnancies at risk. Obstet. Gynecol. *95*, 881–888.

Lilleri, D., Baldanti, F., Gatti, M., Rovida, F., Dossena, L., De Grazia, S., Torsellini, M., and Gerna, G. (2004). Clinically-based determination of safe DNAemia cutoff levels for preemptive therapy of human cytomegalovirus infections in solid organ and hematopoietic stem cell transplant recipients. J. Med. Virol. *73*, 412–418.

Lilleri, D., Gerna, G., Fornara, C., Lozza, L., Maccario, R., and Locatelli, F. (2006). Prospective simultaneous quantification of human cytomegalovirus-specific CD4+ and CD8+ T-cell reconstitution in young recipients of allogeneic hematopoietic stem cell transplants. Blood *108*, 1406–1412.

Lilleri, D., Fornara, C., Furione, M., Zavattoni, M., Revello, M.G., and Gerna, G. (2007a). Development of human cytomegalovirus-specific T-cell immunity during primary infection of pregnant women and its correlation with virus transmission to the fetus. J. Infect. Dis. *195*, 1062–1070.

Lilleri, D., Gerna, G., Furione, M., Bernardo, M.E., Giorgiani, G., Telli, S., Baldanti, F., and Locatelli, F. (2007b). Use of a DNAemia cut-off for monitoring human cytomegalovirus infection reduces the number of preemptively treated children and young adults receiving hematopoietic stem cell transplantation compared with qualitative pp65 antigenemia. Blood *110*, 2757–2760.

Lilleri, D., Fornara, C., Revello, M.G., and Gerna, G. (2008). Human cytomegalovirus-specific memory CD4+ and CD8+ T-cell differentiation after primary infection. J. Infect. Dis. *198*, 536–543.

Lilleri, D., Lazzarotto, T., Ghisetti, V., Ravanini, P., Capobianchi, M.R., Baldanti, F., and Gerna, G. (2009). Multicenter quality control study for human cytomegalovirus DNAemia quantification. New Microbiol. *32*, 245–253.

Limaye, A.P., Bowden, R.A., Myerson, D., and Boeckh, M. (1997). Cytomegalovirus disease occurring before engraftment in marrow transplant recipients. Clin. Infect. Dis. *24*, 830–835.

Locatelli, F., Percivalle, E., Comoli, P., Maccario, R., Zecca, M., Giorgiani, G., De Stefano, P., and Gerna, G. (1994). Human cytomegalovirus (HCMV) infection in paediatric patients given allogeneic bone marrow transplantation: role of early antiviral treatment for HCMV antigenaemia on patients' outcome. Br. J. Haematol. *88*, 64–71.

Lozza, L., Lilleri, D., Percivalle, E., Fornara, C., Comolli, G., Revello, M.G., and Gerna, G. (2005). Simultaneous quantification of human cytomegalovirus (HCMV)-specific CD4+ and CD8+ T-cells by a novel method using monocyte-derived HCMV-infected immature dendritic cells. Eur. J. Immunol. *35*, 1795–1804.

Lurain, N.S., Ammons, H.C., Kapell, K.S., Yeldandi, V.V., Garrity, E.R., and O'Keefe, J.P. (1996). Molecular analysis of human cytomegalovirus strains from two lung transplant recipients with the same donor. Transplantation *62*, 497–502.

Maine, G.T., Stricker, R., Schuler, M., Spesard, J., Brojanac, S., Iriarte, B., Herwig, K., Gramins, T., Combs, B., Wise, J., *et al.* (2000). Development and clinical evaluation of a recombinant-antigen-based cytomegalovirus immunoglobulin M automated immunoassay using the Abbott AxSYM analyzer. J. Clin. Microbiol. *38*, 1476–1481.

Mengelle, C., Sandres-Sauné, K., Pasquier, C., Rostaing, L., Mansuy, J.M., Marty, M., Da Silva, I., Attal, M., Massip, P., and Izopet, J. (2003). Automated extraction and quantification of human cytomegalovirus DNA in whole blood by real-time PCR assay. J. Clin. Microbiol. *41*, 3840–3845.

Nelson, C.T., Istas, A.S., Wilkerson, M.K., and Demmler, G.J. (1995). PCR detection of cytomegalovirus DNA in serum as a diagnostic for congenital cytomegalovirus infection. J. Clin. Microbiol. *33*, 3317–3318.

Novak, Z., Ross, S.A., Patro, S.K., Pati, S.K., Reddi, M.K., Purser, M., Britt, W.J., and Boppana, S.B. (2009). Enzyme-linked immunosorbent assay method for detection of cytomegalovirus strain-specific antibody responses. Clin. Vaccine Immunol. *16*, 288–290.

Pang, X.L., Fox, J.D., Fenton, J.M., Miller, G.G., Caliendo, A.M., and Preiksaitis, J.K. (2009). Interlaboratory comparison of cytomegalovirus viral load assays. Am. J. Transplant. *9*, 258–268.

Razonable, R.R., Van Cruijsen, H., Brown, R.A., Wilson, J.A., Harmsen, W.S., Wiesner, R.H., Smith, T.F., and Paya, C.V. (2003). Dynamics of cytomegalovirus replication during preemptive therapy with oral ganciclovir. J. Infect. Dis. *187*, 1801–1808.

Revello, M.G., and Gerna, G. (2002). Diagnosis and management of human cytomegalovirus infection in the mother, fetus, and newborn infant. Clin. Microbiol. Rev. *15*, 680–715.

Revello, M.G., and Gerna, G. (2010). Human cytomegalovirus tropism for endothelial/epithelial cells: scientific background and clinical implications. Rev. Med. Virol. *20*, 136–155.

Revello, M.G., Percivalle, E., Zavattoni, M., Parea, M., Grossi, P., and Gerna, G. (1989). Detection of human cytomegalovirus immediate early antigen in leukocytes as a marker of viremia in immunocompromised patients. J. Med. Virol. *29*, 88–93.

Revello, M.G., Percivalle, E., Arbustini, E., Pardi, R., Sozzani, S., and Gerna, G. (1998a). *In vitro* generation of human cytomegalovirus pp65 antigenemia, viremia, and leukoDNAemia. J. Clin. Invest. *101*, 2686–2692.

Revello, M.G., Zavattoni, M., Sarasini, A., Percivalle, E., Simoncini, L., and Gerna, G. (1998b) Human cytomegalovirus in blood of immunocompetent persons during primary infection: prognostic implications for pregnancy. J. Infect. Dis. *177*, 1170–1175.

Revello, M.G., Zavattoni, M., Sarasini, A., Baldanti, F., de Julio, C., De-Giuli, L., Nicolini, U., and Gerna, G. (1999). Prenatal diagnostic and prognostic value of human cytomegalovirus load and IgM antibody response in blood of congenitally infected fetuses. J. Infect. Dis. *180*, 1320–1323.

Revello, M.G., Lilleri, D., Zavattoni, M., Stronati, M., Bollani, L., Middeldorp, J.M., and Gerna, G. (2001). Human cytomegalovirus immediate-early messenger RNA in blood of pregnant women with primary infection and of congenitally infected newborns. J. Infect. Dis. *184*, 1078–1081.

Revello, M.G., Zavattoni, M., Furione, M., Lilleri, D., Gorini, G., and Gerna, G. (2002). Diagnosis and outcome of preconceptional and periconceptional primary human cytomegalovirus infection. J. Infect. Dis. *186*, 553–557.

Revello, M.G., Lilleri, D., Zavattoni, M., Genini, E., Comolli, G., and Gerna, G. (2006). Lymphoproliferative response in primary human cytomegalovirus (HCMV) infection is delayed in HCMV transmitter mothers. J. Infect. Dis. *193*, 269–276.

Revello, M.G., Genini, E., Gorini, G., Klersy, C., Piralla, A., and Gerna, G. (2010). Comparative evaluation of eight commercial human cytomegalovirus IgG avidity assays. J. Clin. Virol. *48*, 255–259.

Revello, M.G., Fabbri, E., Furione, M., Zavattoni, M., Lilleri, D., Tassis, B., Quarenghi, A., Cena, C., Arossa, A., Montanari, L., *et al.* (2011). Role of prenatal diagnosis and counseling in the management of 735 pregnancies complicated by primary human cytomegalovirus infection: A 20-year experience. J. Clin. Virol. *50*, 303–307.

Schoppel, K., Kropff, B., Schmidt, C., Vornhagen, R., and Mach, M. (1997). The humoral immune response against human cytomegalovirus is characterized by a delayed synthesis of glycoprotein-specific antibodies. J. Infect. Dis. *175*, 533–544.

Singh, N. (2006). Antiviral drugs for cytomegalovirus in transplant recipients: advantages of preemptive therapy. Rev. Med. Virol. *16*, 281–287.

Singh, N., Wannstedt, C., Keyes, L., Wagener, M.M., Gayowski, T., and Cacciarelli, T.V. (2005). Indirect outcomes associated with cytomegalovirus (opportunistic infections, hepatitis C virus sequelae, and mortality) in liver-transplant recipients with the use of preemptive therapy for 13 years. Transplantation *79*, 1428–1434.

Sylwester, A.W., Mitchell, B.L., Edgar, J.B., Taormina, C., Pelte, C., Ruchti, F., Sleath, P.R., Grabstein, K.H., Hosken, N.A., Kern, F., *et al.* (2005). Broadly targeted human cytomegalovirus-specific CD4+ and CD8+ T-cells dominate the memory compartments of exposed subjects. J. Exp. Med. *202*, 673–685.

Verkruyse, L.A., Storch, G.A., Devine, S.M., Dipersio, J.F., and Vij, R. (2006). Once daily ganciclovir as initial pre-emptive therapy delayed until threshold CMV load > or =10000 copies/ml: a safe and effective strategy for allogeneic stem cell transplant patients. Bone Marrow Transplant. *37*, 51–56.

Waldrop, S.L., Pitcher, C.J., Peterson, D.M., Maino, V.C., and Picker, L.J. (1997). Determination of antigen-specific memory/effector CD4+ T-cell frequencies by flow cytometry: evidence for a novel, antigen-specific homeostatic mechanism in HIV-associated immunodeficiency. J. Clin. Invest. *99*, 1739–1750.

Wang, D., and Shenk, T. (2005). Human cytomegalovirus virion protein complex required for epithelial and endothelial cell tropism. Proc. Natl. Acad. Sci. U.S.A. *102*, 18153–18158.

Wattanamano, P., Clayton, J.L., Kopicko, J.J., Kissinger, P., Elliot, S., Jarrott, C., Rangan, S., and Beilke, M.A. (2000). Comparison of three assays for cytomegalovirus detection in AIDS patients at risk for retinitis. J. Clin. Microbiol. *38*, 727–732.

Yamamoto, A.Y., Mussi-Pinhata, M.M., Boppana, S.B., Novak, Z., Wagatsuma, V., de Frizzo Oliveira, P., Duarte, G., and Britt, W.J. (2010). Human cytomegalovirus reinfection is associated with intrauterine transmission in a highly cytomegalovirus-immune maternal population. Am. J. Obstet. Gynecol. *202*, 297.e1–8.

Antiviral Therapy, Drug Resistance and Computed Resistance Profiling

II.19

Detlef Michel, Meike Chevillotte and Thomas Mertens

Abstract

Three drugs are currently used for treatment of human cytomegalovirus (HCMV) disease or infection: ganciclovir (GCV)/valganciclovir (valGCV), cidofovir (CDV), and foscarnet (FOS). They all target the viral DNA polymerase pUL54, thereby inhibiting viral DNA replication. All current anti-HCMV compounds cause drug-specific and severe side effects and have been reported to select for clinically relevant drug resistant virus variants. Active systemic HCMV infection can be first asymptomatic or symptoms can be non-specific. Therefore, a reliable and fast diagnosis of active systemic viral infection is needed, based on virological markers. This includes early and quantitative detection of drug resistant HCMV variants. Genotyping will become the method of choice for identification of viral drug resistance, but absolutely requires previous quantitative characterization of the effect of individual and combined mutations on the resistance phenotype. A database has been made available on the internet containing all published mutations and according information on the quantitative resistance phenotypes. In view of the many published resistance mutations, the database is a helpful tool to correlate diagnosed mutations in a viral genome with resistance. New compounds to treat HCMV infection and disease are urgently needed. They should ideally combine few adverse effects, good oral bioavailability, the option to treat children *in utero*, as well as novel mechanisms of action, to possibly reduce selection of resistant virus variants by combination therapy. Some substances should also target early steps of viral infection, thereby inhibiting IE- and E-gene expression in infected cells. Modified and new nucleoside analogues, small-molecule compounds, benzimidazoles, indolocarbazole protein kinase inhibitors as well as drugs with originally other indications (leflunomide, artesunate) will soon augment the HCMV antiviral portfolio. They will expand treatment options, but will also increase the number of resistance-conferring mutations, thereby rendering resistance analyses even more complex.

Introduction

Cytomegalovirus poses a well-known major health problem for connately infected children as well as for patients with innate, acquired or iatrogenically induced immunodeficiencies (see also Chapter II.1). More recently, it has been shown that HCMV reactivation does also occur in patients suffering from sepsis (Heininger *et al.*, 2001). The clinical relevance of these reactivations in patients without canonical immunosuppression and with low antigenaemias has still to be established (von Muller *et al.*, 2007). A hypothetic role of HCMV in other disease entities like atherosclerosis or even tumorigenesis and a potential need for antiviral therapy is a matter of debate and research (Stassen *et al.*, 2006; Maussang *et al.*, 2009a; Caposio *et al.*, 2011; see also Chapter II.23). Clinical manifestations of HCMV disease show a very interesting but unexplained variation associated with different patient populations, depending on the type of underlying disease and the kind and intensity of immunosuppression, e.g. retinitis in AIDS patients but interstitial pneumonia in bone marrow transplant recipients (Emanuel *et al.*, 1988). In solid organ transplant (SOT) patients, symptomatic infections occur in approximately 40% of heart–lung, up to 35% of heart, up to 29% of liver and pancreas and 8–32% of renal transplant recipients not receiving antiviral prophylaxis. Antiviral agents have been introduced for treatment with significant benefit for these patients (see Chapters II.13, II.14 and II.16).

All compounds (ganciclovir (GCV)/valganciclovir (valGCV), cidofovir (CDV), and foscarnet (FOS)) currently available for the systemic treatment of HCMV infections target the polymerization reaction of the viral DNA (Chrisp and Clissold, 1991; Crumpacker, 1996; Snoeck *et al.*, 1998) (Fig. II.19.1).

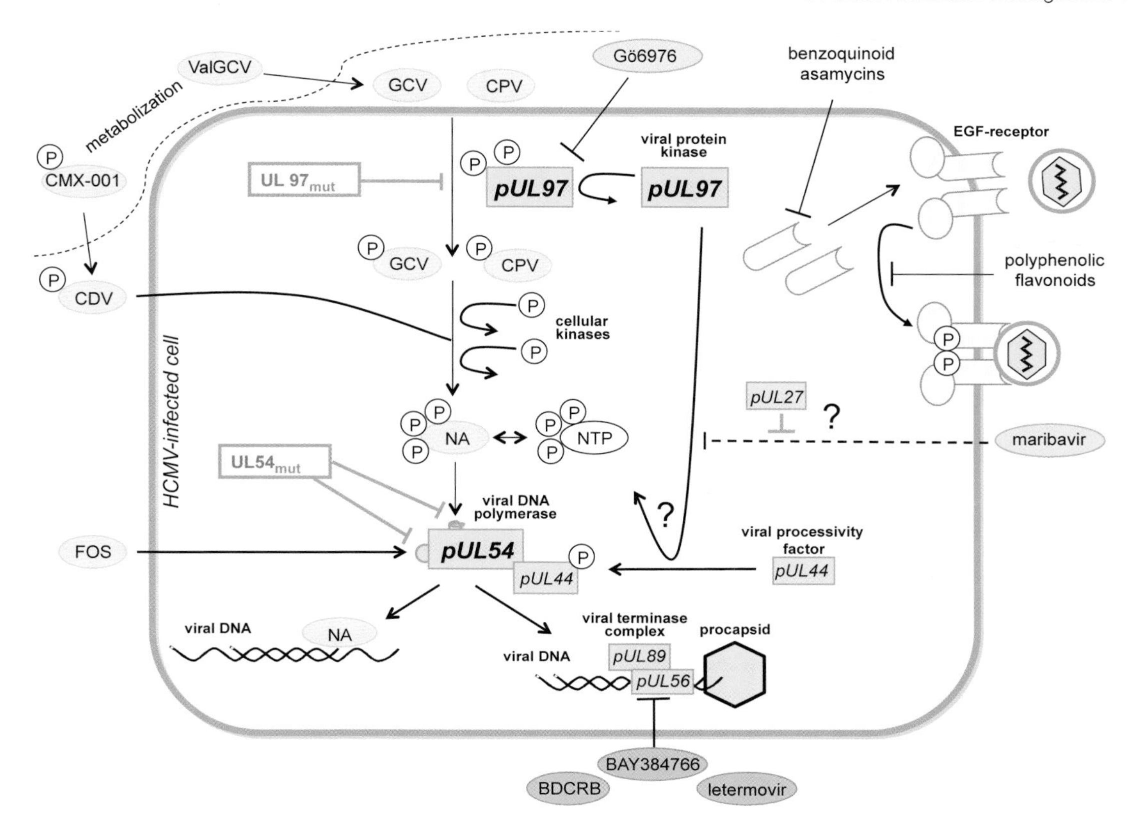

Figure II.19.1 Mechanisms of action of anti-HCMV compounds. The nucleoside analogue (NA) ganciclovir (GCV), can be used directly or is metabolically derived from its precursor compound valganciclovir (valGCV). GCV or cyclopropavir (CPV), another NA, are monophosphorylated by the virally encoded protein UL97 (pUL97). After further phosphorylation, the GCV triphosphate competes with the natural nucleoside triphosphate (NTP) GTP at the viral DNA polymerase pUL54 and is incorporated into the viral DNA. Cidofovir (CDV) can be used directly or is metabolically derived from its precursor compound CMX-001. Both do not require monophosphorylation by pUL97. Foscarnet (FOS) directly inhibits the polymerase at the pyrophosphate binding site. Maribavir (MBV) and Gö6976 inhibit the viral protein kinase pUL97. New drugs such as BDCRB, BAY384766, or letermovir target the viral terminase complex consisting of pUL89 and pUL56, thereby inhibiting viral DNA processing or encapsidation. The first drugs to target HCMV entry inhibit its putative receptor EGFR. Benzoquinoid asamycins hamper proper folding of the receptor, whereas polyphenolic flavonoids inhibit phosphorylation and thereby activation of receptor-associated kinases. Mutations in pUL97 can confer reduced susceptibility to GCV, however, these strains remain sensitive to CDV and FOS. Different mutations in the viral polymerase can confer reduced susceptibility to CDV (often with a cross-resistance to GCV) or FOS. Mutations in UL97 or in UL27 can confer reduced susceptibility to MBV.

An antisense phosphorothioate oligonucleotide, fomivirsen, targeting the immediate early gene of HCMV, was the first approved drug with a different target and mechanism of action (Azad *et al.*, 1993; Perry and Balfour, 1999), but is no longer available. During the past two decades, GCV-resistant isolates have been first recovered mainly from AIDS patients suffering from HCMV retinitis or gastroenteritis (Drew *et al.*, 1991; Jabs *et al.*, 1998a). Several studies reported factors favouring the emergence of drug resistant virus variants, e.g. type of transplantation, severity of immunosuppression, especially lack of specific T-cells. Most factors led to high levels of viral replication, followed by prolonged antiviral therapy with ongoing viral replication in the presence of a drug (Kruger *et al.*, 1999; Limaye *et al.*, 2002; Boivin *et al.*, 2004; Li *et al.*, 2007). The prophylactic administration of GCV in transplant recipients, especially after approval of the orally applicable valine-ester of GCV, valganciclovir (valGCV) (Curran and Noble, 2001) with a better oral bioavailability, may have also led to prolonged drug exposure of some asymptomatic patients with ongoing virus replication. Monitoring viral load in patients at risk and genotypic resistance testing of HCMV antiviral drug susceptibility in patients not responding to antiviral therapy is becoming part of the standard care to treat HCMV infection and disease (Kotton *et al.*, 2010).

In recent years, management of prophylaxis and therapy has been steadily evolving, as did our understanding of selection and dynamic expansion of drug resistant HCMV variants and their impact on clinical outcome in the different patient groups. Sophisticated sequencing techniques and the growing number of data pose additional challenges for the clinical virologist:

1. More than 200 mutations in HCMV pUL54 or pUL97 have been reported in the context of drug resistance (Lurain *et al.*, 1994; Baldanti *et al.*, 1995; Hanson *et al.*, 1995; Chou, 1999; Erice, 1999; Ijichi *et al.*, 2002; Chou and Marousek, 2008; Boutolleau *et al.*, 2011) (Fig. II.19.2). Constant review of the literature is required to base the decision on antiviral medication onto the best available evidence.
2. The diagnosis of virus resistance based on the correlation between genotype and phenotype highly depends on the quality of the phenotypic data. However, a large number of mutations has still not been sufficiently characterized to discriminate mutations reducing drug susceptibility, thus potentially requiring changes in medication, from polymorphisms *not* conferring drug resistance of the respective viruses (Chevillotte *et al.*, 2009b; 2010a).
3. The occurrence of multiple mutations in UL97 and in UL54 has been observed, highlighting the need to investigate such combinatorial genotypes. Reported effects of combined mutations range from synergistically increased resistance levels to compensation of resistance (Mousavi-Jazi *et al.*, 2001; Ijichi *et al.*, 2002; Ducancelle *et al.*, 2007; Scott *et al.*, 2007; Chevillotte *et al.*, 2010b).
4. The virus population in a patient at a given time is always a mixture of different virus subpopulations, which also applies to drug-sensitive and -resistant viruses. New sequencing techniques such as pyrosequencing allow detection and quantification of minor resistant subpopulations (Gorzer *et al.*, 2010; Schindele *et al.*, 2010; Kampmann *et al.*, 2011). Yet our knowledge of the evolution of HCMV subpopulations under drug exposure and its impact on clinical outcome is still scarce.

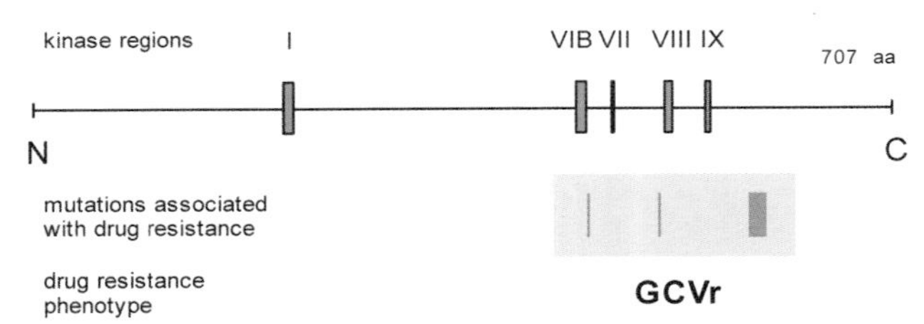

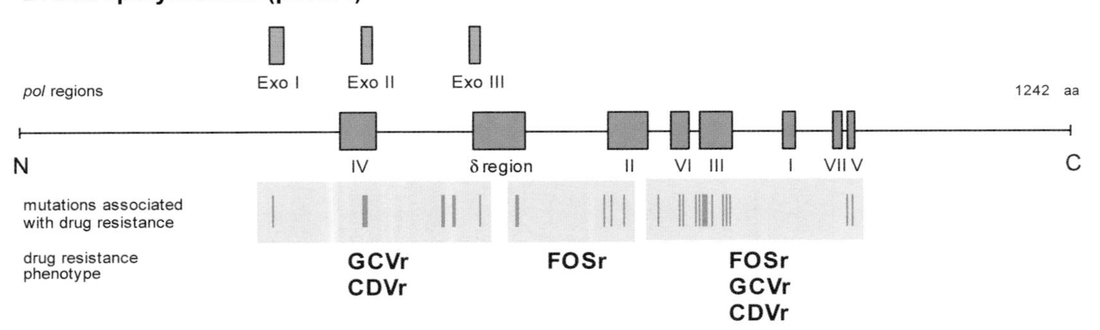

Figure II.19.2 Schematic representation of pUL97 (A) and pUL54 (B) protein structure. Mutations conferring reduced drug susceptibility are indicated by orange bars. GCV = ganciclovir, CDV = cidofovir, FOS = foscarnet. aa, amino acids.

With the use of FOS and CDV as second-line drugs to treat HCMV infection and disease, selection of multiresistant HCMV has been reported (Cihlar *et al.*, 1998; Gilbert and Boivin, 2005; Martin *et al.*, 2006; Chou *et al.*, 2007; Scott *et al.*, 2007). As GCV, CDV and FOS target the same viral protein, the viral DNA polymerase, pUL54, single point mutations in UL54 have been shown to reduce susceptibility to all three drugs. Data from both *in vitro* studies and clinical studies on the above issues have deepened our understanding of many aspects regarding resistant HCMV. Future directions will require more clinical studies to evaluate the predictive value of genotypic resistance testing for patient's outcome as well as the benefit of changes of medication triggered by resistance testing.

Current drugs and therapy

(see Table II.19.1)

Mechanisms of action of available antiviral drugs against HCMV

The nucleoside analogue GCV is mono-phosphorylated by the virally encoded protein kinase pUL97 and subsequently converted to the triphosphate by cellular enzymes, thereby preferentially inhibiting viral DNA synthesis (Fig. II.19.1). Unlike it has been shown for aciclovir (ACV), the incorporation of GCV into the growing viral DNA strand does not lead to an obligatory chain termination, but rather triggers slowing down of DNA polymerization (Biron, 2006). The most relevant adverse effect of GCV, myelotoxicity, results in leucopenia, thrombocytopenia, anaemia, or even bone marrow hypoplasia. CDV (De Clercq *et al.*, 1987; Snoeck *et al.*, 1998) is a phosphonomethoxy analogue of cytosine not requiring viral-mediated phosphorylation. It is directly converted to the active triphosphate by cellular enzymes. The pyrophosphate analogue FOS does also not require chemical modification for antiviral activity, since it binds to and blocks the pyrophosphate binding site of the viral polymerase. As neither CDV nor FOS depends on pUL97-mediated phosphorylation, both compounds are active against HCMV variants with mutated pUL97 in GCV-resistant clinical isolates.

Advantages and limitations of available antiviral compounds

Intravenous GCV has been successfully used in randomized and controlled trials to treat transplant recipients with HCMV disease. Retinitis, several years ago a frequent disease in AIDS patients, responded well to GCV therapy. Fortunately, due to the highly active antiretroviral therapy (HAART) regime, today retinitis is rare in these patients (Baldanti *et al.*, 2002; Kedhar and Jabs, 2007). Interstitial pneumonitis caused by HCMV was not as receptive to GCV monotherapy. For better results, a combination with HCMV-specific immunoglobulins was initiated with the rationale that mechanisms other than viral replication may contribute to HCMV-related pathology in the lung. However, only limited data support an improved efficacy of this approach after lung transplantation (Emanuel *et al.*, 1988; Zamora *et al.*, 2005).

Adverse effects of GCV such as leucopenia, thrombocytopenia, anaemia or even generalized bone marrow hypoplasia, eosinophilia, diarrhoea, renal toxicity, rash and others, occur less frequently in SOT recipients than in bone marrow transplant recipients. Renal toxicity may occur when GCV is administered in combination with other nephrotoxic compounds such as azathioprine, cyclosporine A and amphotericin B. *In vitro* studies suggested that the combination of GCV and FOS could result in synergistic antiviral activity. However, a randomized, controlled study did not support a synergistic effect *in vivo* (Mattes *et al.*, 2004; Drew, 2006; Shereck *et al.*, 2007). FOS and CDV are useful to treat HCMV disease caused by GCV-resistant HCMV strains. However, both substances are associated with nephrotoxicity.

Disease prophylaxis

To date, the two main strategies to prevent HCMV-associated disease are antiviral prophylaxis and pre-emptive therapy. Ideally, prophylaxis should be free of severe adverse effects and interactions with other medications used in transplantation, and should require only a minimum of laboratory tests and drug monitoring. GCV prophylaxis for 100–200 days after transplantation has become a routine in SOT (Eid and Razonable, 2010; Kotton *et al.*, 2010). Nevertheless, pre-emptive therapy – guided by the detection of active systemic HCMV infection by HCMV-DNA PCR or pp65 antigenaemia – remains a valid alternative, as it potentially reduces the amount of drug given to a patient and thereby reduces overtreatment. Although the relatively non-toxic ACV is likewise monophosphorylated by pUL97 (Zimmermann *et al.*, 1997), a randomized, placebo-controlled trial using high-dose oral ACV revealed only a moderate prophylactic effect in preventing HCMV disease following kidney transplantation (Balfour Jr. *et al.*, 1989; Abraham *et al.*, 2008). Concerning liver transplantations, one study suggested a beneficial prophylactic effect of ACV administered orally to HCMV-seropositive recipients while other studies reported decreases of HCMV infections and disease

Table II.19.1 Licensed antiviral drugs and compounds

	Compound	Clinical usage	Dosage/administration
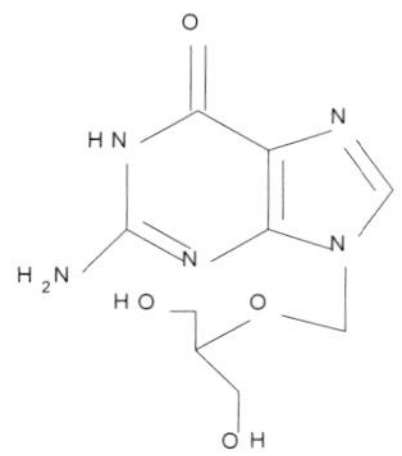	Ganciclovir (INN) (GCV, DHPG) guanosine analogue	Therapy of HCMV diseases Pre-emptive therapy of active HCMV infection Prophylaxis of active HCMV infection and disease in immunosuppressed patients	Initial: 2 × 5 mg/kg/day i.v. (14 days) Maintenance: 5 mg/kg/day i.v. Adjustment of dosage in patients with impairment of renal function Resistance: UL97 mutations (amino acids 460, 520, 590 to 607); UL54 mutations (e.g. aa 408, 412, 413, 501, 513)
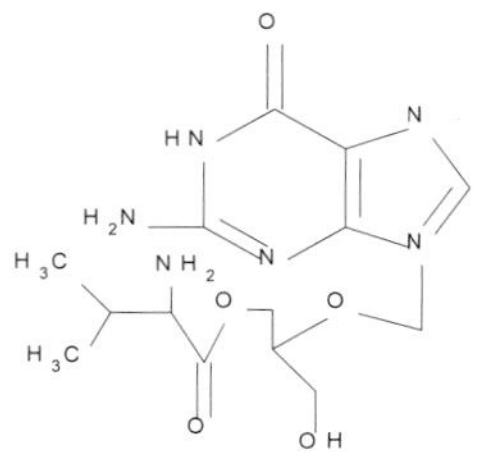	Valganciclovir (ValGCV) guanosine analogue L-valinester of GCV	Maintenance therapy of HCMV diseases Prophylaxis of active HCMV infection and disease in immunosuppressed patients	Initial: 2 × 900 mg/day orally (21 days) Maintenance: 1 × 900 mg/day orally Prophylaxis: 1 × 900 mg/day orally (100 days) Resistance: UL97 mutations (amino acids 460, 520, 590 to 607), UL54 mutations (e.g. aa 408, 412, 413, 501, 513)
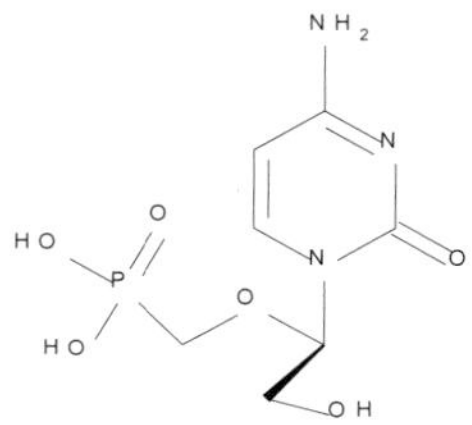	Cidofovir (CDV, HPMPC) acrylic cytosine analogue	Therapy of HCMV diseases Especially for GCV-resistant HCMV (due to UL97 mutation)	Initial: 5 mg/kg i.v. once a week for 2 weeks Maintenance: 5 mg/kg i.v. 1× every 14 days Resistance: UL54 mutations
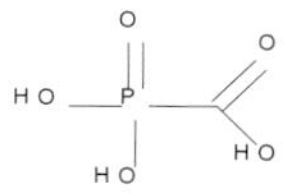	Foscarnet (PFA) pyrophosphate analogue	Therapy of HCMV diseases Especially for GCV-resistant HCMV (due to UL97 mutation) Prophylaxis of active HCMV infection and disease in immunosuppressed patients.	Initial: 3 × 60 mg Foscarnet-natrium-hexahydrate/kg/day i.v. (14 days). Maintenance: 90–120 mg/kg foscarnet-natrium-hexahydrate/kg/day i.v. Adjustment of dosage in patients with impairment of renal function Resistance: UL54 mutations (e.g. aa 419, 578, 773, 781)
5′-G*C*G*T*T*T*G*C*T*C*T*T*C*T*T*C*T*T*G*C*G-3′	Fomivirsen (ISIS2922) antisense phosphorothioate oligonucleotide	Therapy of retinitis. Especially for GCV- and PFA-resistant HCMV	Intravitreal: 330 μg/2 weeks; resistance possible but mechanism uncertain
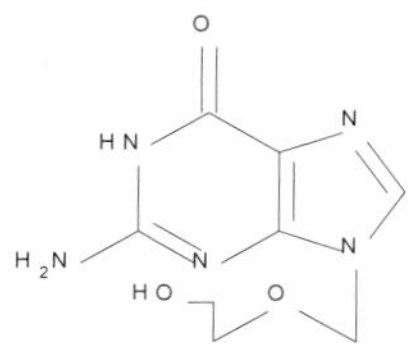	Aciclovir (ACV) guanosine analogue	Therapy of diseases with HSV and VZV Prophylaxis of HSV and VZV infections (Prophylaxis of HCMV infections in solid organ transplantation?)	Not generally established for HCMV Resistance in HCMV: UL97 mutations (amino acids 460, 520, 590 to 607)

but no effect on survival or even no effect of oral ACV at all (Saliba *et al.*, 1993; Singh *et al.*, 1994; Gavaldà *et al.*, 1997).

GCV prophylaxis by intravenous GCV or oral valGCV should be adapted to the risk of the transplant recipient to develop a severe HCMV manifestation and should be given e.g. (1) to all HCMV-seropositive recipients and to HCMV-seronegative recipients receiving a liver or heart transplant from a HCMV-seropositive donor or (2) to HCMV-seronegative renal transplant patients with an organ graft from an HCMV-seropositive donor (Dmitrienko *et al.*, 2007; Kotton *et al.*, 2010).

FOS and CDV are not generally used for prophylaxis in transplantation, particularly not in kidney transplant recipients, because of nephrotoxicity. An open question remains whether antiviral prophylaxis may lead to accelerated selection of resistant virus during subsequent treatment of HCMV-associated disease. Boivin and colleagues investigated the emergence of GCV-resistant HCMV in 301 high-risk SOT recipients after oral prophylaxis with either valGCV or GCV (Boivin *et al.*, 2001). The authors conclude that the prophylaxis did not promote the emergence of resistant virus. Similar results were obtained by Hantz and colleagues in a large multicentre French study (Hantz *et al.*, 2010). In contrast, other authors (Alain *et al.*, 2004; Hantz *et al.*, 2005) reported the detection of GCV-resistant HCMV after valaciclovir-prophylaxis in renal transplant recipients with active HCMV infection. In an *in vitro* study, we showed that ACV selects precisely for the mutations in UL97 that also confer resistance to GCV (Michel *et al.*, 2001). In conclusion, general anti-HCMV prophylactic regimens based on data from clinical studies have still to be established for different patient groups.

Pre-emptive therapy

Rapid and sensitive diagnostic tests allow detection of viral replication even prior to a potential onset of HCMV disease (see Chapter II.18). They provide the opportunity of timely antiviral treatment based on laboratory-proven diagnosis of active systemic infection. This so-called 'pre-emptive therapy' often permits shorter treatment of individuals who are at risk to develop serious HCMV disease, thereby also helping to avoid overmedication. It has been shown to be more efficient and superior to general prophylaxis, especially in patients at risk of interstitial pneumonia (Einsele *et al.*, 2008). Similar to disease prophylaxis, it is important to know whether pre-emptive therapy induces increased selection of resistant HCMV variants, as has been indicated by recent studies (Hantz *et al.*, 2010; Humar *et al.*, 2010; van der Beek *et al.*, 2010; Myhre *et al.*, 2011).

Deferred therapy

Deferred therapy – therapy after onset of clinical symptoms – is generally given to recipients of SOT and to AIDS patients (Spector *et al.*, 1993; Brennan *et al.*, 1997). It has been shown that also postnatal treatment of prenatally infected children may improve the prognosis of hearing loss (Kimberlin *et al.*, 2003).

Viral target genes

Biological function of pUL97

The UL97 protein is a prototype of herpesviral kinases (Michel *et al.*, 1996; Marschall *et al.*, 2003; Michel and Mertens, 2004) and does not share homology with any other known nucleoside kinase; rather it resembles protein kinases and bacterial phosphotransferases (Hanks *et al.*, 1988; Chee *et al.*, 1989). Mark Prichard first generated a UL97-deficient virus, which exhibited a severe replication deficiency in fibroblasts, indicating that pUL97 plays an important role in virus replication *in vitro* (Prichard *et al.*, 1999). This has been confirmed using chemical inhibition of the kinase activity (Marschall *et al.*, 2002; Gilbert *et al.*, 2011). pUL97 is transported into the nucleus via a bipartite nuclear localization signal in the amino terminal region (Michel *et al.*, 1996; Webel *et al.*, 2011). It has been published recently that there exist two different isoforms of pUL97 (Webel *et al.*, 2011), but it has still to be unravelled whether these two isoforms possess specific functions during HCMV replication. pUL97 interacts with the DNA polymerase processivity factor pUL44, an essential component of the replication complex, and might be necessary for nuclear egress.

The main replication steps regulated by pUL97 activity are believed to be the nuclear egress of capsids, viral DNA synthesis and transcription (Prichard *et al.*, 2009; Lee and Chen, 2010; Marschall *et al.*, 2011). However, there exist partially conflicting data concerning pUL97 roles in DNA replication, DNA encapsidation and/or nuclear egress (Wolf *et al.*, 2001; Krosky *et al.*, 2003a,b; Marschall *et al.*, 2003). The region responsible for the interaction with pUL44 has been mapped between aa 366 and aa 459 of pUL97. pUL44 and pUL97 both accumulate in the nucleus and are incorporated into replication centres. Treatment with CDV or with the indolocarbazoles NGIC-I and Gö6976, the latter being specific inhibitors of the pUL97 kinase (Gilbert *et al.*, 2011), prevented co-localization of pUL97 and pUL44 (Marschall *et al.*, 2003). pUL97 has also been shown to directly interact with the major tegument protein of HCMV, pp65 (pUL83) (Kamil and Coen, 2007). Furthermore, blocking pUL97 kinase function using the pUL97 inhibitor maribavir results in nuclear inclusions consisting of pp65 and the minor tegument protein pUL25 (Prichard *et al.*, 2005). In addition, absence of pp65 results in a slightly different autophosphorylation status of pUL97 and the lack of incorporation of pUL97 into virus particles (Chevillotte *et al.*, 2009a). These findings suggest that pUL97 and pp65 may act together during viral infection, which could be important for acquisition of the tegument.

pUL97 has also been shown to interact with cellular proteins, e.g. the cellular protein p32 appears to promote the accumulation of pUL97 at the nuclear lamina (Marschall *et al.*, 2005). Thereby, p32 itself as well as lamin proteins are phosphorylated by pUL97.

The M97 protein (pM97) of mouse cytomegalovirus (MCMV) is the homologue of pUL97. We constructed an MCMV M97 deletion mutant and then replaced the MCMV M97 gene with the HCMV pUL97 gene to construct a gene swap mutant UL97-MCMV (Wagner *et al.*, 2000). The deletion of pM97 in mutant delta M97-MCMV (Wagner *et al.*, 2000) severely affected virus growth. This growth deficit was only partially amended by pUL97 expression in UL97-MCMV.

Phosphorylation and autophosphorylation by pUL97

Biochemical studies have demonstrated protein kinase activity for pUL97 that leads to autophosphorylation and also to phosphorylation of specific exogenous substrates such as histones but also antiviral drugs like GCV (Sullivan *et al.*, 1992; He *et al.*, 1997; Michel *et al.*, 1998, 1999; Baek *et al.*, 2002; Baldanti *et al.*, 2002). The capacity of pUL97 to phosphorylate GCV correlates with viral drug susceptibility: different mutations specifically abolishing GCV phosphorylation *in vitro* (Baldanti *et al.*, 2002) also reduce GCV susceptibility in infection experiments using recombinant vaccinia virus in cell culture. These findings correlate with resistance testing results obtained from patients isolates (Chou *et al.*, 2002).

It has been shown that each of a series of mutations (amino acids) in pUL97 (G340V, A442V, L446R or F523C) resulted in a complete loss of pUL97 autophosphorylation, which was strictly associated with a complete loss of GCV phosphorylation (Michel *et al.*, 1999). Interestingly, although pUL97 mutations found in GCV-resistant HCMV isolates interfere with GCV-phosphorylation, they never reduce pUL97 autophosphorylation, indicating an important role for phosphorylated pUL97 during viral infection. Though GTP can replace ATP as phosphate donor for autophosphorylation, it remains to be determined whether GTP can also replace ATP in the phosphorylation of GCV.

Specificity of pUL97 for nucleosides and nucleoside analogues

The natural nucleosides dA, dC, dT and dG are not substrates for pUL97 as has been shown in the vaccinia virus system (Metzger *et al.*, 1994; Michel *et al.*, 1996). This proves that pUL97 is not a nucleoside kinase. In addition to GCV, pUL97 phosphorylates the nucleoside analogues ACV and penciclovir – both used for therapy of HSV and varicella zoster virus infections – as shown in a quantitative analysis again using recombinant vaccinia viruses (Zimmermann *et al.*, 1997). This is also in line with the observation that ACV selects for GCV cross-resistance of HCMV *in vitro* (Michel *et al.*, 2001).

Biological function of pUL54

The primary protein structure of pUL54, the viral polymerase, contains eight domains that are conserved among members of eukaryotic cells, bacteriophages, and viruses (Wong *et al.*, 1988). No crystal structure of pUL54 does exist, but several 3D models based on the structure of RB69 polymerase have been proposed (Shi *et al.*, 2006; Tchesnokov *et al.*, 2009). These models have been used to retrospectively explain resistance mechanisms due to single aa exchanges resulting in altered tertiary structure of the protein. However, prediction of resistant phenotypes based on primary structure modifications is currently not possible.

Although pUL54 is essential for viral replication, it is a relatively polymorphic protein, and aberrations from 'wild-type' sequences of sensitive laboratory strains can be frequently found in clinical HCMV isolates. However, not all of them confer reduced drug susceptibility (Erice *et al.*, 1997; Chou *et al.*, 2003; Chevillotte *et al.*, 2010a; Lurain and Chou, 2010; Hakki and Chou, 2011). Two different mechanisms of resistance have been proposed for mutations in pUL54 (Gilbert and Boivin, 2005). Some mutations located in the drug binding sites for GCV and CDV, and others at the pyrophosphate binding site for FOS, change the structure of these sites, resulting in a lower affinity to the abovementioned drugs. Other mutations, located in the exonuclease domain of the polymerase, increase its proofreading efficiency. Thus, the non-canonically incorporated nucleoside analogues GCV and CDV are recognized and excised from the growing viral DNA strain more efficiently and lose their inhibitory effect.

Viral resistance

Selection of HCMV resistant to antiviral drugs in clinical settings

The first GCV-resistant virus was reported by Biron and colleagues (Biron *et al.*, 1986). At that time the molecular basis for the resistant phenotype remained unknown, since HCMV does not encode a thymidine kinase-like enzyme. In 1992, Sullivan and colleagues presented data on a viral protein encoded by the ORF UL97 (pUL97, 80 kDa, 707 aa) converting GCV to its monophosphate (Sullivan *et al.*, 1992). GCV phosphorylation was abrogated by deleting aa 590 to 593 (AACR) in the N-terminal region of pUL97. Later, a growing number of mutations in pUL97 has been described conferring GCV resistance in clinical HCMV isolates and often associated with treatment failures in immunosuppressed patients (Baldanti *et al.*, 1995; Erice *et al.*, 1997; Chou, 1999; Erice, 1999; Michel *et al.*, 2003; Schreiber *et al.*, 2009; Lurain and Chou, 2010; Hakki and Chou, 2011). The vast majority (> 90%) of drug-resistant HCMV isolates from patient specimens carry characteristic mutations in pUL97 (Schreiber *et al.*, 2009; Lurain and Chou, 2010). However, none of these isolated viruses carried extended deletions or lacked the UL97 gene. The mutations are not evenly distributed throughout the whole protein but are clustered to aa 460, 520, and a region between aa 590 and 607. Vaccinia virus recombinants (rVV) were generated encoding different mutated pUL97 that had been detected in resistant clinical HCMV isolates at codons 460, 520, 592, 594, 595, 598 and 607 (Baldanti *et al.*, 2002). These rVVs allowed quantification of GCV phosphorylation catalysed by the different mutated pUL97s. When compared to rVV-UL97 wild-type, mean levels of residual intracellular GCV phosphorylation differed by a factor of ~ 10 for the mutated UL97 proteins ranging from 5.2 to 51.8% (Fig. II.19.3). Mutations M460V (located in a pUL97 region homologous to domain VIb of protein kinases) and H520Q (located in a cytomegalovirus-specific, functionally critical domain) were responsible for the lowest levels of residual GCV phosphorylation (9.3% for M460V and 5.2% for H520Q). Mutations in a region homologous to the domain IX had a lower impact on GCV phosphorylation (15.8–51.8%). The restricted location of the mutations suggests that they alter the substrate specificity without abolishing the natural biological function. Recently, mutagenesis has been facilitated by the adoption of bacterial artificial

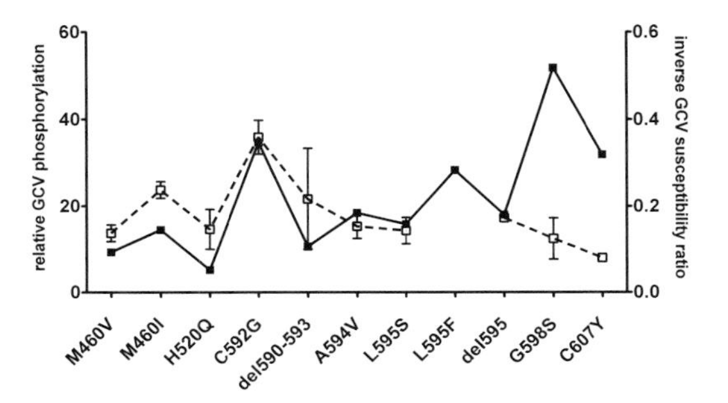

Figure II.19.3 *In vitro* GCV phosphorylation capacity and viral GCV susceptibility of selected UL97 mutants. Relative GCV phosphorylation values (left y-axis; filled squares, solid line) were taken with permission from Baldanti and colleagues (Baldanti *et al.*, 2002). This study used a recombinant vaccinia virus system expressing UL97 mutant proteins, which were tested for GCV phosphorylation capacity *in vitro*. GCV susceptibility data (right y-axis; empty squares) were obtained from the HCMV drug resistance mutations database (http://www.informatik.uni-ulm.de/ni/mitarbeiter/HKestler/hcmv/index.html), containing all published UL97 and UL54 mutations and respective phenotypes. Only marker transfer validated GCV susceptibility ratios between the sensitive parental virus and the respective mutant virus were included. When several entries for the same mutation are present in the database, derived from different publications, mean GCV susceptibilities are shown. For a better comparison to the phosphorylation data, values were inverted. The capacity of GCV phosphorylation from the *in vitro* study inversely correlates with viral GCV susceptibility.

chromosome (BAC) clones of HCMV (Borst *et al.*, 1999; Martin *et al.*, 2006; Gilbert *et al.*, 2011) (see also Chapter I.3). BAC-technology has enabled to validate a large number of mutations conferring resistance as well as polymorphisms not conferring resistance (Chevillotte *et al.*, 2009b; Chou, 2010; 2012; Hakki and Chou, 2011).

Frequencies of phenotypic GCV resistance are difficult to determine since some clinical isolates, especially from treated patients, are difficult to propagate in cell culture, and biological drug sensitivity assays are not well standardized. At first, incidences of GCV resistance have been reported to be approximately 8% in AIDS patients after 3 months of therapy (Drew *et al.*, 1991). At this time only case reports were available describing emergence of GCV resistance in transplant recipients of lung, liver, and bone marrow grafts. Meanwhile, several reports have documented the early emergence of GCV-resistant HCMV in children with severe immunodeficiencies (Eckle *et al.*, 2000; Wolf *et al.*, 2001; Hantz *et al.*, 2010). Studies showed that the percentage of AIDS patients harbouring GCV-resistant HCMV in blood or urine increased (7%, 12% and 28%) after 3, 6, and 9 months of GCV therapy, respectively (Jabs *et al.*, 1998a). Boivin and colleagues detected resistant virus in 2%, 7%, 9%, 13% and 15% of AIDS patients after 3, 6, 9, 12 and 18 months of therapy, respectively (Boivin *et al.*, 2001). Kruger found that GCV-resistant HCMV could be detected in blood or bronchoalveolar lavage from 18 (5.2%) of 348 lung transplant patients after a median GCV therapy of 79 (± 52) days (Kruger *et al.*, 1999). Limaye reported the detection of GCV-resistant HCMV in 4 (9%) of 45 lung-transplant recipients after a median of 4.4 months (range 3.1–6.6 months) after transplantation (Limaye *et al.*, 2002). A prospective study by Boivin (Boivin *et al.*, 2004, 2005) found a lower incidence of GCV resistance in SOT patients (0% in valGCV treated patients, and 6.1% in oral GCV treated patients). Intriguingly, the presence of resistant HCMV strains did not always correlate with disadvantageous clinical outcome (Chmiel *et al.*, 2008). However, some recent studies actually showed a higher incidence of GCV resistance in patients on pre-emptive management than those given GCV prophylaxis (van der Beek *et al.*, 2010; Myhre *et al.*, 2011).

FOS and CDV are less often used for initial therapy. By performing a culture-based study of the frequency of resistance to FOS and CDV in retinitis patients, Jabs and colleagues found baseline frequencies of resistance of < 3% for FOS and < 7% for CDV. During this study, 44 patients received FOS, and 37% of those treated for 9 months had developed at least one FOS-resistant isolate by that time. For CDV only 13 patients could be evaluated and 29% of those treated for 3 months had at least one CDV-resistant isolate (Jabs *et al.*, 1998b).

In addition to pUL97, mutations conferring antiviral resistance have been found in the viral polymerase pUL54 (Cihlar *et al.*, 1998; Chou, 1999; Chevillotte *et al.*, 2010b; Hakki and Chou, 2011). In contrast to some earlier reports, GCV resistance mutations in UL54 do not *per se* confer higher levels of resistance than those in UL97 (Fig. II.19.4A,B). But GCV-resistant strains may carry mutations in both the DNA polymerase and the UL97 gene, and may thus be cross-resistant to CDV and/or FOS. Also, mutations in UL54 may further increase the GCV resistance level conferred by UL97 mutations. Single mutations in the polymerase conferring multi-drug resistance have also been reported. Most of the highly GCV-resistant strains carry mutations in the UL54 and the UL97 gene and consequently are cross-resistant to CDV and/or FOS.

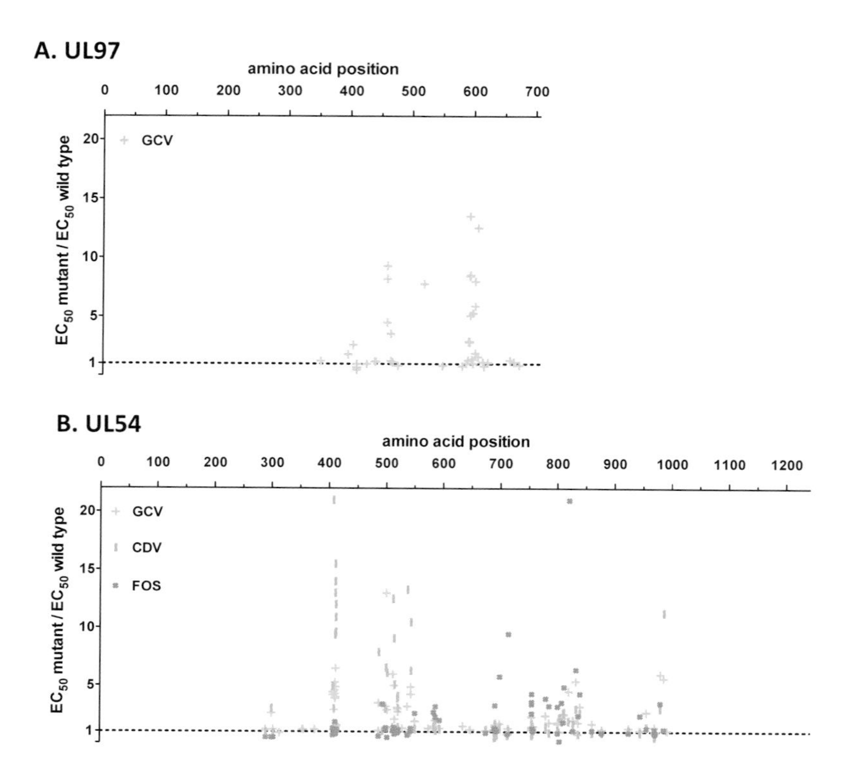

Figure II.19.4 Relative drug susceptibility levels of published, marker transfer validated HCMV pUL97 and pUL54 mutations from the HCMV drug resistance mutations database A. Mutations in pUL97, B. mutations in pUL54.

Summarizing the aspect of the overall sensitivity-phenotype of a mixed virus population occurring in a treated patient, the phenotype depends on the mutated gene (UL97 and/or UL54), the position and kind of the mutated amino acid and the dynamic amount of mutated viruses within the total population. It has also been shown – at least for herpes simplex virus – that the sensitive and resistant virus subpopulations interfere with each other (Okuda *et al.*, 2004). The occurrence of clinically relevant drug-resistant HCMV has raised the need for antiviral susceptibility testing for fast identification of resistant virus populations in treated patients.

Genotypic susceptibility testing: advantages and limitations

Genotypic resistance testing is currently the method of choice, because results can be obtained within short time without time-consuming virus isolation (Drew, 2007). For GCV resistance, the relevant HCMV genes are UL97 and UL54; for CDV and FOS it is UL54 only. Known resistance-conferring mutations in UL97 are located in a relatively small region of the protein, so that only approximately 700 bp have to be sequenced.

The situation in the polymerase pUL54 is more complex, as resistance-conferring mutations can be found almost throughout the protein (Fig. II.19.4B), requiring sequencing of a region of almost 2.5 kbp. This prevailing problem of sequencing larger gene regions will diminish in decent time with the availability of more advanced sequencing technologies. The relevant drug susceptibility mutations in UL54 are located in clusters and resistance phenotypes differ according to their position in the protein: Mutations located between aa 300 and 550 – exonuclease domains – tend to confer resistance against the nucleoside analogues GCV and CDV (Lurain *et al.*, 1992; Cihlar *et al.*, 1998; Chou *et al.*, 2003; Ducancelle *et al.*, 2005; Drew, 2006; Marfori *et al.*, 2007; Scott *et al.*, 2007; Chevillotte *et al.*, 2010a; Gilbert *et al.*, 2011; Chou, 2012). Mutations between aa 495 and 715 have the potential to confer resistance to FOS (Baldanti *et al.*, 1996; Cihlar *et al.*, 1998; Mousavi-Jazi *et al.*, 2001; Ducancelle *et al.*, 2005; Springer *et al.*, 2005; Chevillotte *et al.*, 2010a; Gilbert *et al.*, 2011). Finally, mutations between aa 750 and 1000 potentially confer a mixed resistance phenotype with loss of susceptibility to all of the three drugs (Baldanti *et al.*, 1996; Cihlar *et al.*, 1998; Chou, 1999, 2003; Waldman *et al.*, 1999a; Mousavi-Jazi *et al.*, 2001; Weinberg *et al.*, 2003; Springer *et al.*, 2005; Tchesnokov *et al.*, 2006; Drew, 2006; Scott *et al.*, 2007; Chevillotte *et al.*, 2010a; Gilbert *et al.*, 2011).

Apart from its short processing time, one major advantage of genotypic resistance testing is the possibility to detect subpopulations of resistant variants in mixed HCMV populations of individual patients. With pyrosequencing techniques, it might be possible to detect fractions of resistant populations down to 2%, already in early phases of therapy (Schindele *et al.*, 2010; Kampmann *et al.*, 2011). Recently, the potential of pyrosequencing has been confirmed by Görzer and colleagues, who observed complex mixtures of up to six different HCMV genotypes per patient sample. In all samples, no more than two major genotypes accounted for at least 90% of the HCMV DNA load, often accompanied by up to four low-abundance genotypes at frequencies of 0.1–9% (Gorzer *et al.*, 2010). However, our knowledge on viral population dynamics in patients is still limited. In addition to mixed virus populations found in the blood of one patient at a given time point, it has been observed that viral subpopulations from other body compartments may differ again, further complicating evaluation of resistance (Michel *et al.*, 2003). The host's immunocompetence certainly plays a crucial role in control of viral replication, but probably also viral fitness of the wild-type and the mutant population (Martin *et al.*, 2010). The predictive value of quantitative analyses of the virus subpopulations on clinical outcome has yet to be determined in clinical studies. The major challenge of genotypic resistance testing remains, however, to link the sequencing results to corresponding resistance phenotypes.

From genotype to phenotype

An efficient management of resistant HCMV requires the knowledge of the resistance-conferring mutations to screen for. In view of the many publications reporting UL97 and UL54 mutations, this is not an easy task. We have reported generation of a web-based search tool where a patient's HCMV UL97 or UL54 sequence can be submitted and will be subsequently aligned to that of a sensitive wild-type (Chevillotte *et al.*, 2010b; Martin *et al.*, 2010). The database (http://www.informatik.uni-ulm.de/ni/mitarbeiter/HKestler/hcmv/index.html) is weekly updated, thus providing most current information as an optimal basis for treatment decisions, and contains all published UL97 and UL54 mutations and corresponding phenotypes including the following information:

1. the mutation's qualitative drug resistance phenotype as stated by the authors of the publication with a link to PubMed;
2. the quantitative drug resistance phenotype, expressed as the ratio of EC_{50} values compared to wild-type from cell culture experiments;

3 the assay used in the cited publication to allow estimation of the quality of the phenotypic data; and
4 the viral fitness, a parameter that has been shown to influence clinical outcome.

Obviously, every approach to correlate phenotypic to genotypic results finally depends on the quality of the phenotypic data.

Phenotypic resistance testing: advantages and limitations

For many years, resistance of isolated viruses has been determined phenotypically in an *in vitro* drug sensitivity assay, mostly performed as focus reduction or plaque reduction assay (Landry *et al.*, 2000; Michel *et al.*, 2001). These assays are dependent on virus isolation and passage in cell culture, which requires up to 14 weeks. Comparative studies have been hampered by technical difficulties, lack of standardization of assays, and thus by different definitions of GCV resistance. Resistance has been defined either as EC_{50} values or as ratios of EC_{50} values obtained from mutated viruses and wild-type controls, after definition of mostly arbitrary cut-off values. The results of these biological assays should be interpreted with some caution. Since mixed virus populations are involved, discordant results can be observed in genotypic and phenotypic assays. We and others have observed that a selection bias by passaging the virus *in vitro* before phenotypic susceptibility testing may select for sensitive virus and lead to an underestimation of the impact of viral resistance in patients with failure of antiviral therapy. In all, plaque reduction assays with clinical isolates are nowadays mostly only used for a retrospective determination of resistance.

The main importance of phenotypic resistance testing today lies in the characterization of newly identified mutations after marker transfer (recombinant phenotyping), providing the crucial link between sequence and phenotype for genotypic resistance testing. Mutations suspected to confer resistance are introduced into a well-defined, drug-sensitive laboratory virus strain and then tested for changes in drug susceptibility in cell culture. Different protocols have been used over time for the generation of these recombinant HCMV strains (Cihlar *et al.*, 1998; Wagner *et al.*, 2000; Chou, 2010). The use of clonal BAC technology is currently state-of-the-art (see Chapter I.3). Best protocols are markerless mutagenesis, meaning that they leave no trace of the genetic manipulation in the BAC clonal viral sequence other than the desired mutation (Warming *et al.*, 2005; Tischer *et al.*, 2006; Maussang *et al.*, 2009b). Markerless BAC mutagenesis also allows introduction of multiple mutations in one strain in order to analyse the effects of combinations of mutations. On the one hand, combinations of mutations in UL97 and/or UL54 sometimes show synergistic effects on resistance (Mousavi-Jazi *et al.*, 2001; Ducancelle *et al.*, 2007; Scott *et al.*, 2007; Chevillotte *et al.*, 2010a), but can on the other hand also partially compensate a prior resistant phenotype (Ijichi *et al.*, 2002). Therefore, analysis of such combinations will be a very important task and a major challenge in the future.

The use of recombinant viruses offers another advantage. Most sensitive parental viruses carry a reporter system for detection of viral replication in cell culture, such as secreted alkaline phosphatase SEAP (Chou, 2010), GFP (Marschall *et al.*, 2000), EYFP (Dal Pozzo *et al.*, 2008) or EGFP (Sanchez *et al.*, 2002). These reporter systems enable circumvention of the poorly standardized plaque reduction assay, providing objective, standardized read-outs. It is highly desirable that all putative resistance mutations – and combinations – are characterized by such reporter virus-based marker transfer analyses. Still, all cell culture-based assays lack clinical validation, correlating an increase in EC_{50} values with diminished therapeutic response in the patient. This clinical evaluation of resistance testing results has to be provided by clinical studies.

Marker transfer experiments revealed that many mutations in UL97 and UL54 previously suspected to confer antiviral drug resistance in fact result in very low EC_{50} ratios (Fig. II.19.4A and B) and probably do not confer relevant drug resistance levels, although they all originate from patients with therapy failure. A reason for this could be that immunodeficiency of the host was too severe to control a sensitive virus bearing simple polymorphisms. Nevertheless, it is equally important to distinguish polymorphisms from resistance mutations in order to avoid unnecessary switching of antiviral therapy.

Alternative antiviral compounds

(see Fig. II.19.5)

Substances targeting the viral DNA polymerase pUL54

Cyclopropavir

Cyclopropavir (CPV) is a methylenecyclopropane nucleoside analogue that, like GCV, requires initial phosphorylation by pUL97 (James *et al.*, 2011). The drug has been used in phase I human studies, with the perspective of better *in vitro* potency and cytotoxicity profiles. However, due to the similar mechanism of pUL97 activation, CPV selects several mutations also observed with GCV, thereby conferring partial GCV/CPV cross-resistance. Interestingly, exonuclease and A987G mutations in UL54 at codons associated with dual GCV-CDV resistance in clinical isolates also conferred increased CPV susceptibility. CPV EC_{50} values against several *pol* mutants were increased about 2-fold by adding UL97 mutation C592G (Chou and Bowlin, 2011; Chou *et al.*, 2011b).

CMX-001

This compound is an orally bioavailable ester formulation of CDV. CMX-001 has the same mechanism of action as the parent compound. Clinical trials are on the way for the prevention of infections caused by different DNA viruses as well as HCMV infection after allogeneic stem cell transplantation (ClinicalTrials.gov Identifier: NCT01143181, 2012). Since CMX-001 is reported to have lesser renal toxicity it might be considered for the substitution of CDV.

pUL97-targeting molecules

Maribavir

Maribavir (MBV) exhibits antiviral potency against different HCMV strains *in vitro*, including strains resistant to GCV (mutations at amino acid positions 460, 520, and 594 in pUL97), FOS (mutation T700A in the viral polymerase) and BDCRB (mutations D344E and A355V in pUL89) (McSharry *et al.*, 2001; Biron *et al.*, 2002). The first isolation of a MBV-resistant HCMV strain revealed a mutated pUL97 to be responsible for the resistance. The mutation conferring resistance was mapped to amino acid position 397 (L397R) of pUL97. The today known MBV resistance mutations are located at the pUL97 amino acid positions 353, 397, 409, and 411. This region is next to the kinase ATP-binding domain of the protein (Chou, 2008). However, Komazin and colleagues reported the isolation of an MBV-resistant HCMV for which sequencing identified a single coding mutation in ORF UL27 (L335P) as the one responsible for resistance (Komazin *et al.*, 2003). Also others reported the isolation of MBV-resistant strains carrying mutations in pUL27, whereas no mutation could be found in the pUL97 gene (Chou *et al.*, 2004). The mutations in pUL27 supposed to be responsible for the MBV resistance were R233S, A406V/C415stop, and W362R. The UL27 mutations investigated in this study conferred only low-grade MBV resistance as measured by yield reduction assays.

An *in vitro* study has been presented by Selleseth and colleagues, who analysed the interaction of maribavir with GCV, ACV, CDV, FOS, and the anti-HIV drugs zidovudine, lamivudine, amprenavir, abacavir, indinavir, didanoside, and zalcitabine (Selleseth *et al.*, 2003). These authors found synergistic interaction of MBV only with CDV and abacavir, while additive effects against HCMV were observed with ACV, GCV and FOS. However, the recent phase III clinical trial in allogeneic bone marrow transplant recipients showed that MBV was not significantly better than placebo for the prevention of HCMV disease (Marty *et al.*, 2011; Marty and Boeckh, 2012) (see also Chapter II.16). Proposed explanations include an inadequate dosing regimen and choice of study endpoint. Similarly, preliminary data in a liver transplant population suggests that MBV was inferior to oral GCV for the prevention of HCMV disease (ClinicalTrials.gov Identifier: NCT00497796, 2008). Meanwhile the first patient has been observed developing MBV resistance mutations (Avery *et al.*, 2010a; Strasfeld *et al.*, 2010) (see also Chapter II.13). In conclusion, MBV would deserve further systematic evaluation as treatment for HCMV infection, and its optimal dose and duration of therapy would have to be determined.

Indolocarbazoles

The inhibitory effect of indolocarbazoles was demonstrated for HCMV infection in cultured cells, but the question of whether pUL97 phosphorylation activity is impaired by these compounds was not addressed (Slater *et al.*, 1999). This was first shown in studies by Zimmermann and colleagues (Zimmermann *et al.*, 1997), who showed that indolocarbazoles reduce the phosphorylation activity of pUL97. The compounds were effective *in vitro* against GCV-sensitive and resistant HCMV strains. The compounds Gö6976, K252a, and K252c were found to reduce the virus yield by three orders of magnitude with nanomolar concentrations. The serine/threonine inhibitors Gö6850 and roscovitine were not effective (Marschall *et al.*, 2001).

Different compounds with unknown mechanism of action

Leflunomide

The immunosuppressive drug leflunomide, licensed for the treatment of active rheumatoid arthritis, has been described as an inhibitor of HCMV replication, probably by preventing virion assembly (Waldman *et al.*, 1999b). The activity of leflunomide

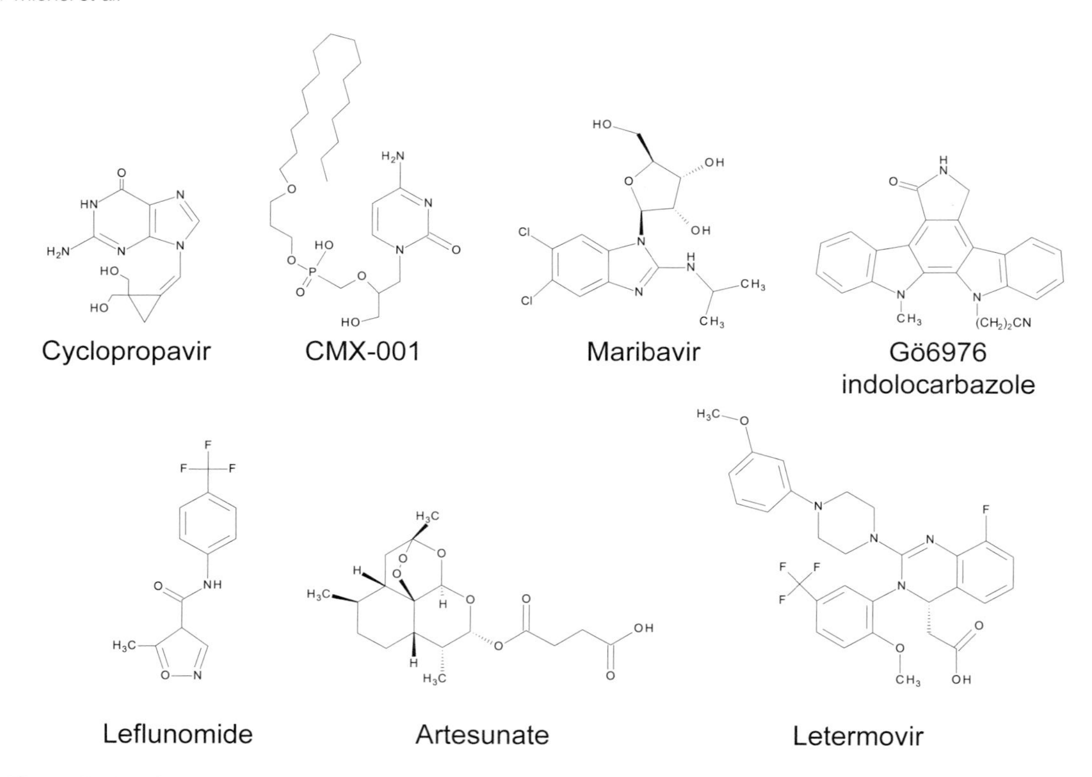

Figure II.19.5 Chemical formula of indicated compounds with antiviral activity against HCMV which are presently not in clinical use.

is attributed mainly to its metabolite termed A771726 [*N*-(4-trifluoromethylphenyl)-2-cyano-3-hydroxycrotoamide]. Leflunomide is considerably less expensive than GCV. Therefore, particularly for transplant patients in developing countries, the compound could be of benefit. Preliminary data suggest that leflunomide therapy for HCMV disease is effective and can be used in allograft recipients provided that a careful monitoring is performed (John *et al.*, 2004).

However, the exact dose for treatment of an HCMV infection still has to be established. It has been reported that the compound was active in clearing the viral load in a bone marrow transplant recipient, but also failure of leflunomide therapy has been observed (Avery *et al.*, 2004; 2010b; Battiwalla *et al.*, 2007).

Artesunate

It has been demonstrated that artesunate, an antimalaria drug, possesses antiviral activity against HCMV strains *in vitro* (Kaptein *et al.*, 2006; Efferth *et al.*, 2008; Li *et al.*, 2008). The substance seems to act on the level of cell signalling as it is believed to down modulate activation of NFκB- or Sp1 pathways by interfering with HCMV replication. The inhibitory effect of artesunate appears to have additive effects with GCV, FOS and CDV. Shapira and colleagues described for the first time the treatment of a stem cell transplant recipient suffering from an infection with a FOS- and GCV-resistant HCMV (Shapira *et al.*, 2008). Artesunate treatment resulted in a significant reduction of the viral load. Recently, it has been reported that artesunate exhibited an effect against a variety of therapy-resistant virus mutants (Chou *et al.*, 2011a).

Viral kinetics of pre-emptive artesunate treatment has been examined in stem cell transplant recipients. Overall, a divergent antiviral efficacy was observed, that appeared to be dependent on the virus baseline growth dynamics (Wolf *et al.*, 2011). Therefore, further dose escalation studies are needed. However, case reports occasionally found that artesunate was ineffective against GCV-resistant HCMV (Lau *et al.*, 2011). *In vitro* studies have shown that more effective modified compounds could be developed. Four artemisinin monomers and two novel artemisinin-derived dimers were tested. The artemisinin dimers most effectively inhibited HCMV replication *in vitro* with no cytotoxicity in fibroblasts at concentrations required for complete HCMV inhibition (Arav-Boger *et al.*, 2010).

Lipophilic alkyl furano pyrimidine dideoxynucleotides (LFPNs)

McGuigan and colleagues showed that the LFPNs are potent and selective antiviral compounds with high specificity for HCMV (McGuigan *et al.*, 2004). Despite their nucleoside structure, the LFPNs do not act as nucleoside analogues but inhibit at a stage in the replication cycle preceding DNA synthesis. However, the exact mechanism of action is open to investigation.

Viral terminase-targeting molecules

Phenylenediaminesulfonamides and benzimidazole derivatives

Using cell culture assays as a primary screening tool for new compounds, two very different classes of HCMV inhibitors, phenylenediaminesulfonamides (e.g. BAY38-4766) and halo-benzimidazoles (e.g. BDCRB, see below) have been identified (Townsend *et al.*, 1995; Reefschlaeger *et al.*, 2001). Benzimidazole-D-ribonucleosides, such as 2,5,6-trichloro-(1-β-D-ribofuranosyl) benzimidazole (TCRB), its 2-bromo homologue (BDCRB) and 5,6-dichloro-2-(isopropylamino)–1,β-L-ribofuranosyl-1-H-benzimidazole (1263W94 or maribavir) are potent and selective inhibitors of HCMV replication (see Fig. II.19.1). Benzimidazole-ribosides inhibit at different stages of HCMV replication (viral DNA maturation). HCMV pUL89, involved in DNA cleavage/packaging, has been identified as a target of both classes of compounds. Inhibition of pUL89 by phenylenediaminesulfonamides or benzimidazoles prevents processing and maturation of head-to-tail concatemeric viral DNA to monomeric genomes (Krosky *et al.*, 1998; Underwood *et al.*, 1998; Buerger *et al.*, 2001). It has been reported that BDCRB inhibits the nuclease activity of HCMV pUL89 and the ATPase activity of pUL56 (Scheffczik *et al.*, 2002; Scholz *et al.*, 2003). HCMV strains resistant to TCRB and BDCRB have been isolated. BDCRB resistance was associated with a mutated UL89 whereas TCRB resistance could be mapped to UL56 and UL89, suggesting specific inhibition of HCMV by these two compounds.

Non-nucleoside terminase inhibitors, letermovir

Hit-to-lead optimization led to the discovery of the substance AIC246 (letermovir) (Lischka *et al.*, 2010). Letermovir targets the viral pUL56 subunit, which is involved in viral DNA processing and/or packaging and belongs to the novel chemical class of 3,4 dihydro-quinazolinyl-acetic acids. As there is no counterpart of the viral terminase in humans, it is expected that compounds targeting this viral enzyme will not show target-related toxicities. In addition, the different mode of action of letermovir should provide new treatment options for patients infected with resistant virus strains for which no effective therapy is currently available (Härter and Michel, 2012; Marschall *et al.*, 2012). *In vitro* data suggested a high antiviral activity without cross-resistance between letermovir and currently available agents (Lischka *et al.*, 2010; Goldner *et al.*, 2011). Phase I data from healthy subjects demonstrated that oral letermovir administration was safe and well tolerated without any drug-specific adverse effects. Letermovir is currently undergoing a clinical phase IIb trial for prophylactic use. Promising perspectives offer the successful treatment of refractory multidrug-resistant HCMV (Kaul *et al.*, 2011). No resistance in patients has been observed until now.

Small interfering RNAs

RNA interference (RNAi) is a natural mechanism of post-transcriptional gene silencing. The mechanism is widely conserved in multicellular organisms and is supposed to be a natural defence strategy against the invasion of viruses and for transposon silencing. The small interfering RNAs (siRNAs) are double-stranded molecules of 21 to 25 base pairs in length that induce the sequence-specific degradation of homologous perfect base pairing RNAs (see Chapter I.5). In recent years the siRNA approach has been shown to be principally effective against HIV, HCV, influenza virus and poliovirus *in vitro*. Wiebusch and colleagues obtained first results that open new possibilities for an antiviral strategy against HCMV (Wiebusch *et al.*, 2004). By designing siRNA against the viral polymerase pUL54 they achieved a reduction in UL54 mRNA levels in infected fibroblasts as well as an inhibition of virus replication. Owing to the high specificity of the siRNA approach these first results are promising to lead to a completely new anti-HCMV strategy.

Epidermal growth factor receptor inhibitors

The epidermal growth factor receptor (EGF-R) has been proposed as a cellular receptor responsible for HCMV entry and HCMV-induced signalling (Wang *et al.*, 2003) (see also Chapter I.8). The viral protein gB appears to be responsible for binding to EGF-R. The putative role of EGF-R as an HCMV receptor may assist the development of anti-HCMV compounds that target very early stages of viral entry. First attempts have been using polyphenolic flavonoids that inhibit EGF-R kinase activity and benzoquinoid asamycins (derivatives of the antibiotic geldanamycin) that inhibit the protein chaperone HSP90 and thus prevent EGF-R expression.

Conclusions and perspectives

Owing to HAART, the incidence of HCMV disease in AIDS patients caused by GCV-resistant HCMV has sharply declined over the last years (see also Chapter II.1). In contrast, GCV-resistant HCMV can be increasingly found in transplant recipients receiving antiviral therapy (see also Chapters II.13, II.14 and II.16). Relevant factors for the emergence of drug-resistant HCMV are severe immunosuppression, high levels of viral replication, prolonged antiviral therapy, and sub-inhibitory drug concentrations at the site of viral replication. During the past years routine administration of antiviral substances for prophylaxis and pre-emptive therapy may also have led to longer GCV exposure of replicating HCMV. A possible differential role of different immunosuppressive drugs such as mycophenolate, tacrolimus, everolimus and cyclosporine A has not been systematically investigated so far, but some data confirm such differences, e.g. everolimus inducing more reactivations than mycophenolate, and clinicians anecdotally report differences in HCMV reactivation and disease depending on the immunosuppressive drugs used. The optimization of the immunosuppressive therapy should be performed also considering potential HCMV reactivations. Antiviral prophylaxis should preferably be restricted to patients known to be at high risk of HCMV disease, and duration of prophylaxis should be matched to the period of severe immunosuppression and to the results of virological monitoring. Concerning prophylaxis it is unclear whether ACV-prophylaxis also promotes GCV resistance in patients. Owing to the growing number of GCV-resistant HCMV strains in transplant patients and the increasing use of CDV and FOS, the incidence of multiresistant HCMV strains will progressively rise with double mutations in pUL97 and the viral polymerase pUL54. Fast routine genotyping of UL54 will become much easier with the advanced sequencing technologies.

The search for new compounds has yielded a wealth of novel antivirals being or behaving as non-nucleosides. Of all these compounds, maribavir, lefluonomid, artesunate and letermovir have so far proceeded into clinical trials. These compounds represent new classes of systemic anti-HCMV drugs with mechanisms of action unrelated to previously available agents. Although resistance has been observed, the gene loci are different from those observed for GCV, CDV and FOS. Owing to high specificity, the siRNA approach enables a completely new anti-HCMV strategy.

A great obstacle for the understanding of HCMV pathogenesis and the preclinical *in vivo* testing of new compounds with anti-HCMV activity is the absence of an appropriate animal model. Hence, one should think about models based on recombinant animal CMVs that exhibit key features of HCMV replication (see Chapters I.23, II.5 and II.22).

It is unknown whether the spread of drug-resistant strains from patient to patient in the clinical setting, as observed for HIV-infected individuals, will become a clinical problem in HCMV, too. Clinicians and virologists should cooperate in characterizing the molecular mechanisms of resistance, in characterizing the relevant mutations for their related phenotypes, in the early detection of emerging resistant virus populations, and finally in optimizing therapy for patients.

Acknowledgements

Part of this work was supported by a grant (RKI 1369–424) from the Robert Koch Institute, Berlin.

References

Abraham, K.A., O'Kelly, P., Spencer, S., Hickey, D.P., Conlon, P.J., and Walshe, J.J. (2008). Effect of cytomegalovirus prophylaxis with acyclovir on renal transplant survival. Ren. Fail. *30*, 141–146.

Alain, S., Hantz, S., Scieux, C., Karras, A., Mazeron, M.C., Szelag, J.C., Imbert, B.M., Fillet, A.M., Gouarin, S., Mengelle, C., *et al.* (2004). Detection of ganciclovir resistance after valacyclovir-prophylaxis in renal transplant recipients with active cytomegalovirus infection. J. Med. Virol. *73*, 566–573.

Arav-Boger, R., He, R., Chiou, C.-J., Liu, J., Woodard, L., Rosenthal, A., Jones-Brando, L., Forman, M., and Posner, G. (2010). Artemisinin-derived dimers have greatly improved anti-cytomegalovirus activity compared to artemisinin monomers. PLoS One. *5*, e10370-doi:10.1371/journal.pone.0010370.

Avery, R.K., Bolwell, B.J., Yen-Lieberman, B., Lurain, N., Waldman, W.J., Longworth, D.L., Taege, A.J., Mossad, S.B., Kohn, D., Long, J.R., *et al.* (2004). Use of leflunomide in an allogeneic bone marrow transplant recipient with refractory cytomegalovirus infection. Bone Marrow Transplantation *34*, 1071–1075.

Avery, R.K., Marty, F.M., Strasfeld, L., Lee, I., Arrieta, A., Chou, S., Tatarowicz, W., and Villano, S. (2010a). Oral maribavir for treatment of refractory or resistant cytomegalovirus infections in transplant recipients. Transpl. Infect. Dis. *12*, 489–496.

Avery, R.K., Mossad, S.B., Poggio, E., Lard, M., Budev, M., Bolwell, B., Waldman, W.J., Braun, W., Mawhorter, S.D., Fatica, R., *et al.* (2010b). Utility of leflunomide in the treatment of complex cytomegalovirus syndromes. Transplantation *90*, 419–426.

Azad, R.F., Driver, V.B., Tanaka, K., Crooke, R.M., and Anderson, K.P. (1993). Antiviral activity of a phosphorothioate oligonucleotide complementary to RNA of the human cytomegalovirus major immediate-early region. Antimicrob. Agents Chemother. *37*, 1945–1954.

Baek, M.C., Krosky, P.M., He, Z., and Coen, D.M. (2002). Specific phosphorylation of exogenous protein and peptide substrates by the human cytomegalovirus UL97 protein kinase. Importance of the P+5 position. J. Biol. Chem. *277*, 29593–29599.

Baldanti, F., Sarasini, A., Silini, E., Barbi, M., Lazzarin, A., Biron, K.K., and Gerna, G. (1995). Four dually resistant human cytomegalovirus strains from AIDS patients: single mutations in UL97 and UL54 open reading frames are responsible for ganciclovir- and foscarnet-specific resistance, respectively. Scand. J. Infect. Dis. *Suppl. 99*, 103–104.

Baldanti, F., Underwood, M.R., Stanat, S.C., Biron, K.K., Chou, S., Sarasini, A., Silini, E., and Gerna, G. (1996). Single amino acid changes in the DNA polymerase confer foscarnet resistance and slow-growth phenotype, while mutations in the UL97-encoded phosphotransferase confer ganciclovir resistance in three double-resistant human cytomegalovirus strains recovered from patients with AIDS. J. Virol. *70*, 1390–1395.

Baldanti, F., Michel, D., Simoncini, L., Heuschmid, M., Zimmermann, A., Minisini, R., Schaarschmidt, P., Schmid, T., Gerna, G., and Mertens, T. (2002). Mutations in the UL97 ORF of ganciclovir-resistant clinical cytomegalovirus isolates differentially affect GCV phosphorylation as determined in a recombinant vaccinia virus system. Antiviral Res. *54*, 59–67.

Balfour Jr., H.H., Chace, B.A., Stapleton, J.T., Simmons, R.L., and Fryd, D.S. (1989). A randomized, placebo-controlled trial of oral acyclovir for the prevention of cytomegalovirus disease in recipients of renal allografts. New Engl. J. Med. *320*, 1381–1387.

Battiwalla, M., Paplham, P., Almyroudis, N.G., McCarthy, A., Abdelhalim, A., Elefante, A., Smith, P., Becker, J., McCarthy, P.L., and Segal, B.H. (2007). Leflunomide failure to control recurrent cytomegalovirus infection in the setting of renal failure after allogeneic stem cell transplantation. Transpl. Infect. Dis. *9*, 28–32.

van der Beek, M.T., Berger, S.P., Vossen, A.C., van der Blij-de Brouwer, C.S., Press, R.R., de Fijter, J.W., Claas, E.C., and Kroes, A.C. (2010). Pre-emptive versus sequential prophylactic-preemptive treatment regimens for cytomegalovirus in renal transplantation: comparison of treatment failure and antiviral resistance. Transplantation *89*, 320–326.

Biron, K.K. (2006). Antiviral drugs for cytomegalovirus diseases. Antiviral Res. *71*, 154–163.

Biron, K.K., Fyfe, J.A., Stanat, S.C., Leslie, L.K., Sorrell, J.B., Lambe, C.U., and Coen, D.M. (1986). A human cytomegalovirus mutant resistant to the nucleoside analog 9-{[2-hydroxy-1-(hydroxymethyl)ethoxy] methyl}guanine (BW B759U) induces reduced levels of BW B759U triphosphate. Proc. Natl. Acad. Sci. U.S.A. *83*, 8769–8773.

Biron, K.K., Harvey, R.J., Chamberlain, S.C., Good, S.S., Smith III, A.A., Davis, M.G., Talarico, C.L., Miller, W.H., Ferris, R., Dornsife, R.E., *et al.* (2002). Potent and selective inhibition of human cytomegalovirus replication by 1263W94, a benzimidazole L-riboside with a unique mode of action. Antimicrob. Agents Chemother. *46*, 2365–2372.

Boivin, G., Gilbert, C., Gaudreau, A., Greenfield, I., Sudlow, R., and Roberts, N.A. (2001). Rate of emergence of cytomegalovirus (CMV) mutations in leukocytes of patients with acquired immunodeficiency syndrome who are receiving valganiclovir as induction and maintenace therapy for CMV retinitis. J. Infect. Dis. *184*, 1598–1602.

Boivin, G., Goyette, N., Gilbert, C., Roberts, N., Macey, K., Paya, C., Pescovitz, M.D., Humar, A., Dominguez, E., Washburn, K., *et al.* (2004). Absence of cytomegalovirus-resistance mutations after valganciclovir prophylaxis, in a prospective multicenter study of solid-organ transplant recipients. J. Infect. Dis. *189*, 1615–1618.

Boivin, G., Goyette, N., Gilbert, C., Humar, A., and Covington, E. (2005). Clinical impact of ganiclovir-resistant cytomegalovirus infections in solid organ transplant. Transpl. Infect. Dis *7*, 166–170.

Borst, E.-M., Hahn, G., Koszinowski, U.H., and Messerle, M. (1999). Cloning of the human cytomegalovirus (HCMV) genome as an infectious bacterial artifical chromosome in *Escherichia coli*: A new approach for construction of HCMV mutants. J. Virol. *73*, 8320–8329.

Boutolleau, D., Burrel, S., and Agut, H. (2011). Genotypic characterization of human cytomegalovirus UL97 phosphotransferase natural polymorphism in the era of ganciclovir and maribavir. Antiviral Res. *91*, 32–35.

Brennan, D.C., Garlock, K.A., Lippmann, B.A., Buller, R.S., Gaudreault-Keener, M., Lowell, J.A., Miller, S.B., Shenoy, S., Howard, T.K., and Storch, G.A. (1997). Control of cytomegalovirus-associated morbidity in renal transplant patients using intensive monitoring and either pre-emptive or deferred therapy. J. Am. Soc. Nephrol. *8*, 118–125.

Buerger, I., Reefschlaeger, J., Bender, W., Eckenberg, P., Popp, A., Weber, O., Graeper, S., Klenk, H.D., Ruebsamen-Waigmann, H., and Hallenberger, S. (2001). A novel nonnucleoside inhibitor specifically targets cytomegalovirus DNA maturation via the UL89 and UL56 gene products. J. Virol. *75*, 9077–9086.

Caposio, P., Orloff, S.L., and Streblow, D.N. (2011). The role of cytomegalovirus in angiogenesis. Virus Res. *157*, 204–211.

Chee, M.S., Lawrence, G.L., and Barrell, B.G. (1989). Alpha-, beta- and gammaherpesviruses encode a putative phosphotransferase. J. Gen. Virol. *70*, 1151–1160.

Chevillotte, M., Landwehr, S., Linta, L., Frascaroli, G., Luske, A., Buser, C., Mertens, T., and v. Einem, J. (2009a). Major tegument protein pp65 of human cytomegalovirus is required for the incorporation of pUL69 and pUL97 into the virus particle and for viral growth in macrophages. J. Virol. *83*, 2480–2490.

Chevillotte, M., Schubert, A., Mertens, Th., and v. Einem, J. (2009b). Fluorescence-based assay for phenotypic characterization of human cytomegalovirus polymerase mutations regarding drug susceptibility viral replicative fitness. Antimicrob. Agents Chemother. *53*, 3752–3761.

Chevillotte, M., Ersing, I., Mertens, T., and v. Einem, J. (2010a). Differentiation between polymorphisms and resistance-associated mutations in human cytomegalovirus DNA polymerase. Antimicrob. Agents Chemother. *54*, 5004–5011.

Chevillotte, M., v. Einem, J., Meier, B.M., Lin, F.-M., Kestler, H.W., and Mertens, Th. (2010b). A new tool linking human cytomegalovirus drug resistance mutations to resistance phenotypes. Antiviral Res. 85, 318–327.

Chmiel, C., Speich, R., Hofer, M., Michel, D., Mertens, T., Weder, W., and Boehler, A. (2008). Ganciclovir/valganciclovir prophylaxis decreases cytomegalovirus-related events and bronchiolitis obliterans syndrome after lung transplantation. Clin. Infect. Dis. 46, 831–839.

Chou, S. (1999). Antiviral drug resistance in human cytomegalovirus. Transpl. Infect. Dis. 1, 105–114.

Chou, S. (2008). Cytomegalovirus UL97 mutations in the era of ganciclovir and maribavir. Rev. Med. Virol. 18, 233–246.

Chou, S. (2010). Recombinant phenotyping of cytomegalovirus UL97 kinase sequence variants for ganciclovir resistance. Antimicrob. Agents Chemother. 54, 2371–2378.

Chou, S. (2012). Phenotypic diversity of cytomegalovirus DNA polymerase gene variants observed after antiviral therapy. J. Clin. Virol. 50, 287–291.

Chou, S., and Bowlin, T.L. (2011). Cytomegalovirus UL97 mutations affecting cyclopropavir and ganciclovir susceptibility. Antimicrob. Agents Chemother. 55, 382–384.

Chou, S., and Marousek, G.I. (2008). Accelerated evolution of maribavir resistance in a cytomegalovirus exonuclease domain II mutant. J. Virol. 82, 246–253.

Chou, S., Waldemer, R.H., Senters, A.E., Michels, K.S., Kemble, G.W., Miner, R.C., and Drew, N.L. (2002). Cytomegalovirus UL97 phosphotransferase mutations that affect susceptibility to ganciclovir. J. Infect. Dis. 185, 162–169.

Chou, S., Lurain, N.S., Thompson, K.D., Miner, R.C., and Drew, W.L. (2003). Viral DNA polymerase mutations associated with drug resistance in human cytomegalovirus. J. Infect. Dis. 188, 33–39.

Chou, S., Marousek, G.I., Senters, A.E., Davis, M.G., and Biron, K.K. (2004). Mutations in the human cytomegalovirus UL27 gene that confer resistance to maribavir. J. Virol. 78, 7124–7130.

Chou, S., Wechel, L.C., and Marousek, G.I. (2007). Cytomegalovirus UL97 kinase mutations that confer maribavir resistance. J. Infect. Dis. 196, 91–94.

Chou, S., Marousek, G., Auerochs, S., Stamminger, T., Milbradt, J., and Marschall, M. (2011a). The unique antiviral activity of artesunate is broadly effective against human cytomegaloviruses including therapy-resistant mutants. Antiviral Res. 92, 364–368.

Chou, S., Marousek,G., and Bowlin,T.L. (2011b). Cyclopropavir susceptibility of cytomegalovirus DNA polymerase mutants selected after antiviral drug exposure. Antimicrob. Agents Chemother. 56, 197–201.

Chrisp, P., and Clissold, S.P. (1991). Foscarnet: A review of its antiviral activity, pharmacokinetic properties and therapeutic use in immunocompromised patients with cytomegalovirus retinitis. Drugs 41, 104–129.

Cihlar, T., Fuller, M.D., and Cherrington, J.M. (1998). Characterization of drug resistance-associated mutations in the human cytomegalovirus DNA polymerase gene by using recombinant mutant viruses generated from overlapping DNA fragments. J. Virol. 72, 5927–5936.

ClinicalTrials.gov Identifier: NCT00497796 (2008). A randomized, double-blind study to assess the efficacy and safety of prophylactic use of maribavir versus oral ganciclovir for the prevention of cytomegalovirus disease in recipients of orthotopic liver transplants. http://clinicaltrials. gov/ct2/show/NCT00497796.

ClinicalTrials.gov Identifier: NCT01143181 (2012). A multicenter, open-label study of CMX001 treatment of serious diseases or conditions caused by dsDNA viruses. http://clinicaltrials. gov/ct2/show/NCT01143181.

Crumpacker, C.S. (1996). Ganciclovir. N. Engl. J. Med. 335, 721–729.

Curran, M., and Noble, S. (2001). Valganciclovir. Drugs 61, 1145–1150.

Dal Pozzo, F., Andrei, G., Daelemans, D., Winkler, M., Piette, J., De Clercq, E., and Snoeck, R. (2008). Fluorescence-based antiviral assay for the evaluation of compounds against vaccinia virus, varicella zoster virus and human cytomegalovirus. J. Virol. Methods 151, 66–73.

De Clercq, E., Sakuma, T., Baba, M., Pauwels, R., Balzarini, J., Rosenberg, I., and Holy, A. (1987). Antiviral activity of phosphonylmethoxyalkyl derivatives of purine and pyrimidines. Antiviral Res. 8, 261–272.

Dmitrienko, S., Yu, A., Balshaw, R., Shapiro, R.J., Keown, P.A., and Genome Canada Biomarkers in Transplantation Group (2007). The use of consensus guidelines for management of cytomegalovirus infection in renal transplantation. Kidney Int. 72, 1014–1022.

Drew, W.L. (2006). Is combination antiviral therapy for CMV superior to monotherapy? J. Clin. Virol. 35, 485–488.

Drew, W.L. (2007). Laboratory diagnosis of cytomegalovirus infection and disease in immunocompromised patients. Curr. Opin. Infect. Dis. 20, 408–411.

Drew, W.L., Miner, R.C., Busch, D.F., Follansbee, S.E., Gullett, J., Mehalko, S.G., Gordon, S.M., Owen Jr., W.F., Matthews, T.R., Buhles, W.C., et al. (1991). Prevalence of resistance in patients receiving ganciclovir for serious cytomegalovirus infection. J. Infect. Dis. 163, 716–719.

Ducancelle, A., Gravisse, J., Alain, S., Fillet, A.M., Petit, F., Pors, M.J., and Mazeron, M.C. (2005). Phenotypic characterisation of cytomegalovirus DNA polymerase: a method to study cytomegalovirus isolates resistant to foscarnet. J. Virol. Methods 125, 145–151.

Ducancelle, A., Alain, S., Petit, F., Sanson Le Pors, M.J., and Mazeron, M.C. (2007). Development and validation of a non-radioactive DNA polymerase assay for studying cytomegalovirus resistance to foscarnet. J. Virol. Methods 141, 212–215.

Eckle, T., Prix, L., Jahn, G., Klingebiel, T., Handgretinger, R., Selle, B., and Hamprecht, K. (2000). Drug-resistant human cytomegalovirus infection in children after allogeneic stem cell transplantation may have different clinical outcomes. Blood 96, 3286–3289.

Efferth, T., Romero, M.R., Wolf, D.G., Stamminger, T., Marin, J.J., and Marschall, M. (2008). The antiviral activities of artemisinin and artesunate. Clin. Infect. Dis. 47, 804–811.

Eid, A.J., and Razonable, R.R. (2010). New developments in the management of cytomegalovirus infection after solid organ transplantation. Drugs 70, 965–981.

Einsele, H., Kapp, M., and Grigoleit, G.U. (2008). CMV-specific T-cell therapy. Blood Cells Mol. Dis. 40, 71–75.

Emanuel, D., Cunningham, I., and Jules-Elysee, K. (1988). Cytomegalovirus pneumonia after bone marrow transplantation successfully treated with the combination of ganciclovir and high-dose intravenous immune globulin. Ann. Intern. Med. 109, 777–782.

Erice, A. (1999). Resistance of human cytomegalovirus to antiviral drugs. Clin. Microbiol. Rev. 12, 286–297.

Erice, A., Gil-Roda, C., Perez, J.L., Balfour Jr., H.H., Sannerud, K.J., Hanson, M.N., Boivin, G., and Chou, S. (1997). Antiviral susceptibilities and analysis of UL97 and DNA polymerase sequences of clinical cytomegalovirus isolates from immunocompromised patients. J. Infect. Dis. *175*, 1087–1092.

Gavaldà, J., de Otero, J., Murio, E., Vargas, V., Rosselló, J., Calicó, I.M.C., and Pahissa, A. (1997). Two grams daily of oral acyclovir reduces the incidence of cytomegalovirus disease in CMV-seropositive liver transplant patients. Transpl. Int. *10*, 462–465.

Gilbert, C., and Boivin, G. (2005). New reporter cell line to evaluate the sequential emergence of multiple human cytomegalovirus mutations during *in vitro* drug exposure. Antimicrob. Agents Chemother. *49*, 4860–4866.

Gilbert, C., Azzi, A., Goyette, N., Lin, S.X., and Boivin, G. (2011). Recombinant phenotyping of cytomegalovirus UL54 mutations that emerged during cell passages in the presence of either ganciclovir or foscarnet. Antimicrob. Agents Chemother. *55*, 4019–4027.

Goldner, T., Hewlett, G., Ettischer, N., Ruebsamen-Schaeff, H., Zimmermann, H., and Lischka, P. (2011). The novel anticytomegalovirus compound AIC246 (Letermovir) inhibits human cytomegalovirus replication through a specific antiviral mechanism that involves the viral terminase. J. Virol. *85*, 10884–10893.

Gorzer, I., Guelly, C., Trajanoski, S., and Puchhammer-Stockl, E. (2010). Deep sequencing reveals highly complex dynamics of human cytomegalovirus genotypes in transplant patients over time. J. Virol. *84*, 7195–7203.

Hakki, M., and Chou, S. (2011). The biology of cytomegalovirus drug resistance. Curr. Opin. Infect. Dis. *24*, 605–611.

Hanks, S.K., Quinn, A.M., and Hunter, T. (1988). The protein kinase family: conserved features and deduced phylogeny of the catalytic domains. Science *241*, 42–52.

Hanson, M., Preheim, L.C., Chou, S., Talarico, C.L., Biron, K.K., and Erice, A. (1995). Novel mutation in the UL97 gene of a clinical cytomegalovirus strain conferring resistance to ganciclovir. Antimicrob. Agents Chemother. *39*, 1204–1205.

Hantz, S., Michel, D., Fillet, A.M., Guigonis, V., Champier, G., Mazeron, M.C., Bensman, A., Denis, F., Mertens, T., Dehee, A., *et al.* (2005). Early selection of a new UL97 mutant with a severe defect of ganciclovir phosphorylation after valaciclovir prophylaxis and short-term ganciclovir therapy in a renal transplant recipient. Antimicrob. Agents Chemother. *49*, 1580–1583.

Hantz, S., Garnier-Geoffroy, F., Mazeron, M.C., Garrigue, I., Merville, P., Mengelle, C., Rostaing, L., Saint, M.F., Essig, M., Rerolle, J.P., *et al.* (2010). Drug-resistant cytomegalovirus in transplant recipients: a French cohort study. J. Antimicrob. Chemother. *65*, 2628–2640.

Härter, G., and Michel, D. (2012). Antiviral treatment of cytomegalovirus infection: an update. Expert. Opin. Pharmacother. *13*, 623–627.

He, Z., He, Y., Kim, Y., Chu, L., Ohmstede, C., Biron, K.K., and Coen, D.M. (1997). The human cytomegalovirus UL97 protein is a protein kinase that autophosphorylates on serines and threonines. J. Virol. *71*, 405–411.

Heininger, A., Jahn, G., Engel, C., Notheisen, T., Unertl, K., and Hamprecht, K. (2001). Human cytomegalovirus infections in nonimmunosuppressed critically ill patients. Crit. Care Med. *29*, 541–547.

Humar, A., Limaye, A.P., Blumberg, E.A., Hauser, I.A., Vincenti, F., Jardine, A.G., Abramowicz, D., Ives, J.A., Farhan, M., and Peeters, P. (2010). Extended valganciclovir prophylaxis in D+/R- kidney transplant recipients is associated with long-term reduction in cytomegalovirus disease: two-year results of the IMPACT study. Transplantation *90*, 1427–1431.

Ijichi, O., Michel, D., Mertens, T., Miyata, K., and Eizuru, Y. (2002). GCV resistance due to the mutation A594P in the cytomegalovirus protein UL97 is partially reconstituted by a second mutation at D605E. Antiviral Res. *53*, 135–142.

Jabs, D.A., Enger, C., Dunn, J.P., and Forman, M. (1998a). Cytomegalovirus retinitis and viral resistance: ganciclovir resistance. J. Infect. Dis. *177*, 770–773.

Jabs, D.A., Enger, C., Forman, M., and Dunn, J.P. (1998b). Incidence of foscarnet resistance and cidofovir resistance in patients treated for cytomegalovirus retinitis. The Cytomegalovirus Retinitis and Viral Resistance Study Group. Antimicrob. Agents Chemother. *42*, 2240–2244.

James, S.H., Hartline, C.B., Harden, E.A., Driebe, E.M., Schupp, J.M., Engelthaler, D.M., Keim, P.S., Bowlin, T.L., Kern, E.R., and Prichard, M.N. (2011). Cyclopropavir inhibits the normal function of the human cytomegalovirus UL97 kinase. Antimicrob. Agents Chemother. *55*, 4682–4691.

John, G.T., Manivannan, J., Chandy, S., Peter, S., and Jacob, C.K. (2004). Leflunomide therapy for cytomegalovirus disease in renal allograft recepients. Transplantation *77*, 1460–1461.

Kamil, J.P., and Coen, D.M. (2007). Human cytomegalovirus protein kinase UL97 forms a complex with the tegument phosphoprotein pp65. J. Virol. *81*, 10659–10668.

Kampmann, S.E., Schindele, B., Apelt, L., Buhrer, C., Garten, L., Weizsaecker, K., Kruger, D.H., Ehlers, B., and Hofmann, J. (2011). Pyrosequencing allows the detection of emergent ganciclovir resistance mutations after HCMV infection. Med. Microbiol. Immunol. *200*, 109–113.

Kaptein, S.J., Efferth, T., Leis, M., Rechter, S., Auerochs, S., Kalmer, M., Bruggeman, C.A., Vink, C., Stamminger, T., and Marschall, M. (2006). The anti-malaria drug artesunate inhibits replication of cytomegalovirus *in vitro* and *in vivo*. Antiviral Res. *69*, 60–69.

Kaul, D.R., Stoelben, S., Cober, E., Ojo, T., Sandusky, E., Lischka, P., Zimmermann, H., and Rubsamen-Schaeff, H. (2011). First report of successful treatment of multidrug-resistant cytomegalovirus disease with the novel anti-CMV compound AIC246. Am. J. Transplantation *11*, 1079–1084.

Kedhar, S.R., and Jabs, D.A. (2007). Cytomegalovirus retinitis in the era of highly active antiretroviral therapy. Herpes *14*, 66–71.

Kimberlin, D.W., Lin, C.Y., Sanchez, P.J., Demmler, G.J., Dankner, W., Shelton, M., Jacobs, R.F., Vaudry, W., Pass, R.F., Kiell, J.M., *et al.* (2003). Effect of ganciclovir therapy on hearing in symptomatic congenital cytomegalovirus disease involving the central nervous system: a randomized, controlled trial. J. Pediatr. *143*, 16–25.

Komazin, G., Ptak, R.G., Emmer, B.T., Townsend, L.B., and Drach, J.C. (2003). Resistance of human cytomegalovirus to the benzimidazole L-ribonucleoside maribavir maps to UL27. J. Virol. *77*, 11499–11506.

Kotton, C.N., Kumar, D., Caliendo, A.M., Asberg, A., Chou, S., Snydman, D.R., Allen, U., and Humar, A. (2010). International consensus guidelines on the management of cytomegalovirus in solid organ transplantation. Transplantation *89*, 779–795.

Krosky, P.M., Underwood, M.R., Turk, S.R., Feng, K.W., Jain, R.K., Ptak, R.G., Westerman, A.C., Biron, K.K., Townsend, L.B., and Drach, J.C. (1998). Resistance of human cytomegalovirus to benzimidazole ribonucleosides maps to two open reading frames: UL89 and UL56. J. Virol. *72*, 4721–4728.

Krosky, P.M., Baek, M.C., and Coen, D.M. (2003a). The human cytomegalovirus UL97 protein kinase, an antiviral drug target, is required at the stage of nuclear egress. J. Virol. *77*, 905–914.

Krosky, P.M., Baek, M.C., Jahng, W.J., Barrera, I., Harvey, R.J., Biron, K.K., Coen, D.M., and Sethna, P.B. (2003b). The human cytomegalovirus UL44 protein is a substrate for the UL97 protein kinase. J. Virol. *77*, 7720–7727.

Kruger, R.M., Shannon, W.D., Arens, M.Q., Lynch, J.P., Storch, G.A., and Trulock, E.P. (1999). The impact of ganciclovir-resistant cytomegalovirus infection after lung transplantation. Transplantation *68*, 1272–1279.

Landry, M.L., Stanat, S., Biron, K., Brambilla, D., Britt, W., Jokela, J., Chou, S., Drew, W.L., Erice, A., Gilliam, B., *et al.* (2000). A standardized plaque reduction assay for determination of drug susceptibilities of cytomegalovirus clinical isolates. Antimicrob. Agents Chemother. *44*, 688–692.

Lau, P.K., Woods, M.L., Ratanjee, S.K., and John, G.T. (2011). Artesunate is ineffective in controlling valganciclovir-resistant cytomegalovirus infection. Clin. Infect. Dis. *52*, 279.

Lee, C.P., and Chen, M.R. (2010). Escape of herpesviruses from the nucleus. Rev. Med. Virol. *20*, 214–230.

Li, F., Kenyon, K.W., Kirby, K.A., Fishbein, D.P., Boeckh, M., and Limaye, A.P. (2007). Incidence and clinical features of ganciclovir-resistant cytomegalovirus disease in heart transplant recipients. Clin. Infect. Dis. *45*, 439–447.

Li, P.C., Lam, E., Roos, W.P., Zdzienicka, M.Z., Kaina, B., and Efferth, T. (2008). Artesunate derived from traditional Chinese medicine induces DNA damage and repair. Cancer Res. *68*, 4347–4351.

Limaye, A.P., Raghu, G., Koelle, D.M., Ferrenberg, J., Huang, M.-L., and Boeckh, M. (2002). High incidence of ganciclovir-resistant cytomegalovirus infection among lung transplant recipients receiving pre-emptive therapy. J. Infect. Dis. *185*, 20–27.

Lischka, P., Hewlett, G., Wunberg, T., Baumeister, J., Paulsen, D., Goldner, T., Ruebsamen-Schaeff, H., and Zimmermann, H. (2010). *In vitro* and *in vivo* activities of the novel anticytomegalovirus compound AIC246. Antimicrob. Agents Chemother. *54*, 1290–1297.

Lurain, N.S., and Chou, S. (2010). Antiviral drug resistance of human cytomegalovirus. Clin. Microbiol. Rev. *23*, 689–712.

Lurain, N.S., Thompson, K.D., Holmes, E.W., and Read, G.S. (1992). Point mutations in the DNA polymerase gene of human cytomegalovirus that result in resistance to antiviral agents. J. Virol. *66*, 7146–7152.

Lurain, N.S., Spafford, L.E., and Thompson, K.D. (1994). Mutation in the UL97 open reading frame of human cytomegalovirus strains resistant to ganciclovir. J. Virol. *68*, 4427–4431.

McGuigan, C., Pathirana, R.N., Snoeck, R., Andrei, G., De Clercq, E., and Balzarini, J. (2004). Discovery of a new family of inhibitors of human cytomegalovirus (HCMV) based upon lipophilic alkyl furano pyrimidine dideoxy nucleosides: action via a novel non-nucleosidic mechanism. J. Med. Chem. *47*, 1847–1851.

McSharry, J.J., McDonough, A., Olson, B., Talarico, C., Davis, M., and Biron, K.K. (2001). Inhibition of ganciclovir-susceptible and -resistant human cytomegalovirus clinical isolates by the benzimidazole L-riboside 1263W94. Clin. Diagn. Lab. Immunol. *8*, 1279–1281.

Marfori, J.E., Exner, M.M., Marousek, G.I., Chou, S., and Drew, W.L. (2007). Development of new cytomegalovirus UL97 and DNA polymerase mutations conferring drug resistance after valganciclovir therapy in allogeneic stem cell recipients. J. Clin. Virol. *38*, 120–125.

Marschall, M., Freitag, M., Weiler, S., Sorg, G., and Stamminger, Th. (2000). Recombinant green fluorescent protein-expressing human cytomegalovirus as a tool for screening antiviral agents. Antimicrob. Agents Chemother. *44*, 1588–1597.

Marschall, M., Stein-Gerlach, M., Freitag, M., Kupfer, R., van den Bogaard, M., and Stamminger, T. (2001). Inhibitors of human cytomegalovirus replication drastically reduce the activity of the viral protein kinase pUL97. J. Gen. Virol. *82*, 1439–1450.

Marschall, M., Stein-Gerlach, M., Freitag, M., Kupfer, R., van den Bogaard, M., and Stamminger, T. (2002). Direct targeting of human cytomegalovirus protein kinase pUL97 by kinase inhibitors is a novel principle for antiviral therapy. J. Gen. Virol. *83*, 1013–1023.

Marschall, M., Freitag, M., Suchy, P., Romaker, D., Kupfer, R., Hanke, M., and Stamminger, T. (2003). The protein kinase pUL97 of human cytomegalovirus interacts with and phosphorylates the DNA polymerase processivity factor pUL44. J. Virol. *311*, 60–71.

Marschall, M., Marzi, A., aus dem Siepen, P., Jochmann, R., Kalmer, M., Auerochs, S., Lischka, P., Leis, M., and Stamminger, T. (2005). Cellular p32 recruits cytomegalovirus kinase pUL97 to redistribute the nuclear lamina. J. Biol. Chem. *280*, 33357–33367.

Marschall, M., Feichtinger, S., and Milbradt, J. (2011). Regulatory roles of protein kinases in cytomegalovirus replication. Adv. Virus Res. *80*, 69–101.

Marschall, M., Stamminger, T., Urbani, A., Wildum, S., Ruebsamen-Schaeff, H., Zimmermann, H., and Lischka, P. (2012). *In vitro* evaluation of the activities of the novel anticytomegalovirus compound AIC246 (letermovir) against herpesviruses and other human pathogenic viruses. Antimicrob. Agents Chemother. *56*, 1137.

Martin, M., Gilbert, C., Covington, E., and Boivin, G. (2006). Characterization of human cytomegalovirus (HCMV) UL97 mutations found in a valganiclovir/oral ganciclovir prophylactic trial by use of a bacterial artificial chromosome containing the HCMV genome. J. Infect. Dis. *194*, 579–583.

Martin, M., Goyette, N., Ives, J., and Boivin, G. (2010). Incidence and characterization of cytomegalovirus resistance mutations among pediatric solid organ transplant patients who received valganciclovir prophylaxis. J. Clin. Virol. *47*, 321–324.

Marty, F.M., and Boeckh, M. (2012). Maribavir and human cytomegalovirus – what happened in the clinical trials and why might the drug have failed? Curr. Opin. Virol. *1*, 555–562.

Marty, F.M., Ljungman, P., Papanicolaou, G.A., Winston, D.J., Chemaly, R.F., Strasfeld,L., Young, J.A., Rodriguez, T., Maertens, J., Schmitt, M., *et al.* (2011). Maribavir prophylaxis for prevention of cytomegalovirus disease in recipients of allogeneic stem-cell transplants: a phase 3, double-blind, placebo-controlled, randomised trial. Lancet Infect. Dis. *11*, 284–292.

Mattes, F.M., Hainsworth, E.G., Geretti, A.M., Nebbia, G., Prentice, G., Potter, M., Burroughs, A.K., Sweny, P., Hassan-Walker, A.F., Okwuadi, S., *et al.* (2004). A randomized, controlled trial comparing ganciclovir to ganciclovir plus foscarnet (each at half dose) for preemptive therapy of cytomegalovirus infection in transplant recipients. J. Infect. Dis. *189*, 1355–1361.

Maussang, D., Langemeijer, E., Fitzsimons, C.P., Stigter-van Walsum, M., Dijkman, R., Borg, M.K., Slinger, E., Schreiber, A., Michel, D., Tensen, C.P., *et al.* (2009a). The human cytomegalovirus-encoded chemokine receptor US28 promotes angiogenesis and tumor formation via cyclooxygenase-2. Cancer Res. *69*, 2861–2869.

Maussang, D., Vischer, H.F., Schreiber, A., Michel, D., and Smit, M.J. (2009b). Pharmacological and biochemical characterization of human cytomegalovirus-encoded G protein-coupled receptors. Methods Enzymol. *460*, 151–171.

Metzger, C., Michel, D., Schneider, K., Lüske, A., Schlicht, H.-J., and Mertens, Th. (1994). Human cytomegalovirus UL97 kinase confers ganciclovir susceptibility to recombinant vaccinia virus. J Virol *68*, 8423–8427.

Michel, D., and Mertens, T. (2004). The UL97 protein kinase of human cytomegalovirus and homologues in other herpesviruses: impact on virus and host. Biochim. Biophys. Acta *1697*, 169–180.

Michel, D., Pavic, I., Zimmermann, A., Haupt, E., Wunderlich, K., Heuschmid, M., and Mertens, Th. (1996). The UL97 gene product of the human cytomegalovirus is an early-late protein with a nuclear localization but is not a nucleoside kinase. J. Virol. *70*, 6340–6347.

Michel, D., Schaarschmidt, P., Wunderlich, K., Heuschmid, M., Simoncini, L., Muhlberger, D., Zimmermann, A., Pavic, I., and Mertens, T. (1998). Functional regions of the human cytomegalovirus protein pUL97 involved in nuclear localization and phosphorylation of ganciclovir and pUL97 itself. J. Gen. Virol. *79 (Pt 9)*, 2105–2112.

Michel, D., Kramer, S., Hohn, S., Schaarschmidt, P., Wunderlich, K., and Mertens, T. (1999). Amino acids of conserved kinase motifs of cytomegalovirus protein UL97 are essential for autophosphorylation. J. Virol. *73*, 8898–8901.

Michel, D., Hohn, S., Haller, T., Jun, D., and Mertens, T. (2001). Aciclovir selects for ganciclovir-cross-resistance of human cytomegalovirus *in vitro* that is only in part explained by known mutations in the UL97 protein. J. Med. Virol. *65*, 70–76.

Michel, D., Lanz, K., Michel, M., Wasner, T., Hauser, I., Just, M., Hampl, W., and Mertens, Th. (2003). Fast genotypic identification and estimation of ganciclovir-resistant cytomegalovirus from clinical specimens. In New Aspects of CMV-related Immunopathology, Prösch, S., Cinatl, J., and Scholz, M., eds. (Karger, Basel), pp. 160–170.

Mousavi-Jazi, M., Hokeberg, I., Schloss, L., Zweygberg-Wirgart, B., Grillner, L., Linde, A., and Brytting, M. (2001). Sequence analysis of UL54 and UL97 genes and evaluation of antiviral susceptibility of human cytomegalovirus isolates obtained from kidney allograft recipients before and after treatment. Transpl. Infect. Dis. 3, 195–202.

von Muller, L., Klemm, A., Durmus, N., Weiss, M., Suger-Wiedeck, H., Schneider, M., Hampl, W., and Mertens, T. (2007). Cellular immunity and active human cytomegalovirus infection in patients with septic shock. J. Infect. Dis. *196*, 1288–1295.

Myhre, H.-A., Dorenberg, D.H., Kristiansen, K.I., Rollag, H., Leivestad, T., Asberg, A., and Hartmann, A. (2011). Incidence and outcomes of ganciclovir-resistant cytomegalovirus infections in 1244 kidney transplant recipients. Transplantation *92*, 217–223.

Okuda, T., Kurokawa, M., Matsuo, K., Honda, M., Niimura, M., and Shiraki, K. (2004). Suppression of generation and replication of acyclovir-resistant herpes simplex virus by a sensitive virus. J. Med. Virol. *72*, 112–120.

Perry, C.M., and Balfour, J.A. (1999). Fomivirsen. Drugs *57*, 375–380.

Prichard, M.N., Gao, N., Jairath, S., Mulamba, G., Krosky, P., Coen, D.M., Parker, B.O., and Pari, G.S. (1999). A recombinant human cytomegalovirus with a large deletion in UL97 has a severe replication deficiency. J. Virol. *73*, 5663–5670.

Prichard, M.N., Britt, W.J., Daily, S.L., Hartline, C.B., and Kern, E.R. (2005). Human cytomegalovirus UL97 kinase is required for the normal intranuclear distribution of pp65 and virion morphogenesis. J. Virol. *79*, 15494–15502.

Prichard, M.N., Quenelle, D.C., Hartline, C.B., Harden, E.A., Jefferson, G., Frederick, S.L., Daily, S.L., Whitley, R.J., Tiwari, K.N., Maddry, J.A., *et al.* (2009). Inhibition of herpesvirus replication by 5-substituted 4′-thiopyrimidine nucleosides. Antimicrob. Agents Chemother. *53*, 5251–5258.

Reefschlaeger, J., Bender, W., Hallenberger, S., Weber, O., Eckenberg, P., Goldmann, S., Haerter, M., Buerger, I., Trappe, J., Herrington, J.A., *et al.* (2001). Novel non-nucleoside inhibitors of cytomegalovirus (BAY 38-4766): *in vitro* and *in vivo* antiviral activity and mechanism of action. J. Antimicrob. Chemoth. *48*, 757–767.

Saliba, F., Eyraud, D., Samuel, D., David, M.F., Arulnaden, J.L., Dussaix, E., Mathieu, D., and Bismuth, H. (1993). Randomized controlled trial of acyclovir for the prevention of cytomegalovirus infection and disease in liver transplant patients. Transplant. Proc. *25*, 1444–1445.

Sanchez, V., Clark, C.L., Yen, J.Y., Dwarakanath, R., and Spector, D.H. (2002). Viable human cytomegalovirus recombinant virus with an internal deletion of the IE2 86 gene affects late stages of viral replication. J. Virol. *76*, 2973–2989.

Scheffczik, H., Savva, C.G.W., Holzenburg, A., Kolesnikova, L., and Bogner, E. (2002). The terminase subunits pUL56 and pUL89 of human cytomegalovirus are DNA-metabolizing proteins with torodial structure. Nucleic Acids Res. *30*, 1695–1703.

Schindele, B., Apelt, L., Hofmann, J., Nitsche, A., Michel, D., Voigt, S., Mertens, T., and Ehlers, B. (2010). Improved detection of mutated human cytomegalovirus UL97 by pyrosequencing. Antimicrob. Agents Chemother. *54*, 5234–5241.

Scholz, B., Rechter, S., Drach, J.C., Townsend, L.B., and Bogner, E. (2003). Identification of the ATP-binding site in the terminase subunit pUL56 of human cytomegalovirus. Nucleic Acids Res. *31*, 1426–1433.

Schreiber, A., Harter, G., Schubert, A., Bunjes, D., Mertens, T., and Michel, D. (2009). Antiviral treatment of cytomegalovirus infection and resistant strains. Expert. Opin. Pharmacother. *10*, 191–209.

Scott, G.M., Weinberg, A., Rawlinson, W.D., and Chou, S. (2007). Multidrug resistance conferred by novel DNA polymerase mutations in human cytomegalovirus isolates. Antimicrob. Agents Chemother. *51*, 89–94.

Selleseth, D.W., Talarico, C.L., Miller, T., Lutz, M.W., Biron, K.K., and Harvey, R.J. (2003). Interactions of 1263W94 with other antiviral agents in inhibition of human cytomegalovirus replication. Antimicrob. Agents Chemother. *47*, 1468–1471.

Shapira, M.Y., Resnick, I.B., Chou, S., Neumann, A.U., Lurain, N.S., Stamminger, T., Caplan, O., Saleh, N., Efferth, T., Marschall, M., *et al.* (2008). Artesunate as a potent antiviral agent in a patient with late drug-resistant cytomegalovirus infection after hematopoietic stem cell transplantation. Clin. Infect. Dis. *46*, 1455–1457.

Shereck, E.B., Cooney, E., van de Ven, C., Della-Lotta, P., and Cairo, M.S. (2007). A pilot phase II study of alternate day ganciclovir and foscarnet in preventing cytomegalovirus (CMV) infections in at-risk pediatric and adolescent allogeneic stem cell transplant recipients. Pediatr. Blood Cancer *49*, 306–312.

Shi, R., Azzi, A., Gilbert, C., Boivin, G., and Lin, S.X. (2006). Three-dimensional modeling of cytomegalovirus DNA polymerase and preliminary analysis of drug resistance. Proteins *64*, 301–307.

Singh, N., Yu, V.L., Mieles, L., Wagener, M.M., Miner, R.C., and Gayowski, T. (1994). High-dose acyclovir compared with short-course preemptive ganciclovir therapy to prevent cytomegalovirus disease in liver transplant recipients. A randomized trial. Ann. Intern. Med. *120*, 375–381.

Slater, M.J., Cockerill, S., Baxter, R., Bonser, R.W., Gohil, K., Gowrie, C., Robinson, J.E., Littler, E., Parry, N., Randall, R., *et al.* (1999). Indolocarbazoles: Potent, selective inhibitors of human cytomegalovirus replication. Bioorgan. Med. Chem. *7*, 1067–1074.

Snoeck, R., Wellens, W., Desloovere, C., Van Ranst, M., Naesens, L., De Clercq, E., and Feenstra, L. (1998). Treatment of severe laryngeal papillomatosis with intralesional injections of cidofovir [(S)-1-(3-Hyrdroxy-2-Phosphonylmethoxypropyl)Cytosine]. J. Med. Virol. *54*, 219–225.

Spector, S.A., Weingeist, T., Pollard, R.B., Dieterich, D.T., Samo, T., Benson, C.A., Busch, D.F., Freeman, W.R., Montague, P., Kaplan, H.J., *et al.* (1993). A randomized, controlled study of intravenous ganciclovir therapy for cytomegalovirus peripheral retinitis in patients with AIDS. J. Infect. Dis. *168*, 557–563.

Springer, K.L., Chou, S., Li, S., Giller, R.H., Quinones, R., Shira, J.E., and Weinberg, A. (2005). How evolution of mutations conferring drug resistance affects viral dynamics and clinical outcomes of cytomegalovirus-infected hematopoietic cell transplant recipients. J. Clin. Microbiol. *43*, 208–213.

Stassen, F.R., Vega-Córdova, X., Vliegen, I., and Bruggeman, C.A. (2006). Immune activation following cytomegalovirus infection: more important than direct viral effects in cardiovascular disease? J. Clin. Virol. *35*, 349–353.

Strasfeld, L., Lee, I., Tatarowicz, W., Villano, S., and Chou, S. (2010). Virologic characterization of multidrug-resistant cytomegalovirus infection in 2 transplant recipients treated with maribavir. J. Inf. Dis. *202*, 104–108.

Sullivan, V., Talarico, C.L., Stanat, S.C., Davis, M., Coen, D.M., and Biron, K.K. (1992). A protein kinase homologue controls phosphorylation of ganciclovir in human cytomegalovirus-infected cells. Nature *358*, 162–164.

Tchesnokov, E.P., Gilbert, C., Boivin, G., and Götte, M. (2006). Role of helix P of the human cytomegalovirus DNA polymerase in resistance and hypersusceptibility to the antiviral drug foscarnet. J. Virol. *80*, 1440–1450.

Tchesnokov, E.P., Obikhod, A., Schinazi, R.F., and Götte, M. (2009). Engineerung of a chimeric RB69 DNA polymerase sensitive to drugs targeting the cytomegalovirus enzyme. J. Biol. Chem. *284*, 26439–26446.

Tischer, B.K., v. Einem. J., Kaufer, B., and Osterrieder, N. (2006). Two-step red-mediated recombination for versatile high-efficiency markerless DNA manipulation in *Escherichia coli*. BioTechniques *40*, 191–197.

Townsend, L.B., Devivar, R.V., Turk, S.R., Nassiri, M.R., and Drach, J.C. (1995). Design, synthesis, and antiviral activity of certain 2,5,6-trihalo-1-(beta-D-ribofuranosyl) benzimidazoles. J. Med. Chem. *38*, 4098–4105.

Underwood, M.R., Harvey, R.J., Stanat, S.C., Hemphill, M.L., Miller, T., Drach, J.C., Townsend, L.B., and Biron, K.K. (1998). Inhibition of human cytomegalovirus DNA maturation by a benzimidazole ribonucleoside is mediated through the UL89 gene product. J. Virol. *72*, 717–725.

Wagner, M., Michel, D., Schaarschmidt, P., Vaida, B., Jonjic, S., Messerle, M., Mertens, T., and Koszinowski, U. (2000). Comparison between human cytomegalovirus pUL97 and murine cytomegalovirus (MCMV) pM97 expressed by MCMV and vaccinia virus: pM97 does not confer ganciclovir sensitivity. J. Virol. *74*, 10729–10736.

Waldman, W.J., Knight, D.A., Blinder, L., Shen, J., Lurain, N.S., Miller, D.M., Sedmak, D.D., Williams, J.W., and Chong, A.S. (1999a). Inhibition of cytomegalovirus *in vitro* and *in vivo* by the experimental immunosuppressive agent leflunomide. Intervirology *42*, 412–418.

Waldman, W.J., Knight, D.A., Lurain, N.S., Miller, D.M., Sedmak, D.D., Williams, J.W., and Chong, A.S. (1999b). Novel mechanism of inhibition of cytomegalovirus by the experimental immunosuppressive agent leflunomide. Transplantation *68*, 814–825.

Wang, X., Huong, S.M., Chiu, M.L., Raab-Traub, N., and Huang, E.S. (2003). Epidermal growth factor receptor is a cellular receptor for human cytomegalovirus. Nature *424*, 456–461.

Warming, S., Costantino, N., Court, D.L., Jenkins, N.A., and Copeland, N.G. (2005). Simple and highly efficient BAC recombineering using galK selection. Nucleic Acids Res. *33*, e36.

Webel, R., Milbradt, J., Auerochs, S., Schregel, V., Held, C., Nöbauer, K., Razzazi-Fazelli, E., Jardin, C., Wittenberg, T., Sticht, H., *et al.* (2011). Two isoforms of the protein kinase pUL97 of human cytomegalovirus are differentially regulated in their nuclear translocation. J. Gen. Virol. *92*, 638–649.

Weinberg, A., Jabs, D.A., Chou, S., Martin, B.K., Lurain, N.S., Forman, M.S., and Crumpacker, C. (2003). Mutations conferring foscarnet resistance in a cohort of patients with acquired immunodeficiency syndrome and cytomegalovirus retinitis. J. Infect. Dis. *187*, 777–784.

Wiebusch, L., Truss, M., and Hagemeier, C. (2004). Inhibition of human cytomegalovirus replication by small interfering RNAs. J. Gen. Virol. *85*, 179–184.

Wolf, D.G., Tan Courcelle, C., Prichard, M.N., and Mocarski, E.S. (2001). Distinct and separate roles for herpesvirus-conserved UL97 kinase in cytomegalovirus DNA synthesis and encapsidation. Proc. Natl. Acad. Sci. U.S.A. *98*, 1895–1900.

Wolf, D.G., Shimoni, A., Resnick, I.B., Stamminger, T., Neumann, A.U., Chou, S., Efferth, T., Caplan, O., Rose, J., Nagler, A., *et al.* (2011). Human cytomegalovirus kinetics following institution of artesunate after hematopoietic stem cell transplantation. Antiviral Res. *90*, 183–186.

Wong, S.W., Wahl, A.F., Yuan, P.M., Arai, N., Pearson, B.E., Arai, K., Korn, D., Hunkapiller, M.W., and Wang, T.S. (1988). Human DNA polymerase alpha gene expression is cell proliferation dependent and its primary structure is similar to both prokaryotic and eukaryotic replicative DNA polymerases. EMBO J. *7*, 37–47.

Zamora, M.R., Davis, R.D., and Leonard, C. (2005). Management of cytomegalovirus infection in lung transplant recipients: evidence-based recommendations. Transplantation *80*, 157–163.

Zimmermann, A., Michel, D., Pavic, I., Hampl, W., Lüske, A., Neyts, J., De Clercq, E., and Mertens, T. (1997). Phosphorylation of aciclovir, ganciclovir, penciclovir and S2242 by the cytomegalovirus UL97 protein: a quantitative analysis using recombinant vaccinia viruses. Antivir. Res. *36*, 35–42.

Cytomegalovirus Vaccine: On the Way to the Future?

II.20

Stanley A. Plotkin and Bodo Plachter

Abstract

Prenatal transmission of CMV is a frequent cause of mental retardation and hearing loss in children. Furthermore, infection with this virus is a severe threat to immunocompromised patients. Consequently, development of a vaccine to prevent CMV disease has been identified as a first rank medical priority. Goals and target populations for such a vaccine have been identified. Antigens to be targeted by vaccine-induced immune responses have been defined indicating that only a subset of the more than 150 viral proteins may be sufficient to induce protective immunity. Using this information, strategies for the development of live virus vaccines as well as subunit vaccines have been developed. At this point at least seven candidate vaccines have been tested clinically and many other approaches are being explored.

Introduction

The human cytomegalovirus is a pathogen for all seasons (Plotkin, 2008). Cytomegalovirus vaccines have been explored since the 1970s, but for many years the field was a wilderness populated by few scientists. Fortunately, in recent years many groups and many manufacturers have developed candidate vaccines and the field is now one of passionate interest. The reasons for the long gestation period are several, but primarily the slow recognition of the importance of the human cytomegalovirus, uncertainty as to the correlates of protection, and doubt about a regulatory pathway were responsible. As this chapter will show, although controversies still exist the probability that a vaccine will emerge has never been higher. The reader is also referred to other recent reviews (Sung and Schleiss, 2010; Bernstein, 2011).

Why a vaccine?

Before describing efforts to develop a vaccine against CMV, it is important to explain why one is needed. The impetus to begin thinking about a vaccine against CMV came from the early studies of congenital CMV infection at the University of Alabama, which showed that fetal infection was common and frequently followed by sequelae (Stagno *et al.*, 1982; Fowler and Boppana, 2006; see also Chapter II.3). A crucial finding also made by the Alabama group was that natural immunity in mothers protected against transplacental infection, suggesting that a vaccine might do the same (Fowler *et al.*, 2003).

The susceptibility of women of child-bearing age to primary infection with CMV varies by country and socioeconomic group, but as an approximation in the USA and Western Europe, about half lack serological evidence of prior infection (Bate *et al.*, 2010; Cannon *et al.*, 2010) and infection during pregnancy occurs at a rate of approximately 2% (Hyde *et al.*, 2010). Sexual exposure, familial exposure and history of breast-feeding all play a role in increased seropositivity, and women in developing countries are likely to be seropositive (Staras *et al.*, 2008a,b). As will be discussed below, being seropositive is not completely protective as reinfection does occur and is in fact responsible for three quarters of fetal infections (Wang *et al.*, 2011). This estimate derives from the relatively low rate of primary infection in pregnancy (1.6%) and the relatively high seropositivity of minority populations (Colugnati *et al.*, 2007), but with frequent transmission in the former and less frequent in the latter case. A meta-analysis of transmission from infected mothers to fetuses showed an overall rate of 0.64% in the USA, with a 32% rate after primary infection and 1.4% after recurrent infection (Kenneson *et al.*, 2007). Approximately 12% of infants born after primary infection will be symptomatic at birth, of whom most will have sequelae such as deafness and mental retardation, whereas about 14% of asymptomatic newborns will eventually have sequelae (Dollard *et al.*, 2007; Rosenthal *et al.*, 2009). The implication of these figures is that each year about 8000 American infants are affected by primary CMV infection, of whom about 20% will have permanent sequelae

(Kenneson *et al.*, 2007) with similar aggregate numbers probable in Europe (Ludwig and Hengel, 2009; Vyse *et al.*, 2009).

The importance of CMV disease in marring the success of both solid organ (see also Chapters II.13 and II.14) and stem cell transplantation (see also Chapter II.16) has also become evident, and although antiviral prophylaxis and treatment are useful in controlling CMV, they are also expensive and imperfect (Boeckh and Geballe, 2011). More speculative reasons for preventing CMV infection have also emerged, including possible implication of the virus in senescence of the immune system, atherosclerosis, stillbirth and gliomas of the brain (Prins *et al.*, 2008; Caposio *et al.*, 2011; Iwasenko *et al.*, 2011; Simanek *et al.*, 2011; see also Chapter II.23).

The importance of congenital CMV and the priority need for a vaccine were recognized by the Institute of Medicine of the US National Academy of Science and the National Vaccine Advisory Committee of the US Public Health Service (Committee to study priorities for vaccine development, Division of health promotion and disease prevention and Institute of Medicine, Washington, 2000; Arvin *et al.*, 2004; see also Chapter II.2 Addendum).

Moreover, aside from the evidence for efficacy of vaccination to be recounted below, administration of immune globulin containing CMV antibodies to infected pregnant women (Nigro *et al.*, 2005; Adler and Nigro, 2009; Maidji *et al.*, 2010) and adoptive T-cell transfer in the treatment of CMV disease in transplant patients (Riddell *et al.*, 1992; Walter *et al.*, 1995; Einsele *et al.*, 2002; Micklethwaite *et al.*, 2007; Horn *et al.*, 2009; Feuchtinger *et al.*, 2010; Peggs *et al.*, 2011) have been successful strategies, underlining the utility of induced immune responses. Thus, there is hope to counter what has been called 'the changeling demon' (Plotkin, 1999) and 'the troll of transplantation' (Balfour, 1979).

Which effector immune responses should be produced by a vaccine?

Antibodies

According to the knowledge gathered in clinical studies and in animal models, successful CMV vaccines will likely have to target multiple branches of the immune system. We will focus here on discussing adaptive immune responses important for control of CMV infection and refer to Chapter II.8 for a review of the interaction of CMVs with the innate immune system.

Humoral immune responses, particularly virus-neutralizing antibodies (NT-Abs), are considered to be essential for protection against CMV infection and, consequently, against congenital disease (Table II.20.1; see also Chapters II.3 and II.10). Transfer of maternal antibodies prevented transfusion-associated infection in newborns in the immediate postnatal period (Yeager *et al.*, 1981). Application of CMV-specific antibodies in multiply transfused preterm infants showed a tendency for reduction of CMV disease in these children (Snydman *et al.*, 1995). As discussed above, CMV antibody seropositivity prior to pregnancy correlates, although incompletely, with protection against prenatal infection and disease (Fowler *et al.*, 1992). The role of antibodies in the prevention of prenatal and perinatal infection has also been emphasized in animal models (see Chapter II.5 and II.6, respectively). However, whether NT-Abs are the sole protective factor in these settings remains uncertain.

More recently, two observations in humans have reinforced the importance of NT-Abs. One was referred to above, namely the apparent utility of giving immune globulin containing CMV NT-Abs to infected women in order to prevent and treat fetal infection (Nigro *et al.*, 2005; Adler *et al.*, 2009; Nigro and Adler, 2011), as well as the apparent protective effect of pre-existing antibody on placental pathology (Maidji *et al.*, 2010).

The second, to be described in detail below, is the reported efficacy of a vaccine containing only the gB glycoprotein in preventing acquisition of CMV by exposed females (Pass *et al.*, 2009). In addition, passive transfer of memory B-cells protected immunodeficient mice from fatal CMV infection, underlining the utility of antibody (Klenovsek *et al.*, 2007). Also in mice, passive transfer of serum antibodies not only prevents infection but reduces spread of an established infection within host organs (Wirtz *et al.*, 2008).

In the transplant setting, the protective role of antibodies is less clear. Solid organ transplant recipients appear to benefit from prophylactic

Table II.20.1 Situations in which antibodies protect against CMV disease

Newborns exposed to white blood cells carrying CMV
Solid organ transplant recipients given passive antibodies
Bone marrow transplant recipients given passive antibodies (equivocal)
Animal models (guinea pigs, mice)
Protection of placenta by maternal antibodies
Protection of fetus by infused antibodies

immune globulin application (Falagas *et al.*, 1997; Valantine *et al.*, 2001; Pereyra and Rubin, 2004). A recent meta-analysis of 11 randomized trials showed that prophylactic administration of CMV immune globulin after solid organ transplant is associated with improved total survival, reduced CMV disease, and reduced CMV-associated deaths (Bonaros *et al.*, 2008). In contrast, an impact of immune globulin on CMV disease and overall survival is still discussed equivocally in the setting of haematopoietic stem cell transplant (Cordonnier *et al.*, 2003; Raanani *et al.*, 2009). Presence of NT-Abs has also been correlated with slower progression towards CMV disease in AIDS patients (Boppana *et al.*, 1995). These findings, although admittedly circumstantial, indicate that CMV specific antibodies may be beneficial in immunosuppressed patients.

Strain differences have to be considered in designing a vaccine. It has been shown that sera from humans infected with one CMV strain may fail to neutralize a heterologous strain (Klein *et al.*, 1999). These experimental data could serve as an explanation for some CMV reinfections. Such infections were seen in the presence of CMV-specific antibodies after exposure to secretions from individuals undergoing acute CMV infection (reviewed in Plotkin, 2002). Severe congenital CMV infections have been reported in children born to antibody seropositive mothers (Ross *et al.*, 2006; 2010). This has been attributed to superinfection with an HCMV strain during pregnancy, which had not been previously present in the mother (Boppana *et al.*, 2001).

Cell-mediated immunity

From the murine model of CMV infection, it became clear that cellular immune responses, in particular cytolytic CD8 T-lymphocytes, are protective against CMV disease and that these cells control viral reactivation from latency (Reddehase *et al.*, 1985; Reddehase, 2000, 2002; Simon *et al.*, 2006; see also Chapters I.22 and II.17). Table II.20.2 lists situations in which cellular immunity is protective. In humans, Reusser and colleagues could show that reconstitution of antiviral CD8 T-cells correlated with protection against CMV disease in bone marrow transplant recipients, a finding that has been confirmed on several occasions thereafter (Reusser *et al.*, 1991; see also Chapter II.16). Making use of that knowledge, the same group provided evidence that *in vitro* expanded, donor-derived CD8 T-cell lines were protective against CMV disease after transfer into bone marrow transplant recipients and that these cells could be sustained by coapplication of CD4 T-cells (Walter *et al.*, 1995). This work was pioneering as it conclusively showed that CD8 T-cells were protective against CMV disease also in humans. Subsequent studies confirmed that combined transfer of CD4 and CD8 T-cells for pre-emptive or prophylactic treatment of CMV reactivation in allogeneic transplant recipients was effective to generate a lasting CD8 T-cell response specific for CMV (Einsele *et al.*, 2002; Peggs *et al.*, 2003; Peggs, 2009). One recent study showed that reconstitution of antiviral immunity in HSCT patients by *ex vivo* selected, pp65-specific T-cells was boosted by CMV infection, indicating that expansion of these cells was antigen driven (Peggs *et al.*, 2011). This implies that transfer of a limited number of antiviral T-cells in combination with a vaccine could be helpful to prevent CMV complications in HSCT patients. In line with this, recent work in the murine model has revealed improved antiviral protection by a therapeutic recombinant dense body (rDB) vaccine (see below), expanding limited numbers of antiviral CD8 T-cells *in vivo* after experimental pre-emptive cell transfer therapy (Gergely *et al.*, 2011; Tracy McGinnis Award, 2011). T-cell expansion by the vaccine was antigen driven and antiviral protection was antigen targeted in this experimental model.

Table II.20.2 Situations in which cell-mediated immunity protects against CMV disease

Primary maternal infection – CD4 T-cells
Recovery of CD8 T-cells after solid organ transplant
Recovery of CD8 T-cells after stem cell transplant
Infusion of CD8 T-cells after transplant
Closure of chronic neonatal infection

Additional evidence for the importance of CD8 T-cells in controlling CMV replication in the transplant setting has been provided by the results of a DNA vaccine study to be described in greater detail below (Wloch *et al.*, 2008; Kharfan-Dabaja *et al.*, 2011). Furthermore, it was recently shown that functional cellular immunity as shown by interferon production correlates with spontaneous clearance of viraemia in solid organ transplant recipients (Lisboa *et al.*, 2012). In addition, in the context of the search for a vaccine against HIV, rhesus cytomegalovirus containing SIV genes was shown to infect seropositive monkeys (Hansen *et al.*, 2011). However, when MHC class-I immunoevasion genes were removed from the CMV vector, the CD8 T-cells of the host were able to control superinfection (Hansen *et al.*, 2010). Thus, it appears that CD8 T-cells may also be important in prevention (see also Chapters II.21 and II.22).

Antigen components for a CMV vaccine

Knowledge about the immune responses to be addressed enables a rational design of the protein composition for a CMV vaccine. Table II.20.3 lists the most obvious candidates. In all instances, genetic complexity and strain diversity of CMV have to be considered (see Chapter I.1). Bioinformatic analyses revealed a genomic coding capacity of the CMV laboratory strain Ad169 for roughly 145 proteins (Davison *et al.*, 2003). Recent CMV isolates encode an additional set of genes (Cha *et al.*, 1996), raising the total coding capacity of the genome of human CMV to at least 165 unique proteins (Murphy *et al.*, 2003; Dolan *et al.*, 2004). Most previous immunological studies focused on viral antigens that were abundantly detectable either in infected cells or in virus particles. Thus, our picture of 'immunodominant' antigens of CMV may be biased to some extent by our lack of knowledge about the immunogenicity of all the putative antigens of the virus.

Target antigens of the humoral immune response

The gB is an important target antigen of the NT-Ab responses against HCMV (Table II.20.3; see Chapter II.10). It is highly conserved between HCMV isolates. All infected individuals develop a humoral response against gB. Several studies have demonstrated that a large portion of the virus-neutralizing capacity found in sera from infected individuals is directed against this protein. Thus, gB is considered to be an essential component of a vaccine against CMV. This is corroborated by the finding that antibodies against the gB homologue of GPCMV protect against prenatal infection in these animals (Schleiss *et al.*, 2003). The recent discovery of the endosomal entry pathway for CMV has, however, raised the issue of the dominant status of gB as vaccine antigen (see below).

The glycoprotein H (gH, gpUL75) is also abundant in the viral envelope. It is included in a complex with gL and possibly gO, although inclusion of the latter protein may be strain dependent (Ryckman *et al.*, 2010). The gH is also part of a complex with gL and gpUL128–131 (Wang and Shenk, 2005).

As with gB, gH is conserved in herpesviruses. Antibodies are detectable in nearly 100% of cases following infection (Britt and Mach, 1996; see also Chapter II.10). NT-Abs are synthesized against gH, and in some individuals this response may be dominant (Urban *et al.*, 1996). Part of the NT-Ab response to gH is strain-specific. Boppana and colleagues detected newly synthesized, strain-specific gH-antibodies in women who encountered a CMV reinfection during pregnancy and transmitted the virus to their offspring. This indicated that superinfection with a CMV strain with a gH-type, different from the endogenous strain had occurred (Boppana *et al.*, 2001). Consequently, different gH-types will have to be considered for vaccine design.

The complex of gN (gpUL73) and gM (gpUL100) is also targeted by NT-Abs (Shimamura *et al.*, 2006). Antibody binding has been shown to be critically dependent on correct interaction of the two components. The gM–gN complex is highly abundant in the viral envelope (Varnum *et al.*, 2004). Although this renders the complex attractive for vaccine development, the known hypervariability of gN may limit the value of this complex for this purpose. Further information is needed about gM-gN to evaluate its applicability in vaccine design.

The relatively simple picture of NT-Abs being

Table II.20.3 Viral proteins possibly to be included in an HCMV vaccine

Viral protein	Function/structure	NT-Abs[a]	CD8 T-cells[b]	CD4 T-cells[c]
gB (gpUL55)	Envelope glycoprotein Tethering/fusion	Yes[d]	Yes	Yes
gH (gpUL75)	Envelope glycoprotein attachment/penetration	Yes[e]	Yes	Yes
gpUL128–131	Envelope proteins in complex with gH/gL, entry into epithelial and endothelial cells	Yes	No	?
pp65 (ppUL83)	Tegument	No	Yes	Yes
IE1 (ppUL123)	Non-structural, regulatory protein	No	Yes	Yes

[a]Target antigen of the neutralizing antibody response against CMV
[b]Target antigen of the CD8 T-cell response against CMV
[c]Target antigen of the Th-cell response against CMV
[d]Binding of NT-Abs highly conformation dependent
[e]NT-Ab response may be strain specific

directed predominantly against the gB glycoprotein and less importantly against gH and gM-gN was overturned by the discovery that entry into epithelial and endothelial cells as well as mononuclear leucocytes is mediated by different proteins (Wang *et al.*, 2005). Whereas gB allows tethering to and penetration into fibroblasts and antibody to gB neutralize infection of these cells, entry into epithelial cells largely is mediated by a pentameric complex of proteins composed of gH/gL/UL128/UL130/UL131. Importantly, culturing of HCMV on fibroblasts leads to the selection of variants that carry deletions or mutations in the UL128–131 gene locus, thereby disrupting the pentameric complex. This means that attenuated CMV strains elicit only low levels of antibody that neutralize entry into epithelial cells and dendritic cells (Hahn *et al.*, 2004; Gerna *et al.*, 2005, 2008; Cui *et al.*, 2008; see also Chapter I.17).

The importance of antibodies that prevent entry into epithelial cells to a vaccine is presently unknown, but it can be suspected that such antibodies would be useful, particularly as they are part of the response to wild virus infection and form the largest portion of antibody in gamma globulin (Fouts *et al.*, 2001). Thus, the inclusion of one or more of the proteins of the gH/gL/UL128–131 complex in a vaccine is of interest, and early indications suggest that the UL128, UL130 and UL131 proteins by themselves can induce NT-Abs against epithelial cell infection (Adler *et al.*, 2006; Macagno *et al.*, 2010; Genini *et al.*, 2011; Saccoccio *et al.*, 2011b; Straschewski *et al.*, 2011). Antibodies to those proteins might also inhibit virus shedding and transmission if infection occurs in spite of vaccination (Oxford *et al.*, 2011). In addition, antibodies to the pentamoric complex appear earlier in mothers who do not transmit to their foetuses (Lilleri *et al.*, 2012).

Targets of the CD8 T-cell response

The tegument protein pp65 (pUL83) and the regulatory IE1 protein (pUL123) have been identified as important CD8 T-cell targets (Reddehase, 2000; Schleiss, 2008a; Herr and Plachter, 2009; see also Chapter II.7). CD8 T-cells specific for pp65 are found in considerable frequencies in CMV-seropositive individuals and in immunocompromised patients (Boppana and Britt, 1996; Wills *et al.*, 1996). Since the protein is highly conserved between viral strains, it appears to be a favourable component to be included in a vaccine. Recent studies using adoptive transfer of pp65-specific T-cells provided evidence that these cells are indeed protective (Peggs *et al.*, 2011).

CD8 T-cells against the IE1 protein are also detectable with a significant prevalence in seropositive individuals (Borysiewicz *et al.*, 1988; Kern *et al.*, 1999; Gyulai *et al.*, 2000; Sylwester *et al.*, 2005). The IE1 was therefore also suggested for a CMV vaccine (Plotkin, 2008). The use of IE1 as a vaccine antigen may be limited, however, by the variability of the protein. This applies in particular to some epitopes that have been classified as immunodominant (Elkington *et al.*, 2003; Prod'homme *et al.*, 2003). It should be noted, however, that the term 'immunodominance', as used in this context reflects frequencies of antigen-specific CD8 T-cells found in PBMCs of infected individuals. It is worthwhile considering that frequency does not necessarily mean functionality in terms of protection. As a matter of fact, it has been shown that subdominant epitopes may be equally protective against mouse CMV in adoptive cell transfer models (Holtappels *et al.*, 2000, 2001, 2008), as well as upon DNA vaccination (Morello *et al.*, 2000; Ye *et al.*, 2004). It remains to be seen whether subdominant, conserved epitopes become relevant inducers of an immune response against human CMV, as has been suggested by data in the mouse model.

An interesting approach to generate CTL responses to pp65 was taken by Paine *et al.* (2010), who generated soluble pp65 in supernatant of cell cultures (human embryonic kidney 293 cells) that stimulated multiple responses in cells from a broad range of individuals.

An additional vaccine approach relating to cellular responses, suggested by studies of mouse CMV and rhesus CMV, consists of the deletion of immunoevasion genes from live viruses in order that they replicate but be rapidly cleared by NK-cell responses (Schleiss, 2010; Slavuljica *et al.*, 2010). It is also relevant to note that in the guinea pig CMV model protection of fetuses can be achieved by induction of either antibodies or CTL responses (Schleiss, 2008b).

Target antigens of the T-helper (Th) lymphocyte response

Infection with CMV is well known to induce a vigorous CD4 T-cell response. Protection conferred by CD4 T-lymphocytes has not been conclusively demonstrated in humans. However, as mentioned above, adoptive transfer experiments have convincingly shown that CD4 T-cell responses are essential for mounting a lasting CTL response. Furthermore, it can be assumed that sustained antibody responses can only be achieved by a vaccine if CD4 T-cell antigens are included. Fortunately, the components that have to be used for vaccine purposes anyway, namely pp65, IE1, gB and gH, are also targeted by CD4 T-cells (Beninga *et al.*, 1995). Interestingly, CD4 and CD8 T-cell epitopes may overlap in pp65, underlining the exceptional properties of that protein for vaccine design (Reiser *et al.*, 2011).

CD4 and CD8 T-cell responses are not exclusively

focused on pp65, IE1, gB and gH. Comprehensive analyses using epitope mapping with overlapping synthetic peptide libraries have identified numerous other viral proteins that are targets of CD4 and CD8 T-cells from seropositive individuals on the population level with high HLA polymorphism coverage (Elkington *et al.*, 2003; Sylwester *et al.*, 2005). The relevance of these antigens for protection against HCMV disease and their usefulness to complement pp65 and IE1 as components of an HCMV vaccine has not been conclusively determined.

Replicating candidate vaccines

Table II.20.4 lists some of the vaccines to be discussed below.

Live, attenuated virus

The 1970s was the decade in which a number of attenuated virus vaccines were licensed (reviewed in Plotkin and Plotkin, 2011) and the first vaccine efforts used that strategy for CMV. Elek and Stern (1974) in the UK and Plotkin and colleagues in the USA (Plotkin *et al.*, 1976) launched studies to attenuate human CMV. The development of a vaccine based on strain Ad169 by Elek and Stern was eventually halted, but the Plotkin group attenuated a strain isolated from a congenitally infected infant by 125 passages in human diploid fibroblast cell strain WI-38. The Towne strain could be distinguished from wild virus by several *in vitro* markers (Plotkin, 1975) and later it was learned that the ULb′ part of the genome was lost during passage in cell culture. The strain was immunogenic in virtually 100%

Table II.20.4 CMV candidate vaccines

	Preclinical testing	Clinical testing (phase I/II)	Efficacy trial	Outcome/perspectives
Live, attenuated (Towne)	Yes	Yes	Yes	Protective against disease in transplant recipients No prevention of infection
Live (Towne–Toledo recombinants)	Yes	Yes	No	Well tolerated in seropositive healthy volunteers Further clinical trials started
Recombinant gB subunit	Yes	Yes	Yes	Induction of NT-Abs and CD4 T-cells in seronegatives Rapid boosting of NT-Abs in seropositives Protection against infection of women Benefit in kidney and liver transplant recipients
Canarypox vectors	Yes	Yes	No	CTL- and Th-responses in seronegative volunteers Poor NT-Ab response Candidate for prime-boost approaches
MVA vectors	Yes	No	No	Phase I trial in healthy volunteers initiated Further trials in haematopoietic stem cell donors and recipients planned
Replication-defective Adenovirus vectors	Yes	No	No	Induction of NT-Abs, CTL and Th responses in animal models Clinical trials pending
Alphavirus replicons	Yes	Yes	No	Induction of NT-Abs, CTL and Th responses
Peptides (combined CD4 and CD8 T-cell epitopes)	Yes	Yes	No	Induction of CTL- and Th-responses in mice Phase Ib safety and immunogenicity trial performed in healthy volunteers with peptides from pp65 combined with a TLR-9 agonist; no serious adverse effects and boosting of CD8 responses
DNA vaccines	Yes	Yes	Yes	Protection against challenge in the MCMV model, using a prime boost approach Induction of NT-Abs and cellular responses in volunteers and tumour patients Reduction of recurrence and occurrence of HCMV viraemia episodes and improved time to event for viraemia episodes in HSCT recipients
Dense bodies	Yes	No	No	Lasting NT-Ab responses in mice CTL- and Th1-type responses in mice Clinical trials pending

of volunteers, including young paediatric nurses, when given subcutaneously, but only a minority seroconverted when given the vaccine intranasally (Fleisher *et al.*, 1982). NT-Abs against several CMV strains were induced in all normal seronegative subjects, and most subjects developed antibodies to IE1, suggesting that some replication had taken place (Kamiya *et al.*, 1982; Plotkin *et al.*, 1983; Quinnan *et al.*, 1984). A minority of subjects developed mucosal IgA antibody (Wang *et al.*, 1996). No virus excretion was found in the throat, urine or blood and systemic symptoms were not seen (Plotkin *et al.*, 1981). Later unpublished studies using PCR confirmed the absence of virus in the blood. Naturally seropositive subjects were not boosted by the vaccine (Plotkin *et al.*, 1984b).

The Towne strain was then tested in patients about to undergo renal transplant, as well as additional normal subjects, including children. Once again, there were no associated symptoms aside from local reaction at the site of injection (Adler *et al.*, 1998). Lymphocyte proliferation, cytotoxic T-cells and delayed type hypersensitivity specific to CMV were demonstrated in nearly all subjects after vaccination (Starr *et al.*, 1981; Gupta *et al.*, 1993; Adler *et al.*, 1998). CD8 cell-mediated CTL responses were seen in about half of subjects. Twelve months after vaccination, CTL responses had waned except for those against immediate-early antigen (Jacobson *et al.*, 2006a). Addition of IL-12 to the vaccine improved cellular responses (Jacobson *et al.*, 2006b). Restriction endonuclease analysis of strains isolated subsequently from urine of vaccinees undergoing kidney transplantation showed them to be always different from the vaccine strain (Glazer *et al.*, 1979; Starr *et al.*, 1981; Plotkin *et al.*, 1985). Thus, there was no evidence for latency of the vaccine virus despite the immunosuppression during transplantation (Plotkin and Huang, 1985). Multiple studies were conducted in renal transplant patients to ascertain efficacy. In summary, and as shown in Table II.20.5 (Plotkin, 1999), vaccination significantly moderated severity of disease in seronegative recipients who received a kidney from a seropositive donor and increased graft survival, but did not significantly reduce rate of infection (Plotkin *et al.*, 1984a,b, 1990, 1991; Brayman *et al.*, 1988). Whether efficacy was due to an antibody or a cellular response could not be ascertained.

As part of the study of the Towne vaccine, a low passage virus called Toledo was used to challenge seronegative, naturally seropositive, and vaccinated volunteers. The results were interesting in that all three groups could be infected, but differed in the challenge dose that successfully induced infection. Whereas seronegative volunteers could be infected with 10 PFU, vaccinees required 100 PFU and naturally seropositives 1000 PFU (Plotkin *et al.*, 1985; 1989; Gonczol *et al.*, 1989). These results presaged the later demonstration by Picker and colleagues, using rhesus CMV that superinfection could be induced in monkeys (Hansen *et al.*, 2009; see also Chapter II.22).

However, when tested as a preventative for infection of mothers whose children were in day care and infected with CMV, Towne vaccine failed to prevent acquisition, although naturally immune mothers were protected (Adler *et al.*, 1995). A possible explanation of the vaccine failure may be that the dose used was small and antibody responses were low. Subsequent theoretical safety concerns have inhibited enthusiasm for further exploitation of this approach. Nevertheless, the studies with Towne demonstrated the feasibility of interfering with the pathogenesis of CMV disease, but also the difficulty of preventing infection.

Towne–Toledo recombinants

It appeared that Towne had been over-attenuated, and therefore an attempt was made to increase its

Table II.20.5 Comparative results of three blinded trials of Towne vaccine in seronegative renal transplant patients who received kidneys from seropositive donors

Trial	*n*	Rate of all CMV disease (%) V	Rate of all CMV disease (%) P	Rate of severe CMV disease (%) V	Rate of severe CMV disease (%) P	Reduction of severe disease in vaccinated compared with placebo (%)
Pennsylvania	67	39	55	6	35	84
Minnesota	35	33	43	5 (10)[a]	36	87
Multicentric	61	38	59	0	17	100
All	163	37	54	0.3	29	89

P, patients given placebo; V, patients given vaccine.
[a]A 10% rate was reported in the original publication, but this includes one case that occurred subsequent to pancreatic transplant after a renal transplant free of CMV disease. Without that case, the incidence was 5%.

immunogenicity by constructing recombinants between Towne and the low passage Toledo isolate, using cosmid technology (Kemble *et al.*, 1996; Fig. II.20.1). Four recombinants were made in different conformations, each containing 70% of the Towne genome and most of the ULb′ region of Toledo (lacking in Towne). These were tested in seropositive subjects (Heineman *et al.*, 2006). None developed symptoms or virus excretion and, not surprisingly, none showed an antibody response. After a long delay due to administrative difficulties, a study in seronegative subjects has begun (S. Adler, Virginia Commonwealth University, Richmond, VA, USA, personal communication).

Vectors

Vectors are in a sense intermediate between live and inactivated vaccines inasmuch as the vehicle is a live virus that expresses foreign antigens encoded by inserted genes, but those genes are not capable of reproducing themselves. Four viral vectors have been applied to CMV vaccination: canarypox, Modified Vaccinia Ankara (MVA) vaccinia mutant, alphavirus replicons, and adenovirus type 5 (both the fully competent virus and an E3 deletion mutant). The replicating adeno 5 contained the gene for gB and was only tested in hamsters, but did induce NT-Abs (Marshall *et al.*, 1990). Canarypox vectors were constructed containing either the gene for gB or for pp65 (Gonczol *et al.*, 1995). The former vector elicited poor antibody responses in humans when used alone, but did serve as a good prime for antibody responses when boosted with the Towne strain (Adler *et al.*, 1999). However, it was a relatively weak prime for boosts with the gB protein in MF59 (see below; Bernstein *et al.*, 2002).

In contrast, the canarypox vector carrying the pp65 gene proved adept at stimulating high levels of long-lasting CTL against CMV in human volunteers after two doses (Berencsi *et al.*, 2001; Fig. II.20.2). So far this lead has not been followed up.

The alphavirus replicon system is one in which genes for foreign antigens are substituted for alphavirus structural proteins. Two replicons were constructed, one that produced gB and a second that produced a pp65–IE1 fusion protein (Reap *et al.*, 2007). They were tested 1:1 in combination at 10^7 and 10^8 replicon particles in a phase I trial, in which they induced NT-Abs as well as CD4 and CD8 T-cell responses as measured by ELISPOT assay and flow cytometry (Bernstein *et al.*, 2009; Fig. II.20.3). After a dose of 10^8 infectious units, 100% of subjects developed neutralizing antibodies, with a mean titre of 218. In addition, at that dose T-cell responses measured by ELISPOT assay were 100% for pp65, 88% for IE1 and 94% for gB (Bernstein *et al.*, 2009). Further clinical trials are awaited.

Recently, Novartis has expressed interest in self-amplifying RNA encapsulated in lipid nanoparticles as a method for immunizing against CMV (Geall *et al.*, 2012). MVA is a spontaneous deletion mutant of vaccinia virus that is widely used as a poxvirus vector. Recombinants were made expressing the gB, pp65 and IE1 proteins of CMV. These were immunogenic in mice (Wang *et al.*, 2006, 2007). An MVA containing a fusion protein of IE1 exon 4 and IE2 exon 5 was particularly good at stimulating cellular responses in mice (Wang *et*

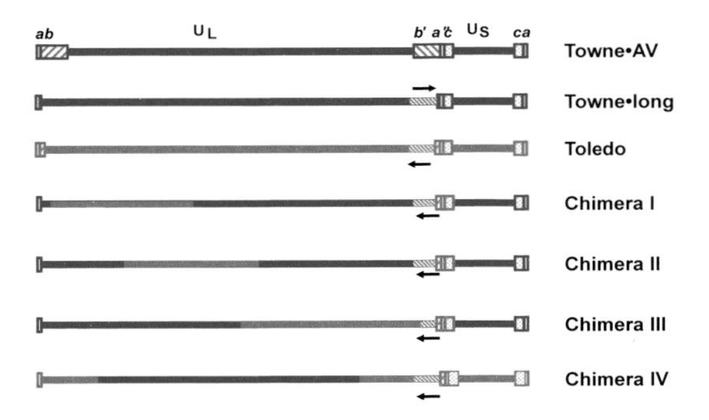

Figure II.20.1 CMV chimeric strains generated by recombination between the Towne attenuated virus (in blue) and the Toledo virulent virus (in red). Hatched sections represent the unique long (UL/*b*′) region, the absence of which may attenuate CMV (Cha *et al.*, 1996, Courtesy of Dr. Richard Spaete and *MedImmune*). Arrows indicate the orientation of UL/*b*′ in a given strain. U_L and U_S, unique long and unique short regions of the HCMV genome, respectively. b, terminal repeat long, b′, internal repeat long, c, terminal repeat short, c′ internal repeat short. a and a′, terminal ‘a’ sequences.

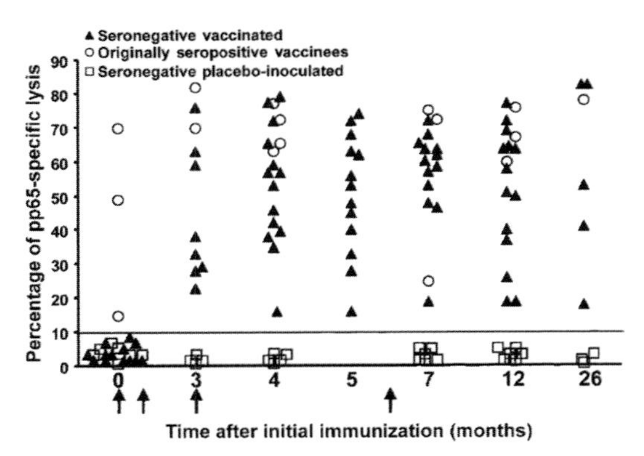

Figure II.20.2 Kinetics of pp65-specific cytolytic activity of 25 subjects after initial immunization with canarypox-CMV-pp65. Specific lysis at an effector to target ratio of 25:1 is shown at all time points. Significant lysis is defined at 10% (Berencsi *et al.*, 2001).

A

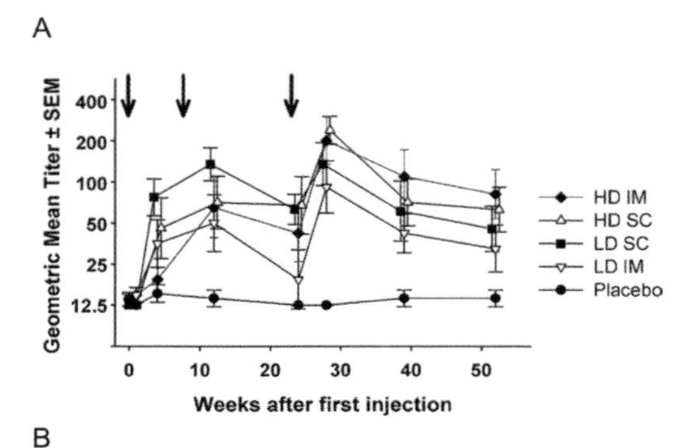

B

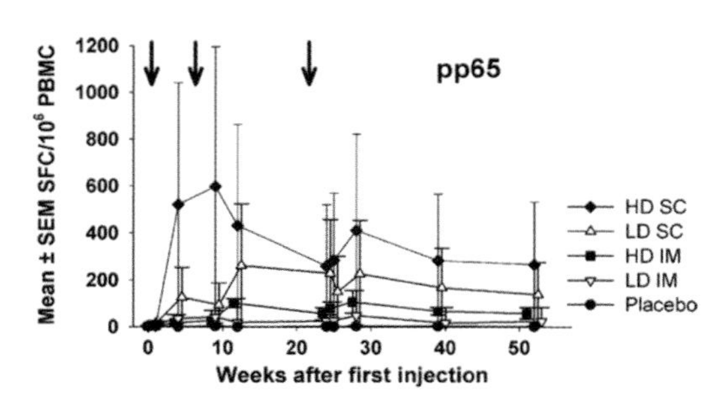

Figure II.20.3 Neutralizing antibody responses and T-cell responses in study participants given alphavirus replicon CMV vaccine. (A) Changes in CMV 50% neutralizing titre (geometric mean ± SEM) over time. Shown are the titres for serum samples assayed at weeks 0, 4, 12, 24, 28, 39 and 52. Titres <25 were assigned a value of 12.5. (B) T-cell responses in study participants, measured by IFN-γ ELISPOT assay. Changes in *ex vivo*, direct IFN-γ ELISPOT assay responses [mean ± SEM spot-forming cells (SFC) per 10^6 PBMCs] over time are shown after stimulation with peptides spanning pp65 (Bernstein *et al.*, 2009). HD, high dose; LD, low dose; IM, intramuscular injection; SC, subcutaneous injection.

al., 2008). MVA was then used to carry the analogous rhesus CMV genes for gB, pp65 and IE1, which proved to be highly immunogenic in monkeys. Although vaccination did not prevent infection, CMV viral loads were decreased (Yue *et al.*, 2008). Priming with DNA plasmids for the same genes followed by MVA vectors gave a similar, although not better result (Abel *et al.*, 2011). An MVA construct expressing pp65, IE1 exon 4 and IE2 exon 5 after a strong vaccinia promoter called mH5 is about to enter clinical trials (Wang *et al.*, 2010; D. Diamond, Beckman Research Institute of the City of Hope, Duarte, CA, USA, personal communication). The first study will be done in normal volunteers and, if safety and immunogenicity are demonstrated, the vaccine would be tested in double-blind trials involving first donors of haematopoietic cells and subsequently recipients of haematopoietic cell transplants. The aim would be to suppress CMV through cellular immune responses.

Replication-deficient adenoviruses have been widely used in experimental vaccinology to produce proteins of vaccine interest, although prior immunity to common adenovirus serotypes inhibits responses. Adenovirus 5 was used to construct a polyepitope vaccine: that is, a vector carrying epitopes from eight different CMV proteins including the three often referred to above, but also including other glycoproteins and targets of cellular responses (Rist *et al.*, 2005). In mice the polyepitope vaccine elicited NT-Abs and both CD4- and CD8 T-cells specific to CMV (Zhong *et al.*, 2008, 2009). It was noted that CD4-deficient mice responded poorly to the vaccine (Zhong *et al.*, 2010). Replicating adenoviruses and lymphocytic choriomeningitis virus are being developed as vectors for CMV genes by biotechnology companies (personal communications by Paxvax and Hookipa biotechs).

Deletion mutants

As mentioned above, CMV depends on immune evasion to replicate in the face of CD8 T-cells and natural killer cells. Therefore, a possible strategy to develop a live attenuated vaccine would be to delete the immune evasion genes so that the virus could not continue to replicate. The success of this strategy was demonstrated with mouse CMV by Cicin-Sain *et al.* (2007) in the group of U.H. Koszinowski. The same group obtained another interesting result through creation of a spread-deficient and therefore attenuated mouse CMV by deletion of a gene crucial for replication (Mohr *et al.*, 2010). Further development and clinical testing of human CMV mutants would be worthwhile.

Non-living candidate vaccines

Glycoprotein B

Several glycoproteins are present in the envelope of CMV, but one that elicits NT-Abs is called gB, by analogy to its *Herpes simplex* glycoproteins (Pereira *et al.*, 1984; Cranage *et al.*, 1986; Britt *et al.*, 1988). Naturally immune and live virus vaccinated individuals develop NT-Abs directed against gB (Gonczol *et al.*, 1989). The gB has been extensively studied and a number of 'neutralizing epitopes' have been defined over the years (see Chapter II.10). The glycoprotein was isolated from the virus using monoclonal antibody and column chromatography (Gonczol *et al.*, 1990). Human volunteers responded with NT-Abs, although, as confirmed later, three doses were necessary for high titres and antibody persistence was not prolonged.

Table II.20.6 Antibody response to three or four doses of subunit CMV glycoprotein B with MF59 adjuvant (Chiron), administered to adults on a schedule of 0, 1, 6 and 18 months

Bleeding schedule (mo.)	Time	Neutralizing antibody titres (geometric mean titre)
7	Post Dose 3	100
18	Pre Dose 4	30
19	Post Dose 4	110
24		65
30		60

However, production of gB was greatly simplified through the development by *Chiron Laboratories* of a Chinese Hamster Ovary cell line containing the gene and producing the protein in quantity (Spaete *et al.*, 1991). The gB was combined with the oil-in-water adjuvant MF59 and shown to be immunogenic in adults (Table II.20.6), and even more so in toddlers (Frey *et al.*, 1999; Pass *et al.*, 1999; Mitchell *et al.*, 2002). The dose that proved most useful was 20 micrograms (mcg) of gB given intramuscularly. Whereas neutralizing titres were in the hundreds in adults, toddlers developed titres in the thousands. Moreover, a single dose in seropositive subjects elicited good anamnestic responses, from which high titre immune globulin could be made (Drulak *et al.*, 2000).

There are several important neutralization epitopes on gB (see Chapter II.10). The immunodominant epitope (AD1) is on the C-terminal portion of the protein. Responses against the other epitopes, including a second epitope (AD2) on the N- terminal portion of gB, may be less prominent and may vary between individuals after natural infection and after gB/MF59 vaccination (Ohlin *et al.*, 1993; Axelsson *et al.*, 2007). The significance of these observations remains to be seen.

Antibody titres after gB vaccination waned considerably in adults after the third dose, but could be restored by a fourth dose (Pass *et al.*, 1999). Unfortunately, persistence of titres was not followed in toddlers. In any case, *Chiron* (now *Novartis*) decided not to pursue the gB vaccine, selling the project to *Sanofi Pasteur*.

A trial of gB/MF59 was launched at the University of Alabama under the leadership of Robert Pass, first in collaboration with *Novartis* and then subsequently with *Sanofi Pasteur*. The subjects were adolescent girls who later might be exposed to CMV through respiratory and sexual contact. The results proved to be important, in that 50% prevention of acquisition was demonstrated during a period of 42 months post-vaccination (Pass *et al.*, 2009). The infection rate was 8% in vaccinees versus 14% in 'placebees', with 95% confidence limits of 7–73%. However, as seen in the Kaplan–Meier curve shown in Fig. II.20.4, the prevention was entirely during the first 12 months. NT-Abs were maximal after the third dose and waned sharply after 12 months, but no direct correlation between level of antibodies and protection against infection was established. However, it did appear that infected vaccinees had a more rapid antibody rise than infected 'placebees', which could be an indication that B-cell responses are important to protection (R. Pass, University of Alabama at Birmingham, personal communication). Three years after vaccination, 60% of vaccinees still had detectable NT-Abs. The same investigators showed that the gB/MF59 vaccine boosted neutralizing antibodies and CD4 T-cells in naturally seropositive women (Sabbaj *et al.*, 2011). Importantly, seropositive women developed higher levels of antibody after vaccination for at least a year (Sabbaj *et al.*, 2011). There were too few congenital infections to judge the ability of the vaccine to prevent transmission, but one infant of 81 vaccinated women who became pregnant was CMV positive versus three of 97 placebo women (Pass *et al.*, 2009).

Another study of gB/MF59 is currently being done

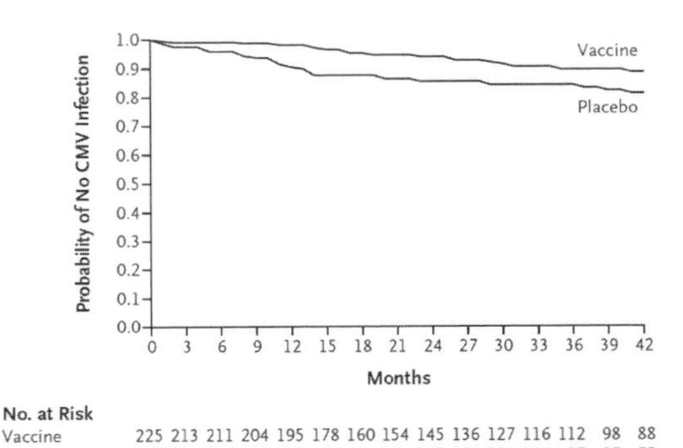

Figure II.20.4 Kaplan–Meier estimates of probability of remaining free of CMV infection in a trial of the gB vaccine (Pass *et al.*, 2009).

in adolescent females by David Bernstein in Cincinnati, and of course it will be crucial to see if confirmation of efficacy is obtained. However, additional evidence for the utility of gB was obtained in recipients of kidney or liver transplantation by Griffiths and colleagues (Griffiths *et al.*, 2011; see also Chapter II.14). Recipients of those organs were randomly distributed to receive three doses of vaccine or placebo and followed for CMV viraemia, which is common after the immunosuppression inherent to transplantation. Antibodies to CMV were markedly increased both in initially seronegative and seropositive recipients. Follow up of approximately 70 patients in each group showed reduction in the duration of viraemia and of the use of ganciclovir to suppress the viraemia (Fig. II.20.5). These were clearly clinically important benefits. Importantly, the duration of viraemia was inversely related to the titre of antibodies at the time of transplantation.

More recently, *Glaxo SmithKline* (GSK) has begun studies with its own gB, combining it with an adjuvant called AS01 that contains QS21 (a saponin derivative) and monophosphoryl lipid A (MPL). The latter is a TLR-4 agonist. In a phase I study, in which 15 mcg were given on a 0, 1 month and 6 months schedule, the vaccine was well tolerated and induced both NT-Abs and CD4 cellular responses (A. Marchant, Université Libre de Bruxelles, Belgium, personal communication). The neutralizing antibody geometric mean titre (GMT) was still higher than average in naturally seropositive individuals at 24 months after the first injection. Conversion of the antibodies from low-avidity to high avidity occurred after the third dose at 6 months. Importantly, gB-specific B-cell memory developed contemporaneously with high avidity. Thus, the adjuvant used by GSK may be superior to MF59, but further data will be needed.

Using a laboratory produced gB, other investigators showed in mice that a CpG oligonucleotide, which is a TLR-9 agonist, also served to adjuvant the glycoprotein (Dasari *et al.*, 2011). The status of gB vaccines is

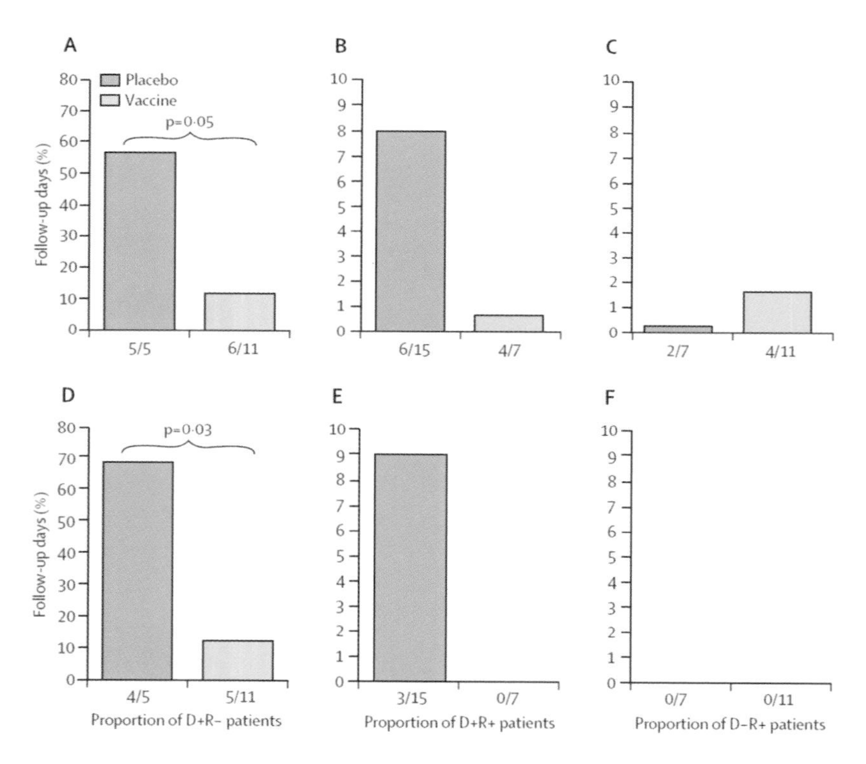

Figure II.20.5 Vaccination of solid organ transplant recipients with a glycoprotein-B vaccine with MF59 adjuvant. Proportion of days that vaccinated transplant patients at risk of CMV infection spent with viraemia or received antiviral treatment. A, B, and C show the duration of viraemia. D, E, and F show the duration of antiviral therapy. D+R–, D+R+, and D–R+ are three groups at risk of primary infection, reinfection, and reactivation, respectively. The numbers below each column indicate the number of patients with viraemia (or treatment) divided by the number in the subgroup. Note the different values on the Y-axes of panels A and D compared with panels B, C, E, and F. **D–**=cytomegalovirus seronegative donor. R–=cytomegalovirus seronegative recipient. D+=cytomegalovirus seropositive donor. R+=cytomegalovirus seropositive recipient (from Griffiths *et al.*, 2011).

certainly promising if persistent protection against infection can be demonstrated.

Other glycoproteins

The glycoprotein now known as gH also induces NT-Abs, and contains three neutralizing epitopes (Rasmussen *et al.*, 1988, Urban *et al.*, 1992; Simpson *et al.*, 1993). There appear to be at least two variants of gH (Urban *et al.*, 1992). Another complex of glycoproteins called gM/gN might also be of interest for a vaccine, as it also elicits NT-Abs and is abundantly present in the virus particle (Gretch *et al.*, 1988; Varnum *et al.*, 2004; Shimamura *et al.*, 2006). The only vaccine-related work done on these proteins thus far are attempts to develop virus-like particles (VLP) containing gH as well as gB and pp65 as well as DNA plasmids coding for gH and gL. The VLP preparations are immunogenic in animals (D. Anderson, *Variations Biotechnologies*, Cambridge, MA, U.S.A., and David Weiner, Inovio Pharmaceuticals, Blue Bell, PA, U.S.A., personal communications).

UL128–131 complex

The discovery of the gH/gL/UL128–131 complex and its role in entry of CMV into epithelial and endothelial cells has given new insight into the biology of CMV but has also provided new candidate antigens for a vaccine (Hahn *et al.*, 2004; Wang and Shenk, 2005). It is obvious that one would wish to prevent CMV entry into macrophages and dendritic cells, and the pentameric complex allows that to happen (Gerna *et al.*, 2005; Ryckman *et al.*, 2008a,b; Sinzger *et al.*, 2008; Straschewski *et al.*, 2011). Antibodies to this complex develop soon after natural primary infection, earlier than antibodies that prevent entry into fibroblasts (Gerna *et al.*, 2008), and rise after reactivated infections (Genini *et al.*, 2011). The pentameric complex also appears to be important for virus shedding and transmission, at least in the rhesus monkey (Oxford *et al.*, 2011).

Thus, the proteins in the pentameric complex have become an additional target of vaccine development, particularly since, not surprisingly, the Towne attenuated virus and the gB subunit vaccine elicit only low levels of antibodies preventing entry into epithelial cells (Cui *et al.*, 2008). Although gH alone does induce NT-Abs, it appears that the ability to block epithelial/endothelial cell entry is better mediated by antibodies to the UL128–131 proteins. Monoclonal antibodies that prevent entry have been developed against all of the proteins (Gerna *et al.*, 2008; Macagno *et al.*, 2010). It should be pointed out, however, that some monoclonal antibodies against gB and gH were capable of blocking epithelial cell entry, albeit more weakly than monoclonal antibodies against the UL128–131 proteins (Macagno *et al.*, 2010). Attempts to include one or more of those proteins in vaccines have already begun, and peptides from pUL130 and pUL131 have already been shown to produce NT-Abs in rabbits when conjugated to Keyhole Limpet Hemocyanin and combined with Freund's adjuvant (Saccoccio *et al.*, 2011b). Interestingly, the same authors showed blockage of entry into airway and genital epithelium.

DNA vaccines

The first attempts to use DNA plasmids to vaccinate against human CMV were published in 1989. Plasmids for both gB and pp65 were shown to be immunogenic in mice for both antibodies and CTL (Endresz *et al.*, 1999, 2001). A later study tested a regimen consisting of gB DNA priming followed by subunit gB and a marked potentiation of antibody response was seen (Endresz *et al.*, 2001; Plotkin, 2001). At that point, DNA plasmids fell out of favour as a vaccine platform but more recently have been restored to favour by the coadministration of adjuvants and delivery by electroporation.

A group at *Vical Inc.* created DNA plasmids for gB and pp65 with the target being recipients of haematopoietic stem cell transplants at risk for primary infection from donor cells or for reactivation of prior infection. They combined the plasmids (5 mg per dose) with CytRx Research Labs (CRL) 1005 poloxamer, a non-ionic lipid that forms nanoparticles of DNA and presumably enhances uptake. The plasmids were first tested in normal volunteers. gB antibodies measured by ELISA developed in 45% of seronegatives but there was no boost in seropositives. ELISPOT responses to pp65 developed in about half of seronegatives and a third of seropositives. The vaccine caused no significant adverse reactions (Wloch *et al.*, 2008). In a phase II trial, vaccine or placebo was randomly given to patients with haematological malignancies once before transplantation and three times after transplantation at approximately 1, 3 and 6 months. ELISA antibody responses against gB were somewhat better in vaccinees at one year but there were high ELISPOT assay responses to gB and particularly to pp65, indicating T-cell responses. The striking observations were a decrease in CMV reactivation, an increase in time to viraemia (Fig. II.20.6) and a decrease in its duration, together with resulting less use of antivirals (Kharfan-Dabaja *et al.*, 2012). This vaccine is moving forward in clinical development by *Astellas Pharma* (Japan) for the transplantation indication and also possibly to prevent congenital infection.

As mentioned previously, DNA plasmids coding for gH/gL stimulate neutralizing antibodies in

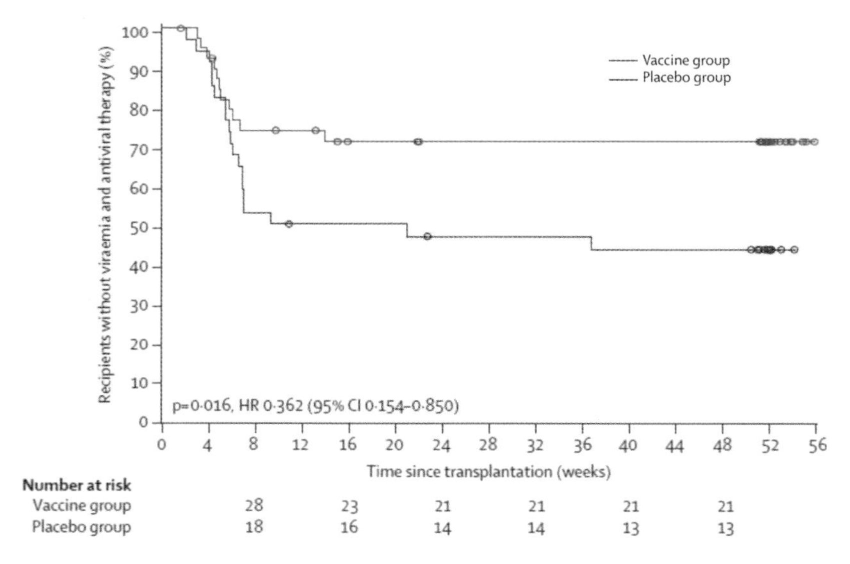

Figure II.20.6 Time to initial viral reactivation in transplant recipients following combined gB/pp65-DNA vaccination. Kaplan–Meier analysis of time to the composite endpoint of viraemia and cytomegalovirus-specific antiviral therapy (circles are censored events, Kharfan-Dabaja *et al.*, 2012). HR=hazard ratio.

mice (Inovio). It is interesting that a DNA prime/MVA boost for the rhesus CMV gB, pp65 and IE1 antigens resulted in decreased salivary excretion of the virus after challenge in monkeys, correlating best with T-cell responses to the pp65 (Abel *et al.*, 2011). Another prime-boost regimen consisting of DNA plasmid for pp65 and IE1, followed by adeno-associated viruses carrying the same genes was immunogenic in mice (Gallez-Hawkins *et al.*, 2004). It has also been shown in normal seronegative subjects that DNA plasmids coding for gB, pp65 and IE1 can prime for a boost with the Towne attenuated virus, indicating that they induced B- and T-cell memory (Jacobson *et al.*, 2009).

The laboratory of Deborah Spector has used the mouse CMV model to develop a prime-boost vaccine strategy that could be applied to HCMV. The prime is a mixture of plasmid DNAs that induce CD8 T-cell responses, followed by a boost with formalin-inactivated MCMV that induces antibodies (Morello *et al.*, 2002). Although additional important CD8 T-cell responses were elicited by using complex mixture plasmids containing DNA for 13 MCMV antigens, the major protection was achieved with plasmids for the proteins analogous to human gB, pp65 and IE1 (Morello *et al.*, 2005). Plasmids for the mouse DNA polymerase and helicase were also found to contribute to protection, but may not be essential (Morello *et al.*, 2007).

Peptides

A defined peptide vaccine would have advantages if it could prevent CMV infection or disease. A laboratory at the City of Hope Hospital in California is attempting to develop one, particularly with transplant patients in mind. The limitations of peptide vaccines are the variety of peptides needed because of the variability of HLA groups in human populations, the lack of a native conformation of antibody epitopes, and their poor immunogenicity without adjuvantation. The first effort used a peptide derived from pp65 that is HLA A*0201 restricted, which gave CD8 T-cell booster responses in peripheral blood cells from seropositive subjects (Diamond *et al.*, 1997). A lipidated version given intranasally induced immune responses in mice (BenMohamed *et al.*, 2000). However, to cover all ethnic groups they found that about 15 epitopes would be necessary for a vaccine to broadly address CD8 T-cell responses (Longmate *et al.*, 2001). Immunogenicity could be increased by coupling the epitopes with the promiscuous Th-cell epitope from tetanus toxin and by adding a CpG oligonucleotide that stimulates TLR-9 (La Rosa *et al.*, 2002). Recently, the City of Hope group used the HLA A*0201 CD8 T-cell epitope at amino acids 495–503 of the pp65 protein coupled with either the PADRE or another tetanus promiscuous Th epitope to immunize healthy adults expressing the same HLA molecule. Volunteers were given four subcutaneous injections at 3 week intervals, either with a CpG oligonucleotide TLR-9 stimulating adjuvant or without the adjuvant. All 30 subjects in the adjuvant group developed T-cell responses to pp65 including some of the previously seronegative subjects. In previously seropositive subjects with low levels of CD8 T-cells specific for pp65 before vaccination, 8 of 17 developed higher reactivity. This approach might be useful to increase immunity in haematopoietic stem cell recipients (La Rosa *et al.*, 2012).

In Germany there has been a focus on vaccination of haematopoietic stem cell recipients with peptide-loaded dendritic cells to stimulate T-cell responses. Dendritic cells were loaded with pp65- and pp150-derived peptides and were administered to stem cell recipients at high risk for CMV disease (Grigoleit *et al.*, 2007). Apparent beneficial effects were seen with respect to the reconstitution of antiviral T-cell responses and long-term CMV control. More recently, T-cells from normal donors were enriched *in vitro* with pp65 peptide-loaded streptamers and were transferred into stem cell recipients. The transferred CD8 cells controlled the viraemia without further antiviral administration (Schmitt *et al.*, 2011; for more discussion of cell transfer therapies in humans and mouse models, see Chapter II.16 and Chapter II.17, respectively).

Subviral particles-dense bodies

CMV-infected fibroblasts release subviral particles, termed Dense Bodies (see also Chapter I.13). They appear attractive for vaccine development, as they contain some of the immunodominant antigenic targets of the CMV-specific immune response. Over 60% of the protein mass of DB consists of pp65 (Varnum *et al.*, 2004; see also Chapter I.6). Using mass spectrometry, these authors also found glycoproteins B, H, L, and M in DB. These polypeptides are obviously inserted into the DB envelope in a functional conformation, as the particles can penetrate host cell membranes in a way comparable to virions (Topilko and Michelson, 1994). Consequently, antibody responses to conformation-dependent epitopes can likely be induced by DB-immunization.

The immunogenicity of DB was demonstrated in mouse models, where NT-Abs, CD8 and CD4 T-cell responses were induced without the addition of adjuvant (Pepperl *et al.*, 2000; Pepperl-Klindworth *et al.*, 2002). More recently, proof could be provided that DB can be modified in their antigenic composition. For this, an HLA-A2 presented IE1-derived peptide was inserted into pp65, thereby generating CMV recombinants that release modified DB. These particles induced CD8 T-cell responses against IE1, showing the potential of DB to accommodate additional antigenic peptides (Mersseman *et al.*, 2008; Becke *et al.*, 2010). A CD8 T-cell response could also be induced against a heterologous peptide contained in modified DB, using the H2-K^b restricted peptide SIINFEKL from Ovalbumin as a model (Fig. II.20.7). Despite all these promising data on DB, clinical testing of the DB approach is still pending.

Lessons from vaccine studies

Although it is still too early to draw firm conclusions, nevertheless some lessons can be extracted from prior work. The most important is that a CMV vaccine is feasible, inasmuch as an attenuated vaccine has reduced disease in solid organ transplant recipients, a subunit gB vaccine has reduced disease in the same population and also prevented acquisition of the virus by previously seronegative women, and a DNA vaccine has reduced viral replication in stem cell transplant recipients.

Thus, the paradigm that antibodies can prevent infection whereas cellular responses enhance recovery is true for CMV (Plotkin, 2010). However, there are still issues to be resolved, of which there are at least four:

1. To prevent congenital infection, should CTL targets be added to an antibody-eliciting vaccine? Certainly one would expect CTL to reduce replication, but each additional antigen in a vaccine must be justified.
2. Although evidence exists for protection afforded by prior infection or vaccination, how can the fact of reinfection be assimilated into CMV vaccine development and strategies for use?
3. Are antibodies protective against entry into epithelial and endothelial cells necessary in addition to those that protect against entry into fibroblasts?
4. Are mucosal antibodies, either IgA or IgG, important for protection, considering that they are present after natural infection (Saccoccio *et al.*, 2011a)?

Target groups for vaccination

Who would be vaccinated, once a CMV vaccine is available? Table II.20.7 gives a list of possible targets. The easiest group to identify is transplant recipients. However, the case differs between solid organ recipients (see also Chapter II.13 and Chapter II.14) and haematopoietic stem cell recipients (Chapter II.16). In the former case, the high risk is seronegative recipients who receive an organ from a seropositive donor. As shown by the studies with the gB vaccine, which in a sense confirm prior studies with hyperimmune globulin,

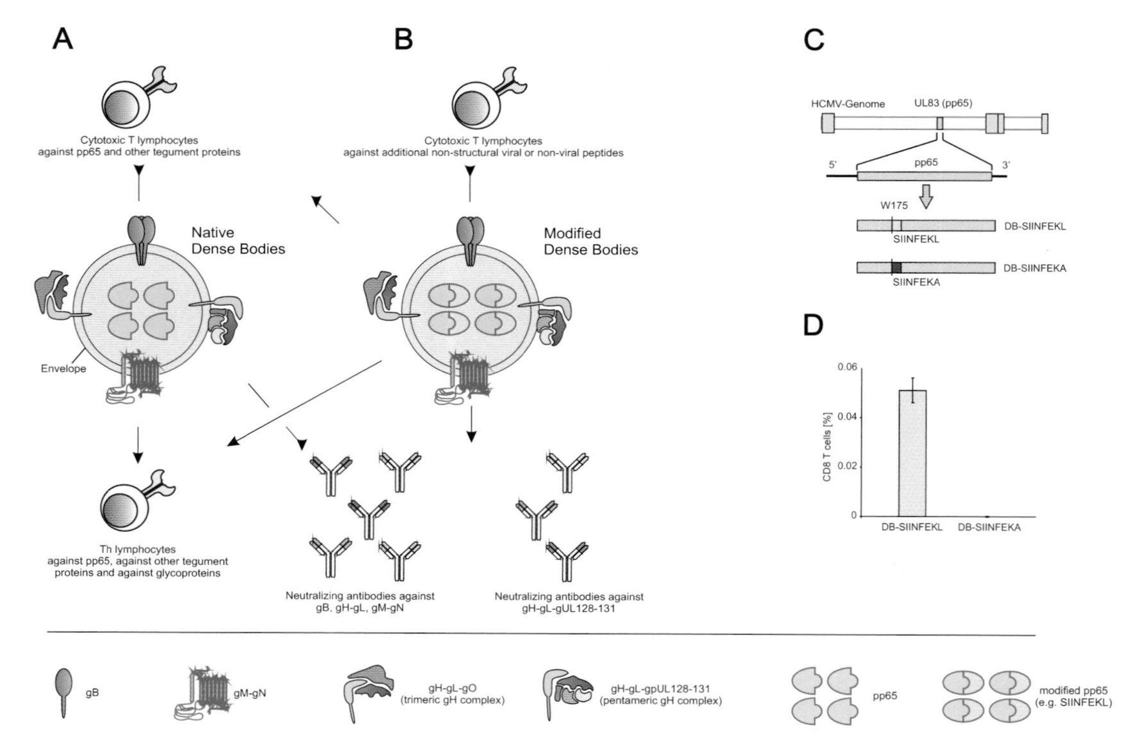

Figure II.20.7 Dense Bodies of HCMV as a basis for vaccine development. (A), Schematic representation of native DB and immune responses induced by these particles. The internal structure of DB mainly consists of tegument proteins, pp65 being the most abundant constituent. The external surface of DB is formed by a lipid bilayer which is comparable to the viral envelope. It contains glycoprotein complexes gB, gM-gN, gH-gL-gO. DB from endotheliotropic strains may also contain the pentameric complex gH-gL-gUL128–131 (see also Chapter I.17). Since the glycoprotein complexes are present in their functional conformation, immunization with DB leads to the induction of a neutralizing antibody response that closely reflects the response after natural infection. In particular, NT-Abs against conformation dependent epitopes, which are considered important for protection (see also Chapter II.10), may be induced by DB. Immunization with these particles, in addition, leads to a distinct CD8 T-cell response (Pepperl *et al*., 2000), which is remarkable for a non-replicating vaccine. Furthermore, Th-cell responses can be induced, as the major CD4 T-cell targets pp65, gB and gH are contained in DB. (B), Strategy to modify the antigenic composition of DB by expressing peptides from non-structural viral proteins or from heterologous proteins in fusion with pp65. HCMV recombinants are generated by bacterial artificial chromosome (BAC) technology to express pp65 and additional peptides in one molecule ('modified pp65'). Expression and packaging of an immunodominant peptide from the HCMV IE1 protein, for example, leads to the formation of recombinant DB that induce a marked CD8 T-cell response against that peptide (Becke *et al*., 2010). This technology also allows inclusion of non-HCMV peptides into recombinant DB (e.g. packaging of the ovalbumin-derived peptide SIINFEKL). (C) and (D), Evaluation of DB-SIINFEKL. (C), Insertion of SIINFEKL or SIINFEKA (negative control) at position W175 of pp65. Recombinant CMVs were generated, using BAC mutagenesis, by inserting the coding sequences for SIINFEKL or SIINFEKA into the UL83 open reading frame. Recombinant viruses were reconstituted on primary human fibroblasts and modified DB were purified from culture supernatant by gradient centrifugation. (D), Interferon-γ ELISPOT analysis of CD8 T-cells from spleens of mice that had been immunized with SIINFEKL- or SIINFEKA-DB, respectively. Stimulator cells were loaded with the SIINFEKL peptide.

antibodies can reduce replication of the virus carried by the transplanted organ. In fact, levels of gB specific antibodies appeared to correlate with improvement of the situation regarding reinfection in seropositive solid organ recipients (Griffiths *et al.*, 2011).

In contrast, the stem cell recipient is at risk from reactivation of endogenous CMV or perhaps reinfection by a virus in the donor cells (for review, see Herr and Plachter, 2009; Thomas and Herr, 2011). T-cell transfer experiments and the studies of DNA plasmids show that a CTL response can reduce the replication of CMV in seropositive stem cell recipients. Thus, depending on the type of vaccine, all recipients of transplants would be candidates for immunization before and in the period after transplantation. This is a sizeable and growing group of patients. Immunization

Table II.20.7 Targets and goals of a prophylactic CMV vaccine

Setting	Target population	Goals
Prepregnancy	CMV-seronegative girls/women	Prevention of primary infection/induction of strain-independent immunity
	CMV-seropositive women	Boosting of pre-existing immunity/induction of strain-independent immunity
Transplantation of haematopoietic cells	D^+/R^- D^-/R^+	Enhancement of CMV-specific immune reconstitution/control of CMV reactivation Donor vaccination to induce antiviral immunity/control of CMV reactivation
Transplantation of solid organs	D^+/R^-	Recipient vaccination Prevention of primary infection/prevention of disease
	D^+/R^+	Recipient vaccination for boosting of pre-existing immunity/ Control of reactivation/prevention of superinfection/prevention of disease
Infants	General population	Induction of a strain-independent response to prevent primary infection of mothers

D^+, donor CMV-antibody seropositive; R^+, recipient CMV-antibody seropositive; D^-, donor CMV-antibody seronegative; R^-, recipient CMV-antibody seronegative.

of donors who are seronegative might be considered in a situation where the recipient is seropositive. The Towne vaccine was used successfully for that purpose in a stem cell transplant situation (Horn *et al.*, 2009).

From a public health point of view, however, the most relevant indication is for the prevention of fetal infection by CMV (see Chapter II.2). There are three groups who potentially could be targeted: infants, adolescents and adults.

In view of the outstanding success of paediatric immunization, an ideal CMV vaccine would be given in infancy to protect females in later pregnancies from transmitting virus to their fetuses. In that case, the vaccine would have to provide about 40 years of protection, although boosters might be considered in adolescence. Although an infant immunization strategy works well with rubella vaccine, it is not yet clear that we have a CMV vaccine with long duration of protection and, in any case, evidence will have to be obtained over many years after vaccination. However, an aspect of CMV epidemiology encourages the use of vaccine in toddlers, and that is the abundant evidence that adults, and particularly mothers, are infected from their young children (Marshall and Adler, 2009). If vaccination of toddlers decreased that transmission, fewer women would be infected during pregnancy.

The most obvious target is the adolescent girl. Vaccination just before puberty would provide protection prior to sexual exposure, but would permit natural infection in childhood to immunize a large proportion of girls. Immunity to HCMV in children would also reduce the risk of infection in their mothers, but as the possibility of infection exists by the sexual route or from other adults by the respiratory route, protection would be imperfect.

Here again boosters might be necessary to prolong protection throughout the entire child-bearing age of approximately 30 years. The problem is that, at least in the USA, coverage of vaccines aimed at adolescents has been mediocre, typified by low uptake of human papillomavirus vaccine.

Finally, the vaccine could be offered to adult women contemplating pregnancy. The difficulty here is the requirement for women to wait until the vaccine regimen is completed, which may be six months, and the possibility that pregnancy may occur during immunization. Vaccination is usually discouraged during pregnancy because of theoretical (or in a few cases, real) concerns about damage to the fetus.

In all of the three possible approaches, vaccination of males against CMV might be considered in order to reduce exposure of females. However, modelling would have to show that vaccination of males is not counterproductive by decreasing circulation of virus in childhood and thus increasing susceptibility of women entering child-bearing years who have themselves not been vaccinated. That is a reason to vaccinate both toddlers and women.

Public health considerations

Clearly, a CMV vaccine would have its initial use in developed countries with large numbers of seronegative women of child-bearing age (Chapter II.2). This is the group whose infants will have the highest proportion of abnormalities if infected *in utero*, as described at the beginning of this chapter as well as in greater detail in Chapter II.3. However, it has been established beyond doubt that reinfection can occur in women who are seropositive, although it has not been established if they differ immunologically from seropositive women who resist reinfection (Boppana *et al.*, 2001; Dar *et al.*, 2008; Mussi-Pinhata *et al.*, 2009). The reinfection of pregnant women also may result in infection and damage to their infants, although at a lower rate than in infants of seronegative women. Wang *et al.* (2011) calculated that although on average 8772 primary CMV infections occurred in American women each year, there were also 29,918 reinfections. Damage from primary infection occurs at a rate of about 20%, whereas damage from reinfection probably occurs only at a rate probably one-tenth of that (Fowler *et al.*, 1992), although a rate as high as 10% has been reported from Brazil (Yamamoto *et al.*, 2011). This means that from a public health point of view, if a vaccine could boost the immune functions of women previously infected, it would be given not only to seropositive women in developed countries but also to girls and women in developing countries. In The Gambia, a 3.9% rate of congenital CMV infection was observed (Kaye *et al.*, 2008). Thus, the prospect is that a CMV vaccination would be justified universally in women, regardless of their serological status.

Issues with respect to licensure of a CMV vaccine

On 10–11 January 2012, the Food and Drug Administration of the United States government, which is the licensing organization for vaccines, held a meeting on cytomegalovirus vaccines in Bethesda, Maryland, attended by manufacturers, regulators from the US and Europe, and academic investigators. The objectives of the meeting were to review vaccine studies thus far, to identify the gaps in knowledge that impede future development, to discuss target groups for vaccination and to determine the best endpoints for efficacy trials.

One principal gap in knowledge revolves around the data showing that reinfection and perhaps reactivation are common in naturally seropositive pregnant women and that at least some infants born to those mothers suffer infection and disease. This phenomenon has been explained by evasion of T-cell responses by CMV immune evasion genes. Nevertheless, prior studies suggest that natural immunity does partially protect against fetal transmission of CMV. It is therefore necessary to obtain more data on the consequences to the fetus of maternal reinfection in comparison with primary infection and to determine whether vaccination can improve the efficacy of natural immunity. Vaccine efficacy studies may require enrolment of both seronegative and seropositive women with evaluation of whether either primary or secondary infections are prevented.

There was extended discussion of target groups for efficacy trials. Women who are contemplating a future pregnancy are the obvious group for vaccination. In enrolling those women, it will be important to choose those already exposed to young children who may transmit CMV to them. Women who have just had a child are statistically more likely to conceive again within several years than women who are childless. Adolescent girls have considerable exposure to CMV and will be the target of a licensed vaccine, but may not be ideal for an efficacy trial because of a lower rate of pregnancy. The identification of an immune correlate of protection in older women will enable bridging efficacy to this population. The definition of such an immune correlate is thus important. Phase II studies must be done with candidate vaccines in such a way that correlates can be identified through analysis of vaccine protection and breakthrough.

Standardization of diagnostic tests will also be necessary before efficacy trials are done, including PCR for the presence of virus infection and serological tests for IgM antibodies, IgG antibodies, and antibody avidity.

Although the primary goal of a CMV vaccine is to prevent congenital disease, the consensus at the meeting was that for an efficacy study the numbers of subjects necessary to show disease prevention is too large, and that infection at birth always precedes disease. Thus, there was agreement that the major endpoint for efficacy studies will be prevention of fetal infection, as demonstrated by absence of virus excretion in the saliva or urine at birth in infants of vaccinated mothers. PCR tests are the most sensitive for this purpose and saliva may be preferable to urine with respect to ease of collection and sensitivity (Boppana *et al.*, 2011).

Although logically prevention of acquisition of virus during pregnancy should protect the fetus, unless efficacy was virtually 100% it would be impossible to know if a high incidence of disease occurred in infants born to women who were infected despite vaccination. On the other hand, a vaccine that protected only some mothers from infection might nevertheless be even more efficacious against transmission to their fetuses or reduce the severity of congenital disease. A possible secondary endpoint is placental thickness. Phase IV follow up studies will be necessary to evaluate disease in infected infants born to mothers in the vaccine and placebo groups.

In addition, because children between 1 and 3 years of age are the major disseminators of CMV infection, paediatric studies should be done at some point in development of vaccines to see if transmission can be reduced, which would be an indication for universal use in children in order to reduce the force of infection and resultant exposure of women.

For an inactivated vaccine, it was said that the numbers of subjects needed for demonstration of safety would not be excessively large, and that probably the population size needed for an efficacy trial would be sufficient to show the absence of serious reactions with a rate of 1/1000 or higher. Thus, one can estimate that the safety data base would involve between 5000 and 20,000 subjects. However, as pregnant women would likely be vaccinated inadvertently after licensure, observations would have to be collected on the health of children born after those pregnancies. Preclinical animal reproductive toxicity tests are mandatory.

At the meeting there was also discussion of the possible use of a CMV vaccine in transplant patients. Two distinct patient groups could profit from vaccination: seronegative solid organ transplant (SOT) recipients who receive an organ from a seropositive donor and seropositive haematopoietic stem cell recipients regardless of donor status, but particularly if the donor is seronegative. In the SOT case, primary infection of seronegative recipients by an organ from a seropositive donor is of major concern and may lead to systemic CMV disease and rejection of the graft. Although antivirals given prophylactically or pre-emptively can control CMV infection, they are expensive and toxic. Thus, a vaccine that prevented or controlled primary infection of SOT recipients would be valuable. Reduction of disease incidence as the endpoint in SOT was proposed. However, because of available antiviral strategies, the sample size would be considerable. Viral parameters such as viral load or time to reactivation appear to correlate well with the risk for disease and were thus suggested as surrogate endpoints.

Although many different types of organs are transplanted, it was said that demonstration of efficacy in a kidney transplant population could lead to extension of vaccination to recipients of other organs, in whom CMV infections are even more serious.

Seropositive stem cell recipients may suffer from serious CMV reactivations that are not always controlled by antivirals. Therefore, in this population, interval to reactivation, number of reactivation episodes, reduction of viral load in the blood, and antiviral use all qualify as endpoints for vaccine studies.

The meeting also heard from NIAID and other US government agencies as well as parent groups, all expressing the need for financial support of CMV vaccine development and for greater awareness of congenital CMV by medical professionals and lay people. The CDC will undertake disease burden modelling and other epidemiological studies of congenital CMV infection. There was palpable enthusiasm at the meeting for CMV vaccine development as a high priority in view of the public health burden of congenital infection, post-transplant disease and other pathologies.

Conclusions

It has been established that vaccination can influence cytomegalovirus infection, both with respect to infection of women and the consequences of infection in transplantation. Much of the effectiveness is mediated by antibodies, although the quality and quantity of antibody might be improved by additional antigens, and the induction of cytotoxic T-cells and helper T-cells could additionally increase control of infection, particularly in transplant patients and possibly also in pregnant women. Despite the here discussed important issues that remain to be resolved, a cytomegalovirus vaccine is now on the horizon. The task now is to move more candidates through early clinical phases so that definitive efficacy trials can be done. The authors of this chapter are optimistic that one day a licensed CMV vaccine will protect unborn children from intrauterine infection and those patients in danger of complications during transplantation.

Acknowledgement

S.P. thanks Grace Smith for secretarial support. B.P. was supported by the Deutsche Forschungsgemeinschaft, Klinische Forschergruppe 183.

References

Abel, K., Martinez, J., Yue,Y., Lacey, S.F., Wang, Z., Strelow, L., Dasgupta, A., Li, Z., Schmidt, K.A., Oxford, K.L., *et al.* (2011). Vaccine-induced control of viral shedding following rhesus cytomegalovirus challenge in rhesus macaques. J. Virol. *85*, 2878–2890.

Adler, B., Scrivano, L., Ruzcics, Z., Rupp, B., Sinzger, C., and Koszinowski, U. (2006). Role of human cytomegalovirus UL131A in cell type-specific virus entry and release. J. Gen. Virol *87*, 2451–2460.

Adler, S.P., and Nigro, G. (2009). Findings and conclusions from CMV hyperimmune globulin treatment trials. J. Clin. Virol. *46(Suppl. 4)*, S54-S57.

Adler, S.P., Starr, S.E., Plotkin, S.A., Hempfling, S.H., Buis, J., Manning, M.L., and Best, A.M. (1995). Immunity induced by primary human cytomegalovirus infection protects against secondary infection among women of childbearing age. J. Infect. Dis. *171*, 26–32.

Adler, S.P., Hempfling, S.H., Starr, S.E., Plotkin, S.A., and Riddell, S. (1998). Safety and immunogenicity of the Towne strain cytomegalovirus vaccine. Pediatr. Infect. Dis. J. *17*, 200–206.

Adler, S.P., Plotkin, S.A., Gonczol, E., Cadoz, M., Meric, C., Wang, J.B., Dellamonica, P., Best, A.M., Zahradnik, J.,

Pincus, S., *et al.* (1999). A canarypox vector expressing cytomegalovirus (CMV) glycoprotein B primes for antibody responses to a live attenuated CMV vaccine (Towne). J. Infect. Dis. *180*, 843–846.

Arvin, A.M., Fast, P., Myers, M., Plotkin, S., and Rabinovich, R. (2004). Vaccine development to prevent cytomegalovirus disease: report from the National Vaccine Advisory Committee. Clin. Infect. Dis. *39*, 233–239.

Axelsson, F., Adler, S.P., Lamarre, A., and Ohlin, M. (2007). Humoral immunity targeting site I of antigenic domain 2 of glycoprotein B upon immunization with different cytomegalovirus candidate vaccines. Vaccine *26*, 41–46.

Balfour, H.H. Jr. (1979). Cytomegalovirus: the troll of transplantation. Arch. Intern. Med. *139*, 279–280.

Bate, S.L., Dollard, S.C., and Cannon, M.J. (2010). Cytomegalovirus seroprevalence in the USA: the national health and nutrition examination surveys, 1988–2004. Clin. Infect. Dis. *50*, 1439–1447.

Becke, S., Aue, S., Thomas, D., Schader, S., Podlech, J., Bopp, T., Sedmak, T., Wolfrum, U., Plachter, B., Reyda, S., (2010). Optimized recombinant dense bodies of human cytomegalovirus efficiently prime virus specific lymphocytes and neutralizing antibodies without the addition of adjuvant. Vaccine *28*, 6191–6198.

Beninga, J., Kropff, B., and Mach, M. (1995). Comparative analysis of fourteen individual human cytomegalovirus proteins for helper T-cell response. J. Gen. Virol. *76*, 153–160.

BenMohamed, L., Krishnan, R., Longmate, J., Auge, C., Low, L., Primus, J., and Diamond, D.J. (2000). Induction of CTL response by a minimal epitope vaccine in HLA A*0201/DR1 transgenic mice: dependence on HLA class II restricted T(H) response. Hum. Immunol. *61*, 764–779.

Berencsi, K., Gyulai, Z., Gonczol, E., Pincus, S., Cox,W.I., Michelson, S., Kari, L., Meric, C., Cadoz, M., Zahradnik, J., *et al.* (2001). A canarypox vector-expressing cytomegalovirus (CMV) phosphoprotein 65 induces long-lasting cytotoxic T-cell responses in human CMV-seronegative subjects. J. Infect. Dis. *183*, 1171–1179.

Bernstein, D.I. (2011). Vaccines for cytomegalovirus. Infect. Disord. Drug Targets *11*, 514–525.

Bernstein, D.I., Schleiss, M.R., Berencsi, K., Gonczol, E., Dickey, M., Khoury, P., Cadoz, M., Meric, C., Zahradnik, J., Duliege, A.M., *et al.* (2002). Effect of previous or simultaneous immunization with canarypox expressing cytomegalovirus (CMV) glycoprotein B (gB) on response to subunit gB vaccine plus MF59 in healthy CMV-seronegative adults. J. Infect. Dis. *185*, 686–690.

Bernstein, D.I., Reap, E.A., Katen, K., Watson, A., Smith, K., Norberg, P., Olmsted, R.A., Hoeper, A., Morris, J., Negri, S., *et al.* (2009). Randomized, double-blind, Phase 1 trial of an alphavirus replicon vaccine for cytomegalovirus in CMV seronegative adult volunteers. Vaccine *28*, 484–493.

Boeckh, M., and Geballe, A.P. (2011). Cytomegalovirus: pathogen, paradigm, and puzzle. J. Clin. Invest. *121*, 1673–1680.

Bonaros, N., Mayer, B., Schachner, T., Laufer, G., and Kocher, A. (2008). CMV-hyperimmune globulin for preventing cytomegalovirus infection and disease in solid organ transplant recipients: a meta-analysis. Clin. Transplant. *22*, 89–97.

Boppana, S.B., and Britt, W.J. (1996). Recognition of human cytomegalovirus gene products by HCMV- specific cytotoxic T-cells. Virology. *222*, 293–296.

Boppana, S.B., Polis, M.A., Kramer, A.A., Britt, W.J., and Koenig, S. (1995). Virus-specific antibody responses to human cytomegalovirus (HCMV) in human immunodeficiency virus type 1-infected persons with HCMV retinitis. J. Infect. Dis. *171*, 182–185.

Boppana, S.B., Rivera, L.B., Fowler, K.B., Mach, M., and Britt, W.J. (2001). Intrauterine transmission of cytomegalovirus to infants of women with preconceptional immunity. N. Engl. J. Med. *344*, 1366–1371.

Boppana, S.B., Ross, S.A., Shimamura, M., Palmer, A.L., Ahmed, A., Michaels, M.G., Sanchez, P.J., Bernstein, D.I., Tolan, R.W., Jr., Novak, Z., *et al.* (2011). Saliva polymerase-chain-reaction assay for cytomegalovirus screening in newborns. N. Engl. J. Med. *364*, 2111–2118.

Borysiewicz, L.K., Hickling, J.K., Graham, S., Sinclair, J., Cranage, M P., Smith, G.L., and Sissons, J.G., (1988). Human cytomegalovirus-specific cytotoxic T-cells. Relative frequency of stage-specific CTL recognizing the 72-kD immediate early protein and glycoprotein B expressed by recombinant vaccinia viruses. J. Exp. Med. *168*, 919–931.

Brayman, K.L., Dafoe, D.C., Smythe, W.R., Barker, C.F., Perloff, L.J., Naji, A., Fox, I.J., Grossman, R.A., Jorkasky, D.K., Starr, S.E., *et al.* (1988). Prophylaxis of serious cytomegalovirus infection in renal transplant candidates using live human cytomegalovirus vaccine. Interim results of a randomized controlled trial. Arch. Surg. *123*, 1502–1508.

Britt, W.J., and Mach, M. (1996). Human cytomegalovirus glycoproteins. Intervirology *39*, 401–412.

Britt, W.J., Vugler, L., and Stephens, E.B. (1988). Induction of complement-dependent and -independent neutralizing antibodies by recombinant-derived human cytomegalovirus gp55–116 (gB). J. Virol. *62*, 3309–3318.

Cannon, M.J., Schmid, D.S., and Hyde, T.B. (2010). Review of cytomegalovirus seroprevalence and demographic characteristics associated with infection. Rev. Med. Virol. *20*, 202–213.

Caposio, P., Orloff, S.L., and Streblow, D.N. (2011). The role of cytomegalovirus in angiogenesis. Virus Res. *157*, 204–211.

Cha, T.A., Tom, E., Kemble, G.W., Duke, G.M., Mocarski, E.S., and Spaete, R.R. (1996). Human cytomegalovirus clinical isolates carry at least 19 genes not found in laboratory strains. J. Virol. *70*, 78–83.

Cicin-Sain, L., Bubic, I., Schnee, M., Ruzsics, Z., Mohr, C., Jonjic, S., and Koszinowski, U.H. (2007). Targeted deletion of regions rich in immune-evasive genes from the cytomegalovirus genome as a novel vaccine strategy. J. Virol. *81*, 13825–13834.

Colugnati, F.A., Staras, S.A., Dollard, S.C., and Cannon, M.J. (2007). Incidence of cytomegalovirus infection among the general population and pregnant women in the USA. BMC Infect. Dis. *7*, 71.

Committee to study priorities for vaccine development, Division of health promotion and disease prevention and Institute of Medicine, Washington. (2000). Vaccines for the 21st Century: a Tool for Decisionmaking, Stratton, K.R., Durch, J.S., and Lawrence, R.S., eds. (National Academy Press, Washington, D.C.).

Cordonnier, C., Chevret, S., Legrand, M., Rafi, H., Dhedin, N., Lehmann, B., Bassompierre, F., and Gluckman, E. (2003). Should immunoglobulin therapy be used in allogeneic stem-cell transplantation? A randomized, double-blind, dose effect, placebo-controlled, multicenter trial. Ann. Intern. Med. *139*, 8–18.

Cranage, M.P., Kouzarides, T., Bankier, A.T., Satchwell, S., Weston, K., Tomlinson, P., Barrell, B., Hart, H., Bell, S.E., Minson, A.C., *et al.* (1986). Identification of the human cytomegalovirus glycoprotein B gene and induction of neutralizing antibodies via its expression in recombinant vaccinia virus. EMBO J. *5*, 3057–3063.

Cui, X., Meza, B.P., Adler, S.P., and McVoy, M.A. (2008). Cytomegalovirus vaccines fail to induce epithelial entry neutralizing antibodies comparable to natural infection. Vaccine *26*, 5760–5766.

Dar, L., Pati, S.K., Patro, A.R., Deorari, A.K., Rai, S., Kant, S., Broor, S., Fowler, K.B., Britt, W.J., and Boppana, S.B. (2008). Congenital cytomegalovirus infection in a highly seropositive semi-urban population in India. Pediatr. Infect. Dis. J. *27*, 841–843.

Dasari, V., Smith, C., Zhong, J., Scott, G., Rawlinson, W., and Khanna, R. (2011). Recombinant glycoprotein B vaccine formulation with Toll-like receptor 9 agonist and immune-stimulating complex induces specific immunity against multiple strains of cytomegalovirus. J. Gen. Virol. *92*, 1021–1031.

Davison, A.J., Dolan, A., Akter, P., Addison, C., Dargan, D.J., Alcendor, D.J., McGeoch, D.J., and Hayward, G.S. (2003). The human cytomegalovirus genome revisited: comparison with the chimpanzee cytomegalovirus genome. J. Gen. Virol. *84*, 17–28.

Diamond, J., York, J., Sun, J.Y., Wright, D.C., and Forman, S.J. (1997). Development of a candidate HLA A*0201 restricted peptide-based vaccine against human cytomegalovirus infection. Blood *90*, 1751–1767.

Dolan, A., Cunningham, C., Hector, R.D., Hassan-Walker, A.F., Lee, L., Addison, C., Dargan, D.J., McGeoch, D.J., Gatherer, D., Emery, V.C., *et al.* (2004). Genetic content of wild-type human cytomegalovirus. J. Gen. Virol. *85*, 1301–1312.

Dollard, S.C., Grosse, S.D., and Ross, D.S. (2007). New estimates of the prevalence of neurological and sensory sequelae and mortality associated with congenital cytomegalovirus infection. Rev. Med. Virol. *17*, 355–363.

Drulak, M.W., Malinoski, F.J., Fuller, S.A., Stewart, S.S., Hoskin, S., Duliege, A.M., Sekulovich, R., Burke, R., and Winston, S. (2000). Vaccination of seropositive subjects with CHIRON CMV gB subunit vaccine combined with MF59 adjuvant for production of CMV immune globulin. Viral Immunol. *13*, 49–56.

Einsele, H., Roosnek, E., Rufer, N., Sinzger, C., Riegler, S., Loffler, J., Grigoleit, U., Moris, A., Rammensee, H.G., Kanz, L., *et al.* (2002). Infusion of cytomegalovirus (CMV)-specific T-cells for the treatment of CMV infection not responding to antiviral chemotherapy. Blood *99*, 3916–3922.

Elek, S.D., and Stern, H. (1974). Development of a vaccine against mental retardation caused by cytomegalovirus infection in utero. Lancet *1*, 1–5.

Elkington, R., Walker, S., Crough, T., Menzies, M., Tellam, J., Bharadwaj, M., and Khanna, R. (2003). Ex vivo profiling of CD8+-T-cell responses to human cytomegalovirus reveals broad and multispecific reactivities in healthy virus carriers. J. Virol. *77*, 5226–5240.

Endresz,V., Kari, L., Berencsi, K., Kari, C., Gyulai, Z., Jeney, C., Pincus, S., Rodeck, U., Meric, C., Plotkin, S.A., *et al.* (1999). Induction of human cytomegalovirus (HCMV)-glycoprotein B (gB)-specific neutralizing antibody and phosphoprotein 65 (pp65)-specific cytotoxic T-lymphocyte responses by naked DNA immunization. Vaccine *17*, 50–58.

Endresz, V., Burian, K., Berencsi, K., Gyulai, Z., Kari, L., Horton, H., Virok, D., Meric, C., Plotkin, S.A., and Gonczol, E. (2001). Optimization of DNA immunization against human cytomegalovirus. Vaccine *19*, 3972–3980.

Falagas, M.E., Snydman, D.R., Ruthazer, R., Griffith, J., Werner, B.G., Freeman, B.G., Rohrer, R., and Boston Center for Liver Transplantation CMVIG Study Group (1997). Cytomegalovirus immune globulin (CMVIG) prophylasis is associated with increased survival after orthotopic liver transplantation. Clin. Transplant. *11*, 432–437.

Feuchtinger, T., Opherk, K., Bethge, W.A., Topp, M.S., Schuster, F.R., Weissinger, E.M., Mohty, M., Or, R., Maschan, M., Schumm, M., *et al.* (2010). Adoptive transfer of pp65-specific T-cells for the treatment of chemorefractory cytomegalovirus disease or reactivation after haploidentical and matched unrelated stem cell transplantation. Blood, *116*, 4360–4367.

Fleisher, G.R., Starr, S.E., Friedman, H.M., Plotkin, S.A. (1982). Vaccination of pediatric nurses with live attenuated cytomegalovirus. Am. J. Dis. Child. *136*, 294–296.

Fouts, A.E., Chan, P., Stephan, J.P., Vandlen, R., and Feierbach, B. (2012). Antibodies against the gH/gL/UL128/UL130/UL131 complex comprise the majority of the anti-cytomegalovirus (ant-CMV) neutralizing antibody response in CMV hyperimmune globulin. J. Virol. *86*, 7444–7447.

Fowler, K.B., and Boppana, S.B. (2006). Congenital cytomegalovirus (CMV) infection and hearing deficit. J. Clin. Virol. 1999, *35*, 226–231.

Fowler, K.B., Stagno, S., Pass, R.F., Britt, W.J., Boll, T.J., and Alford, C.A. (1992). The outcome of congenital cytomegalovirus infection in relation to maternal antibody status. N. Engl. J. Med. *326*, 663–667.

Fowler, K.B., Stagno, S., and Pass, R.F. (2003). Maternal immunity and prevention of congenital cytomegalovirus infection. JAMA *289*, 1008–1011.

Frey, S.E., Harrison, C., Pass, R.F., Yang, E., Boken, D., Sekulovich, R.E., Percell, S., Izu, A.E., Hirabayashi, S., Burke, R.L., *et al.* (1999). Effects of antigen dose and immunization regimens on antibody responses to a cytomegalovirus glycoprotein B subunit vaccine. J. Infect. Dis. *180*, 1700–1703.

Gallez-Hawkins, G., Li, X., Franck, A.E., Thao, L., Lacey, S.F., Diamond, D.J., and Zaia, J.A. (2004). DNA and low titer, helper-free, recombinant AAV prime-boost vaccination for cytomegalovirus induces an immune response to CMV-pp65 and CMV-IE1 in transgenic HLA A*0201 mice. Vaccine *23*, 819–826.

Geall, A.J., Verma, A., Otten, G.R., Shaw, C.A., Hekele, A., Banerjee, K., Cu, Y., Beard, C.W., Brito, L.A., Krucker, T., *et al.* (2012). Nonviral delivery of self-amplifying RNA vaccines. Proc. Natl. Acad. Sci. U.S.A. *109*, 14604–14609.

Genini, E., Percivalle, E., Sarasini, A., Revello, M.G., Baldanti, F., and Gerna, G. (2011). Serum antibody response to the gH/gL/pUL128–131 five-protein complex of human cytomegalovirus (HCMV) in primary and reactivated HCMV infections. J. Clin. Virol. *52*, 113–118.

Gergely, K., Plachter, B., Reddehase, M.J., Holtappels, R., and Lemmermann, N.A.W. (2011). Immunization with recombinant dense bodies protects mice from cytomegalovirus infection. Abstr 17.08; 13. International CMV/Betaherpesvirus Workshop, Nuremberg, Germany, May 14–17, 2011.

Gerna, G., Percivalle, E., Lilleri, D., Lozza, L., Fornara, C., Hahn, G., Baldanti, F., and Revello, M.G. (2005). Dendritic-cell infection by human cytomegalovirus is restricted to strains carrying functional UL131–128 genes and mediates efficient viral antigen presentation to CD8+ T-cells. J. Gen. Virol. *86*, 275–284.

Gerna, G., Sarasini, A., Patrone, M., Percivalle, E., Fiorina, L., Campanini, G., Gallina, A., Baldanti, F., and Revello, M.G. (2008). Human cytomegalovirus serum neutralizing antibodies block virus infection of endothelial/epithelial cells, but not fibroblasts, early during primary infection. J. Gen. Virol. *89*, 853–865.

Glazer, J.P., Friedman, H.M., Grossman, R.A., Starr, S.E., Barker, C.F., Perloff, L.J., Huang, E.S., and Plotkin, S.A. (1979). Live cytomegalovirus vaccination of renal transplant candidates. A preliminary trial. Ann. Intern. Med. *91*, 676–683.

Gonczol, E., Ianacone, J., Furlini, G., Ho, W., and Plotkin, S.A. (1989). Humoral immune response to cytomegalovirus Towne vaccine strain and to Toledo low-passage strain. J. Infect. Dis. *159*, 851–859.

Gonczol, E., Ianacone, J., Ho, W.Z., Starr, S., Meignier, B., and Plotkin, S. (1990). Isolated gA/gB glycoprotein complex of human cytomegalovirus envelope induces humoral and cellular immune-responses in human volunteers. Vaccine *8*, 130–136.

Gonczol, E., Berensci, K., Pincus, S., Endresz, V., Meric, C., Paoletti, E., and Plotkin, S.A. (1995). Preclinical evaluation of an ALVAC (canarypox)--human cytomegalovirus glycoprotein B vaccine candidate. Vaccine *13*, 1080–1085.

Gretch, D.R., Kari, B., Rasmussen, L., Gehrz, R.C., and Stinski, M.F. (1988). Identification and characterization of three distinct families of glycoprotein complexes in the envelopes of human cytomegalovirus. J. Virol. *62*, 875–881.

Griffiths, P.D., Stanton, A., McCarrell, E., Smith, C., Osman, M., Harber, M., Davenport, A., Jones, G., Wheeler, D.C., O'Beirne, J., *et al.* (2011). Cytomegalovirus glycoprotein-B vaccine with MF59 adjuvant in transplant recipients: a phase 2 randomised placebo-controlled trial. Lancet *377*, 1256–1263.

Grigoleit, G.U., Kapp, M., Hebart, H., Fick, K., Beck, R., Jahn, G., and Einsele, H. (2007). Dendritic cell vaccination in allogeneic stem cell recipients: induction of human cytomegalovirus (HCMV)-specific cytotoxic T-lymphocyte responses even in patients receiving a transplant from an HCMV-seronegative donor. J. Infect. Dis. *196*, 699–704.

Gupta, R., Gonczol, E., Manning, M.L., Starr, S., Johnson, B., Murphy, G.F., and Plotkin, S.A. (1993). Delayed type hypersensitivity to human cytomegalovirus. J. Med. Virol. *39*, 109–117.

Gyulai, Z., Endresz, V., Burian, K., Pincus, S., Toldy, J., Cox,W.I., Meric, C., Plotkin, S., Gonczol, E., and Berencsi, K. (2000). Cytotoxic T-lymphocyte (CTL) responses to human cytomegalovirus pp65, IE1-Exon4, gB, pp150, and pp28 in healthy individuals: reevaluation of prevalence of IE1-specific CTLs. J. Infect. Dis. *181*, 1537–1546.

Hahn, G., Revello, M.G., Patrone, M., Percivalle, E., Campanini, G., Sarasini, A., Wagner, M., Gallina, A., Milanesi, G., Koszinowski, U., *et al.* (2004). Human cytomegalovirus UL131–128 genes are indispensable for virus growth in endothelial cells and virus transfer to leukocytes. J. Virol. *78*, 10023–10033.

Hansen, S.G., Vieville, C., Whizin, N., Coyne-Johnson, L., Siess, D.C., Drummond, D.D., Legasse, A.W., Axthelm, M.K., Oswald, K., Trubey, C.M., *et al.* (2009). Effector memory T-cell responses are associated with protection of rhesus monkeys from mucosal simian immunodeficiency virus challenge. Nat. Med. *15*, 293–299.

Hansen, S.G., Powers, C.J., Richards, R., Ventura, A.B., Ford, J.C., Siess, D., Axthelm, M.K., Nelson, J.A., Jarvis, M.A., Picker, L.J., *et al.* (2010). Evasion of CD8+ T-cells is critical for superinfection by cytomegalovirus. Science *328*, 102–106.

Hansen, S.G., Ford, J.C., Lewis, M.S., Ventura, A.B., Hughes, C.M., Coyne-Johnson, L., Whizin, N., Oswald, K., Shoemaker, R., Swanson, T., *et al.* (2011). Profound early control of highly pathogenic SIV by an effector memory T-cell vaccine. Nature *473*, 523–527.

Heineman, T.C., Schleiss, M., Bernstein,D.I., Spaete, R.R., Yan, L., Duke, G., Prichard, M., Wang, Z., Yan, Q., Sharp, M.A., *et al.* (2006). A phase 1 study of 4 live, recombinant human cytomegalovirus Towne/Toledo chimeric vaccines. J. Infect. Dis. *193*, 1350–1360.

Herr, W., and Plachter, B. (2009). Cytomegalovirus and varicella-zoster virus vaccines in hematopoietic stem cell transplantation. Expert. Rev. Vaccines *8*, 999–1021.

Holtappels, R., Thomas, D., Podlech, J., Geginat, G., Steffens, H.P., and Reddehase, M.J. (2000). The putative natural killer decoy early gene m04 (gp34) of murine cytomegalovirus encodes an antigenic peptide recognized by protective antiviral CD8 T-cells. J. Virol. *74*, 1871–1884.

Holtappels, R., Podlech, J., Grzimek, N.K., Thomas, D., Pahl-Seibert, M.F., and Reddehase, M.J. (2001). Experimental preemptive immunotherapy of murine cytomegalovirus disease with CD8 T-cell lines specific for ppM83 and pM84, the two homologs of human cytomegalovirus tegument protein ppUL83 (pp65). J. Virol. *75*, 6584–6600.

Holtappels, R., Simon, C.O., Munks, M.W., Thomas, D., Deegen, P., Kühnapfel, B., Däubner, T., Emde, S.F., Podlech, J., Grzimek, N.K., *et al.* (2008). Subdominant CD8 T-cell epitopes account for protection against cytomegalovirus independent of immunodomination. J. Virol. *82*, 5781–5796.

Horn, B., Bao, L., Dunham, K., Stamer, M., Adler, S., Cowan, M., and Lucas, K. (2009). Infusion of cytomegalovirus specific cytotoxic T-lymphocytes from a sero-negative donor can facilitate resolution of infection and immune reconstitution. Pediatr. Infect. Dis. J. *28*, 65–67.

Hyde, T.B., Schmid, D.S., and Cannon, M.J. (2010). Cytomegalovirus seroconversion rates and risk factors: implications for congenital CMV. Rev. Med. Virol. *20*, 311–326.

Iwasenko, J.M., Howard, J., Arbuckle, S., Graf, N., Hall, B., Craig, M.E., and Rawlinson, W.D. (2011). Human cytomegalovirus infection is detected frequently in stillbirths and is associated with fetal thrombotic vasculopathy. J. Infect. Dis. *203*, 1526–1533.

Jacobson, M.A., Sinclair, E., Bredt, B., Agrillo, L., Black, D., Epling, C.L., Carvidi, A., Ho, T., Bains, R., and Adler, S.P. (2006a). Antigen-specific T-cell responses induced by Towne cytomegalovirus (CMV) vaccine in CMV-seronegative vaccine recipients. J. Clin. Virol. 35, 332–337.

Jacobson, M.A., Sinclair, E., Bredt, B., Agrillo, L., Black, D., Epling, C.L., Carvidi, A., Ho, T., Bains, R., Girling, V., and Adler, S.P. (2006b). Safety and immunogenicity of Towne cytomegalovirus vaccine with or without adjuvant recombinant interleukin-12. Vaccine *24*, 5311–5319.

Jacobson, M.A., Adler, S.P., Sinclair, E., Black, D., Smith, A., Chu, A., Moss, R.B., and Wloch, M.K. (2009). A CMV DNA vaccine primes for memory immune responses to live-attenuated CMV (Towne strain). Vaccine *27*, 1540–1548.

Kamiya, H., Starr, S.E., Arbeter, A.M., and Plotkin, S.A. (1982). Antibody-dependent cell-mediated cytotoxicity against varicella-zoster virus-infected targets. Infect. Immunol. *38*, 554–557.

Kaye, S., Miles, D., Antoine, P., Burny, W., Ojuola, B., Kaye, P., Rowland-Jones, S., Whittle, H., van der Sande, M., and Marchant, A. (2008). Virological and immunological correlates of mother-to-child transmission of cytomegalovirus in The Gambia. J. Infect. Dis. *197*, 1307–1314.

Kemble, G., Duke, G., Winter, R., and Spaete, R. (1996). Defined large-scale alterations of the human cytomegalovirus genome constructed by cotransfection of overlapping cosmids. J. Virol. *70*, 2044–2048.

Kenneson, A., and Cannon, M.J. (2007). Review and meta-analysis of the epidemiology of congenital cytomegalovirus (CMV) infection. Rev. Med. Virol. *17*, 253–276.

Kern, F., Surel, I.P., Faulhaber, N., Frommel, C., Schneider-Mergener, J., Schonemann, C., Reinke, P., and Volk, H.D. (1999). Target structures of the CD8(+)-T-cell response to human cytomegalovirus: the 72-kilodalton major immediate-early protein revisited. J. Virol. *73*, 8179–8184.

Kharfan-Dabaja, M.A., Boeckh, M., Wilck, M.B., Langston, A.A., Chu, A.H., Wloch, M.K., Guterwill, D.F., Smith, L.R., Rolland, A.P., and Kenney, R.T. (2012). A novel therapeutic cytomegalovirus DNA vaccine in allogeneic haemopoietic stem-cell transplantation: a randomised, double-blind, placebo-controlled, phase 2 trial. Lancet Infect. Dis. *12*, 290–299.

Klenovsek, K., Weisel, F., Schneider, A., Appelt, U., Jonjic, S., Messerle, M., Bradel-Tretheway, B., Winkler, T.H., and Mach, M. (2007). Protection from CMV infection in immunodeficient hosts by adoptive transfer of memory B cells. Blood *110*, 3472–3479.

Klein, M., Schoppel, K., Amvrossiadis, N., and Mach, M. (1999). Strain-specific neutralization of human cytomegalovirus isolates by human sera. J. Virol. *73*, 878–886.

La Rosa, C., Wang, Z., Brewer, J.C., Lacey, S.F., Villacres, M.C., Sharan, R., Krishnan, R., Crooks, M., Markel, S., Maas, R., *et al.* (2002). Preclinical development of an adjuvant-free peptide vaccine with activity against CMV pp65 in HLA transgenic mice. Blood *100*, 3681–3689.

La Rosa, C., Longmate, J., Lacey, S.F., Kaltcheva, T., Sharan, R., Marsano, D., Kwon, P., Drake, J., Williams, B., Denison, S., *et al.* (2012). Clinical evaluation of safety and immunogenicity of PADRE-cytomegalovirus (CMV) and tetanus-CMV fusion peptide vaccines with or without PF03512676 adjuvant. J. Infect. Dis. *205*, 1294–1304.

Lilleri, D., Kabanova, A., Lanzavecchia, A., Gerna, A. (2012). Antibodies against neutralization epitopes of human cytomegalovirus gH/gL/pUL1128–131 complex and virus spreading may correlate with virus control *in vivo*. J. Clin. Immunol. Epub ahead of print.

Lisboa, L.F., Kumar, D., Wilson, L.E., and Humar, A. (2012). Clinical utility of cytomegalovirus cell-mediated immunity in transplant recipients with cytomegalovirus viremia. Transplantation *93*, 195–200.

Longmate, J., York, J., La Rosa, C., Krishnan, R., Zhang, M., Senitzer, D., and Diamond, D.J. (2001). Population coverage by HLA class-I restricted cytotoxic T-lymphocyte epitopes. Immunogenetics *52*, 165–173.

Ludwig, A., and Hengel, H. (2009). Epidemiological impact and disease burden of congenital cytomegalovirus infection in Europe. Euro. Surveill. *14*, 26–32.

Macagno, A., Bernasconi, N.L., Vanzetta, F., Dander, E., Sarasini, A., Revello, M.G., Gerna, G., Sallusto, F., and Lanzavecchia, A. (2010). Isolation of human monoclonal antibodies that potently neutralize human cytomegalovirus infection by targeting different epitopes on the gH/gL/UL128–131A complex. J. Virol. *84*, 1005–1013.

Maidji, E., Nigro, G., Tabata, T., McDonagh, S., Nozawa, N., Shiboski, S., Muci, S., Anceschi, M.M., Aziz, N., Adler, S.P., *et al.* (2010). Antibody treatment promotes compensation for human cytomegalovirus-induced pathogenesis and a hypoxia-like condition in placentas with congenital infection. Am. J. Pathol. *177*, 1298–1310.

Marshall, B.C., and Adler, S.P. (2009). The frequency of pregnancy and exposure to cytomegalovirus infections among women with a young child in day care. Am. J. Obstet. Gynecol. *200*, 163.e1–5.

Marshall, G.S., Ricciardi, R.P., Rando, R.F., Puck, J., Ge, R.W., Plotkin, S.A., and Gonczol, E. (1990). An adenovirus recombinant that expresses the human cytomegalovirus major envelope glycoprotein and induces neutralizing antibodies. J. Infect. Dis. *162*, 1177–1181.

Mersseman, V., Besold, K., Reddehase, M.J., Wolfrum, U., Strand, D., Plachter, B., and Reyda, S. (2008). Exogenous introduction of an immunodominant peptide from the non-structural IE1 protein ofhuman cytomegalovirus into the MHC class I presentation pathway by recombinant dense bodies. J. Gen. Virol. *89*, 369–379.

Micklethwaite, K., Hansen, A., Foster, A., Snape, E., Antonenas, V., Sartor, M., Shaw, P., Bradstock,K., and Gottlieb, D. (2007). Ex vivo expansion and prophylactic infusion of CMV-pp65 peptide-specific cytotoxic T-lymphocytes following allogeneic hematopoietic stem cell transplantation. Biol. Blood Marrow Transplant. *13*, 707–714.

Mitchell, D.K., Holmes, S.J., Burke, R.L., Duliege, A.M., and Adler, S.P. (2002). Immunogenicity of a recombinant human cytomegalovirus gB vaccine in seronegative toddlers. Pediatr. Infect. Dis. J. *21*, 133–138.

Mohr, C.A., Arapovic, J., Mühlbach, H., Panzer, M., Weyn, A., Dolken, L., Krmpotic, A., Voehringer, D., Ruzsics, Z., Koszinowski, U., *et al.* (2010). A spread-deficient cytomegalovirus for assessment of first-target cells in vaccination. J. Virol. *84*, 7730–7742.

Morello, C.S., Cranmer, L.D., and Spector, D.H. (2000). Suppression of murine cytomegalovirus (MCMV) replication with a DNA vaccine encoding MCMV M84 (a homolog of human cytomegalovirus pp65). J. Virol. *74*, 3696–3708.

Morello, C.S., Ye, M., and Spector, D.H. (2002). Development of a vaccine against murine cytomegalovirus (MCMV), consisting of plasmid DNA and formalin-inactivated MCMV, that provides long-term, complete protection against viral replication. J. Virol. *76*, 4822–4835.

Morello, C.S., Ye, M., Hung, S., Kelly, L.A., and Spector, D.H. (2005). Systematic priming-boosting immunization with a trivalent plasmid DNA and inactivated murine cytomegalovirus (MCMV) vaccine provides long-term protection against viral replication following systemic or mucosal MCMV challenge. J. Virol. *79*, 159–175.

Morello, C.S., Kelley, L.A., Munks, M.W., Hill, A.B., and Spector, D.H. (2007). DNA immunization using highly conserved murine cytomegalovirus genes encoding homologs of human cytomegalovirus UL54 (DNA Polymerase) and UL105 (Helicase) elicits strong CD8 T-cell responses and is protective against systemic challenge. J. Virol. *81*, 7766–7775.

Murphy, E., Yu, D., Grimwood, J., Schmutz, J., Dickson, M., Jarvis, M.A., Hahn, G., Nelson, J.A., Myers, R.M., and Shenk, T.E. (2003). Coding potential of laboratory and clinical strains of human cytomegalovirus. Proc. Natl. Acad. Sci. U.S.A. *100*, 14976–14981.

Mussi-Pinhata, M.M., Yamamoto, A.Y., Moura Brito, R.M., de Lima Isaac, M., de Carvalho e Oliveira P.F., Boppana, S., and Britt,W.J. (2009). Birth prevalence and natural history of congenital cytomegalovirus infection in a highly seroimmune population. Clin. Infect. Dis. *49*, 522–528.

Nigro, G., and Adler, S.P. (2011). Cytomegalovirus infections during pregnancy. Curr. Opin. Obstet. Gynecol. *23*, 123–128.

Nigro, G., Adler, S.P., La Torre, R., and Best, A.M. (2005). Passive immunization during pregnancy for congenital cytomegalovirus infection. N. Engl. J. Med. *353*, 1350–1362.

Ohlin, M., Sundqvist,V.A., Mach, M., Wahren, B., and Borrebaeck,C.A. (1993). Fine specificity of the human immune response to the major neutralization epitopes expressed on cytomegalovirus gp58/116 (gB), as determined with human monoclonal antibodies. J. Virol. *67*, 703–710.

Oxford, K.L., Strelow, L., Yue, Y., Chang, W.L., Schmidt, K.A., Diamond, D.J., and Barry, P.A. (2011). Open reading frames carried on UL/b′ are implicated in shedding and horizontal transmission of rhesus cytomegalovirus in rhesus monkeys. J. Virol. *85*, 5105–5114.

Paine, A., Oelke, M., Tischer, S., Heuft, H.G., Blasczyk, R., and Eiz-Vesper, B. (2010). Soluble recombinant CMVpp65 spanning multiple HLA alleles for reconstitution of antiviral CD4+ and CD8+ T-cell responses after allogeneic stem cell transplantation. J. Immunother. *33*, 60–72.

Pass, R.F. (2009). Development and evidence for efficacy of CMV glycoprotein B vaccine with MF59 adjuvant. J. Clin. Virol. *46*, 73–76.

Pass, R.F., Duliege, A., Boppana, S., Sekulovich, R., Percell, S., Britt, W., and Burke, R.L. (1999). A subunit cytomegalovirus vaccine based on recombinant envelope glycoprotein b and a new adjuvant. J. Infect. Dis. *180*, 970–975.

Pass, R.F., Zhang, C., Evans, A., Simpson, T., Andrews, W., Huang, M.L., Corey, L., Hill, J., Davis, E., Flanigan, C., and Cloud, G. (2009). Vaccine prevention of maternal cytomegalovirus infection. N. Engl. J. Med. *360*, 1191–1199.

Peggs, K.S. (2009). Adoptive T-cell immunotherapy for cytomegalovirus. Expert. Opin. Biol. Ther. *9*, 725–736.

Peggs, K.S., Verfuerth, S., Pizzey, A., Khan, N., Guiver, M., Moss, P.A., and Mackinnon, S. (2003). Adoptive cellular therapy for early cytomegalovirus infection after allogeneic stem-cell transplantation with virus-specific T-cell lines. Lancet *362*, 1375–1377.

Peggs, K.S., Thomson, K., Samuel, E., Dyer, G., Armoogum, J., Chakraverty, R., Pang, K., Mackinnon, S., and Lowdell, M.W. (2011). Directly selected cytomegalovirus-reactive donor T-cells confer rapid and safe systemic reconstitution of virus-specific immunity following stem cell transplantation. Clin. Infect. Dis. *52*, 49–57.

Pepperl, S., Münster, J., Mach, M., Harris, J.R., and Plachter, B. (2000). Dense bodies of human cytomegalovirus induce both humoral and cellular immune responses in the absence of viral gene expression. J. Virol. *74*, 6132–6146.

Pepperl-Klindworth, S., Frankenberg, N., and Plachter, B. (2002). Development of novel vaccine strategies against human cytomegalovirus infection based on subviral particles. J. Clin. Virol. *25 Suppl 2*, S75-S85.

Pereira, L., Hoffman, M., Tatsuno, M., and Dondero, D. (1984). Polymorphism of human cytomegalovirus glycoproteins characterized by monoclonal antibodies. Virology *139*, 73–86.

Pereyra, F., and Rubin, R.H. (2004). Prevention and treatment of cytomegalovirus infection in solid organ transplant recipients. Curr. Opin. Infect. Dis. *17*, 357–361.

Plotkin, S.A. (1999). Vaccination against cytomegalovirus, the changeling demon. Pediatr. Infect. Dis. J. *18*, 313–325.

Plotkin, S.A. (2001). Vaccination against cytomegalovirus. Arch. Virol. *17 Suppl. S1*, 121–134.

Plotkin, S.A. (2002). Is there a formula for an effective CMV vaccine? J. Clin. Virol. *25 Suppl. 2*, S13-S21.

Plotkin, S.A. (2008). Cytomegalovirus vaccines. In Vaccines, Plotkin, S., Orenstein, W.A., and Offit, P.A., eds. (Saunders, W.B., Philadelphia, PA), pp. 1147–1154.

Plotkin, S.A. (2010). Correlates of protection induced by vaccination. Clin. Vaccine. Immunol. *17*, 1055–1065.

Plotkin, S.A., and Huang, E.S. (1985). Cytomegalovirus vaccine virus (Towne strain) does not induce latency. J. Infect. Dis. *152*, 395–397.

Plotkin, S.A., and Plotkin, S.L. (2011). The development of vaccines: how the past led to the future. Nat. Rev. Microbiol. *9*, 889–893.

Plotkin, S.A., Furukawa, T., Zygraich, N., and Huygelen, C. (1975). Candidate cytomegalovirus strain for human vaccination. Infect. Immun. *12*, 521–527.

Plotkin, S.A., Farquhar, J., and Horberger, E. (1976). Clinical trials of immunization with the Towne 125 strain of human cytomegalovirus. J. Infect. Dis. 2005 *134*, 470–475.

Plotkin, S.A., Friedman, H.M., Starr, S.E., Arbeter, A.M., Furukawa, T., and Fleisher, G.R. (1981). Prevention and treatment of cytomegalovirus infection. In The Human Herpesviruses: an Interdisciplinary Perspective. Nahmias, A.J., Dowdle, W.R., and Schinazi, R.F. eds. (Elsevier, New York), pp. 403–413.

Plotkin, S.A., Friedman, H., Starr, S.E., Smiley, M.L., Grossman, R., and Barker, C. (1983). The prevention of cytomegalovirus disease. In Human Immunity to Viruses, Ennis, F., ed. (Academic Press, New York), pp. 275–276.

Plotkin, S.A., Smiley, M.L., Friedman, H.M., Starr, S.E., Fleisher, G.R., Wlodaver, C., Dafoe, C., Friedman, A.D., Grossman, R., and Barker, C. (1984a). Prevention of cytomegalovirus disease by Towne strain live attenuated vaccine. In CMV: Pathogenesis and Prevention of Human Infection. Birth Defects: Original Article Series *20*, pp. 271–287.

Plotkin, S.A., Smiley, M.L., Friedman, H.M., Starr, S.E., Fleisher, G.R., Wlodaver, C., Dafoe, D.C., Friedman,

A.D., Grossman, R.A., and Barker, C.F. (1984b). Towne-vaccine-induced prevention of cytomegalovirus disease after renal transplants. Lancet *1*, 528–530.

Plotkin, S.A., Weibel, R.E., Alpert, G., Starr, S.E., Friedman, H.M., Preblud, S.R., and Hoxie, J. (1985). Resistance of seropositive volunteers to subcutaneous challenge with low-passage human cytomegalovirus. J. Infect. Dis. *151*, 737–739.

Plotkin, S.A., Starr, S.E., Friedman, H.M., Gonczol, E., and Weibel, R.E. (1989). Protective effects of Towne cytomegalovirus vaccine against low-passage cytomegalovirus administered as a challenge. J. Infect. Dis. *159*, 860–865.

Plotkin, S.A., Starr, S.E., Friedman, H.M., Gonczol, E., and Brayman, K. (1990). Vaccines for the prevention of human cytomegalovirus infection. Rev. Infect. Dis. *990 Suppl. 7*, S827–838.

Plotkin, S.A., Starr, S.E., Friedman, H.M., Brayman, K., Harris, S., Jackson, S., Tustin, N.B., Grossman, R., Dafoe, D., and Barker, C. (1991). Effect of Towne live virus vaccine on cytomegalovirus disease after renal transplant. A controlled trial. Ann. Intern. Med. *114*, 525–531.

Prins, R.M., Cloughesy, T.F., and Liau, L.M. (2008). Cytomegalovirus immunity after vaccination with autologous glioblastoma lysate. N. Engl. J. Med. *359*, 539–541.

Prod'homme, V., Retiere, C., Imbert-Marcille, B.M., Bonneville, M., and Hallet, M.M. (2003). Modulation of HLA-A*0201-restricted T-cell responses by natural polymorphism in the IE1(315–324) epitope of human cytomegalovirus. J. Immunol. *170*, 2030–2036.

Quinnan, G.V.J., Delery, M., Rook, A.H., Frederick, W.R., Epstein, J.S., Manischewitz, J.F., Jackson, L., Ramsey, K.M., Mittal, K., and Plotkin, S.A. (1984). Comparative virulence and immunogenicity of the Towne strain and a nonattenuated strain of cytomegalovirus. Ann. Intern. Med. *101*, 478–483.

Raanani, P., Gafter-Gvili, A., Paul, M., Ben-Bassat, I., Leibovici, L., and Shpilberg, O. (2009). Immunoglobulin prophylaxis in hematopoietic stem cell transplantation: systematic review and meta-analysis. J. Clin. Oncol. *27*, 770–781.

Rasmussen, L., Nelson, M., Neff, M., and Merigan, T.C., Jr. (1988). Characterization of two different human cytomegalovirus glycoproteins which are targets for virus neutralizing antibody. Virology *163*, 308–318.

Reap, E.A., Morris, J., Dryga, S.A., Maughan, M., Talarico, T., Esch, R.E., Negri, S., Burnett, B., Graham, A., Olmsted, R.A., *et al.* (2007). Development and preclinical evaluation of an alphavirus replicon particle vaccine for cytomegalovirus. Vaccine *25*, 7441–7449.

Reddehase, M.J. (2000). The immunogenicity of human and murine cytomegaloviruses. Curr. Opin. Immunol. *12*, 738.

Reddehase, M.J. (2002). Antigens and immunoevasins: opponents in cytomegalovirus immune surveillance. Nat. Rev. Immunol. *2*, 831–844.

Reddehase, M.J., Weiland, F., Münch, K., Jonjic, S., Luske, A., and Koszinowski, U.H. (1985). Interstitial murine cytomegalovirus pneumonia after irradiation: characterization of cells that limit viral replication during established infection of the lungs. J. Virol. *55*, 264–273.

Reiser, M., Wieland, A., Plachter, B., Mertens, T., Greiner, J., and Schirmbeck, R. (2011). The immunodominant CD8 T-cell response to the human cytomegalovirus tegument phosphoprotein $pp65_{495-503}$ epitope critically depends on CD4 T-cell help in vaccinated HLA-A*0201 transgenic mice. J. Immunol. *187*, 2172–2180.

Reusser, P., Riddell, S.R., Meyers, J.D., and Greenberg, P.D. (1991). Cytotoxic T-lymphocyte response to cytomegalovirus after human allogeneic bone marrow transplantation: pattern of recovery and correlation with cytomegalovirus infection and disease. Blood *78*, 1373–1380.

Riddell, S.R., Watanabe, K.S., Goodrich, J.M., Li, C.R., Agha, M.E., and Greenberg, P.D. (1992). Restoration of viral immunity in immunodeficient humans by the adoptive transfer of T-cell clones. Science *257*, 238–241.

Rist, M., Cooper, L., Elkington, R., Walker, S., Fazou, C., Tellam, J., Crough, T., and Khanna, R. (2005). Ex vivo expansion of human cytomegalovirus-specific cytotoxic T-cells by recombinant polyepitope: implications for HCMV immunotherapy. Eur. J. Immunol. *35*, 996–1007.

Rosenthal, L.S., Fowler, K.B., Boppana, S.B., Britt, W.J., Pass, R.F., Schmid, S.D., Stagno, S., and Cannon, M.J. (2009). Cytomegalovirus shedding and delayed sensorineural hearing loss: results from longitudinal follow-up of children with congenital infection. Pediatr. Infect. Dis. J. *28*, 515–520.

Ross, S.A., Fowler, K.B., Ashrith, G., Stagno, S., Britt, W.J., Pass, R.F., and Boppana, S.B. (2006). Hearing loss in children with congenital cytomegalovirus infection born to mothers with preexisting immunity. J. Pediatr. *148*, 332–336.

Ross, S.A., Arora, N., Novak, Z., Fowler, K.B., Britt, W.J., and Boppana, S.B. (2010). Cytomegalovirus reinfections in healthy seroimmune women. J. Infect. Dis. *201*, 386–389.

Ryckman, B.J., Rainish, B.L., Chase, M.C., Borton, J.A., Nelson, J.A., Jarvis, M.A., and Johnson, D.C. (2008a). Characterization of the human cytomegalovirus gH/gL/UL128–131 complex that mediates entry into epithelial and endothelial cells. J. Virol. *82*, 60–70.

Ryckman, B.J., Chase, M.C., and Johnson, D.C. (2008b). HCMV gH/gL/UL128–131 interferes with virus entry into epithelial cells: evidence for cell type-specific receptors. Proc. Natl. Acad. Sci. USA. *105*, 14118–14123.

Ryckman, B.J., Chase, M.C., and Johnson, D.C. (2010). Human cytomegalovirus TR strain glycoprotein O acts as a chaperone promoting gH/gL incorporation into virions but is not present in virions. J. Virol. *84*, 2597–2609.

Sabbaj, S., Pass, R.F., Goepfert, P.A., and Pichon, S. (2011). Glycoprotein B vaccine is capable of boosting both antibody and CD4 T-cell responses to cytomegalovirus in chronically infected women. J. Infect. Dis. *203*, 1534–1541.

Saccoccio, F.M., Gallagher, M.K., Adler, S.P., and McVoy, M.A. (2011a). Neutralizing activity of saliva against cytomegalovirus. Clin. Vaccine Immunol. *18*, 1536–1542.

Saccoccio, F.M., Sauer, A.L., Cui, X., Armstrong, A.E., Habib, E., Johnson, D.C., Ryckman, B.J., Klingelhutz, A.J., Adler, S.P., and McVoy, M.A. (2011b). Peptides from cytomegalovirus UL130 and UL131 proteins induce high titer antibodies that block viral entry into mucosal epithelial cells. Vaccine *29*, 2705–2711.

Schleiss, M.R. (2008a). Cytomegalovirus vaccine development. Curr. Top. Microbiol. Immunol. *325*, 361–382.

Schleiss, M.R. (2008b). Comparison of vaccine strategies against congenital CMV infection in the guinea pig model. J. Clin. Virol. *41*, 224–230.

Schleiss, M.R. (2010). Can we build it better? Using BAC genetics to engineer more effective cytomegalovirus vaccines. J. Clin. Invest. *120*, 4192–4197.

Schleiss, M.R., Bourne, N., and Bernstein, D.I. (2003). Preconception vaccination with a glycoprotein B (gB) DNA vaccine protects against cytomegalovirus (CMV) transmission in the guinea pig model of congenital CMV infection. J. Infect. Dis. *188*, 1868–1874.

Schmitt, A., Tonn, T., Busch, D.H., Grigoleit, G.U., Einsele, H., Odendahl, M., Germeroth, L., Ringhoffer, M., Ringhoffer, S., Wiesneth, M., *et al.* (2011). Adoptive transfer and selective reconstitution of streptamer-selected cytomegalovirus-specific CD8+ T-cells leads to virus clearance in patients after allogeneic peripheral blood stem cell transplantation. Transfusion *51*, 591–599.

Shimamura, M., Mach, M., and Britt, W.J. (2006). Human cytomegalovirus infection elicits a glycoprotein M (gM)/gN-specific virus-neutralizing antibody response. J. Virol. *80*, 4591–4600.

Simanek, A.M., Dowd, J.B., Pawelec, G., Melzer, D., Dutta, A., and Aiello, A.E. (2011). Seropositivity to cytomegalovirus, inflammation, all-cause and cardiovascular disease-related mortality in the USA. PLoS One *6*, e16103.

Simon, C.O., Holtappels, R., Tervo, H.M., Böhm, V., Däubner, T., Oehrlein-Karpi, S.A., Kühnapfel, B., Renzaho, A., Strand, D., Podlech, J., *et al.* (2006). CD8 T-cells control cytomegalovirus latency by epitope-specific sensing of transcriptional reactivation. J. Virol. *80*, 10436–10456.

Simpson, J.A., Chow, J.C., Baker, J., Avdalovic, N., Yuan, S., Au, D., Co, M.S., Vasquez, M., Britt, W.J., and Coelingh, K.L. (1993). Neutralizing monoclonal antibodies that distinguish three antigenic sites on human cytomegalovirus glycoprotein H have conformationally distinct binding sites. J. Virol. *67*, 489–496.

Sinzger, C., Digel, M., and Jahn, G. (2008). Cytomegalovirus cell tropism. Curr. Top. Microbiol. Immunol. *325*, 63–83.

Slavuljica, I., Busche, A., Babic, M., Mitrovic, M., Gasparovic, I., Cekinovic, D., Markova, C.E., Pernjak, P.E., Cikovic, A., Lisnic, V.J., *et al.* (2010). Recombinant mouse cytomegalovirus expressing a ligand for the NKG2D receptor is attenuated and has improved vaccine properties. J. Clin. Invest. *120*, 4532–4545.

Snydman, D.R., Werner, B.G., Meissner, H.C., Cheeseman, S.H., Schwab, J., Bednarek, F., Kennedy, J.L., Jr., Herschel, M., Magno, A., Levin, M.J., *et al* (1995). Use of cytomegalovirus immunoglobulin in multiply transfused premature neonates. Pediatr. Infect. Dis. J. *14*, 34–40.

Spaete, R.R. (1991). A recombinant subunit vaccine approach to HCMV vaccine development. Transplant. Proc. *23*, 90–96.

Stagno, S., Pass, R.F., Dworsky, M.E., Henderson, R.E., Moore, E.G., Walton, P.D., and Alford, C.A. (1982). Congenital cytomegalovirus infection: The relative importance of primary and recurrent maternal infection. N. Engl. J. Med. *306*, 945–949.

Staras, S.A., Flanders, W.D., Dollard, S.C., Pass, R.F., McGowan, J.E., Jr., and Cannon, M.J. (2008a). Cytomegalovirus seroprevalence and childhood sources of infection: A population-based study among pre-adolescents in the USA. J. Clin. Virol. *43*, 266–271.

Staras, S.A., Flanders, W.D., Dollard, S.C., Pass, R.F., McGowan, J.E., Jr., and Cannon, M.J. (2008b). Influence of sexual activity on cytomegalovirus seroprevalence in the USA, 1988–1994. Sex. Transm. Dis. *35*, 472–479.

Starr, S.E., Glazer, J.P., Friedman, H.M., Farquhar, J.D., and Plotkin, S.A. (1981). Specific cellular and humoral immunity after immunization with live Towne strain cytomegalovirus vaccine. J. Infect. Dis. 2005, *143*, 585–589.

Straschewski, S., Patrone, M., Walther, P., Gallina, A., Mertens, T., and Frascaroli, G. (2011). Protein pUL128 of human cytomegalovirus is necessary for monocyte infection and blocking of migration. J. Virol. *85*, 5150–5158.

Sung, H., and Schleiss, M.R. (2010). Update on the current status of cytomegalovirus vaccines. Expert. Rev. Vaccines. *9*, 1303–1314.

Sylwester, A.W., Mitchell, B.L., Edgar, J.B., Taormina, C., Pelte, C., Ruchti, F., Sleath, P.R., Grabstein, K.H., Hosken, N.A., Kern, F., *et al.* (2005). Broadly targeted human cytomegalovirus-specific CD4+ and CD8+ T-cells dominate the memory compartments of exposed subjects. J. Exp. Med. *202*, 673–685.

Topilko, A., Michelson, S., (1994). Hyperimmediate entry of human cytomegalovirus virions and dense bodies into human fibroblasts. Res.Virol. *145*, 75–82.

Thomas, S., and Herr, W. (2011). Natural and adoptive T-cell immunity against herpes family viruses after allogeneic hematopoietic stem cell transplantation. Immunotherapy *3*, 771–788.

Urban, M., Britt, W., and Mach, M. (1992). The dominant linear neutralizing antibody-binding site of glycoprotein gp86 of human cytomegalovirus is strain specific. J. Virol. *66*, 1303–1311.

Urban, M., Klein, M., Britt, W.J., Hassfurther, E., and Mach, M. (1996). Glycoprotein H of human cytomegalovirus is a major antigen for the neutralizing humoral immune response. J. Gen. Virol. *77*, 1537–1547.

Valantine, H.A., Luikart, H., Doyle, R., Theodore, J., Hunt, S., Oyer, P., Robbins, R., Berry, G., and Reitz, B. (2001). Impact of cytomegalovirus hyperimmune globulin on outcome after cardiothoracic transplantation: a comparative study of combined prophylaxis with CMV hyperimmune globulin plus ganciclovir versus ganciclovir alone. Transplantation *72*, 1647–1652.

Varnum, S.M., Streblow, D.N., Monroe, M.E., Smith, P., Auberry, K.J., Pasa-Tolic, L., Wang, D., Camp, D.G., Rodland, K., Wiley, S., *et al.* (2004). Identification of proteins in human cytomegalovirus (HCMV) particles: the HCMV proteome. J. Virol. *78*, 10960–10966.

Vyse, A.J., Hesketh, L.M., and Pebody, R.G. (2009). The burden of infection with cytomegalovirus in England and Wales: how many women are infected in pregnancy? Epidemiol. Infect. *137*, 526–533.

Walter, E.A., Greenberg, P.D., Gilbert, M.J., Finch, R.J., Watanabe, K.S., Thomas, E.D., and Riddell, S.R. (1995). Reconstitution of cellular immunity against cytomegalovirus in recipients of allogeneic bone marrow by transfer of T-cell clones from the donor. N. Engl. J. Med. *333*, 1038–1044.

Wang, C., Zhang, X., Bialek, S., and Cannon, M.J. (2011). Attribution of congenital cytomegalovirus infection to primary versus non-primary maternal infection. Clin. Infect. Dis. *52*, e11-e13.

Wang, D., and Shenk, T. (2005). Human cytomegalovirus virion protein complex required for epithelial and endothelial cell tropism. Proc. Natl. Acad. Sci. USA. *102*, 18153–18158.

Wang, J.B., Adler, S.P., Hempfling, S., Burke, L., Duliege, A.M., Starr, S.E., and Plotkin, S.A. (1996). Mucosal antibodies to human cytomegalovirus glycoprotein B occur following both natural infection and immunization with human cytomegalovirus vaccines. J. Infect. Dis. *174*, 387–392.

Wang, Z., La Rosa, C., Lacey, S.F., Maas, R., Mekhoubad, S., Britt, W.J., and Diamond, D.J. (2006). Attenuated poxvirus expressing three immunodominant CMV antigens as a vaccine strategy for CMV infection. J. Clin. Virol. *35*, 324–331.

Wang, Z., La Rosa, C., Li, Z., Ly, H., Krishnan, A., Martinez, J., Britt, W.J., and Diamond, D.J. (2007). Vaccine properties of a novel marker gene-free recombinant modified vaccinia Ankara expressing immunodominant CMV antigens pp65 and IE1. Vaccine *25*, 1132–1141.

Wang, Z., Zhou, W., Srivastava, T., La Rosa C., Mandarino, A., Forman, S.J., Zaia, J.A., Britt, W.J., and Diamond, D.J. (2008). A fusion protein of HCMV IE1 exon4 and IE2 exon5 stimulates potent cellular immunity in an MVA vaccine vector. Virology *377*, 379–390.

Wang, Z., Martinez, J., Zhou, W., La Rosa, C., Srivastava, T., Dasgupta, A., Rawal, R., Li, Z., Britt, W.J., and Diamond, D. (2010). Modified H5 promoter improves stability of insert genes while maintaining immunogenicity during extended passage of genetically engineered MVA vaccines. Vaccine *28*, 1547–1557.

Wills, M.R., Carmichael, A.J., Mynard, K., Jin, X., Weekes, M.P., Plachter, B., and Sissons, J.G. (1996). The human cytotoxic T-lymphocyte (CTL) response to cytomegalovirus is dominated by structural protein pp65: frequency, specificity, and T-cell receptor usage of pp65-specific CTL. J. Virol. *70*, 7569–7579.

Wirtz, N., Schader, S.I., Holtappels, R., Simon, C.O., Lemmermann, N.A., Reddehase, M.J., and Podlech, J. (2008). Polyclonal cytomegalovirus-specific antibodies not only prevent virus dissemination from the portal of entry but also inhibit focal virus spread within target tissues. Med. Microbiol. Immunol. *197*, 151–158.

Wloch, M.K., Smith, L.R., Boutsaboualoy, S., Reyes, L., Han, C., Kehler, J., Smith, H.D., Selk, L., Nakamura, R., Brown, J.M., *et al.* (2008). Safety and immunogenicity of a bivalent cytomegalovirus DNA vaccine in healthy adult subjects. J. Infect. Dis. *197*, 1634–1642.

Yamamoto, A.Y., Mussi-Pinhata, M.M., Isaac, M.D., Amaral, F.R., Carvalheiro, C.G., Aragon, D.C., Manfredi, A.K., Boppana,S.B., and Britt, W.J. (2011). Congenital cytomegalovirus infection as a cause of sensorineural hearing loss in a highly immune population. Pediatr. Infect. Dis. J. *30*, 1043–1046.

Ye, M., Morello, C.S., and Spector, D.H. (2004). Multiple epitopes in the murine cytomegalovirus early gene product M84 are efficiently presented in infected primary macrophages and contribute to strong CD8+-T-lymphocyte responses and protection following DNA immunization. J. Virol. *78*, 11233–11245.

Yeager, A.S., Grumet, F.C., Hafleigh, E.B., Arvin, A.M., Bradley, J.S., and Prober, C.G. (1981). Prevention of transfusion-acquired cytomegalovirus infections in newborn infants. J. Pediatr. *98*, 281–287.

Yue, Y., Wang, Z., Abel, K., Li, J., Strelow, L., Mandarino, A., Eberhardt, M.K., Schmidt, K.A., Diamond, D.J., and Barry, P.A. (2008). Evaluation of recombinant modified vaccinia Ankara virus-based rhesus cytomegalovirus vaccines in rhesus macaques. Med. Microbiol. Immunol. *197*, 117–123.

Zhong, J., and Khanna, R. (2009). Ad-gBCMVpoly: A novel chimeric vaccine strategy for human cytomegalovirus-associated diseases. J. Clin. Virol. *46 Suppl 4*, S68-S72.

Zhong, J., and Khanna, R. (2010). Delineating the role of CD4+ T-cells in the activation of human cytomegalovirus-specific immune responses following immunization with Ad-gBCMVpoly vaccine: implications for vaccination of immunocompromised individuals. J. Gen. Virol. *91*, 2994–3001.

Zhong, J., Rist, M., Cooper, L., Smith, C., and Khanna, R. (2008). Induction of pluripotent protective immunity following immunisation with a chimeric vaccine against human cytomegalovirus. PLoS One 3, e3256.

Vaccine Vectors Using the Unique Biology and Immunology of Cytomegalovirus

II.21

Michael A. Jarvis, Scott G. Hansen, Jay A. Nelson, Louis J. Picker and Klaus Früh

Abstract

Cytomegaloviruses (CMVs) have a unique immunobiological relationship with their mammalian hosts in which low level persistent infection and/or frequent low level reactivation of latent infection results in the development and life-time maintenance of high frequency, effector memory T-cell (T_{EM}) responses that are characterized by the continuous presence of fully differentiated antiviral T-cells in the periphery of the infected host. This unique immunology has recently been exploited to develop vaccine vectors that are distinguished from other viral vectors in their ability to elicit and maintain T-cells with similar T_{EM}-biased characteristics that are directed against heterologous foreign pathogens. Such responses would be predicted to intercept and control/eliminate such pathogens early in infection (without needing an anamnestic response) prior to full implementation of the pathogen's own immune evasion strategies. In non-human primate models of AIDS such T_{EM}-inducing rhesus CMV (RhCMV) vectors were able to completely control highly pathogenic, mucosally administered SIV infection prior to systemic spread, a pattern of protection that has not been observed with conventional T-cell vaccines. An additional feature that makes CMV particularly suited for use as a vaccine vector is its ability to superinfect and then persist in the healthy CMV immune (CMV-seropositive) host. This remarkable characteristic has enabled use of this vaccine platform regardless of the CMV status of the vaccine recipient. Superinfection, defined as the establishment and maintenance of a secondary infection in an infected (and therefore 'immune') host, requires evasion of pre-existing CMV-specific T-cells. A final distinguishing feature of CMV vectors is the surprising finding that viral latency and persistent immune stimulation does not require spread of the vector beyond the initial infected host cell. The long-term immune stimulatory capacity of such 'spread-deficient' CMVs suggests that CMV-based vectors can be rendered safe whilst maintaining T_{EM} responses to the heterologous target antigen. T_{EM}-inducing CMV vectors have the potential to be useful against many infectious diseases as well as cancer in diverse human and animal populations.

Introduction

Persistent infection by CMV is reflected by an ongoing immune response that is biased towards so-called effector memory T-cells (T_{EM}). T_{EM}-biased immune responses are generally distributed towards effector sites of potential pathogen encounter, such as mucosal and epithelial tissues. T_{EM} also have immediate effector function without the need for time-consuming anamnestic cellular expansion and differentiation. This life-long induction of durable T_{EM} memory – a hallmark of CMV infection – is presumably attributable to many aspects of CMV biology, but probably includes the low level persistent/latent CMV gene expression resulting in continuous viral antigen production, the cellular and tissue sites of CMV persistence within the host, and the multitude of immunomodulatory molecules made by the virus. Even healthy individuals harbouring latent CMV frequently display a large percentage (often more than 10%) of CMV-specific T-cells in tissues such as the lung and liver, as well as in the blood. Thus, while CMV infection is going unnoticed in terms of disease, the host's immune system is maintained in a state of constant vigilance with the capacity to respond rapidly without the usual time delay associated with anamnestic responses of classical T-cell memory (i.e. central memory, T_{CM}). Owing to the persistent nature of CMV infection, this heightened state of immunity is maintained for the lifespan of the host. This unique ability of CMV to induce high levels of T_{EM}-biased immunity in a setting of a low, controlled infection – combined with bacterial-based recombinant DNA technology that allows genetic manipulation of CMV (see Chapter I.3) – creates a unique opportunity to exploit CMV as a vaccine vector to induce lifelong T_{EM} responses to target heterologous pathogens and cancer. Such T_{EM} responses may be particularly useful for situations where a constant and highly vigilant immune system is required within particular tissues of the body. For example, (i) to prevent rapidly replicating pathogens from establishing a systemic infection after breaching mucosal barriers; (ii) to efficiently eliminate infected or cancerous cells from epithelial tissue (i.e. prostate, lung or breast); and (iii) to prevent recurring or reactivating infections. The impact of T_{EM}-induction by CMV vectors has recently been illustrated in the development of AIDS vaccines using the rhesus macaque non-human primate (NHP) model (see also Chapter II.22).

A second hallmark of CMV is its ability to superinfect, and thereby establish a secondary infection in persistently CMV-infected hosts. This remarkable characteristic allows CMV vectors to be used repeatedly without losing efficacy, and also to be used regardless of the CMV-specific immune status of the host. Evasion of CD8$^+$ T-cells has recently emerged as a key mechanism by which CMV bypasses the host immune response to enable superinfection. Given the ability of CMV to superinfect regardless of host immunity, CMV vectors can conceivably be designed in such a fashion as to disseminate through their target host population despite the high prevalence of CMV. While this may not be a desirable feature for CMV vectors to be used in humans, replicating and 'disseminating' CMV vectors may be useful for vaccination of livestock or wild animals. This strategy is currently being developed as a immunocontraceptive vaccine to prevent plagues of house mice (*Mus musculus domesticus*) in Australia, and two vaccines designed to target wild animal species involved in zoonotic transmission of Sin Nombre virus and Ebola virus to humans from deer mice (*Peromyscus maniculatis*) and great apes, respectively.

While the use of replicating CMV vectors in humans who are already CMV-positive is expected to carry no more risk than that associated with their ongoing natural CMV infection, the use of CMV vectored vaccines in CMV-negative individuals will carry the risk associated with normal CMV infection. As reviewed elsewhere in this volume (see Chapter II.1), this risk is largely one that is only realized following immunosuppression. CMV can cause overt clinical disease in a variety of immunosuppressed states, such as during late stages of HIV infection (in the absence of HAART), or following iatrogenic immunosuppression during transplantation. CMV can also cause significant morbidity during congenital infection of the neonate. CMV vectors designed for use in seronegative and immunosuppressed humans will have to meet increased safety standards. Recent studies have indicated that 'spread-deficient' CMV vectors retain their immunogenicity, particularly with respect to T-cell immunity, indicating that CMV vectors may be rendered safe enough for use in all human populations, even those who are CMV seronegative or immunosuppressed (Table II.21.1).

A number of recent discoveries, discussed in more detail below, suggest that safe and immunogenic CMV vectors will be ready for clinical trials in the near future. The first likely applications for these vectors are HIV and cancer.

The unique T-cell immunology of CMV

Persistent infection by animal or human CMV (HCMV) is characterized by large CD4$^+$ and CD8$^+$ T-cell responses that are maintained for the life of the host (Kern *et al.*, 1999; Gillespie *et al.*, 2000; Bitmansour *et al.*, 2001). These CMV-specific T-cell responses to CMV are broadly targeted to many different antigens of the virus (Sylwester *et al.*, 2005), so that T-cell assays measuring single epitope or single antigens likely underestimate the overall percentage of CMV-positive T-cells. During maturation of the immune response in mice, CMV-specific CD8$^+$ T-cells assume a steadily larger percentage of the overall T-cell pool. This characteristic of CMV-specific immunity was initially observed in a mouse bone marrow transplantation (BMT) model (see Chapter II.17), wherein large numbers of L-selectin (CD62L)^{Low} CD8$^+$ 'effector memory' T-cells were shown to accumulate against epitopes of MCMV IE1 and m164 proteins in the lungs, a recognized extralymphoid site of latent infection (Holtappels *et al.*, 2000, 2002). This process, which was also shown to be a characteristic of MCMV infection outside of the BMT setting, was subsequently termed 'memory inflation' (Karrer *et al.*, 2003). Although many factors appear to influence which MCMV epitopes are biased towards development of 'inflationary' responses (discussed in Chapter I.22), the corresponding proteins are expressed during latency or at early stages of reactivation, and it is believed that these frequent antigenic pulses play a critical role in driving these responses (Holtappels *et al.*, 2000, 2002; Karrer *et al.*, 2003, 2004; Munks *et al.*, 2006; Simon *et al.*, 2006). The development of CMV-specific T-cell responses in rhesus macaques differs slightly in pattern; in this species both CD4$^+$ and CD8$^+$ CMV-specific T-cells appear at high frequency during primary infection, and then persist indefinitely at these levels (Price *et al.*, 2008; Hansen *et al.*, 2009).

CMV-specific CD4$^+$ and CD8$^+$ T-cell responses are unusual in that they manifest a strong 'effector memory' bias, which is reflected by their phenotype in blood, functional potential and distribution in the

Table II.21.1 Comparison of CMV vectors with other vectors

Viral vector	Persistence	DNA coding capacity	Immune response to target antigen	Vector seroprevalence in population	Sequential use capacity	Pathogenicity
CMV	Life-long	[i]Large (>50 kb)	High T-cells (T_{EM}-biased). Abs unclear	Moderate to high	Unlimited	Low (spread-deficient); similar to natural infection (spreading).
Pox (NYVAC, MVA, ALVAC)[a,b,c]	Transient	Medium (>25 kb)	Moderate to low T-cells (T_{CM}- biased) and low Abs	Moderate and declining	Single	Low (replication-deficient); moderate (replicating)
Adenovirus[a,b,c,d]	Transient	Moderate (<8 kb)	Moderate to high T-cells (T_{CM}- biased) and Abs	High (some serotypes)	Single	Low (replication-deficient); low (??) (replicating)
AAV[b,e,f]	Transient	Small (<5kb)	Moderate to low T-cells and Abs	High	Unknown	Low (?). *In vitro* indication of integration
Lentiviral[b,g,h]	Life-long	Moderate (<8 kb)	Moderate to high T-cells and Abs	Low	Multiple (?)	Oncogenic potential

[a]Bett *et al.* (2010).
[b]Draper and Heeney (2010).
[c]Casimiro *et al.* (2003).
[d]Shiver and Emini (2004).
[e]Johnson *et al.* (2005).
[f]Mehendale *et al.* (2008).
[g]Hu *et al.* (2011).
[h]Lemiale and Korokhov (2009).
[i]Based on DNA size that can be deleted without affecting virus replication.

body (Kern *et al.*, 1999; Gillespie *et al.*, 2000; Sierro *et al.*, 2005; Munks *et al.*, 2006; Snyder *et al.*, 2008). The effector/central memory paradigm was originally proposed based on the expression of CCR7 (CD197), a receptor for chemokines required for extravasation of lymphocytes into secondary lymphoid tissue via high endothelial venules (Sallusto *et al.*, 1999). CCR7-positive central memory T-cells (T_{CM}) are secondary lymphoid tissue-based memory T-cells that have limited immediate effector functions, but expand rapidly during recall responses and differentiate into effector cells. The long-term memory T-cell responses to transient antigen exposure, such as responses to conventional (non-persistent) vaccines, are characterized by a strong T_{CM} bias. In contrast, CCR7-negative T_{EM} are effector site-based (mucosa, parenchymal organs, splenic red pulp), have potent immediate effector function and limited expansion capacity. Long-term memory responses with these characteristics are mostly observed in chronic viral infections, in particular CMV (Klenerman and Hill, 2005; Reddehase *et al.*, 2008). This is illustrated in Fig. II.21.1, which shows the high levels of T_{EM} responses against multiple CMV-vectored antigens, which are maintained in peripheral tissues of infected rhesus macaques over many years.

Thus, T_{CM} respond to systemic infections by rapidly proliferating before engagement of the invading pathogen, whereas T_{EM} continuously patrol peripheral tissues and mucosal surfaces, and do not need to expand or differentiate for acquisition of effector function. Since T_{EM} maintain an effector state, T_{EM} are more able to rapidly control invading pathogens at the point of initial exposure prior to systemic spread. In contrast, T_{CM} are relatively delayed in acquisition of their effector response, requiring *de novo* exposure to the invading pathogen before differentiation into cells with full effector function.

Paradoxically, all animal CMVs and HCMV express multiple proteins that efficiently inhibit antigen presentation to T-cells by modulating major histocompatibility (MHC) class I surface levels (Powers *et al.*, 2008). Consequently, CMV-infected cells are unable to stimulate CMV-specific T-cells in *in vitro* assays, which seemingly contradicts the high frequency of CMV-specific T-cells in infected individuals. While it was initially thought that these immune evasion molecules would be required for CMV to maintain persistent

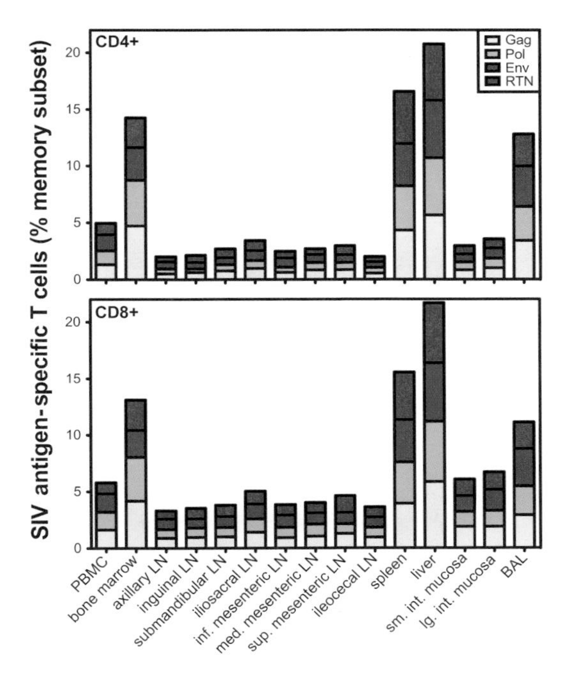

Figure II.21.1 High levels of SIV antigen-specific T-cells in a RhCMV/SIV vaccinated rhesus macaque. SIV Gag, Pol, Env, and Rev/Tat/Nef (RTN)-specific CD4+ and CD8+ total memory T-cell responses were determined in multiple tissues at necropsy from a RhCMV+ rhesus macaque that had received RhCMV/Gag, RhCMV/Pol, RhCMV/Env, and RhCMV/RTN vectors simultaneously >1.5 years earlier. Response frequencies were determined by intracellular expression of TNF-α and/or IFN-γ after stimulation with overlapping SIV Gag, Pol, Env, and RTN peptides.

infection, results from both MCMV and RhCMV studies show that CMV recombinants lacking inhibitors of antigen presentation are, in principle, still able to infect CMV-naïve animals and establish persistent infections (Gold *et al.*, 2004; Hansen *et al.*, 2010; reviewed by Lemmerman *et al.*, 2011). As discussed below, MHC class I evasion mechanisms seem instead to have evolved primarily to facilitate superinfection of hosts that are already CMV infected.

The mechanism by which CMV solves the problem of superinfection – specifically by suppression of the normal MHC I-directed presentation of antigens – leads to certain consequences for the virus. Owing to the absence of direct presentation, priming of immune responses induced by CMV infection appears to result primarily from cross-presentation. Indeed, recent evidence suggests that the majority of CMV-specific T-cells are not induced by direct presentation of viral antigens on the surface of infected cells, but rather by cross-presentation of exogenously acquired antigen by professional antigen-presenting cells (APCs) (Snyder *et al.*, 2010). Consistent with this model, a recent study using Batf3-deficient mice, which are deficient in two main DC subsets believed to be primarily responsible for cross-presentation (lymphoid organ-resident CD8α+ and migratory CD103+ CD11b− DCs) showed a reduced capacity to prime anti-CMV immune responses. Interestingly, these mice were still able to maintain and expand CMV responses against a subset of 'inflationary' CMV epitopes expressed during chronic persistent/latent infection (Torti *et al.*, 2011). This differential requirement for these cross-priming DC populations is consistent with the expression of the MHC class I immunomodulators during acute productive infection, but not latent infection in CMV-infected mice. A recent study suggests that the inability of professional APCs in salivary glands to cross-present CMV antigens may also be responsible for the relative CD4+ T-cell-dependence of CMV control in these tissues, at least in mice (Walton *et al.*, 2011). With respect to development of CMV-based vaccines, the presence of MHC class I inhibitory genes does not appear to prevent the induction of robust immune responses to multiple epitopes within heterologous antigens. Instead, virus expressing these inhibitory genes enhances priming by downmodulating early CD8 T-cell-mediated control of virus infection, apparently by 'feeding' the cross-presentation pathway more efficiently due to increased virus replication and antigen production (Böhm *et al.*, 2008). This heavy bias towards the use of cross-presentation may also explain the apparent minimal requirement for active virus replication for induction of high T-cell responses by CMV after high-dose systemic infections (Mohr *et al.*, 2010; Snyder *et al.*, 2011), since under such conditions most priming is already occurring via the cross-presentation pathway.

Superinfection by CMV

The adaptive immune response is the major block to superinfection, which is defined as the establishment of a secondary infection in a previously infected (and therefore 'immune') host. The efficiency by which adaptive immunity prevents superinfection of many different infectious pathogens can be seen by the large array of highly effective vaccines that exist today – all of which presumably function based on this underlying principle. An inability to superinfect poses major problems for the development of viral and bacterial-based vector platforms for use in populations that already possess high levels of immunity to the platform resulting from natural infection [i.e. human adenovirus serotype 5 (Ad5)], or prior vaccination (i.e. BCG and vaccinia virus). The inability to superinfect also complicates 'prime'/'boost' strategies that are frequently required to achieve high levels of immunity, necessitating the use of immunologically disparate vectors for the 'prime' compared to the 'boost' phase. Consequently, the ability of CMV-based vaccines to superinfect the CMV-seropositive host is a major advantage (especially given the high worldwide seroprevalence of CMV).

The ability to superinfect appears to be a unique quality of CMV, which is all the more remarkable given the high level of CMV-specific immunity present in infected individuals. Indeed, the capacity of CMV to establish superinfection in the healthy seropositive host has long been an area of considerable controversy. It was not until recent findings using the RhCMV infection model in rhesus macaques that the impact of pre-existing CMV immunity on superinfection was finally resolved. Using RhCMV viruses that could be distinguished on the basis of expression of distinct simian immunodeficiency virus (SIV) antigens, RhCMV superinfection was shown to be extremely efficient (Hansen *et al.*, 2009; 2011), with infection being established with ≤ 100 infectious virus particles (Hansen *et al.*, 2010). Superinfection appears to be independent of infection route, since both subcutaneous and oral inoculation of CMV vectors results in comparable immune responses to heterologous target antigens (Fig. II.21.2).

The use of deletion RhCMV recombinants identified four virally encoded down-regulators of MHC class I (HCMV homologues US2, 3, 6 and 11) that were together required for superinfection (Hansen *et al.*, 2010). *In vivo* transient cellular depletion studies showed that these MHC class I modulators were mediating their effect through evasion of the CD8⁺ T-cell response. Interestingly, virus deleted for these MHC class I regulators was able to persist after recovery of CD8⁺ T-cells indicating that CD8⁺ T-cell evasion was required only during the establishment of superinfection, and that once established, virus could persist in these animals without the need for MHC class I modulation. Thus, evasion of pre-existing CD8⁺ T-cell responses by CMV is a key mechanism that enables CMV to overcome CMV-specific immunity during superinfection. However, these results do not rule out the possibility that other CMV-encoded immune modulatory strategies of innate or humoral adaptive immune responses are also involved in mediating this remarkably efficient immune escape.

On a virological level, superinfection appears to be less efficient than primary infection. Studies in the RhCMV infection model have shown that pre-existing immunity can temper the level of secondary infection resulting in lower viraemia in seropositive compared to seronegative animals (Sequar *et al.*, 2002). In the same

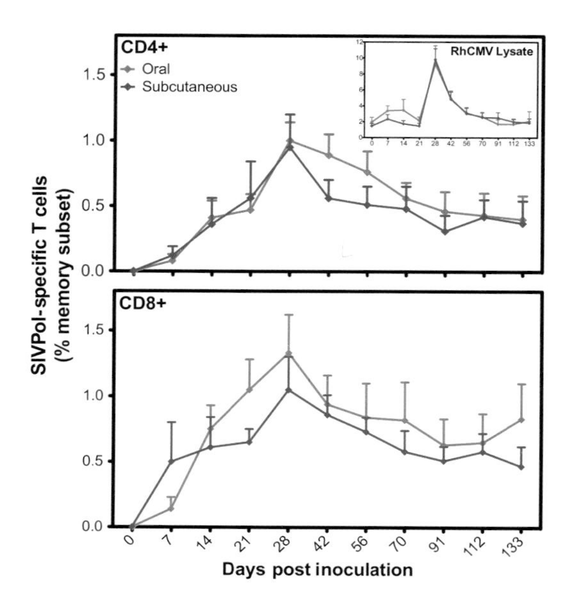

Figure II.21.2 RhCMV/Pol is able to superinfect RhCMV⁺ rhesus macaques via the subcutaneous or oral inoculation route. RhCMV⁺ rhesus macaques were inoculated subcutaneously ($n = 4$) or orally ($n = 4$) with 10^7 PFU of RhCMV/Pol at day 0. PBMC CD4⁺ and CD8⁺ T-cell mean response frequencies (± SEM) to SIV Pol were then determined at indicated times by intracellular expression of TNF-α and IFN-γ after stimulation with overlapping SIV Pol peptides. Note: pre-existing CD4⁺ T-cell responses to RhCMV lysate show a 'boost' after RhCMV/Pol inoculation confirming superinfection via subcutaneous and oral inoculation routes.

model, subunit vaccines were also shown to be able to lower local and systemic viraemia (Abel *et al.*, 2011). These findings suggest that it will probably not be possible to prevent HCMV infection by vaccination, since it seems unlikely that any potential candidate vaccine will induce immune responses, including neutralizing antibody responses, that are substantially better than those induced by natural CMV infection. However, the induction of CMV-specific immunity by vaccination may be able to lower the incidence of congenital infection, similar to the lower incidence reported in seropositive women (Kenneson and Cannon, 2007), and decrease viraemia in seronegative transplant recipients (Emery *et al.*, 2002) (for HCMV vaccine strategies, see also Chapter II.20).

In humans, the question of whether HCMV naturally superinfects CMV seropositives, or whether detectable, patent CMV infection in these individuals results from reactivation of latent virus (with CMV-specific immunity preventing superinfection) was an area of active debate for > 25 years. It is only recently that superinfection has become accepted as a general and relatively common occurrence in humans. Early experimental studies of superinfection in humans were performed as a component of vaccine trials (Quinnan *et al.*, 1984). However, the lack of immunological markers in these clinical studies limited interpretation of the results. Despite this caveat, it seems that a clinical isolate (Toledo strain) was able to cause symptoms in seropositive individuals at high, but not at low dose (Plotkin *et al.*, 1989).

Results from other early studies suggested that superinfection was linked to AIDS- and iatrogenic-mediated immunosuppression (Drew *et al.*, 1984; Spector *et al.*, 1984; Chou, 1986) or high-risk sexual behaviour (Chandler *et al.*, 1987), and that re-activation of the primary infecting virus was the predominant source of virus in seropositives (Adler *et al.*, 1985; 1986; Adler, 1988). The comparable frequency of patent CMV infection (based on anamnestic CMV-specific IgG responses) observed in two CMV-seropositive patient cohorts receiving red blood cells from either CMV negative or random donors (which contained CMV positive donors) was taken as evidence that reactivation was the primary source of most post-transfusion CMV infections in CMV-seropositive patients (Adler *et al.*, 1985). Similarly, a long-term study in hospitalized newborns showed no difference in the number of acquired infections in low birthweight infants receiving blood products from seronegative compared to random donors (Adler *et al.*, 1986).

Reactivation also appeared to be the norm in healthy individuals. In a study of CMV transmission in children attending a day care centre, there was no evidence of superinfection (Adler, 1988). A subsequent study found superinfection to be inefficient, but did identify superinfection in a small number (6%) of children attending day care (Adler, 1991). This was the first indication that healthy seropositive individuals could be superinfected, albeit at low levels – previously thought to be the domain of the immunosuppressed. Although a few studies over subsequent years raised the possibility that superinfection was more common than suggested (Bale *et al.*, 1996; Boppana *et al.*, 2001), reactivation of the primary infecting strain (with immunological prevention of superinfection by a new strain) was the established paradigm as the source of patent virus in healthy CMV seropositives (Adler *et al.*, 1995; Schleiss, 2008). However, the reliance of these early studies on indirect serology and relatively insensitive molecular methods for virus detection left the possibility that superinfection was more frequent in healthy seropositives, but was being overlooked.

Recent epidemiological studies in US and Brazilian women suggest that superinfection is indeed far more common in healthy seropositives than was originally believed (Ross *et al.*, 2010; Yamamoto *et al.*, 2010). In both of these epidemiological studies, superinfection was assessed by detection of antibodies directed against polymorphisms in two essential genes of the virus (gB and gH). Nearly a third of women (59 of 205) were shown to be superinfected with a new CMV strain over a 3 year study period. This rate of superinfection was comparable to the rate of primary infection in the general US population (Colugnati *et al.*, 2007), suggesting that pre-existing CMV-specific immunity has little impact on superinfection. A similarly high frequency of re-infection was found in Brazilian women, where seropositivity approaches 100% (Yamamoto *et al.*, 2010). Together, these studies indicate that superinfection occurs at a high frequency with rates approaching those observed in seronegatives. The frequency of superinfection in these studies may still represent an underestimate of the level of superinfection, as only new strains that differ at the polymorphic regions in gB and gH will be detected. Although highly unlikely, the possibility remains that the acquisition of the new polymorphism reactivity resulted from reactivation of virus strains acquired during primary infection. A definitive assessment of the level of superinfection in healthy human seropositives will have to await phase I clinical trials using CMV vectors carrying heterologous antigens.

Superinfection has also been demonstrated in other animal species. Recent studies using degenerate PCR targeting of conserved CMV genes to identify CMVs unique to wild great ape populations showed that mixed infection with multiple CMVs was a common characteristic of CMV infection in these species (Leendertz *et al.*, 2009). Studies in wild mice similarly suggest that rodent CMVs may share a common capacity for superinfection, as individual mice are infected with multiple genetically distinct MCMV strains. In one PCR-based study over a third of wild mice sampled (40 of 117) were co-infected with multiple CMV strains (Gorman *et al.*, 2006). Whether the high frequency of mixed infections results from co-infection during primary infection or superinfection of seropositive animals is unclear. Interestingly, although superinfection may be common in wild mouse populations, efficient superinfection of laboratory mice has been difficult to achieve (Farroway *et al.*, 2005, Wirtz *et al.*, 2008) (Dr Alec Redwood, personal communication). Superinfection by MCMV in wild mice, and primate CMVs in either the laboratory setting or the field seem to be very similar. In contrast, MCMV immunity has been shown to have differing abilities in preventing superinfection in laboratory strains of mice (Gorman *et al.*, 2006; Mohr *et al.*, 2010). These differences may be due to differences in genotype of wild and laboratory strains of both mice and CMV, methods of detection, or other factors such as stress and behavioural differences.

Immunology and protectivity of CMV-based vectors

Using rodent CMV models (Lloyd *et al.*, 2003; Rizvanov *et al.*, 2003, 2006; Karrer *et al.*, 2004; Smith *et al.*, 2005; Tsuda *et al.*, 2011) and the RhCMV model (Hansen *et al.*, 2009; 2011), a number of laboratories have demonstrated the utility of CMV as a vaccine vector.

Influenza A and lymphocytic choriomeningitis virus

In the first study to investigate the capacity of CMV vectors to induce immunity against infectious target pathogens, recombinant MCMV expressing $CD8^+$ T-cell epitopes derived from influenza A (IA) nucleoprotein (designated MCMV-NP) or lymphocytic choriomeningitis virus (LCMV) glycoprotein fused to the C-terminus of MCMV immediate-early 2 (IE2) protein were shown to induce a CD8 T-cell response against the respective heterologous epitope (Karrer *et al.*, 2004). In contrast to the rapid but transient T-cell responses observed during natural IA or LCMV infection, MCMV infection was associated with slower but steadily increasing induction of epitope-specific T-cell responses, which reached comparable peak levels seen from the natural infection with IA or LCMV, but did not wane (deemed memory inflation, see above). Although protection against the natural pathogens was not investigated, a persistent level of epitope-specific immunity was associated with a more complete and prolonged protection in an IA NP-expressing vaccinia challenge model, compared to a partial and transient protection afforded by immunity from natural IA infection.

Simian and human immunodeficiency viruses

The most advanced studies investigating the capacity for CMV-based vectors to induce protective immunity against an infectious agent have been performed using the SIV NHP model of AIDS (Hansen *et al.*, 2009; 2010). Using a panel of recombinant RhCMV vectors expressing distinct target SIV antigens (designated RhCMV/SIV vectors), these studies have shown CMV-based vectors are capable of inducing a profound level of control against systemic infection using a low-dose mucosal challenge with the highly pathogenic SIV strain, SIVmac239. The low dose mucosal route of challenge is designed to more closely replicate the exposure that would be encountered during sexual exposure to HIV. The results from these studies show that responses elicited by CMV-based vectors can intercept and completely control SIV infection prior to establishment of progressive, systemic SIV infection. This is a new pattern of protection against a pathogenic lentivirus that has not been previously observed except with live attenuated SIV vaccines. Protection against systemic infection was correlated with the ability of the RhCMV/SIV vectors to induce and maintain a T_{EM}-biased SIV-specific $CD8^+$ T-cell response. Notably, and consistent with the ability of CMV to superinfect, all animals used in these studies were CMV seropositive at the time of vaccination. Protection was extremely durable, as vaccinated animals were still protected > 365 days following the last RhCMV/SIV vaccination. Together, these results suggest that human CMV vectors encoding HIV antigens may be at least one component of an HIV vaccine strategy that will be effective at interfering with sexual transmission of HIV in humans. Since vaccinated animals that controlled SIV infection reduced the viral burden to undetectable levels, a CMV-based vaccine may also be efficacious in a therapeutic setting.

Virally vectored immunocontraception

Mouse plagues are a major environmental problem in grain-growing regions of Australia (Brown and Singleton, 2000). A MCMV-based vaccine vector expressing the mouse egg coat protein (murine zona pellucida 3, mZP3) is being developed as a virally vectored immunocontraceptive (VVIC) to induce immune-mediated sterility (immunocontraception) in mice. A recombinant MCMV (MCMV-mZP3) expressing mZP3 was shown to induce sterility in mice (Lloyd *et al.*, 2003; Smith *et al.*, 2005). MCMV-mZP3 induced sterility after a single inoculation, which was long-lasting (> 250 days) and corresponded to a durable high titred antibody response against mZP3 (Lloyd *et al.*, 2003). These MCMV-mZP3 studies further emphasize the immunogenicity of CMV, which even broke tolerance to induce an immune response against a self-antigen.

Sin Nombre hantavirus

In 1993, Sin Nombre hantavirus (SNV) was identified as the causative agent of a severe form of pneumonia called hantavirus pulmonary syndrome (HPS) in the four corners region of the US. Deer mice (*Peromyscus maniculatus*) are the natural reservoir of SNV, and a deer mouse CMV (PCMV) has been engineered to express the SNV viral glycoprotein G1 (Gn). Although T-cell immunity has not been investigated, the G1-expressing

PCMV was shown to induce antibodies to SNV. The recombinant PCMV was also able to superinfect PCMV-infected deer mice, and shared a similar tissue distribution and capacity for reactivation as wild type PCMV (Rizvanov *et al.*, 2003, 2006). The ability of the PCMV-based vector to protect against SNV infection has yet to be determined.

Ebola virus

Great apes represent a major source of transmission of Ebola virus into the human population (Groseth *et al.*, 2007). A MCMV expressing a CD8 CTL epitope from Ebola virus (*Zaire*) nucleoprotein (NP) fused to the C-terminus of the MCMV IE2 protein (designated MCMV/ZEBOV-NP_{CTL}) was shown to induce high levels of NP-specific $CD8^+$ T-cells (Tsuda *et al.*, 2011). These responses were durable, lasting for > 33 weeks following a single vaccination. Importantly, immunity induced by MCMV/ZEBOV-NP_{CTL} was protective against high dose, lethal challenge with mouse-adapted Ebola virus (*Zaire*). These studies serve as a 'proof-of-concept' for the ability of a CMV-based vaccine to protect against a highly lethal human pathogen. Future studies will need to address the level of protection afforded by CMV-based vectors in NHPs – the 'gold-standard' for Ebola virus protection.

'Disseminating' CMV-based vaccines

In addition to the ability of CMV to superinfect, CMV has evolved many characteristics that facilitate its dissemination from individual to individual through its host population. While a capacity for dissemination may not be a desirable feature for CMV vectors to be used in humans, replicating and 'disseminating' CMV-based vectors may be useful for inexpensive vaccination of livestock, to target pathogens in inaccessible wild animals involved in disease transmission to humans (SNV and Ebola virus) or for immunocontraception (mouse plagues). The high host-restriction of CMVs to their target species is expected to prevent spread of a disseminating vaccine 'off-species' to non-targeted animals within the environment. Future studies will need to address the level of protection afforded by 'dissemination' compared to direct inoculation of these vaccines in their target species, as well as examine in more detail the extent of CMV host restriction (for host restriction, see also Chapter I.18).

Safety of CMV vectors

While replicating and disseminating CMV vectors may be desirable for livestock and wild animals, a disseminating vaccine is unlikely to be used in humans. Since HCMV asymptomatically infects a large part of the population an argument can be made for the use of HCMV vectors in healthy HCMV-seropositive subjects, subjecting them to no more risk than that inherent from their pre-existing, naturally acquired HCMV infection. However, HCMV is a pathogen in immune-compromised individuals (Vogel *et al.*, 1997) and the leading infectious cause of congenital birth defects (Ross *et al.*, 2006), (see also Chapter II.2) so that a fully replicative CMV vector is unlikely to be used as a vaccine vector in humans.

Recent work using the MCMV model has shown that persistent immune stimulation by CMV does not require spread of CMV from initially infected cells (Mohr *et al.*, 2010; Snyder *et al.*, 2011). Spread-deficient CMVs were generated by deletion of essential late genes [either glycoprotein L (Snyder *et al.*, 2010) or the tegument protein M94 (Mohr *et al.*, 2010)], using the BAC system and growth of viruses on complementing fibroblasts (see also Chapter I.3). Viruses recovered from the supernatants of complementing cells were able to infect non-complementing fibroblasts, but then unable to spread to neighbouring cells. These 'spread-deficient' viruses induced antibody responses to CMV proteins that were slightly reduced compared to WT-MCMV (Mohr *et al.*, 2010). However, the inability of these viruses to spread did not weaken or shorten the duration of CMV virus-specific T-cell responses compared to WT-MCMV (Snyder *et al.*, 2011). Memory inflation was also observed against MCMV-encoded antigens (Snyder *et al.*, 2010), and the T-cell responses induced by the spread-deficient vectors protected animals against challenge with WT virus one year after vaccination (Mohr *et al.*, 2010). In Snyder *et al.* (2011), normal T-cell responses were only observed upon intravenous inoculation with the spread-deficient virus, whereas Mohr and colleagues employed intraperitoneal or subcutaneous infection.

These data suggest that long-term immunogenicity requires persistent antigen presentation, but not spread of virus from cell to cell. Since CMV-associated disease almost certainly requires viral spread, the spread-deficient vectors would no longer be able to cause disease, even in the absence of a functioning immune system. This critical finding suggests that it may be possible to develop efficacious HCMV vectors with an excellent safety profile. The mechanism by which spread-deficient CMV is able to induce memory inflation and long-term immunity is unknown (for discussion and a hypothesis, see Chapter I.22). In the case of HCMV, one could argue that viral genomes are harboured in self-renewing myeloid lineage stem cells and viral replication and reactivation occurs upon differentiation into APCs. Thus, the viral genome would be passively replicated via cell division and viral protein would be produced in APCs similar to reactivation of latent WT virus (see Chapters I.19 and I.20).

A precedent for the use of persistent herpesvirus-based vaccines is already established based on the use of an attenuated strain of varicella zoster virus (VZV, Oka strain) to prevent VZV disease (chickenpox and shingles). VZV Oka strain, which has been attenuated by multiple passaging, is widely used for vaccination against chickenpox in children and herpes zoster in the elderly (Arvin and Gershon, 1996; Gilden, 2011). The Oka strain is replication-competent, and can cause disseminated infection in immune-deficient individuals (Levy *et al.*, 2003; Banovic *et al.*, 2011). The vaccine can also reactivate and cause herpes zoster (Uebe *et al.*, 2002; Sauerbrei *et al.*, 2003). Furthermore, VZV breakthrough in vaccinated children results in lifelong co-existence of both WT and vaccine strain recipients (Vessey *et al.*, 2001). Yet, given all these negative safety parameters, VZV Oka has still been used with minimal serious complications for worldwide vaccination in children since 1995. Given that CMV vectors can be made spread-deficient by precise genetic engineering they are expected to be far safer than the currently used VZV vaccine.

Future applications for CMV vectors

CMV-based vaccine vectors represent a new paradigm in vaccinology. Unlike other vectors, CMV-based vectors continuously stimulate the immune system at a low level resulting in extraordinary 'effector memory' levels that are long-lived and maintained indefinitely. This permanent immune stimulation, presumably due to low levels of persistent antigen expression, may be a desirable feature for targeting a number of diseases. In addition to chronic viral diseases such as HIV, it is believed that vaccines against recurrent parasitic diseases such as malaria and chronic bacterial disease like *M. tuberculosis* (Mtb) 'will require the generation of durable T-cell responses of sufficient magnitude and quality' (Seder *et al.*, 2008). For Mtb infection, differences in T-cell functional signatures have been observed between subjects who control Mtb infection, compared to those with active Mtb disease (Caccamo *et al.*, 2009; 2010). The loss of effector T-cell signatures and Mtb-specific immunity upon Mtb clearance could conceivably be counteracted by immunization with CMV-based vectors expressing critical Mtb antigens. Sterilizing immunity against malaria parasites has been shown to be achievable by induction of T_{EM} directed against the short-lived, transient liver stage of the parasite (Epstein *et al.*, 2011). Currently, this type of immunity can only be achieved by vaccination with attenuated sporozoites, whilst heterologous prime/boost vaccines fail to induce long-term immunity (Jiang *et al.*, 2009). Conceivably, CMV-based vectors carrying malaria antigens would be able to achieve this desired level of T_{EM} immunity, resulting in a permanent immunological barrier in the liver against sporozoite infection. Persistent low level exposure to antigen that maintains a high frequency of $CD4^+$ T-cells is also thought to be required for protection against Leishmania (Seder *et al.*, 2008).

In addition to prophylactic vaccines, CMV-based vectors may also represent a novel opportunity for therapeutic vaccination against cancer (viral and non-viral). Chronic viral infections that can ultimately lead to cancer such as hepatitis C virus are known to exhaust the cellular immune system (Wherry, 2011). While HCV-specific T-cells are present in chronically infected individuals, these cells are unable to proliferate in response to antigen due to expression of negative regulatory surface molecules (Radziewicz *et al.*, 2007). Expression of negative regulatory molecules is also observed in cancers of non-viral aetiology. Although tumour immunology is extremely complex and multifactorial, exhaustion of memory T-cells is believed to be one factor responsible for cancer vaccine failure despite the induction of tumour-specific T-cells (Klebanoff *et al.*, 2011). For reasons that are not entirely clear, but perhaps relate to the low level of antigen expression, persistent antigen stimulation by CMV does not seem to lead to an exhausted phenotype (Podlech *et al.*, 2000; Snyder *et al.*, 2008; Hertoghs *et al.*, 2010; Cicin-Sain *et al.*, 2011; Richter *et al.*, 2012). This observation suggests that cancer vaccines based on CMV may be able to overcome one of the limitations of cancer immunotherapy. Recently, a CMV-based vector expressing a defined CD8 CTL epitope from a human tumour-associated antigen, prostate-specific antigen (PSA), was shown to decrease tumour progression in a mouse prostate cancer model (Klyushnenkova *et al.*, 2012). The demonstration that CMV-vectors can overcome self-tolerance (Redwood *et al.*, 2005) also indicates that induction of T-cell responses to self-antigens expressed by CMV is possible.

In conclusion, CMV vectors have several unique characteristics: (1) persistent maintenance of non-exhausted, polyfunctional effector memory T-cells that patrol non-lymphoid tissues; (2) the ability to superinfect, raising the possibility to repeatedly use the same vector platform regardless of the CMV serostatus of the recipient; (3) the ability to induce and maintain T_{EM}-biased responses by vectors with increased safety profiles due to their inability to spread. The recent demonstration that CMV-vectored vaccines are protective in several animal models, most notably SIV infection of NHPs and Ebola virus infection of mice, further indicates that CMV vectors may significantly increase our option in developing new vaccines against a series of infectious diseases and possibly cancer.

Acknowledgements

We appreciate Drs. Britt (University of Alabama) and Redwood (University of Western Australia) for insightful discussion during preparation of this chapter. We also appreciate Andrew Townsend for his help with the figures and graphics, and Melissa Brewer for her help with the editing.

Author information: KF, LJP, JAN and SGH have a significant financial interest in TomegaVax Inc., a company that may have a commercial interest in the results of this research and technology. This potential individual and institutional conflict of interest has been reviewed and managed by OHSU. MAJ has no conflicting financial interests.

References

Abel, K., Martinez, J., Yue, Y., Lacey, S.F., Wang, Z., Strelow, L., Dasgupta, A., Li, Z., Schmidt, K.A., Oxford, K.L., *et al.* (2011). Vaccine-induced control of viral shedding following rhesus cytomegalovirus challenge in rhesus macaques. J. Virol. *85*, 2878–2890.

Adler, S.P. (1988). Molecular epidemiology of cytomegalovirus: viral transmission among children attending a day care center, their parents, and caretakers. J. Pediatr. *112*, 366–372.

Adler, S.P. (1991). Molecular epidemiology of cytomegalovirus: a study of factors affecting transmission among children at three day-care centers. Pediatr. Infect. Dis. J. *10*, 584–590.

Adler, S.P., Baggett, J., and McVoy, M. (1985). Transfusion-associated cytomegalovirus infections in seropositive cardiac surgery patients. Lancet *2*, 743–747.

Adler, S.P., Baggett, J., Wilson, M., Lawrence, L., and McVoy, M. (1986). Molecular epidemiology of cytomegalovirus in a nursery: lack of evidence for nosocomial transmission. J. Pediatr. *108*, 117–123.

Adler, S.P., Starr, S.E., Plotkin, S.A., Hempfling, S.H., Buis, J., Manning, M.L., and Best, A.M. (1995). Immunity induced by primary human cytomegalovirus infection protects against secondary infection among women of childbearing age. J. Infect. Dis. *171*, 26–32.

Arvin, A.M., and Gershon, A.A. (1996). Live attenuated varicella vaccine. Annu. Rev. Microbiol. *50*, 59–100.

Bale, J.F., Jr., Petheram, S.J., Souza, I.E., and Murph, J.R. (1996). Cytomegalovirus reinfection in young children. J. Pediatr. *128*, 347–352.

Banovic, T., Yanilla, M., Simmons, R., Robertson, I., Schroder, W.A., Raffelt, N.C., Wilson, Y.A., Hill, G.R., Hogan, P., and Nourse, C.B. (2011). Disseminated varicella infection caused by varicella vaccine strain in a child with low invariant natural killer T-cells and diminished CD1d expression. J. Infect. Dis. *204*, 1893–1901.

Bett, A.J., Dubey, S.A., Mehrotra, D.V., Guan, L., Long, R., Anderson, K., Collins, K., Gaunt, C., Fernandez, R., Cole, S., *et al.* (2010). Comparison of T-cell immune responses induced by vectored HIV vaccines in non-human primates and humans. Vaccine *28*, 7881–7889.

Bitmansour, A.D., Waldrop, S.L., Pitcher, C.J., Khatamzas, E., Kern, F., Maino, V.C., and Picker, L.J. (2001). Clonotypic structure of the human CD4+ memory T-cell response to cytomegalovirus. J. Immunol. *167*, 1151–1163.

Böhm, V., Simon, C.O., Podlech, J., Seckert, C.K., Gendig, D., Deegen, P., Gillert-Marien, D., Lemmermann, N.A., Holtappels, R., and Reddehase, M.J. (2008). The immune evasion paradox: immunoevasins of murine cytomegalovirus enhance priming of CD8 T-cells by preventing negative feedback regulation. J. Virol. *82*, 11637–11650.

Boppana, S.B., Rivera, L.B., Fowler, K.B., Mach, M., and Britt, W.J. (2001). Intrauterine transmission of cytomegalovirus to infants of women with preconceptional immunity. N. Engl. J. Med. *344*, 1366–1371.

Brown, P.R., and Singleton, G.R. (2000). Impacts of house mice on crops in Australia – costs and damage. In Human Conflicts With Wildlife: Economic Considerations, (National Wildlife Research Center, Fort Collins, CO), pp. 48–58.

Caccamo, N., Guggino, G., Meraviglia, S., Gelsomino, G., Di Carlo, P., Titone, L., Bocchino, M., Galati, D., Matarese, A., Nouta, J., *et al.* (2009). Analysis of Mycobacterium tuberculosis-specific CD8 T-cells in patients with active tuberculosis and in individuals with latent infection. PLoS One *4*, e5528.

Caccamo, N., Guggino, G., Joosten, S.A., Gelsomino, G., Di Carlo, P., Titone, L., Galati, D., Bocchino, M., Matarese, A., Salerno, A., *et al.* (2010). Multifunctional CD4(+) T-cells correlate with active Mycobacterium tuberculosis infection. Eur. J. Immunol. *40*, 2211–2220.

Casimiro, D.R., Chen, L., Fu, T.M., Evans, R.K., Caulfield, M.J., Davies, M.E., Tang, A., Chen, M., Huang, L., Harris, V., *et al.* (2003). Comparative immunogenicity in rhesus monkeys of DNA plasmid, recombinant vaccinia virus, and replication-defective adenovirus vectors expressing a human immunodeficiency virus type 1 gag gene. J. Virol. *77*, 6305–6313.

Chandler, S.H., Handsfield, H.H., and McDougall, J.K. (1987). Isolation of multiple strains of cytomegalovirus from women attending a clinic for sexually transmitted disease. J. Infect. Dis. *155*, 655–660.

Chou, S.W. (1986). Acquisition of donor strains of cytomegalovirus by renal-transplant recipients. N. Engl. J. Med. *314*, 1418–1423.

Cicin-Sain, L., Sylwester, A.W., Hagen, S.I., Siess, D.C., Currier, N., Legasse, A.W., Fischer, M.B., Koudelka, C.W., Axthelm, M.K., Nikolich-Zugich, J., *et al.* (2011). Cytomegalovirus-specific T-cell immunity is maintained in immunosenescent rhesus macaques. J. Immunol. *187*, 1722–1732.

Colugnati, F.A., Staras, S.A., Dollard, S.C., and Cannon, M.J. (2007). Incidence of cytomegalovirus infection among the general population and pregnant women in the USA. BMC Infect. Dis. *7*, 71.

Draper, S.J., and Heeney, J.L. (2010). Viruses as vaccine vectors for infectious diseases and cancer. Nat. Rev. Microbiology. *8*, 62–73.

Drew, W.L., Sweet, E.S., Miner, R.C., and Mocarski, E.S. (1984). Multiple infections by cytomegalovirus in patients with acquired immunodeficiency syndrome: documentation by Southern blot hybridization. J. Infect. Dis. *150,* 952–953.

Emery, V.C., Hassan-Walker, A.F., Burroughs, A.K., and Griffiths, P.D. (2002). Human cytomegalovirus (HCMV) replication dynamics in HCMV-naïve and -experienced immunocompromised hosts. J. Infect. Dis. *185,* 1723–1728.

Epstein, J.E., Tewari, K., Lyke, K.E., Sim, B.K., Billingsley, P.F., Laurens, M.B., Gunasekera, A., Chakravarty, S., James, E.R., Sedegah, M., *et al.* (2011). Live attenuated malaria vaccine designed to protect through hepatic CD8 T-cell immunity. Science *334,* 475–480.

Farroway, L.N., Gorman, S., Lawson, M.A., Harvey, N.L., Jones, D.A., Shellam, G.R., and Singleton, G.R. (2005). Transmission of two Australian strains of murine cytomegalovirus (MCMV) in enclosure populations of house mice (Mus domesticus). Epidemiol. Infect. *133,* 701–710.

Gilden, D. (2011). Efficacy of live zoster vaccine in preventing zoster and postherpetic neuralgia. J. Intern. Med. *269,* 496–506.

Gillespie, G.M., Wills, M.R., Appay, V., O'Callaghan, C., Murphy, M., Smith, N., Sissons, P., Rowland-Jones, S., Bell, J.I., and Moss, P.A. (2000). Functional heterogeneity and high frequencies of cytomegalovirus- specific CD8(+) T-lymphocytes in healthy seropositive donors. J. Virol. *74,* 8140–8150.

Gold, M.C., Munks, M.W., Wagner, M., McMahon, C.W., Kelly, A., Kavanagh, D.G., Slifka, M.K., Koszinowski, U.H., Raulet, D.H., and Hill, A.B. (2004). Murine cytomegalovirus interference with antigen presentation has little effect on the size or the effector memory phenotype of the CD8 T-cell response. J. Immunol. *172,* 6944–6953.

Gorman, S., Harvey, N.L., Moro, D., Lloyd, M.L., Voigt, V., Smith, L.M., Lawson, M.A., and Shellam, G.R. (2006). Mixed infection with multiple strains of murine cytomegalovirus occurs following simultaneous or sequential infection of immunocompetent mice. J. Gen. Virol. *87,* 1123–1132.

Groseth, A., Feldmann, H., and Strong, J.E. (2007). The ecology of Ebola virus. Trends Microbiol. *15,* 408–416.

Hansen, S.G., Vieville, C., Whizin, N., Coyne-Johnson, L., Siess, D.C., Drummond, D.D., Legasse, A.W., Axthelm, M.K., Oswald, K., Trubey, C.M., *et al.* (2009). Effector memory T-cell responses are associated with protection of rhesus monkeys from mucosal simian immunodeficiency virus challenge. Nature Med. *15,* 293–299.

Hansen, S.G., Powers, C.J., Richards, R., Ventura, A.B., Ford, J.C., Siess, D., Axthelm, M.K., Nelson, J.A., Jarvis, M.A., Picker, L.J., *et al.* (2010). Evasion of CD8+ T-cells is critical for superinfection by cytomegalovirus. Science *328,* 102–106.

Hansen, S.G., Ford, J.C., Lewis, M.S., Ventura, A.B., Hughes, C.M., Coyne-Johnson, L., Whizin, N., Oswald, K., Shoemaker, R., Swanson, T., *et al.* (2011). Profound early control of highly pathogenic SIV by an effector memory T-cell vaccine. Nature *473,* 523–527.

Hertoghs, K.M., Moerland, P.D., van Stijn, A., Remmerswaal, E.B., Yong, S.L., van de Berg, P.J., van Ham, S.M., Baas, F., ten Berge, I.J., and van Lier, R.A. (2010). Molecular profiling of cytomegalovirus-induced human CD8+ T-cell differentiation. J. Clin. Invest. *120,* 4077–4090.

Holtappels, R., Pahl-Seibert, M.F., Thomas, D., and Reddehase, M.J. (2000). Enrichment of immediate-early 1 (m123/pp89) peptide-specific CD8 T-cells in a pulmonary CD62L(lo) memory-effector cell pool during latent murine cytomegalovirus infection of the lungs. J. Virol. *74,* 11495–11503.

Holtappels, R., Thomas, D., Podlech, J., and Reddehase, M.J. (2002). Two antigenic peptides from genes m123 and m164 of murine cytomegalovirus quantitatively dominate CD8 T-cell memory in the H-2d haplotype. J. Virol. *76,* 151–164.

Hu, B., Tai, A., and Wang, P. (2011). Immunization delivered by lentiviral vectors for cancer and infectious diseases. Immunol. Rev. *239,* 45–61.

Jiang, G., Shi, M., Conteh, S., Richie, N., Banania, G., Geneshan, H., Valencia, A., Singh, P., Aguiar, J., Limbach, K., *et al.* (2009). Sterile protection against Plasmodium knowlesi in rhesus monkeys from a malaria vaccine: comparison of heterologous prime boost strategies. PLoS One *4,* e6559.

Johnson, P.R., Schnepp, B.C., Connell, M.J., Rohne, D., Robinson, S., Krivulka, G.R., Lord, C.I., Zinn, R., Montefiori, D.C., Letvin, N.L., *et al.* (2005). Novel adeno-associated virus vector vaccine restricts replication of simian immunodeficiency virus in macaques. J. Virol. *79,* 955–965.

Karrer, U., Sierro, S., Wagner, M., Oxenius, A., Hengel, H., Koszinowski, U.H., Phillips, R.E., and Klenerman, P. (2003). Memory inflation: continuous accumulation of antiviral CD8+ T-cells over time. J. Immunol. *170,* 2022–2029.

Karrer, U., Wagner, M., Sierro, S., Oxenius, A., Hengel, H., Dumrese, T., Freigang, S., Koszinowski, U.H., Phillips, R.E., and Klenerman, P. (2004). Expansion of protective CD8+ T-cell responses driven by recombinant cytomegaloviruses. J. Virol. *78,* 2255–2264.

Kenneson, A., and Cannon, M.J. (2007). Review and meta-analysis of the epidemiology of congenital cytomegalovirus (CMV) infection. Rev. Med. Virol. *17,* 253–276.

Kern, F., Khatamzas, E., Surel, I., Frommel, C., Reinke, P., Waldrop, S.L., Picker, L.J., and Volk, H.D. (1999). Distribution of human CMV-specific memory T-cells among the CD8pos. subsets defined by CD57, CD27, and CD45 isoforms. Eur. J. Immunol. *29,* 2908–2915.

Klebanoff, C.A., Acquavella, N., Yu, Z., and Restifo, N.P. (2011). Therapeutic cancer vaccines: are we there yet? Immunol. Rev. *239,* 27–44.

Klenerman, P., and Hill, A. (2005). T-cells and viral persistence: lessons from diverse infections. Nature Immunol. *6,* 873–879.

Klyushnenkova, E.N., Kouiavskaia, D.V., Parkins, C.J., Caposio, P., Botto, S., Alexander, R.B., and Jarvis, M.A. (2012). Acytomegalovirus-based vaccine expressing a single tumor-specific CD8+ T-cell epitope delays tumor growth in a murinemodel of prostate cancer. J. Immunother. *35,* 390–399.

Leendertz, F.H., Deckers, M., Schempp, W., Lankester, F., Boesch, C., Mugisha, L., Dolan, A., Gatherer, D., McGeoch, D.J., and Ehlers, B. (2009). Novel cytomegaloviruses in free-ranging and captive great apes: phylogenetic evidence for bidirectional horizontal transmission. J. Gen. Virol. *90,* 2386–2394.

Lemiale, F., and Korokhov, N. (2009). Lentiviral vectors for HIV disease prevention and treatment. Vaccine. *27*, 3443–3449.

Lemmerman, N.A., Bohm., Holtappels, R., and Reddehase, M.J. (2011). *In vivo* impact of cytomegalovirus evasion of CD8 T-cell immunity: facts and thoughts based on murine models. Virus Res. *157*, 161–174.

Levy, M.H., Quilty, S., Young, L.C., Hunt, W., Matthews, R., and Robertson, P.W. (2003). Pox in the docks: varicella outbreak in an Australian prison system. Public Health *117*, 446–451.

Lloyd, M.L., Shellam, G.R., Papadimitriou, J.M., and Lawson, M.A. (2003). Immunocontraception is induced in BALB/c mice inoculated with murine cytomegalovirus expressing mouse zona pellucida 3. Biol. Reprod. *68*, 2024–2032.

Mehendale, S., van Lunzen, J., Clumeck, N., Rockstroh, J., Vets, E., Johnson, P.R., Anklesaria, P., Barin, B., Boaz, M., Kochhar, S., *et al.* (2008). A phase 1 study to evaluate the safety and immunogenicity of a recombinant HIV type 1 subtype C adeno-associated virus vaccine. AIDS Res. Hum. Retroviruses. *24*, 873–880.

Mohr, C.A., Arapovic, J., Muhlbach, H., Panzer, M., Weyn, A., Dölken, L., Krmpotic, A., Voehringer, D., Ruzsics, Z., Koszinowski, U., *et al.* (2010). A spread-deficient cytomegalovirus for assessment of first-target cells in vaccination. J. Virol. *84*, 7730–7742.

Munks, M.W., Cho, K.S., Pinto, A.K., Sierro, S., Klenerman, P., and Hill, A.B. (2006). Four distinct patterns of memory CD8 T-cell responses to chronic murine cytomegalovirus infection. J. Immunol. *177*, 450–458.

Plotkin, S.A., Starr, S.E., Friedman, H.M., Gonczol, E., and Weibel, R.E. (1989). Protective effects of Towne cytomegalovirus vaccine against low-passage cytomegalovirus administered as a challenge. J. Infect. Dis. *159*, 860–865.

Podlech, J., Holtappels, R., Wirtz, N., Steffens, H.-P., and Reddehase, M.J. (2000). Murine model of interstitial cytomegalovirus pneumonia in syngeneic bone marrow transplantation: persistence of protective pulmonary CD8-T-cell infiltrates after clearance of acute infection. J. Virol. *74*, 7496–7507.

Powers, C., DeFilippis, V., Malouli, D., and Fruh, K. (2008). Cytomegalovirus immune evasion. Curr. Top. Microbiol. Immunol. *325*, 333–359.

Price, D.A., Bitmansour, A.D., Edgar, J.B., Walker, J.M., Axthelm, M.K., Douek, D.C., and Picker, L.J. (2008). Induction and evolution of cytomegalovirus-specific CD4+ T-cell clonotypes in rhesus macaques. J. Immunol. *180*, 269–280.

Quinnan, G.V., Jr., Delery, M., Rook, A.H., Frederick, W.R., Epstein, J.S., Manischewitz, J.F., Jackson, L., Ramsey, K.M., Mittal, K., Plotkin, S.A., *et al.* (1984). Comparative virulence and immunogenicity of the Towne strain and a nonattenuated strain of cytomegalovirus. Ann. Intern. Med. *101*, 478–483.

Radziewicz, H., Uebelhoer, L., Bengsch, B., and Grakoui, A. (2007). Memory CD8+ T-cell differentiation in viral infection: a cell for all seasons. World J. Gastroenterol. *13*, 4848–4857.

Reddehase, M.J., Simon, C.O., Seckert, C.K., Lemmermann, N., and Grzimek, N.K. (2008). Murine model of cytomegalovirus latency and reactivation. Curr. Top. Microbiol. Immunol. *325*, 315–331.

Redwood, A.J., Messerle, M., Harvey, N.L., Hardy, C.M., Koszinowski, U.H., Lawson, M.A., and Shellam, G.R. (2005). Use of a murine cytomegalovirus K181-derived bacterial artificial chromosome as a vaccine vector for immunocontraception. J. Virol. *79*, 2998–3008.

Richter, K., Brocker, T., and Oxenius, A. (2012). Antigen amount dictates CD8+ T-cell exhaustion during chronic viral infectionirrespective of the type of antigen presenting cell. Eur. J. Immunol. *42*, 2290–2304.

Rizvanov, A.A., van Geelen, A.G., Morzunov, S., Otteson, E.W., Bohlman, C., Pari, G.S., and St Jeor, S.C. (2003). Generation of a recombinant cytomegalovirus for expression of a hantavirus glycoprotein. J. Virol. *77*, 12203–12210.

Rizvanov, A.A., Khaiboullina, S.F., van Geelen, A.G., and St Jeor, S.C. (2006). Replication and immunoactivity of the recombinant Peromyscus maniculatus cytomegalovirus expressing hantavirus G1 glycoprotein *in vivo* and *in vitro*. Vaccine *24*, 327–334.

Ross, S.A., Fowler, K.B., Ashrith, G., Stagno, S., Britt, W.J., Pass, R.F., and Boppana, S.B. (2006). Hearing loss in children with congenital cytomegalovirus infection born to mothers with preexisting immunity. J. Pediatr. *148*, 332–336.

Ross, S.A., Arora, N., Novak, Z., Fowler, K.B., Britt, W.J., and Boppana, S.B. (2010). Cytomegalovirus reinfections in healthy seroimmune women. J. Infect. Dis. *201*, 386–389.

Sallusto, F., Lenig, D., Forster, R., Lipp, M., and Lanzavecchia, A. (1999). Two subsets of memory T-lymphocytes with distinct homing potentials and effector functions. Nature *401*, 708–712.

Sauerbrei, A., Pawlak, J., Luger, C., and Wutzler, P. (2003). Intracerebral varicella-zoster virus reactivation in congenital varicella syndrome. Dev. Med. Child Neurol. *45*, 837–840.

Schleiss, M.R. (2008). Cytomegalovirus vaccine development. Curr. Top. Microbiol. Immunol. *325*, 361–382.

Seder, R.A., Darrah, P.A., and Roederer, M. (2008). T-cell quality in memory and protection: implications for vaccine design. Nature Rev. Immunol. *8*, 247–258.

Sequar, G., Britt, W.J., Lakeman, F.D., Lockridge, K.M., Tarara, R.P., Canfield, D.R., Zhou, S.S., Gardner, M.B., and Barry, P.A. (2002). Experimental coinfection of rhesus macaques with rhesus cytomegalovirus and simian immunodeficiency virus: pathogenesis. J. Virol. *76*, 7661–7671.

Shiver J.W., and Emini, E.A. (2004). Recent advances in the development of HIV-1 vaccines using replication-incompetent adenovirus vectors. Ann. Rev. Med. *55*, 355–372.

Sierro, S., Rothkopf, R., and Klenerman, P. (2005). Evolution of diverse antiviral CD8+ T-cell populations after murine cytomegalovirus infection. Eur. J. Immunol. *35*, 1113–1123.

Simon, C.O., Holtappels, R., Tervo, H.M., Bohm, V., Daubner, T., Oehrlein-Karpi, S.A., Kuhnapfel, B., Renzaho, A., Strand, D., Podlech, J., *et al.* (2006). CD8 T-cells control cytomegalovirus latency by epitope-specific sensing of transcriptional reactivation. J. Virol. *80*, 10436–10456.

Smith, L.M., Lloyd, M.L., Harvey, N.L., Redwood, A.J., Lawson, M.A., and Shellam, G.R. (2005). Species specificity of a murine immunocontraceptive utilising murine cytomegalovirus as a gene delivery vector. Vaccine *23*, 2959–2969.

Snyder, C.M., Cho, K.S., Bonnett, E.L., van Dommelen, S., Shellam, G.R., and Hill, A.B. (2008). Memory inflation during chronic viral infection is maintained by continuous production of short-lived, functional T-cells. Immunity *29*, 650–659.

Snyder, C.M., Allan, J.E., Bonnett, E.L., Doom, C.M., and Hill, A.B. (2010). Cross-presentation of a spread-defective MCMV is sufficient to prime the majority of virus-specific CD8+ T-cells. PLoS One *5*, e9681.

Snyder, C.M., Cho, K.S., Bonnett, E.L., Allan, J.E., and Hill, A.B. (2011). Sustained CD8+ T-cell memory inflation after infection with a single-cycle cytomegalovirus. PLoS Pathog. *7*, e1002295.

Spector, S.A., Hirata, K.K., and Newman, T.R. (1984). Identification of multiple cytomegalovirus strains in homosexual men with acquired immunodeficiency syndrome. J. Infect. Dis. *150*, 953–956.

Sylwester, A.W., Mitchell, B.L., Edgar, J.B., Taormina, C., Pelte, C., Ruchti, F., Sleath, P.R., Grabstein, K.H., Hosken, N.A., Kern, F., *et al.* (2005). Broadly targeted human cytomegalovirus-specific CD4+ and CD8+ T-cells dominate the memory compartments of exposed subjects. J. Exp. Med. *202*, 673–685.

Torti, N., Walton, S.M., Murphy, K.M., and Oxenius, A. (2011). Batf3 transcription factor-dependent DC subsets in murine CMV infection: differential impact on T-cell priming and memory inflation. Eur. J. Immunol. *41*, 2612–2618.

Tsuda, Y., Caposio, P., Parkins, C.J., Botto, S., Messaoudi, I., Cicin-Sain, L., Feldmann, H., and Jarvis, M.A. (2011). A replicating cytomegalovirus-based vaccine encoding a single Ebola virus nucleoprotein CTL epitope confers protection against Ebola virus. PLoS Negl. Trop. Dis. *5*, e1275.

Uebe, B., Sauerbrei, A., Burdach, S., and Horneff, G. (2002). Herpes zoster by reactivated vaccine varicella zoster virus in a healthy child. Eur. J. Pediatr. *161*, 442–444.

Vessey, S.J., Chan, C.Y., Kuter, B.J., Kaplan, K.M., Waters, M., Kutzler, D.P., Carfagno, P.A., Sadoff, J.C., Heyse, J.F., Matthews, H., *et al.* (2001). Childhood vaccination against varicella: persistence of antibody, duration of protection, and vaccine efficacy. J. Pediatr. *139*, 297–304.

Vogel, J.U., Scholz, M., and Cinatl, J., Jr. (1997). Treatment of cytomegalovirus diseases. Intervirology *40*, 357–367.

Walton, S.M., Mandaric, S., Torti, N., Zimmermann, A., Hengel, H., and Oxenius, A. (2011). Absence of cross-presenting cells in the salivary gland and viral immune evasion confine cytomegalovirus immune control to effector CD4 T-cells. PLoS Pathog. *7*, e1002214.

Wherry, E.J. (2011). T-cell exhaustion. Nature Immunol. *12*, 492–499.

Wirtz, N., Schader, S.A., Holtappels, R., Simon, C.O., Lemmermann, N.A.W., Reddehase, M.J., and Podlech, J. (2008). Polyclonal cytomegalovirus-specific antibodies not only prevent virus dissemination from the portal of entry but also inhibit focal virus spread within target tissues. Med. Microbiol. Immunol. *197*, 151–158.

Yamamoto, A.Y., Mussi-Pinhata, M.M., Boppana, S.B., Novak, Z., Wagatsuma, V.M., Oliveira Pde, F., Duarte, G., and Britt, W.J. (2010). Human cytomegalovirus reinfection is associated with intrauterine transmission in a highly cytomegalovirus-immune maternal population. Am. J. Obstet. Gynecol. *202*, 297 e291–298.

Non-Human-Primate Models of Cytomegalovirus Infection, Prevention, and Therapy

II.22

Klaus Früh, Daniel Malouli, Kristie L. Oxford and Peter A. Barry

Abstract

The last few years have witnessed significant expansion of the Non-Human-Primate (NHP) models of CMV persistence and pathogenesis. Progress in the utilization of the NHP CMV models has been highlighted by a better understanding of natural history, comparative genomic sequence analyses, and *in vivo* studies addressing mechanisms of tissue tropism, immune modulation, vaccine development, and optimization of the use of CMV as a vaccine vector for ectopic expression of heterologous antigens. The earliest observations of CMV infection in NHP during the first part of the twentieth century were remarkable for their prescient descriptions of CMV–host relationships based entirely on microscopic characterization of the protozoan-like (cytomegalic) cells that had been noted in congenitally infected human infants (Ribbert, 1904; Goodpasture and Talbot, 1921) and guinea pigs (Jackson, 1920). In particular, it was noted in the 1920s and 1930s that NHP CMV (1) is a ubiquitous infectious agent, (2) infects multiple cell types, (3) is characterized by low virulence, and (4) modifies host inflammatory responses. In addition, the first use of the term 'latency' to describe the ability of CMV to reactivate may have been used for NHP CMV. In 1935, Cowdry and Scott recognized that treatment of CMV-infected monkeys with irradiated ergosterol stimulated reactivation of CMV in multiple tissues, and they noted that the treatment 'may have activated or intensified a process already latent in the kidneys' (Cowdry and Scott, 1935). The recent progress in the NHP models follows these earliest insights into the hallmarks of CMV infections, and now enables the unique positioning of NHP models to provide a better understanding, treatment, and prevention of HCMV infection and disease in humans. This chapter summarizes the current status of our understanding of NHP CMVs with particular emphasis on viral gene function and viral disease models.

Introduction

Evolutionary relatedness of primate CMVs

Cytomegaloviruses are highly host-restricted and only rarely cross the species barrier even between closely related species. This extreme host-restriction results in cospeciation, i.e. CMVs co-evolving with their hosts (McGeoch *et al.*, 1995). Therefore, the sequence relationships of CMV genomes generally recapitulate the evolutionary relationship of their hosts very closely (see also Chapters I.1, I.2, and I.18). Thus, the closest relatives of HCMV are CMVs of apes, such as chimpanzee (*Pan troglodytes*) CMV (CCMV) (Swinkels *et al.*, 1984; Davison *et al.*, 2003), gorilla (*Gorilla gorilla*) CMV (GgorCMV), and orang-utan (*Pongo pygmaeus*) CMV (PpygCMV) (Leendertz *et al.*, 2009). Owing to the associated inability of HCMV to replicate in other species, animal models have to be employed in which the respective animal-specific CMV is examined. The most commonly used CMV model system is the mouse with the mouse CMV (MCMV) as its pathogen. Although MCMV and HCMV share large parts of their genome, there are substantial differences in their genomic coding contents. While many viral mechanisms are functionally conserved between these two CMV species, especially in immune evasion, the responsible viral genes are not. The same can be said about other established CMV animal models like the rat (RCMV) and the guinea pig (GPCMV) models. The best model would be a laboratory animal with a CMV that closely resembles human CMV in terms of coding content, natural history, pathogenesis, and immune responses of the host. Owing to cospeciation, the closest relatives of HCMV are found in Non-Human-Primates (NHP) comprised of the great apes and monkeys. However, experimental infection of great apes (gorillas, chimpanzees, orang-utans) by their respective CMVs cannot be used as an animal model since apes are protected species, costly, and only available in very small numbers for very limited purposes. The closest evolutionary relatives of HCMV that can be inoculated into experimental animals are derived from lesser NHP (i.e. monkeys; hereafter NHP refers to monkey species), particularly rhesus CMV (RhCMV), which was isolated from rhesus macaques (*Macaca mulatta*) (Asher *et al.*, 1974). Other NHP CMVs that have been isolated are baboon CMV (BaCMV; from different species of the genus *papio, Papio anubis, Papio cynocephalus* and *Papio ursinus*) (Blewett *et al.*, 2001), African Green Monkey (*Cercopithecus aethiops*) CMV (Simian CMV, SCMV) (Black *et al.*, 1963), cynomolgus macaque (*Macaca fascicularis*) CMV (CyCMV) (Ambagala *et al.*, 2011; Marsh *et al.*, 2011), Mandrill (*Mandrillus sphinx*) CMV (MsphCMV) (Leendertz *et al.*, 2009), Drill (*Mandrillus leucophaeus*) CMV (DrCMV) (Blewett *et al.*, 2003), and black-and-white colobus (*Colobus guereza*) CMV (CgueCMV) (Prepens *et al.*, 2007). Full-length and partial genomes of these Old World NHP CMVs have been determined. In addition, CMVs have been isolated from New World monkeys, specifically owl (*Aotus trivirgatus*) (AtriCMV) and squirrel monkeys (*Saimiri sciureus*) (SsciCMV) (Daniel *et al.*, 1971; 1973).

Alignments of the viral glycoprotein B (gB) protein and DNA polymerase of each of these species reveal the much closer evolutionary relationship of old world monkey (OWM) CMVs to HCMV compared to that between new world monkey (NWM) CMVs and HCMV (Fig. II.22.1). This is consistent with the separation of old and new world species occurring prior to the separation of great apes and monkeys. The closest relatives of HCMV that are likely to be most relevant for virus–host interactions and that can be experimentally studied are, therefore, CMVs of old world NHPs. The most commonly used animals in research studies are rhesus macaques as well as cynomolgus macaques. RhCMV was first isolated almost four decades ago (Asher *et al.*, 1974), and the pathogenesis of the virus and the function of multiple viral genes have been extensively studied and are described herein.

Cross-species transmission

Despite the close relationship of CMVs isolated from different primate species, evidence for naturally occurring cross-species infection is very limited. For instance, the recently sequenced cynomolgus CMV (Ambagala *et al.*, 2011; Marsh *et al.*, 2011) shows a very high conservation on the DNA and amino acid level to all published RhCMV sequences. The relationship is so close that CyCMV could be characterized as a sub-strain of RhCMV (RhCMV-Ottawa, see Chapter I.1). An explanation for this phenomenon could be found in new evidence suggesting interbreeding between rhesus and cynomolgus macaques in areas where their natural habitats overlap (Tosi *et al.*, 2002; Otting *et al.*, 2007; Kanthaswamy *et al.*, 2008; Kita *et al.*, 2009; Stevison and Kohn, 2009). This would have opened the opportunity for the virus to cross the species barrier numerous times and could have led to the cospeciation of RhCMV with two different species. On the other hand, it was reported that experimental infection of cynomolgus macaques with RhCMV is not readily apparent suggesting that inter-species transfer of RhCMV might be a rare event (Ambagala *et al.*, 2011; Marsh *et al.*, 2011). Similarly, CMV sequences isolated from wild chimpanzees and gorillas indicated that interspecies transfer occurred after initially separate lineages arose by cospeciation (Leendertz *et al.*, 2009). Thus, it may be that chimpanzees and gorillas can be infected with each other's CMV following natural exposure to infectious virus. There is also only extremely limited and tentative evidence for zoonotic infections of humans by primate CMVs. In one case, CMV was isolated after passage of a brain biopsy sample from a child with clinical encephalopathy on human fibroblasts (Huang *et al.*, 1978). DNA hybridization and sequence analyses identified the virus as a simian CMV (Colburn strain) and not a new strain of HCMV. No study has been reported investigating in further detail the potential for a true zoonotic infection. In a second case, another African green monkey CMV isolate (stealth virus 1) was cultured from the blood of a 43-year-old patient suffering from Chronic Fatigue Syndrome (Martin *et al.*, 1994, 1995). A direct connection between the isolated virus and clinical symptoms either patient suffered from was not established. More interestingly, there are no reports that these patients have had any contact with NHPs, so the source of their putative SCMV infection remains unverified. The circumstances were very different in a third case; here a 35-year-old man with end-stage liver disease secondary to infection with hepatitis B virus received a liver transplant from a BaCMV-positive baboon (*Papio Anubis*) (Michaels *et al.*, 2001). Because the patient was known to be HCMV positive, a prophylactic ganciclovir treatment was administered to avoid viral reactivation and graft rejection. After the ganciclovir administration was lifted, both HCMV and BaCMV could be detected by PCR in the patient's peripheral blood leucocytes (PBL). It is not clear, however, whether BaCMV ever spread to human tissue, or whether the virus was rather detected in baboon leucocytes from the xenotransplant mixed in with the patient's PBLs. Nonetheless, from these anecdotal observations it seems possible that NHP CMVs can infect humans, although this is unlikely to occur naturally. From a biosafety perspective it is worth emphasizing that human fibroblasts and some human

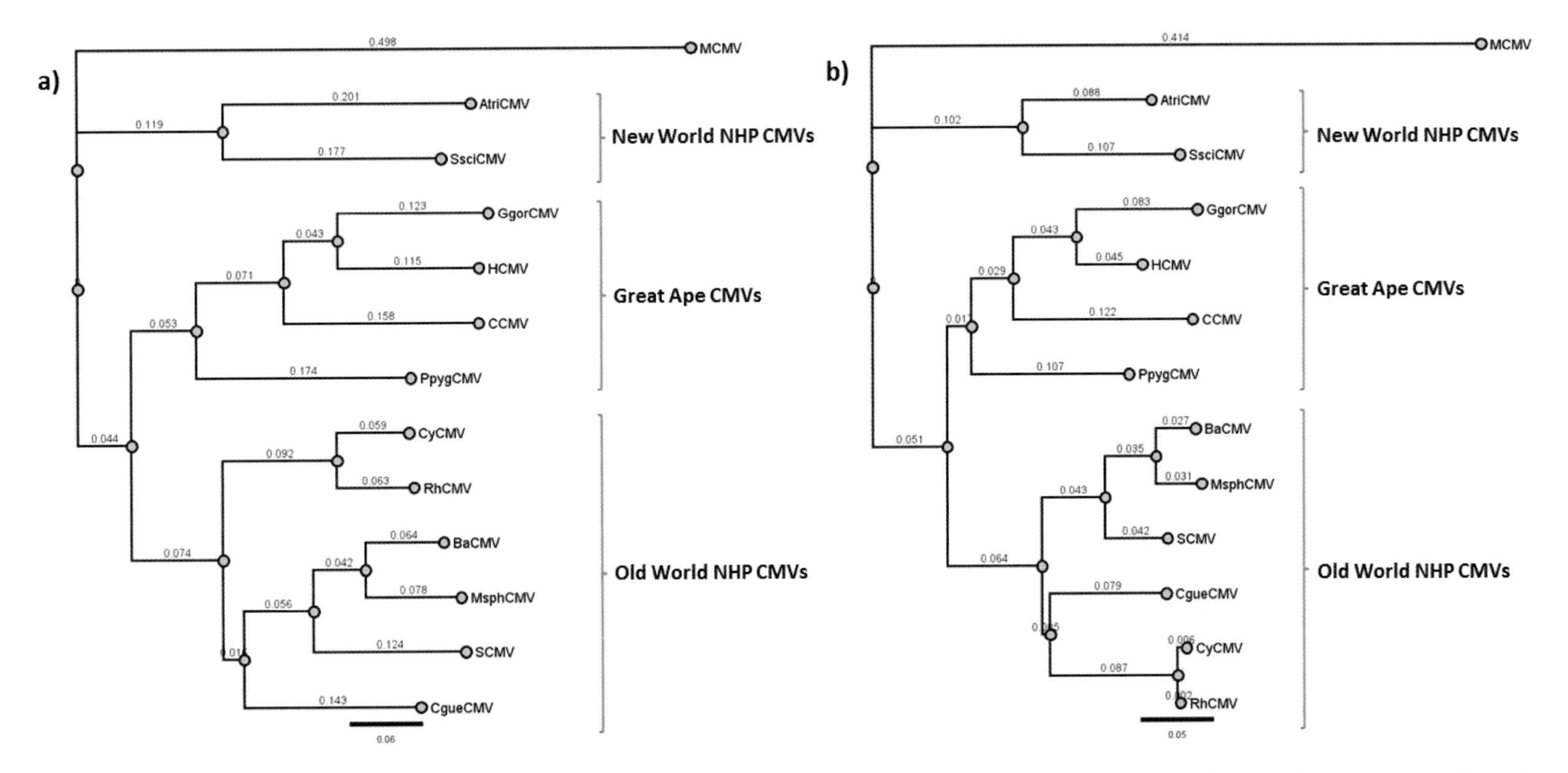

Figure II.22.1 Phylogenetic tree of (A) gB and (B) the viral DNA polymerase for great ape and monkey CMVs. Accession numbers of the protein sequences used to make this figure are HCMV (X17403), CCMV (NC_003521), GgorCMV (FJ538490), PpygCMV (AY129396), RhCMV (AY186194), BaCMV (AC090446), SCMV (FJ483968), CyCMV (AY728171), MsphCMV (AY129399), CgueCMV (AY129397), AtriCMV (FJ483970) and SsciCMV (FJ483967). MCMV (GU305914) was used as an outgroup in both graphs. The phylogenetic trees were made using Geneious Pro 5.5.2.

endothelial cells are in fact fully permissive for NHP CMV replication *in vitro* (Alcendor *et al.*, 1993; Rivailler *et al.*, 2002; Lilja and Shenk, 2008).

While such putative zoonotic infections have been rare, these observations provide suggestive *in vivo* evidence for the conservation of gene function between the viral species, including genes that are non-essential for growth *in vitro* but modulate host immune responses, latency and tissue tropism *in vivo*. However, zoonotic infections seem to be limited to the originally infected host since spreading from the original host to other individuals has never been reported. Thus, while many gene functions are conserved, there are also clear species-specific gene functions and adaptations. In the first part of this chapter, we will thus summarize our current knowledge of conserved and divergent gene functions between NHP CMVs and HCMV, with particular focus on RhCMV. In the second part, we will summarize *in vivo* observations of rhesus macaques naturally exposed to or inoculated with RhCMV.

Conservation of gene sequence and function of NHP CMVs compared to HCMV

Genome content of non-human-primate CMVs compared to HCMV

The recent sequencing of several NHP genomes reveals a high degree of genomic conservation across OWM with many of the genes being more than 90% identical across all species. All NHP CMV genomes are similar in length (about 220,000 bp compared to ~230,000 bp for HCMV) and lack internal repeats. Thus, NHP CMV genomes do not have a unique long (UL) or unique short (US) region, and they do not isomerize. The analysis of the genomes of two fibroblast-adapted strains of RhCMV (68-1 and 180.92) and the coding content of the genomic region corresponding to the UL/b′ region of HCMV from a low passage isolate of RhCMV indicated that the RhCMV genome might encode for over 260 ORFs of more than 100 aa (Hansen *et al.*, 2003; Rivailler *et al.*, 2006; Oxford *et al.*, 2008). Two nomenclature systems have been used to annotate the coding content of the RhCMV genome. The system used for the prototypical 68–1 strain successively numbers the ORFs beginning with the first ORF on the coding or complementary strand at the left end of the genome. RhCMV genes homologous to HCMV genes are listed with uppercase prefixes (e.g. Rh88), whereas ORFs without sequence similarity to HCMV genes are listed with lowercase prefixes (e.g. rh06) (Hansen *et al.*, 2003). Annotation of the 180.92 strain of RhCMV used a system whereby RhCMV ORFs with homology to HCMV ORFs were listed with the prefix 'Rh' in front of the standard HCMV nomenclature for an ORF (e.g. RhUL55) (Rivailler *et al.*, 2006). Here, we adopt the nomenclature established in Chapter I.1.

Of the more than 260 predicted ORFs in the RhCMV genome mentioned above, 172 have orthologues in either OWM CMV or HCMV (Fig. II.22.2). With one exception (Rh91.1), the remaining ORFs most likely do not represent actual genes since all of these ORFs are small (< 300 aa with an average length of 138 aa) and most of them overlap in their entirety with conserved ORFs. This contrasts with conserved ORFs that have an average length of 366 aa and do not overlap completely with other ORFs. This comparative genomics analysis therefore suggests that only very few, if any, truly RhCMV-specific genes exist. One gene family within the RhCMV genome that might contain RhCMV-specific genes is the RL11 family of predicted glycoproteins (Fig. II.22.2). This gene family consists of more than a dozen family members that show a large sequence variation even among different strains of the same virus. Within the 5′ terminal portion and the genomic region corresponding to the UL/b′ region of HCMV, many genes are NHP CMV-specific without apparent orthologues in HCMV. In contrast, almost all of the genes encoded in the central part of the NHP CMV genome, roughly from ORF Rh31 (RhUL13) to ORF Rh160 (RhUL132) in RhCMV, are conserved in all old world primates since they have clear orthologues in the HCMV genome. This includes genes for all structural proteins, including glycoproteins, tegument proteins, and the major immediate-early (MIE) transactivator genes IE1 and IE2. Importantly, even NHP CMV gene products that have very low homology to HCMV can have the same function (Pande *et al.*, 2005).

Conserved genes with conserved functions

Most of the centrally located genes in the NHP CMV genomes are likely functionally conserved given their high level of homology. Examples for this class of genes are structural proteins such as glycoproteins or tegument proteins as well as key transcriptional regulators such as the MIE proteins. Experimentally, conserved functions have been demonstrated for RhCMV gB (RhUL55/Rh89) that is proteolytically processed similarly to HCMV gB (Kravitz *et al.*, 1997; Kropff and Mach, 1997). Another example is the gL protein, which was shown to be essential for RhCMV and HCMV replication (Bowman *et al.*, 2011). Interestingly, although a cell line expressing RhCMV gL (RhUL115/Rh147) was able to complement a RhCMV null mutant, an

Figure II.22.2 Predicted RhCMV ORFs with homology to HCMV and OWM NHP (green), or to OWM NHP only (red). The RL11 family is indicated in yellow. The map presented here is based on a RhCMV sequence combined out of the two fully sequenced RhCMV strains 68–1 (AY186194) and 180.92 (DQ120516) and the RhCMV UL/b′ region sequenced from non-passaged virus (EF990255). Geneious Pro 5.5.2 was used to generate this map.

HCMV null mutant could not be complemented by this cell line. Similarly, a cell line expressing HCMV gL could complement an HCMV gL null mutant but not its RhCMV homologue. The MIE proteins IE1 (IE72) and IE2 (IE86) of HCMV are expressed from the MIE promoter (MIEP) and are splice variants of each other (Stenberg *et al.*, 1984; 1989, 1990, 1996; Stenberg and Stinski, 1985; Pizzorno *et al.*, 1988; Meier and Stinski, 1996). Both IE1 and IE2 play important regulatory roles in viral replication by transactivating early (E) gene expression, and, if IE2 is deleted, no viral gene other than the genes expressed with IE kinetics will be transcribed (Marchini *et al.*, 2001; White *et al.*, 2004). In addition to transactivating functions, IE1 was shown to inhibit interferon signalling by inhibiting the Jak-STAT pathway by binding to STAT 2 (Paulus *et al.*, 2006; Huh *et al.*, 2008; Krauss *et al.*, 2009). IE2 also has a role in innate immune evasion in that it blocks NFκB signalling by preventing the binding of NFκB to the IFN-β promoter (Taylor and Bresnahan, 2005, 2006a,b). The MIE proteins IE1 and IE2 are conserved in all primate CMVs, and their exact origination in five exons homologous to HCMV was described in detail for RhCMV, SCMV and CCMV (Jeang and Gibson, 1980; Jeang *et al.*, 1982; 1984; 1987; Alcendor *et al.*, 1993; Chiou *et al.*, 1993; Chang *et al.*, 1995; Barry *et al.*, 1996; Davison *et al.*, 2003). At present it is not known if the innate immune evasion of IE1 or IE2 is conserved in NHP CMVs.

Also conserved in NHP CMVs is the viral kinase (UL97) that, in HCMV, is responsible for the phosphorylation of a number of viral and host proteins (Kawaguchi *et al.*, 1999; Krosky *et al.*, 2003; Marschall *et al.*, 2003, 2005; Baek *et al.*, 2004; Romaker *et al.*, 2006; Hume *et al.*, 2008; Prichard *et al.*, 2008; Hamirally *et al.*, 2009; Thomas *et al.*, 2009; Becke *et al.*, 2010; Milbradt *et al.*, 2010; Webel *et al.*, 2011) and will also autophosphorylate (Schregel *et al.*, 2007). UL97 is required for the phosphorylation and activation of ganciclovir, which then inhibits viral DNA strand synthesis. RhCMV can function as a perfect animal model here, because the UL97 kinase and the viral DNA polymerase have been shown to be functionally conserved between the human and the rhesus virus (Swanson *et al.*, 1998). This was also shown for other anti-herpesviral drugs directed against conserved CMV genes like UL89 and UL56 (North *et al.*, 2004).

HCMV UL69 is a late nuclear phosphoprotein that facilitates nuclear mRNA export by shuttling between the nucleus and the cytoplasm and recruiting components of the cellular-mRNA export machinery (Winkler *et al.*, 1994; Winkler and Stamminger, 1996; Lischka *et al.*, 2006) (see also Chapter I.11). Another characteristic is that this protein forms a homomultimer (Lischka *et al.*, 2007). All these properties are conserved between primate CMVs, even to the extent that closely related UL69 proteins from different species, including RhCMV, can form functional heteromultimers if they are co-expressed (Zielke *et al.*, 2011). Some conserved genes occur in different copy numbers between HCMV and NHP CMVs. Examples are phosphoprotein 65 (pp65), the chemokine receptor US28, and a cluster of CXC chemokine-like ORFs in the UL/b′ region. In each case, NHP CMVs contain more copies, which are tandemly arrayed, than HCMV. In the case of pp65, HCMV contains one copy whereas all fully sequenced OWM CMVs contain two copies. Interestingly NWM CMVs contain only one copy of the gene and resemble Great Ape CMVs in this regard. Similar to HCMV pp65, Rh-pp65-2 (RhUL83-2/Rh112) is expressed with late (L) kinetics, localizes to the nucleus, is a constituent of the virion tegument, and is a target of both humoral and cellular immunity (Yue *et al.*, 2006). Moreover, unpublished work from our groups suggests that, similar to HCMV, pp65 (both copies) of RhCMV are non-essential for growth *in vitro*. Interestingly, pp65-deleted RhCMV is able to establish a persistent infection in rhesus macaques (our unpublished observations). In the case of US28, RhCMV contains five copies whereas HCMV contains just one (Hansen *et al.*, 2003; Rivailler *et al.*, 2006). The five RhUS28 ORFs (1) exhibit considerable genetic divergence in pair-wise comparisons with each other and with HCMV US28, (2) are expressed with E kinetics, and (3) have distinct functional activities (Penfold *et al.*, 2003). Of the five RhUS28 ORFs, only one (RhUS28.5) has demonstrable binding activity for the chemokine fractalkine, which it internalizes comparable to HCMV US28 (Kledal *et al.*, 1998). Moreover, both US28 and RhUS28.5 were localized to the envelope of infectious virions, suggesting a potential mechanism for targeting cells that are expressing CC chemokines and the surface-bound fractalkine. Finally, HCMV UL146 is a CXC chemokine that strongly stimulates positive chemotaxis of neutrophils (Penfold *et al.*, 1999). Both RhCMV and CyCMV encode six CXC-like ORFs that are located immediately downstream of RhUL146 (Marsh *et al.*, 2011; Oxford *et al.*, 2008). Two of these ORFs, RhUL146a and RhUL146b, may be duplications of RhUL146, and likewise Rh161.1 and Rh161.2, appear to represent a duplication of a single progenitor, Rh161. Finally, RhCMV encodes a CXC-like chemokine Rh158, a homologue of the HCMV CXC-like chemokine UL147. Like RhUL146 and UL146, these other ORFs in RhCMV and CyCMV contain ELRCXC-like motifs. Alpha chemokines contain a hallmark motif (CXC), and target cell activation requires integrity of the three residues (ELR) immediately preceding the CXC motif (Clark-Lewis *et al.*, 1991).

Whereas the preceding examples illustrate amplification of certain ORFs in the NHP genomes, the reverse situation is found for the US6 family (see below) where HCMV contains eight members (US2, 3, 6, 7, 8, 9, 10, 11) and RhCMV contains only five members, namely Rh182(RhUS2), Rh184(RhUS3), Rh185(RhUS6), Rh186(RhUS8) and Rh189(RhUS11). The functional significance of these gene duplications is presently not clear, but it is maybe that they represent a compensatory gene amplification during cospeciation with the NHP host.

Tissue tropism genes

It is now well-established that serial passage of clinical HCMV and RhCMV isolates in fibroblasts can lead to rapid rearrangement, deletions, and accumulation of point mutations within the UL/b′ region of the genome, which encodes ORFs involved with immune modulation and cell tropism, in addition to point mutations within other ORFs outside of UL/b′ (Cha *et al.*, 1996; Davison *et al.*, 2003; Hansen *et al.*, 2003; Murphy *et al.*, 2003; Hahn *et al.*, 2004; Rivailler *et al.*, 2006; Murphy and Shenk, 2008; Oxford *et al.*, 2008; Dargan *et al.*, 2010). Absence of UL/b′ (UL128–UL150) greatly restricts the cell tropism of HCMV to fibroblasts. Loss of tropism for epithelial and endothelial cells, and missing transfer of virus from infected endothelial cells to leucocytes is due to absence of UL128, UL130, and/or UL131A (Hahn *et al.*, 2004; Ryckman *et al.*, 2006, 2008) (see also Chapter I.17). These proteins, together with gH/gL (the UL128 complex), facilitate low pH-dependent endocytic entry into epithelial and endothelial cells (Ryckman *et al.*, 2006, 2008) and transfer of virus from endothelial cells to polymorphonuclear leucocytes (Gerna *et al.*, 2000, 2002, 2003; Revello *et al.*, 2001; Hahn *et al.*, 2004; Wang and Shenk, 2005; Adler *et al.*, 2006).

RhCMV encodes orthologues of HCMV UL128, UL130, and UL131A (RhUL128, RhUL130, and RhUL131A), in addition to gH and gL (Hansen *et al.*, 2003; Bowman *et al.*, 2011). Formation of a quaternary complex of RhUL128, RhUL130, RhUL131A, gH, and gL is presumed, based on the precedent of the HCMV UL128C, but this remains to be formally proven. Individual Basic Local Alignment Search Tool (BLAST) analysis of these RhCMV proteins with their respective HCMV orthologue shows extensive genetic drift from the HCMV proteins (% aa identity: UL128 = 40; UL130 = 36; UL131A = 30). In addition, each of these three RhCMV proteins is larger than its HCMV counterpart. Mutational analyses of HCMV UL130 and UL131A identified clusters of charged amino acids, which when mutated to non-charged amino acids, abrogated infection of human umbilical vein endothelial cells (HUVEC). BLAST alignments of RhUL131A/UL131A and RhUL130/UL130 shows there is generally a strong positional conservation of charged amino acids, including those clusters of charged amino acids identified as being important for HUVEC tropism (Schuessler *et al.*, 2010; 2012) (Fig. II.22.3). In addition, cysteine residues are also strongly conserved, suggesting that the individual proteins of the pentamer may assume similar conformations during formation of the complex.

Similar to HCMV, fibroblast-adapted strains of RhCMV, e.g. RhCMV 68-1, lack a functional UL128–UL131A complex due to genomic rearrangements of the RhCMV UL/b′ region during propagation of the 68-1 strain. Another example is the SCMV strain Colburn, which has a frame shift in its UL128 gene, leading to a premature stop codon after amino acid 89 that shortens the protein to 1/3 of its full length, most likely rendering it non-functional (Aaccession no. FJ483969). Consistent with the phenotype of HCMV strains lacking a functional UL128C, 68-1 is severely impaired for infection of rhesus retinal pigmented epithelial cells (RRPEC) (Lilja and Shenk, 2008). Engineering 68-1 *in vitro* to express intact copies of RhUL128 and RhUL130 (BRh68-1.2) restored infectivity for RRPEC. Notably, BRh68-1.2 also infected and replicated to high copy numbers in primary human retinal pigmented epithelial cells and HUVEC, whereas neither cell type supported productive infection with 68-1. Taken together, the protein alignments and the cross-species infectivity of BRh68-1.2 in human cells suggest that RhUL128C and UL128C form conserved quaternary structures that bind to a conserved receptor.

A systematic mutagenesis analysis of genes conserved between rhesus and human CMV that are non-essential for growth in fibroblasts revealed additional conserved genes that augment epithelial cell tropism, namely Rh01 (HCMV TRL1), Rh159 (HCMV UL148), Rh160 (HCMV UL132), and Rh203 (HCMV US22) (Lilja *et al.*, 2008). These studies were performed using the 68-1 strain, which is impaired for RRPEC infectivity due to loss of RhUL128 and RhUL130. Thus, there may be additional epithelial and endothelial determinants identified using a fully tropic strain of RhCMV. Moreover, it is not known whether the HCMV homologues likewise control infection of epithelial cells. Interestingly, RhCMV lacking R01 was still able to enter epithelial cells but showed reduced spreading, indicating that this gene controls a post-entry checkpoint in epithelial cells.

Unlike HCMV, however, RhCMV lacking UL128

```
RhUL130    34  FCP-MYPSPRQNFGLFTSYKTHPTGPECGNTSLYVFHNRYNQYLIERPSAWSNKLAFYLS  92
UL130      56  YCPFLYPSPPRSPSQFSGFQRVSTGPECRNETLYLLYNREGQTLVERSSTWVKKVIWYLS  115
CyUL130    39  FCP-MYPSPRQNFGLFTSYKTHPTGPECGNTSLYVFHNRYNQYLIERPSAWSNKLAFHYS  97
Consensus       CP  YPSP      F       TGPEC N  LY   NR  Q L ER S W  K     S

RhUL130    93  ALHSPVFQKFSKMATS-STTSMNITEEEKKTFAAHMIPMRSTILRYIVKDGTDVEHCQMR  151
UL130     116  GRNQTILQRMPRTASKPSDGNVQISVEDAKIFGAHMVPKQTKLLRFVVNDGTRYQMCVMK  175
CyUL130    98  QLHSPVFQKISKMATSP-ATSMNITEEEKKTFAAHMIPMRSTILRYIVKDGTDVEHCQMR  158
Consensus             Q+  + A         I  E+ K F AHM P     LR  V DGT    C M+

RhUL130   152  VITWAKTEANF-ISFKVKIELSNAYRRPSSICTRPNLVV  189
UL130     176  LESWAHVFRDYSVSFQVRLTFTEANNQTYTFCTHPNLIV  214
CyUL130   159  VITWAKTEANF-ISFKVKIELSNAYRRPSSICTRPNIFV  196
Consensus         WA        SF V+     A       CT PN  V

RhUL131A    1  MGACRVFITVVTVCMLLCAFTVQGSICTKEFGNWNPNYRPDGYWDICSRQIDAPLRNQLL  60
UL131A      1  MRLCRVWLSVC-----LCAVVL--GQCQRETAEKNDYYRVPHYWDACSRALPDQTRYKYV  53
CyUL131A    1  MGACRVFITV-TVCVLFCAFTVQGSICTKEFGNWNPNYRPDGYWDICSRQIDATLRNQLL  59
Consensus      M  CRV   V      LCA       C +E    N  YR   YWD CSR      R

RhUL131A   61  EKIINASVSYHYATSHNHDDVSLLKRINVTEVALVVNSVQVRPGEIDECLYRQQPEEEIK  120
UL131A     54  ERLVDLTLNYHYDASHGLDNFDVLKRINVTEVSLLISDF-----------RRQNRRGGTN  102
CyUL131A   60  EKIINASVSYHYATSHNHDDVSLLKRINVTEVALVVNSVQVRPGEIDECLHRQQPEEEI-  118
Consensus      E+       YHY  SH  D    LKRINVTEV L                 RQ

RhUL131A  121  KNTKQPQLTRRIGLIKDYISAKKLITFAASGSHTSQSRILTVAIRL  166
UL131A    103  KRT----------------------TFNAAGSLAPHARSLEFSVRL  126
CyUL131A  121  KNTKQPQLTRRIGLIKDYISAKKLFTFAASGSHTSQSRILTIAIRL  166
Consensus      K T                      TF A GS     R L    RL
```

Figure II.22.3 Alignments of the primate CMV UL130 and UL131A proteins. Top three panels: BLAST alignment of HCMV UL130 (Accession #:YP_081565) with the UL130 proteins of RhCMV (RhUL130, Accession #: ABV45241) and CyCMV (CyUL130/cy174, Accession #s: AEQ32264 and AEQ32263). CyUL130 represents the putative spliced product of the CyUL130 and cy174 ORF, based on the RhUL130 protein that is the product of the spliced RhUL130/rh157.4 transcript (Lilja and Shenk, 2008; Oxford *et al.*, 2008). Charged amino acids (D, E, H, K, R) are presented in bold. Double-underlined amino acids represent those amino acids, which, when mutated, exhibit an intermediate reduction of epithelial tropism; solid underlined amino acids represent amino acids critical for epithelial cell tropism (Schuessler *et al.*, 2010; Schuessler *et al.*, 2012). Amino acids with grey background represent potential N-linked glycosylation sites (NXS, NXT). Consensus: Amino acids conserved in all three proteins are indicated; conservation of a charged amino acid at a particular position is indicated by '+'. Only the regions of maximal sequence alignment are presented. Bottom three panels: BLAST alignment of HCMV UL131A (Accession #: YP_081566) with RhUL131A (Accession #: ABV45242) and CyUL131A (based on Accession #: AEQ32265 and a predicted second exon between the complement of nucleotides 158,731 – 158,986 of CyCMV – Accession #: JN227533).

is still able to infect and spread in cultured EC (Rue *et al.*, 2004; Carlson *et al.*, 2005). Thus, RhCMV likely contains additional tropism genes that are not conserved between HCMV and RhCMV. One such gene is a homologue to cellular cyclooxygenase-2 (COX-2) (Rh10) (Rue *et al.*, 2004). This gene is conserved throughout all OWM and is only absent in one published monkey CMV sequence, the closely related CyCMV (Marsh *et al.*, 2011). Since vCOX-2 is dispensable for growth in fibroblasts (Rue *et al.*, 2004), it is conceivable that this gene was lost in CyCMV during passage on fibroblasts. Interestingly, HCMV does not encode a vCOX-2 but up-regulates cellular COX-2 expression and COX-2 inhibitors reduce viral replication (Zhu *et al.*, 2002).

In addition, genes of the RL11 gene family have been associated with cell tropism in HCMV (Stanton *et al.*, 2010). As discussed above, this gene family is highly polymorphic among primate CMVs. Up to now, a role of RhCMV RL11 gene family members in cell tropism has not been established.

Latency-associated genes

As discussed in greater detail in other chapters of this book (Chapters I.19 and I.20), the non-coding transcript LUNA and the protein-encoding transcript UL138 are specifically expressed in cells latently infected with HCMV. The UL138 gene is part of several multicistronic transcripts that also encodes UL133, UL135 and UL136. Both UL133 and UL138 have been identified as critical viral proteins that influence the outcome of infection in different cell types, including haematopoietic progenitor cells (Umashankar *et al.*, 2011). Interestingly, no positional homologues are found in NHP CMVs, apart from CCMV. However, it has been noted that Rh166 exhibits low amino acid identity with HCMV UL133 (27%) and UL138 (35%), and Rh171 is 28% identical with HCMV UL133, suggesting that RhCMV may encode functional homologues of HCMV UL133 and UL138. (Umashankar *et al.*, 2011). It remains to be established whether latency-associated transcripts exist in RhCMV and whether these genes impact the ability to establish and maintain a chronic infection in rhesus macaques.

Immunomodulatory genes

Since host proteins involved in the immune response, e.g. MHC molecules, evolve more rapidly than non-immune proteins, one would expect that the greatest differences between human and NHP CMVs would be observed in immunomodulatory gene products. Therefore, it is probably safe to assume that many of the NHP-CMV-specific genes will be non-essential for growth *in vitro* but play a role in immune evasion *in vivo*. This assumption is supported by the finding that an RhCMV deletion mutant lacking a large region of the genome which encodes for the majority of RhCMV-specific genes was able to replicate *in vitro* (Powers and Früh, 2008). The host immune response to HCMV is multilayered and involves both innate and adaptive arms of the immune system (see Chapters II.7, II.8 and II.10). Moreover, HCMV seems to have developed countermeasures against most of these responses. As discussed next, some of these countermeasures are conserved in NHP CMVs whereas others are not. In addition, NHP CMVs also contain unique immunomodulatory genes not found in HCMV.

Evasion of intracellular innate defence mechanisms

Virus infection of host cells triggers the production of interferon via pattern recognition receptors that activate the IFN gene via a signal transduction cascade involving IRF3 (see also Chapters I.8 and I.16). In case of HCMV, it was shown that this involves the Z-DNA binding protein ZBP-1 and the ER-associated adaptor protein STING (DeFilippis *et al.*, 2010). A major difference between RhCMV and HCMV is the finding that RhCMV does not induce an IFN response in infected fibroblasts whereas HCMV induces IFN and IFN-stimulated genes (ISGs) (DeFilippis and Früh, 2005). This is due to the fact that RhCMV inhibits activation of IRF3 whereas HCMV does not inhibit IRF3 activation (DeFilippis and Früh, 2005). The RhCMV ORF responsible for IRF3 inhibition has so far not been identified, but it is interesting that RhCMV is unable to inhibit IFN induction in human cells (our unpublished observations) suggesting that this inhibitory factor is highly host-adapted.

Another innate cellular response to viral infection is the initiation of apoptosis by infected cells. Since premature apoptosis prevents the virus from completing its replication cycle, viruses counteract these apoptotic mechanisms (see also Chapter I.15). HCMV inhibitors of apoptosis are conserved in RhCMV. Both the UL36, the viral inhibitor of caspase-8-induced apoptosis (vICA), and UL37, the viral Mitochondria-localized Inhibitor of Apoptosis (vMIA), homologues in RhCMV were able to prevent Fas-mediated apoptosis in HeLa cells, whereas the MCMV UL37 homologue M37 was not (McCormick *et al.*, 2003). Interestingly, the BAC-cloned RhCMV 68-1 (Chang and Barry, 2003) has a mutation in rhUL36 that results in a functionally inactive protein (McCormick *et al.*, 2003). Since 68-1 is able to establish persistent infection (see below), it seems that while RhUL36 might be required for optimal replication and spreading, it does not seem to be essential for the establishment and maintenance of persistent infections in rhesus macaques. Presumably, redundant inhibitory mechanisms can compensate for the lack of a functional RhUL36.

Evasion of innate immune cell stimulation

In mice, NK cells play a pivotal role in controlling viral infection (see Chapters II.9 and II.11). For HCMV, multiple genes have been identified that control various aspects of NK cell activation (Wilkinson *et al.*, 2008; see also Chapter II.8). However, the *in vivo* role of these genes is unknown and it would be interesting to use RhCMV as a model to determine the function of homologous genes *in vivo*. However, most NK cell evasion genes of HCMV are not conserved in RhCMV or other NHP CMVs. Humans express at least eight NKG2D ligands (NKG2D-L): MICA/B, ULBP1–3, RAET1E (ULBP4), RAET1G (ULBP5), and RAET1L (ULBP6). All of these ligands are conserved throughout all primate species examined so far, both OWM (Seo *et al.*, 2001; Shiina *et al.*, 2006; Averdam *et al.*, 2007; Doxiadis *et al.*, 2007) and NWM (Shiina *et al.*, 2011). Although all of the NKG2D-L show high polymorphism within the different species, the high overall conservation between species is interesting (Romphruk *et al.*, 2009; Antoun *et al.*, 2010; Naruse *et al.*, 2011). In HCMV, UL16 retains the NKG2D ligands ULBP-1, 2, 6 and MIC-B, and UL142 retains the NKG2D-L MIC-A and ULBP-3 (Wilkinson *et al.*, 2008). However, neither UL16 nor UL142 is conserved in RhCMV. Also not conserved is HCMV UL18, an MHC-I homologue that binds to the inhibitory leucocyte immunoglobulin-like receptor (LIR)-1 expressed on myeloid and lymphoid immune cells (Cosman *et al.*, 1999; Yang and Bjorkman, 2008). The complete absence of RhCMV homologues of HCMV genes inhibiting NKG2D-L is somewhat surprising given the conservation of NKG2D and its ligands. Given the fact that down-regulation of NKG2D-L is also observed in MCMV (Chapter II.9; Lenac *et al.*, 2008) it is safe to assume that NHP CMVs likely contain non-conserved ORFs that prevent NKG2D stimulation. The identity of these genes is currently unknown.

Also, an MICB-targeting miRNA of HCMV, miR-UL112, is not conserved in RhCMV (J.A. Nelson, personal communication), and it remains to be demonstrated whether a similar mechanism of inhibiting MIC-B expression exists for RhCMV. It is also not known whether the observed function of HCMV pp65 in inhibiting NK cell responses by dissociating the CD3ζ signalling chain from NKp30 (Arnon *et al.*, 2005) is conserved in one of the RhCMV pp65 proteins.

However, two NK cell evasion functions seem to be conserved in RhCMV: HCMV UL40 and UL141. HCMV UL40 ensures the expression of HLA-E molecules on HCMV-infected cells (Braud *et al.*, 2002). On uninfected cells, the non-polymorphic HLA-E associates with a highly conserved peptide derived from the signal peptide of classical HLA-A, B, C molecules. Signal-peptide carrying HLA-E is transported to the cell surface where it can engage the inhibitory receptor NKG2A/CD94 on NK cells, thus inhibiting NK cell killing. However, since HCMV inhibits TAP (see below) and thus translocation of HLA-derived signal peptides, HLA-E would be absent from the cell surface unless loaded with a decoy peptide provided by UL40. Mamu-E, the rhesus macaque homologue of HLA-E, is highly conserved and seems to perform a similar function in macaques (Dambaeva *et al.*, 2008). Moreover, RhCMV RhUL40 (Rh67) contains a nonamer peptide in its signal sequence that corresponds to the highly conserved nonamer peptides bound by HLA-E and present in signal peptides of MHC-I molecules (Richards *et al.*, 2011). CyCMV UL40 also contains the same nonamer sequence as RhUL40, and CyUL40 and RhUL40 are highly conserved (Marsh *et al.*, 2011). In contrast, the remaining peptide sequence of RhUL40 is highly divergent from UL40 so that this homology was missed in the initial annotation of RhCMV (Hansen *et al.*, 2003). While it has yet to be demonstrated conclusively that the RhUL40-derived peptide is bound to Mamu-E and triggers inhibition of NK cells, it seems highly likely that this function is conserved in RhCMV.

Another conserved gene, HCMV UL141, has been shown to down-regulate both CD155 and CD112, which represent ligands of the NK cell receptors DNAM-1 (CD226) and TACTILE (CD96) (Tomasec *et al.*, 2005; Prod'homme *et al.*, 2010). The RhUL141 orthologue is encoded by Rh164 (Hansen *et al.*, 2003). It remains to be demonstrated whether this ORF is capable of down-regulating the rhesus macaque homologues of CD155 and CD96, but given the primary structure homology this seems very likely. Interestingly, RhCMV strain 180.92 lacks 13 ORFs, including RhUL141, that are present in low-passage RhCMV strains (Oxford *et al.*, 2008). The loss of coding capacity within 180.92 was apparently due to genomic rearrangements during fibroblast adaptation (Rivailler *et al.*, 2006). There are no published reports describing the growth parameters of 180.92 after experimental inoculation of rhesus macaques with RhCMV 180.92. Therefore, it is not yet known whether the absence of RhUL141 has an impact on growth of RhCMV *in vivo*. The absence of RhUL141 has no apparent defect on replication fitness in cultured fibroblasts (Lilja and Shenk, 2008).

Chemokines, cytokines, and chemokine receptors

Chemokine receptors

The HCMV genome contains four proteins encoding for G protein-coupled receptors (GPCR) that show homology to human chemokine receptors, UL33, UL78, US27, and US28 (Rosenkilde *et al.*, 2001, 2008). The RhUL33 (Rh56) and RhUL78 (Rh107) proteins are highly conserved in RhCMV, whereas the RhUL27 protein is not. Instead, RhCMV contains five homologues to the HCMV chemokine receptor US28, a unique 7-transmembrane domain gene family not found in rodent CMVs. One of these, RhUS28.5, has been shown to have a similar ligand binding profile as HCMV US28 (Penfold *et al.*, 2003). The US12 gene family of seven-transmembrane proteins shows weak homology to GPCR and is found only in NHP CMVs, including OWM CMVs such as RhCMV (Lesniewski *et al.*, 2006) or SCMV (Sahagun-Ruiz *et al.*, 2004). The function of the US12 gene family is unknown.

Chemokines

Clinical HCMV isolates contain two highly divergent genes (Dolan *et al.*, 2004; Arav-Boger *et al.*, 2006) in their UL/b′ region with strong homology to CXCL chemokines (Penfold *et al.*, 1999), UL146 and UL147 (Cha *et al.*, 1996). This gene family is conserved throughout all primate CMVs, but the number of genes within the family varies between different species. Whereas HCMV has two family members, CCMV has three, UL146, UL146b and UL147 (Davison *et al.*, 2009). Within the OWM, RhCMV (Oxford *et al.*, 2008) and BaCMV (Alcendor *et al.*, 2009) both encode for six viral CXCL-like chemokines while SCMV contains eight different genes (Alcendor *et al.*, 2009). Even NWM CMVs encode this gene family (Alcendor *et al.*, 2009), as was shown for AtriCMV containing two viral CXL chemokine genes, which underlines the potential importance of these viral genes. Despite this sequence conservation, only very little is known about the actual *in vitro* or *in vivo* functions of any of these viral chemokine-like ORFs. HCMV UL146 binds to human CXCLR2 and has potent calcium mobilization and chemotactic properties for neutrophils *in vitro* (Penfold *et al.*, 1999; Sparer *et al.*, 2004). The *in vitro* properties suggest that UL146 might have a role in virus spread and dissemination. Similar results where shown for the CCMV UL146 homologue (Miller-Kittrell *et al.*, 2007). The conservation of these proteins in RhCMV opens the possibility to study the role of these proteins *in vivo*.

Cytokines

One notable distinction between primate and rodent CMVs was the transduction of a cellular IL-10 gene into the genomes of primate CMVs. Viral orthologues of IL-10 (vIL-10) have been identified in HCMV, RhCMV, SCMV, BaCMV and CyCMV (Kotenko *et al.*, 2000; Lockridge *et al.*, 2000; Marsh *et al.*, 2011). The absence of a vIL-10 gene in rodent CMVs is consistent with the interpretation that transduction of a progenitor cellular IL-10 (cIL-10) occurred after the evolutionary split of rodents and primates. For unknown reasons, CCMV does not encode a vIL-10, despite the otherwise strong conservation between the HCMV and CCMV genomes (Davison *et al.*, 2003). As each primate CMV cospeciated with its host, the vIL-10 genes underwent extreme genetic drift from the cIL-10 gene of their host such that the vIL-10 proteins share only 25–27% identity with their host's cIL-10 (Lockridge *et al.*, 2000). The extent of genetic drift in the viral orthologues is highlighted by the facts that (1) primate cIL-10 proteins share > 95% identity, and (2) the vIL-10 are as divergent from each other as they are from the cIL-10 of their host. While sharing only 31% amino acid identity, both cmvIL-10 and rhcmvIL-10 are highly stable in sequence (> 98% identity) amongst different strains of HCMV and RhCMV, respectively (Barry and Chang, 2007; Cunningham *et al.*, 2010) (P. Barry, unpublished). The immunosuppressive functionalities of cmvIL-10 and rhcmvIL-10 are almost identical to those of cIL-10 on multiple lymphoid cell types (Kotenko *et al.*, 2000; Spencer *et al.*, 2002; Chang *et al.*, 2004, 2007, 2009; Raftery *et al.*, 2004; Spencer *et al.*, 2008; Slobedman *et al.*, 2009; Chang and Barry, 2010; Logsdon *et al.*, 2011). There is no evidence that rhcmvIL-10 and cmvIL-10 have evolved new IL-10 receptor (IL-10R)-mediated signalling responses. Inter-specific drift of vIL-10 proteins was likely driven as a compensatory selection to some aspect of their hosts' evolution. It has been shown that the binding affinity of cmvIL-10/IL-10R exceeds that of cIL-10/IL-10R (Jones *et al.*, 2002), and maintenance of the higher binding affinity to the host IL-10R was probably critical in shaping the particular vIL-10 sequence. As a result, the progenitor cIL-10, from which vIL-10 arose, has drifted from what was once a 'self' protein, expressed in the context of viral infection, to one that is now highly recognizable by the host immune system. Almost 100% of rhesus macaques infected with RhCMV develop binding antibodies to rhcmvIL-10, although there is a wide range in antibody titres that neutralize rhcmvIL-10 function (Logsdon *et al.*, 2011; Eberhardt *et al.*, 2012). Binding antibodies to cmvIL-10 have been detected in HCMV-infected humans (Cicin-Sain *et al.*, 2010). Non-functional versions of rhcmvIL-10 have been engineered by introduction of two aa changes in the rhcmvIL-10 region critical for binding to IL-10R (Logsdon *et al.*, 2011). Importantly from a vaccine perspective, immunization of RhCMV-infected rhesus macaques with these non-functional versions of rhcmvIL-10 boosts both binding antibody titres and titres of antibodies that neutralize wild-type rhcmvIL-10 function *in vitro*. Induction of high titres of neutralizing antibodies specific to rhcmvIL-10 does not induce any cross-reactivity to cIL-10. The importance of targeting IL-10 in CMV natural history is emphasized by the observation that MCMV has evolved the capacity to exploit the IL-10 signalling pathway to enable the establishment and maintenance of a persistent infection. While cIL-10 is essential for preventing MCMV-induced immunopathologies (Cheeran *et al.*, 2007; Oakley *et al.*, 2008), MCMV-induced up-regulation of cIL-10 impairs MHC class II antigen presentation (Redpath *et al.*, 1999), antigen-specific T-cell expansion (Jones *et al.*, 2010), and clearance of virally infected cells from sites of persistence, such as the salivary glands (SG) (Humphreys *et al.*, 2007; Campbell *et al.*, 2008). Blockade of IL-10R signalling by treatment with a NT-Ab to IL-10R in MCMV-infected mice leads to CD4 T-cell-mediated reduction of SG infection (Humphreys *et al.*, 2007), the expansion of functional MCMV-specific CD8 T-cells and reduction of relative viral loads in the spleen and lung. Taken together, the results suggest a convergence of evolution in that primate CMVs and MCMV use parallel mechanisms to achieve the same end result in enabling persistence. The role of rhcmvIL-10 *in vivo* is discussed below.

Evasion of humoral adaptive immunity

One way CMVs evade the humoral immune response is by expression of virally encoded Fcγ-receptors (Furukawa *et al.*, 1975; Rahman *et al.*, 1976), which are thought to prevent antiviral immunoglobulin G (IgG) from neutralizing free virus and engaging in antibody-dependent cellular cytotoxicity (ADCC) against infected cells (Jenkins *et al.*, 2004). The E phase protein expressed by the identical ORF TRL11/IRL11 (gp34) in HCMV strain AD169 was the first Fcγ-receptor identified in HCMV (Lilley *et al.*, 2001). A second viral Fcγ-receptor encoded by HCMV was discovered in the spliced product of UL119-UL118 (gp68) (Atalay *et al.*, 2002). gp68 binds to the CH2-CH3 interdomain interface of the Fcγ dimer with its Fcγ binding region amino acid residues 71 to 289 (Sprague *et al.*, 2008). Although both viral Fcγ-receptors show a high degree of homology to host Fcγ-receptors, their homology is to different host proteins. UL119-UL118 relates most closely to the third domain of Fcγ-receptor I, whereas TRL11/IRL11 is reminiscent of the second domain of Fcγ-receptor II/III (Atalay *et al.*, 2002). It is not clear whether the gp34 protein (TRL11/IRL11) of HCMV is conserved in NHP CMVs, because the RL11 family is so divergent that no clear homologue can be identified. UL119-UL118, however, is clearly conserved throughout all NHP CMVs and is encoded by Rh152-Rh151 in RhCMV. Whether these RhCMV ORFs function in the same fashion like their HCMV orthologues remains to be experimentally tested. However, a protein of ~68 kDa, the predicted size of the Rh152-Rh151-encoded Fc-receptor, is co-precipitated with antibodies, thus indicating functional conservation (Powers and Früh, 2008).

Evasion of cellular adaptive immunity

CD8^{+} T-cells recognize virally infected cells via MHC-I molecules presenting peptides derived from intracellular virus at the cell surface. HCMV and RhCMV share a group of genes, termed the US6 family (previously also subgrouped into the US2 and US6 family) that prevent MHC-I antigen presentation to T-cells in a multistep fashion. HCMV encodes four proteins that interfere with various steps of the MHC-I/peptide assembly pathway: US2, US3, US6, and US11 (Tortorella *et al.*, 2000). These genes are functionally conserved in RhCMV (Rh182, Rh184, Rh185, Rh189), although the primary structure homology is very low and often missed by standard BLASTP analysis (Pande *et al.*, 2005). Similar to HCMV, it was shown that Rh184 (RhUS3) retains MHC-I, Rh185 (RhUS6) inhibits the peptide transporter TAP, whereas Rh182 (RhUS2) and Rh189 (RhUS11) lead to the degradation of newly synthesized MHC-I proteins. In addition, RhCMV contains an MHC-I targeting gene found in all OWM CMVs: Rh178 or viral inhibitor of heavy chain expression (VIHCE) (Powers and Früh, 2008; Richards *et al.*, 2011). Rh178 seems to take advantage of the fact that all MHC-I proteins contain highly conserved signal peptides and intercept nascent MHC-I heavy chains at a step prior to protein translocation across the ER membrane. Transfer of the MHC-I signal peptide onto other proteins renders them susceptible to inhibition by Rh178 suggesting that this viral protein is able to distinguish MHC-I from unrelated signal peptides. Interestingly, Rh67 (RhUL40) is resistant to VIHCE despite containing the MHC-I-like nonapeptide in its signal peptide (Richards *et al.*, 2011). The role of US6 proteins and VIHCE in establishing and maintaining infection as well as super-infection has been studied extensively and will be discussed below.

NHP CMV natural history

The natural history of RhCMV in immune competent rhesus macaques, the best characterized NHP CMV model, strongly reflects that of HCMV infection in healthy humans. Key elements include the ubiquity of the virus in the population, long-term excretion of virus in bodily fluids, and the importance of horizontal transmission of virus in effectively maintaining the virus in the population. Recent advances in characterizing RhCMV natural history in mixed populations of infected and uninfected animals now enable the design of rigorous vaccination/challenge studies in NHP that accurately recapitulate the conditions that HCMV vaccinees encounter during repeated mucosal exposure to antigenically distinct strains of HCMV excreted by close contacts.

Seroprevalence

Seroprevalence of CMV infection in NHP populations is exceedingly high. In colony-reared rhesus macaques, almost 100% of animals are serologically reactive to RhCMV antigens by 1 year of age (Vogel *et al.*, 1994), comparable to the near universal infection observed in wild populations of rhesus macaques (Jones-Engel *et al.*, 2006). Similarly, CMV is ubiquitous in both wild and captive populations of other NHP species, including African green monkeys (AGM) (*Cercopithecine aethiops*), baboons (*Papio* sp.), Japanese macaques (*Macaca fuscata*), drill monkeys (*Mandrillus leucophaeus*), and new world marmosets (*Callithrix jacchus*) (Black *et al.*, 1963; Minamishima *et al.*, 1971; Swack *et al.*, 1971; Nigida *et al.*, 1975; Swack and Hsiung, 1982; Ohtaki *et al.*, 1986; Eizuru *et al.*, 1989; Kessler *et al.*, 1989; Blewett *et al.*, 2001, 2003; Andrade *et al.*, 2003). Because of the high seroprevalence, CMV is a frequent 'adventitious contaminant' during culture of primary NHP cells and bodily fluids

(Smith *et al.*, 1969). Restriction endonuclease and sequence analyses indicate that there are multiple strains of genetically distinct CMV endemic within different NHP populations (Swinkels *et al.*, 1984; Eizuru *et al.*, 1989; Alcendor *et al.*, 1993; Oxford *et al.*, 2008). Genomic and sequence analyses of CMVs isolated from different NHP species is most consistent with the interpretation that each host species is infected with its own species-specific CMV. Colony management practices in the past sometimes accommodated co-housing and inter-mingling of different NHP species, affording the potential for cross-species transmission. To date, there is no evidence that such a horizontal cross-species transmission event has occurred after co-housing. One report has observed that seronegative rhesus macaques experimentally inoculated i.p. with SCMV seroconvert to SCMV antigens and remain viruric for 6 months post inoculation (Swack and Hsiung, 1982).

Specific pathogen-free (SPF) cohorts

CMV seroprevalence is dependent on housing conditions. If neonatal animals are separated from the dams at, or immediately after, birth and hand-reared under conditions absolutely restricting contact with infected animals, the animals remain CMV-free well past the age of sexual maturity (Minamishima *et al.*, 1971; Nigida *et al.*, 1979). The animals will remain SPF for other endemic infections, such as other simian herpesviruses or simian foamy virus, as long as they are never exposed to animals shedding these endemic viruses. Using this method of nursery-rearing infants, it has been possible to develop self-sustaining, breeding populations of rhesus macaques that are SPF of RhCMV (Barry and Strelow, 2008) and cynomolgus CMV in cynomolgus macaques (Yasutomi, 2010). Because of this, it is now feasible to maintain RhCMV-SPF animals in sufficient numbers to accommodate increased demand for such animals in future studies, removing a potential bottleneck to continued growth of this NHP model.

Viral excretion in bodily fluids

Congenital infection has never been documented in rhesus macaques and other NHP (discussed in detail below), and there is no evidence either way about the presence of CMV in breast milk. The rapid seroconversion of naïve macaques to RhCMV is undoubtedly due to persistent high-level excretion (i.e. shedding) of RhCMV into bodily fluids, such as saliva and urine, of infected animals. Because of this, uninfected cohorts are constantly exposed to horizontally transmitted virus, and, when coupled with the interactive social behaviour of rhesus macaques, the result is an exceedingly high force of viral infection. One study, measuring seroconversion in animals ($N = 25$) born and raised within their large natal group ($N = 142$ animals total), determined that virus was horizontally transmitted to an uninfected animal every ~35 days (Vogel *et al.*, 1994). Another study showed that, when a single RhCMV-infected adult (6 years of age) was introduced into a co-housed group of 15 uninfected juvenile (i.e. non-sexually mature) macaques, RhCMV was horizontally transmitted to an uninfected animal every 56 days. This latter study demonstrated that a 6-year-old animal was still shedding virus ~5 years after it itself had undergone primary infection during its first year of life. Viral shedding was not unique to this particular animal used in this study. Animals naturally exposed to horizontally transmitted virus can shed virus in high copy numbers in saliva and/or urine for extended periods of time (Asher *et al.*, 1969, 1974; Swack and Hsiung, 1982). One study quantified RhCMV DNA in saliva samples collected weekly over the course of 11 weeks from 14 RhCMV-infected animals and showed that RhCMV could be detected on average in 64% of the animals at each time of sample collection (range 45–82%) (Oxford *et al.*, 2011). Since (1) urine was not analysed in this study, and (2) shedding in saliva and urine are not always contemporaneous, the average frequency of animals shedding in some bodily fluid at each time of saliva collection was likely higher than 64%. As with humans, the frequency of animals shedding RhCMV declines with age. A cross-sectional survey of RhCMV-infected animals determined that 75% of juveniles (3–5 years) had detectable RhCMV DNA in saliva, whereas only 25% of older adult animals (≥ 14 years) were positive for viral DNA in saliva (Oxford and Barry, unpublished). Increased knowledge about viral excretion and horizontal transmission is useful for issues related to management of NHP colonies. In addition, these findings highlight the strengths of the NHP to model clinically relevant aspects of HCMV infection, persistence, and pathogenesis (described below).

Mechanisms of viral excretion

As noted above, horizontal transmission is critical for efficient spread of RhCMV to susceptible hosts. Knowledge about viral mechanisms of trafficking to and shedding from certain tissues, such as the salivary glands and the genitourinary tract, could provide novel insights into potential HCMV vaccine candidates. Primary infection of healthy humans with wild-type HCMV results in prolonged viraemia and shedding of virus in saliva and urine (Zanghellini *et al.*, 1999; Arora *et al.*, 2010). In contrast, no shedding of HCMV is detected following experimental inoculation of human volunteers with the tissue culture-adapted AD169 and Towne strains of HCMV (Elek and Stern, 1974; Just *et al.*, 1975; Quinnan *et al.*, 1984; Plotkin *et al.*, 1989; Zanghellini *et al.*, 1999; Arora *et al.*, 2010). Similarly, inoculation with the Toledo strain of HCMV, generally considered more wild-type-like than Towne and AD169, results in a shorter duration of shedding than individuals naturally exposed to wild-type HCMV. Taken together, the studies with different HCMV strains suggests that coding content differences of the Towne, AD169, and Toledo strains from wild-type HCMV may be responsible for the impaired shedding phenotype, particularly absence of a functional UL128 pentamer complex (UL128C), which is essential for epithelial/endothelial cell tropism (Oxford *et al.*, 2011). It should be noted that uncertainties about the exact coding content of the HCMV strains used at the time of inoculation cloud this interpretation of the human studies (see Chapter I.1). Recent studies in rhesus macaques emphasize the potential importance of ORFs encoded in the region corresponding to HCMV UL/b′ in enabling persistent high-level shedding of virus in saliva and urine.

RhCMV-uninfected animals were inoculated (s.c.) with one of three strains of RhCMV that differed in genetic coding content (Oxford *et al.*, 2011). Two strains (UCD52 and UCD59) contained a full-length UL/b′ region of the genome, compared to naturally circulating RhCMV, but were distinguished from each other by multiple genetic polymorphisms. The third strain, 68-1, was the prototypical RhCMV strain (Asher *et al.*, 1974) that had been extensively passaged on fibroblasts. The 68-1 genome has undergone deletions and rearrangements within UL/b′, such that genes coding for two of the five proteins of the RhCMV UL128C (RhUL128 and RhUL130) have been deleted, along with three CXC chemokine-like ORFs (Hansen *et al.*, 2003; Oxford *et al.*, 2008). Notably, only animals inoculated with either UCD52 or UCD59 exhibited the kinetics, magnitude, and duration of shedding in saliva and urine that were comparable to the pattern of shedding exhibited by animals naturally exposed to wild-type RhCMV. In marked contrast, RhCMV 68-1 was profoundly attenuated for shedding in both urine and saliva, as quantified by qPCR. These results implicate a functional role for the RhUL128C and/or the CXC cytokine-like ORF in mediating trafficking from the site of s.c. inoculation to sites of shedding, such as the salivary glands and the genitourinary tract, as well as possibly enabling a state of persistent viral replication in these distal sites. The full-length genomes of UCD52 and UCD59 have not been sequenced, and a causal role for RhUL128C and the CXC-like ORF in facilitating persistent high-level shedding remains to be formally proven. It should be noted that the attenuation of shedding by RhCMV 68-1 is partial, not absolute. Previous studies have noted sporadic, low-level shedding in bodily fluids following experimental inoculation with 68-1 (Yue *et al.*, 2006, 2008; Abel *et al.*, 2008). In addition, RhCMV 68-1 engineered to express SIV antigens can be persistently recovered from urine following s.c. inoculation (Hansen *et al.*, 2009; 2010). Taken together, these studies point to a role for specific ORFs in the shedding of virus in bodily fluids, particularly those involved in epithelial and endothelial cell tropism.

NHP models

Vaccine studies

RhCMV vaccine studies have primarily focused on those HCMV ORF that are being evaluated for HCMV vaccines. These include the glycoprotein B (gB), phosphoprotein 65 (pp65), and immediate-early 1 (IE1) proteins. RhCMV encodes orthologues of the full complement of HCMV glycoproteins. RhCMV gB, like HCMV gB, is the predominant target of NT-Ab that block infection of fibroblasts (Yue *et al.*, 2003). The gB proteins are 60% identical, although sequence identity is higher (74%) in regions of HCMV gB implicated in oligomerization and viral fusion to the cell membrane (Navarro *et al.*, 1993; Kravitz *et al.*, 1997; Britt *et al.*, 2005). Studies in rhesus macaques defining the ORF specificities of $CD4^+$ and $CD8^+$ responses are limited in the number of different viral ORFs that have been analysed. The choice of ORFs analysed has been based largely on precedents of HCMV. Effector responses have been mostly defined by cytokine responses by antigen-specific T-cells. RhCMV pp65 and IE1 are prominent targets of cellular immune responses, in addition to epitopes within other ORFs, including IE2 and the RhCMV IL-10 protein (Yue *et al.*, 2006; Chan and Kaur, 2007; Berger *et al.*, 2008; Cicin-Sain *et al.*, 2011). CTL responses have not been characterized in terms of epitope specificity. One study, using RhCMV-infected autologous fibroblasts as target cells, determined that a RhCMV E-phase protein(s) represents a large portion of the CTL response (Kaur *et al.*, 1996).

Vaccine studies in rhesus macaques have used different vaccine approaches and routes of RhCMV challenge. In short, the results of these studies demonstrate that outcomes following vaccination and RhCMV challenge included reduced (1) local viral replication, (2) RhCMV viraemia and (3) shedding of RhCMV in saliva. Immunization with plasmid expression vectors for RhCMV pp65-2, gB (deleted of the transmembrane region), and +/− rhcmvIL-10 stimulated pp65-specific

IFN-γ responses in PBMCs (as determined by ELISPOT assay), and gB-binding antibodies (Yue *et al.*, 2007). However, no NT-Ab responses were detected following immunization (1:20 minimum dilution), comparable to the absence of NT-Ab following genetic immunization with MCMV gB (Morello *et al.*, 2002). In addition, only weak and transient rhcmvIL-10-binding antibodies were stimulated following five priming and booster immunizations. i.v. challenge with RhCMV strain 68-1 (10^5 PFU) resulted in significantly reduced peak viral loads in the plasma of immunized animals, compared to unimmunized controls. The peak RhCMV copy number in the plasma of controls was ~ 1.4 logs higher than in vaccinated animals. One notable aspect of DNA immunization was observed following challenge. Whereas DNA immunization did not stimulate NT-Ab pre-challenge, a vigorous memory response was noted within 1-week of challenge. These results suggest that DNA immunization effectively primed the immune system for subsequent antigen presented in the context of viral challenge. A subsequent study extended this interpretation by immunizing three times with DNA (gB, pp65–2, and IE1) followed by two booster immunizations with formalin-inactivated RhCMV virions (FI-RhCMV) (Abel *et al.*, 2008). This protocol had been previously shown to induce protective immunity against MCMV challenge (Morello *et al.*, 2005). For the heterologous DNA/FI-RhCMV immunization study, animals were challenged by s.c. inoculation with altogether 2×10^6 PFU of RhCMV strain 68-1, one-fourth of the inoculum delivered into each of four sites. Biopsies of one of the inoculation sites at seven days post RhCMV challenge showed that prior immunization markedly reduced local challenge virus replication at the site of inoculation. Unvaccinated control animals were noted for a marked mononuclear cell infiltrate and numerous cytomegalic, RhCMV IE1 antigen-positive cells. In contrast, the immunized animals had essentially no inflammatory cell infiltrate and no IE1-positive cells. While the results suggest complete local control of the challenge virus, the development of long-term immune responses in the vaccinated/challenged animals is more consistent with a break-through infection and establishment of a persistent infection, or with stochastic episodes of limited viral gene expression during viral latency as shown in a model of MCMV latency in the lungs (Simon *et al.*, 2006) (see also Chapter I.22 and Chapter II.21). Since the kinetics of memory responses at the site of s.c. challenge have not been defined, it is also possible that an inflammatory response in the vaccinees was resolved by the time of the biopsy. RhCMV was detected sporadically at very low copy numbers in the plasma of three of the five control animals and not detected in any of the vaccinees. Shedding in urine and saliva was also not detected in any of the animals in this study. The minimal detection of plasma DNAemia and the absence of shedding of virus greatly restricted analysis of the effect of vaccination on limiting dissemination of challenge virus from the inoculation to distal sites, such as the salivary glands and genitourinary tract. In addition, these studies were based on homologous vaccination and challenge in that the vaccine antigens were based on 68-1, the challenge virus. In addition, the 68-1 strain of RhCMV is severely attenuated for persistent shedding (described above).

To address these issues, and to provide a better model for a vaccine strategy that might be used in human clinical trials, a subsequent study used a single DNA priming immunization and two modified vaccinia Ankara (MVA) booster immunizations followed by a heterologous viral challenge (Abel *et al.*, 2011). RhCMV-uninfected macaques were immunized by this prime/boost strategy with either gB alone (gB DNA/MVA-gB), or gB/pp65/IE1 (gB-DNA + pp65-DNA + IE1-DNA followed by MVA-gB and MVA-pp65/IE1). A third group was also primed with gB/pp65/IE1 DNA expression plasmids followed by two booster immunizations, but with FI-RhCMV in place of MVA. Animals were challenged s.c. with 10^5 PFU of RhCMV strain UCD52, which unlike 68-1, contains a full-length UL/b′ region of the genome and whose gB protein sequence is 88% identical with that of 68-1 (Oxford *et al.*, 2008; 2011). Following immunization, all three vaccine groups developed NT-Ab that neutralized fibroblast infection, and it was observed that boosting with FI-RhCMV stimulated longer-lasting NT-Ab responses than boosting with either MVA/gB or MVA/-gB/pp65/IE1. Notably, the presence of NT-Ab at the time of RhCMV challenge correlated with reductions in both the frequency and magnitude of RhCMV detected in plasma. However, the presence of NT-Ab was not associated with a reduction in the magnitude of RhCMV DNA in saliva. Animals immunized with gB alone or FI-RhCMV exhibited patterns of shedding that were indistinguishable from that of mock-vaccinated controls. However, one-half of the animals that were DNA/MVA immunized with gB/pp65/IE1 showed significant reductions in the magnitude of RhCMV DNA in saliva (1 to 1.9 logs lower than in the controls). In addition, there was a strong association of memory pp65 T-cell responses post challenge in animals exhibiting the greatest reduction in RhCMV DNA in saliva. IE1-specific T-cells were exceedingly low or undetectable. Taken together, the results demonstrate that NT-Ab are critical for limiting viraemia but that cell-mediated immunity, as measured by pp65 responses, plays a critical role in limiting dissemination of challenge virus from the site of inoculation to distal sites, such as the salivary glands. The results of this study highlight that a DNA/MVA vaccination vaccine approach can achieve a significant reduction in a critical parameter of viral replication post challenge (i.e. oral shedding) but also indicate that additional antigens will be required to achieve a greater level of vaccine-induced protective efficacy to viral challenge.

Viral immune modulation *in vivo*

The HCMV gB vaccine clinical trial demonstrated a delay in the occurrence of primary HCMV infection after vaccination (Pass *et al.*, 2009). The result of this clinical trial is encouraging for the prospects for development of an HCMV vaccine (for a detailed discussion of HCMV vaccines, see Chapter II.20). However, one of the challenges facing an HCMV vaccine recipient is the virus' ability to reinfect those with prior immunity. It is now well established that reinfection of seroimmune women with antigenic variants of HCMV also constitutes a significant source of congenital infection and sequelae (Sohn *et al.*, 1992; Boppana *et al.*, 2001; Gaytant *et al.*, 2002, 2003; Gandhoke *et al.*, 2006; Ross *et al.*, 2006; 2010; Yamamoto *et al.*, 2010; Wang *et al.*, 2011). The mechanisms by which HCMV overcomes prior immunity are not established, but studies in rhesus macaques have shed light on the importance of viral proteins that subvert antigen presentation and host cell activation, signalling, and trafficking.

RhCMV encodes multiple proteins that down-regulate MHC class I antigen presentation, including orthologues of the HCMV US2, US3, US6 and US11 proteins (Rh182–Rh189) (Pande *et al.*, 2005; Richards *et al.*, 2011). Functionalities of these proteins have been defined by *in vitro* experiments. The importance of the US2, US3, US6 and US11 proteins in primate CMV natural history is suggested by the fact that the individual protein pairs are < 30% identical at the amino acid level, yet are functionally conserved. Taking advantage of the rhesus model to investigate the role these viral proteins play in a primate host, Hansen and colleagues deleted the RhCMV ORFs encoding the proteins involved in disruption of surface expression of MHC class I, by engineering in an expression cassette for the SIV gag gene into the deletion sites (Hansen *et al.*, 2010). This study unequivocally demonstrated that (1) RhUS2, RhUS3, RhUS6 and RhUS11 are dispensable for establishment of a primary infection in uninfected rhesus macaques, but (2) are essential for reinfection of an animal with prior immunity to RhCMV. Naïve animals inoculated with a variant lacking these ORFs developed gag-specific immune responses in the periphery and bronchioalveolar lavages (BAL) that were equivalent in kinetics, magnitude, and duration as those observed in animals inoculated with the parental virus expressing US2, US3, US6 and US11, in addition to SIV gag (RhCMV-gag). In addition, the deleted RhCMV variant was recovered in the urine of the infected animals. A markedly different result was observed following inoculation in animals with prior immunity to RhCMV. RhCMV variants lacking RhUS2, RhUS3, RhUS6 and RhUS11 were incapable of reinfecting animals previously infected with RhCMV. Immune animals reinfected with RhCMV-gag developed gag-specific T-cell responses in PBMCs and BAL, and RhCMV-gag was cultured from urine. However, repeated s.c. inoculations of RhCMV-immune animals with RhCMV deleted of RhUS2, RhUS3, RhUS6 and RhUS11 failed to stimulate gag-specific T-cell immunity, and the deleted variant was never recovered in the urine. In sum, this study clearly demonstrates that T-cell evasion, mediated by RhUS2, 3, RhUS6 and RhUS11, is not required to establish a primary infection, but these ORFs are essential for evading effector/memory responses in previously infected animals. These data suggest that viral immune evasion mechanisms enable the CMV to repeatedly infect seropositive individuals despite the presence of significant CMV-specific antibody and T-cell responses. Vaccines against CMV are thus unlikely to prevent infection unless the vaccines are able to induce immune responses that are more efficient than those induced by natural infection.

A separate study characterized the viral and immunological patterns of infection following inoculation of naïve animals with an engineered RhCMV variant that lacked the RhCMV IL-10 gene (UL111A) (Chang and Barry, 2010). Based on the ability of cmvIL-10 to modulate immune functions of multiple cell types *in vitro* (discussed above), expression of this viral protein would likely attenuate host immunity during primary infection and play an important role in the establishment of a persistent infection in an immune competent host. To test this hypothesis, naïve monkeys were inoculated s.c. with 400 PFU of either the rhcmvIL-10 deleted variant (RhCMVΔUL111A) or the parental RhCMV 68–1, which contains an intact version of UL111A and expresses the rhcmvIL-10 protein. Infection with RhCMVΔUL111A was notable for (1) increased innate responses at the site of inoculation and (2) increased acute and long-term B and T-cell responses to RhCMV antigens, compared to infection with the 68-1 expressing rhcmvIL-10. A biopsy of one of the inoculation sites obtained 7 days after RhCMVΔUL111A inoculation revealed a markedly increased mononuclear cell infiltrate, compared to the biopsy from animals inoculated with 68-1. In addition to increased response to viral antigens with RhCMVΔUL111A, there was a distinction in the cell types in the vicinity of RhCMV antigen-expressing cells. With 68-1, macrophages comprised a substantial portion of the cell infiltrate. With RhCMVΔUL111A, macrophages represented a minority of the cells, replaced by mononuclear non-T, non-B-cells, which possibly represented NK cells. The absence of rhcmvIL-10 during infection with RhCMVΔUL111A also stimulated acute innate and adaptive changes in the draining axillary lymph node 2 weeks after inoculation. These included significant increases in both CD11b$^-$ CD11c$^+$ MHC-II$^+$ DCs and CD11b$^+$ CD11c$^+$ MHC-II$^+$ myeloid DCs as well as in the frequencies of RhCMV-specific CD4$^+$ T-cells secreting IL-2, IFN-γ, or TNF-α. While the numbers of B-cells in the axillary lymph node were no different between animals infected with RhCMVΔUL111A or 68-1, the titre of RhCMV-specific IgG was higher in RhCMVΔUL111A animals at 2 weeks post inoculation. This early difference in antibody titres to RhCMV in RhCMVΔUL111A animals remained statistically elevated in the periphery out to 40 weeks post inoculation when the experiment was stopped. Acute changes in the frequency of RhCMV-specific T-cells observed in the lymph node at 2 weeks post inoculation were generally maintained in the periphery during the period of observation. RhCMVΔUL111A-infected animals had a higher frequency of RhCMV-specific CD4$^+$ T-cells secreting IFN-γ or IL-2 than animals inoculated with 68-1. In addition, CD8$^+$ T-cell proliferative responses were higher in RhCMVΔUL111A-infected animals. In sum, these studies support a mechanism in which acute interactions between viral IL-10 and innate effector cells at the site of infection, particularly DCs, translate into alterations of early and long-term adaptive antiviral immunity, potentially facilitating the establishment of a persistence infection. Effects on rhcmvIL-10 deletion on RhCMV replication were limited by the sporadic and exceedingly low titres of RhCMV detected in the saliva and urine of the animals inoculated with the parental 68-1 strain.

NHP Models of immune suppression and CMV

Experimental studies of immunosuppressed NHP species receiving allo- or xenotranplants have shown that prolonged and rigorous immune suppression regimens can lead to the frequent reactivation of the host CMV species (Mueller *et al.*, 2002, 2003, 2005; Gollackner *et al.*, 2003; Bielefeldt-Ohmann *et al.*, 2004; Jonker *et al.*, 2004; Mueller and Fishman, 2004; Teotia *et al.*, 2005; Kean *et al.*, 2006, 2007; Barry and Chang, 2007; Haustein *et al.*, 2008; Han *et al.*, 2010). One example is shown in Fig. II.22.4 in which a cynomolgus macaque received a pig heart xenograft (courtesy of D. Bouley, Stanford Univ.). To prevent rejection of the pig tissue, animals were placed on a rigorous anti-rejection drug cocktail (cyclosporine, methylprednisolone, everolimus, and antithymocyte globulin). The animal developed clinical signs of unknown origin (fever, anorexia, diarrhoea) beginning on day 28 post transplant, and the animal was euthanized on day 36. Histopathological lesions consistent with cynomolgus CMV (CyCMV) were confined primarily to the lungs. These included severe oedema in the adventitia of arteries and veins, multifocal medial necrosis, and numerous prominent cytomegalic cells (Fig. II.22.4, left side). Immunohistochemical staining for infected cells, using a polyclonal antibody to RhCMV IE1, demonstrated an abundance of antigen-positive cells within and adjacent to the vasculature (Fig. II.22.4, right side). These results demonstrate that immunosuppression in NHPs can lead to pathogenic CMV reactivation, similar to HCMV reactivation in immunosuppressed humans who are recipients of solid organ or bone marrow allografts.

A recent study compared different immunosuppression regimes in cynomolgus macaques on the frequency of CyCMV reactivation (Han *et al.*, 2010). This study is the first description in NHP describing an immunosuppression regime that efficiently leads to reactivation of endogenous CMV. Recipient animals were treated with different immunosuppressive regimes prior to and following a hepatic infusion of pancreatic islet cells. The frequency and magnitude of CyCMV reactivation was monitored by longitudinally quantifying CMV genome copy numbers in peripheral blood by qPCR. Animals treated with the combination of thymoglobulin and fludarabine exhibited the highest frequency of animals with reactivated CyCMV with 100% of the treated animals having at least one positive sample for CMV. Animals in this group also had higher copy numbers of CyCMV in blood and for longer duration than animals in groups undergoing other immunosuppression treatments. Pre-emptive and therapeutic treatment with valganciclovir was only partially effective in limiting CyCMV reactivation. This study establishes a research foundation for future immunosuppression studies in NHP related to mechanisms of CMV reactivation, as well as prevention and intervention strategies. Another study involving haematopoietic stem cell transplantation and concurrent IL-2 receptor blockade immunosuppression in rhesus macaques similarly observed that RhCMV was partially refractory to prophylactic cidofovir and valganciclovir and therapeutic treatment with ganciclovir (Kean *et al.*, 2007). Since RhCMV exhibits *in vitro* sensitivities to ganciclovir and benzimidazole nucleosides

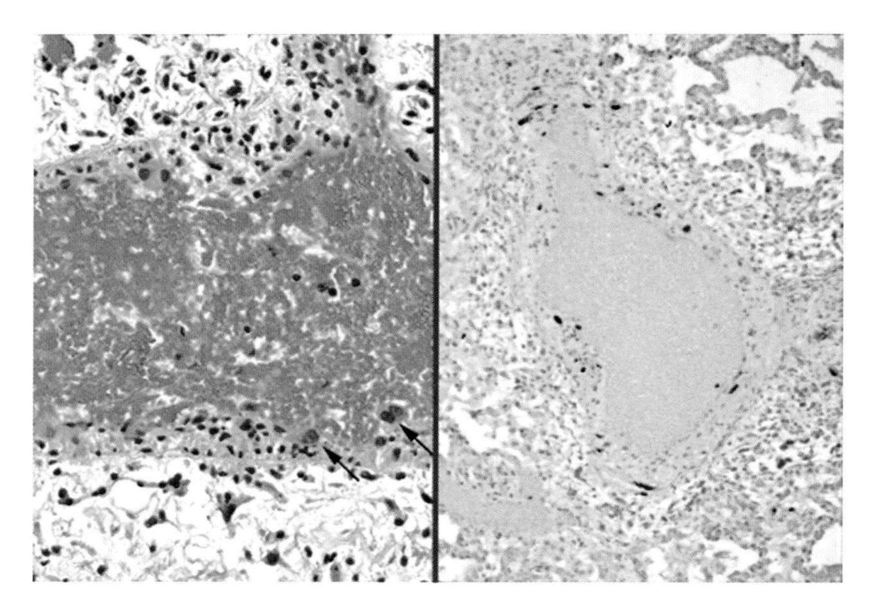

Figure II.22.4 Reactivation of CyCMV in an immunosuppressed cynomolgus macaque. Left: Haematoxylin and eosin staining of the lung of an immunosuppressed cynomolgus monkey 36 days post xenotransplantation. The animal had been naturally exposed with CyCMV prior to the xenotransplantation. The arrows indicate cytomegalic cells. Right: immunohistochemical staining for RhCMV IE1 antigen (brown) in a different area of the lung, demonstrating conservation of IE-1 epitopes between CyCMV and RhCMV. Images courtesy of D. Bouley, Stanford University.

comparable to those of HCMV (Swanson *et al.*, 1998; North *et al.*, 2004), *in vivo* replication of CyCMV and RhCMV in the presence of clinically relevant HCMV antivirals opens the door to future studies in NHP focused on mechanisms of emergence of drug-resistant variants and pre-clinical testing of novel anti-HCMV therapeutics (see also Chapter II.19).

Another study illustrates the expanding utility of NHP to optimize novel therapeutics for chronic diseases using CMV-specific T-cells to track cells adoptively transferred into a recipient. Adoptive transfer of antigen-specific T-cells has been proposed as a treatment for persistent pathogens and various cancers. A central issue in the utility of wide-scale use of adoptive transfer is the long-term self-renewal of the adoptively transferred cells. In studies involving CMV-infected pig-tailed macaques (*Macaca nemestrina*), CMV-specific (IE1, IE2) $CD8^+$ T-cells were isolated, expanded *ex vivo*, antigenically tagged (CD19 or CD20), and infused back into recipient animals to determine whether long-term maintenance of the infused T-cells was due to a particular subset of the T-cells (Berger *et al.*, 2008). The authors of this study determined that the ability for transferred cells to be maintained long-term in the recipient was associated with antigen-specific $CD8^+$T-cells derived from central memory cells ($CD8^+$ $CD28^+$ Fas^{hi}) and not from effector memory cells ($CD8^+$ $CD28^-$ Fas^{hi}).

NHP models of immune deficiency and CMV

While infection of immune competent NHP with species-specific CMV is subclinical, infection in hosts that are immune deficient due to coinfection with an immune deficiency or immune suppressing virus results in CMV sequelae almost equivalent to HCMV sequelae in HIV-infected humans. Simian viruses associated with CMV disease during primary or reactivated RhCMV infection in rhesus macaques include simian immunodeficiency (SIV), simian Type D retrovirus (SRV; genus: *Betaretrovirus*), and measles virus. The details of multiple studies have been provided previously (Barry and Chang, 2007), but, in general, fulminant SIV/RhCMV coinfection with these other viruses results in multiorgan CMV disease, including in the central and peripheral nervous systems, lung, thymus, lymph nodes, liver, stomach, gastrointestinal tract, and arteries (Baskin, 1987; Choi *et al.*, 1999; Kaur *et al.*, 2000, 2003; Sequar *et al.*, 2002). The frequency of animals with evidence of RhCMV cytopathology during SIV- or SRV-induced simian AIDS (SAIDS) can vary, but some reports have noted up to one-third to one-half of the coinfected animals have evidence of RhCMV reactivation at necropsy (Barry and Chang, 2007). Other manifestations of RhCMV reactivation during SAIDS include increased viral loads in blood and tissues, and declining measures of anti-CMV immune functions, such as CTL activity, cytokine secretion, and NT-Ab titres. There is one notable distinction in clinical outcomes associated with RhCMV disease in immune deficient macaques and HCMV disease in HIV-infected people. There have been no conclusively documented cases of RhCMV retinitis during progression of SAIDS. One report noted the presence of herpesvirus-like particles in electron micrographs of the retinas of two SIV-infected, RhCMV-seropositive rhesus macaques (Conway *et al.*, 1990). One simple explanation for the absence of RhCMV-induced retinitis is that the mechanisms for transmission to and infection of the macaque retina are different from HCMV infection in the human retina. A more likely explanation is that the course of SIV infection in rhesus monkeys is probably too short for RhCMV retinitis to develop. *In vitro* studies have documented that rhesus retinal pigmented epithelial cells are fully permissive for supporting infection and replication of a repaired RhCMV 68-1 expressing an intact RhUL128C (Lilja and Shenk, 2008). HCMV retinitis is a late-stage outcome associated with a protracted period of time in patients with severe depletion of $CD4^{+}$T-cells. There are strict criteria for culling study animals to spare them pain and suffering. SIV-infected animals are rarely maintained long-term, and prolonged severe immune deficiency, characterized by massive loss of $CD4^{+}$ T-cells over a long period of time, is never achieved.

NHP models of fetal pathogenesis and CMV

As described in an earlier review (Barry *et al.*, 2006), fetal rhesus macaques are acutely sensitive to the devastating effects of intrauterine RhCMV, and RhCMV sequelae and/or high viral loads can be observed in almost every fetal tissue and bodily fluid (London *et al.*, 1986; Tarantal *et al.*, 1998; Chang *et al.*, 2002; Barry *et al.*, 2006). The developing central nervous system is notably susceptible to infection and RhCMV-associated developmental insults that are comparable to those associated with the subtle to severe cases of congenital infection with HCMV, including damage within the sensorineural structures of the brain (see also Chapter II.3). It should be stressed that all of these studies in fetal rhesus macaques required direct ultrasound-guided inoculation of virus into defined targets of the fetus, such as the peritoneal cavity, the brain, or the amniotic fluid. To date, there has been no confirmed instance of transplacental transmission consistent with congenital infection. As with the absence of RhCMV-associated retinitis observed during SAIDS, one possibility for the absence of congenital transmission in NHP is that NHP CMVs do not traffic to and infect the placenta. Studies comparable to those in humans demonstrating the presence of HCMV in the human placenta (Pereira and Maidji, 2008; see also Chapter II.4) have not been reported for NHP. Another possible explanation for an absence of documented congenital transmission is related to the natural history of NHP CMV in colony-reared cohorts. As described above, NHP CMV is endemic within breeding cohorts of NHP with universal transmission generally around one year of age. Macaque species reach sexual maturity by about 4 years of age. Prior infection with HCMV in pregnant women confers partial, but not absolute, protection against congenital infection. Based on this and the epidemiology of NHP infection in macaques, sexually mature macaques all have robust antiviral immune responses by the time breeding begins. In addition, the presence of multiple genetic CMV variants circulating in breeding populations and the interactive social structure of macaque colonies should elicit immune responses against a wide array of antigenic CMV variants by the time of the animal's first pregnancy. Accordingly, the rate of congenital infection is likely to be exceedingly low ($\leq 1\%$), and the frequency of clinically apparent infections are likely to be even less. No large-scale studies investigating congenital infection in neonatal monkeys have ever been reported. There have been no reports of NHP CMV cytopathology in stillbirths or in early neonatal deaths, and histologically intact fetuses that have been spontaneously aborted are rare. Finally, there have been no reports of idiopathic hearing loss in NHP. Except in the cases of severe or complete hearing loss, subtle hearing loss would likely not be recognized in animals. Since the fetal inoculation studies were done prior to the development of breeding cohorts of macaques that are SPF for RhCMV infection (Barry and Strelow, 2008), all studies have been performed in immune dams. Maternal IgG can be detected in fetal circulation early in the beginning of the second trimester (Barry *et al.*, 2006). Gestation in rhesus macaques is 165 +/- 10 days, with each trimester divided into 55-day increments. Fetal development in humans and macaques are comparable, including organ development and maturation, placental structure, growth characteristics, and haematopoietic and immune system ontogenies (Tarantal and Gargosky, 1995; Barry *et al.*, 2006). Fetal inoculations that are done late in the first trimester (50–55 days) or early in the second trimester (55–60 days) are noted for a higher frequency of fetuses with RhCMV sequelae, greater severity of developmental abnormalities, and greater viral loads in tissues. Although studies are limited, data suggest that the frequency and severity of RhCMV pathology is greater in those fetuses whose dams have lower fibroblast NT-Ab titres.

With development of breeding of self-sustaining cohorts of macaques that are SPF for RhCMV, it should be possible to directly confirm whether or not RhCMV can be vertically transmitted across the placenta following primary infection of a naïve dam. Full utility of this model will depend on development of sufficient numbers of breeding-age, naïve dams so that animals can be permanently removed from the SPF cohort since the study will require their conversion to non-SPF status. However, the intrauterine pathogenesis model has potential utility in addressing clinically relevant issues for HCMV. These include optimizing the passive immune IgG transfer strategy being tested to minimize intrauterine HCMV sequelae following confirmation of congenital infection. The model is also capable of testing novel antiviral drugs that (1) have appropriate pharmacokinetics across the placenta, (2) are effective at limiting intrauterine RhCMV replication, and (3) do not induce any teratological abnormalities in the fetus. Finally, since the fetus is especially susceptible to intrauterine RhCMV, it is possible to sensitively evaluate the attenuation of modified RhCMV vectors being designed as models of attenuated HCMV vectors.

Application of RhCMV natural history to modelling HCMV vaccination and challenge

The natural history of RhCMV in breeding colonies of rhesus macaques and the availability of animals that are SPF for RhCMV now make it possible to improve the translational relevance of vaccination and challenge studies in rhesus macaques. The collective rate of congenital HCMV infection is ~0.6% (including both primary and non-primary maternal infections), indicating that the overwhelming factor for transmission of HCMV in humans is the horizontal transfer of virus in bodily fluids (Kenneson and Cannon, 2007; see also Chapter II.2). Multiple studies have documented that virus can be excreted in saliva and urine long after resolution of primary infection, and in breast milk following successive pregnancies (Stagno *et al.*, 1975; Dworsky *et al.*, 1983; Gautheret-Dejean *et al.*, 1997; Howard *et al.*, 1997; Mansat *et al.*, 1997; Hamprecht *et al.*, 1998, 2001; Kashiwagi *et al.*, 2001; Schleiss, 2006; Britt, 2008). This includes congenitally infected children who can excrete high titres of HCMV for years (Gehrz *et al.*, 1982; Alford and Britt, 1993). The clinical relevance of shedding is manifested by the fact that anyone who sheds virus is an infectious source of virus to those susceptible to horizontally transmitted HCMV, particularly pregnant women without pre-conceptional immunity to HCMV. The probability of primary infection, and by extension the probability of congenital infection, is a direct function of the frequency of close personal contact (i.e. mucosal exposure to infectious bodily fluids) between infected and uninfected individuals (Hamprecht *et al.*, 2001; Jim *et al.*, 2004, 2009). HCMV-infected children in daycare and sexual transmission are two well-described vectors for horizontal transmission to seronegative women (Adler, 1986a,b, 1988a,b; 1989, 1991a,b; Fowler and Pass, 2006; Marshall and Adler, 2009; Cannon *et al.*, 2011). HCMV vaccines, particularly those designed to minimize maternal and/or congenital infection, face a stringent threshold for achieving a significant level of protective efficacy in that the vaccine must elicit immune responses that repeatedly restrict viral replication following multiple exposures to HCMV strains that may encode antigens immunologically variant from the vaccine (see also Chapter II.20).

It is now possible to model this aspect of HCMV vaccine challenge in the rhesus macaque model in ways that cannot be done in other animal models. In particular, it is now possible to factor into vaccination/challenge studies the transmucosal acquisition of virus following repeated exposure to potentially high copy numbers of infectious virus that may be antigenically homologous or distinct from the vaccine antigens. Persistent shedding of RhCMV following either natural exposure to or experimental inoculation with RhCMV (Oxford *et al.*, 2011) enable vaccination/challenge studies to be performed in which vaccinated and control animals are housed with RhCMV-excreting animals, which would serve as a source of challenge virus. Because of the high risk of women acquiring HCMV from their infected children, the proposition has been put forth to immunize young children when a vaccine is approved 'to prevent them acquiring CMV infection and so becoming infectious to their mothers' (Griffiths, 2002). Accordingly, vaccines that prevented shedding of RhCMV in the vaccinated/virus-exposed animals would provide support for the concept of expanding vaccination to include young children.

NHP models of ageing

Advanced age is associated with dysregulated immunity, particularly the impaired development of *de novo* immunity against novel antigens, presented either in the context of vaccination or seasonal infections (Gomez *et al.*, 2005; Nikolich-Zugich, 2005, 2008; Aspinall *et al.*, 2007; Kovaiou *et al.*, 2007; Ostan *et al.*,

2008; Weinberger *et al.*, 2008; Siegrist and Aspinall, 2009). This condition, known as immunosenescence, is hypothesized to contribute to age-related susceptibilities to physical and mental pathologies, inflammatory conditions, and reactivation of latent infections (Krabbe *et al.*, 2004; Fulop *et al.*, 2005; McGeer *et al.*, 2006; Doxiadis *et al.*, 2007; Kovaiou *et al.*, 2007). Work to define the aetiology of immunosenescence implicates age-associated perturbations in both innate and adaptive compartments, with pronounced alterations of lymphocyte populations and inflammatory mediators. Aged humans exhibit decreases in circulating naïve T and B-cells, reduced expression of co-stimulatory molecules, impairment of immune cell proliferation, and constricted T and B-cell diversity, contemporaneously with increases in memory effector T-cell populations and pro-inflammatory cytokines (Cosman *et al.*, 1999; Zanni *et al.*, 2003; Nikolich-Zugich, 2005; Gibson *et al.*, 2009; Haynes and Maue, 2009; Kita *et al.*, 2009). Mechanistic causalities for these changes remain incompletely described. However, a temporal conspiracy of biologically programmed thymic involution and chronic antigenic stimulation from persistent pathogens, particularly HCMV, has become a prime focus of investigation (for a discussion, see also Chapter II.7).

Thymic involution leads to a protracted accumulation of memory lymphocytes, especially $CD8^+$ T-cells, and causes severely restricted diversity within the naïve T-cell pool around the sixth decade of life (Cosman *et al.*, 1999; Nikolich-Zugich, 2005, 2008). In addition to the extraordinarily large devotion of the T-cell repertoire to HCMV (Sylwester *et al.*, 2005), long-term HCMV carriers also have an age-associated increase in clonally expanded HCMV-specific $CD8^+$ $CD28^-$ effector T-cells (Waldrop *et al.*, 1997; Looney *et al.*, 1999; Weekes *et al.*, 1999; Asanuma *et al.*, 2000; Wedderburn *et al.*, 2001; Khan *et al.*, 2002; Sester *et al.*, 2002; Almanzar *et al.*, 2005; Kovaiou and Grubeck-Loebenstein, 2006). Many papers have reported predictive biomarkers of immunosenescence, and these generally focus on terminally differentiated T-cells in HCMV-immune individuals (Koch *et al.*, 2006; Nikolich-Zugich, 2008; Pawelec *et al.*, 2009). These include (1) an inversion of the CD4/CD8 ratio (immune risk phenotype – IRP), which has been associated with a statistically increased risk of mortality in large population studies in Sweden (Olsson *et al.*, 2000; Wikby *et al.*, 2002, 2005, 2008; Huppert *et al.*, 2003; Hadrup *et al.*, 2006); (2) accumulation of $CD8^+$ $CD28^-$ $CD45RA^+$ T-cells, which during ageing has been associated with reduced responsiveness to influenza vaccination (Goronzy *et al.*, 2001; Saurwein-Teissl *et al.*, 2002; Effros, 2007; Targonski *et al.*, 2007; Xie and McElhaney, 2007) and Epstein–Barr virus infection (Khan *et al.*, 2004); and (3) seroimmunity to HCMV, which drives a progressive oligoclonal expansion of HCMV-specific T-cells, usually bearing the $CD8^+$ $CD28^-$ $CD45RA^+$ phenotype (Looney *et al.*, 1999; Weekes *et al.*, 1999; Khan *et al.*, 2002; Vescovini *et al.*, 2004; Almanzar *et al.*, 2005; Sansoni *et al.*, 2008; Chidrawar *et al.*, 2009). One emerging theme is that clonally expanded HCMV-specific CD8 T-cells accumulate during ageing and reduce the naïve TCR repertoire available to face novel antigenic challenges, thereby exacerbating immunosenescence.

Work to characterize the monkey model of ageing has focused on age-related differences in the numbers and percentages of lymphoid populations and in functional capacity of effector T-cells. Rhesus macaques in the wild live on average 15 years, whereas macaques maintained at breeding centres can live >30 years (Otting *et al.*, 2007). With increased lifespans coincident with the absence of predation, disease, and competition for food in the wild, colony-raised aged macaques (≥20 years) develop clinical conditions normally rare or absent in younger animals, such as gastrointestinal adenocarcinomas and age-related changes of the brain and genitourinary tract (Rahman *et al.*, 1976; Bennett *et al.*, 2009; Romphruk *et al.*, 2009; Shiina *et al.*, 2011). As noted above, NHP CMVs are endemic in NHP breeding centres, as are other herpesviruses, such as rhesus rhadinovirus (White *et al.*, 2009). Aged rhesus macaques exhibit alterations of immune cell subsets very similar to those observed in humans demonstrating immune senescent phenotypes, including an age-related reduction of naïve T-cells ($CD95^-$) in both $CD4^+$ and $CD8^+$ subsets concomitantly with a rise in memory T-cells ($CD95^+$), particularly effector memory (EM) T-cells ($CD28^-$) (Jankovic *et al.*, 2003; Messaoudi *et al.*, 2006; Shiina *et al.*, 2006; Cicin-Sain *et al.*, 2007; 2011; Simmons and Mattison, 2011). The loss of naïve T-cells is correlated with increases in naïve cell proliferation and differentiation to the memory compartment (Simmons and Mattison, 2011). Ageing rhesus macaques also exhibit increases in stable T-cell clonal expansions (TCE) that inversely correlate with naïve $CD8^+$ and $CD4^+$ T-cell frequencies and support the hypothesis that these events lead to a temporal decrease in TCR diversity (Hadrup *et al.*, 2006; Cicin-Sain *et al.*, 2007; Simmons and Mattison, 2011). A recent paper showed that, while the activity of antigen-presenting cells was preserved in older macaques, humoral and cellular responses to MVA immunization were diminished in aged (28–25 years) compared to young adult (7–10 years) monkeys (Simmons and Mattison, 2011). CD8 T-cell responses correlated with the frequency of naïve T-cells and were significantly reduced in animals showing TCE compared to those without clonally expanded T-cell

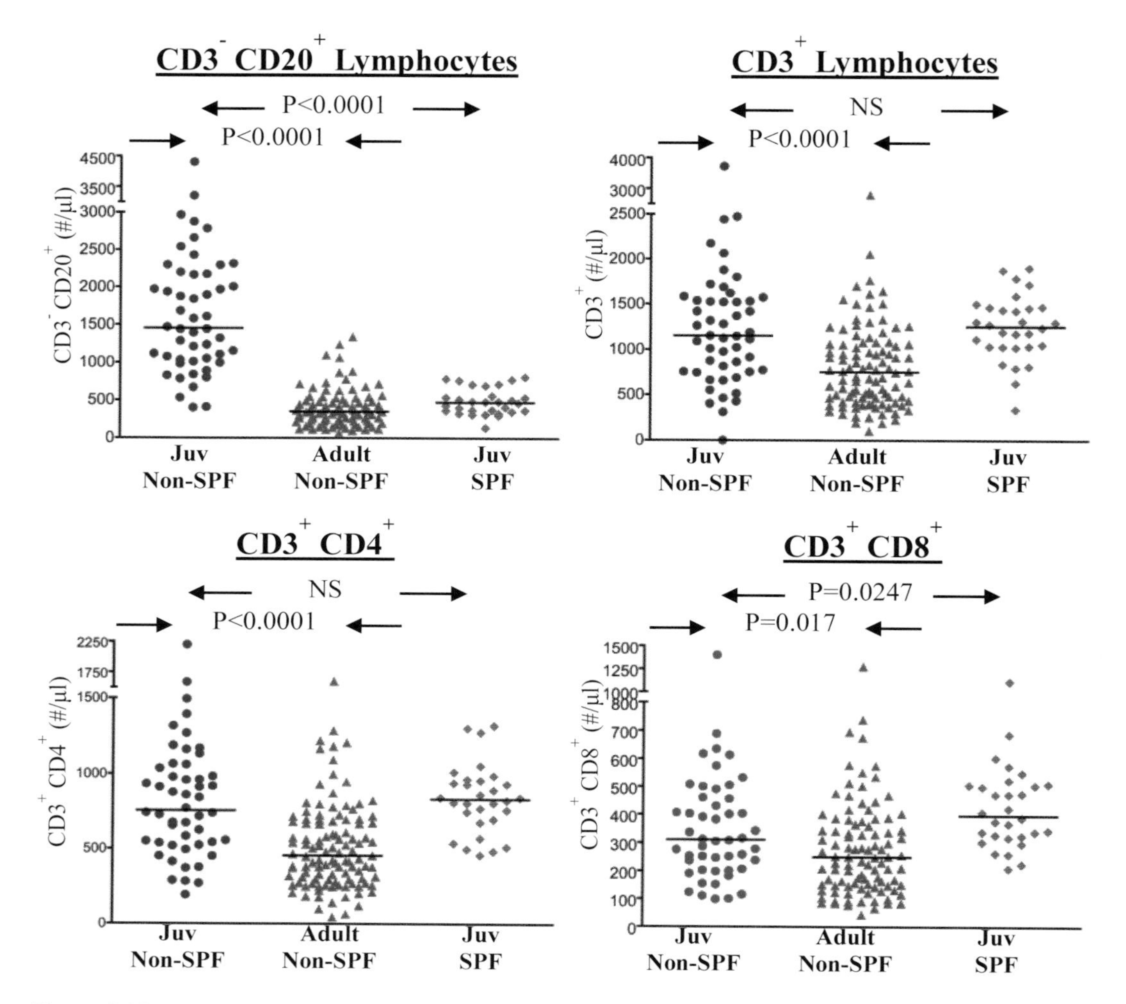

Figure II.22.5 Peripheral lymphoid cell numbers in juvenile (3–5 years) and older adult (≥14 years) RM, non-Specific Pathogen Free (SPF) for RhCMV and other simian herpesviruses, and SPF juvenile RM. The line represents the median value. NS: Not Significant.

populations. These results provide strong evidence that age-related decreases in the naïve T-cell pool directly impact the efficacy and efficiency of immune responses to newly introduced antigens.

A central question is whether shaping of the immune system that occurs concurrently with ageing is inherently acquired or is effected by external factors, especially by exposure to pathogens. A recent study analysed peripheral lymphoid populations of conventionally raised juvenile (3–5 years; $N=50$) and adult (> 14 years, $N=100$) rhesus macaques (i.e. non-SPF), and juvenile macaques ($N=30$) that were SPF for RhCMV and, likely, all simian herpesviruses (Oxford, Krishnan, Barry; unpublished observation). Comparison of cell populations between juvenile and adult non-SPF RM identified age-related changes within specific lymphoid cell populations, and comparison of age-matched non-SPF and SPF juvenile macaques identified changes in lymphoid populations in relation to the presence and absence, respectively, of RhCMV and other simian herpesviruses. Emerging evidence indicates that while the frequency and/or absolute numbers of some lymphocyte populations appear to change as a result of age, other subsets are also influenced by the infectious burden. For example, there are apparent age-related declines in B ($CD3^-$ $CD20^+$), T ($CD3^+$), $CD3^+$ $CD4^+$, and $CD3^+$ $CD8^+$ cell numbers (Fig. II.22.5). However, there are also effects that associate with the presence or absence of RhCMV. Particularly notable, RhCMV and the other simian herpesviruses appear to drive a relative expansion of B-cells in non-SPF juvenile macaques, compared to SPF juveniles

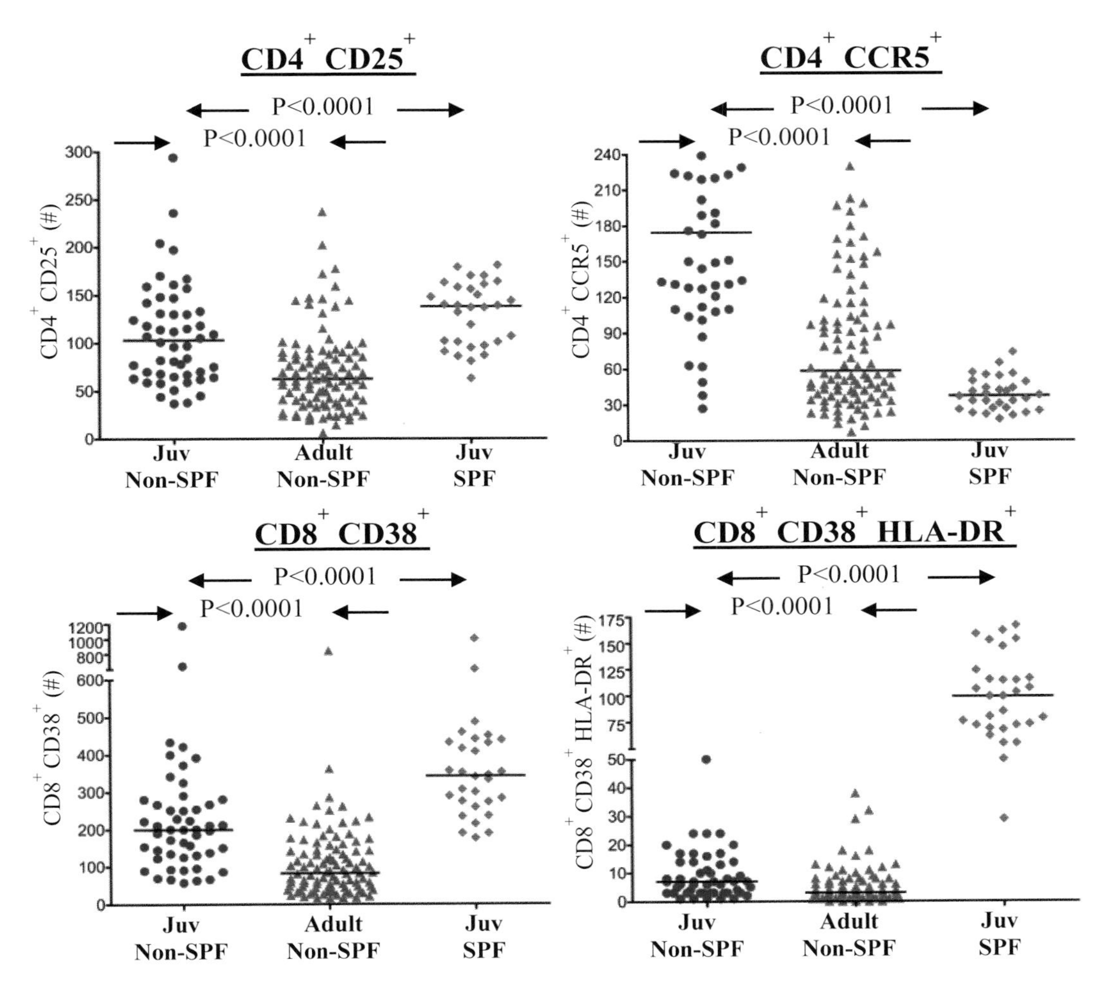

Figure II.22.6 Peripheral cell numbers of CD4+ and CD8+ T-cell subpopulations in non-SPF juvenile and older adult RMs, and SPF juvenile RM.

(Fig. II.22.5). Similar alterations in subsets of CD4 and CD8+T-cells were observed in relation to age and infectious burden (Fig. II.22.6). Significant differences have been observed in the milieu of cytokines secreted by PBMCs from these cohorts following non-specific stimulation with concanavalin A (Oxford, Barry; unpublished observations). These differences suggest that exposure to persistent pathogens alters host signalling patterns and may result in enhanced or reduced immune responses depending on the type of response required. Importantly, since cytokines not only induce immediate immune responses but also orchestrate the proliferation and differentiation of specific immune cell types (see also Chapter II.11), and since differences in cytokine profiles were observed in the early years following primary infection, these results suggest that persistent pathogens may launch an altered course of immune development early and independent of the effects of thymic involution. Since an age-matched adult SPF cohort is unavailable at this time, conclusions cannot be drawn about the specific effect of repeated antigenic stimulation on immune cells in monkeys, but suggest that host exposure to persistently replicating microbes may have a profound and lasting impact on the immune system.

References

Abel, K., Strelow, L., Yue, Y., Eberhardt, M.K., Schmidt, K.A., and Barry, P.A. (2008). A heterologous DNA prime/protein boost immunization strategy for rhesus cytomegalovirus. Vaccine *26*, 6013–6025.

Abel, K., Martinez, J., Yue, Y., Lacey, S.F., Wang, Z., Strelow, L., Dasgupta, A., Li, Z., Schmidt, K.A., Oxford, K.L., *et al.* (2011). Vaccine-induced control of viral shedding

following rhesus cytomegalovirus challenge in rhesus macaques. J. Virol. *85*, 2878–2890.
Adler, B., Scrivano, L., Ruzcics, Z., Rupp, B., Sinzger, C., and Koszinowski, U. (2006). Role of human cytomegalovirus UL131A in cell type-specific virus entry and release. J. Gen. Virol. *87*, 2451–2460.
Adler, S.P. (1986a). Cytomegalovirus infection in parents of children at day-care centers. N. Engl. J.Med. *315*, 1164–1165.
Adler, S.P. (1986b). Molecular epidemiology of cytomegalovirus: evidence for viral transmission to parents from children infected at a day care center. Pediatr. Infect. Dis. J. *5*, 315–318.
Adler, S.P. (1988a). Cytomegalovirus transmission among children in day care, their mothers and caretakers. Pediatr. Infect. Dis. J. *7*, 279–285.
Adler, S.P. (1988b). Molecular epidemiology of cytomegalovirus: viral transmission among children attending a day care center, their parents, and caretakers. J. Pediatr. *112*, 366–372.
Adler, S.P. (1989). Cytomegalovirus and child day care. Evidence for an increased infection rate among day-care workers. N. Engl. J. Med. *321*, 1290–1296.
Adler, S.P. (1991a). Cytomegalovirus and child day care: risk factors for maternal infection. Pediatr. Infect. Dis. J. *10*, 590–594.
Adler, S.P. (1991b). Molecular epidemiology of cytomegalovirus: a study of factors affecting transmission among children at three day-care centers. Pediatr. Infect. Dis. J. *10*, 584–590.
Alcendor, D.J., Barry, P.A., Pratt-Lowe, E., and Luciw, P.A. (1993). Analysis of the rhesus cytomegalovirus immediate-early gene promoter. Virology *194*, 815–821.
Alcendor, D.J., Zong, J., Dolan, A., Gatherer, D., Davison, A.J., and Hayward, G.S. (2009). Patterns of divergence in the vCXCL and vGPCR gene clusters in primate cytomegalovirus genomes. Virology 395, 21–32.
Alford, C.A., and Britt, W.J. (1993). Cytomegalovirus. In The Human Herpesviruses, Roizman, B., Whitley, R.J., and Lopez, C., eds. (Raven Press, Ltd., New York), pp. 227–255.
Almanzar, G., Schwaiger, S., Jenewein, B., Keller, M., Herndler-Brandstetter, D., Wurzner, R., Schonitzer, D., and Grubeck-Loebenstein, B. (2005). Long-term cytomegalovirus infection leads to significant changes in the composition of the CD8+ T-cell repertoire, which may be the basis for an imbalance in the cytokine production profile in elderly persons. J. Virol. *79*, 3675–3683.
Ambagala, A.P., Marsh, A., Chan, J., Pilon, R., Fournier, J., Mazzulli, T., Sandstrom, P., Willer, D.O., and MacDonald, K.S. (2011). Isolation and characterization of cynomolgus macaque (Macaca fascicularis) cytomegalovirus (CyCMV). Virology *412*, 125–135.
Andrade, M.R., Yee, J., Barry, P.A., Spinner, A., Roberts, J.A., Cabello, H., Leite, J.P., and Lerche, N.W. (2003). Prevalence of Antibodies to Selected Viruses in a Long-term Closed Breeding Colony of Rhesus Macaques (*Macaca mulatta*) in Brazil. Amer. J. Primatol. *59*, 123–128.
Antoun, A., Jobson, S., Cook, M., O'Callaghan, C.A., Moss, P., and Briggs, D.C. (2010). Single nucleotide polymorphism analysis of the NKG2D ligand cluster on the long arm of chromosome 6: Extensive polymorphisms and evidence of diversity between human populations. Hum. Immunol. *71*, 610–620.
Arav-Boger, R., Foster, C.B., Zong, J.C., and Pass, R.F. (2006). Human cytomegalovirus-encoded alpha -chemokines exhibit high sequence variability in congenitally infected newborns. J. Infect. Dis. *193*, 788–791.
Arnon, T.I., Achdout, H., Levi, O., Markel, G., Saleh, N., Katz, G., Gazit, R., Gonen-Gross, T., Hanna, J., Nahari, E., *et al.* (2005). Inhibition of the NKp30 activating receptor by pp65 of human cytomegalovirus. Nat. Immunol. *6*, 515–523.
Arora, N., Novak, Z., Fowler, K.B., Boppana, S.B., and Ross, S.A. (2010). Cytomegalovirus viruria and DNAemia in healthy seropositive women. J. Infect. Dis. *202*, 1800–1803.
Asanuma, H., Sharp, M., Maecker, H.T., Maino, V.C., and Arvin, A.M. (2000). Frequencies of memory T-cells ppecific for varicella-zoster virus, herpes simplex virus, and cytomegalovirus by intracellular detection of cytokine expression. J. Infect. Dis. *181*, 859–866.
Asher, D.M., Gibbs, C.J., Jr., and Lang, D.J. (1969). Rhesus monkey cytomegaloviruses: persistent asymptomatic viruses. Bacteriol. Proc. *69*, 191.
Asher, D.M., Gibbs, C.J., Jr, Lang, D.J., Gadjusek, D.C., and Chanock, R.M. (1974). Persistent shedding of cytomegalovirus in the urine of healthy rhesus monkeys. Proc. Soc. Exp. Biol. Med. *145*, 794–801.
Aspinall, R., Del Giudice, G., Effros, R.B., Grubeck-Loebenstein, B., and Sambhara, S. (2007). Challenges for vaccination in the elderly. Immun. Ageing *4*, 9.
Atalay, R., Zimmermann, A., Wagner, M., Borst, E., Benz, C., Messerle, M., and Hengel, H. (2002). Identification and expression of human cytomegalovirus transcription units coding for two distinct Fcgamma receptor homologs. J. Virol. *76*, 8596–8608.
Averdam, A., Seelke, S., Grutzner, I., Rosner, C., Roos, C., Westphal, N., Stahl-Hennig, C., Muppala, V., Schrod, A., Sauermann, U., *et al.* (2007). Genotyping and segregation analyses indicate the presence of only two functional MIC genes in rhesus macaques. Immunogenetics *59*, 247–251.
Baek, M.C., Krosky, P.M., Pearson, A., and Coen, D.M. (2004). Phosphorylation of the RNA polymerase II carboxyl-terminal domain in human cytomegalovirus-infected cells and *in vitro* by the viral UL97 protein kinase. Virology *324*, 184–193.
Barry, P.A., and Chang, W.-L.W. (2007). Primate Betaherpesviruses. In Human Herpesviruses: Biology, Therapy and Immunoprophylaxis, Arvin, A., Campadielli, G., Moore, P., Mocarski, E., Roizman, B., Whitley, R., and Yamanishi, K., eds. (Cambridge University Press, Cambridge), pp. 1051–1075.
Barry, P.A., and Strelow, L. (2008). Development of breeding populations of rhesus macaques (*Macaca mulatta*) that are specific pathogen free for rhesus cytomegalovirus. Comp. Med. *58*, 43–46.
Barry, P.A., Alcendor, D.J., Power, M.D., Kerr, H., and Luciw, P.A. (1996). Nucleotide sequence and molecular analysis of the rhesus cytomegalovirus immediate-early gene and the UL121–117 open reading frames. Virology *215*, 61–72.
Barry, P.A., Lockridge, K.M., Salamat, S., Tinling, S.P., Yue, Y., Zhou, S.S., Gospe, S.M., Jr., Britt, W.J., and Tarantal, A.F. (2006). Nonhuman primate models of intrauterine cytomegalovirus infection. ILAR J. *47*, 49–64.
Baskin, G.B. (1987). Disseminated cytomegalovirus infection in immunodeficient rhesus macaques. Amer. J. Pathol. *129*, 345–352.

Becke, S., Fabre-Mersseman, V., Aue, S., Auerochs, S., Sedmak, T., Wolfrum, U., Strand, D., Marschall, M., Plachter, B., and Reyda, S. (2010). Modification of the major tegument protein pp65 of human cytomegalovirus inhibits virus growth and leads to the enhancement of a protein complex with pUL69 and pUL97 in infected cells. J. Gen. Virol. *91*, 2531–2541.

Bennett, M.W., Dick, E.J., Jr., Schlabritz-Loutsevitch, N.E., Lopez-Alvarenga, J.C., Williams, P.C., Mark Sharp, R., and Hubbard, G.B. (2009). Endometrial and cervical polyps in 22 baboons (*Papio* sp.), 5 cynomolgus macaques (*Macaca fascicularis*) and one marmoset (*Callithrix jacchus*). J. Med. Primatol. *38*, 257–262.

Berger, C., Jensen, M.C., Lansdorp, P.M., Gough, M., Elliott, C., and Riddell, S.R. (2008). Adoptive transfer of effector CD8+ T-cells derived from central memory cells establishes persistent T-cell memory in primates. J. Clin. Invest. *118*, 294–305.

Bielefeldt-Ohmann, H., Gough, M., Durning, M., Kelley, S., Liggitt, H.D., and Kiem, H.P. (2004). Greater sensitivity of pigtailed macaques (*Macaca nemestrina*) than baboons to total body irradiation. J. Comp. Pathol. *131*, 77–86.

Black, P.H., Hartley, J.W., and Rowe, W.P. (1963). Isolation of a cytomegalovirus from African green monkey. Proc. Soc. Exper. Biol. Med. *112*, 601–605.

Blewett, E.L., White, G., Saliki, J.T., and Eberle, R. (2001). Isolation and characterization of an endogenous cytomegalovirus (BaCMV) from baboons. Arch. Virol. *146*, 1723–1738.

Blewett, E.L., Lewis, J., Gadsby, E.L., Neubauer, S.R., and Eberle, R. (2003). Isolation of cytomegalovirus and foamy virus from the drill monkey (*Mandrillus leucophaeus*) and prevalence of antibodies to these viruses amongst wild-born and captive-bred individuals. Arch. Virol. *148*, 423–433.

Boppana, S.B., Rivera, L.B., Fowler, K.B., Mach, M., and Britt, W.J. (2001). Intrauterine transmission of cytomegalovirus to infants of women with preconceptional immunity. N. Engl. J. Med. *344*, 1366–1371.

Bowman, J.J., Lacayo, J.C., Burbelo, P., Fischer, E.R., and Cohen, J.I. (2011). Rhesus and human cytomegalovirus glycoprotein L are required for infection and cell-to-cell spread of virus but cannot complement each other. J. Virol. *85*, 2089–2099.

Braud, V.M., Tomasec, P., and Wilkinson, G.W. (2002). Viral evasion of natural killer cells during human cytomegalovirus infection. Curr. Top. Microbiol. Immunol. *269*, 117–129.

Britt, W. (2008). Manifestations of human cytomegalovirus infection: proposed mechanisms of acute and chronic disease. Curr. Top. Microbiol. Immunol. *325*, 417–470.

Britt, W.J., Jarvis, M.A., Drummond, D.D., and Mach, M. (2005). Antigenic domain 1 is required for oligomerization of human cytomegalovirus glycoprotein B. J. Virol. *79*, 4066–4079.

Campbell, A.E., Cavanaugh, V.J., and Slater, J.S. (2008). The salivary glands as a privileged site of cytomegalovirus immune evasion and persistence. Med. Microbiol. Immunol. *197*, 205–213.

Cannon, M.J., Hyde, T.B., and Schmid, D.S. (2011). Review of cytomegalovirus shedding in bodily fluids and relevance to congenital cytomegalovirus infection. Rev. Med. Virol. *21*, 240–255.

Carlson, J.R., Chang, W.-L.W., Zhou, S.-S., Tarantal, A.F., and Barry, P.A. (2005). Rhesus brain microvascular endothelial cells are permissive for rhesus cytomegalovirus infection. J. Gen. Virol. *86*, 545–549.

Cha, T., Tom, E., Kemble, G., Duke, G., Mocarski, E., and Spaete, R. (1996). Human cytomegalovirus clinical isolates carry at least 19 genes not found in laboratory strains. J. Virol. *70*, 78–83.

Chan, K.S., and Kaur, A. (2007). Flow cytometric detection of degranulation reveals phenotypic heterogeneity of degranulating CMV-specific CD8+ T-lymphocytes in rhesus macaques. J. Immunol. Meth. *325*, 20–34.

Chang, W.L., and Barry, P.A. (2003). Cloning of the full-length rhesus cytomegalovirus genome as an infectious and self-excisable bacterial artificial chromosome for analysis of viral pathogenesis. J. Virol. *77*, 5073–5083.

Chang, W.L., and Barry, P.A. (2010). Attenuation of innate immunity by cytomegalovirus IL-10 establishes a long-term deficit of adaptive antiviral immunity. Proc. Natl. Acad. Sci. U.S.A. *107*, 22647–22652.

Chang, W.L., Tarantal, A.F., Zhou, S.S., Borowsky, A.D., and Barry, P.A. (2002). A recombinant rhesus cytomegalovirus expressing enhanced green fluorescent protein retains the wild-type phenotype and pathogenicity in fetal macaques. J. Virol. *76*, 9493–9504.

Chang, W.L., Baumgarth, N., Yu, D., and Barry, P.A. (2004). Human cytomegalovirus-encoded interleukin-10 homolog inhibits maturation of dendritic cells and alters their functionality. J. Virol. *78*, 8720–8731.

Chang, W.L., Baumgarth, N., Eberhardt, M.K., Lee, C.Y., Baron, C.A., Gregg, J.P., and Barry, P.A. (2007). Exposure of myeloid dendritic cells to exogenous or endogenous IL-10 during maturation determines their longevity. J. Immunol. *178*, 7794–7804.

Chang, W.L., Barry, P.A., Szubin, R., Wang, D., and Baumgarth, N. (2009). Human cytomegalovirus suppresses type I interferon secretion by plasmacytoid dendritic cells through its interleukin 10 homolog. Virology *390*, 330–337.

Chang, Y.-N., Jeang, K.-T., Lietman, T., and Hayward, G.S. (1995). Structural organization of the spliced immediate-early gene complex that encodes the major acidic nuclear (IE1) and transactivator (IE2) proteins of African green monkey cytomegalovirus. J. Biomed. Sci. *2*, 105–130.

Cheeran, M.C., Hu, S., Palmquist, J.M., Bakken, T., Gekker, G., and Lokensgard, J.R. (2007). Dysregulated interferon-gamma responses during lethal cytomegalovirus brain infection of IL-10-deficient mice. Virus Res. *130*, 96–102.

Chidrawar, S., Khan, N., Wei, W., McLarnon, A., Smith, N., Nayak, L., and Moss, P. (2009). Cytomegalovirus-seropositivity has a profound influence on the magnitude of major lymphoid subsets within healthy individuals. Clin. Exp. Immunol. *155*, 423–432.

Chiou, C.-J., Zong, J., Waheed, I., and Hayward, G.S. (1993). Identification and mapping of dimerization and DNA-binding domains in the C terminus of the IE2 regulatory protein of human cytomegalovirus. J. Virol. *67*, 6201–6214.

Choi, Y.K., Simon, M.A., Kim, D.Y., Yoon, B.I., Kwon, S.W., Lee, K.W., and Seo, I.B. (1999). Fatal measles virus infection in Japanese macaques (*Macaca fuscata*). Vet. Pathol. *36*, 594–600.

Cicin-Sain, L., Messaoudi, I., Park, B., Currier, N., Planer, S., Fischer, M., Tackitt, S., Nikolich-Zugich, D., Legasse, A., Axthelm, M.K., *et al.* (2007). Dramatic increase in naïve T-cell turnover is linked to loss of naïve T-cells from old primates. Proc. Natl. Acad. Sci. U.S.A. *104*, 19960–19965.

Cicin-Sain, L., Smyk-Pearson, S., Currier, N., Byrd, L., Koudelka, C., Robinson, T., Swarbrick, G., Tackitt, S., Legasse, A., Fischer, M., *et al.* (2010). Loss of naïve T-cells and repertoire constriction predict poor response to vaccination in old primates. J. Immunol. *184*, 6739–6745.

Cicin-Sain, L., Sylwester, A.W., Hagen, S.I., Siess, D.C., Currier, N., Legasse, A.W., Fischer, M.B., Koudelka, C.W., Axthelm, M.K., Nikolich-Zugich, J., *et al.* (2011). Cytomegalovirus-specific T-cell immunity is maintained in immunosenescent rhesus macaques. J. Immunol. *187*, 1722–1732.

Clark-Lewis, I., Schumacher, C., Baggiolini, M., and Moser, B. (1991). Structure-activity relationships of interleukin-8 determined using chemically synthesized analogs. Critical role of NH2-terminal residues and evidence for uncoupling of neutrophil chemotaxis, exocytosis, and receptor binding activities. J. Biol. Chem. *266*, 23128–23134.

Conway, M.D., Didier, P., Fairburn, B., Soike, K.F., Martin, L., Murphy-Corb, M., Meiners, N., and Insler, M.S. (1990). Ocular manifestation of simian immunodeficiency syndrome (SAIDS). Curr. Eye Res. *9*, 759–770.

Cosman, D., Fanger, N., and Borges, L. (1999). Human cytomegalovirus, MHC class I and inhibitory signalling receptors: more questions than answers. Immunol. Rev. *168*, 177–185.

Cowdry, E.V., and Scott, G.H. (1935). Nuclear inclusions suggestive of virus action in the salivary glands of the monkey, *Cebus fatuellus*. Am. J. Path. *11*, 647–658 641.

Cunningham, C., Gatherer, D., Hilfrich, B., Baluchova, K., Dargan, D.J., Thomson, M., Griffiths, P.D., Wilkinson, G.W., Schulz, T.F., and Davison, A.J. (2010). Sequences of complete human cytomegalovirus genomes from infected cell cultures and clinical specimens. J. Gen. Virol. *91*, 605–615.

Dambaeva, S.V., Bondarenko, G.I., Grendell, R.L., Kravitz, R.H., Durning, M., and Golos, T.G. (2008). Non-classical MHC-E (Mamu-E) expression in the rhesus monkey placenta. Placenta *29*, 58–70.

Daniel, M.D., Melendez, L.V., King, N.W., Fraser, C.E., Barahona, H.H., Hunt, R.D., Garcia, F.G., and Trum, B.F. (1971). Herpes virus aotus: a latent herpesvirus from owl monkeys (*Aotus trivirgatus*) isolation and characterization. Proc. Soc. Exper. Biol. Med. *138*, 835–845.

Daniel, M.D., Melendez, L.V., King, N.W., Barahona, H.H., Fraser, C.E., Garcia, F.G., and Silva, D. (1973). Isolation and characterization of a new virus from owl monkeys: herpesvirus aotus type 3. Amer. J. Phys. Anthropol. *38*, 497–500.

Dargan, D.J., Douglas, E., Cunningham, C., Jamieson, F., Stanton, R.J., Baluchova, K., McSharry, B.P., Tomasec, P., Emery, V.C., Percivalle, E., *et al.* (2010). Sequential mutations associated with adaptation of human cytomegalovirus to growth in cell culture. J. Gen. Virol. *91*, 1535–1546.

Davison, A.J., Dolan, A., Akter, P., Addison, C., Dargan, D.J., Alcendor, D.J., McGeoch, D.J., and Hayward, G.S. (2003). The human cytomegalovirus genome revisited: comparison with the chimpanzee cytomegalovirus genome. J. Gen. Virol. *84*, 17–28.

Davison, A.J., Eberle, R., Ehlers, B., Hayward, G.S., McGeoch, D.J., Minson, A.C., Pellett, P.E., Roizman, B., Studdert, M.J., and Thiry, E. (2009). The order Herpesvirales. Arch. Virol. *154*, 171–177.

DeFilippis, V., and Früh, K. (2005). Rhesus cytomegalovirus particles prevent activation of interferon regulatory factor 3. J. Virol. *79*, 6419–6431.

DeFilippis, V.R., Sali, T., Alvarado, D., White, L., Bresnahan, W., and Früh, K.J. (2010). Activation of the interferon response by human cytomegalovirus occurs via cytoplasmic double-stranded DNA but not glycoprotein B. J. Virol. *84*, 8913–8925.

Dolan, A., Cunningham, C., Hector, R.D., Hassan-Walker, A.F., Lee, L., Addison, C., Dargan, D.J., McGeoch, D.J., Gatherer, D., Emery, V.C., *et al.* (2004). Genetic content of wild-type human cytomegalovirus. J. Gen. Virol. *85*, 1301–1312.

Doxiadis, G.G., Heijmans, C.M., Otting, N., and Bontrop, R.E. (2007). MIC gene polymorphism and haplotype diversity in rhesus macaques. Tissue Antigens *69*, 212–219.

Dworsky, M., Yow, M., Stagno, S., Pass, R.F., and Alford, C. (1983). Cytomegalovirus infection of breast milk and transmission in infancy. Pediatrics *72*, 295–299.

Eberhardt, M.K., Chang, W.-L., Logsdon, N.J., Yue, Y., Walter, M.R., and Barry, P.A. (2012). Host immune responses to a viral immune modulating protein: immunogenicity of viral interleukin-10 in rhesus cytomegalovirus-infected rhesus macaques. PLoS ONE *7*, e37931.

Effros, R.B. (2007). Role of T-lymphocyte replicative senescence in vaccine efficacy. Vaccine *25*, 599–604.

Eizuru, Y., Tsuchiya, K., Mori, R., and Minamishima, Y. (1989). Immunological and molecular comparisons of simian cytomegaloviruses isolated from African green monkey (*Ceropithicus aethiops*) and Japanese macaque (*Macaca fuscata*). Arch. Virol. *107*, 65–75.

Elek, S.D., and Stern, H. (1974). Development of a vaccine against mental retardation caused by cytomegalovirus infection in utero. Lancet *1*, 1–5.

Fowler, K.B., and Pass, R.F. (2006). Risk factors for congenital cytomegalovirus infection in the offspring of young women: exposure to young children and recent onset of sexual activity. Pediatrics *118*, e286–292.

Fulop, T., Larbi, A., Wikby, A., Mocchegiani, E., Hirokawa, K., and Pawelec, G. (2005). Dysregulation of T-cell function in the elderly: scientific basis and clinical implications. Drugs Aging *22*, 589–603.

Furukawa, T., Hornberger, E., Sakuma, S., and Plotkin, S.A. (1975). Demonstration of immunoglobulin G receptors induced by human cytomegalovirus. J. Clin. Microbiol. *2*, 332–336.

Gandhoke, I., Aggarwal, R., Lal, S., and Khare, S. (2006). Congenital CMV infection in symptomatic infants in Delhi and surrounding areas. Ind. J. Pediatr. *73*, 1095–1097.

Gautheret-Dejean, A., Aubin, J.T., Poirel, L., Huraux, J.M., Nicolas, J.C., Rozenbaum, W., and Agut, H. (1997). Detection of human Betaherpesvirinae in saliva and urine from immunocompromised and immunocompetent subjects. J. Clin. Microbiol. *35*, 1600–1603.

Gaytant, M.A., Steegers, E.A., Semmekrot, B.A., Merkus, H.M., and Galama, J.M. (2002). Congenital cytomegalovirus infection: review of the epidemiology and outcome. Obstet. Gynecol. Surv. *57*, 245–256.

Gaytant, M.A., Rours, G.I., Steegers, E.A., Galama, J.M., and Semmekrot, B.A. (2003). Congenital cytomegalovirus infection after recurrent infection: case reports and review of the literature. Eur. J. Pediatr. *162*, 248–253.

Gehrz, R.C., Linner, K.M., Christianson, W.R., Ohm, A.E., and Balfour, H.H., Jr. (1982). Cytomegalovirus infection in infancy: virological and immunological studies. Clin. Exp. Immunol. *47*, 27–33.

Gerna, G., Percivalle, E., Baldanti, F., Sozzani, S., Lanzarini, P., Genini, E., Lilleri, D., and Revello, M.G. (2000). Human cytomegalovirus replicates abortively in polymorphonuclear leukocytes after transfer from infected endothelial cells via transient microfusion events. J. Virol. *74*, 5629–5638.

Gerna, G., Percivalle, E., Sarasini, A., Baldanti, F., and Revello, M.G. (2002). The attenuated Towne strain of human cytomegalovirus may revert to both endothelial cell tropism and leuko- (neutrophil- and monocyte-) tropism *in vitro*. J. Gen. Virol. *83*, 1993–2000.

Gerna, G., Percivalle, E., Sarasini, A., Baldanti, F., Campanini, G., and Revello, M.G. (2003). Rescue of human cytomegalovirus strain AD169 tropism for both leukocytes and human endothelial cells. J. Gen. Virol. *84*, 1431–1436.

Gibson, K.L., Wu, Y.C., Barnett, Y., Duggan, O., Vaughan, R., Kondeatis, E., Nilsson, B.O., Wikby, A., Kipling, D., and Dunn-Walters, D.K. (2009). B-cell diversity decreases in old age and is correlated with poor health status. Aging Cell *8*, 18–25.

Gollackner, B., Mueller, N.J., Houser, S., Qawi, I., Soizic, D., Knosalla, C., Buhler, L., Dor, F.J., Awwad, M., Sachs, D.H., *et al.* (2003). Porcine cytomegalovirus and coagulopathy in pig-to-primate xenotransplantation. Transplantation *75*, 1841–1847.

Gomez, C.R., Boehmer, E.D., and Kovacs, E.J. (2005). The aging innate immune system. Curr. Opin. Immunol. *17*, 457–462.

Goodpasture, E.W., and Talbot, F.B. (1921). Concerning the nature of 'protozoan-like' cells in certain lesions of infancy. Am. J. Dis. Child. *21*, 415–425.

Goronzy, J.J., Fulbright, J.W., Crowson, C.S., Poland, G.A., O'Fallon, W.M., and Weyand, C.M. (2001). Value of immunological markers in predicting responsiveness to influenza vaccination in elderly individuals. J. Virol. *75*, 12182–12187.

Grazia Revello, M., Baldanti, F., Percivalle, E., Sarasini, A., De-Giuli, L., Genini, E., Lilleri, D., Labo, N., and Gerna, G. (2001). *In vitro* selection of human cytomegalovirus variants unable to transfer virus and virus products from infected cells to polymorphonuclear leukocytes and to grow in endothelial cells. J. Gen. Virol. *82*, 1429–1438.

Griffiths, P.D. (2002). Strategies to prevent CMV infection in the neonate. Semin. Neonatol. *7*, 293–299.

Hadrup, S.R., Strindhall, J., Kollgaard, T., Seremet, T., Johansson, B., Pawelec, G., thor Straten, P., and Wikby, A. (2006). Longitudinal studies of clonally expanded CD8 T-cells reveal a repertoire shrinkage predicting mortality and an increased number of dysfunctional cytomegalovirus-specific T-cells in the very elderly. J. Immunol. *176*, 2645–2653.

Hahn, G., Revello, M.G., Patrone, M., Percivalle, E., Campanini, G., Sarasini, A., Wagner, M., Gallina, A., Milanesi, G., Koszinowski, U., *et al.* (2004). Human cytomegalovirus UL131–128 genes are indispensable for virus growth in endothelial cells and virus transfer to leukocytes. J. Virol. *78*, 10023–10033.

Hamirally, S., Kamil, J.P., Ndassa-Colday, Y.M., Lin, A.J., Jahng, W.J., Baek, M.C., Noton, S., Silva, L.A., Simpson-Holley, M., Knipe, D.M., *et al.* (2009). Viral mimicry of Cdc2/cyclin-dependent kinase 1 mediates disruption of nuclear lamina during human cytomegalovirus nuclear egress. PLoS Pathog. *5*, e1000275.

Hamprecht, K., Vochem, M., Baumeister, A., Boniek, M., Speer, C.P., and Jahn, G. (1998). Detection of cytomegaloviral DNA in human milk cells and cell free milk whey by nested PCR. Virol. Meth. *70*, 167–176.

Hamprecht, K., Maschmann, J., Vochem, M., Dietz, K., Speer, C.P., and Jahn, G. (2001). Epidemiology of transmission of cytomegalovirus from mother to preterm infant by breastfeeding. Lancet 357, 513–518.

Han, D., Berman, D.M., Willman, M., Buchwald, P., Rothen, D., Kenyon, N.M., and Kenyon, N.S. (2010). Choice of immunosuppression influences cytomegalovirus DNAemia in cynomolgus monkey (*Macaca fascicularis*) islet allograft recipients. Cell Transplant. *19*, 1547–1561.

Hansen, S.G., Strelow, L.I., Franchi, D.C., Anders, D.G., and Wong, S.W. (2003). Complete sequence and genomic analysis of rhesus cytomegalovirus. J. Virol. *77*, 6620–6636.

Hansen, S.G., Vieville, C., Whizin, N., Coyne-Johnson, L., Siess, D.C., Drummond, D.D., Legasse, A.W., Axthelm, M.K., Oswald, K., Trubey, C.M., *et al.* (2009). Effector memory T-cell responses are associated with protection of rhesus monkeys from mucosal simian immunodeficiency virus challenge. Nat. Med. *15*, 293–299.

Hansen, S.G., Powers, C.J., Richards, R., Ventura, A.B., Ford, J.C., Siess, D., Axthelm, M.K., Nelson, J.A., Jarvis, M.A., Picker, L.J., *et al.* (2010). Evasion of CD8+ T-cells is critical for superinfection by cytomegalovirus. Science *328*, 102–106.

Haustein, S.V., Kolterman, A.J., Sundblad, J.J., Fechner, J.H., and Knechtle, S.J. (2008). Nonhuman primate infections after organ transplantation. ILAR J. *49*, 209–219.

Haynes, L., and Maue, A.C. (2009). Effects of aging on T-cell function. Curr. Opin. Immunol. *21*, 414–417.

Howard, M.R., Whitby, D., Bahadur, G., Suggett, F., Boshoff, C., Tenant-Flowers, M., Schulz, T.F., Kirk, S., Matthews, S., Weller, I.V., *et al.* (1997). Detection of human herpesvirus 8 DNA in semen from HIV-infected individuals but not healthy semen donors. Aids *11*, F15–F19.

Huang, E.S., Kilpatrick, B., Lakeman, A., and Alford, C.A. (1978). Genetic analysis of a cytomegalovirus-like agent isolated from human brain. J. Virol. *26*, 718–723.

Huh, Y.H., Kim, Y.E., Kim, E.T., Park, J.J., Song, M.J., Zhu, H., Hayward, G.S., and Ahn, J.H. (2008). Binding STAT2 by the acidic domain of human cytomegalovirus IE1 promotes viral growth and is negatively regulated by SUMO. J. Virol. *82*, 10444–10454.

Hume, A.J., Finkel, J.S., Kamil, J.P., Coen, D.M., Culbertson, M.R., and Kalejta, R.F. (2008). Phosphorylation of retinoblastoma protein by viral protein with cyclin-dependent kinase function. Science *320*, 797–799.

Humphreys, I.R., de Trez, C., Kinkade, A., Benedict, C.A., Croft, M., and Ware, C.F. (2007). Cytomegalovirus exploits IL-10-mediated immune regulation in the salivary glands. J. Exp. Med. *204*, 1217–1225.

Huppert, F.A., Pinto, E.M., Morgan, K., and Brayne, C. (2003). Survival in a population sample is predicted by proportions of lymphocyte subsets. Mech. Ageing Dev. *124*, 449–451.

Jackson, L. (1920). An intracellular protozoan parasite of the ducts of the salivary glands of the guinea pig. J. Infect. Dis. *26*, 347–350.

Jankovic, V., Messaoudi, I., and Nikolich-Zugich, J. (2003). Phenotypic and functional T-cell aging in rhesus macaques (*Macaca mulatta*): differential behavior of CD4 and CD8 subsets. Blood *102*, 3244–3251.

Jeang, K.-T., and Gibson, W. (1980). A cycloheximide-enhanced protein in cytomegalovirus-infected cells. Virology *107*, 362–374.

Jeang, K.T., Chin, G., and Hayward, G.S. (1982). Characterization of cytomegalovirus immediate-early genes. I. Nonpermissive rodent cells overproduce the IE94K protein form CMV (Colburn). Virology *121*, 393–403.

Jeang, K.-T., Cho, M.-S., and Hayward, G.S. (1984). Abundant constitutive expression of the immediate-early 94K protein from cytomegalovirus (Colburn) in a DNA-transfected mouse cell line. Mol. Cell. Biol. *4*, 2214–2223.

Jeang, K.-T., Rawlins, D.R., Rosenfeld, P.J., Shero, J.D., Kelly, T.J., and Hayward, G.S. (1987). Multiple tandemly repeated binding sites for cellular nuclear factor 1 that surround the major immediate-early promoters of simian and human cytomegalovirus. J. Virol. *61*, 1559–1570.

Jenkins, C., Abendroth, A., and Slobedman, B. (2004). A novel viral transcript with homology to human interleukin-10 is expressed during latent human cytomegalovirus infection. J. Virol. *78*, 1440–1447.

Jim, W.T., Shu, C.H., Chiu, N.C., Kao, H.A., Hung, H.Y., Chang, J.H., Peng, C.C., Hsieh, W.S., Liu, K.C., and Huang, F.Y. (2004). Transmission of cytomegalovirus from mothers to preterm infants by breast milk. Ped. Infect. Dis. J. *23*, 848–851.

Jim, W.T., Shu, C.H., Chiu, N.C., Chang, J.H., Hung, H.Y., Peng, C.C., Kao, H.A., Wei, T.Y., Chiang, C.L., and Huang, F.Y. (2009). High cytomegalovirus load and prolonged virus excretion in breast milk increase risk for viral acquisition by very low birth weight infants. Ped. Infect. Dis. J. *28*, 891–894.

Jones, B.C., Logsdon, N.J., Josephson, K., Cook, J., Barry, P.A., and Walter, M.R. (2002). Crystal structure of human cytomegalovirus IL-10 bound to soluble human IL-10R1. Proc. Natl. Acad. Sci. U.S.A. *99*, 9404–9409.

Jones, M., Ladell, K., Wynn, K.K., Stacey, M.A., Quigley, M.F., Gostick, E., Price, D.A., and Humphreys, I.R. (2010). IL-10 restricts memory T-cell inflation during cytomegalovirus infection. J. Immunol. *185*, 3583–3592.

Jones-Engel, L., Engel, G.A., Heidrich, J., Chalise, M., Poudel, N., Viscidi, R., Barry, P.A., Allan, J.S., Grant, R., and Kyes, R. (2006). Temple monkeys and health implications of commensalism, Kathmandu, Nepal. Emerg. Infect. Dis. *12*, 900–906.

Jonker, M., Ringers, J., Kuhn, E.M., t Hart, B., and Foulkes, R. (2004). Treatment with anti-MHC-class-II antibody postpones kidney allograft rejection in primates but increases the risk of CMV activation. Amer. J. Transplant. *4*, 1756–1761.

Just, M., Buergin-Wolff, A., Emoedi, G., and Hernandez, R. (1975). Immunisation trials with live attenuated cytomegalovirus TOWNE 125. Infection *3*, 111–114.

Kanthaswamy, S., Satkoski, J., George, D., Kou, A., Erickson, B.J., and Smith, D.G. (2008). Interspecies Hybridization and the Stratification of Nuclear Genetic Variation of Rhesus (*Macaca Mulatta*) and Long-Tailed Macaques (*Macaca Fascicularis*). Inter. J. Pathol. *29*, 1295–1311.

Kashiwagi, Y., Nemoto, S., Hisashi, Kawashima, Takekuma, K., Matsuno, T., Hoshika, A., and Nozaki-Renard, J. (2001). Cytomegalovirus DNA among children attending two day-care centers in Tokyo. Pediatr. Int. *43*, 493–495.

Kaur, A., Daniel, M.D., Hempel, D., Lee-Parritz, D., Hirsch, M.S., and Johnson, R.P. (1996). Cytotoxic T-lymphocyte responses to cytomegalovirus in normal and simian immunodeficiency virus-infected macaques. J. Virol. *70*, 7725–7733.

Kaur, A., Rosenzweig, M., and Johnson, R.P. (2000). Immunological memory and acquired immunodeficiency syndrome pathogenesis. Philos. Trans. Roy. Soc. Lond. Biol. Sci. *355*, 381–390.

Kaur, A., Kassis, N., Hale, C.L., Simon, M., Elliott, M., Gomez-Yafal, A., Lifson, J.D., Desrosiers, R.C., Wang, F., Barry, P., *et al.* (2003). Direct relationship between suppression of virus-specific immunity and emergence of cytomegalovirus disease in simian AIDS. J. Virol. *77*, 5749–5758.

Kawaguchi, M., Nakamura, A., and Tsurusawa, M. (1999). The role of mitochondria in apoptosis in U937 and Molt-4 cells: difference in order of mitochondrial membrane potential (delta psi m) reduction and interleukin-1 beta-converting enzyme (ICE) on signal transduction pathway in each cell type. Gan. ToKagaku Ryoho. Canc. Chemother. *26*, 679–685.

Kean, L.S., Gangappa, S., Pearson, T.C., and Larsen, C.P. (2006). Transplant tolerance in non-human primates: progress, current challenges and unmet needs. Amer. J. Transplant. *6*, 884–893.

Kean, L.S., Adams, A.B., Strobert, E., Hendrix, R., Gangappa, S., Jones, T.R., Shirasugi, N., Rigby, M.R., Hamby, K., Jiang, J., *et al.* (2007). Induction of chimerism in rhesus macaques through stem cell transplant and costimulation blockade-based immunosuppression. Amer. J. Transplant. *7*, 320–335.

Kenneson, A., and Cannon, M.J. (2007). Review and meta-analysis of the epidemiology of congenital cytomegalovirus (CMV) infection. Rev. Med. Virol. *17*, 253–276.

Kessler, M.J., London, W.T., Madden, D.L., Dambrosia, J.M., Hilliard, J.K., Soike, K.F., and Rawlins, R.G. (1989). Serological survey for viral diseases in the Cayo Santiago rhesus macaque population. Puerto Rican Health Sci. J. *8*, 95–97.

Khan, N., Shariff, N., Cobbold, M., Bruton, R., Ainsworth, J.A., Sinclair, A.J., Nayak, L., and Moss, P.A. (2002). Cytomegalovirus seropositivity drives the CD8 T-cell repertoire toward greater clonality in healthy elderly individuals. J. Immunol. *169*, 1984–1992.

Khan, N., Hislop, A., Gudgeon, N., Cobbold, M., Khanna, R., Nayak, L., Rickinson, A.B., and Moss, P.A. (2004). Herpesvirus-specific CD8 T-cell immunity in old age: cytomegalovirus impairs the response to a coresident EBV infection. J. Immunol. *173*, 7481–7489.

Kita, Y.F., Hosomichi, K., Kohara, S., Itoh, Y., Ogasawara, K., Tsuchiya, H., Torii, R., Inoko, H., Blancher, A., Kulski, J.K., and Shiina, T. (2009). MHC class I A loci polymorphism and diversity in three Southeast Asian populations of cynomolgus macaque. Immunogenetics *61*, 635–648.

Kledal, T.N., Rosenkilde, M.M., and Schwartz, T.W. (1998). Selective recognition of the membrane-bound CX3C chemokine, fractalkine, by the human cytomegalovirus-encoded broad-spectrum receptor US28. FEBS Lett. *441*, 209–214.

Koch, S., Solana, R., Dela Rosa, O., and Pawelec, G. (2006). Human cytomegalovirus infection and T-cell immunosenescence: a mini review. Mech. Ageing Dev. *127*, 538–543.

Kotenko, S.V., Saccani, S., Izotova, L.S., Mirochnitchenko, O.V., and Pestka, S. (2000). Human cytomegalovirus harbors its own unique IL-10 homolog (cmvIL-10). Proc. Natl. Acad. Sci. U.S.A. *97*, 1695–1700.

Kovaiou, R.D., and Grubeck-Loebenstein, B. (2006). Age-associated changes within CD4+ T-cells. Immunol. Lett. *107*, 8–14.

Kovaiou, R.D., Herndler-Brandstetter, D., and Grubeck-Loebenstein, B. (2007). Age-related changes in immunity: implications for vaccination in the elderly. Expert Rev. Mol. Med. *9*, 1–17.

Krabbe, K.S., Pedersen, M., and Bruunsgaard, H. (2004). Inflammatory mediators in the elderly. Exp. Gerontol. *39*, 687–699.

Krauss, S., Kaps, J., Czech, N., Paulus, C., and Nevels, M. (2009). Physical requirements and functional consequences of complex formation between the cytomegalovirus IE1 protein and human STAT2. J. Virol. *83*, 12854–12870.

Kravitz, R.H., Sciabica, K.S., Cho, K., Luciw, P.A., and Barry, P.A. (1997). Cloning and characterization of the rhesus cytomegalovirus glycoprotein B. J. Gen. Virol. *78*, 2009–2013.

Kropff, B., and Mach, M. (1997). Identification of the gene coding for rhesus cytomegalovirus glycoprotein B and immunological analysis of the protein. J. Gen. Virol. *78*, 1999–2007.

Krosky, P.M., Baek, M.C., and Coen, D.M. (2003). The human cytomegalovirus UL97 protein kinase, an antiviral drug target, is required at the stage of nuclear egress. J. Virol. *77*, 905–914.

Leendertz, F.H., Deckers, M., Schempp, W., Lankester, F., Boesch, C., Mugisha, L., Dolan, A., Gatherer, D., McGeoch, D.J., and Ehlers, B. (2009). Novel cytomegaloviruses in free-ranging and captive great apes: phylogenetic evidence for bidirectional horizontal transmission. J. Gen. Virol. *90*, 2386–2394.

Lenac, T., Arapovic, J., Traven, L., Krmpotic, A., and Jonjic, S. (2008). Murine cytomegalovirus regulation of NKG2D ligands. Med. Microbiol. Immunol. *197*, 159–166.

Lesniewski, M., Das, S., Skomorovska-Prokvolit, Y., Wang, F.Z., and Pellett, P.E. (2006). Primate cytomegalovirus US12 gene family: a distinct and diverse clade of seven-transmembrane proteins. Virology *354*, 286–298.

Lilja, A.E., and Shenk, T. (2008). Efficient replication of rhesus cytomegalovirus variants in multiple rhesus and human cell types. Proc. Natl. Acad. Sci. U.S.A. *105*, 19950–19955.

Lilja, A.E., Chang, W.L., Barry, P.A., Becerra, S.P., and Shenk, T.E. (2008). Functional genetic analysis of rhesus cytomegalovirus: Rh01 is an epithelial cell tropism factor. J. Virol. *82*, 2170–2181.

Lilley, B.N., Ploegh, H.L., and Tirabassi, R.S. (2001). Human cytomegalovirus open reading frame TRL11/IRL11 encodes an immunoglobulin G Fc-binding protein. J. Virol. *75*, 11218–11221.

Lischka, P., Toth, Z., Thomas, M., Mueller, R., and Stamminger, T. (2006). The UL69 transactivator protein of human cytomegalovirus interacts with DEXD/H-Box RNA helicase UAP56 to promote cytoplasmic accumulation of unspliced RNA. Mol. Cell Biol. *26*, 1631–1643.

Lischka, P., Thomas, M., Toth, Z., Mueller, R., and Stamminger, T. (2007). Multimerization of human cytomegalovirus regulatory protein UL69 via a domain that is conserved within its herpesvirus homologues. J. Gen. Virol. *88*, 405–410.

Lockridge, K.M., Zhou, S.S., Kravitz, R.H., Johnson, J.L., Sawai, E.T., Blewett, E.L., and Barry, P.A. (2000). Primate cytomegaloviruses encode and express an IL-10-like protein. Virology *268*, 272–280.

Logsdon, N.J., Eberhardt, M.K., Allen, C.E., Barry, P.A., and Walter, M.R. (2011). Design and analysis of rhesus cytomegalovirus IL-10 mutants as a model for novel vaccines against human cytomegalovirus. PLoS ONE *6*, e28127.

London, W.T., Martinez, A.J., Houff, S.A., Wallen, W.C., Curfman, B.L., Traub, R.G., and Sever, J.L. (1986). Experimental congenital disease with simian cytomegalovirus in rhesus monkeys. Teratol. *33*, 323–331.

Looney, R.J., Falsey, A., Campbell, D., Torres, A., Kolassa, J., Brower, C., McCann, R., Menegus, M., McCormick, K., Frampton, M., *et al.* (1999). Role of cytomegalovirus in the T-cell changes seen in elderly individuals. Clin. Immunol. *90*, 213–219.

McCormick, A.L., Skaletskaya, A., Barry, P.A., Mocarski, E.S., and Goldmacher, V.S. (2003). Differential function and expression of the viral inhibitor of caspase 8-induced apoptosis (vICA) and the viral mitochondria-localized inhibitor of apoptosis (vMIA) cell death suppressors conserved in primate and rodent cytomegaloviruses. Virology *316*, 221–233.

McGeer, P.L., Rogers, J., and McGeer, E.G. (2006). Inflammation, anti-inflammatory agents and Alzheimer disease: the last 12 years. J. Alzheim. Dis. *9*, 271–276.

McGeoch, D.J., Cook, S., Dolan, A., Jamieson, F.E., and Telford, E.A. (1995). Molecular phylogeny and evolutionary timescale for the family of mammalian herpesviruses. J. Mol. Biol. *247*, 443–458.

Mansat, A., Mengelle, C., Chalet, M., Boumzebra, A., Mieusset, R., Puel, J., Prouheze, C., and Segondy, M. (1997). Cytomegalovirus detection in cryopreserved semen samples collected for therapeutic donor insemination. Hum. Reprod. *12*, 1663–1666.

Marchini, A., Liu, H., and Zhu, H. (2001). Human cytomegalovirus with IE-2 (UL122) deleted fails to express early lytic genes. J. Virol. *75*, 1870–1878.

Marschall, M., Freitag, M., Suchy, P., Romaker, D., Kupfer, R., Hanke, M., and Stamminger, T. (2003). The protein kinase pUL97 of human cytomegalovirus interacts with and phosphorylates the DNA polymerase processivity factor pUL44. Virology *311*, 60–71.

Marschall, M., Marzi, A., aus dem Siepen, P., Jochmann, R., Kalmer, M., Auerochs, S., Lischka, P., Leis, M., and Stamminger, T. (2005). Cellular p32 recruits cytomegalovirus kinase pUL97 to redistribute the nuclear lamina. J. Biol. Chem. *280*, 33357–33367.

Marsh, A.K., Willer, D.O., Ambagala, A.P., Dzamba, M., Chan, J.K., Pilon, R., Fournier, J., Sandstrom, P., Brudno, M., and Macdonald, K.S. (2011). Genomic Sequencing and Characterization of Cynomolgus Macaque Cytomegalovirus. J. Virol. *85*, 12995–13009.

Marshall, B.C., and Adler, S.P. (2009). The frequency of pregnancy and exposure to cytomegalovirus infections among women with a young child in day care. Am. J. Obstet. Gynecol. *200*, 163e161–165.

Martin, W.J., Zeng, L.C., Ahmed, K., and Roy, M. (1994). Cytomegalovirus-related sequence in an atypical cytopathic virus repeatedly isolated from a patient with chronic fatigue syndrome. Am. J. Pathol. *145*, 440–451.

Martin, W.J., Ahmed, K.N., Zeng, L.C., Olsen, J.C., Seward, J.G., and Seehrai, J.S. (1995). African green monkey origin of the atypical cytopathic 'stealth virus' isolated from a patient with chronic fatigue syndrome. Clin. Diag. Virol. *4*, 93–103.

Meier, J.L., and Stinski, M.F. (1996). Regulation of human cytomegalovirus immediate-early gene expression. Intervirology *39*, 331–342.

Messaoudi, I., Warner, J., Fischer, M., Park, B., Hill, B., Mattison, J., Lane, M.A., Roth, G.S., Ingram, D.K., Picker, L.J., *et al.* (2006). Delay of T-cell senescence by caloric restriction in aged long-lived nonhuman primates. Proc. Natl. Acad. Sci. U.S.A. *103*, 19448–19453.

Michaels, M.G., Jenkins, F.J., St George, K., Nalesnik, M.A., Starzl, T.E., and Rinaldo, C.R., Jr. (2001). Detection of infectious baboon cytomegalovirus after baboon-to-human liver xenotransplantation. J. Virol. *75*, 2825–2828.

Milbradt, J., Webel, R., Auerochs, S., Sticht, H., and Marschall, M. (2010). Novel mode of phosphorylation-triggered reorganization of the nuclear lamina during nuclear egress of human cytomegalovirus. J. Biol. Chem. *285*, 13979–13989.

Miller-Kittrell, M., Sai, J., Penfold, M., Richmond, A., and Sparer, T.E. (2007). Functional characterization of chimpanzee cytomegalovirus chemokine, vCXCL-1(CCMV). Virology *364*, 454–465.

Minamishima, Y., Graham, B.J., and Benyesh-Melnick, M. (1971). Neutralizing antibodies to cytomegaloviruses in normal simian and human sera. Infect. Immun. *4*, 368–373.

Morello, C.S., Ye, M., and Spector, D.H. (2002). Development of a vaccine against murine cytomegalovirus (MCMV), consisting of plasmid DNA and formalin-inactivated MCMV, that provides long-term, complete protection against viral replication. J. Virol. *76*, 4822–4835.

Morello, C.S., Ye, M., Hung, S., Kelley, L.A., and Spector, D.H. (2005). Systemic priming-boosting immunization with a trivalent plasmid DNA and inactivated murine cytomegalovirus (MCMV) vaccine provides long-term protection against viral replication following systemic or mucosal MCMV challenge. J. Virol. *79*, 159–175.

Mueller, N.J., and Fishman, J.A. (2004). Herpesvirus infections in xenotransplantation: pathogenesis and approaches. Xenotransplant. *11*, 486–490.

Mueller, N.J., Barth, R.N., Yamamoto, S., Kitamura, H., Patience, C., Yamada, K., Cooper, D.K., Sachs, D.H., Kaur, A., and Fishman, J.A. (2002). Activation of cytomegalovirus in pig-to-primate organ xenotransplantation. J. Virol. *76*, 4734–4740.

Mueller, N.J., Sulling, K., Gollackner, B., Yamamoto, S., Knosalla, C., Wilkinson, R.A., Kaur, A., Sachs, D.H., Yamada, K., Cooper, D.K., *et al.* (2003). Reduced efficacy of ganciclovir against porcine and baboon cytomegalovirus in pig-to-baboon xenotransplantation. Am. J. Transplant. *3*, 1057–1064.

Mueller, Y.M., Petrovas, C., Bojczuk, P.M., Dimitriou, I.D., Beer, B., Silvera, P., Villinger, F., Cairns, J.S., Gracely, E.J., Lewis, M.G., *et al.* (2005). Interleukin-15 increases effector memory CD8+ T-cells and NK Cells in simian immunodeficiency virus-infected macaques. J. Virol. *79*, 4877–4885.

Murphy, E., and Shenk, T. (2008). Human cytomegalovirus genome. Curr. Top. Microbiol. Immunol. *325*, 1–19.

Murphy, E., Yu, D., Grimwood, J., Schmutz, J., Dickson, M., Jarvis, M.A., Hahn, G., Nelson, J.A., Myers, R.M., and Shenk, T.E. (2003). Coding potential of laboratory and clinical strains of human cytomegalovirus. Proc. Natl. Acad. Sci. (USA) *100*, 14976–14981.

Naruse, T.K., Okuda, Y., Mori, K., Akari, H., Matano, T., and Kimura, A. (2011). ULBP4/RAET1E is highly polymorphic in the Old World monkey. Immunogenetics *63*, 501–509.

Navarro, D., Paz, P., Tugizov, S., Topp, K., La Vail, J., and Pereira, L. (1993). Glycoprotein B of human cytomegalovirus promotes virion penetration into cells, transmission of infection from cell to cell, and fusion of infected cells. Virology *197*, 143–158.

Nigida, S.M., Jr., Falk, L.A., Wolfe, L.G., Deinhardt, F., Lakeman, A., and Alford, C.A. (1975). Experimental infection of marmosets with a cytomegalovirus of human origin. J. Infect. Dis. *132*, 582–586.

Nigida, S.M., Falk, L.A., Wolfe, L.G., and Deinhardt, F. (1979). Isolation of a cytomegalovirus from salivary glands of white-lipped marmosets (*Saguinus fuscicollis*). Lab. Anim. Sci. *29*, 53–60.

Nikolich-Zugich, J. (2005). T-cell aging: naïve but not young. J. Exp. Med. *201*, 837–840.

Nikolich-Zugich, J. (2008). Ageing and life-long maintenance of T-cell subsets in the face of latent persistent infections. Nat. Rev. Immunol. *8*, 512–522.

North, T.W., Sequar, G., Townsend, L.B., Drach, J.C., and Barry, P.A. (2004). Rhesus cytomegalovirus is similar to human cytomegalovirus in susceptibility to benzimidizole nucleosides. Antimicrob. Agents Chemother. *48*, 2760–2765.

Oakley, O.R., Garvy, B.A., Humphreys, S., Qureshi, M.H., and Pomeroy, C. (2008). Increased weight loss with reduced viral replication in interleukin-10 knock-out mice infected with murine cytomegalovirus. Clin. Exp. Immunol. *151*, 155–164.

Ohtaki, S., Kodama, H., Hondo, R., and Kurata, T. (1986). Activation of cytomegalovirus infection in immunosuppressed cynomolgous monkeys inoculated with varicella-zoster virus. Acta Patholog. Jpn. *36*, 1553–1563.

Olsson, J., Wikby, A., Johansson, B., Lofgren, S., Nilsson, B.O., and Ferguson, F.G. (2000). Age-related change in peripheral blood T-lymphocyte subpopulations and cytomegalovirus infection in the very old: the Swedish longitudinal OCTO immune study. Mech. Ageing Dev. *121*, 187–201.

Ostan, R., Bucci, L., Capri, M., Salvioli, S., Scurti, M., Pini, E., Monti, D., and Franceschi, C. (2008). Immunosenescence and immunogenetics of human longevity. Neuroimmunomod. *15*, 224–240.

Otting, N., de Vos-Rouweler, A.J., Heijmans, C.M., de Groot, N.G., Doxiadis, G.G., and Bontrop, R.E. (2007). MHC class I A region diversity and polymorphism in macaque species. Immunogenetics *59*, 367–375.

Oxford, K.L., Eberhardt, M.K., Yang, K.W., Strelow, L., Kelly, S., Zhou, S.S., and Barry, P.A. (2008). Protein coding content of the U(L)b' region of wild-type rhesus cytomegalovirus. Virology *373*, 181–188.

Oxford, K.L., Strelow, L., Yue, Y., Chang, W.L., Schmidt, K.A., Diamond, D.J., and Barry, P.A. (2011). Open reading frames carried on UL/b' are implicated in shedding and horizontal transmission of rhesus cytomegalovirus in rhesus monkeys. J. Virol. *85*, 5105–5114.

Pande, N.T., Powers, C., Ahn, K., and Früh, K. (2005). Rhesus cytomegalovirus contains functional homologues of US2, US3, US6, and US11. J. Virol. *79*, 5786–5798.

Pass, R.F., Zhang, C., Evans, A., Simpson, T., Andrews, W., Huang, M.L., Corey, L., Hill, J., Davis, E., Flanigan, C., *et al.* (2009). Vaccine prevention of maternal cytomegalovirus infection. N. Eng. J. Med. *360*, 1191–1199.

Paulus, C., Krauss, S., and Nevels, M. (2006). A human cytomegalovirus antagonist of type I IFN-dependent signal transducer and activator of transcription signaling. Proc. Natl. Acad. Sci. U.S.A. *103*, 3840–3845.

Pawelec, G., Derhovanessian, E., Larbi, A., Strindhall, J., and Wikby, A. (2009). Cytomegalovirus and human immunosenescence. Rev. Med. Virol. *19*, 47–56.

Penfold, M.E.T., Dairaghi, D.J., Duke, G.M., Saederup, N., Mocarski, E.S., Kemble, G.W., and Schall, T.J. (1999). Cytomegalovirus encodes a potent a chemokine. Proc. Natl. Acad. Sci. U.S.A. *96*, 9839–9844.

Penfold, M.E., Schmidt, T.L., Dairaghi, D.J., Barry, P.A., and Schall, T.J. (2003). Characterization of the rhesus cytomegalovirus US28 locus. J. Virol. *77*, 10404–10413.

Pereira, L., and Maidji, E. (2008). Cytomegalovirus infection in the human placenta: maternal immunity and developmentally regulated receptors on trophoblasts converge. Curr. Top. Microbiol. Immunol. *325*, 383–395.

Pizzorno, M.C., O'Hare, P., Sha, L., LaFemina, R., and Hayward, G.S. (1988). Trans-activation and autoregulation of gene expression by the immediate-early region 2 gene products of human cytomegalovirus. J. Virol. *62*, 1167–1179.

Plotkin, S.A., Starr, S.E., Friedman, H.M., Gonczol, E., and Weibel, R.E. (1989). Protective effects of Towne cytomegalovirus vaccine against low-passage cytomegalovirus administered as a challenge. J. Infect. Dis. *159*, 860–865.

Powers, C.J., and Früh, K. (2008). Signal peptide-dependent inhibition of MHC class I heavy chain translation by rhesus cytomegalovirus. PLoS Pathog. *4*, e1000150.

Prepens, S., Kreuzer, K.A., Leendertz, F., Nitsche, A., and Ehlers, B. (2007). Discovery of herpesviruses in multi-infected primates using locked nucleic acids (LNA) and a bigenic PCR approach. Virol. J. *4*, 84.

Prichard, M.N., Sztul, E., Daily, S.L., Perry, A.L., Frederick, S.L., Gill, R.B., Hartline, C.B., Streblow, D.N., Varnum, S.M., Smith, R.D., *et al.* (2008). Human cytomegalovirus UL97 kinase activity is required for the hyperphosphorylation of retinoblastoma protein and inhibits the formation of nuclear aggresomes. J. Virol. *82*, 5054–5067.

Prod'homme, V., Sugrue, D.M., Stanton, R.J., Nomoto, A., Davies, J., Rickards, C.R., Cochrane, D., Moore, M., Wilkinson, G.W., and Tomasec, P. (2010). Human cytomegalovirus UL141 promotes efficient down-regulation of the natural killer cell activating ligand CD112. J. Gen. Virol. *91*, 2034–2039.

Quinnan, G.V.J., Delery, M., Rook, A.H., Frederick, W.R., Epstein, J.S., Manischewitz, J.F., Jackson, L., Ramsey, K.M., Mittal, K., Plotkin, S.A., and Hilleman, M.R. (1984). Comparative virulence and immunogenicity of the Towne strain and a nonattenuated strain of cytomegalovirus. Ann. Int. Med. *101*, 478–483.

Raftery, M.J., Wieland, D., Gronewald, S., Kraus, A.A., Giese, T., and Schonrich, G. (2004). Shaping phenotype, function, and survival of dendritic cells by cytomegalovirus-encoded IL-10. J. Immunol. *173*, 3383–3391.

Rahman, A.A., Teschner, M., Sethi, K.K., and Brandis, H. (1976). Appearance of IgG (Fc) receptor(s) on cultured human fibroblasts infected with human cytomegalovirus. J. Immunol. *117*, 253–258.

Redpath, S., Angulo, A., Gascoigne, N.R., and Ghazal, P. (1999). Murine cytomegalovirus infection down-regulates MHC class II expression on macrophages by induction of IL-10. J. Immunol. *162*, 6701–6707.

Ribbert, H. (1904). Über protozoenartige Zellen in der Niere eines syphilitischen Neugeborenen und in der Parotis von Kindern. Centralblatt für allgemeine Pathologie und pathologische Anatomie *15*, 945–948.

Richards, R., Scholz, I., Powers, C., Skach, W.R., and Früh, K. (2011). The cytoplasmic domain of rhesus cytomegalovirus Rh178 interrupts translation of major histocompatibility class I leader peptide-containing proteins prior to translocation. J. Virol. *85*, 8766–8776.

Rivailler, P., Jiang, H., Cho, Y.G., Quink, C., and Wang, F. (2002). Complete nucleotide sequence of the rhesus lymphocryptovirus: genetic validation for an Epstein–Barr virus animal model. J. Virol. *76*, 421–426.

Rivailler, P., Kaur, A., Johnson, R.P., and Wang, F. (2006). Genomic sequence of rhesus cytomegalovirus 180.92: insights into the coding potential of rhesus cytomegalovirus. J. Virol. *80*, 4179–4182.

Romaker, D., Schregel, V., Maurer, K., Auerochs, S., Marzi, A., Sticht, H., and Marschall, M. (2006). Analysis of the structure–activity relationship of four herpesviral UL97 subfamily protein kinases reveals partial but not full functional conservation. J. Med. Chem. *49*, 7044–7053.

Romphruk, A.V., Romphruk, A., Naruse, T.K., Raroengjai, S., Puapairoj, C., Inoko, H., and Leelayuwat, C. (2009). Polymorphisms of NKG2D ligands: diverse RAET1/ULBP genes in northeastern Thais. Immunogenetics *61*, 611–617.

Rosenkilde, M.M., Waldhoer, M., Luttichau, H.R., and Schwartz, T.W. (2001). Virally encoded 7TM receptors. Oncogene *20*, 1582–1593.

Rosenkilde, M.M., Smit, M.J., and Waldhoer, M. (2008). Structure, function and physiological consequences of virally encoded chemokine seven transmembrane receptors. Brit. J. Pharmacol. *153*, 154–166.

Ross, S.A., Fowler, K.B., Ashrith, G., Stagno, S., Britt, W.J., Pass, R.F., and Boppana, S.B. (2006). Hearing loss in children with congenital cytomegalovirus infection born to mothers with preexisting immunity. J. Pediatr. *148*, 332–336.

Ross, S.A., Arora, N., Novak, Z., Fowler, K.B., Britt, W.J., and Boppana, S.B. (2010). Cytomegalovirus reinfections in healthy seroimmune women. J. Infect. Dis. *201*, 386–389.

Rue, C.A., Jarvis, M.A., Knoche, A.J., Meyers, H.L., DeFilippis, V.R., Hansen, S.G., Wagner, M., Früh, K., Anders, D.G., Wong, S.W., *et al.* (2004). A cyclooxygenase-2 homologue encoded by rhesus cytomegalovirus is a determinant for endothelial cell tropism. J. Virol. *78*, 12529–12536.

Ryckman, B.J., Jarvis, M.A., Drummond, D.D., Nelson, J.A., and Johnson, D.C. (2006). Human cytomegalovirus entry into epithelial and endothelial cells depends on genes UL128 to UL150 and occurs by endocytosis and low-pH fusion. J. Virol. *80*, 710–722.

Ryckman, B.J., Rainish, B.L., Chase, M.C., Borton, J.A., Nelson, J.A., Jarvis, M.A., and Johnson, D.C. (2008). Characterization of the human cytomegalovirus gH/gL/UL128–131 complex that mediates entry into epithelial and endothelial cells. J. Virol. *82*, 60–70.

Sahagun-Ruiz, A., Sierra-Honigmann, A.M., Krause, P., and Murphy, P.M. (2004). Simian cytomegalovirus encodes five rapidly evolving chemokine receptor homologues. Virus Genes *28*, 71–83.

Sansoni, P., Vescovini, R., Fagnoni, F., Biasini, C., Zanni, F., Zanlari, L., Telera, A., Lucchini, G., Passeri, G., Monti, D., *et al.* (2008). The immune system in extreme longevity. Exp. Gerontol. *43*, 61–65.

Saurwein-Teissl, M., Lung, T.L., Marx, F., Gschosser, C., Asch, E., Blasko, I., Parson, W., Bock, G., Schonitzer, D., Trannoy, E., *et al.* (2002). Lack of antibody production following immunization in old age: association with CD8(+) CD28(-) T-cell clonal expansions and an imbalance in the production of Th1 and Th2 cytokines. J. Immunol. *168*, 5893–5899.

Schleiss, M.R. (2006). Role of breast milk in acquisition of cytomegalovirus infection: recent advances. Curr. Opin. Pediatr. *18*, 48–52.

Schregel, V., Auerochs, S., Jochmann, R., Maurer, K., Stamminger, T., and Marschall, M. (2007). Mapping of a self-interaction domain of the cytomegalovirus protein kinase pUL97. J. Gen. Virol. *88*, 395–404.

Schuessler, A., Sampaio, K.L., Scrivano, L., and Sinzger, C. (2010). Mutational mapping of UL130 of human cytomegalovirus defines peptide motifs within the C-terminal third as essential for endothelial cell infection. J. Virol. *84*, 9019–9026.

Schuessler, A., Sampaio, K.L., Straschewski, S., and Sinzger, C. (2012). Mutational mapping of pUL131A of human cytomegalovirus emphasizes its central role for endothelial cell tropism. J. Virol. *86*, 504–512.

Seo, J.W., Walter, L., and Gunther, E. (2001). Genomic analysis of MIC genes in rhesus macaques. Tissue Antigens *58*, 159–165.

Sequar, G., Britt, W.J., Lakeman, F.D., Lockridge, K.M., Tarara, R.P., Canfield, D.R., Zhou, S.S., Gardner, M.B., and Barry, P.A. (2002). Experimental coinfection of rhesus macaques with rhesus cytomegalovirus and simian immunodeficiency virus: pathogenesis. J. Virol. *76*, 7661–7671.

Sester, M., Sester, U., Gartner, B., Kubuschok, B., Girndt, M., Meyerhans, A., and Kohler, H. (2002). Sustained high frequencies of specific CD4 T-cells restricted to a single persistent virus. J. Virol. *76*, 3748–3755.

Shiina, T., Ota, M., Shimizu, S., Katsuyama, Y., Hashimoto, N., Takasu, M., Anzai, T., Kulski, J.K., Kikkawa, E., Naruse, T., *et al.* (2006). Rapid evolution of major histocompatibility complex class I genes in primates generates new disease alleles in humans via hitchhiking diversity. Genetics *173*, 1555–1570.

Shiina, T., Kono, A., Westphal, N., Suzuki, S., Hosomichi, K., Kita, Y.F., Roos, C., Inoko, H., and Walter, L. (2011). Comparative genome analysis of the major histocompatibility complex (MHC) class I B/C segments in primates elucidated by genomic sequencing in common marmoset (*Callithrix jacchus*). Immunogenetics *63*, 485–499.

Siegrist, C.A., and Aspinall, R. (2009). B-cell responses to vaccination at the extremes of age. Nat. Rev. Immunol. *9*, 185–194.

Simon, C.O., Holtappels, R., Tervo, H.M., Böhm, V., Däubner, T., Oehrlein-Karpi, S.A., Kühnapfel, B., Renzaho, A., Strand, D., Podlech, J., *et al.* (2006). CD8 T-cells control cytomegalovirus latency by epitope-specific sensing of transcriptional reactivation. J. Virol. *80*, 10436–10456.

Simmons, H.A., and Mattison, J.A. (2011). The incidence of spontaneous neoplasia in two populations of captive rhesus macaques (*Macaca mulatta*). Antiox. Redox Sign. *14*, 221–227.

Slobedman, B., Barry, P.A., Spencer, J.V., Avdic, S., and Abendroth, A. (2009). Virus-encoded homologs of cellular interleukin-10 and their control of host immune function. J. Virol. *83*, 9618–9629.

Smith, K.O., Thiel, J.F., Newman, J.T., Harvey, E., Trousdale, M.D., Gehle, W.D., and Clark, G. (1969). Cytomegaloviruses as common adventitious contaminants in primary African green monkey kidney cell cultures. J. Natl. Canc. Inst. *42*, 489–496.

Sohn, Y.M., Park, K.I., Lee, C., Han, D.G., and Lee, W.Y. (1992). Congenital cytomegalovirus infection in Korean population with very high prevalence of maternal immunity. J. Korean Med. Sci. *7*, 47–51.

Sparer, T.E., Gosling, J., Schall, T.J., and Mocarski, E.S. (2004). Expression of human CXCR2 in murine neutrophils as a model for assessing cytomegalovirus chemokine vCXCL-1 function *in vivo*. J. Interferon Cyto. Res. *24*, 611–620.

Spencer, J.V., Lockridge, K.M., Barry, P.A., Lin, G., Tsang, M., Penfold, M.E., and Schall, T.J. (2002). Potent immunosuppressive activities of cytomegalovirus-encoded interleukin-10. J. Virol. *76*, 1285–1292.

Spencer, J.V., Cadaoas, J., Castillo, P.R., Saini, V., and Slobedman, B. (2008). Stimulation of B-lymphocytes by cmvIL-10 but not LAcmvIL-10. Virology *374*, 164–169.

Sprague, E.R., Reinhard, H., Cheung, E.J., Farley, A.H., Trujillo, R.D., Hengel, H., and Bjorkman, P.J. (2008). The human cytomegalovirus Fc receptor gp68 binds the Fc CH2–CH3 interface of immunoglobulin G. J. Virol. *82*, 3490–3499.

Stagno, S., Reynolds, D., Tsiantos, A., Fuccillo, D.A., Smith, R., Tiller, M., and Alford, C.A., Jr. (1975). Cervical cytomegalovirus excretion in pregnant and nonpregnant women: suppression in early gestation. J. Infect. Dis. *131*, 522–527.

Stanton, R.J., Baluchova, K., Dargan, D.J., Cunningham, C., Sheehy, O., Seirafian, S., McSharry, B.P., Neale, M.L., Davies, J.A., Tomasec, P., *et al.* (2010). Reconstruction of the complete human cytomegalovirus genome in a BAC reveals RL13 to be a potent inhibitor of replication. J. Clin. Invest. *120*, 3191–1208.

Stenberg, R.M. (1996). The human cytomegalovirus major immediate-early gene. Intervirology *39*, 343–349.

Stenberg, R.M., and Stinski, M.F. (1985). Autoregulation of the human cytomegalovirus major immediate early gene. J. Virol. *56*, 676–682.

Stenberg, R.M., Thomsen, D.R., and Stinski, M.F. (1984). Structural analysis of the major immediate early gene of human cytomegalovirus. J. Virol. *49*, 190–199.

Stenberg, R.M., Depto, A.S., Fortney, J., and Nelson, J.A. (1989). Regulated expression of early and late RNAs and proteins from the human cytomegalovirus immediate-early gene region. J. Virol. *63*, 2699–2708.

Stenberg, R.M., Fortney, J., Barlow, S.W., Magrane, B.P., Nelson, J.A., and Ghazal, P. (1990). Promoter-specific trans activation and repression by human cytomegalovirus immediate-early proteins involves common and unique protein domains. J. Virol. *64*, 1556–1565.

Stevison, L.S., and Kohn, M.H. (2009). Divergence population genetic analysis of hybridization between rhesus and cynomolgus macaques. Molec. Ecol. *18*, 2457–2475.

Swack, N.S., and Hsiung, G.D. (1982). Natural and experimental simian cytomegalovirus infections at a primate center. J. Med. Primatol. *11*, 169–177.

Swack, N.S., Liu, O.C., and Hsiung, G.D. (1971). Cytomegalovirus infections of monkeys and baboons. Am. J. Epidem. *94*, 397–402.

Swanson, R., Bergquam, E., and Wong, S.W. (1998). Characterization of rhesus cytomegalovirus genes associated with antiviral susceptibility. Virology *240*, 338–348.

Swinkels, B.W., Geelen, J.L., Wertheim-van Dillen, P., van Es, A.A., and van der Noordaa, J. (1984). Initial characterization of four cytomegalovirus strains isolated from chimpanzees. Brief report. Arch. Virol. *82*, 125–128.

Sylwester, A.W., Mitchell, B.L., Edgar, J.B., Taormina, C., Pelte, C., Ruchti, F., Sleath, P.R., Grabstein, K.H., Hosken, N.A., Kern, F., *et al.* (2005). Broadly targeted human cytomegalovirus-specific CD4+ and CD8+ T-cells dominate the memory compartments of exposed subjects. J. Exp. Med. *202*, 673–685.

Tarantal, A.F., and Gargosky, S.E. (1995). Characterization of the insulin-like growth factor (IGF) axis in the serum of maternal and fetal macaques (*Macaca mulatta* and *Macaca fascicularis*). Growth Reg. *55*, 190–198.

Tarantal, A.F., Salamat, S., Britt, W.J., Luciw, P.A., Hendrickx, A.G., and Barry, P.A. (1998). Neuropathogenesis induced by rhesus cytomegalovirus in fetal rhesus monkeys (*Macaca mulatta*). J. Infect. Dis. *177*, 446–450.

Targonski, P.V., Jacobson, R.M., and Poland, G.A. (2007). Immunosenescence: role and measurement in influenza vaccine response among the elderly. Vaccine *25*, 3066–3069.

Taylor, R.T., and Bresnahan, W.A. (2005). Human cytomegalovirus immediate-early 2 gene expression blocks virus-induced Beta interferon production. J. Virol. *79*, 3873–3877.

Taylor, R.T., and Bresnahan, W.A. (2006a). Human cytomegalovirus IE86 attenuates virus- and tumor necrosis factor alpha-induced NFkappaB-dependent gene expression. J. Virol. *80*, 10763–10771.

Taylor, R.T., and Bresnahan, W.A. (2006b). Human cytomegalovirus immediate-early 2 protein IE86 blocks virus-induced chemokine expression. J. Virol. *80*, 920–928.

Teotia, S.S., Walker, R.C., Schirmer, J.M., Tazelaar, H.D., Michaels, M.G., Risdahl, J.M., Byrne, G.W., Logan, J.S., and McGregor, C.G. (2005). Prevention, detection, and management of early bacterial and fungal infections in a preclinical cardiac xenotransplantation model that achieves prolonged survival. Xenotrans. *12*, 127–133.

Thomas, M., Rechter, S., Milbradt, J., Auerochs, S., Muller, R., Stamminger, T., and Marschall, M. (2009). Cytomegaloviral protein kinase pUL97 interacts with the nuclear mRNA export factor pUL69 to modulate its intranuclear localization and activity. J. Gen. Virol. *90*, 567–578.

Tomasec, P., Wang, E.C., Davison, A.J., Vojtesek, B., Armstrong, M., Griffin, C., McSharry, B.P., Morris, R.J., Llewellyn-Lacey, S., Rickards, C., *et al.* (2005). Down-regulation of natural killer cell-activating ligand CD155 by human cytomegalovirus UL141. Nat. Immunol. *6*, 181–188.

Tortorella, D., Gewurz, B., Schust, D., Furman, M., and Ploegh, H. (2000). Down-regulation of MHC class I antigen presentation by HCMV; lessons for tumor immunology. Immunol. Invest. *29*, 97–100.

Tosi, A.J., Morales, J.C., and Melnick, D.J. (2002). Y-chromosome and mitochondrial markers in *Macaca fascicularis* indicate introgression with indochinese *M. mulatta* and a biogeographic barrier in the Isthmus of Kra. Internatl. J. Primatol. *23*, 161–178.

Umashankar, M., Petrucelli, A., Cicchini, L., Caposio, P., Kreklywich, C.N., Rak, M., Bughio, F., Goldman, D.C., Hamlin, K.L., Nelson, J.A., *et al.* (2011). A novel human cytomegalovirus locus modulates cell type-specific outcomes of infection. PLoS Pathog. *7*, e1002444.

Vescovini, R., Telera, A., Fagnoni, F.F., Biasini, C., Medici, M.C., Valcavi, P., di Pede, P., Lucchini, G., Zanlari, L., Passeri, G., *et al.* (2004). Different contribution of EBV and CMV infections in very long-term carriers to age-related alterations of CD8+ T-cells. Exp. Gerontol. *39*, 1233–1243.

Vogel, P., Weigler, B.J., Kerr, H., Hendrickx, A., and Barry, P.A. (1994). Seroepidemiologic studies of cytomegalovirus infection in a breeding population of rhesus macaques. Lab. Anim. Sci. *44*, 25–30.

Waldrop, S.L., Pitcher, C.J., Peterson, D.M., Maino, V.C., and Picker, L.J. (1997). Determination of antigen-specific memory/effector CD4+ T-cell frequencies by flow cytometry: evidence for a novel, antigen-specific homeostatic mechanism in HIV-associated immunodeficiency. J. Clin. Invest. *99*, 1739–1750.

Wang, C., Zhang, X., Bialek, S., and Cannon, M.J. (2011). Attribution of congenital cytomegalovirus infection to primary versus non-primary maternal infection. Clin. Infect. Dis. *52*, e11–13.

Wang, D., and Shenk, T. (2005). Human cytomegalovirus virion protein complex required for epithelial and endothelial cell tropism. Proc. Natl. Acad. Sci. U.S.A. *102*, 18153–18158.

Webel, R., Milbradt, J., Auerochs, S., Schregel, V., Held, C., Nobauer, K., Razzazi-Fazeli, E., Jardin, C., Wittenberg, T., Sticht, H., *et al.* (2011). Two isoforms of the protein kinase pUL97 of human cytomegalovirus are differentially regulated in their nuclear translocation. J. Gen. Virol. *92*, 638–649.

Wedderburn, L.R., Patel, A., Varsani, H., and Woo, P. (2001). The developing human immune system: T-cell receptor repertoire of children and young adults shows a wide discrepancy in the frequency of persistent oligoclonal T-cell expansions. Immunology *102*, 301–309.

Weekes, M.P., Wills, M.R., Mynard, K., Hicks, R., Sissons, J.G., and Carmichael, A.J. (1999). Large clonal expansions of human virus-specific memory cytotoxic T-lymphocytes within the CD57+ CD28- CD8+ T-cell population. Immunology *98*, 443–449.

Weinberger, B., Herndler-Brandstetter, D., Schwanninger, A., Weiskopf, D., and Grubeck-Loebenstein, B. (2008). Biology of immune responses to vaccines in elderly persons. Clin. Infect. Dis. *46*, 1078–1084.

White, E.A., Clark, C.L., Sanchez, V., and Spector, D.H. (2004). Small internal deletions in the human cytomegalovirus IE2 gene result in nonviable recombinant viruses with differential defects in viral gene expression. J. Virol. *78*, 1817–1830.

White, J.A., Todd, P.A., Yee, J.L., Kalman-Bowlus, A., Rodgers, K.S., Yang, X., Wong, S.W., Barry, P., and Lerche,

N.W. (2009). Prevalence of viremia and oral shedding of rhesus rhadinovirus and retroperitoneal fibromatosis herpesvirus in large age-structured breeding groups of rhesus macaques (*Macaca mulatta*). Comp. Med. *59*, 383–390.

Wikby, A., Johansson, B., Olsson, J., Lofgren, S., Nilsson, B.O., and Ferguson, F. (2002). Expansions of peripheral blood CD8 T-lymphocyte subpopulations and an association with cytomegalovirus seropositivity in the elderly: the Swedish NONA immune study. Exp. Gerontol. *37*, 445–453.

Wikby, A., Ferguson, F., Forsey, R., Thompson, J., Strindhall, J., Lofgren, S., Nilsson, B.O., Ernerudh, J., Pawelec, G., and Johansson, B. (2005). An immune risk phenotype, cognitive impairment, and survival in very late life: impact of allostatic load in Swedish octogenarian and nonagenarian humans. J. Gerontol. Biol. Sci. Med. Sci. *60*, 556–565.

Wikby, A., Mansson, I.A., Johansson, B., Strindhall, J., and Nilsson, S.E. (2008). The immune risk profile is associated with age and gender: findings from three Swedish population studies of individuals 20–100 years of age. Biogerontol. *9*, 299–308.

Wilkinson, G.W., Tomasec, P., Stanton, R.J., Armstrong, M., Prod'homme, V., Aicheler, R., McSharry, B.P., Rickards, C.R., Cochrane, D., Llewellyn-Lacey, S., *et al.* (2008). Modulation of natural killer cells by human cytomegalovirus. J. Clin. Virol. *41*, 206–212.

Winkler, M., and Stamminger, T. (1996). A specific subform of the human cytomegalovirus transactivator protein pUL69 is contained within the tegument of virus particles. J. Virol. *70*, 8984–8987.

Winkler, M., Rice, S.A., and Stamminger, T. (1994). UL69 of human cytomegalovirus, an open reading frame with homology to ICP27 of herpes simplex virus, encodes a transactivator of gene expression. J. Virol. *68*, 3943–3954.

Xie, D., and McElhaney, J.E. (2007). Lower GrB+ CD62Lhigh CD8 TCM effector lymphocyte response to influenza virus in older adults is associated with increased CD28null CD8 T-lymphocytes. Mech. Ageing Dev. *128*, 392–400.

Yamamoto, A.Y., Mussi-Pinhata, M.M., Boppana, S.B., Novak, Z., Wagatsuma, V.M., Oliveira Pde, F., Duarte, G., and Britt, W.J. (2010). Human cytomegalovirus reinfection is associated with intrauterine transmission in a highly cytomegalovirus-immune maternal population. Am. J. Obstet. Gynecol. *202*, 297 e291–298.

Yang, Z., and Bjorkman, P.J. (2008). Structure of UL18, a peptide-binding viral MHC mimic, bound to a host inhibitory receptor. Proc. Natl. Acad. Sci. U.S.A. *105*, 10095–10100.

Yasutomi, Y. (2010). Establishment of specific pathogen-free macaque colonies in Tsukuba Primate Research Center of Japan for AIDS research. Vaccine *28*, 75–77.

Yue, Y., Zhou, S.S., and Barry, P.A. (2003). Antibody responses to rhesus cytomegalovirus glycoprotein B in naturally infected rhesus macaques. J. Gen. Virol. *84*, 3371–3379.

Yue, Y., Kaur, A., Zhou, S.S., and Barry, P.A. (2006). Characterization and immunological analysis of the rhesus cytomegalovirus homologue (Rh112) of the human cytomegalovirus UL83 lower matrix phosphoprotein (pp65). J. Gen. Virol. *87*, 777–787.

Yue, Y., Kaur, A., Eberhardt, M.K., Kassis, N., Zhou, S.S., Tarantal, A.F., and Barry, P.A. (2007). Immunogenicity and protective efficacy of DNA vaccines expressing rhesus cytomegalovirus glycoprotein B, phosphoprotein 65-2, and viral interleukin-10 in rhesus macaques. J. Virol. *81*, 1095–1109.

Yue, Y., Wang, Z., Abel, K., Li, J., Strelow, L., Mandarino, A., Eberhardt, M.K., Schmidt, K.A., Diamond, D.J., and Barry, P.A. (2008). Evaluation of recombinant modified vaccinia ankara virus-based rhesus cytomegalovirus vaccines in rhesus macaques. Med. Microbiol. Immunol. *197*, 117–123.

Zanghellini, F., Boppana, S.B., Emery, V.C., Griffiths, P.D., and Pass, R.F. (1999). Asymptomatic primary cytomegalovirus infection: virologic and immunologic features. J. Infect. Dis. *180*, 702–707.

Zanni, F., Vescovini, R., Biasini, C., Fagnoni, F., Zanlari, L., Telera, A., Di Pede, P., Passeri, G., Pedrazzoni, M., Passeri, M., *et al.* (2003). Marked increase with age of type 1 cytokines within memory and effector/ cytotoxic CD8+ T-cells in humans: a contribution to understand the relationship between inflammation and immunosenescence. Exp. Gerontol. *38*, 981–987.

Zhu, H., Cong, J.P., Yu, D., Bresnahan, W.A., and Shenk, T.E. (2002). Inhibition of cyclooxygenase 2 blocks human cytomegalovirus replication. Proc. Natl. Acad. Sci. U.S.A. *99*, 3932–3937.

Zielke, B., Thomas, M., Giede-Jeppe, A., Muller, R., and Stamminger, T. (2011). Characterization of the betaherpesviral pUL69 protein family reveals binding of the cellular mRNA export factor UAP56 as a prerequisite for stimulation of nuclear mRNA export and for efficient viral replication. J. Virol. *85*, 1804–1819.

Putative Disease Associations with Cytomegalovirus: a Critical Survey

II.23

Ann B. Hill

Abstract

In recent years it has been suggested that CMV may be involved in the pathogenesis of a variety of conditions in which there may not be clear evidence of viral replication. These 'non-traditional' disease associations include glioblastoma and various other cancers, atherosclerotic cardiovascular disease, Alzheimer's disease, and immunosenescence, amongst others. The pathologies fall into two broad groups: tumours, and inflammatory diseases of ageing. In the case of tumours, some groups have used ultrasensitive detection techniques and report finding CMV in the majority of tumour cells. In the case of inflammatory diseases of ageing, the evidence mostly comes from epidemiological studies that have associated CMV serology or CMV-driven alterations in T-cell populations with various outcomes. While CMV's biology provides ready explanations for these putative disease associations, the actual evidence for its being involved remains controversial. This chapter will review the evidence for several putative disease associations.

Introduction

CMV's status as a subclinical chronic infection provides fertile ground for theorists who would like to implicate it in a broad array of diseases, rather like McCarthy era anti-communists in the USA, who could see 'a red under every bed'. CMV's biology readily assists these theories, since the virus encodes genes with the ability to manipulate seemingly every facet of host biology. In addition, the theories are aided by CMV's relationship with inflammation: CMV reactivates and replicates in an inflammatory milieu, actively recruits inflammatory cells to the site of replication and also elicits an enormous T-cell and antibody response. CMV can frequently be detected in sites of inflammation. With growing appreciation that inflammatory processes drive many chronic diseases of ageing that were previously considered degenerative, it is reasonable to ask whether CMV could be involved in their pathogenesis. However, moving from plausible theory to convincing evidence of pathogenesis can be remarkably difficult, and for the majority of these putative disease associations, the jury remains out.

This chapter is concerned with putative disease associations with CMV, i.e. associations that are disputed, but are areas of active research. These fall into two main groups: cancer and chronic diseases of ageing (Soderberg-Naucler, 2006). The latter grouping is quite broad, and includes immunosenescence (which has been covered in Chapter II.7), a systemic state of frailty and overall mortality risk known as the 'immune risk phenotype (IRP)', as well as specific inflammatory diseases such as atherosclerosis and neurodegenerative disease. The term 'inflammaging' has been coined to incorporate all of these concepts (Franceschi *et al.*, 2007). The cancer associations are more specific. Histological or molecular evidence for the presence of CMV has been presented for several tumours. However, a role for CMV in the pathogenesis of the CNS tumour glioblastoma has attracted the most attention, and indeed has resulted in at least three clinical trials targeting CMV with the goal of slowing tumour progression. Since the contentious issues regarding CMV's role in glioblastoma are applicable to other tumours, this chapter's discussion will be restricted to glioblastoma. In what follows I have tried to present both sides of the argument. However, inevitably, my own conservative bias will be evident. I hope at least to highlight the main issues that it seems to me prevent firm conclusions being made at this time.

Glioblastoma

Glioblastoma multiforme (GBM) is a malignant brain tumour affecting both adults and children; it is usually fatal within 12–18 months. It is a tumour of astrocytes, CNS cells that regulate the tissue environment and share with neurons an origin in neuroglial precursor stem cells. GBM is thought to arise through one of two pathways: either a gradual transformation and de-differentiation from a less aggressive astrocytoma, or direct transformation of neuroglial precursor stem cells. Oncogenic mechanisms include inactivation of p53, and amplification of the epidermal growth factor receptor (EGFR) gene, a receptor tyrosine kinase that activates PI3-K/Akt to promote cell cycling. The malignant tumour is thought to be maintained and regenerated by a small subpopulation of cells within the tumour called cancer stem cells, or glioma stem cells.

In 2002, Charles Cobbs and colleagues described the presence of HCMV protein and nucleic acids in virtually all of a series of GBM tumours, but not in normal brain tissue or other tumours, and suggested that HCMV may be involved in the pathogenesis of glioblastoma (Cobbs *et al.*, 2002). This report sparked intense interest and a controversy that is still unresolved. There are three major aspects to the controversy: (a) Is CMV truly present in the majority of tumours?; (b) if/when present, is it present in the majority or minority of cells?; and (c) if/when present, is CMV actually involved in tumorigenesis, or is it just an incidental bystander?

Is CMV present in the majority of gliomas?

Various groups have attempted to replicate the original findings from the Cobbs group, and four of six reports have described similar results. Evidence for the presence of HCMV has been obtained through immunohistochemisty, *in situ* hybridization (ISH), FACS of tissue infiltrating mononuclear cells, and PCR, sometimes with sequencing. The most common technique has been immunohistochemistry (IHC), most commonly staining for IE1. Four studies detected IE1 in virtually all (93–100%) of GBM tumours studied (Cobbs *et al.*, 2002; Mitchell *et al.*, 2008; Scheurer *et al.*, 2008; Slinger *et al.*, 2010). Another study detected IE1 in only 16% of samples, but found pp65 in 51% (Lucas *et al.*, 2011). When HCMV has been detected in virtually all samples, ultrasensitive techniques pioneered by the Cobbs group were generally used. Other groups, however, did not detect CMV in the majority of tumours (Sabatier *et al.*, 2005). Those reporting positive results have emphasized the necessity of these optimal techniques, particularly when staining paraffin-embedded sections, to detect what is considered to be a very low level of infection (Scheurer *et al.*, 2008). These ultrasensitive techniques did not detect CMV in surrounding normal brain tissue, nor in other CNS pathologies, and isotype control antibodies for IHC and irrelevant nucleotide sequences for ISH were negative in the tumours.

PCR has also been used to detect CMV DNA in GBM tumour samples. Using nested PCR on DNA extracted from paraffin-embedded tissue, Cobbs and colleagues were able to amplify a sequence from the gB gene in seven of nine cases; sequencing revealed different CMV strains (Cobbs *et al.*, 2002). Mitchell and colleagues were able to amplify gB in 21/34 (62%) of GBM samples, using DNA extracted from 10mg of freshly resected tumour as template. The product was confirmed as CMV by sequencing in each case, and a high degree of nucleotide variation was taken to indicate that these were unique viral isolates (Mitchell *et al.*, 2008). A recent study reported PCR for 20 different viral genes on DNA extracted from paraffin-embedded tissue (Ranganathan *et al.*, 2012). Each of the genes could be amplified in at least some tumours. Overall, many of the CMV genes were statistically more likely to be amplified from GBM tumours than from non-malignant brain tumours or brain tissue from epileptic patients. However, the signal was very low in many tumours; the question of quantification is discussed further below.

Additional support for the presence of CMV in GBM tumours came from a case report in which a patient displayed a brisk rise in CMV pp65-specific $CD8^+$ T-cells following administration of a tumour vaccine consisting of dendritic cells pulsed with autologous tumour lysate (Prins *et al.*, 2008). Although not a controlled study, this report is consistent with the interpretation that CMV antigen (or infectious virus) was present in the tumour lysate and boosted the patient's CMV-specific T-cell response.

Thus, protagonists for CMV involvement in GBM pathogenesis claim that CMV is present in the vast majority of GBM tumours. They argue that, although ubiquitous, viral infection is at a very low level, and that ultrasensitive techniques are needed to detect it, explaining the lack of detection in some studies. An open meeting on this topic in 2011 issued a consensus report which concluded that 'there is sufficient evidence to conclude that HCMV sequences and viral gene expression exist in most, if not all, malignant gliomas' (Dziurzynski *et al.*, 2012).

Relationship between CMV serostatus and GBM

Currently, only approximately 50% of people in developed countries are CMV seropositive. CMV seroprevalence increases steadily with age (Fig. II.23.1). The median patient age of most adult GBM series is around 50 years, an age at which ~ 70–80% of individuals of the US population are CMV seropositive. For childhood GBM, less than 50% of patients would be expected to

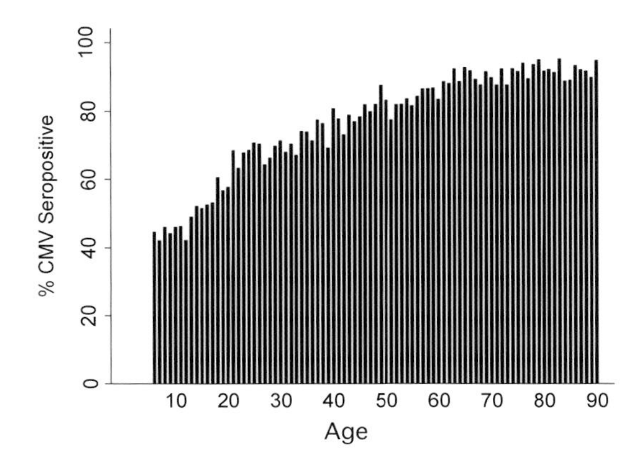

Figure II.23.1 CMV serostatus by age in the US population The data are taken from NHANES III, a survey carried out from 1988 to 1994 on a population selected from census data as representative of the US population.

be CMV seropositive by chance. However, CMV seroprevalence is markedly affected by demographic factors (Dowd *et al.*, 2009) (see also Chapter II.2). It is very high in African Americans, Hispanics, and Asians, and is in addition inversely correlated with socioeconomic status (SES). CMV is also more common in women than men. These demographics do not parallel those of GBM, which is somewhat more common in whites and in males (Lehrer *et al.*, 2012). Thus, if all patients with GBM were CMV seropositive, this would provide strong correlative evidence to support a role for CMV in GBM pathogenesis.

However, CMV seroepidemiology does not provide support for the CMV/glioblastoma hypothesis. In 2001, Wrensch and colleagues reported on serology for four herpesviruses in 134 adult GBM cases from the San Francisco area and 165 controls (Wrensch *et al.*, 2001). CMV seroprevalence was 57% for both GBM cases and controls. However, since controls were slightly older than GBM patients (median age 54 vs. 48), when a statistical adjustment was made for age, there was a slightly higher tendency for GBM patients to be CMV seropositive compared to controls. This is expressed statistically as an odds ratio of 1.2, albeit with a very broad confidence interval (95% CI 0.7–2.0), rendering the association non-significant. It is possible that the association may have been strengthened after adjustment for gender, race, and SES, which was not done. Nevertheless, the fact that 43% of GBM cases were CMV seronegative cannot be avoided. Similarly, in a recent Australian study, only 16 of 27 GBM cases (60%) were CMV seropositive (Crough *et al.*, 2012). Although neither of these studies examined tumours for evidence of CMV, it seems likely that many of the patients whose tumours have been scored positive for CMV in the studies described above will have no serological evidence of infection.

Some investigators believe that such a discrepancy could be accounted for by false negative serology. Evidence has been presented that some patients who are CMV seronegative using standard viral antigens are seropositive to a clinical isolate (Rahbar *et al.*, 2004). Alternately, it has been suggested that some individuals can be infected with CMV without seroconverting, but instead will have T-cell responses to CMV. However, consistent with the accepted position that CMV serology accurately reflects the presence or absence of CMV infection, in the recent GMB series from Australia, T-cell responses were detected in all 16 CMV-seropositive GBM patients but in none of the nine CMV seronegative patients (Crough *et al.*, 2012).

Serum and tumours from the same patients will need to be examined to definitively determine whether CMV is frequently found in the tumours of seronegative patients. However, the current evidence suggests that the seroprevalence of CMV in GBM patients is not dramatically different to that of the general population. This either casts doubt on the validity of finding CMV in 100% of GBM samples, or necessitates a radical rethinking of the meaning of CMV serology.

What percentage of GBM cells are infected with CMV?

Another area of discrepancy in the reports concerns the number of cells actually infected with CMV, and possibly the identity of those cells. Some studies report detecting IE1 in virtually all tumour cells (Cobbs *et al.*, 2002; Mitchell *et al.*, 2008). In contrast, Sabatier and colleagues found CMV only sporadically throughout the tumour, in cells with an astrocytic morphology, which could either have been tumour cells or reactive astrocytes (Sabatier *et al.*, 2005).

Quantitative PCR should be able to shed light on this question. If all GBM cells are making at least IE1 protein, each cell must contain a minimum of one copy of viral DNA, which would result in one copy of the IE1 gene for each two copies of a gene from the diploid host cell. If we assume that the viral genome is intact, there should be the same number of copies of other viral genes. It is likely that the actual viral genome copy number would be higher, since even latently infected cells are reported to carry on average 10 genomes per cell, and productively infected cells carry many times more. Hence, even if the malignant cells constitute only a fraction of nucleated cells within a piece of tumour, there should be a robust PCR signal for HCMV genes, and quantitative PCR should give a ratio to host cell genes proportional to the percentage of tumour cells in the sample. Although very few studies have actually provided quantitative data for CMV genome copies, most reports using PCR suggest that the signal was very low. Two recent papers have been more quantitative. Ranagnathan and colleagues concluded that cellular genomes vastly outnumbered viral genomes in most of these samples, and that CMV was unlikely to promote oncogenesis by acting directly within the tumour cell (Ranganathan *et al.*, 2012). Similarly, Bhattacharjee and colleagues were only able to quantify CMV DNA in 37% of glioma specimens (Bhattacharjee *et al.*, 2012). There was a wide range of copy numbers in the seven tumours in which quantification was successful, from 365 copies per 500 ng of DNA to a high of 10^6, with a median value of 10^4. This was five logs lower than the copy number for experimental fibroblast infection at a MOI of 1. Although copy numbers in GBM would be expected to be logs lower than a productive infection, this degree of discrepancy seems consistent with the conclusions of Ranagnathan and colleagues, i.e. that CMV affects only a minority of malignant cells in most tumours.

In summary: while quantitative data for CMV genome copies in tumour samples is rare in this literature, the data that do exist are not concordant with the supposition that almost all malignant cells in a GMB tumour contain CMV genomes.

Proposed functional consequences of CMV for GBM pathogenesis

HCMV infection of NPCs or glioma cells is believed to promote tumour growth by triggering pathways that are known to be involved in gliomagenesis. HCMV is thus considered to be 'oncomodulatory' rather that oncogenic- i.e. it does not directly transform cells in the manner of classic tumour viruses (Dziurzynski *et al.*, 2012). Investigation of the impact of HCMV on glial cells and glioma cell lines has suggested several ways in which HCMV infection of GBM could promote tumour growth. Three important pathways in gliomagenesis are increased activation of PI3-K/Akt signalling, inhibition of tumour suppressor genes including p53 and Rb, and activation of STAT3. Evidence has been presented for HCMV's ability to perform each of these functions. Glioma cells up-regulate platelet-derived growth factor receptor (PDGFR), a molecule which facilitates HCMV infection of cells (Soroceanu *et al.*, 2008), and is thought to explain HCMV's apparent predilection for infecting gliomas. HCMV gB binding to PDGFR activates PI3K/Akt, which is thought to promote viral entry and replication (Chapter I.8), but could also promote gliomagenesis. Expression of HCMV IE in glioma cell lines also activated PI3K/Akt signalling, and additionally inactivated p53 and Rb (Cobbs *et al.*, 2008). The HCMV gene US28 was proposed to promote tumorigenesis by inducing secretion of IL-6 and VEGF, which led to STAT3 activation (Maussang *et al.*, 2006; Slinger *et al.*, 2010; Soroceanu *et al.*, 2011). Finally, HCMV IE1 was shown to activate telomerase, a common feature of tumours (Straat *et al.*, 2009).

While most of these oncomodulatory properties would require that HCMV be present in the malignant cells, some potential oncomodulatory activities could operate even if CMV were only present in a minority of cells within the tumour. IL-6 and vascular endothelial cell growth factor (VEGF) secreted by HCMV-infected cells promote angiogenesis, and vIL-10 is immunosuppressive. Thus, CMV presence in only part of the tumour, or even in non-malignant cells infiltrating the tumour, could promote tumour growth by its impact on the tumour microenvironment. If CMV acts in either of these ways, clinical interventions that reduce CMV load or activity could have a favourable impact on prognosis in GBM patients.

With this goal, several clinical studies have been initiated (Cobbs, 2011). A prospective, randomized phase II trial of the antiviral drug valgancyclovir in 42 patients began in Sweden in 2006; the results have not yet been reported. Investigators at Duke University, USA, vaccinated GBM patients with autologous dendritic cells transfected with pp65. The patient survival (median 15 months progression free survival) was significantly better than matched historical controls. A second trial using a peptide-based vaccine is planned. An Australian study used adoptive immunotherapy of *in vitro* expanded, pp65-specific $CD8^+$ T-cells in one patient in conjunction with chemotherapy and reported stable clinical improvement (Crough *et al.*, 2012). A clinical trial involving more patients is planned. This last trial will only involve CMV-seropositive patients, since those are the only ones from whom CMV-specific T-cells can be expanded.

CMV and glioblastoma summary

Ten years after the provocative initial publication reporting an association between CMV and glioblastoma, the field remains controversial. The low amount of viral DNA detected in most samples argues against the presence of CMV within all the malignant cells. Further, the fact that published studies report CMV seroprevalence in GBM series that are not radically different from the general population argues against a pivotal role for CMV in the majority of GBM cases. Nevertheless, the data are consistent with CMV being present in some GBM tumours, significantly more so than in surrounding brain tissue and other tumours, although perhaps only in a subset of cells within the tumour. CMV's biology would allow it to play an oncomodulatory role even if only present in a subset of cells, especially in the minority of cases where the viral load in the tumour is substantial. Possibly supporting this idea, a recent case control study from Sweden reported that individuals with higher CMV load in their tumour did worse than those with low CMV load (Rahbar *et al.*, 2012). This result would be consistent with CMV promoting pathology, whether or not those patients that the study identified as having 'low CMV load' are in fact truly CMV-infected. Ongoing studies, with attention to CMV serostatus and to quantitative DNA analysis of CMV burden in the tumour, are likely to lead to clarification.

CMV, immunosenescence and chronic inflammatory disease

Recent years have seen an increased or renewed interest in the idea that CMV is involved in the pathogenesis of a number of chronic inflammatory diseases that are common in ageing populations. Many diseases that used to be classified as degenerative are now considered to have an inflammatory basis. Prominent examples are atherosclerotic cardiovascular disease, neurodegenerative diseases (Alzheimer's and Parkinson's disease), and osteoporosis. In fact, it is now appreciated that the ageing process is accompanied by a general increase in inflammation, manifest by increased serum levels of inflammatory cytokines such as TNF and IL-6, or by downstream markers of their activity, such as C-reactive protein (CRP). The universally familiar phenomenon of frailty, which has both physical and cognitive components, also correlates with increased inflammation, and is a good predictor of mortality in the elderly. The term 'inflammaging' has been coined to as a catch-all term to describe these phenomena (Franceschi *et al.*, 2007). There is an increasing body of literature which suggests that CMV may be involved in inflammaging, frailty, and in some of these chronic inflammatory diseases of ageing.

Over the same time period, there has been a growing interest in the idea that CMV is involved in immunosenescence – the decline in immune function that accompanies ageing (Pawelec *et al.*, 2009). This topic has been dealt with in Chapter II.7 of the current volume. As described there, CMV is clearly responsible for the most dramatic alterations in the T-cell compartment that are considered a hallmark of immunosenescence: the accumulation of large numbers of terminally differentiated 'effector memory' $CD8^+$ T-cells. A variety of phenotypic markers has been used to identify this population, but the most common is lack of the co-stimulatory molecule CD28. Since the size of the $CD8^+CD28^-$ T-cell population correlates in some studies with increased inflammatory cytokines, the phenomena of immunosenescence and inflammaging are probably linked (Wikby *et al.*, 2006).

These considerations have led to the proposition that CMV is causally involved both in immunosenescence and chronic inflammatory diseases of ageing (Wikby *et al.*, 2006; Nikolich-Zugich, 2008). This idea is supported by a series of studies in which either the size of the $CD8^+CD28^-$ T-cell population or titres of anti-CMV antibody have been correlated with various conditions, including frailty, all-cause mortality, poor vaccine response, osteoporosis, cognitive decline, cardiovascular disease, and depression (Table II.23.1).

Biological and statistical considerations in examining the relationship between CMV, inflammation and disease

Given that a significant percentage of the $CD8^+CD28^-$ T-cell population in CMV-infected individuals consists of T-cells specific for CMV, the size of this population is a crude surrogate for the size of the CMV-specific T-cell response. Most of the studies listed in Table II.23.1 therefore report correlations between the size of the immune response to CMV- either T-cell or antibody- and poor outcome. This is usually interpreted as meaning that only a subset of CMV-infected individuals are at increased risk of the outcome- those whose immune system is devoting the most energy to combating the virus, which is presumably those in whom CMV is most active. This would certainly be consistent with CMV being causally involved in these conditions.

However, current knowledge of CMV biology allows for several ways in which these correlations could come about (Fig. II.23.2). CMV could indeed be the 'driver' of the inflammation or immunosenescence which underlie the disease (Fig. II.23.2A). An alternate hypothesis would be that failing immunity is the primary event, which would cause an increased level of CMV activity. This would stimulate antibody and T-cell responses, even if they were less effective in controlling the virus than in younger individuals (Fig. II.23.2B). In this scenario, CMV antibody titres and the $CD8^+28^-$ population size would correlate with immunosenescence, even though CMV was not the cause of the immunosenescence. Yet another possibility is based on the known biology of CMV. Since inflammation provokes CMV reactivation and supports virus replication, increased CMV activity with its attendant immune response could simply be a reflection of the underlying inflammatory milieu, which is the true

Table II.23.1 CMV serology, T-cell subsets and diseases of inflammaging

Outcome	Study results	Reference	Comments
All-cause mortality	OCTO and NONA studies: two cohorts of approx. 100 healthy Swedes aged 86–92, studied longitudinally. An 'immune risk profile' characterized by inverted CD4⁺:CD8⁺ T-cell ratio was found only in CMV seropositives and was strongly correlated with 2-year mortality in successive 2-year periods, in two separate cohorts	Ferguson *et al.* (1995), Wikby *et al.* (1998)	CD4:CD8 < 1 is due to accumulation of CMV-specific $CD8^+CD28^-$ cells >90% of subjects were CMV+ in both studies, but overall mortality rates did not differ between CMV+ and CMV–groups
	Inverted CD4:CD8 ratio < 1 predicted mortality in a study of 497 healthy elderly British individuals; result remained significant after adjusting for age but became (just) non-significant after adjusting for sex	Huppert *et al.* (2003)	CMV status not measured
Frailty	Women's Health and Aging Study, Baltimore, MD. CMV seropositivity associated with prevalent frailty, adjusted odds ratio (AOR) 3.2, $P = 0.03$. High IL-6 significantly increased association: high IL-6 + CMV AOR 20.3 ($P = 0.0007$)	Schmaltz *et al.* (2005)	Positive correlation is with CMV seropositivity *per se*, rather than antibody titres. High number of women with RA in the CMV+ group
	SALSA study: 1559 Latinos in Sacramento area aged 60–101; CMV antibody titre correlated with impaired activities of daily living (ADL); result just non-significant in fully adjusted model	Aiello *et al.* (2008)	CRP level correlated with impaired ADL in fully adjusted model
Cognitive decline	SALSA study: 1204 Sacramento area Latinos, significantly higher rate of decline in cognitive function tests over a 4-year period in individuals with higher CMV antibody titres compared to lowest	Aiello *et al.* (2006)	CRP did not modify relationship between CMV Ab titre and cognitive decline
Flu vaccine response	Mayo Clinic: 153 community dwelling adults aged 65–98; failure to respond to flu vaccine correlated with the size of the $CD8^+CD28^-$ T-cell subset	Goronzy *et al.* (2001)	
	91 Polish nursing home residents aged 65–91; non-responders to vaccine had significantly higher CMV IgG titres and higher numbers of $CD8^+CD28^-$ T-cells	Trzonkowski *et al.* (2003)	
Cardiovascular disease	Atherosclerosis Risk in Communities (ARIC) study: 515 participants; risk of developing coronary artery disease was significantly higher in individuals with the highest levels of CMV antibodies	Sorlie *et al.* (2000)	
	Sweden: 43 men with coronary artery disease; number of $CD8^+CD28^-$ cells significantly higher in patients than healthy controls	Jonasson *et al.* (2003)	$CD8^+$ $CD28^-$ numbers correlated with CMV antibody, but CMV seropositivity did not correlate independently with coronary artery disease
	Germany: 101 patients with abdominal aortic aneurysm compared to 38 healthy controls: patients had higher percentage of CD8 cells that were $CD28^-$ (42% vs. 25%, $P < 0.001$)	Duftner *et al.* (2005)	
	NHANES III analysis for history of CVD in individuals aged over 45. CMV seropositives significantly more likely to have a history of CVD in a fully adjusted model: AOR 1.75	Simanek *et al.* (2009)	Second of two studies that correlate outcome with CMV seropositivity
Depression	137 residents of Birmingham, UK, aged >65; 66% CMV seropositive. CMV seropositivity was not correlated with prevalent depression, but amongst CMV seropositives, antibody titre correlated with depression	Phillips *et al.* (2008)	Depression correlates with inflammation. This result may be an example of CMV acting as a reporter of underlying inflammation, but not mediating disease

The table lists some studies that report positive associations between CMV seropositivity, antibody titres or the size of the $CD8^+CD28^-$ T-cell population and various outcomes within the 'inflammaging' spectrum.

driver of the disease (Fig. II.23.2C). These scenarios are not mutually exclusive. However, it does not seem safe to assume that a correlation between the size of the immune response to CMV and a particular outcome necessarily implicates CMV as causally involved in the outcome.

If CMV actually contributes to the outcome, even if it does so only in a minority of infected individuals,

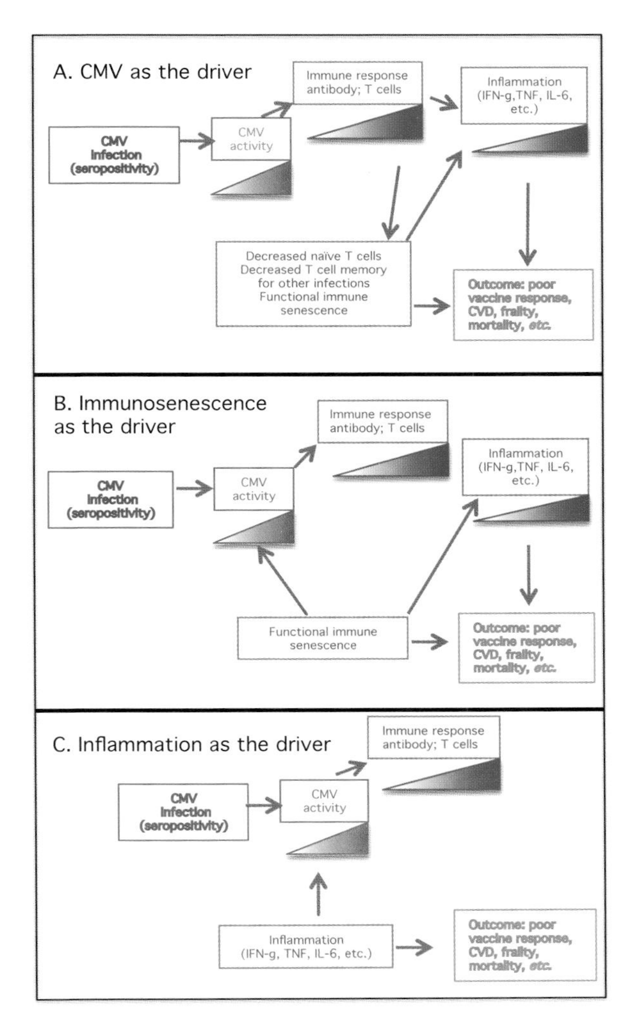

Figure II.23.2 Possible interrelationships between CMV, inflammation and the diseases of 'inflammaging' Many studies have correlated immune responses to CMV – either antibody titres, or the size of the $CD8^+CD28^-$ T-cell population – with systemic inflammation and various diseases of 'inflammaging'. This figure illustrates three scenarios which would result in these correlations, only one of which would involve CMV as a causative agent in pathogenesis.

A. CMV as the driver. Here the immune response to CMV activity increases the level of inflammation at sites of infection (e.g. in blood vessels) and systemically, contributing to the pathogenesis of inflammatory disease. At the same time, the immune response to CMV usurps immunological resources to the extent that the immune system is impaired in its ability to respond to other infections, contributing to chronic inflammation and impaired vaccine responses. If this scenario is correct, CMV is the driver, and there should be a higher incidence of the disease in CMV seropositives.

B. Immunosenescence as the driver. Here, the primary problem is progressive functional decline of the immune system- including impaired functional ability of T-cells. This leads to impaired infection control and increased levels of systemic inflammation, which contributes to the diseases of inflammaging. In individuals who are infected with CMV, CMV activity will increase, with a corresponding increase in numbers of CMV-specific T-cells and antibody, although these may be less effective. However, these are not the cause of the chronic inflammatory disease, merely acting as a litmus paper to report on the underlying problem of a general decline in immunity. CMV seropositives will be no more likely to develop the disease than seronegatives.

C. Inflammation as the driver. Here, systemic inflammation due to other causes is the underlying problem that causes the diseases of inflammaging. Inflammatory conditions favour CMV replication, and CMV activity increases, resulting in increased antibody titres and T-cell responses ($CD8^+CD28^-$ population). However, CMV is not the cause of the inflammation; again, it acts as a litmus paper to report on the degree of inflammation. CMV antibody titres and T-cell responses will be increased in CMV seropositives who have diseases of inflammaging, but CMV seropositive will be no more likely to develop disease than CMV seronegatives.

there should be a greater incidence or severity of the disease in CMV-infected people. There should thus be a correlation between CMV seropositivity *per se* and the outcome, rather than just the antibody titre or the magnitude of T-cell subset alterations. Studies showing such a correlation are rare. CMV seroprevalence is very high in the elderly, so studies need to enrol a very large number of subjects in order to have enough CMV seronegatives to achieve statistical significance. Even when a study is powered to detect this association, major statistical difficulties are imposed by the demographics of CMV infection.

As described in more detail in Chapter II.2, CMV infection is much more common in individuals of lower SES. In addition, in the USA, there is a marked influence of ethnicity on the likelihood of being CMV-infected. African Americans, Hispanics, and Asians all have very high rates of CMV seropositivity. The discrepancy is such that infection rates vary from ~20% to >90% in different communities. Race, ethnicity and socioeconomic status all have profound effects on health outcomes. Low SES, African American and Hispanic ethnicity are all associated with a shorter life expectancy and a higher prevalence of cardiovascular disease than high SES and non-Hispanic whites. Hence, any attempt to correlate markers of CMV infection with these outcomes must take these demographic factors into account. The statistical methodology most commonly used to take into account multiple variables in analysing associations between CMV and disease outcomes is logistic regression analysis.

Logistic regression generates a model that best predicts the outcome (e.g. mortality) based on the risk factor (CMV) and other variables. If the risk factor is binary, e.g. CMV seropositive or seronegative, the outcome is reported as an odds ratio (OR) – the relative odds of a CMV-seropositive person having the outcome (such as cardiovascular disease) compared to a seronegative person. For survival analysis using Cox proportional hazard modelling, the outcome is reported as a hazard ratio (HR). In both cases, the 95% confidence interval (CI) expresses lower and upper limits of the OR (or HR) that would occur 95% of the time if the study sample was selected randomly from the total population. An OR or HR of 1 is a null value, indicating no relationship between the risk factor and outcome. The result is considered statistically significant if the CI does not include the null value.

Variables that are independently associated with both the risk factor and outcome but are not considered part of a causal chain, such as age, gender, race, education level (a proxy for SES) *etc.* are called confounders and need to be included in the model. As an extreme example: since the rate of CMV seropositivity rises so markedly with age, and since old people are far more likely to die than young people, a model that does not adjust for age will give a very high HR of CMV seropositivity for mortality. Including age in the model-adjusting for age- markedly attenuates the association. Other variables which are already known to impact the outcome which may be more or less frequent in the CMV^+ population, are also adjusted for.

In adjusting for the impact of other variables, logistic regression makes the assumption that the impact of each variable is the same for all values of the other variables, i.e. that there is no interaction between the variables. For example, smoking should affect mortality risk equally in men and in women, and across all age strata. Interactions should be tested for in establishing the model, and if present, the dataset should be stratified by the affected variable. If, for example, smoking impacts mortality more in men than in women, then the analysis should be run separately for men and women. Of course, stratifying a dataset reduces the sample size in each analysis and markedly reduces the chance of detecting a significant association. Given the complexity of the biology of CMV as well as the complexity of these disease processes, interactions between variables are almost inevitable, and it is rarely possible to stratify the dataset to completely exclude them. Further, it is clearly difficult to accurately measure some important confounders, particularly SES, rendering it likely that some reported associations with CMV may be the result of inadequate adjustment for confounders.

Allison Aiello and co-workers have raised a provocative question regarding the interrelationship between poor health outcomes, SES, and infection with CMV or other chronic pathogens (Dowd *et al.*, 2009). There are many reasons why SES impacts health, including nutrition, lifestyle, and access to health care. However, if CMV does indeed contribute to inflammaging, they argue that some of the impact of SES on health outcomes could be due to the increased CMV (and other pathogen) burden. If so, adjusting for SES in logistic regression models may remove evidence of an actual causal association. While there is not enough evidence to support this concept at present, it is an interesting and provocative idea.

The immune risk phenotype, CMV, frailty and all-cause mortality

The immune risk phenotype

The combination of immunological measures that came to be called an 'immune risk phenotype' was reported by Ferguson and colleagues in a longitudinal study of very old individuals in Jönköping, Sweden (Ferguson *et al.*, 1995). A panel of baseline immune parameters of the individuals who were alive at the end of a two-year period was compared with those of the deceased, and cluster analysis was used to identify a combination of parameters that predicted mortality. The initial definition of the IRP included increased $CD8^+$ percentages, poor T-cell proliferative responses to concanavalin A, as well as low $CD4^+$ and $CD19^+$ percentages. These studies were continued with the OCTO and NONA longitudinal studies performed in Sweden. As more information accrued over the years, the IRP definition underwent a series of changing descriptions, which can make the literature confusing to read. However, these were basically refinements attempting to generate a robust and simple descriptor. An inverted CD4:CD8 ratio emerged as the most critical component, and, since all individuals with inverted CD4:CD8 ratio in the Swedish Jönköping population were found to be CMV seropositive, compared to 85% of the general age-matched population, CMV seropositivity was included as part of the IRP definition (Ferguson *et al.*, 1995; Wikby *et al.*, 1998, 2002). A more recent follow-up of these individuals found that no individuals who reached 100 years had the IRP, suggesting that only those who never developed the IRP were able to survive to extreme old age (Strindhall *et al.*, 2007). Perhaps the most interesting feature of these studies was the observation that individuals who did not display an IRP at baseline could develop it several years later, reflecting a rather dramatic alteration in their T-cell populations (Wikby *et al.*, 2006). This change seemed to reflect a 'tipping point' in what may have previously been a stable equilibrium and was more predictive of incipient mortality than prevalent disease status. It would be interesting to know whether this rather rapid alteration in T-cell populations was precipitated by an increased level of CMV activity.

The concept of the IRP was moderately supported by a second study from Great Britain, which analysed peripheral blood lymphocytes on 497 healthy elderly individuals, and followed them for 9 years, during which period 36% of the group died (Huppert *et al.*, 2003). In that study, an inverted CD4:CD8 ratio predicted 9-year mortality, although, after adjusting for age and sex, the HR of 1.38 failed to reach statistical significance (95% CI 0.96–1.97). Although this study did not assess CMV seropositivity, current knowledge of the impact of CMV on T-cell populations makes it is reasonable to assume that the inverted CD4:CD8 ratio was associated with CMV infection.

While the IRP was clearly associated with CMV, these studies did not report an association between CMV seropositivity *per se* and all-cause mortality. The overall rate of CMV seropositivity in the Swedish study was too high to make any meaningful conclusion about whether CMV infection causally contributed to earlier mortality, and CMV serology was not performed in the British study. These studies were provocative in linking alterations in CMV-driven T-cell populations with poor ageing outcomes, and they generated the hypothesis that CMV activity could be instrumental in those outcomes. However, by themselves, they could not provide evidence for a causal role for CMV.

Association of CMV with frailty and all-cause mortality

Since the IRP predicted death from a number of causes, it seems to reflect a general state of decline or disease susceptibility, probably due to an increased inflammatory milieu, a decline in immune competence, or perhaps both (Wikby *et al.*, 2006). In this way, the IRP is similar to the notion of frailty, which is well-accepted as a predictor of all-cause mortality. Frailty is a universally recognizable state, considered to represent cumulative declines across multiple physiological systems, distinct from specific morbidities. It is characterized by unintentional weight loss, weakness, exhaustion, slow walking, and impaired ability to carry out the activities of daily living. It is associated with increased serum levels of IL-6 and CRP, and systemic inflammation is considered to underlie to the frail state. Several studies have examined the association between CMV serology and either frailty or all-cause mortality.

A cross-sectional analysis of CMV serostatus, serum IL-6 and prevalent frailty in 700 elderly women in Baltimore, US, found that CMV seropositivity correlated with prevalent frailty: adjusted OR 3.2, $P = 0.03$. IL-6 interacted with CMV to strengthen the association; AOR for CMV+ high IL-6 was 20.3 ($P = 0.007$) (Schmaltz *et al.*, 2005). A prospective study of the effect of CMV on development of frailty and mortality over a 3-year period was performed in women from the same cohort who were not frail at baseline. After adjusting for confounders, women with CMV antibody titres in the highest quartile had an increased risk for mortality (HR 2.79; 95% CI 1.2–6.4). Seropositive women with any level of antibodies had a trend towards higher mortality, but the association was not significant. Similar results were reported for incident frailty in the 3-year follow-up (Wang, 2010). When cardiovascular disease was excluded as a cause of death, CMV was still associated with higher mortality, although the result was not significant.

In the Sacramento Area Latino Study on Aging (SALSA), a study of elderly Latinos, high CMV antibody titres correlated with impaired activities of daily living, a measure of frailty, although the result just failed to reach significance (Aiello *et al.*, 2008). High CMV antibody titres did predict cognitive decline in the same study (Aiello *et al.*, 2006). Furthermore, high CMV antibodies also predicted overall mortality in a 9-year follow-up (HR 1.43; 95% CI 1.14–1.79) in a fully adjusted model (Roberts *et al.*, 2010). When baseline IL-6 and TNF levels were added to the model, the impact of CMV was attenuated, consistent with the interpretation that the effect of CMV on mortality was mediated by these inflammatory cytokines. Because of the very high CMV seroprevalence in elderly Latinos (97% in this study), tests of the risk associated with CMV seropositivity *per se* were not reported.

The Aiello group also investigated the impact of CMV seropositivity on mortality using the 'National Health and Nutrition Examination Survey' (NHANES) III dataset. NHANES III is a survey of a population selected to be representative of the US population, carried out between 1988 and 1994. Mortality data from death certificates is available for approximately a 10 year follow-up period. CMV serology was available on ~ 14,000 participants aged > 25. After adjusting for multiple confounders, CMV seropositivity was significantly although only modestly associated with increased risk for all-cause mortality (HR: 1.19; 95% CI 1.01–1.41). Because Schmaltz and colleagues (2005) had reported a robust association between CMV and frailty in elderly women, Nielson and colleagues analysed the association between CMV and frailty in the NHANES III datasets, but found no association between CMV seropositivity and prevalent frailty in women was seen. However, there was a weak but non-significant association between CMV and frailty in men (C. Nielson, P. Srikanth, A. Simaneck, A. Aiello and A.B. Hill, manuscript in preparation).

These results are subject to the caveats discussed above. Nevertheless, they provide some support for the idea that CMV may contribute causally to inflammation and mortality in the elderly, since seropositivity *per se* contributed a modestly increased mortality risk in the NHANES III study. CMV seropositivity was strongly correlated with frailty in the Women's Health in Aging Study (WHAS), but not in NHANES III. The evidence for high CMV titres been associated with inflammation and poor ageing is stronger, but is subject to the alternative explanations discussed above.

CMV's impact on mortality in NHANES III was largely attributable to an increase in deaths due to cardiovascular disease (CVD). CMV imposed little or no risk for mortality from other causes, including cancer (our unpublished data). The relationship between CMV and CVD has been studied for many years, and is discussed in Chapter II.15. However, because the epidemiological data are contentious, I will briefly review data from prospective studies.

CMV and cardiovascular disease

There is a large literature describing CMV's involvement in the pathogenesis of CVD. In the case of allogeneic organ transplantation, it is clear that CMV accelerates vascular disease in the transplanted organ and also systemically in the recipient. Here, the twin stimuli of allogeneic immune activation and immune suppression apparently create an ideal environment for CMV replication. This pathology is readily replicated in animal models, as reviewed in Chapter II.15. CMV's ability to increase the rate of restenosis after balloon angioplasty also seems clear. However, the question whether CMV plays a role in primary atherosclerosis in non-immunosuppressed patients is much more contentious.

Atherosclerosis is an inflammatory disease of arteries, thought to be mediated primarily by inflammatory macrophages activated by oxidized low density lipoprotein (ox-LDL). In the 1990s, there was significant interest amongst atherosclerosis scientists and clinicians in the idea that chronic pathogens could contribute to the inflammation. CMV and *Chlamydia trachomatis* were the two organisms most frequently considered. However, although the CMV research community frequently continues to consider CMV a player in primary cardiovascular disease, most atherosclerosis scientists and cardiologists appear to have abandoned the idea.

The waning enthusiasm for the hypothesis amongst cardiologists is largely based on epidemiological data. Most evidence that CMV may be associated with CVD has come from small, retrospective studies. Large, prospective studies, in which CMV serology is performed on serum taken at the baseline and controls are nested within the same contemporaneous cohort as cases, are considered a more rigorous approach to this question (Danesh, 1999). Such studies are usually undertaken for a different primary purpose (often interventional), but accurate baseline data are obtained, and cardiovascular outcomes are carefully documented. Sometimes CMV serology has been performed on the entire cohort, but more frequently, case control studies have been performed. After cases (usually subjects with confirmed myocardial infarction, cardiac death or stroke) have been identified, controls are randomly chosen from within the study cohort, matched as closely as possible to cases on criteria such as age, sex, and study site. CMV seropositivity and other risk factors and covariates are then compared between cases and controls. Since it is impossible to match for all covariates, it is still necessary to adjust for covariates before reporting a hazard ratio. In 1999, John Danesh reported a meta-analysis of three prospective studies in the literature, and concluded that there was no evidence that CMV seropositivity was associated with an increased risk of CVD (Danesh, 1999).

A number of studies have reported on the risk of CMV seropositivity for subsequent major cardiac events (myocardial infarction, revascularization or cardiac death, sometimes also thromboembolic stroke) in comparison to subjects who were CMV seronegative at baseline, after adjusting for demographic and other CVD risk factors. The Physicians Health Study (PHS) of male US physicians actually found fewer CMV seropositives amongst 643 cases than among controls matched on age and smoking status, giving an HR for CMV for major cardiac events of 0.77 (95% CI 0.6–0.9) (Ridker *et al.*, 1998). Most studies have reported neutral results. In 213 cases of myocardial infarction or CVD death in the 'Cardiovascular Health Study' (CHS) of US men and women aged > 65, the adjusted OR for CMV seropositivity was 1.2 (95% CI 0.7–1.9) (Siscovick *et al.*, 2000). In the 'Women's Health Study' (WHS) of healthy post-menopausal US female health professionals, 122 cases of a first cardiac even were reported, with CMV seropositivity imparting an HR of 1.0 (95% CI 0.6–1.7). The 'Caerphilly Prospective Heart Disease Study' (CAPS) of men in South Wales, UK identified 195 cases of ischaemic heart disease in a 10 year follow-up, and reported an HR for baseline CMV seropositivity of 1.11 (95% CI 0.64–1.97) (Strachan *et al.*, 1999). Two other studies have reported a modestly increased risk attributable to CMV. The 'Heart Outcomes Prevention Evaluation' (HOPE) Study identified 494 cases amongst high risk Canadian patients, and reported an HR associated with CMV of 1.24 (95% CI 1.01–1.53) (Smieja *et al.*, 2003). Finally, the Aiello group analysed data from NHANES III, in which CMV serology was available on ~ 14,000 individuals chosen to be representative of the US population aged > 25. CMV was associated with a modest increase in risk of cardiovascular mortality which was not quite significant in a fully adjusted model: HR 1.19 (95% CI 0.95–1.49). Both these latter studies had serological data on the entire cohort, whereas the other reports were nested case control studies.

Other prospective studies correlated CMV antibody titres with CVD risk, mostly comparing the risk of individuals with highest levels of CMV antibodies to those with the lowest. In the SALSA study of elderly Latinos, those with CMV antibodies in the highest quartile had an adjusted HR of 1.35 (95% CI 1.10–1.8) compared to those in the lowest quartile (Roberts *et al.*, 2010). Similarly, the Atherosclerosis Risk in Communities (ARIC) study in the USA identified 221 new cases of CVD. Compared to controls, those with the highest CMV antibody titres had an adjusted relative risk of myocardial infarction of 1.89 (95% CI 0.98–3.67) (Sorlie *et al.*, 2000). These studies may suggest that only the subset of individuals with greatest CMV activity are at risk for CVD, but it is also possible that underlying inflammation drives both CMV activity and CVD (see Fig. II.23.2).

Together, these epidemiological studies do not suggest a major role for CMV in CVD pathogenesis in the general population, in clear contrast to the special case of immunosuppressed solid organ transplant patients. However, a modest role cannot be excluded. It is also possible that CMV contributes strongly in a small subset of individuals, presumably those with the highest antibody titres and CMV-specific T-cell responses, in whom CMV reactivation episodes may be common. These studies are of course subject to the methodological caveats discussed above. An additional consideration is the possible influence of study population selection criteria, since many of these studies specifically included or excluded subjects based on health at baseline. This could easily skew results if any increased risk imposed by CMV applies only to a small section of the population. It is interesting that Sorlie and colleagues found that the risk associated with CMV was higher in diabetics (Sorlie *et al.*, 2000). Synergy between CMV and diabetes for CVD has been suggested by some other studies (Visseren *et al.*, 1997; Kalil *et al.*, 2003; Guech-Ongey *et al.*, 2006), and perhaps warrants further investigation.

CMV, 'inflammaging' and mortality: conclusions

CMV's complex interrelationship with inflammation and the demographics of infection make interpreting epidemiological studies particularly challenging. The data do suggest that the more extreme levels of CMV antibodies and CMV-induced T-cell changes correlate with an increase in inflammatory cytokines and impaired health in the elderly. However, the cause and effect of this relationship are difficult to entangle: is CMV driving inflammation, or is inflammation driving CMV?

On the other hand, the data do not suggest that CMV plays a major role in any of these pathologies in the majority of people. Even when positive associations were detected, they were modest, and many studies failed to demonstrate an effect. If CMV does drive pathology, it seems likely that it does so only in a minority of cases. If so, it will be important to identify the characteristics of that minority.

Acknowledgements

This work was supported by grants from the N.I.H. AG039805 and AI047206.

References

Aiello, A.E., Haan, M., Blythe, L., Moore, K., Gonzalez, J.M., and Jagust, W. (2006). The influence of latent viral infection on rate of cognitive decline over 4 years. J. Am. Geriatr. Soc. *54*, 1046–1054.

Aiello, A.E., Haan, M.N., Pierce, C.M., Simanek, A.M., and Liang, J. (2008). Persistent infection, inflammation, and functional impairment in older Latinos. J. Gerontol. A. Biol. Sci. Med. Sci. *63*, 610–618.

Battacharjee, B., Renzette, N., and Kowalik, T.F. (2012). Genetic analysis of cytomegalovirus in malignant gliomas. J. Virol. *86*, 6815–6824.

Cobbs, C.S. (2011). Evolving evidence implicates cytomegalovirus as a promoter of malignant glioma pathogenesis. Herpesviridae *2*, 10.

Cobbs, C.S., Harkins, L., Samanta, M., Gillespie, G.Y., Bharara, S., King, P.H., Nabors, L.B., Cobbs, C.G., and Britt, W.J. (2002). Human cytomegalovirus infection and expression in human malignant glioma. Cancer Res. *62*, 3347–3350.

Cobbs, C.S., Soroceanu, L., Denham, S., Zhang, W., and Kraus, M.H. (2008). Modulation of oncogenic phenotype in human glioma cells by cytomegalovirus IE1-mediated mitogenicity. Cancer Res. *68*, 724–730.

Crough, T., Beagley, L., Smith, C., Jones, L., Walker, D.G., and Khanna, R. (2012). Ex vivo functional analysis, expansion and adoptive transfer of cytomegalovirus-specific T-cells in patients with glioblastoma multiforme. Immunol. Cell Biol. *90*, 872–880.

Danesh, J. (1999). Coronary heart disease, Helicobacter pylori, dental disease, Chlamydia pneumoniae, and cytomegalovirus: meta-analyses of prospective studies. Am. Heart J. *138*, 434–437.

Dowd, J.B., Aiello, A.E., and Alley, D.E. (2009). Socioeconomic disparities in the seroprevalence of cytomegalovirus infection in the US population: NHANES III. Epidemiol. Infect. *137*, 58–65.

Duftner, C., Seiler, R., Klein-Weigel, P., Gobel, H., Goldberger, C., Ihling, C., Fraedrich, G., and Schirmer, M. (2005). High prevalence of circulating CD4+CD28- T-cells in patients with small abdominal aortic aneurysms. Arterioscler. Thromb. Vasc. Biol. *25*, 1347–1352.

Dziurzynski, K., Chang, S.M., Heimberger, A.B., Kalejta, R.F., McGregor Dallas, S.R., Smit, M., Soroceanu, L., and Cobbs, C.S. (2012). Consensus on the role of human cytomegalovirus in glioblastoma. Neuro. Oncol. *14*, 246–255.

Ferguson, F.G., Wikby, A., Maxson, P., Olsson, J., and Johansson, B. (1995). Immune parameters in a longitudinal study of a very old population of Swedish people: a comparison between survivors and nonsurvivors. J. Gerontol. A. Biol. Sci. Med. Sci. *50*, 378–382.

Franceschi, C., Capri, M., Monti, D., Giunta, S., Olivieri, F., Sevini, F., Panourgia, M.P., Invidia, L., Celani, L., Scurti, M., *et al.* (2007). Inflammaging and anti-inflammaging: a systemic perspective on aging and longevity emerged from studies in humans. Mech. Ageing Dev. *128*, 92–105.

Goronzy, J.J., Fulbright, J.W., Crowson, C.S., Poland, G.A., O'Fallon, W.M., and Weyand, C.M. (2001). Value of immunological markers in predicting responsiveness to influenza vaccination in elderly individuals. J. Virol. *75*, 12182–12187.

Guech-Ongey, M., Brenner, H., Twardella, D., Hahmann, H., and Rothenbacher, D. (2006). Role of cytomegalovirus sero-status in the development of secondary cardiovascular events in patients with coronary heart disease under special consideration of diabetes. Int. J. Cardiol. *111*, 98–103.

Huppert, F.A., Pinto, E.M., Morgan, K., and Brayne, C. (2003). Survival in a population sample is predicted by proportions of lymphocyte subsets. Mech. Ageing Dev. *124*, 449–451.

Jonasson, L., Tompa, A., and Wikby, A. (2003). Expansion of peripheral CD8+ T-cells in patients with coronary artery disease: relation to cytomegalovirus infection. J. Intern. Med. *254*, 472–478.

Kalil, R.S., Hudson, S.L., and Gaston, R.S. (2003). Determinants of cardiovascular mortality after renal transplantation: a role for cytomegalovirus? Am. J. Transplant. *3*, 79–81.

Lehrer, S., Green, S., Ramanathan, L., Rosenzweig, K., and Labombardi, V. (2012). No consistent relationship of glioblastoma incidence and cytomegalovirus seropositivity in whites, blacks, and Hispanics. Anticancer Res. *32*, 1113–1115.

Lucas, K.G., Bao, L., Bruggeman, R., Dunham, K., and Specht, C. (2011). The detection of CMV pp65 and IE1 in glioblastoma multiforme. J. Neurooncol. *103*, 231–238.

Maussang, D., Verzijl, D., van Walsum, M., Leurs, R., Holl, J., Pleskoff, O., Michel, D., van Dongen, G.A., and Smit, M.J. (2006). Human cytomegalovirus-encoded chemokine receptor US28 promotes tumorigenesis. Proc. Natl. Acad. Sci. U.S.A. *103*, 13068–13073.

Mitchell, D.A., Xie, W., Schmittling, R., Learn, C., Friedman, A., McLendon, R.E., and Sampson, J.H. (2008). Sensitive detection of human cytomegalovirus in tumors and peripheral blood of patients diagnosed with glioblastoma. Neuro. Oncol. *10*, 10–18.

Nikolich-Zugich, J. (2008). Ageing and life-long maintenance of T-cell subsets in the face of latent persistent infections. Nat. Rev. Immunol. *8*, 512–522.

Pawelec, G., Derhovanessian, E., Larbi, A., Strindhall, J., and Wikby, A. (2009). Cytomegalovirus and human immunosenescence. Rev. Med. Virol. *19*, 47–56.

Phillips, A.C., Carroll, D., Khan, N., and Moss, P. (2008). Cytomegalovirus is associated with depression and anxiety in older adults. Brain Behav. Immun. *22*, 52–55.

Prins, R.M., Cloughesy, T.F., and Liau, L.M. (2008). Cytomegalovirus immunity after vaccination with autologous glioblastoma lysate. N. Engl. J. Med. *359*, 539–541.

Rahbar, A.R., Sundqvist, V.A., Wirgart, B.Z., Grillner, L., and Söderberg-Naucler, C. (2004). Recognition of cytomegalovirus clinical isolate antigens by sera from cytomegalovirus-negative blood donors. Transfusion *44*, 1059–1066.

Rahbar, A., Stragliotto, G., Orrego, A., Peredo, I., Taher, C., Willems, J., and Söderberg-Naucler, C. (2012). Low levels of human cytomegalovirus infection in glioblastoma multiforme associates with patient survival: a case–control study. Herpesviridae *3*, 3.

Ranganathan, P., Clark, P.A., Kuo, J.S., Salamat, M.S., and Kalejta, R.F. (2012). Significant association of multiple human cytomegalovirus genomic loci with glioblastoma multiforme samples. J. Virol. *86*, 854–864.

Ridker, P.M., Hennekens, C.H., Stampfer, M.J., and Wang, F. (1998). Prospective study of herpes simplex virus, cytomegalovirus, and the risk of future myocardial infarction and stroke. Circulation *98*, 2796–2799.

Roberts, E.T., Haan, M.N., Dowd, J.B., and Aiello, A.E. (2010). Cytomegalovirus antibody levels, inflammation, and mortality among elderly latinos over 9 years of follow-up. Am. J. Epidemiol. *172*, 363–371.

Sabatier, J., Uro-Coste, E., Pommepuy, I., Labrousse, F., Allart, S., Tremoulet, M., Delisle, M.B., and Brousset, P. (2005). Detection of human cytomegalovirus genome and gene products in central nervous system tumours. Br. J. Cancer *92*, 747–750.

Scheurer, M.E., Bondy, M.L., Aldape, K.D., Albrecht, T., and El-Zein, R. (2008). Detection of human cytomegalovirus in different histological types of gliomas. Acta Neuropathol. *116*, 79–86.

Schmaltz, H.N., Fried, L.P., Xue, Q.L., Walston, J., Leng, S.X., and Semba, R.D. (2005). Chronic cytomegalovirus infection and inflammation are associated with prevalent frailty in community-dwelling older women. J. Am. Geriatr. Soc. *53*, 747–754.

Simanek, A.M., Dowd, J.B., and Aiello, A.E. (2009). Persistent pathogens linking socioeconomic position and cardiovascular disease in the US. Int. J. Epidemiol. *38*, 775–787.

Siscovick, D.S., Schwartz, S.M., Corey, L., Grayston, J.T., Ashley, R., Wang, S.P., Psaty, B.M., Tracy, R.P., Kuller, L.H., and Kronmal, R.A. (2000). Chlamydia pneumoniae, herpes simplex virus type 1, and cytomegalovirus and incident myocardial infarction and coronary heart disease death in older adults: the Cardiovascular Health Study. Circulation *102*, 2335–2340.

Slinger, E., Maussang, D., Schreiber, A., Siderius, M., Rahbar, A., Fraile-Ramos, A., Lira, S.A., Söderberg-Naucler, C., and Smit, M.J. (2010). HCMV-encoded chemokine receptor US28 mediates proliferative signaling through the IL-6-STAT3 axis. Sci. Signal. *3*, ra58.

Smieja, M., Gnarpe, J., Lonn, E., Gnarpe, H., Olsson, G., Yi, Q., Dzavik, V., McQueen, M., and Yusuf, S. (2003). Multiple infections and subsequent cardiovascular events in the heart outcomes prevention evaluation (HOPE) study. Circulation *107*, 251–257.

Söderberg-Naucler, C. (2006). Does cytomegalovirus play a causative role in the development of various inflammatory diseases and cancer? J. Intern. Med. *259*, 219–246.

Sorlie, P.D., Nieto, F.J., Adam, E., Folsom, A.R., Shahar, E., and Massing, M. (2000). A prospective study of cytomegalovirus, herpes simplex virus 1, and coronary heart disease: the atherosclerosis risk in communities (ARIC) study. Arch. Intern. Med. *160*, 2027–2032.

Soroceanu, L., Akhavan, A., and Cobbs, C.S. (2008). Platelet-derived growth factor-alpha receptor activation is required for human cytomegalovirus infection. Nature *455*, 391–395.

Soroceanu, L., Matlaf, L., Bezrookove, V., Harkins, L., Martinez, R., Greene, M., Soteropoulos, P., and Cobbs, C.S. (2011). Human cytomegalovirus US28 found in glioblastoma promotes an invasive and angiogenic phenotype. Cancer Res. *71*, 6643–6653.

Straat, K., Liu, C., Rahbar, A., Zhu, Q., Liu, L., Wolmer-Solberg, N., Lou, F., Liu, Z., Shen, J., Jia, J., *et al.* (2009). Activation of telomerase by human cytomegalovirus. J. Natl. Cancer Inst. *101*, 488–497.

Strachan, D.P., Carrington, D., Mendall, M.A., Butland, B.K., Sweetnam, P.M., and Elwood, P.C. (1999). Cytomegalovirus seropositivity and incident ischaemic heart disease in the caerphilly prospective heart disease study. Heart *81*, 248–251.

Strindhall, J., Nilsson, B.O., Lofgren, S., Ernerudh, J., Pawelec, G., Johansson, B., and Wikby, A. (2007). No immune risk profile among individuals who reach 100 years of age: findings from the swedish NONA immune longitudinal study. Exp. Gerontol. *42*, 753–761.

Trzonkowski, P., Mysliwska, J., Szmit, E., Wieckiewicz, J., Lukaszuk, K., Brydak, L.B., Machala, M., and Mysliwski, A. (2003). Association between cytomegalovirus infection, enhanced proinflammatory response and low level of anti-hemagglutinins during the anti-influenza vaccination--an impact of immunosenescence. Vaccine *21*, 3826–3836.

Visseren, F.L., Bouter, K.P., Pon, M.J., Hoekstra, J.B., Erkelens, D.W., and Diepersloot, R.J. (1997). Patients with diabetes mellitus and atherosclerosis; a role for cytomegalovirus? Diabetes Res. Clin. Pract. *36*, 49–55.

Wikby, A., Maxson, P., Olsson, J., Johansson, B., and Ferguson, F.G. (1998). Changes in CD8 and CD4 lymphocyte subsets, T-cell proliferation responses and non-survival in the very old: the swedish longitudinal OCTO-immune study. Mech. Ageing Dev. *102*, 187–198.

Wikby, A., Johansson, B., Olsson, J., Lofgren, S., Nilsson, B.O., and Ferguson, F. (2002). Expansions of peripheral blood CD8 T-lymphocyte subpopulations and an association with cytomegalovirus seropositivity in the elderly: the swedish NONA immune study. Exp. Gerontol. *37*, 445–453.

Wikby, A., Nilsson, B.O., Forsey, R., Thompson, J., Strindhall, J., Lofgren, S., Ernerudh, J., Pawelec, G., Ferguson, F., and Johansson, B. (2006). The immune risk phenotype is associated with IL-6 in the terminal decline stage: findings from the swedish NONA immune longitudinal study of very late life functioning. Mech. Ageing Dev. *127*, 695–704.

Wrensch, M., Weinberg, A., Wiencke, J., Miike, R., Barger, G., and Kelsey, K. (2001). Prevalence of antibodies to four herpesviruses among adults with glioma and controls. Am. J. Epidemiol. *154*, 161–165.

Epilogue

Résumé and Visions: From CMV Today to CMV Tomorrow

Ulrich H. Koszinowski

This new book on cytomegalovirus is both a timely update and an extension of the previous book, now giving more room also to clinical observations and studies. '... and what is the use of a book,' thought Alice, 'without pictures or conversation?' (Lewis Carroll, *Alice's Adventures in Wonderland*). This book is written by a selected trustworthy group of world-renowned virologists ... and, yes, it has great pictures!

It is the inevitable fate of all published scientific observations from a given period to be digested and to end up as a state-of-the-art review book. The digest is made for us by other experts and we are usually more than happy to trust their work. The book gives us the advantage to memorize less than we used to or had to. We can just look it up. Such a collection of reviews must be counted, according to Orwell's definition, among the best books: '...those that tell you what you know already'.

I have the additional pleasure to know almost all authors as personalities and not only from their writings. Science is made by humans. It provides pleasure to hear the familiar sound which adds another dimension to their writing. Thus, for me it is not only written but almost spoken text. As anybody else I have colleagues that I favour more than others. How can I control that my trust is not too much controlled by sympathy? Thus, knowing the authors may not be of advantage in all cases. This brings me to the quote of Terentianus Maurus 'pro captu lectoris, habent sua fata libelli' meaning that books have their destinies according to the capacity of the reader. Each of us will take from this attractive thesaurus what she or he is able to. So many presents are laid out and we have many options and choices, indeed.

Such science books in review form often deal with matters we are already pondering, areas we wish to venture in order to ask new questions or we are even working on. Because the answer we gave to a scientific question yesterday may be quite different from the answer to the same questions asked today or in the future such a book must have a limited half-life. Answers offered suffer from systematic oversimplification: a hallmark of science. Already a major oversimplification is what we believe to constitute the properties and functions of a cytomegalovirus protein. We constantly find new functions for old candidates. In addition, some of us have this itch to coin new names whenever describing a new function for a protein of an already known function. But what do we really know for safe about gene functions and how many of such renaming events lie ahead of us until we know? This protein-name-function connection bears difficulties. If we constantly name and rename any of these translation products according to functions we describe to them, even when the observation has not yet been reproduced, we duplicate the maze already presented by CMV as we know it today. Abraham Lincoln said 'How many legs does a dog have if you call the tail a leg? Four. Calling a tail a leg doesn't make it a leg.'

No particular vision is needed to predict that there is serious work coming up. The CMV of today is not the CMV of tomorrow. What will be the state of the gene–protein–function connection in the future? Good enough that this book considers research only until end of 2011. At the recent 2012 Herpesvirus Workshop it was reported that ribosomal footprinting reveals an unexpected complexity of translation products coming from individual ORFs. This demonstrates the value of measuring gene expression at the level of translation. The abundance of footprint fragments in deep sequencing data reports on variability and amount of translation of a gene. In addition, footprints reveal the exact translated regions. This apparently more than doubles the number of CMV proteins known today. How will we confirm this complexity and annotate functions of these translation products?

This problem is not easy to solve. After all, what is THE human cytomegalovirus (HCMV)? Known isolates vary considerably and isolates do not describe a clonal entity but an inhomogeneous group of genomes that, taken together, make a phenotype. In addition, the lack of clonal stability and the difficulty to propagate some of the strains will make it very difficult to study certain CMV strain properties. Only from the genetic side there is hope coming through sequencing and systems biology approaches.

With regard to functional properties of individual translation products there is no easy solution. Genetic engineering for studying specific functions will necessarily get more complex. A more stumbling problem may pose the ignorance-hierarchy in CMV research. We live in an anthropocentric world. Clearly, only HCMV is the relevant pathogen, the subject is funded best, and HCMV research is published, often ignoring work on CMVs from other species. Rhesus CMV is next in line and needs to consider only HCMV research. Then follow the rest, guinea pig CMV, rat CMV and mouse CMV at the end of the 'relevance chain'. This ignorance is justified to some extent: there is high genetic variability already among human CMVs with respect to gene sequences, gene numbers and gene functions. Positional identity is clearly not identity and similarity is just what it is. Yet, this ignorance is not fruitful. Given the flood of data coming up it is perhaps more fruitful to use Occam's razor (entities must not be multiplied beyond necessity), to assume comparable gene functions unless a divergence is clearly proven. Namely, without proposals on potential functions of newly identified proteins by work on animal CMVs there is little hope to unravel the complexity of new HCMV gene products. Work on animal CMVs cannot give right answers for HCMV, yet it can ask right questions.

Subject index

A

B

C

F

G

N

U

V

X

Z

Open reading frame index

See specific chapters for CMVs other than HCMV and MCMV.

HCMV

MCMV

Cluster of differentiation index

CMV strains index

See specific chapters for CMVs other than HCMV and MCMV.

Antiviral drugs index

Frequently discussed molecules

Cytokines and their receptors

Interferons and their receptors

Interleukins and their receptors